OCCUPATIONAL HEALTH & SAFETY

Theory, Strategy & Industry Practice

Fourth Edition

Dianne E.G. Dyck
RN, BN, MSc, COHN(C), CRSP

Occupational Health & Safety: Theory, Strategy & Industry Practice, Fourth Edition
© LexisNexis Canada Inc. 2020
August 2020

All rights reserved. No part of this publication may be reproduced or stored in any material form (including photocopying or storing it in any medium by electronic means and whether or not transiently or incidentally to some other use of this publication) without the written permission of the copyright holder except in accordance with the provisions of the *Copyright Act*. Applications for the copyright holder's written permission to reproduce any part of this publication should be addressed to the publisher.

Warning: The doing of an unauthorized act in relation to a copyrighted work may result in both a civil claim for damages and criminal prosecution.

The publisher, (Author(s)/General Editor(s)) and every person involved in the creation of this publication shall not be liable for any loss, injury, claim, liability or damage of any kind resulting from the use of or reliance on any information or material contained in this publication. While every effort has been made to ensure the accuracy of the contents of this publication, it is intended for information purposes only. When creating this publication, none of the publisher, the (Author(s)/General Editor(s)) or contributors were engaged in rendering legal or other professional advice. This publication should not be considered or relied upon as if it were providing such advice. If legal advice or expert assistance is required, the services of a competent professional should be sought and retained. The publisher and every person involved in the creation of this publication disclaim all liability in respect of the results of any actions taken in reliance upon information contained in this publication and for any errors or omissions in the work. They expressly disclaim liability to any user of the work.

Library and Archives Canada Cataloguing in Publication

Dyck, Dianne E. G.

Occupational health & safety: theory, strategy and industry practice / Dianne E.G. Dyck.

Includes bibliographical references and index.

ISBN 978-0-433-50207-4

1. Industrial hygiene—Canada. 2. Industrial safety—Canada. I. Title.

HD7261.D93 2007   658.3'820971   C2007-904632-0

Published by LexisNexis Canada, a member of the LexisNexis Group
LexisNexis Canada Inc.
111 Gordon Baker Road, Suite 900
Toronto, Ontario
M2H 3R1

Customer Service
Telephone: (905) 479-2665 • Fax: (905) 479-2826
Toll-Free Phone: 1-800-668-6481 • Toll-Free Fax: 1-800-461-3275
Email: customerservice@lexisnexis.ca
Web Site: www.lexisnexis.ca

Printed and bound in Canada.

DEDICATION

This book is dedicated to Anthony (Tony) Roithmayr with whom I had the great fortune to work and collaborate with for the past 20 years.

Tony is a human performance specialist whose ideas and approaches remain beyond his time. He embraced the field of Occupational Health & Safety as an avenue to educate, promote, and facilitate human performance in the workplace. By partnering with leaders, Tony was able to offer his expertise in organizational effectiveness and human performance improvement to develop work environments in which employees willingly hold themselves accountable for great business results.

> *"I firmly believe that great work performance can only be sustained when employees feel energized, satisfied, supported, and secure in their workplace."*

Tony Roithmayr, 2014

Thanks to Tony, I have an appreciation for the magnitude and complexity of the knowledge and effort required to make a workplace a healthy and safe environment.

DEDICATION

This book is dedicated to Anthony (Tony) Rollimaver with whom I had the great fortune to work and collaborate with for the past 20 years.

Tony is a human performance specialist whose ideas and approaches run far beyond his time. He embraced the field of Occupational Health & Safety as an avenue to educate, promote, and facilitate human performance in the workplace. By partnering with leaders, Tony was able to offer his expertise in organizational effectiveness and human performance improvement to develop work environments in which employees willingly hold themselves accountable for great business results.

"I firmly believe that great work performance can only be sustained when employees feel energized, satisfied, supported, and secure in their workplace."

— Tony Rollimaver, 2014

Thanks to Tony, I have an appreciation for the magnitude and complexity of the knowledge and effort required to make a workplace a healthy and safe environment.

ABOUT THE AUTHOR

Dianne Dyck is an Occupational Health and Safety Specialist in Canada and the United States. She has a comprehensive background in the development and management of Occupational Health and Safety (OH&S) services. Her consulting services have included the review, design, development, implementation, and refinement of OH&S programs, disability management programs, and attendance management programs; auditing of OH&S services, disability management programs and employee assistance programs; and the development of OH&S and Disability Management data management systems and software programs. As well, Dianne has provided OH&S and Disability Management instructional services in a number of university settings and has written a well-known textbook, *Disability Management: Theory, Strategy and Industry Practice*, 6th ed. (Toronto: LexisNexis Canada, 2017); and *A Practical Guide to Psychological Health & Safety in the Workplace* (Markham, ON: LexisNexis Canada, 2015).

PREFACE

The evolution of the field of Occupational Health and Safety in Canada and the United States is an exciting area; however, as with any field of practice, it brings with it uncertainty as well as many opportunities. For Occupational Health and Safety Management Systems to be effective, appropriate structures, processes, and outcomes must be in place. To that end, this book was prepared to provide some general guidelines on creating and managing quality Occupational Health and Safety Management Systems and Occupational Health and Safety Programs.

When writing about such a dynamic subject, the process can be somewhat daunting in an attempt to provide the most recent and relevant information to readers. My hope is that *Occupational Health & Safety: Theory, Strategy and Industry Practice, Fourth Edition* will serve as an "everyday guide" to meet your practice and professional needs and that you will find it useful as you make your contributions towards advancing the field of Occupational Health and Safety in Canada, the United States, and New Zealand.

I would like to acknowledge the contributions made by Dr. Ian Arnold, Mary Ann Baynton, Heidi Börner, Jono Everst, Sharon Chadwick, Norm Keith, Randy Kelbert, Sandra Lassowski, Nina Novak, Kristine Robidoux, Tony Roithmayr, Dr. Bonnie Rogers, Dale Rud, Dr. Joti Samra, Dr. Martin Shain, Dr. Steven Simon, Dan Steinke, and Brett Webb. Likewise, valuable assistance and industry examples were received from the New Zealand Electricity Engineers' Association (EEA); ENMAX Corporation; Manitoba Hydro; Network Waitaki Contracting Ltd.; Orange Umbrella®; Petro-Canada (now Suncor Energy Inc.); the Public Services, Health & Safety Association (PSHSA); Trustpower Ltd.; and Unison Contracting Ltd. Without their support, writing this resource would not have been possible.

Lastly, I would like to thank all the Occupational Health and Safety Practitioners/Professionals who encouraged me to write this text. I trust it will meet their expectations and needs.

Dianne Dyck
July 2020

Table of Contents

Dedication iii
About the Author v
Preface vii

Introduction

Introduction — Occupational Health and Safety: Theory, Strategy and Industry Practice

A. Introduction 1
B. Challenges 1
C. Intended Audience 2
D. Format 3
E. Structure 3
   (a) Section 1: Occupational Health and Safety: Theory 3
   (b) Section 2: Occupational Health and Safety: Strategies and Industry Practices 4
   (c) Section 3: Occupational Health and Safety: Supportive Programs ... 8
   (d) Section 4: Occupational Health and Safety: Knowledge and Skills .. 9
   (e) Section 5 - Occupational Health and Safety: The Future 11
   (f) Section 6: Occupational Health and Safety: Glossary and Index ... 11
F. Summary 12
   Chapter References 12

Section 1 — Occupational Health and Safety: Theory

Chapter 1 — Occupational Health and Safety: Historical Perspectives

A. What is Occupational Health and Safety? 15
B. OH&S: The Canadian and American Experience 16
   1. Background History: The 1800s 16
   2. The Industrial Revolution 17
   3. Birth of the Workers' Compensation Systems 18
   4. The Investigational Era 19
   5. The 1950s and 1960s 19
   6. Occupational-based Assistance for Workers 20
   7. Summary 21
   8. History of OH&S in the United States of America 22
   9. History of OH&S in Canada 25
   10. Canadian-American OH&S Systems: Current Difference 26
C. OH&S Management: Theoretical Background 27
   1. Management Theories: Influence on Motivational and OH&S Modern Practices 27
      (a) Classical Management Approaches 28
      (b) Scientific Management Theory 28

Table of Contents

		(c) Administrative or Classical Management Theory	30
		(d) Bureaucratic Management Theory	32
		(e) Behavioural Theory and Approaches	33
		(f) Quantitative Management Approaches	42
		(g) Modern Management Theories	42
	2.	Organizational Designs: Characteristics	57
	3.	Organizational Designs: Relationship with Organizational Elements	60
	4.	Organizational Designs: Structures	63
D.	OH&S: Internal and External Responsibility Systems		67
E.	Canada's OH&S System: Federal and Provincial		69
F.	Summary		70
	Chapter References		70

Chapter 2 — Occupational Health and Safety Management Systems

A.	Introduction		73
B.	Occupational Health and Safety Elements		73
	1.	Management Leadership and Commitment to Occupational Health & Safety	74
	2.	Occupational Health & Safety Communication	76
	3.	OHSMS Structure	76
	4.	Employee Qualifications, Orientation and Training	76
	5.	Hazard Identification and Assessment	77
	6.	Hazard Control	78
	7.	Worksite Inspections	79
	8.	Emergency Preparedness and Response	79
	9.	Incident Investigation and Corrective Actions	80
	10.	System Review and Continuous Improvement	80
C.	Occupational Health and Safety Management Systems		80
	1.	Rationale for Implementation	80
	2.	Purpose and Value	82
	3.	Framework	83
D.	OHSMS Standards: Description and Comparison		84
	1.	OHSAS 18001/02	84
	2.	ISO 45001/Z45001	86
	3.	ILO-OSH 2001	88
	4.	ANSI/AIHA Z10-2005	89
	5.	CSA Z-1000-06	90
	6.	Alberta Partnerships in Injury Reduction Program	91
	7.	Conclusion	92
E.	OHSMS: Typical Evolution Within a Company		92
	1.	The DuPont Bradley Curve	92
	2.	The Fennell Model: ESSO	93
	3.	The Stages of Organizational OH&S Culture Evolution	95
		(a) Stage 1: Barely Conscious	95

TABLE OF CONTENTS

		(b)	Stage 2: Struggling	95
		(c)	Stage 3: Managing	96
		(d)	Stage 4: Consistent and Collaborative	96
		(e)	Stage 5: Continually Improving and Deeply Involved	97

F. OH&S Program: Components 97
G. OH&S Program: Elements 99
    1. Occupational Aspects of an OH&S Program 99
        (a) Workplace/Employee Safety and Incident Prevention 99
        (b) Exposure Monitoring and Health Surveillance 100
        (c) Injury and Illness Management 101
        (d) Emergency Response (ERP) and Disaster Planning 102
        (e) OH&S Legislation Governance and Stewardship 102
    2. Non-occupational Aspects 103
        (a) Disease and Injury Prevention/Health Promotion 103
        (b) Employee Assistance Programs (EAPs) 103
        (c) Attendance Support and Assistance Program (ASAP) 104
        (d) Workplace Wellness Program (WWP) 105
H. OHSMS and OH&S Program: Measurement 105
    1. Evaluation 105
    2. Program Measurement 105
        (a) Structure, Process, and Outcomes: Evaluation 106
        (b) Program Goals, Objectives and Targets: Evaluation 106
        (c) Summative and Formative Evaluation 108
        (d) Impact Evaluation 108
        (e) Why Measure OH&S Programs and Results? 110
    3. Measurement of an OHSMS 112
I. Summary 118
    Chapter References 119

Chapter 3 — Occupational Health and Safety: Key Stakeholders

A. OH&S Key Stakeholders: Who Are They? 123
B. OH&S Key Stakeholders: Roles, Responsibilities and Qualifications ... 125
    1. Employer 125
        (a) Upper Management 125
        (b) Front-line Management 126
    2. Employee 126
    3. Trade Union 127
    4. Joint Health and Safety Committee 128
    5. Occupational Safety Practitioner/Professional 131
    6. Occupational Health Nurse 132
    7. Safety Engineer 133
    8. Conclusion 134
C. Summary 134
    Chapter References 134

TABLE OF CONTENTS

Chapter 4 — Occupational Health Program
A. Introduction 137
B. Occupational Health Program Components and Services 137
C. Occupational Health Professionals 139
   1. Occupational Health Nurses 139
   2. Occupational Health Physicians 143
D. Occupational Health Service: A Business Case 144
   1. The OH Service 144
   2. The OH Risk 145
   3. Going Forward 151
   4. The OH Service: Its Value 151
E. Confidentiality of Medical Information 156
F. Fitness to Work 160
G. Job Demands Analysis (JDA) 162
H. Hearing Conservation Program 164
I. Medical Monitoring 169
J. Respiratory Conservation Program 169
K. First Aid 171
L. Medical Absence Reporting 171
M. OH Program and Service: Program Evaluation 172
N. Summary 172
Appendix 1: Pledge and Oath of Confidentiality 173
Appendix 2: Medical Fitness-to-Work Form 175
Appendix 3: Authorization for the Release of Privileged Information 177
Appendix 4: Job Demands Analysis Form 179
Appendix 5: Absent Reporting: Flowchart (Sample) 181
Appendix 6: Work Absence Form: Non-Medical 183
Appendix 7: Work Absence Form: Medical 185
            Chapter References 186

Section 2 – Occupational Health and Safety: Strategies

Chapter 5 — Occupational Health and Safety Program: Manual Development
A. Introduction 193
B. OH&S Program Manual: Development 193
   1. Policies, Standards and Procedures 193
   2. Manual Components 194
      (a) Mission Statement 194
      (b) Program Goals and Objectives 196
      (c) Administration of the OH&S Program 196
      (d) Stakeholder Roles and Responsibilities 197
      (e) Implementation Strategies 197
      (f) Program Evaluation 197
      (g) Program Promotion 198

3. Industry Example 199
    (a) Preliminary Concerns 200
    (b) Development and Maintenance of the OH&S Program Manual 200
    (c) Formatting the Manual 201
C. Summary 202
Appendix 1:   OH&S Program Manual: An Industry Example 203
Appendix 2:   Visitor Health and Safety 271
              Chapter References 271

Chapter 6 — Occupational Health and Safety: Leadership and Commitment

By Tony Roithmayr and Dianne Dyck

A. Introduction 273
B. Development of an Occupational Health and Safety Culture 274
   1. Corporate Culture 274
   2. Subcultures 278
   3. OH&S (Safety) Culture 281
   4. The Evolution of an OH&S (Safety) Culture 287
C. Key Aspects of OH&S Leadership and Commitment 288
D. Rationale for Strong OH&S Leadership and Commitment 288
E. Great Safety Performance: An Industry Example 289
   1. Corporate OH&S Duty of Care 290
   2. Sustainable High-Performance Workplace Cultures 291
   3. Applying the Performance Maximizer® to Occupational Health and Safety 293
      (a) Project Methodology 293
   4. Future State: Developing the Great Safety Performance™ Model and Methodology 294
   5. Current State: Baseline Measurement of Safety Performance 296
   6. Gap Analysis 299
      (a) Develop Responses and Implement Actions 299
      (b) Measurement and Follow-through 299
   7. Project Results and Findings: Reliability of the Workplace Safety Survey 299
   8. Conditions for Great Safety Performance: Actual Performance of the System for Enabling Safe Work Actions 300
   9. Importance of the Conditions for Great Safety Performance 300
   10. Actual Performance of Safe Work Actions 303
   11. Importance of Safe Work Actions 303
       (a) Predictive Relationship 304
       (b) Extent to which the Conditions for Great Safety Performance are Predictive of Safe Work Actions 304
       (c) Extent to which the Safe Work Actions are Predictive of Safe

Table of Contents

	Work Outcomes	304
F.	Project Conclusion	305
G.	Discussion	305
H.	Measurement of OH&S Leadership and Commitment	309
I.	Summary	312
Appendix 1:	Reliability and Correlation Tables	313
	Chapter References	314

Chapter 7 — Safe Workplace = Great Workplace: Building a Sustainable Culture of Safety

by Tony Roithmayr

- A. Introduction ... 317
- B. New Directions for the Practice of Occupational Health and Safety ... 317
- C. Understanding Workplaces as Systems ... 321
    1. Introduction ... 321
    2. A Guide to Understanding The Performance Maximizer® ... 323
    3. Leading Indicators — The Conditions for Great Safety Performance ... 324
        - (a) Know What to do ... 324
        - (b) Able to do it ... 325
        - (c) Equipped to do it ... 325
        - (d) Want to do it ... 325
        - (e) Interactions ... 326
    4. A Performance System for Building a Safe/Great Workplace ... 326
    5. Transformational Leadership Practices ... 327
    6. The Key Indicators of a Great/Safe Workplace ... 328
- D. Measuring and Analyzing the Performance System ... 329
- E. A New Role for the OH&S Practitioner/Professional ... 330
- F. Defining the New Role – An Industry Example ... 333
    1. Organization Evaluation ... 342
- G. "Taking a Stand": An Industry Experience ... 343
    1. Taking a Stand: An Industry Example ... 343
    2. The Self-Reinforcing Cycle of Culture Development ... 346
        - (a) The Strategy to Shift the Culture ... 348
- H. Workplace Culture Evaluation: The Value Offered ... 355
- I. Summary ... 355
    - Chapter References ... 355

Chapter 8 — Safety Culture: Safety Climate

by Heidi Börner, Dianne Dyck and Brett Webb

- A. Introduction ... 357
- B. Occupational Health and Safety Climate: Assessment ... 358
    1. PSHSA Health and Safety Climate Assessment Project ... 358
        - (a) Methodology ... 359

xiv

		(b)	Findings and Interpretation	360

TABLE OF CONTENTS

       (b) Findings and Interpretation 360
       (c) Implications 361
       (d) Observations Warranting Further Investigation 362
       (e) Transferrable Learning 363
   2. Orange Umbrella® Methodology (New Zealand) 364
       (a) Approach toward Assessing the Safety Climate 364
       (b) Findings and Learnings 367
       (c) Implications of the Safety Climate Assessment Process 368
C. Great Safety Performance: The New Zealand Experience 368
   1. Improvement Trends for the SCP Group 371
   2. Pattern of Improvement 374
   3. Leadership Follow-through on Action Plans: A Success Factor ... 376
   4. The Impact of "Not Following Through" 377
   5. Leader Influence on OH&S (Safety) Culture 378
   6. Observations and Issues from the Safety Climate Project 380
       (a) Comprehensive Reporting 380
       (b) OH&S (Safety) Culture and Climate Improvement Takes Time 381
       (c) Momentum 381
       (d) Industry Involvement and Support 381
   7. Safety Climate Project, Year 4: Emerging Issues 383
       (a) Public Safety 383
       (b) Financial Risk 383
       (c) Business Continuity 383
       (d) Health and Safety at Work Act in 2014 383
D. Discussion 384
E. Summary 386
   Chapter References 386

Chapter 9 — Occupational Health and Safety: Hazard Identification, Assessment, and Control

A. Introduction 389
B. Hazard Management 390
C. Hazard Identification, Assessment, and Control 391
   1. Definitions 391
       (a) Hazard Identification 391
       (b) Risk Assessment 392
       (c) Hazard Control 395
D. New Workplace Hazards 399
E. Hazard Management: Importance 400
F. Hazard Management: Hazard Types 400
G. Hazard Management: Techniques for Hazard Identification, Assessment, and Control 401
H. Hazard Management: Worker Involvement 407
I. Hazard Management: Effectiveness Measurement 409

xv

Table of Contents

J.	Hazards: New Industry Challenges	410
K.	Summary	417
Appendix 1:	Sample Hazard Assessment and Control Form	419
Appendix 2:	Sample Risk Assessment Form	423
Appendix 3:	Sample Worksite Inspection Form	425
Appendix 4:	Sample Ergonomic Assessment for an Office Environment	427
Appendix 5:	Sample Job Safety Analysis	429
	Chapter References	430

Chapter 10 — New Technology: Impact on the Workplace

A. Introduction ... 433
B. New Technology: Terminology ... 433
C. New Technology: What is Happening? ... 435
D. New Technology: Impact ... 438
   1. Change in Hours of Work ... 438
   2. Change in Work Location ... 438
   3. Corporate Security Issues ... 439
   4. Data Safeguarding Issues ... 439
   5. Need for New Corporate Policies and Procedures ... 439
   6. Big Data ... 439
   7. Business Advantages ... 440
   8. Worker Benefits ... 440
E. New Technology: Workplace Hazards ... 440
   1. Automation: Associated Hazards ... 440
   2. Cellphones: Associated Hazards ... 441
   3. Robots: Associated Hazards ... 442
   4. 3-D Printers: Associated Hazards ... 443
   5. Magnetic Resonance Imaging (MRI): Associate Hazards ... 443
F. Employer Response ... 443
G. Summary ... 443
   Chapter References ... 443

Chapter 11 — Occupational Health and Safety: Workplace Standards and Rules

A. Introduction ... 447
B. Workplace Standards and Rules: Importance ... 448
C. Workplace Standards and Rules: Effective Communication Techniques ... 448
D. Workplace Standards and Rules: Industry Examples ... 451
E. OH&S Standards and Rules: Techniques for Worker Conformance ... 452
F. OH&S Standards and Rules: Measurement for Compliance ... 453
G. Summary ... 454
   Chapter References ... 454

Chapter 12 — Occupational Health and Safety: Emergency Preparedness and Response

A. Introduction ... 457

TABLE OF CONTENTS

B. Emergency Response Programs and Plans: Value 458
C. Emergency Response Plan: Design 458
   1. Vulnerability Assessments 459
      (a) Determine Which Hazards Pose a Threat to any Specific
          Enterprise 459
      (b) For Each Potential Emergency Situation, Determine the Associated
          Risk 460
      (c) For Each Potential Emergency Situation, Identify the Potential
          Impact and Develop Control Plans 461
      (d) For Each Emergency Situation, Determine the Necessary Response
          Actions 464
   2. Elements of an Emergency Response Plan 464
D. Emergency Response Plans 470
   1. Company XYZ Emergency Response Plan 470
   2. Fire Emergency Response Plan 475
   3. Chemical Spill/Emission Release Response Plan 479
   4. Major Injury Incident Response Plan 479
   5. Bomb Threat Response Plan 480
   6. Bioterrorism Protection Plan 482
   7. Pandemic Flu Planning 486
      (a) Current Assumptions about a Pandemic Flu Outbreak 486
      (b) Business Continuity Risks 488
      (c) Employee Exposure to Pandemic Influenza at Work:
          Classification 489
      (d) Pandemic Flu: Business Continuity Risk Management 492
      (e) Other Control Measures: Discussion 508
E. Emergency Response Program: Measurement of Effectiveness 509
F. Summary 509
   Appendix 1:   Sample Emergency Response Plan 511
                 Chapter References 516

Chapter 13 — Occupational Health and Safety: Injury Management

A. Introduction 519
B. Injury Management: Importance 519
C. Injury Management: Definitions 520
D. Injury Management: Key Stakeholder Roles and Responsibilities 523
   1   Employer 523
   2   Line Manager 523
   3   Director, OH&S 523
   4   Training Department 523
   5   First Aider 524
   6   Employee 524
E. Injury Management: Preparedness 524
F. Injury Management: Effective Techniques 527
   1. Management of Workplace Injury Incidents 527

TABLE OF CONTENTS

    2. Disability Management 528
    3. Return-to-Work Assistance 528
G. Injury Management: Measurement of Effectiveness 532
H. Summary 537
Appendix 1:   Sample Occupational Health and Safety Incident & Investigation Report Form 539
Appendix 2:   Sample First Aid Record Form 547
    Chapter References 550

Chapter 14 — Occupational Health and Safety: Incident Investigation

A. Introduction 551
B. Definitions 552
C. Incident Investigation 552
D. Importance 552
E. Types of Incidents 553
F. Accident Theory 553
    1. Accident Causation Theories 554
    2. Accident Theory: The Lessons Learned 559
    3. Anatomy of Incidents 561
G. Which Incidents Should be Investigated? 563
H. Causes 563
I. Sequencing 564
    1. Incident Occurs 564
    2. Incident Report Generated 564
    3. Decision to Investigate Made 565
    4. Evidence Collected 565
    5. Investigation Scope Defined 566
    6. Investigation Leader and Team Selected 566
    7. Incident Investigation 567
    8. Investigative Procedures 569
    9. Systematic Root Cause Analysis Conducted 569
    10. Corrective Action(s) Recommended 570
    11. Incident Investigation Report Completed 570
    12. Final Report Presented 570
J. Effective Techniques 571
K. Stakeholder Communication 573
L. Measurement of Effectiveness 574
M. Calculating the Costs of Incidents 576
N. Summary 577
    Chapter References 577

Chapter 15 — Occupational Health and Safety: Risk Management and Communication

A. Introduction 579
B. OH&S Risk Management 579

TABLE OF CONTENTS

    1. Importance 580
        (a) Risk Management 582
        (b) OH&S Risk Management 583
    2. Key Roles and Responsibilities 588
        (a) Senior Management (Employer) 588
    3. Definitions 589
    4. OH&S Practitioner/Professional and Risk Manager 591
    5. Effective Techniques 592
        (a) Risk Estimation 592
        (b) Risk Evaluation 594
        (c) Risk Control 595
        (d) Implement the Plan 597
        (e) Risk Management 597
        (f) Risk Communication 597
    6. Risk Management: Quantifying OH&S Risk 598
C. OH&S Risk Communication 598
    1. OH&S Risk Communication: Importance 599
    2. OH&S Risk Communication: Effective Techniques 600
    3. Managing Hostile Situations 607
    4. Working with the Media 608
    5. OH&S Risk Communication: Additional Considerations 609
    6. OH&S Risk Communication Plan: Evaluation 610
D. Summary 610
    Chapter References 611

Chapter 16 — Marketing Occupational Health and Safety Programs; Communicating the Results

A. Introduction 615
B. Program Marketing 615
    1. Marketing Concepts 615
    2. Marketing Management 617
    3. Key Marketing Concept 617
    4. General Marketing Principles 618
    5. Build a Business Case for the OH&S Program 622
C. OH&S Communication Strategy and Plan 627
    1. OH&S Communication Goal and Objectives 627
        (a) OH&S Program: Sample Communication Goal 627
        (b) OH&S Program: Sample Communication Objectives ... 627
    2. OH&S Communication Plan 628
    3. OH&S Communication Tools 631
        (a) OH&S Program: Website 631
        (b) OH&S Program: Brochures 631
        (c) OH&S Program: Poster 632
        (d) OH&S Program: Outcome Report 633
D. Summary 633

TABLE OF CONTENTS

Appendix 1:   Sample OH&S Communication Plan 635
Appendix 2:   Sample OH&S Program Brochure 639
Appendix 3:   Sample OH&S Poster 643
Appendix 4:   Sample Management Report: Unison Contracting: GSP™
               Journey 645
               Chapter References 650

Chapter 17 — Occupational Health and Safety: Worker OH&S Training
A. Introduction 651
B. Worker OH&S Training 651
C. Worker Competency versus Workplace Complacency 652
D. Worker OH&S Training: Importance 656
E. Worker OH&S Training: Important Elements 658
F. Worker OH&S Training: Developing an OH&S Training Session .. 664
   1. Identify Worker OH&S Training Needs 664
   2. Decide the OH&S Training Priorities 665
   3. Establish the OH&S Training Objectives 665
   4. Develop a Plan for Achieving the OH&S Training Objectives 668
   5. Prepare the Teaching/Lesson Plan and Related Materials 670
   6. Deliver the OH&S Training 676
   7. Evaluate the OH&S Training 677
G. Worker OH&S Training: Measurement of Effectiveness 677
   1. Evaluation of the Entire Worker OH&S Training Program 677
   2. Evaluation of a Specific OH&S Training Course or Program 678
   3. OH&S Training: Measurement Model 680
H. Worker OH&S Training: Making it Fun 681
   1. An Industry Application: ENMAX: Rule Book Madness 681
      (a) The ENMAX Health & Safety Rule Book 681
      (b) The Birth of an Idea 682
      (c) Game Play Rules 682
      (d) Game Finals 684
      (e) What Comes Next? 684
      (f) Safety Can Be Exciting 684
   2. An Industry Application: OH&S Crossword Puzzles Online 685
   3. An Industry Application: Interactive Safety Education 686
   4. An Industry Application: Online Interactive OH&S Training .. 687
I. The Value of OH&S Training Programs 687
J. Summary 689
Appendix 1:   Sample New Employee OH&S Orientation Checklist 691
               Chapter References 692

Chapter 18 — Occupational Health and Safety: Best Practices
A. OH&S Best Practices: Defined 695
B. OH&S Best Practices 695
   1. Demonstrate Strong OH&S Leadership, Commitment, and

xx

Table of Contents

	Passion	695
2.	Strive to Develop a High-performance Work System	698
3.	Build a Strong OH&S (Safety) Culture and Climate	702
4.	Incorporate Occupational Health and Safety into Corporate Business Strategies and Planning	706
5.	Coordinate OH&S Strategies, Processes, and Activities	706
6.	Create OH&S Requirements	710
7.	Create and Communicate OH&S Policies, Procedures and Practice/Performance Standards	711
8.	Protect the Confidentiality of Employee Personal Health Information	712
9.	Clarify OH&S Ownership and Responsibilities	713
10.	Promote Transformational Leadership	713
11.	Invest in Educating Upper and Line Management on Desired Management Practices	715
12.	Coordinate OH&S Staffing and OH&S Budgets	716
13.	Implement an Effective OH&S Data Management System	717
14.	Provide, Coordinate and Integrate Worker OH&S Education and Training	718
15.	Measure OH&S Performance	719
16.	Align Corporate Standards with Recognized OH&S Standards to Uniformly Measure OH&S Programs	721
17.	Develop an OH&S Communication Strategy	722
18.	Strive for Technology-driven OH&S Awareness	724
19.	Implement an Effective Disability Management Program	725
20.	Implement a Graduated Return-to-Work Program	726
21.	Link the OH&S Program and Disability Management Program	728
22.	Link the OH&S Program and Employee Assistance Program	728
23.	Link Prevention to Workplace Illness/Injury Management	729
24.	Adapt to a Changing Workplace/Workforce	730
25.	Address Workplace Health and Safety for Older Workers	731
26.	Address Workplace Health and Safety for Young/New Workers	732
27.	Support/Protect the Contingent Workforce	734
28.	Strive to be Culturally Competent and Proficient	735
29.	Conduct More OH&S Research	735
30.	Get OH&S Research Ideas to Those Who Need It	736
C.	Summary	737
Appendix 1:	OH&S Action Plan and Register	739
	Chapter References	753

Chapter 19 — Occupational Health and Safety: Supporting Disciplines

A.	Introduction	759
B.	Support from Other Disciplines: Value and Importance	759
1.	Occupational Health Professionals	759
2.	Occupational/Industrial Hygienists	759

TABLE OF CONTENTS

    3. Ergonomists 760
    4. Toxicologists 761
    5. Employee Assistance Counsellors 762
    6. Rehabilitation Therapists 763
        (a) Physical Therapists 763
        (b) Occupational Therapists 763
        (c) Vocational Therapists 764
    7. Human Resources Professionals 764
    8. Workplace Wellness Professionals 765
C. Summary 765
    Chapter References 765

Chapter 20 — Occupational Health and Safety: Ergonomics
A. Introduction 767
B. Musculoskeletal Disorders 767
C. Musculoskeletal Disorders: Causes 768
D. OH&S: Value of Ergonomics 769
E. Ergonomic Program: Purpose 771
F. Ergonomic Program: Value 773
G. Ergonomic Programs and Work Accommodation 774
H. Ergonomics and the Aging Worker 775
I. Summary 779
Appendix 1: Educational Materials – Sample #1: General Ergonomic Safety Tips for Employees 781
Appendix 2: Educational Materials – Sample #2: Industrial Ergonomics Program 787
Appendix 3: Educational Materials – Sample #3: Ergonomics Program Prevention 795
    Chapter References 796

Chapter 21 — Occupational Health and Safety: Prevention of Workplace Illness and Injury
A. Introduction 799
B. Role of Occupational Health and Safety 799
C. Role of Employee Assistance Programs in the Prevention of Workplace Illness and Injury 801
D. Role of Workplace Wellness in Disability Management in the Prevention of Workplace Illness and Injury 802
E. Positioning of the Occupational Health and Safety Program with Employee Assistance, Workplace Wellness and Disability Management Programs . 804
F. Links to Corporate Culture, Management Practices, and Human Resource Management Theories 806
    1. Stress Risk Management in the Workplace: Research by Dr. Martin Shain 807
    2. Employee Engagement: Recognition of the Importance of Personal and

Family Life 813
    (a) Family-friendly Policies 814
    (b) Options for the Family-responsive Workplace 817
G. Supportive Infrastructure for the OH&S Program 818
    1. Corporate Culture 818
    2. Policies and Procedures 819
H. Workplace Illness/Injury Prevention Strategy 820
I. Summary 824
    Chapter References 825

Chapter 22 — Toxic Work Environments: Impact on Employee Illness/Injury

by Tony Roithmayr

A. Introduction 829
B. Have Organizations Become Toxic to Human Life? 830
C. The Performance Maximizer® Model 833
D. Organizational Stressors and Health 834
    1. Explanation of the Model 834
E. Are We Treating Symptoms Instead of Root Causes? 840
F. Use of Good Performance Support Practices Is Good Business ... 842
G. Original Research 844
H. Summary 848
    Chapter References 849

Chapter 23 – Psychological Health and Safety in the Workplace

by Ian Arnold, Dianne Dyck and Joti Samra

A. Introduction 853
    1. Psychological Health Disorders: Costs 855
    2. Psychological Health and Performance in the Workplace .. 855
B. Psychological Health and Safety: Legislative Requirement 856
C. Psychological Health and Safety: Definitions 857
D. Psychological Health and Safety: National Standard 859
E. Common Psychological Health Conditions 860
    1. Mental Distress (Psychological Distress) 860
    2. Mental Injury (Psychological Injury) 860
    3. Post-Traumatic Stress Disorder 860
    4. Depression 861
    5. Bipolar Disorder 865
    6. Schizophrenia 865
    7. Alzheimer's Disease 866
F. Available Community Resources 866
    1. A Leadership Framework for Advancing Workplace Mental Health 866
    2. Guarding Minds at Work 866

TABLE OF CONTENTS

		3.	Psychologically Safe Leader Assessment	867
		4.	Workplace Strategies for Mental Health	867
		5.	Mental Health Commission of Canada	868
		6.	Health Canada	868
		7.	Mental Health Works	868
		8.	Canadian Mental Health Association	868
		9.	Centre of Addiction and Mental Health	868
		10.	Mood Disorders Society of Canada	869
		11.	The Schizophrenia Society of Canada	869
		12.	The Depression Center	869
		13.	The Panic Center	869
		14.	Beyondblue: the National Depression Initiative (Australia)	869
		15.	Mental Health Workplace Training Programs	869
		16.	Canadian Centre for Occupational Health and Safety (CCOHS)	870
		17.	Mental Injury Tool Kit	870
		18.	Stress Assess Survey	870
G.	Employer Support			870
H.	Summary			871
	Chapter References			872

Chapter 24 – Psychological Health and Safety: Practical Application in the Workplace

By Mary Ann Baynton, Dianne Dyck, Dan Steinke and Tony Roithmayr

A.	Introduction			877
B.	Psychological Health and Safety: Applied			878
	1.	Strategic Direction		880
C.	Stakeholder Roles and Responsibilities			880
	1.	Senior Management Role		880
	2.	The Front-Line Manager Role		888
		(a)	Approaching Employees	889
		(b)	Implementing Needs-based Problem Solving	892
		(c)	Coaching the Distressed Employee	893
	3.	The Trade Union Role		898
	4.	Role of Joint Health and Safety Committee		899
	5.	The Occupational Safety Professional/Practitioner Role		899
	6.	The Employee Role		899
	7.	The Co-worker Role: Helping the Troubled Employee		901
	8.	The Occupational Health Nurse (OHN) Role		904
		(a)	Promote Training for Understanding	904
		(b)	Consider all the Possibilities	905
		(c)	Medicalizing Performance Issues	905
		(d)	The Rule-out-Rule	906
		(e)	Working for the Best Results	906
	9.	The Role of Employee Assistance Program		907

xxiv

TABLE OF CONTENTS

D. Supporting the Bereaved Employee by Dan Steinke 912
   1. What Is Grief? 913
      (a) Stages of Grief 914
      (b) The New Normal 915
   2. Workplace Wellness Programs and Employee Assistance Programs: Link with Grief Management 915
      (a) Negative Impact of Grieving on the Workplace 916
      (b) Impact of Grieving on the Workplace 917
      (c) Workplace Response 917
   3. Conclusion 918
E. Summary 918
Appendix 1: An Operational Approach to Promoting Good Psychological Health 919
      Chapter References 921

Chapter 25 – Psychological Health and Safety in the Workplace: Measurement

by Mary Ann Baynton, Dianne Dyck and Tony Roithmayr

A. Introduction 923
B. Psychological Health and Safety: Assessment 923
C. Psychological Health and Safety: Workplace Evaluation 929
   1. Structural Evaluation 929
   2. Process Evaluation 932
   3. Outcome Evaluation 934
D. Summary 935
Appendix 1: GM@W Organizational Review 937
Appendix 2: GM@W Initial Scan 941
Appendix 3: Psychological Health and Safety Outcome Measurements .. 943
      Chapter References 945

Chapter 26 — Occupational Health and Safety: Workplace Wellness Strategy

A. Introduction 947
B. Health Promotion 948
C. Workplace Wellness Program: Strategy 955
D. Organizational Wellness: Key Elements 967
E. Development of a Workplace Wellness Program: An Industry Example . 968
F. Business Case for Workplace Wellness Programs 970
   1. Workplace Wellness Program: Business Case Theory 970
   2. The Business Case: From a "Business" Perspective 976
   3. The Business Case: Based on Industry Findings 978
      (a) The Ongoing Business Case 984
G. Human Resources: Role in Workplace Wellness 985
H. Summary 985
Appendix 1 987

Chapter References 988

Chapter 27 — Disability Management: Overview 991
A. What is Disability Management? 991
   1. Management-Labour Commitment and Supportive Policies 993
   2. Policies and Procedures 994
   3. Stakeholder Education and Involvement 994
   4. Supportive Benefit Programs 995
   5. A Coordinated Approach to Injury/Illness Management 995
   6. A Communication Strategy 996
   7. Graduated Return-to-Work 997
   8. Performance Measurement 997
   9. Workplace Wellness 997
B. The Cornerstones of Disability Management Programs 998
   1. Early Intervention 999
   2. Disability Claim Management 1003
   3. Disability Case Management 1005
   4. Return-to-Work Planning 1006
   5. Return-to-work Placement 1007
   6. Confidentiality 1008
   7. Documentation 1008
   8. Program Evaluation and Continuous Improvement 1009
   9. Ethical Disability Management Practice 1010
   10. Legal Compliance 1011
C. Disability Management Models 1012
D. An Integrated Disability Management Program: Value to Stakeholders 1015
   1. For the Corporation 1015
   2. For the Union 1017
   3. For the Employee 1018
E. Integrated Disability Management Program: Hype or Good Business Practice 1019
F. How to Sell an Integrated Disability Management Program to Senior Management 1022
   1. Perceived Barriers 1022
   2. How to Move Forward 1024
     (a) Step 1: Analyze Your Situation 1024
     (b) Step 2: Gather Supportive Disability Data 1025
     (c) Step 3: Demonstrate the Value 1026
G. Evolution of an Integrated Disability Management Program ... 1027
H. Summary 1028
   Chapter References 1029

Chapter 28 — Integrated Workplace Health Management
A. Introduction 1035

TABLE OF CONTENTS

B. Canada: The Current State 1036
C. Integrated Workplace Health Management 1037
    1. IWHM: Explanation 1038
    2. IWHM: Components and Linkages 1039
        (a) Approach 1039
    3. Benefit of Linkages 1040
D. IWHM: Management's Role 1042
E. Marketing the IWHM 1045
F. Summary 1049
Appendix 1:   Performance Metrics: Definitions 1051
              Chapter References 1052

Section 3 – Occupational Health and Safety: Knowledge and Skills

Chapter 29 — Occupational Health and Safety: Ethical Practice
A. Introduction 1057
B. Why Discuss Occupational Health and Safety Ethical Issues? 1057
C. Ethical Considerations 1058
D. Ethics and the Concept of Reasonableness 1060
E. Ethics and Occupational Health & Safety 1060
    1. Common Ethical OH&S Considerations 1061
        (a) Personal Beliefs versus Professional Expectations 1061
        (b) Confidentiality versus Right to Know 1061
        (c) Individual Wishes versus Family or Legal Constraints 1062
    2. Tools for Determining Ethical OH&S Practices 1062
        (a) Discussion with Supervisors 1062
        (b) Legal Counsel 1063
        (c) Self-Appraisal 1063
        (d) Resources 1063
F. Ethical Decision-Making 1063
    1. Ethical Dilemma 1063
        (a) Ethical Fitness™ Model 1064
        (b) Model for Ethical Decision-making in a Professional Situation 1066
G. Ethical Dilemma: Case Study 1067
H. Summary 1070
Appendix 1:   Board of Canadian Registered Safety Professionals: Code of Ethics 1071
Appendix 2:   American Association of Occupational Health Nurses: Code of Ethics 1075
              Chapter References 1076

Chapter 30 — Occupational Health and Safety: Legal Aspects
A. Introduction 1077

TABLE OF CONTENTS

B. Legislation that Influences Workplaces: An Overview 1077
   1. Canadian OH&S Acts 1077
   2. Workers' Compensation Acts 1078
   3. Criminal Code, Section 217 (Commonly Referred to as Bill C-45) 1080
   4. Workplace Hazardous Materials Information Systems Legislation 1081
   5. Transportation of Dangerous Goods (TDG) 1082
   6. Canadian Electrical Safety Regulatory System 1082
   7. National Fire Code of Canada 1083
   8. Building Codes and Regulations 1084
   9. The Highway Traffic Acts 1086
   10. Environmental Legislation 1087
      (a) Historical Background 1087
      (b) Canadian Environmental Law 1088
      (c) Regulatory Instruments 1089
      (d) Relevant Environmental Legislation 1091
      (e) Other Relevant Terms 1093
   11. Human Rights Legislation 1099
      (a) Discrimination 1100
      (b) Duty to Accommodate 1101
      (c) Undue Hardship 1102
      (d) Bona Fide Occupational Requirements 1102
      (e) Human Rights and Substance Abuse Policies 1104
   12. Employment Standards 1104
   13. Freedom of Information and Privacy Legislation 1105
   14. Personal Information Protection and Electronic Documents Act (PIPEDA) 1106
   15. Current Legislation: Summary 1108
C. The Impact of Changing Legislation on the Field of Occupational Health and Safety 1108
   1. Occupational Health and Safety Legislation 1108
   2. Workers' Compensation Legislation 1110
   3. Human Rights Legislation: Recent Developments 1111
   4. Canadian Human Rights and Substance Abuse Policies .. 1111
   5. Legalization of Marijuana 1112
   6. Privacy Legislation: Evolving Legislation 1112
      (a) Request for Employee Personal Information 1112
      (b) Protection of Employee Health Information 1113
      (c) Disclosure of Employee Health Information 1115
D. Psychological Health Safety: Corporate Duty of Care 1117
E. Important Legal Terminology 1118
F. Summary 1136
   Chapter References 1137

TABLE OF CONTENTS

Chapter 31 — Canadian Workplace Safety: Legislation

by Norm Keith

- A. Introduction 1141
- B. The Internal Responsibility System 1146
- C. OH&S General Duty Clauses 1149
- D. The Defence of Due Diligence 1154
- E. OH&S Due Diligence and "Reasonable Steps" 1169
- F. An Effective OH&S Management System (OHSMS) 1172
  1. OH&S Policy 1173
  2. Planning 1173
  3. Implementation and Operation 1174
  4. Evaluation: Inspections and Auditing 1175
  5. Management Review 1176
- G. Conclusion 1176
  - Chapter References 1177

Chapter 32 – OH&S: Legalization of Marijuana: Impact on the Canadian Workplace

- A. Introduction 1181
- B. Marijuana: What is it and the health effects 1181
- C. Marijuana: Detection 1184
- D. Marijuana Use: Management in the Workplace 1187
- E. Marijuana Use: Impact on the Workplace 1189
- F. Employer Response 1193
- G. Legal Response 1196
- H. Summary 1198
  - Chapter References 1198

Chapter 33 – OH&S: The Canadian Workers' Compensation System

- A. Introduction 1203
- B. The Workers' Compensation System: Intent 1203
- C. The Workers' Compensation System: Fundamental Principles 1204
- D. The Workers' Compensation System: Purpose 1204
- E. The Workers' Compensation System: How does it Work? 1205
- F. The Workers' Compensation System: Players 1205
- G. The Workers' Compensation System: Reporting Requirements 1206
- H. Impact of the WCB 1208
- I. How Are We Doing? 1209
- J. Summary 1211
  - Chapter References 1211

Chapter 34 – Occupational Health and Safety: Diversity Considerations

- A. Introduction 1215
- B. Culture 1215
- C. Other Relevant Terms 1217

Table of Contents

- D. Cultural Diversity 1218
- E. Cultural Diversity: What is the Issue? 1219
- F. Transcultural Nursing 1222
- G. Cultural Competence 1222
- H. Development of Cultural Competence 1224
- I. Cultural Competence: It's Importance 1226
- J. Cultural Competence: Other Considerations 1227
- K. Cultural Assessment 1232
- L. Recommendations for Working with People of Diverse Cultures 1234
- M. Diversity Types 1237
  - 1. The Older Worker (Over 60 Years of Age) 1237
    - (a) Recommendations for OH&S/Disability Management Professionals/ Practitioners 1240
  - 2. Women 1240
    - (a) Recommendations for OH&S and Disability Management Professionals/ Practitioners 1242
  - 3. Generations 1242
  - 4. Ethnic Groups 1243
- N. Western Culture 1247
- O. Cultural Viewpoints: Approach to Work and the World of Work 1249
- P. Cultural Diversity Management 1252
  - 1. The Organizational Approach 1253
  - 2. The OH&S Practitioner/Professional Approach 1253
- Q. Summary 1254
  - Chapter References 1255

Chapter 35 – Impact of Five Generations in the Workplace on Occupational Health and Safety Programs

- A. Introduction 1259
- B. What is a Generation? 1260
- C. What is a Generation Gap? 1261
- D. Understanding Generational Differences: The Benefits 1262
- E. The Five Generations: Descriptions 1265
  - 1. Veterans, Traditionalists, or Depression Generation 1265
  - 2. Baby Boomers 1266
  - 3. Generation X or Baby Bust Generation 1269
  - 4. Baby Boomlets (Gen YsEcho Boomers, Nexus, Nexters, and Millennials) 1272
  - 5. GenZ 1276
- F. The Four Generations: Personal and Lifestyle Characteristics 1278
- G. The Four Generations: Impact on the Workplace 1286
- H. The Four Generations: Implications for OH&S Practitioners/Professionals 1304
  - 1. The Organizational Approach 1305
  - 2. Intergenerational Conflicts 1306

TABLE OF CONTENTS

       (a)   Potential ClashPoints® 1306
       (b)   Challenges for OH&S and Disability Management Practitioners/
            Professionals 1307
   3.   Motivational and Performance Reward Differences 1309
   4.   Other Differences 1310
I.   The Five Generations: Generational Integration 1311
J.   The Four Generations: Recommendations for OH&S and Disability
   Management Practitioners/Professionals 1312
K.   Summary 1316
Appendix 1:   Multi-generational Workplaces: Effective Management ... 1317
   Chapter References 1317

Chapter 36 — Internal/External Consulting: Tips for Occupational Health and Safety Practitioners/Professionals

A.   Introduction 1321
B.   OH&S Practitioners/Professionals as Consultants 1321
C.   Consulting: What is Involved? 1322
D.   The Consulting Relationship: Its Characteristics 1323
E.   The Consulting Tool Kit 1324
F.   Six Phases of an OH&S Consulting Project 1325
   1.   Entry Phase 1326
   2.   Contracting Phase 1327
   3.   Data Gathering and Diagnosis Phase 1330
   4.   Feedback and Planning Phase 1333
   5.   Action PHase 1335
   6.   Evaluation and Disengagement Phase 1335
G.   OH&S Consulting: Role of the Internal Consultant 1337
H.   OH&S Consulting: Project Recommendations 1338
I.   OH&S Consulting: Ethical Guidelines 1339
J.   Criteria for Success in a Consulting Situation 1339
K.   For External OH&S Consultants: Suggestions for Building Your OH&S
   Consulting Business 1340
L.   Summary 1342
   Chapter References 1342

Chapter 37 — Occupational Health and Safety: Project Management

A.   Introduction 1343
B.   Request for Proposal: The Response 1343
   1.   Relevant Terms 1343
   2   Proposal Submission Information 1344
   3.   Preparation Time 1344
   4.   Proposal Details 1344
   5.   Proposal Writing 1345
C.   Finalist Interview 1349
D.   Service Agreement Contract Development 1350

TABLE OF CONTENTS

E. Project Management 1351
   1. Project Management Stages 1352
   2. What Can Go Wrong? 1354
   3. What Leads to a Successful Project? 1355
F. Project Report 1356
   1. Report Format 1357
   2. Report Contents 1358
   3. Limitations of Written Reports 1360
G. OH&S Project Report Presentation 1360
   1. Preparing the Presentation 1361
   2. Presenting 1363
H. Summary 1364
Appendix 1: Proposal to Evaluate and Enhance the XYZ Case Management Service 1367
Appendix 2: OH&S Project Presentation (Sample) 1379
   Chapter References 1387

Chapter 38 — Occupational Health and Safety: Organizational Behaviour
A. Introduction 1389
B. Organizational Behaviour: Fundamentals 1390
C. Organizations: Structure and Characteristics 1390
D. Leadership 1392
   1. Leadership: Characteristics of a Leader 1393
   2. Leadership: Determinants of Leadership 1394
   3. Management and Managers 1395
   4. The Leader Role versus the Manager Role 1396
E. Work and Work Performance 1397
   1. Individual Performance Factors 1399
   2. Leadership: Impact on Employees 1404
F. Corporate Culture and Work Attitudes 1404
G. Job Design, Goal Setting and Work Scheduling 1407
   1. Job Design, Goal Setting and Work Scheduling: Impact on Employees 1408
H. Groups and Group Work 1409
   1. Group Norms 1409
   2. Group Dynamics 1410
   3. Team Building 1411
   4. Group Roles and Tasks 1411
   5. Group Communication 1413
I. Relevance to Integrated Workplace Health Management (IWHM) 1413
J. Summary 1414
   Chapter References 1414

Chapter 39 — Occupational Health and Safety: Effective Communication
A. Introduction 1417

TABLE OF CONTENTS

B.	Communication and Its Importance	1417
C.	Communication: Types	1417
D.	Communication: Elements	1417
E.	Communication: General Principles	1418
F.	Communication: Oral Communication	1419
	1. Speaking – Conversations	1419
	2. Occupational Health and Safety – Types of Oral Communication	1424
	3. Oral Presentations	1425
G.	Communication: Written Communication	1426
	1. Technical Writing	1428
	2. Business Writing	1429
	3. Other Types of Written Communication	1431
H.	Communication: Non-Verbal Communication	1435
I.	Summary	1436
Appendix 1: Effective Writing: Tips		1437
	Chapter References	1442

Chapter 40 — Outsourcing Occupational Health and Safety Services

A.	Introduction	1443
B.	Why Outsource?	1443
	1. Decision-making Process	1443
C.	Preparation for Outsourcing	1445
	1. Development of an OH&S Program	1445
	2. Determination of the Preferred Customer-Service Provider Relationship	1446
	3. Determination of the Nature of Services Required	1447
	4. Establishment of Service Criteria	1447
	5. Determination of the Desired Performance Criteria and Measurement Techniques	1447
	6. Establishment of Desired Funding Arrangement	1448
	7. Establishment of a Desired Payment Arrangement	1449
D.	Service Provider Market Search	1449
	1. Steps	1449
	(a) Develop a Request for Proposal	1449
	(b) Select Suitable Service Providers	1450
	(c) Distribute the RFP	1450
	(d) Data Collection and Collation	1450
	(e) Analysis of the RFP Responses	1450
	(f) Selection Process	1451
	(g) Response to Bidders	1452
E.	Service Contract Development	1452
F.	Vendor Management	1453
	1. Recommended Strategies	1453
	(a) Partnering	1453

		(b)	Quality Assurance and Continuous Improvement	1453
		(c)	Performance Measurement	1453
		(d)	Service Provider Reporting	1454
		(e)	Cost-containment	1454
		(f)	Regular Meetings	1454

- G. Vendor Risk Management 1455
  1. Case Law 1456
  2. Balance between Management Rights and Employee's Rights ... 1458
  3. Considerations for Disability Management 1459
  4. Conclusion 1460
- H. Summary 1461
     Chapter References 1461

Chapter 41 — Occupational Health and Safety Practitioners/Professionals: Career Development

- A. Introduction 1463
- B. Career Streaming: Model of Career Development for OH&S Practitioners/Professionals 1463
  1. Career Plateauing 1465
  2. Career Anchors 1465
  3. Career Planning and Development 1466
  4. Career Planning Concepts 1467
     (a) Career Ladder 1467
     (b) Career Modelling 1468
     (c) Career Streaming 1469
  5. Career Streaming: Industry Application 1469
     (a) Career Streaming Principles 1470
     (b) Development Stages 1473
  6. Conclusion 1474
- C. OH&S Job Descriptions 1474
- D. Summary 1475

Appendix 1: Occupational Health Nurses Career Stream 1477
Appendix 2: Occupational Safety Practitioner/Professional Career Stream 1481
Appendix 3: Position Profile: Safety Advisor 1487
Appendix 4: Position Profile: Senior Safety Professional 1491
            Chapter References 1495

Section 4 Occupational Health and Safety: Future Concepts

Chapter 42 Future Challenges in Occupational Health & Safety

- A. Introduction 1499
- B. Professionalism 1499
  1. A Profession 1499
  2. Profession Characteristics: How Does the Field of OH&S Measure

TABLE OF CONTENTS

		Up?	1500
	3.	Professionalization	1504
	4.	The Field of Occupational Health and Safety: Prepared or Not?	1505
	5.	The Occupational Health and Safety Practitioner: Ready or Not?	1506
	6.	Summary	1507
C.	OH&S Management System: Sustainability		1507
	1.	Definitions	1508
	2.	Sustainability Model Process	1509
D.	The Contingent Workforce		1512
E.	Increased Reliance on Contractors: Implications for Employers		1513
F.	Use of OH&S Programs, Attendance Support Programs and Disability Management Programs		1514
	1.	How Big is the Problem?	1514
	2.	Why are Workers Taking Time Off Work?	1514
	3.	Presenteeism	1517
	4.	Impact on Available Labour Supply	1518
	5.	What Can Organizations/Companies Do?	1518
	6.	What Do Organizations/Companies Need to Achieve Success?	1520
G.	Measurement of Leading versus Lagging Indicators of OH&S Performance		1520
H.	Summary		1521
		Chapter References	1521
Glossary			1527
Index			1569

Introduction

OCCUPATIONAL HEALTH AND SAFETY: THEORY, STRATEGY AND INDUSTRY PRACTICE

A. INTRODUCTION

The field of Occupational Health and Safety (OH&S) is an asset to employees, employers, and unions. It is commonly cited as being "required for legislative compliance", "a good business practice", and "the right thing to do". However, the true intent of OH&S is worker health and safety — physical and psychological health and safety.

The field of Occupational Health and Safety grew out of the recognized need to keep workers safe, healthy, and productive. However, that aim was not readily embraced by industry. It involved a lengthy process of "learning by trial and error" on the part of industry, government, and labour; hence the saying: "Occupational Health and Safety law is written in blood — the blood of workers."

As well, the workplace and the nature of the work done, tend to create health problems and/or exacerbate some pre-existing worker health conditions. It took years for industry leaders to accept this fact, along with the associated social and legal responsibilities and accountabilities for worker health, safety, and well-being. Fortunately, today there is great interest in OH&S interventions and practices. In fact, those approaches have been demonstrated to be directly related to organizational health, productivity, and strong financial results.

B. CHALLENGES

Many challenges still exist in the field of Occupational Health & Safety. They stem from the evolving legislation; differences in the interpretation of the legislative decisions; gaps between employer activities and employee well-being; recognition of new "hot illnesses";[1] various ethical considerations; the misperception that "Safety" just means "physical health and safety"; the lack of accreditation and OH&S practice standards; and poorly understood methods for program evaluation.

The challenge for employers is to create a workplace system designed to keep workers psychologically and physically healthy and safe. To achieve this goal, employers must create a work system that informs, energizes, motivates, equips, enables, and rewards workers to consistently work safely. Working safely should be the "path of least resistance" and the "norm" within a workplace.

---

[1] Many illnesses, which in the past were not compensable, have now, become recognized as being compensable under employee benefit plan coverage. The result is that employees can be off work with illnesses that are ill-defined and poorly understood in terms of case management techniques. Some examples are chronic fatigue syndrome, fibromyalgia, multiple chemical sensitivities, Epstein-Barr syndrome, and post-COVID-19 syndrome.

The outcome is cultural change — a cultural change that enables workers and the organization to realize positive returns on their efforts. Once achieved, the employer must monitor the workplace conditions to ensure that the work climate and work environment continues to support a strong safety culture. For more information on a systems approach to workplace health and safety, refer to Chapter 6, "Occupational Health and Safety: Leadership and Commitment" and the Great Safety Performance Model.

This book initially was the result of the encouragement received from OH&S Practitioners/Professionals to produce an "OH&S reference book" that contains the information and tools needed for them to practise ethically, legally, and competently. Having picked up that gauntlet, this text is now in its fourth edition and is viewed as a valuable educational resource.

C. INTENDED AUDIENCE

My intention in writing this book was to create an everyday resource for the various players in the OH&S field. Those who may find this book relevant to their practice or area of work include:

- employers;
- Occupational Safety Practitioners/Professionals;
- Occupational Health Nurses;
- Occupational Health Physicians;
- Human Resources Practitioners/Professionals;
- corporate legal counsel;
- Employee Assistance Program counsellors;
- Return-to-work Coordinators;
- union leaders;
- physiotherapists;
- occupational therapists;
- vocational therapists;
- management personnel;
- insurance carrier staff;
- educators;
- students; and
- arbitrators and judges.

In Canada, few comprehensive OH&S books have been written. The books that do exist focus on the various aspects of OH&S law or on the specific aspects of Occupational Health and Safety, rather than on the practice techniques and tools for competent OH&S practice. This book describes how to design and implement an

INTRODUCTION

effective OH&S Management System (OHSMS) and OH&S Program inclusive of a Psychological Health and Safety Management System (PHSMS); who should be involved; the role of each player; the legalities/ethics involved; how to market the program; the relationship with other corporate programs; and how program success or failure can be measured and communicated. Additionally, OH&S practice techniques such as hazard management; worker training; emergency preparedness and response planning; incident investigation; and injury management are provided. One unique aspect is the inclusion of the "tools of the trade", namely program marketing; internal/external consulting; career development; outsourcing OH&S services; effective communication; and OH&S project management.

As a result, care has been taken to craft the contents of this book so that they can be widely understood and broadly applied. As with any dynamic field, what is written today will quickly become dated; but this textbook is an attempt at keeping OH&S Practitioners/Professionals abreast on the OH&S basics.

D. FORMAT

The reader will note that in this text, when a key OH&S term is used for the first time, it is presented in bold print. This denotes that a definition of that term is available in the Glossary.

The text has been arranged so that the reader is first presented with a "macro-level" view of the OH&S theories and concepts; followed by many OH&S strategies and practical applications of the various constructs and supportive programs; and ending up with the knowledge and skills that OH&S Practitioners/Professionals require to effectively practise now and in the future.

Attention is paid to proactive as well as reactive OH&S strategies. To illustrate how these strategies can be operationalized, many industry examples and practices are provided.

Lastly, specific OH&S skills, tools, and challenges are discussed. The intent is to provide OH&S Practitioners/Professionals with career development and career enhancement knowledge and tools.

E. STRUCTURE

The content of this book is organized so that OH&S theory precedes the delivery of various OH&S strategies and industry practices. The last few chapters are dedicated to the development of OH&S Practitioner/Professional skills and careers. The book is divided into four distinct sections:

(a) Section 1: Occupational Health and Safety: Theory

Chapter 1 provides an historical perspective of the field of Occupational Health and Safety in Canada and the United States (U.S.). In addition to defining the term, "Occupational Health and Safety", it describes how the field evolved; the differences between the Canadian and U.S. OH&S systems; the various accident theories and models; organizational design features and structures; and the concepts of external and internal responsibility systems. As well, Canada's current OH&S system — federal and provincial, is explained.

Chapter 2 addresses the rationale for companies/organizations having a robust and highly effective Occupational Health and Safety Management System (OHSMS) in terms of the value offered and the business return on investment. It describes the OHSMS elements and their contributions to worker safety; explains the OHSMS framework and infrastructure; identifies and compares the various OHSMS Standards; describes how an OHSMS tends to evolve; and discusses how to evaluate an OHSMS through the use of techniques such as auditing, benchmarking, trending, establishing the return on investment, and using a balanced score card and perception surveys.

The roles and responsibilities of the various stakeholders in an OH&S Program are presented in Chapter 3. Because companies use unique approaches to meet their OH&S requirements, no two OH&S Programs are structured the same. Although they may contain the same elements, how each element is implemented differs. As a result, role confusion occurs. Questions like: "Who does what?"; "Where do my responsibilities start and stop?"; "How can we work successfully together to deal with all the relevant issues involved?"; "Who should be responsible and accountable for workplace health and safety?"; and many more exist. Chapter 3 attempts to deal with many of these questions. However, one caveat is that each organization has different stakeholders and players. Their exact roles and responsibilities in an OH&S Program vary depending on the available organizational resources and internal/external expertise. These variations are to be expected — the key is to clarify in each situation, who does what, when, and how.

Chapter 4 introduces the Occupational Health (OH) Program, its components and services. The value of an OH Program is explained along with the legal requirement for upholding employee personal health and medical information confidential. The concepts of fitness-to-work, fit-for-duty, job demands analysis, medical monitoring, health surveillance, and program evaluation are presented, along with programs such as hearing conservation, respiratory conservation, and first aid. Having an awareness of the OH Program and what it offers, assists the OH&S Practitioner/Professional with their mandate in the workplace.

(b) Section 2: Occupational Health and Safety: Strategies and Industry Practices

The importance of a supportive infrastructure for an OH&S Program — structure, process, and recognizable outcomes — is fully described in Chapter 5. As well, some examples of the recommended elements are included. To make this chapter even more useful, an industry example of some aspects of an OH&S Program manual is provided. The intent is to assist the reader in exploring some of the areas that should be addressed in an OH&S Program manual.

Chapter 6 deals with OH&S leadership and commitment. It describes the key aspects of OH&S leadership and commitment, and the rationale for strong OH&S

leadership and commitment. The Great Safety Performance Model is a unique approach to the development of a strong safety culture and to the measurement of OH&S leadership and commitment.

Prepared by Tony Roithmayr, Chapter 7, "Safe Workplace = Great Workplace: Building a Sustainable Culture of Safety", addresses the importance of the "software" of Safety: the essential elements that support the effectiveness of Safety "hardware". New directions in OH&S practice are presented, as are the measurement and analysis of a workplace performance system. Inherent in this approach is to present and define a new role for OH&S Practitioners/Professionals.

OH&S (Safety) culture and OH&S (Safety) climate are two terms described in Chapter 8. Their application from an industry perspective is described thanks to the Ontario Public Services Health and Safety Association (PSHSA); and the New Zealand Electricity Engineers' Association (EEA), Orange Umbrella®, Trustpower Ltd., and Unison Contracting Ltd. Special thanks is extended to Sandra Lassowski for allowing her thesis material to be presented.

Chapter 9 addresses hazard identification, assessment, and control (hazard management). It explains the importance of hazard management; some relevant techniques; the use of specific tools such as worksite inspections; the value of worker involvement; and the measurement of hazard management.

The hazards and benefits associated with new technology in the workplace, are addressed in Chapter 10. Automation, artificial intelligence (AI), machine learning, digitization, digitalization, robotics, 3-D printers, mobile devices, CCTV cameras, GPS systems, magnetic resonance imaging (MRI) biometric devices, and Segway scooters are discussed.

Chapter 11 outlines OH&S Program practice/performance standards and rules, their value, and the need for enforcement of these standards and safety rules. Effective techniques for communicating workplace standards and rules as well as the measurement of compliance with workplace standards and rules are also discussed.

Emergency preparedness and response planning, a legally required element, is dealt with in Chapter 12. The aspects that are described include:

- the importance of emergency response programs and plans;
- the design and elements of an Emergency Response Plan (ERP);
- the key roles and responsibilities;
- effective techniques for emergency response;
- sample Emergency Response Plans (ERPs);
- a sample Fire ERP;
- a sample Chemical Spill/Emission Release Response Plan;
- a sample Bomb Threat Response Plan;

- a sample Terrorism Protection Plan;
- a sample Pandemic Flu Plan; and
- measurement of the effectiveness of the ERP.

Injury management, another legally required element of an OH&S Program, is discussed in Chapter 13. The topics covered include:

- importance of injury management;
- injury management preparedness;
- key roles and responsibilities;
- effective techniques for injury management, including Disability Management and return-to-work programs; and
- measurement of the effectiveness of injury management.

A fourth legally required element of an OH&S Program is incident reporting and investigation. Chapter 14 deals with the importance of incident reporting and investigation; the types of incidents; Accident Theory; the key roles and responsibilities; the effective techniques for incident investigation (systematic analysis); and the measurement of the effectiveness of a company's incident investigation techniques.

OH&S Risk Management and Communication is a relatively hot topic. In Chapter 15, the author discusses the importance of OH&S risk management and communication; the key stakeholder roles and responsibilities; some effective techniques; how to quantify OH&S risk; and how to measure the effectiveness of a company's OH&S risk management and communication approaches.

Once an OH&S Program is developed, it is important to "tell and sell" stakeholders about and on the program. Chapter 16 addresses effective program marketing and communication techniques, along with industry examples of communication tools.

Worker OH&S education and training are covered in Chapter 17. Topics such as the importance of worker training; worker competency; worker complacency; the psychology of worker training; worker training programs and an example of a training matrix; effective techniques for worker training; ways to make safety training fun; and the measurement of the effectiveness of worker training programs are covered. The provision of worker OH&S training is legally required under the OH&S Acts in all Canadian jurisdictions.

Over the years, there has been increased interest in the current best practices in the OH&S field, which are summarized in Chapter 18. The concepts presented include the need to:

- demonstrate strong OH&S leadership, commitment, and passion;
- strive for a high-performance work system;
- build a strong OH&S culture and climate;

Introduction

- incorporate OH&S planning into corporate business strategies and planning;
- coordinate OH&S strategies, processes and activities;
- create OH&S requirements;
- create and communicate OH&S policies, procedures and practice/performance standards;
- protect the confidentiality of employee personal health information;
- clarify OH&S ownership and responsibilities;
- promote transformational leadership;
- invest in educating Upper and Line Management on desired management practices;
- coordinate OH&S staffing and OH&S budgets;
- implement an effective OH&S data management system;
- provide, coordinate, and integrate worker OH&S training;
- measure OH&S performance;
- align company standards with recognized OH&S standards to uniformly measure OH&S Programs;
- develop an OH&S communication strategy;
- strive for technology-driven OH&S awareness;
- implement effective Disability Management and return-to-work programs;
- implement an effective Disability Management Program;
- implement a Graduated Return to Work Program
- link the OH&S Program with the Disability Management Program;
- link the OH&S Program with the Employee Assistance Program (EAP);
- link prevention to workplace illness/injury management;
- adapt to a changing workplace/workforce;
- address workplace health and safety for older workers;
- address workplace health and safety for young/new workers;
- support and protect the contingent workforce;
- strive to be culturally competent and proficient: appreciate the impact cultural diversity and four generations in the workplace have on OH&S programming;
- conduct more OH&S research; and
- get OH&S research to those who need it.

For an OH&S Program to be effective, it needs competent Practitioners/

Professionals, as well as support from various disciplines, including:
- OH Professionals (*i.e.*, nurses and physicians);
- industrial hygienists;
- ergonomists;
- Environmental Practitioners/Professionals;
- toxicologists;
- Workplace Wellness Practitioners/Professionals;
- Human Resources Practitioners/Professionals; and
- physiotherapists/massage therapists.

(c)   Section 3: Occupational Health and Safety: Supportive Programs

The field of OH&S relies on the technical expertise of OH Professionals, Occupational/Industrial Hygienists, Ergonomists, Toxicologists, Employee Assistance Counsellors, Rehabilitation Therapists, and Human Resources and Workplace Wellness Practitioners/Professionals. Chapter 19 addresses the importance of this multidisciplinary support.

With musculoskeletal disorders and injuries ranking the major cause of Workers' Compensation claims, companies must include ergonomics in their OH&S services. It is part of their legal commitment to undertake hazard assessment and manage identified hazards. Chapter 20 explains musculoskeletal disorders and their causes; the value of an ergonomic program; what an ergonomic program entails and how to use ergonomics when accommodating an employee post-illness/injury, or when supporting aging workers.

Chapter 21, "Occupational Health and Safety: Prevention of Workplace Illness and Injury", examines the role of Employee Assistance Programs (EAPs) in the prevention of workplace illness and injury; the function of a Disability Management Program; the role of a Workplace Wellness Program; the links between workplace Occupational Health and Safety, and management practices and human resource management theories; stress risk management in the workplace; and research by Dr. Martin Shain. By examining the interrelationship between the OH&S Program and other company programs such as the Disability Management Program, Employee Assistance Program, Workplace Wellness Program, and Human Resources Program, opportunities for prevention are identified. Examples of possible synergies are provided, along with a sample Workplace Wellness model.

The impact that management practices can have on employee health and wellness is discussed in Chapter 22. The term "toxic workplace" is used by Tony Roithmayr to depict those workplaces in which a high number of organizational stressors exist — a place where employee performance is compromised. The outcome is poor employee morale, increased employee absenteeism, high staff turnover, and reduced productivity. This chapter was co-authored with Tony Roithmayr, a specialist in helping organizations develop high performance cultures.[2]

Psychological health and safety is a legal requirement that Canadian employers are bound to uphold under the provincial OH&S Acts. In Chapter 23, this topic is introduced by Dr. Ian Arnold and Dr. Joti Samra. The definition of psychological health and safety is provided along with practical suggestions for workplace implementation. Further to the topic of psychological health and safety, practical application tips are provided in Chapter 24 by Mary Ann Baynton, Dianne Dyck, Tony Roithmayr and Dan Steinke. Additionally, the effective measurement of a psychological health and management system is described with the assistance of Mary Ann Baynton and Tony Roithmayr, and the support of Guarding Minds @ Work (Chapter 25).

Workplace Wellness Programs (WWPs) are industry-based health promotion programs designed to promote and foster employee health and well-being. Using a comprehensive approach, WWPs can be linked with other company programs like the OH&S Program, Disability Management Program, and Employee Assistance Program. Chapter 26 provides details on the development of a workplace wellness strategy and implementation of a WWP.

In Chapter 27, the topic of disability management is covered. It provides an overview in terms of what disability management is, why it is important, the current models in place, and the value that a Disability Management Program can offer various stakeholders. As with any business program, a Disability Management Program must add value to an organization. Chapter 24 discusses how to demonstrate that value to Senior Management and union leaders in order to garner their support and endorsement.

An integrated approach to managing health in the workplace is described in Chapter 28. This upstream approach goes a long way towards positioning the organization/company to implement an enlightened approach to employee health, safety, and well-being. The best practice is to prevent, instead of mitigating workplace losses.

(d)   Section 4: Occupational Health and Safety: Knowledge and Skills

The ethical and legal aspects involved in OH&S are presented in Chapters 29, 30, and 31. Inevitably, when dealing with people and organizational issues, ethical considerations need to be addressed. In Chapter 29, Dr. Bonnie Rogers begins by discussing the key ethical theories and their implications. Jane Hall examines ethics in terms of disability management. The topics of ethical dilemmas and ethical decision-making are described. In Chapter 30, information on the legislation relevant to the field of OH&S and the issues surrounding changing legislation are presented by contributors such as Sharon Chadwick and Kristine Robidoux. To assist OH&S Practitioners/Professionals, important legal terms are included at the

---

[2] Online: http://www.performance-bydesign.com/.

end of that chapter. The key concepts in Canadian workplace legislation are presented by Norm Keith (Chapter 31) — the Internal Responsibility System, the External Responsibility System, General Duty Clauses, due diligence, and reasonable steps that employers can take.

Chapter 32 explores the impact of legalizing marijuana on the workplace. Marijuana, its health effects, detection, and impact on the workplace one-year post legalization are addressed. As well, suggestions on how Canadian employers can best address worker impairment are provided.

The Canadian Workers' Compensation System is often confusing for employers, employees, and union leaders to understand and navigate. Chapter 33 explains the intent, fundamental principles, purpose, functioning, and reporting requirements of the various provincial Workers' Compensation Boards.

The impact of cultural diversity in the workplace is evident in the area of Occupational Health and Safety. Chapter 34 explores cultural diversity and cultural competence. Related to cultural diversity is the impact of five generations in the workplace — a phenomenon that is impacting today's workplaces (Chapter 35).

On a daily basis, OH&S Practitioners/Professionals provide internal/external consulting services to their customers/clients and external agencies. In Chapter 36, consulting principles, tools, and tips are described.

OH&S project management is addressed in Chapter 37. Project management is a discipline unto itself. Many OH&S Practitioners/Professionals tend to undertake OH&S projects with little or no formal project management training or experience. Given that over 60% of projects tend to "miss the intended" performance targets,[3] this chapter is included to increase knowledge levels of effective project management techniques.

Having a working knowledge of organizational behaviour enables the OH&S Practitioners/Professionals to effectively function within the workplace system. The topics covered in Chapter 38 include:

- the fundamentals and characteristics of organizational behaviour;
- leadership;
- work and work performance;
- corporate culture and work attitudes;
- job design, goal setting and work scheduling;
- groups and group work; and
- relevance to Integrated Workplace Health Management.

Effective communication, a critical skill for OH&S Practitioners/Professionals, is

---

[3] CompuCom Systems (2003). Project Management Best Practices. CompuCom Systems Inc.

described in Chapter 39. Oral, written, and non-verbal communication are defined along with industry examples of how each skill can be applied.

Chapter 40 offers information on outsourcing OH&S services and how this approach to OH&S servicing can be achieved. It explores issues such as the reasons for outsourcing; what internal preparation is required; conducting a market search; and establishing performance measures, contract development, and vendor management. This is a more recent aspect of OH&S programming and an area that remains poorly understood.

For the OH&S Practitioners/Professionals, Chapter 41 on career development, is designed to be informative and thought-provoking. Few OH&S Practitioners/Professionals ever purposely plan out their career paths. Rather, they tend to end up in the field of Occupational Health and Safety by "happenstance". This chapter provides career planning and development information for the OH&S Practitioner/Professional, along with an appreciation of the importance of position/job descriptions. The material presented can assist OH&S Practitioners/Professionals in explaining their value to the companies/organizations they serve.

(e) Section 5: Occupational Health and Safety: The Future

The last chapter, Chapter 42, provides a discussion on some of the current and future challenges in the field of Occupational Health and Safety, namely:

- OH&S professionalism;
- enhanced legal expectations of employers;
- sustainability of the OH&S Management System;
- the contingent workforce;
- increased reliance on contractors and the implications for employers;
- use of Occupational Health and Safety, Attendance Support and Assistance, and Disability Management Programs to mitigate employee absenteeism; and
- the need to measure leading versus lagging indicators of Safety performance.

The intent of this chapter is to foster thought about the future of the field of Occupational Health and Safety in Canada, the U.S., and New Zealand; and how to effectively address these critical issues.

(f) Section 6: Occupational Health and Safety: Glossary and Index

To assist the reader, a Glossary containing relevant OH&S definitions has been included at the end of this book. As mentioned earlier, each term within the Glossary is denoted by the wording/phrase being bolded when it is first used within this textbook.

F. SUMMARY

As noted, this OH&S programming resource is designed for "front-line use". The hope is that readers will find the information useful and readily applicable to their practice areas. As well, if this book can, in any way, clarify the cloudy topic of OH&S Program, then this author's aim will be achieved.

In essence, the goal of this book is to promote the field of Occupational Health and Safety, and the practitioners/professionals working within it. The message to these OH&S Practitioners/Professionals is to:

> *Simply become who you are and remember from where you came. Become more than anyone dreamed, and be an inspiration to many. . .*
>
> Author Unknown

CHAPTER REFERENCES

CompuCom Systems (2003). *Project Management Best Practices*. CompuCom Systems Inc.

T. Roithmayr, *Performance by Design* (2000), online: http://www.performance-bydesign.com/.

Section 1
Occupational Health and Safety: Theory

Section I
Occupational Health and Safety: Theory

Chapter 1

OCCUPATIONAL HEALTH AND SAFETY: HISTORICAL PERSPECTIVES

A. WHAT IS OCCUPATIONAL HEALTH AND SAFETY?

Occupational Health and Safety (OH&S) is a cross-disciplinary area focused on protecting the health and safety of people engaged in work or employment. As a secondary effect, OH&S may also protect employers, customers, suppliers, and members of the public who may be impacted by the workplace environment.[1]

According to the International Labour Organisation (ILO), **Occupational Health** aims at:[2]

> . . . the promotion and maintenance of the highest degree of physical, mental and social well-being of workers in all occupations; the prevention amongst workers of departures from health caused by their working conditions; the protection of workers in their employment from risks resulting from factors adverse to health; the placing and maintenance of the worker in an occupational environment adapted to his physiological and psychological capabilities; and, to summarize, the adaptation of work to man and of each man to his job.

Occupational Safety is devoted to the identification, evaluation, and control of those hazards and stressors arising in, and from, the workplace which may cause losses, *i.e.*, property/equipment damage, product loss, worker injury/illness, intellectual losses, and/or financial losses. Loss control is vital to enhancing worker productivity and organizational profitability. In short, "minimizing loss is as much an improvement as is the maximization of profit".[3]

In this book, the term "Occupational Health and Safety" is used to include both disciplines — Occupational Health and Occupational Safety. Hence, an **Occupational Health and Safety Program** is a complete system that ensures high safety

---

[1] Wikipedia, Occupational Health and Safety entry, online: http://en.wikipedia.org/wiki/Occupational_safety_and_health (date accessed: February 28, 2020).

[2] *Ibid.*; see also the ILO/WHO (1950). *Joint ILO/WHO Expert Committee on Industrial Hygiene Report*, 1st Session, August 28 to September 2, 1950 (Geneva: International Labour Office).

[3] L. Allen, quoted in F. Bird and G. Germain, *Loss Control Management: Practical Loss Control Leadership* (Loganville, GA: DNV, 1996), at 27.

standards throughout the organization's/company's operations and:

- reflects a strong commitment from management towards workplace health and safety — both physical and mental;
- encourages worker commitment towards workplace health and safety;
- helps workers understand their responsibility in preventing workplace incidents;
- promotes a work environment characterized by the elements required to work safely, namely, ensures that employees know how to work safely, are able to work safely, are equipped to work safely and are motivated to work safely;
- values strong management-employee interactions (respect, honesty, trust, cooperation, open communication and cultural acceptance);
- provides the necessary infrastructure; and
- enables program evaluation and continuous improvement.

An **Occupational Health and Safety (OH&S) Program** is a defined action plan designed to prevent incidents and occupational diseases. Some form of an OH&S Program is required under OH&S legislation in most Canadian and American jurisdictions. At a minimum, an OH&S Program must include the elements required by the applicable OH&S legislation. Because organizations/companies differ, an OH&S Program developed for one organization/company cannot necessarily be expected to meet the needs of another.

B. OH&S: THE CANADIAN AND AMERICAN EXPERIENCE

How did workplace health and safety come into being? In the sections that follow, the history of OH&S in Canada and the United States will be discussed, as well as the influence that the various management theories had on the modern field of Occupational Health and Safety.

1. Background History: The 1800s

Prior to the 1800s, human losses due to workplace injuries/illness were recognized, but the economic and health effects of the infectious diseases of the day overshadowed industrial efforts to address workplace issues. As a result, the prevention efforts of disease or conditions such as Coal Miner's pneumoconiosis (Black Lung), Chalicosis (Stone Cutter's Lung), Byssinosis (Brown Lung), chimney sweep's Sooty Warts (cancer of the scrotum), Silicosis (Grinder's Disease and Potter's Rot), *etc.*, were limited or non-existent.

In the 1800s, people began to be aware of the relationship between alcohol abuse and workplace incidents and injuries. Alcohol abuse was viewed as a self-inflicted state, and the victim of a related workplace incident was perceived as being responsible for his/her injury. This concept of "victim blaming" was pervasive because it aligned well with many religious doctrines, the military chain of command principle, and industrial incident investigation findings of the day. The military chain-of-command principle holds that workplace incidents occur when people fail to follow orders; the industrial incident investigation findings of the day indicated that more than 90 per cent of incidents were due to human error, with the rest being viewed as an Act of God.[4]

2. The Industrial Revolution

The Industrial Revolution brought with it a huge movement of workers from rural settings and occupations (cottage industries), to urban settings and factories (industrialized environments). The use of mechanical energy, machines, and new production methods took over the traditional work practices that involved human skills and strength. Workplaces were organized into larger work settings such as mills, factories, forgeries, mines, and railroads. Workers were arranged into workgroups and their work supervised.

Along with the Industrial Revolution, occupational injury and disease flourished. Unskilled labourers were thrown into a highly mechanized economy in which they had little control over their working conditions and work environments. Mechanization led to the creation of many new workplace hazards. The introduction of new machines and processes far exceeded any advancement in workplace safety. Few, if any, hazard controls existed, or were even sought. As usual, there was a considerable lag between the presence and recognition of the problem, and the development of a workable solution.

Deplorable and wretched by today's workplace health and safety standards, people accepted their working conditions and rationalized that workplace incidents and work-related illness would occur based on:

- the belief that the work was dangerous and workers/employers would have to expect that incidents were inevitable — in essence, worker injury/illness was part of the job;
- employer disregard for human life;
- the uncontrolled dangers within the workplace; or
- carelessness by workers who failed to adhere to the known safety precautions.

However, with time, people began to question the needless loss of human life, viewing it as morally and socially unjustified. Management and workers were identified as being accountable for the poor state of workplace affairs. From an economic standpoint, work incidents severely and negatively impacted company operational efficiency, productivity, and profitability. They also led to negative social effects because an injured worker could not take care of himself or his family and would have to rely on charitable groups to survive or would go without.

---

[4] M. Guarnieri, "Landmarks in the History of Safety" (1992) 23(3) J. Safety Res. 151-158.

In terms of obtaining income replacement from the employer, the injured worker had little recourse. To resolve disputes, the employer and employee relied on the legal system which was a lengthy and costly venture that rarely ended in justice for either party. The employee seldom succeeded in being compensated for his/her injury or illness. The main reason for this fact was the existence within the law of what is known as the **Holy Trinity**, which upheld that:

1. The employer was not responsible for employee injury if the employee contributed in any way to the incident or event.
2. The employer was not responsible for employee injury if a co-worker contributed in any way to the incident that injured the employee.
3. When accepting a job, an employee also accepted the risks of the job and so should have planned for them.[5]

Should the employee successfully sue the employer, the chance of receiving payment was minimal. Many an employer would become financially insolvent, be bankrupted by the court costs, or bankrupted by other related legal costs.

3. Birth of the Workers' Compensation Systems

Between the late 1800s and early 1900s, the concept of a workers' compensation system began in Germany, Great Britain, and the U.S. In Germany, Chancellor Otto Von Bismarck introduced a compulsory, state-run, incident compensation system (1884-1886). This initial workers' compensation system was collectively financed by employers with the employees providing individualized financial contributions.

In Britain, the workers' compensation system was developed because the sheer number of lawsuits related to workplace injury-related disputes between workers and their employers were overtaxing the courts. The new system addressed the Holy Trinity and lessened the burden of proof required from workers to prove their employers liable.

In the U.S., between 1908 and 1915, several American states introduced compensation legislation. The State of Washington enacted an exclusive mandatory system based on collective liability. As workers' compensation was given state jurisdiction, the United States developed a variety of Workers' Compensation Boards (WCBs), mandatory insurance, self-insurance, and various hybrids of each.

Workers' compensation in Canada had its beginnings in the province of Ontario in 1886. Ontario adopted the British Workers' Compensation system, which unfortunately did not last long. By 1897, a new compensation law, based on "no-fault" principles was introduced. Although it placed compensation responsibilities on the individual employers, employees that previously had no protection if

---

[5] W. Fox-Decent, "Workers' Compensation: Past, Present and Future – An Historical Overview" (2002).

their employer was financially unable to pay, were financially assisted.[6] Then in 1910, Justice William Meredith was appointed to the Ontario Royal Commission to study workers' compensation. His final report, known as the Meredith Report, was produced in 1913.

The Meredith Report outlined a unique trade-off in which workers relinquished their right to sue their employer in exchange for injury compensation benefits. Meredith advocated for "no-fault" insurance, collective liability, independent administration, and exclusive jurisdiction. The proposed system would exist at arm's length from the government and be shielded from political influence, allowing only limited powers to the Minister responsible. The outcome was a workers' compensation system that remains in effect today. For more information on the Canadian Workers' Compensation System, refer to Chapter 33, "Occupational Health & Safety: The Canadian Workers' Compensation System".

4. The Investigational Era

Between 1915 and 1930, much was done in workplace safety within the time known as the Investigational Era. The belief that incidents could be prevented, began to emerge.

As insurers, the various Canadian and American Workers' Compensation Boards (WCBs) were responsible for and interested in workplace injuries and their related costs. They funded research in this area, the focus of which was to determine how to reduce workplace injuries and their costs, and to determine, "What was wrong with people that caused them to injure themselves?"[7] Some safety themes surfaced:

- Incidents can be prevented.
- Education is vital because people's behaviours cause incidents.
- The injured worker is usually at fault ("victim blaming").

During this time, much was written on the topic of workplace safety and published by groups like the National Safety Council, the Safety Institute of America, and individual researchers. By individualizing the problem and focusing attention on how to reshape employees who were deemed "accident prone" or "troubled", researchers began investigating the psychological factors of incident prevention. However, with time, these theories failed to translate into a reduction in the rate of workplace incidents.

5. The 1950s and 1960s

In the 1950s and 1960s, a new viewpoint on workplace safety developed. Individual researchers independently arrived at the conclusion that the practice of "blaming the victim" and the various definitions of the term "accident" were stumbling blocks to

---

[6] *Ibid.*

[7] M. Guarnieri, "Landmarks in the History of Safety" (1992) 23(3) J. Safety Res. 151-158.

understanding the real causes of workplace losses. Gibson and Haddon[8] proposed that the focus of accident causation should be on the outcome of the injury. Stapp and Dehaven independently started to investigate the effects of energy forces on the human body.[9] Haddon (1957) proposed that the prevention of injuries depended on the control of energy forces and their negative impact(s).[10] Gibson (1961) described the relationship of mechanical, chemical, thermal, and electrical energy to human injury. He also proposed that safety education should concentrate on children — a very novel concept, even by today's standards.[11]

The outcome was a shift in research focus from behavioural psychology approaches to those from the fields of engineering and epidemiology. As well, the fields of engineering and medicine became linked, allowing for a merge of ideas about engineering controls and a reduction of human injury. The result was that the principles of physics were applied to injury prevention. Test dummies and computer modelling were used to simulate the negative effects of energy releases on the human body. Most importantly, the term "accident" started to disappear from modern scientific and engineering use (1985).[12] A focus on the mechanism of injury took hold, instead of "blaming the victim".

6. Occupational-based Assistance for Workers

Although the provision for occupational-based assistance for workers started in the 1800s with the growth of Welfare Capitalism in North America, occupational interventions gained a strong foothold in the 1940s and 1950s with the Occupational Alcoholism Programs (OAPs).[13] As policies and procedures changed in Canada and the U.S. in the 1970s, an increase in occupational-based assistance for workers occurred. The focus was on the "troubled or maladjusted workers". Welfare Capitalism, OAPs, and later, Employee Assistance Programs (EAPs), held that workers needed to "be fixed", or "moulded to some specific conventional form". Little attention was paid to the impact of work and workplace dynamics on the person.[14] For many years, this viewpoint was sustained, and only now, does an understanding of the reciprocal relationship between man, the family, work, and the workplace exist.[15]

---

[8] *Ibid.*
[9] *Ibid.*
[10] *Ibid.*
[11] *Ibid.*
[12] *Ibid.*
[13] R. Csiernik, *Wellness and Work* (Toronto: Canadian Scholars' Press, 2005).
[14] *Ibid.*
[15] L. Duxbury and C. Higgins, "Voices of Canadians: Seeking Work-life Balance", Human Resources and Social Development of Canada Report (2003), online: http://publications.gc.ca/collections/Collection/RH54-12-2003E.pdf (date accessed: February 28, 2020).

During the 1980s and 1990s, workplace health promotion appeared. From that field of endeavour came the concept of workplace wellness and its business advantages. Although both concepts promote a proactive approach to workplace health and safety, the same old viewpoints remained, namely:

- individualization of problems;
- perception of the "troubled worker" and the "accident-prone person";
- the injured worker being perceived as "flawed in some way"; and
- the need to mould workers to better fit the workplace.[16]

7. Summary

Over the last century, some of the notable advancements in Occupational Health and Safety include a growth in workplace health and safety knowledge; increased workplace hazard controls; more factual and objective information on loss-control and incident causation; more emphasis on analyzing potential losses; and a gradual decline in workplace incidents. With progress, came many key common standards and regulations in the areas of:

- hazard assessment, identification and control;
- hazard communication;
- confined space entry;
- lock out/tag out procedures;
- provision, use, maintenance, and replacement of personal protective equipment (PPE);
- respiratory protection;
- hearing protection;
- exposure-prevention to blood-borne pathogens;
- exposure-prevention to harmful agents like asbestos, silica, radiation, lead, mercury, formaldehyde, *etc.*;
- industrial hygiene standards;
- chemical exposure limits;
- mining standards;
- electrical safety standards;
- control of workplace emergencies;
- worker exposure and medical record standards;
- injury management standards;
- whistleblower regulations;

---

[16] R. Csiernik, *Wellness and Work* (Toronto: Canadian Scholars' Press, 2005).

- working alone protection;
- workplace violence protection;
- support for injured workers;
- return-to-work provisions for injured workers; and
- psychological health and safety in the workplace.

This list is by no means exhaustive. Rather, it was included to indicate the nature and amount of progress made. In Chapter 42, "Future Challenges in Occupational Health and Safety", the future challenges discussed provide an indication of how the "OH&S finish line" keeps changing, and how far the field of Occupational Health and Safety has yet to go.

8. History of OH&S in the United States of America

In the U.S., OH&S reforms advanced and were often triggered by major disasters such as the 1911 fire in New York's Triangle Waist Factory (Sidebar). The Triangle Waist Fire tragically illustrated that the fire inspections and precautions of the day were woefully inadequate. Workers recounted helpless efforts to open the exit doors, only to learn that they were deliberately locked. Owners frequently locked the factory's exit doors to discourage workers from stealing materials. Other workers waited at the windows for rescue workers to arrive, only to discover that the firefighters' ladders were inadequate and the water from the hoses could not reach the top floors. Many workers chose to jump to their deaths rather than be burned alive.[17]

Sidebar: Triangle Shirtwaist Factory Fire[18]

In New York City, the fire at the Triangle Waist Company that killed 146 young immigrant workers was one of the worst workplace disasters since the beginning of the Industrial Revolution. It is still viewed as significant because it highlights the inhumane working conditions to which industrial workers can be subjected. To many, its horrors epitomize the extremes of industrialism. The tragedy remains in the collective memory of the nation and of the international labour movement. The victims are still celebrated as martyrs at the hands of industrial greed.

Typical of the Manhattan "sweat factory" of the day, the Triangle Waist Company offered workers low wages, excessively long hours, and unsanitary

---

[17] ILR School, Cornell University, "The Triangle Waist Factory Fire", online: http://www.ilr.cornell.edu/trianglefire/story/introduction.html (date accessed: February 28, 2020).

[18] ILR School, Cornell University, "The Triangle Waist Factory Fire", online: http://www.ilr.cornell.edu/trianglefire/story/introduction.html (date accessed: February 28, 2020).

and dangerous working conditions. Even though many workers toiled under one roof in the Asch building, the owners, Max Blanck and Isaac Harris, claimed they were unaware of the inhumane working conditions and low pay offered by subcontractors to these workers.

Near closing time on Saturday afternoon, March 25, 1911, a fire broke out on the top floors of the Asch Building in the Triangle Waist Company. Within minutes, the fire spread, killing 146 of the 500 employees. Survivors were left to live and relive the agonizing moments of fear and pain, and the sights of people leaping from the ninth-floor windows.

"Many of the Triangle Waist Factory workers were women, some as young as 15 years old. They were, for the most part, recent Italian and European Jewish immigrants who had come to the United States with their families to seek a better life. Instead, they faced lives of grinding poverty and horrifying working conditions. As recent immigrants struggling with a new language and culture, the working poor were ready victims for the factory owners."[19]

Speaking out against the inhumane working conditions could end with the loss of desperately needed jobs for workers, a prospect that forced them to endure personal indignities and severe exploitation.

As time passed, employers, employees, and trade unions were made legally accountable for developing and maintaining safe and healthy workplaces. The Occupational Safety and Health Administration (OSHA) has regulated Occupational Health and Safety in the U.S. since its official introduction on April 28, 1971, when the American OSH Act came into effect.

For several decades before that, a limited number of OH&S regulations for specifically defined industries existed. As well, some broad regulations by some of the individual American states were in place. There was no such thing as standardization of the application of the OH&S codes and standards from state to state. As well, there was limited, to no, enforcement of the regulations that did exist.[20]

In passing the OSH Act, the U.S. Congress declared "its purpose and policy is to assure so far as possible every working man and woman in the Nation safe and healthful working conditions . . .".[21] The Act is a landmark in the history of American labour and public health legislation, for it was the first major piece of legislation that governed Occupational Health and Safety across a great breadth of industries within the U.S. Despite its lofty intentions, from the outset, the Act was

---

[19] *Ibid.*

[20] B. Plog, *Fundamentals of Industrial Hygiene*, 3d ed. (Chicago, Illinois: National Safety Council, 1988) at 677.

[21] *Ibid.*

not intended to protect the millions of employees of public agencies (Federal, State, county, and municipal), or private establishments with less than ten employees.

The general duty and obligations set by the OSH Act included:
- Employers must provide employees with a safe and healthy workplace — free of recognized hazards that could cause death or serious injury or illness. This is known as the **General Duty Clause**.
- Employers must comply with the terms of the OSH Act.
- Employees must comply with the OH&S standards, rules, regulations, and orders stated in the OSH Act, and that are relevant to the employee's actions and behaviours.[22]

The main provisions of the OSH Act include:
- Ensure, as far as is reasonably practicable, that every employee has a safe and healthy place in which to work.
- Require employers to document and inform employees of exposures to toxic materials, or the harmful physical agents identified by OSHA standards as dangerous substances.
- Provide for employee input in worksite tours/inspections.
- At the request of any employee, investigate alleged workplace health and safety violations.
- Empower the OSHA to enact OH&S regulations and standards.
- Establish the National Institute for Occupational Safety and Health (NIOSH) to research workplace hazards and to develop new OSH standards.
- Fund on a 50/50 cost-sharing basis, state-operated OSH programs with interested states.[23]

The Occupational Safety and Health Administration (OSHA), the agency primarily responsible for enforcing the OSH Act, develops and enacts standards, rules, and regulations by which employers are expected to conduct business. However, not all potential hazardous exposures can be conceived of. The General Duty Clause covers all possible hazards and assigns the responsibility and accountability for eliminating or minimizing hazardous conditions to employers.

Enforcement of the OSH Act is the responsibility of the OSHA. The OSHA's mandate is to:
- Conduct worksite inspections following workplace incidents that result in fatalities or multiple serious injuries, or following employee complaints, or

---

[22] B. Plog, *Fundamentals of Industrial Hygiene*, 3d ed. (Chicago, Illinois: National Safety Council, 1988) at 678.

[23] *Ibid*.

based on the degree of hazard associated with the work, or as follow-up to an earlier inspection.

- Be able to inspect, unannounced, any worksite and review injury incident records.
- In the instance of perceived imminent danger, OSHA can request a "shutdown" of the operations, or seek legal injunction to force a shutdown should the company refuse the original request.
- Seek criminal charges/penalties through the courts.
- Uphold a system of appeal for employers who believe they are being unduly charged.

In summary, the development of OH&S legislation in the U.S. tended to be:

- part of broader social reforms that occurred;
- impacted by coalition politics and political compromise between industry and labour;
- spurred on by tragic events; and
- championed by heroic individuals.

9. History of OH&S in Canada

Canadian OH&S developments were slower than in the U.S. As well, the paths taken and the progress made, differed.

In Canada, workers are covered by provincial or Federal labour codes depending on the sector in which they work. Workers covered by Federal legislation (including those in mining, transportation, communications, and Federal employment) are covered by Part II of the *Canada Labour Code*; all other workers are covered by the OH&S legislation of the province in which they reside and work.

The Canadian Centre for Occupational Health and Safety (CCOHS), an agency of the Government of Canada, was created in 1978 by an Act of Parliament. The Act was based on the belief that all Canadians had "a fundamental right . . . to a healthy and safe working environment".[24] The CCOHS is mandated to promote safe and healthy workplaces by providing information and advice about occupational health and safety.

There are 14 legal jurisdictions in Canada — one Federal, 10 provincial, and three territorial, each having its own OH&S legislation. Federal legislation covers employees of the Federal government and Crown agencies and corporations across Canada. The *Canada Labour Code*, Part II, also applies to employees of companies

---

[24] R. Robinson, D. Smith, G. Golder and B. Sharper, "Program Evaluation and Performance Measurement Study", CCOHS (Jan. 26, 2006), online: http://www.ccohs.ca/ccohs/reports/studies/performanceJan06.html (date accessed: February 28, 2020).

or sectors that operate across provincial or international borders. These businesses include:
- airports;
- banks;
- canals;
- exploration and development of petroleum on lands subject to Federal jurisdiction;
- ferries, tunnels, and bridges;
- grain elevators licensed by the Canadian Grain Commission, and certain feed mills and feed warehouses, flour mill, and grain seed cleaning plants;
- highway transport;
- pipelines;
- radio and television broadcasting and cable systems;
- railways;
- shipping and shipping services; and
- telephone and telegraph systems.

In all Canadian jurisdictions, the basic elements, such as worker/employer/supervisor rights and responsibilities, are similar. However, the details of the OH&S legislation and how the laws are enforced, vary.

10. Canadian-American OH&S Systems: Current Difference

Despite similar inceptions, the outcomes of the Canadian and the U.S. OH&S systems differ. Besides the obviously different structural arrangements of the jurisdictions and laws, there are several process and philosophical differences, such as:
- The U.S. has one overriding government body — the OSHA, whose mandate is to develop, enact, monitor and enforce OH&S legislation. In Canada, each province or territory and the Federal government is responsible for the OH&S legislation in its domain.
- In the U.S., the philosophy of OH&S appears to be "command and control" in nature; whereas in Canada, it is more of a "self-regulation" philosophy.
- In the U.S., the OSHA cannot readily "shut down" an operation when imminent danger is perceived to exist, or levy penalties without court action. In Canada, OH&S officers can stop work without requiring court action and can also levy penalties.
- In the U.S., provisions for national OSH research exist with NIOSH; not so in Canada. OH&S research occurs in universities, technical schools, colleges, government-funded agencies, and industry. Funding tends to come

from various sources and there is no one central body tasked to collect and disseminate OH&S research information.

- In the U.S., each state has its own regulatory body in terms of a Workers' Compensation Act and Workers' Compensation Board (WCB). The mandate of these state-run WCBs is to provide compensation for employees who are injured in the course of employment. While schemes differ between jurisdictions, provisions are made for weekly payments in place of wages (functioning in this case as a form of disability insurance); compensation for economic loss (past and future); reimbursement or payment of medical and like expenses (functioning in this case as a form of health insurance); general damages for pain and suffering; and benefits payable to the dependants of workers killed during employment (functioning in this case as a form of life insurance). Cash benefits are established by state formulas with maximum benefit levels. The benefits are administered on a state level, primarily by the state department of labour. In some states, the financial coverage of the Workers' Compensation (WC) claim is adjudicated and paid by the WCB with independent WC insurance carriers covering the injury healthcare costs. However, there are a number of variations in the delivery of WCB claims coverage in the U.S.

In Canada, WCBs exist in every province and employer participation is mandatory for all medium and high-risk industries. Canadian employers with at least one employee must have a valid WCB account. The Canadian WCBs provide injury claim adjudication, case management, claim payment, and prevention research and education. Although there are some differences in how the administration and the services are handled, Canadian WCBs tend to be more homogeneous in their approach and provision of services than the WCBs in the U.S.

C. OH&S MANAGEMENT: THEORETICAL BACKGROUND

To further understand the historical evolution of the field of Occupational Health and Safety in Canada and the U.S., OH&S Practitioners/Professionals must gain an appreciation of the management theories and the organizational designs and structures that helped shape today's organizations.

1. Management Theories: Influence on Motivational and OH&S Modern Practices

Although safety-focused individuals worked in the area of workplace safety since the early 1900s, the concept of an Occupational Health and Safety Management System (OHSMS) only emerged in the late 1950s. Prior to that, workplace safety was viewed as an employer initiative for improving working conditions.

Over the years, four dominant management theories impacted workplace safety and the evolution of the OHSMS. The purpose of this section is to examine these four major management approaches and their influence on worker motivation and modern OH&S practices. These management approaches are: (1) Classical Management Approaches; (2) Behavioural Approaches; (3) Quantitative Management

Approaches; and (4) Modern Management Theories. They are discussed below and later summarized in Table 1.1.

(a)   Classical Management Approaches

Traditional or classical management focuses on efficiency. It is the grouping of similar ideas on the management of organizations that evolved in the late 1800s and early 1900s. The Classical Management Approach contains three general branches: (1) the **Scientific Management Theories** which focus on the "one best way" to do a job; (2) the **Administrative Management Theories** which emphasize the flow of information in the operation of the organization; and (3) the **Bureaucratic Management Theories** which rely on a rational set of structuring guidelines, such as rules and procedures, hierarchy of authority, and a clear division of labour. The predominant characteristic of all three branches is an emphasis on the economic rationality of the individual employee at work.

(b)   Scientific Management Theory

The **Scientific Management Theory** focuses on universal principles and operates on the premise that people are rational and economically driven to work, and that workers will seize work opportunities that enable them to achieve economic gains.

Frederick Taylor, known as the "Father of Scientific Management Theory", published *The Principles of Scientific Management* in 1911.[25] He believed that employees typically perform below their capacity, and a scientific management approach that provided proper direction and financial incentives to employees would correct this performance problem. In Taylor's day, work was perceived as nothing more than an impersonal exchange of labour for money.

Taylor believed that:

- Scientific methods, such as time and motion studies, could be used to enhance work practices and increase overall worker productivity.
- Employees would work harder and be more productive if they were offered proper financial incentives.
- Management and worker prosperity through increased production and greater profitability, could be attained through:
    - standardizing jobs, tasks, work tools, procedures, processes, and conditions (***production lines***);
    - using rules of motion for each job;
    - implementing work efficiencies;
    - selecting the right workers for the job (***right skills and abilities, fitness to work, person-job-fit***);

---

[25] NETMBA, "Frederick Taylor and Scientific Management", Management section, online: http://www.netmba.com/mgmt/scientific/ (date accessed: February 28, 2020).

- training workers;
- planning work (organizing, **managing, and directing**); and
- supporting workers and processes (**job descriptions, industrial hygiene practices, occupational safety practices, etc.**).

According to Taylor, the drivers (motivators) for working hard were perceived to be:

- enhancing profits through streamlining and standardizing work; and
- defining tasks because people work better when tasks are well-defined.

The limitations of this theory are that:

- workers are not primarily driven by making money;
- it focuses mainly on improving workers and neglects enhancements of the management system;
- the proposed production-efficiency techniques resulted in monotony and boring jobs — a condition known today to result in reduced worker alertness (inattentiveness) and incidents;
- it can lead to repetitive-strain and over-use injuries;
- it failed to recognize or promote cross-functional work experiences or abilities;
- it does not provide any guidance to Management on how to support workers;
- it created a "caste-like" hierarchy within the workplace — the "worker/Management perception";
- it provided no room for career advancement; and
- it may inhibit self-actualization, a higher state of human performance.

The influences that this theory had on the modern field of Occupational Health and Safety are:

- acknowledgement of the value of worker training;
- the beginnings of "person-job fit", or the science of ergonomics;
- the use of standardized work procedures;
- the introduction of task analysis, or job task analysis;
- the use of results-based compensation to drive worker performance (incentive wage plan);
- designing job and work methods to increase worker performance;
- the introduction of worker recruitment and selection based on their ability to do the job; and
- the idea of providing workers with supervisory support and job aids.

Two other contributors to this management theory were Frank and Lillian Gilbreth, who are recognized as the "Pioneers of Motion Studies". They used time and motion studies to enhance the work Taylor began. Their contribution was towards the value that ergonomics and human factors could add to organizational efficiencies, productivity, and profitability.

(c) Administrative or Classical Management Theory

The **Administrative or Classical Management Theory**, which focuses on the flow of information within an organization, was introduced by Henri Fayol, the "Father of Modern Management Theory", in his publication ***Administration Industrille et Generale*** (1917).[26] Fayol provided the business world with the **Five Duties of Management**,[27] which are still used today and include:

1. foresight (planning);
2. organization;
3. command (leadership);
4. coordination of work; and
5. controlling.

The other concepts that Fayol brought forth, and the influences he had, include:

- The idea of systematically documenting the experiences of successful managers so that greater understanding could be gained of what skills/abilities they possessed over those skills held by the less capable managers.
- A change in management focus from worker attributes to manager attributes.
- The key duties of a manager as being the planning, organizing, leading/directing, coordinating, and controlling work.
- The introduction of ***Fourteen Principles for Management*** as methods for implementing his rules:
  - *Division of Labour* — reduces the span of attention and/or effort of any one group. It enables the development of expertise through practice and familiarity.
  - *Authority* — the right to give an order which is supported by assigned responsibility. It aligns well with *span of control*, or the area of influence that a manager possesses or is assigned.
  - *Discipline* — the outward marks of respect in accordance with the formal and informal agreements between employer and employee.

---

[26] Wikipedia, Henri Fayol entry, online: http://en.wikipedia.org/wiki/Henri_Fayol (date accessed: February 28, 2020).

[27] *Ibid.*

- *Unity of Control* — having one "boss" or top-authority figure.
- *Unity of Direction* — having one person manage all the business activities that relate to a singular business performance objective. The concept of unity of direction results in greater efficiency when the work is planned and directed by a supervisor.
- *Scalar Chain of Command* — the unbroken line of communication from top to bottom in an organization.
- *Subordination of Individual Interests to Group Interests* — the belief that the interests of one person should not overrule the interests of the group.
- *Remuneration* — compensation should be fair for the employee and the employer.
- *Centralization* — like gravity, it is always present to some degree depending on the size of the organization and quality of the managers.
- *Initiative* — within the limits of authority and discipline, all levels of employees and Management should demonstrate initiative.
- *Order* — a place for everything and everything in its place. The concept of the right person for the job.
- *Equity* — a combination of kindness and justice towards employees.
- *Stability of Tenure* — employees need to be given time to settle into their jobs.
- *Esprit de Corps* — harmony is a great strength for an organization. Teamwork should be encouraged.

The major motivational influence of this theory is that, by encouraging Management direction, workers had a clear understanding of the work expectations. This provides the worker with job clarity and the resulting job satisfaction.

In terms of the limitations of the Fayol Theory, it:

- is unifocal in its approach — it deals solely with the manager and management system, neglecting the contributions that workers make;
- promotes psychological failure when workers are not able to define their own work goals;
- fails to account for instances when managers do not take ownership for worker health and safety;
- fails to account for instances when workers do not take ownership for their own health and safety, or for that of their fellow workers; and
- does not address situations when workers defer the "control of the work environment" to their manager.

The contributions made by Fayol to the modern field of Occupational Health and Safety are that he identified the:

- key roles for managers;
- fact that managers need safety training; and
- **Fourteen Principles for Management** as ways of implementing his rules. These are still widely used today.

In the 1930s, Mary Parker Follett[28] proposed that by using workgroups and teamwork, organizations could use the combined worker talent to get work done. The principle of power "with", and not power "over", workers was proposed. According to Follett, the manager's role was to help employees cooperatively achieve the needed integration of interests. She advocated the idea of "reciprocal relationships" in terms of understanding the dynamic aspects of the "individual" in relationship to others.

As well, Follett believed in making each employee "an owner in the business" so that he/she would "own" a share of the responsibility for the company's success or failure. Follett advocated the principle of integration — "power sharing". Her ideas on negotiation, power, and employee participation were influential in the development of organizational studies.

In terms of motivational factors, Follett believed that her approach would encourage management and workers to work in harmony; and that by making each employee "an owner in the business", feelings of collective responsibility for business operations and success, would ensue. This signalled the beginning of ideas like **profit sharing**[29] and **gain sharing**.[30]

Follett's contribution to the field of Occupational Health and Safety includes the introduction of concepts like teamwork, negotiation, and collective responsibility for business success or failure.

(d)   Bureaucratic Management Theory

In the 1930s, Max Weber[31] introduced his **Bureaucratic Theory** which is founded

---

[28] A. Marriner, *Guide to Nursing Management*, 2nd ed. (St. Louis, MO: The C.V. Mosby Company, 1984) at 159-160.

[29] **Profit sharing** refers to various company incentive plans that provide direct or indirect payments to employees based on the company's profitability in addition to the employees' regular salary and bonuses.

[30] **Gain sharing** is a system of reward for increased employee productivity: a system in a company by which employees' pay is increased in line with the gains in productivity or reductions in costs achieved by the company that are the direct result of the employees' cooperation.

[31] A. Marriner, *Guide to Nursing Management*, 2nd ed. (St. Louis, MO: The C.V. Mosby Company, 1984) at 158.

on logic, order, and the concept of legitimate authority. He proposed that:

- greater productivity could be achieved by using a proper organizational structure;
- a well-defined hierarchy of authority needs to exist;
- a clear division of labour is essential;
- formal rules and procedures are required;
- increased bureaucracy leads to greater operational efficiency and fairness for employees;
- career growth ought to be based on merit; and
- the result would be an expedient way to deal with urgent situations and multiple worksites.

This model is most effective when collective action needs to be taken expediently, quickly, and simultaneously at multiple geographic locations or sites. It is no wonder that the bureaucratic management approach has been the model of choice for the military, government, emergency response services, and hospitals.

The motivational impacts of this theory have proven to be both positive and negative in nature. Although with the bureaucratic management approach, workers clearly know what they are to do and how they will do it, the impersonality and multiple levels of hierarchy are viewed as cumbersome, impersonal, and career limiting. So, with this management approach, the task functions of the job tend to be efficiently addressed, while the relationship aspects suffer.

The limitations of Weber's theory and the bureaucratic management approach are that they promote the development of a bureaucratic system which:

- is slow to change and to adapt to change;
- is rigid when handling problems;
- can be inefficient due to the amount of "red tape" to go through when workers are trying to address fluctuations in customer needs, change, handling problems, and countering employee apathy; and
- can lead to the belief that workers have little control over their work. This can be manifested as employee apathy or learned helplessness.

For the field of modern Occupational Health and Safety, this management approach is typically adopted for Emergency Response Plans (ERP) where a single point of command is used to orchestrate actions efficiently and effectively in times of crisis.

(e) Behavioural Theory and Approaches

The **Human Resources and Behavioural Management Theory** and approaches focus on human needs, workgroups, and the social factors that impact work. It upholds that people are naturally social and self-actualizing beings.

An example of this theory is the work done by Elton Mayo (1924). He conducted the famous Hawthorne Studies that centred on the human aspects of work and set the stage for a field of study called **Organizational Behaviour**.[32] The **Hawthorne Effect**, which he described, is the initial improvement in a process of production caused by the obtrusive observation of that process. This effect was first noticed in the Hawthorne plant of Western Electric. In this instance, worker production increased not because of actual changes in working conditions that were introduced by the plant's Management, but because Management demonstrated an interest in such improvements and in them as workers.[33]

The tendency for workers, singled out for special attention at work, to perform as anticipated merely because of Management expectations, has far-reaching ramifications. It points to the impact of the human elements of work (the relationship aspects) in that they have a greater effect on worker productivity than do the job's technical and physical aspects.

The limitations of this theory are that it has been poorly understood. Instead of recognizing that workers positively respond when recognized by Management, many managers and Human Resources Practitioners viewed Mayo's work as indicating that the outcome of workplace initiatives are often confounded by a manager's interest in the workers. The work by Dyck and Roithmayr (2002) indicates that Management's interest in workers is a very powerful factor in promoting and enabling workgroups to maximize their safety performance.[34]

For OH&S Practitioners/Professionals, this theory helps to explain employee reactions to Management's interest, recognition, and investment in them.

In 1943, Abraham Maslow produced his well-known model, the **Hierarchy of Needs**.[35] The main assumptions held are:

1. **The Deficit Principle**: a "satisfied need" is not a motivator of behaviour: rather, people act to satisfy deprived needs.

2. **The Progression Principle**: a need at any level only becomes activated

---

[32] **Organizational Behaviour (OB)** is the study of individual and group dynamics in an organizational setting, as well as the nature of the organizations themselves. Refer to Chapter 38, "Occupational Health and Safety: Organizational Behaviour" for more detail.

[33] A. Weber, "The Hawthorne Works" (2002) *Assembly Magazine*, online: http://www.assemblymag.com/CDA/Archives/0cdaaa2e0d5c910VgnVCM100000f932a8c0 (date accessed: February 28, 2020).

[34] Chapter 6, "Occupational Health and Safety: Leadership and Commitment", provides details on the Great Safety Performance research.

[35] G. Norwood, "Maslow's Hierarchy of Needs", Deeper Minds, online: http://www.deepermind.com/20maslow.htm (date accessed: February 28, 2020); see also A. Marriner, *Guide to Nursing Management*, 2nd ed. (St. Louis, MO: The C.V. Mosby Company, 1984) at 161-162; and F. Bird and G. Germain, *Loss Control Management: Practical Loss Control Leadership*, revised ed. (Loganville, GA: DNV, 1996) at 46.

once the next lower-level need has been satisfied.

These two principles indicate that people strive to sequentially satisfy the five human-need levels.

Maslow's **Hierarchy of Needs** proposes that people bring "their individual needs" to work and are driven to satisfy those human needs:

- Physiological Needs — the needs to stay alive.
- Security Needs — the needs to feel secure.
- Social Needs — the needs to belong.
- Ego Needs — the needs to be "somebody".
- Self-actualization — the needs to develop one's potential as a human being.

The OH&S management applications associated with the **Hierarchy of Needs** are:

- the theory helps managers understand employee needs and how to address them at work, thereby making their management duties and life easier; and
- managers who satisfy employee-human needs achieve better worker respect, "buy-in", support, and operational productivity.

The major critique of this model is that life does not work that way — it is far more complex.

For the field of OH&S, Maslow's Hierarchy of Needs offers:

- an explanation for employee actions at work and how to motivate them to attain desired OH&S attitudes and perform the desired OH&S behaviours;
- an explanation as to why employees expect more than a paycheque from work; and
- anticipatory guidance on how to effectively respond to the workgroup after a traumatic event.

In the 1940s, McClelland, in his **Basic Needs Theory**, put forth the idea that there are three levels of human needs:

- Need for Achievement.
- Need for Power.
- Need for Affiliation.[36]

According to McClelland, people acquire or develop these needs over time as a result of their individual life experiences. Each need is associated with a distinct set of work preferences. The message is that managers must recognize the strength of each of these human needs in themselves and in others, and to create a work

---

[36] A. Marriner, *Guide to Nursing Management*, 2nd ed. (St. Louis, MO: The C.V. Mosby Company, 1984) at 195.

environment that is responsive to those needs as well as conducive to enabling workers to meet those needs.

Clayton Alderfer introduced the **ERG Theory**[37] which built on Maslow's Hierarchy of Needs, but holds that the basic human needs are:

1. Existence needs.
2. Relatedness needs.
3. Growth needs.

Alderfer proposed that people do not have to satisfy a lower need before addressing a higher-order need. He also assumes that satisfied needs can be motivators when a higher-order need is unmet (**Frustration-Regression Principle**).

J. Stacy Adams[38] (1963) introduced the **Equity Theory** in which perceived inequity is a motivator. It holds that perceived inequity is based on worker comparisons with each other regarding rewards, job opportunities, pay increases, and other benefits. Comparisons can be internal or external, and workers respond to the perceived inequities by:

- changing their work input (effort);
- trying to change rewards received;
- using different comparators;
- rationalizing the inequities; or
- leaving the situation.

It is proposed that when a worker perceives positive inequity, he/she tends to balance the situation by increasing their work effort and input.

In 1960, Douglas McGregor in his book, *The Human Side of Enterprise*, introduced **Theory X and Theory Y**.[39] Influenced by Maslow and the Hawthorne Studies, he believed that managers should pay more attention to the social and self-actualization needs of workers. He held that managers believe and use either

---

[37] Value Based Management, ERG Theory by Clayton P. Alderfer, online: http://www.value basedmanagement.net/methods_alderfer_erg_theory.html (date accessed: February 28, 2020).

[38] A. Marriner, *Guide to Nursing Management*, 2nd ed. (St. Louis, MO: The C.V. Mosby Company, 1984) at 167; Wikipedia, J. Stacy Adams entry, online: http://en.wikipedia.org/wiki/Equity_ theory (date accessed: February 28, 2020); J. Stacy Adams, "Toward an understanding of inequity" (1963) 67 J. Abnorm. Psychol. 422-436; and J. Stacy Adams, "Inequity in social exchange", in L. Berkowitz, ed., *Advances in Experimental Social Psychology*, Vol. 2 (New York, NY: Academic Press, 1965), at 267-299.

[39] NETMBA, "Theory X and Theory Y", Management section, online: http://www.netmba.com/ mgmt/ob/motivation/mcgregor/ (date accessed: February 28, 2020); A. Marriner, *Guide to Nursing Management*, 2nd ed. (St. Louis, MO: The C.V. Mosby Company, 1984), at 163-165.

Theory X or Theory Y when managing their business functions (Table 1.1).

Table 1.1: Manager Beliefs and Characteristics

Theory X Manager	Theory Y Manager
• Employees hate work and are inherently lazy • Employees lack ambition and avoid responsibility • Employees are irresponsible • Employees hate change • Employees seek security • Employees prefer to be led by others versus lead themselves; • Employees are motivated by punishment • Employees are self-centred and indifferent to the organization's needs	• Employees want to work • Employees are capable of self-direction • Employees are responsible • Employees are capable of self-control • Employees are capable of imagination and creativity • Employees are motivated by achievement

Managers that adhere to either the Theory X or the Theory Y assumptions create a **self-fulfilling prophecy**[40] with their employees. That is, they reap what they profess/demonstrate to believe. Workers mirror the expectations that their manager holds for them. This concept added a whole new dynamic to the employee management practices of the day.

The learning from this theory is that whether or not an OH&S system will be implemented, depends on whether the leader of the organization is a Theory X or Theory Y proponent.

For OH&S Practitioners/Professionals, the messages that the Theory X and Theory Y provide are:

- Management should keep their behavioural expectations, in terms of safe work practices, high.
- This theory aligns with the current management emphasis on leadership, empowerment and self-managed teams.
- Managers holding either the Theory X or the Theory Y philosophies can create self-fulfilling prophecies: subordinates tend to act in accordance with their manager's expectations.

Chris Argyris' **Immaturity/Maturity Theory**[41] (1957) is counter to Taylor's and Fayol's beliefs in that it argues that certain principles used in the Classical Management Approaches that were intended to increase worker efficiency are

---

[40] **Self-fulfilling Prophecy** is a prediction that, in being made, actually causes itself to become true. In terms of managing people, managers net what they profess/demonstrate to believe about workers: workers mirror the expectations their manager holds for them.

[41] A. Marriner, *Guide to Nursing Management*, 2nd ed. (St. Louis, MO: The C.V. Mosby Company, 1984), at 196; C. Argyris, "The Organizational Environment and Productivity, Human Relations Contributors", online: http://www.accel-team.com/human_relations/hrels_06ii_argyris.html (date accessed: February 28, 2020).

inconsistent with the mature-adult personality. Argyris proposes that if an organization's management style is in conflict with employee personalities, it will negatively impact their work performance. In fact, the greater the disparity between the individual's needs and the organization's needs, the more tension, conflict, and dissatisfaction that will result. Argyris recommended that managers focus on helping workers to achieve self-actualization by using a Theory Y philosophy and expanding job requirements to include more task variety and responsibility.

The motivational message gained from Argyris' work is: To encourage the development of responsible, mature employees, adjust supervisory styles to include more employee participation.

A limitation of this theory is that despite all that a manager may do to enrich the work experience, talented employees may still leave.

The messages offered to OH&S Practitioners/Professionals by this theory are:

- employee involvement in OH&S issues/problems leads to ownership of the problem; and
- it can help to explain the lack of worker "buy-in" to a proposed change.

Robert Blake and Jane Mouton (1964), when they introduced the idea of team management styles, upheld two important leadership dimensions — concern for people and concern for production. These two dimensions are independent, so a manager can score high on one measure and score low on the other; or high/low on both. Blake and Mouton's **Managerial Grid** described the various management styles as follows:

- **Country Club Management** — attention to worker needs for a satisfying work relationship leads to a comfortable, friendly organizational atmosphere, and work tempo. This translates to a low regard for production and a high regard for people and their needs/relationships.
- **Team Management** — work is accomplished by people being committed to a common goal, thereby creating relationships of trust and respect. This translates to a high regard for both production and people relationships.
- **Organization Man Management** — adequate performance is possible through balancing "the necessity to get work done" with "maintaining worker morale at a satisfactory level". This translates into a balance of concern for production and for people.
- **Impoverished Management** — minimal effort is exerted to get the work done or to sustain organizational membership. This translates to a low regard for both production and people.
- **Authority-Obedience Management** — efficiency in operations results from arranging work conditions so that the interference of human elements is kept to a minimum. This translates to a high regard for production and

low regard for relationships.[42]

According to Blake and Mouton, the ideal managerial style is **Team Management**. By integrating concern for people and production, this style of manager can confront problems directly while fostering mutual trust, respect, and interdependence.[43]

Based on Victor Vroom's **Expectancy Theory**[44] (1964), worker motivation stems from factors such as:

- what employees value;
- the employee's perceived performance/reward ratio; and
- employee expectations of success or failure at work.

The Expectancy Theory of motivation is used to help us understand how individuals make decisions regarding various behavioural alternatives. The model deals with the directional aspect of motivation; that is, once behaviour is energized, what behavioural alternatives are individuals likely to pursue.

When deciding among behavioural options, individuals select the option with the greatest motivational drivers. The motivational force for a behaviour, action or task is a function of three distinct perceptions, namely:

1. **Expectancy** is the belief that one's effort will result in the attainment of desired performance goals. This belief, or perception, is generally based on an individual's past experience, self-confidence (often termed self-efficacy), and the perceived difficulty of the performance standard or goal.

2. **Instrumentality** is the belief that if one does meet performance expectations, he/she will receive a greater reward. This reward may come in the form of a pay increase, promotion, recognition, or sense of accomplishment. It is important to note that the reward must be appropriate to the level of performance, or the instrumentality will be low.

3. **Valance** refers to the value the individual personally places on the rewards. This is a function of personal needs, goals, values and sources of motivation.

Expectancy and instrumentality are attitudes. As such, they represent a worker's perception of the likelihood that working hard will lead to good work performance, and in turn, good work performance will lead to advancement. These perceptions represent the worker's subjective reality and may or may not align with workplace

---

[42] A. Marriner, *Guide to Nursing Management*, 2nd ed. (St Louis, MO: The C.V. Mosby Company, 1984), at 168.

[43] *Ibid*.

[44] Value Based Management, Expectancy Theory by Victor Vroom, online: http://www.valuebasedmanagement.net/methods_vroom_expectancy_theory.html (date accessed: February 28, 2020).

reality: they are influenced by the person's experiences, by what they observe and by their self-perceptions.

The message regarding motivating the employee is that managers should:

- tell employees what is expected of them;
- make the work valuable/meaningful;
- make the work doable;
- provide feedback on the employee's progress; and
- reward success.

The overall message to OH&S Practitioners/Professionals is that organizations should focus more on the use of the leading indicators of safety performance than the lagging indicators of safety performance. This idea is reinforced throughout this book.

The **Goal-setting Theory**, introduced by Edwin Locke in the mid-1960s,[45] upholds the belief that task goals can be highly motivating if they are well set up and well-managed. Likewise, worker ownership of organizational goals depends on the degree of worker participation in setting those goals. Thus, to elicit worker interest, involve workers in establishing the work goals that they are going to be expected to work towards.

Thorndike and B.F. Skinner (mid-1960s) contributed to the foundation of management styles through the creation of the **Reinforcement Theory of Motivation**.[46] The Reinforcement Theory implies that if a person has a need, then he/she will be motivated to work hard to achieve the need desired. Furthermore, there is a continuous pattern within reinforcement, because consequences influence behaviour and behaviour influences consequences. It is important to recognize that people constantly need to be rewarded (reinforcement) for work that is done. Reinforcement is what motivates and drives people to push themselves.[47]

According to this theory, if hard work is rewarded and positively reinforced, then the outcome increases the probability of the employee demonstrating more hard work and dedication. When an employee is negatively rewarded (reinforced), then the probability of the occurrence of that behaviour or work diminishes.

Based on Thorndike's **Law of Effect**,[48] worker behaviour that results in a pleasant

---

[45] Wikipedia, Goal-setting Theory entry, online: http://en.wikipedia.org/wiki/Goal-Setting_Theory (date accessed: February 28, 2020).

[46] C. Wortman, E. Loftus & M. Marshall, *Psychology*, 3d ed. (New York, NY: Knopf, 1988), at 130-140.

[47] University of West Virginia, Reinforcement Theory, online: http://www.managementstudy guide.com/reinforcement-theory-motivation.htm (date accessed: February 28, 2020).

[48] The **law of effect** states that responses that are closely followed by satisfying

outcome is likely to be repeated: that which does not, will not be repeated. This theory aligns well with Skinner's **Operant Conditioning Theory**[49] and the principles of Organizational Behaviour modifications.

The associated motivational messages include:
- positive reinforcement and negative reinforcement can lead to the performance of desired behaviours;
- punishment eliminates undesired behaviours; and
- extinction eliminates behaviours — positive or negative in nature.

Frederick Herzberg (1959) developed the **Maintenance Factors/Motivating Factors Theory** which holds that two types of work factors exist in the work environment:

1. *Maintenance (Hygiene) Factors* (*i.e.*, physical conditions, pay, status, benefits) — these factors prevent dissatisfaction and are neutral in regards to worker motivation.

2. *Motivating Factors* (*i.e.*, responsibility, achievement, growth, and recognition) — these factors help employees attain their ego and self-actualization needs, and are therefore, motivational.

The limitations of the **Maintenance Factors/Motivating Factors Theory** are that the new generations of workers like Generation X and Baby Boomlets do not come to work merely for a pay cheque. These workers want to be involved and participatory in work design and decision-making. Also, organizations are advised to avoid the use of generalized OH&S programs; rather, tailor worker education to meet specific employee learning needs.

The message for OH&S Practitioners/Professionals is that Herzberg's theory:
- aligns with continuous improvement, performance excellence, Total Quality Management, and worker performance;
- has an impact regarding OH&S training/educational approaches and techniques; and

---

consequences become associated with the situation and are more likely to recur when the situation is subsequently encountered. Conversely, if the responses are followed by aversive consequences, associations to the situation become weaker.

Wikipedia, Edward Thorndike entry, online: http://en.wikipedia.org/wiki/Edward_Thorndike (date accessed: February 28, 2020).

[49] **Operant Conditioning Theory** is defined as "the use of consequences to modify the occurrence and form of behavior. Operant conditioning is distinguished from Pavlovian conditioning in that operant conditioning deals with the modification of voluntary behavior through the use of consequences, while Pavlovian conditioning deals with the conditioning of behavior so that it occurs under new antecedent conditions". See Wikipedia, Operant Conditioning entry, online: http://en.wikipedia.org/wiki/Operant_conditioning (date accessed: February 28, 2020).

- impacts Occupational Safety recognition programs.

(f) Quantitative Management Approaches

Quantitative Management Approaches use mathematical techniques for problem solving. Their primary focus is towards decisive decision-making using economic decision-making criteria, formal mathematical models, and computer modelling. The assumptions associated with this approach are that, the mathematical approaches listed below can enhance managerial decision-making and problem solving. For example, managers can use:

- trend analysis (identifying trends based on past experiences);
- mathematical forecasting (making future projections useful in planning);[50]
- linear programming (calculating how to best allocate scarce resources among competing uses);[51]
- queuing theory models (allocating organizational resources to minimize customer waiting times);[52]
- network models (breaking down the large tasks in a complex project into smaller ones for better analysis and control);[53]
- inventory modelling (determining the inventory needed through mathematical means);[54] and
- simulations (computerized modelling problems to test various solutions).[55]

For OH&S Practitioners/Professionals, the management tools that emerged from this approach are the use of OH&S auditing techniques to evaluate the status of the OHSMS; incident investigation approaches and techniques; the collection and analysis of lagging indicators of safety performance; the techniques for trend analysis; and the use of metrics regarding the leading indicators of safety performance. It also provides insight into effective ways for communicating OH&S results and outcomes in business jargon.

(g) Modern Management Theories

Modern Management Theories focus on total systems thinking, contingency thinking and an awareness of global developments in Management.

---

[50] D. Sayers & Associates, *Management and Organizational Behaviour*, BCRSP Self-Study Guide (Mississauga, ON: Board of Certified Safety Professionals, 2000), at 27.

[51] *Ibid.*

[52] *Ibid.*

[53] *Ibid.*

[54] *Ibid.*

[55] *Ibid.*

In the 1940s and 1950s, Bird[56] and Ludwig von Bertalanffy[57] put forth their Safety System Theories. This theoretical approach views worker errors to be a function of flaws in the work system that creates an "error-prone" work environment as opposed to worker neglect or incompetence. In this theory, error is viewed not as a cause, but as a consequence or a symptom of latent conditions that originate at an organizational level and are displayed by Line Management. Latent conditions stem from deficiencies in organizational functions such as developing policies and procedures, budgeting, staffing, maintaining equipment, and managing processes. These latencies create work conditions that exacerbate worker fallibility and stress the limits of human performance. In fact, the latent conditions can also originate even further "upstream" beyond the sphere of individual organizations, to include the activities of external bodies such as OH&S legislative agencies, equipment manufacturers, and even an industry as a whole.

Safety System Theories view organizations as closed systems that function towards a common purpose. They propose that when one part of the system undergoes change, all the other parts are affected. This theory also recognizes that people are complex with multiple talents and abilities, and that people are variable. Well-accepted in the field of engineering, System Theory was poorly understood and not used in the field of Occupational Health and Safety until recently.

The motivational aspect of this theory centres on the belief that high-performing organizations offer a variety of managerial styles and job opportunities.

The message to OH&S Practitioners/Professionals is that an impact — positive or negative — that affects one part of the system, is felt within the other parts as well. This concept applies to OH&S program inputs, throughputs, and outcomes, both positive and negative.

Total Quality Management (TQM)[58] was introduced in the 1990s. It encompasses concepts such as:

- visionary leadership;
- management commitment;
- corporate culture;
- customer focus;
- statistical process control;
- benchmarking;

---

[56] F. Bird and G. Germain, *Loss Control Management: Practical Loss Control Leadership*, revised ed. (Loganville, GA: DNV, 1996).

[57] Wikipedia, Ludwig von Bertalanffy entry, online: http://en.wikipedia.org/wiki/Ludwig_von_Bertalanffy (date accessed: February 28, 2020).

[58] Wikipedia, Total Quality Management entry, online: http://en.wikipedia.org/wiki/Total_Quality_Management (date accessed: February 28, 2020).

- continuous improvement;
- worker training;
- cross-functional teams;
- worker empowerment;
- self-managed teams;
- learning organizations;
- change management; and
- managing the total system.

The key players in TQM were Deming, Juran, Tom Peters, Conway, Senge and Drucker. The TQM motto is "Commitment to Quality". The theory hinges on a management philosophy that focuses on satisfying customers. The objective is **continuous improvement** (CI) using fundamental concepts such as:

- employee involvement;
- new culture;
- new tools; and
- use of measures like "best practices" and "upstream and downstream measures".

The impact that this theory has on the area of worker motivation is dependent on worker heredity, worker needs, the work environment, and work culture. It considers the "cognitive", "behavioural", and "affective" aspects that are also at play.

For OH&S Practitioners/Professionals, **loss control** became integrated into the vision, mission, strategic plans, policies, and procedures of organizations. Concepts such as an Occupational Health and Safety Management System; the **Plan-Do-Check-Act Model** (P-D-C-A); standardized formats for OH&S auditing; and continuous improvement were introduced and remain dominant elements today.

Woodward (1958) originally introduced the **Contingency Theory**,[59] which is similar to Hersey and Blanchard's **Situational Leadership Theory**[60] in that it puts forth the assumption that there is no simple, one right way to lead. The main difference is that Situational Leadership Theory tends to focus more on the behaviours that the leader should adopt, given situational factors (often about follower behaviour); whereas Contingency Theory takes a broader view that includes contingent factors about leader capability and other variables within the situation.

---

[59] Wikipedia, Contingency Theory entry, online: http://en.wikipedia.org/wiki/Contingency_theory (date accessed: February 28, 2020).

[60] Wikipedia, Situational Leadership Theory entry, online: http://en.wikipedia.org/wiki/Situational_leadership_theory (date accessed: February 28, 2020).

Situational Leadership Theory, developed by Hersey and Blanchard (1960s), suggests that leaders adjust their leadership style based on the "readiness" of the followers to perform in certain situations. It puts forth the belief that depending on follower readiness, leaders use a management approach of either:

- delegating;
- participating;
- selling; or
- telling.

From a motivational standpoint, managers can enhance worker motivation by how they lead. For OH&S leaders, there emerged the belief that there is no "one best way" to lead, rather "it depends" on the situation and the related circumstances.

The **Path-Goal Theory**, developed by Robert House (1971),[61] holds that an effective leader is one who clarifies the paths for followers to take so they can achieve both the task-related and personal work goals.

The Path-Goal Theory of Leadership proposes that the leader can affect the work performance, job satisfaction and motivation of a group in different ways, such as:

- offering rewards for achieving performance goals;
- clarifying paths towards these goals; and
- removing obstacles to performance.

A person may achieve the leadership goals listed above by adopting a certain leadership style that is appropriate to the situation. For example:

- *Directive leadership*: specific advice is given to the group and ground rules and structure are established, for example, clarifying expectations, specifying or assigning certain work tasks to be followed, *etc*.
- *Supportive leadership*: good relations are promoted within the group and sensitivity to the subordinates' needs is shown.
- *Achievement-oriented leadership*: challenging goals are set and high performance is encouraged while confidence is shown in the group's ability to achieve.
- *Participative leadership*: decision-making is based on consultation with the group and information is shared with the group.

Good leaders help employees progress, remove barriers, and provide opportunities for employee growth and development. The Career Streaming Model described in Chapter 41, "Occupational Health and Safety Practitioners/Professionals: Career Development", is a prime industry-applied example of this theory. It is designed to

---

[61] R.J. House, "A Path-Goal Theory of Leader Effectiveness" (1971) 16(3) Adm. Sci. Q. 321-339.

promote practitioner/professional growth and development.

Transformational Leadership, defined by Burns (1978),[62] is inspired leadership that influences the beliefs, values, and goals of followers so that they can perform in an extraordinary manner.[63] Transformational leaders offer a "purpose" that transcends short-term goals and focuses on higher-order intrinsic needs. This results in followers identifying with the leader and his/her directions.

The four dimensions of transformational leadership are:

- *Charisma or Idealized Influence*: the degree to which the leader behaves in ways that followers view as admirable and which cause followers to identify with the leader.
- *Inspirational Motivation*: the degree to which the leader articulates a vision that is appealing and inspiring to followers.
- *Intellectual Stimulation*: the degree to which the leader challenges assumptions, takes risks, and solicits the ideas of followers. Leaders with this trait stimulate and encourage follower creativity.
- *Individualized attention*: the degree to which the leader attends to each follower's needs, acts as a mentor or coach to the follower, and listens to the follower's concerns and needs.

Transformational leaders motivate followers through:

- providing a clear vision;
- their charisma;
- using symbolism;
- empowerment;
- intellectual stimulation; and
- having integrity.[64]

Charismatic Leadership, although similar and complementary to Transformational Leadership, has been the basis of its own distinct development.[65] The

---

[62] Changing Minds, Leadership Theories: Transformational Leadership, online: http://changingminds.org/disciplines/leadership/styles/transformational_leadership.htm (date accessed: February 28, 2020).

[63] D. Hofmann and F. Morgeson, "The Role of Leadership in Safety", in J. Barling and M. Frone, (eds.) *The Psychology of Workplace Safety* (Washington, DC: American Psychological Association, 2004), at 176.

[64] For more information on Transformational Leadership and leaders, refer to Chapter 7, "Safe Workplace = Great Workplace: Building a Sustainable Culture of Safety", Chapter 24, "Psychological Health and Safety: Practical Application in the Workplace", and Chapter 18, "Occupational Health and Safety: Best Practices".

[65] Max Weber, "The Nature of Charismatic Authority and its Routinization" in *Theory of*

assumptions of Charismatic Leadership are that followers are attracted to leaders who believe in themselves, possess charm and grace, and are admired.

In their book, *In Search of Excellence*, Tom Peters and Robert Waterman (1982)[66] propose an approach towards performance excellence which encompasses eight attributes:

- bias for action;
- closeness to customer;
- autonomy and entrepreneurship;
- productivity through people;
- "hands on" and value-driven;
- simple form and a lean staff complement;
- "sticking to the knitting"; and
- simultaneous "loose-tight properties".

This concludes the description of the management theories and approaches that helped to shape today's organizations. As an everyday tool, or quick-reference guide, Table 1.2 summarizes the history of the above management approaches and theories, and their influences on worker motivation and the modern OH&S practices.

Social and Economic Organization, translated by A.R. Anderson and Talcot Parsons (1947), originally published in 1922 in German under the title *Wirtschaft und Gesellschaft* (chapter III, § 10); R.J. House, "A 1976 theory of charismatic leadership", in J.G. Hunt and L.L. Larson (eds.), *Leadership: the cutting edge* (Carbondale, IL: Southern Illinois University Press, 1977) at 189-207; J.A. Conger and R.N. Kanungo (eds.), *Charismatic Leadership in Organizations* (Thousand Oaks, CA: Sage Publications, 1998); and P. Sellers, "What exactly is Charisma?", *Fortune* (January 15, 1996).

[66] T. Peters and R. Waterman, *In Search of Excellence* (New York, NY: HarperCollins Publishers, 1982).

Table 1.2: Management Strategies and their Impact on Modern OH&S Practices

Management Theory	Time frame	Theorist	Concepts	Impact on the Concept of Worker Motivation	Limitation of the Theory	Impact on Modern OH&S Practices
I. CLASSICAL APPROACHES						
I.) Scientific Management Theory (*Focus is on universal principles*)	Early 1900s (1911)	**Frederick Taylor**, "Father of Scientific Management Theory". Published "*The Principles of Scientific Management*", 1911. Believed that workers perform below capacity and developed a scientific approach to management that provided proper direction and financial incentives would correct the problem. In Taylor's day, work was an impersonal exchange of labour for money.	• Use scientific methods to enhance work practices (time studies) and productivity of people at work • Offer proper pay incentives • Management and worker prosperity can be attained by: • Standardizing jobs • Standardizing work tools, procedures, processes and conditions (production lines) • Planning work (managing and directing) • Selecting the right workers for the job (right abilities, fitness to work) • Training workers • Supporting workers and work processes (job descriptions and industrial hygiene) • Implementing work efficiencies • Having rules of motion for each job	• Profits can be enhanced by streamlining work and standardizing work • People work better when their work tasks are well-defined	Faulty because: • Workers are not driven by money alone • Primary focus on improving the worker system, not the management system • Provided no guidance to management on how to support workers at work • Resulted in the creation of monotonous repetitive tasks and boring jobs • Created a "caste-like" hierarchy – "Supervisor-worker" perception • May inhibit self-actualization • Set up condition for repetitive strain and overuse injuries • Allowed for no cross-functional work experiences or abilities • Provided no room for career advancement	• Worker training is important • Introduced standardized work procedures • Beginning of job task analysis • Used results-based compensation to drive performance (incentive wage plan) • Recommended design jobs and standardized work methods to increase employee performance • Introduced worker recruitment and careful selection of employees based on their skills and ability to do the job • Provided employees with supervisory support
II.) Administrative or Classical Management Theory (*Focus is on the flow of information within an organization*)	1900s 1916	**Frank and Lillian Gilbreth** (Pioneers of Motion Studies) **Henri Fayol**, Father of Modern Management Theory. Published "*Administration Industrielle et Generale*". 1916. Changed the management focus from workers to managers. The duties of a manager: 1. Planning 2. Organization 3. Command 4. Coordination 5. Controlling Introduced the 14 Principles for Management as methods for implementing his rules, e.g.: o Division of Labour	• Use time and motion studies to enhance the above • Systematic attempts to document and understand the experiences of successful managers	• By encouraging management direction, workers would have a clear line of sight regarding work	• Unifocal in its approach – it deals solely with the manager and management system, neglecting the contributions that workers can make • Can lead to psychological failure when employees are not able to define their own work goals • Managers may not take ownership for safety • Employees may not take ownership for safety • Employees defer 'control of	• Beginning of the science of ergonomics and human factors • Identified the manager role • Identified that managers need safety training • Identified the 14 principles for management as ways of implementing rules. These are still in use today.

CH. 1: OCCUPATIONAL HEALTH AND SAFETY: HISTORICAL PERSPECTIVES

Management Theory	Time frame	Theorist	Concepts	Impact on the Concept of Worker Motivation	Limitation of the Theory	Impact on Modern OH&S Practices
II.) Administrative or Classical Management Theory (cont'd)			<ul><li>Authority</li><li>Discipline</li><li>Unity of Control</li><li>Unity of Direction</li><li>Scalar Chain of Command</li><li>Subordination of Individual Interests to Group Interests</li><li>Remuneration</li><li>Centralization</li><li>Initiative</li><li>Order</li><li>Equity</li><li>Stability of Tenure</li><li>Esprit de Corps</li></ul><ul><li>Assumed that unity of direction would result in greater efficiency when the work is planned and directed by a supervisor</li></ul>		the environment" to the manager	
	1930s	Mary Parker Follett	<ul><li>By using groups, organizations can use the combined worker talent to get work done (the beginnings of team work)</li><li>Manager's role is to help employees co-operatively achieve the needed integration of interests</li><li>Believed that by making each employee "an owner in the business", it would create feelings of collective responsibility (the beginning of ideas like profit sharing, gain sharing)</li></ul>	<ul><li>Encouraged management and workers to work in harmony (team work)</li><li>Believed that making every employee an owner in the business would create feelings of collective responsibility for company success/failure</li></ul>		<ul><li>Team work came into being</li><li>Promoted a collective responsibility for business success/failure (Profit Sharing Plans, Gain Sharing Plans, etc.)</li></ul>
III.) Bureaucratic Management Theory *(Focus is on the organizational structure and processes)*	1930s	Max Weber	<ul><li>The concept of bureaucracy holds that greater productivity can be achieved by using a proper organizational structure that is founded on:<ul><li>Logic, order, and legitimate authority</li><li>Clear division of labour</li><li>Well-defined hierarchy of authority</li></ul></li></ul>	<ul><li>Demonstrates a plan for action. This is comforting to workers and is viewed as positively motivating</li><li>This management theory can be very "career limiting" — it is a negative motivator for workers</li></ul>	<ul><li>Can be inefficient in some business responses due to the amount of "red tape", lack of flexibility in meeting the fluctuations in customer needs, problems addressing change, handling problems and a challenge in countering employee apathy</li><li>Tends to be slow to</li></ul>	<ul><li>The typical set-up for Emergency Response Plans and responses (command centres)</li><li>The typical set up of governments and the military</li></ul>

OCCUPATIONAL HEALTH & SAFETY: THEORY, STRATEGY AND INDUSTRY PRACTICE

Management Theory	Time frame	Theorist	Concepts	Impact on the Concept of Worker Motivation	Limitation of the Theory	Impact on Modern OH&S Practices
III.) Bureaucratic Management Theory (cont'd)			• Formal rules and procedures • Impersonality • Career growth based on merit • Viewed as an expedient way to deal with urgent situations and multiple worksites • Proposed that increased bureaucracy leads to greater efficiency and fairness for employees		• Can lead to worker passivity believing that they have little control over their work (worker apathy) • Demonstrates rigidity in handling problems "change"	
2. BEHAVIOURAL APPROACHES						
Human Resources and Behavioural Management Theory *(Focus on human needs, work groups, and social factors that impact work)*	1924	Elton Mayo (Hawthorne Studies)	• Focuses on human aspects of work • Set the stage for a field called "Organizational Behaviour" *The Hawthorne Effect* is the tendency for people who are being singled out for special attention to perform as anticipated merely because of the expectations created	• Human elements of work have a greater effect on productivity than the job's technical and physical aspects	• The Hawthorne Effect has been poorly understood over the years. Instead of recognizing that workers respond positively when recognized, many HR practitioners and managers viewed Mayo's work as indicating that the outcome of workplace initiatives are often confounded by manager interest in workers • Life doesn't quite work this way....	• Helps to explain employee reactions to a manager's interest in the work being done. Usually what interests a worker's supervisor, tends to fascinate the worker

	1943	**Abraham Maslow**
		(Hierarchy of Needs)
		Holds that people are naturally social and self-actualizing
		Deficit Principle – a satisfied need is not a motivator of behaviour, rather people act to satisfy deprived needs
		Progression Principle – individuals have to satisfy a lower need before moving to a higher order of need (step-by-step progression)
			• People bring "needs" to work and are driven to satisfy those needs which are: *(Five Human Needs)*:
			  ○ Physiological Needs – to stay alive
			  ○ Security Needs – to feel secure
			  ○ Social Needs – to belong
			  ○ Ego Needs – to be somebody
			  ○ Self-actualization – to develop potential
			• A need at any level only becomes activated once the next lower level need has been satisfied. These two principles mean that people strive to sequentially satisfy the five need levels.
				• Helps managers understand employee needs and how to address them through work
				• Managers who satisfy employee human needs will attain good operational productivity
						• Helps to explain employee actions at work and how to motivate them to perform at the desired level
						• Explains why employees expect more than a paycheck from work
						• Provides guidance on how to effectively respond to the workgroup after a traumatic event

50

CH. 1: OCCUPATIONAL HEALTH AND SAFETY: HISTORICAL PERSPECTIVES

Management Theory	Time frame	Theorist	Concepts	Impact on the Concept of Worker Motivation	Limitation of the Theory	Impact on Modern OH&S Practices

Human Resources and Behavioural Management Theory (cont'd)	1940s	David McClelland (Basic Needs Theory)	• Three human needs are:
			  1. Need for Achievement
			  2. Need for Power
			  3. Need for Affiliation
			
			People acquire or develop these needs over time as a result of individual life experiences. Each need is associated with a distinct set of work preferences.
				• Managers are encouraged to recognize the strength of each need in themselves, and in others, and to create work environments responsive to those basic human needs		• Links workplace influences with worker motivation

	(1969)	Clayton Alderfer (ERG Theory)	• Built on Maslow's work, but holds that needs are:
			  1. Existence Needs
			  2. Relatedness Needs
			  3. Growth Needs
			• Believes that people do not have to satisfy a lower need before addressing a higher-order need
			• Assumes that satisfied needs can be motivators
			• A satisfied need can become a motivator when a higher-order need is unmet (Frustration-Regression Principle)
				• Helps managers to better understand motivational drivers and worker responses		• Enhances knowledge on worker motivation

	1963	J. Stacy Adams (Equity Theory)	• Perceived inequity is a motivator
			• Perceived inequity is based on worker's comparison with others regarding rewards, job opportunities, pay increases, etc.
			• Comparisons can be internal/external
			• People respond by:
			  1. Changing their work inputs (their efforts)
			  2. Trying to change the rewards received
			  3. Using different comparators
				• Managers need to understand this concept and how to use this type of information wisely		• Furthers the field of knowledge on what motivates workers to perform well at work
						• Explains the impact of work rewards, compensation, benefits, and work conditions on worker motivation

Management Theory	Time frame	Theorist	Concepts	Impact on the Concept of Worker Motivation	Limitation of the Theory	Impact on Modern OH&S Practices
Human Resources and Behavioural Management Theory (cont'd)						
	1960	McGregor *(Theory X and Theory Y)* • Influenced by Maslow and Hawthorne Studies • Believed that managers should pay more attention to the social and self-actualization needs of workers	4. Rationalizing the inequities 5. Leaving the situation • When a worker perceives a positive inequity, he/she tends to balance the situation by increasing their work input • Managers either use: *Theory X:* • Employees hate work and are inherently lazy • Lack ambition/avoid responsibility • Irresponsible • Hate change • Seek security • Prefer to be led • Motivated by punishment • Self-centred and indifferent to organization's needs *or* *Theory Y:* • Employees want to work • Capable of self-direction • Are responsible • Capable of self-control • Capable of imagination, creativity • Motivated by achievement • Managers holding either the Theory X or Y assumptions can create "self-fulfilling prophecies" with their employees and their work groups	• Managers tend to adhere to either theory and employees respond accordingly — *"Self-fulfilling Prophecy"* • Added a new dimension to employee management	• Whether OH&S systems will be implemented or not depends on whether the leader of the organization is a Theory X or Y proponent	• Advises management and OH&S practitioners/professionals to keep expectations high in terms of safe work practices • Aligns with current emphasis on leadership, empowerment and self-managed teams • Managers holding either Theory X or Y philosophies can create "self-fulfilling prophecies"; subordinates tend to act in accordance with the manager's expectations
	1957	Chris Argyris *(Immaturity/Maturity Theory)*	• Argues that certain principles used in the Classical Management Approaches (Taylor and Fayol) that are intended to increase worker	• To encourage the development of responsible, mature employees, there is a need to adjust	• Talented employees may leave regardless of how much encouragement is provided	• More employee involvement in OH&S issues leads to ownership of the problem and the solution • Helps to explain the lack of "buy-in" to changes proposed

CH. 1: OCCUPATIONAL HEALTH AND SAFETY: HISTORICAL PERSPECTIVES

Management Theory	Time frame	Theorist	Concepts	Impact on the Concept of Worker Motivation	Limitation of the Theory	Impact on Modern OH&S Practices
Human Resources and Behavioural Management Theory (cont'd)			efficiency, are inconsistent with the mature adult personality • The organization's management style can be in conflict with employee personalities • Recommended that managers expand job requirements to include more task variety and responsibility	supervisory styles to include more employee participation		without employee input/involvement
	1964	Robert Blake and Jane Mouton (Managerial Grid)	Described the team management styles seen in organizations. The Managerial Grid includes styles like: • Country Club Management; • Impoverished Management; • Organization Management; • Authority-Obedience Management; and • Team Management. Also addressed Relationship Management versus Task Management			• Helps OH&S practitioners/professionals to understand team management styles and their characteristics so that they gain insight on how to deal with the various managers encountered in the workplace • Promotes team management as being the most effective approach
	1964	Victor Vroom (Expectancy Theory)	Worker motivation stems from: • What employees value; • Perceived performance/reward ratio; and • Employee expectations at succeeding	• Tell workers what you expect them to do • Make the work valuable • Make the work achievable • Provide feedback on the worker's progress • Reward success		• Provides valuable insight on how to motivate workers to adopt OH&S practices • Encourages leading indicators of safety versus the lagging indicators of safety • Supports the concepts of goal-oriented modified work and the use of performance appraisals
	mid-1960s	Edwin Locke (Goal-setting Theory)	• Task goals can be highly motivating if they are well-set and well-managed • Worker ownership of the goals depends on the degree of worker participation in setting the goals	Involve workers in setting work goals		• Provides valuable insight on how to motivate workers to adopt OH&S practices • Supports worker involvement in the development of safe work practices, development of OH&S training programs, and design of work

Management Theory	Time frame	Theorist	Concepts	Impact on the Concept of Worker Motivation	Limitation of the Theory	Impact on Modern OH&S Practices
Human Resources and Behavioural Management Theory (cont'd)	mid-1960s	Edward Thorndike and B.F. Skinner (*Reinforcement Theory of Motivation*)	• Based on *Thorndike's Law of Effect* behaviour that results in a pleasant outcome is repeated; that which does not, is not • Fits with Skinner's operant conditioning • Aligns with the concept of Organizational Behaviour modifications	• Positive Reinforcement and Negative Reinforcement can lead to desired behaviours • Punishment eliminates undesired behaviours • Behavioural extinction, eliminates behaviours		• Positive Reinforcement and Negative Reinforcement can lead to desired behaviours • Punishment eliminates undesired behaviours • Behavioural extinction eliminates behaviours – positive and negative in nature
	1959	Fredrick Herzberg (*Maintenance Factors/Motivating Factors*)	Two types of factors exist: 1. *Maintenance (Hygiene) Factors* (physical conditions, pay, status, benefits). These prevent dissatisfaction and are neutral re: motivation 2. *Motivating Factors* (responsibility, achievement, growth and recognition)	• These help employees meet ego and self-actualization needs and are therefore motivating	• The new generations of workers do not necessarily come to work for a paycheck – they want to be involved and participatory • Cannot use packaged OH&S training programs for all employee groups. OH&S training programs need to be tailored to meet worker needs	• Aligns with continuous improvement, excellence, TQM and performance • The theory impacts training approaches and techniques • The theory impacts Safety recognition programs
3. QUANTITATIVE MANAGEMENT APPROACHES						

Quantitative Management Approaches
(Focus on mathematical techniques for problem solving)
		The common characteristics are:
		• Primary focus is towards decisive decision-making
		• Based on economic decision-making criteria
		• Uses formal mathematical models
		• Uses computers
			• Assumes that mathematical approaches can enhance managerial decision-making and problem solving, e.g.,
			  ○ Trend Analysis
			  ○ Mathematical forecasting
			  ○ Linear programming
			  ○ Queuing Theory
			  ○ Network models
			  ○ Inventory modeling
			  ○ Simulations
						Promotes the use of:
						• OH&S audits
						• Incident investigation approaches
						• Collection and analysis of the lagging indicators of Safety
						• Trend analysis techniques
						• Metrics regarding leading indicators of safety
						• Business Jargon: Ways to speak about Occupational Health and Safety in business terms

4. MODERN MANAGEMENT THEORIES						
Modern Management Theories *(Focus on total systems thinking, contingency thinking and an awareness of global developments in management)*	1940-50s	Frank Bird; Ludwig von Bertalanffy (*Safety System Theory*)	• Organizations are systems that function toward a common purpose • Change one part of the system and all parts are affected • Recognizes that people are complex with multiple talents and abilities, and that people are variable	• High performing organizations offer a variety of managerial styles and job opportunities	Poorly understood and used until recently.	• Negative impact in one part of the system is felt within the other parts – applies to OH&S

CH. 1: OCCUPATIONAL HEALTH AND SAFETY: HISTORICAL PERSPECTIVES

Management Theory	Time frame	Theorist	Concepts	Impact on the Concept of Worker Motivation	Limitation of the Theory	Impact on Modern OH&S Practices

Modern Management Theories (cont'd) *(Focus on use of standards, measurement and continuous improvement)*	1990	Deming, Juran, Tom Peters, Conway, Peter Senge, Peter Drucker *(Total Quality Management)*
		• Management philosophy that focuses on satisfying customers
		• The objective is continuous improvement
		• Fundamental concepts:
		  ○ Employee involvement
		  ○ New culture
		  ○ New tools
		  ○ Continuous improvement and use of measures like "best practices, upstream and downstream measures"
			"Commitment to Quality"
			
			Total Quality Management includes:
			• Visionary leadership
			• Management commitment
			• Corporate culture
			• Customer focus
			• Statistical process control
			• Benchmarking
			• Continuous improvement
			• Training
			• Cross-functional teams
			• Empowerment
			• Self-managed teams
			• Learning organizations
			• Change management
			• Managing the total system
				• Dependent on heredity, needs, environment, culture
				• Considers "cognitive", "behavioural" and "affective" aspects
						• Loss control became integrated into the organization's vision, mission, strategic plans, policies and procedures
						• Safety Management System concept
						• Use of the Plan, Do, Check, Act Model
						• Format for OH&S auditing
						• Concept of continuous improvement (CI)

	1958	James Woodward *(Contingency Theory)*	• Assumes that there is no one right way to lead people • Holds a broad view on leader capability and other variables in the situation			• Introduces that concept of situational factors playing a role in leadership style

	1960s	Hersey and Blanchard *(Situational Leadership Theory)*	The concept of Situational Leadership holds that:
			• Leaders adjust their leadership style based on the "readiness" of the followers to perform in certain situations
			• Depending on follower readiness, leaders:
			  ○ Delegate
			  ○ Participate
			  ○ Sell
			  ○ Tell
				• Managers can enhance motivation in workers by how they lead		• Belief that there is no one best way to lead – The style selected depends on the situation and circumstances

	1971	Robert House *(Path-goal Theory)*	• An effective leader is one who clarifies the paths for followers to take to achieve both task-	• Managers can enhance motivation in workers by how they lead		• Provides insight into effective techniques of OH&S leadership

Management Theory	Time frame	Theorist	Concepts	Impact on the Concept of Worker Motivation	Limitation of the Theory	Impact on Modern OH&S Practices
Modern Management Theories (cont'd)			related and personal goals (career streaming) • Good leaders help employees progress, remove barriers and provide opportunities for employee growth and development	1. Directive Leadership 2. Supportive Leadership 3. Achievement-oriented Leadership 4. Participative Leadership		
	1978	**James MacGregor Burns** *(Transformational Leadership)*	• Inspired leadership that influences beliefs, values and goals of followers so that they can perform in an extraordinary manner	• Leader motivates followers by: 1. Clear vision 2. Charisma 3. Symbolism 4. Empowerment 5. Intellectual stimulation 6. Integrity		• Provides further insight into effective techniques of OH&S leadership
	1947-1987	**Charismatic Leadership**	• Followers are attracted to leaders who believe in themselves, possess charm and grace, and are admired	• Leader motivates followers by: 1. Self-confidence 2. Charm and grace 3. Admired		• Provides further insight into effective techniques of OH&S leadership
	1982	**Tom Peters and Robert Waterman** *(In Search of Excellence)*	• The drive for performance excellence is based on eight attributes: 1. Bias for action 2. Closeness to customer 3. Autonomy and entrepreneurship 4. Productivity through people 5. "Hands on" and value-driven 6. Simple form and lean staff complement 7. "Sticking to the knitting" 8. Simultaneous loose-tight properties			• Demonstrates how Occupational Health and Safety performance can be enhanced

2. Organizational Designs: Characteristics[67]

Given that OH&S Practitioners/Professionals will spend most of their careers operating in some sort of organizational system, the characteristics of four primary organizational designs are worthy of discussion. Critical to the effectiveness of the OH&S function is the OH&S Practitioner's/Professional's knowledge of, and ability to understand how, the organization is designed and operates.

The **Bureaucratic Design (B)** is typically encountered in government, military and hospital settings. It is characterized by:

- a clear-cut division of labour;
- strict hierarchy of authority;
- staffing by technical competency;
- existence of formal rules and procedures; and
- an impersonal approach to workplace decision-making.

In terms of the **Mechanistic Design**, it is an organizational structure characterized by a rigidly and tightly controlled structure that is aimed at minimizing the impact of differing human traits. Most large organizations have some elements of the mechanistic organizational design. The variations include:

1. **Machine Bureaucracy (MB)**

    Exemplified by McDonalds Restaurants, this organizational design has:

    - a clear hierarchy of authority with a large middle-management group;
    - highly specialized and standardized tasks;
    - functional departmentalization; and
    - top-management decision making.

2. **Professional Bureaucracy (PB)**

    Typical of the organizational design of law firms and universities, the professional bureaucracy:

    - is staffed with highly-educated professionals;
    - demonstrates a high level of autonomy;
    - has a decentralized structure;
    - possesses a large number of support staff; and
    - has few middle-management personnel.

---

[67] Johnson Safety Management, *Occupational Health, Safety and Environmental Competencies, Module Two: OHS and E Systems* (Edmonton, AB, 2001); D. Sayers & Associates, *Management and Organizational Behaviour, BCRSP Self-Study Guide* (Mississauga, ON: Board of Certified Safety Professionals, 2000), at 41-51.

3. **Divisionalized Bureaucracy (DB)**

   This organizational design demonstrates the traits of:
   - a hybrid form of departmentalization;
   - a number of relatively autonomous internal units operating within a common organizational umbrella;
   - divisions that are formed based on product/service, customer, or geographic location; and
   - top-management coordination assisted by a large staff complement.

The **Adaptive Organization Design (A)** is more modern in approach; thus, adaptive organizations:
- operate with minimal bureaucratic features;
- focus on developing cultures that value worker empowerment and participation; and
- use team and network structures.

The **Organic Design Alternatives (O)** operate with a more decentralized authority structure in which there are few rules and procedures, less division of labour, wider spans of control and a more personal means of coordination. There are two types:

1. *Simple Structure Organizational Design:*
   - has little or no staff complement; and
   - has the presence of a single top manager who works directly with other personnel and performs a variety of tasks.

2. *Adhocracy Organizational Design:*
   - has little staff-line distinction;
   - relationships develop based on knowledge and expertise;
   - has highly-skilled personnel; and
   - includes cross-functional teams.

Table 1.3 summarizes the characteristics of these four primary organizational designs.

Table 1.3: The Characteristics of the Primary Organizational Designs

DESIGN	DESCRIPTION	EXAMPLES
Bureaucratic (B)	• Clear-cut division of labour • Strict hierarchy of authority • Staffing by technical competency • Formal rules and procedures • Impersonal approach to decision-making	• Military • Hospitals

Mechanistic Design Alternatives	**Machine Bureaucracy: (MB)** • Clear hierarchy of authority with a large middle-management group • Highly specialized and standardized tasks • Functional departmentalization • Top-management decision making	• McDonalds restaurants
	Professional Bureaucracy: (PB) • Staffed with highly educated professionals • A high level of autonomy • Decentralized structure • Large number of support staff • Few middle management	• Legal firms • Universities
	Divisionalized Bureaucracy: (DB) • Hybrid form of departmentalization • Made up of a number of relatively autonomous internal units operating within a common organizational umbrella • Divisions formed based on product/service, customer, or geographic location • Coordinated by top management assisted by a large staff complement	Many national and international companies such as Petro-Canada Inc.
Adaptive Organizations (A)	• Operates with minimal bureaucratic features • Focuses on developing cultures that value worker empowerment and participation • Uses team and network structures	
Organic Design Alternatives (O)	• Operates with a more decentralized authority • Few rules and procedures • Less division of labour • Wider span of control • A more personal means of coordination	Many software development companies
	• Two types: 1. Simple Structure • Has little or no staff • Has a single, top manager who works directly with other personnel doing a variety of tasks 2. Adhocracy • Has little "staff-line" distinction • Relationships develop based on knowledge and expertise • Has highly skilled personnel • Involves cross-functional teams	

3. Organizational Designs: Relationship with Organizational Elements[68]

Each of the organizational designs described above impacts the organizational elements of:

- **Hierarchy of Authority** — a chain of command; a superior-subordinate or leader-follower relationship. The hierarchy is that the subordinate reports to the superior in a linear manner. It allows for quick decisions and action because the manager has complete authority for the operation and only has to consult with his or her direct supervisor to make decisions.
- **Rules and Procedures** — the work and safety standards to which workers are expected to perform.
- **Division of Labour** — work is divided up so that workers can focus on specific work duties. Enabling workers to specialize, so they can become proficient and efficient in the performance of their work tasks.
- **Span of Control** — the number of subordinates that reports to a superior.
- **Organizational Coordination** — how work is organized and how organizations operate. It includes the structure, processes, and nature of the work relationships.

Figure 1.1 demonstrates organizational design in relation to the elements of an organization. By way of explanation, for each element there are two poles. For example, one can characterize **Hierarchy of Authority** as centralized or decentralized. In between, there is a continuum. The same can be said about **Rules and Procedures**, **Division of Labour**, **Span of Control**, and **Organizational Coordination**.

Each organizational design is identified symbolically. The code is:

- Bureaucratic Design (B)
- Machine Bureaucracy (MB)
- Professional Bureaucracy (PB)
- Divisionalized Bureaucracy (DB)
- Adaptive Organization Design (A)
- Organic Design Alternatives (O)

Each organizational design can be classified on each of these elements. For example, the Bureaucratic Design has a centralized hierarchy of authority, many formalized rules and procedures, a precise division of labour, narrow span of control, and formal and impersonal organizational coordination. In contrast, the

---

[68] Johnson Safety Management, *Occupational Health, Safety and Environmental Competencies, Module Two: OHS and E Systems* (Edmonton, AB, 2001); D. Sayers & Associates, *Management and Organizational Behaviour, BCRSP Self-Study Guide* (Mississauga, ON: Board of Certified Safety Professionals, 2000), at 41-51.

Organic Design Alternatives features a decentralized hierarchy of authority, few formalized rules and procedures, a precise division of labour, an open span of control, and informal and personal organizational coordination. The rest of the organizational designs are in-between.

Figure 1.1: Organizational Design in Relationship to the Elements of an Organization

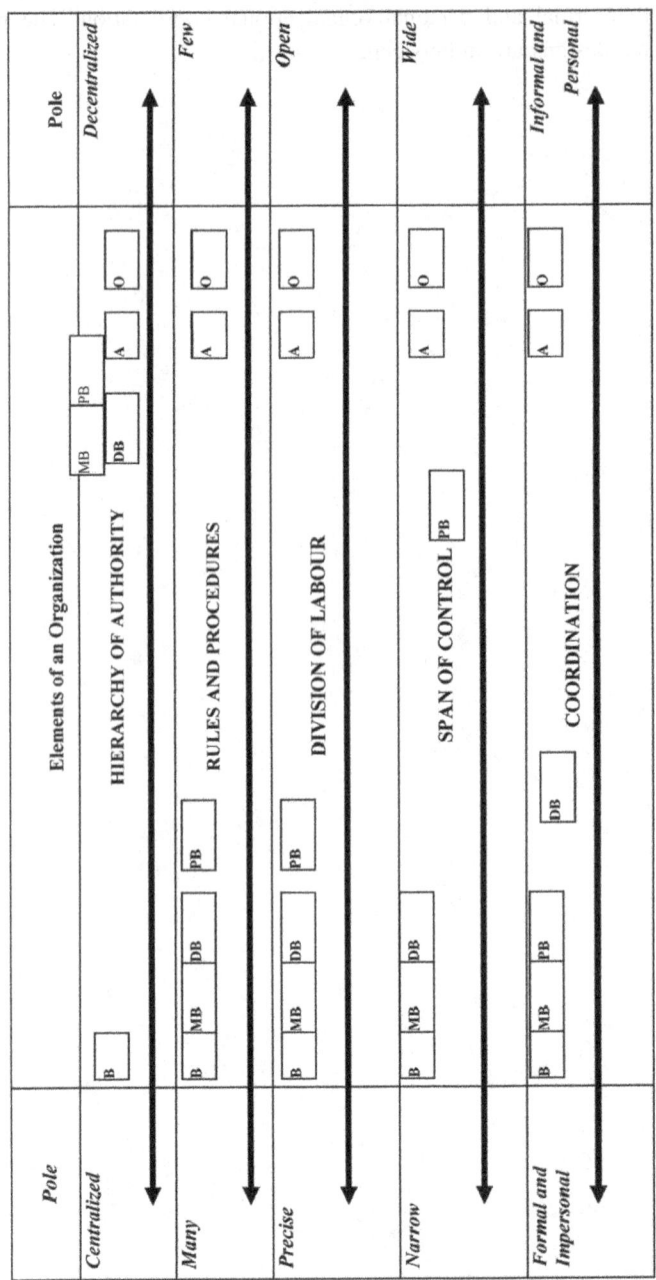

4. Organizational Designs: Structures[69]

Organizations choose to group people and jobs in ways that align with their business strategies and functions. OH&S Practitioners/Professionals need to recognize and appreciate why these structures were adopted, along with the advantages and limitations of each design structure.

To begin with, **departmentalization** is the process whereby people and jobs are grouped into various work units or departments to create a well-functioning whole. Workers can be grouped according to function, division, a mix of function and division, a cross-functional matrix, teams, and networks.

In an approach that uses a functional structure (**Functional Structure Approach**), workers are grouped based on similar skills and common functions. For example, workers doing accounting and payment functions would be in the financial department; workers assigned with the legal and legal-related functions would be in the legal department; and employee recruitment, selection, compensation, human resource management, and other human resource-related functions would be in the Human Resources department. The advantages of this approach are that it tends to:

- work well for small organizations in a stable environment with one or a few products or services to market;
- offer technical strengths;
- capitalize on the economies of scale and the efficient use of resources;
- use task assignments consistent with worker training/education;
- support high-quality technical problem-solving;
- enable in-depth training and skill development; and
- afford clear-cut career paths for employees.

The limitations of the Functional Structure Approach are that it promotes a "silo-effect" because of the limited opportunities for cross-functional activities and communication barriers. Overspecialization and a narrow view of the corporate objectives tend to occur. There is the tendency for too many decisions being referred upward in the hierarchy, as well as a lack of clear responsibility for product/service delivery, to be exhibited. The organization tends to be slow to demonstrate an adaptive response to change in the market, industry and social environments.

Grouping workers based on similar product/services, customers in the same geographic area, or on the same time schedule, is known as a **Divisional Structure Approach**. For example, workers who are dealing with a product like natural gas may be the Natural Gas Marketing Group; or workers located in a particular part of the country might be identified as being the Central Division or Eastern Division. A Night Crew would be those workers assigned to working the night shift doing a specific set of duties. The strength of a divisional structuring approach is that it

---

[69] *Ibid.*

works well in complex situations where organizations are pursuing diverse business strategies and allows for more integration of business functions and workers.

The **Hybrid Structure Approach** of grouping workers by functions and divisions is often used by larger organizations. It helps them to address different operating contingencies, and to match sub-unit structures with specific challenges. It is also characteristic of organizations seeking the benefits of global operations.

A **Matrix Structure Approach** uses permanent cross-functional teams to blend the technical strengths of the functional structures with the integrating potential of divisional structures. Team members have dual alliances to a functional manager as well as to a project manager. The advantage of this approach is that it:

- facilitates the management of several projects operating at any given time;
- offers greater flexibility;
- promotes more cross-functional co-operation;
- enables greater customer satisfaction;
- promotes better accountability;
- improves decision making;
- improves strategic management; and
- enables the pursuit of growth strategies in dynamic and complex market environments.

The limitations of the matrix structure are that it is a challenge for workers to answer to two bosses — team members may become confused as to who to take work directions from. As well, power struggles in a two-boss system are not uncommon. "Groupitis" is another challenge: team members often become too focused on themselves and their beliefs, thereby losing sight of the bigger corporate goals. Lastly, it can result in additional costs to the organization because it adds an extra level of managers to the system.

The approach that uses a **Team Structure** is related to the Matrix Structure Approach, but in this structure, permanent and temporary teams are created to improve lateral relations and to solve problems throughout the organization. The team structure is superimposed over the functional structure. Cross-functional teams are used to supplement, not replace, the normal mode of operation. They work together to solve problems and to seek out opportunities. For example, a team might exist with a mandate to seek operational efficiencies throughout an organization. This structural approach tends to harness the intellectual and problem-solving potential of the workforce.

The last structural approach to be presented is the **Network Structure Approach**. It consists of a central-core workforce that is linked through networks with external suppliers of essential business services. These networks, in the form of strategic alliances and business contracts, allow the organization to operate without having to "own" all its supporting functions. For example, an organization might link with a contracted Disability Management service and an Employee Assistance Program (EAP) service to provide more robust Disability Management Program within the organization. The advantages of this approach are that it tends to work well for small companies and helps to keep overhead costs down. However, some of the limitations are that the success of the outsourcing arrangement is reliant on sound supply-chain practices and services, and good vendor management skills (see Chapter 40, "Outsourcing Occupational Health and Safety Services"). It can also be negatively impacted by a variety of market pressures.

Table 1.4 provides a summary of the traditional approaches to organizational structuring.

Table 1.4: Traditional Approaches to Organizational Structuring

APPROACH	DESCRIPTION	ADVANTAGES	LIMITATIONS
Functional Structure Approach	Employees grouped based on similar skills and common functions, *e.g.*, Financial Department, Legal Department, OH&S Department	• Works well for small organizations in a stable environment with one or a few products or services being marketed • Offers technical strengths • Offers economies of scale and an efficient use of resources • Task assignment is consistent with training • Promotes high-quality technical problem-solving • Enables in-depth training and skill development • Affords clear-cut career paths for employees	• Results in a "silo-effect" due to limited cross-functional activities and communications • Tends to lead to overspecialization • Leads to a narrow view of corporate objectives • Too many decisions get referred up the corporate hierarchy • Can result in the unclear responsibility for product/service delivery • The organization tends to be slow to demonstrate an adaptive response to change in the marketplace, industry and social environments
Divisional Structure Approach	Employees are grouped based on similar product, customers in same geographic area, or on the same time schedule, *e.g.*, Western Division, Night Crew	• Works well in complex situations where organizations are pursuing diverse business strategies • Allows for more integration of functions and employees	• Can result in a "silo-effect" due to limited cross-functional activities and communications • Tends to lead to a narrow view of corporate objectives
Hybrid Structure Approach	Blends the functional and divisional approaches described so far	• Used by larger organizations • Helps to address different operating contingencies and match subunit structures with specific challenges • Characteristic of organizations seeking the benefits of global operations	• Can result in a "silo-effect" due to limited cross-functional activities and communications • Tends to lead to a narrow view of corporate objectives

Matrix Structure Approach	• Uses permanent cross-functional teams to blend the technical strengths of the functional structures with the integrating potential of divisional structures • Team members have dual allegiances to their functional managers and to their project managers	• Useful when organizations have several projects operating at any given time • Offers great flexibility • Affords more interfunctional co-operation • Leads to greater customer satisfaction • Promotes better accountability • Results in improved decision-making capabilities • Enables improved strategic management • Characteristic of organizations pursuing growth strategies in dynamic and complex business environments	• Can result in "power struggles" due to the use of a "two-boss system" • Can lead to anarchy: team members may become confused as to who to take work directions from • Can lead to "groupitis": team members become too focused on themselves and lose sight of the corporate goals • May result in excessive overhead costs: this approach tends to add extra managers
Team Structure Approach	• Related to the Matrix approach, but here permanent and temporary teams are created to improve lateral relations and to solve problems throughout the organization • A team structure is superimposed over the functional structure • Cross-functional teams are used to supplement, not replace, the normal mode of operation. They work together to solve problems and seek out opportunities	• Harnesses intellectual and problem-solving potential of workforce	• Can result in "power struggles" due to the use of the "two-boss system" • Can be perceived by workers as "taking on two jobs", but being paid for just one

Network Structure Approach	• Consists of a central core workforce that is linked through networks with outside suppliers of essential business services • These networks, in the form of strategic alliances and business contracts, allow the organization to operate without having to own all its supporting functions (*e.g.*, contracted disability management services, industrial hygiene services, EAP services)	• Works well for small and medium-sized companies • Keeps overhead costs down	• Reliant on sound supply chain practices and services • Requires the use of good vendor management skills • Can be negatively impacted by market pressures

D. OH&S: INTERNAL AND EXTERNAL RESPONSIBILITY SYSTEMS

Under Canadian and U.S. OH&S legislation, companies are legally required to develop an Internal Responsibility System for ensuring high quality workplace health and safety. An **Internal Responsibility System** (IRS) is the "people framework" within an organization. It maintains that:

- everyone within a company is responsible for workplace health and safety;
- everyone must do their part to identify and solve health and safety issues or problems and to look for ways to improve the work processes with which they are involved; and
- by capturing the creativity, leadership, experience, and knowledge of workers, companies can improve workplace health and safety.

When examined further, the term *Internal* means "within the workplace" and "responsibility is within the job". *Responsibility* refers to the personal duty of care that everyone must ensure workplace health and safety. Workers have three essential rights:

1. the right to know about workplace hazards;
2. the right to refuse unsafe work; and
3. the right to participate in health and safety decision-making activities directly or through representatives.

The term *System* indicates that the workplace is a closed system with individual parts that all work together for the good of workers and the whole organization. The purpose of the system is to eliminate/reduce workplace injuries/illness. Everyone operating within the Internal Responsibility System must take reasonable precautions to prevent harm to themselves and others.

Within an organization, OH&S accountability in the form of authority and responsibility is delegated downward and accountability for workplace health and safety move upward (Figure 1.2). Here, accountability is defined as "measured commitment".

Figure 1.2: Internal Responsibility System: Structure and Accountability

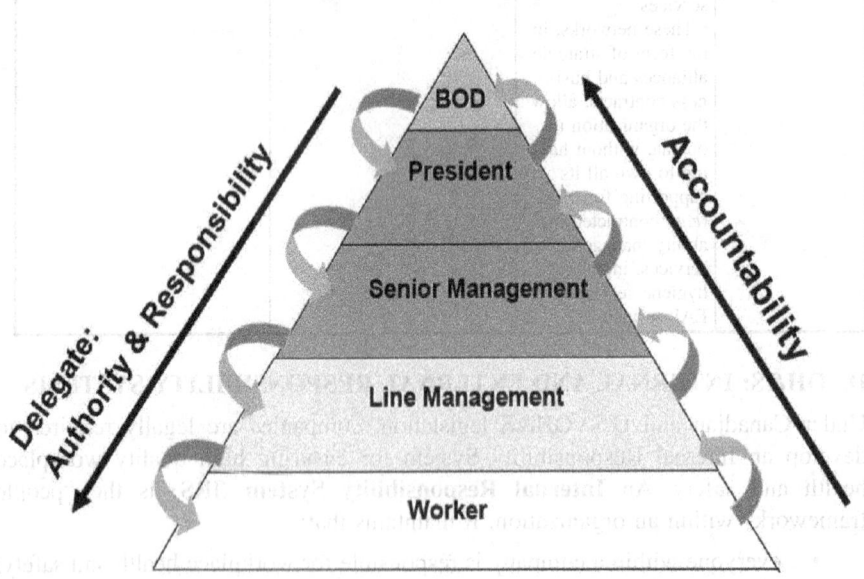

To operationalize this concept further, each player in the system has a role to play towards minimizing unsafe acts/conditions and the resultant work-related illness/injury (Figure 1.3).

---

[70] Reproduced courtesy of *Workplace Environment, Health & Safety Reporter*, "Minimizing Unsafe Acts" (Toronto, ON: Templegate Information Services Inc., 1998). For subscription information, phone: (416) 920-0768.

Figure 1.3: Workplace Contributions to the Minimization of Unsafe Acts[70]

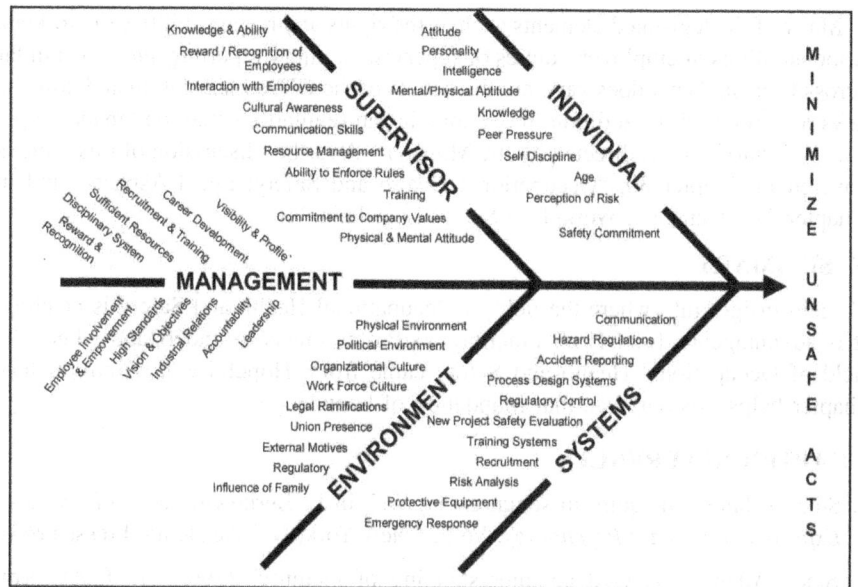

In summary, responsibility for Occupational Health and Safety is internal to the workplace and is internal to the job of everyone within the organization. Actions taken to fulfil this responsibility are also referred to as OH&S due diligence.

In comparison, the **External Responsibility System** (ERS) comes into play when an organization's IRS fails. The ERS translates to government intervention in the form of enforcement action by a provincial or State OH&S governmental agency. Hence, when an organization/company fails to uphold its internal responsibilities, the government steps in.

E. CANADA'S OH&S SYSTEM: FEDERAL AND PROVINCIAL

In Canada, the provinces and territories are accountable for the development and operation of the External Responsibility System (ERS). Each province has a government department whose mandate is to protect the well-being of workers.

The general responsibilities of the Federal and provincial governments regarding Occupational Health and Safety include:

- enforcement of OH&S legislation;
- workplace inspections;
- dissemination of OH&S information;
- promotion of worker training, education and OH&S; and

- resolution of OH&S disputes.[71]

Many of the legislated elements such as the rights and responsibilities of workers, responsibilities of employers, duties of supervisors, injury reporting, *etc.*, are similar across Canada. What does vary, are the details of the OH&S legislation and how the laws are enforced. In addition, provisions in the regulations may be "mandatory", "discretionary" or "as directed by the Minister". A further discussion of this topic is covered in Chapter 30, "Occupational Health and Safety: Legal Aspects" and in Chapter 31, "Canadian Workplace Safety Legislation".

F. SUMMARY

To gain insight into where the field of Occupational Health and Safety is heading, it is advantageous for OH&S Practitioners/Professionals to understand where the field of Occupational Health and Safety came from. Hopefully, this introductory chapter helps to strengthen that foundation of knowledge.

CHAPTER REFERENCES

J. Stacy Adams, "Inequity in social exchange", in L. Berkowitz, ed., *Advances in Experimental Social Psychology*, Vol. 2 (New York, NY: Academic Press, 1965).

J. Stacy Adams, "Toward an understanding of inequity" (1963) 67 J. Abnorm. Psychol. 422-436.

C. Argyris, "The Organizational Environment and Productivity, Human Relations Contributors", online: http://www.accel-team.com/human_relations/hrels_06ii_argyris.html (date accessed: February 28, 2020).

F. Bird and G. Germain, *Loss Control Management: Practical Loss Control Leadership*, revised ed. (Loganville, GA: DNV, 1996).

Changing Minds, Leadership Theories: Transformational Leadership, online: http://changingminds.org/disciplines/leadership/styles/transformational_leadership.htm (date accessed: February 28, 2020).

J.A. Conger and R.N. Kanungo (eds.), *Charismatic Leadership in Organizations* (Thousand Oaks, CA: Sage Publications, 1998).

R. Csiernik, *Wellness and Work* (Toronto: Canadian Scholars' Press, 2005).

L. Duxbury and C. Higgins, "Voices of Canadians: Seeking Work-life Balance", Human Resources and Social Development of Canada Report (2003), online: http://publications.gc.ca/collections/Collection/RH54-12-2003E.pdf (date accessed: February 28, 2020).

W. Fox-Decent, "Workers' Compensation: Past, Present, and Future — An Historical Overview" (2002).

---

[71] OH&S Legislation in Canada — Basic Responsibilities, Legislation (OH&S Answers), online: http://www.ccohs.com/oshanswers/> (date accessed: February 28, 2020).

M. Guarnieri, "Landmarks in the History of Safety" (1992) 23(3) J. Safety Res. 151-158.

D. Hofmann and F. Morgeson, "The Role of Leadership in Safety", in J. Barling and M. Frone (eds.), *The Psychology of Workplace Safety* (Washington, DC: American Psychological Association, 2004).

R.J. House, "A Path-Goal Theory of Leader Effectiveness" (1971) 16(3) Adm. Sci. Q. 321-339.

R.J. House, "A 1976 theory of charismatic leadership", in J.G. Hunt and L.L. Larson (eds.), *Leadership: the cutting edge* (Carbondale, IL: Southern Illinois University Press, 1977).

ILR School, Cornell University, "The Triangle Waist Factory Fire", online: http://www.ilr.cornell.edu/trianglefire/story/introduction.html (date accessed: February 28, 2020).

Johnson Safety Management, *Occupational Health, Safety and Environmental Competencies, Module Two: OHS and E Systems* (Edmonton, AB, 2001).

ILO/WHO (1950). *Joint ILO/WHO Expert Committee on Industrial Hygiene Report*, 1st Session, August 28 to September 2, 1950 (Geneva: International Labour Office).

A. Marriner, *Guide to Nursing Management*, 2d ed. (St. Louis, MO: The C.V. Mosby Company, 1984).

NETMBA, "Fredrick Taylor and Scientific Management", Management section, online: http://www.netmba.com/mgmt/scientific/ (date accessed: February 28, 2020).

NETMBA, "Theory X and Theory Y", Management section, online: http://www.netmba.com/mgmt/ob/motivation/mcgregor/ (date accessed: February 28, 2020).

G. Norwood, "Maslow's Hierarchy of Needs", Deeper Minds, online: http://www.deepermind.com/20maslow.htm (date accessed: February 28, 2020).

OH&S Legislation in Canada — Basic Responsibilities, Legislation (OH&S Answers), available online: http://www.ccohs.com/oshanswers (date accessed: February 28, 2020).

T. Peters and R. Waterman, *In Search of Excellence* (New York, NY: HarperCollins Publishers, 1982).

B. Plog, *Fundamentals of Industrial Hygiene*, 3d ed. (Chicago, Illinois: National Safety Council, 1988).

R. Robinson D. Smith, G. Golder and B. Sharper, "Program Evaluation and Performance Measurement Study", *CCOHS* (Jan. 26, 2006), online: http://www.ccohs.ca/ccohs/reports/studies/performanceJan06.html (date accessed: February 28, 2020).

D. Sayers & Associates, *Management and Organizational Behaviour, BCRSP*

Self-Study Guide (Mississauga, ON: Board of Certified Safety Professionals, 2000).

P. Sellers, "What exactly is Charisma?", *Fortune* (January 15, 1996).

University of West Virginia, Reinforcement Theory, online: http://www.managementstudyguide.com/reinforcement-theory-motivation.htm (date accessed February 28, 2020).

Value Based Management, ERG Theory by Clayton P. Alderfer, online: http://www.valuebasedmanagement.net/methods_alderfer_erg_theory.html (date accessed: February 28, 2020).

Value Based Management, Expectancy Theory by Victor Vroom, online: http://www.valuebasedmanagement.net/methods_vroom_expectancy_theory.html (date accessed: February 28, 2020).

A. Weber, "The Hawthorne Works" (2002) *Assembly Magazine*, online: http://www.assemblymag.com/CDA/Archives/0cdaaa2e0d5c910VgnVCM100000f932a8c0 (date accessed: February 28, 2020).

Max Weber, "The Nature of Charismatic Authority and its Routinization" in *Theory of Social and Economic Organization*, translated by A.R. Anderson and Talcot Parsons (eds.) (1947). Originally published in 1922 in German under the title *Wirtschaft und Gesellschaft* (chapter III, § 10).

Wikipedia entries cited in this chapter are available online: http://en.wikipedia.org (date accessed: February 28, 2020).

Workplace Environment, Health & Safety Reporter, "Minimizing Unsafe Acts" (Toronto: Templegate Information Services Inc, 1998).

C. Wortman, E. Loftus and M. Marshall, *Psychology*, 3rd ed. (New York, NY: Knopf, 1988).

Chapter 2

OCCUPATIONAL HEALTH AND SAFETY MANAGEMENT SYSTEMS

A. INTRODUCTION

An **Occupational Health and Safety Management System** (OHSMS) can be defined as the part of the overall management system which includes the organizational structure; planning activities; responsibilities; practices; procedures; processes; and resources for developing, implementing, achieving, reviewing, and maintaining the Occupational Health and Safety Policy.[1]

An **Occupational Health and Safety (OH&S) Policy** is a statement made by an organization of its intentions and principles in relation to its overall OH&S performance. It provides a framework for action and for setting its OH&S goals, objectives, and performance targets.

An **Occupational Health and Safety (OH&S) Program** is designed to promote, protect, and restore worker health within the context of a safe and healthy work environment. The effectiveness of an OH&S Program depends on many factors, most notably Management, union, and employee endorsement and support. A well-designed OH&S Program provides a framework for compliance with the applicable legislation. In addition, it can result in decreased injuries and lost workdays; significant cost savings; increased productivity; enhanced employee morale and commitment; and good business outcomes.

In this chapter, the elements of an OHSMS are described as well as the rationale for having an OHSMS; the purpose and value of an OHSMS; the structure of an OHSMS; OHSMS standards; and the evaluation of the effectiveness of an OHSMS. The latter part of this chapter is dedicated to a discussion of an OH&S Program, followed by the evaluation of an OHSMS and OH&S Program.

B. OCCUPATIONAL HEALTH AND SAFETY ELEMENTS

The elements of an OHSMS are:

1. Management leadership and commitment to Occupational Health and Safety (usually expressed as an OH&S Policy).

---

[1] BSI Management Systems, "OHSAS 18001 Awareness" (Presented at the City of Calgary Informational Seminar, January 2003), at 4-3.

2. Occupational Health and Safety communication.
3. OHSMS structure, usually expressed as an OH&S Program.
4. Employee qualifications, orientation, and training.
5. Hazard identification and assessment.
6. Hazard control.
7. Worksite inspections.
8. Emergency preparedness and response.
9. Incident investigation and corrective actions.
10. System review and continuous improvement (see Figure 2.1).

A more in-depth discussion of the elements of an OHSMS follows Figure 2.1.

Figure 2.1: An Effective Occupational Health & Safety Management System: The Basic Elements of an OHSMS

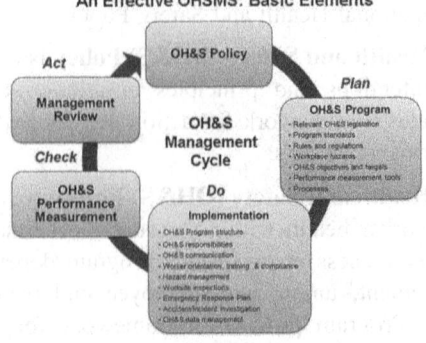

1. Management Leadership and Commitment to Occupational Health & Safety

At every level within an organization, **loss control** is an important part of the Management leader's role. The rationale for this is that Management is:

- responsible for the health and safety of subordinates, the public, and the environment;
- responsible for controlling business costs through managing health, safety and environmental risks; and
- accountable for upholding operational strategy through loss control management that can improve overall management, including managing property damage, process loss, and service/product quality.[2]

---

[2] F. Bird and G. Germain, *Loss Control Management: Practical Loss Control Leadership*, revised ed. (Loganville, GA: DNV, 1996), at 27.

The management functions recognized as critical in today's workplace include planning, organizing, leading/directing, and controlling. The last function — controlling — warrants further discussion.

Controlling is defined as comparing results with predetermined standards of performance and taking corrective action when performance deviates from these standards.[3] It involves establishing performance standards, determining how performance will be measured, evaluating performance, and providing performance feedback so that behaviours/culture can be changed.[4] Given that OH&S legislation is a performance-based standard, it is vital for employers to be able to demonstrate effective and compliant OH&S performance.

There are five steps Management can use for controlling losses. According to the Det Norske Veritas (DNV), they can be summarized with the acronyms: **I** (Identification); **S** (Standards); **M** (Measurement); **E** (Evaluation); and **Cm** (Commending Compliance) (see Figure 2.1).[5]

Identification describes the work that needs to be done to achieve the desired loss control objectives. ***Standards*** establish the terms for safe work performance. ***Measurement*** of performance is done as compared to established standards. ***Evaluation*** of performance and stakeholder communication occurs. ***Commending compliance*** refers to correcting deficiencies in performance standards (see Figure 2.2).

Figure 2.2: Leadership Activities for Managing Control[6]

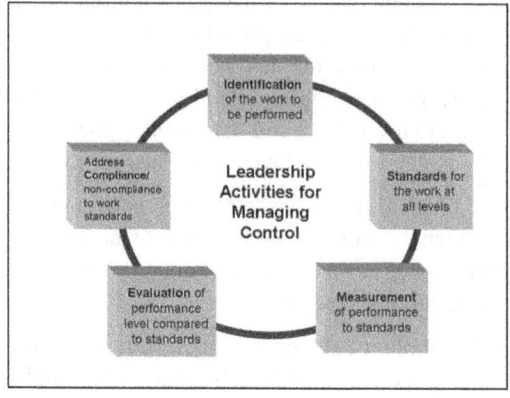

---

[3] E. Sullivan and P. Decker, *Effective Management in Nursing*, 2nd ed. (Menlo Park, CA: Addison-Wesley Publishing Company, 1988), at 56.

[4] *Ibid*.

[5] F. Bird and G. Germain, *Loss Control Management: Practical Loss Control Leadership*, revised ed. (Loganville, GA: DNV, 1996), at 27.

[6] Adapted from F. Bird and G. Germain, *Loss Control Management: Practical Loss Control Leadership*, revised ed. (Loganville, Georgia: DNV, 1996), at 32.

A short, a well-managed loss control system provides an operational strategy to improve overall management.

2. Occupational Health & Safety Communication

Effective Occupational Health & Safety (OH&S) communication is vital to marketing and gaining "buy-in" for an OH&S Program and its initiatives. By developing an OH&S Communication Plan and using the available company communication vehicles, Management can distribute the desired OH&S information to all stakeholders. Further discussion of OH&S communication is provided in Chapter 15, "Occupational Health and Safety: Risk Management and Communication" and Chapter 16, "Marketing Occupational Health and Safety Programs; Communicating the Results".

3. OHSMS Structure

The typical operational structure for an OHSMS is an OH&S Program which is described later in this chapter.

4. Employee Qualifications, Orientation and Training

Ensuring that employees are properly qualified is the first step to ensuring that the employee can competently and safely do the job. However, to determine if the employee is competent, the employer needs to observe the employee's work practices.

One way to achieve this outcome is by using new employee orientation sessions. An employee orientation should be designed to be a two-way process of reaching mutual understanding,[7] welcoming the new person to the organization,[8] socializing the employee to the workplace culture, helping the new employee to become familiar with the work environment,[9] and determining the skills and nature of the information required for the new employee to do his/her job safely.

Another purpose of the new employee orientation is to enable the employee to be able to meet the company's work standards and to do the job safely. Research indicates that the provision of new employee orientations is associated with a 25 per cent lower incident rate among new hires.[10]

Once on the job, the employee requires ongoing training to remain competent in the job. **Training** is a process of continually instructing employees on the skills necessary to do a job competently and safely. Managers provide training that stresses safety, work methodologies, and hazard assessments. According to the

---

[7] F. Bird and G. Germain, *Loss Control Management: Practical Loss Control Leadership*, revised ed. (Loganville, GA: DNV, 1996), at 228.

[8] *Ibid.*

[9] *Ibid.*

[10] *Ibid.*

DNV, workplace training significantly reduces incident costs.[11] Chapter 17, "Occupational Health and Safety: Worker OH&S Training" provides more detail on employee education and training.

5. Hazard Identification and Assessment

Hazard identification involves doing an inventory of the workplace hazards. The most effective way to do this is to observe routine work operations as they are being performed. These observations include the:

- general condition of the worksite;
- condition of the equipment/tools;
- presence of workplace hazards, *i.e.*, noise, electricity, mechanical, chemical, gravity, thermal, pressure, radiant energy, biological, and body mechanics;[12]
- presence, quality and effectiveness of the hazard controls, *i.e.*, engineering controls, administrative controls, and personal protective equipment;
- availability, condition, and employee use of personal protective equipment;
- presence and condition of access equipment, elevated surfaces, walkways, aisles, and means of egress/exit;
- quality of the hazard communication available to employees;
- presence of available safety measures such as safe work practices, codes of practice, rescue drills, and emergency procedures;
- presence of fire and spill-prevention equipment; and
- workstation/job design.

Employees, supervisors, and union representatives are well-positioned to assist with the identification of workplace hazards and hazardous conditions.

Hazard assessment is the evaluation of the risks posed by these identified hazards to the people, products, plant, processes, and environment. The assessments deal with the frequency, severity, and potential for loss that each hazard poses. The hazards are then ranked according to risk potential, and other factors such as legislative requirements, corporate standards, industry standards, and "best practices". Based on these risk assessments, companies can establish the required resources, activities, processes, and programs to control the identified hazards.

The topic of hazard identification, assessment and control is discussed further in Chapter 9, "Occupational Health and Safety: Hazard Identification, Assessment and Control", and in Chapter 10, "New Technology: Impact on the Workplace".

---

[11] *Ibid.*, at 249.

[12] These hazard types are described in greater detail in Chapter 9, "Occupational Health and Safety: Hazard Identification, Assessment and Control".

6. Hazard Control

Since it is impossible to eliminate all workplace hazards, organizations must make provisions for controlling these hazards if they wish to manage their losses. According to William Haddon,[13] there are ten general hazard control measures that organizations can use to prevent or mitigate the negative effects posed by a specific hazard:

- prevent the hazard from occurring in the first place;
- reduce the amount of the hazard present;
- prevent exposure to the hazard by controlling the release;
- modify the rate or spatial distribution of the release of the hazard;
- separate workers from the hazard;
- use a barrier to isolate the hazard from the workers;
- modify the chemical components of the hazard to make it less toxic;
- increase worker resistance against the hazard's negative effects;
- act to counter any negative effects that may occur; and
- stabilize, repair, and rehabilitate the injured worker.[14]

To achieve these measures, there are four well-recognized, hazard-control techniques. They are presented here in order of priority for use:

1. **Substitution** — the replacement of a hazardous substance, process or piece of equipment with a less problematic option. For example, replacing isocyanates with less hazardous paint hardeners.

2. **Engineering Controls** — tangible controls that change or modify worker exposure to a hazard is using enclosures, shielding, buffers, different/revised processes, or ventilation. For example, the use of baffles or enclosures on a piece of equipment to dampen the noise. The advantage of engineering controls is that they do not rely on human activation or implementation to be effective.

3. **Administrative Controls** — behavioural or personal management controls designed to lessen or limit the risk of exposure to a specific hazard. For example, the use of job reassignments, or policies and procedures to control worker exposure to a hazard such as noise; or scheduling hazardous work like painting during "off" hours.

4. **Personal Protective Equipment** — equipment worn by workers that is designed to protect the worker from specific hazards or hazardous

---

[13] F. Bird and G. Germain, *Loss Control Management: Practical Loss Control Leadership*, revised ed. (Loganville, GA: DNV, 1996), at 427-433.

[14] *Ibid.*

conditions. For example, the use of steel-toed boots to protect the worker from foot injuries when working with heavy equipment. This is the worker's "last line of defence" and is not a "true" hazard management strategy.

The topic of hazard identification, assessment, and control is discussed further in Chapter 9, "Occupational Health and Safety: Hazard Identification, Assessment and Control".

7. Worksite Inspections

An effective way to identify workplace hazards and unsafe conditions is by doing worksite inspections. The strength of this practice is that workplace hazards and hazardous conditions can be identified *before* they result in a loss.

The goal of the worksite inspection is to:

- identify potential problems;
- identify equipment that is substandard, inadequate or improperly used;
- identify improper work practices;
- identify what is going on in terms of work processes and use of raw products;
- identify issues that stem from actions taken to remedy a problem and implementation of new process components;
- provide leaders with a means of self-appraisal as a leader; and
- demonstrate leadership commitment to workplace safety.[15]

The topic of worksite inspections is discussed further in Chapter 9, "Occupational Health and Safety: Hazard Identification, Assessment and Control".

8. Emergency Preparedness and Response

Appropriate and adequate emergency response preparedness is a business requirement. It reduces the losses and costs associated with a disaster situation and it meets the legal requisite to respond to contain the emergency situation, protect employees, and assist injured employees. The extent and complexity of emergency preparedness and disaster response plans depends on the nature of the industry and the work done, the size of the organization, the hours of operation, and the community services available. At a minimum, the following elements should be part of a company's emergency response and disaster planning:

- an adequate emergency disaster preparedness and response plan;
- identification of the response team's equipment needs and training requirements;
- an injury response mechanism;

---

[15] *Ibid.*, at 123-124.

- procedures for incident investigation;
- incident management;
- documentation of the plan and its components;
- response data collection and analysis;
- process/procedural evaluation, including doing trend analysis to identify areas for performance/plan improvement; and
- a proactive approach to prevent, or enhance response to, future disasters that threaten the organization, its facilities, its business, and/or its human resources.

The topic of Emergency Preparedness and Response is discussed in Chapter 12, "Occupational Health and Safety: Emergency Preparedness and Response".

9. Incident Investigation and Corrective Actions

Incident investigation is a critical element to understanding why an incident occurred; to identify the potential for future incidents; and to identify jobs that result in a high frequency of incidents. With any workplace incident, it is important to document accurately and completely the incident and responses; and to identify the root causes and necessary corrective actions/measures needed to prevent the recurrence of a similar incident.

Chapter 13, "Occupational Health and Safety: Injury Management" and Chapter 14, "Occupational Health and Safety: Incident Investigation", provide in-depth information on this subject.

10. System Review and Continuous Improvement

A system review of an OHSMS allows Management to understand how effective or ineffective the OHSMS is and whether it is meeting the set objectives and targets. The intent is to ensure that the conditions put in place to support workplace health and safety are functional, that they are being followed and that they are achieving the desired goals. A system review is designed to identify continuous improvement opportunities as well as to demonstrate the organization's commitment to its duty of health and safety care to employees and the public. The latter part of this chapter addresses OHSMS and OH&S Program evaluation.

C. OCCUPATIONAL HEALTH AND SAFETY MANAGEMENT SYSTEMS

1. Rationale for Implementation

An Occupational Health and Safety Management System (OHSMS) consists of the elements of an organization (including its resources, systems, planning, implementation, and monitoring processes, culture, structure, and tasks), that taken together, support employees in the achievement of a company's business objectives. These objectives tend to fall into one or more of the following categories:

- operational effectiveness and efficiency;

- reliability of the OH&S program and services; and/or
- legislative compliance.

The Ontario Workplace Safety & Insurance Board (WSIB) in a document titled, ***Business Results Through Health & Safety (2001)***, explains how OH&S Practitioners/ Professionals can build a business case for implementing an OH&S Program.[16] The premise is that besides making good common sense, OH&S Practitioners/ Professionals must be able to demonstrate to Management the financial and business benefits of the OH&S Program. The cited business justification is:

- compliance with the law;
- cost reduction;
- business interruption protection;
- employee relations improvement;
- reliability and productivity improvement;
- building public trust; and
- building organizational capability.[17]

By implementing an effective OHSMS, companies can realize superior OH&S results, which in turn will:

- lower costs;
- improve employee relations, morale and trust;
- improve company reliability and productivity;
- lessen business interruptions;
- increase public trust and hence, the corporate image; and
- increase organizational capability, competitiveness, and profitability.

The relationship between company values, the OHSMS, and the occurrence of incidents/injuries is depicted in Figure 2.3. For OH&S practitioners/ professionals, this resource is extremely valuable and readily available free of charge at http://www.mtpinnacle.com/pdfs/Biz.pdf.

---

[16] WSIB/CSPAAT, "Business Results Through Health & Safety" (2001), online: http://www.mtpinnacle.com/pdfs/Buisness-Results.pdf (date accessed: February 28, 2020).

[17] *Ibid.*

[18] WSIB/CSPAAT, "Business Results Through Health & Safety" (2001), online: http://www.mtpinnacle.com/pdfs/Biz.pdf (date accessed: February 28, 2020). Reprinted with permission in 2009.

Figure 2.3: Model on the Management of Occupational Health & Safety Outcomes[18]

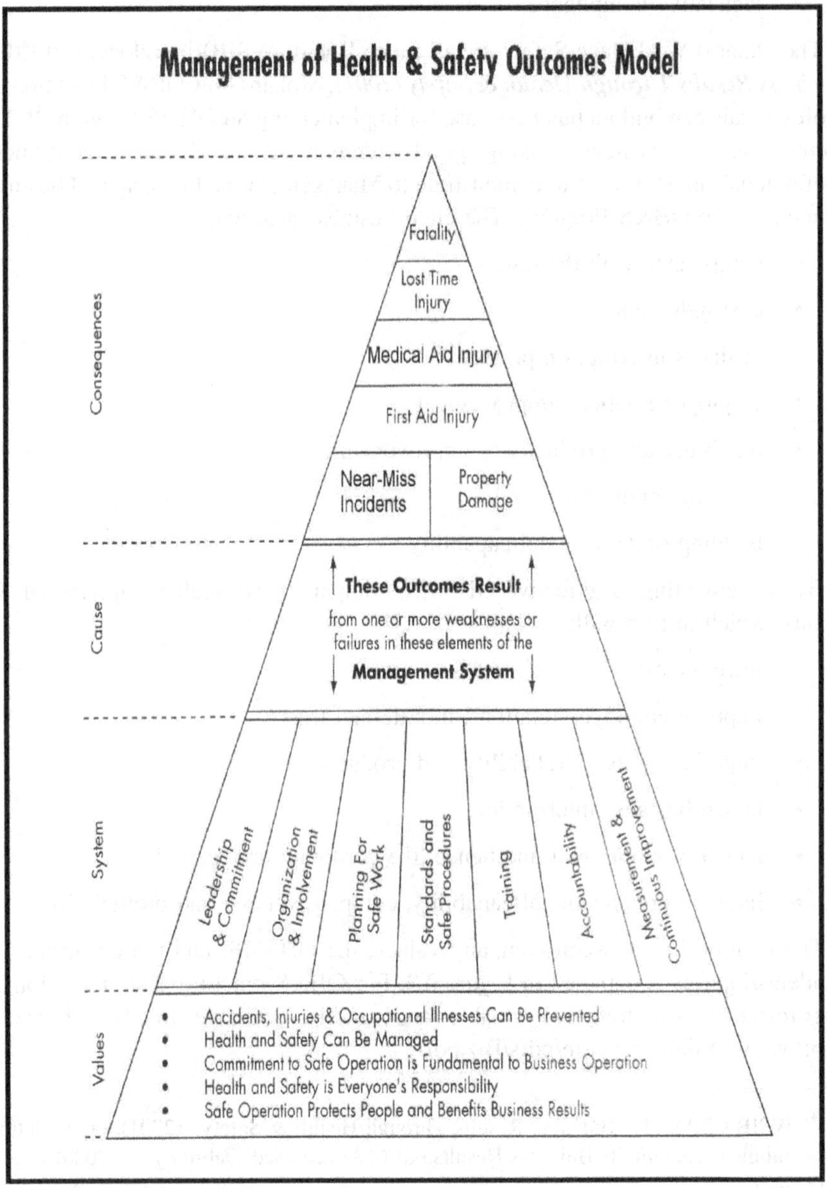

Management of Health & Safety Outcomes Model

Consequences:
- Fatality
- Lost Time Injury
- Medical Aid Injury
- First Aid Injury
- Near-Miss Incidents | Property Damage

Cause:
↑ These Outcomes Result ↑
from one or more weaknesses or failures in these elements of the
↓ Management System ↓

System:
- Leadership & Commitment
- Organization & Involvement
- Planning For Safe Work
- Standards and Safe Procedures
- Training
- Accountability
- Measurement & Continuous Improvement

Values:
- Accidents, Injuries & Occupational Illnesses Can Be Prevented
- Health and Safety Can Be Managed
- Commitment to Safe Operation is Fundamental to Business Operation
- Health and Safety is Everyone's Responsibility
- Safe Operation Protects People and Benefits Business Results

2. Purpose and Value

An OHSMS provides a framework for effective Occupational Health and Safety management and outcomes in terms of:

- compliance with the applicable legislation, company, and industry standards

that demonstrate an OH&S duty of care and corporate due diligence;
- assurance that workplace health and safety hazards and risks are identified and effectively managed;
- application of sound business management processes to the management of Occupational Health and Safety;
- continuous improvement of the OH&S Program and services through regular evaluation and corrective actions;
- being a means for demonstrating conformity; and
- combining the quality, OH&S, and environment management systems into an integrated management system.

The value that an OHSMS adds to an organization is that it enables an organization to:
- identify system weaknesses and focus improvement efforts in critical areas;
- identify system strengths and leverage its abilities in other areas, or to influence external groups (such as government agencies or third-party insurers);
- link upstream successes (**leading indicators of safety performance**) with the downstream (**lagging indicators of safety performance**) performance indicators;
- analyze OH&S performance trends and assess the effectiveness of the safety improvement efforts;
- compare the success of more discrete safety performances and their end results;
- validate the organization's best practices;
- leverage change within the organization using proven evidence from the tracked OH&S performances;
- demonstrate a correlation between OH&S practices, OH&S outcomes, and the related business costs/savings;
- demonstrate the organization's OH&S duty of care and due diligence in the provision of a healthy and safe workplace; and
- respond to marketplace and societal changes in a timely manner, thereby promoting business continuity.

3. Framework

An OHSMS is a systematic approach to managing Occupational Health and Safety within an organization. It contains elements that interrelate and provides a road map for action and promotes continuous improvement through linkages between the elements. The linkages between the elements are as follows:

1. **Standard Setting** — development of the OH&S Policy.
2. **Planning** — activities such as risk management, identification of OH&S legal requirements, and business planning that are associated with developing plans for the control and management of safety hazards and risk mitigation.
3. **Implementation and Operation** — actions associated with hazard control and risk mitigation such as accountabilities and leadership activities, worker training, OH&S communication, emergency preparedness, operational control, employee involvement, and documentation.
4. **Checking and Corrective Action** — activities associated with checking to determine if hazards are controlled, risks are mitigated, and problems are corrected such as performance monitoring, recognition of desired performance, incident management, auditing, and records management.
5. **Review** — activities associated with Management review to ascertain that OH&S policies, objectives, and performance targets are met.

D. OHSMS STANDARDS: DESCRIPTION AND COMPARISON

There are several OHSMS standards available to which organizations can voluntarily subscribe. The purpose of this section is to describe and compare those standards and to introduce the Alberta Partnerships in Injury Reduction Program.

1. OHSAS 18001/02

The **OHSAS 18001/02**, introduced in 2007, is an Occupational Health and Safety Management System (OHSMS) specification developed through the concerted effort of many of the world's leading national standards bodies, certification bodies, and specialist consultancies. The main driver for the OHSAS 18001 was to try to remove confusion in the workplace from the proliferation of certifiable OH&S specifications.

More specifically, OHSAS 18001/02 is an Occupational Health and Safety Assessment Series for Occupational Health and Safety Management Systems. It is intended to help organizations control OH&S risks. It was developed in response to the widespread demand for a recognized standard against which to be certified and assessed.

The OHSAS specification is applicable to any organization or company seeking to:

- establish an OH&S management system aimed at eliminating or reducing risk to employees and other interested parties who may be exposed to OH&S risks associated with its activities;
- assure alignment and compliance with its stated OH&S policy;
- demonstrate due diligence in terms of OH&S duty of care;
- implement, maintain, and continually improve its OH&S management system;

- make a self-determination and declaration of compliance with this OHSAS specification; and
- seek certification/registration of its OH&S management system by an external organization.

OHSAS 18001 and 18002 are two standards which assist in the implementation of an OH&S Management System certification system (Table 2.1). Essentially, OHSAS 18001/02 can substantially help organizations to minimize their OH&S risks, continually improve an existing OH&S Management System, demonstrate OH&S diligence and attain assurance that their OH&S Management System is effective and reliable.

Table 2.1: OHSAS OH&S Management System Specification

OH&S MANAGEMENT SYSTEM (OHSAS 18002) STRUCTURE

The elements consist of:
1. **Occupational Health and Safety Policy**
2. **Program Planning:**
  • Legal risk assessment
  • Setting objectives
  • Planning for hazard identification, risk assessment and risk control
  • Health & Safety Management Program for maintaining the objectives
3. **Implementation and Objectives:**
  • Outline structure and responsibilities
  • Provide training and awareness
  • Communicate the program (marketing)
  • Documentation
  • Data collection and control
  • Operational control
  • Emergency response
4. **Checking and Corrective Action:**
  • Performance measurement
  • Incidents, corrective and preventative actions
  • Records management
  • Auditing
  • Management review
5. **Management Review and Leadership**

PROCESS

For each element, one needs to:
  • Outline the requirement
  • Describe the intent of the requirement
  • Consider the type of inputs required
  • Describe the process involved
  • State the output (what is expected to happen)
The crucial aspects for any OH&S Program are:
  • Senior Management commitment
  • Responsibility for the ownership of the OH&S Program by line managers
  • A supportive work culture as fostered by Management's leadership for OH&S
  • OH&S engrained in all aspects of the work
  • Employee involvement

OUTCOMES

Companies that implement the OH&S Program described above experience:
  • Fewer incidents and injuries
  • Healthier and happier employees
  • Lower Workers' Compensation Board assessment rates and claims costs
  • Lower business costs
  • Better shareholder return on investment

2. ISO 45001/Z45001-19

ISO 45001, released in 2018, it is an upgrade to OHSAS 18001/02, which will be withdrawn by March 2021. It is an international standard on OHSMS specifications. It provides a framework for the management of workplace risks, the prevention of occupational illness and injury, and the development of a safe and healthy workplace.

This new standard offers a number of benefits namely:

- Its process and structure make the integration of other management systems (ISO 9001 and 14001) easier;
- It uses the plan-do-check-act (PDCA) model — a model that organizations are familiar with and can be used to implement a safe and healthy workplace regardless of their size and industry;
- It positions organizations to support worker health, safety, and well-being, by effectively identifying OH&S hazards, risks, and opportunities; and
- It requires management to integrate OH&S issues as part of their business strategies, create a corporate culture of worker involvement in the OHSMS, and integrate the OHSMS with the business strategies.

The main changes in ISO 45001 centre on:

- The introduction of new clauses for systematic determination and monitoring of the business context;
- An inclusive OH&S culture whereby employees and other stakeholders are involved in the OHSMS;
- Risk and opportunity management to deliver enhanced OH&S outcomes;
- Increased leadership accountability to deliver an effective OHSMS;
- Stronger focus on OHSMS improvement and performance evaluation;
- Prescriptive communication in terms of what, when, and how to communicate; and
- Procurement of outsourced services and contractors.[19]

For details on the migration to ISO 45001 from OHSAS 18001, refer to the *ISO 45001 Occupational Health and Safety Management Systems Migration Guide, 2018*.[20] Of particular interest is the table on the differences between the two standards (pages 14-16).

In 2019, the Canadian Standards Association (CSA) adopted and amended ISO 45001 and produced CSA Z45001. The rationale for this action was to modify it to align with Canadian law, culture, language, work environments, and social context. In Canada, the concepts of worker involvement in the OHSMS and the hierarchy of hazard control are important; hence, they needed to be incorporated into CSA Z45001. Likewise, the standard had to be connected to existing Canadian standards, such as the National Standard on Psychological Health & Safety in the Workplace, CSA Z1004 Workplace Ergonomics, CSA Z1005 Incident Investigation, CSA Z1006 Confined Spaces, CSA Z1007 Hearing Loss Prevention, *etc.* Lastly, the Canadian

---

[19] NSF (2018). *ISO 45001 Occupational Health and Safety Management Systems Migration Guide*, online: https://www.nsf.org/newsroom_pdf/isr_dis45001_guide.pdf (date accessed: February 28, 2020).

[20] *Ibid*.

version focuses on constructive management change.

Using CSA Z1000 as the foundation and the ISO 45001's structure, processes and outcomes, CSA Z45001 was developed to enable the integration of the OHSMS with the other company resources. It specifies the requirements of an OHSMS and provides guidance on how to prevent occupational injuries and illness, thereby creating a safe and healthy workplace. There is a strong focus on performance measurement, evaluation, and continuous improvement. The OHSMS outcomes are continuous OH&S performance improvement, legal compliance, and the achievement of OH&S objectives.[21]

Being a voluntary standard, organizations of varying sizes, can use it whole or in part, to systematically improve their OHSMS and its effectiveness.

3. ILO-OSH 2001

The International Labour Organisation (ILO) hosted an international workshop on OSH-MS standardization in 1996. From that initiative, an OSH management standard for national and organizational use was released (December 2001). The **ILO-OSH 2001** is an international model that is designed to be compatible with other management system standards and guides. By virtue of its design and focus, it encourages integration of OH&S Management System elements into overall policy and business management. It also stresses that OH&S management should be a Line Management responsibility as opposed to an OH&S Department function.[22]

ILO-OSH 2001 provides guidelines for implementation at national and organizational levels. Nationally, countries can adopt the standard and use it as the basis for certification of the country's employers.[23] From the perspective of an organization, it has five main sections which align with Deming's Plan-Do-Check-Act Model, the basis to a system approach for management (Figure 2.4).

The main sections and elements of the ILO-OSH 2001 are:

- *Policy*: The OH&S Policy is the basis for the OH&S Management System — it provides direction for the organization to follow in terms of its OH&S performance. The elements in this section are the OH&S policy and the level of employee involvement.

- *Organizing*: It ensures that the required OH&S structure is in place and that the OH&S roles and responsibilities are defined. The related elements include stakeholder responsibility and accountability, employee competence and training, OH&S documentation, and OH&S communication.

---

[21] CSA (2019). *CSA Z45001:19*, online:www.scc.ca/en/standardsdb/standards/29765 (date accessed: February 28, 2020).

[22] ILO, "Guidelines on Occupational Safety and Health Management Systems" (2001), online: http://www.ilo.org/public/libdoc/ilo/2001/101B09_287_engl.pdf (date accessed: February 28, 2020).

[23] *Ibid.*

- *Planning and Implementation*: Through an initial review, the organization's level of OH&S performance is determined. This section includes the elements of system planning, development and implementation; OH&S objectives and performance targets; and hazard prevention and management.

- *Evaluation*: Assesses how the organization's OH&S Management System is functioning and identifies system limitations. Included are the elements of performance monitoring and measurement; investigation of occupational incidents and injuries/illness; auditing; and Management review.

- *Action for Improvement*: Implements the recommended preventative and corrective actions and emphasizes the need for continuous improvement.

Figure 2.4: ILO-OSH 2001: Main Sections and Elements[24]

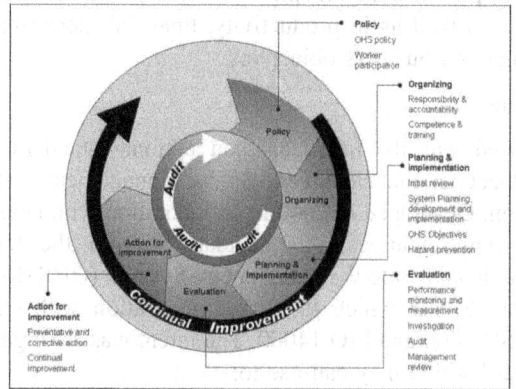

This OH&S Management System standard aligns very well with the environment management standard, ISO 9000 and ISO 14000.

4. ANSI/AIHA Z10-2005

ANSI/AIHA Z10-2005 is the American National Standard for Occupational Health and Safety Management Systems. This standard provides the critical management systems requirements and guidelines for the improvement of Occupational Health and Safety. Experts from labour, government, professional organizations, and industry formulated this valuable standard in 2005 after extensive examination of current national and international standards, guidelines, and practices.

ANSI/AIHA Z10-2005 is made up of the following sections:

- Management Leadership and Employee Participation;
- Planning;

---

[24] Figure 2.4 has been adapted for public and educational use; online: http://www.ilo.org (date accessed: February 28, 2020).

- Implementation and Operation;
- Evaluation and Corrective Action; and
- Management Review.

Contained in the appendices, the standard addresses roles and responsibilities; policy statements; assessment and prioritization; audit information; and much more.

This broad-reaching standard enables organizations of all sizes and types to integrate OH&S management into their overall business management systems. It is compatible with relevant OH&S, environmental, and quality management standards, such as International Organization for Standardization (ISO) 9000 and 14000, and with approaches to OH&S management commonly used in the United States and Canada.

ANSI/AIHA Z10 provides the blueprint for widespread benefits in Occupational Health and Safety, as well as in productivity, financial, performance, quality and other organizational and business objectives.

5. CSA Z1000-06

The **CSA Z1000-06** was the first Canadian national standard that defined the requirements for Occupational Health and Safety Management System development and implementation. Developed in 2006 in collaboration with labour, Management, and government representatives, this standard set out the framework for an Occupational Health and Safety Management System (OHSMS). It follows the Plan-Do-Check-Act Model which facilitates integration with other management systems such as ISO 9000 and ISO 14000. The intent was for organizations to have a Canadian standard which they can use to:

- improve their OHSMS;
- encourage a systematic approach for meeting defined OH&S objectives;
- define individual roles and accountabilities within the OHSMS;
- increase awareness of Occupational Health and Safety;
- clarify the need for Management and worker commitment and involvement;
- address hazard- and risk-based prevention and control measures;
- identify desired outcomes with concrete goals;
- prevent workplace injuries/illness;
- enhance their OH&S performance; and
- enable compliance with OH&S legislation.[25]

---

[25] Canadian Standards Association, Product Highlights: "CSA Z1000-06: New Occupational Health and Safety Management (OHSM) Standard Garners the Media Spotlight" (2006), online: http://www.csagroup.org/ca/en/home or http://www.shop.csa.ca (date accessed: February 28, 2020).

CH. 2: OCCUPATIONAL HEALTH AND SAFETY MANAGEMENT SYSTEMS

6. Alberta Partnerships in Injury Reduction Program[26]

The **Alberta Partnerships in Injury Reduction Program** (the Partnerships Program) initially set a performance-based standard for a basic health and safety program within the Canadian province of Alberta. Introduced in 1989, the Alberta Partnerships in Injury Reduction Program is a voluntary program designed to create a partnership between the Alberta government and Alberta employers to promote and develop strong workplace cultures within Alberta businesses.

The Alberta Partnerships in Injury Reduction Program's intent is to encourage employers and workers to build effective Occupational Health and Safety Management Systems, and ultimately, strong safety cultures. The scope and complexity of the sought-after OHSMS varies depending on the type of workplace and the nature of the work done.

Under the Partnerships Program, the elements of the OHSMS are:

- clearly stated employer policy and commitment;
- identification and analysis of workplace hazards;
- hazard control measures;
- worker competency and training;
- worksite inspection program;
- incident reporting and investigation;
- emergency response planning; and
- program administration.[27]

Although a voluntary program, financial incentives exist for Alberta employers who participate in the Partnerships Program. Employers who earn a Certificate of Recognition (COR) for having a basic health and safety program are entitled to a financial rebate on their Workers' Compensation Board (WCB) insurance premium. Additional financial incentives exist for companies that reduce the amount of their WCB claims over the rates of the previous two years.

Based on the success of this program, other Canadian provinces (*e.g.*, British Columbia, Manitoba, Ontario, New Brunswick, Nova Scotia and the Yukon Territories) have adopted the approach. For example, the Partners in Injury and Disability Prevention Program (Partners), in British Columbia (BC), has been under development since September 2015.[28] Of the 1,135 COR-certified construction

---

[26] Alberta Workplace Health and Safety, "The Partnerships in Health and Safety Program: Infosheet 1" (2014), online: http://www.employment.alberta.ca/documents/WHS/WHS-PS-InfoSheet1.pdf (date accessed: February 28, 2020).

[27] *Ibid.*

[28] WorkSafe BC, "The Partners in Injury and Disability Prevention Program (Partners Program)" (2018) Discussion Paper, online: https://www.worksafebc.com/en/resources/law-

companies surveyed in BC (2005-2012), they averaged 12 per cent fewer short-term disabilities, long-term disabilities and fatalities, and fewer serious occupational injuries (17% reduction).[29]

7. Conclusion

Of all the OHSMS specifications described, the Partnerships in Injury Reduction Program has proven to be very effective in reducing workplace injuries. The difference is that although all the standards are voluntary in nature, the Partnerships Program is the only one that links the concept of an OH&S Management Standard with financial incentives for implementing the recommended standard. This phenomenon parallels the observation that, knowing what to do does not translate into "doing" unless motivation for action exists. The message is that if organizations and governments really want employers to manage workplace health and safety through the implementation of OHSMS, then, reward them to do so in a meaningful way.

E. OHSMS: TYPICAL EVOLUTION WITHIN A COMPANY

The development of an OHSMS tends to occur at different rates. For example, certain areas like OH&S governance, policy, programs, and senior Management commitment might be deemed mature; yet, other aspects like line management, OH&S leadership, and worker commitment remain in development stage.

1. The DuPont Bradley Curve

The maturity scales typically found in industry and how they relate to injury/

---

policy/discussion-papers/partners-in-injury-and-disability-prevention-program?lang=en (date accessed: February 28, 2020).

[29] P. Caulfield, "WorkSafeBC said 'no changes' to COR program in 2018" (2018) Journal of Commerce, online: https://canada.constructconnect.com/joc/news/associations/2018/03/worksafebc-said-no-changes-cor-program-2018 (date accessed: February 28, 2020).

incident rates were described by Bradley in the DuPont Bradley Curve (see Figure 2.5).

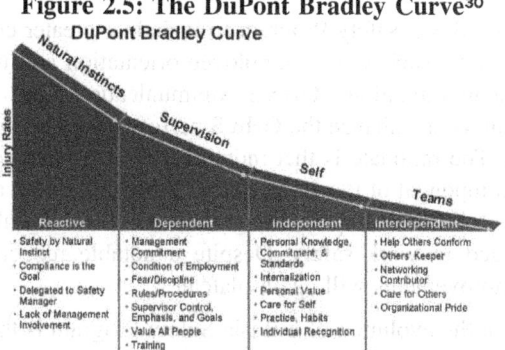

Figure 2.5: The DuPont Bradley Curve[30]

In the ***Reactive Stage***, the champions for the OH&S Program are the OH&S Practitioners/Professionals: Management commitment is limited or missing. Compliance with legislative requirements (minimum OH&S standards) is the main goal. Safety practices occur by instinct. Zero injury/incident rates are unrealistic.

In the ***Dependent Stage***, there is evidence of Management OH&S leadership and commitment. The message that "all people are valued", is evident. Worker training is provided, and supervisors are responsible for ensuring that safe work practices are upheld. Fear of discipline tends to motivate compliance because observance to OH&S rules and standards, and the use of safe work practices by employees is a condition for employment. Safety occurs through supervision. Zero injury/incident rates are difficult to achieve.

In the ***Independent Stage***, personal knowledge and commitment to workplace safety and standards are evident. In essence, Safety is internalized as opposed to being an "add-on" effort to the work being done. Employees take responsibility and accountability for their safety and are recognized for demonstrating safe work practices and behaviours. People work together as teams to care for themselves and each other. Organizational pride for working safety is evident. At this stage, zero injury/incident rates are attainable.

2. The Fennell Model: ESSO

Another way to look at OHSMS evolution is to measure the growth of the OH&S system in terms of injury frequency (Figure 2.6). According to Dave Fennell (2001), a Basic Safety Program demonstrates the fundamentals of incident reporting, worksite inspections, maintenance plans, and awareness plans. The framework and foundation provide for further safety progress. The goal is regulatory compliance

---

[30] DuPont Safety Resources, "Dupont Bradley Curve" (Newark, DE: DuPont Safety Resources, 1994), online: http://www.dupont.com/safety (date accessed: February 28, 2020). Reprinted with permission.

which aligns with the inherent desire to prevent workplace injury. The introduction of a Basic Safety Program results in a reduction in the injury frequency rate. However, this usually takes 9-12 months to occur, and the improvement will plateau.

By enhancing the Basic Safety Program to include a greater commitment to the fundamentals of safety, such as new employee orientation and training, proactive reporting, incident investigations, OH&S communications, and clear supervisory roles, an organization can enhance the OH&S performance outcomes (*e.g.*, reduced injury frequency). The rationale is that more energy is applied so that supervisors understand each component of the Basic Safety Program as well as their respective roles and responsibilities. Safety meetings become meaningful. Employees are emotionally engaged to work safely. Despite a notable reduction in the injury frequency rate, improvements will again plateau.

The next stage in the evolution of a Basic Safety Program is the implementation of Advanced Approaches with Supporting Management Systems. Performance measurement and analysis, supervisor/employee accountability for safety, employee involvement, and development of safety values are the elements evident. In short, the Safety Management System that has management commitment and support emerges. Safety becomes an "approach" as opposed to a "program". Safety becomes a personal value and employees willingly participate in safe work practices. Again, the injury frequency rates drop, but plateau.

Figure 2.6: Growth of Effective Safety Systems[31]

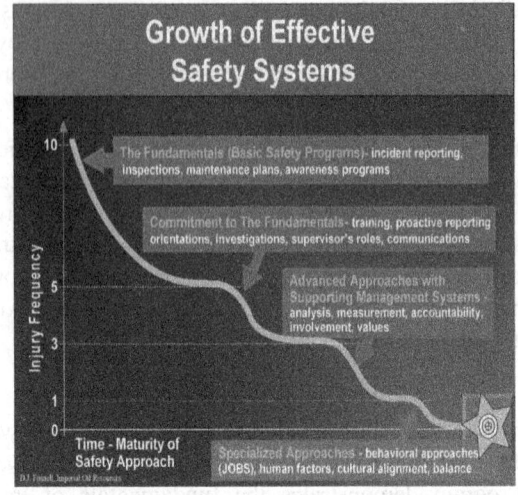

The final stage is a Basic Safety Program that is augmented with Specialized Approaches. It includes behavioural safety approaches, management of human

---

[31] D. Fennell, "Growth of Effective Safety Systems" (Presented at the Canadian Society of Safety Engineers luncheon, January 10, 2005).

factors, alignment of the safety culture, use of leading-edge safety techniques, alignment with the safety and corporate cultures, and an understanding of the positive impact of safety culture on safety outcomes. Employees understand and embrace safe work practices and concepts. The use of peer support and feedback strengthens the system. A Safety Program at this stage of evolution generates sustainable, low injury frequency rates.

3. The Stages of Organizational OH&S Culture Evolution

A third way to look at the evolution of an organization's OHSMS is in terms of organizational behaviours. Having reviewed the research conducted by Lardner,[32] Hudson[33] and Kelbert,[34] the following model has been developed to encompass the work of all three researchers.

The Stages of Organizational OH&S Culture Evolution model is comprised of five developmental stages (Figure 2.7):

(a) Stage 1: Barely Conscious

During this stage, Management does not perceive Occupational Health and Safety as a "key business risk". Although the organization has OH&S standards, regulations, and rules, these elements tend to be ignored when workloads increase. The prevailing attitude is *"everyone for themselves"*. Accidents are viewed as "unavoidable" — "part of the job". Workers are not aware that they are involved in risky behaviours. Workplace incidents may not even be acknowledged or counted.

(b) Stage 2: Struggling

Through the development of management commitment to Occupational Health and Safety, the company can mature to Stage 2. Companies at this stage of development experience an average rate of incidents, but still experience severe injury incidents. OH&S rules and regulations exist but are not always followed. Management tends to blame workers for their "unsafe" acts and attitudes. The success of the OHSMS is measured in terms of the lagging indicators of safety performance, namely injury frequency and severity rates.

---

[32] R. Lardner, M. Fleming & P. Joyner, "Towards a Mature Safety Culture", Symposium Series #148 (Edinburgh, UK: Keil Centre, 2001), online: http://www.keilcentre.co.uk/media/1064/towards-a-mature-safety-culture-lardner-2002.pdf (date accessed: February 28, 2020).

[33] P. Hudson, "Safety Culture: The Ultimate Goal" (Transport Canada, Aviation Safety Letter, 2002), Vol. 2, at 6-7.

[34] R. Kelbert, "Steps Towards Organizational Safety Maturity model" (Presented as part of the presentation on Creating a World Class Safety Culture, Western Conference on Safety, Vancouver, British Columbia, 2005).

Figure 2.7: Stages of Organizational OH&S Culture Evolution Model[35]

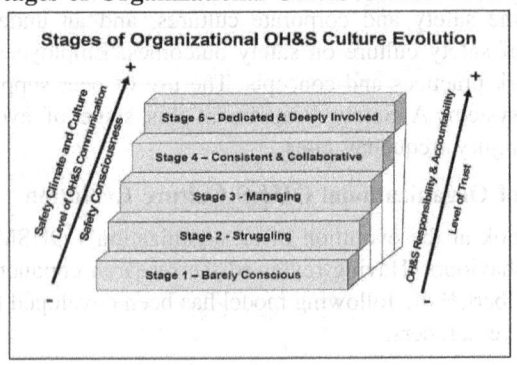

(c) Stage 3: Managing

By realizing the importance of frontline staff and by developing personal responsibility for OH&S, companies can move to Stage 3. At this stage, accident rates are below industry averages, but have plateaued. OH&S standards, regulations and rules are the primary driving force for workplace health and safety. The focus of the OH&S Program is to make employees aware of unsafe behaviours and the personal risks involved. Management promotes employee involvement but believes that very few of the employees are interested in seeking ways to improve workplace health and safety. Incident frequency and severity rates continue to be used as measures of the success of the OHSMS.

(d) Stage 4: Consistent and Collaborative

Maturity occurs as a result of engaging staff to co-operate and commit to improving workplace safety. Organizations at this stage of development, experience good to excellent rates of incident frequency and severity. OH&S standards, regulations, and rules are not the driving force for safe work performance. Rather, Management recognizes their management contribution in terms of the root causes of incidents. Employee input is perceived as valuable: employees appreciate that fact. The focus of the OH&S Program is to make the "safe way to work, the first choice". Task analyses are continually done to improve jobs. Some preventative activity measures occur. The OH&S measurement systems that are used to determine the success of an OHSMS are sophisticated.

---

[35] Developed by D. Dyck (2007) as an adaptation from P. Hudson, "Safety Culture: The Ultimate Goal" (Transportation Canada, Aviation Safety Letter, 2002), Vol. 2, at 6-7; R. Lardner, M. Fleming and P. Joyner, "Towards a Mature Safety Culture", Symposium Series #148 (Edinburgh, UK: Keil Centre, 2001), online: http://www.keilcentre.co.uk/media/1064/towards-a-mature-safety-culture-lardner-2002.pdf (date accessed: February 28, 2020); and R. Kelbert, "Steps Towards Organizational Safety Maturity model" (Presented as part of the presentation on Creating a World Class Safety Culture, Western Conference on Safety, Vancouver, British Columbia, 2005).

(e) Stage 5: Continually Improving and Deeply Involved

To move forward, companies must promote the consistent use of safe work practices and fight complacency. This final evolutionary stage is characterized by the occurrence of few, if any, incidents. The prevailing belief is that worker injuries, illness, or poor health are unacceptable. All stakeholders are committed to workplace health and safety. At this point, cultural change has occurred — safe work behaviours have become part of daily work activities. A wide range of performance measures are used to evaluate the success of the OHSMS.

According to Kelbert,[36] when the OH&S system becomes sophisticated enough, the greatest degree of hourly worker involvement is possible. A smoothly operating OHSMS allows employees to take over aspects of the daily practices, *e.g.*, inspection, correction, near-miss reports, safety sampling, work observations, *etc*. A well-functioning OHSMS has checks and balances built in so that it does not require precious management time and resources to maintain them.

Managers often want this level of worker involvement at the lower evolutionary stages, and become frustrated when workers are unwilling to contribute much beyond the minimum requirements, such as attending Safety meetings, committee meetings, *etc*. However, clear standards for preventive safety activities (*e.g.*, set agendas for crew safety meetings, formats for recording suggestions, well-defined incident investigation criteria, *etc*.) must be consistently applied throughout an operation.

Workers respect a system that is clear, straightforward, and produces good results. Expecting discretionary effort from employees who observe inconsistently applied management standards, and arbitrariness (*e.g.*, one department's crew meeting takes five minutes while another takes 45 minutes and some departments are allowed to skip theirs), is unreasonable. For this reason, Kelbert entitled the last stage, **Continually Improving and Deeply Involving**, not **Involved**.

Although the descriptions of these evolutionary stages of safety maturity appear "black and white"; in reality, organizations can demonstrate a mix of the above features.

F. OH&S PROGRAM: COMPONENTS

For an OH&S Program to be successful, it should contain eight components, commonly referred to as the "Factor of 8". They are:

1. ***Involvement*** — Everyone in the company should be involved in the OH&S Program. The use of an OH&S Steering Committee is one way to promote this involvement.[37] The mandate of this committee is to ensure that the OH&S Program priorities are met; that the compliance require-

---

[36] Comments submitted by Randy Kelbert (May 2007).

[37] C. Lawson, "Developing a Corporate Safety Program" (Presented at the American Occupational Health Nurses' Conference, April 20-27, 2001, San Francisco, CA).

ments are defined; that the program is visible, supported and adequately resourced; and that qualified OH&S practitioners/professionals are in place. At a minimum, all stakeholders should be made aware of their respective OH&S roles and responsibilities.

2. ***Accountability*** — Through the development of an OH&S Policy, organizational and labour OH&S accountability can be defined.[38] Accountability can also be defined as part of the corporate Code of Conduct, vision, and OH&S philosophy. OH&S accountability should also be built into organizational and employee performance appraisals and viewed as measured commitment.

3. ***Communication*** — Company expectations for OH&S performance must be defined and articulated to all involved.[39] To that end, an OH&S Safety Communication Plan should be developed (refer to Chapter 16, Appendix 1, for an industry example of an OH&S Communication Plan). In addition to the spoken and written message, leaders must model the messages they wish to convey. There should be avenues for employee involvement in identifying workplace health and safety issues, such as safety, meetings, hazard report forms, communication sessions with leaders, *etc.*

4. ***Education*** — All stakeholders should be provided instruction on the company's OH&S Plan, its intent and implementation, and the expected OH&S performance standards. Education should also encompass worker OH&S training, regular OH&S educational sessions, brochures, online communication, *etc.*

5. ***Evaluation*** — To measure the performance of the OH&S Program, regular program evaluation must occur. Through the use of audits, program enhancement reviews and collection of OH&S data, a company can monitor and evaluate the effectiveness of its OH&S Program and its compliance with the applicable legislation.

6. ***Reinforcement*** — Working safely must be viewed as valued and integrated into the corporate culture.[40] Regular reinforcement of this component is vital to the achievement of sound and consistent workplace OH&S performance.

7. ***Continuous Improvement*** — The goal of OH&S Program evaluation is continuous improvement — *i.e.*, ways to continually improve the OH&S Program.

8. ***Return on Investment*** — To sustain any business initiative or program, the related benefits must be demonstrable and recognized. For OH&S Pro-

---

[38] *Ibid.*
[39] *Ibid.*
[40] *Ibid.*

grams, the returns on the company's investment in OH&S include:
- a proactive approach to workplace health and safety;
- OH&S diligence;
- a healthier and safer place for people to work;
- high quality work performance;
- a high level of employee morale;
- employee loyalty;
- low employee absenteeism rates;
- good staff retention levels;
- good employee productivity and company profitability; and
- compliance with the applicable legislation.

A proper amount of attention to these eight components of an OH&S Program can assist companies in meeting and exceeding their business goals, as well as their OH&S due diligence and duty of care obligations.

G. OH&S PROGRAM: ELEMENTS

A discussion of the occupational and non-occupational elements of an OH&S Program follows.

1. Occupational Aspects of an OH&S Program

(a) Workplace/Employee Safety and Incident Prevention

Every Canadian/American employer is responsible for providing an OH&S Program that ensures the health and safety of employees, including the development of strategies for workplace incident/injury/illness prevention. The overall objectives for workplace/employee safety and incident prevention are to:

- establish an OHSMS that includes management and leadership; competence, selection and training; communication, risk assessment and control; measuring performance and corrective actions; and system evaluation;
- identify workplace hazards and risks;
- provide OH&S training in accordance with the applicable legislation and industry standards;
- establish communication strategies to ensure that employees receive accurate and timely OH&S information, and to allow for ease of employee reporting of worksite hazards;
- establish procedures to promote and ensure employee/operational compliance;
- ensure prompt reporting and investigation of injury/illness to determine root cause(s);

- plan and implement corrective actions with ongoing evaluation; and
- regularly evaluate the OH&S Program with a view to continuous improvement.

Management and employee involvement in, ownership of, and accountability for, workplace health and safety are critical to the success of any OH&S Program. This can be promoted through the implementation of employee/management committees such as Joint Health & Safety Committees (JHSCs).

In terms of determining "person-job" fit, job demands analyses (JDAs) and employee fitness-to-work assessments are two important tools. By having an inventory of JDAs for all job positions, employers can establish the physical and psychological demands for each job position. Pre-placement and periodic fitness-to-work assessments help to determine the degree of "person-job fit" — an important measure when determining the employee's ability to do the job safely.

Risk-based, periodic fitness-to-work assessments such as hearing conservation, respiratory conservation, ergonomic evaluations, and back assessments are some other effective safety and incident/injury/illness prevention approaches.

(b) Exposure Monitoring and Health Surveillance

Prevention of employee exposure to hazardous agents involves:
- hazard elimination through substitution or elimination of the noxious agents;
- exposure reduction through engineering controls;
- administrative controls to reduce the duration of hazard exposure for the employee;
- worker training to ensure employee awareness of hazardous products/ substance such as WHMIS 2015 training; and
- personal protective equipment to deal with hazards that cannot be addressed by the other preferred control methods.

Critical to exposure management is **exposure monitoring** — the continual evaluation of the workplace and employees for potential exposure. In terms of the workplace, this involves worksite inspections and surveys, environmental monitoring, evaluation of the results, and interpretation of the findings. In terms of the employee, health surveillance and medical monitoring are commonly used approaches.

Health Surveillance is the systematic collection and evaluation of employee data to identify instances of illness or health trends suggesting adverse workplace exposures coupled with actions to reduce hazardous workplace exposures. It involves:
- determination of potential for hazard exposure;
- provision of employee education and training;

- selection of medical tests and frequency of testing;
- identification of triggers for monitoring;
- identification of employee population(s) to be monitored;
- determination of employee occupational health history;
- baseline assessment;
- employee exposure assessment;
- interpretation of individual and aggregate results;
- employee notification of the results;
- development of a results-driven action plan;
- documentation and record keeping; and
- quality control.

(c) Injury and Illness Management

Occupational injury/illness does occur, thereby necessitating injury/illness management. This includes:

- injury/illness reporting;
- assessment, treatment and referral for injury/illness;
- incident investigation;
- claim management;
- case management;
- worker accommodation and gradual return to work;
- documentation and record keeping; and
- statistical reporting.

More information on injury management is provided in Chapter 13, "Occupational Health and Safety: Injury Management".

Incident investigation is performed whenever there is a major injury incident, or the potential for the same. This involves:

- investigating the incident circumstances and details;
- identifying the root cause(s);
- making recommendations for corrective actions;
- documenting the incident and findings; and
- reporting the findings and recommended controls.

More information on incident investigation is provided in Chapter 14, "Occupational Health and Safety: Incident Investigation".

The **Workers' Compensation Board (WCB)** is a third-party, government-operated insurance system designed to protect the injured/ill worker and afford employers litigation protection. It is a no-fault, industry-funded insurance that is mandatory for certain industry groups. Employers pay all the premiums and claims costs, while employees forfeit their right to take legal action against the employer. Claim administration includes claim submission, claim adjudication, claim appeal (if required), and claim termination. For more information on the Canadian Workers' Compensation System, refer to Chapter 33, "Occupational Health and Safety: The Canadian Workers' Compensation System".

In most instances, employers are legally required to manage their WCB claims and to offer work accommodation. This is best achieved through Disability Claims Management and Disability Case Management endeavours. Having an integrated Disability Management Program in place allows the employer to combine both essential elements in all disability situations (occupational illness/injury and non-occupational illness/injury); and in both short-term disability (STD) and long-term disability (LTD) situations. More information on disability management is provided in Chapter 27, "Disability Management: Overview".

(d) Emergency Response (ERP) and Disaster Planning

Emergency Response Plans (ERP) require an evaluation of the potential emergency situations and the available on-site resources. When developing these plans, it is important to consider potential emergency situations such as:

- medical emergencies;
- hazardous materials releases;
- events like fire, explosions, bomb threats, workplace violence and terrorism; and
- natural disasters.

In all cases, there is a need for documented ERPs. These plans should be built in consultation and collaboration with employees and the community.

In general, ERPs include:

- stakeholder roles and responsibilities;
- a system of communication;
- posted evacuation plans and routes;
- regular practice drills;
- follow-up assessment of the planned process(es); and
- feedback to the participants on their level of performance.

More information on emergency preparedness and response is provided in Chapter 12, "Occupational Health and Safety: Emergency Preparedness and Response".

(e) OH&S Legislation Governance and Stewardship

Numerous guidelines, regulations, and laws exist to protect the worker and the environment. Employers must understand the applicable legislation. Failure to do so will result in occupational injuries/illness, financial penalties, and even criminal charges. By being aware of, and monitoring the applicable legislation, employers can positively impact legislation development and enforcement. Chapters 30, 31, 32 and 33 address the legal aspects related to Occupational Health and Safety.

2. Non-occupational Aspects

(a) Disease and Injury Prevention/Health Promotion

Disease and injury prevention are the focus of OH&S Programs. Three levels of prevention exist:

i. **Primary Prevention** — preventing problems before they exist, such as education on heart health, smoking cessation, cancer awareness, sun safety, nutrition, and off-the-job safety.

ii. **Secondary Prevention** — the early detection of disease and the initiation of early treatment programs such as screening for vision disorders, hearing deficits, cholesterol, diabetes, tuberculosis, lung disorders, *etc.*

iii. **Tertiary Prevention** — intervention and/or treatment of disease and/or prevention of further health deterioration as a result of disease such as rehabilitation and restoration related to chronic diseases and conditions (*e.g.*, substance abuse, psychological health conditions, *etc.*).

Health promotion encompasses efforts like those of the Primary Prevention activities. Health promotion activities differ in that they are more generalized and less focused on addressing a particular health problem. Some examples are physical/emotional fitness, stress management, and effective parenting programs. Chapter 21, "Occupational Health and Safety: Prevention of Workplace Illness and Injury", and Chapter 26, "Occupational Health and Safety: Workplace Wellness Strategy" further address this topic.

(b) Employee Assistance Programs (EAPs)

Employee Assistance Programs are a support system for employees and management. The goal of the EAP is to help employees resolve concerns that affect their well-being and/or job performance. The EAP combines access to confidential counselling, referral to community healthcare professionals, and the expertise of professional external counsellors. Counselling is available for virtually any situation that is causing undue stress and concern (Table 2.2).

Table 2.2: Stressful Situations

• stress/lifestyle	• harassment	• parenting
• emotional issues	• trauma/crisis	• elder care
• physical or sexual abuse	• bereavement	• coping with chronic illness
• substance abuse/dependency	• career concerns	
	• personal legal issues	• sleeping disorders
• family/marital concerns	• financial concerns	• substance abuse treatment/rehabilitation
• relationship issues	• health concerns	
	• eating disorders	

The intent of the EAP is to offer a helping hand before a concern reaches a serious or crisis situation. It must be recognized however, that resolution of a problem is the responsibility of the employee and to be successful, requires individual motivation and active participation. More information on EAPs is provided in Chapter 19, "Occupational Health and Safety: Supporting Disciplines", Chapter 21, "Occupational Health and Safety: Prevention of Workplace Illness and Injury", and Chapter 26, "Occupational Health and Safety: Workplace Wellness Strategy".

(c) Attendance Support and Assistance Program (ASAP)

An **Attendance Support and Assistance Program** is a proactive approach to promoting and supporting employee attendance at work, and to prevent "work presenteeism".[41] To be successful, the program implementation requires:

- internal partnerships to address employee attendance;
- increased employee understanding of the impact lost work time has on the organization and its employees;
- standards involving regular work attendance;
- employee accountability for work attendance;
- fair and consistent treatment of employees;
- a focus on supporting and helping the employee attend work;
- the capacity to work with individual employees who are having difficulty maintaining regular work attendance; and
- workplace solutions to maintain regular employee attendance and reduce absenteeism.

The ASAP is custom-made to fit a company and its attendance rates. The key functions of this program are:

- Identifying the importance to the organization of employee dependability and responsibility, as demonstrated through good attendance.
- Indicating the concern of the organization regarding excessive absenteeism for any reason.
- Identifying the relationship between absence and performance management.
- Defining culpable or "blameworthy" absences and non-culpable or "innocent" absences, and the measure for dealing with these separately; that is,

---

[41] **Presenteeism** is the phenomenon of employees being at work, but because of wasted time, failure to concentrate, sleep deprivation, distractions, poor health, and/or lack of training, they may not be working at all.

progressive discipline for culpable absence and counselling or resource assistance for non-culpable absence.

- Clearly outlining the rules of the organization on reporting absences, for example:

    (a) the frequency and direction of reporting;

    (b) when, and if, a medical certificate is required; and

    (c) the nature and frequency of any additional information required by the employer during a period of absence from work.

- Being consistently enforced, while at the same time flexible enough to allow for some discretion on the part of the employer in the case of emergencies or unusual circumstances.

- Providing guidance to managers on what information is required from the absent employee, and what type of information is necessary for tracking purposes.

- Ensuring that there is a method for documentation and follow-up in the management of absenteeism.

The concept of an Attendance Support and Assistance Program, is discussed in D. Dyck, *Disabilility Management: Theory, Strategy & Industry Practice* (Markham, ON: LexisNexis, 2013), Chapter 9.

(d) Workplace Wellness Program (WWP)

A WWP is aimed at proactively promoting organizational and worker well-being. The focus is on managing both psychological and physical needs of both parties in response to environmental stress so that they are more resilient to environmental pressures. For an in-depth discussion on health promotion, health protection, and workplace wellness refer to Chapter 26, "Occupational Health and Safety: Workplace Wellness Strategy".

H. OHSMS AND OH&S PROGRAM: MEASUREMENT

1. Evaluation

Once an OHSMS is established, the next step is to measure its actual performance and compare the results against established standards. The effectiveness of an OHSMS can also be demonstrated in terms of achievement of stated program goals, objectives, and performance targets.

2. Program Measurement

Before proceeding with the evaluation of an OHSMS, the topic of program measurement should be addressed. Program measurement identifies the gaps between **the current state** and **the desired state** of a program, indicates whether the program goals/objectives are met or not, and enables improvements both along the way and periodically.

(a) Structure, Process, and Outcomes: Evaluation

Inherent to program evaluation is the establishment of a program that has an infrastructure which includes structure, processes, and outcomes. For an OH&S Program, the **Structure** would include the make-up of the team; its position within the organization; its functions; the qualifications and career paths of the OH&S Practitioners/Professionals; the OH&S Program Manual that houses the OH&S standards, the OH&S forms and other tools; OH&S equipment; OH&S communication plan; and OH&S communication tools (*e.g.*, webpage, brochures, worker tips, safety promotional items, the company's safe work practices, etc.).

The **Process** involves hazard identification, assessment, and control; worker OH&S training; OH&S communication; OH&S services and activities; workplace inspections; incident reporting and investigation; data management; audits and other program evaluation; and trend analyses and other analyses of the OH&S outcomes.

The **Outcomes** are comprised of the degree of hazard control, level of worker OH&S knowledge, degree of implementation of safe work practices, the number and quality of worksite inspections, the identification of injury trends, and the degree of proactivity demonstrated in terms of workplace safety (Figure 2.8). In this model, the OH&S structure contributes to the processes, and both, result in the nature of the outcomes achieved.

Figure 2.8: A Typical OH&S Program Structure

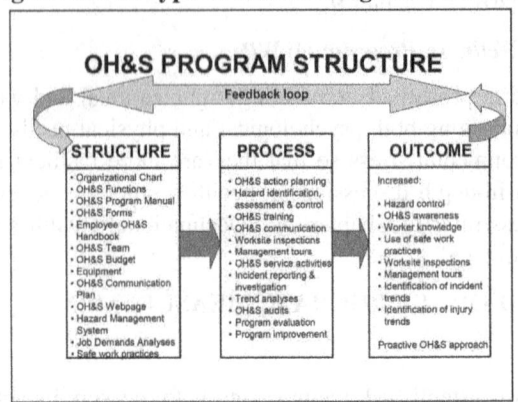

In addition to these components, the system must include a feedback mechanism for continuous improvement of the individual components.

(b) Program Goals, Objectives and Targets: Evaluation

OH&S Programs require program goals, objectives, and performance targets. These are the elements that are examined during program evaluation. A **goal** is a broad statement about what the program is designed to accomplish.

An **objective** is a specific aim set to achieve a desired goal. When developing objectives, it is important to be "SMART": that is, they must be:

- **S**pecific — Should be clearly defined.
- **M**easurable — Can be measured.
- **A**ttainable — Within one's power to do and willingness to do.
- **R**ealistic — Can be achieved.
- **T**ime-specific — Attainable within a specific time frame.

Targets are levels of performance that the company wants to attain. Typically, a company sets their targets as improvements of 5 to 10 per cent over the previous year's results. Since OH&S targets are usually linked to performance appraisals and incentive pay and other performance-related bonuses, it is important to set targets so that they are realistic and attainable through diligent and conscientious performance of desired work and safety behaviours.

Sample goals for an OH&S Program might be stated as:

1. *To ensure a safe and healthy workplace.*
2. *To promote illness/injury prevention.*
3. *To minimize the personal and economic costs of injury and disability.*

Based on the above sample goals, the OH&S Program objectives might be:

1. *To have an effective OH&S Program in place as proof that 100 per cent of the program targets were met or exceeded in 2021.*
2. *To promote illness/injury prevention as evidenced by the achievement of 90 per cent of the OH&S Worker Training schedule in 2021.*
3. *To promote illness/injury prevention as evidenced by the implementation of an ergonomic program in 2021.*
4. *To minimize the personal and economic costs of injury and disability as evidenced by a reduction of 10 per cent in the number of occupational injury incidents in 2021.*

A typical measurement plan for an OH&S Program would include:

1. *Establish OH&S Program goals.*
2. *Set the OH&S Program objectives.*
3. *Determine the OH&S targets.*
4. *Plan a schedule/action plan for the achievement of the objectives and targets and for the measurement of each.*
5. *Implement the plan.*
6. *Check progress.*
7. *Determine the results.*
8. *Analyze the results.*
9. *Communicate the outcomes.*

(c) Summative and Formative Evaluation

Two other evaluation terms and techniques that are commonly used include:

- **Summative Evaluation** provides information on a program's efficacy; that is, the ability to do what it was designed to do.[42] For example, did the employees learn what they were supposed to learn from a training course/program. In a sense, it lets the employee and the trainer know "how they did". More importantly, looking at the employee's level of achievement enables the trainer to determine if the training course/program teaches what it is supposed to teach.

    Summative evaluation is typically quantitative in nature. It uses numeric scores or letter grades to assess learner achievement. It is usually conducted after a certain period (*e.g.*, the first year, third year) to determine if the program has met established standards, goals, objectives, or performance targets. The findings and reports can be distributed internally or externally.

- **Formative Evaluation** is a little more complex. It is conducted during the development, delivery or improvement of a program/product/training course.[43] Typically, it is done with small groups of people who are asked to evaluate, or "test run", various aspects of a program/product/training course or materials. For example, an OH&S Practitioner/Professional might ask a group of employees to comment on a draft, worksite inspection form prior to its publication for use in the workplace. At times, feedback from a target audience might be sought. For example, when developing content for an OH&S webpage, comments should be sought from the stakeholders for whom the webpage content is being designed.

    The purpose of formative evaluation is to validate or ensure that the goals of the program/product/training course are being achieved, and to make improvements, if necessary, by means of identification and subsequent remediation of problematic aspects.[44] Other distinguishing features are that a formative evaluation tends to be conducted repeatedly with a view to improvement, and the findings/reports are intended primarily for internal use.

(d) Impact Evaluation

An **Impact Evaluation** measures the degree of change in the OH&S behaviours/ well-being/attitudes of workers, workgroups, business units, or organizations that

---

[42] M. Scriven, *Evaluation Thesaurus*, 4th ed. (Newbury Park, CA: Sage Publications, 1991).

[43] *Ibid.*

[44] C. Weston, L. McAlpine and T. Bordonaro, "A model for understanding formative evaluation in instructional design" (1995) 43(3) Educational Technology Research and Development 29-46.

can be attributed to an OH&S initiative, program or policy. The central impact evaluation question is: "What would have happened to those receiving the intervention if they had not in fact been exposed to the OH&S initiative, program or policy?"

Since evaluation of workers/workgroups/business units/organizations with and without an intervention is not possible, the use of a control group is adopted for comparison purposes. A **control group** is a group which is as similar as possible (in observable and unobservable dimensions) to those receiving the intervention (the study group). This comparison allows for the establishment of **definitive causality**, *i.e.*, attributing observed changes to the program, while removing confounding factors.

Impact evaluation is aimed at providing feedback to help improve the design of OH&S Programs and policies. In addition to providing for improved accountability, impact evaluations are a tool for dynamic learning, allowing policymakers to improve ongoing OH&S Programs and ultimately better allocate funds across OH&S Programs.

Table 2.3 provides a summary of the evaluation types, the techniques that are typically used, and the level of resources required to conduct the evaluation.

Table 2.3: Program Evaluation: Types, Resources and Techniques

Evaluation Type	Resources: Minimal	Resources: Modest	Resources: Substantial
Structure	• Document review • Comparison against a standard	Audits	Comparison against "best practices"
Process	Record keeping (*e.g.*, monitoring activity timetables)	Program checklist (*e.g.*, review of adherence to programs plans)	Management audit (*e.g.*, external management review of activities)
Outcome	Activity assessments (*e.g.*, numbers of health screenings and outcomes or program attendance and audience response)	Progress in attaining objectives monitored (*e.g.*, periodic calculation of percentage of aware, referred, participating)	Assessment of target audience for knowledge gain (*e.g.*, pre-test and post-test of change in knowledge)
Summative	Injury statistics	Trend analysis	Performance comparisons between groups, between years, or between industry groups
Formative	Readability/ "eyeballing" or "snicker test"	• Central location • Participant interviews	• Focus groups • Individual in-depth interviews
Impact	Print media review (*e.g.*, monitoring of content of articles appearing in newspapers)	Public surveys (*e.g.*, telephone surveys of self-reported behaviour)	Studies of public behaviour/health change (*e.g.*, data on physician visits or changes in public's health status)

(e) Why Measure OH&S Programs and Results?

One question that is always raised is: "Why continuously monitor and check OH&S Programs and results?" The rationale is to:

1. Promote stakeholder awareness for the OH&S Program goals, objectives, and performance targets.

2. Help the company's OH&S performance remain "on target".

CH. 2: OCCUPATIONAL HEALTH AND SAFETY MANAGEMENT SYSTEMS

3. Know how much further there is to go to realize "success".
4. Enhance the alignment of the OH&S Program goals, objectives, and performance targets with other business strategies and needs.
5. Correct identified problems.
6. Provide feedback to those involved.
7. Promote accountability for workplace OH&S.
8. Increase the likelihood of accomplishing the OH&S Program goals, objectives and performance targets.

To improve the OH&S Program result, there are a few recommended strategies that can be adopted:

- To begin with, be clear on the company's OH&S Program's vision, goals, objectives, and performance targets.
- Regularly monitor and measure the OH&S objectives and targets.
- Use an integrated bottom line.
- Learn from the results.
- Be persistent and diligent.
- Let go of what is not working.
- Recognize, communicate and celebrate progress.
- Learn from the experts.
- Aim for continuous improvement.

With an OH&S Program, there are barriers or challenges that can prevent the attainment of desired OH&S results. They include:

- Being afraid to dream or "think outside the box".
- Adoption of the "It didn't work before" attitude. That may be so, but with time and change, the proposed approach may now work.
- Discomfort with looking at the OH&S results. The ostrich approach (*i.e.*, denial and fear) gets a company nowhere: Face the results and determine how to address them.
- An established OH&S goal that was not a priority and so it was not addressed.
- Loss of focus.
- Seeing only problems and roadblocks, not ways to get to workable solutions.
- Choosing someone else's goals (but not your own). To attain a goal, it requires ownership of that goal, as well as passion.
- Reluctance to ask for help.

- Not keeping the OH&S commitments that were made.
- The presence of competing work demands, pressures, and stressors.

Knowing these are possible detractors, OH&S Program designers can build drivers into their program that counteract these barriers.

3. Measurement of an OHSMS

There are several ways to measure the functionality and effectiveness of an OHSMS:

- **Auditing** — using this measurement technique, the OHSMS is measured against a predetermined protocol. **Auditing** involves documentation reviews, interviews, and observations to verify the existence and functionality of various programs, policies, and procedures.

  This measurement technique allows for the identification of the gaps between the current and ideal state of the OHSMS. For example, Figure 2.9 demonstrates the OHSMS audit results as compared to the OHSMS audit protocol. It indicates that this organization's OHSMS is strong in management leadership and commitment for safety; worker qualifications, orientation and training; health and safety communication; and incident investigation. The OHSMS components that warrant attention are hazard identification and assessment; worksite inspections; hazard control; and emergency response. The OHSMS audit results provide direction for the organization to make OHSMS improvements. Typically, the audit process includes recommendations for reaching the ideal state.

  According to Dr. Peter Strahlendorf, School of Environmental Health, Ryerson Polytechnic University, an OHSMS audit "is a means to an end. It allows or enables individuals in the organization to meet their own due diligence requirements, as well as ensuring due diligence for the organization as a whole".[45]

---

[45] P. Strahlendorf, "The Occupational Health and Safety System: Tools for Due Diligence" (Workplace Environment Health & Safety Reporter, Special Report, April 1996), at 11.

Figure 2.9: OHSMS Audit Results (example)

Element	Total Points Scored	Total Points Possible	Percent-age
1. Management Leadership and Commitment to Safety	114	125	91%
2. Hazard Identification and Assessment	102	170	60%
3. Hazard Control	133	160	83%
4. Worksite Inspections	82	105	78%
5. Worker Qualifications, Orientation and Training	91	100	91%
6. Emergency Response	78	90	86%
7. Incident Investigation	114	125	91%
8. Health and Safety Communication	116	125	93%
OHSMS Survey Score	829	1000	83%

- **Benchmarking** — benchmarking is a continual and collaborative discipline that involves measuring and comparing the results of the key process with "best performers", or with one's own previous achievements. Internally, benchmarking can be used to compare OHSMS Review results against previous performance results. This can be done in terms of the whole company, or among the various divisions within the company. Externally, benchmarking involves comparing an organization's OHSMS and/or results against that of another organization.

By way of an example, Figure 2.10 demonstrates the benchmarking of OHSMS of various companies. The findings are that Company 1 and Company 11 scored strongly on the OHSMS audit, as compared to Companies 6 and 14.

Figure 2.10: Benchmarking Example of a Number of OHSMS

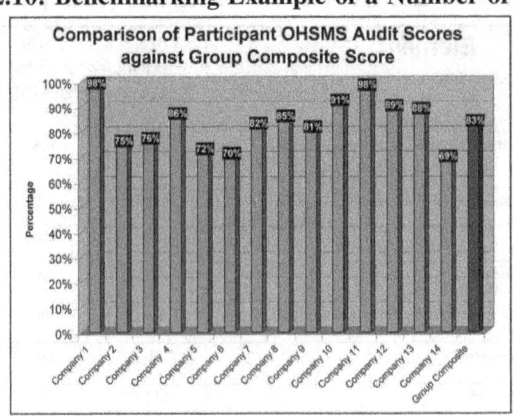

Upon examining the data more closely (Figure 2.11), the benchmarking indicates that the group of companies is strong at *Management Leadership and Commitment*; *Worker Qualifications*; *Orientation and Training*; *Emergency Response;* and *Incident Investigation* but needs to address *Hazard Identification and Assessment*; *Worksite Inspections*; and *Hazard Control*. Knowing this, the group can target its efforts towards improvement of those elements.

Figure 2.11: Benchmarking Example of OHSMS Components

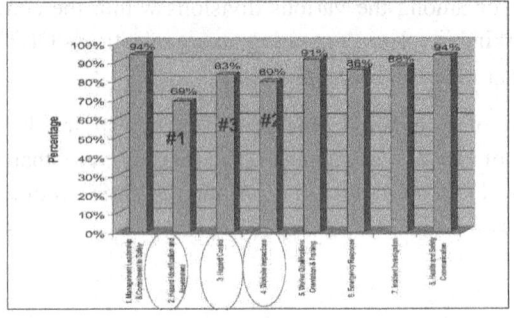

- **Performance Metric** — is a measurement of the performance of OH&S performance. Using the *Institute for Work & Health Organizational Performance Metric*, organizations/companies can assess and improve their OH&S performance.[46] This evidence-based, eight-time questionnaire measures the percentage of time each OH&S practice, as based on the CSA Z1000 standard, occurs. Interpretation of the results is explained. Results can then be compared with the performances of other workplaces; refer to

---

[46] IWH, *Institute for Work & Health Organizational Performance Metric* (2016), online: www.iwh.on.ca (date accessed: February 28, 2020).

at www.iwh.on.ca/iwh-opm for more information and resources.

- **Trend Analysis** — is an analysis technique that is based on the idea that what has happened in the past will provide the OH&S Practitioner/Professional with an idea of what will happen in the future. The term refers to the concept of collecting and evaluating information and attempting to spot a pattern, or *trend*, in the information.

  Effective trend analysis is highly dependent on the availability of sound OH&S data. It also requires that the data sets being examined are similar in nature. That is, each data set:
  - covers a comparable period of time (*e.g.*, for one year or six months);
  - covers the same data elements (*e.g.*, number of workplace incidents within a specific workgroup, or injury frequency rates for a specific workgroup); and
  - was collected and analyzed using the same methods.

  This technique can be simple or complex, depending on the information sought. Of importance, is to ensure that the trend analysis techniques used, are well understood in terms of their strengths and limitations.

- **Cost Effectiveness and Cost-Benefit Analysis** — evaluation of OHSMS often includes some measurement of cost. This may range from a requirement to demonstrate the cost effectiveness of the OHSMS to analyzing the cost benefit of a specific aspect of the program, such as worker training, worker involvement in safety, perception of the safety culture, *etc*.

  Cost containment refers to keeping costs to a minimum. **Cost effectiveness** is demonstrating the results of the OHSMS from a financial perspective. **Cost benefit** is weighing costs of the OHSMS against the benefits provided.

  Effective analysis of costs requires planning, data collection and the appropriate analysis tools. The first step is to decide what is to be measured. The second step is to decide what data is required to provide the information, and how the data will be collected and analyzed.

  Data to demonstrate cost effectiveness usually needs to be collected over time; both to provide meaningful information for analysis and to provide the necessary information for measurement of improvement.

  To demonstrate a **return on investment** (ROI) for having a sound OHSMS, the formula is:

$$\text{ROI} = \frac{(\text{Costs without an OHSMS}) - (\text{Costs with an OHSMS})}{\text{Cost of developing/operating the OHSMS}}$$

Typically, the return on investment for having a sound OHSMS in place is $3 for every $1 spent on safety.[47] When determining the ROI, companies tend to deal with **direct safety management costs**.[48] However, if the indirect costs[49] of safety management are included, the ROI increases three to fivefold the direct safety costs.[50] In addition, the non-financial returns can include the provision of a safe work environment, compliance with regulations, the reduction in work-related injuries/illness, a healthier workforce, good employee morale, retention of workers, and reputation and corporate image management.

- **Balanced Score Card** — measurement of costs and injury rates (**lagging safety performance indicators**) has been the approach used by organizations to justify the effectiveness, or lack thereof, of their OHSMS. However, there is a growing amount of research indicating that cultural and leadership traits drive employee well-being and sustainable high performance.

  A **Balanced Score Card** is a measurement tool that focuses on financial measures, customer service indicators, internal work efficiencies, innovation, learning, growth measures, and system measures. It stipulates the organization's strategic goal(s) and the related business objectives. It quantifies the capabilities of the system elements to produce positive financial results, to enhance customer loyalty, to result in smooth cost-effective operations, to promote leading edge business endeavours, and to demonstrate strong health and safety outcomes. This means organizations need to create and use measures that are valid, and which reliably quantify the above values. By tracking the Balanced Score Card Results, organizations can determine the areas that warrant improvement and leverage those areas in which they do well in order to facilitate the desired improvements.

- **Perception Surveys** — measurement of employee experience and perception of the quality and effectiveness of the OHSMS can be achieved using a perception survey. Employees are well-positioned to provide feedback on the quality and effectiveness of the corporate safety culture, as well as the individual components of the OHSMS (such as OH&S leadership and commitment, OH&S communications, worker training, employee involve-

---

[47] Liberty Mutual Research Institute, reported in "Proving the Return on Investment on Safety" (2001), *Safety Compliance Insider*, Vol. 1(2), at 15.

[48] **Direct safety management costs** are the costs directly identifiable in terms of management of the safety system such as the OH&S budget, training costs, safety equipment costs and associated program costs.

[49] **Indirect safety management costs** include the "hidden" costs such as the management time and effort to promote workplace safety, worker time to do things properly the first time, data management time and costs, *etc.*

[50] *Ibid.*

ment, physical working conditions, and compliance with safe work practices).

An example of this OHSMS measurement technique is provided in Chapter 6, Figure 6.6, "Great Safety Performance™ Model: Measurement of Leading Indicators". It enables a gap analysis of the current OHSMS in comparison with the desired state for the OHSMS. From there, gaps can be identified and an action plan to eliminate or reduce the identified gaps, developed. The strength of the Great Safety Performance Survey Tool is that it identifies what continuous improvement actions employees perceive as valuable for the organization to pursue.[51]

There are many reasons for conducting system/program evaluations, namely to:

- Determine whether the system/program has been successful and/or effective in achieving its stated goals, objectives, and performance targets.
- Determine whether the system/program reached the target population(s).
- Determine the cost of the system/program, and if it was cost-effective or not.
- Justify the past and future costs and the challenges faced by the system/program.
- Justify continuation, continuation with modifications, or termination of the system/program.
- Contribute to the field of OH&S knowledge.
- Determine the strengths and weaknesses of the system/program elements for continuous improvement actions.
- Demonstrate that the system is meeting the organization's expectations and that it aligns with the overall business strategies.
- Promote confidence in the quality, value, and benefits of the system/program.
- Assess and comment on the current governance, accountability, and performance measurement systems within the organization.

The value obtained from conducting a system/program evaluation is that it:

- Creates greater stakeholder awareness of the system/program, its goals, elements, functions, and outcomes.

---

[51] D. Dyck & T. Roithmayr, "Great Safety Performance: An Improvement Process Using Leading Indicators" (2004) 52:12 AAOHN 511-520.

- Identifies opportunities for system/program improvement (Figure 2.12).
- Provides direction for enhancement of the OHSMS elements, standards, and procedures.
- Increases stakeholder appreciation that "upstream" activities (**leading indicators of safety performance**) can positively impact "downstream" outcomes (**lagging indicators of safety performance**).
- Promotes greater focus on inducing long-term behavioural and organizational cultural change.
- Can be used as a performance measurement for organizational incentive programs.
- Provides "real" data for organizational marketing initiatives; for enhancing the organization's image as a responsible player in Occupational Health and Safety; and for leveraging system/program improvements, and for worker training programs.
- Demonstrates corporate OH&S due diligence.

Figure 2.12: OHSMS Cycle[52]

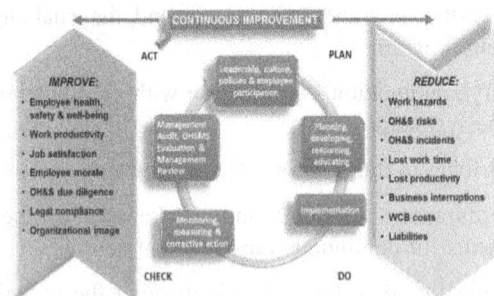

I. SUMMARY

An OHSMS is a valuable management tool. It integrates organizational elements involved in the continuous cycle of planning, implementing, evaluating, and continually improving the loss control processes within an organization. It is a systematic approach to managing workplace safety and contains elements that interrelate and provide a roadmap for action. Companies that possess a sound OHSMS tend to experience fewer workplace incidents/injuries/illness and related costs. This was attributed to the creation of good safety cultures, more employee

---

[52] Adapted from American National Standards Institute, ANSI 10-2005, Washington, D.C.; and Government of Alberta & Reliance OHS Inc., *Overview of Best Practices in Occupational Health and Safety in the Healthcare Industry* (Edmonton, AB: Government of Alberta, 2009-2011), at 41.

participation in safety activities, increased hazard reporting, better Management response to safety issues, and enhanced employee appreciation of both the physical and psychological working environments.[53] In short, having an OHSMS makes good business sense.

"Safety, like quality, improves when we improve the system; not when we hire more specialists to find defects or remove hazards."[54]

CHAPTER REFERENCES

Alberta Workplace Health and Safety, "The Partnerships in Health and Safety Program: Infosheet 1" (2014), online: http://www.employment.alberta.ca/documents/WHS/WHS-PS-InfoSheet1.pdf (date accessed: February 28, 2020).

American National Standards Institute, ANSI 10-2005, Washington, D.C.

BSI Management Systems, "OHSAS 18001 Awareness" (Presented as a City of Calgary Informational Seminar, January 2003).

F. Bird and G. Germain, *Loss Control Management: Practical Loss Control Leadership*, revised ed. (Loganville, GA: DNV, 1996).

Canadian Standards Association (CSA), Product Highlights: "CSA Z1000-06: New Occupational Health and Safety Management (OHSM) Standard Garners the Media Spotlight" (2006), online: http://www.csagroup.org/ca/en/home or http://www.shop.csa.ca (date accessed: February 28, 2020).

Canadian Standards Association (CSA), *CSA Z45001:19* (2019), online:www.scc.ca/en/standardsdb/standards/29765 (date accessed: February 28, 2020).

P. Caulfield, "WorkSafeBC said 'no changes' to COR program in 2018" (2018), Journal of Commerce, online: https://canada.constructconnect.com/joc/news/associations/2018/03/worksafebc-said-no-changes-cor-program-2018 (date accessed: February 28, 2020).

Comments submitted by Randy Kelbert (May 2007).

W.E. Deming, *Minerva News* (March 1993).

DuPont Safety Resources, "Dupont Bradley Curve" (Newark, DE: DuPont Safety Resources, 1994), online: http://www.dupont.com/safety (date accessed: February 28, 2020). Reprinted with permission.

D. Dyck, *Disability Management: Theory, Strategy and Industry Practice*, 6th ed. (Toronto: LexisNexis Canada, 2017).

D. Dyck and T. Roithmayr, "Great Safety Performance: An Improvement Process

---

[53] L. Robson, J. Clarke, K. Cullen, *et al.*, "The Effectiveness of Occupational Health and Safety Management Systems: A Systematic Review" (Toronto, ON: Institute for Work & Health, 2005).

[54] W.E. Deming, *Minerva News* (March 1993), at 3.

Using Leading Indicators" (2004) 52:12 AAOHN 511-520.

D. Fennell, "Growth of Effective Safety Systems" (presented at the Canadian Society of Safety Engineers luncheon, January 10, 2005).

Government of Alberta & Reliance OHS Inc., *Overview of Best Practices in Occupational Health and Safety in the Healthcare Industry* (2011), Edmonton, AB: Authors, at 41.

P. Hudson, "Safety Culture: The Ultimate Goal" (Transport Canada, Aviation Safety Letter, 2002), Vol. 2.

IWH, *Institute for Work & Health Organizational Performance Metric* (2016), online: www.iwh.on.ca (date accessed: February 28, 2020).

ILO, "Guidelines on Occupational Safety and Health Management Systems" (2001), online: http://www.ilo.org/public/libdoc/ilo/2001/101B09_287_engl.pdf (date accessed: February 28, 2020).

R. Kelbert, "Steps Towards Organizational Safety Maturity model" (Presented as part of the presentation on Creating a World Class Safety Culture, Western Conference on Safety, Vancouver, British Columbia, 2005).

R. Lardner, M. Fleming and P. Joyner, "Towards a Mature Safety Culture", Symposium Series #148 (Edinburgh, UK: Keil Centre, 2001), online: http://www.keilcentre.co.uk/media/1064/towards-a-mature-safety-culture-lardner-2002.pdf (date accessed: February 28, 2020).

C. Lawson, "Developing a Corporate Safety Program" (Presented at the American Occupational Health Nurses' Conference, April 20-27, 2001, San Francisco, CA).

Liberty Mutual Research Institute, reported in "Proving the Return on Investment on Safety" (2001), *Safety Compliance Insider*, Vol. 1(2).

NSF, *ISO 45001 Occupational Health and Safety Management Systems Migration Guide* (2018), online: https://www.nsf.org/newsroom_pdf/isr_dis45001_guide.pdf (date accessed: February 28, 2020).

L. Robson, J. Clarke, K. Cullen, *et al.*, "The Effectiveness of Occupational Health and Safety Management Systems: A Systematic Review" (Toronto: Institute for Work & Health, 2005).

M. Scriven, *Evaluation Thesaurus*, 4th ed. (Newbury Park, CA: Sage Publications, 1991).

P. Strahlendorf, "The Occupational Health and Safety System: Tools for Due Diligence", *Workplace Environment Health & Safety Reporter, Special Report (April 1996)*.

E. Sullivan and P. Decker, *Effective Management in Nursing*, 2d ed. (Menlo Park, CA: Addison-Wesley Publishing Company, 1988).

WSIB/CSPAAT, "Business Results Through Health & Safety" (2001), online: http://www.mtpinnacle.com/pdfs/Buisness-Results.pdf (date accessed: February 28, 2020).

C. Weston, L. McAlpine and T. Bordonaro, "A model for understanding formative evaluation in instructional design" (1995), 43:3 Educational Technology Research and Development 29-46.

WorkSafe BC, Discussion Paper: The Partners in Injury and Disability Prevention Program (Partners Program) (2018), online: https://www.worksafebc.com/en/resources/law-policy/discussion-papers/partners-in-injury-and-disability-prevention-program?lang=en (date accessed: February 28, 2020).

Chapter 3

OCCUPATIONAL HEALTH AND SAFETY: KEY STAKEHOLDERS

A. OH&S KEY STAKEHOLDERS: WHO ARE THEY?

Occupational Health and Safety (OH&S) Programs involve many stakeholders and highly technical functions. These OH&S functions vary depending on the industry; nature of the work, size, and location of the organization; available community resources; and the applicable legislative requirements. The most prevalent components of an OH&S Program include:

- **Occupational Health Management** providing elements such as fitness-to-work and fitness-for-duty assessments; medical monitoring and surveillance; respiratory protection; hearing conservation; health education; ergonomic assessments; human factors management; psychological risk management; emergency medical response; workplace wellness; and quality assurance audits.

- **Occupational Safety Management** providing elements such as hazard identification and management; general safety rules; safe work procedures; work permits; worker training; worksite inspections; equipment inspection and maintenance; product safety; fleet safety; pre-job planning; work observations; personal protection equipment provision, use and maintenance; incident investigation; a system for occupational safety communication; off the job safety; and program evaluation such as auditing.

- **Occupational Hygiene Management** providing elements such as hazard inventories; measurement of workplace hazards; walk-through surveys; ventilation assessments; respiratory protection; worker education; and audits.

- **Environmental Management** addressing elements such as hazard products inventories; waste management; emission control; pollution prevention; worker training; and audits.

- **Process Safety Management** providing elements such as process hazard information and knowledge; process hazard analysis; process equipment integrity; process design considerations and facility set-up; pre-start-up reviews and compliance audits; sharing of process safety information; and incident findings.

- **Work System** that includes elements such as an environment in which workers know how to work safely; are able to work safely; are equipped to work safely; and are motivated to want to work safely. It includes employee interactions and relationship aspects such as treating each other with respect; being honest; dealing with conflicts directly and fairly; listening; adapting to diverse working and communication styles; working in unison; freely communicating; taking pride in their work; and being accountable for their own conduct and results.

- **Corporate Safety Culture** that impacts elements such as the corporate values and ethics; beliefs regarding safety; degree of compliance with safety rules/procedures; willingness to accept safety responsibilities and accountabilities; and a belief that continuous improvement is necessary and possible.

Due to the technical expertise required to deliver these functions, a multi-disciplinary team can often be part of an OH&S Department. These professionals along with other key players, often make up the roster of stakeholders involved in the operation of a robust OH&S Program:

- employers;
- employees;
- trade union leaders (where applicable);
- supervisors (line managers, team leaders, foremen);
- Joint Health and Safety Committee (JHSC) members;
- OH&S Practitioners/Professionals;
- Human Resources Practitioners/Professionals;
- government OH&S officers; and
- workers' compensation personnel, *i.e.*, industry managers, adjudicators, case managers.

Occupational Health and Safety involves internal stakeholders functioning at various levels of the organization as well as external stakeholders. To give their endorsement and support, these stakeholders need to understand the company's OH&S Program goals, objectives, benefits, and desired outcomes. Only by understanding the magnitude of the issues within the workplace can stakeholders provide support and input towards common solutions.

Clarity of stakeholder roles is an essential element towards garnering stakeholder support and ensuring the existence of a successful OH&S Program. This chapter focuses on the roles of the specific personnel involved in workplace health and safety.

B. OH&S KEY STAKEHOLDERS: ROLES, RESPONSIBILITIES AND QUALIFICATIONS

1. Employer

From a legislative perspective, an employer may be a company, manager, or supervisor; it may even be an employee. The employer is anyone directing the work that is to be done by others.

The general duties of the employer/employer-representative are to ensure, as far as reasonably practicable, the health and safety of workers, and that these workers are aware of their responsibilities and duties under the relevant OH&S legislation. Given that organizations tend to have a variety of levels of Management, and that their responsibilities and duties differ, the following discussion will focus on two levels of Management: Upper Management and Front-line Management.

(a) Upper Management

The levels of Upper Management can be defined as:

- **Senior Management** — those individuals who have a number of operating managers reporting to them. This includes senior managers not generally at the worksite, but who may be periodically on-site (CEOs, COOs, VPs, Directors, Senior Management, *etc.*).
- **Operations Management** — the site-based personnel who have a number of supervisors reporting to them (Superintendents, Coordinators, Team Leaders, Operations Manager, General Manager, *etc.*).

The responsibilities and duties of Upper Managers are to:

- be cognizant of their legal OH&S and workers' compensation responsibilities;
- establish the OH&S Program vision and philosophy;
- set the OH&S Program objectives;
- develop a visual image of the OH&S Program in terms of a design/model;
- map out the process flow for the OH&S Program;
- develop OH&S policies and procedures;
- develop a discipline policy and clearly communicate the consequences for non-conformance with organizational safety values and practices;
- develop and approve the skill sets required for the OH&S professional positions;
- communicate the OH&S roles and responsibilities of the various corporate stakeholders;
- explore the needs of the various corporate stakeholders;
- increase the general awareness of the OH&S Program goals, benefits, and outcomes throughout the company;

- model the OH&S behaviours valued by the company;
- oversee the implementation of the OH&S program;
- manage the work being done;
- provide a forum for dealing with "refusal to work" situations;
- participate in the monitoring and evaluation of the OH&S Program's performance and effectiveness;
- consider worker health and safety issues when acquiring or procuring new equipment/tools, or designing new facilities, or adopting new processes;
- support the efforts to mitigate losses due to employee injury by supporting modified work options;
- identify and recommend needed support systems and other program elements; and
- focus on preventative strategies, such as strengthening the safety culture and the use of leading indicators of safety performance.

(b) Front-line Management

Front-line Management encompasses those individuals who assume dual functions of directing work and doing the work. They can be defined as:

- *Supervisor* — the site-based personnel who direct the work of hourly employees (Foremen, Shift Supervisors, *etc.*).
- *Foreman/Lead Hand/Team Leader* — those individuals who direct work at the front lines.

In addition to supporting the duties of Upper Management, Front-line Management should:

- uphold and model the OH&S behaviours valued by the company;
- correct the behaviour of employees who are behaving in an unsafe manner;
- stop work when there is an unsafe act or condition evident. Failure to do so indicates that the Front-line Manager condones the unsafe behaviour. This sets a poor example for employees and contractors, and makes it difficult to control future unsafe behaviours; and
- initiate discipline if repeated interventions with the employee in question are unsuccessful, or if the unsafe behaviour is extreme.

2. Employee

From a legal perspective, employees, while engaged in an occupation, must take reasonable care to protect their own health and safety as well as that of their co-workers (OH&S Act). Employees should cooperate with their employer to protect the health and safety of those involved in the organization.

The general duties and responsibilities of the employee are to:
- get OH&S information about the job from the employer including access to the appropriate safety manual, procedures, specifications, *etc.*;
- cooperate with and participate in the employer's OH&S Program;
- comply with the organization's safe work practices;
- participate fully in training programs/opportunities;
- use, care for, and maintain the personal protective equipment provided;
- assume work duties for which they are competent to safely undertake and complete;
- protect the health and safety of him/herself and others;
- display a positive attitude towards work and workplace health and safety;
- report unsafe work conditions or substandard equipment;
- refuse unsafe work by telling the supervisor if imminent danger is perceived. The employer is responsible for investigating any arising disputes. Employees cannot be disciplined or fired for refusing unsafe work;
- establish good working relations with the supervisor and co-workers;
- exercise self-discipline in terms of work practices and OH&S practices;
- be accountable for their own actions and behaviours;
- understand the consequences of non-compliance with OH&S standards; and
- uphold the organization's OH&S values and commitments.

3. Trade Union

Trade union leaders play an influential role in workplace safety practices, including involvement and support in:
- promoting a broader understanding of the OH&S Program;
- addressing the corporate cultural issues that impact workplace health and safety, employee attendance, and employee attitudes towards work;
- actively promoting a trusting and positive environment in which safe work practices can successfully be undertaken;
- complying with the applicable legislation: *i.e.*, OH&S Acts, Human Rights and Duty to Accommodate legislation, and Freedom of Information Acts;
- assisting with the development of prevention strategies required to lower the incidence of employee illness/injury;
- enhancing the OH&S Program's communication capabilities within the organization;
- assisting with the marketing and explanation of the OH&S Program to their members;

- assisting with the identification of workplace hazards and their control; and
- working collaboratively with Management on health and management issues that tend to co-exist in incident investigations and disability management situations and can be difficult to resolve.

4. Joint Health and Safety Committee

A **Joint Health and Safety Committee** (JHSC) is a "forum for bringing the internal responsibility system into practice",[1] and as such has a mandate to identify, evaluate, and participate in the resolution of workplace health and safety issues/concerns. It consists of worker (employee) and Management representatives.

It is Management's responsibility to establish the JHSC.[2] Equal representation of Management and Labour is required. Selection of the committee members is defined in Canadian OH&S legislation. Generally, Management representatives are selected by the employer, while the worker representatives are selected by employees or by the union (if there is one).

In deciding the ideal committee size, organizations must consider:

- the total number of employees in the organization;
- the number of different trades or unions involved;
- the complexity of the operation(s);
- the types and number of work hazards;
- whether or not all segments of the workforce are represented (*i.e.*, Management, supervisors, male employees, female employees, field staff, office staff); and
- whether or not the committee encompasses adequate knowledge of the work conditions, processes and/or practices.

A review of the legislative requirements for JHSCs is recommended.[3]

JHSC members must be adequately trained in Occupational Health and Safety for them to fully contribute to all the committee's activities. In Canadian jurisdictions, the employer is legally required to ensure that JHSC members receive training on elements such as:

- company expectations;

---

[1] CCOHS, "Joint Health and Safety Committee – What is a Joint Health and Safety Committee?" CCOHS (2018), online: https://www.ccohs.ca/oshanswers/hsprograms/hscommittees/whatisa.html (date accessed: February 28, 2020).

[2] *Ibid*.

[3] Refer to CCOHS, "When are health and safety committees required, how many people are on the committee, and who are the committee members?" CCOHS (2018), online: https://www.ccohs.ca/oshanswers/hsprograms/hscommittees/whatisa.html.

- JHSC scope of responsibilities and authority;
- the applicable OH&S legislation;
- the principles of incident causation;
- hazard recognition;
- job safety analysis;
- basic principles of Occupational/Industrial Hygiene;
- effective methods of raising OH&S awareness;
- worksite inspections;
- incident investigation; and
- effective techniques of oral communication.

In some provinces, like Ontario, JHSC certification is available and required.[4]

The roles of the OH&S Practitioner/Professional and the JHSC members must be clearly defined to prevent misunderstanding and conflict within the organization. The OH&S Practitioner/Professional attends JHSC meetings as a technical specialist. His/her role at these meetings is that of resource person, advisor, or guest. Whatever the role, the JHSC should be assisted, not controlled (or seen as controlled), by the OH&S practitioner/professional.

In general, the JHSC member is responsible for:

- attending all JHSC meetings;
- promoting the organization's OH&S Policy and Program;
- assisting in the development of the organizational OH&S rules;
- assisting in the development of safe work procedures;
- advising on personal protective equipment (PPE);
- assisting the workplace to raise OH&S standards above minimal requirements;
- encouraging adequate worker education and training programs so that workers are knowledgeable about their rights, responsibilities, duties, and restrictions under the applicable OH&S legislation;
- assisting with the training of new employees (*e.g.*, Kimberly-Clark's JHSC conducts Safety Orientations for new or transferred employees[5]);
- conducting health and safety education programs (*e.g.*, Kimberly-Clark's

---

[4] Refer to JHSC Regulations, online: https://osg.ca/jhsc-regulation/.

[5] R. Bryson & C. Opas, "Best Practices for a Successful Joint Health & Safety Committee" (Presented at the Industrial Accident Prevention Association Conference, April 26-28, 2004, Toronto, ON).

JHSC assists in the communication and explanation of safety policies, practices and procedures[6]);
- promoting and monitoring the organization's degree of compliance with the applicable OH&S regulations;
- participating in the identification and control of workplace hazards by doing physical condition inspections with a senior member of management for the respective areas;
- participating in assessments and development of control programs for hazardous substances (Workplace Hazardous Materials Information System — WHMIS 2015);
- assisting the employer to resolve employee OH&S issues/complaints;
- providing feedback on employee issues/complaints/suggestions;
- assisting with the resolution of work-refusal situations;
- maintaining records of workplace incidents and injuries;
- participating in incident investigations;
- reviewing and monitoring corrective actions that result from incident investigations;
- initiating other activities as indicated by incident experience;
- monitoring the effectiveness of the organization's OH&S Program;
- studying OH&S programs of other companies with a view to enhancing the company's own OH&S program; and
- making health and safety improvement recommendations.[7]

In jurisdictions such as Ontario where JHSCs are legally mandated, the powers and duties of the JHSC include:

- ***Obtain information from the employer about Health and Safety in the workplace:*** The employer is legally obligated to cooperate with and assist the JHSC in undertaking its legal functions.
- ***Provide input into existing and proposed workplace health and safety programs.***
- ***Recommend health and safety improvements*** in the workplace.
- ***Be consulted on worker training programs.***
- ***Identify actual and potential hazards*** in the workplace.
- ***Conduct regular worksite inspections.***

---

[6] *Ibid.*
[7] *Ibid.*

- *Participate in any health and safety-related testing in the workplace.*
- *Take part in the investigation of a work refusal.*
- *Investigate a serious injury incident* or fatality.
- *Accompany a government official* during a tour of the workplace.
- *Receive worker concerns, complaints, and recommendations.*
- *Address worker issues and recommend solutions.*
- *Hold regular JHSC meetings* at least every three months in a location provided by the employer — guidelines exist on how to conduct an effective JHSC meeting.[8]

5. Occupational Safety Practitioner/Professional

Depending on the size of the organization, Occupational Safety Practitioners/Professionals can assume a variety of roles from a front-line Safety Advisor to a Safety Specialist who tends to assume more of a corporate role. Chapter 41, "Occupational Health and Safety Practitioners/Professionals: Career Development" describes in detail, the possible levels of Occupational Safety roles that can be found in organizations.

By way of an example, the Career Streaming Model presents the Safety Advisor/Safety Specialist as playing a key role in advising and guiding the various Business Lines in terms of their OH&S responsibilities. He/she acts as an expert resource on OH&S issues for employees and managers within that Business Line. This position:

- assists assigned business lines to plan and promote the OHSMS within its respective operation;
- advises on compliance with, and interpretation of, the *Occupational Health and Safety Act* and other relevant legislation;
- ensures that all responsible persons are aware of the legal requirements of notifying the enforcing authorities of workplace serious incidents;
- encourages effective hazard management;
- assists the business line to conduct incident investigations;
- ensures that the appropriate incident reports are sent to the enforcing authorities in a timely manner;
- conducts diagnostic investigations as required;
- participates in the evaluation of the OHSMS;
- gathers and manages OH&S data;

---

[8] R.S.O. 1990, c. O.1, Amended 2019. "Part II, Section 9: JHSC Regulations", online: https://www.ontario.ca/laws/statute/90o01 (date accessed: February 28, 2020).

- analyzes OH&S data; and
- generates and distributes relevant reports to Management.

The following (Table 3.1) is an example of what the Occupational Safety Advisor role entails:

Table 3.1: Occupational Safety Advisor

Skill Types	Advisor Role
Technical Specialist Skills:	• Regulatory compliance and governance • Legislation monitoring • OH&S Program development for a business line • Standard setting (advise, research, review) • Loss control (operational) • Hazard identification, assessment and control • Risk management (Safety) • Emergency preparedness and response • Incident investigation (assist and advise) • Technical communication • Proactive/preventative approach • Safety leadership • Quality assurance • Auditing/program evaluation
Relationship Skills:	• OH&S communication to business lines • Interpersonal relationship skills • Advocacy • Team building • Coaching • Community relationship • Reputation management
Business Skills:	• Business communication • Decision-making • Project management • Business/operational advice • Process facilitation • Issue management • Safety marketing • Customer-focus servicing

Often Occupational Safety Practitioners/Professionals function within an organization without a position or job description. This is not a recommended practice, especially when legal and insurance coverage tends to centre on whether the Occupational Safety Practitioner/Professional is functioning within the scope of the position. As a means of protection, the Occupational Safety Practitioner/Professional can use the information provided in the relevant Career Stream model to develop a position or job description. An industry example is provided in Chapter 41, "Occupational Health and Safety Practitioners/Professionals: Career Development" at Appendix 1.

6. Occupational Health Nurse

Occupational Health Nurses (OHNs) are registered nurses who possess certification in Occupational Health Nursing, and often, possess additional education and skills in the areas of Occupational Health and Safety, relationship building, program development, human resources management and business management. Their mandate is to promote healthy working environments, protect the health of the worker, and prevent work-related injuries/illnesses.

Although many companies have OHNs on staff, most do not. Their services are often purchased in the marketplace on a contractual or part-time basis. As a result, the discussion on the OHN and Occupational Health Nursing is provided in Chapter 4, "Occupational Health Program".

7. Safety Engineer

Safety engineering is an applied science that assures that a life-critical system behaves as needed, even when individual system parts fail. Safety engineering refers to any act of incident prevention by a person qualified in the field. The multidisciplinary nature of safety engineering means that an array of professionals is actively involved in incident prevention or safety engineering.

Safety engineering is often reactionary to workplace incidents. This arises largely because of the complexity and difficulty of collecting and analyzing data on near-miss incidents.[9] Safety reviews are now recognized as an important risk-management tool. Failure to identify safety risks, and the inability to control these risks, can result in massive loss costs, both human and economic, to an organization.

Safety engineers define potential operational defects. A **failure** is the inability of a system or component to perform its required functions within specified performance requirements; while a **fault** is a defect in a device or component, *e.g.*, a short circuit or a broken wire. System-level failures are caused by lower-level faults, which are ultimately caused by basic component faults. The unexpected failure of a device that was operating within its design limits is a **primary failure**, while the expected failure of a component stressed beyond its design limits is a **secondary failure**. A device which appears to malfunction because it has responded as designed to a bad input is suffering from a **command fault**. A **critical fault** is one that endangers one or a few people. A **catastrophic fault** is one that endangers, harms or kills a significant number of people.

Safety engineers also identify the different modes of safe operation: A **probabilistically safe system** has no single point of failure, and enough redundant sensors, computers, and effectors so that it is very unlikely to cause harm ("very unlikely" means, on average, less than one human life lost in a billion hours of operation). An **inherently safe system** is a clever mechanical arrangement that cannot be made to cause harm — obviously the best arrangement, but this is not always possible. A **fail-safe system** is one that cannot cause harm when it fails. A **fault-tolerant system** can continue to operate with faults, though its operation will be substandard/less efficient.

These terms combine to describe the degree of safety needed by systems: for

---

[9] A **near-miss incident** is an incident that has the potential to cause damage or injury.

example, most biomedical equipment is only critical, and often another identical piece of equipment is nearby, so it can be merely probabilistically fail-safe. Train signals can cause catastrophic incidents (imagine chemical releases from tank-cars) and are usually inherently safe. Aircraft failures are catastrophic, so aircraft are usually probabilistically fault tolerant. Without any safety features, nuclear reactors might have catastrophic failures, so real nuclear reactors are required to be at least probabilistically fail-safe, and some such as pebble bed reactors are inherently fault tolerant.[10]

8. Conclusion

In an organization, although each of these stakeholders has different roles and responsibilities, in a strong OHSMS, they are positioned to work together towards a common goal — the minimization of unsafe work acts (Figure 3.1).

Figure 3.1: Stakeholder Roles in Minimizing Unsafe Acts[11]

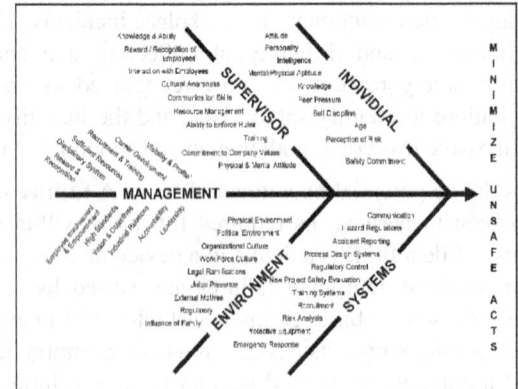

C. SUMMARY

The purpose of this chapter is to describe the roles and responsibilities of the various stakeholders involved in an OH&S Program. As you continue to read the remaining book chapters, these roles will come alive.

CHAPTER REFERENCES

R. Bryson and C. Opas, "Best Practices for a Successful Joint Health & Safety Committee" (Presented at the Industrial Accident Prevention Association Conference, April 26-28, 2004, Toronto, Ontario).

---

[10] Wikipedia, "Safety engineering entry", online: http://www.wikipedia.org (date accessed: February 28, 2020).

[11] Reproduced courtesy of *Workplace Environment, Health & Safety Reporter*, "Minimizing Unsafe Acts" (Toronto, ON: Templegate Information Services Inc., 1998). For subscription information, phone: (416) 920-0768.

CCOHS, "Joint Health and Safety Committee – What Is a Joint Health and Safety Committee?" CCOHS (2018), online: https://www.ccohs.ca/oshanswers/hsprograms/hscommittees/whatisa.html (date accessed: February 28, 2020).

R.S.O. 1990, c. O.1, Amended 2019. "Part II, Section 9: JHSC Regulations", online: https://www.ontario.ca/laws/statute/90o01 (date accessed: February 28, 2020).

S. Glasbeek & F. Campbell, *Occupational Health and Safety in Ontario Education: A Risk and Compliance Manual*, c. 6 (Markham, ON: Butterworths, 2005).

Ontario Ministry of Labour, "Joint Health & Safety Committees" (April 2014), online: http://www.labour.gov.on.ca/english/hs/pubs/jhsc/jhsc_jhsc.php.

Wikipedia, Safety Engineering entry, online: http://www.wikipedia.org (date accessed: February 28, 2020).

Workplace Environment, Health & Safety Reporter, "Minimizing Unsafe Acts" (Toronto, ON: Templegate Information Services Inc., 1998).

Chapter 4

OCCUPATIONAL HEALTH PROGRAM

A. INTRODUCTION

An Occupational Health Program, which can be operated internally or externally to an organization, provides appropriate health services to workers as well as monitors the work environment. It is designed to implement nursing and medical skills and knowledge in the workplace so that:

- workers are protected against health hazards resulting from work and/or the work environment;
- the workplace can be adapted to fit the worker's physical and psychological needs;
- the workplace is a safe and healthy place to work;
- workers can enjoy an optimal level of workplace health; and
- rehabilitation services are made accessible to ill/injured workers.

Prevention is the goal. The focus is on promoting a safe and healthy work environment in which workers can function without experiencing any negative impacts from the assigned work.

Occupational Health (OH) involves a diverse team of people who have experience in occupational health nursing, occupational health medicine, psychology, ergonomics, and industrial hygiene. One of their roles, termed **governance**, is to provide strategic and technical advice on occupational health issues to Management.

B. OCCUPATIONAL HEALTH PROGRAM COMPONENTS AND SERVICES

The components of an Occupational Health Service fall within five major areas:

1. Work and Health Interactions:

- medical surveillance of workplace factors that negatively impact worker health;
- ergonomic job analyses;
- periodic health monitoring;
- disability management programs;
- work accommodation; and

- data collection, management, interpretation and dissemination.

2. **Health Impacts on the Capacity to Work:**
   - pre-placement screening;
   - return-to-work assessment;
   - periodic fitness-to-work assessments; and
   - identification of substance or gambling addictions.

3. **Effects of Work on Health:**
   - education on work factors that could negatively impact human health;
   - evaluation and recommendations for appropriate hazard control;
   - diagnosis and treatment recommendations of occupational disease;
   - emergency treatment and medical management;
   - individual health and wellness counselling for employees;
   - disaster planning for the workplace and, when appropriate, the community;
   - interpretation of government OH&S legislation;
   - assistance with the development of government OH&S legislation; and
   - advice on the health impacts of organization/company operations/products on environment and public health.

4. **Health and Wellness Support:**
   - health education;
   - promotion of physical fitness;
   - promotion of psychological counselling;
   - medical support related to the control of employee medical absenteeism;
   - provision of immunizations;
   - screening for early detection of non-occupational disease;
   - promote the treatment of non-occupational illness; and
   - occupational health research.

5. **Occupational Health Technical Support:**
   - development, maintenance and evaluation of the Occupational Health Program;
   - development, maintenance and evaluation of the Disability Management Program;
   - assistance with Workers' Compensation Board issues/claims;
   - oversee/management of the Employee Assistance Program;

- oversee/management of the Employee Physical Fitness Program; and
- oversee/management of the Ergonomics Program.

C. OCCUPATIONAL HEALTH PROFESSIONALS

Occupational Health professionals are nurses and physicians who have attained education in their respective fields and then pursued advanced education in Occupational Health.

1. Occupational Health Nurses

What can **Occupational Health Nurses** (OHNs) offer to an organization? They are qualified to:

- implement and manage OH Programs and the related services;
- facilitate occupational health and safety, disability management and workplace wellness efforts with internal and external players;
- provide assessment and surveillance of the worker;
- perform assessments and surveillance of the workplace;
- coordinate and/or perform emergency response services;
- treat, counsel, and refer workers;
- develop and manage/coordinate Disability Management Programs;
- coordinate effective treatment and/or referrals;
- facilitate safe and timely return-to-work opportunities for recovering ill/injured workers;
- liaise with management, the worker and/or external resources such as:
    - physicians
    - therapists
    - rehabilitation specialists
    - Workers' Compensation Board(s)
    - insurance providers
    - Employee Assistance Program counsellors
    - other allied health professionals;
- engage in workplace health promotion activities;
- provide workplace education and training;
- act as a vital link between workplace stakeholder groups such as the employees, employer, and union(s); and
- build strong relationships with employees and their families.

In many organizations, the OHN manages and operates the OH&S Program, including the Disability Management Program and oversees the Employee Assistance Program and Workplace Wellness Program. Given their education as generalists in the field of Occupational Health, they are valuable professionals for organizations and their employee populations.

The Occupational Health Nurse provides value to an organization by assisting with *operational efficiency, loss control, injury/illness management and disability management*. Occupational health nursing is a specialty that provides for and delivers healthcare services to workers, work populations, and organizations. Graphically depicted, the scope of Occupational Health Nursing Practice (Figure 4.1), can positively impact the organization, the workplace environment, employees, and the community at large.

Figure 4.1: Scope of Occupational Health Nursing Practice

SCOPE OF OHN PRACTICE

- Community at Large
- Employees
- Workplace Environment
- Organization

OHN Practice
- Promote healthy work environment
- Protect employee health
- Prevent work-related injuries/illness

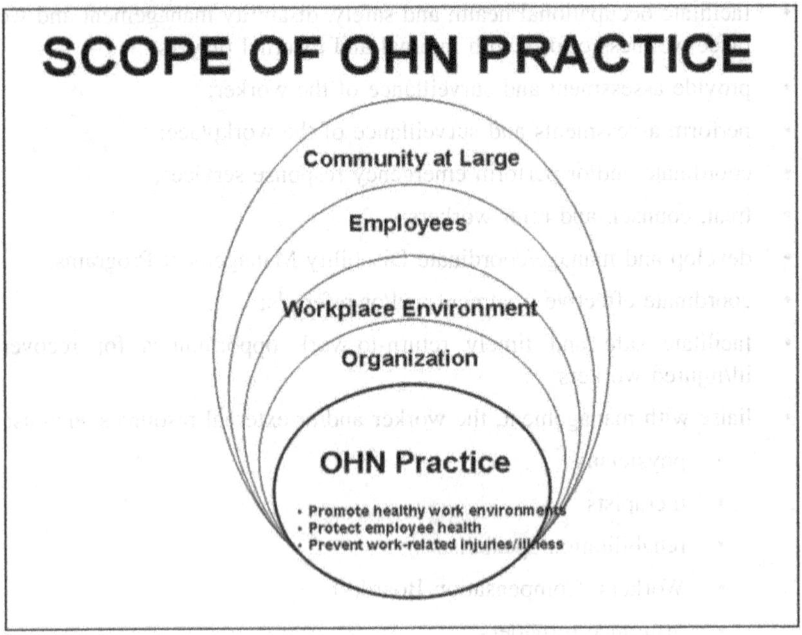

In terms of *operational efficiency*, OHNs:
- provide customized, effective OH&S Programs;
- promote the integration of employee services into organization/company strategic plans and operations;
- align OH&S Program goals with corporate/operational goals and objectives;
- promote worker wellness and fitness-to-work, thereby improving worker productivity, reducing expenses, and improving organization/company profits;

- assist new workers to integrate into the workplace;
- monitor the work environment in terms of hazard control, thereby reducing operational downtime, reducing expenses, and improving organization/company profits;
- provide anticipatory guidance on potential workplace problems and offer counselling so as to enable the organization/company to avoid/mitigate the negative impacts;
- monitor worker well-being and workplace injury/illness with a view to loss control and prevention;
- strive to enhance employee morale by working closely with employees and epitomizing the message that the organization/company cares for them personally; and
- promote the organization/company as a "responsible corporate citizen".

Loss control means minimizing loss due to people, property, process, plant, or profit damages/threats. According to Allen, "[m]inimizing loss is as much of an improvement as is the maximization of profit."[1] In terms of occupational health nursing, prevention involves the activities associated with "health-protecting behaviour". The emphasis is on guarding or defending an individual or group against specific illness/injury.

Regarding operational production and service demands, OHNs can support the workplace by:

- ensuring workers are **fit-to-work**[2] and **fit-for-duty**;[3]
- ensuring the worksite is free of uncontrolled hazards;
- medically monitoring workers exposed to known hazards;
- conducting risk assessments;
- communicating the nature and severity of identified risks;
- promoting worker well-being;
- conducting human factor/ergonomic assessments and identifying suitable remedial actions;
- participating in emergency response activities;

---

[1] L. Allen, cited in "Managing Control of Loss" in *Loss Control Management: Practical Loss Control Leadership*, revised ed., by F. Bird & G. Germain (Loganville: GA: DNV, 1996).

[2] **Fitness-to-work** is defined as the physical and psychological ability to safely undertake stated physical and psychological job demands.

[3] **Fit-for-duty** is defined as a condition in which an employee's physical, physiological, and psychological state enables them to continuously perform assigned tasks safely. It encompasses physical requirements, psychological conditions, and psychological status.

- facilitating critical incident stress debriefing post-incident; and
- assisting with the management of strategic OH&S issues.

Hence, OHNs contribute to promoting and maintaining worker health and safety, as well as workplace safety. By controlling losses, OHNs contribute to the enhancement of organization/company profits.

OHNs have a major role to play in the areas of injury/illness management and disability management. Through **client advocacy** — the activity associated with pleading or representing an employee's or organization's cause, OHNs act as a **client liaison** — the position of responsibility within an organization for maintaining communication links with external individuals, agencies, or organizations.

OHNs are educated, skilled, experienced, and ideally positioned within an organization to facilitate injury/illness management and disability management. They are competent at:

- mitigating the workplace illness/injury through timely response and referral for medical treatment;
- determining worker fitness to work and assessing perceived worker impairment (fitness for duty);
- managing injury/illness cases;
- co-managing insurer (government/private insurers) responsibilities and actions;
- coordinating disability management assistance;
- assisting workers to successfully return to work in a safe and timely manner;
- negotiating service provider contracts and activities;
- evaluating the outcomes and determining the return on investment for the organization/company; and
- conducting trend analyses with a view to illness/injury prevention and the introduction of suitable loss control measures.

In terms of managing a Disability Management Program, the OHN is positioned and capable of:

- supporting all the operations with the development and implementation of the Disability Management Program and the related practices;
- assisting the operations and employees with disability management and related labour relations issues and resolutions that may affect employee employability;
- investigating and assisting with the preparation of WCB reports on behalf of the organization and its employees;
- supporting the operations and employees with their individual disability

management and work accommodation performances;
- mentoring and coaching line management on attendance and disability management-related issues and practices that can affect employee employability;
- coordinating disability management services internally and externally with insurers and other healthcare providers;
- ensuring that employee personal health information is maintained in a confidential manner;
- providing disability management education;
- assisting the organization to comply with the applicable legislation (Workers' Compensation Acts, *Canada Labour Code*,[4] Human Rights legislation) and industry standards for disability management;
- maintaining the organization's disability management data management system (workers' compensation and non-occupational disabilities);
- conducting disability management performance audits and preparing the resulting reports;
- providing strong attendance support and disability management leadership; and
- representing the organization's business interests on government, industry, and professional committees/forums.

In summary, OHNs are professionals that have much to offer an organization and its employees.

2. Occupational Health Physicians

Occupational Health Physicians (OHPs) are medical physicians who have chosen to specialize in the field of occupational medicine. They are knowledgeable about medicine, the workplace and stakeholder interests, as well as the areas of ergonomics, human factors, industrial hygiene, toxicology, Workers' Compensation System, and disability management.

Occupational medicine is a fairly new field of medicine. In 1984, it became recognized as a medical specialty in Canada. Since its inception, OHPs have played a major role in promoting workplace health and safety. Today, the Occupational Medicine Specialists of Canada (OMSOC) is the only Canadian organization dedicated to the specialty practice of occupational medicine. Its membership comprises occupational health physicians certified as specialists by the Royal College of Physicians and Surgeons of Canada, as well as specialists from other medical and surgical specialties with an interest in occupational medicine. OMSOC provides a forum for advancing the practice of occupational medicine by facilitating

---

[4] *Canada Labour Code*, R.S.C. 1985, c. L-2.

dialogue amongst physician specialists and between occupational medicine practitioners and members of allied fields, notably government, industry, management, and the law.[5]

The OHP is concerned both with the effect of work on human health and with the effects of health on an employee's ability to work. Thus, OHPs function as consultants to management, employees, unions, and the public when required. They are well-positioned to advise managers, workers, and union leaders about health-centred problems that concern people and the work environment. As well, OHPs can explain health-related problems in terms of the workplace, their occurrence, and the significance of hazards and statutory requirements.

OHPs are also able to interpret the conditions of the workplace and the nature of the work to external healthcare providers. They may be involved in studies and research on occupational injuries and diseases, and how to prevent them.

In the area of Disability Management, they liaise with the employee's attending physician(s) to get a better understanding of the nature of the employee's disabling condition and how it impacts his/her ability to work. They also interpret the job demands and conditions to the attending physician(s) and the workplace accommodations available to the employee. In this way, they are a valuable member of the Disability Management Team.

OHPs, like other medical professionals, adhere to professional ethics. As a professional advisor, the OHP is legally bound to maintain confidentiality of the employee's health information. However, given the OHP's knowledge and skills, he/she can provide Management with information on the employee's fitness for work and work limitations.

D. OCCUPATIONAL HEALTH SERVICE: A BUSINESS CASE

An Occupational Health Service (OH Service), like any business function, is expected to justify from a business standpoint, its position within, and the value offered to the workplace. Developing a business plan, working the plan, and then, measuring and reporting on the performance of the OH Service, is critical. Unfortunately, this approach is often overlooked. The lack of meaningful performance measurements hampers the OH Service from demonstrating the value it adds to an organization. Consequently, the senior management team is left questioning the value of the OH Service and its personnel.

1. The OH Service

Occupational Health is defined as "the promotion and maintenance of the highest degree of physical, mental and social well-being of workers in all occupations by preventing departures from health, controlling risks and the adaptation of work to

---

[5] Occupational Medicine Specialists of Canada, *OMSOC Home Page*, online: http://www.OMSOC.org (date accessed: February 28, 2020).

people, and people to their jobs".[6] Hence, an OH Service entails the

> (a) . . . services entrusted with essentially preventive functions and responsible for advising the employer, the workers and their [union] representatives in the undertaking on:
>
>> (i) the requirements for establishing and maintaining a safe and healthy working environment which will facilitate optimal physical and mental health in relation to work;
>>
>> (ii) the adaptation of work to the capabilities of workers in the light of their state of physical and mental health.[7]

Additionally, the OH Service includes:

> 1. The maintenance and promotion of workers' health and working capacity;
>
> 2. The improvement of working environment and work, to become conducive to safety and health; and
>
> 3. The development of work organization and working cultures in a direction which supports health and safety at work, and in doing so also, promotes a positive social climate and smooth operation, and may enhance the productivity of the undertaking.[8]

2. The OH Risk

An Occupational Health (OH) risk is the probability of a loss — typically, static risk means a loss in which there can be no hope of any gain. A risk analysis entails identifying and evaluating real and potential risks.

To qualify the health risks evident in the general Canadian population, the following known data are used:

- smokers cost Canadian employers $4,256 in lost productivity and increased absenteeism;[9]

---

[6] World Health Organization, Reported in Agius, "What is Occupational Health?" (1950), online: http://www.agius.com/hew/resource/ohsilo.htm (date accessed: February 28, 2020).

[7] World Health Organization, *C161 - Occupational Health Services Convention, 1985 (No. 161)* (1985), online: http://www.ilo.org/dyn/normlex/en/f?p=NORMLEXPUB:55:0::NO::P55_TYPE,P55_LANG,P55_DOCUMENT,P55_NODE:CON,en,C161,/Document (date accessed: February 28, 2020).

[8] World Health Organization, *Good Practice in Occupational Health Services: A Contribution to Workplace Health* (1995) at 2-3.

[9] Conference Board of Canada, *Smoking Cessation and the Workplace: Benefits of Workplace Programs* (2013), online: https://www.quitnow.ca/files/QN/files/library/Smoking_Cessation_and_the_Workplace_Briefing_3_Benefits_of_Workplace_Programs.pdf (date accessed: February 28, 2020).

- depression costs the Canadian economy at least $32.3 billion annually;[10]
- anxiety costs the Canadian economy $17.3 billion a year;[11]
- drug users are anticipated to cost their employers around $7,000 per user per year;[12]
- work stress contributes to 40 per cent of the staff turnover costs[13] and staff turnover is priced at 1.5-2 times the employee's annual salary;
- almost a quarter of Canadians living with a mental illness are unable to work because of their symptoms.[14] This translates to 500,000 employed Canadians being unable to work due to mental health problems lending to 355,000 disability cases plus approximately 175,000 full-time workers absent from work due to mental illness;[15]
- musculoskeletal disorders (MSDs): In Canada, the economic burden of MSDs is estimated to be upwards of $22 billion annually (2014).[16] Musculoskeletal injuries cost an additional $15 billion annually;[17]
- physical disorder claims cost on average $9,000 and psychological disability claims cost $18,000 per case;[18]
- the direct costs of oral disease treatment in Canada are $11.6 billion

---

[10] Conference Board of Canada, *Canadian Alliance for Sustainable Health Care* (2016), online: http://www.conferenceboard.ca/press/newsrelease/16-09-01/unmet_mental_health_care_needs_costing_canadian_economy_billions.aspx (date accessed: February 28, 2020).

[11] *Ibid.*

[12] National Safety Council, "Marijuana at Work: What Employers Need to Know", online https://nsc.org/membership/training-tools/best-practices/marijuana-at-work (2018) (date accessed: February 28, 2020).

[13] R. Tangri, *Stress Costs Stress Cures* (Victoria, BC: Trafford Publishing, 2003).

[14] Conference Board of Canada, *Canadian Alliance for Sustainable Health Care* (2016), online: http://www.conferenceboard.ca/press/newsrelease/16-09-01/unmet_mental_health_care_needs_costing_canadian_economy_billions.aspx (date accessed: February 28, 2020).

[15] CAMH, Mental Illness and Addictions: Facts and Statistics (2016), online: http://www.camh.ca/en/hospital/about_camh/newsroom/for_reporters/Pages/addictionmentalhealthstatistics.aspx (date accessed: February 28, 2020).

[16] Institute of Musculoskeletal Health and Arthritis, "Strategic Plan 2014-2018" Canadian Institutes of Health Research (Winnipeg, Manitoba, 2014), online: http://www.cihr-irsc.gc.ca/e/48830.html (date accessed: February 28, 2020).

[17] *Ibid.*

[18] Centre for Addiction and Mental Health (CAMH), "The numbers add up, so make your vote count", *camhconnexions (Fall 2011) 11(3)* online: http://www.camh.ca/en/hospital/about_camh/newsroom/connexions_newsletter/Documents/4515Connexions_Fall2011EN.pdf (date accessed: February 28, 2020).

CH. 4: OCCUPATIONAL HEALTH PROGRAM

(2011);[19]

- Canadian work-related injuries cost an average of $39,534 (2018) for assessable employers;[20]
- Canadian work absence (2019) was an average of 10.3 days per full-time employee, which equates to $2,386.40 per full-time employee in lost productivity[21] and $37 billion for all Canadian full-time employees;[22]
- work-related fatalities (2017) in the U.S.,[23] cost an average of $1.15 million per death; these are the direct costs such as wage losses, medical expenses, administrative expenses, vehicle damage, and employer costs, excluding property damage.[24] Missing are the indirect costs such as increased WCB insurance premiums, worksite investigation costs, fines (averaging $97,500/

---

[19] Health Canada, *Report on the Findings of the Oral Health Component of the Canadian Health Measures Survey 2007–2009* (2010), online: http://publications.gc.ca/site/archivee-archived.html?url=http://publications.gc.ca/collections/collection_2010/sc-hc/H34-221-1-2010-eng.pdf (date accessed: February 28, 2020).

[20] Calculated using Association of Workers' Compensation Boards of Canada, "2018 Injury Statistics", *Statistics*, online: http://awcbc.org/?page_id=14 (date accessed: February 28, 2020). Involves the addition of IR5 – the Current Year Average Benefit Cost/Claim and IR6 – Administration Cost per claim.

[21] Calculated by determining the daily wage from the average hourly wage for full-time Canadian employees as of December 2019 ($28.96 per hour) and multiplying that by eight hours/day and by the average number of lost time days per full-time Canadian worker (10.3 days). Data obtained from Statistics Canada, *Table 14-10-0190-01 Work Absence of full-time employees by geography, annual* (Ottawa: Statistics Canada, January 31, 2020), online:https://www150.statcan.gc.ca/t1/tbl1/en/tv.action?pid=1410019001; and Statistics Canada, *Table 14-10-0287-01, Labour force characteristics, monthly adjusted and trend-cycle, last 5 months* (2019), online: https://www150.statcan.gc.ca/t1/tbl1/en/tv.action?pid=1410028701 (date accessed: January 31, 2020).

[22] Calculated by determining the daily wage from the average hourly wage for full-time Canadian employees as of October 2019 ($28.96 per hour) and multiplying that by eight hours/day and by the average number of lost time days per Canadian worker (10.3 days). That value is then multiplied by the number of full-time workers in December 2019 (15,536,200 FTEs). Data obtained from Statistics Canada, *Table 14-10-0190-01 Work Absence of full-time employees by geography, annual* (Ottawa: Statistics Canada, January 31, 2020), online: https://www150.statcan.gc.ca/t1/tbl1/en/tv.action?pid=1410019001; and Statistics Canada, *Table 11, Average Usual Hours and Wages of Employees by Selected Characteristics*, online: https://www150.statcan.gc.ca/n1/daily-quotidien/191108/t011a-eng.htm; and Statistics Canada, *Table 14-10-0287-01, Labour force characteristics, monthly adjusted and trend-cycle, last 5 months* (2019), online: https://www150.statcan.gc.ca/t1/tbl1/en/tv.action?pid=1410028701 (date accessed: January 31, 2020).

[23] Note: USA data are used because Canada does not provide workplace fatality costs.

[24] National Safety Council, "Costs" *Injury Facts* (2019), online: https://injuryfacts.nsc.org/work/costs/work-injury-costs/ (date accessed: February 28, 2020).

case[25]), loss of an employee (1.5 times their annual wage), business interruption costs, and the price of negative media for the organization/ company. These indirect costs equate to 1.5[26]-4[27] times the direct costs, making the total cost of worker fatality $2.9 million - $5.8 million. The Injury Cost Calculator tool,[28] indicates that the organization/company would have to generate revenue of $57.5 million to cover this cost (assuming a 10 per cent margin);[29]

- preventable injuries cost Canadians more than $26.8 billion a year (2010);[30]
- sedentary adults living in Australia, are estimated to cost the healthcare system $1.5 billion per year.[31] In Canada, physical inactivity costs $10 billion (2012).[32] Every sedentary employee, costs the employer $488 per year;[33]
- obesity costs between $4.6 and $7.1 billion annually, and contributes to the development of diabetes, high blood pressure, and cancer. Obese employees use 35 per cent more health services and 77 per cent more on medications

---

[25] K. Annable, J. Marcoux & V. Kubinec, "The Price of Death" *CBC News* (November 30, 2017), online: https://www.cbc.ca/news2/interactives/workplace-death-injury-labour-laws/ (date accessed: January 31, 2020).

[26] Statistics Canada, reported in A. Carofano, "Controlling the Costs of Absenteeism – The CFO Perspective" *Workplace Medical Corp.* (2014), online: http://www.workplacemedical.com/wp-content/uploads/2015/04/Absence-Managment-Whitepaper-The-CFO-Perspective-.pdf (date accessed: January 31, 2020).

[27] Safety Management Group, *Injury Cost Calculator* (2019), online: https://safetymanagementgroup.com/resources/injury-cost-calculator/ (date accessed: January 31, 2020).

[28] Safety Management Group, *Injury Cost Calculator* (2019), online: https://safetymanagementgroup.com/resources/injury-cost-calculator/ (date accessed: January 31, 2020).

[29] Safety Management Group, *Injury Cost Calculator* (2019), online: https://safetymanagementgroup.com/resources/injury-cost-calculator/ (date accessed: January 31, 2020).

[30] Public Health Agency of Canada, *The Cost of Injury in Canada* (2015), online: http://www.parachutecanada.org/downloads/research/Cost_of_Injury-2015.pdf (date accessed: February 28, 2020).

[31] K. Cogdon, "Couch potatoes cost health system $1.5B a year" *The Cairns Post,* (October 12, 2007) at 21.

[32] P. Katzmarzyk, "Tallying the global economic burden of physical inactivity" *Alberta Centre for Active Living* (October 1, 2016), online: https://www.centre4activeliving.ca/news/2016/10/global-economic-burden-physical-inactivity/ (date accessed: February 28, 2020).

[33] Health & Safety Ontario, *The Business Case for a Healthy Workplace* (2008), online: https://www.wsps.ca/WSPS/media/Site/Resources/Downloads/BusinessCaseHW_Final.pdf?ext=.pdf (date accessed: January 31, 2020).

than do non-obese employees. By 2019, it is estimated that approximately 21 per cent of the Canadian adult population will be obese;[34]
- presenteeism, the phenomenon of employees being at work, but not productive, costs employers nine times more than absenteeism. The cause of presenteeism is often health-related and can be addressed with support from OHNs;
- fatigue costs Canadian employers $330 million annually in lost productivity, and is associated with absenteeism, presenteeism, workplace injuries and health conditions such as high blood pressure, depression, stroke, and obesity;[35]
- substance abuse: the total societal cost of addiction is $40 billion or $1,267 per Canadian. Tobacco accounts for 43 per cent ($17 billion); while alcohol accounts for 37 per cent of this cost.[36] In Alberta alone, the total cost of substance abuse was $4.4 billion.[37] The cost per employee who drinks an excessive amount of alcohol is $597 per year; and
- unhealthy lifestyles: for each noted health risk, Telus Mobility estimates the cost per employee to be $2,000 per year.[38]

When the financial costs of these risks are added, the overall "risk price tag" is more than $219 billion. Through OH programming, OHNs and Occupational Health Physicians can significantly assist organizations to prevent and mitigate these staggering risks and costs. The benefit is that most OH Services lower illness/injury rates and costs by between 10-50 per cent.[39] Hence, the financial losses can be managed, and significant savings realized (Table 4.1):

---

[34] K. Twells, "Current and predicted prevalence of obesity in Canada: A trend analysis" CMAJ Open, Mar 3;2(1):E18-26. doi: 10.9778/cmajo.20130016. eCollection 2014 Jan.

[35] C. Baglien, "Fighting Workplace Fatigue" *PEO Canada Employee Management*, online: https://www.peocanada.com/peo-blog/fighting-workplace-fatigue/ (date accessed: February 28, 2020).

[36] Canadian Centre on Substance Abuse, *The Costs of Substance Abuse in Canada, 2002* (2003).

[37] AADAC, "The Cost of Substance Abuse in Alberta, 2002", *Profile* (2007), online: http://www.assembly.ab.ca/lao/library/egovdocs/2007/alad/159248.pdf (date accessed: January 31, 2020).

[38] Health & Safety Ontario, *The Business Case for a Healthy Workplace* (2008), online: https://www.wsps.ca/WSPS/media/Site/Resources/Downloads/BusinessCaseHW_Final.pdf?ext=.pdf (date accessed: January 31, 2020).

[39] Workplace Safety and Prevention Services Ontario, *The Business Case for a Healthy Workplace* (2011) at 8-11, online: https://www.wsps.ca/WSPS/media/Site/Resources/Downloads/ BusinessCaseHW_Final.pdf?ext=.pdf (date accessed: February 28, 2020).

Table 4.1: Risk Reduction through OH Service

Risk ($)	Rate of Reduction	Loss Avoidance ($)
$219B	10%	$21.9B
$219B	20%	$43.8B
$219B	30%	$65.7B
$219B	50%	$109.5B

The above approach of calculating the financial and human costs that could negatively impact the workplace, uses a mix of population and occupational data, as opposed to strictly occupational-specific data. However, Health & Safety Ontario presented a similar approach in its 2011 report, *The Business Case for a Healthy Workplace*.[40] Using occupational-specific data, it advocated action by employers to prevent and mitigate employee illness/injury. To sell its premise, examples of successful industry interventions; of known costs of "doing nothing"; and of financial benefits realized by industry, were provided. They concluded that Canadian employers can annually save:

- $700 million in stress-related absences;
- $2.22 billion in lost productivity due to mental illness;
- $6.6B in lost productivity due to all forms of clinical and sub-clinical mental illness; and
- $1.1 billion in absenteeism due to work-family conflict[41]

If according to Fries, 20-30 per cent of health conditions are preventable,[42] then an OH Service could offer an organization at least a 20-30 per cent cost reduction in these known risks, regardless of which of the above cost-models is used. That amount of *loss avoidance* would vastly outweigh the much smaller price-tag for funding a corporate OH Service. In support of that hypothesis, research indicates that the return on investment of an OH Service is $3 to $5 for every $1 spent on the OH service.[43],[44],[45]

---

[40] WSIB/CSPAAT, "Business Results Through Health & Safety" (2001), online: http://www.mtpinnacle.com/pdfs/Buisness-Results.pdf (date accessed: February 28, 2020).

[41] Workplace Safety and Prevention Services Ontario, *The Business Case for a Healthy Workplace* (2011) at 9, online: https://www.wsps.ca/WSPS/media/Site/Resources/Downloads/ BusinessCaseHW_Final.pdf?ext=.pdf (date accessed: February 28, 2020).

[42] J.F. Fries, "Beyond Health Promotion: Reducing Need and Demand for Medical Care" (1998) 17:2 Health Affairs 70-84.

[43] C. Smith, "Down with Costs, Up with Safety" OH&S (July 10, 2015), online: https://ohsonline.com/Blogs/The-OHS-Wire/2015/07/Down-with-Costs-Up-with-Safety.aspx (date accessed: February 28, 2020).

[44] Bongarde, *Selling Safety to Your CEO* (Penticton, BC: MMV Bongarde, 2005).

[45] Workplace Safety and Prevention Services Ontario, *The Business Case for a Healthy*

3. Going Forward

Risk management and loss reduction are particularly important to an organization given that the Canadian workforce is aging. With 36 per cent of Canadian employees being over the age of 55 years,[46] the risks of chronic health conditions, musculoskeletal disorders, and the complications of injury, are high.

Likewise, 20 per cent of Canadian employees are foreign-born;[47] some of whom have faced significant political strife, social upheaval, and economic hardship along with health challenges and lack of adequate healthcare, before coming to Canada.

The legalization of marijuana; limitations on the employer's right to seek proof of employee fitness-to-work related to casual absenteeism in Ontario; and the newly recognized medical conditions as per the *Diagnostic and Statistical Manual of Mental Disorders (DSM5)*,[48] make it difficult for the employer to differentiate between culpable and non-culpable employee behaviours.

Employers must factor these realities into their approach to occupational health and safety. Having an OH Service and an OHN who is educated in health risks and suitable interventions, positions the employer to be duly diligent in meeting the applicable Occupational Health & Safety General Duty Clause: *To provide a safe and healthy workplace.*[49]

4. The OH Service: Its Value

An OH Service, whether delivered in-house or externally, provides the technical expertise needed by an employer, to establish and maintain a safe and healthy working environment. Optimal physical and psychological health is critical to maximizing human performance on the job. This translates to a work environment that supports the employee to know how to do their job safely, to be able to do the job safely, to be equipped to do the job safely, and to be motivated (want) to do the job safely.[50]

Workplace (2011), 8-11, online: https://www.wsps.ca/WSPS/media/Site/Resources/Downloads/BusinessCaseHW_Final.pdf?ext=.pdf, (date accessed: February 28, 2020).

[46] Statistics Canada, *Insights on Canadian Society: The impact of aging on labour market participation rates*, Cat. No. 75-006-X, online: http://www.statcan.gc.ca/pub/75-006-x/2017001/article/14826-eng.htm (date accessed: February 28, 2020).

[47] Statistics Canada, "Immigration and Ethnocultural Diversity" *National Household Survey* (2011), online: https://www12.statcan.gc.ca/nhs-enm/2011/as-sa/99-010-x/99-010-x2011001-eng.cfm#a1 (date accessed: February 28, 2020).

[48] American Psychiatric Association (2013). *Diagnostic and Statistical Manual of Mental Disorders version* (5th ed.). Arlington, VA: Author.

[49] The General Duty Clause is stated in every provincial OH&S Act, as well as in Canada Labour Code II.

[50] T. Roithmayr in D. Dyck, *Disability Management: Theory, Strategy & Industry Practice*, 6th ed. (Toronto: LexisNexis Canada, 2017).

An OH Service can counsel, coach, and mentor the employer on the value of creating a respectful work environment in which honesty, trust, loyalty, cultural diversity, and respect are evident. Ensuring a good "person-job fit" and facilitating suitable work accommodation when needed, are two strengths of an OH Service. Table 4.2 provides specific clinical evidence of the value offered by an OH Service and the contributions that an OHN can make.

Table 4.2: Value of an OH Service and the Contributions made by OHNs

Contributions	Value to Company	Qualifiers
Manage Internal OH&S Program	42% saving over external OH Service	Internal OH&S Programs are 42% less costly to operate than are external OH&S services.[51]
Pre-placement Assessments	Right people for the job	Having the right person/job fit means lower absenteeism and staff turnover costs. The estimated saving is the replacement cost of the "misplaced" employee's salary (1.5-2 times the annual salary[52]). For 2018, the average Canadian annual salary was $55,800,[53] making this cost-avoidance measure worth $83,700 - $111,600 per new employee.
Periodic Risk-based Monitoring	No fines/penalties	Provincial OH&S legislation dictate that hearing conservation, respiratory conservation and monitoring for some chemical exposures occur. Fines for non-compliance, although rare, can be levied.

---

[51] G. Lantos, "Cost-effectiveness of in-house Occupational Health Programs" presented at the Toronto Occupational Health Conference, 1992; Public Health Agency of Canada, *The Cost of Injury in Canada* (2015), online: http://www.parachutecanada.org/downloads/research/Cost_of_Injury-2015.pdf (date accessed January 31, 2020).

[52] E. Skronski, "Is Your Company Healthy? 7 Healthy Strategies for Employee Retention" *Canada One* (2008), online: https://www.canadaone.com/ezine/march08/strategies_for_employees_retention.html (date accessed January 31, 2020).

[53] Statistics Canada, *Average Canadian Salary, 2018* (2019), online: https://www150.statcan.gc.ca/t1/tbl1/en/tv.action?pid=1410032002 (date accessed January 31, 2020).

Contributions	Value to Company	Qualifiers
Emergency Preparedness	Fewer injuries Injury management when needed No fines/penalties	Emergency preparedness is a legislated requirement. An OH Service can oversee and enhance this effort thereby not only complying with the legislation, but also mitigating the risk of further injury or death. Fines in Canada can be as high as $500,000 for a first-time offence.
Ergonomic Support	Increased productivity: Lower WCB costs for musculoskeletal injuries	Management of ergonomic-related health conditions can result in reduced WCB claims and increased worker productivity.[54]
Workers' Compensation Reporting	No fines/penalties	Failure to report WCB claims in accordance with provincial legislation can result in substantial fines ranging from $100 per day late to a $25,000 fine.[55]
Wellness Programs/ Services	Reduced absenteeism Reduced presenteeism Increased productivity	OH Services are well-positioned to manage Workplace Wellness Programs that reportedly offer a savings of $3.27 per dollar spent on medical costs and $2.73 per dollar spent on absenteeism.[56] Based on a literature review, the return on investment per dollar spent ranged from $1.50 to $6.18.[57]

---

[54] Workers Health & Safety Centre Federation of Ontario, Ergonomic Training Resources (2019), online: https://www.whsc.on.ca/Resources/Publications/Ergonomic-Resources (date accessed: January 31, 2020).

[55] AWCBC, Legislation and Links (2019), online: www.awcbc.org (date accessed: January 31, 2020).

[56] K. Baicker, D. Cutler & Z. Song, "Workplace Wellness Programs Can Generate Savings" (2010) 29:2 Health Affairs 304-311.

[57] Health & Safety Ontario, *The Business Case for a Healthy Workplace* (2008), online: https://www.wsps.ca/WSPS/media/Site/Resources/Downloads/BusinessCaseHW_Final.pdf?ext=.pdf (date accessed: January 31, 2020).

Contributions	Value to Company	Qualifiers
Attendance / Disability Management	30-50% cost avoidance 9-25% savings with an integrated disability management program	The management of workplace absenteeism and disabilities can result in a 30-50% reduction in related costs.[58] In 2018, Canada spent $16.6 billion in lost productivity due to casual absence.[59] The average cost per employee for all disability-related absences was $2,386.[60] A 30-50% reduction would lower this cost to between $716-$1,193 per employee per year. For company with 1000 employees, the saving would equate to $716K - $1,193K. Put another way, an OH Service can save companies 2.8% of payroll through an integrated disability management program. According to Marsh Risk, the total cost of employee absence and the operation of disability management programs equals 15% of payroll; WBGH and Watson Wyatt claim that Integrated Disability Management Programs save companies 19%-25% in disability costs. By having documented return-to-work plans in place, 81% of survey participants reported that they are effective in reducing disability costs.[61]
Oversight of EFAP Services	65% increase in productivity	Effectively managing an EAP can result in a 65% reduction in stress and improvement in worker productivity. Mental health issues cost businesses almost $1,500 per employee per year.[62] OHNs can help employees identify and manage their distress and reduce the related costs. Otherwise, a psychological disability can cost on average $18,000.[63]
Employee Support	Priceless!	The relationship an OHN forges with an employee has been shown to enhance compliance with treatment regimens, rehabilitation plans, and lifestyle changes.[64]

Contributions	Value to Company	Qualifiers
Employee Loyalty	Return on investment	By addressing psychosocial issues, the OHN can help the employee and increase employee satisfaction and health. "For every 5-unit increase in employee satisfaction in a [business] quarter, there is a 1.3-unit increase in customer satisfaction in the next [business] quarter, and a 0.5-unit increase in revenues above the national average in the following quarter".[65]

Investing in an OH Service and employing the technical expertise of an OHN can save organizations a considerable amount of money and position it to retain its valuable human capital.

---

[58] National Institute of Disability Management and Research (NIDMAR), *Disability Management Success: A Global Corporate Perspective* (Port Alberni, BC: NIDMAR, 2005), online: http://www.nidmar.ca (date accessed: January 31, 2020).

[59] Mercer, "How Much Are You Losing to Absenteeism?", Mercer Website (2019), online: https://www.mercer.ca/en/our-thinking/how-much-are-you-losing-to-absenteeism.html (date accessed: January 31, 2020).

[60] Calculated by determining the daily wage from the average hourly wage for full-time Canadian employees as of December 2019 ($28.96 per hour) and multiplying that by eight hours/day and by the average number of lost time days per full-time Canadian worker (10.3 days). Data obtained from Statistics Canada, *Table 14-10-0190-01* Work *Absence of full-time employees by geography, annual* (Ottawa: Statistics Canada, January 31, 2020), online:https://www150.statcan.gc.ca/t1/tbl1/en/tv.action?pid=1410019001; and Statistics Canada, *Table 14-10-0287-01, Labour force characteristics, monthly adjusted and trend-cycle, last 5 months* (2019), online: https://www150.statcan.gc.ca/t1/tbl1/en/tv.action?pid=1410028701 (date accessed: January 31, 2020).

[61] Watson Wyatt Worldwide, *Staying @ Work Survey* (2005), online: www.watsonwyatt.com (date accessed: January 31, 2020).

[62] Benefits Canada (2013). "Workplace Mental Health". *Benefits Canada*, online: http://www.benefitscanada.com/benefits/health-wellness/workplace-mental-health-44885 (date accessed: January 31, 2020).

[63] Centre for Addiction and Mental Health (CAMH), *Mental Illness and Addictions: Facts and Statistics* (2016), online: http://www.camh.ca/en/hospital/about_camh/newsroom/for_reporters/Pages/addictionmentalhealthstatistics.aspx (date accessed: January 31, 2020).

[64] D. Anderson, *Effects of Blood Pressure Screening on Lifestyle Behaviours* (Calgary: University of Calgary, 1989).

[65] Health & Safety Ontario, "The Business Case for a Healthy Workplace" *WSPS* (2008), online: http://www.oxfordcounty.ca/Portals/15/Documents/WorkplaceHealth/fd_business_case_healthy_workplace.pdf at 6 (date accessed: January 31, 2020).

E. CONFIDENTIALITY OF MEDICAL INFORMATION

An employee's personal health information is privileged, and the employee has the right to control this information and communicate or retain it as deemed fit. However, access to relevant medical information is required to determine the employee's fitness to work and eligibility for disability coverage. In such instances, the organization has the legal right to request that the ill/injured employee provide documentation on his/her fitness to work.

All employee health information must be used only for the purpose for which it was collected, and not released to a third party without the employee's informed consent, unless the organization is legally required to do so.

Informed Consent

For employee consent to be valid, it must be an informed consent. The requirements for informed consent include:

- the provision of adequate information about the nature and consequence of the intended action to allow the employee to come to a reasoned decision;
- ensuring that the employee is mentally competent, and can understand and appreciate the nature and consequences of the procedure;
- consent being freely given;
- consent being obtained without misrepresentation or fraud;
- consent cannot be given for the performance of an illegal procedure; and
- consent is often in relation to the specific act contemplated unless the employee's life is immediately endangered, and it is impractical to obtain consent.

Procedure

The practices upheld by an OH Service are that:

- All documents containing employee health information are retained by the organization or its designate, which may be an external OH Service Provider.
- The medical information in these documents is not made available for use in employment decisions.
- The OH professionals handling employee medical health information on XYZ's behalf are charged with maintaining the confidentiality of this information. This includes written guidelines for dealing with the collection, retention, storage, security, access, disclosure, and destruction of identifiable employee health information.
- Employees have the right to access their medical health information held by the OH Services, including a right to request that corrections be made, if necessary, or a notation of objections.

Principles of Confidentiality

All persons who collect, maintain, handle, and use health information related to employees, must protect the confidentiality of that information. Organizations must recognize the employee's right to privacy in relation to health information collected by the organization or it's designates.

Many principles govern this area and include:

- personal health information is only distributed on a "need-to-know" basis;
- personal health information is restricted to OH Services staff that sign off on a *Pledge and Oath of Confidentiality* (Appendix 1) and are subject to a recognized professional code of ethics;
- employees have the right to access all information regarding his/her health and fitness;
- documented health information is the property of XYZ entrusted to OH Service Provider staff for safeguarding and protection of confidentiality; and
- information will not be released without the written consent of the employee.

Definitions

Confidentiality is the maintenance of trust expressed by an individual verbally, or in writing, and the avoidance of an invasion of privacy through accurate reporting and authorized communication.

Health Information is an accumulation of data relevant to the past, present, and future health status of an individual that includes all that OH Services staff learn in the exercise of their responsibilities.

Privacy is the claim of individuals, groups, or institutions to determine for themselves when, how, and to what extent, information about them is communicated to others.

Designated Representative is any individual or organization to whom an employee gives written authorization to exercise a right to access.

Collection of Health Information

The primary purpose of collecting and retaining health information is to monitor the health status of employees, and thus, to protect the rights of both employees and the employer using a factual employee health database.

Health information is collected through various methods including, but not limited to, interviewing, examination, discussion, written documentation, and electronic data processing, all of which are subject to confidentiality.

The classes of health information collected relate to health assessments, EAP treatment reports, illness and injury reports, personal and family history, consultant reports, and laboratory tests.

Health information collected and documented on employee health records, medical logs, and data processing files are subject to confidentiality.

Retention and Storage

All health information on employees working with designated hazards is maintained as part of the employee health record for a period of 40 years from the time such records were first made, or for a period of 20 years from the time the last entry was made to such records, whichever is longer.

All health information is stored separately from other employee information. The storage location is to be checked regularly and safeguarded from fire, water and other potential disasters.

All computerized health information is to be secured using passwords and access codes.

The medical records of terminated employees are to be archived according to an established protocol by the OH Service Provider.

Accessibility

Employees, former employees, or their properly designated representatives, have the right to inspect, and copy all or in part, their health records. All such written requests are to be honoured within a reasonable time that should not normally exceed 15 business days. To the extent practicable, inspection of a health record is to be made in the presence of an OH professional who will endeavour to explain the meaning of the content of the record to the employee. Rebuttal of the information contained in the health record by the employee must be included in the record, signed and dated by the employee. The OH professional may add a note of explanation to the file that addresses any agreement or disagreement.

It is recommended that employees:

- accept a summary of material facts and opinions in lieu of copies of the records requested; or
- accept a release of the requested information only to the family physician or other qualified healthcare professionals.

With respect to health information whose disclosure to the employee may have an adverse impact upon the health of the employee, access should only be provided to a designated physician of the employee.

No other XYZ personnel have the right to access employee health information unless the following disclosure obligations have been met:

- **Management Disclosure** is having health information released to Management and is limited to the following:
    - report of employee fitness-to-work status;
    - determination that a medical condition exists, and that the employee is under medical care;

- time that the employee has been or is expected to be off work;
- medical limitations, if any, to carry out work in a safe and timely manner; and/or
- medical restrictions, if any, regarding specific tasks.

However, if disclosure is necessary because of a **clear danger** to the employee, the co-workers, the workplace, or the public and:
- the employee concerned consistently refuses to give consent; and
- a second opinion is obtained from the employee's personal physician when the concern is for the health of the employee, or to fellow employees, or from the Medical Officer of Health when the risk is to the public,

the OH Service Provider may make the disclosure to the appropriate manager after giving notice in writing to the employee, indicating that confidential information will be disclosed.

- **External Disclosure** is disclosure of employee health information to external sources; however, this must not occur, unless the individual has authorized such a release by providing a signed and dated consent form for release of medical information.
- **Routine Request for Release of Medical Information** pertains to a written request by a physician, medical institution, another health agency, or insurance company for abstracts or copies of part or all of the employee's health record. It is honoured when the *Authorization for Release of Privileged Information Form* (Appendix 1), or its equivalent, has been signed by the employee.
- **Medical Emergencies:** In the event of a medical emergency (*i.e.*, an imminent threat to life or limb), information contained in the employee's health record may be released upon the request of a responsible family member, or the attending physician.
- **Release of Pertinent Medical Data to Appropriate Public Health Authorities:** When it is determined that a public health issue or risk exists, as in the case of a reportable communicable disease, appropriate notification to provincial or municipal health authorities must be made in accordance with the statutory requirement.
- **Disclosure to Designated Representative:** Upon presentation of a written consent by an employee or former employee, copies of the employee's medical record will be released to the designated representative. However, again, when medical information is deemed to have a detrimental impact upon the health of the employee, medical information must be provided only to the employee family physician.

Destruction
When it becomes appropriate to dispose of health information pertaining to an

individual employee, including formal health records, notes and messages, they must be rendered completely and permanently unidentifiable through destruction by burning, shredding or automated erasure.

Misuse of Health Information

Any individual aware of an abuse of confidentiality of health information will document and report the incident to the VP, Human Resources or designate for investigation and further action.

F. FITNESS TO WORK

Work-related illness/injury is prevented by the control of exposure to health hazards in the workplace. Pre-placement and periodic fitness-to-work (FTW) assessments supplement these efforts. All **FTW assessments** are "risk-based" and dependent on the demands and related hazards of the employee's assigned job.

Typically, the Director, OH&S is responsible and accountable for organizing the OH Program be it in-house or by a designated OH Service Provider; and for ensuring that the OH records are maintained in accordance with legislated standards.

FTW Assessments

The program is comprised of the following elements (refer to Table 4.3 for the scheduled times):

(a) Determination of the job demands using various tools like a Job Demands Analysis.

(b) A risk-based occupational health examination to include:
- height
- weight
- pulse
- blood pressure
- baseline hearing assessment (includes hearing test examination of ear canal, ear drums and outer ear; hearing and symptom history; noise exposure history; and hearing education)
- basic eyesight
- colour vision
- urine test
- body system review
- assessment of overall psychological status.

(c) Bi-annual audiometric test and education for employees exposed to noise levels of 85 dBA.

(d) Bi-annual respiratory assessments for all employees exposed to atmospheric hazards, and/or required to use respiratory gear.

(e) Regular vision testing for employees whose visual acuity is important to their job.

(f) Review of lifestyle and health status with appropriate advice and guidance to the employee.

(g) Back Assessment.

(h) Review of immunization status.

(i) Biological testing, as required.

Using these elements, employees are medically examined according to their job demands, hazards, and risks.

Table 4.3: Occupation and FTW Elements

Job Type	A	B	C	D	E	F	G	H	I
Electricians	X	X	X	X	X	X	X	X	X
Crane Operators	X	X	X	X	X	X	X	X	X
Welders	X	X	X	X	X	X	X	X	X
Labourers	X	X	X	X		X	X	X	X
Professional Drivers	X	X	X		X	X	X		X
Fork Lift Operators	X	X	X			X	X		
Manual Handlers	X	X			X	X		X	X
Mechanics	X	X	X		X	X	X		X
International Travellers	X					X		X	X
Executives	X	X				X			X

The extent of the assessments will be altered to accommodate any changes in legislation.

The frequency of subsequent FTW assessments and follow-up action(s) depend on employee age and test results.

Pre-placement FTW Assessment

Pre-placement assessments are "risk-based" — that is, those candidates who are being placed within the field, or high-risk, jobs are required to undergo a pre-placement assessment.

The objectives of the pre-placement assessments are:

- To ensure the candidate is fully informed of the health risks associated with the job.
- To ensure the candidate's health status does not expose co-workers, the public, or the candidate to any increased risks.
- To attempt to ensure that the candidate is fit for the requirements of the job, and not impaired in anyway.

- To establish a base line data for subsequent occupational health and wellness assessments.
- To attempt to ensure that XYZ does not inherit occupational health problems from previous employment.
- To identify existing health conditions that may require job accommodations or support at XYZ.

The FTW assessment should be arranged as early as possible within the employment probationary period.

Periodic FTW Assessments

The following is a description of:

- **Executive Medicals:** Executives are offered annual medical examinations that include the elements noted in the above table.
- **Professional Driver Medicals:** Employees who possess a Class 1 driver's licence and drive commercial vehicles are regularly provided Department of Transport (DOT) Medicals.
- **Pilots:** Employees who fly corporate or commercial planes are regularly provided Department of Transport (DOT) Medicals.

FTW Assessment Results

Information and test results remain confidential in accordance with the organization's Confidentiality Policy and are maintained in the employee health records housed within the OH Service.

Reports to Management are limited to employee fitness to work using the *OH Medical Fitness-to-Work Form* (Appendix 2). If there are instances where discussion with Management is warranted, the employee will be advised by the OH Service and permission to release the required information will be obtained using the *Authorization for Release of Privileged Information Form* (Appendix 3).

Annual reports on the FTW Assessments are limited to aggregate data reported as program statistics.

G. JOB DEMANDS ANALYSIS (JDA)

Every job is made up of many tasks. The Job Demands Analysis (JDA) shows a breakdown of job tasks and determines the physical and psychological demands involved. This information is used for:

- conducting Fitness-to-Work (FTW) Assessments;
- determining modified work opportunities;
- setting rehabilitation goals for injured/ill employees; and
- identifying "high-risk" job tasks that warrant modification.

A JDA for every job can be completed by the organization's OH&S Department or designate. These JDAs are retained in the organization's JDA Inventory and that inventory should be housed on the organization's OH&S website.

Procedure for a Job Demands Analysis

To use the *Job Demands Analysis* form (Appendix 4), complete the form using data collected through observation, interviews, and direct measurement of the various job tasks/demands involved for a specific job.

The following is a description of the terms used in this process and of how to respond to each term:

Conducting a Job Demands Analysis

Description:	Information explaining the answers to the other columns which helps in getting a good idea about the demands of a job. Write a brief explanation or notation as to how a task is carried out.
Fine Finger Movements:	Refers to jobs where finger dexterity is needed to complete the job. Examples are typing, assembly of electronics and writing mechanics.
Handling:	This refers to jobs where the hands and manual dexterity are needed to carry out the job. An example is material handling.
Gripping:	Indicate the number of hours a worker may spend grasping or holding material during their work tasks. Examples are welding, drilling, pulling apart to dissemble, or holding materials in place while assembling.
Hearing, Smelling, Vision, Reading and Writing, Speech:	Approach these from a safety point of view. Does the worker have to be able to smell noxious fumes? Does the worker have to be able to hear an alarm? Do they have to distinguish between colours to do their job safely?
Controlled Substances:	Are any of the substances on-site WHMIS controlled? If yes, which ones do these workers have to handle?
Operate Equipment/ Machinery/Vehicles:	What equipment or machinery is required to be operated as part of the job? Which vehicles must be operated as part of the job?
Comments:	Write a brief summary of what the job is, and what the worker is required to do. Note which training courses, certifications, or licences are required by the company before the worker can do this job and operate equipment, machinery or vehicles.

Responsibilities

The related responsibilities of the various stakeholders tend to vary by organization, yet typically, the:

<u>Line Manager</u>

- Maintains copies of JDAs for their respective work area with the Hazard Identification and Control Records.

- Ensures that the JDAs are accessible to employees for review.
- Contacts the OH/OH&S Departments to complete a new JDA if jobs change due to modifications in processes, procedures, or the worksite.
- Reviews the JDAs every two years and notifies the OH/OH&S Departments of any changes in job tasks so that the relevant JDA can be reviewed.
- Modifies "high-risk" job tasks as per Risk Assessment Control (refer to Chapter 9) process and using the OH&S Department's support as required.

OH&S Department
- Completes a JDA form for each job/position.
- Maintains the JDA Inventory.
- Retains a copy for use for Pre-placement FTW Assessments and Disability Claim/Case Management (refer to Chapter 27 "Disability Management Overview").

H. HEARING CONSERVATION PROGRAM

Occupational noise is the most common "on-the-job" health hazard. Prolonged exposure to excessive noise can damage hearing and result in a reduced quality of life for the employee. It may also result in psychological and physical stress and can contribute to workplace incidents. Once hearing is lost, it cannot be replaced or repaired. Fortunately, work-related hearing loss is preventable.

The provincial noise regulations are designed to protect employees from exposure to excessive noise and any resulting permanent hearing loss. Employers are responsible for minimizing the noise hazard at their workplaces; employees are legally required to participate in noise control programs to prevent hearing loss. The costs related to hearing loss claims for work-related hearing loss can and are assessed against the organization by the provincial Workers' Compensation Board.

Objectives

The objectives of the Hearing Conservation Program are to:
- Reduce noise exposure levels to the lowest level reasonably practicable.
- Monitor noise levels in the workplace.
- Post signage in accordance with the relevant legislation that denotes work environment noise levels and acceptable work practices.
- Provide employees with suitable noise protection.
- Provide employee training on the hazards of noise and noise control measures.
- Implement a Hearing Conservation Program in accordance with the relevant legislation.
- Annually review the effectiveness of the education and training program noise abatement.

Noise Control

There are three main noise control techniques:

- **Engineering Controls** Noise control at the source is the preferred method of control; its results are usually permanent, are applicable to new and old installations, and generally meet with greater employee acceptance. Interference with the path of noise is an alternate means of noise control and includes, but is not limited to, such action as:
    - noise-source isolation;
    - noise-source enclosure;
    - separation of noisy equipment; and
    - use of sound-absorbing materials.
- **Administrative Noise Control**
  This will be used when and where appropriate, as an effective control method. Examples of such controls are:
    - scheduling maintenance during operational downtime;
    - limiting exposure to excessively noisy work environments; and
    - increasing distance from noise source to personnel.
- **Personnel Hearing Protection**
  Canadian Safety Association standard Z94.2-94 (R2011) is used in designated areas where engineering or administrative controls are not possible (Table 4.4).

Table 4.4: Selection of Hearing Protection

SELECTION OF HEARING PROTECTOR	
Maximum – Minimum Noise Level Level (dBA Lex)	CSA Class of Hearing Protection
90	C, B or A
95	B or A
100	A
105	A
110	A earplug + A or B earmuff
110	A earplug + A or B earmuff and limited exposure time to keep sound reaching the worker's eardrum below 85 dBA Lex

Occupational Exposure Limits (OEL)

OELs define a worker's maximum permitted daily exposure to noise without hearing protection. OELs take into consideration the loudness of the noise —

measured in decibels (dBA) — and the duration of the exposure. Noise is classified as either continuous or impulse. Impulse noise is very short in duration. Employers are responsible for making sure that employees are not exposed to noise exceeding the OELs (Table 4.5).

Table 4.5: Occupational Exposure Limits (OEL)

OELs for Continuous Noise	
Sound Level (dBA)	Maximum Permitted Duration of Exposure(hrs/day)
82	16
83	12 hrs + 41 min
84	10 hrs + 4 min
85	8
88	4
91	2
94	1
97	30 min
100	15 min
103	8 min
106	4 min
109	2 min
112	56 sec
115 and greater	0

OELs for Impulse Noise	
Peak Sound Level (dBA)	Maximum Number of Impulses Permitted per 8-hr Day
120	10,000
130	1,000
140	100
150	0

Assessment

When any employee is likely to be exposed to noise exceeding the OEL, then a Noise Level Survey must be conducted by a competent person in accordance with ANSI standards. The assessment must include the following:

- the noise level in terms of the OEL;
- methods of noise reduction; and
- recommended hearing protection.

The noise-level testing equipment must be used according to ANSI standards. The assessment results and recommendations are dealt with as noted in Table 4.6:

Table 4.6: Noise Level Survey Results

Result	Recommended Action
Below OEL	• monitor
Above OEL	• bi-annual audiometric testing
	• provide and enforce use of hearing protection
	• provide education regarding hearing conservation
	• designate the area as an "Ear Protection Zone" and identify it with the appropriate signage.

Noise Level Surveys are initially carried out (baseline), and then every two years, or when there is a change in processes or equipment, they should be repeated to determine if any change exists.

Records of Noise Level Surveys and/or personal dosimetry testing of the facility areas must be maintained and:

- made readily available to an affected employee or to an officer; and
- permanently retained.

Audiometric Testing

Testing is conducted by the OH professionals in accordance with the applicable legislation. Noise-exposed employees should be tested within six months of joining the organization (baseline audiogram), and then, 12 months after the baseline testing and every two years after that as part of the Medical Monitoring Program. As a rule, testing is arranged and paid for by the organization.

Employee Hearing Results

Hearing assessment results are typically managed as follows:

- The documentation includes hearing history, current hearing status, noise exposures (present and past), hearing protection used, any signs/symptoms of hearing problems, and non-occupational noise exposure.
- A copy of the test is provided to the employee.
- If the results indicate an abnormal shift, a relevant medical history is requested, and the employee is advised and referred to a designated physician along with the test results and medical data.
- The designated physician and audiologist will assess and refer the employee as is clinically indicated and required by the applicable legislation.
- The hearing test records must be maintained on each noise-exposed employee for a period of at least 10 years by the company's OH Service.

Hearing Conservation Program: Program Evaluation

The Hearing Conservation Program must be evaluated annually to verify its effectiveness. Analysis of audiometric test results can provide an indication of the effectiveness of the Program. Trends for specific occupations, processes, departments and worksites can be established, and prevention opportunities identified.

Purchasing

When work equipment is being purchased, consideration must be given to the noise generated by that equipment and the total noise level in the area. This is to ensure that unnecessarily noisy equipment is not purchased. This is a prudent and cost-effective preventative approach.

Responsibilities

The related responsibilities of the various stakeholders can vary by organization, yet typically:

Management

- Ensures that whenever economically feasible, new work equipment with the least noise emission, is procured.
- Ensures that unprotected exposure of the organization's personnel will not exceed the OELs.
- Evaluates noise levels and exposure.
- Implements noise control measures.
- Makes audiometric testing and hearing conservation education available.
- Posts appropriate hearing protection signage where noise levels exceed the OEL.
- Ensures that the Hearing Conservation Program is administered, and the records are kept appropriately.
- Ensures that the Hearing Conservation procedure is followed.

Line Manager

- Ensures that the specified hearing protection is provided and replaced when defective or worn.
- Ensures that hearing protection is worn by employees.
- Provides information and training to their workforce on the need for hearing protection and how to use it.
- Allows employees to attend hearing testing when needed.

Employees

- Report to Line Manager, or designate, any significant situation that may have an adverse effect on employee hearing.
- Wear the provided hearing protection in all posted areas.
- Report any defects in the supplied hearing protection.

- Report for hearing testing appointment.

Director, OH&S
- Reviews noise surveys and assessments.
- Recommends methods for noise reduction.
- Specifies the appropriate hearing protection.
- Advises on equipment purchasing.
- Ensures that audiometric testing is conducted in accordance with professional standards.
- Administers Hearing Conservation Program.

I. MEDICAL MONITORING

Medical monitoring involves the health surveillance of employees who are exposed to specific workplace hazards. The biological effects of exposure to toxic substances such as asbestos, silica, lead, cadmium, nickel, arsenic, cobalt, chromium, tellurium, chlorine, *etc.*, are measured and monitored using medical examinations and laboratory testing. Other hazards such as noise exposure, VDT use, respiratory inhalants, international travel, *etc.*, are assessed. The focus is on prevention as well as the provision of employee and management education.

The provincial OH&S Acts stipulate the regulated substances and the minimal medical monitoring required. For example, according to the Alberta OH&S Act (2019), Schedule 1, employers must develop a code of practice for: arsenic and arsenic compounds, asbestos, benzene, beryllium, 1,3-butadiene, cadmium, coal tar pitch volatiles, 1,2-dibromoethane (Ethylene dibromide), ethylene oxide, hexachlorobutadiene, hydrazines, hydrogen sulfide, isocyanates, lead and lead compounds, ethyl bromide, methyl hydrazine, perchlorates, silica-crystalline, respirable styrene in styrene resin fabrication, vinyl chloride (chloroethylene), and zinc chromate.

J. RESPIRATORY CONSERVATION PROGRAM

Employees are often required to wear respiratory protection when working around/with toxic airborne hazardous substances. Employers must ensure that a respiratory protection policy and procedure exists, and that employees are medically fit to safely wear respiratory protection.

Exposure to dusts (flour, grain, wood working); metal fumes (due to welding, cutting, and smelting); solvent vapors (paints, adhesives, strippers, and cleaning agents); gases (benzene, chlorine, bromine); pharmaceuticals; biohazards; sensitizing vapors (isocyanates, some epoxies, and beryllium); and infectious agents all warrant the use of respiratory protection.

Respiratory equipment must be selected in accordance with the nature of the known hazard. For example, a filtering facepiece respirator may protect against particulate hazards (dusts), but not against gases and vapors. Rather, an air-purifying respirator with chemical cartridges or an atmosphere-supplying respirator, such as an airline respirator or a self-contained breathing apparatus — also known as Self-contained Breathing Apparatus (SCBA), would be appropriate. In addition to selecting the right respirator, it is critical to ensure that the employee is fit to wear it. This involves "Fit Testing".

The Canadian Law on respiratory protection states:

> **12.7 (1)** Where there is a hazard of an airborne hazardous substance or an oxygen deficient atmosphere in a work place, the employer shall provide a respiratory protective device that is listed in the NIOSH Certified Equipment List published on February 13, 1998 by the National Institute for Occupational Safety and Health, as amended from time to time, and that protects against the hazardous substance or oxygen deficiency, as the case may be.
>
> **(2)** A respiratory protective device referred to in subsection (1) shall be selected, fitted, cared for, used and maintained in accordance with the standards set out in CSA Standard Z94.4-M1982, Selection, Care and Use of Respirators, the English version of which is dated May, 1982, as amended to September, 1984 and the French version of which is dated March, 1983, as amended to September, 1984, excluding clauses 6.1.5, 10.3.3.1.2 and 10.3.3.4.2(c).
>
> **(3)** Where air is provided for the purpose of a respiratory protective device referred to in subsection (1),
>
>> **(a)** the air shall meet the standards set out in clauses 5.5.2 to 5.5.11 of CSA Standard CAN3-Z180.1-M85, Compressed Breathing Air and Systems, the English version of which is dated December 1985 and the French version of which is dated November 1987; and
>>
>> **(b)** the system that supplies air shall be constructed, tested, operated and maintained in accordance with the CSA Standard referred to in paragraph (a).[66]

Respiratory fit testing as per the CSA Standard Z94.4-02, is generally required under the provincial OH&S Acts. Respiratory safety is dependent on tight-fitting face pieces, be it a full-faced mask, half-faced mask, or filtering face piece. The employee must be clean shaven or free of facial hair to attain a tight seal. The quantitative fit test involves testing the wearer with a smell to determine if it can be detected or not. This type of testing is quick and easy to administer. The result is either pass/fail.

A second test, a quantitative fit test, gives a number that is referred to as the fit factor (FF). It determines how well a face piece seals against the wearer's face. A high FF indicates that a good seal with the face has been achieved, and that the mask provides protection against harm.

In combination with Fit Testing, the employer should require respiratory

---

[66] Canada Occupational Health and Safety Regulations (SOR/86-304), 2018, Respiratory Protection, Section 12.7, SOR/88-68, s. 14; SOR/94-263, s. 45; SOR/99-151, s. 1; SOR/2002-208, s. 43(F).

protection wearers to first undergo a health assessment, respiratory assessment (pulmonary function testing — spirometry) and an electrocardiogram (ECG) to rule out any heart defects. These tests must be administered by qualified personnel, such as an OHN.

Employee education is the third element of a respiratory protection program. The employee must understand the related hazards and the use of the protective actions and equipment to ensure their safety. Equipment care and maintenance is a critical part of the education

The last element of a respiratory protection program is the documentation of the testing, its results and the provision of employee education.

K. FIRST AID

This topic is covered in Chapter 13, "Occupational Health and Safety: Injury Management".

L. MEDICAL ABSENCE REPORTING

The reasons for employee absenteeism vary and include chronic medical conditions, personal problems, job dissatisfaction, lack of awareness of the attendance expectations, medical appointments, and frequent illness/injury. The causes of absenteeism are numerous, to name a few: an irresponsible work attitude, abuse of company sick leave/short-term disability insurance programs, substandard supervision, and/or poor employee-employer relations.

For employers, absenteeism is costly. To engage employees, Management should encourage regular work attendance by offering challenging and interesting jobs, opportunities for personal advancement, recognition for accomplishments, involvement in decision making, and clear communication that regular work attendance is expected, as well as what to do if ill/injured.

As part of the organization's attendance control efforts, absence reporting must be clearly explained so that the employee and supervisor can effectively deal with work absences and a plan for a prompt return to work by the employee. This reporting process is specific to the organization. However, a sample flowchart of absence reporting is provided on in Appendix 5, along with a sample absence report form (Appendix 6). The forms used to report employee absence are organization-specific and are worded in accordance with the OH resources available. If an OH Service exists, then the absence report form can ask for the absent employee's medical diagnosis (Appendix 7); if not, then, the form must be limited to the employee's fitness-to-work capabilities (Appendix 6).

A major part of managing employee attendance is monitoring employee work attendance. With that data, the supervisor can then recognize those employees who regularly attend work and assist those employees who demonstrate high rates of absenteeism or excessive absenteeism. The OH Service and professionals can assist the Operations with that obligation.

M. OH PROGRAM AND SERVICE: PROGRAM EVALUATION

Evaluation is recognized as a critical component to a successful OH Program and its services. Annual program evaluation is recommended along with ongoing program monitoring throughout the year.

Reports on activities undertaken and performance outcomes should be provided to Management and employees. The data used must be aggregate data, not individual employee data.

N. SUMMARY

The purpose of this chapter is to explain the purpose and related services of the corporate OH Program, as well as its business value to the organization.

APPENDIX 1

PLEDGE AND OATH OF CONFIDENTIALITY

- All personal health information related to an identified employee will be treated as confidential, regardless of format (written, oral, electronic, *etc.*).
- Confidentiality extends to everything the OH&S Practitioner/Professional learns in the exercise of their responsibilities. It extends to both obviously important and apparently trivial information and includes the nature of the employee's contact with the staff, all information an employee discloses, and all the information learned from external caregivers.
- Personal health information can be shared with occupational health professionals employed by Company XYZ *in privacy* to enhance continuity of care and a coordinated approach.
- The dissemination of personal health information will be considered a breach of confidentiality and will be reported to Director, Human Resources and the CEO, Company XYZ. Disciplinary action will be taken up to and include immediate termination of employment *with cause*.
- Senior Management is responsible for ensuring the OH&S staff are aware of and sign the Pledge of Confidentiality acknowledging this awareness.
- To acknowledge and emphasize the serious responsibility in safeguarding employee health information, all Company XYZ staff (permanent or temporary), or contract staff involved with disability management will sign a pledge of confidentiality on the first day of work and annually thereafter, which will be worded as follows:

Pledge of Confidentiality

I have read and reviewed Company XYZ's Standard of Practice on Confidentiality of Personal Health Information. I understand that all employee personal health information, to which I may have access, is confidential and will not be communicated except as outlined above.

Signed	Witness	Date
Signed	Witness	Date

Note: This pledge is to be signed annually and the original form sent to the employee's file. Copies are to be retained by the area Manager and the employee.

APPENDIX 2

MEDICAL FITNESS-TO-WORK FORM

OH&S Service Provider
Address: _____
Telephone: (XXX) XXX-XXXX   Fax: (XXX) XXX-XXXX

MEMORANDUM

TO: XXXXXXXXXXXXXXXX   cc: Employee Medical File
      Director, Occupational Health & Safety

FROM: Occupational Health Service Provider

DATE:

SUBJECT: **Fitness to Work: Pre-placement**

A pre-placement health assessment for the position of _____ has been
                                                     (Job Title)
completed for _____ who is
considered to be:
                    (Name of Employee)

____   Fit for the job

____   Fit for the job with the restrictions listed below

____   Further investigation or job modification required, as noted below

____   Unfit for the job

Comment: _____

_____

_____

_____   _____
Examiner Name (Signature)   Examiner Name and Credentials (Print)
Service Provider
Occupational Health Services

APPENDIX 3

AUTHORIZATION FOR THE RELEASE OF PRIVILEGED INFORMATION

TO: _____

ADDRESS: _____

_____

_____

I AUTHORIZE YOU TO RELEASE HEALTH INFORMATION IN STRICT CONFIDENCE

TO: _____

ADDRESS: _____

_____

_____

INFORMATION TO BE PROVIDED CONSISTS OF:

_____

_____

_____

_____

_____

FOR THE PERIOD (DATES): _____

_____	_____
EMPLOYEE NAME (PRINT)	WITNESS NAME (PRINT)
_____	_____
EMPLOYEE NAME (SIGNATURE)	WITNESS (SIGNATURE)
_____	_____
DATE	DATE

APPENDIX 4

JOB DEMANDS ANALYSIS FORM

APPENDIX 5

ABSENT REPORTING: FLOWCHART (SAMPLE)

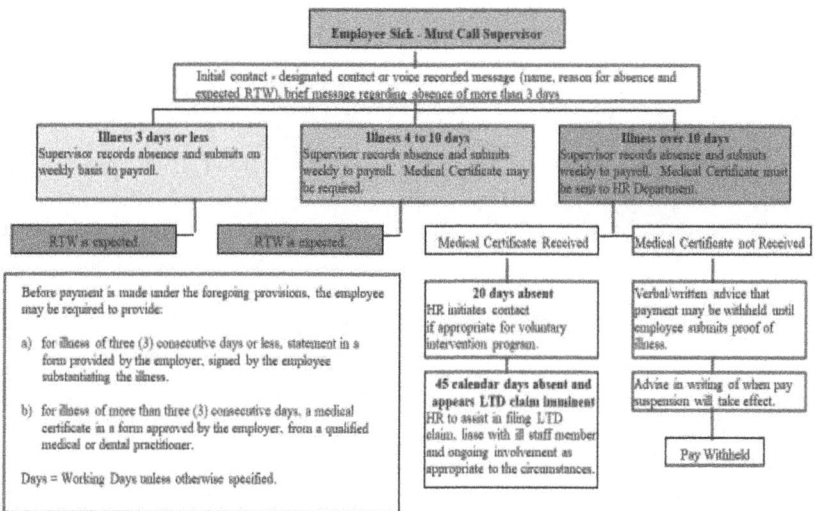

Source: Dyck, 2000-2018

APPENDIX 6

WORK ABSENCE FORM: NON-MEDICAL

For Organizations without Occupational Health Professional Support

Return-to-Work Certificate

This must be completed and signed by an employee returning to work after an absence of 3 (three) or more days	
Non-Work Related: ☐ *i.e.*, the flu, sports injury	**Work Related:** ☐ *i.e.*, a possible Workers' Compensation claim
1.(a) ☐ I have seen a physician who advised me that I would be medically fit to work on this date: _____	
(b) ☐ To the best of my knowledge I am fit for work. *(check one)*	
2. Specify any work restrictions recommended by your physician: _____ _____ _____	
Physician's Name and Address: _____ _____	
Employee Name: (please print) _____	*Employee Signature:* _____
Employee Number: _____	*Date:* _____
Return-to-work date: _____	

APPENDIX 6

WORK ABSENCE FORM: NON-MEDICAL

Used in conjunction with and Occupational Health Departmental Support

Return-to-Work Certificate

This is to confirm that I am prepared and agree to have the employee returning to work after an absence of _____ shifts or _____ days.

Non-Work Related: ☐	Work Related: ☐
Vacation, flu, sports injury	i.e. a possible Workers' Compensation claim

1. (a) ☐ I have seen a physician who advises me that I would be medically fit to work on this date.

   (b) ☐ To the best of my knowledge I am fit for work. I plan to start

2. ☐ Specify any work restrictions recommended by your physician

_____

_____

_____

Physician's Name and Address

_____

Physician's Name (please print)	Employee Signature
_____	_____

Employee Number	Date
_____	_____

Return-to-work date _____

APPENDIX 7

WORK ABSENCE FORM: MEDICAL[67]

---

[67] Adapted from: L. Ydreos, "Medical Status Report Forms" (Presented at the Absenteeism and Disability Management Seminar, Toronto, May 9-10, 1995) [unpublished].

CHAPTER REFERENCES

AADAC, "The Cost of Substance Abuse in Alberta, 2002", *Profile* (2007), online: http://www.assembly.ab.ca/lao/library/egovdocs/2007/alad/159248.pdf (date accessed: January 31, 2020).

L. Allen, cited in "Managing Control of Loss" in *Loss Control Management: Practical Loss Control Leadership*, revised ed., by F. Bird & G. Germain (Loganville: GA: DNV, 1996).

American Psychiatric Association (2013), *Diagnostic and Statistical Manual of Mental Disorders version* (5th ed.). Arlington, VA: Author.

D. Anderson, *Effects of Blood Pressure Screening on Lifestyle Behaviours*. (Calgary: University of Calgary, 1989).

K. Annable, J. Marcoux & V. Kubinec, "The Price of Death", *CBC News* (November 30, 2017), online: https://www.cbc.ca/news2/interactives/workplace-death-injury-labour-laws/ (date accessed: January 31, 2020).

Association of Workers' Compensation Boards of Canada, "2017 Injury Statistics", *Statistics*, online: http://awcbc.org/?page_id=14 (date accessed: February 28, 2020).

Association of Workers' Compensation Board, "2018 Financial Measures". online: http://awcbc.org/?page_id=9759 (date accessed April 2020).

AWCBC, *Legislation and Links* (2019), online: www.awcbc.org (date accessed: January 31, 2020).

C. Baglien, "Fighting Workplace Fatigue" *PEO Canada Employee Management* (2015), online: https://www.peocanada.com/peo-blog/fighting-workplace-fatigue/ (date accessed: February 28, 2020).

K. Baicker, D. Cutler & Z. Song "Workplace Wellness Programs Can Generate Savings", (2010) 29:2 Health Affairs 304-311.

Benefits Canada, "Workplace Mental Health" *Benefits Canada* (2013), online: http://www.benefitscanada.com/benefits/health-wellness/workplace-mental-health-44885 (date accessed: January 31, 2020).

Bongarde, *Selling Safety to Your CEO* (Penticton, BC: MMV Bongarde, 2005).

Canada Labour Code, R.S.C. 1985, c. L-2.

Canada Occupational Health and Safety Regulations (SOR/86-304), 2018, Respiratory Protection, Section 12.7, SOR/88-68, s. 14; SOR/94-263, s. 45; SOR/99-151, s. 1; SOR/2002-208, s. 43(F).

Canadian Centre on Substance Abuse, *The Costs of Substance Abuse in Canada, 2002* (2003).

Centre for Addiction and Mental Health (CAMH), "The numbers add up, so make your vote count", 11:3 camhconnexions (Fall 2011), online: http://www.camh.ca/en/hospital/about_camh/newsroom/connexions_newsletter/Documents/4515Connexions_

Fall2011EN.pdf (date accessed: February 28, 2020).

CAMH, *Mental Illness and Addictions: Facts and Statistics* (2016), online: http://www.camh.ca/en/hospital/about_camh/newsroom/for_reporters/Pages/addictionmentalhealthstatistics.aspx (date accessed: February 28, 2020).

K. Cogdon, "Couch potatoes cost health system $1.5B a year", *The Cairns Post* (October 12, 2007) at 21.

Conference Board of Canada, *Smoking Cessation and the Workplace: Benefits of Workplace Programs* (2013), online: https://www.quitnow.ca/files/QN/files/library/Smoking_Cessation_and_the_Workplace_Briefing_3_Benefits_of_Workplace_Programs.pdf (date accessed: February 28, 2020).

Conference Board of Canada, *Canadian Alliance for Sustainable Health Care* (2016), online: http://www.conferenceboard.ca/press/newsrelease/16-09-01/unmet_mental_health_care_needs_costing_canadian_economy_billions.aspx (date accessed: February 28, 2020).

J.F. Fries, "Beyond Health Promotion: Reducing the need and demand for medical care" (1998) 17:2 *Health Affairs* 70-84.

Health Canada, *Report on the findings of the oral health component of the Canadian Health Measures Survey 2007–2009* (2010), online: http://publications.gc.ca/site/archivee-archived.html?url=http://publications.gc.ca/collections/collection_2010/sc-hc/H34-221-1-2010-eng.pdf (date accessed: February 28, 2020).

Health & Safety Ontario, *The Business Case for a Healthy Workplace* (2008), online: https://www.wsps.ca/WSPS/media/Site/Resources/Downloads/BusinessCaseHW_Final.pdf?ext=.pdf (date accessed: January 31, 2020).

Institute of Musculoskeletal Health and Arthritis, "Strategic Plan 2014-2018" Canadian Institutes of Health Research (Winnipeg, Manitoba, 2014), online: http://www.cihr-irsc.gc.ca/e/48830.html (date accessed: February 28, 2020).

P. Katzmarzyk, "Tallying the global economic burden of physical inactivity", *Alberta Centre for Active Living* (October 1, 2016), online: https://www.centre4activeliving.ca/news/2016/10/global-economic-burden-physical-inactivity/ (date accessed: February 28, 2020).

Mercer, "How Much Are You Losing to Absenteeism?" *Mercer Website* (2019), online: https://www.mercer.ca/en/our-thinking/how-much-are-you-losing-to-absenteeism.html (date accessed: January 31, 2020).

National Safety Council, "Costs" *Injury Facts* (2019), online: https://injuryfacts.nsc.org/work/costs/work-injury-costs/ (date accessed: February 28, 2020).

National Safety Council, "Marijuana at Work: What Employers Need to Know" (2018), online https://nsc.org/membership/training-tools/best-practices/marijuana-at-work (date accessed: February 28, 2020).

National Institute of Disability Management and Research (NIDMAR), *Disability Management Success: A Global Corporate Perspective* (Port Alberni, BC:

NIDMAR, 2005), online: http://www.nidmar.ca (date accessed: January 31, 2020).

Occupational Medicine Specialists of Canada, OMSOC Home Page, online: http://www. OMSOC.org (date accessed: February 28, 2020).

Public Health Agency of Canada, *The Cost of Injury in Canada* (2015), online: http://www.parachutecanada.org/downloads/research/Cost_of_Injury-2015.pdf (date accessed: February 28, 2020).

T. Roithmayr in D. Dyck (2017), *Disability Management: Theory, Strategy & Industry Practice*, 6th ed. (Markham, ON: LexisNexis Canada, 2017).

Safety Management Group, *Injury Cost Calculator* (2019), online: https://safetymanagementgroup.com/resources/injury-cost-calculator/ (date accessed: January 31, 2020).

E. Skronski, "Is Your Company Healthy? 7 Healthy Strategies for Employee Retention", *Canada One* (2008), online: https://www.canadaone.com/ezine/march08/strategies_for_employees_retention.html (date accessed January 31, 2020).

C. Smith, "Down with Costs, Up with Safety", *OH&S* (July 10, 2015), online: https://ohsonline.com/Blogs/The-OHS-Wire/2015/07/Down-with-Costs-Up-with-Safety.aspx (date accessed: February 28, 2020).

Statistics Canada, Average Canadian Salary, 2018 (2019), online: https://www150.statcan.gc.ca/t1/tbl1/en/tv.action?pid=1410032002 (date accessed January 31, 2020).

Statistics Canada, *Table 14-10-0190-01* Work *Absence of full-time employees by geography, annual* (Ottawa: Statistics Canada, January 31, 2020), online:https://www150.statcan.gc.ca/t1/tbl1/en/tv.action?pid=1410019001 (date accessed January 31, 2020).

Statistics Canada, *Table 14-10-0287-01, Labour force characteristics, monthly adjusted and trend-cycle, last 5 months* (2020), online: https://www150.statcan.gc.ca/t1/tbl1/en/tv.action?pid=1410028701 (date accessed January 31, 2020).

Statistics Canada, *Table 11, Average Usual Hours and Wages of Employees by Selected Characteristics* (2020), online: https://www150.statcan.gc.ca/n1/daily-quotidien/191108/t011a-eng.htm (date accessed: January 31, 2020).

Statistics Canada, *Insights on Canadian Society: The impact of aging on labour market participation rates, Cat. No. 75-006-X*, online: http://www.statcan.gc.ca/pub/75-006-x/2017001/article/14826-eng.htm (date accessed: February 28, 2020).

Statistics Canada, reported in A. Carofano, "Controlling the Costs of Absenteeism - The CFO Perspective" (2014), *Workplace Medical Corp.*, online: http://www.workplacemedical.com/wp-content/uploads/2015/04/Absence-Managment-Whitepaper-The-CFO-Perspective.pdf (date accessed: January 31, 2020).

Statistics Canada, "Immigration and Ethnocultural Diversity" *National Household*

Survey (2011), online: https://www12.statcan.gc.ca/nhs-enm/2011/as-sa/99-010-x/99-010-x2011001-eng.cfm#a1 (date accessed: February 28, 2020).

R. Tangri *Stress Costs Stress Cures* (Victoria, BC: Trafford Publishing, 2003).

E. Twells, "Current and predicted prevalence of obesity in Canada: a trend analysis", *CMAJ Open*, Mar 3;2(1):E18-26. doi: 10.9778/cmajo.20130016. eCollection 2014 Jan.

Watson Wyatt Worldwide, *Staying @ Work Survey*, online: www.watsonwyatt.com (date accessed: January 31, 2020).

Workplace Safety and Prevention Services Ontario, *The Business Case for a Healthy Workplace* (2011) at 8-11, online: https://www.wsps.ca/WSPS/media/Site/Resources/Downloads/ BusinessCaseHW_Final.pdf?ext=.pdf, (date accessed: February 28, 2020)

World Health Organization, *Good Practice in Occupational Health Services A Contribution to Workplace Health* (1995) at 2-3.

World Health Organization, *C161 - Occupational Health Services Convention, 1985 (No. 161)* (1985), online: http://www.ilo.org/dyn/normlex/en/f?p=NORMLEXPUB:55:0::NO::P55_TYPE,P55_LANG,P55_DOCUMENT,P55_NODE:CON,en,C161,/Document (date accessed: February 28, 2020).

World Health Organization, Reported in Agius, "What is Occupational Health?" (1950), online: http://www.agius.com/hew/resource/ohsilo.htm (date accessed: February 28, 2020).

Workers Health & Safety Centre Federation of Ontario, Ergonomic Training Resources (2019), online: https://www.whsc.on.ca/Resources/Publications/Ergonomic-Resources (date accessed: January 31, 2020).

L. Ydreos, "Medical Status Report Forms" (Presented at the Absenteeism and Disability Management Seminar, Toronto, May 9-10, 1995) [unpublished].

Section 2
Occupational Health and Safety: Strategies

Chapter 5

OCCUPATIONAL HEALTH AND SAFETY PROGRAM: MANUAL DEVELOPMENT

A. INTRODUCTION

An **Occupational Health and Safety (OH&S) Program Manual** is the documented evidence of Management's endorsement and commitment to OH&S and is designed to serve as a standard for OH&S practice within the organization/company. The format of the manual can be hardcopy, electronic, or both. The key aspect is to ensure that the OH&S Program Manual is readily accessible to all employees and is in an acceptable format, so employees are willing and able to use it.

The value of having an OH&S Program Manual is that it:

- articulates the organization's OH&S standards and practices;
- standardizes the OH&S practices within the organization;
- demonstrates to all stakeholders the organization's leadership and commitment to a certain level of OH&S performance;
- explains how the organization plans to administer, implement, evaluate, and continuously improve its OH&S Program; and
- demonstrates compliance with the OH&S Act and the other applicable legislation.

B. OH&S PROGRAM MANUAL: DEVELOPMENT

When developing an OH&S Program Manual, it is imperative to consider many elements. The following is a discussion of these elements, as well as the provision of suggestions for action:

1. Policies, Standards and Procedures

Corporate policies, standards, and procedures reflect Management's attitude regarding workplace health and safety, and the organization's commitment to worker health and safety. They are designed to facilitate the achievement of the goals established by the OH&S Program.

Management develops policies as guidelines for employees during their employment activities. Ideally, policies are consistent with the company's corporate goals and business strategies. One of their purposes is to prevent, or help to resolve, problems. For that reason, policies should be comprehensive in their scope, clear in their intent, fair to all, documented, and readily available.

A performance **Standard** is a stated approach to a specific practice. Standards tend to be based on evidence-based practices, or if they are not available, then they are developed as guidelines and rules for practice, as well as to set boundaries on practice activities. In essence, a standard serves to clarify roles and responsibilities and can be used as a benchmark performance metric.

Procedures are defined actions that serve to standardize the OH&S Program. They provide a basis for stakeholder education, clarify the process, and facilitate the smooth functioning of the OH&S Program.

OH&S policies and procedures need to be current and appropriate for the organization. For this reason, a periodic review of their applicability is critical to any OH&S Program's success.

The following sections of this chapter address the various components of a typical OH&S Program Policy Manual. Industry illustrations are provided to clarify the presented concepts.

2. Manual Components

(a) Mission Statement

The OH&S Program **mission statement**, or **policy**, describes Management's commitment to workplace safety and the provision of a safe work environment. It presents the high-level program objectives and describes the organization's values and beliefs towards maximizing workplace health and safety and minimizing the impact of workplace incidents on stakeholders. It must:

- be documented, implemented, and regularly reviewed and updated;
- be appropriate to the nature and scale of the organization's OH&S risks;
- state the company's commitment to allocate resources to OH&S;
- include a commitment to continuous improvement;
- state compliance with the applicable legislation and industry practices;
- be communicated to employees so that they are aware of their respective OH&S obligations and duties; and
- be readily available and accessible.

A well-composed OH&S mission statement, or policy, emphasizes a collaborative approach towards achieving workplace health and safety. This means that all the stakeholders have a responsibility and role to play in promoting workplace health and safety. It emphasizes that they will be held accountable to uphold their respective OH&S roles and responsibilities.

An example of a mission statement or policy is as follows:

CH. 5: OCCUPATIONAL HEALTH AND SAFETY PROGRAM: MANUAL DEVELOPMENT

Policy: Occupational Health and Safety Policy	Approved: January 2, 2007
Effective Date: January 2, 2007	Reconfirmed: January 2, 2020

Company XYZ (XYZ) is committed to the protection of its employees, contractors and physical assets and to maintaining the safety of the public.

This OH&S Policy is an integral part of the daily work activities of XYZ management and employees. The goal and objective of this policy is to prevent injury and personal suffering, minimize losses associated with incidents, and comply with the applicable legislation. Everyone is responsible for their own health and safety and that of co-workers.

Managers and supervisors will encourage and promote health and safety excellence by:
- demonstrating the goals and objectives of the XYZ OH&S Program through actions, participation and follow-up;
- communicating the roles and responsibilities for OH&S to all employees and contractors;
- developing, updating, and implementing safe work practices and procedures;
- ensuring workers have safe and appropriate equipment/tools with which to work;
- developing training, educational, and certification programs to ensure employees are able to perform their duties to acceptable safety standards and safe work practices;
- providing employees with suitable training and monitoring their work competence; and
- ensuring employees are aware of their OH&S responsibilities and duties under all applicable legislation and regulatory standards.

Employees will work in partnership by:
- maintaining their awareness and fulfilling their OH&S responsibilities and duties under all applicable legislation and regulatory standards;
- actively participating in company health and safety programs; and
- participating in the development, updating, and implementation of safe work practices and procedures.

Company XYZ believes excellence in OH&S performance is an essential requirement to the successful operation of our company. We are dedicated to providing the leadership support and management systems required for XYZ to be a leader in all our endeavours. Our accident and loss prevention goals and objectives will be achieved through the cooperation, involvement, awareness, and actions of each employee.

Policy Approved:

_____

Company President
Date: January 2, 2020

(b) Program Goals and Objectives

OH&S Program goals are the broad statements that are central to the OH&S Program. **Objectives** are the specific aims for the OH&S Program. They state the desired outcomes in a manner that is meaningful, relevant, realistic, actionable, sustainable, useful, measurable, and results-oriented.

For the successful functioning of the OH&S Program, the program objectives should be mutually agreed upon by Management and labour. By accepting these objectives, both parties agree to work together towards workplace health and safety. An example of a program goal is as follows:

- *To develop an integrated and comprehensive OH&S Program.*

Examples of OH&S Program objectives are:

By December 2021,

- *To obtain Management support for OH&S Program as evidenced by adequate manpower and resources for the OH&S Program.*
- *To determine current corporate and employee views on the current status of the OH&S Program as evidenced by positive results from the perception survey.*
- *To increase stakeholder knowledge and use of OH&S Program concepts, programs, and services as evidenced by an increased level of awareness.*

(c) Administration of the OH&S Program

This section includes the formal duties of the people (committee members or individuals) responsible for the administration of the OH&S Program. It includes the lines of responsibility and accountability.

The specific policies and procedures related to the OH&S Program are also presented. These policies and procedures clarify and standardize how key OH&S Program activities occur, such as the:

- OH&S Policy
- Respectful Workplace Policy (Prevention of Workplace Violence Policy, Harassment Policy)
- Non-smoking Policy
- Substance Abuse Policy
- Working Alone Policy
- Seat-belt Use Policy
- Cellphone Use Policy
- Employee Assistance Program Policy
- Ergonomics Program
- Disability Management Policy

- Return-to-work Policy

(d) Stakeholder Roles and Responsibilities

The policy and procedure manual identifies the primary stakeholders in the OH&S Program. These include the Senior and Middle Management, supervisors, employees, unions, OH&S Practitioners/Professionals, and other relevant stakeholders.

The roles for each stakeholder are described, as well as the specific lines of communication and reporting. An effective method of describing the roles and lines of communication involves using a simple organization chart.

(e) Implementation Strategies

Implementation strategies define methods of assessment, referral points, intervention options, potential job accommodations, and alternative jobs.

Implementation strategies also involve gathering information about the various services and resources available to support the OH&S Program. These include Employee Assistance Programs, Human Resources services, Disability Management Program, union services, industry groups/associations, government or municipal agencies (OH&S agencies, Workers' Compensation Boards), insurer (government and private) benefits, in-house training programs, and local caregivers (*i.e.*, physicians, hospitals, rehabilitation specialists, *etc.*). More information on the supporting disciplines and programs can be found in Chapter 19, "Occupational Health and Safety: Supporting Disciplines".

Lastly, the establishment of a database to collect information on the OH&S Program is required. This database can be manual or electronic. The key aspect is that the data on workplace health and safety is maintained.

(f) Program Evaluation

Program evaluation is critical; measurement enables organizations to manage and control losses. As well, what gets measured within an organization/company tends to get attention. This section of the manual should define the desired methods of measuring OH&S performance and the expected outcomes of the OH&S Program.

There are a few terms that need to be defined here. The first is **Occupational Health and Safety Audit**, which is a systematic, documented, verification process of objectively obtaining and evaluating evidence to determine whether specific OH&S activities, events, conditions, management systems, or information conform to the audit criteria. The process includes effective communication of the results to the client(s).[1]

Occupational Health and Safety performance is the measurable results of the Occupational Health and Safety Management System, related to an organization's

---

[1] BSI Management Systems, "OHSAS 18001 Awareness" (Presented at an Informational Seminar, April 2002, Calgary, AB), at 4-3.

control of its OH&S hazards and risks, based on its OH&S Policy, goal(s), and targets.[2]

An **Occupational Health and Safety goal** is the overall, quantifiable aim of the OH&S Program that the organization sets out to achieve. An OH&S target is a detailed performance requirement, quantified where practicable, that is applicable to the organization or parts thereof, that arises from an OH&S goal and that needs to be set and met to achieve that goal.[3]

The point of OH&S Program evaluation is continuous improvement — the process of enhancing the OH&S Management System to achieve improvements in overall OH&S performance in accordance with the organization's OH&S Policy.[4]

(g) Program Promotion

Within the OH&S Program Policy Manual, it is important to describe the lines of communication for the OH&S Program. The "who, what, when, how and where" of program promotion should be described in the manual. For example:

Policy

The OH&S Program should be accessible and made easy for employees to identify their need for assistance, or for supervisors to refer employees when required.

Procedures

i. The OH&S Program should be promoted no less than three times per year at all worksites. Promotion may be accomplished through the distribution of promotional materials, information sessions, email announcements and/or posters.

ii. There should be some form of targeted promotional material made available to eligible employees at least every 12 months.

iii. Educational sessions on the OH&S Program are to be presented as a component of the standard employee orientation and supervisory training packages.

iv. Key OH&S Program education/training is to be provided to Human Resources professionals, OH&S personnel, Joint Health & Safety Committee members, union representatives, and similar personnel.

This section also explains the strategies that have been developed to inform the

---

[2] *Ibid.*, at 4-4.

[3] *Ibid.*

[4] *Ibid.*, at 4-2.

internal and external stakeholders of the OH&S Program of any changes that may develop.

Lastly, the policy section creates and maintains an awareness of the benefits of the OH&S Program. This can involve regular feedback on the OH&S audit results, number and nature of workplace incidents, modified work initiatives, related costs, other program outcomes, and the goals of the future OH&S Program. Using a graphical representation of the results can have a positive impact (Figure 5.1). It depicts that although the total audit scores for the company were similar over the three years, performance of the OH&S Program declined over the three years in the areas of organizational commitment, hazard identification, hazard control, worksite inspections, and incident investigation. However, program improvements were achieved in the areas of worker qualifications, orientation, and training; emergency response planning; and program administration.

Figure 5.1: Graphic Presentation of the Company Occupational Health and Safety Audit Results, 2017-2019

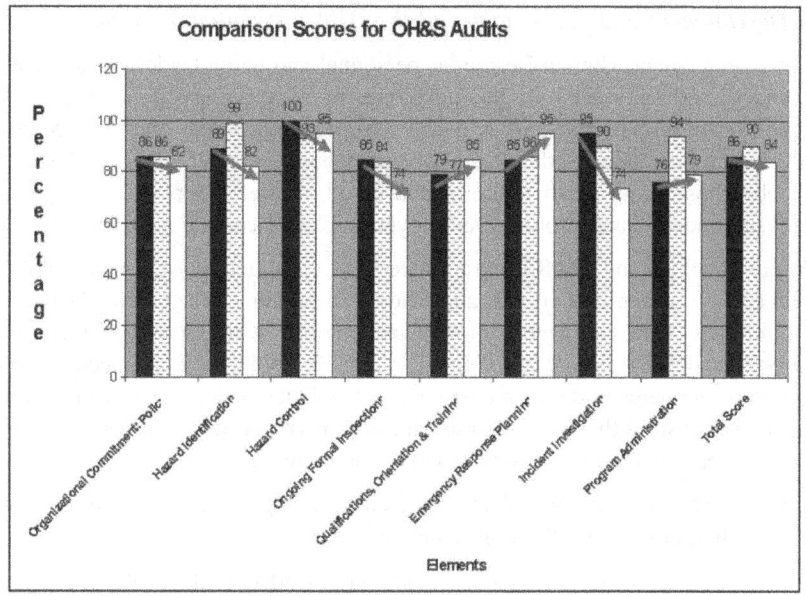

3. Industry Example

For an OH&S Program to operate effectively, it requires policies and procedures. As discussed above, policies and procedures function to treat employees fairly and equitably; to communicate the rules and regulations that are in place; and to facilitate the functioning of the OH&S Program. The sections that follow provide some examples of the issues and concerns that a company needs to consider when developing its OH&S Program policies and procedures. Appendix 1 provides the reader with an industry example of an OH&S Program Manual.

(a) Preliminary Concerns

The following OH&S-related policies are only useful to a company if they are accepted, endorsed, and supported by Senior Management, union leaders, and employees. To create acceptance, a company should:

- have a committee regularly review and modify the policies to meet stakeholder and business needs;
- involve representatives from all stakeholders in the planning and implementation processes;
- communicate the policies to all employees;
- implement the policies in a unified manner with all employees;
- evaluate their impact on the company operations and employees; and
- modify and readjust the policies as required.

(b) Development and Maintenance of the OH&S Program Manual

Senior Management, Human Resources personnel and union leaders (if applicable) must be directly involved in the development of company policies and their presentation to employees. This is one way to get stakeholder "buy in". Management, Human Resources, and union involvement ensure greater understanding of the policy by those required to interpret and to administer it; provides insight into employee concerns; and promotes employer-employee communication.

Where practical, an OH&S Program Steering Committee should be selected from employees, union representatives, and Middle to Senior Management. The size of the committee varies depending on Management's willingness to broaden the process. Broader representation brings different viewpoints and perspectives to the discussions and helps to develop a better set of policies. The involvement of Senior Management ensures that the corporate philosophies regarding business, production, and employee satisfaction are found within the manual.

Some suggestions that the OH&S Program Steering Committee may choose to consider when developing the policy manual are to:

- meet with Senior Management and union leaders to delineate what is to be accomplished by the guidelines;
- select a person on the committee to coordinate the development of the OH&S Program Manual;
- determine which of the company's current policies are to be included in the manual, and whether or not they need revision;
- have supervisors/managers and union representatives respond to a checklist of tentative policies and an outline of instructions for implementing policies; and/or

- set deadlines for completion of the OH&S Program Manual.

Another important responsibility of the committee is the introduction of the OH&S Program Manual to all employees. One option is to consider an employee-orientation meeting during which:

- Management and union can jointly comment on the importance of the OH&S Program Manual and its purposes;
- one or more of the steering committee members involved in developing the manual presents a brief discussion on the set-up of the OH&S Program Manual, each of its policies and the reasons for including them in the manual; and
- time is allotted for questions and comments from the employees.

If the company has revised any of its current policies, or is adding new policies, a more informal meeting may be appropriate.

Note: An important part of the process, prior to introducing the manual to employees, is to have legal counsel specializing in human resources, labour and employment law, and OH&S legislation review the contents.

The manual's publication does not finish the project. As the company grows and its workforce gets larger and more diverse, new issues will emerge and new policies will have to be developed to address them. Similarly, changes in laws, regulations, employee benefits, and other areas necessitate revisions. Plans for periodic reviews of the developed policy and procedure manual should be made and documented for due diligence purposes.

(c) Formatting the Manual

There are several formats available for organizing an OH&S Program Manual. When customizing the guidelines, the following questions should be considered:

- What image does the company want to portray to employees?
- Will the manual be presented in hardcopy, or in an electronic format using a company intranet system? Or, will it be a combination of both?
- How often are major revisions to the manual anticipated?
- What will be the maintenance method for the manual? Will manuals be returned to one location for updating, or distributed to employees who will be expected to maintain their own manuals, or will it be electronically updated and distributed?

Here are some general points to consider when developing the content of the manual policies and procedures:

- Keep in mind for which departments/employee groups the policies have been written.
- Organize each policy in a logical operational sequence.

- Stay on topic.
- Make the sentences and paragraphs brief and succinct.
- Avoid rigid formality unless that degree of strictness is desired.
- Check for user understanding. Try to be flexible and to avoid vague, unclear, or indirect statements.
- Establish the terminology for key words such as gender, organization, location, department, division, and positions. These should be used consistently throughout the manual.

If the loose-leaf binder format for the manual is selected, the company name or logo should be prominently presented on each page. The section name and policy title on each page helps in organizing and referencing the manual. The use of "page __ of __ pages" makes it easy for everyone to determine if he/she has the complete policy in their version of the manual. The use of "issue effective date" and "revision date" assists in researching the history and changes to each policy. The inclusion of "Approved by" legitimizes the policy and signifies that Senior Management has reviewed the manual.

C. SUMMARY

Having a functional infrastructure for an OH&S Program ensures that the program continues to operate despite changes in the personnel responsible for the program; that all employees are treated fairly and equitably; that outcome measures are identified; and that the program's success can be measured and reported. An OH&S Program Manual can be used to make the OH&S Program's infrastructure visible and accepted.

Appendix 1

OH&S Program Manual: An Industry Example

The remainder of this chapter has been used to illustrate some components of an OH&S Program Manual template that could be found in industry.

OH&S Program Manual
Company XYZ
January 2, 2020

Table of Contents

SECTION 1	**INTRODUCTION**	
	1.0	XYZ Occupational Health and Safety Policy
	2.0	Manual Introduction
	3.0	Manual Issue Control
	4.0	Orientation
SECTION 2	**OH&S LEADERSHIP AND MANAGEMENT**	
	1.0	OH&S Leadership and Management
	2.0	Worker Competence, Selection and Training
	3.0	OH&S Communication and Feedback
	4.0	Risk Assessment and Control *(materials provided in Chapter 15)*
	5.0	Performance Measurement and Corrective Actions
	6.0	Audit Procedures
SECTION 3	**RELATED POLICIES AND PROCEDURES**	
	1.0	Employee Assistance Program
	2.0	Substance Abuse Policy
	3.0	Non-Smoking Workplace Policy
	4.0	Workplace Hazardous Materials Information System (WHMIS)
	5.0	Emergency Response Plan *(materials provided in Chapter 12)*
	6.0	Incident Investigation *(materials provided in Chapter 14)*
	7.0	Visitors and Contractor Control Policy
	8.0	Ergonomics Program *(materials provided in Chapter 20)*
	9.0	Disability Management Policy *(materials provided in Chapter 27)*
SECTION 4	**OCCUPATIONAL HEALTH PROGRAM**	
	(materials provided in Chapter 4)	

1.0	Occupational Health Program *(materials provided in Chapter 4)*
2.0	Confidentiality of Health Information Policy
3.0	Fitness to Work
4.0	Job Demands Analysis
5.0	Hearing Conservation Program
6.0	Respiratory Conservation Program
7.0	First Aid *(materials provided in Chapter 13)*
8.0	Medical Absence Reporting
9.0	Medical Monitoring Program
10.0	Medical Monitoring (for applicable substances)

SECTION 5 OCCUPATIONAL SAFETY PROGRAM

1.0	General Safety Rules
2.0	Work Equipment Safety Rules
3.0	Access Equipment Safety Rules
4.0	Crane and Hoist Safety Rules
5.0	Electrical Equipment Safety Rules: Portable and Transportable
6.0	Forklift Safety Rules
7.0	Rigging and Slinging Equipment Safety Rules
8.0	Manual Handling Safety Rules
9.0	Grinding Wheel Safety Rules
10.0	Personal Protective Equipment (PPE) Safety Rules
11.0	Working Alone Safety Policy
12.0	Workplace Violence Safety Policy

SECTION 6 ENVIRONMENTAL PROGRAM
(not addressed)

1.0	Environmental Plan
2.0	Control of Substances Hazardous to Health (COSHH)
3.0	Storage and Handling of Hazardous Substances
4.0	Waste Management Plan

SECTION 7 CONTROL OF CONTRACTORS PROGRAM
(not addressed)

1.0	Selection, Assessment and Evaluation of Contractors/Sub-Contractors
2.0	List of Approved Contractors
3.0	Procedures for Contractors
4.0	General Conditions of Work in/on Site
5.0	Code of Practice for Contractors Working in/on Company Premises
6.0	Contractors Records

(The underlining indicates available links to sections of the manual.)

SECTION 1: INTRODUCTION

1.0 XYZ OCCUPATIONAL HEALTH AND SAFETY POLICY

This OH&S Program Manual is designed to clearly state the policies and procedures that balance employee and employer rights and expectations at Company XYZ (XYZ).

Policy: **Occupational Health and Safety Policy**
Effective Date: January 2, 2007
Approved: January 2
Reconfirmed: January 2, 2020

Company XYZ (XYZ) is committed to the protection of its employees, contractors, and physical assets and to maintaining the safety of the general public. This OH&S policy is an integral part of the daily work activities of XYZ Management and employees. The goals of this policy are to prevent injury and personal suffering, minimize losses associated with incidents and comply with the applicable legislation. Everyone is responsible for their own health and safety and that of co-workers.

Managers and supervisors will encourage and promote health and safety excellence by:

- demonstrating the goals and objectives of the XYZ OH&S Program through actions, participation and follow-up;
- communicating the roles and responsibilities for Occupational Health and Safety to all employees and contractors;
- developing, updating, and implementing safe work practices and procedures;
- ensuring workers have safe and appropriate equipment/tools with which to work;
- developing training, educational, and certification programs to ensure employees can perform their duties to acceptable safety standards and safe work practices;
- providing employees with suitable training and monitoring their work competence; and
- ensuring employees are aware of their OH&S responsibilities and duties under all applicable legislation and regulatory standards.

Employees will work in partnership with Management by:

- maintaining their awareness and fulfilling their OH&S responsibilities and duties under all applicable legislation and regulatory standards;
- actively participating in the OH&S Program; and
- participating in the development, updating, and implementation of safe work practices and procedures.

CH. 5: OCCUPATIONAL HEALTH AND SAFETY PROGRAM: MANUAL DEVELOPMENT

Policy: **Occupational Health and Safety Policy**
Effective Date: January 2, 2007
Approved: January 2
Reconfirmed: January 2, 2020

Company XYZ believes excellence in OH&S performance is an essential requirement to the successful operation of our company. We are dedicated to providing the leadership support and management systems required for XYZ to be a leader in all our endeavours. Our accident and loss prevention goals and objectives will be achieved through the cooperation, involvement, awareness, and action of each employee.

Policy Approved:

_____

Company President
Date: January 2, 2020

2.0 MANUAL INTRODUCTION

This manual explains XYZ's OH&S Management System (OHSMS) and contains the relevant OH&S policies and procedures. XYZ employees are expected to review this manual, along with the XYZ *Employee OH&S Handbook* at the new employee orientation and annually thereafter. Acknowledgment of the completion of this action is to be recorded in the employee training log which is then maintained in the employee's personnel file.

Suggestions for improvements to the manual, or requests for clarification of details in the manual should be sent to Management or to the XYZ OH&S Department.

3.0 MANUAL ISSUE CONTROL

The OH&S Program Manual is reviewed annually and forms part of the Annual OH&S Action Plan.[5] Revisions to the manual are made when required and controlled by the Director, OH&S Department. The date of revision is shown on each page and recorded on the Record of Revisions. Employees are notified of revisions to this manual via email and the manual's Record of Revisions.

The OH&S Program Manual is the property of XYZ. No part of the OH&S Program Manual, procedures or practices are to be updated or disclosed to unauthorized persons without prior written approval of the Director, OH&S Department.

3.1 List of Manual Holders

Hard copies of the OH&S Manual will be held by the:

1. Director, OH&S Department (master copy)

---

[5] An example of an OH&S Action Plan is provided in Chapter 18, "Occupational Health and Safety: Best Practices", Appendix 1.

2. XYZ Safety Specialists
3. XYZ Disability Management Coordinator
4. XYZ's Occupational Health Service Provider

All XYZ employees have access to the OH&S Manual via the XYZ intranet system.

3.2 Record of Revisions

Rev. #	Amended By	Date	Details

4.0 ORIENTATION

Line Management is responsible for ensuring that new employees are aware of the XYZ OH&S policies and procedures; and that they receive an OH&S orientation and the appropriate OH&S training prior to undertaking new job responsibilities.

4.1 OH&S Orientation Process

The orientation to the XYZ OH&S Program includes an orientation to the following elements:

- the XYZ OH&S Policy;
- the OH&S Leadership and Management section of the OH&S Program Manual;
- Occupational Health procedures and practices;
- Occupational Safety procedures and practices;
- the XYZ OH&S Website; and
- the *Employee OH&S Handbook* — hardcopy or electronic version.

4.2 Documentation

It is the responsibility of Line Management to ensure that the OH&S orientation for new employees is documented on the OH&S Orientation Checklist and is signed and placed in the employee's personnel file.

4.2.1 XYZ OH&S Orientation Checklist

Line Management is responsible for explaining each item on XYZ's Orientation Checklist (*found in Chapter 17, Appendix 1*) to the new or transferred employee. As each item is covered, the employee and Line Manager should date and initial the checklist. The completed checklist is then kept in the employee's personnel file.

4.2.2 Ongoing OH&S Knowledge

Employees are responsible for remaining current regarding XYZ's OH&S practices and procedures, the OH&S Program Manual and the *Employee OH&S Rule Book*.

SECTION 2: OH&S LEADERSHIP AND MANAGEMENT

1.0 OH&S Leadership and Management

1.1 Introduction

XYZ is committed to business excellence and being an industry leader, providing outstanding value to our customers; a healthy, safe, and enjoyable work environment for our employees; and superior returns on investment to our shareholder(s).

XYZ is committed to:

- providing high quality services and products to our customers;
- maintaining a high standard of integrity, ethics, and excellence in all aspects of our business;
- providing a healthy, safe, challenging, and rewarding environment for our employees;
- fulfilling our OH&S responsibilities to our industry, communities, and to the protection of the environment;
- continually improving our services, products, and profitability; and
- providing corporate stewardship regarding public and industry safety.

The XYZ OH&S Philosophy is that ***"Workplace health and safety is priority #1"***.

1.2 Guiding Principles

- *Compliance*: We will endeavour to remain in compliance with all regulations, guidelines, and codes of practice for health, safety, and the protection of the environment.
- *Internal monitoring (auditing)*: We will conduct regular and routine evaluations of our operations to ensure we are meeting our objectives for OH&S stewardship.
- *Training*: We will provide adequate training opportunities for our employees, so they can perform their work safely, efficiently, and for the protection of the environment.
- *Research*: Where appropriate, we will conduct or fund research programs to provide healthier, safer, and more environmentally sound methods for doing business.
- *Communications*: We will maintain ongoing communications with our clients, employees, suppliers, contractors, and the community so they are aware of our commitment and intentions to protect employee health, safety, and the environment.

- *Emergency response:* We will provide an Emergency Response Plan that will be able to respond efficiently to all foreseeable emergencies.
- *Waste management*: We will tailor our operations to produce the lowest volume and least toxic waste achievable within responsible levels of financial and administrative control. We will insist that our waste management contractors share this responsibility and use accepted practices.

1.3 Applicable Legislation

Every employer and employee (an individual who works under a contract of employment) has legal duties/responsibilities according to the applicable OH&S legislation.

For example, the Alberta Occupational Health and Safety Act[6] states:

3 (1) Every employer shall ensure, as far as it is reasonably practicable for the employer to do so,

(a) the health and safety and welfare of

(i) workers engaged in the work of that employer,

(ii) those workers not engaged in the work of that employer but present at the worksite at which that work is being carried out, and

(iii) other persons at or in the vicinity of the worksite who may be affected by hazards originating from the worksite,

(b) that the employer's workers are aware of their rights and duties under this Act, the regulations and the OHS code and of any health and safety issues arising from the work being conducted at the worksite,

(c) that none of the employer's workers are subjected to or participate in harassment or violence at the worksite,

(d) that the employer's workers are supervised by a person who

(i) is competent, and

(ii) is familiar with the Act, the regulations and the OHS code that apply to the work performed at the worksite,

(e) that the employer consults and cooperates with the joint worksite health and safety committee or the health and safety representative, as applicable, to exchange information on health and safety matters and to resolve health and safety concerns,

(f) that health and safety concerns raised by workers, supervisors, self-employed persons and the joint worksite health and safety committee or health and safety representative are resolved in a timely manner, and

(g) that on a worksite where a prime contractor is required, the prime contractor is advised of the names of all of the supervisors of the

---

[6] R.S.A. 2017, c. O-2, s. 3(1), (2), (3), (4).

workers

(2) Every employer shall ensure that workers are adequately trained in all matters necessary to protect their health and safety, including before the worker

    (a) begins performing a work activity,

    (b) performs a new work activity, uses new equipment or performs new processes, or

    (c) is moved to another area or work site.

(3) Every employer shall cooperate with any person exercising a duty imposed by this Act, the regulations and the OHS code.

(4) Every employer shall comply with this Act, the regulations and the OHS code.

1.4 Responsibility, Authority, and Resources

XYZ considers the most effective way of implementing its OH&S Policy, procedures and practices, is through the established company management structure. Senior Management commitment and leadership is vital for successfully meeting our OH&S objectives.

Every employee at XYZ is responsible and accountable for the health and safety of themselves, fellow employees, the public, and the surrounding environment.

The responsibilities and lines of authority within XYZ are as follows:

Company President

The Company President is responsible and accountable for ensuring that all activities under his/her control are carried out in accordance with the OH&S policy. This entails ensuring that:

- All the statutory OH&S requirements relevant to XYZ operations are identified and the XYZ OH&S Policy and practices take these fully into account.
- The XYZ OH&S Management System, Psychological Health & Safety (PH&S) Management System, Annual Management Review, Annual OH&S Action Plan, and OH&S Program Manual are prepared in accordance with company policy.
- Individual OH&S responsibilities and objectives are clearly established and monitored.
- The OH&S Management System is allocated sufficient resources to attain the high standards of performance that the XYZ policy demands.
- The general Occupational Health and Safety of the workforce and the environment are maintained.

Senior Management

Each Director or Senior Manager at XYZ is responsible and accountable for ensuring that the operations under their span of control are carried out in accordance with the policies and procedures laid down in XYZ's OH&S Program Manual.

This entails ensuring that:

- all company and statutory OH&S requirements of direct relevance to the activities under their control are identified, adhered to, and monitored;
- the XYZ's Annual OH&S Action Plan is developed and implemented;
- an annual OH&S Management System review, inclusive of a PH&S Management System review, is conducted and the required changes in the OH&S Management System or procedures are communicated to all personnel;
- roles and responsibilities for OH&S at Company XYZ are clearly established and performance against expectations, monitored;
- line management objectives are established and performance against these objectives monitored;
- adequate safe systems of work and procedures are documented and implemented for activities under their control;
- arrangements are in place for consulting and briefing the workforce on OH&S issues;
- employee OH&S training needs are assessed and provided;
- a system of communication is in place to provide the workforce with current and future OH&S information;
- OH&S hazards are identified, risk assessments conducted, and recommendations implemented;
- all incidents are investigated, and prevention strategies implemented;
- contractors adhere to XYZ's OH&S standards;
- adequate programs of OH&S inspection, review, and auditing are carried out;
- emergency response resources and procedures are in place and emergency response drills are regularly held; and
- a clear leadership example is set by the promotion of OH&S awareness.

Line Management

Line Managers have a key role to play in the day-to-day maintenance of a safe and healthy workplace and should:

- Be aware of the XYZ OH&S philosophy, policies, standards, and procedures.
- Set a clear leadership example by promoting OH&S awareness.
- Be responsible and accountable for the health and safety of themselves and

- others (including visitors and contractors) within their areas of responsibility.
- Be responsible and accountable for ensuring the facility and equipment are maintained and controlled under the OH&S Management System.
- Work with the XYZ Training department; Director, OH&S; and Human Resources to plan the OH&S training of all employees from new employee orientation to specific job training.
- Ensure that work under their control is:
    (a) adequately supervised;
    (b) conducted by trained and competent staff who are familiar with the hazards and risks associated with the work;
    (c) carried out in a safe manner, in accordance with approved procedures and instructions;
    (d) conducted with adequately maintained work equipment and process systems suitable for the task;
    (e) conducted by personnel wearing specified personal protective equipment for the task; and
    (f) conducted in accordance with applicable OH&S legislation.
- Assist in the preparation, implementation, and maintenance of the XYZ Safe Work Practices.
- Conduct periodic worksite inspections to identify unsafe working conditions and practices, report the findings, and recommend corrective actions.
- Monitor all maintenance schedules and emergency procedures according to XYZ OH&S Policy and procedures.
- Provide leadership during an emergency by following the Emergency Response Plan and carrying out responsibilities as designated.
- Promote OH&S awareness to develop positive attitudes at all levels using current incident prevention techniques. All safety incidents, environmental events, and the circumstances leading to them, should be investigated. Line Managers should report and record incidents and ensure that action is taken to eliminate or control the immediate and root causes of the incident.
- Encourage employee involvement and participation to improve OH&S protection by means of pre-job meetings, group meetings, personal contacts, and safety meetings.
- Monitor the use of personnel protective equipment and protective clothing.
- Maintain contact with the XYZ OH&S Department, the Environmental Departments, the Disability Management Coordinator, and the Employee Assistance Program (EAP) to promote OH&S compliance.

Director, OH&S Department

The Director, OH&S Department, acts as a resource on OH&S issues for employees and managers at XYZ. This position:

- Advises on compliance with, and interpretation of, the applicable statutory and OH&S legislation.
- Assists all levels of Management in the planning and promotion of the OH&S Management System and OH&S Program.
- Manages the XYZ OH&S Department.
- Ensures all responsible persons are aware of their legal obligation to notify the enforcing authority of incidents, and that incident reports are sent to the enforcing authority(ies) in a timely manner.
- Assists with the provision of OH&S services when required.
- Represents XYZ on OH&S issues on government/industry task forces and at public forums.
- Reports to the VP, Human Resources.

Safety Practitioners/Professionals

The Safety Practitioner/Professional plays a key role in advising and guiding the various XYZ business lines in terms of their OH&S responsibilities. He/she acts as an expert resource on OH&S issues for employees and managers within that business line. This position:

- Assists the assigned business line to plan and promote the OH&S Management System within its respective operation.
- Advises on compliance with, and interpretation of, the OH&S and other relevant legislation.
- Ensures that all responsible persons are aware of the legal requirements of notifying the enforcing authorities of workplace incidents.
- Assists the business line to conduct incident investigations.
- Ensures that the appropriate incident reports are sent to the enforcing authorities in a timely manner.
- Conducts diagnostic incident investigations as required.
- Reports to the Director, OH&S Department.

Joint Health and Safety Committee Members

The function of Joint Health & Safety Committee (JHSC) members does not diminish in any way the line responsibilities of managers, immediate supervisors, and all the other employees for promoting effective OH&S standards.

JHSC members have a duty to represent the views of the workforce in their area, on

CH. 5: OCCUPATIONAL HEALTH AND SAFETY PROGRAM: MANUAL DEVELOPMENT

OH&S matters and to assist in communicating the information from the JHSC Committee meetings.

The JHSC members will:

1. Be invited both formally through JHSC meetings, and on an informal basis, to discuss the implications of any proposed new modifications to plant, processes, equipment, and work routines.
2. Covariance. Have access to all statutory OH&S/PH&S[7] documentation and copies of company policies, procedures, and standards.
3. Be kept informed of the company's plans and programs and be given details of the company OH&S/PH&S performance.
4. Receive appropriate training to enable them to fulfil their roles effectively.
5. Be invited to form part of incident investigation teams depending on the nature of the incident and whether the interests of their constituents are affected.
6. Take part in worksite inspections and OH&S/PH&S audits with Management and employees.
7. Assist in preparation of procedures and advise on the implementation of company codes of practice and Safe Work Practices.
8. Assist in conducting periodic inspections to identify unsafe plant and working conditions and assist in reports containing findings and recommendations for correction of defects.
9. Assist in monitoring trials of personal protective equipment and protective clothing.
10. Encourage employees to report workplace incidents.

Employees

Employees are always responsible and accountable for the health and safety of themselves and others (including visitors and contractors), and for ensuring that their actions do not cause adverse OH&S effects.

Each XYZ employee has a responsibility to:

- be familiar with the XYZ OH&S policy and OH&S/PH&S Management System through the regular review and use of the OH&S Program Manual, *Employee OH&S Handbook*, and XYZ Safe Work Practices;
- follow the XYZ OH&S standards and Safe Work Practices;
- handle all hazardous materials and equipment in a safe and responsible manner;

---

[7] PH&S – Psychological Health & Safety.

- protect the health and safety of themselves and other employees, the environment, and the public;
- report unsafe acts and conditions as well as possible environmental problems;
- report all workplace incidents and environmental events; and
- seek help from their Line Manager if problems/concerns/issues arise.

Fire Wardens

Fire Wardens have been appointed for each department and trained to deal with fires and the evacuation of buildings.

First Aiders

First Aiders are specially trained and appointed to deal with illness and injuries sustained on the company's premises. They are available to deal with visitors and employees who become ill while at XYZ worksites. They are expected to render First Aid to visitors and employees within the scope of their First Aid training level.

2.0 STRUCTURE AND RESPONSIBILITY/ACCOUNTABILITY

2.1 Organizational Chart

(List the applicable legislation and industry standards)

2.2 XYZ Annual OH&S Action Plan

The purpose of the Annual OH&S Action Plan is to ensure the continuous improvement of XYZ's OH&S activities and ultimately the OH&S Management System. XYZ's goal is to establish a "best-in-class" OH&S Program (refer to Chapter 18, Appendix 1).

2.2.1 Timing

The OH&S Action Plan is produced annually and timed to be prepared for submission to the company leaders prior to the development of the annual business plan. The intent is to incorporate OH&S issues into XYZ's business strategies.

2.2.2 Development

The Director, OH&S, with guidance from Senior Management and Line Management develops the OH&S Action Plan.

The OH&S Action Plan is based on the company's previous OH&S Action Plan, XYZ's current business strategy, the results from the Management Review of the OH&S Management System, applicable changes in legislation and/or industry practices, and safety audit.

The OH&S Action Plan identifies individual responsibilities along with target dates for completion.

2.2.3 Approval

The OH&S Action Plan is approved by the Company President and Board of Directors.

2.2.4 Monitoring

Progress of the OH&S Action Plan is monitored by the Director, OH&S Department.

3.0 TRAINING, AWARENESS, AND COMPETENCE

OH&S training needs are to be identified and training provided to all XYZ personnel as required.

3.1 Identification of Training Requirements

The Director, OH&S, in conjunction with the Line Managers, the XYZ Training Centre and the Human Resources personnel identify the OH&S training requirements for XYZ employees and contractors.

3.2 Provision of Training

In most cases, the Line Manager arranges for employee training according to the XYZ OH&S Training Matrix (Table 4.1).

3.3 Refresher Training

Identification of refresher training needs is the joint responsibility of the Director, OH&S, the XYZ Training Centre, Human Resources personnel, and XYZ employees.

3.4 Employee Competency

The Management of the various XYZ business lines ensures that staff are adequately qualified, suitably trained, and sufficiently experienced to safely carry out their job tasks. This is achieved through the provision of the identified training and completion of vocational qualifications. Competency also considers relevant experience.

4.0 OH&S COMMUNICATION

To ensure the smooth running of the OH&S Management System, it is important that an effective OH&S communication strategy is developed and maintained (Section 2.3).

5.0 ANNUAL MANAGEMENT REVIEW

5.1 Introduction

The Annual Management Review involves an assessment of the overall effectiveness of XYZ's OH&S/PH&S Management System in achieving the stated objectives and policy requirements.

The Review is conducted annually and is dependent on the availability of Upper Management.

5.2 The Review Team

The Review Team consists of the following personnel:

- XYZ Company President
- Chief Financial Officer
- Senior Management
- Director, OH&S
- Line Management

5.3 Agenda

The following topics of the OH&S Management System are reviewed:
- Results of the previous Management Review
- OH&S Policy
- Status of the OH&S Management System
- Status of the PH&S Management System
- Annual OH&S Action Plan achievements
- Incident experience (present and past)
- Audit program results
- Legislative changes
- OH&S Integration with the XYZ business strategy.

5.4 Follow-up Action(s)

Minutes of the Annual Management Review are to be circulated to the Management team and an action plan implemented.

The Review is to be used to augment the XYZ Annual OH&S Action Plan.

Records of the Annual Management Review are to be retained by the Director, OH&S.

2.0 WORKER COMPETENCE, SELECTION AND TRAINING

2.1 Introduction

Developing and maintaining a good OH&S culture is dependent on the availability of adequate staffing levels and competent, well-trained employees.

A **competent worker** means a worker who is adequately qualified, suitably trained, and has enough relevant experience to safely perform work with only a minimal degree of supervision.

This procedure details the requirements for the selection of staff, ensuring that acceptable levels of competence are achieved, and that suitable training is provided to maintain this level of competence.

2.2 References

(List the applicable legislation and industry standards)

2.3 Selection

When recruiting and selecting new staff for high-risk activities, the following must be determined:

- level of expertise;
- competence;
- capabilities;
- attitude;
- physical capabilities; and
- awareness and understanding of OH&S regulations.

2.4 Competence

Competency is a method of ensuring that the workforce can carry out the expected work from a technical, quality, and OH&S perspective related to qualifications, training, and experience.

2.4.1 *Competency of Trainers*

To become competent trainers, Managers, Line Managers, the Director, OH&S, and others involved in training need to complete instructor training. This may include leadership training, "Train the Trainer" courses on specific OH&S topics (WHMIS 2015, confined space entry, hazard awareness and control, *etc.*), and/or on adult learning principles.

Peer review of training sessions is encouraged to provide support and constructive feedback on content, training methods used, and effectiveness of the presentation.

2.5 Training

1. Employees are to be trained as indicated in the XYZ OH&S Training Matrix (Table 4.1).
2. OH&S orientation and training are provided.
3. Training gaps are to be identified and the necessary training provided.
4. Refresher OH&S training is provided.
5. Appropriate training when new processes, products, or work equipment are introduced to the workplace.

2.6 Responsibilities

<u>Senior Management</u>

- Recruit and select staff possessing the appropriate skills for the job tasks.
- Ensure that the workforce is competent for job tasks they are assigned.
- Ensure adequate funding is available to allow the annual training plan to be completed.

- Identify training needs for inclusion in the XYZ Annual OH&S Action Plan.
- Review the OH&S Training Matrix (Table 4.1) annually with Line Managers.

Line Management
- Provide on-the-job training, as required.
- Conduct OH&S orientation and training related to the operation.
- Identify employee OH&S training needs and arrange for training opportunities.
- Identify OH&S training requirements for new processes and work equipment.
- Provide in-house OH&S awareness training courses.
- Include worker OH&S training needs in the XYZ Annual OH&S Action Plan.
- Review the XYZ OH&S Training Matrix (Table 4.1) annually.

Director, OH&S
- Produce the Annual OH&S Training Plan and ensure that the requirements are met.
- Assist in the organization of employee training as stipulated in the OH&S Training Matrix.
- Identify the OH&S training gaps and furnish reports to Management.
- Review the OH&S Training Matrix (Table 4.1) annually.

XYZ Training Department
- Assist in determining the OH&S training needs for XYZ employees.
- Co-ordinate the OH&S Training Program, including the planning, implementation, and evaluation of the content and delivery of the various training offerings.
- Make recommendations regarding future employee OH&S training needs and enhancements.
- Maintain adequate records of all the OH&S training provided.

CH. 5: OCCUPATIONAL HEALTH AND SAFETY PROGRAM: MANUAL DEVELOPMENT

Table 4.1: XYZ OH&S Training Requirements

SAFETY TRAINING TOPICS	COO/CFO/DIRECTOR/GM	MANAGERS	LINE MANAGE-MENT	OH&S SECTION	TECHNICAL & OFFICE EMPLOYEES	FIELD EMPLOYEES
EMPLOYEE/CONTRACTOR ORIENTATION	✓	✓	✓	✓	✓	✓
OH&S MANAGEMENT	✓	✓	✓	✓	✓	✓
RISK/HAZARD ASSESSMENT TRAINING	As Req'd	✓	✓	✓	As Req'd	As Req'd
ELECTRICAL SAFETY	✓	✓	✓	✓	As Req'd	✓
WHMIS 2015	✓	✓	✓	✓	✓	✓
TDG TRAINING	Awareness	Awareness	✓	✓		As Req'd
ERGONOMIC AWARENESS	✓		✓	✓	✓	✓
CONTRACTOR MANAGEMENT	Awareness	✓	✓	✓	As Req'd	As Req'd
FIRST AID/CPR/AED	Awareness	✓	✓	✓	✓	✓
WORKING ALONE	Awareness	As Req'd	As Req'd	As Req'd	As Req'd	As Req'd
NOISE AWARENESS & HEARING CONSERVATION	Awareness	Awareness	✓	✓	As Req'd	As Req'd
OFFICE SAFETY	✓	✓	✓	✓	✓	As Req'd
MANUAL HANDLING AWARENESS		✓	✓	✓	As Req'd	As Req'd
CONFINED SPACE ENTRY		Awareness	As Req'd	✓	As Req'd	As Req'd
FIRE WARDEN TRAINING		✓	As Req'd	✓	As Req'd	As Req'd
INCIDENT REPORTING & INVESTIGATION	Awareness		✓	✓	As Req'd	As Req'd
RESPIRATORY CONSERVATION & FITNESS TESTING			As Req'd	✓	As Req'd	As Req'd
RESPIRATORY PROTECTION EQUIPMENT TRAINING			As Req'd	✓	As Req'd	As Req'd
SCBA TRAINING			As Req'd	Awareness	As Req'd	As Req'd

SAFETY TRAINING TOPICS	COO/ CFO/ DIRECTOR/ GM	MANAGERS	LINE MANAGE-MENT	OH&S SECTION	TECHNICAL & OFFICE EMPLOYEES	FIELD EMPLOYEES
EMERGENCY RESCUE TRAINING			As Req'd	√	As Req'd	As Req'd
PSYCHOLOGICAL HEALTH & SAFETY (PH&S) IN THE WORKPLACE	√	√	√	√	√	√

Definitions:

As Required (As Req'd) means that if the employee has any contact with this relevant section, he/she must be trained on all aspects pertaining to that section.

Awareness means that all employees who have contact with this section should be aware of all risks associated with that section.

3.0 OH&S COMMUNICATION AND FEEDBACK

3.1 Introduction

Good communication is essential if the company is to have a well-motivated workforce and hence, an effective OH&S Management System. This section covers communication, suggestions, and complaint procedures.

3.2 Company OH&S Program Manual

All XYZ staff will have access to an electronic copy of the OH&S Program Manual. Employees are expected to read the OH&S Program Manual in its entirety at orientation, and then, annually. When revisions are distributed, or when processes change, employees are to review the relevant OH&S Program Manual sections.

3.3 XYZ OH&S Communication Strategy

3.3.1 Internal Communication

OH&S Communication within the organization is achieved via:

1. **XYZ Intranet System:** The OH&S Website contains regular OH&S information and regular updates, the current OH&S Action Plan, applicable legislation, the XYZ OH&S Program Manual, *Employee OH&S Handbook* and the XYZ Office Safety Program.
2. **Bulletin Boards:** Relevant OH&S information, articles, bulletins, and meeting minutes are posted on company bulletin boards, both within the various worksites and on the XYZ Intranet System.
3. **Safety Meetings:** Regularly scheduled safety meetings are held with the business units to identify and address OH&S issues.
4. **Management Reports:** Senior Management is regularly notified of the ongoing OH&S activities and of any critical OH&S issues/risks.
5. **Employee OH&S Handbook:** This document exists in both soft and hard copy versions. Employees and Management, who are without ready access to the XYZ Intranet System, or who require a hard copy version for a business reason, are to be provided with a hard copy version of the *Employee OH&S Handbook*.
6. **OH&S Suggestions:** Employees are encouraged to submit OH&S ideas via this manner of communication.
7. **OH&S Communications:** The OH&S Department produces and distributes OH&S information.

3.3.2 External Communication

Communications received from interested parties relevant to the XYZ OH&S Management System are documented.

If a response is required, the Director, OH&S will reply and/or forward the communication to the personnel qualified to respond. All responses will be copied and appended to the original correspondence and retained within the XYZ OH&S Department.

3.4 Regulatory Information

All company bulletin boards, namely the XYZ Intranet System or some other prominently positioned announcement board, should display the following information:

1. First Aid information.
2. Action in the event of a fire, including evacuation plans.
3. Action in the event of a chemical spill.
4. Action in the event of a major incident.
5. Names and contact numbers of the OH&S personnel.
6. Names and contact numbers of Fire Wardens.
7. Workers' Compensation Board information on injury/illness claim reporting.
8. XYZ Disability Management Coordinator name and contact number.
9. Employee Assistance Program (EAP) contact name and number.
10. OH Service Provider contact name and number.

3.5 Non-regulatory Information

The company bulletin boards, the XYZ Intranet System, or some other communication means, should also prominently display:

1. The XYZ OH&S policy.
2. The XYZ PH&S policy.
3. The XYZ Annual OH&S Action Plan.
4. The minutes of the relevant safety meetings.
5. Any relevant incident statistics.
6. Incident reports.
7. Building plans showing the location of fire equipment and emergency evacuation routes.
8. Any other appropriate OH&S information or posters.

All personnel are required to remain current on this OH&S information.

3.6 OH&S-Related Alerts/Bulletins/Memos

OH&S alerts/bulletins/memos may come from the OH&S Department. When one is initiated, the Director, OH&S determines if it is appropriate to the various XYZ business units. If it is, then it will be sent to the appropriate Directors and Managers for distribution.

3.7 Occupational Health and Safety Literature

The OH&S Department maintains OH&S informational resources available for use.

3.8 OH&S Suggestions

XYZ operates in a manner that encourages OH&S suggestions from employees. Suggestions forwarded directly to the OH&S Department, or to the Line and Senior Management are used to make contributions towards the improvement of XYZ's OH&S standards and practices.

Questions from employees or Management are encouraged through the XYZ Intranet System: *Safety/Questions and Answers*.

3.9 Issues and Concerns

All OH&S related feedback is recorded and followed in a manner that builds trust and demonstrates commitment to our OH&S Management System.

Issues or concerns about OH&S issues are to be addressed by Line Management, or through the completion of the Incident & Investigation Report Form. If the matter is not resolved in a satisfactory manner, then the Operational Director is to be informed.

Employees are strongly encouraged to report and pursue OH&S concerns.

If the Manager/Director is unable to resolve a problem or address a concern, he/she may contact the Director, OH&S or designate for assistance.

If an issue or concern is related to an activity/work that is determined to have a "high risk" for injury, then the activity/work must stop until the problem is resolved.

Non-urgent concerns or questions should be raised at the respective Departmental Safety Meeting.

3.10 Meetings

Various meetings are held throughout the organization via:

(a) ***Safety Meetings:*** OH&S concerns can be raised and addressed at these departmental meetings.

(b) ***Regular Management Meetings:*** OH&S is a regular agenda item where information, safety concerns, incidents, and corrective action status may be reviewed.

(c) ***Annual Management Review Meeting:*** Management meets annually to review OH&S matters. The intent is to monitor the progress of the OH&S Management System. At this time, relevant communications, suggestions, and complaints may be addressed.

4.0 RISK ASSESSMENT AND CONTROL

(Available in Chapter 9, "Occupational Health and Safety: Hazard Identification, Assessment, and Control")

5.0 PERFORMANCE MEASUREMENT AND CORRECTIVE ACTIONS

(*not provided*) This topic is addressed in Chapter 2, "Occupational Health and Safety Management Systems", and partially addressed in Chapter 42, "Occupational Health and Safety: Future Challenges", HS^{3TM} and program sustainability.

6.0 AUDIT PROCEDURES

(*not provided*) An audit tool is provided for evaluating the organization's Psychological Health & Safety Management System, Chapter 25, "Psychological Health and Safety in the Workplace: Measurement", Appendix 3.

SECTION 3: RELATED POLICIES AND PROCEDURES

1.0 EMPLOYEE ASSISTANCE PROGRAM (EAP)

1.1 Introduction

XYZ is mindful of the fact that everyone experiences personal difficulties from time-to-time, and that these situations can be emotionally, physically, and psychologically disruptive to an otherwise well-balanced and fulfilling life. In recognition of these circumstances, and as a means of minimizing any potential adverse effect on the employee job performance, XYZ has established an Employee Assistance Program (EAP).

1.2 Policy

Permanent employees, and/or their spouses and families, may contact XYZ's EAP either as "self-referrals" or as "assisted referrals" for short-term professional counselling to resolve personal issues or treatment/rehabilitation for substance abuse. **All referrals are voluntary and confidential.**

1.3 Purpose of the EAP

The EAP is designed to assist, at the earliest opportunity, any employee of XYZ who has a personal problem that may adversely affect job performance. An employee suffering from personal problems will be given similar consideration, assistance and medical coverage as that provided to employees suffering from a physical illness/injury.

For the purpose of this program, personal problems are defined as *physical illness, emotional and stress-related problems, family problems, alcohol or drug abuse, or similar related matters which may adversely affect the employee's job performance and/or health.*

XYZ recognizes that alcoholism and drug dependency/addiction are health problems that are as treatable as are many other psychological and emotional problems. XYZ is concerned with the employee's use of alcohol or drugs in so far as it affects the employee's job performance, health, and job safety.

The goal of the EAP is to help employees resolve concerns that may affect their well-being and/or job performance. The program combines access to in-house counselling, referral to community health care professionals and the expertise of professional external counsellors, which the program offers to employees on a confidential basis. Counselling is available for:

- stress/lifestyle issues
- emotional issues
- physical or sexual abuse
- substance abuse/dependency
- family/marital concerns
- relationship issues
- harassment

- trauma/crisis
- bereavement
- career concerns
- personal legal issues
- financial concerns
- health concerns
- eating disorders

- parenting issues
- elder care issues
- coping with chronic illness
- sleeping disorders
- substance abuse treatment rehabilitation

The intent of the EAP is to offer a helping hand before a concern reaches a serious or crisis situation. It must be recognized, however, that XYZ only offers short-term assistance. Successful resolution of a problem is the responsibility of the employee and requires employee motivation and active participation. As well, this is a brief-counselling model: long-term counselling when warranted, can be arranged through other community sources.

1.4 Funding

The program is 100 per cent company-funded for counselling through the arranged EAP service provider. A client-coding invoice system is in place to guarantee the confidential handling of the costs for this program.

1.5 When to Access the EAP

Employees may experience work-related difficulties or personal problems in their day-to-day activities. In most cases, these concerns or problems can be resolved. There are times, however, when a personal issue becomes too difficult to manage and is detrimental to families, relationships, and personal health. Without assistance in reaching a solution, difficult issues can often get progressively worse. Professional assistance can identify options and strategies to help to resolve issues and thus prevent a life crisis.

Workplace-based counselling services are effective in providing easy access to professional counselling. It is hoped that employees will have easier access to these services and will seek help earlier than if they had to find help on their own.

An employee can access the EAP in one of four ways:

1. *Self-Referral:* Employees or spouses may contact, without prior approval, their EAP Service Provider at (XXX) XXX-XXXX (local office) or 1-800-XXX-XXXX.

2. **Disability Management Coordinator:** The XYZ Disability Management Coordinator may, as part of the Disability Management Program, recommend EAP assistance to an employee.
3. **OH&S Department:** XYZ's OH&S Department may, as part of the Emergency Response Plan and Incident Investigation process, recommend EAP assistance to an employee, or group of employees.
4. **Director/Line Manager/Union Referral:** Director/Line Managers/Union may encourage employees to use the EAP when a personal problem is believed to be related to poor job performance. This type of referral is part of the Director/Line Manager's coaching role in the performance management process. The Director/Line Manager/Union would be informed only of the employee's effort to access the EAP and general progress. All details and discussions between the employee and the external counsellor are strictly confidential and will not be divulged to the Director/Line Manager.

When the employee contacts the EAP, a qualified professional determines the nature of the concern and arranges to have a counsellor contact the employee within 24 hours to set up an appointment. In an emergency, the call will be immediately handled.

Time off work for employees seeking assistance through the EAP is considered the same as for any health-related issue. Inform your Line Manager in advance of your appointment and, to maintain confidentiality, merely disclose that you are taking time from work for a medical appointment.

1.6 Confidentiality

Confidentiality is critical to the success of an EAP. XYZ's EAP is managed by a firm specializing in Employee Assistance Programs. All client information is maintained off-site and only non-identifying, statistical information is provided to XYZ's Director, OH&S.

Services provided through the EAP are invoiced by the EAP Service Provider to XYZ without disclosing client names. Under no circumstances are related records kept on personnel files. Personal information will only be released upon the employee's written approval. Accessing the EAP has no impact on one's present job or future with XYZ. Those decisions are based solely on employee performance and capabilities.

1.7 Responsibilities

Line Managers
- Be aware of the EAP services available to XYZ employees.
- Objectively observe employee behaviour.
- Inform employees that if personal problems are in some way responsible for work performance problems, EAP assistance is available.

CH. 5: OCCUPATIONAL HEALTH AND SAFETY PROGRAM: MANUAL DEVELOPMENT

- Document work performance issues clearly and focus strictly on substandard performance or observable behaviours.
- Respect the employee's right to privacy and confidentiality when discussing substandard work performance.
- Be ready to cope with the employee's emotional reaction, resistance, defensiveness, and/or hostility.
- Indicate concern if the employee begins to talk about personal problems by saying that although concerned, knowledge of personal details is not expected.

Union
- Be aware of the EAP services available to union members.
- Assist union members to attain EAP support if required or sought.

Director, OH&S
- Oversee the administration of the XYZ EAP.
- Set performance expectations for the EAP Service Provider.
- Evaluate the performance of the EAP Service Provider.
- Assess the value of the EAP to XYZ and its employees. Design and implement enhancements as required.
- Link the EAP to other XYZ programs such as the OH&S, Workplace Wellness and Disability Management Program.

Disability Management Coordinator
- Provide confidential assistance, guidance to employees/Line Management by way of referral services and case coordination.
- Assist the Director, OH&S as required.

OH&S Department
- Be aware of the EAP services available.
- Assist employee to attain EAP support if required or sought.

2.0 SUBSTANCE ABUSE POLICY

2.1 Introduction

XYZ recognizes that the misuse of alcohol and drugs, as well as the side-effects of prescription and over-the-counter drugs, can have a detrimental effect on employee health and welfare, and can lead to addiction. This, in turn, can have an adverse effect on the employee's personal health and safety, and the health and safety of others, as well as on work quality and productivity and customer service.

2.2 Policy

Alcohol/drugs are not permitted for any purpose in any XYZ offices, buildings, worksites, or vehicles. Alcohol at XYZ-sponsored events will be served at licensed premises only. Post-function transportation will be arranged for employees and paid for by XYZ.

Employees taking prescription or over-the-counter (OTC) medications, or who are addicted to a substance, must notify the XYZ Occupational Health (OH) Service Provider if the substance does or might impact their work performance.

XYZ employees who are under the influence of alcohol/drugs (impaired) while at work may be subject to disciplinary action. It is illegal to enter or stay in workplace when impaired.

2.3 Definitions

Workplace includes any company or client premise, yard, or parking area. This policy also applies when the employee is "on call" or working off company premises.

Alcohol means any alcoholic beverage.

Substance misuse in this policy refers to the **use** of illegal/legal drugs; and the misuse, whether deliberate or unintentional, of prescribed drugs and substances such as cannabis, solvents, or gases.

Prohibited substances for this policy, refers to the following definition of prohibited substances:

- any substance (See Note (1) below) which an individual may not possess, sell or use under applicable Canadian legislation;
- any legal, but unlawfully used substance, *e.g.*, prescription drugs obtained or used without medical supervision; and
- any solvent or gas used, or intended to be used, for the purposes of intoxication.

Note (1): *For the sake of clarity most, but not all, prohibited substances fall into the following categories:*

- *Stimulants (including, but not limited to, amphetamines and cocaine).*
- *Depressants (including, but not limited to, narcotics and barbiturates).*
- *Hallucinogens (including, but not limited to, cannabis, hashish, and LSD).*
- *Deliriants (including, but not limited to, glue, solvents, and aerosol sprays).*

2.4 Mandatory Requirements

2.4.1 The following actions are <u>strictly prohibited:</u>

- Possessing or consuming alcohol and/or a prohibited substance while in the workplace.
- Reporting to work, working, or attempting to work impaired.

CH. 5: OCCUPATIONAL HEALTH AND SAFETY PROGRAM: MANUAL DEVELOPMENT

- Reporting to work, working, or attempting to work while under the influence of *any drug/medication* that affects the person's ability to safely perform his/her job or affect the safety and well-being of others.

2.4.2 The following are <u>strictly required</u>:

2.4.2.1 Prescription Drug Use

Any employee/contractor who is using prescription, or over-the-counter drugs, which may impair the ability to perform the job safely, or which may affect the safety or well-being of others, is to contact the XYZ OH&S Department. If it is determined that drug use is, or may negatively impact the employee's ability to perform, the employee/contractor will be referred to the XYZ Occupational Health (OH) Service Provider for a fitness-to-work assessment.

2.4.2.2 Medical Marijuana (Cannabis) Use

Employees who are prescribed medical marijuana must notify their employer. The use of medical marijuana (cannabis) is strictly prohibited in the workplace. Failure to comply with this requirement may result in dismissal.

2.4.2.3 Notice of Conviction

Any employee/contractor who is convicted of a drug/alcohol offence must inform the company (Director of OH&S) within five days of the conviction. Failure to comply with this requirement may result in dismissal.

2.5 Policy Administration

XYZ acknowledges that alcohol or drug dependency is a disease state. Employees who feel they may have developed a dependency on alcohol/substance, or an addiction to either, are encouraged to voluntarily seek help to deal with their problem. It is important that such problems be identified so that help can be offered.

To maintain the company policy, alcohol and/or drug testing may be carried out when required by a job contract; or after a serious incident; or when there is "reasonable cause". If warranted, testing will be conducted by a certified laboratory and consist of urinalysis or blood analysis.

The company will respect the rights of its employees to privacy and confidentiality and ensure that strict standards are maintained in the implementation and subsequent administration of this policy.

2.6 Recognition of the Problem

Alcohol/drug abuse by employees can have an adverse impact on XYZ and has the potential to jeopardize the safety of the employee and co-workers.

The following characteristics, especially when arising in combinations, may indicate the presence of an alcohol or drug-related problem. Some of the signs associated with alcohol/drug abuse may be caused by other factors such as stress and should be regarded only as indications that an employee may be misusing alcohol or drugs.

2.6.1 Presenteeism
- at work, but not performing;
- distracted with personal issues, illness, or injury;
- negativity impacting group processes.

2.6.2 Absenteeism
- instances of unauthorized leave;
- frequent Friday and/or Monday absences;
- leaving work early;
- lateness (especially on returning from lunch);
- excessive level of sickness absence;
- strange and increasingly suspicious reasons for absence; and/or
- unusually high level of sickness for colds, flu, and stomach upsets.

2.6.3 High Incident Level
- at work; and/or
- elsewhere, *e.g.*, driving or at home.

2.6.4 Work Performance
- difficulty in concentrating;
- work requires increased effort;
- individual tasks take more time; and/or
- problems with remembering instructions.

2.6.5 Mood Swings
- irritability;
- depression; and/or
- general confusion.

2.7 Intervention

Line Managers who feel an employee's/contractor's unsatisfactory performance may be alcohol or drug-related should discuss the matter with their Senior Manager. If necessary, the Senior Manager, in consultation with Human Resources, will arrange a meeting with the employee/contractor to discuss the unsatisfactory performance. The required standards of performance will be stated and employee understanding of them confirmed. Efforts will be made to establish why job performance is unsatisfactory.

The employee should be informed of the assistance XYZ is prepared to give those who have personal problems. Referral to the XYZ OH Service Provider, and/or EAP, must be strongly encouraged to the employee. The employee should also be informed of community agencies where help can be obtained, such as Alcoholics Anonymous and local treatment centres.

2.8 Voluntary Disclosure and Treatment

At any time prior to being selected for an alcohol or drug test, an employee may request assistance through the company EAP without jeopardizing their continued employment. The caveat is that they undertake and complete an appropriate recovery program. This disclosure will be treated in confidence.

Every effort will be made to ensure that, upon completion of the recovery program and testing negative on random alcohol/drug screens for up to one-year post-completion of a rehabilitation program, employees are able to return to the same or equivalent work.

However, where such a return-to-work would jeopardize either a satisfactory level of job performance, safety of the employee or co-workers, or the employee's recovery, the Manager/Line Manager with XYZ's Director, OH&S; OH Service Provider; and EAP support; will review the circumstances surrounding the case and agree on a suitable course of action.

2.9 Relapse

Where an employee, having undergone treatment, tests positive on a random alcohol/drug screen, or suffers a relapse, the company will consider the case on its individual merits. Medical advice will be sought to ascertain how much more treatment/rehabilitation time is likely to be required for a full recovery. At XYZ's discretion, further treatment or rehabilitation time may be given to help the employee fully recover.

However, if recovery seems unlikely after the additional treatment, dismissal may result.

2.10 Roles and Responsibilities

Management

- Set a good example for their staff and others.
- Be familiar with the relevant policy and the **Under the Influence Procedure** (Figure 4.2).
- Ensure, through the provision of training seminars, that staff understand the policy and their individual responsibilities.
- Be alert to, and monitor changes in, employee work performance, attendance, sickness, and incident patterns.
- Seek guidance and support from Human Resources or OH&S personnel.
- Take an objective and non-judgmental approach when counselling or interviewing employees.

- Remain current on the EAP services available and how to access them.
- Refer employees for EAP support when appropriate.
- Identify any aspects of the working environment that could lead to alcohol or drug misuse problems and, if possible, change them.
- Intervene early when there are signs of problems.
- Intervene immediately when the safety of the employee or co-workers may be threatened due to an employee showing signs of impairment.
- Act upon alcohol/drug abuse as indicated in this policy.

Figure 4.2: Under the Influence

UNDER THE INFLUENCE PROCEDURE

Should a manager or line manager either suspect or observe that an employee/contractor/visitor is under the influence of alcohol or drugs (impaired), then the following procedure is to be adopted:

- Since it is useful to have a witness to events, where possible call for assistance (*e.g.*, another line manager).
- Ask the person to accompany you to an office or quiet location.
- State clearly to the person the reasons you have for suspecting they are **unfit for work (impaired)** (*e.g.*, smell of alcohol/drug odor, unsteadiness, unclear speech, uncharacteristic behaviour, personal appearance, *etc.*).

No attempt at diagnosis should be made, nor should the individual be accused of being intoxicated. Describe only the actions or behaviour observed.

- Ask the individual to explain his/her behaviour or actions (there may be other possible explanations for their behaviour or actions, *e.g.*, the person may be unwell).
- If it is believed that the individual is ill, or in need of medical attention, then seek the advice of an occupational health professional.
- Decide, given the evidence, whether to suspend the individual from duties. If there is substantive evidence, *e.g.*, strong smell of alcohol/drug odor, then testing is not necessary. However, if there is no satisfactory explanation for the behaviour or if drug use is suspected, then the employee should be requested to undergo an OH Services assessment.
- The person must be accompanied and driven in a company vehicle or a taxi to the OH Service Provider's office. Do not allow the employee to drive.
- If the decision is taken to suspend the individual, send the person home by the most convenient method, with a letter requesting them to attend a disciplinary meeting.
- If the person becomes belligerent and refuses to leave, or insists upon driving, do not attempt to restrain physically; rather, decide if it is appropriate to inform the police regarding a possible drinking and driving, or driving under the influence of drugs offence.

CH. 5: OCCUPATIONAL HEALTH AND SAFETY PROGRAM: MANUAL DEVELOPMENT

UNDER THE INFLUENCE PROCEDURE

- As soon as possible, complete a report documenting all observations, responses from person, and the actions taken.
- Carry out further investigations prior to the person returning for disciplinary hearing (*e.g.*, sickness/absence record).
- Conduct the disciplinary meeting.

Note: There is no facility to carry out a test during the night shift or on the weekend. On these occasions, if a test is considered necessary, then it should be conducted the morning following the night shift, or as early as possible on a Monday morning.

Employee

- Find out about alcohol and drugs, and the related social, health, and employment effects.
- Avoid covering up or colluding with colleagues.
- Urge colleagues to seek help if they have problems larising from alcohol or drug misuse.
- Seek help where they themselves have problems from alcohol or drug misuse.
- Be familiar with the relevant policy and procedures.
- Use alcohol responsibly and avoid illegal drug or risky prescription medication use.

Director, OH&S

- Support Line Management as required, with the management of "troubled" employees.
- Support employees seeking assistance with substance abuse.
- Assist with the management of alcohol or drug testing results.
- Keep this policy current.

OH Service Provider

- Arrange alcohol/drug tests when appropriate.
- Obtain consent from the employee to release the alcohol/drug test results to XYZ's Director, OH&S.
- Undertake appropriate drug testing procedures.
- Receive the test results, disseminate them, and ensure the confidential handling of the results.
- Release test results to the Director, OH&S directly by telephone or confidential fax and documented in the employee's medical file.

3.0 NON-SMOKING WORKPLACE POLICY

3.1 Introduction

XYZ promotes a smoke-free working environment; and offers a comprehensive program that includes education, drug treatment to reduce physical withdrawal and cravings, and provides counselling to deal with the behavioural and psychological triggers.

3.2 Policy

XYZ maintains a safe and healthy work environment by providing a smoke-free work environment for our employees. Out of concern for the health and comfort of all employees, XYZ provides a smoke-free workplace.

Smoking of any kind (*e.g.*, cigarettes, cigars, pipes, or electronic cigarettes) is not permitted anywhere, or at any time, on company premises, or in company vehicles. This policy applies to all persons while on company premises including visitors.

Failure to comply with this policy will result in disciplinary action.

3.3 Environment

All employees must maintain an acceptable level of productivity during working hours whether they are non-smokers or smokers. Immediate supervisors and managers are responsible for administering this policy and are expected to ensure that an equitable working environment is maintained. To this end, specific guidelines will not be implemented under this policy for "smoke breaks".

Management must insist on strict adherence to a smoke-free work environment. Complaints regarding violations of this policy should be directed to the immediate supervisor or manager. Policy enforcement will be dealt with under the disciplinary process.

During the recruiting process, candidates will be advised of this policy. New employees will be required to agree to compliance with this policy as a condition of employment.

3.4 Smoking Cessation Assistance

Financial assistance for smoking cessation will be provided under the following guidelines:

1. Reimbursement for the cost of enrollment in a recognized smoking cessation program to a maximum of $600 on a one-time only basis for:
    (a) Employee.
    (b) Smoking employee's dependant (optional).
2. Reimbursement is subject to providing proof of payment and completion of a cessation program and will be handled by submitting an expense statement with necessary receipts to the Human Resources Department.

4.0 Workplace Hazardous Materials Information System[8]

4.1 Introduction

The Workplace Hazardous Materials Information System (**WHMIS 2015**) is Canadian legislation that addresses workers' "right-to-know" about health and safety hazards of hazardous products used in the workplace. It provides Canadian employees, employers and suppliers with information on the handling, use, storage, and disposal of hazardous products used in the workplace that could be dangerous to the safety of people and the environment. Some examples of hazardous products used at XYZ are acids, caustics used for cleaning and compressed gases.

WHMIS 2015 ensures that information is readily available to employees handling controlled products in three ways:

- Any hazardous product present in the workplace must have a supplier label attached to it.
- The manufacturer/supplier of a hazardous product must provide a Safety Data Sheet (SDS) which explains the precautions necessary for the safe handling, storage, and disposal of the hazardous product.
- Employers are responsible for ensuring WHMIS 2015 is used and understood by employees; that all hazardous products are labelled; that SDSs are available; and that all employees receive WHMIS 2015 education and site-specific training.

4.2 Objectives

- To reduce illness and injury to all employees resulting from hazardous products used in the workplace.
- To ensure that employers and employees know how to identify, label, use, store, and handle hazardous products.
- To comply with Canadian legislation.

4.3 Key Elements

Labelling — designed to alert workers to the identity and dangers of products, and to the basic safety precautions.

Safety Data Sheets (SDS) — technical bulletins that are prepared by the supplier and which provide detailed hazard and precautionary information.

Worker education and training programs — instruction for all employees in both general information about WHMIS 2015; and in specific training for safe handling, use, and storage of hazardous products.

---

[8] Canada introduced amendments to the WHMIS legislation in 2015-2016 to align with the Globally Harmonized System (GHS) for identifying and communicating workplace hazards.

4.4 Hazardous Products

The materials that are regulated by WHMIS 2015 are called **controlled products**. They are classified as:

- *Physical hazards* that relate to the product's physical or chemical properties
- *Health hazards* that relate to the product's negative health effects.

The above hazard groups are further broken down into hazard classes that group similar products. For information refer to: http://www.worksafebcmedia.com/media/WebBooks/whmis2015workers/?_ga=2.248989959.1082315295.1557280769-600751087.1557280769#/chapter/03/page/1/3.

Pictograms are a visual representation of the product's hazard type. The ten WHMIS hazard symbols are:

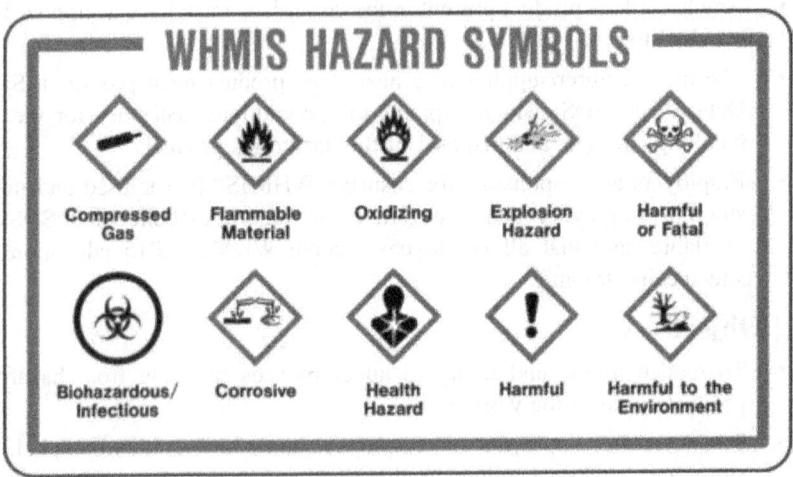

WHMIS Pictograms (WHMIS 2015), https://www.ccohs.ca/oshanswers/chemicals/whmis_ghs/pictograms.html, *OSH Answers*, Canadian Centre for Occupational Health and Safety (CCOHS), October 9, 2019. Reproduced with the permission of CCOHS, 2019.

CH. 5: OCCUPATIONAL HEALTH AND SAFETY PROGRAM: MANUAL DEVELOPMENT

The WHMIS 2015 symbols are defined as:

Symbol	Name	Symbol	Name	Symbol	Name
	Exploding bomb (for explosion or reactivity hazards)		Flame (for fire hazards)		Flame over circle (for oxidizing hazards)
	Gas cylinder (for gases under pressure)		Corrosion (for corrosive damage to metals, as well as skin, eyes)		Skull and Crossbones (can cause death or toxicity with short exposure to small amounts)
	Health hazard (may cause or suspected of causing serious health effects)		Exclamation mark (may cause less serious health effects or damage the ozone layer*)		Environment* (may cause damage to the aquatic environment)
	Biohazardous Infectious Materials (for organisms or toxins that can cause diseases in people or animals)				

* The GHS system also defines an Environmental hazards group. This group (and its classes) was not adopted in WHMIS 2015. However, you may see the environmental classes listed on labels and Safety Data Sheets (SDSs). Including information about environmental hazards is allowed by WHMIS 2015.

WHMIS Pictograms (WHMIS 2015), https://www.ccohs.ca/oshanswers/chemicals/whmis_ghs/pictograms.html, *OSH Answers*, Canadian Centre for Occupational Health and Safety (CCOHS), October 9, 2019. Reproduced with the permission of CCOHS, 2019.

4.4.1 Introducing Hazardous Products into XYZ

All hazardous products must be reviewed before being introduced into XYZ. The intent is to make certain that the product is safe to use and that the appropriate SDS exists onsite. The required procedure is explained in the XYZ *Hazardous Product Acceptance Process* and involves obtaining the required SDS and completion of the *XYZ HazardousProduct Acceptance/Rejection Request* form.

4.5   Labelling

As hazardous products are introduced into the marketplace, various types of labels, or other identifiers are necessary to alert workers to the hazards and safe handling procedures.

4.5.1   Manufacturer/Supplier's Role

When a manufacturer/supplier produces, or imports, a product for distribution and sale in Canada, that supplier must prepare a supplier label in English and French. A supplier label must be attached or printed on any hazardous product or its container. The supplier labels must be replaced with workplace labels if they have been accidentally removed, or become unreadable, or the product is decanted. Suppliers must also prepare *Transportation of Dangerous Goods* labels when the product is being shipped.

Although the format of the WHMIS 2015 supplier label is flexible, it must contain the following information:

- product identification;
- initial supplier identifier;
- pictogram(s);
- signal word - *Danger* or *Warning*;
- hazard statement(s) — specific to the hazard class and category; describe the most important hazard(s) of the product;
- precautionary measures; and
- supplemental label information — addresses precautions, physical state, route of exposure, and any hazards not already noted.

A sample Supplier Label is:

Product K1 / Produit K1

Danger
Fatal if swallowed.
Causes skin irritation.

Precautions:
Wear protective gloves.
Wash hands thoroughly after handling.
Do not eat, drink or smoke when using this product.

Store locked up.
Dispose of contents/containers in accordance with local regulations.

IF ON SKIN: Wash with plenty of water.
If skin irritation occurs: Get medical advice or attention.
Take off contaminated clothing and wash it before reuse.
IF SWALLOWED: Immediately call a POISON CENTRE or doctor.
Rinse mouth.

Danger
Mortel en cas d'ingestion.
Provoque une irritation cutanée.

Conseils :
Porter des gants de protection.
Se laver les mains soigneusement après manipulation.
Ne pas manger, boire ou fumer en manipulant ce produit.

Garder sous clef.
Éliminer le contenu/récipient conformément aux règlements locaux en vigueur.

EN CAS DE CONTACT AVEC LA PEAU : Laver abondamment à l'eau.
En cas d'irritation cutanée : Demander un avis médical/consulter un médecin.
Enlever les vêtements contaminés et les laver avant réutilisation.
EN CAS D'INGESTION : Appeler immédiatement un CENTRE ANTIPOISON ou un médecin.
Rincer la bouche.

Compagnie XYZ, 123 rue Machin St, Mytown, ON, N0N 0N0 (123) 456-7890

WHMIS Supplier Label (WHMIS 2015), https://www.ccohs.ca/oshanswers/chemicals/whmis_ghs/labels.html, OSH Answers, Canadian Centre for Occupational Health and Safety (CCOHS), October 9, 2019. Reproduced with the permission of CCOHS, 2019.

4.5.2 The Employer's Role

Supplier Labels

The employer is responsible for checking that supplier labels have been provided and applied to all hazardous products received in the workplace. Hazardous products with missing, altered, or illegible labels must not be handled or used, except in some provinces (*e.g.*, Alberta), where they can be held in storage while actively seeking an appropriate label.

Workplace Labels

Workplace labels are required when decanting or replacing damaged supplier labels on hazardous products:

- produced in the workplace for use there;
- transferred from the supplier's container to another container where it will not be used right away; and/or
- transferred by bulk shipment with a supplier label only on the outside package.

Workplace labels:

1. have identifiers (product name that matches the SDS product name);
2. have Safe Handling Information — hazard information, pictograms, precautionary measures, First Aid measures; and
3. should refer to the SDS.

Workers must be instructed about the information contained on labels and identifiers. The employer must take steps to ensure labels are not defaced and are always easy to read.

Identifiers

There are a few circumstances where hazardous products are required only to be clearly identified. A mark, signs, stamp, sticker, tag, wrapper, or colour codes/labels may be used. These are used when a hazardous product is a:

- bulk shipment;
- pipe, piping system, process or reaction vessel, conveyer belt, or storage vessel; and/or
- container holding small amounts where a label would compromise product recognition.

4.6 Safety Data Sheets (SDS)

The SDS provides Canadian workers with the information they need to work safely with hazardous products. They are technical bulletins that provide detailed hazard, precautionary, and emergency information on a product, in English and French.

All SDS must be current — no more than three years old. Where information is "Not Available" or "Not Applicable" under a heading, this must be indicated. Blank space under a heading is not acceptable.

At XYZ, employees can access the required MSDSs through the company's Intranet System.

The SDS should be read and understood prior to XYZ employees working with hazardous products.

4.6.1 Content

All SDSs must provide:

- product identifier and information;
- hazardous classification;
- composition;
- first Aid measures;
- fighting measures;
- accidental release measures;
- handling and storage;
- toxicological properties (health effects);
- exposure controls/PPE;
- physical and chemical properties;
- stability and reactivity data;
- toxicology data;
- ecological data;
- disposal consideration;
- transportation information;
- regulatory information; and
- other information, like the recent date of the SDS.

4.6.2 Purpose

To provide workers with information on:

- product properties;
- potential hazards, *e.g.*, health, fire, explosion, reactivity;
- how to use, handle, store, and dispose of safely; and
- how to deal with emergencies (First Aid, spill, fire) involving the product.

4.6.3 Manufacturer/Supplier Responsibility

Manufacturers/Suppliers must develop or obtain SDSs in both official languages for each hazardous product they sell or import. Information must be current and prepared no more than three years before sale or importation. A copy of the current SDS must be sent to the purchaser on, or before, the date of sale at the time of first purchase.

4.6.4 Employer Responsibility

The employer must:

- Ensure that SDSs are received for all hazardous products supplied to the workplace.
- Contact the supplier for an updated sheet when the MSDS at the workplace is three years old.
- Store and label or identify hazardous products that are received without an SDS.
- Prepare SDSs for hazardous products **if produced** in the workplace according to WHMIS requirements.
- Ensure that copies of SDSs are readily available, at all times, to all employees who may be exposed to these products. **At XYZ, employees can access the required SDSs through the XYZ Intranet System.**
- Provide SDS information to medical professionals, as needed, in cases of exposures requiring medical treatment.
- Provide the appropriate resources to ensure required engineering, administrative, and personal protective equipment controls are in place prior to use of the chemical(s); and
- Train workers in the content, purpose, and significance of information contained in the SDS and hazard control measures.

4.7 Employee Education

According to the legislation, Canadian workers (employees, and contractors) must receive education if they use, handle, store, or dispose of a hazardous product, or supervise workers performing these duties; or, work near the hazardous product such that their health and safety could be at risk during the normal storage, handling, use, or disposal, during maintenance operations, cleaning or housekeeping, or in emergencies.

At XYZ, education and training programs for employees are provided to all employees so that they are able to **apply** the information and protect their health and safety on the job. Generic WHMIS 2015 education is provided to employees in both classroom and self-paced learning environments, and occurs:

- prior to starting work as per the XYZ new employee orientation practices;
- immediately if working conditions change, or if new hazard information becomes available;
- if worksite inspections show deficiencies in WHMIS 2015 requirements; and
- as part of post-incident corrective action(s).

Generic WHMIS 2015 education covers:

- what WHMIS 2015 is;
- how WHMIS 2015 works;
- what a hazardous product is;
- the hazards associated with hazardous products;
- what a WHMIS 2015 label is;
- how to read a WHMIS 2015 label;
- what anSDS is;
- how to read anSDS; and
- other generic WHMIS 2015 details.

Site-specific WHMIS 2015 training involves hands-on, job-specific training aimed at teaching workers how to work safely with the hazardous products in a particular workplace. It involves:

- worker training in XYZ's Safe Work Procedures for the handling, use, storage, and disposal of specific hazardous products; and
- worker training in XYZ's emergency response procedures should a chemical spill or emission occur.

5.0 EMERGENCY RESPONSE POLICY (refer to Chapter 12, "Occupational Health and Safety: Emergency Preparedness and Response").

6.0 INCIDENT INVESTIGATION (refer to Chapter 14, "Occupational Health and Safety: Incident Investigation").

7.0 VISITOR AND CONTRACTOR CONTROL POLICY

7.1 Introduction

The company is responsible for the health and safety of contractors, visitors, and to a lesser extent, trespassers. Additionally, the company must ensure that employees are not harmed by the activities of a contractor. This procedure has been produced to facilitate compliance with the duties described.

7.2 Definitions

Contractor — a contractor is any person, or persons, who are invited onto company premises in order to carry out work under a contract for employment.

Visitor — any person from any company, or member of the public, who is invited onto company premises.

Trespasser — any uninvited person who is on company premises either intentionally or unintentionally.

7.3 Control of Contractors

1. Contractors must be approved and have adequate provincial Workers'

Compensation and business insurance coverage. Contractors work to XYZ's code of practices while on the company premises.
2. A list of approved contractors will be issued to managers and supervisors. The list will be updated as required.
3. Contractors who are not approved must not be used.
4. Contractors must sign in and out at XYZ's Main Reception.
5. Contractors must be given information about any site hazards that may affect them while they are on company premises. Completion of the *Control of Contractors Form* is evidence that this information has been supplied to the contractor.
6. The safety performance of the contractor must be monitored while on site. This is particularly important if the work involves working at heights, conducting hot work, or doing construction/demolition activities.
7. If additional contractors are required, then the XYZ OH&S Department must be informed.

7.4 Control of Visitors

1. Visitors must sign in and out at the XYZ Main Reception.
2. Basic safety rules have been compiled for visitors and should be drawn to the attention of first-time visitors (Manual, Appendix 2).

7.5 Control of Trespassers

1. The most likely trespassers are children, and they are particularly at risk of injury due to their inquisitiveness and poor recognition of hazards. They are also the persons most likely to cause property damage through arson.
2. Preventing access to this group is clearly the best option for dealing with this potential problem. This can be achieved by:
   (a) always locking the yard gates on completion of work as well as the buildings; and
   (b) challenging the trespasser(s) as to their presence and inviting them to leave.
3. Where possible, keep the workplace and storage areas free of hazards.
4. Place warning signs on gates to deter trespassers.

7.6 Responsibilities

Management
1. Ensure that this procedure is followed within their areas of responsibility.

Immediate Supervisors (normally the person requesting the work)
1. Ensure that the method of work proposed by the contractor is satisfactory.

2. Monitor contractor work activities and ensure that any unsafe practices are reported to both the contractor and the OH&S Department.

3. Complete the Contractor Form and send it to the OH&S Department upon completion of the work.

Note: In many instances, the above requirements will be carried out by Plant Engineering.

Plant Engineering

1. Assess contractor eligibility and approve/reject.
2. Prepare and circulate the contractor list, updating it as required.
3. Maintain all records as required by company Contractor Procedures.

Receptionists

1. Ensure all contractors and visitors sign in and sign out of the facility.
2. Ensure all new visitors are notified of the basic safety rules.

8.0 ERGONOMICS PROGRAM

(*not provided*)

Available in Chapter 20, "Occupational Health and Safety: Ergonomics" in this publication.

9.0 DISABILITY MANAGEMENT POLICY

(*not provided*)

Available in Chapter 27, "Disability Management: Overview", in this publication.

SECTION 4: OCCUPATIONAL HEALTH PROGRAM

1.0 OCCUPATIONAL HEALTH (OH) PROGRAM

(*not provided*)

Available in Chapter 4, "Occupational Health Program", in this publication.

SECTION 5: OCCUPATIONAL SAFETY PROGRAM

1.0 GENERAL SAFETY RULES

1.1 General Safety Rules

General safety rules are standards and work practices developed to protect the well-being of XYZ employees from the hazards in the XYZ work environments. These general safety rules apply to all XYZ employees and contracted staff.

It is the responsibility of XYZ employees and contracted staff to be knowledgeable of these general safety rules.

1.2 Housekeeping[9]

- Good housekeeping is fundamental to creating safe work areas. It can bring greater efficiency, eliminate hazards, prevent fires, improve employee morale, and create a good impression on customers and visitors.

- Housekeeping includes keeping work areas neat and orderly; maintaining halls and floors free of slip and trip hazards; and removal of waste materials (*e.g.*, paper, cardboard) and other fire hazards from work areas. It also requires paying attention to important details such as the layout of the whole workplace, aisle marking, the adequacy of storage facilities, and maintenance. Good housekeeping is also a basic part of accident and fire prevention. Effective housekeeping is an ongoing operation.

- Poor housekeeping can be a cause of accidents, such as:
  - tripping over loose objects on floors, stairs, and platforms;
  - being hit by falling objects;
  - slipping on greasy, wet, or dirty surfaces;
  - striking against projecting, poorly stacked items, or misplaced material;
  - cutting, puncturing, or tearing the skin of hands or other parts of the body on projecting metal, nails, or file cabinet drawers.

- All worksites and areas under the authority of XYZ will therefore maintain the highest possible standard of housekeeping. The list below is not exhaustive but is the minimum that is required.

1.3 Office Work Areas

For information on Office Safety, refer to the *Employee OH&S Handbook*, and/or the XYZ Office Safety Program posted on the OH&S Website.

1.4 Shop Areas

- Employees entering a shop must stay within the designated walkways.
- Work areas must be kept clear and clean for the efficient movement of materials and people.
- Oil leaks and spills must be wiped away as soon as they appear.
- Ensure adequate working space around machines.

---

[9] **Housekeeping:** More information is available online: https://www.ccohs.ca/oshanswers/hsprograms/house.html and http://www.slideshare.net/mohsin61292/housekeeping-ppt?related=6.

- Position equipment so as to avoid "striking against" incidents.
- Prevent objects from protruding into walkways and creating tripping dangers.
- Place caution cones at the corners of walkways to prevent injury and damage by vehicles "cutting" corners.
- Put barriers in place while testing electrical equipment.

1.5 General Storage

- Store and stack all materials so that they are easy to get at.
- Stored materials must not interfere with work; yet be readily available when required. Stored materials must allow at least one meter of clear space under overhead sprinkler heads.
- Use stacking equipment as designed.
- Make sure objects cannot fall from the tops of shelving.
- Employees should use proper lifting techniques (refer to the *Employee OH&S Handbook*).
- Store flammable liquids in approved storage cabinets.
- Compressed gas cylinders should be stored according to the TDG and WHMIS 2015 legislation.
- Controlled products must be stored according to WHMIS 2015, TDG and fire regulations.

1.6 Scrap and Garbage

- Use suitable containers for garbage and scrap.
- Make sure garbage containers or waste materials are never left in walkways.
- Never allow garbage to accumulate on floors where it can cause tripping incidents.
- Always keep oil rags in approved containers.
- Place items for recycling in the appropriate recycle bins.

2.0 WORK EQUIPMENT SAFETY RULES

2.1 Introduction

Work equipment should not give rise to unacceptable risks to health and safety regardless of its age or place of origin. This procedure must be followed to ensure that suitable work equipment is procured, used safely, and maintained in a safe condition.

2.2 References

(Add in the applicable legislation)

2.3 Definitions

Work Equipment includes any machinery, appliance, apparatus, or tool used or available for use at work. This covers all equipment ranging from simple hand tools, to items of machinery and complete plants.

Suitable means suitable in any respect, which it is reasonably foreseeable, will affect the health and safety of any person. Suitability applies to three areas:

1. the initial integrity;
2. the place where it will be used; and
3. the purpose for which it will be used.

Use means any activity involving work equipment including starting, stopping, erecting, installing, dismantling, programming, setting, transporting, repairing, modifying, maintaining, servicing, and cleaning.

2.4 Applicability

This applies to the following classes of equipment:

- new;
- secondhand;
- hired or leased; and
- modified.

2.5 Procurement

2.5.1 Pre-acquisition Process

The Pre-acquisition Process at XYZ involves the following steps:

- A needs assessment is conducted which includes a cost/benefit analysis of purchasing the product/service and a risk determination (financial, efficiency, and safety) of the service/product.
- A market research and product/service search are undertaken.
- A decision is made whether to proceed/not proceed with the acquisition.
- Criteria for product/vendor selection are developed.

2.5.2 Evaluation of the Proposed Work Equipment

Once the decision to purchase and the acquisition is made, a product/service evaluation is undertaken. This includes the assessment of whether the product/service:

- performs the desired tasks;
- meets the safety requirements;
- functions fairly maintenance free; and

- meets the expected product life span.

2.5.3 Responsibilities

Management
- Ensure that an effective procurement process exists.
- Ensure that adequate resources are available.

XYZ Fleet Services
- Ensure that the equipment meets the current OH&S and CSA requirements.
- Ensure that the equipment is tested pre-delivery and meets the manufacturer and CSA specifications.
- Maintain a record of the specifications and the pre-delivery acceptances.

2.6 Use

2.6.1 Training/Provision of Information

- Personnel, who use or work on machines that have a specific health and safety risk, must be adequately trained and competent.
- They must also be provided with adequate information and instructions about the safe work methods and applicable hazard identification and safeguards.

2.6.2 General Precautions

- Prevent access to any dangerous parts of machinery or equipment, or stop the movement of any dangerous part, by provision of suitable guards or safety devices where there is a risk of contact with any dangerous part.
- Where practicable, provide means of protection against any likely failure of the equipment, such as overheating, fire, ejected objects, premature release of articles or substances, or possible explosion.
- Safeguard against any persons coming into contact with any high or low temperatures that may be associated with the use of the plant or apparatus.
- Provide suitable, clearly identified and located controls to enable the equipment to be safely started or stopped or initiate changes of speed or other operating conditions and enable emergency stop when appropriate.
- Ensure that equipment and any other work components are stable and safe to use.
- Provide a safe and appropriate means of isolating the equipment from sources of energy.
- Ensure that any area where work equipment is being used has adequate lighting levels which are suitable for the operations being performed.

- Ensure that all plant and equipment have suitable markings and, where applicable, appropriate devices (*e.g.*, audible warnings) to warn of any health and safety hazards.

2.6.3 Responsibilities Related to Equipment Use

Line Manager
- Ensure that employees are trained and competent in the safe use of the equipment.
- Ensure the equipment is used safely.
- Ensure that equipment is inspected on a regular basis.
- Equipment is tagged and removed from service if safety devices become defective.
- Equipment defects are rectified in a timely manner.
- Ensure that daily equipment inspections are undertaken as per manufacturer's guidelines.

Employee
- Use equipment as trained and instructed.
- Wear the appropriate PPE, as required.
- Report equipment and PPE defects to the Line Manager.
- Have a working knowledge of location and usage of safety equipment such as fire extinguishers, eyewash stations, showers, *etc*.
- Complete daily equipment inspections as per manufacturer's guidelines.

2.7 Maintenance

Employers have a legal duty to provide safe facility, equipment, and working conditions. Refer to the legal requirements applicable to the jurisdiction.

Employees are legally responsible for identifying defective work equipment before use, and for removing it from service ("tagging it") until it is repaired or replaced.

2.7.1 Work Equipment Register

An Equipment Register is maintained by XYZ Fleet Services.

2.7.2 Routine Maintenance

- Routine maintenance is carried out on work equipment based on the manufacturers' recommendations.
- If no recommendations exist for routine maintenance, then these are to be developed with employee input.
- The maintenance is to be carried out by either competent "in-house" personnel, or approved contractors.

- All maintenance carried out is to be recorded on maintenance logs and retained for auditing purposes.
- Defects that are discovered during maintenance are to be recorded.
- Repairs must be carried out in a timely manner.

2.7.3 Machinery/Equipment Repairs
- Equipment that is to be serviced or repaired is to be "tagged" or "locked out" in accordance with the XYZ Lock out/Tag Out Safe Work Practice.
- Equipment needing repair, servicing, testing, or adjustment must be performed according to manufacturer's specifications and the applicable XYZ Safe Work Practice(s).

2.7.4 Responsibilities

Senior Management
- Ensure that maintenance schedules exist, and that the appropriate maintenance is carried out on equipment for which they have responsibility.

Line Manager
- Ensure that only qualified persons conduct maintenance activities.
- Ensure Safe Work Practices are developed and used for lock out/tag out.
- Ensure Safe Work Practices are developed and used for related confined space entry.
- Ensure that noted defects are fixed in a timely manner.
- Maintain equipment repair records.

XYZ Fleet Services
- Ensure that all the required inspections occur at the required intervals.
- Ensure that all defects are reported and addressed.
- Maintain equipment maintenance records.

Employees
- Perform and log daily equipment inspections.
- Report defects.
- File the equipment inspection reports.

3.0 ACCESS EQUIPMENT SAFETY RULES

3.1 Introduction

The provision of safe access equipment (*e.g.*, ladders, stepstools, aerial devices, and manlifts) and a system for its use are essential elements in controlling the high risk of injuries associated with equipment use.

3.2 References
(Add in the applicable legislation)

3.3 Definition
Access equipment in this instance is:
- portable ladders regardless of size;
- stepladders;
- stools;
- aerial devices; and
- manlifts.

3.4 Safe Use
- Access equipment shall be used in accordance with the manufacturer's specifications and maintained to current standards (ANSI, CSA, ULC).
- Ladders are to be used in accordance with the *Employee OH&S Handbook*.
- Aerial devices are to be used in accordance with the *Employee OH&S Handbook*.

3.5 Inspection
Access equipment will be inspected by a competent person. Records of inspection and comments must be kept, and that any defects repaired by a qualified person according to the established maintenance standards and schedules.

3.6 Responsibilities

Management
- Ensure that this standard is maintained and followed in their area(s) of responsibility.

Line Manager
- Ensure that any person who is instructed to use access equipment is aware of and understands this procedure.
- Any new access equipment purchased is suitable for its use.
- Any defective equipment is identified and removed from use.

Employees
- Use access equipment in accordance with this standard.
- Report, tag, and take out of service any defective access equipment.

4.0 CRANE AND HOIST SAFETY RULES

4.1 Introduction
This procedure has been produced to ensure the safe use of cranes and hoists. It also

covers the design, installation, inspection, and certification.

4.2 References

(Add in the applicable legislation)

4.3 Applicability

This procedure applies to cranes and hoists that have a capacity of 2,000 kilograms or more.

4.4 Definitions

Check means to determine whether the equipment meets the relevant code, standards, or manufacturer's or professional engineer's specifications and is operating properly.

Examine means a visual inspection carried out by a competent person to ensure the equipment is in a condition that will not compromise worker safety.

Inspect and/or test refers to carrying out a procedure, including operating the equipment which will determine whether the equipment is correctly assembled and functioning, and is likely to continue to do so.

4.5 Installation/Testing/Certification

1. All cranes will be designed, installed, erected, checked, examined, inspected, operated, maintained, and repaired in accordance with the appropriate CSA standards.
2. Cranes will not be brought into use until they have been tested and certified by a competent person and the certificate of test and examination sent to XYZ.
3. All major structural, mechanical, and electrical components must be permanently and legibly identified as being component parts of a specific make and model of hoist.
4. All hoists must be equipped with positive pressure controls.
5. Only competent workers are to operate the hoist.
6. Signals are to be given by a designated signaller where necessary to ensure a safe hoisting operation.
7. If an operator has any doubt as to worker safety, he/she will not move any equipment or load until:
   (a) safe working conditions have been assured; or
   (b) orders to proceed have been issued by a designated signaller.
8. Where there is a danger created by the movement of a load being moved, raised, or lowered by a crane/hoist:
   (a) a tag line of an adequate length to ensure that the worker

9. The bridge crane and other similar hoists operating on rails, tracks, or trolleys will:

   (a) have a positive stop or limiting device to prevent actions beyond safe operating limits or contact with other equipment that is on the same rail, track, or trolley;

   (b) be equipped with an over-speed limiting device;

   (c) have a means of ensuring that the rails, tracks, or trolleys cannot spread or misalign;

   (d) have sweepguards installed to prevent materials on the rail, track, or trolley from causing dislodgment of the hoist; and

   (e) have a bed designed to carry all anticipated loads.

10. Each crane, or similar hoist, will have permanently affixed to it, in accordance with the appropriate CSA Standard, a plate or weatherproof chart which is legible and shows:

    (a) the manufacturer's rated capacity load for the machine;

    (b) the manufacturer's name;

    (c) the model, serial number, and year of manufacture or shipment date;

    (d) hand signals for control of hoisting operations; and

    (e) the load ratings for all possible boom angles and boom radii.

11. The manufacturer's rated capacity load for each crane boom or mast section will be permanently and legibly marked on the boom or section, as the case may be.

12. Blocking procedures are developed for the installation, removal, or replacement of a derrick or the mast or boom section of a crane, so that the collapse or upset of any part of the derrick or crane is prevented; and where outriggers are installed on a crane or similar hoist, they are extended and supported by solid footings before being used.

13. The Engineering Department will maintain records of certificates and test examinations.

4.6 Safety Precautions

4.6.1 *Electrically Operated Cranes*

Do

1. Make certain that the crane structure, including rails, are free of obstructions and personnel before operating the Crane.

Controlling the tag line cannot be struck by any movement of the load is to be used.

2. Ensure that the path over which the load will travel is free of personnel and obstructions before operating the crane.
3. Stand near the load to obtain the best possible view of the movement of the load but not sufficiently close to be struck by the load.
4. Ensure that the load is correctly slung (a) before taking up the slack; and (b) before raising the load clear of the ground. Avoid stretch lifting.
5. When lifting the load, do not raise it higher than is necessary.
6. Before lifting the load ensure that the crane rope is correctly spooled on the winch drum and correctly received through the crane block.
7. Ensure the load suspension point is located securely in the bowl of the crane hook and retained by the safety catch before lifting.
8. Perform all operations smoothly and carefully. Haste will result in a swinging and uncontrolled load. If the load starts to swing, stop and allow it to settle before resuming.
9. Having completed the lift, leave the crane parked with the crane block and/or slings at an adequate height to clear personnel and vehicles.

Do Not
1. Operate a crane/hoist unless competent to do so.
2. Work on the crane unless the electrical supply is isolated, the tag and/or lockout procedures are in place, and the fuses are removed.
3. Place materials on, or near, conductor rails, junction boxes, or motors.
4. Rely on limit switches or stops to arrest movements of the crane.
5. Allow the hoist rope to become slack, except during inspection/servicing (in which case, correct revving and spooling must be ensured before returning the crane to service).
6. Rapidly depress and release the pendant buttons (this will cause overheating of the contractors and premature failure).
7. Operate the crane if you see or suspect a defect of any kind. If there is a defect, without delay, have the crane isolated and report the fault to the Immediate Supervisor.
8. Leave suspended loads unattended.
9. Work under a suspended load.

4.6.2 Engine Build/Strip Stands

These must be operated in accordance with their safe system of work detailed in the associated risk assessment.

4.7 Maintenance and Repairs

In addition to XYZ's equipment maintenance standards, the following applies to the operation of cranes/hoists:

> Structural repairs or modifications to components of a crane or similar hoist are to be:
> 1. performed only under the direction and control of a professional engineer; and
> 2. tested and certified by the professional engineer to the effect that the workmanship and quality of materials used are such that the capacity of the components is not less than their original capacity.
>
> Following the repair work, sections that are repaired or modified are individually and uniquely identified, and the engineer's certification refers to those sections.

4.8 Examination

1. All cranes and hoists will be thoroughly examined by a competent person every 12 months.
2. The Engineering Department will maintain records of these examinations.

4.9 Log Books

XYZ will ensure that there is a logbook for each crane, or hoist, operating at the facility. The logbook will contain details on date and time of:

1. inspections, examinations, checks, and tests, including those specified in the manufacturer's specifications;
2. defects or deficiencies;
3. repairs;
4. sizes and types of wire ropes in use, including rigging information;
5. hours of service;
6. any matter or incident that may affect the safe operation of the crane/hoist;
7. a record of certification;
8. any other operational information that has been specifically identified by the employer; and
9. in the case of a tower crane, whether the weight testing device was lifted for that working day before the lifting of materials began.

All log entries are to be signed by the operator.

Operators are to be familiar with all recent log entries before starting to operate his or her machinery.

4.10 Training

All personnel who use these cranes and hoists must be trained in their safe operation.

CH. 5: OCCUPATIONAL HEALTH AND SAFETY PROGRAM: MANUAL DEVELOPMENT

5.0 Electrical Equipment Safety Rules: Portable and Transportable[10]

5.1 Introduction

Failure to use and maintain portable electrical equipment correctly can result in severe injury to the user through electrical shock and burns. Failure of the equipment may also cause fires.

5.2 Reference

(Add in the applicable legislation)

5.3 Definition

Electrical equipment — equipment that is connected to a fixed installation by means of a flexible cable by a plug/socket or spur box. It includes equipment that is either handheld, or hand-operated, moveable, or intended to be moved while connected to the electrical supply.

5.4 Voltage

Whenever practicable, workshop handheld power tools should be 110V and grounded.

5.5 Use of Equipment

The following basic precautions must be followed when using electrical equipment:

- use only CSA/ULC approved equipment;
- use the equipment for the purpose for which it was designed;
- only use the equipment if you are trained to do so;
- do not overload the equipment;
- wear the correct personal protective equipment — refer to the *Employee OH&S Handbook*;
- do not use the equipment if it is damaged;
- ensure that cables and extension leads do not create a tripping hazard; and
- flexible cables must not be used for raising or lowering the equipment.

For more details, refer to the *Employee OH&S Handbook*.

5.6 Before Use Inspection

Electrical equipment must be visually inspected before use. In particular, watch for damaged insulation and cables. Any defects are to be removed from service for repair/replacement.

---

[10] **Electrical Equipment:** For more information, refer to: https://www.ccohs.ca/oshanswers/safety_haz/electrical.html and https://www.ccohs.ca/topics/hazards/safety/electrical/.

5.7 Responsibilities

Line Manager
- Coordinate the repair or replacement of damaged equipment.

Employees
- Use the equipment as directed in this standard.
- Report any defects found on the equipment to Line Management.

6.0 FORK LIFT SAFETY RULES

(Not covered)

7.0 RIGGING AND SLINGING EQUIPMENT SAFETY

(Not covered)

8.0 MANUAL HANDLING SAFETY RULES[11]

8.1 Introduction

Manual handling includes pushing, pulling, carrying, as well as lifting. Many incidents, mainly sprains, fractures, abrasions, and cuts are caused due to unnecessary or incorrect manual handling. Furthermore, back injuries account for many of the more serious and chronic problems resulting in pain for the employee and loss to the employer.

8.2 Assessment

To assess the hazards associated with manual handling:

(a) identify the hazards;

(b) evaluate the hazard in terms of risk;

(c) identify suitable controls; and

(d) implement controls.

8.3 Lifting Guidelines

Back injuries can be extremely painful and can lead to a serious injury. Always practice safe lifting techniques when lifting heavy objects. The preferred practice is the use of cranes, hoists, or a forklift to move or lift heavy objects.

Safe lifting practices are discussed in the *Employee OH&S Handbook*.

8.4 Training

Employees are to receive training based on their assigned tasks. Refer to Chapter 20,

---

[11] **Manual Handling:** For more information, refer to: http://www.slideshare.net/vtsiri/manual-handling-ppt-presentation?related=1; and https://www.ccohs.ca/oshanswers/ergonomics/mmh/hlth_haz.html.

"Occupational Health and Safety: Ergonomics" in this publication for suggested training materials.

8.5 Records

As detailed in this manual, refer to the section on Risk Assessment and Control.

8.6 Responsibilities

Line Management
- Perform manual handling risk assessments as required within their areas of responsibility.
- Develop and implement the appropriate control measures.
- Ensure manual handling training is provided as required.

Employees
- Identify and report manual handling concerns to Line Management.
- Use the proper manual handling techniques.

9.0 Grinding Wheel Safety Rules

(Not covered)

10.0 Personal Protective Equipment (PPE) Safety Rules[12]

10.1 Personal Protective Equipment

Personal protective equipment (PPE) should be regarded as the last resort in protecting employees from the hazards associated with their work.

XYZ's strategy is to minimize the hazards and introduce engineering and administrative controls to reduce the risk of injury. However, there are many instances when PPE plays a vital part in protecting the employee.

10.2 References

(Add in the applicable legislation)

10.3 PPE Assessment

Hazard assessments, worksite inspections, and incident investigations can be used to identify and specify the PPE required.

Some work activities may require a formal assessment and the following assessment process is to be used to:

- identify hazards to employee health or safety;

---

[12] **Personal Protective Equipment:** For more information, refer to: https://www.ccohs.ca/oshanswers/prevention/ppe/.

- evaluate the hazard against available PPE controls;
- select an appropriate PPE; and
- evaluate the PPE's effectiveness to control the identified hazard.

For further information, refer to the *Employee OH&S Handbook*.

10.4 Standard Issue PPE

At XYZ, PPE is issued to employees in accordance with specific job hazards. Refer to the *Employee OH&S Handbook* for more information.

10.5 Prescription Safety Glasses

XYZ provides an allowance towards the purchase of prescription safety glasses to employees who normally require safety glasses as per the respective collective agreements.

10.6 Footwear

XYZ provides an allowance for the purchase of approved safety footgear by field employees as per their respective collective agreements.

10.7 Maintenance

PPE must be maintained in a workable condition to perform the function for which it was designed. This involves regular inspection by Line Management and employees. Employees are legally responsible to maintain their PPE in good working order every day and to replace as necessary.

10.8 Responsibilities

Line Manager

- Be aware of PPE requirements and ensure compliance.
- Provide employees with the appropriate PPE.
- Ensure that employees receive information, instruction, and training on the following:
  - the hazards that PPE is protecting against and the related limitations; and
  - PPE selection, use, maintenance, storage, and defect reporting.
- Provide suitable accommodation for the safe storage of the PPE.

Employee

- Use the PPE as instructed.
- Report any equipment limitations, defects or loss to Line Management.

11.0 Working Alone Safety Policy[13]

11.1 Working Alone

A significant number of job functions at XYZ have been, and will continue to be, performed by employees working alone. XYZ recognizes that working alone in certain circumstances or environments can be deemed unsafe and requires special arrangements to minimize potential health and safety risks. For this reason, XYZ has made provisions to protect its employees who are working alone.

Working alone is defined as those circumstances in which the employee is working without radio or audible contact with another worker or the worksite and/or where assistance is not readily available.

11.2 Reference

(Add in the applicable legislation)

11.3 Policy

Certain jobs or circumstances may require an XYZ employee to perform job duties without the benefit of observation by, or immediate contact with, another employee who could assist in the event of an injury, illness, or emergency.

While XYZ endeavours to ensure that all employees work with a high level of protection against workplace hazards, and other risks, a particular level of care is required when employees are working alone.

XYZ will:

- Identify and assess existing and potential hazards that may arise where an employee is working alone.
- Make every reasonable effort to eliminate or minimize such hazards.
- Provide an effective means of communication between the employee and persons able to respond to the employee's needs.
- Ensure that employees are trained and educated so that they can perform their jobs safely.

11.4 Responsibilities

<u>Line Management</u>

Business unit management, in consultation with the XYZ Safety Section, is responsible for:

- Identifying and assessing workplace hazards with follow-up evaluations at appropriate intervals.

---

[13] **Working Alone:** For more information, refer to: http://www.ccohs.ca/oshanswers/hsprograms/workingalone.html.

- Providing adequate means of communication for employees working alone to ensure compliance with this policy.
- Where effective communication is not practical for a specific job or worksite, maintaining regular contact with the employee through visits or scheduled phone-ins to ensure the employee's safety is supervised.
- Implementing safety measures to reduce the risk to employees from the identified hazards.
- Establishing a Safe Work Practice for Working Alone.
- Providing adequate employee training so they can perform their jobs safely.

OH&S Department

The OH&S Department is responsible for:

- Creating and maintaining appropriate awareness by Management and employees of the risks and procedures when working alone.
- Monitoring XYZ's safety standards and performance where employees are working alone.
- Reporting to Management on XYZ's performance and compliance with applicable regulations.

Employees

All employees who are working alone are responsible for:

- Reporting hazards and incidents to their supervisor.
- Complying with the applicable Safe Work Practice for Working Alone for their area.

11.5 Procedures

XYZ business lines will consult with employees, Line Management, and the operations to assess the inherent risks associated with working alone and develop a specific plan of action for their department. This is done using the following approach:

11.5.1 Assessment

- Conduct a hazard assessment to identify existing or potential safety hazards in the workplace associated with working alone using the *Risk Assessment Form* (see Chapter 9, "Occupational Health and Safety: Hazard Identification, Assessment, and Control", Appendix 2).

11.5.2 Control Measures and Documentation

- implement control measures to reduce or eliminate the risk for employees associated with the identified hazards;
- document the corrective actions taken on the *Risk Assessment Form* (see

Chapter 9, "Occupational Health and Safety: Hazard Identification, Assessment, and Control", Appendix 2);

- ensure that XYZ employees have an effective way of communicating with their employer, immediate supervisor, or another designated person in case of an emergency situation; and
- monitor the hazard identification, hazard control measures and communication practices implemented.

11.5.3 Employee Training

- ensure that XYZ employees are trained so they can perform their jobs safely; and
- monitor the effectiveness of the training.

11.6 Control Measures

- identify lone-worker situations in the various business units at XYZ;
- develop specific Safe Work Procedures for these lone workers (Figure 11.1);
- provide appropriate training associated with lone work;
- institute the practice of periodic radio/telephone contact, or a system of central monitoring; and
- provide radio alarms, panic buttons, and/or personal pendant alarms where appropriate.

Figure 11.1: Development of Working Alone Safe Work Practice

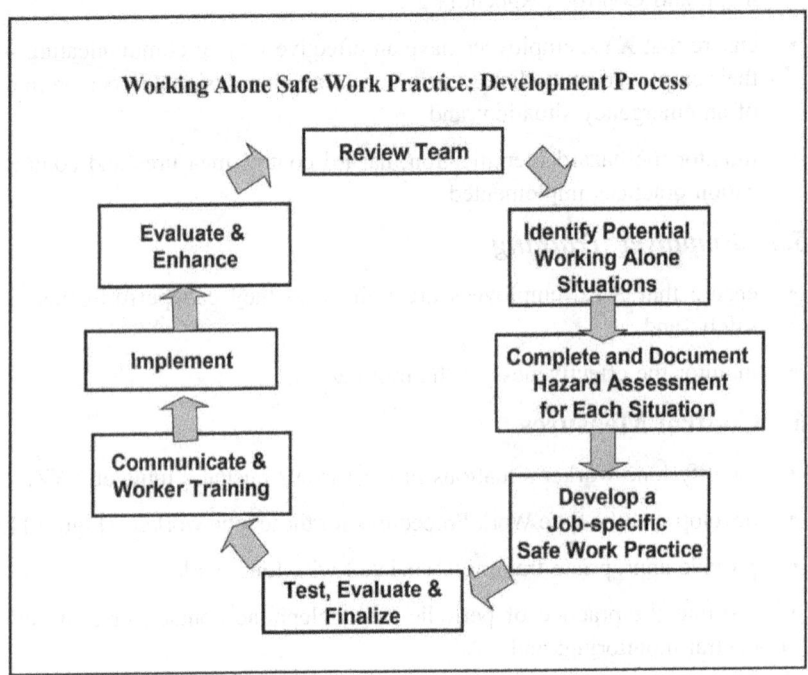

11.7 Standard Review

This Working Alone standard will be reviewed annually and updated as required by the CEO, XYZ, or designate.

12.0 PREVENTION OF WORKPLACE VIOLENCE POLICY[14]

12.1 Introduction

Most people think of violence as a physical assault. However, workplace violence is a much broader problem. It includes:

- **Threatening behaviour** — such as shaking fists, destroying property, or throwing objects.
- **Verbal or written threats** — any expression of intent to inflict harm.
- **Harassment** — any behaviour that demeans, embarrasses, humiliates, annoys, alarms, or verbally abuses a person and that is known or would be expected to be unwelcome. This includes words, gestures, intimidation, bullying, or other inappropriate activities.

---

[14] **Workplace Violence:** For more information, refer to: http://www.ccohs.ca/oshlinks/subject/workplaceviolence.html.

- **Verbal abuse** — swearing, insults, or condescending language.
- **Physical attacks** — hitting, shoving, pushing, or kicking.

Workplace violence is defined as any act in which an employee is abused, threatened, intimidated, or assaulted in his or her employment.

Workplace violence is not limited to incidents that occur within a traditional workplace. Work-related violence can occur at off-site business-related functions (*i.e.*, conferences or trade shows), at social events related to work, in clients' homes, or away from work, but resulting from work (*e.g.*, a threatening telephone call to your home from a client).

Certain work factors, processes, and interactions can put people at increased risk from workplace violence. Examples include:

- working with the public;
- handling money (*e.g.*, cashiers);
- carrying out inspection or enforcement duties; or
- working alone.

12.2 Reference

(Add in the applicable legislation)

12.3 Application

XYZ recognizes the potential for violent acts or threats directed against staff. Every effort will be made to identify possible sources of violence and implement procedures to eliminate or minimize risks. XYZ management will ensure that all employees are aware of violence hazards and are properly trained to protect themselves.

Workers are to follow the procedures to reduce the risk of violence and immediately report all incidents to their manager. XYZ employees who commit acts of violence towards other XYZ employees will face disciplinary action, up to and including termination.

12.4 Control Measures

- Identify potential workplace violence situations at XYZ.
- Develop Safe Work Procedures for managing and reporting workplace violence.
- Provide appropriate training associated with workplace violence.

12.5 Procedure

XYZ business lines will consult with employees and Line Management to assess the inherent risks associated with workplace violence and develop a specific plan of action for their department. This is done using the attached process flowchart

(Figure 12.1) and the following approach:

12.5.1 Assessment

- Conduct a hazard assessment to identify potential workplace violence situations using the *Risk Assessment Form* (see Chapter 9, "Occupational Health and Safety: Hazard Identification, Assessment, and Control", Appendix 2).

12.5.2 Control Measures and Documentation

- implement control measures to reduce or eliminate the risk of workplace violence for employees or customers;
- document the hazard controls on the attached *Risk Assessment Form* (see Chapter 9, "Occupational Health and Safety: Hazard Identification, Assessment, and Control", Appendix 2);
- ensure that XYZ employees have an effective way of managing workplace violence with their employer, immediate supervisor, or another designated person in case of an emergency;
- ensure that XYZ employees have an effective way of managing and reporting workplace violence situations; and
- monitor the hazard identification, hazard control measures, and workplace violence management practices.

12.5.3 Worker Training

- Ensure that XYZ employees are trained so they can identify and reduce or eliminate potential workplace violence.
- Ensure that XYZ employees are trained in techniques for managing and reporting workplace violence.
- Monitor the effectiveness of the training.

12.6 Responsibilities

Management

- conduct a hazard assessment to identify existing or potential safety hazards in the workplace associated with workplace violence;
- implement safety measures to reduce the risk to workers from the identified hazards;
- ensure that workers have an effective way of managing workplace violence; and
- ensure that workers are trained in addressing workplace violence.

OH&S Department

The OH&S Department is responsible for:

CH. 5: OCCUPATIONAL HEALTH AND SAFETY PROGRAM: MANUAL DEVELOPMENT

- Creating and maintaining appropriate awareness of workplace violence risks by Management and employees.
- Reporting to Management on XYZ's performance and compliance with applicable regulations.

Employee
- Take reasonable care to protect themselves and the public from workplace violence.
- Participate in workplace training and follow the procedures for managing and reporting workplace violence.

Main Receptionist
- In the event of a violent act, initiate Emergency Response Plans.

12.7 Standard Review

This standard will be reviewed annually and updated as required by the Director, OH&S, or designate.

Figure 12.1: Development of Workplace Safe Work Practice

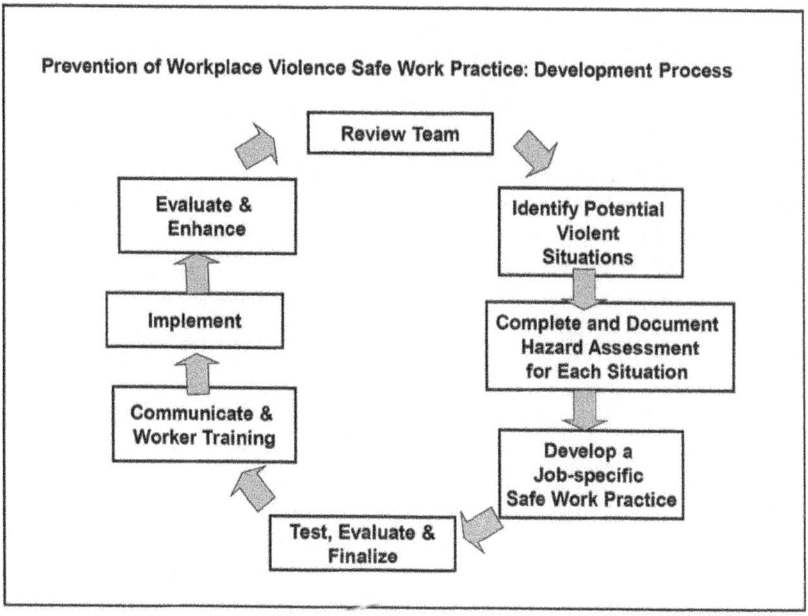

```
Prevention of Workplace Violence Safe Work Practice: Development Process

            Review Team
    ↗                        ↘
Evaluate &              Identify Potential
Enhance                 Violent Situations
    ↑                        ↓
Implement               Complete and Document
                        Hazard Assessment
                        for Each Situation
    ↑                        ↓
Communicate &           Develop a
Worker Training         Job-specific
                        Safe Work Practice
    ↖                        ↙
            Test, Evaluate &
            Finalize
```

Although not covered in this text, the following two sections could also be included in the manual:

SECTION 6: ENVIRONMENTAL PROGRAM

1.0   Environmental Plan

2.0	Control of Substances Hazardous to Health
3.0	Storage and Handling of Hazardous Substances
4.0	Waste Management Plan

SECTION 7: CONTROL OF CONTRACTORS PROGRAM

1.0	Selection, Assessment and Evaluation of Contractors/Sub-Contractors
2.0	List of Approved Contractors
3.0	Procedures for Contractors
4.0	General Conditions of Work in/on Site
5.0	Code of Practice for Contractors Working in/on Company Premises
6.0	Contractors Records

Appendix 2

Visitor Health and Safety

Welcome to COMPANY XYZ.

As a visitor, you are owed a duty of care while you are on our premises. We therefore ask you to read and follow some basic precautions designed to ensure your SAFETY.

Basic Rules

- Please sign in and sign out at the Main Reception.
- Do not enter buildings unescorted.
- While you are in the workplace, please do not enter a hazardous area unless you are escorted (Personal Protective Equipment is required in hazardous areas).
- Please obey all the posted SAFETY signs.
- Familiarize yourself with directional maps located throughout the building.

Protective Clothing

Not normally required.

Smoking

COMPANY XYZ employees maintain a NO SMOKING workplace. Please observe this practice.

Alcohol and Drugs

The possession and/or consumption of drugs or alcohol are strictly prohibited at COMPANY XYZ.

Emergencies

FIRE: When you hear the fire alarm, please leave the building immediately by the nearest exit and go with a COMPANY XYZ employee to that building's assembly point.

BOMB THREAT: If this occurs, the fire alarm sounds — a short audio burst followed by an announcement. Please follow the instructions given and take any personal belongings, such as your coat or briefcase, with you.

CHAPTER REFERENCES

Alberta, *Occupational Health and Safety Act*, R.S.A. 2017, c. O-2, s. 3(1), (2), (3), (4).

BSI Management Systems, "OHSAS 18001 Awareness" (Presented as an Informational Seminar, April 2002, Calgary, AB).

CCOHS website, online: http://www.ccohs.ca (date accessed: January 31, 2020).

D. Dyck, *Disability Management: Theory, Strategy and Industry Practice*, 6th ed. (Toronto: LexisNexis Canada, 2017).

WHMIS Pictograms (WHMIS 2015), online: https://www.ccohs.ca/oshanswers/chemicals/whmis_ghs/pictograms.html, *OSH Answers*, Canadian Centre for Occupational Health and Safety (CCOHS), October 9, 2019.

WHMIS Supplier Label (WHMIS 2015), online: https://www.ccohs.ca/oshanswers/chemicals/whmis_ghs/labels.html, *OSH Answers*, Canadian Centre for Occupational Health and Safety (CCOHS), October 9, 2019.

Chapter 6

OCCUPATIONAL HEALTH AND SAFETY: LEADERSHIP AND COMMITMENT

By Tony Roithmayr and Dianne Dyck

A. INTRODUCTION

Leadership is defined as an art that liberates people to do what is required of them in the most effective and humane way possible. Leadership is "an art, something to be learned over time, not simply by reading books. Leadership is more tribal than scientific; more a weaving of relationships than an amassing of information".[1] Leadership is both about getting results and, about how those results are obtained. **Leaders** are individuals who excite, stimulate, and drive other people to work towards a vision, making it a reality.

The essence of good leaders is that they possess the:

- *Vision* to spell out clearly what they will do for those who depend on them.
- *Drive* to share their vision with those who have the greatest stake in the leader's success.
- *Courage* to change what is, initiate change, and make strategic decisions to move forward.
- *Ability* to inspire others to achieve their goals.
- *Foresight* to empower others to learn new skills and to achieve to their functional potential.
- *Wisdom* to listen, learn, and translate that knowledge into value and added performance.
- *Integrity* to set a positive example and be a strong role model.
- *Willingness* to recognize accomplishments and celebrate individual and team successes.[2]

According to the Great Place to Work Institute Canada, "organizational leaders

---

[1] M. DePree, *Leadership is an Art* (New York, NY: Dell Publishing, 1989).

[2] J.T. O'Rourke, "The Essence of Leadership" (1993) 6:2 Drake Business Review 16-17.

who develop a level degree of trust with their employees see stronger worker performance".[3]

This chapter discusses leadership, corporate culture, safety culture, Occupational Health and Safety (OH&S) leadership, the key aspects of OH&S leadership, the importance of OH&S leadership, and provides an industry example of an effective OH&S leadership model. Good OH&S leadership and commitment leads to a strong safety culture. Inherent in successful OH&S leadership and the maintenance of a strong OH&S culture is regular OH&S performance measurement.

B. DEVELOPMENT OF AN OCCUPATIONAL HEALTH AND SAFETY CULTURE

According to Ryan, "[t]he corporate culture is dictated by what Management does; what Management pays attention to; what Management condones or ignores; and what Management measures. (Add to that,) management controls the resources necessary to effect change".[4] To best position the concept of an OH&S culture, let us first discuss corporate culture in general terms.

1. Corporate Culture

All organizations have cultures, whether they know it or not.[5] **Corporate culture** is defined as the system of shared beliefs and values that develops within an organization and guides behaviour of its members.[6] Culture is to an organization what memory is to a person.[7] It is the way things are done within an organization. Just as no two individuals or families are the same, neither are two corporate cultures the same. Each is unique and has a profound impact on the performance of the organization and the quality of the work-life experience of employees.

The corporate culture imparts Management's beliefs, attitudes, values, and approaches to employees. It lets employees know whether they are trusted, valued, or respected. In addition, they learn the priorities that Senior Management holds dear; which procedures to follow; and which workplace rules can be ignored. They

---

[3] G. Lowe & J. Wetherow, "High Trust Cultures = Health and Performance" (Presented at the International Accident Prevention Association Conference, April 2007, Toronto, ON).

[4] D. Ryan, "Moving off Your OH&S Plateau: Cutting Edge Strategies for Fostering and Measuring a Dynamic OH&S Culture" (Presented at the Occupational Health & Safety Amendment Act, 2002 — What's New, What's Changing and What You Need to Do to Comply, Insight Conference, May 27-28, 2003, Edmonton, AB).

[5] T. Rutledge, "Culture – Not Just a Plaque on the Wall" (2007) *Canadian Occupational Safety*, April/May 2007 Issue, at 14.

[6] J. Schermerhorn, J. Hunt & R. Osborn, *Managing Organizational Behavior*, 4th ed. (Toronto, ON: John Wiley & Sons, 1991) at 340-341.

[7] International Atomic Energy Agency, "Safety Culture in Nuclear Installations: Guidance for Use in Enhancement of Safety Culture" (December 2002), online: http://www-pub.iaea.org/ MTCD/publications/pdf/te_1329_web.pdf (date accessed: February 28, 2020).

learn who is really in charge; how decisions get made; how problems are handled; how conflicts are resolved; how much support exists for employees; and who shares responsibility/accountability for what. Corporate culture helps employees figure out how to get around workplace challenges or barriers — officially and unofficially — and which values, attitudes, and behaviours will or will not be tolerated.

Although corporate culture is the man-made part of the work environment,[8] it is not a conscious element, even for Senior Management. Corporate culture includes formal policies that are laid out in black and white, but it is much more than that. It is the "personality" that differentiates one organization from another. A strong, clear corporate culture has three characteristics:

- *Consensus* — The beliefs, values, attitudes, habit, and traditions that are shared by all organizational members. This includes certain key issues that the company holds dear. For example, a company leadership that cherishes the health and well-being of their employees (*a value*), may adopt zero injuries as their fundamental safety goal.

- *Consistency* — Actions and symbols within the organization that are congruent. The organization's formal and informal practices are aligned with the company's objective of managing safety.

- *Clarity* — The company members know what is expected of them and why.[9]

There are two elements of corporate culture: the **observable culture** and the **core culture**.[10] The observable culture is what can be seen and heard within an organization. It is the way people behave, dress, talk about customers, and arrange their offices. It includes the rites, rituals, norms, symbols, stories, and the heroes. It is made up of the readily observable aspects of culture and those that are less obvious. The core culture is composed of the corporate values, beliefs, and acceptable behaviours. It is the reason things are done the way they are, and it drives the observable culture elements as depicted in Figure 6.1.

---

[8] *Ibid.*

[9] B. Bain, "Expanding the Traditional View of Safety Culture" (Presented at the Industrial Accident Prevention Association Health & Safety Conference, Toronto, ON, April 2004)

[10] E. Schein, *Organizational Culture and Leadership*, 2d ed. (San Francisco, CA: Jossey-Bass, 1992)

[11] S. Hazzard, "Management and Organizational Behaviour", *BCRSP Self-Study Guide* (Mississauga, ON: Board of Certified Safety Professionals, 2000) at 47.

Figure 6.1: Elements of Corporate Culture (adapted from OHS&S Systems Domain, BCRSP Study Guide, 2007[11])

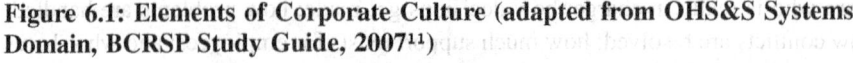

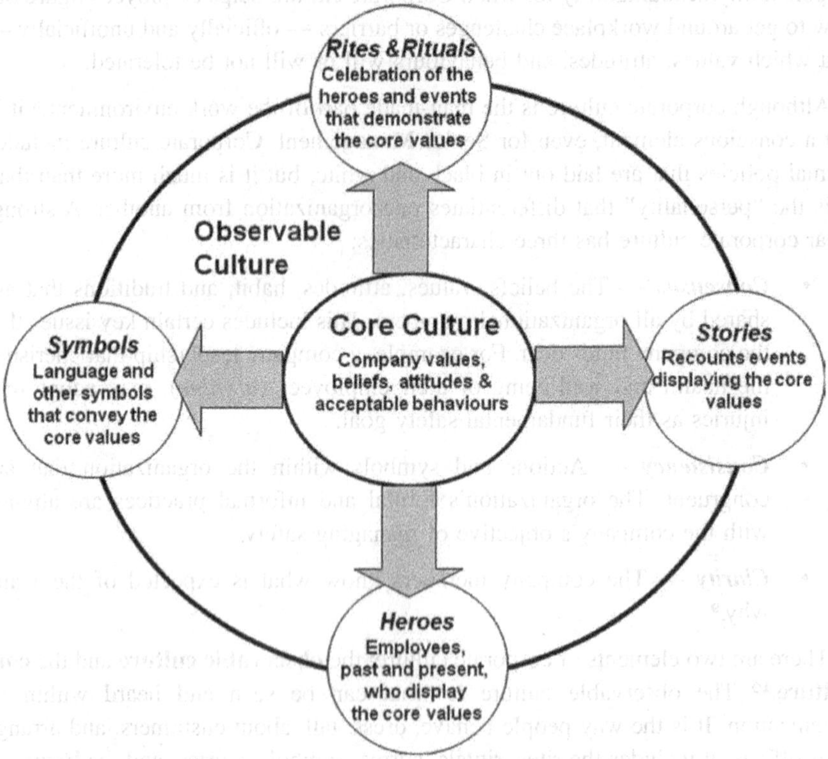

Using Schein's concepts on corporate culture, the International Atomic Energy Agency (IAEA)[12] put forth a similar construct when identifying three levels of culture. The first and most obvious level is the ***artifacts***. They parallel Schein's observable culture. The second level is the ***espoused values*** — the philosophies, goals, and strategies held dear by the organization and its people. These are the more visible aspects of Schein's core culture. The third level is the ***assumptions***. They are the fundamental beliefs that people take for granted and uphold in a subconscious manner. This is the deepest level of culture — one which can be difficult to identify. For a more comprehensive view of the IAEA's construct refer to Table 6.1.

---

[12] International Atomic Energy Agency, "Safety Culture in Nuclear Installations: Guidance for Use in Enhancement of Safety Culture" (December 2002), online: http://www-pub.iaea.org/ MTCD/publications/pdf/te_1329_web.pdf (date accessed: February 28, 2020).

Table 6.1: IAEA's Three-Level Model of Culture

Level	Level of Culture	Description	Manifestation
1	Cultural Artifacts	The architecture of facilities, personal greetings, rituals, mode of dress, and stated values	• Easy to see, feel, and identify • Most difficult to understand
2	Espoused Values of Culture	Philosophies, values, and strategies adopted by the organization	***Observable Aspects:*** • Identifiable by asking people what they believe and support • Include the attitudes and values towards equal opportunity, teamwork, empowerment, and safety ***Hidden Aspects:*** • The "talk may not match the walk" • Workers may verbalize support for safety, but in reality, they take short cuts and demonstrate a complacent approach to work

3	Cultural Assumptions	Reflects human nature, societal cultural beliefs, and organizational values	This level is very deep and hidden. It:
			• Resides at the very heart/ core of the organization
			• Encompasses many fundamental beliefs about people and human nature
			• Reflects the history, values, beliefs, and philosophies of the organizational leaders (formal and informal)
			• Impacts how workers are treated and supervised
			• Deals with assumptions about things like:
			  • Nature of time and space
			  • Why humans behave as they do
			  • Moral values
			  • Relationship values
			  • Acceptable mode of interaction
			  • Nature of truth
			  • Nature of trust

To understand an organization's culture, all three of these levels of culture need to be examined and understood. Knowledge and use of the Management Theories presented in Chapter 1, "Occupational Health and Safety: Historical Perspectives", and the ethical constructs described by Dr. Bonnie Rogers in Chapter 29, "Occupational Health and Safety: Ethical Practice", can assist with this process.

Lastly, it is important to recognize that leadership drives corporate culture. The Canadian Corporate Culture Study indicates that 90 per cent of the respondents believed that leadership drives corporate culture while 85 per cent reported that corporate culture in turn, drives performance and hence, profitability.[13] This is supported by the finding that organizations with strong corporate cultures function collaboratively and demonstrate strong corporate earnings. According to a recent Canadian study, organizations with the most admired cultures outperformed the Toronto Stock Exchange (TSX 60) by an average of almost 600 per cent.[14]

2. Subcultures[15]

Within any corporate culture, many subcultures exist. These are "mini cultures" that have their own unique beliefs, values, attitudes, work habits, and traditions. They may or may not align with or support the corporate culture. For example, the "Us/Them" subculture is common within organizations. The members of this subculture are driven to protect jobs and its members even if their actions conflict with the organization's objectives and goals. One manifestation is that they tend to disbelieve Management's intention and motives. This can have a negative impact on Management's efforts. For example, they tend to be suspicious of Management's safety initiatives, believing that workplace safety is being managed solely to control financial losses.

Some other subcultures include:

- *Occupational Subcultures* — Various occupational groups hold unique beliefs, values, attitudes, and work preferences and practices. For example, professionals have their own professional work ethics and need for autonomy and independence. Professional groups tend to work towards their ethical values, which may be in conflict with the values held by the organization. If ignored, this subculture can make it difficult to integrate these professionals into the organizational culture.

- *Ethnic and Racial Subcultures* — Some of the cultural dimensions that impact the workplace include:

---

[13] Waterstone Human Capital, Canada's "10 Most Admired Corporate Cultures Program and the Canadian Corporate Culture Study", online: http://www.waterstonehc.com/cmac/canadas-10 (date accessed: February 28, 2020).

[14] S. Klie, "Culture drives performance" *Canadian HR Reporter* (December 13, 2010) at 1 and 15.

[15] Excerpts from D. Dyck, "Health & Wellness Domain" *BCRSP Study Guide* (Mississauga, ON: Board of Certified Safety Professionals, 2015). Reprinted with permission from BCRSP.

- *Language*

    Language reflects a group's culture and the vocabulary reveals the history of the society and the aspects that are important to it. The structure of the language used can influence how one understands the surrounding environment.[16]

- *Time Orientation*

    Different cultures hold different attitudes about time. A traditional view of time is that it is circular, suggesting repetition. If something does not happen today, that is okay, because the opportunity will return tomorrow. A modern viewpoint is that time is linear. The past is gone; the present is here; and the future is almost upon us. Rather than measuring time with recurring natural events, time is measured with the precise movement of a clock.[17]

    Another aspect is the difference between monochronic (do one thing at a time) and polychronic (do many things at once) cultures. Monochronic cultures separate work and rest; polychronic cultures do not.[18]

- *Use of Space*

    Personal space is that "distance of comfort" to which we have adapted. We feel uncomfortable if others invade that space or if they are too far away for ready communication. The size of the personal space zone varies with cultures. For example, South Americans and Arabs are comfortable at closer distances than are North Americans.[19]

    Organization of space also differs. Spanish and Italian towns are set up around central squares, whereas North American towns are structured linearly along Main Street. In the workplace, North Americans prefer individual offices, whereas the Japanese prefer an open floor plan.[20]

- *Religion*

    Religion is a major component of culture. Rituals, religious days, icons, and foods are some of the visual manifestations. Religion also influences codes of ethics and moral behaviour.

---

[16] J. Schermerhorn, J. Hunt & R. Osborn, *Managing Organizational Behavior*, 4th ed. (Toronto: John Wiley & Sons, 1991), at 79-80.

[17] *Ibid.*, at 80.

[18] *Ibid.*

[19] *Ibid.*, at 80-81.

[20] *Ibid.*, at 81.

- *Values and Attitude Foundations*

  Values and attitudes regarding achievement and work, wealth and material gain, risk and change vary by culture and impact how people view employment and organizations. Four main values or attitudes that are worth consideration are:

  1. **Power Distance** — The degree to which a society accepts a hierarchical or unequal distribution of power within organizations. An organization with a small power distance is characterized by superiors viewing subordinates as "people like me". Superiors are accessible and the general belief is that all the organizational members have equal rights.
  2. **Uncertainty Avoidance** — The degree to which a society perceives unequal and ambiguous situations as threatening and to be avoided. An organization with strong uncertainty avoidance is characterized by the belief that time is money, security is paramount and documented rules and regulations are critical.
  3. **Individualism-Collectivism** — The degree to which a society focuses on individuals or groups as resources for work and social problem solving. Collectivism holds a "we" consciousness: a socialistic outlook and that decision-making is a group responsibility.
  4. **Masculinity-Femininity** — The degree to which a society emphasizes the so-called "masculine" traits such as assertiveness, independence, and insensitivity to human feelings. A masculine-oriented society has sex roles clearly differentiated, values independence, and perceives the big and fast as beautiful.[21]

  For more details, refer to Chapter 34, "Occupational Health and Safety: Diversity Considerations".

- *Generational Subcultures* — Having five generations of workers in the workplace brings with it at least five generational subcultures, each with their own beliefs, values, wants and needs. For more details on this phenomena and each of the subcultures, refer to Chapter 35, "Impact of Five Generations in the Workplace on Occupational Health and Safety".
- *Gender Subcultures* — Aspects such as different communication styles, relationship styles, primary caregiver's responsibilities, absence of mentoring programs, and the need for flexibility on work assignments, set women apart from men.

---

[21] G. Hofstede, "Motivation, Leadership and Organization: Do American Theories Apply Abroad?" *Organizational Dynamics*, Vol. 9, Summer 1980, pp. 46-49.

These are but a few of the subcultures that can exist within a workplace. For employers, Human Resources, and OH&S practitioners/ professionals, it is important to recognize which subcultures exist within an organization and to determine if they conflict with the corporate core and observable culture.

3. OH&S (Safety) Culture

An **OH&S culture** is the moral, social, and behavioural norms of an organization that are based on the shared beliefs, values, attitudes, habits, and traditions on safety that give meaning to an organization's employees and provides them with the accepted safety behaviours within their organization. *It is the way safety gets done within an organization when no one is looking.* Newcomers learn the organization's safety patterns when they join the organization, and eventually teach them to others. It is what employees believe the organization really wants and dictates whether employees will or will not adhere to the safety rules when Line Management is not around. A safety culture is what makes a workplace a safe or unsafe place to work. It is crucial for organizations to have a strong OH&S culture.[22]

An organization's corporate culture drives its OH&S practices and culture (Figure 6.2).

Figure 6.2: The Impact of Corporate Culture on OH&S and other Important Elements of Business

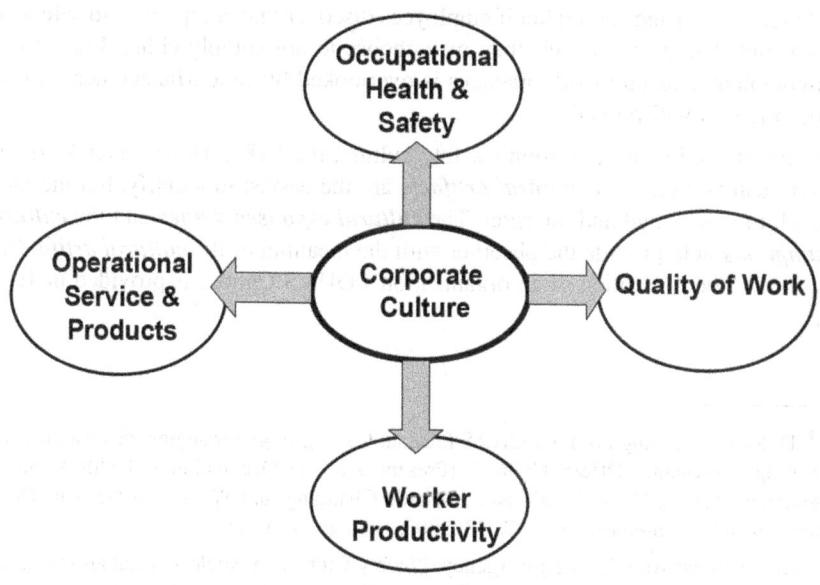

---

[22] International Labour Office, "Safety in Numbers: Pointers for Global Safety Culture at Work" (Geneva: ILO, 2003) at 21.

The OH&S culture is determined by:

- *Trust* — Does Management "walk the talk"? Do workers "walk the talk"?
- *Credibility* — Has Management demonstrated credibility by its words and actions?
- *Leadership* — What is the leadership style and does Management live by its stated values?
- *Commitment* — Does Management back its beliefs and values with appropriate support and adequate resources?
- *Care* — Do employees believe that Management and the company care about them as people?
- *Communication* — Are employees able to raise safety concerns without fear of reprisal?
- *Recognition* — Are employees recognized and rewarded for working safely?
- *Norms* — Are the expected safety behaviours known and supported?[23]

These cultural determinants are important to recognize because they shape the organization's OH&S culture. For example, culture is learned through observation and experiences within an organization. The written work rules may support workplace health and safety; but if employees discover that compliance to safe work rules is not shared by all staff, they may choose to not comply either. Likewise, if non-compliance to safe work practices is overlooked by Line Management, unsafe work practices will prevail.

To understand an organization's safety culture, the IAEA Three-Level Model of Culture can be used. The *cultural artifacts* are the easiest to identify, but the most difficult to understand and interpret. The *cultural espoused values* and the *cultural assumptions* help provide the observer with the meaning of the *cultural artifacts*.[24] Insight into the evaluation of an organization's OH&S Culture is provided in Table 6.2.

---

[23] D. Ryan, "Moving off Your OH&S Plateau: Cutting Edge Strategies for Fostering and Measuring a Dynamic OH&S Culture" (Presented at the Occupational Health & Safety Amendment Act, 2002 — What's New, What's Changing and What You Need to Do to Comply, Insight Conference, May 27-28, 2003, Edmonton, AB).

[24] International Atomic Energy Agency, "Safety Culture in Nuclear Installations: Guidance for Use in Enhancement of Safety Culture" (December 2002), online: http://www-pub.iaea.org/MTCD/publications/pdf/te_1329_web.pdf (date accessed: February 28, 2020).

Ch. 6: Occupational Health & Safety: Leadership and Commitment

Table 6.2: Insight into an Organization's OH&S (Safety) Culture

Level	IAEA Three-Level Model	Aspects	Evidence	Characteristics
1	OH&S Cultural Artifacts	• OH&S Objects • Language • Stories • Rituals • Celebrations • Behaviours	• OH&S Policy and Manual • Zero injury philosophy • Past worker injuries • Safety awards • Safety banquets/BBQs • Use of safety equipment, conformance with safe work practices and procedures	• Top Management commitment to workplace safety • Visible OH&S leadership • Use of a systems approach to OH&S • Worker self-assessment of workplace safety and safety practices • Business drivers for workplace safety • A balance between the use of sound safety practices and production demands • A sound relationship between government regulators and the company • Use of a proactive and long-term perspective to OH&S • Effective management of change • Quality management of OH&S documents and procedures • Compliance with legal regulations and company standards

Level	IAEA Three-Level Model	Aspects	Evidence	Characteristics
				- Sufficient and competent staff
- A positive worker interest in identifying safety improvements
- Strong understanding of the interaction between man, technology, and the organizational goals
- Clear stakeholder OH&S roles and responsibilities
- Positive work motivation and job satisfaction
- Worker involvement and commitment to OH&S
- Reasonable workloads, time pressures, and workplace stressors
- Regular measurement of OH&S performance
- Appropriate resource allocation
- Collaboration and teamwork
- Effective management of workplace conflict
- Positive working relationships between supervisors and workers
- Awareness of work processes
- Good housekeeping

Level	IAEA Three-Level Model	Aspects	Evidence	Characteristics
2	Espoused OH&S Cultural Values	• OH&S Philosophy	• "Safety is #1" philosophy	• High priority and time for safety
		• OH&S Goals	• Need to identify workplace hazards	• A belief that safety can always be enhanced
		• OH&S Strategies	• Zero tolerance for workplace incidents	• A propensity for openness and strong communication
			• Blame-free environment	• Organizational learning is valued
			• Errors are "learning-takes"	
			• Desire to go home safe and sound	
			• Worker safety is paramount	
3	OH&S Cultural Assumptions	• Fundamental beliefs that people take for granted and uphold in a subconscious manner	• Incidents are caused by carelessness	• A belief that a balance of past, present, and future is required
			• Some people are accident prone	• Errors are viewed as a learning opportunity — a "learning-take" not a "mistake"
			• Risks have to be taken in this business	• Safety is everyone's responsibility

Level	IAEA Three-Level Model	Aspects	Evidence	Characteristics
			• Incidents are avoidable • Properly designed facility promotes workplace safety • Use of safe work procedures prevents incidents	• Views OH&S from a systems perspective • Use of a transformational leadership style of management by all levels of management • A view that people are interested, trustworthy, motivated to work, and value personal growth and development

4. The Evolution of an OH&S (Safety) Culture

As mentioned, all organizations have cultures, whether they recognize this fact or not.[25] Some cultures are supportive; some are not. Both impact the organization's achievement of its business goals — either positively or negatively. Culture will drive strategy or drag strategy.[26] Strong OH&S cultures are characterized by good communication between Management and the company stakeholders. It enhances workplace health and safety, nurtures positive employee morale, and enables profitability.[27]

OH&S cultures tend to fit along a continuum (Figure 6.3). They are also impacted by the nature of the company's OH&S leadership and commitment.

Figure 6.3: Continuum of the Evolution of an OH&S (Safety) Culture: Distinguishing Features[28]

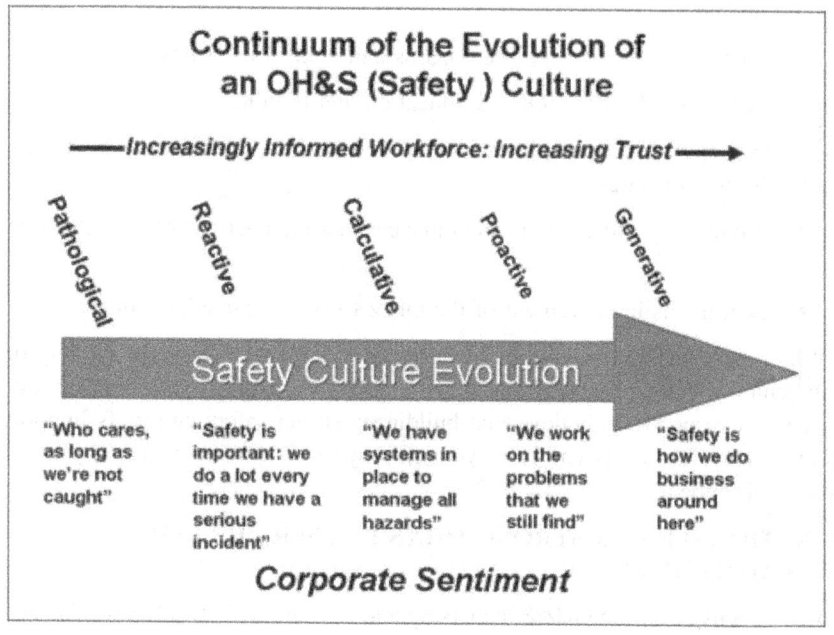

---

[25] T. Rutledge, "Culture — Not Just a Plaque on the Wall" *Canadian Occupational Safety*, April/May 2007 Issue, at 14.

[26] G. Lowe & J. Wetherow, "High Trust Cultures = Health and Performance" (Presented at the International Accident Prevention Association Conference, April 2007, Toronto: ON).

[27] P. Hudson, "Safety Culture: The Ultimate Goal" (Transport Canada, Aviation Safety Letter, 2002) at 6-7.

[28] Adapted from P. Hudson, *ibid.*, at 6.

C. KEY ASPECTS OF OH&S LEADERSHIP AND COMMITMENT

OH&S leadership is defined as the ability to enable and drive workplace safety. It includes:

- development of an OH&S vision;
- demonstrating commitment to workplace health and safety;
- requiring the development of a system that facilitates workplace health and safety, including an OH&S structure, as well as the related processes and meaningful outcome measurements;
- ensuring the OH&S system's infrastructure clearly defines stakeholder responsibility and accountability;
- provision of appropriate resources to allow the system to successfully operate;
- identification of workplace hazards and control measures;
- provision of worker OH&S education and training;
- provision of the right information, tools, and supports to enable workers to do the job safely;
- monitoring and regular performance measurement of the OH&S system; and
- continuous improvement of the OH&S system and safety culture.

Who is responsible for building a strong safety culture? That responsibility rests with Senior Management and leadership team.[29] According to a Canadian survey, 93 per cent of respondents believe that building a strong safety culture is important. "How better to show that you value your employees than creating a healthy and safe work environment?"[30]

D. RATIONALE FOR STRONG OH&S LEADERSHIP AND COMMITMENT

Why the focus on strong OH&S leadership and commitment? Leadership shapes the corporate culture and hence, the organization's safety culture and practices. In turn, strong OH&S leadership and commitment are associated with organizational productivity and profitability. Organizations with sound OH&S leadership and commitment enjoy better employee morale; more employee participation and investment in the organization; lower staff turnover rates; increased hazard reporting; and a significant reduction in recordable incident rates and costs. A strong

---

[29] T. Phillips, "Reader Panel: Building a Safety Culture" *Canadian Occupational Safety* (2007), at 12.

[30] *Ibid.*

safety culture is associated with good business and organizational success.[31] According to Steven Simon: "[s]tart with the culture. Without a good safety culture, you're planting a good seed in the desert."[32]

To operationalize the OH&S leadership and commitment concepts discussed so far, an example of an industry-tested, safety performance model follows which addresses the measurement and building of a safety culture.

E. GREAT SAFETY PERFORMANCE: AN INDUSTRY EXAMPLE[33]

In the U.S., 900,400 of the 2.8 million workplace injuries and illness among workers in 2018 resulted in recuperation time from work.[34] In Canada, 2.8 workers die from an occupational injury or disease every day,[35] and more than 264,438 workers are injured; one worker is seriously injured or killed every 2.2 minutes of work.[36]

Unsafe work practices have severe consequences for individuals and organizations such as:

- personal grief and hardship for the affected workers and their families;[37]

---

[31] D. Pratt, "How to Put Employee Health on the Executive Radar Screen" (Keynote Speech presented to Alberta Occupational Health Nurses Conference, Calgary, AB, May 25, 2000).

[32] S. Simon, Co-founder, Culture Change Consultants, quoted by D. Ryan in "Moving off Your OH&S Plateau: Cutting Edge Strategies for Fostering and Measuring a Dynamic OH&S Culture" (Presented at the Occupational Health & Safety Amendment Act, 2002 – What's New, What's Changing and What You Need to Do to Comply, Insight Conference, May 27-28, 2003, Edmonton, AB), at 15.

[33] D. Dyck & T. Roithmayr, "Great Safety Performance: An Improvement Process Using Leading Indicators" (December 2004), 52:12 AAOHN Journal at 511-520. Copyright held by the American Association of Occupational Health Nurses, Inc. Used with permission. All rights reserved.

[34] U.S. Department of Labor, "Injuries, Illnesses and Fatalities, 2018" (2020), online: http://www.bls.gov/iif/ (date accessed: February 28, 2020).

[35] According to Association of Canadian Workers' Compensation Boards, "2018 Injury Statistics", Statistics, 1027 work-related fatalities occurred in 2018. Using 365 days per year, 2.8 workers are killed per day. Statistics available online at: http://awcbc.org/?page_id=14 (date accessed: February 28, 2020).

[36] According to the Association of Workers' Compensation Boards of Canada, there were a total of 265,465 lost-time injuries (264,438) and fatalities (1027) in 2018. Using 250 days and assuming an average of an eight-hour workday, the number of work minutes is 120,000 minutes. See Association of Workers' Compensation Boards of Canada, "2017 Injury Statistics", *Statistics 2019*, online: http:// awcbc.org/?page_id=14 (date accessed: February 28, 2020).

[37] D. Dyck, *Disability Management. Theory, Strategy and Industry Practice*, 6th ed. (Toronto: LexisNexis Canada, 2017); and National Institute of Disability Management and Research, *Disability Management in the Workplace: A Guide to Establishing a Joint*

- financial liabilities for workers, corporations, and government agencies;[38] and
- potential criminal prosecution for persons who direct work and demonstrate wanton disregard of worker or public safety, along with Senior Management and directors.[39]

These facts indicate the need for a more proactive approach to the management of workplace safety — an approach that enables organization leaders to monitor all the relevant risks and take corrective action before incidents happen.

Great Safety Performance™ is an improvement process that describes how organizations can improve their safety outcomes by maximizing the conditions for safety within their workplaces. In keeping with the accepted hierarchy of workplace hazard controls, the Great Safety Performance™ places greater emphasis on the "antecedents" of behaviour as a way of identifying, monitoring, and managing the leading indicators of safe work performance. The Great Safety Performance™ also enables companies to demonstrate their level of compliance with the applicable OH&S legislation.

1. Corporate OH&S Duty of Care

In Canada, **OH&S Duty of Care** means:

> Everyone who undertakes, or has the authority, to direct how another person does work or performs a task is under a legal duty to take reasonable steps to prevent bodily harm to that person, or any other person, arising from that work or task.[40]

Organizations, corporations, and individuals who direct others to perform work, or have the authority to do so, must take reasonable and practicable steps to provide a safe and healthy workplace, and to protect workers and the public from potential harm because of the work. They must also be able to provide evidence of their actions to ensure people are protected from harm.[41]

Lapses in corporate OH&S duty can have severe consequences for organizations and their leaders. In addition to the risk of sizable financial penalties and social embarrassment, front-line supervisors, managers, executives, and directors of corporations — any individual who directs work and workers — can be held legally accountable for preventable workplace injuries and death. Leading-edge companies know that the "ideal defence" is a "sound offence". By establishing a robust OH&S

Workplace Program (Port Alberni, BC: NIDMAR, 2003).

[38] National Institute of Disability Management and Research, *Disability Management in the Workplace: A Guide to Establishing a Joint Workplace Program* (Port Alberni, BC: NIDMAR, 2003).

[39] *Criminal Code*, R.S.C. 1985, c. C-46.

[40] *Ibid.*, s. 217.1.

[41] N. Keith, *Workplace Health and Safety Crimes* (Markham, ON: Butterworths, 2004).

Management System, corporations can "bullet-proof" themselves.[42] The key elements of an OH&S Management system are:

- Management Leadership and Commitment;
- Hazard Identification, Assessment, and Control;
- Worker Qualifications, Orientation, and Training;
- OH&S Communication;
- Incident Investigation and Reporting; and
- Program/Process Evaluation and Continuous Improvement.

Safety professionals and researchers not only agree that this is the ideal OH&S management structure,[43] but recognize that 85 per cent of the safety failures in the workplace stem from system problems that only Management can address.[44]

Leading-edge companies deal with system problems by identifying and addressing the leading indicators of safety performance to maximize the effectiveness of their OH&S Management Systems. A **leading indicator of safety performance** is defined as an index designed to anticipate or forecast the Safety outcomes of current trends. This means identifying, monitoring, and continuously improving all the system conditions that are necessary to enable employees to work safely. It translates into leaders creating a workplace culture that supports a high degree of commitment to workplace safety; having a process for measuring, monitoring, and managing the leading indicators of safety performance; and demonstrating the effectiveness of the established OH&S Management System. This is the type of evidence required of company leaders to prove that they are upholding their OH&S duty.

2. Sustainable High-Performance Workplace Cultures

The Great Safety Performance™ model is particularly relevant to the leadership, communication, measurement, and continuous improvement elements of an OH&S Management System. Leadership is addressed by providing Management with data, a focus and an agenda for safety improvement and culture change efforts. Communication is focused on creating dialogue between front-line workers and Management; and involving front-line personnel in improvement planning. As for measurement, a specific set of leading indicators are defined, measured, and monitored to demonstrate the predictors of safe work in a specific job function. In terms of continuous improvement, a process to identify system performance gaps, implement improvements, monitor progress, and manage cultural shift is implemented.

---

[42] *Ibid.*

[43] G. Germain, R. Arnold, R. Rowan and J. Roane *Safety, Health, Environment and Quality Management: A Practitioner's Guide*, 2d ed. (Loganville, GA: International Risk Control America, Inc., 1998)

[44] F. Bird & G. Germain, *Practical Loss Control Leadership* (Georgia: DNV, 1996).

Strengthening these elements of an OH&S Management System is accomplished by using a model of organization performance systems to enable and support worker performance in a variety of ways. The Performance Maximizer® (Figure 6.4) illustrates the nature of human performance in the workplace by describing all the factors that exist when successful human performance pertaining to any function occurs in the workplace. The model asserts that leaders and workers need to jointly create conditions whereby everyone will:

- know *What* to do;
- be *Able* to do it;
- be *Equipped* to do it;
- Want to do it and have the experience; and
- experience *Interactions* that foster trust, respect, integrity, collaboration, and accountability.

Figure 6.4: The Performance Maximizer®[45]

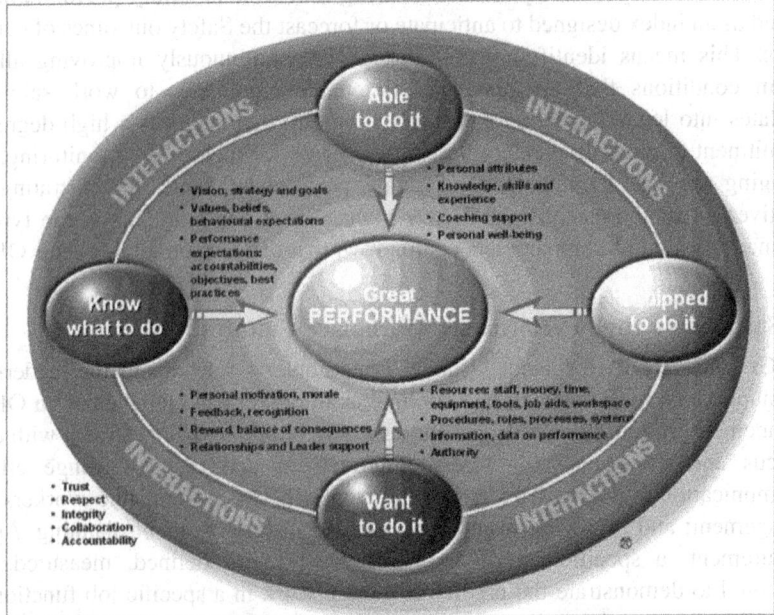

These factors have been proven to support people in performing at their best and are referred to as the "Conditions for Great Performance". Interactions refer to the quality and effectiveness of interpersonal interactions among people in the work-

---

[45] Reprinted with permission from ENMAX Corporation and Performance! . . . by design® (1999-2004); T. Roithmayr, *The Performance Maximizer®* (2000), online: http://www.performance-bydesign.com (date accessed: February 28, 2020).

place.[46] The absence of these conditions constitute obstacles and barriers to successful worker performance.

3. Applying The Performance Maximizer® to Occupational Health and Safety

The premise implicit in *The Performance Maximizer®* is that great human performance is enabled by using a performance system approach to create the right overall work environment — the *Conditions for Great Performance*. With that premise in mind and with the guidance of *Performance! . . . by design®*, an external performance consulting firm, a Canadian electrical utility embarked on the design of the first Great Safety Performance™ Model in April 2000. The company was determined to be "best in class" in regard to safe work and safe workplaces. The corporate OH&S Team identified the desired OH&S and business outcomes, hypothesized that the *Conditions for Great Performance* are the "leading indicators" of occupational safety, and formulated a set of goals for the proposed Great Safety Performance initiative. The project goals were to:

1. shift the focus of safety management from lagging to leading indicators of safety performance;
2. demonstrate a predictive relationship between leading and lagging indicators of safety performance;
3. use the leading indicators to provide data to monitor system risks and to drive injury/illness prevention;
4. provide leaders in the company with a focus and an agenda for their safety improvement and culture change efforts; and
5. "raise the bar" on safety performance by significantly improving the company's safety results.

(a) Project Methodology

To guide the project, a continuous improvement process was incorporated into the Great Safety Performance™ Model. This process consists of steps designed to systematically maximize the *Conditions for Great Performance* and create a high-performance culture that can sustain effective work practices. This process provides the structure for the following description of the methodology used in the pilot project.

(i) Explore the Situation

The Great Safety Performance™ Project began in April 2000. The initiation stage included an exploration of high-performance workplaces, defining the scope of the work to be done, and developing the goals provided above.

---

[46] The category, Interactions, was added to *The Performance Maximizer®* after the work reported in this article was begun and is therefore not included in the results. However, in our current work, Interactions have also proven to be a critical part of the performance system.

(ii) Partner with Clients and Manage Change

At various stages of the project, the partnering and change Management activities included:

- working with Management to set goals and make decisions about their participation, resource commitments, and improvement actions;
- working with front-line workers and leaders to tailor the Great Safety Performance™ Model to meet the group needs;
- educating stakeholders about the benefits of adopting the Great Safety Performance™ Model to them; and
- working with front-line workers and leaders to validate and understand data, recommend improvements, and implement actions.

In general, these involvement activities served to ensure that Management were active leaders in the process, and that all the stakeholders felt ownership of the initiative.

4. Future State: Developing the Great Safety Performance™ Model and Methodology

The OH&S Team began defining all the required elements of the Great Safety Performance™ Model by addressing two questions:

1. What does great safety performance look like?
2. What are all the specific factors that would make up the enabling *Conditions for Great Safety Performance* in the workplace?

The Great Safety Performance™ Model indicates that great safety performance consists of *Safe Work Actions* that lead to *Safe Work Outcomes* that produce positive *Safety Results* and *Individual Outcomes* which, in turn, improve the impact on *Company Business* (Figure 6.5).

Ch. 6: Occupational Health & Safety: Leadership and Commitment

Figure 6.5: Great Safety Performance: An Illustration of the Leading and Lagging Indicators of Safety

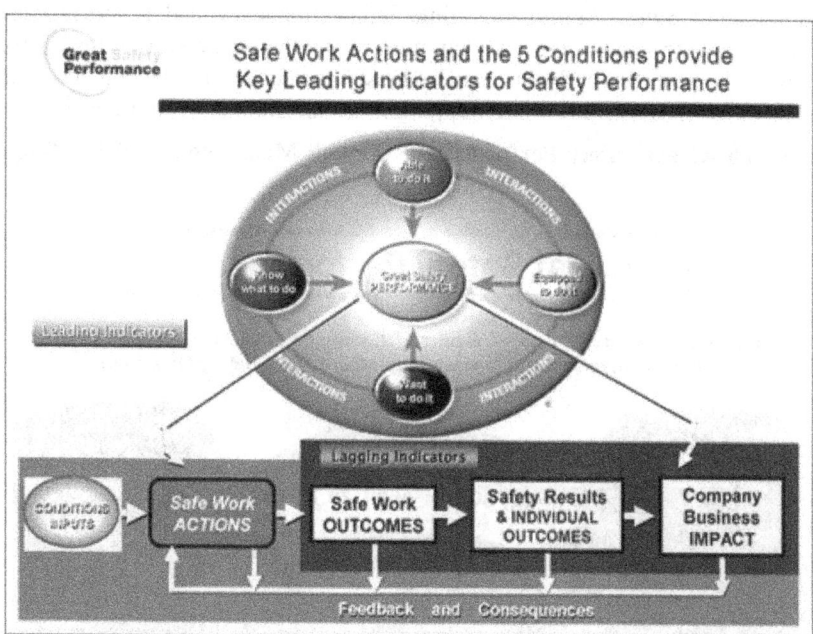

The OH&S Team identified 10 practices for the Safe Work Actions and 26 specific factors necessary to enable those Safe Work Actions. These factors are the Conditions for Great Safety Performance which specify what is required to:

- *Know What to do* to work safely (five factors).
- Be *Able to* work safely (five factors).
- Be *Equipped to* work safely (10 factors).
- *Want to* work safely (six factors).

The OH&S Team also proposed that the *Safe Work Actions* and the four *Conditions for Great Safety Performance* are the *Leading Indicators of Safety*. In the model, the *Safe Work Outcomes* and *Safety Results* elements are the *Lagging Indicators of Safety* (Figure 6.5).

The OH&S Team operated on the premise that a predictive relationship exists between the leading and lagging indicators of safety performance. That is, given the right supportive environment (*Conditions for Great Safety Performance*), workers would be enabled to perform *Safe Work Actions*, and then, would produce the desired *Safe Work Outcomes* and *Safety Results* (**Lagging Indicators of Safety Performance**). It was believed that by monitoring both the leading and lagging indicators of safety performance, the company would have the information needed to continuously improve its safety performance.

To help explain and market this novel concept, a communication metaphor using the dashboard of a car was developed (Figure 6.6). The traditional focus on predominantly *Safety Results* (lost-time injury frequency and severity) can be thought of as the view in the rearview mirror. The landscape seen in the mirror is a view of the past. Similarly, an incident must have occurred to be counted (it is in the "rearview mirror"). Prevention opportunities are limited to future actions.

Figure 6.6: Great Safety Performance™ Model: Measurement of Leading Indicators

The instruments on the automobile's dashboard monitor its performance in "real time" and allow the driver to take appropriate corrective action before problems occur. Similarly, the Great Safety Performance™ Model and methodology monitors a "dashboard" of leading indicators of safety performance and provides real-time data to guide leaders in proactively maximizing the performance environment for safety.

5. Current State: Baseline Measurement of Safety Performance

The OH&S Team determined that the model elements could be measured using a combination of "hard" and "soft" (perception) data for leading and lagging indicators of safety performance. For lagging indicators, the typical safety results were available: incident and injury records, frequency and severity rates, as well as business data. In terms of the leading indicators of safety performance, the company already tracked a variety of "hard" data such as worker training records, worksite inspection reports, equipment inspection and maintenance logs, vehicle inspection logs, and safety audit results. However, significant factors in the *Conditions for Great Safety Performance* are rooted in the actual workplace experience of workers themselves and for these there were no practical "hard" data possible. The OH&S team decided to measure these factors using a worker perception tool.

The *Workplace Safety Survey*, a 36-item survey instrument, was developed to measure all the leading indicators of safety performance as well as to collect the needed perception data. Using a six-point frequency scale ranging from "0" (almost never) to "5" (almost always), workers were asked to rate how frequently they engage in the *Safe Work Actions* — a measure of their own performance (10 items). In the second part of the survey, workers used a six-point agreement scale ranging from "0" (strongly disagree) to "5" (strongly agree) to rate 26 items on the degree to which they actually experienced a variety of factors required to support their safe work actions (*i.e.*, the Conditions for Great Safety Performance — a measure of how well the performance system supports them (26 items)). Figure 6.7 illustrates the design of this workplace safety survey tool.

All the survey items were then rated again on how important the workers felt it is to:

- perform the listed *Safe Work Actions* to ensure their safety; and
- have the *Conditions for Great Safety Performance* in place to enable them to perform the identified *Safe Work Actions*.

Importance was rated on a six-point importance scale from "0" (Very Unimportant) to "5" (Very Important).

Figure 6.7: Design of Workplace Safety Survey

Overview of the Workplace Safety Survey

Scale = 0 1 2 3 4 5

Part 1: My EXPERIENCE ~ What actually happens			Part 2: My Belief about IMPORTANCE to be safe		
Part 1A	Safe Work Actions Almost Never ⟷ Almost Always	Page 2	Part 2A	Safe Work Actions Very Unimportant ⟷ Very Important	Page 6
	Conditions for Great Safety Performance: Strongly Disagree ⟷ Strongly Agree			Conditions for Great Safety Performance: Very Unimportant ⟷ Very Important	
Part 1B	• Know What to do • Able to do it • Equipped to do it • Want to do it	Page 3	Part 2B	• Know What to do • Able to do it • Equipped to do it • Want to do it	Page 7
Part 1C	• Interactions	Page 5	Part 2C	• Interactions	Page 9

The purpose of administering the survey was to:

- establish the reliability of the survey tool;
- determine whether a predictive relationship between *Safe Work Actions* and the *Conditions for Great Safety Performance* could be established;
- determine if a predictive relationship between the leading and lagging indicators of safety performance could be established; and
- determine the overall impact of the Great Safety Performance™ Model on improving workplace safety.

The research group was comprised of 65 Meter Readers that were new to the company. Their safety results (lost-time injury frequency and severity rates) were five times higher than that of the rest of the company, and they believed that workplace injuries were an inherent part of meter reading.

Meter Readers are utility workers who travel about the city reading residential, commercial, and industrial utility meters located at customer sites. While driving to and walking about their assigned routes, they encounter numerous hazards such as traffic, slippery sidewalks, decks or stairs, agitated dogs, insect nests, poorly lit stairwells, violent people, cluttered walkways, temperature extremes, and the like. Although the company appeals to its customers to keep access to the meters clear and safe, many hazards remain.

Prior to implementing the pilot project, the utility Meter Readers were educated on the Great Safety Performance™ Model, the planned measurement techniques, and how the appropriate remedial actions would be determined. They were also advised that their individual responses would be kept confidential because all the results would be reported in aggregate form. They were then invited to participate in the pilot project.

Before any intervention took place, the Meter Readers completed a baseline *Workplace Safety Survey* which was tailored to the meter-reading situation (December 2001). The purpose of administering the surveys with Meter Readers specifically was to:

- establish a level of current safety performance;
- identify performance gaps and their causes;
- develop recommendations for improving safety performance; and
- track changes in worker perceptions of the conditions for performance and their safety performance over time (quarterly during the first year).

The survey data were processed, analyzed, and interpreted by *Performance!. . . by design*® in collaboration with a researcher at the Department of Psychology, University of Calgary, Alberta. The survey data were analyzed using special statistical software (SSPS) suited for this type of data. The reliability and validity of the *Workplace Safety Survey* tool were also analyzed.

Means were calculated for all items and a constant value of 20 was used as a multiplier for each. This was done to facilitate both the interpretation of the results and their presentation as "gauge scores" in the Great Safety Performance "dashboard" display. The highest possible mean score (or gauge score) for each leading indicator was 100 (as opposed to 5, the highest rating on the survey scale).

6. Gap Analysis

The analysis of baseline results indicated a variety of relatively lower mean scores in both *Safe Work Actions* and the *Conditions for Great Safety Performance*. Based on the researchers' hypothesis, low mean scores in the *Conditions for Great Safety Performance* are linked to low mean scores in *Safe Work Actions*. To understand and validate the data, the results of the *Workplace Safety Survey* were compared against the "hard" safety data as well as other available measures such as results of the company Employee Relationship Survey (soft data). Likewise, discussions were held with Meter Readers and their management to gain insight into the data.

(a) Develop Responses and Implement Actions

The OH&S Team worked with Management and employee groups to identify appropriate improvement actions. Through collaboration, company Management, the Meter Readers, and the OH&S Team initiated a variety of improvements actions in order to reduce or eliminate the identified barriers and obstacles to improved safety performance. Table 6.3 provides a sampling of weaknesses in the *Conditions for Great Safety Performance* — actions that were taken and the results achieved.

(b) Measurement and Follow-through

Quarterly *Workplace Safety Surveys* were administered between December 2001 and December 2002. One more was done, July 2003, at the end of the 18-month pilot period.

In July 2002, the Meter Readers were asked to record the number of occurrences they personally had in each of five injury/incident categories: First Aid, Medical Aid, Lost-time Injury, Motor Vehicle Accident, and Property Damage. This data would be used to determine if a predictive relationship between the leading and lagging indicators of safety performance could be established.

Data for all surveys were again processed, analyzed, and the results communicated to Management and the Meter Readers as described above. With this information on progress, Management and the OH&S Team were able to make decisions about the success of their improvement efforts and take further corrective action as needed.

7. Project Results and Findings: Reliability of the Workplace Safety Survey

The *Workplace Survey* proved to be an impressively reliable measurement tool for the leading indicators of safety performance. Method 2 (Covariance Matrix) was used for this analysis and the results are reported in the Reliability Analysis Table in the Appendix. This analysis determines the extent to which the items in each

Condition for Performance are related to each other; and measures the internal consistency of each scale.

8. Conditions for Great Safety Performance: Actual Performance of the System for Enabling Safe Work Actions

The degree to which Meter Readers experienced the four *Conditions for Great Safety Performance* during the 18-month pilot period is shown graphically in Figure 6.8 and numerically in Table 6.3. The July 2003 results indicate that since December 2001 (baseline), there had been a modest improvement in the workplace system to enable workers to *Know What to do* (7.5 per cent) and be *Able to do it* (10.3 per cent). Stronger improvement took place in the *Equipped to do it* (15.2 per cent) gauge score. The *Want to do it* gauge score measured feedback, recognition, Management's responsiveness to unsafe conditions and practices, and motivation to perform well. It is in the *Want to do it* category that the most dramatic improvement occurred (42.9 per cent). Improvement in a variety of factors in the conditions for great safety performance drove these results. Table 6.3 provides a sampling of the specific improvements noted.

9. Importance of the Conditions for Great Safety Performance

For each of the individual items in the *Conditions for Great Safety Performance*, the Meter Readers were asked to indicate how important each item is for them to work safely every day. The Meter Readers steadfastly reported the individual items as being "important" or "very important" for them to work safely (Figure 6.8). Overall, the Importance Ratings for the *Conditions for Great Safety Performance* have increased approximately 9 per cent from baseline.

Table 6.3: Sample Results of Specific Improvements in Conditions for Great Safety Performance

Issues identified	Actions implemented	Survey score increase after 18 months
Some lack of awareness regarding expectations and standards as well as priorities and direction.	Clarify and strengthen the communication of expectations, Management's commitment to safety and the importance of both Safe Work Actions and the Conditions for Great Safety Performance.	13%
Meter Readers were unclear about how their safety performance is measured and received little information in that regard.	More information provided at monthly safety meetings; quarterly progress updates using Great Safety Performance Survey data.	14%

CH. 6: OCCUPATIONAL HEALTH & SAFETY: LEADERSHIP AND COMMITMENT

Issues identified	Actions implemented	Survey score increase after 18 months
There were serious issues with hazards at customers' meter reading sites; insufficient signage, guards and barriers.	Gain customers' cooperation in controlling or eliminating hazards at meter sites.	40%
Some deficiency in properly designed tools/equipment.	Improve some tools; Meter Readers select own footwear.	14%
Quick access to assistance or guidance was not readily available when needed.	Provide cell phones to maintain contact while en route.	14%
Meter Readers received infrequent (meaningful) recognition for doing their jobs safely.	Enhance front line leadership practices to increase the quantity and quality of performance feedback and recognition.	56%
Meter Readers received insufficient specific feedback when they did their jobs safely.		56%
Meter Readers had insufficient helpful corrective feedback when doing their jobs in an unsafe manner.		23%
Management was not perceived to be quick and decisive in responding to unsafe conditions and practices.	Improve Management's responsiveness to safety issues and communication of actions and results.	23%
It was not absolutely clear to Meter Readers that they could refuse to enter hazardous sites.	Clarify authority to make individual decisions regarding safety at work and Management then supporting their decisions.	17%

Figure 6.8: Mean Scores for the *Conditions for Great Safety Performance*

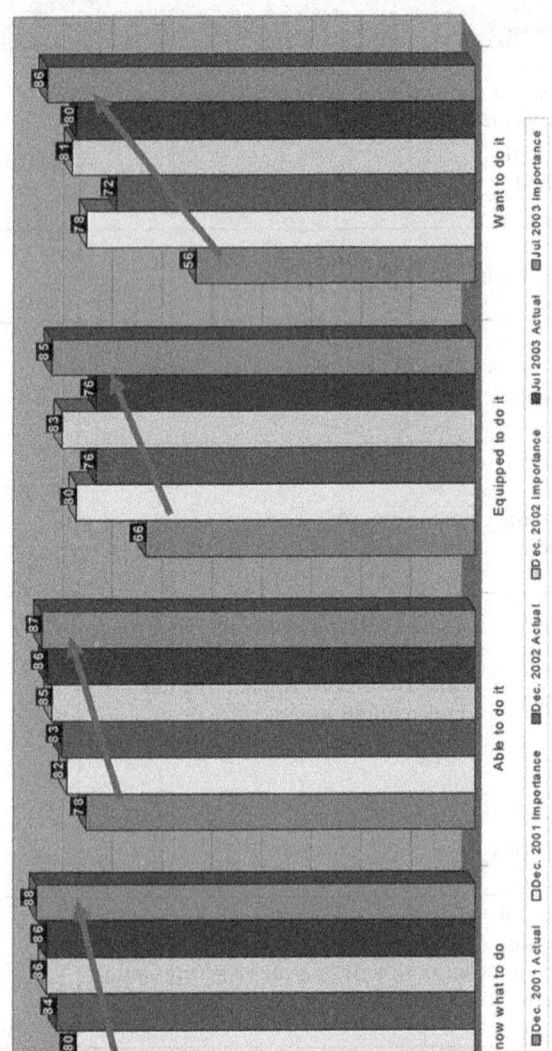

Actual = the degree to which Meter Readers reported that these Conditions were in place to support their safety.

Importance = the degree to which Meter Readers reported that these Conditions are important for their safety.

CH. 6: OCCUPATIONAL HEALTH & SAFETY: LEADERSHIP AND COMMITMENT

10. Actual Performance of Safe Work Actions

The mean scores for the 10 *Safe Work Actions* measured by the Workplace Safety Survey are presented graphically in Figure 6.7. The degree to which the Meter Readers reported that they engage in *Safe Work Actions* improved 7.5 per cent from baseline (Table 6.4). This indicated that the Meter Readers perceived their performance of the *Safe Work Actions* to be higher than "very frequently" and approaching the "almost always" level (A score of 80 equates to "very frequently" and 100 to "almost always" on the survey instrument, and the individual items as being "important" or "very important" for them to work safely (Figure 6.7)). This finding aligned with the dramatic decrease in the group's actual safety results (injury frequency and severity rates) over the 18-month pilot period (Table 6.4). Overall, the Importance Ratings for the *Conditions for Great Safety Performance* increased approximately nine per cent from baseline.

Table 6.4: Great Safety Performance "Dashboard"

Leading Indicators	DEC. 2001	DEC. 2002	JULY 2003	% Change
Know What to do	80	84	86	7.5
Able to do it	78	82	86	10.3
Equipped to do it	66	76	76	15.2
Want to do it	56	72	80	42.9
Safe Work Actions	80	80	86	7.5
Lagging Indicators	DEC. 2001	DEC. 2002	JULY 2003	
Frequency of Injuries	33.08	24.95	0.0	
Severity of Injuries	288.87	61.26	0.0	

11. Importance of Safe Work Actions

For each of the individual items in Safe Work Actions, Meter Readers were asked, "[t]o work safely everyday, how important it is to engage in each of the ten individual *Safe Work Actions*?" The average ratings of importance increased approximately five per cent over the 18-month period (Figure 6.9). By July 2003, the Meter Readers were rating the individual items as "important" to "very important" for working safely on the job.

The correlation analysis done on all the surveys conducted in 2002 and July 2003 indicates that all the Actual Frequency and Importance Rating scores strongly correlate. For the 10 *Safe Work Actions* the analysis produced Pearson Correlations ranging from 0.601 to 0.778 with significance (1-tailed) at 0.000 for all 10. The more important the Meter Readers rated a *Safe Work Action*, the more likely they were to engage in that *Safe Work Action*. This meant that increasing Meter Reader perceptions about the importance of *Safe Work Actions* would be a very strong influence on the *Safe Work Actions* being performed with more frequency. This significant finding should encourage Management and industry trainers to accelerate and make more effective their efforts to help workers appreciate, value, and commit to acting on corporate safety standards and practices.

(a) Predictive Relationship

This project sought to establish a predictive relationship between the *Conditions for Great Safety Performance* and *Safe Work Actions*, and subsequently, between *Safe Work Actions* and *Safe Work Outcomes*. Experience, logic, and data from other areas tell us that this relationship is intrinsic to worker safety performance.

(b) Extent to which the Conditions for Great Safety Performance are Predictive of Safe Work Actions

Bivariate Correlation Analysis (Pearson; 1-tailed) was done on the data for each round of surveys administered during the 18-month period. The data support a strong predictive relationship from the *Conditions for Great Safety Performance* to *Safe Work Actions*. Correlation Analysis (Independent variables = *Conditions for Great Safety Performance;* dependent variable = *Safe Work Actions*) yielded Pearson Correlations of 0.447 to 0.726 with significance at $=/ .001$.

It is also very important to note that the *Conditions for Great Safety Performance* are highly correlated with each other (see Correlation Analysis Table in the Appendix). This finding supported the researchers' assertion that *The Performance Maximizer®* illustrates a "performance system" and that the *Conditions for Great Safety Performance* were a set of performance enablers that were characterized by high degree of interdependence. The important implication for how leaders manage safety performance is addressed in the discussion that follows.

(c) Extent to which the Safe Work Actions are Predictive of Safe Work Outcomes

Once the methodology for identifying and measuring the leading indicators of safety performance proved reliable and predictive (valid), the OH&S Team sought to correlate the leading indicators of safety performance (*Conditions for Great Safety Performance* and *Safe Work Actions*) with the actual lagging indicators of safety performance (*Safe Work Outcomes; Safety Results* — workplace injury frequency and severity rates). Both logic and experience tell us that *Safe Work Outcomes* almost always follow consistent performance of *Safe Work Actions*. By establishing a predictive relationship to *Safe Work Outcomes*, the OH&S Team hoped to understand the degree to which some, or all, of the *Conditions for Great Safety Performance* needed to improve in order to achieve better *Safe Work Outcomes* and *Safety Results*.

In July 2002, the OH&S Team began tracking the predictive relationship between *Safe Work Actions* and *Safe Work Outcomes*. The data collected indicated that *Safe Work Actions* negatively correlated to the occurrence of First Aid and Medical Aid injury incidents (Pearson Correlations and Significance respectively: -.370, .022; -.374, .021). That is, the higher the *Safe Work Actions* score, the less likely it is for First Aids and Medical Aids to occur. Since then, statistical testing of the predictive relationship between *Safe Work Actions* and *Safe Work Outcomes* could not be achieved due to a relatively small target population and a dramatic reduction in the number of workplace injury incidents (see Table 6.4). Correlations cannot be established when the injury/incident numbers are at, or close to, "0". This idea was abandoned in favour of finding a larger population with potentially more workplace incidents.

F. PROJECT CONCLUSION

Great Safety Performance™ combines a model of organization performance systems with a continuous improvement process. It requires the integration of perception ("soft") data and objective ("hard") data for the *Conditions for Great Safety Performance*. The available "hard data" supported the trend noted in the relevant perception data.

The results achieved through the application of the Great Safety Performance™ Model with the pilot group (Meter Readers) supported several significant conclusions:

- The data collection and performance measurement methods proved to be impressively reliable. The four *Conditions for Great Safety Performance* were statistically significant predictors of on-the-job safety performance.
- By strengthening the *Conditions for Great Safety Performance*, obstacles and barriers to the performance of the *Safe Work Actions* were reduced or removed.
- Managing and improving the *Conditions for Great Safety Performance* led to improvements in *Safety Results*.
- Workers' beliefs about the importance of *Safe Work Actions* strongly influenced their performance of those practices.
- The *Conditions for Great Safety Performance* is an interdependent set of variables that must be managed as a system.

G. DISCUSSION

The Great Safety Performance™ Model is intuitively sound and easy to understand and use. The *Workplace Safety Survey* tool, which was found to be valid and reliable, allows an organization to measure the degree to which workers practise *Safe Work Actions* and the degree to which they are enabled to do so by the workplace performance environment. By doing repeated surveys, safe work practices and their enabling conditions can be tracked and intervention efforts evaluated. This is one way in which a company can demonstrate its commitment to workplace safety and provide evidence of its "OH&S duty of care" for workers.

Knowing that there is a predictive relationship between the five[47] *Conditions for*

---

[47] In subsequent Great Safety Performance™ projects *Interactions* also demonstrated strong correlations to the performance of *Safe Work Actions* on the job.

Great Safety Performance and the *Safe Work Actions*, company leaders who want to improve safety outcomes, can focus on the leading indicators of safety performance as opposed to dwelling on the lagging ones. Using this approach, there is an opportunity to change the safety outcomes, as opposed to merely observing and reporting on them. As Charles E. Gilmore asked the National Safety Congress, "[w]hat is the sense of measuring, if the loss must occur, before you can act? That is reaction, not control."[48]

It is very valuable to know that as worker perceptions about the importance of *Safe Work Actions* increase, the frequency with which the desired *Safe Work Actions* are performed increase as well. Instead of dictating to workers that certain behaviours must be practised, Management and trainers can be more effective if they explain the benefits of the desired *Safe Work Actions* to workers and their families. In essence, focus on "what is in it" for each player to practise the proposed *Safe Work Actions*. This approach is consistent with current research on effective occupational safety training techniques.[49]

Although the predictive relationship between *Safe Work Actions* and Safe Work Outcomes could not be conclusively proven in this pilot project, there is a strong indication that there is a direct relationship. Once the Great Safety Performance Model™ and its elements were initiated, the target group began to demonstrate a dramatic reduction in workplace Safety Results. The injury frequency rate[50] for the group (Table 6.4) dropped from 33 (December 2001) to 0 (July 2003) and the injury severity rate[51] decreased from 289 (December 2001) to 0 (July 2003).

More importantly, a significant cultural change occurred. Line Management and the Meter Readers moved from a state of "learned helplessness" to one of "being empowered". For example, when they joined the company in December 2000, they voiced the belief that workplace injuries were "just part of meter reading". By July 2003, the group realized that they could work safely without experiencing any workplace injuries. Their experience changed their beliefs.

Two additional important findings emerged. First, the research confirmed that the *Conditions for Great Safety Performance* are highly interdependent. This is a crucial message to leaders and OH&S practitioners/ professionals. They must recognize that in managing safety, they are dealing with an interconnected system. Investigations, analyses and remedies that have a singular focus may be oblivious to other factors in the safety "performance system" which are influencing safety practices and

---

[48] F. Bird & G. Germain, *Practical Loss Control Leadership* (Georgia: DNV, 1996).

[49] M. Colligan & A. Cohen, "The role of training in promoting workplace safety and health" in J. Barling & M.R. Frone, eds., *The Psychology of Workplace Safety* (Washington, DC: American Psychological Association, 2004) at 223-264.

[50] **Injury Frequency rate** is the number of lost-time injury incidents per 200,000 work hours.

[51] **Injury Severity rate** is the number of lost-time workdays due to a work injury per 200,000 work hours.

results. This means that safety performance can only be maximized when all the enabling conditions are considered, planned for, measured, and managed as interdependent elements.

Second, Great Safety Performance™ can serve as a vehicle to quantify, document, and demonstrate the efforts a company invests to create a safe workplace with safe work practices. Such information is valuable evidence which can attest to the company's due diligence in promoting and providing a safe workplace as well as its compliance in meeting its OH&S duty.

Lastly, using the Great Safety Performance™ Model, organizations can design and implement a variety of high-leverage improvement initiatives specific to their business situations. These include:

- identifying the leading indicators for safety performance;
- assessing the gaps between ideal and actual safety performance;
- developing workable solutions to strengthen the *Conditions for Great Safety Performance* and deliver the desired safety results;
- establishing a monitoring mechanism to measure the effectiveness of the implemented initiatives and interventions; and
- creating a system that demonstrates organizational/corporate commitment to workplace safety.

Figure 6.9: Mean Scores for Safe Work Actions — Actual Performance and Importance Scale: 0-100

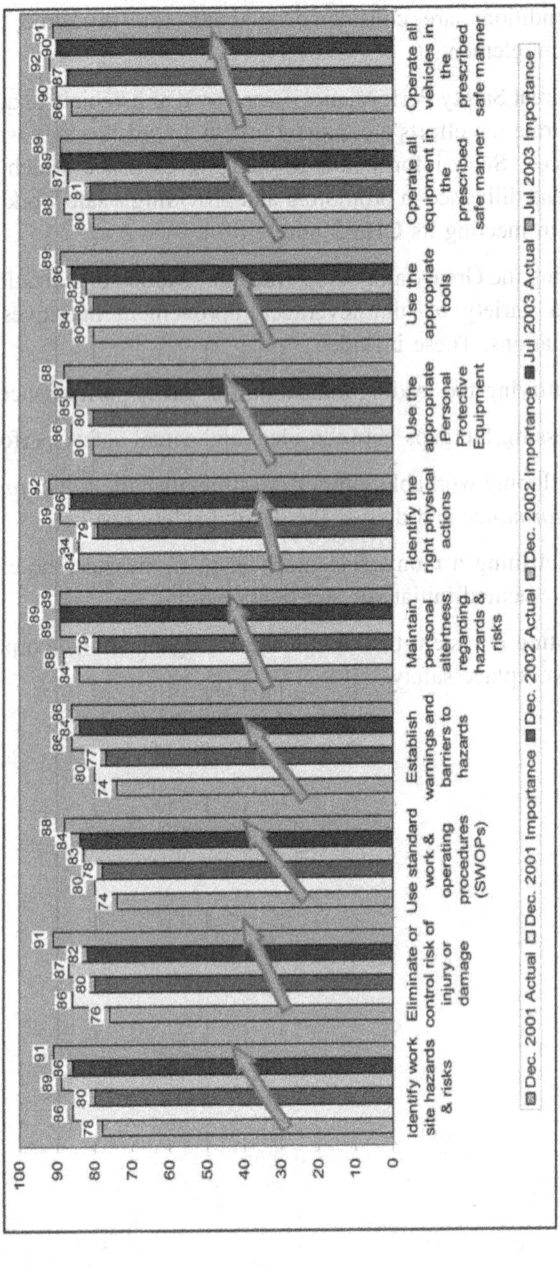

Safe Work Action	Dec. 2001 Actual	Dec. 2001 Importance	Dec. 2002 Actual	Dec. 2002 Importance	Jul 2003 Actual	Jul 2003 Importance
Identify work site hazards & risks	78	80	86	89	86	91
Eliminate or control risk of injury or damage	76	80	86	87	82	91
Use standard work & operating procedures (SWOPs)	74	78	80	83	84	88
Establish warnings and barriers to hazards	74	77	80	86	84	86
Maintain personal alertness regarding hazards & risks	79	84	89	89	89	—
Identify the right physical actions	79	84	84	86	92	—
Use the appropriate Personal Protective Equipment	80	85	86	87	88	—
Use the appropriate tools	80	82	84	86	89	—
Operate all equipment in the prescribed safe manner	80	81	87	88	89	—
Operate all vehicles in the prescribed safe manner	86	87	90	92	91	—

Actual = how frequently Meter Readers perform the *Safe Work Actions*.

Importance = how Important Meter Readers believe it is to perform the *Safe Work Actions* for their own safety.

H. Measurement of OH&S leadership and commitment

An organization's culture is a critical determinant of its longevity and production, service, quality, and health and safety success.[52] However, the development of a strong safety culture requires regular measurement of the status of an organization's OH&S leadership and commitment, as well as taking actions towards continuous improvement.

Measurement of OH&S leadership and management commitment can be challenging. Although there are many available measurement approaches, it is important to closely examine them to determine if they are appropriate to the task:

- **Safety Statistics** — Comparison of organizational safety statistics (lagging indicators of safety performance) has been, and is, the dominant measurement approach used to establish the functionality of an OH&S program and its leadership. However, this approach is fraught with problems. Lagging indicators of safety performance do not predict future safety success. As well, incident statistics do not measure the quality of Line Management supervision or the quality of the safety program. There are too many confounders involved in the occurrence of an incident and the outcome data can be misleading. As well, fluctuations in year-to-year incident statistics can be difficult to explain.[53]

- **Auditing** — OH&S audits measure the organization's OH&S program against recommended standards for OH&S management systems. These audit protocols include performance measures of OH&S leadership and commitment. The intent of an audit is to identify system strengths and weaknesses known to prevent or cause incidents. However, few audits have a system focus and are not designed to identify "why" deficiencies exist or how well the OH&S system is functioning.[54]

- **Perception Surveys** — OH&S perception surveys are used to garner employee opinions and perceptions of how well the organization's OH&S system is working. Many researchers believe that perception surveys are a much better predictor of an organization's future safety performance than any of the other indicators.[55] It is a common belief that an organization's

---

[52] D. Ryan, "Corporate Safety Culture: A Determinant of Program Effectiveness" (*CSSE Contact*, Winter 2002), at 10-11.

[53] D. Ryan, "Moving off Your OH&S Plateau: Cutting Edge Strategies for Fostering and Measuring a Dynamic OH&S Culture" (Presented at the Occupational Health & Safety Amendment Act, 2002 — What's New, What's Changing and What You Need to Do to Comply, Insight Conference, May 27-28, 2003, Edmonton, AB).

[54] S. Simon, "Safety Culture Assessment as a Transformative Process" (Culture Change Consultants, Larchmont, NY, 1998).

[55] N. Novak, "Evaluation of Safety Indicators", Masters Thesis (Department of Occupational Health, McGill University, September 22, 2006).

OH&S program is only as good as employees perceive it to be. However, the results of perception surveys are a measurement of soft data only and to be valid, they must be well designed and carefully implemented.

- **Safety Culture Assessment** — Dr. Steve Simon recommends the use of a transformative approach when conducting a safety culture assessment. By assessing the safety culture, organizations can gain insight into the status of other performance areas, such as productivity, quality, cost control, and customer service. The intent is to create a cultural database from which information on the organizational climate,[56] as well as the cultural processes that enable the OH&S program to function, can be obtained. The assessment measures the leading indicators that can make or break an OH&S program and can be a valuable resource for cultural change.[57]

Dr. Simon's transformative cultural assessment is more than a perception survey. It is a management and leadership tool that uses a change intervention process based on the collection of qualitative and quantitative data. It is best used in conjunction with a model that ensures a comprehensive approach to evaluating the entire organization and that provides a shared framework for interpreting the data and making recommendations.

Dr. Steven Simon developed the Simon Open System (S.O.S.) Culture Change Model™ for this purpose. It "views safety performance as an integral part of organizational work, technology, systems, people and culture. The S.O.S. Model embraces a whole systems perspective rather than focusing on individual, fix-it strategies".[58] The intent is to capture all the elements that impact safety performance.

The S.O.S. Model evaluates the safety process through an examination of the leading indicators of safety performance. It develops a profile of all the barriers and supports that impact how an organization manages safety. This includes the structural, technological, social, incentives, and measurement systems (Figure 6.10).

---

[56] **Organizational Climate** is the manifestation of the organization's culture. This concept is discussed in Chapter 8, "Safety Culture: Safety Climate".

[57] S. Simon, "Safety Culture Assessment as a Transformative Process" (Culture Change Consultants, Larchmont, NY, 1998).

[58] *Ibid.*, at 3-4.

[59] Printed with permission from S. Simon, "Safety Culture Assessment as a Transformative Process" (Culture Change Consultants, Larchmount, NY, 1998).

Figure 6.10: S.O.S. Culture Change Model™: Framework for Diagnosis and Action Planning[59]

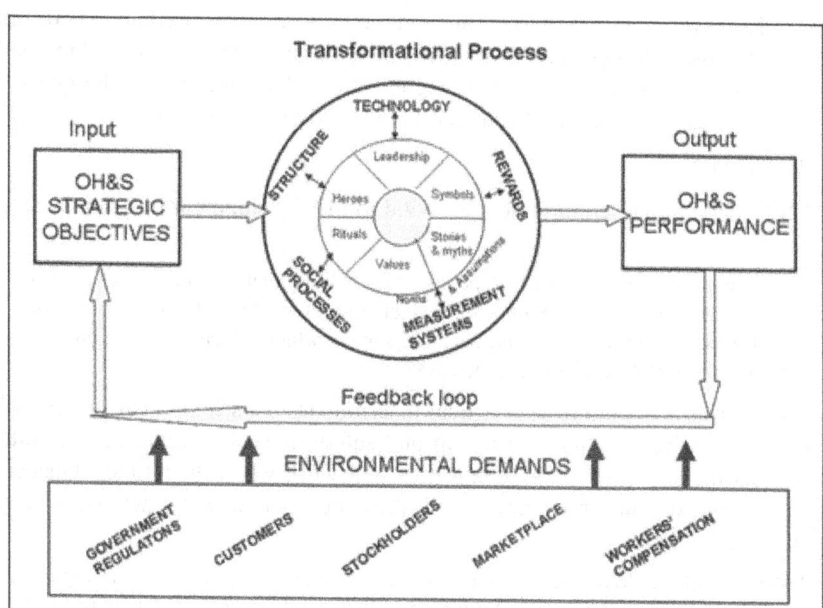

A company's OH&S strategic objectives influence and contribute to the development of corporate culture, which in turn impacts the company's OH&S performance. This model and its concepts parallel the concepts of the Great Safety Performance™ Model, which was presented earlier and will be discussed further in Chapter 7, "Great Safety Performance: A Management Approach".

- **Measurement of the OH&S Performance System** — One of the unexpected benefits of the Great Safety Performance™ methodology is that it not only measures the nature of the workplace environment in which employees function, it also permits performance measurement of the organization's OH&S management system both at initial implementation (baseline) and as time progresses. In this manner, organizations can identify areas of improvement and the issues that warrant attention. As well, they can track and demonstrate the improvements made to the OH&S management system over time. This can attest to their OH&S Duty of Care and to their due diligence (Figures 6.8 and 6.9).

The strengths of this approach is that it measures the OH&S management system, safety performance, employee perceptions and values, and corporate culture, as well as identifying what areas need to be addressed and where employees perceive improvements can be made. As well, it can result in sustainable changes to safety performance, the safety culture and ultimately, to the corporate culture.

- **Comparison with "Best in Class" Companies** — Benchmarking can be used to compare OH&S leadership and commitment among organizations. However, it is critical to ensure that the measurements used are comparable, and the findings may not be relevant to the organization. Additionally, noting "differences" between organizations does not answer the questions such as "why?", and "what can be done to evoke the desired changes?"

I. SUMMARY

OH&S leadership and commitment are vital to the development of a sound Safety culture:

> Leadership is about people, and people who believe that executives really care about what happen to them will be more likely to follow when other initiatives arise . . . Reducing injuries is a visible accomplishment which affects everyone and one which, employees genuinely appreciate.[60]

An organization's culture is a critical determinant of its long-term success with production, service, quality, and health and safety. It is the organization's Safety culture that dictates how safely employees will work when their Line Manager is absent. It also dictates how new employees are socialized to the ways of the organization:

> Safety is culture-driven, and management establishes the culture . . . an organization's culture translates into a system of expected behaviour . . . The injury and illness experience that results are a direct reflection of an organization's safety culture . . . Major improvements in safety will be achieved only if a change in culture takes place — only if major changes occur in the system of expected behavior.[61]

A strong positive culture is required if an organization is striving to become a leading-edge organization. According to D. Pratt, Healthy Business Inc., "[g]ood health is good leadership, is good business."[62]

---

[60] T. Pattenden, Presentation made at the 1994 OSH Conference, as reported by D. Pratt in "How to put Employee Health on the Executive Radar Screen" (Keynote Speech, presented to the Alberta Occupational Health Nurses Conference, Calgary, AB, May 25, 2000).

[61] F.A. Manuele, *Advanced Safety Management: Focusing on Z10 and Serious Injury Prevention* (New York: John Wiley & Sons, 2008) at 82.

[62] D. Pratt, "How to put Employee Health on the Executive Radar Screen" (Keynote Speech presented to the Alberta Occupational Health Nurses Conference, Calgary, AB, May 25, 2000).

Appendix 1

RELIABILITY AND CORRELATION TABLES

Reliability Coefficients — Standardized item alpha (=/ .7 = Very Good)

Leading Indicators	# of Items	April 2002	July 2002	October 2002	December 2002	July 2003
Know What to do	5	.85	.78	.85	.81	.85
Able to do it	5	.82	.70	.85	.76	.81
Equipped to do it	10	.81	.80	.82	.86	.85
Want to do it	6	.93	.85	.89	.88	.86
Safe Work Actions	10	.91	.92	.95	.87	.86
N — number of cases		49	45	57	37	48

Method 2 (Covariance Matrix) was used for this analysis.

Bivariate Correlation Analysis: July 2003 — Meter Readers (total population = 65)

		Know What to do	Able to do it	Equipped to do it	Want to do it
Know What to do	Pearson Correlation	1	.762(**)	.697(**)	.644(**)
	Sig. (1-tailed)		.000	.000	.000
	N	48	48	48	48
Able to do it	Pearson Correlation	.762(**)	1	.643(**)	.648(**)
	Sig. (1-tailed)	.000		.000	.000
	N	48	48	48	48
Equipped to do it	Pearson Correlation	.697(**)	.643(**)	1	.881(**)
	Sig. (1-tailed)	.000	.000		.000
	N	48	48	48	48
Want to do it	Pearson Correlation	.644(**)	.648(**)	.881(**)	1
	Sig. (1-tailed)	.000	.000	.000	
	N	48	48	48	48
Safe Work Actions	Pearson Correlation	.516(**)	.726(**)	.484(**)	.447(**)
	Sig. (1-tailed)	.000	.000	.000	.001
	N	48	48	48	48

CHAPTER REFERENCES

Association of Canadian Workers' Compensation Boards, "2017 Injury Statistics" *Statistics*, online: http://awcbc.org/?page_id=14 (date accessed: February 28, 2020).

B. Bain, "Expanding the Traditional View of Safety Culture" (Presented at the Industrial Accident Prevention Association Health & Safety Conference, Toronto, ON, April 2004).

F. Bird & G. Germain, *Practical Loss Control Leadership* (Georgia: DNV, 1996).

M. Colligan & A. Cohen, "The role of training in promoting workplace safety and health" in J. Barling & M.R. Frone, eds., *The Psychology of Workplace Safety* (Washington, DC: American Psychological Association, 2004).

Criminal Code, R.S.C. 1985, c. C-46.

M. DePree, *Leadership is an Art* (New York, NY: Dell Publishing, 1989).

D. Dyck, *Disability Management: Theory, Strategy and Industry Practice*, 6th ed. (Toronto: LexisNexis Canada, 2017).

D. Dyck, "Health & Wellness Domain", *BCRSP Study Guide* (Mississauga, ON: Board of Certified Safety Professionals, 2015).

D. Dyck & T. Roithmayr, "Great Safety Performance: An Improvement Process

Using Leading Indicators" (December 2004), 52:12 AAOHN.

G. Germain *et al.*, *Safety, Health, Environment and Quality Management: A Practitioner's Guide*, 2d ed. (Loganville, GA: International Risk Control America, Inc., 1998).

S. Hazzard, "Management and Organizational Behaviour", *BCRSP Self-Study Guide* (Mississauga, ON: Board of Certified Safety Professionals, 2000).

G. Hofstede, "Motivation, leadership and organization: Do American theories apply abroad?" *Organizational Dynamics*, Vol. 9, Summer 1980.

P. Hudson, "Safety Culture: The Ultimate Goal" (Transport Canada, Aviation Safety Letter, 2002).

International Atomic Energy Agency, "Safety Culture in Nuclear Installations: Guidance for Use in Enhancement of Safety Culture" (December 2002), online: http://www-pub.iaea.org/MTCD/publications/pdf/te_1329_web.pdf (date accessed: February 28, 2020).

International Labour Office, "Safety in Numbers: Pointers for Global Safety Culture at Work" (Geneva: ILO, 2003).

N. Keith, *Workplace Health & Safety Crimes* (Markham, ON: Butterworths, 2004).

S. Klie, "Culture drives performance" *Canadian HR Reporter* (December 13, 2010) at 1 and 15.

G. Lowe & J. Wetherow, "High Trust Cultures = Health and Performance" (Presented at the International Accident Prevention Association Conference, April 2007, Toronto, ON).

F.A. Manuele, *Advanced Safety Management: Focusing on Z10 and Serious Injury Prevention* (New York: John Wiley & Sons, 2008) at 82.

National Institute of Disability Management and Research, *Disability Management in the Workplace: A Guide to Establishing a Joint Workplace Program* (Port Alberni, BC: NIDMAR, 2003).

N. Novak, "Evaluation of Safety Indicators", Masters Thesis (Department of Occupational Health, McGill University, September 22, 2006).

J.T. O'Rourke, "The Essence of Leadership" (1993) 6:2 Drake Business Review.

T. Pattenden, Presentation made at the 1994 OSH Conference, as reported by D. Pratt in "How to put Employee Health on the Executive Radar Screen" (Keynote Speech presented to the Alberta Occupational Health Nurses Conference, Calgary, AB, May 25, 2000).

T. Phillips, "Reader Panel: Building a Safety Culture" (2007) *Canadian Occupational Safety*.

D. Pratt, "How to put Employee Health on the Executive Radar Screen" (Keynote Speech presented to the Alberta Occupational Health Nurses Conference, Calgary, AB, May 25, 2000).

T. Roithmayr, "The Performance Maximizer®" (2000), online: http://www.performancebydesign.com (date accessed: February 28, 2020).

T. Roithmayr, "Workplace Safety Survey Report", ENMAX Power Meter Reader Group (2003), unpublished.

T. Rutledge, "Culture — Not Just a Plaque on the Wall" (2007) *Canadian Occupational Safety*, April/May 2007 Issue, at 14.

D. Ryan, "Moving off Your OH&S Plateau: Cutting Edge Strategies for Fostering and Measuring a Dynamic OH&S Culture" (Presented at the Occupational Health and Safety Amendment Act, 2002 — What's New, What's Changing and What You Need to Do to Comply, Insight Conference, May 27-28, 2003, Edmonton, AB).

D. Ryan, "Corporate Safety Culture: A Determinant of Program Effectiveness" (*CSSE Contact*, Winter 2002).

D. Sayers & Associates, "Management and Organizational Behaviour", *BCRSP Self-Study Guide* (Mississauga, ON: Board of Certified Safety Professionals, 2007) at 47.

E. Schein, *Organizational Culture and Leadership*, 2d ed. (San Francisco, CA: Jossey-Bass, 1992).

J. Schermerhorn, J. Hunt & R. Osborn, *Managing Organizational Behavior*, 4th ed. (Toronto: John Wiley & Sons, 1991).

S. Simon, "Safety Culture Assessment as a Transformative Process" (Culture Change Consultants, Larchmont, NY, 1998).

U.S. Department of Labor, "Injuries, Illnesses and Fatalities, 2018" (2020), online: http://www.bls.gov/iif/ (date accessed: February 28, 2020).

Waterstone Human Capital, Canada's "10 Most Admired Corporate Cultures Program and the Canadian Corporate Culture Study", online: http://www.waterstonehc.com/cmac/canadas-10 (date accessed: February 28, 2020).

Chapter 7

SAFE WORKPLACE = GREAT WORKPLACE: BUILDING A SUSTAINABLE CULTURE OF SAFETY

By Tony Roithmayr[1]

A. INTRODUCTION

This chapter is dedicated to the unthinkably large number of employees whom we have failed — albeit, unknowingly and unintentionally — yet failed nonetheless and too often, fatally. Our failure as organization leaders, Occupational Health and Safety (OH&S) Practitioners/Professionals and OH&S consultants has not been due to a lack of caring; we care deeply. It has been due to a lack of awareness; not knowing what we don't know, and yes, benign neglect that does not recognize and challenge the development within our workplaces of a "lethal cocktail"[2] of beliefs, attitudes, and habituated behaviours that, coupled with "system-induced" obstacles to safe performance, are a breeding ground for disaster.

B. NEW DIRECTIONS FOR THE PRACTICE OF OCCUPATIONAL HEALTH AND SAFETY

There is a growing realization that there is more to creating and maintaining a safe workplace than what was previously thought. For a long time, the field of Occupational Health and Safety focused on the technical aspects of "how to do tasks safely", *e.g.*, the use of policies and procedures; management of workplace hazards and risks; adoption of safe work practices; use of personal protection equipment; use of appropriate tools and equipment; conducting regular worksite inspections; doing

---

[1] Tony Roithmayr is President of Performance by Design® Inc. and originator of *The Performance Maximizer®*; co-developer of the *Great Safety Performance*™ model; and co-founder of the Orange Umbrella®. Tony firmly believes that injury-free workplaces are possible and can be developed through committed leadership that fosters an environment in which employees feel energized, satisfied, and healthy, and are fully supported and enabled to perform efficiently, effectively, and safely; see online: http://www.performance-bydesign.com (date accessed: January 31, 2020).

[2] P. Hudson, *et al.*, *Meeting Expectations: A New Model for a Just and Fair Culture* (2008), at 3, online: http://www.onepetro.org (date accessed: January 31, 2020).

root cause analysis of incidents; and measurement through OH&S audits. These OH&S efforts are very important indeed and the depth and sophistication of the knowledge and practice in these areas has grown considerably. The technical OH&S aspects — the "hardware" of Safety — have become a well-researched science.

OH&S Practitioners/Professionals and Line Managers are just now beginning to realize that a multitude of things influence what people do on the job. Knowing the traditional preventive measures, or how to do specific tasks safely, does not necessarily mean that things get done that way. Many OH&S Practitioners/Professionals and Line Managers are sensing a need to expand their awareness and level of understanding of how the performance of work in organizations happens, and what it means in terms of the management of workplace safety. Indeed, "managing safety is by far the most difficult task in any hazardous endeavour".[3]

In-depth study of disastrous events in a variety of industries has brought much needed attention to the subtle complexities present within organizations. For example, valuable lessons have been learned, albeit the hard way, from the disasters at Chernobyl (nuclear), Zeebrugge (ferry), Clapham Junction (railway), Texas City (refining), and this list is increasing. These and many other spectacular events, make it

> clear that those people at the "sharp end" in direct contact with each **system** were not so much the instigators of bad events as the inheritors of an "accident in waiting" that had, in some cases, been lying dormant within the system for years.[4]

In his comments, James Reason points out that ***system understanding*** is crucial because accidents, in even well-defended systems,

> arose from a concatenation[5] of many different factors arising from all levels of the organization . . . [and] that these latent systemic conditions, in combination with local triggers, opened up a brief window of accident opportunity.[6]

When looking outside the traditional boundaries of OH&S, and by examining the knowledge available in Human Performance Technology and Organization Development, as well as the lessons learned during the Quality movement of the 80s and 90s, we broaden our understanding of the effective functioning of organizations. A fundamental lesson is that **all organizations are systems**. They are collections of employees, processes, and things that are interconnected and interdependent, as well as "organic"[7] in their functioning. These disciplines teach us that there are a

---

[3] J. Reason, "Foreword", in A. Hopkins, *Safety, Culture and Risk* (CCH Australia Ltd., 2005).

[4] *Ibid.* [Emphasis added].

[5] ***Concatenation*** — a series of interconnected or interdependent things or events.

[6] J. Reason, "Foreword", in A. Hopkins, *Safety, Culture and Risk* (CCH Australia Ltd., 2005).

[7] **Organic** in this sense means viewing or explaining something as having a growth and development analogous to that of living organisms.

multitude of influences on an employee's performance that are not adequately explained and dealt with in the traditional approach to the field of OH&S.

In their definition of Human Performance Technology (HPT), the International Society for Performance Improvement (ISPI), describes the significance of understanding organizations as systems:

> A system implies an interconnected complex of functionally related components. The effectiveness of each unit depends on how it fits into the whole and the effectiveness of the whole depends on the way each unit functions. A systems approach [to organization performance] considers the larger environment that impacts processes and other work.[8]

With respect to workplace Safety, organizational systems have been designed with careful consideration to processes, workflow, job design, procedures, engineering of safety features, policies, physical barriers, programs and training, rules and regulations, and the management and measurement systems — much of this is the "hardware" of Safety. Major incidents happen only when many, or all, of the built-in defence mechanisms fail. "But there is something else at work — something that reaches all parts of the system for good or ill . . . safety culture."[9]

As Reason and many others have concluded, a very powerful component in the organizational system is the Safety culture — a largely unexamined set of beliefs, values, attitudes and habituated behaviours that are imbedded deep within the fabric of the organization. The Safety culture has a huge impact on worker performance, often more than the practices, knowledge, and skills that employees have been taught. In this chapter, we identify the psychosocial factors that are an integral part of the organizational system.

These psychosocial factors can be viewed as the Safety "software" elements of the system that facilitate the functioning of the system's Safety "hardware". The Safety "software" includes the aforementioned beliefs, values, attitudes, and habits, as well as the level of functioning of performance support[10] provided by the leadership team and the interactions that occur among the employees within the organization.

Recognizing organizations as whole systems of interconnected hardware and software elements leads to a profound and critical conclusion:

> safe performance on the job (indeed, all performance) can only be maximized when all the system elements are considered, measured, planned for, and managed as the interdependent elements that they are.

---

[8] ISPI, "What is HPT?" online: http://www.ispi.org (date accessed: February 28, 2020).

[9] J. Reason, "Foreword", in A. Hopkins, *Safety, Culture and Risk* (CCH Australia Ltd., 2005).

[10] **Performance support** means communication and interactions with workers regarding safety; information sharing; performance expectations and feedback; recognition; coaching and skill development; and dealing with task and employee obstacles and problems.

Lessons from the gurus in Quality Improvement (Deming – Quality) and Human Performance Technology (Gilbert, G. Rummler, and Robinson and Robinson – Human Performance Technology) have significant implications for the management and improvement of Occupational Health and Safety:

- More than 85 per cent of performance problems are not about the employee; they are management problems (Deming, 1982).[11] This is supported by the Safety literature (Bird and Germain, 1996).[12]

- When the system is not performing well, it is seldom that only a single component has failed. Therefore, the most effective solutions are rarely the single-point solutions. We are often asked, "What one thing should an organization do to improve performance?" The response: "Stop looking for the one thing."[13]

- When you pit a good performer against a bad system, the system will win each and every time.[14]

- In over 90 per cent of incidents, there are multiple causes of performance deficiency. If performance improvement is to be achieved, all the significant causes of performance deficiency must be resolved.[15]

- The absence of performance support in the work environment, and not the absence of knowledge and skill, is the single greatest block to exemplary performance.[16]

Going forward, if OH&S Practitioners/Professionals need to understand and deal with the systemic issues surrounding workplace Safety, they must broaden their practice scope and adopt multidisciplinary approaches to their professional practices. Increasingly, their professional practice will need to:

- include the concepts, tools, and processes found in Human Performance Technology in order to better understand their organization systems; and

- access the methodologies in Organization Development to develop and sustain more supportive work environments. This approach is critical to the

---

[11] W.E. Deming, *Out of the Crisis* (MIT Press, 1986).

[12] F. Bird & G. Germain, *Loss Control Management: Practical Loss Control Leadership*, revised ed. (Loganville, GA: DNV, 1996).

[13] Anatomy of Performance.pdf, online: http://www.performancedesignlab.com (date accessed: February 28, 2020).

[14] G. Rummler & A. Brache, *Improving Performance*, 2d ed. (San Francisco, CA: Jossey-Bass, 1995) at 64.

[15] D.G. Robinson & J.C. Robinson, *Performance Consulting* (San Francisco, CA: Berrett-Kohler, 1995) at 180.

[16] Tom Gilbert, quoted in D.G. Robinson & J.C. Robinson, *Moving from Training to Performance* (San Francisco, CA: Berrett-Kohler, 1998), at 7.

development of a sustainable culture of Safety.

This chapter addresses:

- the concept of workplaces as "systems" and hence, an industry application of systems theory;
- the measurement and analysis of a performance system;
- a new role for the OH&S Practitioner/Professional;
- the use of a performance model in terms of the new OH&S Practitioner/Professional role; and
- the evaluation of the workplace culture through the use of a Safety culture continuum.

C. Understanding Workplaces as Systems

Organizations are systems according to the **System Safety Theory**. This theory recognizes the inseparable connection between individuals, their tools and machines, and the general work environment. Changes to one part of the work system have an impact on the remaining parts and the "whole" system. Recognizing and understanding this fact has a profound influence on the understanding of the functioning of organizations and how workplace systems factor into improving Safety in the workplace.

In this section, we will take a deeper look at the components of workplace systems. The functioning of these components (referred to as "Conditions for Performance") has a direct impact on performance. We will establish that these conditions, the Conditions for Performance are leading indicators of Safety behaviour and are thus predictive of the trend in Safety outcomes.

Lastly, we will stress the implications for leadership in managing and improving the performance of employees within the workplace system. A Transformational Leadership model will be introduced as a tool for developing a robust culture of Safety.

1. Introduction

Great Workplaces energize, support, and satisfy employees. They are safe and healthy — physically, emotionally, and psychologically. They enable performance at its best. And, they enable organizations to retain good employees. In a Great Workplace, employees willingly hold themselves accountable for working efficiently, effectively, and safely while achieving great results.

Despite the great intentions and sincere efforts of all involved, workplace safety often ends up as an "add-on program" which holds employees accountable for workplace Safety as opposed to building a system in which working safely is "a way of life". This typically results in continued sub-optimal level of Safety performance; that outcome is of great concern to front-line workers and Management alike.

The Great Safety Performance™ model and methodology that was introduced in

Chapter 6, "Occupational Health and Safety: Leadership and Commitment", enables organizations to advance beyond the traditional Safety procedures and training and target the development of a total culture of Safety. *The Performance Maximizer®* (Figure 7.1) is a foundation concept for understanding human performance that is an essential element to building a Great Workplace. Great Safety Performance™ is built upon *The Performance Maximizer®*. We must begin by understanding this concept and its role in enabling great performance.

The Performance Maximizer® takes a workplace **systems perspective** in providing practical insights into the nature of workplace performance, its management and the areas for improvement. The resulting Great Safety Performance™ model incorporates safe work behaviours, states safe outcomes and business results. Most importantly, it defines a comprehensive set of leading indicators for working safely that include psychosocial factors.

Figure 7.1: The Performance Maximizer

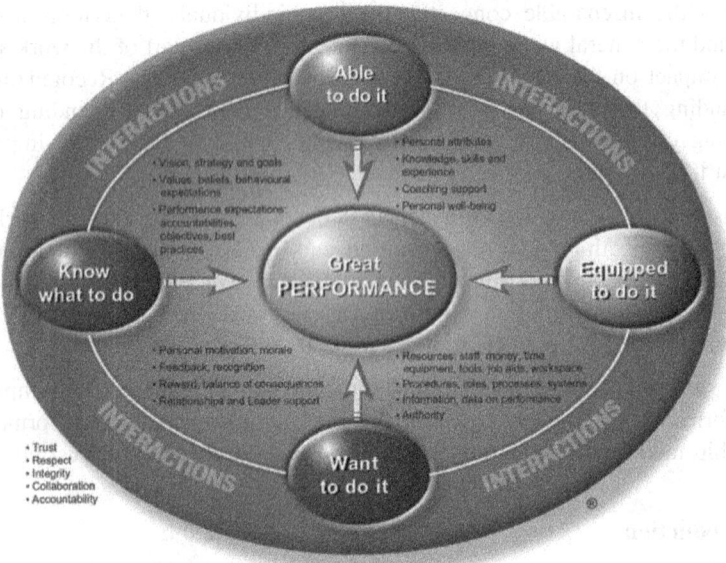

The Performance Maximizer® provides a picture of what safe performance looks like and what is needed to achieve it. It is used to:

- define/clarify desired safety performance;
- identify what is needed in each of five conditions to enable great safety performance;
- assess what enables (and what prevents) working safely; and

- implement actions to better manage or improve safety performance.

The Performance Maximizer® brings into sharp focus all the factors that impact working safely, simply, and clearly.

These factors are clustered into what is termed the "Conditions for Great Safety Performance". That is, **Know What to do**, **Able to do it**, **Equipped to do it**, **Want to do it** and **Interactions** that foster trust, respect, integrity, collaboration and accountability. Having these conditions, or not having them, in an organization is not a matter of choice. Like gravity, they are a fact of life — they exist! Choice is a factor in terms of what the organization does about it. How well the overall "system" of leading indicators is managed can be illustrated through The Great Safety Performance™ model. It can enable organizations to determine progress towards the goal of an injury-free workplace.

2. A Guide to Understanding the Performance Maximizer®

Performance is about what employees do, achieve, and contribute to their organization. Great safety performance is defined as superior results achieved in an exemplary manner. To attain this level of performance, it is important to begin by addressing these questions:

- To what organizational business goals is Safety performance expected to contribute? How are they measured within the organization?
- What Safety outcomes are required to make that contribution and how can they be measured?
- What has to be done well to produce those results (safe work actions and interaction behaviours)?

Defining great safety performance specifies the safe work actions (both task and interactions), the safe outcomes and business results (Figure 7.2). Once great safety performance has been defined and aligned with the organization goals, identify the specific factors in the five conditions for great safety performance that need to be in place to enable success.

Figure 7.2: Performance System for Building Great Workplaces

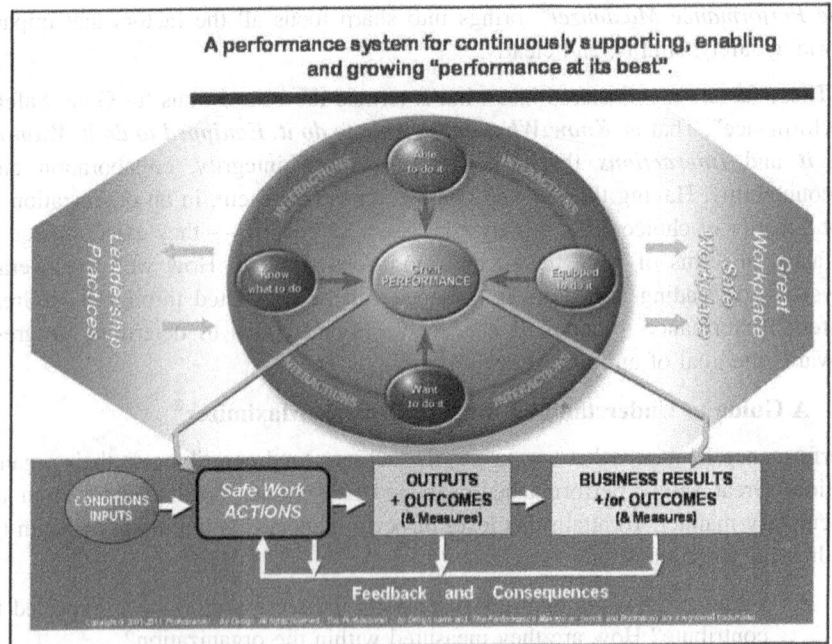

3. Leading Indicators — The Conditions for Great Safety Performance

The Conditions for Great Safety Performance, which were introduced in Chapter 6, "Occupational Health and Safety: Leadership and Commitment", are described in more detail here:

(a) Know What to do

- Vision, strategy, and goals
- Values, beliefs, and behavioural expectations
- Performance expectations: accountabilities, objectives, and best practices

The *Know What to do* condition emphasizes clarity and alignment in terms of what needs to be accomplished — that everyone is crystal clear about the "Great Safety Performance" that is expected of them, *i.e.*, there are no assumptions or guessing.

Performance expectations for workplace safety and their measures have been discussed, clarified, agreed upon, communicated, and understood. These expectations are aligned to organization vision, strategy, and goals for creating a safe workplace. Also, clear and valued are the most effective activities and practices that will accomplish desired work results in a safe manner. The *Know What to do* condition also includes behavioural expectations (*e.g.*, conduct/interactions), that is, not only what work gets done, but also how it gets done.

(b) Able to do it

- Personal attributes
- Knowledge, skills, and experience
- Coaching support
- Personal well-being

The *Able to do it* condition refers to the capability to perform safely from a variety of perspectives. It means that the personal attributes, traits, and characteristics should "fit" the requirements of the work — including a person's psychological, physical, and emotional capacities.

The *Able to do it* condition is about competence — the knowledge, skills, and experience necessary to achieve the required work results in a safe manner. On-the-job coaching by peers, a supervisor or a mentor, is also an essential factor in enabling safe performance. The physical and emotional health necessary to be safely productive and personally fulfilled is part of the *Able to do it* condition.

(c) Equipped to do it

- Resources: staff, money, time, equipment, tools, job aids, workspace, *etc.*
- Procedures, roles, processes, and systems
- Information and the data on performance
- Authority

The *Equipped to do it* condition refers to resources, business/work processes and procedures, organizational and IT systems, and the physical environment in which work is done. It is also about the importance of having the information required to do the job and the measurement data that lets people know how they are doing. Employees need to be clear about the scope and authority of their roles so that there is no confusion about who is responsible for each specific task and how they are expected to work together to achieve the organization's goals. These elements — resources, tools, processes, environment, information, role clarity, *etc.* — need to be geared towards the ability to work safely.

(d) Want to do it

- Personal motivation and morale
- Feedback and recognition
- Reward and balance of consequences
- Relationships and leader support

The *Want to do it* condition is about an employee's willingness to hold him/herself accountable for quality work and achieving agreed upon results in a safe manner. It is about knowing that the contribution he/she makes is valued and rewarded; that he/she is recognized for "doing the right things right"; and that he/she is held accountable for performance shortfalls, unsafe practices, and inappropriate behaviour. Leader support is crucial. A large part of employees feeling challenged, valued, and recognized is up to the leaders. Poor interactions among employees, as well as obstacles within other conditions for great safety performance, can often erode motivation, or the *Want to do it*.

(e) Interactions
- Trust
- Respect
- Integrity
- Collaboration
- Accountability

Interactions are about the way employees work. They are the behaviours and personal conduct that create the emotional climate of the workplace. Positive, healthy Interactions support personal well-being, develop a respectful workplace, foster productive relationships and have a profound impact on performance outcomes, including Safety in the workplace. Interactions that enable great performance are characterized by trust, respect, integrity, responsiveness, fairness, collaboration, and accountability. Interactions constitute the behavioural performance expectations of the organization.

The five conditions for great safety performance are connected to each other; they are not independent variables. They are part of a system and impact each other. A change in one condition will impact the status of one or more of the other conditions. Safety performance can only be maximized when all the five enabling conditions are considered, planned, and systematically managed together, as a system.

Like gravity, the five conditions for great safety performance are always present and inescapable. The arrows indicate that these conditions are always active and impact performance simultaneously and continuously. This reflects the "organic" nature of organizations.

The large oval background represents the total performance environment in which an organization's employees operate. This total environment fosters great safety performance when all the interdependent conditions for performance are in a strong, positive, and enabling state.

4. A Performance System for Building a Safe/Great Workplace

Research has confirmed that the Conditions for Great Performance enable effective performance on the job. Furthermore, when applied to Workplace Safety, statistical analysis has identified a predictive relationship between the five Conditions for Great Performance and working safely on the job. In other words, the Conditions for Great Performance are leading indicators of Safety performance which are directly correlated to the safe work practices of employees and can forecast the trend in the safety-related outcomes of their actions on the job. Research has also demonstrated that the Conditions for Great Performance are highly interdependent which confirms that we are dealing with a system.

This is a crucial message for leaders. Great Workplace cultures do not just happen. Leaders create and sustain them. Leaders must recognize that in managing the Safety of employees, they are dealing with an interconnected system. This reinforces the earlier assertion that safe performance on the job can only be maximized when all the enabling conditions are considered, planned, measured, and managed as interdependent elements. Continuously maximizing the conditions for great safety performance is therefore the crucial leadership task that will surround employees with a strong, enabling performance environment for Safety.

Figure 7.2 illustrates how leaders can support, enable, and grow a culture of Safety. The Performance Maximizer® implicitly points to a required set of leadership practices which, when skillfully practised, produce outcomes that strengthen and sustain the Conditions for Great Safety Performance. In this way, leaders can make Safety a central theme in their management and improvement of workplace performance. They model the change they would like to see; they inspire shared purpose and personal growth; they challenge the status quo; they enable exemplary performance; and they create a focus on the Interactions that foster trust, respect, integrity, collaboration, and accountability.

The presence of the Conditions for Great Safety Performance tends to strengthen as a result of the leaders' intentional and persistent performance of these leadership practices. When they reach a critical mass, their impact "ripples" through the organization. The outcome is a Great Workplace culture in which employees are energized, supported, and enabled to safely perform at their best.

Summarized below are the leadership practices (Transformational Leadership) that foster a Great/Safe Workplace and the outcomes that are indicators of success in building it.

5. Transformational Leadership Practices

Transformational leadership is inspired leadership that influences the beliefs, values, and goals of followers so that they can perform in an extraordinary manner. In Chapter 1, "Occupational Health and Safety: Historical Perspectives" and Chapter 18, "Occupational Health and Safety: Best Practices", transformational leadership as a concept, is discussed. In this chapter, a model of transformational leadership practices is provided. It is a leadership skill set that OH&S Practitioners/ Professionals should strive to develop.

Transformational leaders:

- **Show the Way**
    - Declare personal values and principles;
    - Take a stand;
    - Pursue small wins to achieve big goals;
    - Set the example for others to follow.

- *Inspire Shared Purpose*
  - Imagine exciting and meaningful possibilities;
  - Discover and appeal to shared aspirations;
  - Build commitment to a common vision through personal passion and belief.
- *Challenge to Change*
  - Challenge the "way it's always been done";
  - Fix what's wrong;
  - Seek innovative opportunities to improve and grow;
  - Experiment, take risks and learn from mistakes.
- *Enable Exemplary Performance*
  - Strengthen others by sharing power and discretion;
  - Foster collaboration and involvement around common goals;
  - Remove obstacles to performance;
  - Build a climate of trust, respect, integrity, and accountability.
- *Hearten the Spirit*
  - Recognize and reward individual and collaborative contributions;
  - Provide genuine, "open-hearted" encouragement;
  - Reinforce spirited collaboration by celebrating values and victories.
- *Foster Courageous Growth*
  - Encourage and challenge others to stretch beyond their "comfort zones";
  - Create opportunities for learning and growth;
  - Inspire confidence and self-esteem in others.

These practices are measured by 33 leadership behaviours. When skillfully and persistently applied, they produce the outcomes listed below.

6. The Key Indicators of a Great/Safe Workplace

The outcomes include:

- Employees understand the organization's vision, values, and goals for Safety.
- Employees embrace the performance results expected of them and know how their success will be measured.
- Employees have adequate and appropriate resources to achieve what is expected.

- Employees are supported to build knowledge and skill and overcome difficulties.
- Effective work groups are fostered and maintained.
- Barriers to safe and effective performance that are beyond the control of individuals and teams are removed.
- Employee progress and contribution to achievement of goals is measured and communicated.
- Employees are held accountable for delivering agreed upon results in a safe manner.
- Ongoing communication maintains focus and commitment to working safely, efficiently and effectively in an environmentally responsible manner.
- Progress, development, and the achievement of desired goals is recognized and celebrated.
- Employees feel energized, enabled, and inspired to contribute their personal best.

D. MEASURING AND ANALYZING THE PERFORMANCE SYSTEM

The reality that workplaces are interconnected systems, coupled with the fact that a comprehensive model of all its components is available and that this model also includes the "software" described above, allows for a complete evaluation of the system and powerful analysis of its functioning.

The measurement process and analysis tools can accomplish the following:

1. Assess all the drivers of Great Safety Performance and generate workplace specific solutions ("one size" does not fit all).
2. Measure a statistically proven set of Leading Indicators - provide a "dashboard" to monitor the predictive relationship to the trend in Safety Performance.
3. Determine if what management believes they are communicating to their employees is actually what is heard and understood.
4. Investigate "the stuff that's really hard to get at", *i.e.*, culture — the people side of the enterprise.

    Employees' experiences in their workplace form their perceptions of the workplace culture. These perceptions shape their assumptions and beliefs about what is okay and not okay. The beliefs in turn govern what employees do on the job which becomes a habituated set of behaviours that comply with and form part of the culture. (See Figure 7.9)

    The evaluation accesses and quantifies perceptions to identify the safety and productivity risks in the workplace culture.

5. Provide data that identifies the heavy "drag" on a Key Business Driver —

Employee Engagement — *and*, it directs management to the appropriate solutions.

Research findings across North America indicate that barely more than 30 per cent of employees are really fully engaged at work. An equal percentage are merely present — a serious drain on productivity, a huge frustration for their managers and much more likely to experience incidents and injury. Engagement levels are a function of the culture.

6. Provide a methodology in which *High Involvement* produces *High "Buy-in"* to the process and the resulting Improvement interventions.

   The methodology takes the issues directly to those most affected and most familiar with them — employees. They are involved from their initial orientation, through surveying, assessment, solution development, implementation, and subsequent follow-through.

7. Positions leadership as the key to the workplace transformation that will be required to achieve the safety performance goals. The process will also provide insight into the organization's readiness to practice Transformational Leadership.

8. The "bottom line": this adds up to a very substantial reduction in the overall margin for error. Safety performance and productivity move in the same direction. Improvements in Safety performance also create significant productivity gains.

E. A NEW ROLE FOR THE OH&S PRACTITIONER/ PROFESSIONAL

In this section, the role of the OH&S Practitioner/Professional will be reframed based on the indisputable fact that Occupational Health and Safety is entering a "new frontier" — the management of the workplace culture.

The focus of Safety in most organizations has been on the aforementioned Safety "hardware" and the associated Safety "hard" skills which many OH&S Practitioners/Professionals are extremely good at. At the same time, OH&S Practitioners/Professionals have been largely unaware of (or neglected) the Safety "software" and the associated "soft" skills that ultimately determine the effectiveness of the Safety "hardware" and Safety results.

The following is a summary of much of the Safety "hardware": the Safety policies, OH&S training, Safety rules and regulations, OH&S management and measurement processes, work flow, job design and the engineering of Safety features, procedures, safe work practices, root cause analysis, audits, inspections, tools and equipment, personal protective equipment, signage, physical barriers, safety meetings, *etc*. All of which are supported by, or are part of, a well-documented OH&S Management System.

- Included in the Safety "software" are employee values, beliefs, attitudes, habituated behaviours ("the way things are done around here"), and the kind

- and quality of interactions among the people in the organization, all contributing to the culture of the organization.

- Added to that are some very critical drivers of culture: the quality and frequency of performance support from organizational leaders — positive feedback regarding what employees do well; helpful corrective feedback on what is not working well; coaching on skills and dealing with obstacles; and recognition for and celebration of "good results".

- Research has shown that the Safety "software" — the organization's Safety culture — has a powerful impact on Safety outcomes. It is only recently that "forward-looking organizations" began to factor these cultural determinants into their Corrective Action Plans. In general, however, they are overlooked.

- Provided the Safety "hardware" present in most organizations is effective, persistent, and consistent attention to the cultural factors can lead to the development of a Safe Workplace.

- Research has also shown that the same cultural factors that lead to excellent Safety performance are also "the drivers" of excellent business performance.[17] Hence, the assertion: A Great Workplace = A Safe Workplace.

- These insights into the role of the Safety culture in terms of managing and improving workplace safety have important implications for OH&S Practitioners/Professionals. They invite OH&S practitioners/professionals to step into a realm of new possibilities; to go where few have gone before; and to understand, participate in, facilitate and ultimately, to shape cultural change initiatives within their organizations.

- Most OH&S Practitioners/Professionals are unaware of, or are not skilled in the processes needed to help transform workplaces. An analogy of a bicycle is offered to explain the role of Safety "hardware" and Safety "software" in organizations (Figure 7.3).

---

[17] J.P. Kotter & J.L Heskett, *Corporate Culture and Performance* (New York: Free Press, 1992); D. Kravetz, *People Management Practices and Financial Success: A Ten-Year Study* (1996), online: http://www.kravetz.com (date accessed: February 28, 2020).

Figure 7.3: Bicycle Analogy

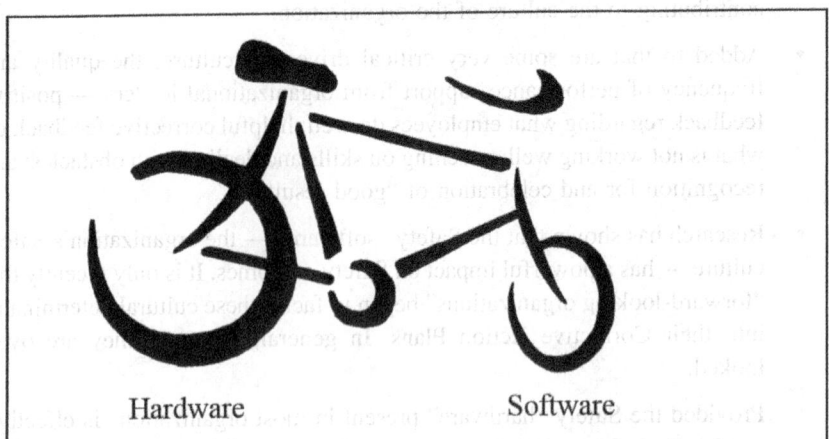

The back wheel is used to provide the power for forward motion — without it, we can't go anywhere. The front wheel is used to get the rider to a destination — the function of the front wheel enables navigation on the route. In this analogy, the "hardware" of Safety is like the back wheel — without it, Safety is going nowhere. The Safety "hardware" provides all the power; everything we know, everything we have learned, and everything we have created to keep people safe is in this back wheel. It offers tremendous power.

The Safety "software" is like the front wheel: The Safety culture and the "soft skills". Like the front wheel of the bike, it is the Safety "software" that gets us to where we want to go with workplace Safety, namely the attainment of injury-free workplaces. We must use this "front wheel" to foster, guide, facilitate, and direct our efforts in building stronger, better and safer workplaces. It is the way we deliver the Safety hardware. Unfortunately, in many of today's organizations, the Safety "front wheel" is currently inadequate for the task at hand. As shown in Figure 7.3, an underdeveloped "front wheel" component of many OH&S systems poorly prepares organizations to venture into the "new frontier of Safety".

The additional knowledge and skills that OH&S Practitioners/ Professionals will need include:

- The effective use of tools like Safety perception surveys to measure and interpret both the effectiveness of the Safety "hardware" and the degree to which employees are experiencing the Safety "software" support they need to work effectively, efficiently, and safely every day.
- Monitoring the leading and lagging indicators of Safety and correlating them to the business Key Business Performance Indicators.
- Recognizing the improvement actions that are needed to build a safer workplace that includes a strong and supportive Safety culture.

- Collaborative performance consulting with managers and exercising influence skills — "non-authority leadership".[18]
- Excellent facilitation skills to lead management meetings and guide the processes that will improve the workplace and Safety performance.
- Understanding and appreciating the importance of employee engagement.
- The ability to support/guide managers and supervisors in engaging employees.
- Understanding the process and practices for shifting beliefs, attitudes, and behaviours.
- The ability to support/guide managers and supervisors in shifting their beliefs, attitudes, and behaviours.

OH&S Practitioners/Professionals need to be both technical OH&S experts, as well as Cultural Change Agents. The new Safety frontier requires a significant repertoire of people skills.

> [S]pecifically, the safety professional of the future will need to be an agent of behavioural and cultural change. It will be his mission to partner with line management in leading the organization toward the adoption of new norms, beliefs, and values. . . .
>
> It is imperative for the safety professional to understand how culture affects organizations and to understand its impact on safety performance . . . . The deepest root cause accident investigation will be stopped in its tracks by a mistrustful, cover your back, blame-the-other-guy culture dominated by norms of concealment.[19]

F. DEFINING THE NEW ROLE — AN INDUSTRY EXAMPLE

Recognizing that the "new Safety frontier" is here, the Plant Programs and Environmental Support Department (Generation South Division) at Manitoba Hydro significantly redefined the role of the OH&S Practitioner/Professional.[20] The Manager and his OH&S Practitioners/Professionals had prior experience using the Performance System model described in Chapter 6, "Occupational Health and Safety: Leadership and Commitment". Their experience includes exposure to the Performance Maximizer®, and the Great Safety Performance tools and assessment methodology derived from it.

---

[18] **Non-authority leadership** has no formal hierarchical authority over others; relies on personal credibility; and leads through the authenticity of personal passion, a sense of purpose and the skills to influence change.

[19] S. Simon, "On the Future of the Safety Profession" (2001) at 2-3, online: http://culturechange.com/wp-content/uploads/2012/08/futuresafeprofpaper.pdf (date accessed: February 28, 2020).

[20] R.G. McKinnon, Manager, Plant Programs and Environmental Support Department, Generation South Division, Manitoba Hydro, Manitoba.

A performance model was developed for the OH&S practitioner/professional role that clearly defines what successful performance should look like. This model specifies what is to be accomplished in the role, how it is best accomplished, and what measures or indicators will be used to determine whether "it" was accomplished in a quality manner. A critical feature of the performance model is that it describes the conditions required to enable and support desired performance — "performance at its best".

The performance model is organized as follows:

1. **Contribution Statement** : describes the purpose of the OH&S Practitioner/Professional role (why it exists), and how it adds value to the business of the organization. The Contribution Statement for the role is:

    Partner with Plant organizations to develop a workplace culture in which everyone demonstrates accountability for Safety; a culture in which Safety is an integral part of the ongoing Business Activities that lead to the achievement of our Business Goals in an Injury-Free Workplace.

2. *A Summary* of the model (does not list measures) is provided in Figure 7.4.

3. **Business Results (& Measures)** : explains the business results to which the OH&S Practitioner/Professional are expected to contribute. These results provide "line of sight" for the OH&S Practitioner/Professional role. Three of the Business Results and Measures are shown in Figure 7.5.

4. **Performance Results** :

    a. <u>Operational Results (& Measures)</u>: are the main, ongoing outputs and/or outcomes, that the OH&S Practitioner/Professional is accountable for achieving. Two Operational Results and Measures are presented in Figure 7.6.

    b. <u>Leadership Results (& Measures)</u>: personal, interaction or leadership outcomes that facilitate achievement of Operational Results. Two Leadership Results and Measures are shown in Figure 7.7.

5. Detailed *Actions* : what OH&S Practitioners/Professionals need to do to achieve the results specified in their Performance Results. Detailed Actions are included in Figures 6.6 and 6.7.

6. **Enabling Conditions** : these are the specific conditions that OH&S Practitioners/Professionals and their leaders need to jointly create so that they each:

    √ Know *What to do;*

    √ are *Able to do it;*

    √ are *Equipped to do it;*

    √ *Want to do it*; and

    √ experience *Interactions* that foster trust, respect, integrity, collaboration and accountability.

CH. 7: SAFE WORKPLACE = GREAT WORKPLACE

This performance model is clear, specific and comprehensive; far more so than is the usual "job description". It places the OH&S Practitioner/ Professional role in the context of the business; positions it to challenge the "new Safety frontier"; and provides direction for the development of the OH&S Practitioner/Professional to excel at meeting that challenge.

Parts of the performance model are reproduced below in Figures 7.4, 7.5, 7.6 and 7.7. In this industry example, which has been provided by the Plant Programs and Environmental Support Department (Generation South Division), at Manitoba Hydro, the OH&S Practitioner/Professional is titled the Field Safety Officer (FSO).

Each of the Performance Results listed on the Summary page (Figure 7.4) are detailed in the manner shown below. Fifty-seven specific factors that support and enable FSO performance are itemized within the "Conditions for Performance" (*Know What to do*, *Able to do it*, *Equipped to do it*, *Want to do it* and *Interactions*).

Figure 7.4: Overall Summary of the Performance Model

FIELD SAFETY OFFICER (FSO) → **KEY ACTIONS** → **PERFORMANCE RESULTS (& Measures)** → **STATION BUSINESS RESULTS (& Measures)**

Contribute to and/or directly impact...

Partner, Lead, Enable...

KEY ACTIONS:
1. Support the Station Management in establishing, operating, and maintaining an effective safety program that meets ALL regulations, policies, and rules.
2. Develop a set of key messages and implement a communication plan to raise Safety culture awareness, develop Safety culture understanding, and foster commitment by all employees to the Journey to an Injury-Free Workplace.
3. "Track Positives": Implement a methodology to help management and employees to identify cultural norms, assess whether they foster an injury-free workplace, strengthen those norms that do, and work to remove/replace those that impede it.
4. Support management in developing employee capability to do the "right things in the right way" to safely achieve the required results.
5. Enable Workplace Safety & Health Committees to be effective in advancing the development of a safer workplace.
6. Provide the mechanism and information to keep management and staff informed about their progress toward achieving an Injury-Free Workplace.
7. Support and enable management's operation of their Safety Management System as a valuable tool to achieve Key Business Results.
8. Contribute to enhancing Manitoba Hydro's reputation among industry peers, regulators, and within our communities.
9. Ensure I am skilled in continuously enabling the journey to an injury-free workplace.
10. Maintain myself as an engaged, productive employee.

PERFORMANCE RESULTS:
1. I am "healthy and whole" at the end of every day.
2. Plant activities and the operation of the Safety Management System comply with all federal, provincial, and corporate regulations, policies, and rules.
3. Consistent Key Safety Culture Messages are frequently delivered to, and understood by, our customers (management and employees).
4. Regular Site Safety visits foster learning about safe work practices and the role of culture, and identifies appropriate corrective actions.
5. Management identifies, communicates, and applies the "software", as well as the "hardware", components required to build a positive Safety culture.
6. Current employees, new employees, and contractors are set up for success: they demonstrate capability to do the right things in the right way to achieve desired results safely, efficiently, and effectively.
7. Promote the existence of actively engaged and proactive Workplace Safety & Health Committees.
8. Corporate Safety & Health initiatives are effectively implemented (e.g., policies, regulations, programs, etc.).
9. Performance data regarding lagging and leading indicators of Safety performance and selected business performance indicators.
10. Management and staff know where they stand regarding their progress in Safety and their capabilities for making improvements.
11. Management view the Safety Management System as a valuable tool to achieve Key Business Results.
12. Promote community awareness about Safety at home and in recreation.
13. Manitoba Hydro is recognized by the industry and regulators as a model of workplace Safety.
14. Is current and proficient in the competencies required for success in the Field Safety Officer role.
15. Is a trusted, respected, engaged, productive employee.

BUSINESS RESULTS:
1. A safe workplace
2. Cost-effective optimization of Operations and Maintenance functions
3. Budget performance
4. Environmentally responsible
5. A great workplace with a skilled, engaged, and productive workforce
6. Positive image/reputation within our communities

Ch. 7: Safe Workplace = Great Workplace

Figure 7.5: Examples of Business Results and Measures

Station Business Results:
the business results to which the FSO is expected to contribute provide "line of sight" for the role.

BUSINESS RESULTS	MEASURES
1. Safe workplace	▲ *Lagging indicators* • Frequency - Number of lost time injuries • Severity - Number of lost time days due to injuries • Number of high risk incidents • Number of major near misses ▲ *Leading indicators* (GSP Survey Results) • Safe Work Actions • Know what to do • Avle to do it • Equipped to do it • Want to do it • Interactions
2. Cost - Effective optimization of Operations and Maintenance	▲ Unit Availability (%)[21] ▲ Forced Outage rate (%)[22] ▲ WMS Performance Index (%)[23]

---

[21] *Unit Availability* is the percentage of time a generating unit is capable of producing power.

[22] *Forced Outage Rate* is percentage unplanned downtime of a generating unit (due to mechanical or other operational problems).

[23] **WMS Performance Index** is a productivity measure produced through the Manitoba Hydro Work Management System.

Station Business Results: the business results to which the FSO is expected to contribute provide "line of sight" for the role.	
5. Great Workplace with skilled, engaged, productive workforce	▲ Performance Results ▲ Survey results (GSP/GW)[24] ▲ Assessment by SWubject Matter Experts ▲ Absenteeism ▲ Retention ▲ Presenteesm

[24] *GSP* is GSP™ Workplace Safety Survey; *GW* refers to the Key Indicators of a Great Workplace.

CH. 7: SAFE WORKPLACE = GREAT WORKPLACE

Figure 7.6: Examples of FSO Performance Results (Operational)

OPERATIONAL RESULTS	MEASURES
3. Consistent Key Safety Culture Messages are frequently delivered to, and understood by, our customers (Management and front-line employees).	▲ Communication plan and materials are in place. ▲ Management and front-line employees are using the language of the messages.
ACTIVITIES	
a. Develop a set of Key Messages and implement a communication plan to raise the Safety culture awareness, develop Safety culture understanding, and foster commitment by all employees to the Journey to an Injury-Free Workplace.	
b. Consistently model *Interactions* that foster trust, respect, integrity, collaboration, and accountability.	
c. Reinforce appropriate behaviour (task & interactions) by acknowledging people that are "doing things right".	
d. Take a stand by challenging inappropriate behaviour in a constructive manner – unsafe actions and *Interactions*.	
e. Implement methodologies that raise awareness about how we interact and ways to improve our interactions.	
5. Management identifies, communicates, and applies Safety "software"*** as well as the "hardware"** components required to build a positive Safety culture.	▲ Corrective actions deal with both "software" and "hardware" elements. ▲ Observations of management interactions with staff demonstrate: - a no-blame approach; and - attending to positive culture building.

OPERATIONAL RESULTS	MEASURES
ACTIVITIES	

a. "Track Positives": Implement a methodology to help management and employees to identify cultural norms.*** Assess whether they foster an injury-free workplace, strengthen those that do, and work to remove/replace those that impede it.
b. Enable management to use actual incidents to work with employees to identify current negative cultural norms and strategize shifts to positive norms.
c. Support management to challenge the cultural myth of "too much to do", and enable accountability for effective time management.
d. Look for, and act on, the opportunities to recognize and reinforce workers' safe work actions – *catch people doing things right!*
e. Create opportunities for management to model on how to "catch people doing things right":
   ✓ giving feedback,
   ✓ providing recognition, and
   ✓ reinforcing desired behaviour.
g. "Catch management doing things right" – reinforce effective feedback and recognition practices.	

* *"Hardware"* refers to the Technical elements: policies, procedures, Safety Management Systems, work processes, equipment, tools, job design, measurement systems, management system, etc.

** *"Software"* refers to psychosocial factors, the "social" (or interactive) elements of employees' work experience that influence their minds (perceptions, assumptions, beliefs) and, in turn, their behaviour: employees' interactions/communication with management and each other – evidence of trust, respect, integrity, collaboration and accountability; clarity of role and expectations; leadership, direction and guidance; recognition and reward, being valued; employee involvement/participation in matters that concern them, etc.

*** *Cultural Norms* = habituated behaviours and the "hidden elements" - underlying beliefs and/or assumptions - that drive those behaviours.

CH. 7: SAFE WORKPLACE = GREAT WORKPLACE

LEADERSHIP RESULTS	MEASURES
14. Is Current and Proficient in the Competencies required for success in the FSO role	▲ Skill self-assessment of required competencies ▲ Applicable Degree, Diplomas, Certificates ▲ Training matrix (required versus completed courses) ▲ Leadership Practices Assessment
ACTIVITIES	

a. Identify and communicate the training needed to achieve the results described in the FSO role.
b. Agree on and implement a learning action plan that ....
  1) enables me to achieve expected performance results, and
  2) prepares me for opportunities that can help realize my potential.
a. Demonstrate performance improvement from the implementation of the learning plan.	

15. Is a trusted, respected, engaged, productive employee	▲ Manager's evaluation ▲ Status of FSO's.... Know what to do, Able to do it, Equipped to do it, Want to do it, Interactions
ACTIVITIES	
a. Discuss with Manager and seek help with obstacles to performance that are outside of his/her control. b. Keep Manager informed and provide best advice when work results can and/or will differ from agreed expectations. c. Deliver on commitments and hold self-accountable. d. Conduct him/herself in ways that foster trust, respect, integrity, collaboration, and accountability. e. Perform daily activities in accordance with our Operating Principles.	

1. Organization Evaluation

The field of Occupational Health and Safety is evolving. Today, Occupational Health and Safety has moved to another stage of maturation: some organizations are on the path of understanding what the new Safety frontier will look like, while others are still stuck in the past.

Using the concepts presented so far in this chapter, examine your organization's position on the following Safety Culture Continuum (Figure 7.7).

Figure 7.7: Safety Culture Continuum: Where is Your Organization on the Continuum of Where We Were to Where We're Going?

The OH&S Evolutionary Continuum	
WHERE WE ARE/WERE	**WHERE WE ARE GOING**
1. Safety is an expense. Safety is something we *must* do (compliance).	Safety is an investment in the welfare of our people, and in our productivity.
2. If we get the technical stuff right, we'll be safe, *i.e.*, Safety engineering; error analysis.	Psychosocial factors and organizational factors account for >85% of incidents and accidents, *i.e.*, workplace culture is a very powerful driver.
3. The Safety department is responsible for safety. Safety is not an integral part of operational functioning.	Management from top to bottom is accountable for Safety. Employees are highly involved in the responsibility for Safety and its improvement.
4. The "Command and Control by Policy and Procedures" is the way we get things done.	Beliefs/Assumptions drive behaviours; "the way we do things around here" *i.e.*, The way we behave is the essence of our culture. Learn to understand it and shift it in the right direction.
5. OH&S practitioners/professionals are Doers with Supervisors acting as the enforcers.	The new OH&S practitioner/ professional is an Agent of Behavioural and Culture change - facilitator, consultant, and leader.
6. Managerial processes and skills are the predominant style and approach to leadership.	Leadership that connects to both the "mind and the heart"; creating willing participants in a transformation that enables great Safety performance and truly embraces the journey to an injury-free workplace.

CH. 7: SAFE WORKPLACE = GREAT WORKPLACE

G. "TAKING A STAND": AN INDUSTRY EXPERIENCE

The conclusion from the foregoing material is that:

> The *new* work of the OH&S Practitioner/Professional (and the management they support) is Shifting Beliefs and Behaviour!

However, this task is easier said than done.

How does one go about changing deep-rooted, system-induced beliefs/ values, and their accompanying risky behaviours? The answer is:

> You "draw a line in the sand" and take a stand!

1. Taking a Stand: An Industry Example

A small contracting company in New Zealand's Electricity Supply Industry took a bold step towards effecting positive culture change. The company is contracted by the Operator of the local Electricity Network to maintain the Network within their geographic area.

This company had used the GSP™ Workplace Safety Survey offered by Orange Umbrella®[25] to measure the effectiveness of their OH&S Management System. The company wanted to understand the strengths and weaknesses of their operation because they believed the evaluation would guide them in improving their Safety performance. A discussion of the findings helped to identify the current factors that created risk for them. Here is an example of what was learned:

- ***System-induced Risk***: (An obstacle to safe work in the *Equipped to do it* Condition for performance support)

    The company was experiencing a poorly functioning work process that did not consistently provide them with timely information. At times, the information about the jobs they are being contracted to do was inaccurate. This resulted in an inability to do forward planning, hindered the proper preparation for jobs and ultimately, eroded the time available to do the work during scheduled electrical outages. For the work crews, this resulted in frustration, "running out of time", risking not meeting deadlines in returning power to customers and potentially, exceeding the budget allotted to the work. This situation is depicted in Figure 7.8.

- ***Culture of "Risk-denial"***:

    As a result of frequently facing the issues described above, employees engage in practices that "push the limits" to get the job done within the allotted time for an outage and the contracted budget (and often just to get the power back on for "the little old lady waiting to put her cookies in the oven!"). They hurry, they cut corners, and they take short cuts. These

---

[25] Orange Umbrella® is an international provider of GSP™ Safety perception surveys and services based on the Great Safety Performance model and methodology described in Chapter 6, "Occupational Health and Safety: Leadership and Commitment".

behaviours get repeated over and over again. Because nothing "bad happens" (so far), the belief that "nothing will happen" develops. In this way, the situation, and the associated level of risk, becomes normal and goes unchallenged. However, at some unpredictable time and place, a "brief window of accident opportunity" may open.

Figure 7.8 Current State Work Process

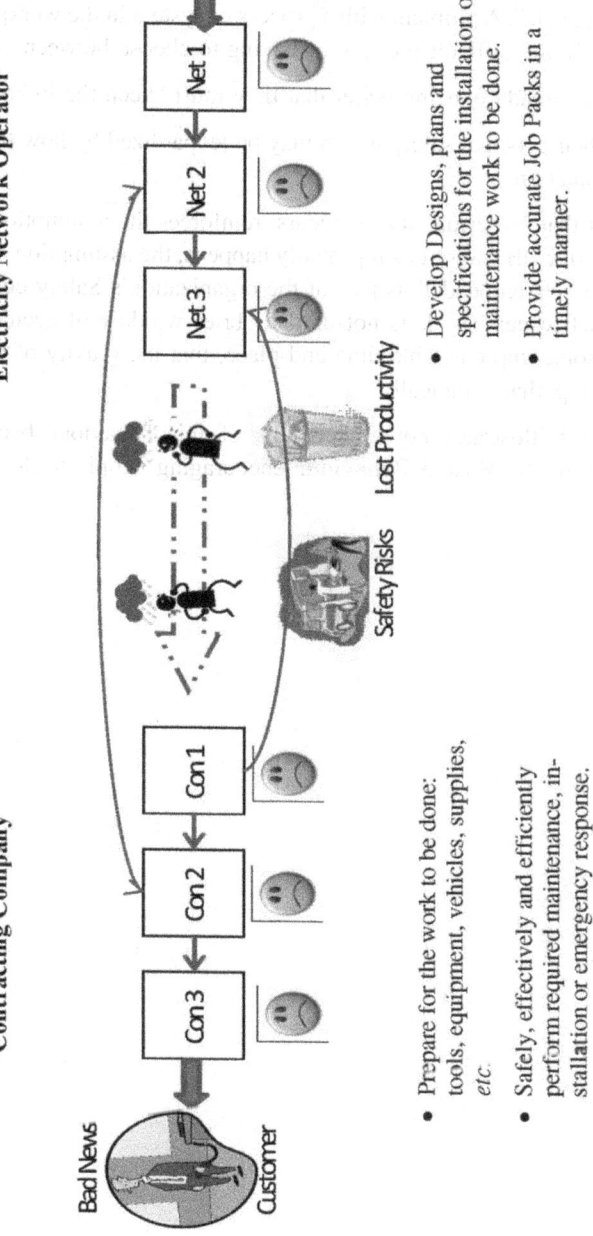

2. The Self-Reinforcing Cycle of Culture Development

The above example illustrates how the work culture develops through a "self-reinforcing cycle". A problem with a process or system in the workplace tends to put employees in the difficult position of having to choose between:

1. an already existing belief that they must "keep the lights on", and
2. their personal safety which may be jeopardized by how they deal with the situation.

The fact that no serious injury occurs, reinforces the assumption that it is *"okay to take the risk"*. Because this repeatedly happens, the assumption becomes a belief, which then gets embedded as part of the organization's Safety culture. Hence, the unsafe practice persists. It is not until a "brief window of accident opportunity" opens, at some unpredictable time and place, that the gravity of the risk becomes apparent — perhaps tragically.

Figure 7.9 illustrates how the results of this behaviour become a **Positive Reinforcer** of the **Wrong Behaviour** encouraging people to do it over and over again.

Figure 7.9: Self-Reinforcing Cycle
The Work Experience

Behaviour
- Took chances
- Cut corners
- 2 – repeated
- 3 – repeated
- Nth time repeat.

Results
- Saved time
- Got the lights on
- Nothing bad happened

1, 2, 3, 4 …. N repetitions of the same behaviour

Perception
1. "That worked out alright"
2. "That worked out alright again"
3. "Wow this is working well"
4. Ditto

Nth time……..

Belief
1. ???
2. "Hmmm, seems to be OK"
3. "That seems to be working"

Nth time: "This method really works well; it's a great time-saver. This is how we do things around here."

Assumption
1. "That method can come in handy"
2. "It worked last time; I'll do it again next time"
3. "It should work again"
4. "And again"

Nth time "…looking good"

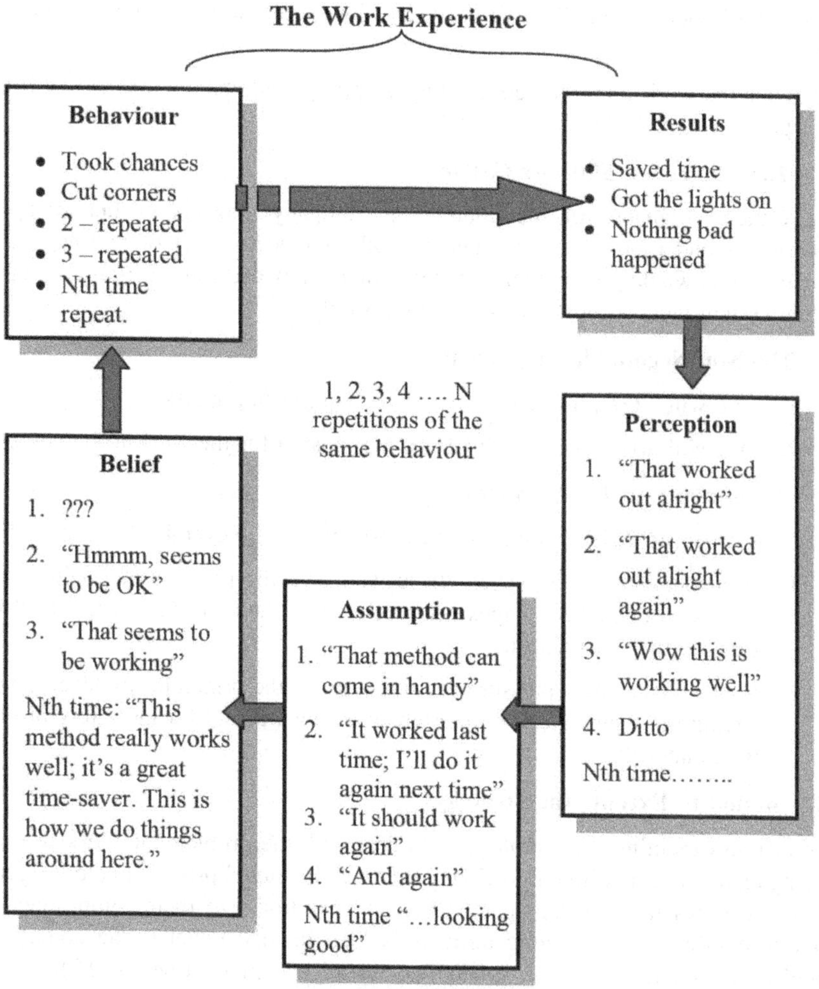

In this scenario, the *Beliefs* must change in order to change the *Behaviour*.

We can also work on the Behaviour aspect; new desirable outcomes will reinforce that new behaviour and create a new belief.

The Belief Shift that *must happen*:

Move From: Productivity is more important than Safety;

Move To: A shared Belief that: Productivity and Safety are not only equally important, but also mutually supportive (poor Safety erodes productivity: a strong Safety culture lifts productivity).

The accompanying Behavioural Shift *must move*:

From: "Let's Get 'er Done'"; "We've always done it this way"; "Nothing ever went wrong";

To: When the work cannot be done safely, stopping, and making it safe, and if that isn't possible, *Not Doing it*.

(a) The Strategy to Shift the Culture

The General Manager of the contracting company and his or her group of Supervisors and Line Mechanics unanimously agreed to a set of non-negotiable principles that would govern their behaviour on the job, and cause action to be taken on the system-induced obstacles to working safely.

(i) The Non-Negotiable Agreement

1. We will *NOT* put ourselves, and our equipment, at risk.
2. We will arrive on the job site fully prepared for the work to be done.

(ii) Agreement: To Take a Stand

Their agreement to *Take a Stand* on these principles has several effects, namely:

- It encourages, even pushes, those who own the poorly functioning work processes to make changes that will enable the Line Mechanics to "stay true" to their principles.
- It also creates peer pressure to "stay true" to the principles by leveraging a strength in their Safety culture which is to watch out for the safety of their workmates.[26]

(iii) Action to Execute the Strategy

One task in executing their strategy is to deal with the problematic work process. Having made the commitment to *Taking a Stand*, the model provided below (Figure 7.11), illustrates how the company plans to operationalize their commitment to workplace safety over the long term. Imbedded in this model is the concept of "*catching people doing things right*". This model is explained below (#2).

1. Redesign the Work Process (Figure 7.8) that is a root cause in this scenario.

    To achieve high levels of performance within both the Network and Contracting organizations in which the business process must operate smoothly, efficiently, and effectively, the inputs to the process and the outputs from the process must be clearly defined. As well the interfaces

---

[26] The company in the example participated in the (2010) Safety Culture Pilot Project sponsored by the Electrical Engineers Association of New Zealand involving 376 employees in 11 companies. Two items in the GSP™ Workplace Safety Survey identified this as a strength: 81 per cent of electrical workers agreed that "my coworkers encourage me to work safely"; 90 per cent said "*I encourage my work mates to do tasks safely.*"

and handoffs need to function seamlessly in order to meet the requirements that both organizations have of each other in order to maintain the essential service in a safe manner.

2. Manage the Expectations set by the Non-Negotiable Agreement through Consequences to Behaviour.

Culture is expressed through patterns of behaviour — "what we do and say". Our behaviour can be encouraged or discouraged. Both good patterns of behaviour, as well as poor behaviours, can be encouraged or discouraged. Such influence can be intentional or inadvertent. These influences are nonetheless the consequences that shape human behaviour. Both encouragement and discouragement come from people and the workplace performance system.[27] The scenario presented above is a classic example of a workplace system discouraging the right safe work practices. As well, the positive outcomes (consequences) shown in Figure 7.9, actually encourage the type of risky behaviour that the company wants to discourage.

The chart in Figure 7.11 shows a continuum of behaviour from exceptional performance, to the deliberate violation of known rules, regulations or Safety practices. The responses to those behaviours range from tracking and recognizing positive behaviours to discipline and dismissal from the organization. An appropriate consequence to the individual and/or team is recommended. These responses include an examination of work environment/system obstacles that inadvertently drive poor, risky practices.

A very powerful part of this chart is that it also leads into the examination of the Supervisor/Foreman's behaviour. The front-line employee does not, and should not, perform work in isolation. Front-Line leadership, through what they do and say, supports and enables the kind of safe, effective and efficient performance that leads to excellent performance outcomes.

Managing expectations also means providing consequences to the Supervisor/Foreman to support and enable him/her. An appropriate consequence across the continuum from left to right is recommended. The consequences to Supervisor/Foreman originate from more Senior Management.

The chart also specifies the accountability Management has to support and enable all employees to work in a manner that ensures their personal safety and the safety of their co-workers. In this way, the chart specifies the roles of everyone in the operational hierarchy in developing a robust

---

[27] A.C. Daniels & J. Agnew, *Safe by Accident* (Performance Management Publications, 2010).

culture of Safety.

The action is to apply the recommended consequences when an individual has demonstrated exemplary behaviour or when performance is simply as required; and to use it to assist in determining the appropriate action following an error or violation.

3. Catch people doing things right! or What's love got to do with it?

Too often employees only get attention when they do things wrong. That negative response from supervisors and managers is by far more frequent than "catching people doing things right". If we are to shift counter-productive beliefs and the dysfunctional risky behaviour those beliefs drive, the balance of consequences must shift dramatically toward positive reinforcement: "[p]ositive reinforcement is any consequence that follows a behaviour that increases that behaviour's frequency . . . *If critical safe behaviour is not increasing, it is not being reinforced.*"[28] This last statement by Daniels and Agnew is very important. It should be one we think about whenever we see someone doing anything that is outside of the parameters of our safe work practices, rules and regulations.

Regarding the chart in Figure 7.11, the left side of the continuum is by far the most important for building a positive culture of safety over the long term. The number of things that go right is by far greater than the amount of things that go wrong. That means there is a huge opportunity for intentional positive reinforcement in the form of positive feedback about the specific behaviour that an employee does efficiently, effectively and safely. When things go right, there is also the opportunity for positive reinforcement through recognition to ensure good work practices continue. Recognition must be appropriate to the task or job being recognized. It must also be meaningful to those being recognized. Let's not be bashful about it, let's give them "some love". It goes a long way to make people feel valued!

Supervisors and managers tend to raise the issue of the frequency of feedback and recognition: "[d]o we have to go around and constantly tell them that they are doing a good job? Doing a good job is what they get paid for." If that is the prevailing attitude, refer back to the two paragraphs above.

> Most OH&S Practitioners/Professionals greatly underestimate the number of reinforcers required to develop (automatic, consistent, and effective) performance. *The latest research indicates that the number (of Reinforcers) is actually in the hundreds.*[29]

---

[28] *Ibid.*
[29] *Ibid.*

4. Using the Chart to respond to Unsafe Acts

When dealing with unsafe behaviour, it is extremely important to distinguish between human error and violation. We must identify the underlying cause or intention for the action before deciding on a "just and fair" consequence. Figure 7.10 illustrates the logic for dealing with unsafe acts that is built into the organization of information in the chart in Figure 7.11.

Figure 7.10: Understanding Unsafe Actions

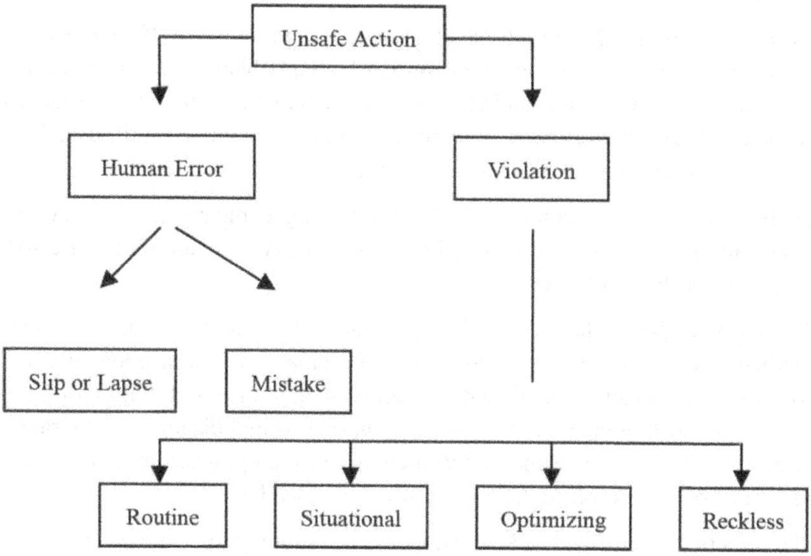

```
                    Unsafe Action
                   /            \
             Human Error      Violation
              /      \         /  |  |  \
        Slip or   Mistake  Routine Situational Optimizing Reckless
        Lapse
```

The definitions of the types of violation are provided in Figure 7.11.

In general, the response to unsafe actions by an individual or team should be:

- Provide coaching when dealing with human error.
- Coaching is usually also appropriate for routine, situational, and company-optimizing violation.
- Invoke formal discipline in cases of deliberate or reckless violation.

Human error can be a mistake, a slip, or a lapse. Mistakes are often because of lack of skill, training, or knowledge. Slips and lapses can be a momentary lapse in memory, temporary loss of focus, or distraction. In these cases, disciplinary procedures are rarely appropriate. Coaching is the most appropriate response.

As Figures 7.10 and 7.11 indicate, violations can occur over a range of severity. Coaching is likely appropriate for the individual/team in cases of routine and situational violations as well as violations intended for company benefit. The individual employee or team would rarely face formal discipline. However, if the

Supervisor/Foreman failed to act regarding routine and situational violations, early-stage discipline is appropriate in addition to coaching. If the Supervisor/Foreman's behaviour is as described in the chart under violation for company or personal benefit or reckless violation, then formal discipline is most likely warranted. The Supervisor/Foreman should be "taking a stand" by following through on the appropriate consequence to the individual or team.

In the case of the final two violations listed on the chart (violation for personal benefit; reckless violation), formal discipline is quite appropriate for both the individual employee and the Supervisor/Foreman.

Unsafe acts are usually symptoms of an underlying root cause. If someone makes an error, or breaks a rule they were not aware of, simply blaming the person is unjust and unfair. This does not address any organizational weaknesses that could contribute to other incidents. It is important to uncover the system obstacles that make desired performance difficult to achieve.

In the case of the situation with the contracting company in our example, a problem in the system — the ineffective work process — is the root cause for the risky behaviour by work crews.

As indicated above, the chart always gets used twice in any situation so that the individual or team as well as the Supervisor/Foreman are held accountable for the actions they have taken or condoned. As can be seen in the chart, failure on the part of the Supervisor/Foreman is more likely to attract formal discipline. The rationale is that a failure by the Supervisor/Foreman puts employees at risk. It is the Supervisor/Foreman's job to "show the way" as a leader.

The practices described above support the strategy of *Taking a Stand* by reinforcing the desired safe work actions and supportively turning around those practices that embody risk to life and limb. Figure 7.11: Culture Change: *Taking a Stand* by Managing Expectations through Consequences to Behaviour, is adapted from Saunders.[30]

---

[30] Inspired by and adapted from "Just and Fair Culture", a presentation by Dave Sanders, A B Consulting (November 2, 2010, EEA Safety Workshop, Wellington New Zealand), online: http://www.eea.co.nz/Section?Action=View&Section_id=416 (date accessed: February 28, 2020). Dave Sanders' model and this adaptation are based on the work of Patrick Hudson, Leiden University.

CH. 7: SAFE WORKPLACE = GREAT WORKPLACE

Figure 7.11: Culture Change: *Taking a Stand* by Managing Expectations through Consequences to Behaviour

● Track & Recognize Positives ● ● ● ● Coach ● ● ● ● Discipline ● ● ●

Type of Performance	Exceptional Performance	Expected Performance	Unintentional Error (slip, lapse, mistake)	Routine Violation	Situational Violation	Violation Optimizes for Company Benefit	Violation Optimizes for Personal Benefit	Reckless Violation
Individual or Team Behaviour	Individual or Team went above and beyond the call of duty in safe, efficient, effective performance? Yes ⇩ No ⇨	Individual or Team follows all procedures to achieve safe, efficient, effective performance? Yes ⇩ No ⇨	Did the Individual *think* he/she was doing things the right way? Yes ⇩ No ⇨	Do other team members normally complete the task in the same way? Yes ⇩ No ⇨	Did the Individual *think* the prescribed procedure was an obstacle to getting the job done? Yes ⇩ No ⇨	Did the Individual *think* it was better for the company to do the job his/her way? Yes ⇩ No ⇨	Did the Individual do it that way to make it easier for him/herself? Yes ⇩ No ⇨	Did the Individual intentionally ignore or not care about the consequences of unsafe actions? Yes ⇩ No
Consequence to Individual or Team	Reward/Recognize at discretion of management: private/public praise, recognition or award.	Recognize and encourage through positive feedback about what is being done well.	Acknowledge what was right about the situation. Train and/or coach the individual in the use of correct procedures.	Coach the Team on the importance of understanding and following correct procedures, *i.e.*, not taking shortcuts.	Assure the Team that they can speak up when procedures cannot be followed and stop work until the job can be done safely.	Assure the Team that they can balance production pressure with HSE requirements. Consider a "first warning" to the individual.	Consider formal discipline according to company policy, procedures and workplace agreements.	Take formal disciplinary action according to company policy, procedures, and workplace agreements.

353

	● Track & Recognize Positives ● ● ● ● ● ● Coach ● ● ● ● ● ● Discipline ● ● ●							
Type of Performance	Exceptional Performance	Expected Performance	Unintentional Error (slip, lapse, mistake)	Routine Violation	Situational Violation	Violation Optimizes for Company Benefit	Violation Optimizes for Personal Benefit	Reckless Violation
Supervisor/ Foreman Behaviour (S/F)	Did the Supervisor/ Foreman (S/F) also perform in an exceptional manner?	S/F consistently focuses the Team on safe, effective work practices and procedures?	Did the S/F fail to supervise work to ensure tasks were done in the required manner?	Did the S/F allow non-compliant, poor work practices and/or work environment/ "system" obstacles to develop without correction?	Did the S/F know the procedure and/or work environment/ "system" obstacles were barriers to getting the job done, AND, do nothing about it?	Did the S/F permit shortcuts, unsafe actions for the sake of getting the job done?	Did the S/F overlook this behaviour on this or previous occasions?	Did the S/F condone the actions of the Individual?
	Yes ⇩ No ⇨	Yes ⇩ No ⇨	Yes ⇩ No ⇨	Yes ⇩ No ⇨	Yes ⇩ No ⇨	Yes ⇩ No ⇨	Yes ⇩ No ⇨	Yes ⇩ No
Consequence to Supervisor/ Foreman (S/F)	If behaviour was in support of the whole Team reward/ recognize the S/F: private or public praise, recognition, or award	If whole Team is working this way or for specific improvements, recognize and encourage through positive feedback about what S/F is doing well.	Coach the S/F on the accountabilities of the Supervisor role. Coach the S/F on identifying and managing errors.	Coach on how to monitor & enforce procedures. Coach on how to identify and correct work environment/ "system" obstacles to performance. Consider discipline e.g. first warning.	Consider formal disciplinary action. Coach as in the previous section; monitoring, enforcement, correcting work environ-ment/system obstacles to performance	Take formal disciplinary action according to company policy, procedures, and workplace agreements Coach as in the previous section.	Take formal disciplinary action as in previous section. Coach on how to recognize such behaviour earlier and how to deal with it.	Take formal disciplinary action as in previous section. Coach on how to recognize such behaviour earlier and how to deal with it.
	⇧	⇧	⇧	⇧	⇧	⇧	⇧	⇧
Management Accountability	Ensure all employees work in a manner that ensures their personal safety and assists the safety of coworkers. Ensure Supervisors/Foremen have the knowledge and skills to fully enable their teams so that they all *Know What* to do, are *Able* to do it, are *Equipped* to do it, *Want* to do it, and are supported by *Interactions* that foster trust, respect, integrity, collaboration and accountability. Ensure Supervisors and Foremen can identify and correct obstacles to performance that are embedded in their overall work environment							

H. WORKPLACE CULTURE EVALUATION: THE VALUE OFFERED

Organizations that elect to conduct an evaluation of their workplace culture can unearth and address deep-rooted labour-management issues that tend to "poison the Safety culture and work environment". As noted above, *"love does have a lot to do with it"*.

Likewise, evaluation of the workplace culture brings to light issues such as:
- the presence and impact of counterproductive systemic processes;
- certain leadership practices that are not supported by the workplace culture;
- the existence of long-standing, dysfunctional workplace relationships; and
- situations where "true leadership" are absent.

Importantly, it forces management to "do something" — the evaluation provides data that cannot be ignored. As well, it provides data/information that facilitates positive action.

So, how do we measure Safety culture in a holistic manner? Although there are a variety of approaches to measuring workplace culture, it is more efficacious[31] to assess Safety culture in a comprehensive manner. That is, to assess the Safety culture in the context of the entire workplace system, not just from one aspect of it. The approach described in this chapter recognizes that the elements of a workplace system are so interconnected and interdependent that they need to be evaluated in the context of a whole workplace system. This concept is reinforced in Chapter 8, "Safety Culture: Safety Climate".

I. SUMMARY

The OH&S skills and Safety hardware have taken organizations a long way towards improving worker safety. However, the results being achieved with this approach have plateaued. To move to the "new Safety frontier", organizations must embrace the concept of Safety software, *i.e.*, the presence of a Safety culture and the soft skills required to develop and maintain it. They will need efficacious methods to measure and analyze the impact of their workplace system on their safety performance. It will take organizations to the next level of Safety functioning and enable them to enjoy an injury-free workplace. If an organization does everything it needs to do to have a truly safe workplace, it will enjoy the additional benefits of a highly engaged workforce as well as greater productivity and profitability.

CHAPTER REFERENCES

Anatomy of Performance.pdf, online: http://www.performancedesignlab.com (date accessed: February 28, 2020).

---

[31] **Efficacy** is what happens in the "real world" as opposed to what is expected to happen. Efficiency and effectiveness are measures of what is expected to happen. Efficacy is the capacity to produce a desired effect within a "real life" situation and is hence impacted by "real-life" factors.

F. Bird & G. Germain, *Loss Control Management: Practical Loss Control Leadership*, revised ed. (Loganville, GA: DNV, 1996).

A.C. Daniels & J. Agnew, *Safe by Accident* (Performance Management Publications, 2010).

W.E. Deming, *Out of the Crisis* (MIT Press, 1986).

P. Hudson, *et al.*, *Meeting Expectations: A New Model for a Just and Fair Culture*, at 3, online: http://www.onepetro.org (date accessed: February 28, 2020).

International Society for Performance Improvement (ISPI), "What is HPT?", online: http://www.ispi.org (date accessed: February 28, 2020).

J.P. Kotter & J.L Heskett, *Corporate Culture and Performance* (New York: Free Press, 1992).

D. Kravetz, *People Management Practices and Financial Success* (1996), online: http://www.kravetz.com (date accessed: February 28, 2020).

R.G. McKinnon, Manager, Plant Programs and Environmental Support Department, Manitoba Hydro, Manitoba.

Orange Umbrella®, online: http://www.orangeumbrella.co/.

J. Reason, "Foreword" in A. Hopkins, *Safety, Culture and Risk* (CCH Australia Ltd., 2005).

D.G Robinson & J.C. Robinson, *Performance Consulting* (San Francisco, CA: Berrett-Kohler, 1995) at 180.

D.G Robinson & J.C. Robinson, *Moving from Training to Performance* (San Francisco, CA: Berrett-Kohler, 1998) at 7.

T. Roithmayr, "The Performance Maximizer® ~ A Powerful Tool for Understanding and Maximizing Performance", online: http://www.performance-bydesign.com (date accessed: February 28, 2020).

G. Rummler & A. Brache, *Improving Performance*, 2d ed. (San Francisco, CA: Jossey-Bass, 1995), at 64.

D. Sanders, "Just and Fair Culture", a presentation by Dave Sanders, A B Consulting (November 2, 2010, EEA Safety Workshop, Wellington, New Zealand), online: http://www.eea.co.nz/Section?Action=View&Section_id=416 (date accessed: February 28, 2020).

S. Simon, "On the Future of the Safety Profession" (2001), online: http://culturechange.com/wp-content/uploads/2012/08/futuresafeprofpaper.pdf (date accessed: February 28, 2020).

Chapter 8

SAFETY CULTURE: SAFETY CLIMATE

By Heidi Börner, Dianne Dyck and Brett Webb

A. INTRODUCTION

The constructs, Safety culture, and Safety climate, evolved from the broader concepts of organizational culture and organizational climate. However, there is considerable confusion as to what exactly each term means.

An Occupational Health & Safety (Safety) (**OH&S (Safety)**) **culture** is defined as the moral, social, and behavioural norms of an organization that are based on the shared beliefs, values, attitudes, habits, and traditions of health and safety that give meaning to an organization's employees and provides them with the accepted health and safety behaviours within their organization. It is what "defines the organization to employees".[1] The OH&S (Safety) culture has a huge impact on worker performance, often more than the practices, knowledge, and skills that employees have been taught. According to the Health & Safety Commission (1993), it determines the degree of commitment to, and the style and proficiency of, an organization's OH&S Management System, program, and practices.

The **OH&S (Safety) climate** is "the prevailing atmosphere of the organization, the socio-psychological environment that profoundly influences behaviour, and is typically measured by employee perceptions".[2] It is the factually, or interpretive, outputs or results of an organization's higher-level safety culture.[3] A positive OH&S (Safety) climate is associated with low occupational injury and illness rates and costs, as well as other desirable business outcomes,

> . . . climate is commonly associated with terms such as "superficial" . . . "snapshot," "quantitative," and "state," whereas culture with "deep," "stable," "qualitative," and "trait."[4]

---

[1] D. Spitzer, *Transforming Performance Measurement: Rethinking the Way We Measure and Drive Organizational Success* (New York, NY: AMACOM, 2007).

[2] *Ibid.*

[3] K. Mearns et al., *Human and Organisational Factors in Offshore Safety (OTH 543)* (Suffolk: Offshore Safety Division, HSE books, 1997).

[4] D. Seo et al., "A cross-validation of safety climate scale using confirmatory factor analytic approach" (2004) 35:4 Journal of Safety Research 427-445.

It is widely recognized that corporate culture and hence, the OH&S (Safety) culture, are primary drivers and predictors of improving Safety performance:

> Safety is culture-driven, and Management establishes the culture . . . an organization's culture translates into a system of expected behaviours (norms). The injury and illness experience that results, is a direct reflection of an organization's Safety culture. Major improvements in Safety will only be achieved if a change in culture takes place.[5]

Leaders have a profound impact on the culture and climate of an organization; the reason being, employees judge a business initiative by the individual(s) leading it. What leaders say is interesting, but what they do speaks volumes. In essence, "[y]ou change the culture of a company by changing the behavior of its leaders."[6]

The intent of this chapter is to present two industry examples of health and safety climate assessments — one conducted in Ontario, Canada, and one in New Zealand; and then, to illustrate how many electricity utility organizations in New Zealand implemented and sustained a positive health and safety culture and climate.

B. OCCUPATIONAL HEALTH AND SAFETY CLIMATE: ASSESSMENT

1. PSHSA Health and Safety Climate Assessment Project[7]

By Dianne Dyck and Brett Webb[8]

In 2013, as a means for improving workplace health and safety outcomes, the Public Services Health and Safety Association (PSHSA) assessed the Health & Safety climate of four Ontario health care organizations. The rationale for this move is that Ontario's health care sector experiences the fifth highest rate of workplace injuries and illnesses in the Province. Given the size of this industry sector (787,000 employees) and the cost of work-related injuries (approximately $2.35B in direct costs related to premiums, benefits, surcharges, and Workplace Safety and Insurance Board (WSIB) costs, and the indirect costs due to the long-term impacts of injuries

---

[5] F. Manuele, *Advanced Safety Management: Focusing on Z10 and Serious Injury Prevention* (New York. NY: John Wiley & Sons, 2008).

[6] L. Bossidy & R. Charan, "Execution: The Discipline of Getting Things Done" in D. Spitzer, *Transforming Performance Measurement: Rethinking the Way We Measure and Drive Organizational Success* (New York, NY: AMACOM, 2007) at 128.

[7] Thanks, is extended to the Public Services Health and Safety Association for their assistance with this chapter. The PSHSA staff's extensive knowledge of Occupational Health and Safety was integral to developing this chapter section, and I am grateful for their guidance and input.

[8] B. Webb, Bbus, DipOSH, CMIOSH, CRSP, is a consultant with PSHSA in Northern Ontario. Brett's safety work experience includes the following industries: health care, education, government, claim management, manufacturing, security, and social housing. He also lectured at the University of Western Ontario in Occupational Health and Safety Management.

on workers), this project was deemed important and funded by the Ministry of Health and Long-Term Care.

The PSHSA mandate is "to influence and support improvements in health and safety outcomes for all public sector organizations".[9] As such, it undertook a climate assessment at four Ontario health care organizations — an acute care, long-term care, community services, and public health organization. Additionally, the assessment enabled PSHSA to "test and validate a health and safety climate assessment tool in the Ontario healthcare setting and determine opportunities to improve the assessment's efficiency, effectiveness, and scalability"[10] for future use.

In addition to the PSHSA mandate, the project was designed to:

- incorporate research and leading practices, where possible;
- build on and leverage ongoing related efforts and initiatives, as well as expand on existing tools and assessments;
- shift focus from lagging to leading indicators of health and safety performance;
- continue to build on knowledge translation opportunities by bridging the gap between research-generated knowledge and practice;
- expand the communication channels and networks between the system (*i.e.*, prevention system) and front-line providers (*i.e.*, individual hospitals); and
- indirectly influence and improve organizational performance in areas such as patient care and patient safety.

For this pilot project, the working definition of **Safety Climate** is "a tangible output, or indicator, of an organization's health and safety culture as perceived by individuals or groups at a point in time".[11] The overview states that

> Assessing an organization's health and safety climate identifies the current values, attitudes, and patterns of behaviour of employees with respect to health and safety. The assessment explains in detail why there may be a gap between the desired outcomes of the organization and the current reality.[12]

(a) Methodology

The intent of the Climate Assessment Tool was to measure the health and safety climate of each organization across 17 distinct areas or dimensions including priority safety areas such as commitment, communication, values, compliance, risk,

---

[9] PSHSA, *2014 Health and Safety Climate Assessment Project Report*, online: http://www.pshsa.ca/wp-content/uploads/2014/07/HFO-climate-project-report.pdf at 3 (date accessed February 28, 2020).

[10] *Ibid.*

[11] *Ibid.*

[12] *Ibid.*

environment, involvement, management style, and safety procedures.

The methodology for this pilot project involved the use of surveys, focus groups sessions, observation, documentation review, as well as the administration of the PSHSA Climate Assessment Tool. The data also included participant characteristics and demographics (level in the organization, profession, age, gender, *etc.*).

The results were subsequently analyzed and provided to the respective participating health care organizations.

(b) Findings and Interpretation

The pilot project determined that:

- The participating health care organizations had an average health and safety climate score of 6.8/10. Without an established threshold value between good and poor performance for the Climate Assessment Tool, it is difficult to state whether this is a high or low score. However, the interpretation was that, there is a significant need for improvement.

- The scores were stronger in the dimensions of *Personal Need for Safety*, *Safety Rules and Procedures*, *Supportive Environment*, *Safe Behaviours*, and *Management*. This finding is viewed as "workers want to be safe and work in an environment where safety is [a] priority".[13]

- Lower scores were noted in the dimensions of *Work Environment*, *Cooperation*, *Shared Values*, *Systems Compliance*, and *Accidents and Incidents*. The interpretation is that "the workplace is not as safe as it should be and can be improved if all workplace parties focus on health and safety processes",[14] workplace parties have a clear, shared vision of health and safety; comply with processes and safe work practices; and a visual commitment from leadership.

- Within all the health care organizations, the leadership and management personnel responded significantly higher on all the assessment dimensions than did the front line employees. The indication appears to be that there is a need to improve employee participation in health and safety, enhance communication, and ensure training results in knowledge translation.[15]

- PSHSA employees are not engaged in the health and safety initiatives, nor are they empowered to make Occupational Health & Safety improvements. To address this situation, a participatory approach must be incorporated.

- Incidents that a supervisor may link to employee behaviour are in fact the result of other system failures such as ineffective communication or ineffective/non-existent training and knowledge transfer.

---

[13] *Ibid.*, at 6.
[14] *Ibid.*
[15] *Ibid.*

Ch. 8: Safety Culture: Safety Climate

- Incident reporting is not encouraged through a mechanism of "blame-free" reporting. Root Cause Analyses are not being completed nor are the appropriate hazard controls implemented. This leaves workplaces "open" for ongoing incidents and potential losses and liabilities.

- Training programs need to be reviewed to ensure that they are reaching the right people, at the right time, using the right delivery, and achieving the right outcomes.

- The feedback from the participating organizations on the assessment process was positive (Figure 8.1).

Figure 8.1: An Example of Participation Feedback on the Climate Assessment Process[16]

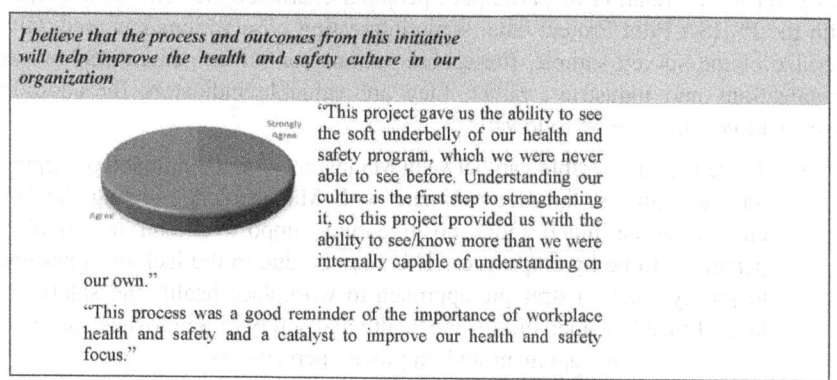

I believe that the process and outcomes from this initiative will help improve the health and safety culture in our organization

"This project gave us the ability to see the soft underbelly of our health and safety program, which we were never able to see before. Understanding our culture is the first step to strengthening it, so this project provided us with the ability to see/know more than we were internally capable of understanding on our own."

"This process was a good reminder of the importance of workplace health and safety and a catalyst to improve our health and safety focus."

(c) Implications

The positive feedback from the participating organizations and the pilot project findings provided PSHSA with needed information to assist other Ontario health care organizations to create strong Health and Safety (H&S) cultures, and organizational cultures. The project recommendations that are expected to be operationalized by the PSHSA over the next three to four years are as follows:

1. Undertake PSHSA Health and Safety Climate Assessments at other Ontario health care organizations.

2. Develop products and services to support Ontario health care organizations in improving their health and safety culture, and organizational culture.

3. Pilot the culture improvement products and services at four to five Ontario health care organizations.

4. Conduct research and additional statistical analysis of the climate assessment findings.

---

[16] *Ibid.*

5. Promote and market the PSHSA Health and Safety Climate Assessment approach.
6. Engage Ontario health care organization leadership in culture improvement initiatives.
7. Support the Ministry of Health and Long-Term Care, through HealthForce-Ontario (HFO), in the inclusion of health and safety leading and lagging indicators as reporting requirements.
8. Continue to support and investigate new opportunities for system-wide health and safety improvements in the Ontario health care sector.

(d) ***Observations Warranting Further Investigation***

Based on the correlation of participant personal characteristics and demographics with the PSHSA Pilot Project data, some interesting observations emerged. Given the size of the survey sample, these observations cannot be generalized to other organizations and industries; rather, they are valuable indicators for additional investigation. They are as follows:

- There is a measurable gap between Management and employee perceptions on the status of the H&S climate with Management viewing the H&S climate to be much more positive and supportive than it actually is perceived to be by employees. This may be due to the lack of opportunity to jointly create a strategic approach to workplace health and safety, or a lack of visible leadership for the health and safety of employees; hence, the gap between management and employee perceptions.
- Employees under age 25 years had more positive perceptions towards health and safety than did the older workers. The exceptions include their perceptions on the dimensions of *Involvement*, and *Safety Rules and Procedures*. This finding is consistent with the research on the younger generation workers who are characterized as possessing a sense of entitlement and the need to be involved in workplace decisions.
- The dimension scores were consistently lower among permanent staff compared to temporary staff with the exception of the *Involvement* dimension.
- Employees with one year or less of service, scored higher in all dimensions as compared to employees with more years of service. This finding indicates that there is a certain level of societal perception towards workplace health and safety; but the fact that this perception worsens the longer the employee is with the organization is most concerning and points to problems with the organization's culture. Further insight could be gained by comparing the "less-than-one-year" employees against other provinces, industries, or countries to determine their social perceptions towards workplace health and safety.
- The Joint Health and Safety Committee (JHSC) Management Members

demonstrated higher scores than did the JHSC Worker Members or Non-members on all the dimensions except for *Safety Rules and Procedures*. Given the large difference between worker and management members, there appears to be a need to build a working relationship between the two groups so as to create a more effective cooperative committee.

- Nurses scored lower in all dimensions except *Personal Need for Safety* than did the other professions. The largest gap was in the *Shared Values* and *Work Environment* dimensions. Given the fact that nurses make up the majority of the workforce in the health care industry and that their perception towards health and safety was consistently lower than other demographics, cultural improvement strategies should be a priority for this group.

- Physicians scored lower on all dimensions and the belief is that this finding is linked to their "contractor" relationship and the "sense of not belonging" to the organization's culture.

- In general, workers do want to be safe and to work in a safe environment where safety is a priority.

(e) Transferable Learning

The PSHSA Pilot Project unearthed many aspects that lend to transferable learning, namely:

- Performance measurement is critical to understanding what is currently happening within an organization and to making informed decisions on how to improve the situation.[17]

- Health care organizations continue to lag behind other industry sectors in their OH&S structures, processes, and outcomes, which correlates with them having the highest work absence days (15.9 days/FT employee) and rates.[18]

- Establishing an OH&S Management System is critical to positive organizational health and safety performance outcomes.[19]

- Transformational leadership and leaders can enable organizations to retain talented human capital and maximize their Occupational Health and Safety

---

[17] D. Spitzer, *Transforming Performance Measurement: Rethinking the Way We Measure and Drive Organizational Success* (New York, NY: AMACOM, 2007).

[18] Statistics Canada, *Work Absence rates of full-time employees by Industry, Annual, 2019*, Table: 14-10-0191-01 (formerly CANSIM 279-0030) (2020), online: https://www150.statcan.gc.ca/t1/tbl1/en/tv.action?pid=1410019101 (date accessed February 28, 2020).

[19] For more information on the Occupational Health and Safety Management System, refer to Chapter 2, "Occupational Health and Safety Management Systems".

Management System.[20]

- A positive H&S culture is one where everyone from the CEO to receptionist, are brought together around "a common set of beliefs". The development of "a common set of beliefs" should be embraced as an opportunity; the development of a positive H&S culture is not something that develops naturally, but rather, something that can be strategically-driven and shaped by all involved.

2. Orange Umbrella® Methodology (New Zealand)[21]
By Heidi Börner[22] and Dianne Dyck

The Orange Umbrella® methodology evolved from the Performance Maximizer model and the Great Safety Performance model.[23] Through collaboration between specialist knowledge and experience in workplace health and safety, human performance, mediation, facilitation, restorative principles, communication, research, information technology design and development, and industry users, this methodology was developed. It was designed to be a simple, user-friendly tool that companies could use to evaluate and address their occupational health and safety risks related to OH&S (Safety) climate and culture.

This section explains OH&S (Safety) climate assessment using the Orange Umbrella® process; and then, in Section C, a New Zealand project from both the industry and individual-company perspectives will be discussed.

(a) Approach towards Assessing the Safety Climate

The primary driver in the Orange Umbrella® methodology is the Great Safety Performance (GSP™) Survey data. This data provides companies with a picture of the perceptions of front-line staff/contractors, support staff, supervisors, and middle managers. However, surveys alone do not bring about change. Using them without follow-up and subsequent actions can cause more damage than good. The GSP™ Survey is, therefore, part of a four-stage "progression" called NewHeights™. This

---

[20] For more information on Transformational Leadership, refer to Chapter 7, "Safe Workplace = Great Workplace: Building a Sustainable Culture of Safety".

[21] Thanks, is extended to Orange Umbrella® Inc., online: http://www.orangeumbrella.co for the submission of their work currently underway in New Zealand.

[22] Heidi Börner, RN, BN, MN, COHN(C) is the co-creator of the Orange Umbrella® tools which help employers diagnose and quantify the hidden risks in their organization, so they can proactively manage health and safety outcomes. She currently coaches and mentors employers to proactively evaluate and reduce their health and safety system risks using the Orange Umbrella® tools. Jon Everest, MSc, works in the area of conflict management as a mediator, conflict coach, facilitator, trainer, and professional supervisor. Heidi and Jon are the Directors of Orange Umbrella® Inc., online: http://www.orangeumbrella.co.

[23] For more information on the Great Safety Performance (GSP) model, refer to Chapter 7, "Safe Workplace = Great Workplace: Building a Sustainable Culture of Safety".

is a facilitated process designed to position company staff and senior management to successfully develop and deliver their own *Action Strategy* aimed at preventing and mitigating workplace OH&S risks.

NewHeights™ acts as a catalyst for creating a workplace culture of collaboration, participation, engagement, and respect, while minimizing OH&S risks and maximizing human and corporate performance. The process involves activities and actions that are fully transparent and measurable.

The steps in the NewHeights™ methodology include:

1. **Commitment by Management.** This step is so critical: without it, the GSP™ Survey is not undertaken. Senior Managers and Board members must understand the journey that they are embarking on so that they can fully support it.

2. **Establish the Process Leaders.** These are "in-house" staff that are trained to assume the project's Process Leader role. This role ensures participant engagement, as well as strong GSP™ Survey response rates and workshop attendance.

3. **Develop Project Communications and Logistics.** Process Leaders prepare the project communications and logistics.

4. **Administer the GSP™ Survey.** The GSP™ Survey data is collected via an online system. An algorithm is applied to the data so that it can be clearly presented and easily analyzed. Respondent confidentiality and anonymity is critical in the collection of perception data.

    The GSP™ Survey tool measures seven pivotal workplace leading indicators of Safety as per the GSP Model (Figure 8.2), that are subsequently used for informed decision-making on how to evaluate and manage risk for workers doing hazardous work and for providing them with ongoing support to work safely.

Figure 8.2: The GSP™ Model

The company needs to know whether workers:

- Are carrying out the *Safe Work Actions* required by the job.
- *Know what to do*: Do they have the knowledge base to do the work? Do they know what their role is in a job?
- Are *Able to do it*: Do they have the right skills? Are they physically fit to do the work?
- *Equipped to do it*: The right equipment, PPE, policies and procedures.
- *Want to do it*: What system motivators are there to get the best performance?
- Have supportive *Interactions* to support safe work actions.
- Have company *Leadership* that sets the workplace health and safety of its employees as a top priority and demonstrates that by ensuring that operational decisions are consistent with safety.

These GSP™ Survey elements are statistically checked for reliability and validity during the NewHeights™ process.

5. *Facilitated workshops*: The GSP™ Survey data are analyzed by the Process Leaders and the Senior Management Team during a facilitated workshop.

6. *Facilitated workshops* are conducted for all involved staff. The purpose is

to validate the GSP™ Survey results, to discover why workers responded as they did, and to provide an opportunity for staff to provide additional feedback and make recommendations for improvement.

7. The information collected from these two workshops is formatted into a *Perspectives Report* and distributed to the participants.
8. The *Action Strategy* process is facilitated for Senior Management and worker collaboration on salient issues. Issues are prioritized and an *Action Plan* is developed to ensure that adequate resources are allotted for its completion.
9. The company then continues on its own, adjusting and completing their *Action Plan*.
10. Tracking the GSP™ journey involves ensuring that the *Action Plan* is completed. NewHeights™ Rounds are repeated annually to monitor the effectiveness of the interventions and provide leading indicators to measure ongoing risk as the company changes over time.

(b) Findings and Learnings

The NewHeights™ Methodology has resulted in a number of findings and learnings, namely:

- **Senior Management Commitment**: The information generated through the Orange Umbrella® Process involves all levels of the company. It can have far-reaching effects that need to be managed by the company leaders. Senior Management therefore, must be fully "on board" with the process, right from the outset. The Orange Umbrella® Process is not a process that should be led by someone who does not have the positional authority, mandate, and accountability for making the needed changes to address the identified OH&S risks. If Senior Management does not support the process, the process will derail and result in even greater OH&S risks for employees and contractors.

- **Time and Resources**: Changing the OH&S (Safety) culture and climate is a complex process. It can be a long-term, slow, and resource-intensive process. It is important that Senior Management has realistic expectations in regards to the speed of change and the resources required to make that change.

- **Communication**: It is equally important for workers to receive regular communication about progress of the stated *Action Plan*.

- **Follow-through**: As a collaborative process for generating improvement action, it is important for everyone to follow-through on their respective commitment to the Orange Umbrella® Process. Each has a role to play, as well as an expectation for improvement, whether it is personal growth or the introduction of a large piece of equipment. Companies that successfully reduce their OH&S risks are the ones that followed through on their stated *Action Plans*.

(c) Implications of the Safety Climate Assessment Process

From experience with facilitating the Orange Umbrella® methodology, several implications related to the Safety Climate Assessment Process have been noted:

- The Safety Climate Assessment tool provides baseline data on the perceived status and manifestation of the company's OH&S (Safety) culture. It is from this point that all subsequent outcomes are measured.
- The baseline Safety Climate Assessment information serves as the basis for assisting the company leaders to bring about change in the OH&S (Safety) culture. It also serves as a motivational driver for taking action.
- The Safety Climate Assessment Process helps to clarify noted OH&S (Safety) climate and culture issues.
- Tracking results demonstrates the areas that have seen OH&S (Safety) culture and climate change, and quantifies the direction and degree of change. This provides the needed information for reporting company safety culture improvement over time.
- The Safety Climate Assessment provides a documented and transparent method for collecting ongoing OH&S risk data, and for identifying the improvements made.

The interaction between Orange Umbrella® and its clients highlights the complexity of measuring and managing OH&S (Safety) culture and climate and has consistently shown that the power of measuring and monitoring the OH&S (Safety) culture and climate is not in the development of the measurement tool; rather, the power lies in the improvements made, and how they are developed and implemented.

C. GREAT SAFETY PERFORMANCE: THE NEW ZEALAND EXPERIENCE[24]

The New Zealand Electricity Engineers' Association (EEA) was formed in 1925. It represents the New Zealand electricity supply industry (electricity industry) in engineering technical and safety matters. The EEA advocates for the industry on legislative and regulatory matters, and it liaises with electricity supply jurisdictions internationally. Electricity industry participants collaborate through the EEA for unified safety rules for work on electricity assets, and for supporting electrical engineering technical and work practice safety guides. This collaboration through committee structures, produces common policy and practice around significant asset and work practice risks. The EEA monitors and measures electricity industry safety performance. It also leads and coordinates research and other initiatives in support of a collective industry approach to resolving safety-critical issues.

---

[24] Gratitude is extended to New Zealand Electricity Engineers' Association (EEA), and Unison Contracting Ltd., and Trustpower Ltd. for permitting their experience with Orange Umbrella® to be presented in this chapter.

CH. 8: SAFETY CULTURE: SAFETY CLIMATE

The more unified electricity industry approach to safety and the introduction of performance based OH&S legislation since the early 1990s, supported better safety performance as compared to the preceding two decades. By 2005, although the incidence of serious injury and fatality had fallen, the improved electricity industry safety performance had plateaued. The 2008 electricity industry safety strategy particularly recognized leadership and workplace safety culture as important keys to reducing OH&S risk and further improving safety performance. In keeping with the collective approach and the goals to improve leadership and workplace safety culture, the EEA coordinated a trial common approach among industry participants. Ten electricity network companies or service providers joined a trial Safety Climate Project (SCP) in 2010. Its goal was to evaluate safety climate and to integrate leadership planning and focus on substantial safety issues out of the evaluation process. This trial led to an overwhelming conclusion in 2011 that the SCP process should continue as an industry collaborative project under the auspices of the EEA.

The SCP utilizes Tripartite Agreement[25] among the EEA, participating companies and the Orange Umbrella®. The Agreement outlines the roles and responsibilities of each party and ensures the confidentiality of company data. However, the arrangement also enables project findings on common significant issues to be evaluated through the EEA for wider electricity industry application.

Since 2010, the SCP has been applied across 28[26] electricity industry companies. This has enabled insights across the participants into safety issues previously hard to quantify or define objectively, but which impact upon the level of risk and performance. The result has been OH&S improvements both at the participating company level, and at the wider industry level. The EEA has also used the SCP findings to support submissions to government regulators to initiate new programs and suppliers for the industry, and to track progress in OH&S risk reduction.

The SCP is unique as a collaborative industry-based initiative to support better OH&S performance, leadership, and culture improvement across an industry sector. Using the Orange Umbrella® methodology, the SCP integrates measurement of worker OH&S experience with a view to proactive continuous improvement. This is achieved through worker engagement and a leadership focus on the OH&S issues that affect worker OH&S performance.

A consistent outcome experienced by the participating companies (including those with high OH&S audit scores and "zero lost time" injuries) is the generation of over 150 ideas for improvements to lower OH&S risks and improve productivity per round. Many of these OH&S (Safety) climate and culture issues were felt and

---

[25] The tripartite approach meant that the participating companies, the EEA, and Orange Umbrella® as the service provider, were project partners.

[26] The data from the 28 companies was amassed from over 4,000 survey responses, an average response rate of 91 per cent, and over 12,000 feedback items from the facilitated workshops.

feedback through the evaluation process indicated that they were impacting company risk and performance. However, such issues are often not visible or objectively understood by Senior Management. The SCP process crystallized these issues to the extent that Senior Management and staff were able to collaborate on improvement plans according to their capacity and available resources. Companies that followed through on their planned improvements (*Action Plans*), experienced valuable gains in workplace safety, worker wellness, work productivity, and service quality.

Figure 8.3 illustrates SCP application from the front line to the wider industry. It is noteworthy that the SCP process supports strategic, industry, and company-level improvements. The SCP does not replace OH&S management systems. Rather, it evaluates the climate in which OH&S management systems are operating, how well these systems are delivering, and what improvements will help them function more effectively. The SCP process is an outstanding means for management/employee engagement, ongoing indication of progress around proactive themes, and reporting of the "state of the nation" and leading measures around positive action to stakeholders such as at the board level. The following sections develop these matters.

Figure 8.3: The Safety Climate Project (SCP)

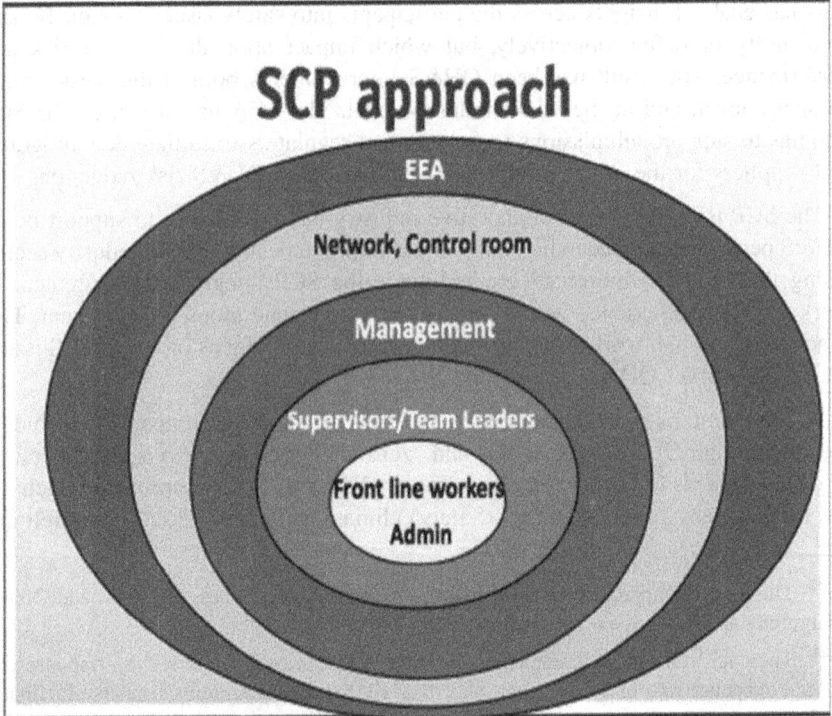

The SCP engagement and evaluation process offers each participating company improvement opportunities through:

- quantitative evaluation, benchmarks, and improvement tracking;
- qualitative detail about Safety risks to the company and staff, and staff input on making improvements;
- a safe and managed process facilitated by a neutral third-party facilitator;
- industry support through the EEA; and
- peer support and sharing of strategies through the SCP community.

1. Improvement Trends for the SCP Group

For the SCP group of companies, the noted trends in the aggregate employee perceptions about the magnitude of change in their OH&S experience across the first four survey and improvement rounds (2010 to 2013) is presented in Figure 8.4.

Figure 8.4: Safety Climate Group Experience: Round 1 to Round 4

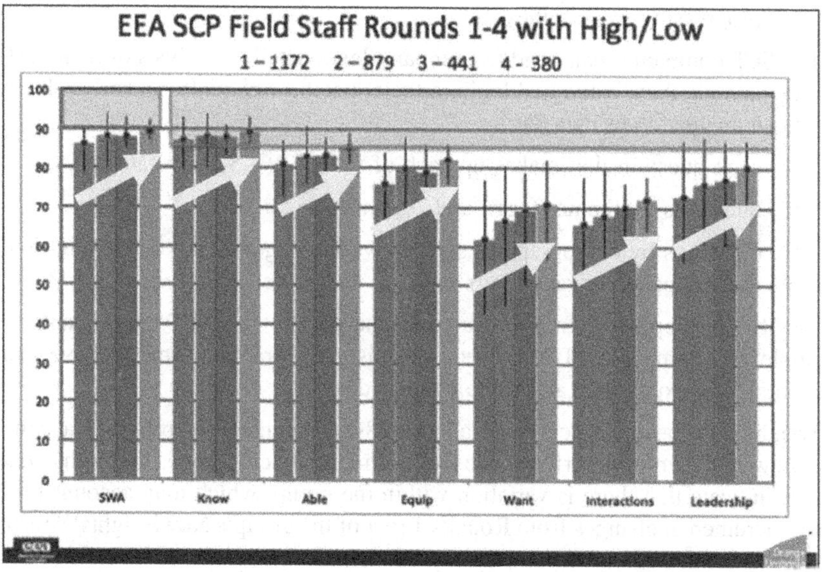

Note: Some of the companies started their first SCP survey later in this period and have yet to complete their third and fourth rounds.

This data indicates:

- An upward trend in the averages of all GSP™ Survey categories, with the greatest increases observed in the *"Want to do it"*, *"Interactions"*, *"Leadership"*, and *"Equipped to do it"* categories. This reflects the impact of the company *Action Plan*s that focused on improving these specific areas in order to reduce the OH&S risks.

- Although there has been improvement, the degree of risk in the "*Want to do it*", "*Interactions*", and "*Leadership*" categories remains. The highest-scoring companies are outside the range of "Lower OH&S Risk". Please note that the lowest scores in the range are trending upwards as the companies move from Rounds 1 to 4 of the NewHeights™ Approach, which indicates improvement for even the lowest-scoring companies.
- High scores (over 85) in the "*Know what to do*" category starts from Round 1 of the NewHeights™ Approach and is maintained across all the Rounds. This reflects the impact of existing industry training and safety rules, and the company training programs. They have been instrumental in establishing a foundation for worker OH&S behaviour.
- The ranges of "highest company average" versus "lowest company average" decrease with subsequent NewHeights™ Rounds; this indicates improvement across the entire SCP Group.
- The high and low ranges for "*Safe Work Actions*" and "*Know what to do*" are small, which increases the level of confidence that the SCP Group is progressing well in these areas.

The SCP companies can conduct company-level detailed analysis of their GSP™ Survey data via their individual login to their specific web platform and dashboard. They can analyze their data for:

- each question that makes up each of the GSP™ Survey categories;
- questions by theme (tags); and
- questions by demographics such as work category, time at the job, time at the company, average hours worked, and more.

As well, this approach enables outcome-sharing by the SCP participants. For example, two companies, Trustpower and Unison Contracting Limited, have shared their company experience and some of their data.[27]

Trending the average scores of all the 28 SCP companies provides a general overview of the group's performance and the improvements made. The high and low ranges indicate that there is variation within the group, which may account for the small incremental changes from Rounds 1 to 4 of the group's NewHeights™ results. It is also indicative of the time and persistence required to achieve the desired changes in OH&S (Safety) culture and OH&S risks.

The survey data for workers were also classified according to the "soft skills" required to manage staff, their performance, and effect culture change, as compared to the "traditional health and safety" elements that are typically measured in OH&S audits (Figure 8.5).

Interestingly in Round 1, the SCP Group started out with "soft" skills that were

---

[27] Online: http://www.orangeumbrella.co/case-studies/.

considerably lower than the measured levels of the traditional OH&S audit elements. This is not surprising because most companies focused on those traditional elements to ensure that their OH&S Management Systems were intact. Strengthening of the OH&S (Safety) culture and climate as an improvement approach, is a relatively new concept; so, it makes sense that these "soft" skills would measure lower.

As the companies progressed through their respective Rounds of the NewHeights™ Approach, their "soft" elements improved, which seems to have had the greatest influence on the survey questions that relate to workplace incident reporting. This finding is consistent with the feedback from workers who say that when a worker reports an incident, is treated respectfully, and the process does not end up being punitive in nature, then workers are more likely to report workplace incidents. This is but one illustration of the positive impact that a strong OH&S (Safety) climate, and hence, OH&S (Safety) culture, has on the OH&S Management System operations.

Figure 8.5: Comparison of "Soft" and "Traditional" OH&S Skills

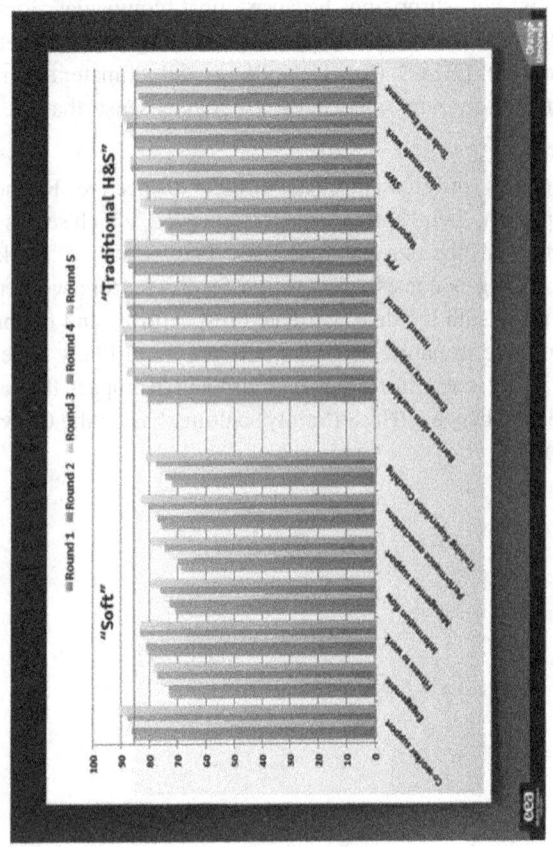

2. Pattern of Improvement

Through the SCP, the participating companies realized the "determining role" of Senior Management in the improvement of the OH&S (Safety) culture. The most visible improvement occurred when Senior Management:

- supported and empowered their employees;
- took a genuine interest in their worker and Supervisors' work;
- provided positive and constructive feedback;
- acted decisively and within a reasonable time frame, on issues;
- provided structure and coaching; and
- maintained visible leadership and positive commitment to the "GSP™ Journey".

A "learning and discovery trend" was evident among the SCP companies. In general, the initial worker GSP™ Survey and the issues analysis process were intense and challenging experiences for them. In a short time frame, the process objectively identified and labelled real and sometimes difficult issues that were significantly impacting the work experience, and worker health and safety. However, Management and employee participants quickly became "encouraged and inspired" when they reached the *Action Strategy* phase and could identify a positive path forward — their own GSP™ Journey.

In the SCP, the second Round of GSP™ Surveys and improvement processes happens soon after the first Round. The rationale for delivering the second survey before completing of all the planned remedial actions quickly becomes evident when the feedback about the project progress, communications, and remedial actions are discussed. In most instances, because of doing Round 2, some "course correcting" is needed. Round 3 is the point at which the SCP participants *really start* to see some process improvements. For example, they report that there is more visible involvement and "buy-in" from Senior Management, and especially the CEO. This result is evident in the pattern of improvement seen from the GSP™ survey scores (Figure 8.6).

Figure 8.6: Pattern of Improvement: Noted Participant Reactions

- Intense
- Skepticism about capacity to change
- Steep learning curve
- Surprise
- Resistance
- Emotive
- Action Planning builds bridges and starts the healing

- What? Next round already?
- Course corrections – "Oh. That's what you meant."
- Improved understanding of the power of the process.
- Increased faith in possibility of improvement.
- "This is not going away".
- Survey scores may not yet reflect changes.

- Culture shift to welcoming the opportunity to address issues.
- Better quality communications.
- Looking forward to seeing survey scores and feedback.
- Survey scores and feedback more positive as Actions get followed through.
- Most survey score improvement seen in perception of manager follow through and commitment.
- Easy and less complex issues are completed.

- Survey scores and feedback significantly improve as work teams see easier issues being resolved.
- Ready now to tackle hard and complex issues.
- Health & Safety risks lower.
- Productivity improving.
- Embrace NewHeights™ round as opportunity to give positive feedback and celebrate successes.
- Working groups taking on issues in-between surveys.

- Maintaining communications and positivity.
- Checking on progress of really tough or complex issues.
- Frequent communications and working groups embedded as new way of operating.
- Starting to look outside the company for ideas, open dialogue on industry issues.
- More interest in full dashboard of performance including GSP™ indicators, turnover, absenteeism, productivity measures, audit scores.

3. Leadership Follow-through on Action Plans: A Success Factor

A company's success in the SCP seems largely determined by how committed and open the CEO and Senior Management were listening, understanding, and acting on worker concerns. As members of the EEA and participants in the SCP, Senior Management Executives of the individual SCP companies tended to support each other, as well as to receive EEA support and encouragement. This phenomenon has been an important factor in the success of the SCP. It has proven to be an excellent opportunity for growth and strengthening of the support mechanism for Senior Management Executives in the SCP.

The SCP evaluation tools and process helped companies to generate their own improvement strategies. Improvement actions are determined and developed collaboratively between company leaders, supervisors, and workers. Collaboration and commitment to act are clearly stated in the company's SCP Agreement. At the commencement of the SCP, workers are briefed on this undertaking. Key to employee engagement is the level of trust established — the belief that Senior Management will follow through on their undertaking. Subsequently, the SCP findings correlate change in worker perception about their workplace OH&S experience with the perceived level of follow-through by Senior Management on agreed-upon actions. This has proven to be a key determinant for maintaining worker engagement and hence, the success of the SCP.

Different relative success rates at improving workplace OH&S (Safety) culture and addressing OH&S risks have been noted among the SCP companies. Where the company leaders have ensured follow-through on their *Action Plan*s, the changes in their GSP™ Survey scores have been much greater relative to the companies that have not done so. Both the qualitative and quantitative SCP data underscore this finding.

Some common features about these successful SCP companies are that they:

- used the information from the SCP to focus on improving their OH&S Management Systems;
- fast-tracked improvements;
- collaborated with workers to ensure that the planned improvements are appropriate;
- followed through on their improvement and *Action Plan*s, and promises made;
- communicated regularly on the progress of their GSP™ Journey;
- recognized successes;
- continued to measure and track performance; and
- had Senior Management leaders that demonstrate tangible commitment to improvement and *Action Plan*s. They are interested, participate, follow up, and remove barriers to progress.

4. The Impact of "Not Following Through"

The SCP data supports the observation that when there is follow-through on *Action*

*Plan*s, the GSP™ Survey scores increase, and the OH&S risks decrease. For example, seven companies that followed through on their *Action Plan*s were compared with six companies that demonstrated limited follow-through. Each company's latest survey scores were compared to their baseline scores, and the magnitude and direction of the performance change determined. The outcome is that: companies that followed through showed greater GSP™ Survey score improvements and more scores in the lower OH&S risk range.

The Safe Work Action (SWA) scores for companies that "followed through" exhibited a greater change (increase); and because those are the desired work behaviours (the ones that keep workers safe), an increase in this measure indicates a lower risk of worker injury. The SWA increases are achieved through increases in the other work environment conditions; that is, by ensuring that people "*Know what to do*", "are *Able to do it*", "are *Equipped to do it*", "*Want to do it*", and have the positive "*Interactions*" and "*Leadership*". In companies that did "little or no follow-through in their *Action Plan*s", the SWA scores remained the same, or worsened.

In terms of "*Leadership*" improvement, the companies that followed through on their *Action Plan*s earned significant benefits. The importance of Senior Management leadership is consistent with the feedback received from the worker workshops.

The SCP data indicate that the leadership demonstrated as part of the SCP and following through with improvements and *Action Plan*s, directly affects OH&S risks. For example, having work actions designed to ensure worker safety, *e.g.*, safer harnesses, better traffic management, complete information in job packs, training, equipment, better planning so there is less rushing to do the job, improved work systems, *etc.*, promotes mutual trust which in turn, supports better workplace communication. This strongly correlated with enhanced reporting of OH&S incidents and risks. Following through on improvements and *Action Plan*s is consistent with the foundation promise of the SCP — it is "walking the talk".

5. Leader Influence on OH&S (Safety) Culture

Lassowski demonstrates the influence that Senior Management leaders have on the OH&S (Safety) culture (Figures 8.7 and 8.8).[28]

---

[28] S. Lassowski completed a six-month internship at Orange Umbrella® in 2013 as a Master student from the Berlin University of Applied Sciences. She analyzed the SCP quantitative and qualitative data, observed the SCP workshops, and interviewed key personnel. Her focus was on the factors that enhance and inhibit the development of a supportive OH&S (Safety) culture. Her thesis is available through Orange Umbrella®.

CH. 8: SAFETY CULTURE: SAFETY CLIMATE

Figure 8.7: Factors *Inhibiting* Development of a Positive OH&S (Safety) Culture:

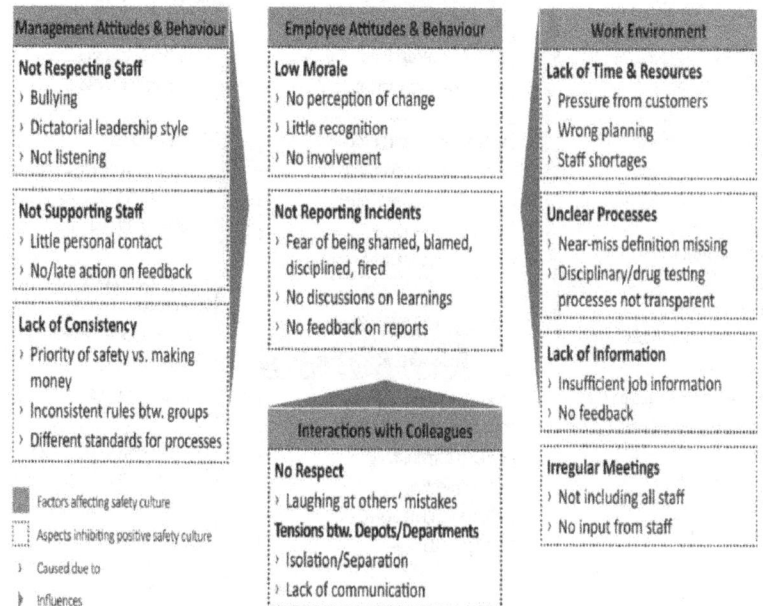

By understanding what inhibits the development of a positive OH&S (Safety) culture, Senior Management can make the necessary changes towards creating a work culture and environment that fosters an OH&S (Safety) culture and climate.

Figure 8.8: Factors *Enhancing* Development of a Positive OH&S (Safety) Culture:

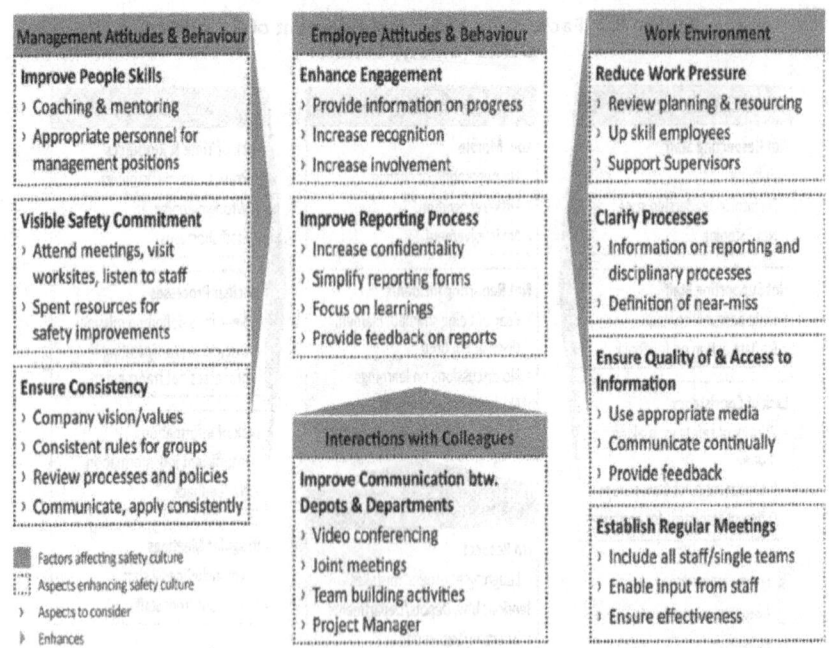

Lassowski's findings, which are consistent with those of other researchers as well as with the related theories on Safety culture, highlights the specific positive actions that Senior Management can immediately take and do so at a low cost.

6. Observations and Issues from the Safety Climate Project

(a) *Comprehensive Reporting*

Companies that commit to safety climate assessment and improvement use the "new" measures to round out their health and safety reporting to their boards and to their staff. Where they previously would have reported only incident severity and frequency, and perhaps audit scores, they are now also reporting related efficiency, quality, and performance metrics. Their reporting includes items specific to their business such as rework rate, variation rate, job completions, restart rate (for electricity generation), number of "skips" for safety reasons, environmental issues, number of action improvements completed, alongside the GSP™ leading indicators and the traditional health and safety lagging indicators. This gives a much clearer view of the impact of health and safety performance on the business and is a reminder to company leaders that health and safety is not an "add-on", but simply "good business".[29]

Other industry groups can benefit from this "group approach" to tackle wider-system issues that affect large numbers of businesses. These groups could be geographic, such as a "safe city" or "safer community" projects, Chambers of Commerce providing learning opportunities for their members, or industry safety associations. In the SCP, there is considerable power in collective problem-solving of common issues. In 2010, companies were concerned that meeting to discuss common health and safety issues would transgress anti-competition laws. But once they recognized that it did not, and that the SCP was sanctioned by the regulators via the EEA, the move to address historic and bad industry habits was greeted enthusiastically.

(b) OH&S (Safety) Culture and Climate Improvement Takes Time

It is important for companies to recognize that changing OH&S (Safety) culture and climate does not happen overnight. The analogy that we use is that it is "a long-distance event and not a sprint". As part of starting the change process and demonstrating Senior Management commitment, we encourage companies to address the smaller items to get some "quick wins". It is important that Senior Management recognizes this so that they continue supporting the *Action Plan* development and completion. For this reason, it is helpful to measure and monitor progress — even small progress — by repeating the NewHeights™ process.

(c) Momentum

The Safety Climate Project has maintained its momentum because of the information flow within each company, and within the industry. The New Zealand Electricity Industry and the Health and Safety Industry have several awards evenings where initiatives that were born from the SCP process, are recognized and celebrated. Now that some SCP companies have been on their improvement "GSP™ Journey" for a number of years, they are also starting to win awards for their OH&S (Safety) culture improvement, *e.g.*, Unison Contracting Services Limited (2014).

The recognition of the effort that companies have made, is not only a celebration for the winning company internally, but also a celebration for the industry because of the learning that has occurred. Taking time to acknowledge effort and success is important to keep momentum going.

(d) Industry Involvement and Support

The SCP gave the participating companies reliable and valid tools to evaluate their leading indicators of OH&S risks, track the effectiveness of process improvements, and to benchmark with their SCP Group peers about the proactive measures and improvements. It also enables an industry-level overview of common issues and

---

[29] For more details on the "Good Business" rationale, refer to Chapter 16, "Marketing Occupational Health and Safety Programs; Communicating the Results", Appendix 4: The Unison Contracting Ltd. Report.

themes among the SCP Group, which are then fed back to the EEA Board and the Safety Strategy and Policy Group (SSPG) and assessed for new or improved OH&S interventions to benefit the wider-industry sector.

The common issues include:

- perceived SAIDI[30] influence on OH&S risks for workers;
- quality and completeness of incident investigation, analysis techniques, and data reporting;
- the need to harness suitability, condition, and usage;
- problems with traffic management;
- the need for enhanced, and more, available industry training;
- retention and development of skilled staff;
- Senior Management leadership effectiveness;
- communication;
- work planning and design processes;
- network and contractor relationships;
- application of drug and alcohol programs;
- supervisory or management styles short of just and fair standard;
- fatigue management; and
- supervisory skill development.

This feedback has been used by the EEA to focus an industry-level response. Some examples include:

- Liaison with regulators concerning perceived regulatory conflicts and submissions.
- Clarifying policy and guidance around live-line work decisions.
- Initiating supervisory model competencies development.
- Arranging specialist training in incident investigation and analysis.
- Improving safety performance indicators.
- Promoting industry awareness and improved knowledge about the above issues and solutions through forums and industry events, including the EEA Conference and Safety Workshops.

These activities support the SCP companies and others in the electricity industry to

---

[30] SAIDI: The System Average Interruption Duration Index (SAIDI) is a reliability indicator used by electric power utilities. SAIDI is the average outage duration for each customer served. Utility regulators often use SAIDI as a network power quality performance measure.

focus improvement on common areas aimed at reducing OH&S (Safety) risks.

7. Safety Climate Project, Year 4: Emerging Issues

(a) Public Safety

The related public safety concerns include traffic management, dangerous dog management, public contacting powerlines or approaching them too closely, emergency response (car-versus-pole), safety of the public when the power is off, re-energizing and reconnecting powerlines, reversed polarity on house connections, and other issues. The actions taken by these companies have included system improvements, more worker training, more signage, and increased awareness across the wider industry.

(b) Financial Risk

Management and workers express concerns about the costs of addressing some of the expensive financial issues raised during the SCP process, such as Information Technology solutions, new plants and vehicles, or network-asset improvements. All parties understand that for people to stay employed, the company does need to maintain good financial health. There can be some difficult decisions where financial drivers are incompatible with safety. These drivers are identified and discussed in the SCP Group, for example, the SAIDI requirements and the impact on workers.

(c) Business Continuity

At the start of the SCP, it was identified that there is a shortage of qualified technical, supervisory, and planning-work people in the industry. Other identified problems were the quality of the competencies earned during qualification, and the limited opportunity for trainees to practice trade skills with proper coaching and supervision. The identified factors were focused on capacity, *i.e.*, having enough people in the company with the right experience to provide coaching, and having the coaching time. Both management and workers said that it was difficult to replace people on holidays and weekends leading to the common occurrence of "on-call duty" while on vacation. As well, there are a significant number of experienced people who are on the verge of retirement leading to the concern of "institutional knowledge walking out the door".

(d) **Health and Safety at Work Act** *in 2014*

A new *Health and Safety at Work Act* came into force in New Zealand. The SCP has positioned companies to look beyond the Hazard Register to evaluate the risks that those hazards pose. With this new legislation, there is clearer and more enforced accountability for directors, senior managers, and principals to create fewer health and safety risks for their employees and contractors. The new legislation makes it clear in its due diligence requirement that everyone, including company officers, must understand the business and the potential OH&S (health and safety) risks. Those with governance and management responsibilities are required to know as much about the OH&S status of the company as they do about the company's financial status.

The Safety Climate Project supports the requirements outlined in the proposed *Health and Safety at Work Act* by:

- providing leading OH&S indicators for use by senior leaders, company officers, and boards of directors;
- providing leading indicators of OH&S to reduce OH&S risks before an incident occurs;
- providing transparency and evidence of how OH&S responsibilities are carried out within an organization and by company officers;
- generating documentation via the Orange Umbrella® tools, to provide evidence of how OH&S is being managed, and how the OH&S risks are being reduced. In instances where companies do not follow through on improvement actions, this can also be clearly evident;
- providing a platform for safe, open, honest, and collaborative consultation about improving OH&S within a company; and
- providing an industry initiative for addressing issues common to and driven by, industry regulations and traditions.

D. DISCUSSION

Historically, the focus has been on the definition and theoretical underpinnings of the OH&S (Safety) culture. Measurement of the OH&S (Safety) *culture* involved the use of case studies, direct observation and interviews, as opposed to organizational outcomes such as productivity, efficiency, and employee satisfaction;[31] whereas measurement of the OH&S (Safety) *climate* was composed of empirical methods such as cross-sectional questionnaires, and analyses that linked a positive OH&S (Safety) climate and culture with outcomes like increased production,[32] and greater employee job satisfaction.[33]

It can be argued though, that OH&S (Safety) culture affects worker attitudes, behaviours, and work performance. Because the OH&S (Safety) climate is accepted to reflect those OH&S (Safety) attitudes,[34] it seems reasonable that measuring the OH&S (Safety) climate addresses the elements of the OH&S (Safety) culture. Hence, even though the OH&S (Safety) culture and OH&S (Safety) climate are viewed as independent concepts, they are interdependent and influence each other.

---

[31] S. Cox & R. Flin, "Safety Culture: Philosopher's Stone or Man of Straw?" (1998) 12:3 Work and Stress 202-216

[32] R. Pritchard & B. Karasick, "The Effect of Organizational Climate on Managerial Job Performance and Job Satisfaction" (1973) 9:1 Organ. Behav. Hum. Perf. 126-146.

[33] B. Schneider, "Organizational Climates: An Essay" (1975) 28 Pers. Psychol. at 147–179.

[34] K. Mearns et al., *Human and Organisational Factors in Offshore Safety (OTH 543)* (Suffolk: Offshore Safety Division, HSE books, 1997).

The belief is that the OH&S (Safety) climate manifests the OH&S (Safety) culture,[35] and is strongly influenced by it.

The value in measuring OH&S (Safety) culture and climate is their potential as *leading indicators of safety performance*. Roithmayr demonstrated that organizational culture, Management leadership, and the conditions of the work environment, strongly correlated with OH&S (Safety) outcomes.[36]

By designing the OH&S (Safety) climate assessment to qualify Management commitment to occupational health and safety, Management practices towards developing a supportive work environment, along with employee performance and OH&S outcomes, a comprehensive picture of the OH&S (Safety) climate and OH&S (Safety) culture can be obtained. In essence, it is a combination of the leading and lagging indicators of Safety.

As indicated by both the PSHSA and Orange Umbrella® clinical research, Senior Management leadership plays a critical role in quality of the OH&S (Safety) culture.

> Leadership is the communication of the actions and standards you expect by words, deeds and silence. Reason, 2006

For an organization to exhibit a strong OH&S (Safety) culture, Senior Management must take the lead and make it happen. Having solid leading and lagging indicators of safety helps Management make the informed decisions necessary to be able to move the organization forward.

Finally, it is important for Senior Management to recognize the five important components of OH&S (Safety) culture, namely:

1. **Informed culture** — an OH&S (Safety) system that collates data from all types of workplace incidents and safety audits, and then, combines that with information from proactive measures such as climate surveys. In many ways, an informed culture is an OH&S (Safety) culture.

2. **Reporting culture** — the active and honest workforce participation in reporting all types of incidents, completing attitude surveys, and becoming involved in how OH&S (Safety) is managed within the organization. It is characterized by an organizational climate in which the workers feel free to contribute to the informed culture.

3. **Just culture** — an atmosphere of trust in which response to incidents is to determine what *actually happened* as opposed to seeking to lay blame; yet, not to turn a blind eye to unsafe acts.

---

[35] E. Schein, *Organizational Culture and Leadership*, 4th ed. (San Francisco, CA: John Wiley & Sons, 2010).

[36] T. Roithmayr, Chapter 5, "Occupational Health and Safety: Leadership and Commitment", originally written in 2004 and published in D. Dyck, *Occupational Health & Safety: Theory, Strategy & Industry Practice* (Markham, ON: LexisNexis Canada, 2007).

4. **Flexible culture** — rewards are offered for reporting "near miss" incidents much like they are given for safe work behaviours.

5. **Learning culture** — is the collection of the information needed to enhance the OH&S (Safety) performance of the organization along with the desire to make the needed changes.[37]

E. SUMMARY

Organizations can attain a positive OH&S (Safety) climate and culture by first determining the state of its current OH&S (Safety) climate, and then, working towards the development and sustainability of a positive OH&S (Safety) culture. The "carrot" is lower staff turnover, lower injury and illness rates and costs, as well as other positive business outcomes. Over the years, one of the most powerful life lessons one can learn is that:

> You have to be the change that you want to see in the world.
>
> Gandhi

CHAPTER REFERENCES

L. Bossidy & R. Charan, *Execution: The Discipline of Getting Things Done*, cited in D. Spitzer, *Transforming Performance Measurement: Rethinking the Way We Measure and Drive Organizational Success* (New York, NY: AMACOM, 2007) at 128.

S. Cox & R. Flin, "Safety Culture: Philosopher's Stone or Man of Straw?" (1998) 12:3 *Work and Stress* 202-216.

F. Manuele, *Advanced Safety Management: Focusing on Z10 and Serious Injury Prevention* (New York, NY: John Wiley & Sons, 2008).

K. Mearns, R. Flin, M. Fleming, & R. Gordon *Human and Organisational Factors in Offshore Safety* (OTH 543) (Suffolk: Offshore Safety Division, HSE books, 1997).Orange Umbrella® Inc., online: http://www.orangeumbrella.co.

R. Pritchard & B. Karasick, "The Effect of Organizational Climate on Managerial Job Performance and Job Satisfaction" (1973) 9:1 Organ. Behav. Hum. Perf. 126-146.

Public Services Health & Safety Association (PSHSA), *PSHSA Health and Safety Climate Assessment Project* (Toronto, ON: PSHSA), online: http://www. pshsa.ca/wp-content/uploads/2014/07/HFO-climate-project-report.pdf (date accessed February 28, 2020).

J. Reason, "Safety Culture" *Safety Management Systems* (2006), online: http://www.airsafety.aero/safety_development/sms/safety_culture/ (date accessed February 28, 2020).

---

[37] J. Reason, "Safety Culture", *Safety Management Systems* (2006), online: http://www.airsafety.aero/safety_development/sms/safety_culture/ (date accessed February 28, 2020).

T. Roithmayr, Chapter 5, "Occupational Health and Safety: Leadership and Commitment", originally written in 2004 and published in D. Dyck, *Occupational Health & Safety: Theory, Strategy & Industry Practice* (Markham, ON: LexisNexis Canada, 2007).

E. Schein, *Organizational Culture and Leadership*, 4th ed. (San Francisco, CA: John Wiley & Sons, 2010).

B. Schneider, "Organizational Climates: An Essay" (1975) 28 Pers. Psychol. 147-179.

D. Seo, M. Torabi, E. Blair, and N. Ellis, "A cross-validation of safety climate scale using confirmatory factor analytic approach" (2004) 35:4 Journal of Safety Research 427-445.

D. Spitzer, *Transforming Performance Measurement: Rethinking the Way We Measure and Drive Organizational Success* (New York, NY: AMACOM, 2007).

Statistics Canada, *Work Absence rates of full-time employees by Industry, Annual, 2019*, Table: 14-10-0191-01 (formerly CANSIM 279-0030) (2020), online: https://www150.statcan.gc.ca/t1/tbl1/en/tv.action?pid=1410019101 (date accessed February 28, 2020).

Chapter 9

OCCUPATIONAL HEALTH AND SAFETY: HAZARD IDENTIFICATION, ASSESSMENT, AND CONTROL

A. INTRODUCTION

A **hazard** is a condition or practice with the potential for leading to accidental loss. It can be the source of danger, or the element of an organization's activities, products, or services that can negatively interact with the health and safety of workers or the public.[1] **Risk** is the chance of loss occurring. It is a measure of the probability and the potential for severity of harm.

Hazards exist in the workplace due to conditions caused by:

- new materials/procedures introduced into the workplace without adequate training;
- unsafe work procedures;
- lack of maintenance on vehicles and equipment;
- failure to conduct vehicle and equipment pre-use inspections;
- tools, machinery, and equipment that wear out;
- tools that get damaged or are misused;
- use of improper tools;
- safeguards that are not used;
- materials/tools being placed in unsafe positions;
- safe working loads that are exceeded;
- worker distractions;
- worker fatigue;
- worker psychological distress;
- worker substance use;

---

[1] BSI Management Systems, "OHSAS 18001 Awareness" (Presented at the City of Calgary Informational Seminar, April 2002, Calgary AB) at 4-3.

- worker complacency;
- exposure to biological or chemical agents; and
- external threats (natural or man-made disasters).

B. HAZARD MANAGEMENT

All Canadian OH&S Acts require employers to identify, assess, control, and evaluate the risks within their respective workplaces. The hazard management process involves the:

1. identification and appraisal of hazardous conditions and practices;
2. development of hazard control methods, procedures, and programs;
3. communication of hazard control information; and
4. measurement of effectiveness of control measures.

The steps involved in hazard management include:

- hazard identification;
- risk assessment;
- hazard control; and
- mitigation through emergency preparedness and response.

The goal of hazard management is to organize and control work activities so that hazards are eliminated/reduced, and the related risks to the employees, public, and organizational assets minimized. These organizational assets are:

- *People* — examples of loss in this category are injury or illness to personnel; short- or long-term disability due to injury; financial loss incurred in compensation payments; payment for replacement personnel; extra workload for co-workers; *etc.*
- *Equipment/Property* — loss may range from minor damage requiring repair to major damage where total replacement or rebuilding may be necessary.
- *Business Process* — examples are loss of product resulting from a process interruption or loss of project time due to damage to personnel, equipment, or materials.
- *Environment* — environmental pollution and impact result in long- or short-term damage, reversible and irreversible effects, as well as adverse publicity.

OH&S legislation requires employers to assess the various and environmental hazards inherent in their operations. This involves carrying out a systematic, general examination of the work activities, recording significant findings, and acting to eliminate/reduce the risk(s) to an acceptable level.

The OH&S hazard identification and control process is covered in this section.

The aim is to reduce the risk of injury, damage to health, equipment/property, business processes, and the environment by introducing appropriate controls. Control selection should follow a system of priority ranking.

C. HAZARD IDENTIFICATION, ASSESSMENT, AND CONTROL

1. Definitions

A **hazard** is defined as something with the potential to cause harm and can include substances, equipment, machines, method of work, or the work environment. **Harm**, in this context, covers injury to people, ill health, equipment/property damage, or environmental leak/spill. **Risk** is the likelihood that harm will occur due to a hazard and the severity of its consequences. **Hazard control** is any action that reduces the risk of loss. It includes the prevention/reduction of loss exposure, reduction of loss-producing agents, and determination/avoidance of risk.

Critical tasks are tasks that have the potential to produce major loss to people, property, business processes, and/or environment when not properly performed.

(a) Hazard Identification

Hazard identification is a systematic approach that involves:
- assessment of each department/area;
- preparing and reviewing a schedule of job tasks, activities, or work operations;
- discussions with Management, Line Managers, and workers;
- reference to incident records, pre-job meetings, incident investigation reports, safety meeting minutes, and worksite inspections;
- observation of the activities under review;
- review of applicable regulations, codes of practice and industry practices to assist with the identification of hazards associated with particular equipment or processes;
- consideration of hazards arising from reasonably foreseeable changes in processes, equipment, circumstances, or work conditions;
- documentation of identified hazards;
- identification of workers through job demands analyses (JDAs) that are potentially "at risk"; and
- identification of others (*e.g.*, visitors, contractors, the general public) potentially "at risk".

There are many hazard identification tools. They include:
- hazard assessments;
- worker training/experience and reporting;
- pre-job meetings;

- worksite inspections;
- OH&S codes and standards;
- OH&S legislation;
- worker training;
- near-miss incident reporting;
- the findings from incident investigations and reports; and
- lagging indicators of safety performance (occupational injury/illness claims).

It is critical for employers to identify the hazards in their workplaces so that they can meet their legal obligation of providing a safe and healthy workplace, and their legal obligation to educate workers about the workplace hazards that exist. It is even more critical to identify workplace hazards so the workforce talent can be protected and maintained.

(b) Risk Assessment

The risk assessment procedure is designed to address existing work and planned maintenance activities. However, for more dynamic activities and unique tasks, such as modifications to buildings, the worksite and/or equipment; *ad hoc* repairs; and complex tasks involving contractors, the risk assessment must be tailored to the specific work involved.

A risk assessment entails:
- identifying the hazards in the work activities. This can be done through the use of a Hazard Assessment and Control Form (Appendix 1);
- rating the risks by determining the likelihood (probability) that harm will occur; and
- establishing the hazard severity.

A team comprising employees, Line Managers, Management, and OH&S technical specialists carry out risk assessments. The results of the risk assessments are recorded on a Risk Assessment Form (Appendix 2) and approved by Management before they can be considered as complete.

(i) Assessment of Hazard Severity

The **hazard severity (harm)** is the extent of the injuries, ill health, and/or damage to equipment, property, process or the environment that may be sustained if the hazard is realized. The steps for doing this include:
1. identify the severity using Table 9.1; and
2. enter the result on the Risk Assessment Form (Appendix 2).

Table 9.1: Hazard Severity Table

Severity (S)	Injury Potential	Potential Environmental Impact	Property/ Equipment Damage
i	Trivial injury/ies (scratch, small cut, burn, bruise, or abrasion)	Slight leak/spill — minimal effect contained locally	$100
ii	Minor injury/ies (laceration requiring stitches, moderate bruises or burns)	Minor leak/spill — no lasting effect	$1,000
iii	Serious injury (broken bone, severe burns or bleeding, eye injury)	Minor leak/spill — public concern	$10,000
iv	Major injury to one or more than one person (amputation, permanent disability)	Localized leak/spill. Non-conformance with Regulations (considerable public concern)	$100,000
v	Single fatality	Massive leak/spill — Non-conformance with Regulations (considerable public concern)	$1,000,000
vi	Multiple fatalities	Massive leak/spill — major public concern. Major cleanup required	$10,000,000

(ii) Assessment of Probability

The assessors should make a judgement on the **probability** (likelihood) that the potential harm will be realized. It may help to carry this out in two steps. Ask:

1. Is it possible that an unsafe act or condition that could cause harm could occur?
2. What is the probability of the harm being realized?

Probability is affected by:

1. *The frequency of exposure to the hazard.* For example, is a piece of equipment in constant use, that is, used two to three times per week, or just used once a week?
2. *How effectively is the hazard controlled by existing or precautionary measures?* For example, does the machine have an effective safety guard?
3. *The way the work is organized.* For example, is the work set up such that it potentiates exposure to the hazard?

Safety audits, worksite inspection reports, incident investigation data, as well as company and national OH&S statistics, affect the evaluation of the probability that a hazard could cause harm. The steps involve:

1. taking all of the above into consideration, assessing the likelihood of the hazard severity being realized;
2. classifying the probability of the hazard severity being realized using one of the five categories in an ascending scale of likelihood listed in the *Probability Scale* (Table 9.2); and

3. entering the Probability on the Risk Assessment Form (Appendix 2).

Table 9.2: Hazard Probability Scale

Probability Scale	
A	Very unlikely to occur
B	Unlikely to occur
C	May happen
D	Likely to occur
E	Highly likely

(iii) Determination of the Risk Rating

The **risk rating** is a combination of the **hazard severity** and the **probability** of the occurrence. The risk rating is primarily a method of comparing risks so that they could be:

1. prioritized in terms of those requiring action; and
2. evaluated in terms of the anticipated improvements due to the introduction of suitable hazard controls.

The steps for determining the risk rating include:

1. Plot both the severity and probability on the Risk Rating, Table 9.3, to determine the risk rating.

   RISK RATING = HAZARD SEVERITY X PROBABILITY OF OCCURRENCE

2. Enter the risk rating onto the Risk Assessment Form (Appendix 2).

Table 9.3: Risk Rating Table

S				Risk Rating Table		
E	i	1	1	1	1	2
V	ii	1	2	2	3	3
E	iii	1	2	4	5	5
R	iv	1	2	4	5	5
I	v	1	3	4	5	5
T	vi	2	3	4	5	5
Y		A	B	C	D	E
				PROBABILITY		

(c) Hazard Control

Hazard control, the systematic approach to risk reduction associated with hazards, includes:

1. Determination of a timeline for the implementation of hazard control measures.
2. Determination of engineering, administrative, or personal protective equipment control measures. Refer to Table 9.4 for the Priority Ranking for Risk Control.
3. Re-evaluation of hazards after hazard control measures have been implemented.
4. Development of the resultant standard Safe Work Procedure.
5. Employee training in the use of the new Safe Work Procedure.
6. Review of hazard control measures when:
   - changes in work processes occur;
   - recommendations from incident investigations are made;
   - recommended as part of a Worksite Inspection; and/or
   - hazard controls are annually reviewed.

(i) Hazard control mechanisms

The field of Occupational Health and Safety uses a hierarchy of controls to manage workplace hazards. This hierarchy of controls consists of:

1. **Elimination/Substitution/Engineering controls**
    - eliminating the hazard;
    - substituting a less hazardous process/product;
    - isolation or enclosure;
    - local exhaust;
    - general ventilation;
    - wetting-down process;
    - shielding;
    - shock or vibration mounting; and
    - machinery or workplace redesign.
2. **Administrative and work practice controls:**
    - worker education and training;
    - good housekeeping;
    - product labelling, use, storage, handling, and disposal;
    - personal hygiene;
    - rules compliance;
    - behaviour re-enforcement;
    - scheduling workers to minimize exposure; and
    - warning and alarm systems.
3. **Use of personal protective equipment (PPE):**
    - eye protection;
    - head protection;
    - foot/toe/instep protection;
    - hand protection;
    - hearing protection;
    - respiratory protection; and
    - protection from the environment.
4. **Mitigation of consequences through emergency response planning:**
    - create awareness of emergency situations;
    - develop clearly defined roles and responsibilities;

- develop and define procedures;
- provide resources (equipment, people);
- provide training and drills; and
- regularly review, evaluate, and improve the response process.

Table 9.4: Priority Ranking for Risk Control

1. **Engineering Controls** such as: • Hazard elimination • Hazard substitution by something less hazardous and risky • Engineering (hazard enclosure, separation from people, mechanical aids, ventilation, *etc.*)
2. **Administrative Controls** such as: • Use of work practices that reduces the risk to an acceptable level • Use of Safe Work Procedures that are followed • Use of adequate supervision • Identification of training needs • Use of information/instruction (i.e., signs, handouts)
3. **Personal Protective Equipment Controls**
This ranking applies to risk control in relation to all types of harm — injury to people, ill health and damage to assets or to the environment. In many cases, a suitable combination of control methods is necessary.

Refer to the CCOHS: Hazard Control Fast Facts Card, online: https://www.ccohs.ca/products/boutique/hazardcontrol/.

(ii) Timing for Risk Reduction

Having determined the risk rating (1-5), refer to the Time Scale, Table 9.5, for the timing of the implementation of the control measures.

Table 9.5: Time Scale

Risk Rating	Risk Reduction Policy
5 Unacceptable	Stop activity. Fix immediately
4 High	Stop activity, Fix within one week
3 Medium	Reduce within one month
2 Low	Reduce within six months
1 Acceptable	Monitor

However, even if risk assessments indicate an acceptable level of risk rating, Management must act to control risks where legal requirements apply.

(iii) Risk Control Measures

The Risk Assessment process helps determine if the hazard controls are adequate.

When additional hazard control measures are required, an action plan is to be documented on the Risk Assessment Form (Appendix 2). This form provides a system for priority ranking of the hazard control.

Prior to the implementation of hazard control measures, the Risk Assessment process for new controls is to be reviewed to establish a forecast risk rating with the new measures in place. The revised probability, severity and forecast risk rating is then entered on the Risk Control Follow-up Form (Table 9.6).

Table 9.6: Risk Control Follow-up Form

Risk Control Follow-Up Form				Hazard #:			Date:
Action Plan							
Date	Identified Risk	Corrective Action(s)	Date Assigned	Person(s) Accountable	Date Completed		Management Signature

This follow-up process is necessary to:

1. ensure that the control measures introduced do not introduce another hazard(s). For example, the provision of personal protective equipment for hearing protection may prevent warning alarms from being heard by the employee; and

2. choose the most cost-effective hazard control option where more than one option with a similar risk rating exists.

(iv) Risk Control Documentation and Follow-up

Both the Risk Assessment Form and the Risk Control Follow-up Form are retained by the manager and OH&S Practitioners/Professionals; a copy is provided to the Line Manager responsible for the activity. The Line Manager coordinates and controls the follow-up action and implementation and ensures employees receive appropriate training.

(v) Review

There is a requirement to review and revise any risk assessment if:

1. there is any reason to believe that the risk assessment is invalid. This may be due to the occurrence of incidents, complaints, or recommendations from inspections;

CH. 9: OH&S: HAZARD IDENTIFICATION, ASSESSMENT, AND CONTROL

2. there are significant changes in the method of working or the equipment used; and/or
3. a period of one year has elapsed since the risk assessment was last reviewed.

D. NEW WORKPLACE HAZARDS

Over the years, most workplace hazards were physical or chemical in nature. This has changed. Today, workers face hazards such as:

- Long periods of stationary/sedentary work.
- Repetitive strain and over-use injuries such as finger, wrist, and elbow injuries due to excessive computer use. Terms like "roller-ball thumb" and references to texting injuries like "text neck", have emerged.
- Excessive work stressors due to high-paced work, increased production expectations, constant changes in work expectations, lack of role clarity regarding assigned work tasks, and uncertain performance measures.
- Worker complacency resulting in reduced alertness or a conscious decision to take unsafe "short cuts".
- Motor vehicle collisions due to distractions from the illegal use of hand-held devices and other distracted driving practices.
- Forklift incidents due to improper equipment use, or use of hand-held devices while operating the forklift.
- Hazards associated with the use of new technologies such as robotic equipment, medical equipment, automated devices, and computers.
- Workplace harassment, bullying, and violence.
- Remote work.
- Animal attacks.
- Terrorism (*i.e.*, violent acts as well as exposure to smallpox, plague, anthrax, toxins, and other biological agents).
- Biohazards (*e.g.*, exposure to Norovirus, West Nile virus, E. Coli, hepatitis, influenza, SARS, salmonella, Ebola, COVID-19, *etc.*).
- Laser burns.
- Pandemic illness.[2]

Most OH&S legislation requires employers to conduct hazard assessments when new processes and equipment are developed or acquired. The intent is for employers to have suitable hazard controls in place to protect workers. The challenge is the

---

[2] Refer to Chapter 12, "Occupational Health and Safety: Emergency Preparedness and Response", for more on the pandemic flu hazard and planning.

recognition of workplace hazards before an incident occurs.

E. HAZARD MANAGEMENT: IMPORTANCE

Hazard management is important because:

- **Loss Prevention**: Hazard identification, assessment, and control enables organizations to eliminate/reduce workplace incidents that can lead to damage, property loss, business interruptions, productivity reduction, and lost profits.

- **Injury Prevention**: Hazard identification, assessment, and control are critical to preventing incidents and enabling workplace safety.

- **Legally Required**: In at least one Canadian province, Alberta, hazard identification, assessment, and control are legally required. Alberta employers are required to assess the worksite and identify existing and potential hazards before work begins at the worksite, or prior to the construction of a new worksite. The employer must prepare and date a report of the results and methods used to control or eliminate the identified hazards. The hazard assessment must be repeated at reasonable intervals, when a new work process is introduced, when a work process changes or before construction of any significance.

Over and above all that, Canadian and U.S. OH&S legislation obligates employers to provide safe and healthy workplaces, and to educate their employees on the work hazards relevant to their workplaces. Conducting hazard assessments is a sound mechanism for meeting this duty of care.

F. HAZARD MANAGEMENT: HAZARD TYPES

Workplace hazards fall under the following hazard types:

- **Gravity**: the potential sources for workplace incidents include falls from heights, falls on the same level, falling items, and collapsed structures.

- **Mechanical**: the potential sources for workplace incidents include the worker getting caught in, on, or between something; getting cut or scraped; getting entangled in moving equipment; and friction.

- **Body Mechanics**: the potential sources for workplace incidents include slips and trips; loss of balance; awkward positioning; reaching; striking against something; sustained posture for a duration; over-exertion from lifting, pulling, or pushing; repetitive movements; and work in cramped areas.

- **Chemical**: the potential sources for workplace incidents include inhalation, ingestion, absorption, burns, asphyxiation, fire, explosion, and dust.

- **Pressure**: the potential sources for workplace incidents include bottled gases, steam, air systems, water systems, chemicals, and wind.

- **Thermal**: the potential sources for workplace incidents include hot surfaces, flames, humidity, hot/cold liquids, cold surfaces, and wind chill.

- *Electrical*: the potential sources for workplace incidents include high voltage, low voltage, exposed conductor, underground conductor, overhead conductor, limits of approach, short circuit, back feed, induction, and grounding.
- *Noise*: the potential sources for workplace incidents include ambient noise, noise from tools, and impact/steady-state noise.
- *Biological*: the potential sources for workplace incidents include sewage, medical/human waste, animal contact/bites, insect stings/bites, water contamination, airborne contaminants, and allergic reaction.
- *Radiant Energy*: the potential sources for workplace incidents include sun exposure, sunburn, glare/brightness, welding arc, laser, and radiation source.
- *Psychological*: the potential sources for workplace incidents include fatigue, distraction due to personal issues, harassment, violence, life-threatening attacks, disasters, injury/death of co-workers, *etc.*

G. HAZARD MANAGEMENT: TECHNIQUES FOR HAZARD IDENTIFICATION, ASSESSMENT, AND CONTROL

The techniques for hazard identification, assessment, and control include:

- *Hazard Assessment Tools*

    Using a tool like the one provided in Appendix 1, employers can identify and assess all the worksite hazards present in their operation, and determine whether or not they are adequately controlled.

- *Job Safety Analysis*

    A **Job Safety Analysis** (JSA) is a tool for breaking the job down into basic steps and identifying the hazards associated with each step. JSAs are used to review work methods and uncover hazards that might result in incidents. As a result, a JSA should be performed on all jobs that involve critical tasks, a change in procedure or equipment, or resulted in an incident. They provide data for developing Safe Operating Procedures and serve as a framework for incident analysis.

    The Job Safety Analysis Form (Appendix 5) can be used by the supervisor or OH&S practitioner to conduct the JSA in collaboration with the employees who regularly perform the task. The job of interest should be broken down into a sequence of steps that describe the process in detail. As a rule, the JSA should contain less than 12 steps. If more steps are needed, the job should be broken into separate tasks.

    Through observation and discussion, the hazards associated with each step are identified and documented on the Job Safety Analysis Form.

The JSA also prescribes controls for each hazard. In totality, a JSA is a chart listing all of these steps, hazards and controls. They are part of many existing OH&S programs.

- *Job Demands Analysis*

    A **Job/Position Demand Analysis** (JDA/PDA) is designed to identify all the physical and psychological demands of the job/position. Although it is typically used as part of the fitness-to-work process, a JDA is an ideal way to identify workplace hazards and their potential effect on worker well-being. A sample JDA (or PDA as they are also termed) is provided in Chapter 4, "Occupational Health and Safety Program: Manual Development", Appendix 4.

- *Equipment Inspections/Pre-trip Inspections*

    The purpose of both these tools is to assess the equipment/vehicle prior to its use/operation. In some instances, companies will require workers to complete an inspection form that is designed to draw their attention to critical factors requiring attention.

- *General Inspections/Walk-through Surveys*

    General worksite inspections/walk-through surveys are valuable tools for identifying hazards before they become problematic. They help to identify hazards that tend to get overlooked in the course of working in the area.

- *Housekeeping Evaluations*

    Housekeeping evaluations include an assessment of both cleanliness and order of the workplace. Equipment and materials should be in a specific location for maximum productivity, quality, safety, and cost control.

- *Worker Reports*

    Hazard reports and near-miss incident reports submitted by workers are critical tools in identifying workplace hazards.

- *Specific Worksite Inspections*

    Worksite inspections are valuable tools for identifying problems and assessing the related risk before accidents or other losses occur. A sample worksite inspection tool for an industrialized setting is provided in Appendix 3.

The goals of worksite inspections are to:

- identify potential problems;
- identify equipment deficiencies;
- identify substandard methods or practices and "at risk" behaviours;
- identify the effects of changes in materials or processes;
- identify equipment problems and deficiencies not anticipated during job design or task analysis;

CH. 9: OH&S: Hazard Identification, Assessment, and Control

- identify inadequacies in remedial action taken for identified problems;
- demonstrate Management commitment to Occupational Health and Safety through visible action;
- provide information for management self-appraisals; and
- address Critical Parts, Critical Tasks, General Inspections, Management Tours and Housekeeping Evaluations.

The typical approach for conducting a worksite inspection is depicted in Figure 9.1.

Figure 9.1: Worksite Inspection: Flow Chart

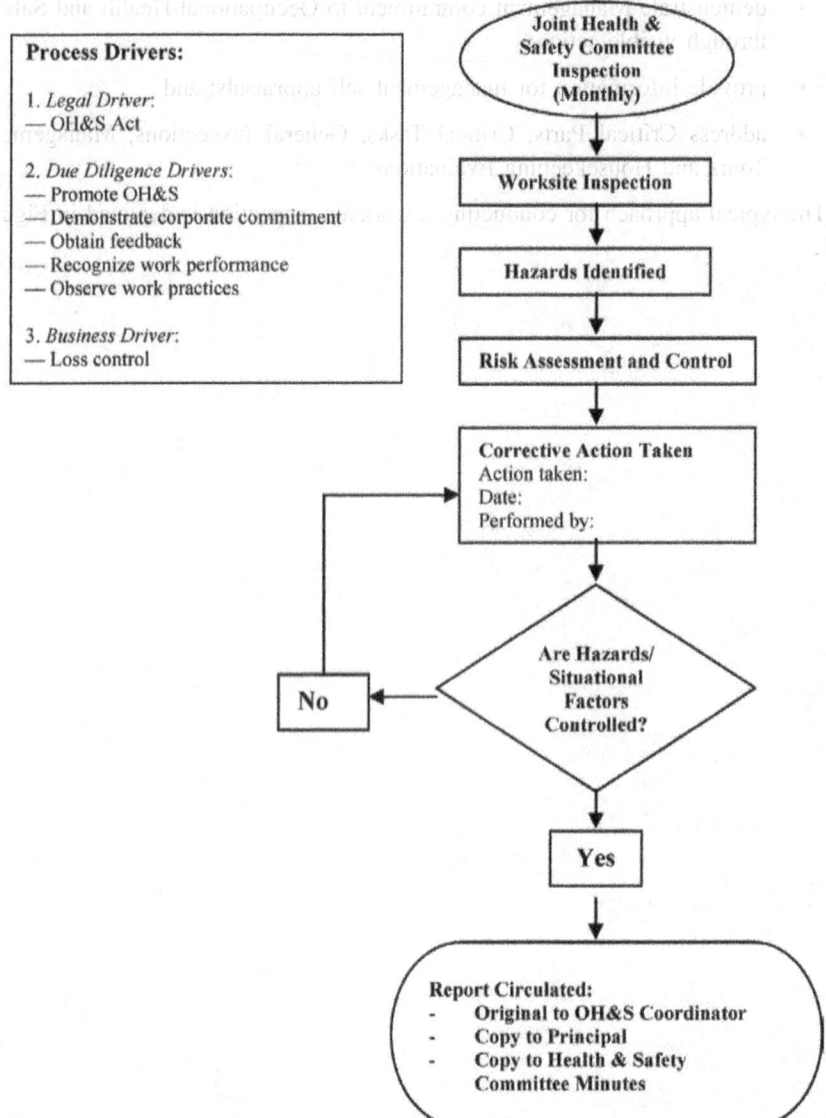

Inspection reports also provide a documented tracking process of a company's efforts to identify and control hazards.

- **Ergonomic Assessments**

    Ergonomic assessments allow for the identification of ergonomic hazards. By using a tool like the one provided in Appendix 4, a systematic assessment of the worker's workstation and body mechanics can be done.

- **Hazard Controls/Safeguards**

    The most effective controls/safeguards are the ones that require low human reliance for success. To apply adequate controls/safeguards, one needs to ask:

    - Can the hazard be eliminated?
    - Can the energy be eliminated?
    - Can physical barriers or safety equipment be used?
    - Can the chance of human error be reduced?
    - Can warning devices be implemented?
    - Can protective procedures such as safe work procedures, permits, confined space entry permits, *etc.*, be used?
    - Are the workers involved competent, motivated, and properly supervised to perform safely?
    - Can the residual risk be accepted?

The principal solutions for minimizing identified hazards are:

- **Find a new way to do the job:** To find an entirely new way to perform a task, determine the goal of the operation and analyze feasible ways to achieve this goal. Select the safest method.
- **Change the physical conditions:** If a new way to perform the job cannot be developed, change the physical conditions (such as tools, materials, equipment, layout, location) to eliminate or control the hazard.
- **Change the work procedure:** Investigate changes in the job procedure that would enable employees to perform the task without being exposed to the hazard.
- **Reduce the frequency of its performance:** Reducing the number of times a job is performed contributes to safer operations only because the frequency of exposure to the hazard is reduced.

Solutions should clearly state what to do and how to do it. A good recommendation explains both "what" and "how" to do the task. For example:

- Set wrench jaws securely on the bolt.
- Test its grip by exerting slight pressure on it.

- Brace yourself against something immovable, or take a solid stance with feet wide apart before exerting slow steady pressure.

This recommendation reduces the possibility of a loss of balance if the wrench slips. These processes are depicted in Figure 9.2.

Figure 9.2: Hazard Controls/Safeguards

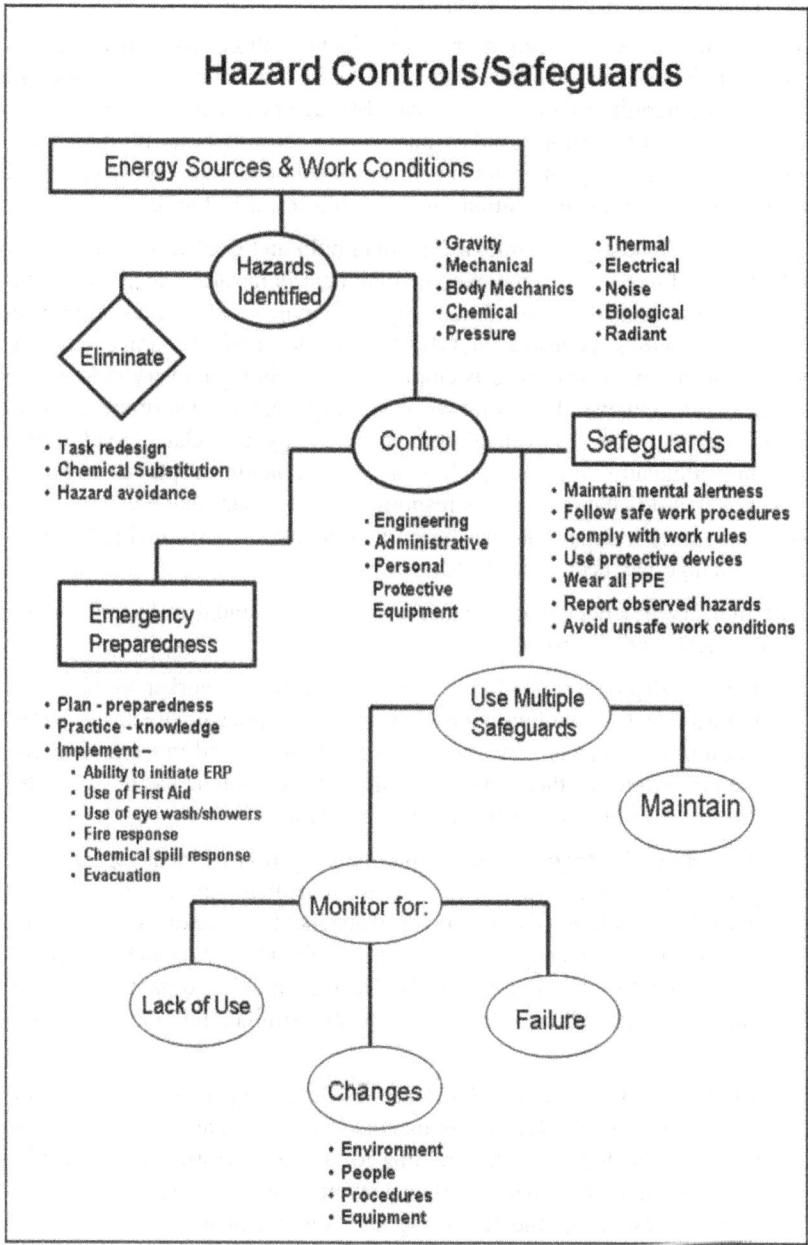

H. HAZARD MANAGEMENT: WORKER INVOLVEMENT

Hazard identification is best achieved by the people doing the work — the employees. Workers have a legal responsibility to identify unsafe work conditions, unsafe work actions, and unsafe equipment.

By providing worker education on hazard identification, assessment, and management, employers can enable employees to be instrumental in hazard identification. The accountability to do so comes from Management's OH&S communication on the importance of workplace safety and worker accountability to assist with the creation of a strong safety environment. This desired behaviour should be rewarded either through employee recognition, or through a reward of sorts.

By way of an industry example, a Canadian pulp and chemical plant developed a guide designed to promote employee involvement in hazard management and the adoption of safe work practices. This company requires and enables employees to be leaders in OH&S. A leader in OH&S promotes and encourages safe work practices. The company also expects employees to ensure that every task/activity is done safely. To achieve this, employees are expected to determine their work readiness. This translates to evaluating the task/activity, the related conditions, and the existing safeguards to decide if they are acceptable/unacceptable (Figure 9.2). The result is that it is the employee's responsibility to ensure that he/she undertakes a task/activity only when the risk has been identified, safeguards are in place and the impact of change has been considered.

Some industry practices that promote worker involvement in hazard identification, assessment, and control are:

- *Hazard Reporting* — Companies that encourage worker involvement in hazard management introduce and promote the practice of hazard reporting. Where the company sets hazard reporting targets and rewards workers for their contribution, there has been a direct correlation between the number of hazards reports and the reduction in workplace incidents.

- *Near-miss Reporting* — To encourage near-miss incident reporting, companies must first clarify with workers that their report will not result in disciplinary action. Given that workplace incidents are due to an 85 per cent failure in the management system,[3] this is a reasonable stance for companies to take. The benefit is that the knowledge gained about these "nearly happened" incidents can prevent similar, but more serious and costly incidents.

- *Worker involvement on Joint Health & Safety Committees* — In all Canadian provinces JHSCs are mandatory; workers are provided time from their regular duties and the training to take part. However, to get full and eager worker participation, Management should reward worker involvement by respecting and rewarding their involvement.

---

[3] F. Bird & G. Germain, *Loss Control Management: Practical Loss Control Leadership*, revised edition (Loganville, GA: DNV, 1996) at 27.

CH. 9: OH&S: HAZARD IDENTIFICATION, ASSESSMENT, AND CONTROL

- *Worker involvement in tool/equipment/machinery/vehicle inspections* — Before using tools/equipment/machinery, or driving vehicles, workers are expected to inspect them for hazards in terms of defects, misuse, substandard condition, or being unsuitable for the task. Noted hazards are to be reported so the equipment/vehicle can be taken out of use and serviced.
- *Worker involvement in worksite inspections* — By inspecting the worksite, workers can identify unsafe conditions/practices that could result in a workplace incident.
- *Good housekeeping practices by workers* — Part of doing the job, should be the expectation that workers leave the worksite in a suitable condition.
- *Worker involvement in incident investigations* — When investigating incidents, workers should be involved because they know the work and the processes, and they can learn from the experience.

I. HAZARD MANAGEMENT: EFFECTIVENESS MEASUREMENT

There are various levels of hazard management measurement that can be used to determine the degree of effectiveness of the hazard management processes:

- *Rate of Hazard Reporting:*

    The number of hazards reported through hazard reports or near-miss incident reports can be counted and compared with the company's past record of hazard reporting. This year-over-year reporting can indicate if workers are reporting workplace hazards as expected. It can also be correlated with the corresponding lagging safety performance indicator (e.g., Injury Frequency Rate) to demonstrate the positive relationship between the two.

    A number of Canadian companies that track and correlate near-miss incident rates with their injury frequency rates have identified a strong inverse relationship; that is, as the rate of near-miss incident reports increases, the incidence of lost-time incidents decreases.

- *System Hazards and the OH&S Risk:*

    In each organization, there are inherent and emerging hazards. **Inherent hazards** are the hazards that are associated with the nature of the work being done. **Emerging hazards** are the hazards that develop with time. To determine the amount of OH&S risk within a given company system, each inherent hazard and emerging hazard must be examined in terms of the controls currently in place. If the hazard is eliminated, or exists at an acceptable level, the risk level is graded as low. If the risk is at an unacceptable level, the risk level is graded as high. Through the use of numerical values, a cumulative risk level can be calculated.

    Again, using this value and comparing it year over year, the OH&S practitioner can enable the company to evaluate the effectiveness of the company's hazard management system.

- **Risk Control Monitoring:**

    When uncontrolled hazards are identified, they must be tracked until suitable controls have been put in place. Through the use of a tracking mechanism, companies can measure how quickly the system responds to eliminate/reduce the associated OH&S risks, as well as the related costs.

    Given the importance of OH&S risk management, the company may set targets for hazard control and measure performance against those targets. This is one way to demonstrate the company's level of reactive due diligence.[4]

    Due diligence means taking "every precaution reasonable" to avoid loss or harm. It means taking action that is beyond minimal legislative requirements. **Reactive due diligence** is properly responding when a workplace hazard is identified.[5] **Proactive due diligence** is developing an OH&S Management System that ensures worker health and safety.[6]

J. HAZARDS: NEW INDUSTRY CHALLENGES

Some emerging workplace hazards are

- **Fatigue**

    Many people work two or more jobs to survive. Our society is recognized as being chronically sleep-deprived: "Canada is the third most sleep-deprived country, with nearly a third (31%) of the Canadians feeling like they don't sleep enough".[7] In particular, a third of the working males get four to six hours of sleep each night.[8]

    Fatigue has long been recognized as a workplace hazard, however, with an increasingly "sleep deprived" society, the prevalence of this hazard is high. Fatigue is strongly associated with functional and cognitive impairment:[9]

---

[4] **Proactive due diligence** is developing an OH&S Management System that ensures worker health and safety.

[5] P. Strahlendorf, "Proactive versus Reactive Due Diligence, Unit 12, Prosecutions" *Occupational Health & Safety Law: Study Guide* (Toronto, ON: Ryerson University, 1998) at 63-66.

[6] *Ibid*.

[7] Aviva, *Health Check: UK Report* (2016), online: file:///C:/Users/degdy/Downloads/Aviva_Health_Check_Report_SECURED_13.10.16.pdf (date accessed: February 28, 2020).

[8] Canadian Men's Health Foundation, "Study Finds a Third of Canadian Men Are Sleep Deprived" *News, Canadian Men's Health Foundation* (July 2016), online: https://menshealthfoundation.ca/study-finds-third-canadian-men-sleep-deprived (date accessed: February 28, 2020).

[9] I. Nijrolder, D. van der Windt, & H. van der Horst, "Prognosis of fatigue and functioning

vigilance,[10] cognitive performance,[11] memory, problem-solving, planning, and even the use of language (verbal fluency) can be negatively impacted.[12] It is also associated with motor vehicle incidents,[13] injuries, human error, medical errors,[14] reduced quality of care,[15] and errors in judgement.

WorkSafeBC claims that sleep deprivation parallels the effects of alcohol consumption as follows:

- 17 hours awake is equivalent to a blood alcohol content of 0.05;
- 21 hours awake is equivalent to a blood alcohol content of 0.08 (the legal limit in Canada);
- 24-25 hours awake is equivalent to a blood alcohol content of 0.10.[16]

These impairments also reduce worker productivity: attention to detail, learning, and creativity are all negatively impacted.

Some industries report worker fatigue as being a major problem. For example, a five-year study at the Centre for Research in Occupational Safety and Health, Laurentian University, Sudbury, Ontario, determined that many Ontario miners (79%) were fatigued. Their definition of fatigue was overwhelming exhaustion, not relieved by rest. Of the 2,224 participating miners, 30% reported that they slept 6.2 hours per night and the quality of their sleep was poor. Contributing factors were worker fear for their psychological safety, burnout, anxiety, depression, post-traumatic stress disorder, and alcohol use.[17]

---

in primary care: a 1-year follow-up study" (2008) 6:6 Annuals Family Medicine 519-527. doi: 10.1370/afm.908.

[10] P. Alhola & P. Polo-Kantola, "Sleep deprivation: Impact on cognitive performance" (2007) 3:5 Neuropsychiatr Dis Treat 553-567.

[11] J.S. Durmer & D.F. Dinges, Neurocognitive consequences of sleep deprivation (2005) 25:1 Semin Neurol 117-129.

[12] M. Miller *et al.*, "Chapter 1: Sleep and Cognition" University of Warwick, Warwick Medical School, Coventry, UK (2014), online: http://www2.warwick.ac.uk/fac/med/research/mhwellbeing/sleep/sleeppublications/sleep_and_cognition_2014.pdf (date accessed: February 28, 2020).

[13] J. Wise, K. Heaton & P. Patrician, "Fatigue in Long-Haul Truck Drivers" (2019) 67:2 WH&S 68-77.

[14] M. Miller *et al.*, "Chapter 1: Sleep and Cognition" University of Warwick, Warwick Medical School, Coventry, UK, 10 (2014), online: http://www2.warwick.ac.uk/fac/med/research/mhwellbeing/sleep/sleeppublications/sleep_and_cognition_2014.pdf (date accessed: February 28, 2020).

[15] *Ibid.*

[16] WorkSafeBC, reported in CCOHS, "Fatigue" *OHS Answers* (2014), online: http://www.ccohs.ca/oshanswers/psychosocial/fatigue.html (date accessed: February 28, 2020).

[17] M. Larivière *et al.*, *Mining Mental Health Report: Executive Summary* (April 2018),

To effectively address this hazard, companies should employ the principles of hazard assessment, control, management, and evaluation as pointed out in the above report. Workers and Management should be educated on the effects of fatigue and the related dangers. Management must take control, focus on worker mental health, and establish a work culture, climate and environment that assists workers (refer to Chapter 6, "Occupational Health and Safety: Leadership and Commitment", Section E). This should include the introduction of a mental health policy, a mental health strategy, open communication about work hazards and concerns, manageable workloads, and adequate resources to do the job safely. For example, some worker aids could be education on sleep management techniques, the importance of a healthy diet, EAP services, and avoiding driving or operating heavy equipment when tired.

- **Remote Work**

    Working in remote locations can lead to:
    - exposure to weather extremes, such as sunshine, heat, as well as low temperatures, high wind chills, blizzards, ice and snow-covered surfaces, treacherous ice craters, *etc.*;
    - exposure to wildlife (*e.g.*, bears, cougars, wolves, and wild dogs);
    - exposure to insects, snakes, poisonous plants (*e.g.*, poison ivy, poison oak) and harmful biological organisms (*e.g.*, guardia in streams);
    - operation of all-terrain vehicles;[18]
    - social isolation and the negative impacts on the worker's psychological health; and
    - delayed medical treatment for illness/injuries due to the lack of accessible health care services.

Hazard control measures include engaging employees in the hazard assessment process, and the implementation of safe health and safety procedures. A complementary approach is to educate workers, namely on:
- the impact of weather extremes and how to identify and assess this hazard and hence, avoid unnecessary exposures. If exposure is unavoidable, then the worker needs to be educated on how to work safely within that environment;
- wildlife awareness and safety training, along with the provision of personal protective equipment;
- prevention of exposure to poisonous plants, insect bites, snake encounters, and water-borne bacteria, and how to address each;

---

online: http://www.vale.com/canada/EN/aboutvale/communities/health-and-safety/Documents/Mining%20Mental%20Health%20Summary%20Report.pdf (date accessed: February 28, 2020).

[18] A. Silliker, "Joy Ride" Canadian Occupational Safety, April/May 2019 issue, at 16-17.

- safe off-road vehicular travel.; and
- an awareness of the related signs and symptoms of mental health problems, and effective intervention measures and positive coping skills.

Of course, First Aid Training, Wilderness Survival education, machine safety, and an emergency preparedness and response plan must be implemented and regularly evaluated.

- ***Worker Impairment***

Impairment is defined as experiencing a state of diminished, reduced, or damaged state of being.[19] According to the Meriam-Webster Dictionary, *impairment* is

> [being in] an imperfect or weakened state or condition: such as [being] diminished in function or ability: lacking full functional or structural integrity; [or] being unable to function normally or safely (as when operating a motor vehicle) because of intoxication by alcohol or drugs.[20]

Impairment stems from many sources. Most people relate impairment with a disease state, or an injury, or a substance-related condition. Of course, those are the obvious sources; yet, there are more to consider.

The World Health Organization (WHO) recognizes *impairment* as being:

> a problem in body function or structure; an activity limitation is a difficulty encountered by an individual in executing a task or action; while a participation restriction is a problem experienced by an individual in involvement in life situations.[21]

This broader definition encompasses more than the traditional sources of impairment, and includes the negative effects of fatigue, shift work, prescription medication use, over-the-counter medication use, medication interactions, and aging.

Illness and injury can result in worker disability, that is, the inability to uphold a functional role in society. This form of impairment can manifest as activity impairment and/or participation restrictions;[22] workers are away from the workplace or functioning in a reduced capacity. In either situation, worker productivity is reduced.

As the result of substance use, the worker might be working but is functionally incapacitated. The degree of impairment varies depending on the dosage, the

---

[19] Dictionary.com, "Impairment" *Dictionary.com.* (2017), online: http://www.dictionary.com/browse/impairment (date accessed: February 28, 2020).

[20] Merriam-Webster, "Impairment", *Merriam-Webster Dictionary* (2017), online: https://www.merriam-webster.com/dictionary/impaired (date accessed: February 28, 2020).

[21] World Health Organization, Disabilities, *Health Topics, WHO* (2017), online: http://www.who.int/topics/disabilities/en/ (date accessed: February 28, 2020).

[22] *Ibid.*

properties of the substance, the gender and age of the user, the length and time since taking the substance, the worker's level of tolerance for the substance, and/or the nature of the activity. Activities involving concentration, fine motor skills, information processing, deductive reasoning, quick response time, and memory are often the most impacted.

With over 26% of today's full-time workforce being over the age of 60 years of age,[23] workplaces are witnessing the effects of aging. Aging is manifested by:

- increased morbidity due to sleep deprivation, chronic health conditions, increased use of medication; and
- functional inability to undertake their previous job tasks.

For example, the older worker is more likely to experience changes in vision, hearing, muscle strength, mobility, motor performance, fine motor skills, perceptual abilities,[24] and response times. Likewise, with aging, comes sleep changes and the related cognitive decline.[25]

To effectively manage impairment, organizations are advised to use a combination of risk management approaches. The employer has a legal duty to provide a safe and healthy workplace, as well as meet their business strategies and obligations. The employer must have policies and procedures addressing situations when worker impairment occurs. For example, an OH&S policy; OH&S program; safe work and operating procedures; and enforcement of each. The employer needs to ensure that the work conditions such as work hours, work demands, pace of work, shiftwork, and travel requirements are evaluated in terms of their health impacts. If deemed hazardous, then the employer is obligated to remedy the situation as part of their OH&S legal obligation. Likewise, the employer must ensure that adequate resources and expertise exist to enable the identification and management of worker impairment.

An OHN is qualified and if positioned to do so, can take a leadership role in the management of worker impairment. In terms of health and safety promotion, and disease prevention, the OHN can support the workplace by:

- ensuring workers are fit-to-work;

---

[23] Statistics Canada, *Labour Statistics at a Glance: Reasons for Working at 60 and beyond*, 71-222-X (2018), online: https://www150.statcan.gc.ca/n1/pub/71-222-x/71-222-x2018003-eng.htm (date accessed: February 28, 2020).

[24] C. Voelcker-Rehage, "Motor-skill learning in older adults—a review of studies on age-related differences" (2008) 5:30 European Review of Aging and Physical Activity. DOI:10.1007/s11556-008-0030-9.

[25] M. Miller *et al.*, "Chapter 1: Sleep and Cognition" University of Warwick, Warwick Medical School, Coventry, UK, 15 (2014), online:: http://www2.warwick.ac.uk/fac/med/research/mhwellbeing/sleep/sleeppublications/sleep_and_cognition_2014.pdf (date accessed: February 28, 2020).

- ensuring the worksite is free of uncontrolled hazards;
- medically monitoring workers exposed to known hazards;
- conducting risk assessments;
- communicating the nature and severity of identified risks;
- conducting human factor/ergonomic assessments and identifying suitable remedial actions;
- participating in planning emergency response activities;
- facilitating critical incident stress debriefing post-incident; and
- assisting with the management of strategic OH&S issues.

OHNs contribute to promoting and maintaining worker health and safety, workplace safety, controlling losses, and enhancing the organization's profits. More details are provided in Chapter 4, "Occupational Health Program".

- ***Workplace Violence***

    Workplace violence can include:

    - Physical violence, which involves the use of physical force by a person against a worker in a workplace that causes or could cause physical injury.
    - Psychological violence, which includes harassment, bullying, intimidation and demeaning treatment of a worker.
    - Sexual harassment, which is a form of harassment that includes repeated unwanted sexual behaviour that has harmful consequences for the victim.
    - Financial violence, which is defined as actions taken to prevent advancement or promotion of an individual, which may have a financial impact.[26]

Workplace harassment, bullying, and violence are genuine threats to the health and safety of workers. These violent acts can happen anywhere and in any workplace; however, some workplaces are more at risk than others. For example, school, hospital, and healthcare settings; police services; industry and community first responders; Emergency Medical Service workers; home-health workers; along with those employed in money-handling businesses, are at high risk for workplace violence.

Between 2006-2015, the number of violence-related workplace injuries among

---

[26] Canadian Nurses Association, "Violence Faced by Health-Care Workers in Hospitals, Long-Term Care Facilities and in Home Care Settings," written submission to the Parliament of Canada, House of Commons, May 16, 2019.

healthcare workers increased by 66%.[27] Over 60% of nurses report experiencing violence at work,[28] with those working in the Emergency Department, psychiatric units[29] and remote/rural areas being the most at risk.[30] Likewise, physicians (33%), paramedics (74%), personal support works (89%) experienced violent incidents in the workplace.[31]

In response to this data, the House of Commons Standing Committee on Health, brought forth the following recommendations for Canadian healthcare settings:

- Adoption of best practices in violence prevention and evaluate their effectiveness;
- Enhanced public awareness on violence against healthcare workers;
- Collection and standardization of violence in healthcare;
- Support research on the prevention of gender-based violence in healthcare settings;
- Address staffing shortages;
- Fund upgrading to long-term care facilities;
- Amendment of the Criminal Code to consider assault on a healthcare worker as an aggravating circumstance when sentencing;
- Establish CCOHS as the repository for sharing best practices in violence prevention.[32]

These External Responsibility System responses force employers, in many cases the provincial/territory/municipal governments, to enhance their Internal Responsibility System.

It is an employer obligation to create a workplace in which worker health and safety are maintained. This means using the principles of hazard assessment, control, management, and evaluation. It also requires the development of a respectful workplace, and a corporate culture in which it is everyone's responsibility to remain vigilant of high-risk situations and to report them. Specifically, it involves

---

[27] Canadian Nurses Association, "Violence Faced by Health-Care Workers in Hospitals, Long-Term Care Facilities and in Home Care Settings" written submission to the Parliament of Canada, House of Commons, May 16, 2019.

[28] *Ibid.*

[29] *Ibid.*

[30] S. Jahner, Exploring the Distressing Events and Perceptions of Support Experienced by Rural and Remote Nurses: A Thematic Analysis of National Survey Data, pending publication in *WH&S Journal*.

[31] HESA Committee Report, Violence Facing Health Care Workers in Canada, House of Commons, Standing Committee on Health, June 2019, online: https://www.ourcommons.ca/DocumentViewer/en/42-1/HESA/report-29/page-5.

[32] *Ibid.*

the use of best practices such as violence response protocols, safety alarm and response systems, violence risk assessments, de-escalation training, patient flagging, and specialized procedures for lone workers.

For details on how to manage workplace violence, refer to Chapter 5, "Occupational Health and Safety Program: Manual Development", Appendix 1, Section 5, Subsection 12, Prevention of Workplace Violence Policy.

- *Foreign/Overseas Work*

Expatriates working in certain countries, risk illness, occupational injury, incarceration, being a victim of a crime or scam, kidnapping, torture, death, exposure to natural disasters, or simply vanishing. This risk is the greatest in poverty-stricken countries where the foreign worker is viewed as wealthy and easy targets. Although foreign travellers can also be at-risk, the main occupations to be impacted are journalists, engineers, medical professionals, aid workers, military, miners, and teachers.[33]

In preparation for foreign travel, employers must ensure that the expat knows the hazards, and are educated on how to deal with them. Support systems and resources like healthcare facilities, security resources, access to the nearby Canadian Embassy, and International Travel Insurance, must be in place. The main message should be, "you are a guest and keep an eye on the larger context. Build positive and respectful relationships with local peers and workers."[34] And above all, be vigilant, avoid needless risks, and keep the exit plan in mind.

K. SUMMARY

Hazard management is a management system. Legally, employers must conduct initial and repeated worksite hazard assessments; and then, control the identified hazards. With that duty of care, employers are advised to proactively address the hazards in their workplace and continue to seek innovative ways to eliminate/reduce the inherent hazards:

> To look is one thing;
>
> To see what you look at is another;
>
> To understand what you see, is another;
>
> To learn from what you understand, is something else;
>
> But to act on what you learn, is all that really matters.

Winston Churchill

---

[33] J. Lian, "Expatriate Safety, Out of the Comfort Zone" *OHS Canada* (March/April 2019) at 20-21.

[34] R. Melles (2019). Reported in J. Lian "Expatriate Safety, Out of the Comfort Zone" *OHS Canada* (March/April 2019) at 20-21.

Appendix 1

SAMPLE HAZARD ASSESSMENT AND CONTROL FORM[35]

Step 1: On the Hazard Identification checklist, check off all the hazards or potential hazards that are present at your worksite. Add any identified hazards specific to your worksite to the list:

---

[35] Alberta Employment and Immigration, *Health and Safety Tool Kit* (2006) at 38-39, available for public use online: http://work.alberta.ca/ (date accessed: February 28, 2020).

HAZARD IDENTIFICATION			
Physical Hazards		**Chemical Hazards**	
Lifting and handling loads	☐	Chemicals (identify types)	☐
Repetitive motion	☐	Type:	☐
Slipping and tripping	☐	Type:	☐
Moving parts of machinery	☐	Type:	☐
Working at heights	☐	Type:	☐
Pressurized systems	☐	Type:	☐
Vehicles	☐	Dusts	☐
Fire	☐	Fumes (identify types)	☐
Electricity	☐	Type:	☐
Noise	☐	Type:	☐
Lighting	☐	Type:	☐
Temperatures	☐	Mists and Vapors (identify types)	☐
Vibration	☐	Type:	☐
Ionizing Radiation	☐	Type:	☐
Workplace Violence	☐	Type:	☐
Other:	☐	Other:	☐
Other:	☐	Other:	☐
Other	☐	Other:	☐
Biological Hazards		**Psychological Hazards**	
Viruses	☐	Working conditions	☐
Fungi (mould)	☐	Fatigue	☐
Bacteria	☐	Stress	☐
Blood and Body Fluids	☐	Other:	☐
Sewage	☐	Other:	☐
Other:	☐	Other:	☐
Other:	☐		

NOTE: If you work in a high hazard industry, an industry-specific checklist may be required.

Ch. 9: OH&S: Hazard Identification, Assessment, and Control

Step 2: Hazard Assessment and Control Sheet (Sample — Page 2)

- Take the hazards identified on the checklist above and list them on the Hazard Assessment and Control Sheet
- Identify the controls that are in place: engineering, administrative, PPE or combination for each hazard

Company: _____ Location: _____

Date of assessment: _____ Completed by: _____

Hazard	Controls in Place (list)			Follow-up Action Required	Date/Person Responsible
	Engineering	Administrative	PPE		

Appendix 2

SAMPLE RISK ASSESSMENT FORM

Risk Assessment Form

Risk Assessment Form		Hazard #:	Date:
RECORD OF ASSESSMENT			

Hazard Type:
Location: TYPE OF HARM: Injury: Y N
Activity/Task: Damage to Property: Y N
Work Group: Damage to Environment: Y N
Hazard Management Team: **RATINGS:** Severity:
Signature: Probability:
Date of Assessment: Review Date: Risk Rating:
Who/what is at risk? Time Line:
Safe Operating Procedures Available: Y N Accidents/Incidents Y N Date:
Hazards Identified:

Situational Factors:

Existing Control Measures:

Are additional control measures required? Y N
If Yes, summarize action plan below:

CONTROL MEASURES REQUIRED:	E/I/M	Target Date	Action By (Name)	Completed By: (Name & Date)
1) Engineering Controls				
Monitoring Plan:				
2) Administrative Controls				
Monitoring Plan:				
3) Personal Protective Equipment Controls				
Monitoring Plan:				

Forecasted Ratings for new control: Severity___ Probability:___ Risk Rating:___ Time Line:___
Date for Full Implementation of Control Measures:
Assessment Accepted By:___ Designation:___ Date:___

Risk Assessment Form

Step 1) Assessment of Hazard Severity: The hazard severity (harm) is the extent of the injuries, ill health, or damage to the environment that may be sustained if the hazard is realized. Identify the severity from Table 1 below and enter the result on the Hazard Assessment and Management Form.

Table 1 Severity

Severity (S)	Injury Potential	Potential Environmental Impact	Property, Equipment Damage
i	Trivial injury/ies (scratch, small cut, burn, bruise or abrasion)	Slight leak/spill - minimal effect contained locally	$100
ii	Minor injury/ies (laceration requiring stitches, moderate bruises or burns)	Minor leak/spill - no lasting effect	$1,000
iii	Serious injury, (broken bone, severe burns or bleeding, eye injury)	Minor leak/spill public concern	$10,000
iv	Major injury to one or more than one person, (amputation, permanent disability)	Localized leak/spill. Non conformance with Regulations (considerable public concern)	$100,000
v	Single fatality	Massive leak/spill. Non conformance with Regulations (considerable public concern)	$1,000,000
vi	Multiple fatalities	Massive leak/spill major public concern. Major clean up required	$10,000,000

Step 2) Assessment of Probability: Probability is split into five categories in an ascending scale of likelihood: Determine the probability of occurrence. Enter the result on the Hazard Assessment and Management Form

Table 2 Probability Scale

	Probability Scale
A	Very unlikely to occur
B	Unlikely to occur
C	May happen
D	Likely to occur
E	Highly likely

Step 3) Determination of Risk Rating: Determine the Risk Rating by entering the Severity and Probability ratings on Table 3. The resulting number is the Risk Rating.

Table 3 Risk Rating Table

			PROBABILITY			
S	i	1	1	1	1	2
E	ii	1	2	2	3	3
V	iii	1	2	4	5	5
E	iv	1	2	4	5	5
R	v	1	3	4	5	5
I	vi	2	3	4	5	5
T/Y		A	B	C	D	E

Step 4) Determination of Time Line for Action
Use Table 4 to determine the Time Line for action required

Table 4 Risk Reduction Policy

Risk Rating	Time Line
5 Unacceptable	Stop activity. Fix immediately
4 High	Stop activity. Fix within 1 week.
3 Medium	Reduce within 1 month
2 Low	Reduce within 6 months
1 Acceptable	Monitor

NOTE: 1) *Engineering Controls* – e.g., elimination, substitution, enclosure, guarding.
2) *Administrative Controls* – e.g., Safe Operating Procedures, Supervision, Training, Information (flow charts, signs, handouts).
3) *Personal Protective Equipment Controls* – e.g., hat, eye and face protection, hearing protection, respiratory protection, suit, apron, gloves, boots.

Last revised: August 1, 2002   ©H. Börner & D. Dyck 2002

Appendix 3

SAMPLE WORKSITE INSPECTION FORM

Company Logo

APPENDIX 3

SAMPLE INDUSTRIAL WORKSITE INSPECTION

13. Lighting/Heating	15. Fire Exits	16 Fire Extinguishers and Small Hose (1 1/2 inches):
☐ OK ☐ Needs Attn	☐ OK ☐ Needs Attn	☐ OK ☐ Needs Attn
Adequate heat ☐Yes ☐ No ☐ NA	Exits clearly marked ☐Yes ☐ No ☐ NA	Are all extinguishers properly charged and pressurized?
Explosion proof lights ☐Yes ☐ No ☐ NA	Doors locked from outside only ☐Yes ☐ No ☐ NA	☐Yes ☐ No ☐ NA
Emergency lighting ☐Yes ☐ No ☐ NA	Exit path clear ☐Yes ☐ No ☐ NA	Are all extinguishers and small hoses in good condition and readily accessible?
Fuel leaks detected ☐Yes ☐ No ☐ NA		☐Yes ☐ No ☐ NA
Adequate distance from combustible materials ☐Yes ☐ No ☐ NA		
Obstructions ☐Yes ☐ No ☐ NA		

14. Bulletin Board	17 Other:	
☐ OK ☐ Needs Attn		
Safety Meeting minutes ☐Yes ☐ No ☐ NA		
Readily seen ☐Yes ☐ No ☐ NA		
Updated/maintained ☐Yes ☐ No ☐ NA		
Current Safety alerts ☐Yes ☐ No ☐ NA		

HAZARD CLASSIFICATION
A work site or equipment condition or employee action that has the potential to cause:

1. A. Permanent injury fatality
   B. Extensive Damage
   C. Permanent Loss
2. A. Temporary, lost time injury
   B. Repairable damage
   C. Temporary loss of use
3. A. First aid, medical aid
   B. No lost time due to injury
   C. Damage without loss of use

Item	Hazard Classification	Specify what needs attention or corrective action	Action By	Date	Completed

Remarks

Inspector(s) Name (Printed)	Supervisor Name (Printed)
Inspector(s) Signature	Supervisor (Signature)

© H. Börner & D. Dyck 2002

Appendix 4

SAMPLE ERGONOMIC ASSESSMENT FOR AN OFFICE ENVIRONMENT

Company Logo

APPENDIX 4
SAMPLE ERGONOMIC ASSESSMENT FOR AN OFFICE WORKSTATION

	Yes	No
1. VDT stations are arranged so that lighting does not reflect directly off the screen.		
2. The seat and backrest of the chair support comfortable posture permitting occasional variation in the sitting position.		
3. Seat height is adjustable so that the entire sole of the foot rests on the floor or footrest, and the back of the knee is slightly higher than the seat of the chair.		
4. Backrest height is adjustable.		
5. Backrest angle is adjustable.		
6. Footrest provided if desired by individual.		
7. The height of the surface on which the keyboard rests is adjustable allowing the person's forearms, with fingers resting on the keyboard, to be nearly horizontal or inclined slightly upward..		
8. The workstation is adjusted so that the wrist is in a straight line (i.e., not bent up or down).		
9. The topmost line of the screen is slightly below eye level.		
10. Screen position can be tilted.		
11. Document holder is positioned at the same height and at the same distance from the viewer as the screen.		
12. Work surface is large enough to hold all needed reference material (at least 35 inches wide).		
13. Paper can be easily and conveniently loaded into printers without the need for lifting heavy boxes in awkward postures.		
14. Screen has colour, brightness, and contrast satisfactory with the operator.		
15. Characters on the screen are clear and free of flicker or jitter.		
17. There is adequate room under the work table to permit movement of operator's legs and a foot rest where necessary.		
18. Task schedules allow the operator to perform duties not requiring use of the VDT at least 15 minutes during each 2-hour period.		
19. Are all adjustments easy to make with single lever or know controls? Equipment that is difficult to adjust will probably not be adjusted properly.		

Score (count all "No" answers)

Comments: _____

_____

_____

Prepared by: _____ Date: _____

Area/Task Identification: _____ Operator's Name: _____

Evaluation: When a group of workstations are evaluated using this check list by the same individual, the workstations with the higher scores should be the ones most likely to cause ergonomic stress. It is not necessary for each workstation to achieve a "zero", or perfect score, on this check list.

©H. Börner & D. Dyck 2002

Appendix 5

SAMPLE JOB SAFETY ANALYSIS[36]

Job Safety Analysis	Job:	Date:
	Title of Person Who does Job:	Analysis By:
Department:		Reviewed:
		Safety Office Review:
Required Recommended PPE:		

Job Steps	Hazards	Safety Procedures

I have reviewed the above JSA and am familiar with the required job tasks.

Print Name	Date	Sign Name	Supervisor

---

[36] University of Louisiana at Munroe, "Job Safety Analysis Form", online: http://www.ulm.edu/safety/manual/djobsafetyanalysis.htm (date accessed: February 28, 2020).

CHAPTER REFERENCES

Alberta Employment and Immigration, *Health and Safety Tool Kit* (2006), online: http://work.alberta.ca (date accessed: February 28, 2020).

Alberta Human Resources and Employment, "Fatigue, Extended Work Hours, and Safety in the Workplace" *Workplace Health and Safety* (June 2004, Reformatted August 2010), online: http://work.alberta.ca/documents/ WHS-PUB-ERG015.pdf (date accessed: February 28, 2020).

P. Alhola & P. Polo-Kantola, "Sleep deprivation: Impact on cognitive performance" (2007) 3:5 Neuropsychiatr Dis Treat. 553-567.

Aviva, *Health Check: UK Report* (2016), online: file:///C:/Users/degdy/Downloads/Aviva_Health_Check_Report_SECURED_13.10.16.pdf (date accessed: February 28, 2020).

F. Bird & G. Germain, *Loss Control Management: Practical Loss Control Leadership*, revised ed. (Loganville, GA: DNV, 1996).

H. Börner & D. Dyck, "Ergonomic Assessment Form" (Calgary, AB: Progressive Health Consulting, 2002).

H. Börner & D. Dyck, "Worksite Inspection Form" (Calgary, AB: Progressive Health Consulting, 2002).

H. Börner & D. Dyck, "Risk Assessment Form" (Calgary, AB: Progressive Health Consulting, 2002).

BSI Management Systems, "OHSAS 18001 Awareness" (Presented at the City of Calgary Informational Seminar, April 2002, Calgary AB).

CCOHS, *CCOHS: Hazard Control Fast Facts Card* (2019), online: https://www.ccohs.ca/products/boutique/hazardcontrol/

Canadian Men's Health Foundation, "Study Finds a Third of Canadian Men Are Sleep Deprived", *News, Canadian Men's Health Foundation*, (July 2016), online: https://menshealthfoundation.ca/study-finds-third-canadian-men-sleep-deprived (date accessed: February 28, 2020).

Canadian Nurses Association, "Violence Faced by Health-Care Workers in Hospitals, Long-Term Care Facilities and in Home Care Settings" written submission to the Parliament of Canada, House of Commons, May 16, 2019.

Dictionary.com, "Impairment", *Dictionary.com*. (2017), online: http://www.dictionary.com/browse/impairment (date accessed: February 28, 2020).

J.S. Durmer & D.F. Dinges, Neurocognitive consequences of sleep deprivation (2005) 25:1 Semin Neurol 117-129.

S. Jahner (2020), Exploring the Distressing Events and Perceptions of Support Experienced by Rural and Remote Nurses: A Thematic Analysis of National Survey Data, pending publication in WH&S Journal.

M. Larivière., Z. Kerekes, C. Lessel, A. Smith, A. Sinclair, M. Tiszberger & C.Leduc

Mining Mental Health Report: Executive Summary (April 2018), online: http://www.vale.com/canada/EN/aboutvale/ communities/health-and-safety/Documents/Mining%20Mental%20Health %20Summary%20Report.pdf (date accessed: February 28, 2020).

J. Lian, "Expatriate Safety, Out of the Comfort Zone" *OHS Canada* (March/April 2019) at 20-21.

R. Melles (2019). Reported in J. Lian, "Expatriate Safety, Out of the Comfort Zone" *OHS Canada* (March/April 2019) at 20-21.

Merriam-Webster, "Impairment" *Merriam-Webster Dictionary* (2017), online: https://www.merriam-webster.com/dictionary/impaired (date accessed: February 28, 2020).

M. Miller *et al.*, "Chapter 1: Sleep and Cognition" University of Warwick, Warwick Medical School, Coventry, UK. (2014), online: http://www2.warwick.ac.uk/fac/med/research/mhwellbeing/sleep/sleeppublications/sleep_and_cognition_2014.pdf (date accessed: February 28, 2020).

I. Nijrolder, D. van der Windt & H. van der Horst, "Prognosis of fatigue and functioning in primary care: a 1-year follow-up study" (2008) 6:6 Annuals Family Medicine 519-527. doi: 10.1370/afm.908.

A. Silliker, "Joy Ride" *Canadian Occupational Safety* (April/May 2019 issue, at 16-17.

P. Strahlendorf, "Proactive versus Reactive Due Diligence, Unit 12, Prosecutions" *Occupational Health & Safety Law: Study Guide* (Toronto, ON: Ryerson University, 1998).

Statistics Canada, *Labour Statistics at a Glance: Reasons for Working at 60 and beyond*, 71-222-X (2018), online: https://www150.statcan.gc.ca/n1/pub/71-222-x/71-222-x2018003-eng.htm (date accessed: February 28, 2020).

University of Louisiana at Munroe, "Job Safety Analysis Form", online: http://www.ulm.edu/safety/manual/djobsafetyanalysis.htm (date accessed: February 28, 2020).

C. Voelcker-Rehage, "Motor-skill learning in older adults—a review of studies on age-related differences" (2008) 5:30 European Review of Aging and Physical Activity. DOI:10.1007/s11556-008-0030-9.

J. Wise, K. Heaton & P. Patrician, "Fatigue in Long-Haul Truck Drivers" (2019) 67:2 WH&S 68-77.

WorkSafeBC, reported in CCOHS, "Fatigue" *OHS Answers* (2014), online: http://www.ccohs.ca/oshanswers/psychosocial/fatigue.html (date accessed: February 28, 2020).

World Health Organization, Disabilities, *Health Topics, WHO* (2017), online: http://www.who.int/topics/disabilities/en/ (date accessed: February 28, 2020).

Chapter 10

NEW TECHNOLOGY: IMPACT ON THE WORKPLACE

A. INTRODUCTION

Over the last 30 years, the use of technology in the workplace has greatly increased and evolved. Initially, the use of computers and the Internet changed how the organization gathered data, communicated corporate information, organized work, conducted business, reached out to clients and global markets, and assessed business performance. Robotics designed to conduct the more repetitive and basic job tasks emerged along with the use of various mobile devices like smartphones, tablets, and hand-held data trackers. Artificial intelligence (AI) technology made simple office tasks like doing a Google search or using the Smartphone, easy and efficient.

This Age of Technology is part of our daily lives, including the way in which we work and interact. It pushes organizations to do more, faster, better, and in different ways. It also is used to protect workers from exposure to hazardous work and conditions (*e.g.*, the use of gas detectors[1]). The purpose of this chapter is to explore the various types of new technology, their advantages and disadvantages, and the associated workplace hazards.

B. NEW TECHNOLOGY: TERMINOLOGY

With new technology comes new terminology that OH&S Professionals/Practitioners need to recognize, for example:

Automation involves the use of machines or technology to automatically do tasks faster, better, and at a reduced cost, with minimal to no manpower. Some examples of automation include ordering foods/beverages using a self-serve computer screen, or self-checkout at grocery and other retail stores. In general, automation:

- Substitutes for labour;
- Substitutes for tasks, not jobs;
- Complements labour;
- Can increase demand thereby creating jobs;

---

[1] L. Johnson, "Gas Up" *Canadian Occupational Safety* (2019), online: www.cos-mag.com at 20-21, (date accessed: February 28, 2020).

- Frees up people to create new products and do new tasks; and
- May/may not result in desired outcomes.[2]

Ubiquitous now, automated services will continue to improve and become far-reaching. It is expected that in Canada over the next 10 years, automatization will impact more than 50 per cent of the current jobs.[3]

Artificial Intelligence (AI) is defined as "the theory and development of computer systems able to perform tasks that normally require human intelligence, such as visual perception, speech recognition, decision-making, and translation between languages".[4] It is a powerful mean for finding information and problem solving; however, the worker has to come up with the questions to be answered and know what they are looking for.

Machine learning is an arm of computer science in which machines learn a task without being programmed to do so.[5] It involves pattern recognition which enables computers to accomplish tasks like diagnosing medical conditions quickly and reliably. **Neural network** refers to computers modelling the way the human brain problem solves through interconnections with other computers (a network).[6]

Digitization is the process of converting a variety of forms of information into a digital format. This a commonly done in business, for example scanning incident reports.

Digitalization is "the use of digital technologies to change a business model and provide new revenue and value-producing opportunities; it is the process of moving to a digital business".[7] It lends to the use of behavioural analytics — by understanding how people/workgroups function, technology can be used to enhance their activities. Through the use of AI and machine learning, companies can analyze their service/product demands and optimize their response and available resources.

---

[2] M. Muro, R. Maxim & J. Whiton, "Automation and Artificial Intelligence: How machines are affecting people and places" *Brookings* (2019), online: https://www.brookings.edu/wp-content/uploads/2019/01/2019.01_BrookingsMetro_ Automation-AI_Report_Muro-Maxim-Whiton-FINAL-version.pdf (date accessed February 28, 2020).

[3] J. Vomiero, "Half of Canadian jobs will be impacted by automation in the next 10 years" *Global News* (March 26, 2018), online: https://globalnews.ca /4105713/automation-workforce-canada-human/ (date accessed February 28, 2020).

[4] Dictionary, "Artificial Intelligence", *Dictionary* (2019), online at p.1.

[5] S. Lazaruk, "Automated future: Computer and robotics already changing retail and the workplace" *Vancouver Sun* (January 28, 2017), online: https://vancouversun.com/author/susanlazarkukprov (date accessed February 28, 2020).

[6] S. Lazaruk, "Automated future: Computer and robotics already changing retail and the workplace" *Vancouver Sun* (January 28, 2017), online: https://vancouversun.com/author/susanlazarkukprov (date accessed February 28, 2020).

[7] Gartner IT, *Glossary IT: Digitalization* (2019), online: https://www.gartner.com/it-glossary/digitalization/, p. 1, (date accessed February 28, 2020).

Robotics is an interdisciplinary branch of engineering and science that includes mechanical engineering, electronic engineering, information engineering, computer science, and others. Robotics deals with the design, construction, operation, and use of robots, as well as computer systems for their control, sensory feedback, and information processing.[8] Today, robots are widely used in warehouses, manufacturing, and hospitals. For example, alongside its workers, Amazon uses more than 100,000 robots in its U.S. warehouses.[9]

3-D Printers involve any of various processes in which material is joined or solidified under computer control to create a three-dimensional object, with material being added together (such as liquid molecules or powder grains being fused together), typically layer by layer.[10]

Mobile devices encompass cellphones, tablets, data-collection devices, GPS devices, *etc*. These are commonplace in the workplace and have drastically changed how work is organized and done.

CCTV cameras, GPS systems, or **biometric devices** are often used in the workplace. Although they are used for business purposes, they can compromise the employee's privacy rights if not properly set-up by the employer.

Magnetic Resonance Imaging (MRI) is a noninvasive medical test that provides pictures of organs, soft tissues, bone, and virtually all other internal body structures through the use of a powerful magnetic field, radio frequency pulses, and a computer.[11]

Segway scooters are two-wheeled, self-balancing personal transporters used in many facilities to enhance swift movement from one area to another. User error often results in face, head, shoulder, and multiple-trauma injuries, some of which are serious in nature.[12]

C. NEW TECHNOLOGY: WHAT IS HAPPENING?

The new and emerging technology offers employers many advantages, namely enhanced business efficiencies, more new job opportunities, better business com-

---

[8] Wikipedia, "Robotics: definition" *Wikipedia* (2020), online: https://en.wikipedia.org/wiki/Robotics (date accessed February 28, 2020).

[9] T. Simonite, "Robots will take jobs from men, the young and minorities" *Business* (January 24, 2019) at 1-13.

[10] Wikipedia, "3-D Printers", *Wikipedia* (2019), online: https://en.wikipedia.org/wiki/3D_printing (date accessed February 28, 2020).

[11] Radiologyinfo.org, "Magnetic Resonance Imaging (MRI) – Body" (2019), online: https://www.radiologyinfo.org/en/info.cfm?pg=bodymr (date accessed February 28, 2020).

[12] M. McKay, (2010). Reported in *EHSToday*, "Injuries Sustained While Riding Segway Transporter 'Severe' " (September 23, 2010), online: https://www.ehstoday.com/safety/management/injuries-sustained-riding-segway-severe-0923 (date accessed February 28, 2020).

munication, and incredible medical care.[13] It has changed the way in which people work and interact with the world in general, for example:

- Most adults (90%) and youth have a smartphone within reach at all times and are "lost" without them;[14]
- Computer use is the norm in Canada and is a vehicle for most business interactions;
- They have ready access to the Internet and vast amounts of information. For example, Internet activity is highly used for information searches, business activities, data collection, and data interpretation;
- Global marketing is enhanced by better access to clients/customers and marketplaces;
- Innovation is achievable and supported by robust technology;
- Access to customers is available 24/7 and can be tailored to meet their specific communication needs;
- Performance measurement and monitoring is made effective and efficient;
- It enables and supports remote work: collaborative technology allows participants throughout the world to interact and work together on projects. This technology has wide-application especially in the field of medicine;[15]
- Continuous monitoring of the human body allows for the early detection of disease (*e.g.*, cancers, diabetes, Parkinson's disease, *etc.*)[16] and for encouraging the adoption of healthy lifestyles (*e.g.*, devices such as Fitbits, iWatches); and
- It facilitates staff recruitment and retention.

Along with the "good aspects" of new technology, come challenges ("bad aspects"). For example, the impact of automation on the workplace is predicted to be that:

1. "Automation and AI will affect [job] tasks in virtually all occupational groups in the future, but the effects will be of varied intensity and drastic for only some [occupations]";[17] and

---

[13] Futureofworking.com, "7 Biggest pros and cons of technology" (2017), online: https://futureofworking.com/7-biggest-pros-and-cons-of-technology (date accessed February 28, 2020).

[14] *Ibid.*

[15] N. Harris-Briggs, "Is a virtual workplace really possible?" (2018), online: https://martechseries.com/mts-insights/guest-authors/vitrual-workplace-really-possbile/ (date accessed February 28, 2020).

[16] National Geographic, *The Future of Medicine*, January 2019 issue, at 27-35; 56-57.

[17] M. Muro, R. Maxim & J. Whiton, "Automation and Artificial Intelligence: How

2. "The impacts of automation and AI in the coming decades will vary especially across occupations, places, and demographic groups."[18]

In Canada, the world of work has certainly changed, and automation has resulted in job loss in a number of occupations — trades; transportation and equipment operation; sales and service; utilities; office support and administration; and health, natural and applied sciences technical occupations. Occupations that involve routine, administrative, and service-oriented tasks are at high-risk for automation given today's current technological capabilities. This includes:

- retail salespersons;
- administrative assistants;
- food-counter attendants;
- cashiers;
- kitchen helpers;
- transport truck drivers.[19]

It is predicted that those to be affected the most will be men, young workers, and minority workers.[20,21]

The occupations at low risk for automation are the ones that require human interaction, communication skills, decision-making and judgement, and creativity. These include education, nursing, medicine, allied health, natural sciences and resources, management and supervisory positions, recreation and sport, arts, and culture.[22,23]

---

machines are affecting people and places" *Brookings* (2019), online: https://www.brookings.edu/wp-content/uploads/2019/01/2019.01_BrookingsMetro_Automation-AI_Report_Muro-Maxim-Whiton-FINAL-version.pdf at 5 (date accessed February 28, 2020).

[18] *Ibid.*

[19] A. Zwack, "The Impacts of Technology in the Canadian Workplace" prepared for the 2017 Midwinter Meeting of the Employment Rights and Responsibilities Committee American, Bar Association (Vancouver, BC: Glgmlaw, 2017).

[20] T. Simonite, "Robots will take jobs from men, the young and minorities" *Business* (January 24, 2019) at 1-13.

[21] M. Muro, R. Maxim & J. Whiton, "Automation and Artificial Intelligence: How machines are affecting people and places" *Brookings* (2019), online: https://www.brookings.edu/wp-content/uploads/2019/01/2019.01_BrookingsMetro_Automation-AI_Report_Muro-Maxim-Whiton-FINAL-version.pdf (date accessed February 28, 2020).

[22] A. Zwack, "The Impacts of Technology in the Canadian Workplace" prepared for the 2017 Midwinter Meeting of the Employment Rights and Responsibilities Committee American, Bar Association (Vancouver, BC: Glgmlaw, 2017).

[23] H. Siu (2019). Reported in S. Lazaruk, "Automated future: Computer and robotics already changing retail and the workplace" *Vancouver Sun* (January 28, 2017), online: https://vancouversun.com/author/susanlazarkukprov (date accessed February 28, 2020).

D. NEW TECHNOLOGY: IMPACT

In the workplace, the impact of new technology is dramatic and widespread. Although so far in this chapter, some impacts have been alluded to, in this section, a more in-depth discussion of some key aspects of work will be discussed.

1. Change in Hours of Work

Computers, mobile phones, other data-gathering devices, and their connectivity with corporate network servers, enable employees to work in accordance with their work and personal demands. On one hand, this allows the employee to balance their work and personal demands; on the other, this technology can be intrusive making the workday seem "never-ending".[24]

Today, the workday is on average 40.2 hours;[25] yet, it tends to be much longer due to the amount of unaccounted work done by the worker at home. This phenomenon is driven by the employer's ready access to the employee, the universal expectation for immediate response to emails/texts, and corporate rewards/recognition for doing so.[26]

2. Change in Work Location

Employees no longer have to be at a set work location; they can and do, work from a number of locations. Working from home is a most attractive option for many employees; it can eliminate the need for:

- childcare and eldercare support;
- lengthy commutes;
- travel/commuting costs;
- business wear costs; and
- other business-related costs.

However, as noted in Chapter 20, "Occupational Health and Safety: Ergonomics", this practice can result in musculoskeletal problems. Home and remote locations tend to require the worker to adapt to the job; not have the job set up to fit the worker.

---

[24] A. Zwack, "The Impacts of Technology in the Canadian Workplace" prepared for the 2017 Midwinter Meeting of the Employment Rights and Responsibilities Committee American, Bar Association (Vancouver, BC: Glgmlaw, 2017).

[25] Statistics Canada, "Average usual and actual hours worked in a reference week by type of work (full-and part-time), annual" *Statistical Canada* (2019), online: https://www150.statcan.gc.ca/t1/tbl1/en/tv.action?pid=1410004301&pickMembers%5B0%5D=1.1&pickMembers%5B1%5D=3.1&pickMembers%5B2%5D=5.1&pickMembers%5B3%5D=6.6 (date accessed February 28, 2020).

[26] A. Zwack, "The Impacts of Technology in the Canadian Workplace" prepared for the 2017 Midwinter Meeting of the Employment Rights and Responsibilities Committee American, Bar Association (Vancouver, BC: Glgmlaw, 2017).

CH. 10: NEW TECHNOLOGY: IMPACT ON THE WORKPLACE

3. Corporate Security Issues

The use of mobile devices that house, or can remotely access, corporate information, introduces security risks and liabilities for the organization. This risk is heightened if the employee is using a personal device which can be used/accessed by others — family members, friends, social-media followers, hackers and/or co-workers. The practice of Bring Your Own Device (BYOD) has been pointed out to be a very risky practice for organizations to adopt.[27] This has been recognized by Telus and the Privacy Commissioner, Ontario who jointly created a white paper on the risks of BYOD.[28]

4. Data Safeguarding Issues

In addition to security risks, Canadian corporations are legally required to safeguard employee and client/customer information from destruction and unauthorized accessed or dissemination.[29] Failure to uphold this legal obligation can result in legal sanctions and fines, along with the potential negative impacts on the company's business reputation.[30]

5. Need for New Corporate Policies and Procedures

Based on the white paper on the use of personal devices developed by the Telus Corporation and the Privacy Commissioner of Ontario (2013),[31] companies must develop clear policies and procedures for the access and transmittal of corporate information, along with the consequences of their violation.

6. Big Data

With computers tracking and collating various forms of data obtained from iPhones, personal computers, Ring doorbells, Amazon Echos, credit cards, loyalty cards, body-tracking devices, Google, Facebook, and various apps, data can be measured and used to make sense of what is happening within a population and prepare for managing human behaviour. For example, mapping data can be used to shape buying patterns, plan for adequate supplies of consumer products, influence voting

---

[27] *Ibid.*

[28] A. Cavoukian & Telus, *BTOD: (Bring Your Own Device) Is Your Organization Ready?* Office of the Information and Privacy Commissioner, Ontario, Canada (2013), online: http://www.ontla.on.ca/library/repository/mon/27012/325215.pdf (date accessed February 28, 2020).

[29] A. Zwack, "The Impacts of Technology in the Canadian Workplace" prepared for the 2017 Midwinter Meeting of the Employment Rights and Responsibilities Committee American, Bar Association (Vancouver, BC: Glgmlaw, 2017).

[30] *Ibid.*

[31] A. Cavoukian & Telus, *BTOD: (Bring Your Own Device) Is Your Organization Ready?* Office of the Information and Privacy Commissioner, Ontario, Canada (2013), online: http://www.ontla.on.ca/library/repository/mon/27012/325215.pdf (date accessed February 28, 2020).

practices, detect weather threats, identify hot spots of political unrest, and quantify societal issues. Although technology is helpful, it can also be a threat to human privacy and well-being – a threat that society must grapple with.

7. Business Advantages

New technology has been shown to complement work tasks; streamline processes; enhance data accuracy and completeness; lend to greater creativity and innovation; expand markets; and increase productivity and profitability. Robots can work 24/7 without fatigue or incurring injuries or Workers' Compensation claims and costs. Automated processes can operate non-stop and be remotely managed. However, this eliminates many "entry-level" labour positions for workers and heightens the health risks for workers functioning alongside them.

8. Worker Benefits

"For some workers, digitalization is associated with increased pay and job resiliency."[32] In the high-digital occupations like computer and finance occupations, there has been job growth; likewise, in the low-digital occupations like personal care and food preparation. Not so in the middle-digital occupations, like administration and educational occupations.[33] Correspondingly, wage increases occurred in the high-digital occupations; but not in the middle- and low-digital occupations.[34]

Robots, which have been in the workplace for years, are often used for the tedious, dangerous, and dirty jobs that are physically demanding and high-risk for injury. This can improve the quality of life for workers, thereby freeing them up to undertake jobs that involve higher-order job skills.

E. NEW TECHNOLOGY: WORKPLACE HAZARDS

As with the introduction of any new process, product, or piece of equipment into the workplace, Canadian employers must identify, assess, control, and manage the related hazards (refer to Chapter 9, "Occupational Health and Safety: Hazard Identification, Assessment, and Control" for more details). To identify and address the related hazards, the following information that is based on current knowledge, is offered:

1. Automation: Associated Hazards

With automation, comes less employee physical activity and interaction with the work equipment. The noted hazards are:

- *Boredom and inattentiveness* – with reduced physical activity and comes

---

[32] A. Zwack, "The Impacts of Technology in the Canadian Workplace" prepared for the 2017 Midwinter Meeting of the Employment Rights and Responsibilities Committee American, Bar Association (Vancouver, BC: Glgmlaw, 2017) at 17.

[33] *Ibid.*, at 24.

[34] *Ibid.*

CH. 10: NEW TECHNOLOGY: IMPACT ON THE WORKPLACE

monotony. There is a strong association between inattentiveness and the occurrence of workplace incidents.[35]

- *Fear of job loss or displacement* – due to automation comes worker distress due to fear of job loss or displacement.[36]

2. Cellphones: Associated Hazards

Cellphones changed how people communicate with each other and the workplace; a boon in many ways; a curse in others. The recognized hazards are:

- *Distracted driving:* Research has repeatedly shown that cellphone use while driving is the number one cause of workplace fatalities. Additional distraction occurs due to text messaging, talking, and game-playing while operating a vehicle.[37]
- *Inattentiveness:* Cellphone distractions can negatively impact the employee's degree of spatial awareness, recognition of hazards, and operation of dangerous equipment.[38,39]
- *Incivility in the workplace*: Cellphones are associated with incivility in the workplace. This can range from the inconsiderate use of cellphones by visitors/clients/customers within the workplace, to worker misuse. Incivility ranges from eye-rolling, to sarcastic comments, to taunting, to racial/ethnic slurs, to intimidation, or to physically violent behaviours.[40]
- *Privacy violations:* Cellphones allow for data sharing, distribution, and

---

[35] J. Hewitt, "This is the hidden risk of automation that no one is talking about" *World Economic Forum* (2017), online: https://www.weforum.org/agenda/2017/11/automation-automated-job-risk-robot-bored-boredom-effort-fourth-industrial-revolution/ (date accessed February 28, 2020).

[36] Encyclopedia Britannica (2019). "Advantages and disadvantages of automation", online: https://www.britannica.com/technology/automation/Advantages-and-disadvantages-of-automation (date accessed February 28, 2020).

[37] OSHA, "Distracted Driving: No Texting" (2012) *OSHA 3416-09R*, online: https://www.osha.gov/Publications/3416distracted-driving-flyer.pdf (date accessed February 28, 2020).

[38] M. Lies & A. Young, "Cell Phones at the Workplace: Protecting Employee Safety", *OSHA Compliance* (October 2016), online: https://www.laborandemploymentlawcounsel.com/2016/10/cell-phones-at-the-workplace-protecting-employee-safety/ (date accessed February 28, 2020).

[39] N. Keith, "Forklift drivers convicted for cellphone use in warehouse", *Canadian Occupational Safety* (January 2019) at 1, online: https://www.cos-mag.com/ohs-laws-regulations/columns/forklift-drivers-convicted-for-cellphone-use-in-warehouse/ (date accessed February 28, 2020).

[40] S. Weinsier, "Workplace incivility and the connected workplace" presented at the 2019 *AAOHN Conference*, Jacksonville, FL, April 9, 2019.

photographing. This is a major risk for the employer and employees.[41]

- *Productivity Losses*: Worker use of the cellphone to make phone calls, surf the Internet, send emails, or play games, results in lost work productivity and constitutes a loss for the employer.

- *Musculoskeletal Injuries*: Heavy use of cellphones and other mobile devices is associated with injuries such as "text-neck", shoulder pain, "selfie" elbows, texting thumb,[42] *etc*. Do refer to Chapter 20, "Occupational Health & Safety: Ergonomics" for more details.

3. Robots: Associated Hazards

OSHA identifies seven main potential hazards to humans who are working with robots:[43]

- *Control errors* – a faulty control system, or software issues, or electromagnetic/radio frequency interference can result in erratic behaviour robotic behaviour and human injury.

- *Environmental sources* – sudden environmental causes of electromagnetic or radio-frequency interference, or power surges/loss, can influence the robot's performance.

- *Human errors* – can cause serious incidents because of faulty programming, interfacing, maintenance, or the incorrect activation of the "teach pendant" or control panel. Failing to recognize "line-of-fire" risks associated with the robot's motions can also lead to injury.

- *Improper installation* – can result in different hazards depending on the degree of variance from the design, installation requirements, and layout of equipment and utilities of the robot or automation system.

- *Mechanical failures* – tend to be unpredictable, dangerous, and result from the robot operating unexpectedly. These can result in a pinch injury or being physically struck by the robot.

- *Power systems* – pneumatic, hydraulic, or electrical power sources with malfunctioning control or transmission elements in the power system can be disrupted and lead to malfunctions. The result can be electrical shocks or fires due to electrical overloads.

---

[41] M. Graham, "Four potential hazards in the workplace" *SafetyPro Resources* (2014), online: https://www.safetyproresources.com/blog/four-potential-hazards-of-cell-phones-in-the-workplace (date accessed February 28, 2020).

[42] OH&S Canada, "Avoiding selfie elbows and texting thumbs" Dispatches *OHS Canada* (March/April 2019) at 15.

[43] OSHA, "Section IV: Chapter 4 Industrial Robots and Robot System Safety" *OSHA Technical Manual* (2019), online: https://www.osha.gov/dts/osta/otm/otm_iv/otm_iv_4.html (date accessed February 28, 2020).

- *Unauthorized access* – a worker unfamiliar with the area and the safety procedures is at high-risk for injury.

4. 3-D Printers: Associated Hazards

Once limited, the availability of 3-D Printers is now greater. From a hazard perspective, these desk-top 3-D Printers have been shown to emit small particles from the toner and chemicals that are associated with asthma.[44]

5. Magnetic Resonance Imaging (MRI): Associate Hazards

The nurses and technologists that work with MRIs have reported vertigo, nausea, dizziness, the presence of a metallic taste, and visual phosphenes.[45] This has been explored without any resolution or leads on how to address this hazard other than administration controls.

F. EMPLOYER RESPONSE

According to the OH&S legislation, the Canadian employer has a general duty clause to uphold, namely, to provide a safe and healthy workplace. With the emergence of new technology, employers are challenged to explore, understand, and assess the related hazards with a view to their management and effective control. However, given the newness to the technology, it also means, monitoring the clinical research on each new technology in regard to emergent hazards.

Based on the available recommendations, employers must:

1. Investigate the magnitude and aspects of the problem;
2. Develop policies, procedures, and practices to address the problem;
3. Control the hazard and monitor the effectiveness of the control measures;
4. Amend if required; and
5. Regularly review their actions.

G. SUMMARY

New technology will continue to emerge; in response, companies and organizations need to position themselves to respond and address the related workplace risks to workers and business endeavours. The tools provided in Chapter 9, "Occupational Health and Safety: Hazard Identification, Assessment, and Control" is available to assist with this effort.

CHAPTER REFERENCES

A. Cavoukian & Telus, *BTOD: (Bring Your Own Device) Is Your Organization*

---

[44] OSHA, "Control Measures for 3-D Printers" *OSHA* (2019), online: https://www.cdc.gov/niosh/research-rounds/resroundsv1n12.html (date accessed February 28, 2020).

[45] A. Gorlin, J. Hoxworth & J. Mueller, "Occupational Hazards of Exposure to Magnetic Resonance Imaging" (2015) 123:10 Anesthesiology 976-977, online: http://anesthesiology.pubs.asahq.org/article.aspx?articleid=2441365 (date accessed February 28, 2020).

Ready? Office of the Information and Privacy Commissioner, Ontario, Canada (2013), online: http://www.ontla.on.ca/library/repository/mon/27012/325215.pdf (date accessed February 28, 2020).

Dictionary, "Artificial Intelligence" *Dictionary* (2019), online, at p.1.

Encyclopedia Britannica, "Advantages and disadvantages of automation" (2019), online: https://www.britannica.com/technology/automation/Advantages-and-disadvantages-of-automation (date accessed February 28, 2020).

Futureofworking.com (2017). *7 Biggest pros and cons of technology*, online: https://futureofworking.com/7-biggest-pros-and-cons-of -technology (date accessed February 28, 2020).

Gartner IT, *Glossary IT: Digitalization* (2019), online: https://www.gartner.com/it-glossary/digitalization/ at 1, (date accessed February 28, 2020).

M. Graham, "Four potential hazards in the workplace" *SafetyPro Resources* (2014), online: https://www.safetyproresources.com/blog/four-potential-hazards-of-cell-phones-in-the-workplace (date accessed February 28, 2020).

N. Harris-Briggs, "Is a virtual workplace really possible?" (2018), online: https://martechseries.com/mts-insights/guest-authors/vitrual-workplace-really-possbile/ (date accessed February 28, 2020).

J. Hewitt, "This is the hidden risk of automation that no one is talking about" *World Economic Forum* (2017), online: https://www.weforum.org/agenda/2017/11/automation-automated-job-risk-robot-bored-boredom-effort-fourth-industrial-revolution/ (date accessed February 28, 2020).

L. Johnson, "Gas Up" *Canadian Occupational Safety* (2019), online: www.cos-mag.com at 20-21.

N. Keith, "Forklift drivers convicted for cellphone use in warehouse" *Canadian Occupational Safety* (January 2019) at 1, online: https://www.cos-mag.com/ohs-laws-regulations/columns/forklift-drivers-convicted-for-cellphone-use-in-warehouse/ (date accessed February 28, 2020).

S. Lazaruk, "Automated future: Computer and robotics already changing retail and the workplace" *Vancouver Sun* (January 28, 2017), online: https://vancouversun.com/author/susanlazarkukprov (date accessed February 28, 2020).

M. Lies & A. Young, "Cell Phones at the Workplace: Protecting Employee Safety" *OSHA Compliance* (October 2016), online: https://www.laborandemploymentlawcounsel.com/2016/10/cell-phones-at-the-workplace-protecting-employee-safety/ (date accessed February 28, 2020).

M. McKay. Reported in *EHSToday*, "Injuries Sustained While Riding Segway Transporter 'Severe' " (September 23, 2010), online: https://www.ehstoday.com/safety/management/injuries-sustained-riding-segway-severe-0923 (date accessed February 28, 2020).

M. Muro, R. Maxim & J. Whiton, "Automation and Artificial Intelligence: How machines are affecting people and places" *Brookings* (2019), online: https://www.brookings.edu/wp-content/uploads/2019/01/2019.01_BrookingsMetro_Automation-AI_Report_Muro-Maxim-Whiton-FINAL-version.pdf (date accessed February 28, 2020).

National Geographic, *The Future of Medicine*, January 2019 issue, at 27-35; 56-57.

OH&S Canada, "Avoiding selfie elbows and texting thumbs" Dispatches *OHS Canada* (March/April 2019) at 15.

OSHA, "Section IV: Chapter 4 Industrial Robots and Robot System Safety, *OSHA Technical Manual* (2019), online: https://www.osha.gov/dts/osta/otm/otm_iv/otm_iv_4.html (date accessed February 28, 2020).

OSHA, "Control Measures for 3-D Printers" *OSHA* (2019), online: https://www.cdc.gov/niosh/research-rounds/resroundsv1n12.html (date accessed February 28, 2020).

OSHA, "Distracted Driving: No Texting", (2012) *OSHA 3416-09R*, online: https://www.osha.gov/Publications/3416distracted-driving-flyer.pdf (date accessed February 28, 2020).

H. Siu (2019). Reported in S. Lazaruk, "Automated future: Computer and robotics already changing retail and the workplace" *Vancouver Sun* (January 28, 2017), online: https://vancouversun.com/author/susanlazarkukprov (date accessed February 28, 2020).

J. Vomiero, "Half of Canadian jobs will be impacted by automation in the next 10 years", *Global News* (March 26, 2018), online: https://globalnews.ca/4105713/automation-workforce-canada-human/ (date accessed February 28, 2020).

Wikipedia, "Robotics: definition" *Wikipedia* (2020), online: https://en.wikipedia.org/wiki/Robotics (date accessed February 28, 2020).

Wikipedia, "3-D Printers" *Wikipedia* (2019), online: https://en.wikipedia.org/wiki/3D_printing (date accessed February 28, 2020).

A. Zwack, "The Impacts of Technology in the Canadian Workplace" prepared for the 2017 Midwinter Meeting of the Employment Rights and Responsibilities Committee American, Bar Association (Vancouver, BC: Glgmlaw, 2017).

Chapter 11

OCCUPATIONAL HEALTH AND SAFETY: WORKPLACE STANDARDS AND RULES

A. INTRODUCTION

Leading edge organizations/companies manage their workplace health and safety obligations by creating an Occupational Health and Safety Management System (OHSMS) that sets the OH&S practice/performance standards so:

- workplace hazards are identified, assessed, and controlled;
- safe work procedures are developed, implemented, and complied with;
- worker OH&S training is provided and evaluated;
- everyone knows they will be held accountable for their actions; and
- provisions for continuous improvements are made.[1]

Practice/performance standards are stated approaches to practice/performance based on industry-recognized and accepted principles of practice for planned processes. They form the guidelines and rules for work practices, provide the boundaries for work activities, clarify stakeholder roles and responsibilities, and serve as a benchmark.

In any organization/company, there are two types of practice/performance standards: the formal and the informal versions. **Formal practice/performance standards** are the documented organizational/company OH&S expectations, rules, and safe work practices. They are evident in OH&S Program manuals, employee handbooks, training materials, and OH&S communication materials. They form the structure of the organization's/company's OH&S Program, practices, and safe work procedures.

The **informal OH&S practice/performance standards** are what get done in the workplace when the employee thinks "no one is watching". These are the actual practices/work performances that go on within the field, worksites, and offices. In terms of workplace safety, they are the greatest predictor of workplace OH&S outcomes.

---

[1] WSIB/CSPAAT, "Business Results Through Health & Safety" (2001), online: http://www.mtpinnacle.com/pdfs/Biz.pdf (date accessed: February 28, 2020).

The greater the gap between the formal and informal work practice/performance standards, the weaker the OHSMS and the higher the potential for failures. The intent of this chapter is to discuss workplace standards and rules.

B. WORKPLACE STANDARDS AND RULES: IMPORTANCE

Workplace OH&S standards and rules are the evidence of an organization's/ company's commitment to their OH&S duty of care and due diligence in action. Good leaders monitor their own work performance and that of the people who work with them. Key to this is being able to measure OH&S performance in quantifiable, objective terms.

Workplace OH&S practice/performance standards establish corporate values and expectations. They must be specific, clear, and aligned with OH&S government and industry standards. They are the "bar" against which organizational/company, Management and employee OH&S performance are measured. According to Bird (1996), Management control is dependent on regular performance evaluation — evaluation is dependent on the existence of performance standards against which to measure performance.[2]

Workplace OH&S standards are a test of performance. They enable continuous improvement through the identification of gaps between the observed and the desired OH&S levels of performance. Often these performance gaps are the reason workplace incidents occur. By identifying them, action can be taken to prevent unnecessary losses.

Practice/performance standards are beneficial because they:

- promote a consistent approach to OH&S practices;
- provide meaningful direction to the practice in question; and
- promote effectiveness and efficiency of practice through a reduction of errors, complications, and costs.

To remain current and credible, practice/performance standards must be regularly reviewed, updated, and enforced.

C. WORKPLACE STANDARDS AND RULES: EFFECTIVE COMMUNICATION TECHNIQUES

Employers are legally responsible for communicating workplace practice/ performance standards and rules to employees and other workers. The typical vehicles for doing this are:

- **Management Communication and Behaviours** — Organizational/ company leaders set the practice/performance standards and therefore should be the role models of the desired OH&S behaviours. For employees, the

---

[2] F. Bird & G. Germain, *Loss Control Management: Practical Loss Control Leadership*, revised ed. (Loganville, GA: DNV, 1996) at 34.

impact of what they observe their leaders demonstrate far outweighs what they hear their leaders say.

Transformational leaders[3] behave as OH&S role models. They are respected because they do what is right for the right reasons, and not because it is easier or more profitable. They value workplace health, safety, and the well-being of workers and inspire workers to work safely.

Supervisors, who have great influence on worker behaviour, must demonstrate the desired OH&S behaviours if the organization expects workers to do likewise. When supervisors actively monitor and recognize/reward sound OH&S practices/performance, their subordinates engage in more safety-related behaviours.[4] This results in positive OH&S outcomes.

Leading-edge organizations recognize the importance of having supervisors that demonstrate the transformational leadership qualities. In addition to eliciting positive OH&S performance and outcomes, transformational leadership is associated with stronger employee loyalty to the organization, increased productivity, lower employee absenteeism, and staff turnover.[5]

- **Management Theories and Practices** — Management theories and practices regarding workplace health and safety are an extension of the organization's/company's vision, mission, culture, and values. They are concrete evidence of Management's commitment to Occupational Health and Safety. Organizations/companies that truly care about the health, safety, and well-being of its workers subscribe to human resource theories and practices that support and demonstrate those beliefs. In Chapter 18, "Occupational Health and Safety: Best Practices", the best practices of developing a high-performance work system and strong safety culture are discussed. It includes the top ten ways to create high-performance work systems and OH&S cultures. The most critical factor is that the management theories and practices used must promote employee trust in Management's leadership.

- **OH&S Communication** — Organizations/companies should as part of their OHSMS, include an OH&S communication strategy (Chapter 16, "Marketing Occupational Health and Safety Programs; Communicating the Results"). The OH&S communication strategy defines the corporate OH&S message, how it is to be delivered, to whom, through what means and by

---

[3] **Transformational leaders** — refer to Chapter 1, "Occupational Health and Safety: Historical Perspectives", for information on transformational leadership and leaders.

[4] D. Hofmann & F. Morgeson, Chapter 8, "The Role of Leadership in Safety" in J. Barling & M. Frone, eds., *The Psychology of Workplace Safety* (Washington, DC: American Psychological Association, 2004) at 161.

[5] *Ibid.*, at 171.

when. This powerful tool is dynamic and should be carefully monitored to ensure it aligns well with the organization's/company's OH&S standards.

Some industry-proven OH&S communication techniques include:

- the practice of conducting Chief Executive Officer (CEO) "coffee-pot" sessions — a forum for getting direct feedback on OH&S issues from employees. To get feedback on the organization's/company's OH&S performance, CEOs are advised to ask the employees who do the work and who have the most to lose should the OHSMS fail them;
- the practice of having the CEO welcome new employees and state the importance of workplace health and safety, as well as upholding organization's/company's OH&S standards during the New Employee Orientation;
- implementation of Management worksite tours/visits designed to show respect for employees and the work they do for the organization;
- the practice of educating supervisory staff on the importance of regularly communicating OH&S standards;
- the use of an OH&S website to provide employees easy access to OH&S standards and rules, and the tools to uphold them;
- the development of brochures/written statements explaining the importance of, and supporting the use of OH&S practice/ performance standards;
- the use of OH&S promotional events such as OH&S fairs/expos, special Safety days, OH&S instructional games, celebration of the organization's/company's OH&S successes, *etc.*;
- the distribution of items like pens, notebooks, truck visors, reminder cards, coffee mugs, calendars, *etc.*, that reinforce messages related to the organization's/company's OH&S standards;
- posting of OH&S information on bulletin boards, walls, or electronic boards;
- issuing Safety Bulletins or Alerts when OH&S hazards or unsafe conditions are identified;
- regularly presenting OH&S educational sessions at Safety Meetings;
- requiring supervisory staff to reinforce OH&S practice/ performance standards and rules at pre-job meetings;
- informally teaching workers about OH&S standards and rules whenever possible.

- **New Employee Orientation** — When an employee begins a new job, employers have a legal obligation to ensure that OH&S practices/performance standards are fully explained and understood.
- **Initial Job Training** — New employees are impressionable. The "talk and walk" of their direct supervisor must align with what has been communicated in the New Employee Orientation. When it does, the OH&S messages come through loud and clear. This reinforces the point made earlier that it is important to ensure that supervisors understand and are willing to communicate the organization's/company's OH&S practice/performance standards and rules.
- **OH&S Training** — Training sessions are prime opportunities to ensure that employees have access to, and awareness of, the organization's/company's OH&S standards and rules. By discussing them, by explaining why they exist, by providing examples of the relevancy of the rule to their work, trainers can operationalize these standards/rules. They can test the employee's understanding and practice of such, reinforcing the importance of these standards and the organization's/company's expectations. Trainers can also use this opportunity to demonstrate the desired OH&S performance/practice standards and rules. For example, during an off-site training session, the OH&S Trainer should ensure that the vehicle being used is safely operated; that means, demonstrating safe driving practices, strict observance of traffic laws, use of seat belts, avoidance of cell phone use while driving, *etc*.
- **Worker Annual Performance Reviews** — This is a prime time for supervisory staff to reinforce the organization's/company's commitment to OH&S, and the values it upholds. By comparing the organization's/company's OH&S practice/performance standards against the employee's demonstrated work and OH&S performance, the supervisor can identify areas of success and areas for improvement.
- **Rule Changes** — When new workplace rules or rule changes occur, the employer is legally required to inform workers of those developments. In addition to being an "update", this is an opportune time to reaffirm Management's commitment to workplace health and safety, as well as to clarify the expectations that:
    - everyone is responsible and accountable for workplace health and safety;
    - this new/revised rule must be upheld; and
    - that any noted shortfalls with the new rule should be communicated to Management.

D. WORKPLACE STANDARDS AND RULES: INDUSTRY EXAMPLES

The following are industry samples of workplace standards and rules that are provided in Chapter 5, "Occupational Health and Safety Program: Manual Development":

- **NON-SMOKING WORKPLACE POLICY** — refer to Chapter 5, Sample Program Manual, Section 3, Related Policies and Procedures, Part 3.
- **WHMIS 2015** — refer to Chapter 5, Sample Program Manual, Section 3, Related Policies and Procedures, Part 4.
- **SUBSTANCE ABUSE POLICY** — refer to Chapter 5, Sample Program Manual, Section 3, Related Policies and Procedures, Part 2.
- **GENERAL SAFETY RULES** — refer to Chapter 5, Sample Program Manual, Section 5, Occupational Safety Program, Part 1.
- **WORK EQUIPMENT SAFETY** — refer to Chapter 5, Sample Program Manual, Section 5, Occupational Safety Program, Part 2.
- **ACCESS EQUIPMENT SAFETY** — refer to Chapter 5, Sample Program Manual, Section 5, Occupational Safety Program, Part 3.
- **CRANE AND HOIST SAFETY RULES** — refer to Chapter 5, Sample Program Manual, Section 5, Occupational Safety Program, Part 4.
- **ELECTRICAL EQUIPMENT SAFETY RULES** — refer to Chapter 5, Sample Program Manual, Section 5, Occupational Safety Program, Part 5.
- **MANUAL HANDLING SAFETY RULES** — refer to Chapter 5, Sample Program Manual, Section 5, Occupational Safety Program, Part 8.
- **PERSONAL PROTECTIVE EQUIPMENT SAFETY RULES** — refer to Chapter 5, Sample Program Manual, Section 5, Occupational Safety Program, Part 10.
- **WORKING ALONE POLICY** — refer to Chapter 5, Sample Program Manual, Section 5, Occupational Safety Program, Part 11.
- **WORKPLACE VIOLENCE POLICY** — refer to Chapter 5, Sample Program Manual, Section 5, Occupational Safety Program, Part 12.
- **INJURY RESPONSE FIRST AID POLICY** — refer to Chapter 13, "Occupational Health and Safety: Injury Management".
- **EMERGENCY RESPONSE PLAN** — refer to Chapter 12, "Occupational Health and Safety: Emergency Preparedness and Response", Part D.
- **CONFIDENTIALITY OF HEALTH INFORMATION POLICY** — refer to Chapter 4, "Occupational Health Program".
- **PSYCHOLOGICAL HEALTH & SAFETY STANDARD POLICY** — refer to Chapter 23 "Psychological Health and Safety: In the Workplace", Chapter 24 "Psychological Health and Safety: Practical Application in the Workplace" and Chapter 24 "Psychological Health and Safety in the Workplace: Measurement".

E. **OH&S STANDARDS AND RULES: TECHNIQUES FOR WORKER CONFORMANCE**

Compliance with an organization's/company's OH&S practice/performance stan-

dards and rules can be promoted by the following principles:
- develop and implement clear policies, standards, and rules that align with the organization's/company's vision, mission, and values;
- educate workers on the organization's/company's policies, standards, and rules;
- review the importance of the organization's/company's policies, standards, and rules;
- explain the consequences for non-conformance;
- consistently enforce the policies; and
- document actions taken.

F. OH&S STANDARDS AND RULES: MEASUREMENT FOR COMPLIANCE[6]

Compliance with an organization's/company's OH&S practice/performance standards should be measured at many levels. The critical questions to be answered are:
- Does Senior Management uphold the OH&S standards that they set?
- Does Line Management uphold the OH&S standards that they have been assigned to uphold?
- Do employees observe the OH&S standards as per management's stated expectations?
- How does Management respond when OH&S standards are violated?

Compliance measurement can take on many forms, but at a minimum it should involve:
- A **Review of Management OH&S Practices/Communications** — Comparing what Management does against what it states it values and wants done;
- **Regular Monitoring of Employee Performance** — Observing worker behaviours and comparing them against the expected OH&S practice/performance standards and rules;
- **Worksite Inspections** — Monitoring worksite conditions and employee behaviours through a regular examination of the worksite, the presence and condition of the required equipment, and the practices of the employees present;
- **Worker Perception Surveys** — Measurement of employee perception of the organization's/company's OH&S leadership, practice/performance standards, and status of workplace safety;

---

[6] Editorial, "Recommendations to Adopt and Enforce Appropriate Standards of Workplace Behaviour" *Workplace News* (March/April Issue, 2007) at 9.

- **Regular OH&S Performance Monitoring and Reporting** — Daily/weekly/monthly measurement and reporting of OH&S performance through the OH&S performance measurement systems implemented;
- **OHSMS/PHSMS Compliance Audit** — Regular auditing of the existing OHSMS against a recognized OHSMS standard such as ISO 45001/Z45001-19; OHSAS 18001/02; ANSI/AIHA Z10; ILO-OSH 2001; CSA Z-1000-06; Alberta Partnerships Audit Protocol; Psychological Health and Safety Standard; *etc.*; and
- **Incident Investigations** — The findings from an incident investigation can be a measure of the OHSMS, Management, and/or employee level of compliance with the organization's/company's OH&S practice/performance standards and rules.

G. SUMMARY

OH&S practice/performance standards and rules are part of the management system for ensuring that the organization's/company's vision, mission, culture, and values are operationalized. Legally required, these OH&S practice/performance standards and rules are the tangible evidence of the organization's/company's OH&S duty of care, and that reasonable steps are being taken to protect workers. However, simply having OH&S practice/performance standards and rules is not enough. CEOs need to monitor and measure the level of compliance with the organization's/company's OH&S practice/performance standards and rules. Organizational/company leaders need to be able to demonstrate that they know:

- the organization's/company's OH&S practice/performance standards and rules;
- what is actually going on in terms of the degree of observance of OH&S practice/performance standards and rules within the organization;
- what their OH&S Practitioners/Professionals are doing;
- that worker OH&S training is being provided and whether it is effective/ineffective;
- what the contractors and contingent workers are doing; and
- that evidence (records) exists supporting the organization's/company's OH&S due diligence.

CHAPTER REFERENCES

F. Bird & G. Germain, *Loss Control Management: Practical Loss Control Leadership*, revised ed. (Loganville, GA: DNV, 1996).

Editorial, "Recommendations to Adopt and Enforce Appropriate Standards of Workplace Behaviour" *Workplace News* (March/April Issue, 2007).

D. Hofmann & F. Morgeson, Chapter 8, "The Role of Leadership in Safety" in J. Barling & M. Frone, eds., *The Psychology of Workplace Safety* (Washington, DC:

American Psychological Association, 2004).

WSIB/CSPAAT, "Business Results Through Health & Safety" (2001), online: http://www.mtpinnacle.com/pdfs/Biz.pdf (date accessed: February 28, 2020).

Chapter 12

OCCUPATIONAL HEALTH AND SAFETY: EMERGENCY PREPAREDNESS AND RESPONSE

A. INTRODUCTION

Organizations/companies strive to eliminate or reduce risks that may arise from the nature of work and work activities to an acceptable level. Despite measures to manage workplace risks, emergency situations tend to occur. In preparation, companies must demonstrate their emergency preparedness and develop Emergency Response Plans (ERPs). To quote Dwight D. Eisenhower: "Plans are nothing, planning is everything."

Legally, employers are required to develop plans designed to deal with emergency situations that might occur within a workplace. This includes responses to fire, explosions, bomb threats, chemical emissions, incident situations, natural disasters, and such. The goal is for the organization to ensure that its people are prepared to respond to the emergency in a manner that prevents or mitigates worker injury/illness and loss.

According to the Canadian Centre for Occupational Health & Safety (CCOHS), the objective of emergency response planning is to reduce the potential consequences of an emergency by:

- containing the emergency situation;
- preventing worker/public fatalities and injuries;
- reducing damage to buildings, stock, and equipment; and
- accelerating the resumption of normal operations.[1]

Emergency Response Plans provide specific instructions for responding to, and dealing with, emergencies. They must be well communicated and regularly tested to ensure that workers are able to respond as planned. All users of the ERP must be familiar with the contents of the plan, as well as their individual responsibilities in an emergency situation. Employees assigned specific duties within the ERP must be

---

[1] CCOHS, "Emergency Planning, Health & Safety Programs" *OSH Answers*, online: http://www.ccohs.ca/oshanswers/hsprograms/planning.html (date accessed: February 28, 2020).

trained and kept familiar with the related hazards and response equipment. The ERP must be reviewed and updated annually, and as required.

B. EMERGENCY RESPONSE PROGRAMS AND PLANS: VALUE

Besides providing guidance during an emergency, developing the ERP has other advantages. During the design and planning phase, unrecognized hazardous conditions that could aggravate an emergency situation may be identified, allowing planners to integrate their elimination into the ERP. The planning process may "bring to light" operational deficiencies, such as the lack of adequate resources (equipment, trained personnel, or supplies), or workplace situations that need to be rectified before an emergency occurs. In addition, an ERP promotes safety awareness and demonstrates Management's leadership and commitment to the health and safety of workers. This is a due diligence requirement for employers.

Other benefits of an ERP are:

- An effective ERP can prevent severe losses such as multiple worker/ public casualties and the possible financial collapse of the organization.
- An effective ERP demonstrates that organizational/company leaders are upholding their duty of care.
- ER planning forces people to face the fact that disasters can happen and that they need to be prepared to manage them.
- ERP drills reinforce worker awareness of the potential for disasters and the responses required.
- ERP procedures should be used in the orientation of new employees, worker training, and in practice sessions to develop the response skills of employees to an emergency situation.
- ERP planning and drills enable responders to respond quickly and efficiently.
- ERP drills can identify shortfalls such as the lack of resources, worker training deficits, response time issues, the need for external supports, lack of equipment, communication problems, and the like.
- ERP planning and drills tend to reinforce an organization's/company's message that all personnel are responsible for reporting and initiating a response to any hazardous condition (*e.g.*, fire, explosion, spill/release of chemicals or gas, workplace violence, medical emergency, or natural disaster).
- ERP planning can identify potential impact to the environment, community, and public.

C. EMERGENCY RESPONSE PLAN: DESIGN

To design a comprehensive ERP, planners should begin by conducting a needs assessment or vulnerability assessment, to help them determine:

- How likely is a situation to occur?
- What means are available to stop or prevent the situation?
- What actions are necessary for a given emergency situation?[2]

Based on the results, appropriate emergency plans, procedures, and external supports can be established.

As part of the Needs Assessment, it is critical to determine which internal and external groups should be involved in the ERP design and planning. Typically, stakeholders such as operational leaders, technical experts, OH&S Practitioners/Professionals, risk management professionals, environmental professionals, Joint Health and Safety Committee (JHSC) members, and security professionals can provide valuable input and a means of wider worker involvement.

The best ERPs include employees in the planning process, specify what employees should do during an emergency, and ensure that employees receive proper training for emergencies. Employees are excellent organizational/company resources in terms of offering suggestions about potential hazards, worst-case scenarios, and proper emergency responses.

Since many emergencies involve or impact external agencies, appropriate municipal officials should also be consulted.

1. Vulnerability Assessments

Companies need to examine the potential threats that exist for them, their employees, and their work facilities. Although emergencies by definition are "sudden events", their occurrence can be predicted with some degree of certainty.

(a) Determine Which Hazards Pose a Threat to any Specific Enterprise

Create an inventory of actual and potential hazards based on worker hazard reports, the organization's/company's incident history, worksite evaluation, recognized external hazards, and known technological and natural hazards. Some well-recognized hazards are:

- fire
- explosion
- building collapse
- major structural failure
- spills of flammable liquids
- accidental release of toxic substances
- deliberate release of hazardous biological agents or toxic chemicals
- computer sabotage

---

[2] *Ibid.*

- airplane crashes
- train derailments
- terrorist activities
- exposure to radiation
- electrical power failures
- loss of water supply
- communication failures
- environmental aspects
- industry-specific incidents.[3]

Natural hazards include:

- tornadoes
- hurricanes
- other severe windstorms
- fallen trees
- floods
- earthquakes
- snow or ice storms
- severe extremes in temperature (cold or hot)
- pandemic diseases

By no way exhaustive, these two lists provide "food for thought" regarding hazards that could potentially impact Canadian and U.S. companies.

The possibility of one disaster event triggering others must be considered. For example, a fire may result in a chemical release and lead to structural failures; while a tornado might initiate a number of the events noted in the list of chemical and physical hazards.

(b) For Each Potential Emergency Situation, Determine the Associated Risk

Assess each emergency situation in terms of the severity of the emergency event, the potential for the emergency situation to occur and the resultant risk. Using the definitions for severity, potential and risk as presented in Chapter 9, "Occupational Health and Safety: Hazard Identification, Assessment, and Control", Tables 9.1, 9.2, and 9.3; Table 12.1 provides examples of how an organization/company can evaluate and prioritize potential emergency situations.

---

[3] *Ibid.*

Table 12.1: Organization/Company Needs Assessment/Vulnerability Assessment (Sample)

Emergency Situation	Injury Severity	Environmental Severity	Business Severity	Potential	Risk	Priority
Fire	High	Minor	High	C	4	High
Explosion	Extreme	Medium	High	C	5	Top
Building Collapse	High	Medium	High	B	3	Medium
Structural Failure	High	Medium	High	A	1	Lowest
Chemical Spill	Low	Extreme	High	D	5	Top
Toxic Substance Release	Low	Extreme	High	B	3	Medium
Deliberate Release of Toxic Agents	High	Medium	Low	A	1	Lowest
Computer Sabotage	Low		Extreme	D	5	Top
Airplane Crash	Extreme	High	Extreme	A	2	Low
Train Derailment	High	High	Extreme	A	2	Low
Exposure to Radiation	Low		Medium	A	1	Lowest
Electrical Power Failure	Minor		Medium	C	3	Medium
Loss of Water Supply	Minor		Medium	C	3	Medium
Communication Failures	Minor		High	B	3	Medium
Environmental Aspects	Minor	Extreme	High	D	5	Top
Industry-specific Incidents	Extreme	Medium	High	D	5	Top

This type of examination enables organizational/company leaders to identify which emergency situations warrant their attention. Based on the above example, developing ERPs for explosion, computer sabotage, environmental aspects, and industry-specific incidents would be the most critical, followed by plans to deal with fire and other potential emergencies.

(c) For Each Potential Emergency Situation, Identify the Potential Impact, and Develop Control Plans

One way to undertake this task is to use the "What if" approach. For each potential emergency situation, ask "what if" an explosion was to happen? What would be the sequential events? What would be the likely damage — to plant, equipment, and environment? What would be the expected number of casualties? How many people could get injured? How would evacuation occur? What is the expected loss of business documentation? What would be the anticipated length of business disruption? Could legal liabilities ensue? What impact could the event have on the organization's/company's reputation or image?

The possible major impacts of each emergency situation should be itemized, such as:

- sequential events (for example, fire after explosion);
- damage to plant infrastructure, equipment, and environment;
- casualties (number and nature — that is, fatalities, serious injuries, and minor injuries, as well as the type of injuries, like fractures, burns, debris impalement, falls from heights, asphyxiation, and such);
- evacuation (nature of evacuation required — that is, self-enacted by employees, facilitated by rescue teams, remote rescue, air rescue, sea rescue, and such);
- business losses (loss of product, vital records/documents, and equipment);
- disruption of work;
- legal liabilities; and
- organization's/company's reputation/image.

By examining each step of the event using the "what if" approach, planners can identify the potential impact of a disaster situation (Table 12.2).

Table 12.2: Potential Impact of Disaster Situations

Disaster Type	Sequential Events	Damage (infrastructure, equipment, environment)	Casualties (number, nature)	Evacuation (nature, necessary provisions)	Business Disruptions	Business losses (product, records, equipment)	Legal Liabilities	Impact to Company Reputation
Fire	• Explosion of oil tanks • Collapse of catwalks and other related structures • Chemical emissions	• Tank damage • Infrastructure collapse • Destruction of equipment • Air pollution • Ground pollution	*Number:* • 10 injuries • 2 deaths *Nature:* • Asphyxiation • Impact injuries • Burns • Falls from heights • Strains and sprains	*Nature:* • Self-enacted • Assisted rescue *Provisions:* • Fire-fighting equipment and in-house personnel • Spill containment kits • Fire-arrest systems	• Shut down for nine months of repairs • No product sales	• Loss of product for nine months • Loss of equipment • Destruction of production, maintenance and training records • Loss of business income	• OH&S infractions • Potential criminal charges under *Criminal Code* s. 17 (Bill C-45) (*Act to amend the Criminal Code criminal liability of organizations*), S.C. 2003, c. 21) • Potential environmental penalties	• Negative press • Negative image

(d) For Each Emergency Situation, Determine the Necessary Response Actions

Although there will be differences with each type of emergency situation, there are some common response actions, for example:

- declare an emergency;
- sound an alert;
- evacuate the danger zone;
- close the relevant main shutoffs;
- call for external help (fire, police, emergency medical services, hazardous spill clean-up, *etc.*);
- initiate rescue operations;
- attend to casualties; and
- control the associated hazards (fire) and energy sources.[4]

2. Elements of an Emergency Response Plan

An ERP describes all the possible emergencies, consequences, required actions, written procedures, and the resources available. It provides the actions employees should take to ensure their safety if a fire or other emergency situation occurs. In most circumstances, immediate evacuation is the best policy, especially if professional firefighting services are available to respond quickly. There may be situations where employee firefighting is warranted to give other employees time to escape, or to prevent danger to others by spread of a fire.

A basic ERP (Appendix 1) is adequate for offices, small retail shops and small manufacturing settings where there are few or no hazardous materials/processes, and employees can readily evacuate when alarms sound or when notified by public address systems. More complex plans are required in workplaces containing hazardous materials or workplaces where employees fight fires, perform rescue and medical maneuvers, or delay evacuation after alarms sound to shut down critical equipment.

It is essential that the ERP be site-specific with respect to emergency conditions evaluated, evacuation policies and procedures, emergency reporting mechanisms, and alarm systems.

As part of emergency preparedness and planning, the organization should:

- determine the conditions under which an evacuation would be necessary and when shelter-in-place is warranted;

---

[4] *Ibid*; see also U.S. Department of Labor, "OSHA Principle Emergency Response and Preparedness: Requirements and Guidelines" (OSHA 3122-06R - 2004), online: https://www.osha.gov/Publications/osha3122.pdf (date accessed: February 28, 2020).

- define personnel roles, lines of authority and communication, and site security and control;
- create a list of emergency responders and emergency contacts;
- develop a written ERP that includes medical and emergency alert procedures;
- use an incident command system;
- dedicate adequate response resources and equipment;
- provide employee orientation to, and ongoing training on, the ERP;
- undertake regular ERP drills;
- evaluate the success of the ERP drills and identify areas for improvement; and
- review the ERP on a regular and "as needed" basis.

When writing an ERP, it is important to define:

- a clear chain of command and designation of the person authorized to order an evacuation or shutdown;
- the emergency reporting procedures such as dialing 911, or an internal emergency number or pulling a manual fire alarm;
- the alarm system to be used to notify employees (including disabled employees) to evacuate and/or take other actions. The alarms used for different actions should be distinctive and might include horn blasts, sirens, or even public address systems;
- the procedures for sheltering-in-place;
- the procedures for employees who remain on-site after the evacuation alarm sounds, if required, before evacuating. Employees may be required to operate fire extinguishers or shut down gas and/or electrical systems and other special equipment that could be damaged if left operating or create additional hazards to emergency responders (such as releasing hazardous materials);
- the specific evacuation procedures, including escape routes and exits. These include the actions employees should take before and while evacuating, such as shutting windows, turning off equipment, and closing doors behind them. Exit diagrams are typically used to identify the escape routes to be followed by employees from each specific facility location;
- the specific evacuation procedures for high-rise buildings;
- the procedures for assisting visitors and employees to evacuate, particularly those with disabilities or who do not speak English;
- the procedures to account for employees after the evacuation to ensure that everyone got out. This might include procedures for designated employees

to sweep areas and check offices and rest rooms, before being the last to leave a workplace or conducting a roll call in the assembly area. Many employers designate a Fire Warden to assist others in an evacuation and to account for personnel;
- the designation of which employees, if any, that will remain after the evacuation alarm to shut down critical operations or perform other duties before evacuating;
- a means of accounting for employees after an evacuation;
- the special equipment (PPE) for employees; and
- the appropriate respirators.

As well, it is vital to identify, document, and locate needed resources such as:
- facility/plant/floor maps;
- large scale maps showing evacuation routes and service conduits (such as gas and water lines);
- a list of trained personnel including their contact phone numbers, duties and responsibilities;
- emergency contact personnel;
- access to ERP documents;
- medical supplies;
- auxiliary communication equipment;
- power generators;
- respirator equipment;
- chemical and radiation detection equipment;
- mobile equipment;
- emergency protective clothing;
- firefighting equipment;
- ambulance or ambulance services; and
- rescue equipment.

Since a sizable document will likely result, the plan should provide specific staff members with individualized written instructions about their specific emergency duties.

The following are descriptions of the various ERP elements, which are presented as guidelines:

- **The Objective**

    The objective is a brief summary of the purpose of the plan — for example, *to reduce human injury and damage to property in an emergency*. It also specifies those staff members who may put the plan into action. The objective identifies clearly who these staff members are since the normal chain of command cannot always be available on short notice. At least one of them must always be on the site when the premises are occupied. The extent of authority of these personnel must be clearly indicated.

- **Coordination of the Emergency Response**

    One individual should be appointed and trained to act as Emergency Coordinator as well as a "back-up" coordinator. However, personnel on the site during an emergency play an essential role in ensuring that prompt and efficient action is taken to minimize loss. In some cases, it may be possible to recall off-duty employees to help, but the critical initial decisions usually must be made immediately by those onsite.

- **Duties, Responsibilities, Authority, and Resources**

    The duties, responsibilities, authority, and resources for the ERP must be clearly defined and assigned:

    - reporting the emergency;
    - activating the emergency plan;
    - assuming overall command of the ERP response;
    - establishing ERP communication;
    - alerting staff;
    - ordering evacuation or shelter-in-place;
    - alerting external agencies;
    - confirming the evacuation is complete;
    - alerting the community of possible risk;
    - requesting external aid;
    - coordinating the activities of various groups;
    - providing medical aid;
    - advising the families of the casualties;
    - ensuring emergency shutoffs are closed;
    - sounding the all-clear signal; and
    - advising the media.

    This list of responsibilities should be completed, considering that in organizations that operate on reduced staff during some shifts, some personnel must assume extra responsibilities during emergencies. Sufficient alternates for each responsible position must be named to ensure that someone with authority is available on-site at all times.

External organizations that may be available to assist (with varying response times) include the:

- fire department;
- emergency medical services;
- hazardous materials clean-up service;
- police department;
- telephone company;
- hospital(s);
- utility companies;
- industrial neighbours (mutual aid); and
- government agencies.

Pre-planned coordination is necessary to avoid conflicting responsibilities. For example, the police, fire department, emergency medical service, company firefighters, and the company first aid responders may be on the scene simultaneously. A predetermined chain of command in such a situation is required to avoid organizational difficulties. Under certain circumstances, an outside agency may assume command.

- **Communication**

During an emergency, communication is vital, especially between key personnel such as the overall commander, on-scene commander, engineering, fire brigade, medical, rescue, and outside agencies. Efforts should be made to provide for alternate means of communication should communication system failures occur. Depending on the size of the organization and the physical layout of the premises, it may be advisable to plan for an emergency control centre with alternate communication facilities. All personnel with alerting or reporting responsibilities must be provided with a current list of telephone numbers and addresses of those people they may have to contact.

- **Procedures**

Many factors determine what procedures are needed in an emergency, such as the:

- nature and degree of the emergency situation;
- size and complexity of the organization;
- capabilities/resources of the organization in an emergency situation;
- availability and immediacy of external support services;
- physical layout of the premises;
- location of the plant/facility;

- proximity and nature of neighbouring businesses;
- availability of emergency response support from neighbouring businesses; and
- number of plant/facility structures.

The common elements to be considered in all emergencies include pre-emergency preparation and provisions for alerting and evacuating staff, for handling casualties, and for containing of the emergency.

Natural hazards, such as floods or severe storms, often provide prior warning. The plan should take advantage of such warnings by strategically locating workers and equipment; providing instructions on sandbagging; moving equipment to needed locations; providing alternate sources of power, light or water; obtaining extra equipment; and relocating personnel with special skills. Phased states of alert allow such measures to be initiated in an orderly manner. This approach is further explained in the section on Pandemic Flu Planning presented later in this chapter.

The evacuation order is of greatest importance in alerting staff. To avoid confusion, only one type of signal should be used for the evacuation order. Commonly used for this purpose are sirens, fire bells, whistles, flashing lights, paging system announcements, or word-of-mouth in noisy environments. The "all-clear signal" is less important since time is not such an urgent concern at that point.

The following actions are "musts":

- Identify and teach staff about the evacuation routes and alternate means of escape.
- Ensure evacuation routes remain unobstructed.
- Specify safe locations (muster areas) for staff to gather for head counts to ensure that everyone has left the danger zone.
- Assign individuals to assist handicapped employees in emergencies.
- Carry out treatment of the injured.
- Search for the missing individuals simultaneously with efforts to contain the emergency.
- Provide alternate sources of medical aid when normal facilities may be in the danger zone or unavailable.
- Ensure the safety of all staff and neighbours at risk, before containing the extent of the property loss.

- **Testing and Enhancement**

Completing a comprehensive plan for handling emergencies is a major step toward preventing disasters. However, it is difficult to predict all of the problems that may happen unless the plan is tested. Exercises and drills must be conducted to practise all, or the critical portions (such as evacuation), of the ERP.

A thorough and immediate debriefing and review after each exercise and drill or after an actual emergency will help planners identify areas that require improvement. Additionally, the knowledge level of individual ERP responsibilities can be evaluated through paper tests, interviews, or mock rescue demonstrations.

The ERP should be revised when shortcomings become known and should be reviewed at least annually. Changes in plant infrastructure, processes, products, and key personnel are occasions for updating the plan.

It should be stressed that provision must be made for the training of both individuals and teams, if they are expected to perform adequately in an emergency. An annual full-scale exercise can maintain a high level of proficiency.

D. EMERGENCY RESPONSE PLANS

The following are several industry-tested ERPs that have been provided to clarify the aforementioned ERP concepts. Some are:

1. Company XYZ Emergency Response Plan

(A sample of an overall introductory to an ERP)

1.0 Introduction

Company XYZ strives to eliminate/reduce the risks that arise from work activities to an acceptable level. Despite measures to manage workplace risks, Company XYZ recognizes that an emergency situation may occur. In preparation, Company XYZ has developed an Emergency Response Plan (ERP) for a few emergency/disaster situations.

All personnel are responsible for reporting and initiating a response to any hazardous condition (*e.g.*, fire, explosion, spill/release of chemicals or gas, workplace violence, medical emergency, or natural disaster).

Specific instructions for responding to and dealing with emergencies involving fires, explosions, major incidents, chemical spills and releases, threats (personal/bomb), and natural or manmade disasters are covered within this ERP. Although the ERP is maintained in the *OH&S Program Manual*, step-by-step ERP procedures are available through various media such as the company Safe Work Procedures, *Employee OH&S Handbook*, and Company XYZ internal website. The *Employee OH&S Handbook* is available to company employees in hard or electronic copy.

These ERP procedures are to be used in the orientation of new employees, in employee training sessions, and in practice sessions to develop and hone the response skills of employees to an emergency situation.

All users of the ERP must be familiar with the contents of the plan, as well as their individual duties and responsibilities in an emergency situation.

Employees assigned specific duties within the plan will be trained and kept current with the related hazards and response equipment.

2.0 Scope

This ERP addresses:

- Fire Emergency Response Plan (*Provide the location of this document*)
- Chemical Spill/Release Response Plan (*Provide the location of this document*)
- Major Injury Incident Response Plan (*Provide the location of this document*)
- Bomb Threat Response Plan (*Provide the location of this document*)
- Bioterrorism Protection Plan (*Provide the location of this document*)

3.0 References

- *Relevant Occupational Health and Safety Act*
- *Relevant Fire Code*
- *Relevant Environmental Protection and Enhancement Act*

4.0 General Requirements

1. The various company facility layouts will be prominently displayed and indicate the location of:

    (a) emergency exits;

    (b) the main gas isolation valve;

    (c) the main electrical isolator;

    (d) main water stopcocks;

    (e) chemical spill kits;

    (f) fire extinguishers;

    (g) emergency eye-wash stations;

    (h) PPE stations;

    (i) First Aid area and kits; and

    (j) Automatic External Defibrillators (AEDs).

2. Company vehicles will prominently display and indicate the location of:
   (a) first aid kits;
   (b) fire extinguishers;
   (c) spill kits; and
   (d) emergency rescue equipment.
3. The above items will be clearly identified with appropriate signage and their access unimpeded.

5.0 Responsibilities

Director, OH&S or Designate
1. Act as the Planning Coordinator, responsible for the development and administration of comprehensive and effective emergency response plans.
2. Maintain the local fire department approval rating, as required.
3. Approve use of outside expertise and resources, such as training, plan development, cleanup, decontamination or other services, as needed.
4. Ensure that contact telephone lists, contractor agreements, flow charts and maps contained in the company OH&S Program Manual and/or posted throughout the company are current and complete.
5. Ensure that an effective communication system is in place.

Line Managers
1. Ensure that responsible persons are trained to assume control in the event of an emergency.
2. Ensure that the chosen personnel are competent to manage an emergency.
3. Ensure that all personnel understand the emergency procedures.
4. Ensure that their workgroup(s) are prepared to effectively handle an emergency.
5. Review the ERP drills to evaluate performance of employees during the response and make any needed changes to the process(es).

Employees
1. Understand what action to take in the event of an emergency and if in any doubt, consult the Line Manager.
2. Ensure emergency exits, sprinklers, fire extinguishers, spill stations, First Aid area, and first aid kits are accessible.

3. Participate in ERP training.

4. Participate fully in emergency response drills.

6.0 External Communications

Company XYZ's Communications Department will handle any related media relation activities and questions. No communication with the press or media is permitted without the prior approval of this group.

7.0 ERP Program Evaluation

The ERP will be reviewed by the Director, OH&S, and updated annually and as required. Changes will be approved by senior management and communicated to the relevant stakeholders.

Each time an ERP is practised or enacted, a documented evaluation is completed using the *Emergency Response Plan Evaluation form*. These forms are retained by the OH&S Department for evaluation and due diligence purposes. Recommendations for process improvement are forwarded to the various company business lines and the proposed changes monitored by the OH&S Department.

8.0 Telephone Numbers

Emergency Facility	Telephone Number
Local Fire Department	911
Emergency Medical Services (EMS)	911
Local Police Department	911
Hazardous Materials Spills Service	911
Poison Centre Calgary	XXX-XXXX (24 hour)
Local Gas Utility Company	XXX-XXXX (24 hour)
Electrical Utility Company	XXX-XXXX (24 hour)
Water Utility	XXX-XXXX (24 hour)
Municipal Solid Waste Department	XXX-XXXX (24 hour)
Company XYZ Contact Numbers	**Telephone Number**
Company Control Centre	XXX-XXXX (24 hour)
Director, OH&S	XXX-XXXX (24 hour)
Director, Environment	XXX-XXXX (24 hour)
Facility Operations	XXX-XXXX (24 hour)
Security	XXX-XXXX (24 hour)
Regulatory Agencies	**Telephone Number**
Environmental Protection Office	XXX-XXXX (24 hour)
Dangerous Goods Incidents (TDG)	XXX-XXXX (24 hour)
Disaster Services	XXX-XXXX (24 hour)
Company Health Services	**Telephone Number**
Occupational Health Services	XXX-XXXX (24 hour)
Employee Assistance Program	XXX-XXXX (24 hour)
Hospital(s)	XXX-XXXX (24 hour)
Monitoring Agencies	XXX-XXXX (24 hour)

2. Fire Emergency Response Plan

(A sample Fire ERP that links the overall Fire ERP with other company documents.)

1.0 Fire Safety

Fire Safety ensures that employees know what action to take in the event of a fire and that detection equipment, first aid equipment, fire-fighting equipment, and evacuation aids function as planned.

2.0 Reference

Local Fire Code

3.0 Responsibilities

Management

1. Designate one Fire Warden and alternate in each work area.
2. Have a Fire Evacuation Plan in place for each company facility.
3. Ensure an adequate number of Fire Wardens are available to implement the Fire Evacuation Plan.
4. Have the Fire Evacuation Plan available for reference by the Fire Department, Fire Wardens, and company staff.
5. Uphold the company Fire Safety Plans.
6. Review the Fire Safety Plans at staff meetings bi-annually.

Director, OH&S

1. Ensure Fire Evacuation Plan training is available for all personnel.
2. Ensure the Fire Evacuation Plan is put into effect.
3. Contact local fire authorities and familiarize them with the company's facility, products, and potential fire hazards.
4. Arrange and assist with fire drills, keeping a written record of the drill date, evacuation time, and comments/recommendations.
5. Ensure that the Fire Evacuation Plan for each work area is posted in a prominent location — electronically and hardcopy.

Facility Operations

1. Provide fire evacuation training to Fire Wardens.
2. Test fire protection/fighting equipment using the standardized approaches developed by Company XYZ.
3. Report any identified deficiencies to the Fire Wardens or OH&S Department.
4. When the fire alarm and/or detection system is inoperative for more than one hour, notify the local fire authorities.

5. Ensure that company staff at leased sites and buildings receive an orientation to the Fire Evacuation procedures for that site.
6. Maintain ERP training records.

Employee
1. Know company's Fire Safety Plan and procedures.
2. Familiarize oneself with location of fire extinguishers, fire alarm pull stations and exits.
3. Know the Fire Wardens.
4. Participate in fire drills.
5. Follow the Fire Warden's directions.
6. Remain at the Assembly Point until the "All Clear" signal is given.
7. Report fire hazards and malfunctioning firefighting/protection equipment to the Facility Operations.

Fire Wardens
1. Fire Wardens and alternates have been assigned to each work unit/areas.
2. Their duties consist of:
   - notifying the OH&S Department if relocated to another work area;
   - ensuring that an up-to-date list of personnel who normally work in the department exists;
   - knowing the evacuation plan procedures;
   - awareness of locations of fire extinguishers, exits, fire alarm-pull stations, and identification and location of incapacitated employees and their mobility aids;
   - alertness to and reporting of unsafe conditions in their area of responsibility to the Chief Fire Warden;
   - being trained to a standard level of first aid, CPR and AED;
   - attendance at Fire Warden training sessions;
   - reporting any noted fire, if required;
   - taking control during a fire or an evacuation ensuring that:
     - the alarm is raised;
     - the fire services are called;
     - the elevator is not used;
     - all people leave the building;

CH. 12: OCCUPATIONAL HEALTH AND SAFETY: EMERGENCY PREPAREDNESS AND RESPONSE

- personnel are checked at the Assembly Point; and
- the Chief Fire Warden is advised of the zone status.
- liaising with the emergency services on arrival (Chief Fire Warden); and
- advising the Chief Fire Warden if they are for any reason unable to fulfil their duties.

3. They will be given training on the above duties.

Have all the responsibilities of a Fire Warden, plus:

Chief Fire Wardens

1. Appoint at least two Alternate Fire Wardens.
2. Update and remain familiar with emergency procedures.
3. Remain familiar with the locations of all emergency exits.
4. Are familiar with the fire alarm system and its operation.
5. Maintain a current "confidential" list of persons with disabilities.
6. Wear a fire hat for identification.
7. Act as the building liaison for all fire- and emergency-related matters.
8. Conduct annual meetings with all Fire Wardens/Alternates and the Fire Inspector.
9. Keep the issued fire kit current.
10. Check and test in accordance with regulations all fire suppression/detection equipment within the assigned facility/building.
11. Advise the Alternate Fire Wardens of any planned absence and during times of absence, vacation, courses, *etc.*
12. Have absolute authority over building evacuations, drills or emergencies and cannot be overruled or disputed except by the Fire Department.
13. In the event of a fire, meet the Fire Department at the main fire panel located at the front entrance.

4.0 Firefighting Equipment

1. Fire extinguishers and sprinkler systems are throughout the various company facilities and in the company vehicles.
2. Fire extinguishers and sprinkler systems are tested and maintained on a regular basis.
3. Fire extinguishers are to be visible, well-marked, wall mounted, and readily accessible.

5.0 Fire Alarm Testing

1. The fire alarm system is tested annually on company-owned buildings. Facility Operations is responsible for conducting these tests and documenting the results.
2. Any deficiencies must be reported and rectified as soon as possible.

6.0 Signs

The following signage is required:

- to indicate fire extinguisher points;
- to indicate emergency exit doors both internally and externally;
- to indicate fire doors; and
- instructions on what action to take in the event of a fire (refer to the *Employee OH&S Handbook*).

7.0 Emergency Exits

1. These must be clearly marked.
2. They must not be obstructed at any time.
3. They must be capable of being opened without the use of a key.
4. They must be checked regularly to ensure that they can be easily operated.

8.0 Fire Drills

These will be carried out at least bi-annually. They are initiated by Facility Operations and coordinated through the OH&S Department. Any problems identified during the exercise are addressed.

9.0 Training

Personnel are given instructions and training on what to do when discovering a fire and/or hearing the fire alarm. A list of the duties for employees and Fire Wardens is posted in a number of workplace locations. This training is conducted during the orientation sessions for new employees and thereafter at annual intervals.

Training records are maintained by the Human Resources Department.

10.0 Maintenance/Testing and Records

The maintenance/testing of fire extinguishers and other firefighting equipment will be carried out annually and the results documented by Facility Operations. Defects will be addressed.

11.0 Specific Response

The actions to be taken in the event of a fire are described in the *Employee OH&S Handbook*.

CH. 12: OCCUPATIONAL HEALTH AND SAFETY: EMERGENCY PREPAREDNESS AND RESPONSE

3. Chemical Spill/Emission Release Response Plan

(A sample ERP that points to a Safe Work Procedure for identifying, assessing, and responding to a chemical spill/emission.)

1.0 Introduction

In the event of a chemical spill, the safety of company employees and the general public is of the utmost importance. For this reason, Company XYZ has developed a Chemical Spill/Emission Release Response Procedure that is to be followed. By way of reinforcement, response actions must always consider the safety of all personnel involved.

2.0 Specific Response

1. Refer to Company XYZ Chemical Spill/Emission Release Response Procedure.
2. Review the Safety Data Sheets for the related control products.

4. Major Injury Incident Response Plan

(A sample ERP that points to a Safe Work Procedure for injury management.)

1.0 Introduction

Fortunately, serious injury incidents that require the urgent attendance of the emergency services are rare. However, it is important that, if they do occur, the procedure followed ensures actions that minimize the negative impact of the incident.

2.0 Specific Response

In the event of a workplace injury incident, these steps are to be taken:

1. Assess the situation.
2. Contact external assistance (911).
3. Control or remove hazards to self and the victim.
4. Provide first aid and/or initiate the ERP for the specific work area. This initiates a cascade of events in accordance with the severity of the injury.
5. Ensure that the scene of the incident is left undisturbed except to:
    (a) attend to injured workers;
    (b) prevent further injuries; and/or
    (c) protect the public/property that is endangered as a result of the incident.
6. Ensure that a company management representative accompanies the injured employee to the medical facility.
7. Take the necessary actions to notify and assist the injured employ-

ee's family. In the case of all fatalities, the Chief Operating Officer and/or Company XYZ/President will notify next of kin.

8. Notify the appropriate government agencies of the incident.
9. Complete (employee and line manager) the required company incident report form and Workers' Compensation Forms.
10. All statements to the press are to be issued through the company's Communications Department.
11. Co-operate with government OH&S officers and the police in investigating the incident.
12. Arrange for critical incident debriefing for employees/dependants.
13. Prepare and issue a preliminary report within 24 hours, followed by a final investigation report.

3.0 Transportation

Ill/injured employees are to be transported to the nearest healthcare facility by ambulance. To summon an ambulance, dial 911, providing the incident location, situation and nature of injury(ies).

5. Bomb Threat Response Plan

(A sample ERP that is more detailed.)

1.0 Introduction

This procedure details what action is to be taken in the event of the receipt of a bomb threat. Any information suggestive of an explosive device having been placed in a company facility must be regarded as a threat until an investigation proves otherwise.

The most likely method for this type of threat is telephone communication. The call may be genuine or a hoax: however, treat it as genuine until proven to be a hoax.

2.0 Bomb Threat

2.1 Taking the Call

The person who receives the call is to record as much information as possible about the package and the caller. Experienced switchboard operators accustomed to sifting questionable calls, have been known to panic when the word "bomb" is heard. However, the call will probably be the only contact that the "threatener" will make and is the only opportunity to obtain vital information, such as location of a device, time of detonation, sex of the caller, background noises, *etc.*

Company XYZ's *Bomb Threat Data Record* is found on the

company OH&S website and provides guidance on handling the caller and is a mechanism for gathering data on the details of the call. It outlines the:

1. procedure for taking and recording a bomb threat made by a telephone caller;
2. roles of the Emergency Coordinators;
3. search teams;
4. telephone procedures; and
5. evacuation procedures.

The majority of bomb threats are received at a publicly advertised telephone local such as a main switchboard. The most likely recipients of a bomb call placed to a company will probably be the main switchboard, system operations and customer services. These areas should keep copies of Company XYZ's *Bomb Threat Data Record*.

2.2 Getting HELP

Once the information has been taken from the "threatener" and the bomb threat is appraised, a decision is made to contact the police and the premises and grounds are searched. To get assistance with this process, contact the Emergency Contacts listed.

2.3 Evaluation of the Bomb Threat

Evaluating a bomb threat requires assessing the credibility of the threat and selecting one of three possible alternatives:

1. take no action to search or evacuate;
2. search without evacuation; or
3. evacuate and search.

3.0 Receipt of a Suspicious Package

If a suspicious package or container is identified, everyone in the facility is to be evacuated from the danger area in accordance with the Company XYZ's Bomb Threat Procedure.

4.0 General Precautions

1. Do not leave parcels or baggage lying around in odd corners or unattended.
2. Do not touch or disturb suspicious packages.
3. In the event of an explosion, stay clear of windows; flying glass can travel up to 200 metres.
4. Employees should evacuate to another building away from the risk

of injury from flying glass or debris.

5.0 Key Players

The Emergency Team members are made up of available individuals such as OH&S personnel and Fire Wardens.

The Emergency Search Team comprises representatives from:

(a) Corporate Security

(b) Facility Operations

(c) Occupational Health and Safety

(d) Environmental Department

(e) Operations

Contact names and numbers are listed on Company XYZ's OH&S website.

6.0 Responsibilities

Emergency Team Coordinator

- In the event of a bomb threat, call the Emergency Team to assemble at reception areas.

Emergency Team

1. Immediately upon notification of a bomb threat, the Emergency Team will:

    (a) set up an Emergency Coordination Centre in the Reception Area of the building where the caller said the bomb has been placed.

    (b) contact the Emergency Search Team.

    (c) assess the nature of the threat and determine:

    i. to initiate a search without evacuation, or

    ii. if a full evacuation is required and then a search, or

    iii. if a selective evacuation is sufficient along with a search.

2. For serious situations, notify the Company/President's office, and then proceed as determined and in accordance with Company XYZ's Bomb Threat Procedure.

6. Bioterrorism Protection Plan

(A sample ERP)

1.0 Introduction

A **bioterrorism attack** is the deliberate release of viruses, bacteria or other germs (agents) used to cause illness/death in people, animals or plants.[5] Typically, these agents are found in nature, but they can also be altered to cause disease, to resist current medicines and/or to spread into the environment via air, water, or food chain. Terrorists may use biological agents because they can be extremely difficult to detect and do not cause illness for several hours to several days. Some bioterrorism agents, like the smallpox virus, can be spread from person to person; and some, like anthrax, cannot, but can cause death.

One significant component of emergency preparedness is the ability to counter the risk of injuries, illnesses, and deaths from attacks on occupational settings.

2.0 Bioterrorism Agents

Bioterrorism agents are classified by the ease with which they can be spread, and the severity of illness caused. There are three categories of bioterrorism agents:

- *Category A* — high-priority agents

    Organisms or toxins that are easily spread or transmitted from person-to-person; result in high illness and death rates; cause public panic and social disruption; and require special action for public health preparedness. These include anthrax, smallpox, botulism toxins, tularemia, and viral hemorrhagic fevers.

- *Category B* — second highest priority agents

    Organisms or toxins that are moderately easy to spread or transmit from person-to-person; result in moderate illness and low death rates; and require enhancements for public health preparedness and disease monitoring.

- *Category C* — third highest priority agents

    Includes organisms/toxins and emerging pathogens that could be engineered for mass spread in future because they are readily available; easily produced and spread; and have the potential for high morbidity, mortality rates and public health impacts.

3.0 Signs of a Bioterrorist Attack

The best defence is awareness of the signs of a bioterrorist attack. Given that the signs and symptoms are vague (headache, fever, malaise, and cough), the following indicators are more reliable for detecting an attack:

- existence of illness in a close-knit group of people (a cluster);

---

[5] Centers for Disease Control and Prevention (CDC), "Bioterrorism", online: http://www.bt.cdc.gov/bioterrorism/overview.asp (date accessed: February 28, 2020).

- a high infection rate within the cluster;
- illness occurring in an unusual location;
- illness occurring at an unusual time of year;
- illness that demonstrates odd symptoms and signs; and/or
- death of animals.[6]

Once detected, the company OH&S Department should contact the local public health authorities for guidance and assistance.

4.0 Preparedness

Bioterrorism is like any other workplace threat, with the exception that the hazardous agent is difficult to detect, has a significant negative impact on employee health and causes public fear and outrage. However, it is a threat that is somewhat invisible, hard to detect, quick to spread, results in high mortality and morbidity rates, and causes panic.

Many of the emergency response plans and security measures that companies have in place can protect the company against a bioterrorist attack. For example:

- company security measures aimed at protecting the company from intruders;
- company reception and mailroom practices that screen visitors and mail;
- worker training programs that can be tailored to include information on bioterrorism and appropriate responses;
- business continuity plans for pandemic flu and other worker health threats;
- emergency evacuation plans; and
- emergency risk communication plans.

In the event of a bioterrorist emergency event, the company would be prepared to respond accordingly:

- *Airborne Attack* — In the event of an airborne release, company emergency evacuation plans would prove useful in getting workers out of the building to safety. Emergency responders equipped with SCBA could shut down ventilation systems and assist First Responders to get victims unable to exit on their own to safety.

---

[6] Minnesota Department of Health, "Bioterrorism: Background and Significance", online: http://www.health.state.mn.us/bioterrorism/hcp/btbackground.ppt (date accessed: February 28, 2020).

- ***Water or Foodborne Attack*** — Employees affected by organisms/agents spread through water or food need to be evacuated to local health facilities. The suspect agents must be isolated for investigation purposes.

5.0 General Precautions

- Educate workers on bioterrorism — what it is, the signs of an attack and the company's preparedness.
- Monitor visitor and contractor entry to the company facility(ies).
- Monitor mail delivery and handling.
- Be alert to the signs of a bioterrorist event.
- Be prepared with the proper equipment (masks and handwashing materials) and procedures to respond.

6.0 Key Players

The Emergency Team comprises representatives from:

(a) Corporate Security

(b) Facility Operations

(c) Occupational Health & Safety

(d) Environmental Department

(e) Operations

Contact names and numbers are provided on the Company XYZ OH&S website.

7.0 Responsibilities

Emergency Team Coordinator

- In the event of a bioterrorism event, will call the Emergency Team to assemble at Reception areas.

Emergency Team

(a) Set up an Emergency Coordination Centre in the Reception Area of the building.

(b) Assess the situation and determine the appropriate response given the nature of the attack (airborne, foodborne, or waterborne).

(c) Contact public health department, emergency medical response, and police for assistance and guidance.

(d) Advise the Senior Management team of the situation and the actions planned.

(e) Respond to protect workers and the facility.

Additional resources on emergency preparedness and planning are available at:

- **Government of Alberta, Employment and Immigration:** They developed in 2011 an Occupational Health and Safety: Tool Kit for Small Business. It contains a sample ERP, which has been included as an industry example in Appendix 1 of this book. Access to an electronic version of this document is available at: http://work.alberta.ca/documents/ohs-best-practices-BP018-greyscale.pdf.
- **U.S. Department of Labor, OSHA:** They provide manuals, software,[7] and publications[8] on emergency preparedness and response at:
  - http://www.osha.gov/dep/etools/ehasp/index.html
  - http://www.osha.gov/Publications/osha3088.pdf.

7. Pandemic Flu Planning

On a different vein, companies also need to develop contingency plans for emergency situations that could negatively impact the workforce, such as an outbreak of pandemic flu. The difference with this type of emergency is that it is neither acute nor short term in nature. A pandemic flu outbreak is expected to be an extended event, with multiple waves of outbreaks in the same geographic area; each outbreak could last from six to eight weeks. Waves of outbreaks may occur over a period of one year or longer and are expected to negatively impact many areas of Canada, the United States, and other countries simultaneously.

A response to a pandemic flu outbreak cannot be fully addressed using the traditional ERP techniques. Contingency plans need to involve organizational/company personnel from areas like Risk Management, Business Continuity, Communication, Human Resources, Security, Operations, and Occupational Health and Safety.

Lastly, the risk here is not to the organization's/company's equipment, facilities or work processes; the risk is to the people involved in doing the work.

(a) Current Assumptions about a Pandemic Flu Outbreak

The following assumptions concerning a pandemic outbreak are important for organizations/companies to know and consider when doing their emergency response planning:

- Susceptibility to the pandemic influenza virus will be universal.

---

[7] U.S. Department of Labor, "OSHA e-HASP Software — Version 2.0 (e-HASP2), March 2006", online: http://www.osha.gov/dep/etools/ehasp/index.html (date accessed: February 28, 2020).

[8] U.S. Department of Labor, "Guidance on Preparing Workplaces for an Influenza Pandemic", OSHA-3327-05R, online: https://www.osha.gov/Publications/influenza_pandemic.html (date accessed: February 28, 2020).

- Efficient and sustained person-to-person transmission signals an imminent pandemic.
- The clinical disease attack rate will likely be 30 per cent or higher in the overall population during the pandemic. Illness rates will be highest among school-aged children (about 40 per cent) and decline with age. Among working adults, an average of 20 per cent will become ill during a community outbreak.
- Some persons will become infected but not develop clinically significant symptoms. Asymptomatic, or minimally symptomatic, individuals can transmit infection and develop immunity to subsequent infection.
- Of those who become ill with influenza, 50 per cent will seek outpatient medical care.
- The number of hospitalizations and deaths will depend on the virulence of the pandemic virus.
- The risk groups for severe and fatal infection cannot be predicted with certainty, but are likely to include infants, the elderly, pregnant women and persons with chronic medical conditions.
- The rates of employee absenteeism will depend on the severity of the pandemic.
- In a severe pandemic, absenteeism attributable to illness, the need to care for ill family members, and fear of infection, may reach 40 per cent during the peak weeks of a community outbreak, with lower rates of absenteeism during the weeks before and after the peak.
- Certain public health measures (*e.g.*, closing schools, quarantining household contacts of infected individuals, and closing daycares) are likely to increase rates of employee absenteeism.
- The typical incubation period (interval between infection and onset of symptoms) for influenza is approximately two days.
- Persons who become ill may shed the virus and can transmit infection for up to one day before the onset of illness. Viral shedding and the risk of transmission will be greatest during the first two days of illness. Children usually shed the greatest amount of virus and therefore, are likely to pose the greatest risk for transmission.
- On average, infected persons will transmit infection to approximately two other people.
- In an affected community, a pandemic outbreak will last about six to eight weeks.
- Multiple waves (periods during which community outbreaks occur across the country) of illness could occur with each wave lasting two to three months. Historically, the largest waves have occurred in the fall and winter,

but the seasonality of a pandemic cannot be predicted with certainty.[9]

It is very difficult to estimate what might occur in the event of an influenza pandemic. Experts believe that pandemic influenza could reach Canada within three months of being detected anywhere in the world. People would be expected to be ill for about seven to 10 days. The pandemic would have its maximum effect on the Canadian population within five to seven months and could last up to 18 months.

According to Health Canada, predicting the impact of a pandemic is challenging because no one knows how the virus will behave or how serious the pandemic will be. During a severe flu season, as many as 8,000 Canadians die from influenza and its complications; on average there are 4,000 deaths from annual flu. In a moderately severe pandemic, it is estimated that between 11,000 and 58,000 deaths may occur in Canada. These numbers assume that the virus would cause illness in 15 per cent to 35 per cent of the population.[10,11]

Although the actual physical properties of the offending flu are unknown, the premise is that influenza will be primarily spread through large droplets (droplet transmission) that directly contact the nose, mouth or eyes. These droplets are produced when infected people cough, sneeze or talk, sending the relatively large infectious droplets and very small sprays (aerosols) into the nearby air for contact with other people. Because large droplets travel a limited range (less than six feet), people are advised to limit close contact with others.

To a lesser degree, the transfer of human influenza material to nose, mouth and eyes through touching objects contaminated with influenza viruses can occur. Influenza may also be spread by very small airborne infectious particles. The contribution of each route of exposure to influenza transmission is uncertain at this time and may vary based upon the characteristics of the offending influenza strain.

(b) Business Continuity Risks

Additional to the health risks, a pandemic flu outbreak brings with it, business continuity risks. For example, in the event of pandemic flu, would the organization/ company survive? Many countries were relatively unaffected by the SARS epidemic of 2003 and so have not experienced the high degree of business disruption resulting

---

[9] U.S. Department of Homeland Security (DHS), *Pandemic Influenza: Preparedness, Response, and Recovery Guide for Critical Infrastructure and Key Resources*, at 13, online: https://www.dhs.gov/sites/default/files/publications/cikrpandemicinfluenzaguide.pdf (date accessed: February 28, 2020).

[10] Health Canada, "Pandemic Flu Preparedness", online: https://www.canada.ca/en/public-health/services/flu-influenza/canadian-pandemic-influenza-preparedness-planning-guidance-health-sector.html (date accessed: February 28, 2020).

[11] Public Health Agency of Canada, "The Canadian Pandemic Influenza Plan for the Health Sector" (2011), online: https://www.canada.ca/en/public-health/services/flu-influenza/canadian-pandemic-influenza-preparedness-planning-guidance-health-sector.html (date accessed: February 28, 2020).

from employee absenteeism that could occur. However, being a member of a global community, Canadian and American companies face a real risk of a major pandemic or epidemic. Some potentially negative business impacts are:

- **Communication System Failures** — Due to the lack of critical information people/contacts and the absence of back-up information systems, failures in how work is to be done can occur. Likewise, a high volume of people seeking information or accessing the Internet to conduct business from a remote site, can overload communication systems.
- *Loss of Key Personnel* — Key personnel may be unavailable to organizational/company operations for lengthy periods or permanently. A pandemic could affect in excess of 40 per cent of the workforce during periods of peak influenza illness. Employee absenteeism could result from sickness/death, caregiving of family members, and childcare due to closed schools or daycare centres. One of the greatest risks is panic. The prediction is that people will adopt a "siege mentality", *i.e.*, wanting to stay in their own homes, surround themselves with their loved ones, and avoid any non-essential interaction with other people. Work duties become insignificant when life and family are in jeopardy.
- *Change in Consumer Patterns* — During a pandemic flu outbreak, consumer demand for items related to infection control is likely to increase dramatically, while consumer interest in other goods may decline.
- *Customer Avoidance of Marketplaces* — Consumers may also change the ways in which they shop as a result of the pandemic. To limit/reduce person-to-person contact, consumers may try to shop online or at "off-peak" hours to reduce contact with other people; show increased interest in home delivery services; or prefer other options such as drive-through services.
- *Business Closures* — Companies may be voluntarily closed or shut down by public health authorities. Shipments of items from companies severely affected by the pandemic may be delayed or cancelled. This can reduce or prevent access to supplies or services needed to conduct or maintain normal business operations.
- *Transportation Disruptions* — Bus, rail, and airline transportation may be disrupted due to lack of available manpower, or due to lack of consumer demand for fear of contracting the flu.
- *Breakdown in Police/Fire/Emergency Medical Services.*
- *Panic-based Evacuations.*
- *Overwhelmed Medical Facilities* — Hospitals/Acute Care Clinics are not expected to be able to cope if the numbers of sick and dying exceed expectations.

(c) Employee Exposure to Pandemic Influenza at Work: Classification

The risk of occupational exposure to influenza during a pandemic may vary from "very high", to "high", to "medium" or to "lower (caution)" risk. The level of risk depends in part on whether the jobs require close proximity to people potentially infected with the pandemic influenza virus or whether they are required to have either repeated or extended contact with known or suspected sources of pandemic influenza virus such as co-workers, the general public, outpatients, schoolchildren, or other such individuals or groups.

To help employers determine appropriate work practices and precautions, OSHA has divided workplaces and work operations into four risk zones, according to the likelihood of employees' occupational exposure to pandemic influenza (Figure 12.1).

Figure 12.1: Occupational Risk Pyramid for Pandemic Influenza[12]

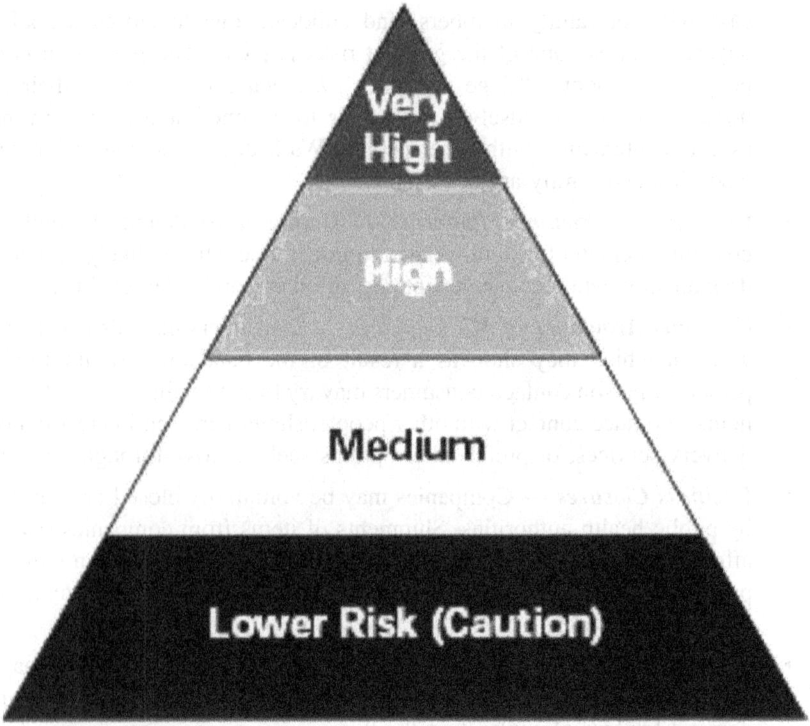

- **Very High-Exposure Risk**

    Very high-exposure risk occupations are those with high potential exposure to high concentrations of known or suspected sources of pandemic influenza

---

[12] U.S. Department of Labor, "Guidance on Preparing Workplaces for an Influenza Pandemic", OSHA-3327-05R, at 11, online: https://www.osha.gov/Publications/OSHA3327pandemic.pdf (date accessed: February 28, 2020).

during specific medical or laboratory procedures. For example, healthcare workers (physicians, nurses, dentists, and laboratory personnel) performing aerosol-generating procedures on known or suspected pandemic patients.

- **High-Exposure Risk**

    High-exposure risk occupations are those with high potential for exposure to known or suspected sources of pandemic influenza virus. For example, healthcare delivery and support staff exposed to known or suspected pandemic patients (physicians, nurses, and other hospital staff that must enter patients' rooms); emergency medical service technicians involved with medical transport of known or suspected pandemic patients in enclosed vehicles; and morgue and mortuary employees performing autopsies on known or suspected pandemic patients.

- **Medium-Exposure Risk**

    Medium-exposure risk occupations include jobs that require frequent, close contact exposures (within six feet) to identified or suspected sources of pandemic influenza virus such as co-workers, the general public, outpatients, schoolchildren, or other such individuals or groups. For example, school staff, high-population-density work environments and some high-volume retail workplaces.

- **Lower-Exposure Risk (Caution)**

    Lower-exposure risk occupations are those that do not require contact with people known to be infected with the pandemic virus, nor frequent close contact (within six feet) with the public. However, employers should be cautious and develop preparedness plans to minimize employee infections, even for those at lower risk levels, *i.e.*, office employees.

    Employers of critical infrastructure and key-resource employees (such as law enforcement, emergency response, or public utility employees) may consider upgrading protective measures for these employees beyond what would be suggested by their exposure risk due to the necessity of such services for the functioning of society, as well as the potential difficulties in replacing them during a pandemic.

    Canadian and U.S. employers have an important role in protecting employee health and safety, and in limiting the impact of an influenza pandemic. It is important to work with community planners to integrate industry pandemic plan into local and provincial/state planning, particularly if the organization's/company's operations are part of the nation's critical infrastructure or key resources. Integration with local community planners allows businesses to access resources and information promptly to maintain business operations and protect employees.

(d) Pandemic Flu: Business Continuity Risk Management

The following are guidelines for action that organizations can use to manage the risk of widespread illness that removes people from the world of work. It is worth noting that many of the issues potentially arising from a pandemic would also apply to a chemical, biological, radiological, or nuclear (CBRN) incident; and to a major flood incident that would remove people from the work equation. A pandemic and a flood would both include warning periods. In the case of a CBRN incident, there would be no warning.

(i) Business Continuity Risk Management

What can organizations do in terms of business continuity risk management? **Business resilience**, that is, reducing the likelihood of an interruption occurring and reducing its impact if and when an incident occurs, is the key. It involves:

- *Implement Effective Knowledge Management:* Ensure there is an effective communication system in place so that in the absence of key information people from the workplace, business operations can continue. Actions like the following are critical:
    - undertake a critical records analysis to identify what documentation currently exists versus what will be needed;
    - capture and manage critical information;
    - use off-site storage of copies of all critical information;
    - develop comprehensive Safe Work Procedures for critical job tasks;
    - provide cross-training on critical job positions/tasks;
    - ensure that knowledge and skills are distributed corporate-wide using reliable training systems; and
    - develop and implement succession planning for critical positions.

- *Treat Communications Issues as a Priority:* The most effective way to control panic is to keep people well-informed and prepared. Likewise, to keep business operations running smoothly, people need to know what is expected and how to do the job. To those ends, some recommendations are:
    - Ensure that all contact lists are up-to-date and maintained.
    - Evaluate all communication options to keep personnel informed — management communications, internal communication systems (newsletters, pay cheque inserts), intranet, bulletins, mobile networks, radio, Internet, *etc.*
    - Establish means to communicate with suppliers.

- Establish means to communicate with your customer base.
- Establish an authorized and competent team to communicate with key stakeholders and the media.
- Consider remotely hosted e-mail and website options.
- Consider alternate telephone providers as a back-up to current systems.
- ***Implement Flexible Work Practices Options:*** Based on best advice, employers need to plan for high (40 to 60 per cent) employee absenteeism. One way to address manpower shortages is to enable employees to work from alternate offices or from home. This may involve:
    - moving personnel to other locations (may need to accommodate families);
    - leasing remote facilities (safe areas) and relocating personnel; and
    - developing key resource requirements for each of the above options.
- ***Provide Secure Transport:*** To enable employees to safely get to and from work, companies may wish to contract with a bus company to provide private transport. This would be especially important for plant/refinery situations. Another option is corporate-sponsored car-pooling for employees.
- ***Ensure Adequate Supplies/Services:*** Lack of supplies and services are probably going to be the second biggest challenge for organizations/companies. With widespread use of "just-in-time" warehousing and contracting of non-essential services, organizations/companies are very dependent on outside resources for their business operations. To ensure adequate resources, some recommendations are:
    - contract multiple suppliers so that contingency arrangements can be made;
    - monitor for geographic (even international) dispersion of goods/services;
    - identify other key customers and know where you stand in the "pecking order"; and
    - ensure that suppliers have effective business continuity plans which have been thoroughly tested.
- ***Take Care of Business:*** Review the organization's/company's customer base and other income-producing opportunities in terms of protecting current business and identifying emerging business oppor-

tunities, such as expanding the customer base and moving into new markets. Customers will be inquiring about business continuity plans — make sure the organization/company can provide them and that they are viewed as adequate. Review the distribution and delivery options to determine if the organization/company can reduce the need for travel during a pandemic. Examine possible hedging opportunities that may be available to spread the organization's/company's risk.

- ***Prepare for Security Issues:*** Given the expected strain on public services and the anticipated manpower shortage, companies will need to plan for self-management of security issues. Ask if:
  - arrangements are in place to secure the organization's/company's premises for short- and long-term periods of service withdrawal or manpower vacancy;
  - provisions exist should the security provider receive multiple demands for increased resources;
  - provisions for the security of personnel — in the office, travelling between home and office, security at home — are in place;
  - secure provision of consumer staples — food, drink, essential household items — has been planned; and
  - service limitations of police, fire and emergency medical services have been addressed.

  These are all security issues that need to be understood and managed.

- ***Address Legal, Regulatory and Insurance Options:*** In terms of ways to transfer the risk associated with a pandemic flu outbreak, carefully examine insurance policies to ensure that coverage exists for all cases of business cessation, including voluntary and mandatory closure and loss of income. As well, ensure that the organization/company is exercising a duty of care to employees by regularly inspecting the status of facility ventilation systems and all shared washroom facilities to guarantee health standards. Ensure that employee illness insurance protection (Workers' Compensation Insurance) for both the worker and organization/company is in place.

(ii) Emergency Preparedness and Response Plan (*Administrative Control Measures*)

- Develop a staged approach to emergency response planning (Table 12.4) that corresponds with the World Health Organization phases of a pandemic flu outbreak (Table 12.3), including pandemic preparedness; review it and conduct regular drills. A Pandemic Planning

Checklist is provided in Table 12.5.

Table 12.3: WHO Stages of a Pandemic Flu Outbreak[13]

WHO Phases		Federal Government Response Stages	
INTER-PANDEMIC PERIOD			
1	No new influenza virus subtypes have been detected in humans. An influenza virus subtype that has caused a human infection may be present in animals. If present in animals, the risk of human disease is considered to be low.	0	New domestic animal outbreak in at-risk country
2	No new influenza virus subtypes have been detected in humans. However, a circulating animal influenza subtype poses a substantial risk of human disease.		
PANDEMIC ALERT PERIOD			
3	Human infection(s) with a new subtype, but no human-to-human spread, or at most rare instances of spread to a close contact.	0	New domestic animal outbreak in at-risk country
		1	Suspected human outbreak overseas
4	Small cluster(s) with limited human-to-human transmission but spread is highly localized, suggesting that the virus is not well adapted to humans.	2	Confirmed human outbreak overseas
5	Larger cluster(s) but human-to-human spread still localized, suggesting that the virus is becoming increasingly better adapted to humans, but may not yet be fully transmissible (substantial pandemic risk).		
PANDEMIC PERIOD			
6	Pandemic phase: increased and sustained transmission in general population.	3	Widespread human outbreaks in multiple locations overseas
		4	First human case in North America
		5	Spread throughout United States
		6	Recovery and preparation for subsequent waves

---

[13] Homeland Security Council, *Implementation Plan for the National Strategy for Pandemic Flu*, at 104, online: http://www.flu.gov/professional/federal/pandemic-influenza-implementation.pdf (date accessed: February 28, 2020).

Table 12.4: Staged Approaches to Pandemic Flu Preparedness Planning for a Domestic Company[14]

WHO Phase	Description	Government Response Stage	Government Goals/Actions	Company Action
1	**Inter-Pandemic Period** No new virus strains in humans, but possibly present in animals. Risk to humans is considered low.	0 New domestic animal outbreak in an "at-risk" country	• Monitor situation, increasing surveillance as the situation heightens • Prepare for a pandemic flu outbreak • Urge similar preparation at all levels of society • Facilitate the desired level of preparation	• Learn about pandemic flu outbreaks and the recommended actions for emergency preparedness and response • Undertake emergency response planning • Conduct business continuity planning
2	**Inter-Pandemic Period** No new virus strains in humans, but present in animals. Pose substantial risk to humans			• Communicate pandemic events to workers/families • Conduct ERP drills • Conduct business continuity evaluation • Provide and encourage general flu immunization
3	**Pandemic Alert Period** Human infection(s), but no/rare instances of human-to-human transmission	0 New domestic animal outbreak in "at-risk" country		

[14] *Ibid.*, at 105-107.

CH. 12: OCCUPATIONAL HEALTH AND SAFETY: EMERGENCY PREPAREDNESS AND RESPONSE

WHO Phase	Description	Government Response Stage	Government Goals/ Actions	Company Action
		1 Suspected human outbreak overseas		
4	**Pandemic Alert Period** Small cluster(s) of limited human-to-human transmission: localized spread only	2 Confirmed human outbreak overseas	• Monitor situation • Prepare to limit entry to country • Prepare specific flu vaccine	• Monitor employee health in foreign business locations • Conduct ERP drills • Conduct business continuity evaluation • Provide and encourage general flu immunization
5	**Pandemic Alert Period** Larger cluster(s): localized spread only			

WHO Phase	Description	Government Response Stage	Government Goals/Actions	Company Action
6	**Pandemic Period** Increased and sustained transmission in a general population	3 Widespread human outbreaks in multiple locations overseas	• Delay outbreak in North America • Monitor closely for first case in North America • Prepare domestic containment and response mechanisms • Activate domestic emergency medical plans • Use layered screening at borders • Concentrate efforts on specific vaccine production • Put hospitals on surveillance for "first case"	• Prioritize preparedness efforts • Monitor public health information and bulletins • Have planned resources (OH&S and medical supports) on "standby" • Communicate the company's pandemic flu plan to employees and family members • Eliminate employee travel to affected areas • Segregate rest of staff from employees travelling in from affected areas • Monitor employee absenteeism

WHO Phase	Description	Government Response Stage	Government Goals/ Actions	Company Action
		4 First human case in North America	• Contain first case in North America • Provide antiviral treatment prophylaxis • Activate pandemic plans • Limit non-essential domestic travel • Continue developing the specific vaccine • Deploy diagnostic reagents for pandemic virus to all laboratories	• Communicate the company's pandemic flu plan to employees and family members • Eliminate employee travel to affected areas • Segregate rest of staff from employees travelling in from affected areas • Monitor employee absenteeism

		5
		
		Spread in Canada and United States
			• Support community response
			• Preserve critical infrastructure
			• Mitigate illness and death
			• Mitigate impact to economy
				• Aim at flu containment — promote messages like:
				  • Stay at home if ill
				  • Good hand hygiene
				  • Good cough etiquette
				• Provide infection control supplies
				• Preserve critical infrastructure
				• Modify face-to-face contact
				• Implement flexible work practices

WHO Phase	Description	Government Response Stage	Government Goals/Actions	Company Action
		6 Recovery and preparation for subsequent waves		• Plan for next wave by learning from experiences of the first wave • Use that portion of the workforce that recovered from the flu

- Be aware of and review national, provincial, and municipal health department pandemic influenza plans. Incorporate appropriate actions from these plans into workplace disaster plans.
- Prepare and plan for running the organization's/company's operations with a reduced workforce.
- Organize and identify a central team of people or a focal point to serve as a communication source so that employees and customers can have accurate information during the crisis.
- Provide employees and customers with easy access to infection control supplies, such as soap, hand sanitizers, personal protective equipment (such as gloves or surgical masks), tissues, and office-cleaning supplies.
- Understand and develop work practices and engineering controls that provide additional protection to employees and customers, such as: drive-through service windows, clear plastic sneeze barriers, ventilation, and the proper selection, use, disposal of personal protective equipment.
- Identify possible exposure and health risks to employees. For example, are employees potentially in contact with people with influenza such as those working in a hospital or clinic? Are employees expected to have a lot of contact with the general public?
- Minimize exposure to co-workers or the public, by enabling employees to work from home.
- Provide worker training, education, and informational material about business-essential job functions and employee health and safety, including proper hygiene practices (refer to Table 12.6) and the use of any personal protective equipment to be used in the workplace.
- Educate and train employees in proper hand hygiene, coughing etiquette, and social distancing techniques.
- Ensure that informational material is available in a usable format for individuals with sensory disabilities and/or limited English proficiency. Encourage employees to take care of their health by eating right, getting plenty of rest, and getting a seasonal flu vaccination.
- Make sure the disaster plan protects and supports employees, customers and the general public. Be aware of concerns about pay, sick leave, safety, and health. Informed employees who feel safe at work are less likely to be absent.
- Regularly exercise the ERP to determine if the ERP will work, or simply make the situation worse. It will identify opportunities for

improvement. As has been aptly stated,

> No plan ever survives contact with the enemy.
>
> Helmuth von Moltke

Table 12.5: Business Pandemic Influenza Planning Checklist[15]

Some recommended planning actions are:

1.1 Plan for the impact of a pandemic on your business:

Action	Completed	In Progress	Not Started
Identify a pandemic coordinator and/or team with defined roles and responsibilities for preparedness and response planning. The planning process should include input from labour representatives.			
Identify essential employees and other critical inputs (*e.g.*, raw materials, suppliers, subcontractor services/ products and logistics) required to maintain business operations by location and function during a pandemic.			
Train and prepare ancillary workforce (*e.g.*, contractors, employees in other job titles/ descriptions, retirees).			
Develop and plan for scenarios likely to result in an increase or decrease in demand for your products and/or services during a pandemic (*e.g.*, effect of restriction on mass gatherings, need for hygiene supplies).			
Determine potential impact of a pandemic on company business financials using multiple possible scenarios that affect different product lines and/or production sites.			
Determine potential impact of a pandemic on business-related domestic and international travel (*e.g.*, quarantines, border closures).			
Find up-to-date, reliable pandemic information from community public health, emergency management, and other sources and make sustainable links.			

---

[15] Adapted from U.S. Department of Health & Human Services, "Business Pandemic Influenza Planning Checklist", online: http://www.pandemicflu.gov/professional/pdf/businesschecklist.pdf (date accessed: February 28, 2020).

CH. 12: OCCUPATIONAL HEALTH AND SAFETY: EMERGENCY PREPAREDNESS AND RESPONSE

Action	Completed	In Progress	Not Started
Establish an emergency communications plan and revise periodically. This plan includes identification of key contacts (with backups), chains of communication (including suppliers and customers) and processes for tracking and communicating business and employee status.			
Implement an exercise/drill to test your plan and revise periodically.			

1.2 Plan for the impact of a pandemic on your employees and customers:

Action	Completed	In Progress	Not Started
Forecast and allow for employee absences during a pandemic due to factors such as personal illness, family member illness, community containment measures and quarantines, school and/or business closures and public transportation closures.			
Implement guidelines to modify the frequency and type of face-to-face contact (*e.g.*, handshaking, seating in meetings, office layout, shared workstations) among employees and between employees and customers (refer to CDC recommendations).			
Encourage and track annual influenza vaccination for employees.			
Evaluate employee access to and availability of healthcare services during a pandemic and improve services as needed.			
Evaluate employee access to and availability of psychological health and social services during a pandemic, including corporate, community and faith-based resources and improve services as needed.			
Identify employees and key customers with special needs and incorporate the requirements of such persons into your preparedness plan.			

1.3 Establish policies to be implemented during a pandemic:

Action	Completed	In Progress	Not Started
Establish policies for employee compensation and sick-leave absences unique to a pandemic (*e.g.*, non-punitive, liberal leave), including policies on when a previously ill person is no longer infectious and can return to work after illness.			
Establish policies for flexible worksite (*e.g.*, telecommuting) and flexible work hours (*e.g.*, staggered shifts).			
Establish policies for preventing influenza spread at the worksite (*e.g.*, promoting respiratory hygiene/ cough etiquette and prompt exclusion of people with influenza symptoms).			
Establish policies for employees who have been exposed to pandemic influenza, are suspected to be ill, or become ill at the worksite (*e.g.*, infection control response, immediate mandatory sick leave).			
Establish policies for restricting travel to affected geographic areas (consider both domestic and international sites), evacuating employees working in or near an affected area when an outbreak begins and guidance for employees returning from affected areas (refer to CDC travel recommendations).			
Set up authorities, triggers, and procedures for activating and terminating the company's response plan, altering business operations (*e.g.*, shutting down operations in affected areas) and transferring business knowledge to key employees.			

1.4 Allocate resources to protect employees and customers during a pandemic:

Action	Completed	In Progress	Not Started
Provide sufficient and accessible infection control supplies (*e.g.*, hand cleansing agents and products, tissues, and receptacles for their disposal) in all business locations.			

Action	Completed	In Progress	Not Started
Enhance communications and information technology infrastructures as needed to support employee telecommuting and remote customer access.			
Ensure availability of medical consultation and advice for emergency response.			

1.5 Communicate with and educate employees:

Action	Completed	In Progress	Not Started
Develop and disseminate programs and materials covering pandemic fundamentals (*e.g.*, signs and symptoms of influenza, modes of transmission), personal and family protection and response strategies (*e.g.*, hand hygiene, coughing/sneezing etiquette, contingency plans).			
Anticipate employee fear and anxiety, rumours and misinformation and plan communications accordingly.			
Ensure that communications are culturally and linguistically appropriate.			
Disseminate information to employees about your pandemic preparedness and response plan.			
Provide information for the at-home care of ill employees and family members.			
Develop platforms (*e.g.*, hotlines, dedicated Web sites) for communicating pandemic status and actions to employees, vendors, suppliers and customers inside and outside the worksite in a consistent and timely way, including redundancies in the emergency contact system.			
Identify community sources for timely and accurate pandemic information (domestic and international) and resources for obtaining countermeasures (*e.g.*, vaccines and antivirals).			

1.6 Coordinate with external organizations and help your community:

Action	Completed	In Progress	Not Started
Collaborate with insurers, health plans and major local healthcare facilities to share your pandemic plans and understand their capabilities and plans.			

Action	Com-pleted	In Prog-ress	Not Started
Collaborate with federal, state and local public health agencies and/or emergency responders to participate in their planning processes, share your pandemic plans and understand their capabilities and plans.			
Communicate with local and/or state public health agencies and/or emergency responders about the assets and/or services your business could contribute to the community.			
Share best practices with other businesses in your communities, chambers of commerce and associations to improve community response efforts.			

Table 12.6: Pandemic Flu Prevention: Recommended Personal Hygiene[16]

Individuals can take preventative actions, stay healthy, and avoid spreading influenza, whether it is the seasonal flu that circulates each winter or pandemic influenza, by:

- **Getting an annual flu shot:** Ensure family members do likewise.
- **Washing hands frequently:** Twenty seconds of handwashing with warm water and soap helps remove bacteria and viruses. Remember to wash before and after eating, after using the bathroom, after coughing or sneezing, and after touching surfaces that may have been contaminated by other people.
- **Keep your hands away from your face:** In most cases, the flu virus enters the body through the eyes, nose, and mouth.
- **Covering up when you cough or sneeze:** Use a tissue or raise your arm/elbow up to your face to cough or sneeze into your sleeve. If you use a tissue, dispose of it as soon as possible and wash your hands immediately.
- **Keeping shared surface areas clean:** Regularly clean doorknobs, light switches, telephones, keyboards, and other surfaces that can become contaminated with all kinds of bacteria and viruses.
- **Staying home when sick!** If you go out when sick, you may spread your illness to co-workers, classmates, neighbours, or others. It may take you longer to get better if you are not well rested. So, wait until

---

[16] Adapted from Public Health Agency of Canada, "Flu Prevention Checklist", online: http://www.phac-aspc.gc.ca/influenza/pdf/lang/english_flu_prevention_factsheet.pdf (date accessed: February 28, 2020).

you no longer have a fever and your cough is improving.
- **Talking about staying healthy.** Encourage others to follow these simple steps. If you have children, be a good role model. Teach them to count to 20 while washing their hands and show them how to cover up when they cough or sneeze.

It also promotes a business continuity management culture and a level of skill in the participants that will be critical in the management of a real pandemic.

(iii) Provisions for Care of Employees and Contract Workers (*Administrative Control Measures*)
- Recognize that employees with ill family members may need to stay home to care for them: plan for this eventuality.
- Work with insurance companies and provincial/local health agencies to provide information to employees and customers about medical care in the event of a pandemic.
- Work with employees and union(s) to address sick leave, disability pay, transportation, travel, childcare, employee non-medical absence, and other human resource issues.
- Develop a sick leave policy that does not penalize sick employees, thereby encouraging employees who have influenza-related symptoms (*e.g.*, fever, headache, cough, sore throat, runny, or stuffy nose, muscle aches, or upset stomach) to stay home and not infect other employees.
- Develop policies and practices that distance employees from each other, the customers, and the general public. Consider practices to minimize face-to-face contact between employees such as e-mail, Web sites, and teleconferences. Policies and practices that allow employees to work from home or to stagger their work shifts may be important as absenteeism rises.
- Recognize that, in the course of normal daily life, all employees will have non-occupational risk factors at home and in community settings that should be reduced to the extent possible. Some employees will also have individual risk factors that should be considered by employers as they plan how the organization will respond to a potential pandemic (*e.g.*, immuno-compromised individuals and pregnant women).
- Assist employees in managing additional stressors related to the pandemic. These are likely to include distress related to personal or family illness; life disruption; grief related to loss of family, friends or co-workers; loss of routine support systems; and similar challenges. Assuring timely and accurate communication will also be important throughout the duration of the pandemic in decreasing fear

or worry. Employers should provide opportunities for support, counselling and psychological health assessment, and referral should these be necessary. If present, EAPs can offer training and provide resources and other guidance on psychological health and resiliency before and during a pandemic.

(iv) Business Continuity Planning (*Administrative Control Measures*)

- Identify business-critical positions and people required to sustain business-necessary functions and operations. Prepare to cross-train or develop ways to function in the absence of these positions. It is recommended that employers train three or more employees to be able to sustain business-necessary functions and operations; and communicate the expectation for available employees to perform these functions if needed during a pandemic.
- Identify the "critical number of people and resources" required to maintain business operations;
- Plan for downsizing services but also anticipate any scenario which may require a surge in your services.
- Work with suppliers to ensure that the necessary products/equipment/services can be obtained.
- Stockpile items such as soap, tissue, hand sanitizer, cleaning supplies, and recommended personal protective equipment. When stockpiling items, be aware of each product's shelf life and storage conditions (*e.g.*, avoid areas that are damp or have temperature extremes) and incorporate product rotation (*e.g.*, consume oldest supplies first) into your stockpile management program.

These actions are not comprehensive in nature. The most important part of pandemic planning is to work with employees, provincial/local public health agencies, and other employers in the community to develop co-operative pandemic plans to maintain organizational/company operations and to keep employees and the public safe. Communicate what is known, be open to ideas from employees, then, identify and share effective health practices with other community members.

(e) Other Control Measures: Discussion

(i) Hierarchy of Controls

OH&S Practitioners/Professionals use a framework called the **hierarchy of controls** to select ways of dealing with workplace hazards. (Refer to Chapter 9, "Occupational Health and Safety: Hazard Identification, Assessment, and Control" for more details.) The hierarchy of controls prioritizes intervention strategies based on the premise that the best way to control a hazard is to systematically remove it from the workplace, rather than relying on employees to reduce their exposure. In the setting of a pandemic, this hierarchy should be used in concert with current public health recommendations. The types of measures that may be used are engineering controls, administrative controls, personal protective equipment (PPE), and the mitigation of consequences through emergency response planning. The control measures listed in the previous section were administrative control measures; however, employers tend to use a combination of these four control measures.

In terms of pandemic flu planning, there are advantages and disadvantages to each type of control measure when considering ease of implementation, effectiveness, and costs. For example, hygiene and social distancing can be implemented relatively easily and with little expense, but this control method requires employees to modify and maintain their behaviours, which may be difficult to sustain. On the other hand, installing clear plastic barriers or a drive-through window will be more expensive and take a longer time to implement, although in the long run may prove to be more effective at preventing transmission during a pandemic. Employers must evaluate their workplace to develop a plan for protecting their employees that may combine both immediate actions as well as longer-term solutions.

E. EMERGENCY RESPONSE PROGRAM: MEASUREMENT OF EFFECTIVENESS

It is not enough for an organization/company to develop an ERP; it needs to test the completeness and effectiveness of the plan. This involves:

- benchmarking the structure of the ERP against relevant standards, industry best practices, and "best in class" companies;
- assessing the various ERP processes;
- evaluating the outcome of various ERP drills; and
- conducting a *post-mortem* on actual emergency response activities.

The goal of ERP program evaluation is to continually improve the ERP so that future workplace/community losses are prevented or reduced. Key to this effort is to document the ERP program evaluation results so that the noted deficiencies and the related remedial interventions can be tracked. In short, the enhancements can be demonstrated.

F. SUMMARY

Although no organization/company wants to face an emergency or disaster situation, planning for an emergency and the required responses is vital to mitigate workplace losses. Emergency planning and preparedness enables an organization/company to responsibly take charge of the situation and limit the potential negative outcomes.

Appendix 1

Sample Emergency Response Plan[17]

Company Name: _____

Location: _____

POTENTIAL EMERGENCIES (Based on Hazard Assessment)	The following are identified potential emergencies: _____ _____ _____ _____
EMERGENCY PROCEDURES	In the event of an emergency (type or general) _____ occurring within or affecting the worksite, the (designated person) _____ _____ makes the following decisions and ensures the appropriate key steps are taken: • • •
LOCATIONS OF EMERGENCY EQUIPMENT	Emergency equipment is located at: • Fire Alarm: _____ • Fire Extinguisher: _____ • Fire Hose: _____ • Panic Alarm Button: _____ • Other _____

---

[17] Government of Alberta, Employment and Immigration, *Occupational Health and Safety: Tool Kit for Small Business* (Edmonton, AB: Government of Alberta, Employment and Immigration, 2011) at 44-47, online: http://work.alberta.ca/documents/OHS-Tool-Kit-Small-Business.pdf (date accessed: February 28, 2020).

EMERGENCY RESPONSE EQUIPMENT TRAINING & REQUIREMENTS (List of names of workers trained to use each type of equipment)	Name:	Training Received:	Frequency:
LOCATION AND USE OF EMERGENCY FACILITIES	The nearest emergency services are located: • Fire station: _____ • Ambulance: _____ • Police: _____ • Hospital: _____ • Other: _____		
FIRE PROTECTION REQUIREMENTS	• _____ are located_____ _____		
ALARM AND EMERGENCY COMMUNICATION REQUIREMENTS	• _____ _____ _____		

FIRST AID	First Aid Kit Type: _____
	Location: _____
	Other Supplies: _____
	First Aiders are: _____
	Workstation & Shift: _____
	Transportation Arrangements: _____

DESIGNATED RESCUE AND EVACUATION WORKERS	The following workers are trained in rescue and evacuation: (Name and area of expertise) _____ _____ _____ _____ _____

COMPLETED ON: _____
                        Date

SIGNED: _____

CHAPTER REFERENCES

CCOHS, "Emergency Planning, Health & Safety Programs" *OSH Answers*, online: http://www.ccohs.ca/oshanswers/hsprograms/planning.html (date accessed: February 28, 2020).

Centers for Disease Control (CDC), "Bioterrorism", online: http://www.bt.cdc.gov/bioterrorism/overview.asp (date accessed: February 28, 2020).

Government of Alberta, Employment and Immigration, *Occupational Health and Safety: Tool Kit for Small Business* (Edmonton, AB: Government of Alberta, Employment and Immigration, 2011), at 44-47, online: http://work.alberta.ca/documents/OHS-Tool-Kit-Small-Business.pdf (date accessed: February 28, 2020).

Health Canada, "Pandemic Flu Preparedness", online: http://www.hc-sc.gc.ca/hc-ps/ed-ud/prepar/flu-pandem/index-eng.php (date accessed: February 28, 2020).

Homeland Security Council, *Implementation Plan for the National Strategy for Pandemic Flu*, online: http://www.flu.gov/professional/federal/pandemic-influenza-implementation.pdf (date accessed: February 28, 2020).

Minnesota Department of Health, "Bioterrorism: Background and Significance", online: http://www.health.state.mn.us/bioterrorism/hcp/btbackground.ppt (date accessed: February 28, 2020).

Public Health Agency of Canada, "The Canadian Pandemic Influenza Plan for the Health Sector" (2011), online at: http://www.phac-aspc.gc.ca/cpip-pclcpi/ (date accessed: February 28, 2020).

Public Health Agency of Canada, "Flu Prevention Checklist", online: http://www.phac-aspc.gc.ca/influenza/flupc_e.html (date accessed: February 28, 2020).

U.S. Department of Health & Human Services, "Business Pandemic Influenza Planning Checklist", online: http://www.pandemicflu.gov/professional/pdf/businesschecklist.pdf (date accessed: February 28, 2020).

U.S. Department of Homeland Security (DHS), *Pandemic Influenza: Preparedness, Response and Recovery Guide for Critical Infrastructure and Key Resources*, online: http://www.flu.gov/professional/pdf/cikrpandemicinfluenzaguide.pdf (date accessed: February 28, 2020).

U.S. Department of Labor, "Guidance on Preparing Workplaces for an Influenza Pandemic", *OSHA-3327-05R*, online: http://www.osha.gov/Publications/OSHA3327pandemic.pdf (date accessed: February 28, 2020).

U.S. Department of Labor, "How to Plan for Workplaces Emergencies and Evacuation", *OSHA-3088*, online: http://www.osha.gov/Publications/osha3088.pdf (date accessed: February 28, 2020).

U.S. Department of Labor, "OSHA e-HASP Software — Version 2.0 (e-HASP2), March 2006", online: http://www.osha.gov/dep/etools/ehasp/index.html (date accessed: February 28, 2020).

U.S. Department of Labor, "OSHA Principle Emergency Response and Preparedness: Requirements and Guidelines" (OSHA 3122-06R - 2004), online: https://www.osha.gov/Publications/osha3122.pdf (date accessed: February 28, 2020).

U.S. Department of Labor, "OSHA Requirements for Emergency Response and Preparedness in Construction Industry 29 CFR 1926", online: http://www.osha.gov/dcsp/alliances/swri/swri_jc_030205bb/index.html (date accessed: February 28, 2020).

Chapter 13

OCCUPATIONAL HEALTH AND SAFETY: INJURY MANAGEMENT[1]

A. INTRODUCTION

Employers are legally required to respond to workplace injuries or illness.[2] This requirement covers first aid, emergency transportation, documentation, reporting, and assistance to the ill/injured employee for a safe and timely return-to-work. Although there are some application differences among Canadian provinces and U.S. states, this legal requirement for injury management remains similar.

This chapter is designed to address the importance of injury management; the preparatory measures for effective injury management; the role and responsibilities of the various stakeholders; the effective techniques; and the measurement of the effectiveness of the organization's/company's injury management.

B. INJURY MANAGEMENT: IMPORTANCE

Besides being legally required, **injury management** is an important management tool for:

- reducing the negative impact(s) of workplace illness/injury;
- reducing the potential for illness/injury to other workers;
- controlling the related costs; and
- expediting a safe and timely return-to-work by the ill/injured employee.

---

[1] Excerpts from D. Dyck, Disability Management: Theory, Strategy & Industry Practice, 6th ed. (Toronto: LexisNexis Canada, 2017).

[2] The Canadian OH&S Acts all require employers to initially respond to workplace injuries. For more details, refer to Chapters 30 "Occupational Health and Safety: Legal Aspects" and 31 "Canadian Workplace Safety Legislation", which deal with the related OH&S law.

C. INJURY MANAGEMENT: DEFINITIONS[3]

Using the Alberta OH&S Act definitions as examples, the injury management terms are:

Acute illness/injury means a physical injury or sudden occurrence of an illness, which results in the need for immediate temporary care.

First aid means the application of accepted principles of immediate and temporary treatment to sustain life, to prevent a condition from becoming worse, and to promote recovery using available equipment, supplies, facilities, and services to provide immediate and temporary care to an injured or ill worker.

First Aider means a person who is designated by an employer to provide first aid to workers at a worksite and who is an Emergency First Aider, Standard First Aider or Advanced First Aider.

Emergency First Aider means a person who holds a certificate in Emergency First Aid from a training agency.

Standard First Aider means a First Aider who holds a certificate in Standard First Aid from a training agency.

Nurse means a graduate of an approved registered nursing program who maintains membership and good standing with the provincial nursing association and who is an advanced first aider.

Healthcare facility means a hospital, medical clinic, or physician's office that has the capability of dispensing emergency medical treatment 24 hours a day.

Close worksite means a worksite that is not more than 20-minute travel time from a healthcare facility under normal travel conditions using the available means of transportation.

Distant worksite means a worksite that is more than 20 minutes, but less than 40 minutes, travel time from a healthcare facility under normal travel conditions using the available means of transportation.

Isolated worksite means a worksite that is more than 40 minutes of travel time from the worksite to a healthcare facility under normal travel conditions using the available means of transportation.

Low hazard work means work described in Figure 13.1.

Medium hazard work means work that is neither low hazard work nor high hazard work.

High hazard work means work described in Figure 13.1.

---

[3] Alberta *Occupational Health and Safety Code 2009* (revised 2018; current as of January 1, 2019), "Part 1: Definitions and General Information" (Edmonton, AB: Alberta Government, 2013), online: http://www.qp.alberta.ca/documents/OHS/OHS.pdf (date accessed: February 28, 2020).

Occupational injury is any "work-related" injury or illness suffered by an employee. It is work-related if an event or exposure in the work environment either caused or contributed to the resulting condition or aggravated a pre-existing condition. This includes psychological trauma.

First Aid Injury is an occupational injury that is treated by a physician/nurse or other healthcare provider and which includes:

- a visit limited to observation;
- diagnostic procedures, including the use of prescription medications solely for the purpose of diagnostic purposes;
- use of non-prescription medications, including antiseptics;
- simple administration of oxygen;
- administration of tetanus or diphtheria or booster shots;
- cleaning, flushing or soaking wounds on skin surface;
- use of wound coverings such as bandages, gauze pads;
- use of any hot/cold therapy except for musculoskeletal disorders;
- use of any totally non-rigid, non-immobilization means of support;
- drilling of a nail to relieve pressure;
- use of eye patches;
- removal of foreign bodies not embedded in the eye if only irrigation or removal with a cotton swab is required; and
- removal of a splinter/foreign material from areas other than eyes.

Medical Aid Injury is an occupational injury that results in medical care that is beyond first aid treatment, but that does not result in lost time from work.

Lost-Time Injury is an occupational injury that results in the worker losing time from work beyond the date of the incident or is likely to lose time in the future.

Fatality is any death resulting from a work injury regardless of the time intervening between injury and death.

Occupational disease is "a disease or ill health arising out of and directly related to an occupation".[4]

---

[4] *Occupational Health and Safety Act*, R.S.A. 2000, c. O-2, s. 1(ii).

Figure 13.1: Degree of Hazard[5]

LOW HAZARD WORK		Operation and maintenance of:	
Low hazard work means work at the following:		(a)	food packing or processing plants;
1.	Administrative sites where the work performed is clerical or administrative in nature;	(b)	beverage processing plants;
2.	Dispersal sites:	(c)	electrical generation and distribution systems;
	(a) where a worker is based;	(d)	foundries;
	(b) where a worker is required to report for instruction; and	(e)	industrial heavy equipment repair and service facilities;
	(c) from which a worker is transported to a worksite where the work is performed.	(f)	sawmills and lumber processing facilities;
HIGH HAZARD WORK		(g)	machine shops;
High hazard work means work involving the following activities:		(h)	metal fabrication shops;
1.	Construction or demolition, including:	(i)	gas, oil, and chemical process plants;
	(a) industrial and commercial process facilities;	(j)	steel and other base metal processing plants; and
	(b) pipelines and related gas/oil transmission facilities;	(k)	industrial process facilities not elsewhere specified.
	(c) commercial, residential and industrial buildings;	Woodlands operations;	
	(d) roads, highways, bridges, and related installations;	Gas and oil well drilling and servicing operations;	
	(e) sewage gathering systems;	Mining and quarrying operations;	
	(f) utility installations; and	Seismic operations; and	
	(g) water distribution systems.	Detonation of explosives.	

---

[5] Alberta *Occupational Health and Safety Code 2009* (revised 2018; current as of January 1, 2019), "Schedule 2" (Edmonton, AB: Alberta Government, 2013), online: http://www.qp.alberta.ca/documents/OHS/OHS.pdf (date accessed: February 28, 2020).

D. INJURY MANAGEMENT: KEY STAKEHOLDER ROLES AND RESPONSIBILITIES

The roles of the various stakeholders in injury management should be clearly defined by the employer. For example:

1. Employer

It is the employer's responsibility to ensure that:

- the required First Aid supplies and facilities are present, adequate, and in good working order; and
- at least one worker is certified in First Aid to a required level; that level is dependent on the number of employees present at the workplace and the level of hazard.

2. Line Manager

It is the Line Manager's duty to:

- ensure First Aider coverage at the worksite including coverage for breaks, vacation, and sick leave;
- ensure that an employee seeking outside medical aid or missing time from work due to a work-related illness/injury has the necessary forms for gathering information to enable a modified work opportunity; and
- ensure that the appropriate Workers' Compensation Board (WCB) forms are completed and submitted in a timely manner.

3. Director, OH&S

It is the OH&S manager's responsibility to:

- oversee the First Aid Policy and its administration;
- assess and monitor the various worksites in terms of their First Aid requirements;
- advise management of the various worksites' First Aid requirements;
- ensure that an incident investigation is started immediately post-incident occurrence;
- ensure that the requisite WCB forms and government notification are completed within the required timelines; and
- ensure that First Aid supplies and equipment are current, adequate, and readily accessible.

4. Training Department

It is the company/organization's Training Department's responsibility to:

- coordinate and monitor the quality of the First Aid Training provided to employees;

- retain First Aid Training logs for a period of seven years from the time the training was completed; and
- evaluate the First Aid Training program regularly.

5. First Aider

It is the First Aider's responsibility to:

- participate in First Aid training;
- ensure that an employee (or contractor/customer/visitor) receives medical aid if required and, if necessary, arrange transportation to receipt of medical aid; and
- report the incident on the organization's/company's incident form and submit to Line Management and the organization's/company's OH&S Department.

6. Employee

It is the employee's responsibility to:

- report any work-related injury/illness immediately to the site First Aider and Line Manager/Foreperson/Coordinator; and
- inform the Line Manager if the injury/illness results in a Lost-Time Injury, requires medical aid, or if the incident had the potential for serious injury (*i.e.*, near-miss incident).

The Canadian Workers' Compensation system is designed to compensate ill/injured workers for work-related diseases and injuries. It is a no-fault insurance system funded entirely by the employers. The provincial Workers' Compensation Acts govern the functioning of the WCB. It is important to remember that WCB benefits are a worker's statutory right. An effective claim management system never attempts to prevent workers from receiving benefits to which they are legitimately entitled.

E. INJURY MANAGEMENT: PREPAREDNESS

Companies/organizations are legally required to prepare for the eventuality of a workplace injury incident. To do this, they must attend to the following:

- **First Aid Policy**

    The company **First Aid Policy** states the company's commitment to the provision of injury management. A sample is:

    Company XYZ is committed to effectively manage workplace illness/injury through the provision of first aid services and maintenance of first aid equipment and supplies at the various company worksites in accordance with the applicable legislation.

- **First Aid Training**

    Companies must have individuals trained and available to provide first aid services if required. Many companies choose to train their workers to a basic or Standard Level of First Aid including Automatic External Defibrillator (AED) training. First aid training needs to be updated on a regular basis.

- **First Aid Supplies and Equipment**

    To effectively manage workplace illness/injury, companies must provide first aid services and maintain first aid equipment and supplies at their various worksites. To determine the level of first aid services and equipment required, each worksite should be assessed in terms of the degree of hazard (*e.g.*, high, medium, low), the proximity to the nearest healthcare facility, and the number of employees present. The greater the degree of hazard, the more distant the nearest healthcare facility, and the higher the number of employees, the more preparation and first aid supplies are needed. For example, the Alberta OH&S Code defines degree of hazard (Figure 13.1).

- **First Aid Services**

    First aid Treatment is to be provided within the scope of the First Aider qualifications. At high-hazard worksites, occupational health professionals may be on-site. In either case, the company should establish a protocol/standard for the treatment of ill/injured persons. A sample protocol is:

Treatment of Ill/Injured Persons

The First Aider provides first aid to the ill/injured person until help arrives using the following procedure:

- *Assess the situation and control any hazards.*
- *Call for help if necessary.*
- *Direct the nearest person to initiate the company Emergency Response Plan.*
- *Provide First Aid until help arrives.*
- *Keep the incident scene intact and secure.*
- *Report the incident to your Line Manager who will initiate an incident investigation.*
- *Employee and Line Manager to report the injury incident internally and externally.*

Exposure to Bloodborne Pathogens

- *Use of a CPR Mask is expected to promote personal protection when administering CPR.*
- *If in the course of providing first aid, the First Aider is exposed to blood or body fluids, immediately notify the direct supervisor and OH&S Practitioner/Professional.*

Administration of Medications

Prescription drugs or over-the-counter remedies are not to be administered by employees. They may assist an ill/injured employee to reach and use medication that has been prescribed for him/her, but not provide the employee with medication from other sources.

Transportation

- *Ill/injured employees are to be transported at company expense to the nearest healthcare facility by ambulance. To call an ambulance, dial 911 or 9-911.*

- *In those rare instances when an ambulance is unavailable, or deemed unnecessary, the employee is to be transported in a suitable vehicle that is equipped with a means of communication with the worksite and healthcare facility; that affords protection against the weather; and that is large enough to accommodate a stretcher.*

In either scenario, if the employee is acutely ill/injured, or needs to be accompanied during transport to a healthcare facility, a First Aider, other than the operator of the vehicle, must accompany the employee.

- **Reporting, Retention, Access, and Confidentiality**

  A workplace injury/illness must be reported using an appropriate incident report form and submitted to Management and the OH&S Department. This record is to be retained for a period of time commensurable with the applicable legislation. A sample Incident Report Form is provided in Appendix 1, and a sample First Aid Record developed by Alberta Jobs, Skills, Training and Labour[6] is provided in Appendix 2.

  Access to First Aid records is restricted to individuals requiring access, use, and disclosure of the information for the purpose of:

  - medical treatment;
  - worksite inspections;
  - incident investigation;
  - evaluation of health and safety programs and statistics; and
  - Workers' Compensation Board for the purposes of worksite health and safety audit programs.

  Because of the nature of the information obtained, confidentiality of First

---

[6] Alberta Jobs, Skills, Training and Labour, *Occupational Health and Safety Tool Kit* (Edmonton, AB: Alberta Jobs, Skills, Training and Labour, 2011) at 52, 53, online: http://work.alberta.ca/documents/OHS-Tool-Kit-Small-Business.pdf (date accessed: February 28, 2020).

Aid information is legally required. Persons with access to the First Aid records must keep the information confidential in accordance with the provincial OH&S Act and the applicable privacy legislation.

F. INJURY MANAGEMENT: EFFECTIVE TECHNIQUES

There are three well-recognized techniques for injury management:

1. Management of Workplace Injury Incidents

In the event of a workplace injury incident, the organization/company should have an ERP developed for the safe and timely management of workplace injuries.[7] The following is an example of the recommended actions:

- *Assess the situation.*
- *Contact 911.*
- *Control or remove the hazards to self and victim.*
- *Provide First Aid and/or initiate the Emergency Response Plan for the area. This initiates a cascade of events in accordance with the severity of the injury.*
- *Ensure that the scene of the incident is left undisturbed except to:*
    - *attend to persons injured;*
    - *prevent further injuries; and/or*
    - *protect the public/property that is endangered as a result of the incident.*
- *Ensure that a management representative accompanies the injured employee to the medical facility.*
- *Take the necessary actions to notify and assist the injured employee's family.*
- *The Director, OH&S, reports the incident to government authorities.*
- *The Line Manager completes the company incident form (A sample Incident and Investigation Report is provided in Appendix 1) and WCB forms.*
- *All statements to the press are to be issued through the company Communications Department.*
- *Co-operate completely with government OH&S Officers and the police in investigating the incident.*
- *Initiate critical incident debriefing if warranted (psychological first aid).*
- *The Director, OH&S, or designate, will issue a preliminary report within 24 hours followed by a final investigation report, and arrange for critical*

---

[7] Refer to Figure 16.4: First Aid Response Steps, at Chapter 16 "Marketing Occupational Health and Safety Programs; Communicating the Results".

incident debriefing for employees/dependants through the Employee and Family Assistance Program.

As part of the preparedness planning for a workplace injury, responders should be provided with guidelines on which injuries/illnesses warrant transportation to medical service, for example, electrical shock, exposure to chemicals, radiation exposure, head injuries, eye injuries, falls, hand injuries, and the like.

2. Disability Management

Disability management is a coordinated approach to injury/illness management. It includes:

- implementing claim management processes;
- undertaking early intervention;
- implementing case management practices;
- providing employee guidance towards responsible healthcare services (case management);
- facilitating medical/vocational fitness-to-work evaluations (*i.e.*, medical forms, job demands analyses, functional capacity assessment referrals, *etc.*);
- implementing multi-disciplinary interventions into the disability management process;
- developing alliances/linkages with external resources (*i.e.*, healthcare providers, EAPs, vocational rehabilitation, insurers, *etc.*);
- promoting early rehabilitation/retraining for the recovering employee;
- gathering injury/illness data using disability management information systems; and
- regularly evaluating the claim and case management processes with a view to continuous improvement. For example, a report by the Institute for Work and Health found that one month after injury, almost half the workers had high levels of depressive symptoms:

    > Six months after injury, almost 40 per cent of workers who were not back at work, still had high levels of depressive symptoms. Most injured workers with ongoing symptoms of depression did not seem to be getting treatment.[8]

Chapter 27, "Disability Management: Overview", provides more information on Disability Management Programs.

3. Return-to-Work Assistance

Graduated return-to-work opportunities are intended to assist recovering employees in safely returning to the workplace and, ultimately, to regular full-time employ-

---

[8] *Impact*, Institute for Work and Health Annual Report (2008) at 5-6.

ment. Often, a Return-to-Work Coordinator is involved to facilitate the process. However, regardless of who is involved, a successful graduated return-to-work outcome depends on a cooperative and collaborative approach between the employee, union representative and Management.

A Graduated Return-to-Work Plan is designed to achieve the following objectives:

1. ensure fair and consistent treatment for all employees who are returning to work;
2. promote shared responsibility for effective return-to-work plans and placements among supervisors, union representatives, ill/injured employees, Occupational Health Nurses, and the Disability Management Coordinator, if applicable;
3. provide coordinated claim and case management services for the ill/injured employee; and
4. mitigate medical absence costs associated with disability claims.

A Graduated Return-to-Work Plan is based on a number of principles. Some examples are as follows:

1. *A safe and timely return to work is in the best interest of the ill/injured employee and the organization.* The employee benefits from having meaningful employment, gradual work conditioning, and the social supports associated with being at work, when deemed appropriate. The organization is able to mitigate the costs associated with lost production, hiring and training replacement workers, and re-scheduling of other workers. Supporting the recovering employee to return to productive work minimizes the direct and indirect costs associated with disability.

2. *Early intervention is critical to achieving a positive return-to-work experience.* It can:
   - help the employee to receive appropriate and timely care;
   - help with the physical, social, psychological, vocational, and financial implications of illness/injury;
   - increase the likelihood of successful rehabilitation;
   - facilitate the process of coping and adjustment for the employee, family, and workgroup;
   - promote a safe, timely and successful return to work; and
   - be cost-effective for the employee, family, and employer.

3. *A positive approach to disability is advantageous.* This means focusing on the person's capabilities and the contributions that he/she can make to the workplace. By bringing ill/injured employees back into the workplace, the organization and unions can demonstrate the belief that each employee,

regardless of disability, has abilities that can be valuable. This approach can enhance employee morale.

4. *Graduated Return-to-Work Plans should include meaningful, goal-oriented work that matches the employee's capabilities.* Modifications consider the type of work to be performed and the hours to be worked.

5. *Employees should be compensated in accordance with the work performed.*

6. *Crossing union jurisdictional issues must be addressed and resolved for a Graduated Return-to-Work Program to function successfully.*

7. *A Graduated Return-to-Work Plan must recognize the employee's diminished capability and not compromise the employee's recovery or safety.*

8. *A Graduated Return-to-Work Plan must ensure that the general workplace safety is not compromised.*

9. *The Graduated Return-to-Work Plan is not a disciplinary tool.* Performance issues are to be resolved through the appropriate administrative processes and collective agreements.

10. *A Graduated Return-to-Work Plan may include a return to:*

    - the employee's own job with reduced hours;
    - a portion of the employee's own job duties with full-time or part-time hours;
    - a different job within the employee's department on a full-time or part-time basis;
    - an unrelated job in another department on a full-time or part-time basis; and/or
    - a new job outside of the organization on a full-time or part-time basis.

The Hierarchy of Return-to-Work Options is provided in Figure 13.2.

---

[9] D. Dyck, *Disability Management: Theory, Strategy & Industry Practice*, 6th ed. (Toronto: LexisNexis Canada, 2017) at 200.

CH. 13: OCCUPATIONAL HEALTH AND SAFETY: INJURY MANAGEMENT

Figure 13.2: Hierarchy of Return-to-Work Options[9]

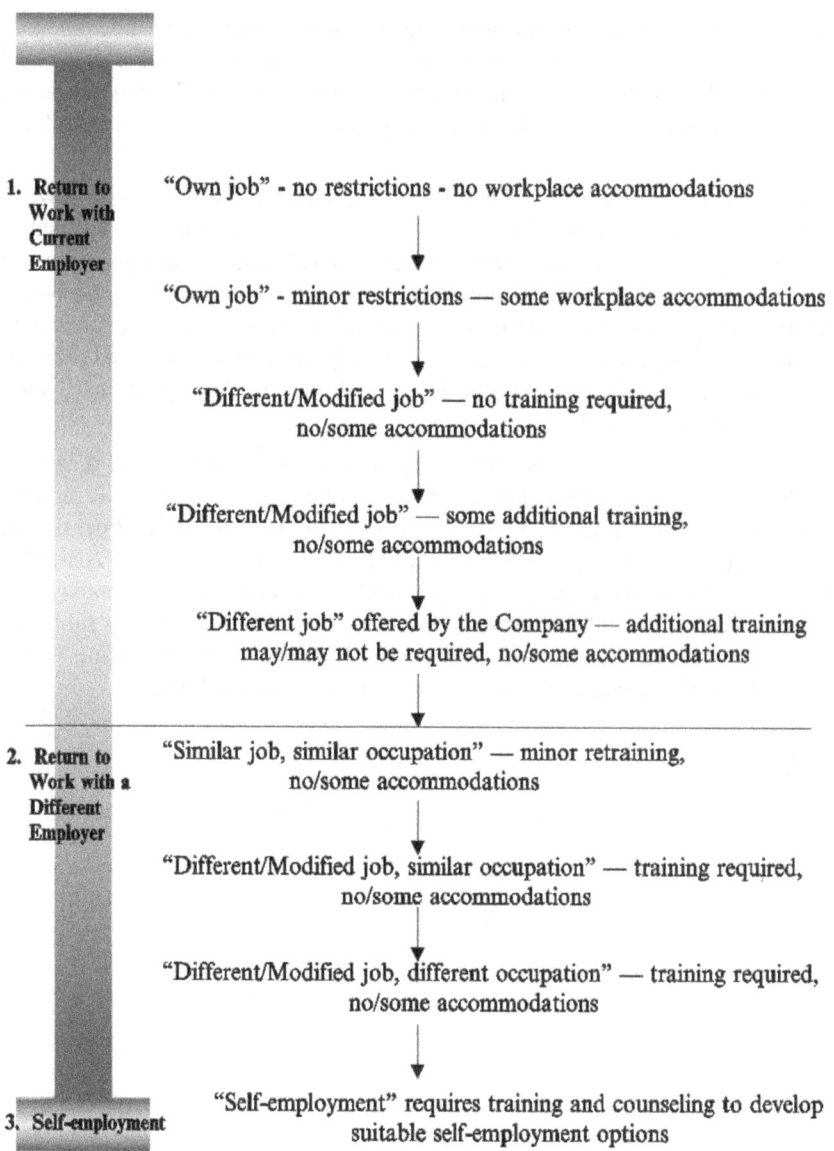

G. INJURY MANAGEMENT: MEASUREMENT OF EFFECTIVENESS[10]

What benefits can be realized by injury management? Realistically, a decline in the number of WCB claims and a reduction in lost-time hours and costs, can be expected. Additionally, the information gathered from illness/injury reports can help identify emerging problems or the development of trends that warrant further investigation.

The evaluation process can occur at many levels. *At the individual case level*, the process and results are continually reviewed throughout the course of the disability and improvements are sought. *At the program level*, injury management results, costs, system concerns, and recommendations are analyzed and reported periodically to management. Confidentiality of individual information is maintained in accordance with medical or nursing confidentiality codes of practice and provincial privacy legislation. *At the process level*, auditing of the injury management practices is recommended.

An example of a spreadsheet for setting up an injury management database is provided in Table 13.1, which indicates that 21 employees were on WCB for a total of 1,196 days at an average of 57 workdays per case. At a rate of $233 per day, this totals $278,668 and averages $13,269.90 per case. As well, 13 (62 per cent) of the employees returned to modified/alternate work for a total of 503 workdays. The dollar saving by the modified/alternate work initiative is $117,199. By subtracting the "days saved" from the total short-term disability days, the "time lost" is calculated. In this scenario, that is 693 days at a cost of $161,469.

---

[10] *Ibid.*

CH. 13: OCCUPATIONAL HEALTH AND SAFETY: INJURY MANAGEMENT

Table 13.1: Injury Management Spreadsheet

Name	WCB Start	WCB End	WCB Length	Nature of Injury	First Aid	MWP	Days Saved
X	19.08.06	19.09.20	32	Leg fracture	Y	N	0
A	19.08.12	19.10.01	35	Arm fracture	Y	N	0
C	19.08.26	20.01.10	92	Back	N	Y	41
B	19.09.03	20.03.03	124	Back	N	Y	66
D	19.09.03	20.03.03	124	R.S.I.	N	N	0
I	19.07.02	19.11.13	90	R.S.I.	N	Y	52
Y	19.09.16	19.10.25	29	Ankle sprain	Y	Y	10
M	19.09.09	20.02.03	119	Knee sprain	N	Y	72
J	19.08.01	19.08.23	15	Elbow laceration	Y	N	0
N	19.09.12	19.09.20	7	Eye injury	Y	N	0
T	19.09.03	19.10.15	30	Foot laceration	Y	Y	15
U	19.09.03	19.09.13	9	Hand laceration	Y	N	0
W	19.09.24	19.11.22	42	Back injury	N	N	0
O	19.09.18	19.09.25	6	Knee sprain	Y	N	0
P	19.09.19	19.10.07	11	Neck sprain	N	Y	5
L	19.09.12	19.11.01	36	Back sprain	N	Y	5
V	19.10.22	20.04.20	124	Back injury	Y	Y	38
E	19.10.22	19.12.02	29	Fractured ribs	Y	Y	19
F	19.10.07	20.04.07	125	Back	N	Y	78
K	19.10.16	20.03.11	100	Back injury	N	Y	97
Q	19.10.30	19.11.22	17	Burn	Y	Y	5
21 Employees			1196 Days		11 First Aid	13 Employees	503 Days

Legend: R.S.I. = Repetitive Strain Injury

Using injury management data, many types of reports can be generated. Table 13.2 is an example of an annual report.

Table 13.2: Annual Results of the Injury Management
Summary Statistics: Injury Management

	Results for 1 year	
Number on WCB	103 employees	
Total WCB days	3309 days	$ 770,997*
Average total WCB time	32 days	$ 7,456
% on MWP**	44%	
Days saved	1382 days	$ 322,006
Actual WCB time	2124 days away	$ 494,892
Average WCB time	21 days away	$ 4,893

* WCB Cost = $770,997: ** MWP = Modified/alternate Work Program

The cost to have 103 employees on WCB was $770,997 for the year. The average time off was 32 days. This meant an average of $7,456 per case. By placing 44 per cent of the recovering employees on modified/alternate work, 1,382 days were saved — a saving of $322,006.

There are other ways to evaluate success. The following are a few techniques that may be employed.

1. Determine the "difference" that injury management has made:
   - Calculate the total number of employees on WCB before the program began and after it had been operational for one year. Compare the differences.
   - Quarterly and annual comparisons can also be made as shown in the example in Table 13.3.

Table 13.3: Comparison of Results for Four Quarters, 2019

	1st Q	2nd Q	3rd Q	4th Q
Number on WCB	30	31	50	41
Total WCB days	882	823	955	960
Average WCB time (days)	29.4	26.5	19.1	23.6
% on MWP	20%	48%	64%	58%
Days saved	204	318	414	429

2. Calculate the cost of the lost-time claims:
   - Establish the total number of days on WCB and then multiply that number by the average, or actual employee salary including the burden factor. Some companies use a set cost, such as an average salary of $233 per day as the "sick leave cost" (Table 13.4). This is increased to $633 per day if a replacement worker is required for that period of time. In this scenario, the "replacement cost" is $400 per day.

Table 13.4: Quarterly Report on Injury Management
July 1, 2019 to September 30, 2019

Number of employees on WCB	30 employees
Total WCB days	882 days
Average WCB time per employee	29.4 days
Total WCB costs	$205,506*
Average WCB cost per employee	$ 6,850.20
WCB costs per month	$ 68,502

* Total WCB costs = WCB Cost (CDN $233 per day) × Total Lost-time days

3. Calculate the total "days saved":
   - This can be done two ways, or by combining both:
     (a) determine the difference between the predicted and actual return-to-work time; or

(b) determine the number of days each employee is on modified/alternate work.

Either of these methods yields the "days saved" by the injury management efforts.

- To place a dollar-savings on this figure, multiply the number of "days saved" by the "sick-leave cost", with or without the "replacement cost", as the actual case might be.

4. Calculate the causes of the WCB claims:
   - Determine the cause of each claim (Table 13.5).

Table 13.5: Causes of Workplace Injuries

n = 99 WCB cases

Cause of WCB	Per cent
Strains/sprains	47%
Fractures	20%
Lacerations	17%
Eye injuries	8%
Repetitive Strain disorders	6%
Burns	2%

5. Calculate the cost of each cause to the organization/company:
   - This involves determining the "time lost" for each category of causes and assigning a dollar figure to the cause (Table 13.6).

Table 13.6: Savings Realized by Modified/Alternate Work Initiatives

n = 48 modified/alternate work opportunities

Condition	Employees Returned to Work	Average Days Saved	Savings per Claim
Back	3/3	51	$ 11,883
Sprains/Strains	7/11	46	$ 10,718
Lacerations	3/5	28	$ 6,524
Eye injuries	2/5	24	$ 5,592
Fractures	8/16	21	$ 4,893
Burns	1/3	93	$ 21,669
Repetitive Strain disorders	1/3	43	$ 10,019

6. Calculate the success of modified/alternate work initiatives:
   - Examine the data on all the WCB claims that worked modified/alternate work. Subdivide the group by the disability causes. Total the number of "days saved" for each case in each disability subgroup and

then, add a dollar figure to the "days saved" for each. This will indicate where the biggest differences were made by modified/alternate work (Table 13.7).

Table 13.7: Causes of WCB Per cent of Total WCB Injury Days (3,208 days)

Cause	WCB Days	Per cent of Total WCB Claims
Strains/sprains	1050	33%
Fractures	603	19%
Lacerations	320	10%
Eye injuries	252	8%
Repetitive Strain disorders	232	7%
Burns	177	6%

- Table 13.7 shows that all the employees with back disabilities went to modified/alternate work and an average of 51 days were saved for each case. This translates to a savings of $11,883 per case. In comparison, only eight out of 16 of the fracture-claims returned early to modified/alternate work. An average of 21 days per case was saved, $4,893 per case.
- This type of data helps OH&S professionals to decide where and how they can make the biggest difference to the organization's/company's "bottom line".

7. The final way of evaluating the injury management efforts is to do an audit. An example of some auditing questions is provided in Table 13.8.

Table 13.8: Injury Management Audit Questions

Audit Questions	Comments
Were first aid services provided at the worksite?	
Was the initial injury management protocol followed?	
Were the appropriate forms provided to the injured worker?	
Did injured employees receive the right injury management services at the right time?	
Is the Disability Case Manager notified of the injury on day #1?	
Is the standardized disability management approach used in all cases?	
Who are the usual members of the disability management team?	
How often is modified/alternate work usually available?	
Is the modified/alternate work considered "gainful employment"?	
Are supervisors pro-active in the injury management process?	
How does senior management demonstrate support for the various injury management efforts?	

How do union leaders demonstrate support for the injury management efforts?	
What percentage of employees return to full duties within one day? One week? One month? Three months? Six months?	
How many Workers' Compensation claims move to a long-term disability status?	
What types of workplace accommodations are made?	

H. SUMMARY

Injury management is an effective approach to manage workplace injury claim numbers, durations, negative outcomes, and injury costs. Injury management as part of a Disability Management Program can help organizations/companies cope with the rising disability costs, the challenges of an aging and changing workforce, the effects of uncertain economic times on disability costs, the intricacies of a complex healthcare system, and the administration of disability cases.

By working with employees, their medical practitioners, community agencies, and Employee Assistance Program, Human Resources, Disability Management, and OH&S Practitioners/Professionals can help to rehabilitate injured employees, thereby getting them back to work earlier. Return-to-Work Programs can facilitate that outcome.

Appendix 1

SAMPLE OCCUPATIONAL HEALTH AND SAFETY INCIDENT & INVESTIGATION REPORT FORM

| OH&S Incident # | |

APPENDIX 1
OCCUPATIONAL HEALTH AND SAFETY INCIDENT & INVESTIGATION REPORT

Incident - an unplanned event resulting in or having a potential for injury, ill health, damage, or other loss, including a near-miss event.

Near-miss - an incident or event that in different circumstances may have caused injury, ill health, damage or other loss.

First Aid Incident – an incident that requires first aid treatment only.

Medical Aid Incident – an incident that results in medical care that is beyond first aid treatment, but that does not result in lost time from work.

Lost-time Incident - any injury or illness that results in lost-time workdays.

Serious Injury – an incident that results in hospitalization or death.

**Incident investigations are not intended to assign blame.
They are undertaken to prevent a recurrence of an incident.**

Part A – Incident Report
This part of the form to be completed by person taking charge of the incident scene as soon as possible and handed to the supervisor/manager of the area.

1. Incident Details
(Tick relevant boxes ☐ or write in relevant details)

Incident type	Severity of incident	Number injured/ill	Exact location of incident
☐ Near-miss ☐ First Aid ☐ Medical Aid ☐ Lost-time Injury ☐ Fatality	☐ Potential to cause injury ☐ Minor (First Aid less than one day off work) ☐ Moderate (Doctor and/or 1-5 days off work) ☐ Major (Hospitalization and/or 5+ days off work) ☐ Fatality	How many people were injured?	

Date/Time of incident
 	
Date of incident:
 	
Time of incident:
 	
Shift: ☐ day
      ☐ afternoon
      ☐ night
	Date reported:
	
	Reported by:
	
	Reported to:
		Management Informed
		
		Which Manager was first informed?
		
		Time and date informed:	

OHS Notification Details	*[must be filled out for Serious Injury/Fatality]*		
Has OHS been advised? ☐ Yes ☐ No		☐ No, not a Serious Injury	

Attach copy of completed OH&S report to this form

Copyright Jon Everest, Heidi Börner and Dianne Dyck 2001. Printed with permission.

| OH&S Incident # | |

2. Description of Incident

Describe activities immediately **prior to** the incident:

Describe the incident (include individuals involved and witnesses):

(Continue on a separate sheet of paper)

Draw a diagram or take photos of the accident scene

(Attach on separate sheets if necessary)

3. Eyewitness(es)
(Complete this information for anyone who saw the incident happen)

Name	Employee/Patient/Contractor/Visitor	Contact Tel. No	Date statement taken

Copyright Jon Everest, Heidi Börner and Dianne Dyck 2001. Printed with permission.

| | | OH&S Incident # | |

4. Injured Person Details
(Photocopy this page and complete one for each person injured in this incident)

Name:	Contact Tel:
Address:	Date of Birth:
	Sex: ☐ Male ☐ Female
Employee: Patient: Contractor: Visitor:	**Complete for Injured Employee** Employee number: Job Title: Start date at this location: Start date on this job: Scheduled length of shift: Number of hours at work when event occurred:

Body Parts Affected
(tick relevant boxes ☐ or write in relevant details)

Head	Body	Arm	Leg	
☐ Head	☐ Neck	☐ Upper arm	☐ Hip/upper leg/thigh	☐ Multiple body parts
☐ Face	☐ Shoulder	☐ Elbow	☐ Knee	
☐ Eye	☐ Upper back	☐ Lower arm	☐ Lower leg	
☐ Nose	☐ Lower back	☐ Wrist	☐ Ankle	
☐ Ear	☐ Chest	☐ Hand including fingers	☐ Foot	
	☐ Abdomen		☐ Toes	

Injury Type
(tick relevant boxes ☐ or write in relevant details)

☐ Aches/pain (gradual)	☐ Chemical reaction	☐ Fatal	Poisoning
☐ Amputation	☐ Choking/suffocation	☐ Foreign Body	Psychological
☐ Blood/body fluid exposure	☐ Cut	☐ eye ☐ nose ☐ ear	☐ Strain/sprain
☐ Broken bone	☐ Dental injury	☐ Head injury	☐ Multiple injuries
☐ Bruising incl. crushing	☐ Dermatitis	☐ Hearing loss	☐ Other (state what)
☐ Burn/scald	☐ Dislocation	☐ Inhalation disease (asbestos/lead)	

Treatment Details
(tick relevant boxes ☐ or write in relevant details)

None ☐ First Aider ☐ Nurse   ☐ Doctor   ☐ Hospital (show name- _____)

Name of Doctor/Nurse/First Aider treating	

Copyright Jon Everest, Heidi Börner and Dianne Dyck 2001. Printed with permission.

CH. 13: OCCUPATIONAL HEALTH AND SAFETY: INJURY MANAGEMENT

| OH&S Incident # | |

5. Hazard Information

Object/Equipment/Substance inflicting injury/damage:	Person in control of object/equipment/substance at time of incident:
Was this company equipment? Yes No	Is the hazard involved in this incident included in the Hazard Register and Hazard Control Log? Yes No
External equipment involved in incident	Name and address of owner of external equipment

6. Equipment/Property Damage

Item(s) Damaged	Nature of Damage	Estimated Cost	Actual Cost
1.			
2.			
3.			
4.			

Form completed to this point by:

Name	Position	Contact Tel.	Date	Signature

Thank you for completing this form.
Please give it to your team leader / supervisor for investigation.

Form received by:

Name	Position	Contact Tel.	Date	Signature

Copyright Jon Everest, Heidi Börner and Dianne Dyck 2001. Printed with permission.   4

	OH&S Incident #

Part B – Investigation Report
This part of the form to be completed by the assigned incident investigator.

Investigation assigned to	
Date investigation opened	

Before proceeding, check that all details for Part A have been completed. Has all the necessary information been obtained?   Yes   No

The Injury Factors Worksheet can be used to determine the factors that contributed to this incident. Once the factors have been identified, complete the tables below.

7. Injury Factors
(tick relevant boxes ☐ or write in relevant details)

Substandard Acts/Practices:	Substandard Conditions:
☐ Operating equipment without training/authority	☐ Inadequate guards or barriers
☐ Failure to warn	☐ Inadequate or improper protective equipment
☐ Failure to secure	☐ Defective tools, equipment or materials
☐ Operating at improper speed	☐ Congestion or restricted action
☐ Failure to follow procedures	☐ Inadequate warning system
☐ Removing or making safety devices inoperable	☐ Fire and explosion hazards
☐ Using defective equipment	☐ Poor housekeeping/disorder
☐ Using equipment improperly	☐ Hazardous environmental conditions (gases, dusts, smokes, fumes, vapours)
☐ Failing to use personal protective equipment properly	☐ Noise exposures
☐ Improper loading	☐ High or low temperature exposures
☐ Improper placement	☐ Inadequate or excess illumination
☐ Improper lifting	☐ Inadequate ventilation
☐ Improper position for task	
☐ Servicing equipment in operation	
☐ Horseplay	
☐ Under influence of alcohol /drugs	

8. Root Causes
(tick relevant boxes ☐ or write in relevant details)

Personal Factors:	Job/System Factors:
☐ Inadequate capability	Inadequate leadership/supervision
☐ Lack of knowledge	Inadequate staffing
☐ Lack of skill	Inadequate engineering
☐ Stress	Inadequate purchasing
☐ Improper motivation	Inadequate maintenance
	Inadequate tools/equipment
	Inadequate work standards/procedures
	Wear and tear
	Abuse or misuse

Copyright Jon Everest, Heidi Börner and Dianne Dyck 2001. Printed with permission.   5

CH. 13: OCCUPATIONAL HEALTH AND SAFETY: INJURY MANAGEMENT

OH&S Incident #

9. Type of Incident
(tick relevant boxes)

Type of Contact:	Contact with:
Struck against	Electricity
Struck by	Heat
Caught in	Cold
Caught on	Radiation
Caught between	Caustics
Slip	Noise
Fall on same level	Toxic or noxious substances
Fall to below	
Overexertion	

9. Risk rating
(Use the risk matrix to determine the risk rating for this incident)

		Probability of recurrence			
		Likely to occur immediately	Probable in time	Possible in time	Remotely possible
S E V E R I T Y	Fatality or permanent total disability	Critical	Critical	Serious	Moderate
	Lost time injury	Critical	Serious	Moderate	Minor
	Reportable Injury (no lost time)	Serious	Moderate	Minor	Negligible
	First Aid treatment	Moderate	Minor	Negligible	Negligible

RISK RATING FOR THIS INCIDENT

10. Does Hazard Register and Hazard Control Log address this incident?	Yes	No

If yes, how and why was the Hazard Control Measure ineffective?

Copyright Jon Everest, Heidi Börner and Dianne Dyck 2001. Printed with permission.   6

OH&S Incident #

11. What corrective actions should be taken to prevent recurrence?

Corrective action	Assigned to (name)	Scheduled completion date	Date completed	Sign-off action completed (signature)
1.				
2				
3.				
4.				
5.				
6.				

12. Investigation Log

Action	Name	Job Title	Date completed	Signature
OSH informed by				
Incident investigated by				
Reviewed by Health and Safety Committee				
Reviewed by Senior Management				
Hazard Register and Control Log updated by				
All remedial actions completed				
Further comments:				

Copyright Jon Everest, Heidi Börner and Dianne Dyck 2001. Printed with permission.   7

Appendix 2

Sample First Aid Record Form[11]

Date of injury or illness: _____ Time: _____ AM ☐
                                Day   Month   Year   PM ☐

Date of injury or illness: _____ Time: _____ AM ☐
Reported to First Aider: Day   Month   Year   PM ☐

Full name of injured or ill worker: _____

Description of the injury or illness:

_____

_____

_____

Description of where the injury or illness occurred/began:

_____

_____

_____

Cause of the injury or illness:

_____

_____

_____

First aid provided? Yes ☐ (If yes, complete the rest of this page) No ☐

Name of first aider: _____

First aid qualifications:

---

[11] Alberta Jobs, Skills, Training and Labour, Occupational Health and Safety Tool Kit (Edmonton, AB: Alberta Jobs, Skills, Training and Labour, 2011) at 52, 53, online: http://work.alberta.ca/documents/OHS-Tool-Kit-Small-Business.pdf (date accessed: February 28, 2020).

Emergency First Aider	☐	Emergency Medical Technician–Paramedic	☐
Standard First Aider	☐	Emergency Medical Technician–Ambulance	☐
Advanced First Aider	☐	Emergency Medical Technician	☐
Registered Nurse	☐	Emergency Medical Responder	☐

First Aid provided:

_____

_____

_____

CONFIDENTIAL

Keep this record for at least 3 years from the date of injury or illness

CHAPTER REFERENCES

Alberta Occupational Health and Safety Code 2018, "Part 1: Definitions and General Information" (Edmonton, AB: Alberta Government, 2018), online: http://www.qp.alberta.ca/documents/OHS/OHS.pdf (date accessed: February 28, 2020).

Alberta Occupational Health and Safety Code 2018, "Part 11 — First Aid" (Edmonton, AB: Alberta Government, 2018), online: http://www.qp.alberta.ca/documents/OHS/OHS.pdf (date accessed: February 28, 2020).

Alberta Occupational Health and Safety Act 2018, "Schedule 2 — First Aid, Table 2" (Edmonton, AB: Alberta Government, 2018), online: http://www.qp.alberta.ca/1266.cfm?page=O02P1.cfm&leg_type=Acts&isbncln=9780779800865&display=html (date accessed: February 28, 2020).

Alberta Jobs, Skills, Training and Labour, *Occupational Health and Safety Tool Kit* (Edmonton, AB: Alberta Jobs, Skills, Training and Labour, 2011) at 52, 53, online: http://work.alberta.ca/documents/OHS-Tool-Kit-Small-Business.pdf (date accessed: February 28, 2020).

Alberta Jobs, Skills, Training and Labour, *Occupational Health and Safety Tool Kit* (Edmonton, AB: Alberta Jobs, Skills, Training and Labour, 2011) at 52, 53, online: http://work.alberta.ca/documents/OHS-Tool-Kit-Small-Business.pdf (date accessed: February 28, 2020).

D. Dyck, *Disability Management: Theory, Strategy & Industry Practice*, 6th ed. (Toronto: LexisNexis Canada, 2017).

Impact, Institute for Work and Health Annual Report (2008). *Occupational Health and Safety Act*, R.S.A. 2000, c. O-2.

Chapter 14

OCCUPATIONAL HEALTH AND SAFETY: INCIDENT INVESTIGATION

A. INTRODUCTION

Thousands of accidents/incidents occur in Canada and the United States (U.S.) every day. In the U.S., the lost-time injury/illness rate is 2.8 claims per 100 full-time workers.[1] In Canada, 29 workers experience a lost-time injury claim per hour.[2] This is a decrease from 2005 when 38 workers sustained a lost-time injury claim per hour.[3] Most incidents (85 per cent) are the result of management system failures.[4] Incidents are usually complex and tend to have multiple causes. Incident investigations determine how and why these failures occur and are aimed at unearthing the root cause(s). By using the information gained, similar, or perhaps more disastrous, incidents can be prevented.

This chapter is arranged so that incident investigation terms and the value provided to the organization/company are first explained, followed by a discussion of the various accident causation theories and their implications for incident investigations and the field of Occupational Health and Safety (OH&S) practical industry applications.

---

[1] U.S. Department of Labor, Bureau of Labor Statistics, "Employer-reported Workplace Injuries and Illnesses - 2018" (2020), online: https://www.bls.gov/news.release/pdf/osh.pdf (date accessed: February 28, 2020).

[2] In 2018, there were 264,438 lost-time injury/disease claims accepted by provincial Workers' Compensation Boards throughout Canada. Dividing that by 365 days and then by 24 hours, the result is 29 lost-time injury claims per hour. Source: Association of Workers' Compensation Boards of Canada, "2018 Injury Statistics; Statistics", online: http.//awcbc.org/?page_id=14 (date accessed: February 28, 2020).

[3] In 2005, there were 337,930 lost-time injury claims accepted by provincial Workers' Compensation Boards throughout Canada. Dividing that by 365 days and then 24 hours, the result is 38 lost-time injury claims per hour. Source: Association of Workers' Compensation Boards of Canada, "Key Statistical Measures for 2005", Workers' Compensation Board/Commission Financial and Statistical Data, online: http://www.awcbc.org/english/en/index.asp (date accessed: February 28, 2020).

[4] F. Bird & G. Germain, *Loss Control Management: Practical Loss Control Leadership*, revised ed. (Loganville, GA: DNV, 1996).

B. DEFINITIONS

Incident is an event or set of circumstances that could, or does, result in an unintended harm or damage. It includes near-miss, damage, chemical spills, and injury incidents.

Accident is an unplanned event, or set of circumstances, which results in harm or damage. It includes ill health/injury; damage to property, plant, products or the environment; production losses; or increased liabilities.

For the purposes of this book, the term "incident" will be used because of its universal application.

C. INCIDENT INVESTIGATION

Incident investigations should be carried out promptly post-incident using a systematic process. The purpose of the investigation is not to find fault, but rather to identify the root causes of the incident and to develop suitable corrective actions to prevent a recurrence. The benefit to companies and their employees is the maintenance of safe work environments through effective hazard management and mitigation of risk, and a healthy future for all involved.

The rationale for incident investigation is:

1. *Moral Logic* — Working or permitting unsafe work to be done unsafely is unethical and morally unacceptable in today's society.
2. *Legal Requirements* — The law requires work to be done safely and with due diligence.
3. *Financial Responsibility* — The human and equipment costs associated with incidents are unacceptable.

The intent of this chapter is to provide OH&S Practitioners/ Professionals with a framework to use, as opposed to explaining how to conduct an incident investigation.

D. IMPORTANCE

Employers are legally required to conduct incident investigations following an injury incident. In addition to that requirement, other benefits of doing an incident investigation are:

- *Demonstration of due diligence:* Companies that seek to provide a safe and healthy workplace want to know about hazards that are uncontrolled, or about unsafe situations, or about unsafe work practices. By investigating an incident, the organization/company demonstrates its OH&S due diligence.
- *Fact-finding:* Through fact-finding, the organization/company can learn exactly what went wrong and why. The process also can identify imminent danger, incident causes, unsafe work practices/situations, and uncontrolled hazards. It may also provide important information on other unrelated weaknesses in the OHSMS and opportunities for improvement.

CH. 14: OCCUPATIONAL HEALTH AND SAFETY: INCIDENT INVESTIGATION

- *Prevention:* By conducting an investigation, organizations/companies can implement corrective actions so that a similar incident can be avoided in future.
- *Information:* Can provide the needed information for insurance claims.
- *Protection:* Might prevent criminal charges against the organization/company. It can identify third-party liability situations.

E. TYPES OF INCIDENTS

Near-miss Incident is an incident that has the potential to cause damage or injury if the events were slightly different. Some individuals refer to it as a "freebie" incident from which the organization/company needs to learn.

Damage Incident is an incident that results in equipment or property damage only.

First Aid Incident is an injury incident that is treated by a physician/nurse or other healthcare provider that involves:

- a visit limited to observation;
- diagnostic procedures, including the use of prescription medications solely for diagnostic purposes;
- use of non-prescription medications, including antiseptics;
- simple administration of oxygen;
- administration of tetanus or diphtheria or booster shots;
- cleaning, flushing, or soaking wounds on skin surface;
- use of wound coverings such as bandages or gauze pads;
- use of any hot/cold therapy except for musculoskeletal disorders;
- use of any totally non-rigid, non-immobilization means of support;
- drilling of a nail to relieve pressure;
- use of eye patches;
- removal of foreign bodies not embedded in the eye if only irrigation or removal with a cotton swab is required; or
- removal of a splinter/foreign material from areas other than eyes.

Medical Aid Incident is an injury incident in which the medical care or treatment extends beyond first aid as described above.

Lost Time Injury Incident is an injury incident that results in the worker experiencing lost work time beyond the day of the incident.

Fatality Incident is an injury incident that results in death of the worker.

F. ACCIDENT THEORY

Accident theory is the science that explains how incidents are caused. An accident theory provides the categories of causes, and then, links them together in an understandable manner. It helps distinguish between immediate and underlying basic or root causes. It can shape the organization's/company's and incident investigator's thinking about its management systems, the conditions of the work environment, work design, worker behaviour and motivation, and other contributing factors. In fact, accident theory serves as the basis for the Internal Responsibility System and is the underpinning of the Canadian and U.S. OH&S legislation.

Accident theory is an important concept because it:

- offers assumptions about why incidents happen, which in turn, can be used to develop suitable corrective actions;
- can help with the incident investigation process; and
- can help get more information and value from worksite inspections.

Preventing incidents is extremely difficult in the absence of an understanding of the causes of incidents. Many attempts have been made to develop a prediction theory of accident causation, but so far none has been universally accepted. Researchers from different fields of science and engineering have been trying to develop a theory of accident causation that will help to identify, isolate, and ultimately remove, the factors that contribute to, or cause incidents.

1. Accident Causation Theories

The belief that all incidents are caused and can be prevented implies that by examining the factors that lead to incident occurrence, prevention knowledge will ensue. Thus, by identifying the root causes of incidents, necessary steps to prevent the recurrence of the incidents can be taken.

The root causes of incidents can be grouped as immediate and contributing. The **immediate incident causes** are unsafe acts of the worker and unsafe working conditions. The **contributing incident causes** could be management-related factors, environment factors, and/or the physical and psychological condition of the worker. A combination of causes must converge in order to result in an incident.

- **Management Theories, or The Domino Theories** — These theories assume that if unsafe practices and unsafe work conditions that lead to near-miss incidents are eliminated, then the probability of the occurrence of more serious incidents are reduced. Management Theories include Heinrich's Theory, Bird's Theory, and Adam's Theory.

    According to W.H. Heinrich (1931), who developed the **Domino Theory**, the ratio of incident causes demonstrates that 88 per cent are caused by unsafe acts by workers, 10 per cent by unsafe conditions, and two per cent by "Acts of God". He proposed a *Five-factor Accident Sequence* in which each factor would initiate the next step in the manner of toppling dominoes lined up in a row. The accident sequence occurs as:

    1. presence of worker ancestry and social environment factors;
    2. occurrence of a worker error;

3. unsafe acts together with mechanical and physical hazards;
4. incident; and
5. damage or injury.

In the same way that the removal of a single domino in the row would interrupt the sequence of toppling, Heinrich suggested that removal of one of the above factors would prevent the incident and resultant injury; with the key domino to be removed from the sequence being, unsafe acts together with mechanical and physical hazards.

Bird's *Loss Causation Theory* broadened the concept to include damage to property. Bird started the use of the term "incident" versus "accident". He proposed that fault results from failures in the management system, such as an inadequate system, inadequate standards, and inadequate compliance. Bird also upheld the need to address near-miss and minor incidents as well as the major ones. A comparison of the Heinrich and Bird Theories is provided in Table 14.1.

Adam's Theory prompts an examination of the individual people within a management system, not just defective policies and procedures.

Table 14.1: Comparison of the Bird and Heinrich Theories

Bird's Theory of Loss Causation	Aspect	Heinrich's Domino Theories
Management lack of control leading to failure	**Ultimate Cause of Incidents**	The personal characteristics of workers
Incorporates the concept of "Loss Control Management" (ISMEC)	**Loss Control**	Not addressed
Contact with released energy or substance	**Incident Description**	Not addressed
The concept of investigating a "near-miss incident" or minor incident was added	**Incident Investigation**	Dealt only with incidents that caused damage(s)
Includes "occupational disease"	**Damage/ Loss Type**	Deals only with "occupational injury"
Multi-causal in nature	**Cause**	Uni-causal in nature
Incident cause is based on failure within the management system. The theory perceived substandard acts and conditions as symptoms of that failure.	**Incident/ Accident Cause**	Incident due to substandard acts and conditions
Addressed damage to people and property	**Damage**	Dealt with damage to people only

Broader scope	**Scope of the Theory**	Narrow scope
Assigned fault to a weakness in the management system	**Fault**	Blamed the victim/other workers for the incident

- **Multiple Causation Theory** — is an outgrowth of the Domino Theory, but it postulates that for a single incident there may be many contributory factors, causes, and subcauses; and that certain combinations of these give rise to incidents. According to this theory, the contributory factors can be grouped into the following two categories:
    1. *Behavioural* — This category includes factors pertaining to the worker, such as improper work/safety attitude, lack of knowledge, lack of skills, and an inadequate physical and psychological condition. It assumes that these negative states occur when employees are not receiving the correct mix of positive rewards and negative sanctions to reinforce safe work behaviours. By reversing the situation, the theory holds that new physical habits will lead to the development of positive attitudes about the work experience and themselves. The ***Behaviour-based Safety*** approach and the ***People-based Safety Model*** emerged out of this incident theory.
    2. *Environmental* — This category includes the improper guarding of other hazardous work elements and the degradation of equipment through use and unsafe procedures.

    The major contribution of this theory is the belief that rarely, if ever, is an incident the result of a single cause or act.

- **The Pure Chance Theory** — holds that with any given set of workers, each worker has an equal chance of being involved in an incident. It further implies that there is no single, discernible pattern of events that leads to an incident. In this theory, all incidents are treated according to Heinrich's "Acts of God", and it is held that there exist no possible interventions to prevent them.

- **The Biased-liability Theory** — is based on the view that once a worker is involved in an incident, the chances of the same worker becoming involved in future incidents are either increased or decreased as compared to the rest of workers. This theory contributes very little, if anything at all, towards the prevention of future similar incidents.

- **Psychological Theories** — emphasize the effect of workplace social factors on employee motivation and attitude. One example of the Psychological Theory is ***Kerr's Goals Freedom Alertness Theory***. It proposes that a worker with the freedom to set his/her own work goals, usually engages in high-quality work performance, including working safely. According to this theory, an incident is due to worker inattentiveness and not because of external causes. Incidents can be prevented by encouraging workers to remain attentive and interested in their work.

Another example of the Psychological Theory is the **Accident Proneness Theory** which maintains that within a given set of workers, there exists a subset of workers who are more liable to be involved in incidents. In the past, researchers were not able to prove this theory conclusively, but based on current research in which clumsiness has been identified as being a genetic disorder that affects up to five per cent of the general population, there may be renewed interest in this theory.

- **The Decision Theory** — focuses on the sequence of events that has to happen for an incident to be avoided, or for an incident to occur. It addresses an operator's memory, perception, and problem-solving abilities.

- **The Energy Transfer Theory** — focuses on the concept of energy release as a necessary part of incident causation. Accordingly, a worker is injured or equipment damaged when energy is released and the protective barriers fail. So, for every exchange of energy, there is a source, a path, and a receiver. This theory is useful for determining injury causation and evaluating energy hazards and control methodology. It supports the development of strategies that are preventive, limiting or ameliorating with respect to the energy transfer.

Dr. Leslie Ball is a major proponent of this theory. His theory assumes that all hazards involve some form of energy and that all incidents are caused by hazards. Incident prevention means controlling those hazards.

Control of energy transfer at the source can be achieved by the following means:

- elimination of the energy source;
- changes made to the design or specification of elements of the work station; and
- preventive maintenance.

The path of energy transfer can be modified by:

- enclosure of the path;
- installation of barriers;
- installation of absorbers; and
- positioning of isolators.

The receiver of the energy transfer can be assisted by limiting the exposure to the energy, and using personal protective equipment. This model is excellent for HAZOP[5] and risk identification.

---

[5] **HAZOP** is an analysis technique that uses a systematic process to (1) identify possible

- **The Ergonomic Theory** — addresses the relationship between the worker and his/her surroundings. This theory incorporates the physical and psychological factors that impact the worker internally and externally.

- **The "Symptoms versus Causes" Theory** — is not so much a theory, as an admonition to be heeded, if incident causation is to be understood. Usually, when investigating incidents, we tend to accept the obvious or immediate causes of the incident and neglect the role of the root causes. Unsafe acts and unsafe conditions are the symptoms — the proximate causes — and not the root causes of the incident.

- **The System Safety Theory (Engineering Theories)** — views the organization/company as a closed system. It recognizes the inseparable connection between individuals and their tools and machines, and the general work environment. Changes to one part of the work system have an impact on the remaining parts and the "whole" system. Proponents of a System Safety Theory are Bob Firenze (*Firenze Model*) and Kjellen and Larsson (*The CSA Model*).

- **The Modern Loss Causation Theory** — maintains that incidents are the result of many causes. This model encompasses the concept of multi-linear interaction of causes and effects, and involves multiple opportunities for control. It proposes that worker behaviour is motivated by a desire to reach a certain goal and that workers vary in their ability and motivation to do things. The motivational principles presented are:

    - *Goal Power* — People perform more efficiently when they have meaningful work goals, objectives, and standards. The motivation to accomplish results increases when workers have meaningful goals.

    - *Participation* — Workers have a tremendous need to be involved in work, to participate and to be part of a team. Meaningful involvement of workers in work tasks and work conditions increases motivation and support.

    - *Feedback* — Workers need timely, tangible, focused, and frequent feedback. They need to know what their jobs are,

---

deviations from normal operations; and (2) ensure that appropriate safeguards are in place to help prevent accidents. HAZOP is a systematic, highly structured assessment used to generate a comprehensive operational review and to ensure that appropriate safeguards against accidents are in place. It is used primarily to identify safety hazards and operability problems of continuous process systems, especially fluid and thermal systems, and to review procedures and sequential operations.

what is expected in terms of their performance, how their performance measures up, and what/how to improve. Effective communication increases worker motivation.
- *Recognition* — Behaviour that results in recognition, reward, or some form of positive reinforcement tends to be repeated. Behaviour that results in negative effects tends to decrease or stop.

This theory proposes that worker attitudes can be positive or negative in nature, and that they are the precursors of behaviour.

2. Accident Theory: The Lessons Learned

The knowledge gained through a review of the various accident theories includes:
- The terms "theory", "model", and "working model" are defined.[6]
- The realization that investigators must look beyond the "direct causes" of incidents and focus on the root causes. In many situations, the root cause is a failure in the management system.
- Workers conducting worksite inspections should look beyond the "direct causes" when a problem/unsafe situation is identified. Once more, the root cause is a failure in the management system.
- Energy theories identify the risk of high energy sources. Workers should be trained to identify hazards in which high energy sources are present and be aware of the potential severity of outcomes should those energies be released.
- Energy theories support a systematic review of the workplace and provide guides on how to establish priorities for action.
- Energy theories help set priorities for incident investigations in that it is not the nature of the incident, but rather the potential for harm, that needs to be considered.
- Decision models and the science of ergonomics help OH&S Practitioners/ Professionals to tease apart problems associated with workstation design.
- Ergonomic models address internal and external stressors that interfere with operator decision-making processes. To reduce operator error, checklists of common weaknesses in displays and controls have been developed.
- Many accident theories are based on Human Performance Factors, which support the need for stress reduction programs, fitness programs, and

---

[6] **Theory** is defined as a set of ideas or principles that work together to explain a concept or something more tangible. **Model** is an approach created to explain a theory. **Working model** is a graphic depiction of a theory. It must be general in nature and be intuitive to promote understanding.

workplace wellness programs to promote worker well-being.
- In relation to new processes/equipment, accident theories support the anticipation of hazards and controls, thereby avoiding a "trial and error" approach to workplace safety.
- Accident theories enable employers to understand the impact that human relationships can have on workplace health and safety.
- Adams's revised domino model helps people understand the Internal Responsibility System and the legal duties assigned. It also purports that the person who knows the job is the best person to know how to do it safely. It reinforces the concept that safety is not an extra; rather, it must be integrated into the work being done.
- Accident theory holds that it is possible to be in compliance with OH&S legislation and still have a high rate of workplace incidents. This is in sharp contrast to OH&S legislation that encourages companies to have well-designed and maintained workplaces so that they experience fewer incidents.
- In terms of OH&S Policy development, accident theory guides employers by acknowledging the role of the employer in incident prevention; by reinforcing employer responsibility for incidents; by recommending the inclusion of health promotion and wellness programs; and by providing wording around the concepts of responsibility, authority, and accountability.
- Accident theory, in concert with the Quality Management Movement, indicates that management focus should be on the "upstream events", or the activities that lead to the outputs. The implication is that companies should focus their efforts on the leading indicators of safety performance if they wish to attain a high-performance Occupational Health and Safety Management System and solid OH&S outcomes.
- Behavioural theory provides information on worker behaviour. To obtain the desired OH&S behaviours, positive behavioural principles can be applied. They include:
  - *Pinpoint the desired OH&S behaviours* — make sure the behaviours are specific, clearly defined, and measurable. For example, the supervisor/line manager should state: "This is what I want to see when I come on this site."
  - *Record baseline data* — measure and keep baseline data on the current OH&S behaviours and performance.
  - *Reinforce specific desired behaviour* — provide positive recognition/ rewards for demonstration of the desired OH&S behaviours.
  - *Evaluate the impact of reinforcement on performance* — measure and track evidence and the frequency of the desired OH&S behav-

iours. Compare this data against the baseline data to identify if a difference has been made.

- *Follow through with appropriate action* — appropriate actions should be taken based on factual data.[7]

- Psychological and Behavioural Theories provide insight into human motivation such as worker coping behaviour, cognitive dissonance, and frustration.

  - *Coping behaviour* — A motive tends to decrease if the drive is satisfied or blocked from being satisfied. The "need" does not go away; rather, the worker copes. **Coping** is an attempt to overcome the obstacle by adopting a variety of human behaviours.

  - *Cognitive Dissonance* — There exists a relationship between people's perceptions and reality. When there is a difference between the two states, tension develops and the worker feels uncomfortable. This discomfort drives the worker into action to better align the worker's perceptions with reality (the state of consonance).

  - *Frustration* — is an irrational human behaviour that occurs when the accomplishment of a worker's goal(s) is blocked in some way. Frustration gets expressed as:
    - aggression — rage, hostility, violence;
    - rationalization — blaming others, intellectualizing the situation, making excuses for why things cannot be accomplished ("It was not meant to be", "It was not my time", "It didn't work last time either");
    - regression — resorting to behaviours of a previous chronological age, *e.g.*, "acting childish" is one type of regression response;
    - fixation — appearing "stuck" by exhibiting the same behaviour(s) repeatedly although those actions gain nothing for the person;
    - resignation — giving up, losing hope of ever reaching the desired goal, "learned helplessness"; and/or
    - withdrawal — removing oneself from social interaction.[8]

3. Anatomy of Incidents

The anatomy of incidents, including the details of the antecedent conditions,

---

[7] J. Johnson, "Module Eight — Accident Theory" *CRSP Exam Preparation Workshop Manual* (Edmonton, AB, 2001) at 13–14.

[8] *Ibid.*, at 14.

immediate causes, types of incidents, incident response, and outcomes of incidents is depicted in Figure 14.1.

Figure 14.1: Anatomy of an Incident

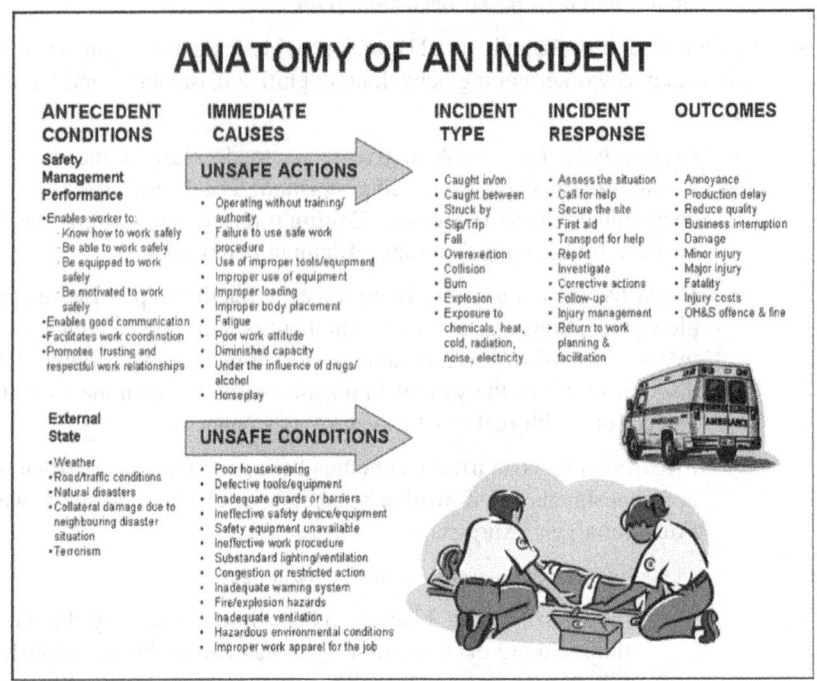

By way of explanation, the quality of the safety management performance system is the most dominant predictor of incident occurrence. A management system that enables workers to know how to work safely, to be able to work safely, to be equipped to work safely, and to be motivated to work safely is the ideal model.[9] Using an agriculture analogy, it is the nature and quality of the soil in which plants grow. Soil rich in nutrients and well-watered, yields much better crops than a nutrient-lacking soil. Likewise, a workplace with a safety management system that sets the stage for safe work attitudes and practices yields fewer incidents. The key is the management of these environments.

Further, the graphic recognizes that external environments can impact workplace safety. Although they are beyond the scope of the individual workplace, it is important to recognize their impact and to prepare a response for the potential occurrence of threat.

Incident Response is added to this graphic because depending on the quality of

---

[9] Refer to Chapter 7, "Occupational Health and Safety: Leadership and Commitment" and the Great Safety Performance Model for additional details.

that response, the outcomes differ. For example, in instances when the incident response is timely and safely enacted, the risk of additional workers being injured is significantly reduced. Likewise, post-incident injury management can mitigate the negative impacts of the incident in terms of human and financial losses. Lastly, return-to-work efforts can shorten the number of lost-time days and reduce the worker-productivity loss.

This graphic is not exhaustive; rather it is designed to demonstrate the "cause and effect" relation of antecedent conditions, and their impact on incident occurrence.

G. WHICH INCIDENTS SHOULD BE INVESTIGATED?

Common sense tells us that any serious loss should be investigated promptly and thoroughly. This includes injury, occupational illness, damage, chemical releases, and fire. Each organization/company must define what losses, and potential losses, are significant to its resources, its people, and its public relations.

However, near-miss incidents that have the potential for serious or catastrophic outcomes should also be investigated. Save for the "Grace of God", a slight change in timing, a different distance and/or the presence of other personnel or conditions, the near-miss incident could have gone from a "close call" to a serious incident. Think of them as "second-chance" or "freebie" incidents — that is, the employer and worker(s) are lucky this time, but given a repeat of the same incident, the outcomes could be much different. Opportunities like that need to be cherished, understood, and heeded.

H. CAUSES

In spite of their complexity, most incidents are preventable by eliminating one or more causes. Incident investigations determine not only what happened, but also how and why it happened. The information gained from these investigations can prevent recurrence of similar, or perhaps more disastrous incidents.

Incident investigators are interested in each event as well as in the sequence of events and conditions surrounding each event that led to an incident. The incident type is also important to the investigator. The recurrence of incidents of a particular type or those with common causes shows areas needing special incident prevention emphasis. A detailed analysis of an incident normally unearths three levels of causation: basic causes, indirect causes, and direct causes (Figure 14.2).

---

[10] U.S. Department of Labor, OHSA, "Safety & Health Management Systems e-Tool: Module 1 — Safety & Health Payoffs", online: http://www.osha.gov/SLTC/etools/safetyhealth/mod1.html (date accessed: February 28, 2020).

Figure 14.2: Analysis of an Incident[10]

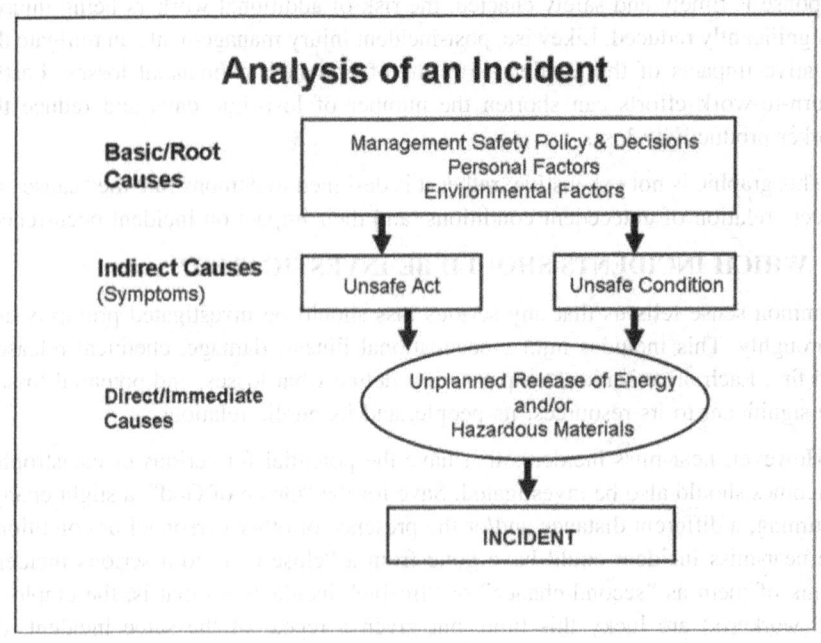

I. SEQUENCING

When investigating an incident, the typical sequence of events is as follows:

1. **Incident Occurs**
2. **Incident Report Generated**

Incident reports capture, in the words of those who were there, what happened, how, when, where, and why. Typically, they include details like:

- worker's name and personal contact details;
- business details such as work location, occupation, length of service, supervisor, *etc.*;
- date, time, and location of the incident;
- who the incident was reported to;
- names of witnesses;
- what the worker(s) was/were doing;
- how the incident happened;
- the nature of the injury(ies) and/or property damage;
- if first aid was provided at the scene;
- if medical aid was required; and

- if time from work will be required and if so, for how long?

This data is combined with a site analysis and further evidence preservation.

3. Decision to Investigate Made

Based on the organization/company's incident investigation standard/protocol, the decision to investigate the incident is made.

4. Evidence Collected

Evidence preservation must be timely and purposeful. According to Ron Rosser (2007), **evidence preservation** is the key building block to a successful incident investigation and involves taking action to keep all the position, people, parts, and paper evidence intact.[11] It must begin promptly post-incident and should start with the gathering of the position evidence. Incident scene management is critical to the preservation of evidence.

Position evidence is determining where everyone and everything was prior to and at the time of the incident.[12] It also includes information on the weather, wind direction, terrain, soil/air/water conditions, and other nearby/existing hazards; additional physical details like sounds, smells, and signs of damage; location of equipment and damaged parts; condition of the equipment/parts; *etc.*

Given that an incident scene can change quickly, this is fragile evidence and therefore needs to be quickly collected. Photographs/videos, witness statements, position maps, and measurements help with this process. Not unlike a forensic investigation, this collection of the data needs to be systematically done and clearly documented.

People evidence is the next time-sensitive type of information. It includes witness statements; the incident report; worker/witness perceptions of what happened, how and why; technical expert opinions; and targeted interviews with key people.[13] The purpose of the interviews is to find out:

- what was going on just prior to the incident;
- what tasks were being done;
- what hazards existed;
- what controls were in place;
- what pieces of equipment/tools were in use;
- who was involved;
- what were each of them doing and why;

---

[11] R. Rosser, "Root Cause Analysis" HATSCAN Training (Presented in Edmonton, AB, January 29, 2007).

[12] *Ibid.*

[13] *Ibid.*

- where were each of them located and why;
- what was the sequence of events that led up to the incident;
- what were the conditions surrounding each event;
- what happened once the energy source was released and the incident was under way;
- how did people respond;
- where did they go;
- whether there were events or conditions that occurred that were out of the norm, such as smells, sounds, sights, or other indications of a problem;
- what each player believes caused the incident and why.

This is by no means an exhaustive list of interview questions — rather, it is presented to evoke thoughts on what else an investigator may wish to find out. The intent is to elicit from the witnesses and technical experts what actually happened.

The **parts evidence** is the collection of pieces of the puzzle.[14] Each piece has the potential to unlock the secret(s) of what actually transpired. This means the retrieval, labelling, and protection of each piece of equipment and the worksite so that a reliable analysis can occur. Some parts will be too massive, or too damaged to be moved, and therefore, remain "in situ" for on-site examination.

When collecting and examining parts, each piece must be identified as to what it is, what it did, how it works, what it normally looked like, its intended function or purpose, its current condition, how it might have become damaged, its previous condition, its inspection and maintenance history, and its manufacturer/supplier.

The **paper evidence** is the least fragile of all the evidence types, and therefore is less time-sensitive than the other three types of evidence.[15] It includes the incident report(s), safe work procedures for critical tasks, pre-job planning meeting document(s), equipment/vehicle inspection reports, equipment/plant maintenance records, loss control records, non-compliance reports, worker training logs, computer data logs, control centre communication records, relevant OH&S rules and regulations, and such.

5. Investigation Scope Defined

A decision is made as to what will, and will not, be included in the investigation.

6. Investigation Leader and Team Selected

A good Incident Investigation Leader possesses specialized training in the area of incident investigation. Often this role is assumed or shared by an OH&S Practitioner/Professional.

---

[14] *Ibid.*

[15] *Ibid.*

The skilled and able leader must be:
- knowledgeable in specialized incident investigation techniques;
- experienced in conducting an incident investigation;
- aware of the necessary elements of a strong incident investigation team;
- a team builder and player;
- skilled in good project management skills;
- a good facilitator;
- a respected and credible member of the organization/company or resource;
- approachable, friendly, and engaging;
- organized;
- an analytical and logical thinker;
- objective;
- focused;
- a communicator;
- able to provide clear direction;
- knowledgeable about industry operations;
- politically savvy; and
- committed to upholding the incident investigation data as confidential.

The make-up of the investigation team should take into consideration the need for operational expertise; technical expertise in terms of the equipment and processes; facilitator skills; OH&S technical expertise; environmental expertise; *etc*. As well, there must be an adequate number of investigators to undertake and complete the needed investigation tasks.

7. Incident Investigation

Companies should develop a procedure to ensure that a standardized and systematic incident investigation process is used. The value of doing so is that it:
- ensures that the incident investigation is conducted in a responsible manner;
- ensures consistency in quality, content, and process;
- provides an auditable tracking trail for examination and review;
- demonstrates due diligence given that incident investigation is a legal requirement; and
- enables them to support the claim that the process used was valid.

An example protocol for an incident investigation follows (Table 14.2):

Table 14.2: Incident and Investigation Protocol

Incident Investigation Protocol

Investigation Team

All incidents are to be investigated by qualified personnel as soon as the injured employee(s) has(ve) been attended to and all hazards have been eliminated or controlled. The investigation team could include the:
- Line Manager of the work area;
- Union representative;
- Other workers;
- OH&S Practitioner/Professional;
- Technical or professional specialists; and
- Senior management.

The company provides incident investigation training for selected employees and management personnel.

Incident Investigation Procedure

For details on performing an incident investigation, refer to Company XYZ's online OH&S tools:
- Incident and Investigation Flow Chart;
- Incident and Investigation Procedure;
- Incident Report Form (sample offered in Chapter 13, "Occupational Health and Safety: Injury Management, Appendix 1); and
- Investigation Report form.

Key Roles and Responsibilities

Management
- Receives notification of the incident directly from Line Management.
- Reviews the report with Line Management, employees and the OH&S Department.
- Provides the necessary resources to implement corrective actions.
- Ensures that the Investigation Team is competent in incident investigation.
- Ensures that the focus of the investigation is to identify root causes and corrective actions, not to assess blame.
- Demonstrates management commitment to workplace health and safety through the timely implementation of corrective actions.

Company XYZ OH&S
- Notifies as required, Senior Management, union leaders and the applicable government agencies of the incident.

- Issues the preliminary incident report.
- Assists Line Management with the incident investigation.
- Assists with incident investigations when required.
- Issues the final incident investigation report.
- Maintains records of the incident investigation report.

Line Management
- Promotes worker reporting of all incidents.
- Ensures that injured/ill persons receive first aid and treatment as required.
- Ensures the incident scene is secured and all hazards are controlled.
- Ensures the appropriate reports are forwarded in a timely manner.

Employee
- Reports all incidents to Line Management.
- Seeks first aid or medical treatment as needed.
- Completes the necessary reports.
- Assists with the incident investigation as required.

8. Investigative Procedures

The Investigation Team Leader:

- presents a preliminary briefing to the Investigation Team. This should cover all the collected evidence; and
- assigns specific tasks to each of the team members, such as the specific tasks of evidence preservation.

The Investigation Team should:

- inspect the incident site;
- interview each victim, witness, people present before the incident, and those who arrived shortly after; and
- determine what was "normal" before the incident, where the abnormality occurred, when it was first noted, and how the abnormality occurred.

9. Systematic Root Cause Analysis Conducted

There are a number of systematic approaches and tools available on the commercial market. Each promotes its strengths in helping companies determine the root cause of incidents. Companies should examine what is available and select the investigative system or tool that best fits their philosophy, resources, nature of industry, and business needs. The investigation system or tool should serve the needs of the investigation and be able to complement a high-level of investigation if required.

Regardless of the system or tool of choice, the steps to be undertaken include:

- analyzing the data obtained and ensuring that all the relevant data has been collected and recorded through a chronological sequence of events and conditions;
- determining why the incident occurred: what the likely sequence of events and probable causes were, and what could have been the alternative sequence of events;
- testing each sequence option against the available data;
- validating the most likely sequence of events and the root causes; and
- developing the appropriate corrective actions.

10. Corrective Action(s) Recommended

The recommendations are the most important part of the investigation process. The investigation fails if the report merely states facts and draws conclusions.[16]

Specific (S.M.A.R.T.)[17] recommendations should be made to address each identified root cause. Actions should be assigned to the persons accountable for their completion. Corrective action logs should contain and track the agreed upon actions.[18]

11. Incident Investigation Report Completed

At this point, conduct a post-investigation briefing with the team, and prepare a final report, including the recommended actions to prevent a recurrence. Distribute the report for review and comments as per the established organizational/company protocol.

12. Final Report Presented

An incident investigation is not complete until a final report is prepared and submitted to the proper internal and external authorities. In many instances, special report forms are available. However, there are times when a more extensive report is required. Such reports are often elaborate and include a cover page, a title page, an abstract, a table of contents, a commentary or narrative portion, a discussion of probable causes, and a section on conclusions and recommendations.

The following outline has been found especially useful in developing the information to be included in the formal report:

A. Background Information
  i. where and when the incident occurred;
  ii. who and what were involved; and

---

[16] *Ibid.*

[17] Refer to Chapter 17, "Occupational Health & Safety: Worker OH&S Training", Section F, item #3. "Establish the OH&S Training Objectives".

[18] *Ibid.*

CH. 14: OCCUPATIONAL HEALTH AND SAFETY: INCIDENT INVESTIGATION

    iii.   operating personnel and other witnesses.

B. **Account of the Incident (what happened?)**
  i. sequence of events;
  ii. extent of damage;
  iii. incident type;
  iv. agency or source (of energy or hazardous material); and
  v. the point of loss of control.

C. **Discussion (Analysis of the incident — How? Why?)**
  i. direct causes (energy sources; hazardous materials);
  ii. indirect causes (unsafe acts and conditions);
  iii. basic causes (management policies, personal/environ-psychological factors); and
  iv. recommendations (to prevent a recurrence) for immediate and long-range action to remedy:
   - basic causes;
   - indirect causes; and
   - direct causes (such as reduced quantities of protective equipment or structures).

J. EFFECTIVE TECHNIQUES

Some effective incident investigation techniques include:

- During an investigation, gather evidence from many sources: get information from witnesses and reports, as well as through observation.
- Inspect the incident site before changes occur. Take photographs and measurements and make sketches of the incident scene. Record all pertinent data on site maps.
- Get copies of all the associated incident reports.
- Review documents containing normal operating procedures, flow diagrams, maintenance charts, or reports of difficulties or abnormalities including previous incident reports.
- Keep complete and accurate investigation notes. Record pre-incident conditions, the incident sequence, and post-incident conditions. In addition, document the location of victims, witnesses, machinery, energy sources, and hazardous materials.
- In some investigations, a particular physical or chemical law, principle, or property may explain a sequence of events. Include laws in the notes taken during the investigation or in the later analysis of data. Gather data during

the investigation that may lend itself to analysis by these laws, principles, or properties. An appendix in the final investigation report can include an extended discussion.
- Assign experienced personnel to conduct interviews. If possible, the team given to this task should include an individual with a legal background. In conducting interviews, the team should:
  - Get preliminary statements as soon as possible from all witnesses.
  - Locate the position of each witness on a master chart (including the direction of view).
  - Arrange for a convenient time and place to talk to each witness (separately).
  - Explain the purpose of the investigation (incident prevention) and attempt to put each witness at ease.
  - Listen; let each witness speak freely; and be courteous and considerate.
  - Take notes without distracting the witness. Use a tape recorder only with the expressed, written consent of the witness.
  - Use sketches and diagrams to help the witness.
  - Emphasize areas of direct observation. Label hearsay accordingly.
  - Be sincere and do not argue with the witness.
  - Record the exact words used by the witness to describe each observation. Do not "put words into a witness' mouth".
  - Word each question carefully and be sure the witness understands what is being asked.
  - Use open-ended questions.
  - Identify the qualifications of each witness (*e.g.*, name, address, occupation, years of experience, *etc.*).
  - Have the witness sign the statement.
  - Supply each witness with a copy of his/her statements. Signed statements are desirable.
- After interviewing all witnesses, the Investigation Team should analyze each witness' statement. They may wish to re-interview one or more witnesses to confirm or clarify some key points. While there may be inconsistencies in witnesses' statements, investigators should assemble the available testimony into a logical order. Analyze this information along with data from the incident site.
- Not all people react in the same manner to a particular stimulus. For example, a witness within close proximity to the incident may have an

entirely different story to tell than does the person who saw it at a distance. Some witnesses may also change their stories after they have discussed it with others. The reason for the change may provide additional clues as to what really happened.

- A witness who has had a traumatic experience may not be able to recall the details of the incident. A witness who has a vested interest in the results of the investigation may offer biased testimony. Finally, eyesight, hearing, reaction time, and the general condition of each witness may affect his/her powers of observation. A witness may omit entire sequences because of a failure to observe them or because their importance was not realized. The use of sound interview techniques can be very useful at getting at the facts.

- Use *Change Analysis* for problem solving. This technique emphasizes change. To solve a problem, an investigator must look for deviations from the norm. Consider all problems that might result from some unanticipated change. Analyze the noted change to determine its causes. Use the following steps in this method:

  1. define the problem (what happened?);
  2. establish the norm (what should have happened?);
  3. identify, locate, and describe the change (what, where, when, to what extent);
  4. specify what was and what was not affected;
  5. identify the distinctive features of the change;
  6. list the possible causes; and
  7. select the most likely causes.

- Use *Job Safety Analysis* (JSA) to break the job into basic steps. JSAs are part of many organization/company incident prevention programs. In general, a JSA breaks down a job into basic steps, and identifies the hazards associated with each step. The JSA also prescribes controls for each hazard. A JSA is a chart listing these steps, hazards, and controls. Review the JSA during the investigation if a JSA has been conducted for the job involved in an incident. Perform a JSA if one is not available. Perform a JSA as a part of the investigation to determine the events and conditions that led to the incident.

- Incident re-enactment can help the investigators to better understand the sequencing of events. Note that care and control during the re-enactment are very important to prevent incident recurrence.

- Finally, the incident report results should be shared with all the relevant stakeholders.

K. STAKEHOLDER COMMUNICATION

Incident investigations can help companies learn from their misfortunes, or vicariously, from the mistakes made by others. They can highlight unique or new information about the system of interest and of other existing systems that could have prevented the incident.

Incident investigation findings — both the process and outcome findings — must be shared with workers and other relevant stakeholders. The information obtained is part of an organization's/company's learning process. Incidents that occur in one part of the organization/company can be avoided in other organizational/company departments provided Management and workers are apprised of the situation and the suitable controls.

As noted, employers have a legal duty of care to advise workers of unsafe work conditions and acts, as well as of suitable controls. However, there are a number of challenges that need to be addressed, such as:

- the confidentiality of the worker information needs to be protected;
- the message should deal with the opportunities to improve the management system: not on blaming the worker(s);
- the message should not just deal with the facts: It must also explain what has been learned and what is going to be done to prevent a similar incident in future;
- the message must be tailored to the audience and relevant to their workplace experience; and
- the message must motivate workers to adopt good work and OH&S behaviours in future.

L. MEASUREMENT OF EFFECTIVENESS

Given the importance of incident investigation, it behooves OH&S Professionals/ Practitioners to regularly evaluate the quality and effectiveness of incident investigations. To begin with, decide the elements to be assessed, for example:

Completion Rate
- Are all incidents investigated? If not, why not?

Completeness
- Are the incident report and investigation forms completed as required by organizational/company protocol? If not, why not?
- Are incident reports and investigations completed in a timely manner? If not, why not?
- Were the basic and immediate causes identified and analyzed? If not, why not?
- Are the corrective actions S.M.A.R.T.? If not, why not?
- Are all the relevant forms signed? If not, why not?

Quality
- Are the forms completed haphazardly? If so, why?

- Are the incident descriptions comprehensive? If not, why not?
- Is the documentation clear and comprehensive? If not, why not?
- Was all the data considered and analyzed? If not, why not?
- Are the corrective actions reasonable, appropriate, and adequate? If not, why not?
- Will they address the uncontrolled hazard(s) identified? If not, why not?

Compliance
- Are the incidents being investigated as per the organization's/company's protocol criteria and government regulations? If not, why not?

Effectiveness
- Are the corrective actions addressed and completed? If not, why not?
- What is the frequency of recurrent incidents?

Other elements that warrant assessment are:

- **The organization's/company's incident protocol** — How does it compare with that of "best in class" organizations/companies?
- **The organization's/company's incident tools** — How do they compare with those tools used by "best in class" companies?
- **The incident training** — Is it meeting the needs of the Investigation Leader and Team members?
- **The system used to analyze the root cause(s)** — Is the approach/system meeting the organization's/company's needs?

Regular evaluation of the effectiveness of the incident investigation system should be undertaken by senior management in concert with the OH&S Practitioners/Professionals. Identified shortfalls should be corrected and their appropriateness reassessed as time goes on. As well, the evaluation should focus on whether the investigation process and outcomes strengthened the organization's/company's ability to prevent similar incidents (losses) in future. For example, were the findings incorporated into management systems such as the operational safe work procedures and/or worker training modules?

The intent of this evaluation process is to determine:
- if the incident investigation process was valid and reliable;
- if it is defensible and repeatable;
- if it reached reasonable and practicable conclusions;
- if all the relevant stakeholders were apprised of the findings; and
- if the findings made a difference to the organization's/company's loss control efforts or not.

M. CALCULATING THE COST OF INCIDENTS

Few companies appreciate the costs associated with workplace incidents and the related incident reporting and investigation. The reason for this is that most companies do not capture these costs. Companies that wish to determine the incident costs would have to identify and tabulate all the activities and costs associated with the incident, from start to finish. That exercise includes activities and costs associated with:

- the incident;
- the reporting and investigation process;
- the resultant damages;
- the replacement of personnel, materials and/or equipment;
- the disability costs; and
- all related business and human resource costs.

To assist companies, a number of software tools have been developed, namely:

- **Workplace Incident Cost Calculator:** WorkSafeBC hosts an online tool designed to enable employers to calculate the cost of workplace incidents. It includes the incident costs, incident investigation costs, damage costs, worker replacement costs, and productivity losses. The user can customize the tool using the organization's/company's profit margin, so that time to recover the calculated costs can be determined.[19]

- **Smart Planner:** The Institute for Work and Health (IWH) has developed a software tool called the Smart Planner that can be readily downloaded and used.[20] The Smart Planner enables the organization/company to calculate the impact of a workplace incident, as well as to undertake a cost-benefit analysis of the organization's/company's OH&S initiatives.

In addition to the financial costs, there are the legal sanctions. Companies can be levied fines, penalties, incarceration, and negative reputation outcomes which in Alberta are publicly posted (Table 14.3). As well, descriptions of the prosecutions and penalties are provided.[21] Over the years, the number of convictions remains to be of concern: as well, the cost of legal sanctions has greatly increased. High fines do not appear to be much of a deterrent.

---

[19] WorkSafeBC, *The Workplace Incident Cost Calculator*, online: https://www.worksafebc.com/en/resources/health-safety/interactive-tools/workplace-incident-cost-calculator?lang=en (date accessed: February 28, 2020).

[20] IWH, *The Smart Planner*, online: http://www.iwh.on.ca/smart-planner (date accessed: February 28, 2020).

[21] Online: http://work.alberta.ca/occupational-health-safety/6750.html (date accessed: February 28, 2020).

This type of press can be very damaging to an organization's/company's reputation and future prosperity.

Table 14.3: Alberta OHS Prosecutions Penalties Summary 2004–2017[22]

Calendar Year	Number of Prosecutions Completed	Total Penalties (fines, alternate penalties & victim surcharge)	Range of Court Penalties (fines & alternate penalties)
2017	21	$3,787,100	$34,500 - $650,100
2016	19	$2,050,625	$11,500 - $325,000
2015	11	$2,001,249	$28,750 - $450,000
2014	14	$1,677,050	$4,000 - $325,000
2013	10	$2,615,750	$35,750 - $1,500,000
2012	10	$3,332,500	$70,000 - $1,250,000
2011	20	$3,457,750	$10,000 - $400,000
2010	11	$1,737,250	$10,000 - $400,000
2009	7	$457,225	$4,025 - $100,750
2008	22	$5,083,000	$45,750 - $425,000
2007	12	$1,720,000	$70,750 - $350,000
2006	10	$1,534,500	$40,000 - $500,000
2005	12	$554,050	$2,000 - $100,000
2004	9	$597,500	$10,000 - $120,000

N. SUMMARY

Given the frequency of workplace incidents, and the associated human and financial costs, incident prevention is critical. By understanding the root/basic causes of incidents, prevention strategies can be identified and implemented.

CHAPTER REFERENCES

Alberta Jobs, Skills, Training and Labour, *Alberta OHS Prosecutions Penalties Summary 2004–2018* (2019), online: http://https://www.alberta.ca/summary-convictions.aspx (date accessed: February 28, 2020).

Association of Workers' Compensation Boards of Canada, "Key Statistical Measures for 2005", online: http://www.awcbc.org/en/index.asp (date accessed: February 28, 2020).

Association of Workers' Compensation Boards of Canada, "2018 Injury Statistics; Statistics", online: http://awcbc.org/?page_id=14 (date accessed: February 28, 2020).

F. Bird & G. Germain, *Loss Control Management: Practical Loss Control Leadership*, revised ed. (Loganville, GA: DNV, 1996).

---

[22] Alberta Jobs, Skills, Training and Labour, *Alberta OHS Prosecutions Penalties Summary 2004–2017* (2019), online: https://www.alberta.ca/summary-convictions.aspx (date accessed: February 28, 2020).

IWH, *The Smart Planner*, online: http://www.iwh.on.ca/smart-planner (date accessed: February 28, 2020).

J. Johnson, "Module Eight — Incident Theory" *CRSP Exam Preparation Workshop Manual* (Edmonton, AB, 2001).

R. Maurer, "2012 Workplace Injury Rate Continues Downward Trend" *SHRM*, 2013, online: http://www.shrm.org/hrdisciplines/safetysecurity/articles/pages/2012-workplace-injury-rate-decline.aspx (date accessed: February 28, 2020).

R. Rosser, "Root Cause Analysis" HATSCAN Training (Presented in Edmonton, AB, January 29, 2007).

U.S. Department of Labor, Bureau of Labor Statistics, "Employer-reported Workplace Injuries and Illnesses - 2012", online: http://www.bls.gov/news.release/archives/osh_11072013.pdf (date accessed: February 28, 2020).

U.S. Department of Labor, OHSA, "Safety & Health Management Systems e-Tool: Module 1 — Safety & Health Payoffs", online: http://www.osha.gov/SLTC/etools/safetyhealth/mod1.html (date accessed: February 28, 2020).

WorkSafe Alberta, *Calculator: Incident Cost Summary*, online: http://work.alberta.ca/smallbusiness/calculator/calculator/summary.html (date accessed: February 28, 2020).

WorkSafeBC, *The Workplace Incident Cost Calculator*, online: http://www.worksafebcmedia.com/media/calculators_html5/WICC/index.html (date accessed: February 28, 2020).

Chapter 15

OCCUPATIONAL HEALTH AND SAFETY: RISK MANAGEMENT AND COMMUNICATION

A. INTRODUCTION

Occupational Health and Safety (OH&S) Practitioners/Professionals are involved in OH&S risk management situations and provide OH&S communications on a daily basis. The purpose of this chapter is to discuss the principles, importance, and effective techniques of OH&S risk management and OH&S risk communications.

B. OH&S RISK MANAGEMENT

Risk management is the process of making and implementing decisions to minimize adverse effects of accidental and business losses to an organization/company. It is a systematic application of the planning, organizing, leading, and controlling functions to:

- anticipate and identify accidental loss exposures;
- evaluate the risk;
- work to avoid or eliminate hazards; and
- attain an acceptable level of risk.

Risk, in this sense, is a state in which losses are possible. It can be defined as the probability of loss of that which we value.[1] Risk can also be perceived as an opportunity that can present as either a gain or a loss:

> Risk is the potential for realization of unwanted, negative consequences of an event.[2]

The field of **Risk Management** is a practice with processes, methods, and tools for managing business risks. The value offered by managing risk is that organizations/companies can take actions to avoid catastrophes by developing and implementing

---

[1] V. Covello, "Risk Communication Slides", online: http://www.centerforrisk communication.org (date accessed: February 28, 2020).

[2] F. Manuele, *On the Practice of Safety*, 3rd ed. (New Jersey: Wiley-Interscience, 2003) at 59.

governance and financial guidelines that afford a level of stability to earnings and minimize potential losses due to a variety of risk exposures.[3]

OH&S is risk management operationalized. Synonymous with every aspect of an Occupational Health and Safety Management System (OHSMS), risk management is aimed at minimizing the costs of **"pure risk"**[4] at a reasonable cost. It is an administrative, managerial function that is similar to standard hazard identification and loss control practices, but more sophisticated and focused on potential risk.

1. Importance

Why OH&S risk management? What value does it add to an organization/company? To better understand this concept and answer these questions, we must first review the following facts:

In Canada, in 2018:[5]

- the work injury rate was 1.4 per 100 workers (787,790 claims);[6]
- workers experienced 264,438 serious injuries requiring recuperation and time away from work;[7]
- 1,027 work-related fatalities occurred, or 2.8 per calendar day,[8] or 4 per workday;[9]
- one worker is seriously injured or killed every 2.2 minutes of work;[10]

---

[3] S. Meltzer, "Risk Management: 2010" (Presented at University of Calgary, Calgary, AB, March 31, 2005).

[4] **Pure risk** refers to risks in which there can be no hope of gain.

[5] At the time of writing this chapter, the AWCBC.org has not finalized the 2018 workplace injury/illness rates.

[6] Association of Workers' Compensation Boards of Canada, "2018 Injury Statistics" *Statistics 2020*, online: http://awcbc.org/?page_id=14 (date accessed: February 28, 2020).

[7] *Ibid.*

[8] Using 365 days per year, 2.8 workers were killed per calendar day. See Association of Workers' Compensation Boards of Canada, "2018 Injury Statistics" *Statistics 2020*, online: http://http://awcbc.org/?page_id=14#fatalities (date accessed: February 28, 2020).

[9] Using 250 workdays per year, 4 workers were killed per workday. See Association of Workers' Compensation Boards of Canada, "2018 Injury Statistics" *Statistics 2020*, online: http://http://awcbc.org/?page_id=14#fatalities (date accessed: February 28, 2020).

[10] According to the Association of Workers' Compensation Boards of Canada, there were a total of 265,465 lost-time injuries (264,438) and fatalities (1,027) in 2018. Using 250 days and assuming an average of an eight-hour workday, the number of work minutes is 120,000 minutes. See Association of Workers' Compensation Boards of Canada, "2017 Injury Statistics" *Statistics 2020*, online: http://awcbc.org/?page_id=14 (date accessed: February 28, 2020).

CH. 15: OCCUPATIONAL HEALTH & SAFETY: RISK MANAGEMENT & COMMUNICATION

- 1.3 per cent of Canadian full-time and part-time workers are seriously injured or killed yearly;[11] and
- the average cost of a Workers' Compensation claim was CDN $39,500 (2018) for assessable employers.[12]

In the United States (U.S.), in 2018:

- the number of recordable situations in the private industry sector were 2.8 million with a total of 900,400 cases requiring recuperation time from work (an incidence rate of 2.8 per 100 full-time equivalent workers); this was the same as in 2017;[13]
- the median days away equalled 8 days off work;[14]
- 5,250 American workers died due to work-related injuries, resulting in a fatality rate of 3.5 per 100,000 full-time workers — a 2% increase from 2017 (5,147 fatalities);[15]
- of the 5,250 fatalities, 1,276 deaths were due to motor vehicle accidents and 453 were due to homocides in the workplace;[16]
- the causes and costs (nearly $47 billion) of the top 10 most costly disabling workplace injuries in 2016 are provided in Figure 15.1 (2019 Liberty Mutual Workplace Safety Index): *"This translates into over a billion dollars*

---

[11] According to Statistics Canada there were 19,933,500 Canadian workers in 2018, and in the same year, according to the Association of Workers' Compensation Boards of Canada, there were a total of 265,465 lost-time injuries and fatalities in 2018: Association of Workers' Compensation Boards of Canada, "2018 Injury Statistics", online: http://awcbc.org/?page_id=14 (date accessed: February 28, 2020); see also Statistics Canada, Labour force characteristics by age and sex (December 2018), online: https://www150.statcan.gc.ca/n1/daily-quotidien/180105/t001a-eng.ht (date accessed: February 28, 2020).

[12] Calculated using Association of Workers' Compensation Boards of Canada, "2018 Injury Statistics" *Statistics 2020*, online: http://awcbc.org/?page_id=14 (date accessed: February 28, 2020). Involves the addition of IR5 – the Current Year Average Benefit Cost/Claim, and IR6 – Administration Cost per claim and the average OH&S cost per claim.

[13] U.S. Department of Labor, "Injuries, Illnesses and Fatalities, 2018" (2020), online: http://www.bls.gov/iif/ (date accessed: February 28, 2020).

[14] U.S. Department of Labor, "Injuries, Illnesses and Fatalities, 2018" (2020), online: http://www.bls.gov/iif/ (date accessed: February 28, 2020).

[15] U.S. Department of Labor, "Census of Fatal Occupational Injuries Summary, 2018" (2020), online: http://www.bls.gov/news.release/cfoi.nr0.htm (date accessed: February 28, 2020).

[16] U.S. Department of Labor, "Census of Fatal Occupational Injuries Summary, 2018" (2020), online: http://www.bls.gov/news.release/cfoi.nr0.htm (date accessed: February 28, 2020).

a week spent by businesses on the most disabling injuries."[17] The top five injury causes are overexertion, fall on the same level, struck by object, falls to a lower level, and other exerctions or bodily reactions.

Figure 15.1: The Top 10 Causes of the Most Disabling Workplace Injuries in 2016 ($ in billions)[18]

Injury Cause	Over-exertion	Fall: same level	Struck by	Fall: lower level	Other exertions/ body reactions	Road incidents	Slip/ trip	Caught in	Repetitive strain	Struck against
Percentage	24%	19%	9%	9%	7%	5%	4%	3%	3%	2%
Cost	$13.11	$10.38	$5.22	$4.98	$3.69	$2.70	$2.18	$1.93	$1.59	$1.15

These workplace injuries account for 85 per cent of the overall occupational disability costs in the U.S.[19]

(a) Risk Management

The interest in risk management is higher now than it has ever been.[20] Each year organizations/companies experience extensive psychological, security, product, people, property, and reputation losses. As a result, organizations/companies are seeking ways to prevent such losses. They look to the field of risk management, the intent of which is to manage these business risks, thereby minimizing real and potential losses.

Risk management is ultimately the responsibility of the Board of Directors or Senior Management of an organization/company. They have to decide the level of their risk appetite, their preferred system for risk controls, and how they will make management accountable for risk management within the organization/company and risk communication.[21]

---

[17] Liberty Mutual Research Institute for Safety, "2019 Workplace Safety Index: The Top 10 Causes of Disabling Injuries at Work", *Viewpoint*, online: https://viewpoint.libertymutualgroup.com/article/top-10-causes-disabling-injuries-at-work-2019/ (date accessed: January 31, 2020).

[18] Liberty Mutual Research Institute for Safety, "2019 Workplace Safety Index: The Top 10 Causes of Disabling Injuries at Work" *Viewpoint*, online: https://viewpoint.libertymutualgroup.com/article/top-10-causes-disabling-injuries-at-work-2019/ (date accessed: January 31, 2020).

[19] Liberty Mutual Research Institute for Safety, "2019 Workplace Safety Index: The Top 10 Causes of Disabling Injuries at Work" *Viewpoint*, online: https://viewpoint.libertymutualgroup.com/article/top-10-causes-disabling-injuries-at-work-2019/ (date accessed: January 31, 2020).

[20] S. Meltzer, "Risk Management: 2010" (Presented at University of Calgary, Calgary, AB, March 31, 2005).

[21] *Ibid.*

The corporate governance issues that Management needs to address are:

- "compliance" versus "due diligence";
- whether to use a checklist approach to risk management;
- the potential penalties for directors and officers;
- the potential impact that specific investment decisions might have on the organization/company; and
- the separation between risk management audit and risk management consulting functions.[22]

(b) OH&S Risk Management

The field of Occupational Health and Safety (OH&S) plays an important role in risk management efforts. OH&S risk management is a framework for gathering information on organizational/company risks and their potential or real impact. The objective is to prevent any serious impact on the organization's/company's financial structure from accidental losses. Figure 15.2 demonstrates the magnitude of the real and potential losses that an organization/company could face as the result of an incident at work.

Workplace incidents are expensive — more expensive than most organizations/companies realize. The total cost of incidents includes the direct and indirect costs. According to the Liberty Mutual 2019 Workplace Safety Index, disabling workplace injuries and illnesses cost U.S. employers more than $1 billion per week in direct workers' compensation costs.[23] There was no estimate of the indirect costs provided.

---

[22] *Ibid.*

[23] Liberty Mutual Research Institute for Safety, "2019 Workplace Safety Index: The Top 10 Causes of Disabling Injuries at Work" *Viewpoint*, online: https://viewpoint.libertymutualgroup.com/article/top-10-causes-disabling-injuries-at-work-2019/ (date accessed: January 31, 2020).

[24] F. Bird & G. Germain, *Practical Loss Control Leadership*, revised ed. (Loganville, GA: DNV, 1996) at 8. Reprinted with permission from Det Norske Veritas (DNV).

Figure 15.2: Cost of Workplace Incidents[24]

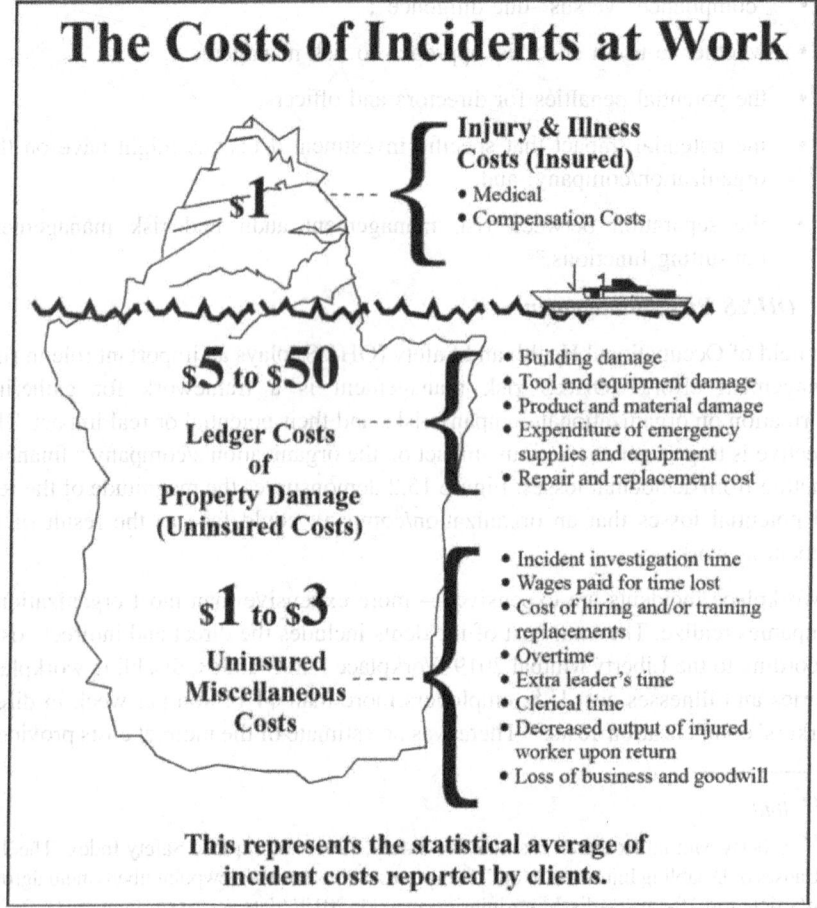

When an incident occurs, an organization/company faces direct and indirect incident costs. According to Occupational Safety and Health Information (OSHA), the ratio of the indirect to direct costs varies greatly, from a high of 20:1 to 1:1.[25]

The direct incident costs include:

- ambulance costs;
- worker injury costs (*e.g.*, diagnostic, treatment, and rehabilitation costs) in the form of worker's compensation claim costs;

---

[25] U.S. Department of Labor, "Cost of Accidents, Safety & Health Management Systems e-Tool", online: http://www.osha.gov/SLTC/etools/safetyhealth/mod1_costs.html (date accessed: February 28, 2020).

CH. 15: OCCUPATIONAL HEALTH & SAFETY: RISK MANAGEMENT & COMMUNICATION

- indemnity payments;
- cost of independent medical examinations;
- fitness-to-work assessment costs;
- return-to-work efforts and costs;
- job accommodation efforts and costs;
- third party injury costs, if applicable;
- third party business interruption costs, if applicable;
- vehicle/equipment damage;
- damage to the building/facility;
- new equipment costs; and
- cleanup costs.

These are the costs that are directly associated with the occurrence of the incident. Often organizations/companies are able to transfer these costs through the use of various forms of insurance.

The indirect incident costs tend to be far greater than the direct costs, and include:

- business interruptions;
- lost productivity;
- overtime costs;
- the value of staff time doing incident investigation and reporting;
- the value of staff time in recruiting, selection, and placement of replacement workers;
- training of replacement workers;
- costs due to inefficiency of replacement workers;
- administrative overtime;
- product damage;
- increased insurance premium rates;
- OH&S fines/penalties/incarceration;
- loss of revenue;
- loss of sales;
- lost customers/clients;
- value of staff time doing public relations work and media relations; and
- damage to the corporate image.

To calculate the indirect costs of injuries and illness, multiply the direct costs by a factor as shown in Table 15.1.

Table 15.1: Direct to Indirect Cost Calculation[26]

Direct Illness/Injury Costs	Indirect Cost Illness/Injury Ratio
$0-$2,999	4.5
$3,000-$4,999	1.6
$5,000-$9,999	1.2
$10,000 or more	1.1

Indirect-incident costs are uninsured and come directly from the organization's/company's profits, thereby negatively impacting its profitability.

Then, follow with an explanation on the cost of incidents. For example, the average cost of a workplace lost-time injury in Canada, 2018, is calculated to be $39,534 per claim for assessable employers.[27] Using the following calculation (Table 15.2):

Table 15.2: Calculation of Required Revenue to Offset Loss[28]

Element	Value	Explanation
Direct Cost of Injury	$39,534	Average DIRECT cost of injury
Indirect Cost of Injury	$43,487	Direct cost X 1.1 (Table 15.1)
Total Cost of Injury	$83,021	Direct Cost + Indirect Cost
Profit Margin of Business (%)	5%	A company-specific value
Revenue Needed	$1,660,428	Total Cost/Profit Margin

For an average lost-time injury claim, Canadian employers have to generate between $7.3 million (one per cent profit margin), $3.7 million (two per cent profit margin), $2.4 million (three per cent profit margin), $1.8 million (four per cent profit margin), or $1.5 million (five per cent profit margin).

For ease of calculation, the U.S. Department of Transportation provides a similar, "quick-to-use" table on how to estimate the amount of revenue required to pay for the costs of a motor vehicle incident (Table 15.3).

---

[26] U.S. Department of Labor, OSHA $afety Pays Program, "Background of the Cost Estimates" $afety Pays Program, 2010, online: https://www.osha.gov/dcsp/smallbusiness/safetypays/background.html (date accessed: February 28, 2020).

[27] Refer to footnote 11 in this chapter for the details on that calculation.

[28] Adapted from Safety Management Group, "Injury Cost Calculator" (2015), online: http://www.safetymanagementgroup.com/injury-cost-calculator.aspx; and U.S. Department of Transportation, "Accident Cost Table — Federal Motor Carrier Safety Administration" U.S. Department of Transportation (2011).

CH. 15: OCCUPATIONAL HEALTH & SAFETY: RISK MANAGEMENT & COMMUNICATION

Table 15.3: Revenue/Sales Required to Cover Losses[29]

Annual Incident Costs	Profit Margin				
	1%	2%	3%	4%	5%
$1K	$100,000	$50,000	$33,000	$25,000	$20,000
$5K	$500,000	$250,000	$167,000	$125,000	$100,000
$10K	$1,000,000	$500,000	$333,000	$250,000	$200,000
$25K	$2,500,000	$1,250,000	$833,000	$625,000	$500,000
$59K	$5,000,000	$2,500,000	$1,667,000	$1,250,000	$1,000,000
$100K	$10,000,000	$5,000,000	$3,330,000	$2,500,000	$2,000,000
$150K	$15,000,000	$7,500,000	$5,000,000	$3,750,000	$3,000,000
$200K	$20,000,000	$10,000,000	$6,666,000	$5,000,000	$4,000,000

As noted in Chapter 4, "Occupational Health Program", the average cost of a workplace fatality in the USA, ranges between $2.9 million - $5.8 million.[30] The Injury Cost Calculator tool[31] that uses the above process, indicates that the organization/company would have to generate revenue of $57.5 million cover this cost (assuming a 10% margin).[32]

OH&S risk management is a decision-making process used to:

- address timely decision-making under conditions of uncertainty and time urgency;
- provide worker involvement in decision-making; and
- ensure the existence of an effective means of risk communication.

Although OH&S risk management works best in the design and "what if stage", it is an "everyday activity" that includes security, product liability, and the safety of employees, customers and public.[33] As such, OH&S risk management focuses on personnel, property, anticipated income, and liability with the objective of preventing accidental loss that negatively impacts the organization's/company's financial structure.

---

[29] U.S. Department of Transportation, "Accident Cost Table — Federal Motor Carrier Safety Administration" U.S. Department of Transportation (2011).

[30] Refer to Chapter 4, "Occupational Health Program", footnotes 24–27.

[31] Safety Management Group, *Injury Cost Calculator* (2019), online: https://safetymanagementgroup.com/resources/injury-cost-calculator/ (date accessed: January 31, 2020).

[32] *For a 5% profit margin, the organization/company would have to generate $115 million to offset the cost of the fatality.* Safety Management Group, *Injury Cost Calculator* (2019), online: https://safetymanagementgroup.com/resources/injury-cost-calculator/ (date accessed: January 31, 2020).

[33] Board of Canadian Registered Safety Professionals (BCRSP), *Risk Management Domain, CRSPEX Study Guide* (Mississauga, ON: BCRSP, 2002).

2. Key Roles and Responsibilities

(a) Senior Management (Employer)

OH&S risk management begins with Senior Management who are tasked with ensuring that:

- all risk exposures are identified;
- risk exposures are analyzed and evaluated;
- combinations of funding and preventing exposures to risk are evaluated; and
- decisions for funding and preventing losses are implemented.

Losses can manifest in many forms. For example, an employer's legal risk reduction measures depend on:

- ***The Applicable Regulations***: Employers need to be aware of the jurisdiction's minimum OH&S standards, and to focus on exceeding the minimum standard thereby assuring a duty of OH&S care and due diligence.
- ***The Industry Standards***: Industry standards emerge when companies within an industry perform to a certain performance level thereby setting a performance expectation.
- ***The Applicable Best Practices***: Best practices are recognized through the approach of evaluating the OH&S practices of top performing companies. Benchmarking, or comparing OH&S management techniques, is a common approach to establishing best practices.
- ***The Company's Own Standards, Policies, and Procedures***: It is important for the organization/company to determine if it is living up to its own standards, policies, and procedures. After an incident, OH&S inspectors, prosecutors, lawyers, and judges tend to ask if these standards, policies, and procedures were followed.
- ***The Employer's Experience***: Employers must learn from their mistakes. Being aware of past mistakes, and not learning from a similar incident, is negligence. In OH&S legislation, penalties for second offences are usually twice that of the first offence.
- ***Worker OH&S Vulnerability:*** Employers are expected to ensure that employees uphold corporate approved work behaviours and OH&S practices. The Institute for Work & Health developed an *OHS Vulnerability Measure* which assesses employee vulnerability to OH&S risks.[34] This tool allows Management to recognize the current level of vulnerability and take actions to prevent workplace inuries. Follow-up measures can be taken to

---

[34] IWH, *OHS Vulnerability Measure*, online: www.iwh.on.ca/ohs-vulnerability-measure (date accessed: January 31, 2020).

determine if the remedial measures actually worked.

3. Definitions

Reasonable care is defined as actions that would be taken by "the reasonable person" — one who plans and takes actions in accordance with general and approved practices. Employers are legally expected to demonstrate reasonable care in terms of worker health and safety.

Due diligence means taking all reasonable steps as part of "due care" to prevent the occurrence of an incident or mishap and having contingency plans in place to control an incident and limit consequential damage. Reasonable care encompasses the three elements of due diligence:

1. *Foreseeability* — recognizing the potential for harm. This can be achieved through the use of hazard assessments.

2. *Establishing workplace systems* — using systems to address the hazards identified including:

    - systems of work (*e.g.*, workplace culture, safe work procedures, work and communication processes, hazard assessments, workplace inspections, OH&S program standards, joint health and safety committees, incident investigations, injury management, *etc.*);
    - worker training;
    - adequate and maintained equipment;
    - personal protective equipment;
    - clear directions and instructions;
    - adequate and clear information; and
    - system monitoring and continuous improvement.

3. *Ensuring the designed systems work* — monitoring and modifying systems to ensure effectiveness through audits, *etc.*

Due diligence is the level of judgement, care, prudence, determination, and activity that a person/company would reasonably be expected to do under particular circumstances. It is "risk-based"; that is, reasonable care activities increase with "high-risk" occupations and decrease with "low-risk" occupations. Good risk assessment processes ensure due diligence.[35] Table 15.4 provides a due diligence checklist for employers.

---

[35] Hatscan, *Handi-guide to Alberta's OH&S Act, Regulation and Code* (Edmonton, AB: Hazard Alert Training Inc., 2004) at c. 2-2.

Table 15.4: Due Diligence Checklist

	Element
☐	Does the company have an OH&S Policy?
☐	Does the company have a formalized OHSMS?
☐	How are OH&S goals, objectives and targets determined?
☐	Do stakeholders know their OH&S responsibilities and accountabilities?
☐	How are stakeholder OH&S responsibilities and accountabilities communicated?
☐	How does management demonstrate its OH&S commitment?
☐	How are managers/workers held accountable for workplace safety?
☐	Are workplace safety behaviours/outcomes included in performance appraisals?
☐	Is safety integrated into the company's business endeavours?
☐	Are the resources allocated to OH&S adequate?
☐	How are OH&S action plans implemented?
☐	What does worker orientation and training look like?
☐	How are workers instructed on the use, care, and replacement of PPE?
☐	How is worker competence measured? Documented?
☐	How do workplace hazards get reported? Tracked?
☐	What does the hazard identification, assessment, and control program look like? Is it documented?
☐	How do workplace incidents get reported? Recorded?
☐	How are workplace incidents investigated? Documented?
☐	How do corrective actions get implemented? Documented?
☐	How are workplace injury situations handled? Recorded?
☐	How are workers involved in the investigation process?
☐	How are "refusal to work" situations handled? Documented?
☐	Are safety factors considered when purchasing equipment or designing new processes? Recorded?
☐	What does the equipment maintenance program look like?
☐	How is damaged/substandard equipment identified and removed from the system? Recorded?
☐	What are the elements of the Emergency Preparedness and Response Plan? How are they communicated to stakeholders? How are they tested?
☐	How frequently is the OHSMS reviewed and by whom?
☐	Do the OHSMS Review results get communicated?
☐	What happens when workers are non-compliant with OH&S standards/procedures/practices?
☐	What does the OH&S Data Management System look like?

In summary, the employer is responsible for:

- ensuring the workplace is healthy and safe;
- ensuring performance standards are developed and followed;
- identifying, assessing, and controlling workplace hazards;
- training workers (*e.g.*, employees and contractors);
- establishing and maintaining a Joint Health & Safety Committee (where applicable);
- managing workplace injuries;

- reporting all critical injuries to the respective provincial Workers' Compensation Boards;
- investigating workplace incidents; and
- seeking ways to continuously improve workplace health and safety.

For more information on due diligence, refer to Chapter 30, "Occupational Health & Safety: Legal Aspects".

4. OH&S Practitioner/Professional and Risk Manager

Employers are typically supported by the:

- **OH&S Practitioner/Professional** — a specialist in preventing workplace losses who strives to eliminate, reduce, and/or control them.
- **Risk Manager** — a specialist in funding business losses who works to ensure that losses can be absorbed in normal cash flow, or buffered by reserves, or transferred to others through legal means, including insurance.

Both these stakeholders deal with **"pure risk"** or **"static risk"** — risks in which there can be no hope of gain. Both deal with unplanned events that cause loss. Both interface with every operation within the organization/ company.

Given the complexity of the relationship between the two functions, Risk Management and Safety Management, the following graphic (Figure 15.3) has been developed to clarify these two functions and their related roles.

Three other risk management functions are:

- **Risk Financing** — an approach for paying for losses through the operating budgets; by borrowing against assets; or through the use of commercial insurance.
- **Security** — an approach for protecting people, property, products, business continuity, intellectual property, and assets from deliberate acts of threat.
- **Insurance** — an approach for legally transferring risk to a third party.

---

[36] Adapted from Johnston Safety Management, "Module Ten: Risk Management" in *CRSP Exam Preparation Workshop Manual* (Edmonton, AB, 2001) at 2. Work originally produced by H.W. Heinrich, D. Petersen & N. Roos, *Industrial Accident Prevention*, 5th ed. (Columbus, OH: McGraw Hill, 1980) (out of print). Reprinted with permission from Dan Petersen.

Figure 15.3: Relationship between Risk Management and OH&S Management[36]

5. Effective Techniques

Although there are a number of models of OH&S Risk Management available, all tend to use a similar approach and to include similar components. The following is a discussion of one OH&S Risk Management model — the Loss Control Model (IEDIM) and its elements.

OH&S Risk Management is a systematic process and the Loss Control Model uses five steps:

- identification of the "pure risks" (**Risk Estimation**);
- evaluation of each risk (**Risk Evaluation/Analysis**);
- development and selection of suitable control measures (**Risk Control**);
- implementation of risk control techniques (**Plan Implementation**);
- monitoring, review, and change as required (**Risk Management**); and
- stakeholder engagement and consultation (**Risk Communication**).[37]

(a) Risk Estimation

The first step involves the identification of all loss exposures — real and potential

---

[37] Board of Canadian Registered Safety Professionals (BCRSP), *Risk Management Domain, CRSPEX Study Guide* (Mississauga, ON: BCRSP, 2002) at 30.

in nature. Risk determination is a two-part process — ***Risk Assessment*** and ***Risk Evaluation*** (Figure 15.4).

Risk assessment entails ***Risk Analysis*** — hazard identification and risk estimation for each hazard, as well as ***Option Evaluation*** — the development of options and analysis of the feasibility of each identified option.

As part of the risk assessment process, risk levels and their consequences are used. For example, when a risk is determined to be intolerable, it is recommended that work not be started or continued. In a substandard risk situation, work should not be started until the risk is reduced or continued until effective controls are in place. Under a moderate risk situation, efforts should be made to reduce the risk provided the prevention costs are measured and limited in scope. Tolerable risks require few additional controls, and trivial risks warrant no action.[38]

---

[38] S. Rodriques & J. Lancelotte, "Risk Management in a Due Diligence World" (Presented at the Industrial Accident and Prevention Association Conference, Toronto, ON, April 14, 2003).

Figure 15.4: OH&S Risk Determination[39]

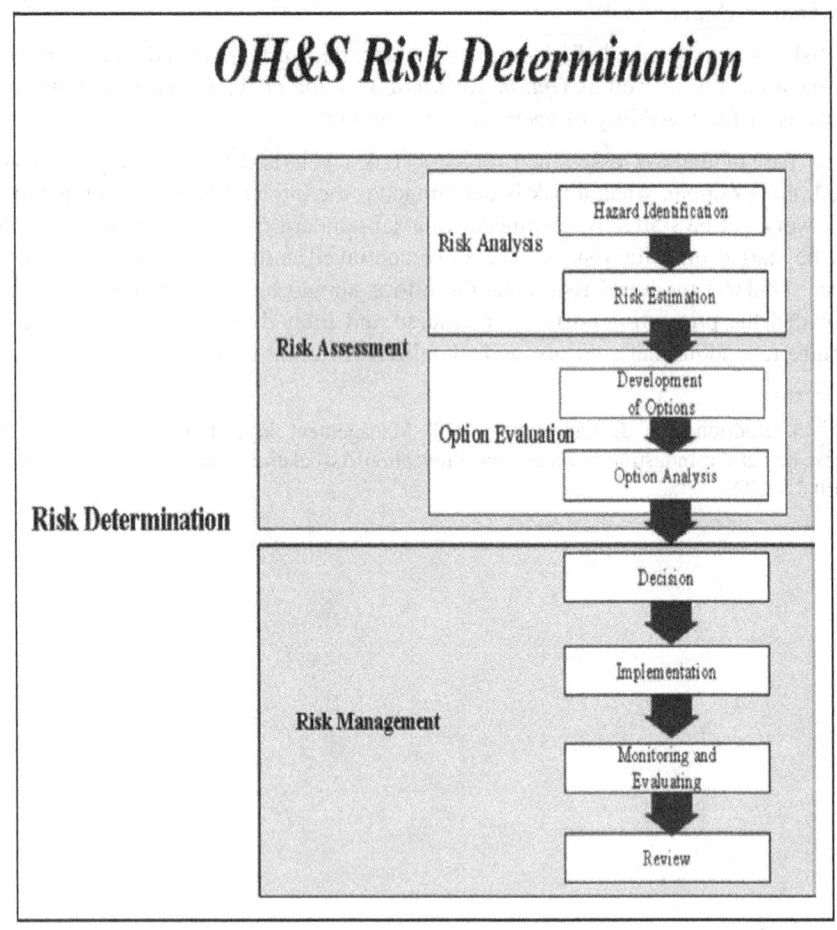

(b) Risk Evaluation

The second step is to determine the level of risk for each of the identified losses using three variables:

- *Severity* — the extent of the loss that may be sustained if an incident/incidental loss occurs.
- *Frequency* — the rate of occurrence of a loss.

---

[39] Adapted from Canada Health Protection Branch, "Health Risk Determination: The Challenge of Health Protection" (Ottawa, ON: Health and Welfare Canada, 1990). Reproduced with the permission of the Minister of Public Works and Government Services Canada, 2007.

- *Probability* — the chance of a loss occurring.

(c) Risk Control

The third step involves using the "four T's" to manage the risk (Figure 15.5). These four recognized techniques for handling risk factors can be used separately or in combination.

Figure 15.5: Risk Management Techniques: Four "T's"

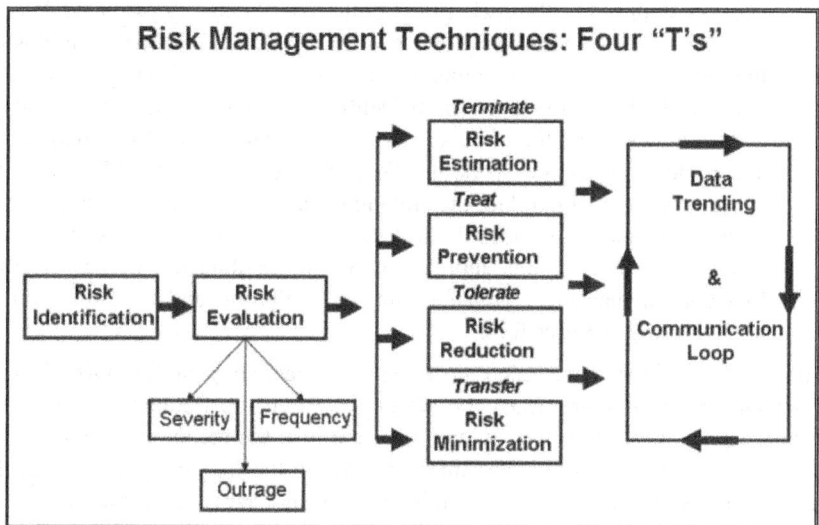

- *Terminate* (**Risk Estimation**) — eliminate or avoid the risk by not undertaking the activity, or by exiting the business or activity. A good illustration is faulty equipment that is removed from service before anyone is injured by its substandard operation. This is the only risk management technique that is designed to be used alone.[40]

- *Treat* (**Risk Prevention**) — reduce the risk by substituting the hazard agent for a safer one; by using engineering controls; by using administrative controls; and by using worker personal protection equipment. Treating the risk involves the use of safety techniques for loss control or loss prevention. Education is a cornerstone of successful risk management programs. It involves the use of well-designed OH&S orientation, training and ongoing communication programs. A second cornerstone is planned maintenance, an aspect that requires stringent documentation to ensure that corporate policies, procedures, and practice routines are upheld. In this way, many risks can be prevented.

---

[40] Johnston Safety Management, "Module Ten: Risk Management" in *CRSP Exam Preparation Workshop Manual* (Edmonton, AB, 2001) at 3.

- *Tolerate* **(Risk Reduction)** — some risks cannot be eliminated or prevented. As such, we accept the risk within the context of prescribed risk tolerances. It involves financing the identified risks through operating budgets, reserves, borrowing money, and some insurance agreements.
- *Transfer* **(Risk Minimization)** — transfer the risk through contracts, leases and/or insurance plans.[41] This technique is similar to having a "safety net" positioned below a high-wire circus performer. It puts in place the tools necessary for controlling what might otherwise be a very extensive loss. It is a technique for controlling the severity of a concurrent loss. It builds on the trends and patterns found in the field of Risk Management. The placement of Automatic External Defibrillators (AEDs) in the workplace and training staff on their use is a good illustration. An employee may suffer a cardiac event while working. Through the use of the AED and the administration of First Aid, the situation can be mitigated. In this way, the transfer technique provides the last measure of control for concurrent risk situations. However, this approach is not a substitute for the use of other loss control measures. Transfers are not foolproof, and tend to result in some loss by the transferor.

The conditions for using each technique are described in the OH&S Risk Assessment Matrix (Figure 15.6). For example, when the frequency of the risk occurrence is high, but the severity and costs are low, the recommended approach is to self-insure or manage the cost internally. When the frequency is low, but the severity and cost are high, the use of third-party insurance is recommended.

---

[41] Board of Canadian Registered Safety Professionals (BCRSP), *Risk Management Domain, CRSPEX Study Guide* (Mississauga, ON: BCRSP, 2002).

Figure 15.6: OH&S Risk Assessment Matrix[42]

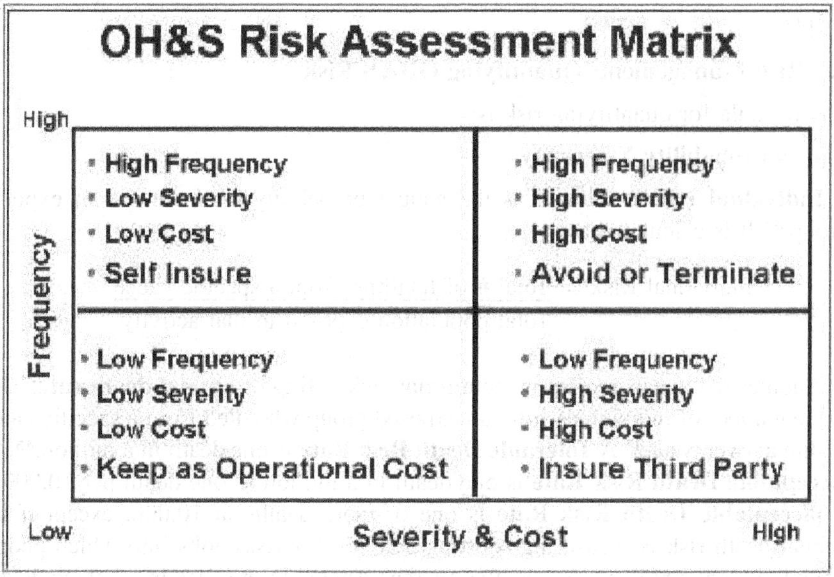

OH&S Risk Assessment Matrix

Frequency	Severity & Cost	
High	• High Frequency • Low Severity • Low Cost • Self Insure	• High Frequency • High Severity • High Cost • Avoid or Terminate
	• Low Frequency • Low Severity • Low Cost • Keep as Operational Cost	• Low Frequency • High Severity • High Cost • Insure Third Party
	Low ———————————— **High**	

(d) Implement the Plan

The fourth step focuses on implementing the risk management plan by:

- converting the plan into action;
- allocating the appropriate and adequate resources;
- educating people on the plan and their respective roles and responsibilities in bringing the plan to fruition; and
- making the necessary equipment available.

(e) Risk Management

The fifth step focuses on the process of risk management. It involves:

- measurement of the performance of the risk management plan;
- evaluation of its effectiveness; and
- actions to commend or correct the risk management plan performance.

(f) Risk Communication

The last step informs the relevant stakeholders of the risk management plan, the

---

[42] Adapted from CSSE, "Competency 7 — Risk Management Module" *BCRSP Exam Preparation Course — Course Participant's Guide (2002)* (Toronto, ON: CSSE, 2002) at 9. Original source: Det Norske Veritas (DNV). Reprinted with permission from DNV.

status of its performance and the anticipated enhancement or continuous improvement plans. Risk communication is discussed in detail in the latter part of this chapter.

6. Risk Management: Quantifying OH&S Risk

The formula for quantifying risk is:

Risk = probability X severity

Individual risk is defined as the annual probability of death for an exposed group.[43] It is calculated as:

$$\text{Individual Risk} = \frac{\text{Total \# of fatalities from a specific cause}}{\text{Total population exposed to that activity}}$$

Another risk measurement is the **annual risk rating** or **annual death rate**. That is the number of individuals from an exposed group who died from a specific cause within a given year.[44] A **Tolerable Death Risk Rate** is one death in a million.[45] An **Acceptable Death Risk Rate** is one death in a million to one death per 10,000.[46] **Unacceptable Death Risk Rate** is one or more deaths in 10,000, except if the annual death risk is measuring sporting activities or risky jobs into which people voluntarily enter.[47] For example, the Average Annual Death Risk in a Vehicle is one per 10,000.

C. OH&S RISK COMMUNICATION

OH&S risk communication is a science-based approach for communicating effectively in high-concern, high-stress situations, emotionally charged situations, and controversial situations.[48] According to the U.S. Department of Health and Human Services:

> Risk communication [is an] interactive process of exchange of information and opinion among individuals, groups, and institutions; often involves multiple messages about the nature of the risk or expressing concern, opinions, or reactions to risk messages or to legal and institutional arrangement for risk management.[49]

---

[43] Johnston Safety Management, "Module Ten: Risk Management" in *CRSP Exam Preparation Workshop Manual* (Edmonton, AB, 2001) at 3.

[44] *Ibid.*

[45] *Ibid.*

[46] *Ibid.*

[47] *Ibid.*

[48] V. Covello R. Peters and J. Wojtecki, *et al.*, "Risk Communication, the West Nile Virus Epidemic and Bioterrorism: Responding to the Communication Challenges Posed by the Intentional or Unintentional Release of a Pathogen in an Urban Setting" (2001) 78:2 J. Urban Health, N.Y. Bull. Acad. Med. 382-391.

[49] U.S. Department of Health and Human Services, *Communicating in a Crisis: Risk*

The objective of risk communication is to increase audience knowledge and understanding; enhance their trust and your credibility; and promote dialogue to resolve disagreements of opinion and belief on what needs to be done. It is an ongoing process that helps to define the problem and solicit involvement and action before an emergency occurs.

Effective OH&S risk communication involves:

1. telling people about the situation at hand;
2. developing an open dialogue with the target group;
3. enhancing the target group's understanding of the situation;
4. getting commitment for action from members of the target group; and
5. resulting in action.

Risk communication differs from crisis communication in that **crisis communication** encompasses those messages delivered to stakeholders during a crisis period.[50]

1. OH&S Risk Communication: Importance

Risk communication is part of everyday life, whether you are at work, at home, or in the community.[51] As already noted, risk is defined as the probability of loss of that which we value. In terms of risk communication, risk is composed of the nature of the hazard, the probability of the risk occurrence, the severity of the risk, as well as outrage.

Risk = the nature of the hazard x probability x severity + *outrage.*

When a "high-risk" situation occurs, the human values that come into play are concerns about safety, security, finances, human health and well-being, quality of life, and protection of family/community members. These can lead to **outrage** — a strong human reaction to "high-risk" situations. When people are upset, they experience difficulty in hearing, understanding, and/or recalling information. As well, when stressed, people tend to distrust that "the people in the know" can listen, care, or empathize with them. When in such a severe emotional state, people seek honest and open communication from competent experts.

Outrage stems from threats posed to people's values and to the items that people "hold dear to their hearts". It is important for risk communicators to recognize the nature of outrage factors (Table 15.5), so they can be prepared to manage them. For

---

Communication Guidelines for Public Officials (Washington, U.S. Department of Health and Human Services, 2002) at 4, online: http://www.hhs.gov/od/documents/RiskCommunication.pdf (date accessed: February 28, 2020).

[50] K. Fearn-Banks, *Crisis Communications: A Casebook Approach* (New Jersey: Lawrence Erlbaum Associates Inc., 2007).

[51] V. Covello, "Risk Communication Slides", online: http://www.nwcphp.org/docs/pdf/april02.ppt (date accessed: February 28, 2020).

example, instances that evoke high outrage are messages delivered by untrustworthy sources; proposed remedial actions that net few benefits or are deemed to be inequitable or unfair; uncontrollable or dreaded disasters; man-made disasters; catastrophic events; disasters that result in children being the main victims; disasters that involve identifiable victims; disasters that result from unethical or immoral practices; disasters that result in irreversible effects; and/or disasters that are not scientifically understood.

Dr. Vincent Covello notes that: "[t]here is virtually no correlation between the ranking of hazards by experts and the ranking of those same hazards by the public."[52]

Table 15.5: Outrage Factors[53]

Low	High
• Trustworthy sources	• Untrustworthy sources
• Substantial benefits	• Few benefits
• Voluntary	• Involuntary
• Controllable	• Uncontrollable
• Fair/equitable	• Unfair/inequitable
• Natural origin	• Manmade
• Familiar	• Unfamiliar/exotic
• Not dreaded	• Dreaded
• Certain	• Uncertain
• Children not victims	• Children are victims
• Not memorable	• Memorable
• Moral/ethical	• Immoral/unethical
• Clear, non-verbal message	• Mixed, non-verbal message
• Responsive	• Unresponsive
• Random/scattered	• Catastrophic
• Little media attention	• Lots of media attention
• Victims statistical	• Victims identifiable
• Immediate efforts	• Delayed effects
• Effect reversible	• Irreversible effects
• Scientifically well-understood	• Not scientifically understood

2. OH&S Risk Communication: Effective Techniques

There are four models of effective risk communication that warrant consideration. They are:[54]

---

[52] *Ibid.*

[53] *Ibid.*

[54] V. Covello *et al.*, "Risk Communication, the West Nile Virus Epidemic and

- *Risk Perception Model*

    Risk perception stems from the magnitude of the outrage factors provided above, each of which can alter perceptions to varying degrees. These factors determine the public's level of concern, fear, hostility, and outrage.

    Given all the above variables, OH&S Practitioners/Professionals need to develop and demonstrate good risk communication skills.

- *Mental Noise Model*

    As the level of public anxiety increases, the ability to process information decreases. The creation of mental noise blocks the receipt of information. Risks with the least amount of control create the most mental noise.

    OH&S Practitioners/Professionals need to keep this model in mind when delivering risk communication messages. Keep the messages simple and direct.

- *Negative Dominance Model*

    In a high-concern situation, people process negative and positive information asymmetrically. That is, they tend to focus more on the negative aspects, downplaying the positive messages. The result is that they place more value on their losses than their gains.

    Hence, for OH&S Practitioners/Professionals, it is important to focus on the wording of the risk communication message. Provide positive messages that explain actionable activities: "[m]ore specifically, risk communications are most effective when they focus on what is being done, rather than on what is not being done."[55]

- *Trust Determination Model*

    In high-concern situations, risk communicators need to build credibility and earn the trust of the target group. To do so, it is critical for the target group to perceive that the risk communicator is:

    - *Empathic and caring.* According to Dr. Vincent Covello: "[i]n high-concern situations, people want to know that *you care*, before they care what you know about the situation."[56]

    - *Competent and a technical expert.* Make sure the risk communicator is the "best person" to deliver the message and to respond to the questions posed.

Bioterrorism: Responding to the Communication Challenges Posed by the Intentional or Unintentional Release of a Pathogen in an Urban Setting" (2001) 78:2 J. Urban Health, N.Y. Bull. Acad. Med. 382-391.

[55] *Ibid.*

[56] *Ibid.*

- *Honest and open.*
- *Dedicated and committed.*

In summary: "[T]rust and credibility are difficult to achieve: if lost, they are even more difficult to regain."[57]

To assist OH&S risk communicators, a four-step Emergency OH&S Risk Communication Model (Table 15.6) is provided.

Table 15.6: Emergency OH&S Risk Communication: Four-step Model[58]

1) Empathy Identify with your audience: • Recognize who has been hurt • Assume responsibility for solving the problem • Define yourself and your values • Acknowledge risk	2) Solution Advance the story: • Describe what you are going to do to SOLVE the problem • Provide action that the audience can do
3) Facts Be clear and straightforward: • Explain what happened • Explain your record on this issue • Explain the benefits of your solution	4) Process Demonstrate how the decision-making, or investigation process is credible: • Explain the process • Explain when more information will be available • Demonstrate openness/accountability • Use third parties to lend credibility to your message

Step One — Empathy is designed to enable the OH&S risk communicator to identify with the audience by recognizing who has been hurt, assuming responsibility for solving the problem, defining oneself and one's values, and acknowledging the risk.

Step Two — The Solution is aimed at advancing the story by describing what the risk communicator is going to do to solve the problem and providing "action(s)" that the audience can do.

Step Three — The Facts enable the risk communicator to be clear and

---

[57] V. Covello, "Risk Communication, Trust and Credibility" (1993) 35 J. Occup. Med. 18.

[58] Center for Disease Control (CDC), "CDC Responds: Risk Communication and Bioterrorism" *CDC Public Health Training Network* (Atlanta, GA: Public Health Foundation, 2001). Available for public use.

straightforward by explaining what happened, explaining the organization's/company's record on this issue, and describing the benefits of the proposed solution.

Step Four — The Process demonstrates how the decision-making or incident investigation process is credible. During this step, the process is explained and the details on when more information will be available are provided. The risk communicator needs to ensure that a feeling of openness and accountability is projected to the target audience. If necessary, a third-party expert can be used to lend credibility to the message being provided.

Risk communicators need to uphold the principles of risk communication, namely:

- Anticipate questions and concerns by closely listening to the audience and recorgnizing that people want to know:
    - Are my family and myself safe?
    - What have you found out that may affect myself and my family?
    - How can I protect myself and my family?
    - Who (or what) caused this situation?
    - Can you fix it?[59]
- Prepare an accurate, well-crafted, relevant, ethical, consistent, and understandable message — the media and the public want to know:
    - Who is in charge?
    - Are those injured, getting help?
    - Is this situation contained?
    - What can be expected going forward?
    - What should we the public/media do?
    - Why did this happen?
    - Were there warning signs that this might happen?
    - Why wasn't this situation prevented?
    - What else can go wrong?
    - When were you notified about this situation?
    - What does the data being presented actually mean to us?
    - What good/bad things are you not telling us about?[60]
- Make sure to be well-briefed on the pertinent details and use the communication materials and visual aids in a professional manner.

---

[59] *Ibid.*

[60] *Ibid.*

- Practice the delivery of the risk communication message so as to appear prepared and "on schedule".
- Ensure the message is short in duration and accurate.
- Limit the message content.
- Engage the audience by involving them in the communication.
- Avoid apologizing for oneself or the organization/company.
- Aim to enhance audience knowledge and trust.

To achieve the above tasks, the risk communicator needs to be aware of some common myths and incorrect actions:

- ***There is not enough time and resources for a risk communication program*** — *False*: Organizations/Companies can ill afford to not have a risk communication program and plan. Disasters occur without warning and developing an OH&S Communication Plan during an emergency is not advisable. Train OH&S personnel to communicate effectively. Plan for projects to involve the public.
- ***Telling people about a risk is more likely to unduly alarm them*** — *False*: The organization's/company's representatives need to partner with employees/clients/the public during a high-concern situation. Partnering requires open, honest, and well-crafted communication as noted earlier. Giving people the chance to air their feelings and concerns can decrease the potential for alarm.
- ***Communication is less important than education*** — *False*: Well-designed communication is vital if the target audience is going to be able to understand the magnitude of the situation without the message being clouded by outrage. Pay as much attention to the process of dealing with people as is paid to explaining the technical data.
- ***Don't go to the public until solutions exist*** — *False*: It is vital for the organization's/company's risk communicators to approach the public in a timely manner. By not doing so, more fear is generated. Barry Johnson, Assistant Surgeon General, U.S. Public Health Service (1987), recommends that the risk communicator "[g]et the receiver [the audience] involved up front."[61]
- ***These issues are too difficult for people to understand*** — *False*: To gain trust and credibility, risk communicators need to ensure that people receive relevant information in a timely manner. In times of high concern, the information that people need to hear is that the company/risk communicator

---

[61] B. Johnson (1987), quoted in "A Primer on Health Risk Communication Principles and Practices" Agency for Toxic Substances and Disease Registry, online: http://www.atsdr.cdc.gov/risk/riskprimer/index.html (date accessed: February 28, 2020).

cares and that their fears are being heard. Tease apart public disagreements with company policies from the lack of knowledge about the technical data. Address each separately.

- *Technical decisions should be made by technical people* — *False*: Stakeholders impacted by the high-concern situation need to provide input into the problem-solving process. The decisions being made need to include the interests of these stakeholders.
- *Risk communication is not my job* — *False*: Risk communication at some level, is part of every OH&S Practitioner's/Professional's job.
- *If we give them an inch, they'll take a mile* — *False*: This statement is definitely a myth. If you listen to people when they are seeking "inches", they are less likely to ask for "miles".
- *If we listen to the public, we'll devote scarce resources to issues that are not important to public health* — *False*: Listening to the stakeholders can actually protect scarce resources. Listen early to people so that controversy can be avoided, and the potential for disproportionate attention to lesser items, reduced.
- *Activists are responsible for stirring up issues* — *True and false*: Activists can help focus public anger. However, outrage tends to stem from those who have the most at stake. Work with the various groups as opposed to against them.[62]

In summary, for risk communicators to overcome these myths and incorrect actions, these cardinal rules of risk communication can be very helpful.

- *Accept and involve the public as a partner in dealing with high-concern situations* — the goal is to produce an informed public, not to defuse public concerns. Involve the public at the earliest possible stage and determine what type of involvement they prefer or want. Clarify their role and acknowledge situations where the public will have only limited decision-making power.
- *Know your audience* — know their characteristics and types of concern.
- *Identify and respond to the needs of different audiences* — different audiences and media require different goals and actions.
- *Plan your message carefully* — merely disseminating information without regard for communicating the complexities and uncertainties of the risk does not equate to effective risk communication. Well-managed risk communication efforts ensure that messages are constructively formulated,

---

[62] C. Chess, B.J. Hance & P.M. Sandman, *Improving Dialogue with Committees: A Short Guide to Government Risk Communication* (New Jersey: New Jersey Department of Environmental Protection, 1988).

transmitted, and received; and that they result in meaningful actions.

- **Listen to the public's specific concerns about the situation** — people tend to care more about trust, credibility, competence, fairness, and empathy than about statistics and event details.
- **Be honest, frank, and open** — honesty and openness begin with being aware of one's feelings and values, and how they affect you. The risk communicator's ability to effect constructive communication is determined by whether or not the audience perceives him/her as a trustworthy and credible person.
- **Speak clearly and with compassion** — avoid having the risk communication message downplay or ignore the "tragedy of the disaster". People can understand risk information even if they do not agree with your take on the situation. However, they will not listen, if the risk communicator fails to acknowledge their pain and fears.
- **Acknowledge that people's values and feelings are a legitimate aspect of a disaster** — provide a forum for people to vent their feelings. Listen to their concerns and recognize their values. When people speak emotionally about their feelings, respond to their emotions versus providing more data.
- **Use ethical communication and avoid the use of "spin" techniques** — people can see through "a spin", and will perceive it as an "insult to their intelligence". As well, the risk communicator's credibility and honesty will be jeopardized.
- **Work with other credible sources to help the target audience grasp the nature of the situation and how to address it** — conflicts and disagreements among organizations/companies on what actions to take, make communication with the public more difficult.
- **Evaluate your risk communication efforts** — measure the effectiveness of your message in terms of whether or not it met your goals and the audience needs. Did it result in the desired changes? What went well? What could have gone better? What did you learn from the experience?
- **Meet the needs of the media** — the media are looking for emotionally charged, newsworthy stories.
- **Avoid the common pitfalls of risk communication** (Table 15.7).

Table 15.7: Common Pitfalls[63]

Pitfall	Pitfall
Use of jargon	Use of humour
Use of negative allegations	Use of negative words and phrases
Use of a great reliance on words	A display of temper
Use of abstract concepts	Lack of clarity
Demonstrating inconsistency between the verbal and non-verbal messages	Attacks on members of the target audience
Making hollow promises	Use of speculation
Focusing on the cost of disaster or remediation actions	Use of organizational identity/fame
Blaming others	Making risk comparisons
Use of "Off the Record" statements	Use of risk/benefit/cost comparisons
Use of health risk numbers	Use of numbers
Use of technical details and debates	Use of long presentations

3. Managing Hostile Situations

As has been mentioned, OH&S issues can arouse strong emotions and hostility. The OH&S Practitioner/Professional is advised to recognize that hostility tends to be directed at the organization/company or the OH&S position, not towards him/her personally.

When faced with a hostile situation, the OH&S Practitioner/Professional is advised to:

- *Acknowledge the existence of the audience's/person's hostility and outrage* — anger, fear, guilt, and denial are all normal reactions to an abnormal situation.

- *Practice self-management* — it is tempting to return hostility with anger. But, that worsens the situation. The use of self-management allows for hostile feelings to be aired and disagreements to be discussed.

- *Be prepared* — recognize and be ready to address hostility. Turn hostility into a collaborative effort for managing the high concern situation.

- *Communicate empathy and caring* — hostile people are expressing fear, anger, grief, and denial. They need understanding to cope with the unexpected situation. Expressions of empathy and caring can go a long way in assisting them.

- *Track your messages* — have someone document or record the events of the meeting, and how hostile audience members were managed. This is particularly important when the high-concern situation is expected to be

---

[63] V. Covello, "Risk Communication, Trust and Credibility" (1992) 6:1 Health and Environmental Digest 1, also published in (1993) 35 J. Occup. Med. 18.

lengthy in duration, or if great controversy exists.[64]

4. Working with the Media

Working with the public can be an important opportunity for public communication — positive relationships with the media are critical. The media seek "human-interest" stories, bad news, people's varied perspectives, "yes or no" or unsafe answers, and front-page stories.[65] They tend to be more interested in politics than risk, simplicity than complexity and danger than safety.[66] According to Peter Hunt, former Vice President Public Affairs, ENMAX Corporation, "[n]ews is anything someone does not want printed; the rest is just advertising."[67]

In preparation for an interview with the media, recognize that the media is looking for information on the disaster event as to Whom? What? When? Where? Why? and How? Ensure that your message is succinct and accurate. For national news, design the communication to cover up to three messages, to air seven to nine seconds (21-27 words), and to allow for a nine-second knowledge/trust window.[68] For broadcast media, limit the message to a 10- to 12-word "soundbite". For print media, prepare a one- to three-line quote.[69]

Some other preparation tactics are:

- ***ask who will be doing the interview*** — avoid asking for a preferred reporter;
- ***determine what issues they want to discuss*** — avoid asking for specific questions in advance or insisting that certain subjects be covered;
- *acknowledge when the topics being covered are not your area of expertise*;
- *ask about the format and planned duration of the interview*;

---

[64] ATSDR, "A Primer on Health Risk Communication Principles and Practices" *Agency for Toxic Substances and Disease Registry (ATSDR)*, online: http://www.atsdr.cdc.gov/risk/riskprimer/index.html (date accessed: February 28, 2020).

[65] *Ibid.*, at 23.

[66] V. Covello & F. Allen, "Seven Cardinal Rules of Risk Communication" U.S. Environmental Protection Agency, Office of Policy Analysis in "A Primer on Health Risk Communication Principles and Practices", Agency for Toxic Substances and Disease Registry, online: http://www.atsdr.cdc.gov/risk/riskprimer/vision.html#cardinal (date accessed: February 28, 2020).

[67] P. Hunt, "Role of Public Affairs within ENMAX Corporation" (Internal presentation at ENMAX Corporation, Calgary, AB, 2004).

[68] V. Covello, "Risk Communication Slides", online: http://www.nwcphp.org/docs/pdf/april02.ppt (date accessed: February 28, 2020).

[69] E. Donovan & V. Covello, *Risk Communication: Student Manual* (Washington, DC: Chemical Manufacturers' Association, 1989) in "A Primer on Health Risk Communication Principles and Practices" Agency for Toxic Substances and Disease Registry, online: http://www.atsdr.cdc.gov/risk/riskprimer/references.html (date accessed: February 28, 2020).

- *find out who else will be interviewed* — avoid insisting that an adversary NOT be interviewed; and
- *prepare and practice the message* — avoid assuming that the interview will be easy.[70]

During the interview, some tips for the OH&S Practitioner/Professional are:
- be honest and accurate;
- stick to your key message(s) and avoid the use of improvisations or negative allegations;
- state your conclusions first and then provide the supporting data;
- be open with your message;
- provide a reason for the subjects that you cannot discuss;
- offer to get any needed information that you may not have with you;
- avoid guessing or speculation as to why things are the way they are;
- stress the facts; and
- correct mistakes by stating that you would like an opportunity to clarify the subject.[71]

Once the interview is over, remember that you are still "on the record". Be helpful and volunteer to get information or to make yourself available. Should inaccuracies in the report occur, politely call and inform the interviewer as opposed to calling the reporter's supervisor to complain.

5. OH&S Risk Communication: Additional Considerations

OH&S practitioners need to be able to effectively deal with erroneous information. Given the ubiquitous social media, stakeholders have ready access to all kinds of information, including what Dr. G. Pennycock terms, "bullshit". As an OH&S Practitioner, it can be very challenging to deal with widely dispersed, erroneous information that stakeholders believe.

Bullshit is defined as "something that is designed to impress but that was constructed absent direct concern for the truth".[72] Pennycock proposes that some people have a propensity or strong bias towards believing something — hence, they are biased towards believing bullshit. Gilbert supports this by proposing that to comprehend an abstract, humans must first believe it. To counter erroneous information, the OH&S practitioner must first help the person to recognize the

---

[70] *Ibid.*

[71] *Ibid.*

[72] H. Frankfurt (2005), reported in G. Pennycock, "On the reception and detection of pseudo-profound bullshit" (2015) 10:6 *Judgment and Decision Making* online: http://journal.sjdm.org/15/15923a/jdm15923a.pdf (date accessed: January 31, 2020) at 1.

conflict of information and carry on from there by:

- Reviewing the presented information;
- Checking its accuracy;
- Determine why people choose to believe it;
- Be empathetic and caring, but clearly explain the facts;
- Assess if a change in belief occurs; and
- Support the person's comprehension of the facts.

6. OH&S Risk Communication Plan: Evaluation

In establishing a risk communication plan, measurable performance objectives need to be established. For each component of the risk communication plan, assess what went well, what could have gone better and why.

The evaluation can be:

- Formative — measurement of the process as it unfolds; or
- Summative — measurement of the plan once it has been completed.

The critical factor is to learn from the experience.

According to Baruch Fischhoff, Department of Engineering & Public Policy, Carnegie-Mellon University:

> If we have not gotten our message across, then we ought to assume that the fault is not with our receivers.[73]

D. SUMMARY

Risk management and risk communication are part of everything Management does. In terms of OH&S issues, the OH&S practitioner/ professional is expected to take a leadership role. According to Bob Nicolay, former CEO and President, ENMAX Corporation:

> Leadership is about relationships. Relationships are built on communication. It's all about communication.[74]

This chapter was designed to provide information and techniques for effective OH&S risk management and risk communication. Remember:

> Life is not about how fast you run, or how high you climb, but how well you bounce. Anonymous

---

[73] B. Fischhoff (1985), quoted in "A Primer on Health Risk Communication Principles and Practices" Agency for Toxic Substances and Disease Registry, online: http://www.atsdr.cdc.gov/risk/riskprimer/index.html (date accessed: February 28, 2020).

[74] B. Nicolay, "Leadership Forum" (Internal presentation at ENMAX Corporation, Calgary, AB, 2004).

CHAPTER REFERENCES

ATSDR, "A Primer on Health Risk Communication Principles and Practices" *Agency for Toxic Substances and Disease Registry*, online: http://www.atsdr.cdc.gov/risk/riskprimer/index.html (date accessed: February 28, 2020).

Association of Workers' Compensation Boards of Canada, "2018 Injury Statistics" *Statistics*, online: http://awcbc.org/?page_id=14 (date accessed: February 28, 2020).

F. Bird & G. Germain, *Practical Loss Control Leadership*, revised ed. (Loganville, GA: DNV, 1996).

Board of Canadian Registered Safety Professionals (BCRSP), *Risk Management Domain, CRSPEX Study Guide* (Mississauga, ON: BCRSP, 2002).

Canada Health Protection Branch, "Health Risk Determination: The Challenge of Health Protection" (Ottawa, ON: Health and Welfare Canada, 1990).

Center for Disease Control (CDC), "CDC Responds: Risk Communication and Bioterrorism" *CDC Public Health Training Network* (Atlanta, GA: Public Health Foundation, 2001).

C. Chess, B.J. Hance & P.M. Sandman, *Improving Dialogue with Committees: A Short Guide to Government Risk Communication* (New Jersey: New Jersey Department of Environmental Protection, 1988).

V. Covello, "Risk Communication, Trust and Credibility" (1992) 6:1 Health and Environmental Digest 1, also published in (1993) 35 J. Occup. Med. 18.

V. Covello, "Risk Communication Slides", online: http://www.nwcphp.org/docs/pdf/april02.ppt (date accessed: February 28, 2020).

V. Covello & F. Allen, "Seven Cardinal Rules of Risk Communication" *U.S. Environmental Protection Agency, Office of Policy Analysis* in "A Primer on Health Risk Communication Principles and Practices" Agency for Toxic Substances and Disease Registry, online: http://www.atsdr.cdc.gov/risk/riskprimer/vision.html#cardinal (date accessed: February 28, 2020).

V. Covello, R. Peters & J. Wojtecki, *et al.*, "Risk Communication, the West Nile Virus Epidemic and Bioterrorism: Responding to the Communication Challenges Posed by the Intentional or Unintentional Release of a Pathogen in an Urban Setting" (2001) 78:2 J. Urban Health, N.Y. Bull. Acad. Med. 382-391.

E. Donovan & V. Covello, *Risk Communication: Student Manual* (Washington, DC: Chemical Manufacturers' Association, 1989) in "A Primer on Health Risk Communication Principles and Practices" *Agency for Toxic Substances and Disease Registry*, online: http://www.atsdr.cdc.gov/risk/riskprimer/references.html (date accessed: February 28, 2020).

K. Fearn-Banks, *Crisis Communications: A Casebook Approach*, 3rd ed. (New Jersey: Lawrence Erlbaum Associates Inc., 2007).

B. Fischhoff (1985), quoted in "A Primer on Health Risk Communication Principles and Practices" *Agency for Toxic Substances and Disease Registry*, online: http://www.atsdr.cdc.gov/risk/riskprimer/index.html (date accessed: February 28, 2020).

H.G. Frankfurt, On Bullshit (Cambridge: Cambridge University Press, 2005).

Government of Canada, "Occupational Injuries and Diseases in Canada, 1996-2008" Labour Program, online: http://www.labour.gc.ca/eng/health_ safety/pubs_hs/oidc.shtml (date accessed: February 28, 2020).

Hatscan, *Handi-guide to Alberta's OH&S Act, Regulation and Code* (Edmonton, AB: Hazard Alert Training Inc., 2004) at c. 2-2.

P. Hunt, "Role of Public Affairs within ENMAX Corporation" (Internal presentation at ENMAX Corporation, Calgary, AB, 2004).

Institute for Work & Health (IWH), *OHS Vulnerability Measure* (2016), online: www.iwh.on.ca/ohs-vulnerability-measure (date accessed: January 31, 2020).

B. Johnson (1987), quoted in "A Primer on Health Risk Communication Principles and Practices" *Agency for Toxic Substances and Disease Registry*, online: http://www.atsdr.cdc.gov/risk/riskprimer/index.html (date accessed: February 28, 2020).

Johnston Safety Management, "Module Ten: Risk Management" in *CRSP Exam Preparation Workshop Manual* (Edmonton, AB, 2001).

Liberty Mutual Research Institute for Safety, "2019 Workplace Safety Index: The Top 10 Causes of Disabling Injuries at Work" *Viewpoint*, online: https://viewpoint.libertymutualgroup.com/article/top-10-causes-disabling-injuries-at-work-2019/ (date accessed: January 31, 2020).

F. Manuele, *On the Practice of Safety*, 3rd ed. (New Jersey: Wiley-Interscience, 2003).

S. Meltzer, "Risk Management: 2010" (Presented at University of Calgary, Calgary, AB, March 31, 2005).

B. Nicolay, "Leadership Forum" (Internal presentation at ENMAX Corporation, Calgary, AB, 2004).

H.G. Frankfurt (2005) reported in G. Pennycock, "On the reception and detection of pseudo-profound bullshit" (2015) 10:6 Judgment and Decision Making, online: http://journal.sjdm.org/15/15923a/jdm15923a.pdf (date accessed: January 31, 2020).

S. Rodriques & J. Lancelotte, "Risk Management in a Due Diligence World" (2015) 10:6 Judgment and Decision Making 549-563 (Presented at the Industrial Accident and Prevention Association Conference, Toronto, ON, April 14, 2003).

Safety Management Group, "Injury Cost Calculator" (2015), online: http://www.safetymanagementgroup.com/injury-cost-calculator.aspx (date accessed: February 28, 2020).

Statistics Canada, Labour force characteristics by age and sex (December 2018), online: https://www150.statcan.gc.ca/n1/daily-quotidien/180105/t001a-eng.htm (date accessed: February 28, 2020).

U.S. Department of Health and Human Services, *Communicating in a Crisis: Risk Communication Guidelines for Public Officials* (Washington, U.S. Department of Health and Human Services, 2002), online: http://store.samhsa.gov/product/Risk-Communication-Guidelines-for-Public-Officials/SMA02-3641 (date accessed: February 28, 2020).

U.S. Department of Labor, "Cost of Accidents, Safety & Health Management Systems e-Tool", online: http://www.osha.gov/SLTC/etools/safetyhealth/mod1_costs.html (date accessed: February 28, 2020).

U.S. Department of Labor, "Census of Fatal Occupational Injuries Summary, 2018" (2020), online: http://www.bls.gov/news.release/cfoi.nr0.htm (date accessed: February 28, 2020).

U.S. Department of Labor, "Injuries, Illnesses and Fatalities, 2018" (2020), online: http://www.bls.gov/iif/ (date accessed: February 28, 2020).

U.S. Department of Labor, "Occupational Illness and Injuries (Annual) News Release, 2018" (2020), online: http://www.bls.gov/news.release/archives/osh_11072013.htm (date accessed: February 28, 2020).

U.S. Department of Labor, OSHA $afety Pays Program, "Background of the Cost Estimates", $afety Pays Program, 2010, online: https://www.osha.gov/dcsp/smallbusiness/safetypays/background.html (date accessed: February 28, 2020).

U.S. Department of Transportation, "Accident Cost Table – Federal Motor Carrier Safety Administration", online: http://www.truckbrakesafety.com/. . ./fmcsa-accident-cost-revenue.pdf (date accessed: February 28, 2020).

Chapter 16

MARKETING OCCUPATIONAL HEALTH AND SAFETY PROGRAMS; COMMUNICATING THE RESULTS

A. INTRODUCTION

The shortfall of most Occupational Health and Safety (OH&S) Programs is that although they exist and function well, few are effectively marketed in terms of the value they offer to employees, contractors, management, and the organization/company. This chapter addresses the concepts of program marketing and how they can be applied to promote an OH&S Program. In addition, it addresses how to effectively communicate the intent of the OH&S Program, its benefits, processes, and program results. The marketing principles and practices, discussed in this chapter, are transferable to other workplace programs.

B. PROGRAM MARKETING[1]

Marketing is defined as a social and managerial process through which individuals and groups obtain what they need and want by creating and exchanging products/services and value with others.[2]

1. Marketing Concepts

The components of marketing include:

Customer Needs, Wants, and Demands

Customer needs, wants, and demands must be considered when producing products or developing services to be delivered. This means that program developers need to know what each stakeholder needs, wants, and expects from the program. In terms of marketing an OH&S Program, the program leaders must demonstrate how the program and its services meet those specific stakeholder needs, wants, and expectations.

---

[1] Adapted from the concepts in D. Dyck, Disability Management: Theory, Strategy & Industry Practice, 5th ed. (Markham, ON: LexisNexis Canada, 2013).

[2] *Ibid.*

Products or Services

Products and services involve anything that can be offered to a market to satisfy a customer need or want.[3] This includes both tangible (products) and intangible items (services). OH&S services provided to employees and workplace operations fall into this category (Figure 16.1).

Figure 16.1: An Example of Some OH&S Products and Services[4]

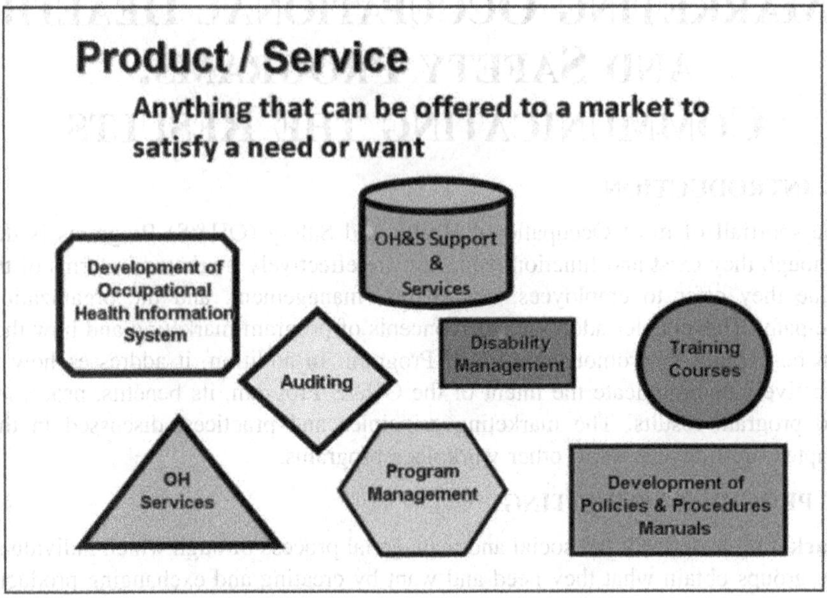

Product / Service
Anything that can be offered to a market to satisfy a need or want

- Development of Occupational Health Information System
- OH&S Support & Services
- Auditing
- Disability Management
- Training Courses
- OH Services
- Program Management
- Development of Policies & Procedures Manuals

Product Value, Consumer Satisfaction and Product Quality

Consumers buy based on their **needs** and the perception of the value that the product/service offers, *i.e.*, its **features**. Satisfaction depends on the product's/service's perceived performance relative to buyer needs and expectations. When the product or service quality meets or exceeds customer expectations, the customer is satisfied. This means that for employees and managers to view an OH&S Program and its services as valuable, they must perceive that they have received more value than what they initially expected. For OH&S Professionals/Practitioners, the message is *to not over-promise and under-deliver*, but rather to keep promises realistic in what can be done, and then, over-deliver.

---

[3] *Ibid.*
[4] *Ibid.*

Product/Services Exchange

The **exchange** is the actual delivery of the product or service.[5] How the product/service is made available to the customer is important. It can be either a negative or a positive experience. This is where customer-service principles come into play. The strategy is to provide the right product/service to the right person, at the right time, for the right reasons, and at the right price.

Market

The **market** is the potential customers for whom the product/service has been designed.[6] For an OH&S Program, this would include the employees, management, unions, and various external stakeholders like government authorities, industry associations, potential customers, and family members.

2. Marketing Management

Effective marketing can be equated to effective marketing management. **Marketing management** involves managing customer demand which, in turn, involves managing customer relationships. Marketers need to build long-term relationships with valued customers, distributors, dealers, and suppliers.

OH&S Professionals/Practitioners can promote their programs by nurturing strong relationships with program champions and those employees who have benefitted from the services. Union leaders, Line Managers, and Human Resources personnel can be valuable allies. They are well-positioned to help market the OH&S program and its services.

3. Key Marketing Concept

Marketing management involves a thorough understanding of a key marketing concept. Successful marketing depends on determining customer ***needs and wants*** and delivering the desired services more effectively and efficiently than one's competitors. The sales equation is:

Features + Needs = Benefits[7]

What does this mean to the promoters of an OH&S Program? Simply stated: *"Know the customer's needs and wants, and then, demonstrate what the OH&S program and its services can offer to each of the stakeholders involved."*

Ensure the marketing message describes "what is in it for them". For example, employees care about what the OH&S Program can offer them and their families. How the OH&S Program is going to improve the organization's/company's bottom line or prevent legal sanctions against the organization/company does not matter to

---

[5] *Ibid.*

[6] *Ibid.*

[7] R. Buttenshaw, *Ready, Set, Sell: How to Succeed in Selling* (Wellington, NZ: Generator, 2008) at 27.

employees. Thus, the marketing message to employees needs to focus on how the OH&S Program can promote the creation of a safe work environment, employee well-being and protection from injury or death. As well, the employee message needs to have a personal touch. It should be directed to the employee and family members.

The core marketing concepts are summarized in Figure 16.2.

Figure 16.2: Core Marketing Concepts

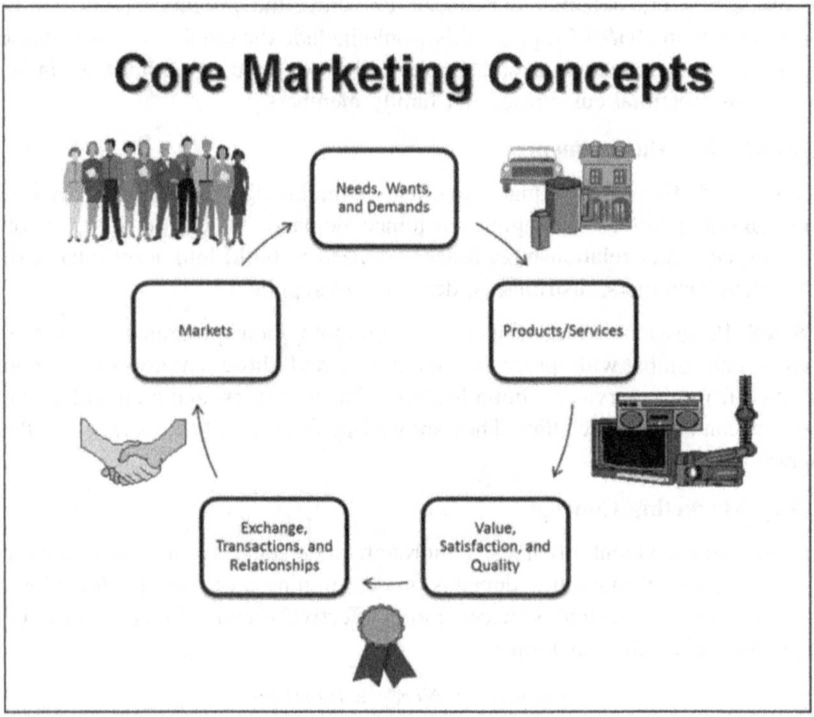

4. General Marketing Principles

The following are some general principles that can be followed so that this key marketing concept can be operationalized:

- **Promote the Benefits of the OH&S Program**

    Articulate to stakeholders the benefits that the OH&S Program and its services can offer to each of them. For example:

    For a company, some of the drivers for having an OH&S Program are to:
    - *meet the applicable legislative requirements;*
    - *address business needs of worker health and safety, business continuity, productivity, and profitability;*

- *promote employee and public safety;*
- *enable employee motivation for work through facilitating a healthy and safe workplace;*
- *work with the operations to attain productivity efficiencies;*
- *facilitate improved communication and understanding between management and staff in regards to workplace health and safety and the applicable legislation;*
- *attain improved relationships with provincial government agencies and industry; and*
- *control the cost of Workers' Compensation Board claims costs and insurance premiums.*

These benefits can be broadly stated, or specifically defined. Some general statements that can be made are:

Corporate Benefits

For the company, the OH&S Program provides a structure and function in which the following can occur:

- *support of a healthy workforce;*
- *hazard management;*
- *monitoring of process changes and their impact on worker health and safety;*
- *worker education about workplace health and safety;*
- *incident prevention;*
- *fewer lost workdays and lower Workers' Compensation claims;*
- *cost-containment of approximately 19 per cent through:*
  - *lower disability cost;*
  - *lower insurance premiums and rates; and*
  - *lower Workers' Compensation rebates;*
  - *reduced costs associated with hiring replacement workers;*
  - *higher level of employee productivity;*
  - *greater profitability; and*
  - *compliance with the applicable legislative obligations.*

Union Benefits

For the Union, the OH&S Program affords union members an opportunity to:

- problem-solve addressing areas of worker and workplace health and safety;
- interact and build relationships;
- protect the health, wellness and employability of members;
- maintain labour rights and principles;
- promote member well-being; and
- contribute to company profitability and competitiveness.

Employee Benefits

The employee and his or her family, benefits from the OH&S Program through:

- *provision of a healthy and safe workplace;*
- *fewer risks of worker injury/illness;*
- *ability to work safely;*
- *ability to earn a regular income; and*
- *ability to return home healthy and safe at the end of the day, and able to enjoy a personal life.*

- **The Message must Fit the Target Audience and the Culture**

  The message must fit the "culture". This means the message sender must know the target audience and the culture(s) within which that audience operates. Doing a "mini" culture assessment is advisable. For example, what are the group's beliefs and values? What are the symbols, rituals, and rites that they observe? Who do they recognize as the group's heroes? Who are their leaders — informal or formal?

  Information on culture and cultural assessment is contained in Chapter 6, "Occupational Health and Safety: Leadership and Commitment"; Chapter 7, "Safe Workplace = Great Workplace: Building a Sustainable Culture of Safety"; and Chapter 8, "Safety Culture: Safety Climate". Once that is determined, design the appropriate marketing message.

- **Selling Techniques**

  Success means using effective marketing techniques. In the world of Occupational Health and Safety, that might look like:

  - marketing OH&S services, internally and externally, through the use of:
    - OH&S training sessions;
    - OH&S brochures/posters;
    - OH&S educational presentations;

- media boards (bulletin boards, electronic information boards, computer message boards, *etc.*); and/or
- articles in the organization/company newsletter;
* use of OH&S web pages;
* promoting public safety awareness and campaigns through Safety Expos, school education programs, radio/television programs, *etc.*);
* use of regular Safety Meetings;
* use of seasonal OH&S educational topics;
* provisions of one-on-one coaching of employees and management;
* use of rewards for demonstrated positive safety behaviours;
* use of rewards for reporting hazards/near-miss incidents;
* provision of regular OH&S reporting to all levels;
* use of incident trend analysis data to identify issues and reinforce the message(s) presented; and
* use of approaches that make OH&S educational/training sessions fun and exciting.

- **Be Prepared to Answer Tough Questions**

Promoters of the OH&S Program must know how to answer the following questions:
- Why is an OH&S Program needed in our organization/company? Is it legally required? Is it necessary for business continuity and success?
- What are the benefits to the organization/company and the various stakeholders?
- What data are required to build a business case that promotes the existence of an OH&S Program?
- What are the real costs related to workplace incidents and employee injury/illness?
- What savings can be gained from having an OH&S Program?
- What return on investment (ROI) can be realized?

To effectively answer these questions, the promoters of the OH&S Program have to do some internal and external research, and legwork. To assist with this process, the following facts may prove helpful (Table 16.1):

Table 16.1: Relevant Facts on Workplace Safety

Workplace Fatalities in Canada (2018):[8]

- In 2018, 1027 work-related fatalities occurred, or 2.8 per calendar day, or 4 per workday;[9]
- The workers with the highest fatality rates in the ages of 65 or more years of age (61%). Workers aged 55-64 years of age ranked second at 17%, with younger workers (15-39 years of age) experiencing 11% of the fatalities;
- Men experience more workplace fatalities (997) than women (30).
- Most worker fatalities occurred in the construction (199[19%]) and manufacturing (182[18%]) industries.

Workplace Lost-time Injuries in Canada (2018):[10]

- Canada reported 264,438 serious injuries requiring recuperation and time away from work.
- Workers 45-59 years of age accounted for 34% (90,404) of the lost-time injuries.
- Men had more lost-time injuries (158,969) than did women (105,443).
- Most lost-time claims occurred in Health and Social Services Industry (47,014), Manufacturing (35,910), and Retail Trade (28,203).
- Work injuries in Canada cost approximately $12.8 billion per year.[11]

5. Build a Business Case for the OH&S Program

A business case for an OH&S Program must demonstrate the benefits and return on investment that could be realized by having an OH&S Program. This is important when initiating the OH&S Program and should be regularly repeated. Stakeholders need to know the rationale for initially supporting the program; and then, why they should continue to offer that support. In essence, what is in it for them to support the OH&S Program?

---

[8] Association of Workers' Compensation Boards of Canada, "2018 Injury Statistics", *Statistics*, online: http:// http://awcbc.org/?page_id=14#fatalities (date accessed: January 28, 2020).

[9] According to Association of Workers' Compensation Boards of Canada, 1,027 work-related fatalities occurred in 2018. Using 365 days per year, 2.8 workers are killed per day. See Association of Workers' Compensation Boards of Canada, "2018 Injury Statistics", *Statistics*, online: http://awcbc.org/?page_id=14#fatalities (date accessed: January 28, 2020).

[10] Association of Workers' Compensation Boards of Canada, "2018 Injury Statistics", Statistics, online: http://awcbc.org/?page_id=14 (date accessed: January 28, 2020).

[11] D. Dyck (2019). Refer to the following pages for the details.

Some suggestions that can be made to garner support from the various stakeholders are:

For the Company:

The OH&S Program is a management tool for:

- *managing workplace hazards;*
- *promoting workplace safety and safe work practices;*
- *identifying the reasons for workplace incidents through data collection and trend analysis;*
- *preventing employee illness/injuries; and*
- *containing occupational injury/illness-related costs.*

It enables the company to comply with applicable legislation, thereby upholding their OH&S Duty of Care and due diligence. A safe workplace is attractive to workers; an important consideration in today's tight labour market. It is also attractive to investors and customers. Finally, there is a profit chain between creating a safe workplace, having good business practices, being productive, and realizing a sound profit. Selling the program as such can help Boards of Directors and Senior Management to recognize the value an OH&S Program can provide to a company.

Add to this "personal-hook message", with an explanation of the costs of workplace incidents. For example, the average cost of a workplace lost-time injury claim, in Canada in 2016-2017, was calculated at $34,900 per claim. In comparison, in the U.S., the average costs of workplace injuries in 2016-2017 were estimated at $40,051.[12] A further analysis of the U.S. data shows:

- Motor vehicle accidents:
  - Non-fatal Disabling Injury = $93,800 per claim;
  - Fatal Injury = $1,615 per claim; and
  - Property damage = $4,400 per vehicle.[13]
- Work injuries (fatal and non-fatal):
  - Without employers' costs = $1,150 per death; $39,000 per disabling injury; and
  - With employers' costs = $1,265 per death; $42,900 per disabling injury.[14]

---

[12] National Safety Council, "Workers' Compensation Costs" (2017), online: https://injuryfacts.nsc.org/work/costs/workers-compensation-costs/ (date accessed: January 28, 2020).

[13] National Safety Council, "Costs (2017)", available at: https://injuryfacts.nsc.org/all-injuries/costs/guide-to-calculating-costs/data-details/ (date accessed: January 31, 2020).

[14] Calculated by multiplying CDN $1,150,000 per death and CDN $39,000 per disabling

Note: This data does not include any estimate of property damage or of the non-disabling injury costs.

In comparison, in Canada in 2017, the average cost of a workplace lost-time injury was calculated at $34,900 per claim for assessable employers. Given that there were 251,625 lost-time injury claims in 2017, the total cost of disabling workplace injuries was $8.8 billion. *These are the direct costs only.* If the indirect costs are included ($8.8 billion x a factor of 1.1 then the annual total cost of workplace injuries is $9.7 billion. For details on the Canadian 2018 costs ($12.8B) refer to Table 15.2.

To put this into perspective for the individual organization/company, state this information in recognizable terms for the organization/company. For example, using a **Workplace Incident Cost Calculator** (Chapter 14, "Occupational Health and Safety: Incident Investigation"), tally the cost of recent organizational/company incidents. In doing that, the organization/company will gain an appreciation of how long it will take them to recover from the identified loss.

Lastly, determine what the organization's/company's total costs for workplace health and safety would be with and without an OH&S Program (Figure 16.3). Then, calculate the savings realized by having an OH&S Program. The return on investment (ROI) for having an OH&S Program is the savings realized divided by the cost of having an OH&S Program.

---

injury by the factor of 1.1 (refer to Table 15.1 for details).

Figure 16.3: Model for a Cost/Benefit Analysis of an OH&S Program[15]

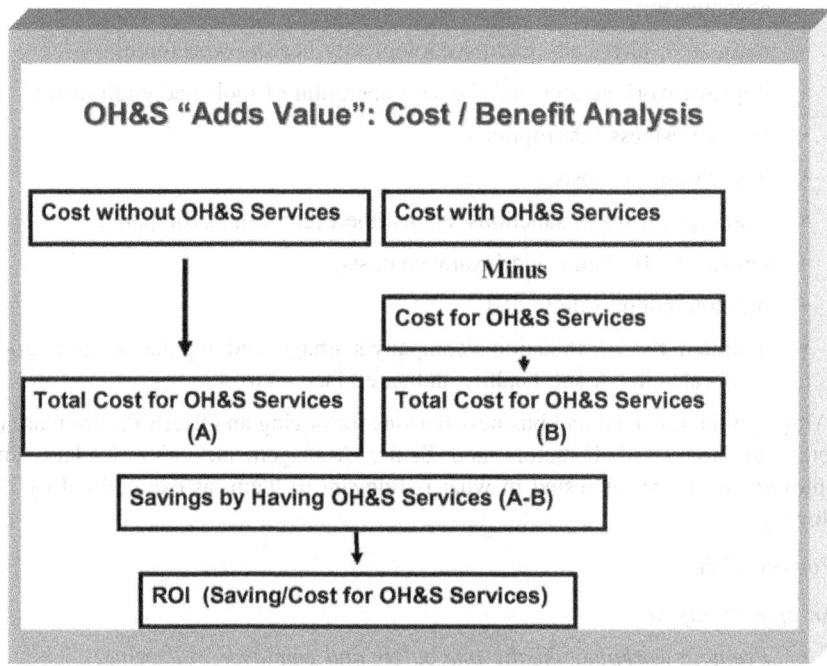

Now, explain the financial benefits of having a sound OH&S Program. For example, a Stanford Study by Levitt and Samuelson indicates safety costs to be 2.5 per cent of direct labour costs;[16] the Business Roundtable estimated safety costs to be 0.625 per cent of total project costs;[17] and another study set safety costs at 8 per cent of payroll.[18] However, the safety return on investment (ROI) has been demonstrated to be $3-$10 per dollar spent on safety.[19]

In addition to the financial benefits, it is important to note that an OH&S Program can add value to an organization/company and its business operations by supporting efforts that:

---

[15] D. Dyck, "Fundamentals of Disability Management" (Presented as a pre-conference offering at the Industrial Accident Prevention Association Conference in Toronto, April 14-15, 2007.)

[16] R. Prichard, "The Cost of Safety: Expert Commentary", AON Worldwide Resources, online: http://www.irmi.com/expert/articles/2002/Prichard10.aspx (date accessed: January 28, 2020).

[17] *Ibid.*

[18] *Ibid.*

[19] J. Phillips, P. Phillips & A. Pulliman, *Measuring ROI in Environment, Health & Safety* (John Willey and Sons, 2014); K. Morrison, "The ROI of Safety", *Safety & Health Magazine* (May 23, 2014).

- improve employee morale, and in turn, reduce employee presenteeism and absenteeism;
- maintain a physically and psychologically healthy workforce;
- improve work actions and the safe operation of tools and equipment;
- lessen business interruptions;
- lower business costs;
- lessen government sanctions against the organization/company;
- reduce WCB claims and insurance costs;
- promote public safety; and
- maintain the organization's/company's image and reputation as a good corporate citizen and healthy and safe place to work.

Although the financial and business reasons for having an OH&S Program are of interest to Boards of Directors and Senior Management, union leaders, and employees are more interested in what is relevant to them, as exemplified by the following:

For the Union:

An opportunity to:

- *Promote workplace health and safety and employee well-being;*
- *Maintain labour rights and principles;*
- *Protect member well-being and the employability of members;*
- *Demonstrate the union's "due diligence" in terms of preventing workplace injuries/illness and helping injured workers to return to work;*
- *Interact and build relationships;*
- *Problem-solve mutual concerns and issues; and*
- *Contribute to the company's profitability and competitiveness.*

For Employees:

The OH&S Program enables the employee to:

- *Function in a healthy and safe work environment;*
- *Have safe equipment and adequate work information with which to operate;*
- *Use safe work practices;*
- *Be exposed to fewer risks of injury/illness;*
- *Be able to work safely;*
- *Report unsafe work conditions or equipment;*

- *Receive recognition for a job safely completed;*
- *Earn a regular income;*
- *Return home safe at the end of the day able to enjoy a personal and family life; and*
- *Witness their co-workers go home safely at the end of the day as well.*

C. OH&S COMMUNICATION STRATEGY AND PLAN

An OH&S Communication Strategy defines the corporate OH&S message, how it is to be delivered, to whom, through what means, and by when.

1. OH&S Communication Goal and Objectives

To begin with, determine what the communication goal or "the message" for the OH&S Program should be. This goal will vary as the OH&S Program evolves. Initially, it may be as follows:

(a) OH&S Program: Sample Communication Goal

All stakeholders are to be knowledgeable about the OH&S Program, the applicable legislation and of their roles and responsibilities regarding both.

From that point, strategies to enable the realization of this goal can be developed. These are the steps that the program promoters would take to move towards the completion of the goal.

(b) OH&S Program: Sample Communication Objectives

By January 2021, 100 per cent of the stakeholders will know about the OH&S Program and their specific roles and responsibilities as evidenced by their level of program awareness and/or their level of participation in the OH&S Program.

By June 2021, all the key stakeholders will have an in-depth knowledge about the OH&S Program and be able to enact their specific roles and responsibilities, as evidenced by their level of program awareness and their participation in the OH&S Program.

These objectives can be operationalized by:

1. Achieving the overall endorsement of the OH&S Program and its services.
2. Providing general education about the OH&S Program and its services.
3. Providing specific education for key players in the OH&S Program (Senior Management, Supervisors/Managers, Union Representatives, and OH&S staff).
4. Offering introductory and regular communication about the OH&S Program.
5. Providing ongoing education about the OH&S Program to all.
6. Offering introductory and regular communication with external stakehold-

ers (Government agencies and Workers' Compensation Board(s)).
7. Developing and distributing regular reports on the OH&S Program and service outcomes (quarterly/annual).

2. OH&S Communication Plan

The next step is to create an OH&S Communication Plan and the accompanying tools to use to tell the key stakeholders about the OH&S Program. Using the organization's/company's OH&S Program Manual (Chapter 5, "Occupational Health and Safety Program: Manual Development", Appendix 1), and OH&S Action Plan (Chapter 18, "Occupational Health and Safety: Best Practices", Appendix 1), determine what messages are to be provided to each stakeholder group. The following is a graphic representation of a sample communication plan for an OH&S Program.

(a) OH&S Program: Sample Communication Plan

Stakeholder	What?	How?	When?	Accountability?

For each stakeholder group, determine the desired message(s) to be delivered, the appropriate approach and timing, the best medium, and who will be accountable for delivering the message.

Different stakeholder groups have different informational needs. For example, Senior Management will want to know about OH&S Program performance measures such as hazard management efforts, incident rates, injury rates and costs, effectiveness of corporate prevention activities, and the degree of corporate compliance with the applicable legislation. The financial officer for the organization/company will want to know the objective measures about the program and the expected return on investment (ROI) from the OH&S Program. Line Managers are interested in outcomes such as reduced business interruptions, lowered incident rates and costs, and increased employee productivity. Employees will want to know how the OH&S Program facilitates workplace and worker health and safety. OH&S Professionals/Practitioners will value information on how the program enables them to enhance employee health, safety, and wellness; and how workplace injuries can be reduced. Lastly, union personnel will want to know what the OH&S Program offers to their members in terms of worker health and safety, and job security.

CH. 16: MARKETING OCCUPATIONAL HEALTH AND SAFETY PROGRAMS

The following are some sample communication plans that were industry-created:

For Employees

Stakeholder	What?	How?	When?	Account-ability?

Employees	• History of workplace injuries and the WCB experiences; the impact of doing nothing
	 • Overview of the OH&S Program
	 • Benefits (direct and indirect) of the OH&S Program for workers
	 • Future plans re: workplace safety and injury prevention
		• Overview of the OH&S Program
		 • Use of presentations using company examples
		 • Explanation of their role in the OH&S Program
		 • Training regarding workplace safety and the OH&S Program:
		   • Introductory
		   • Ongoing
		 • Scheduled union meetings (shift workers)
		 • Handouts
			• Scheduled meetings (Introductory, Follow-up, New Employee Orientation)	• Union Representatives
				 • Management
				 • HR
				 • Home Department

629

For Senior Management and Corporate Directors

Stakeholder	What?	How?	When?	Account-ability?
Senior Management	• Impart knowledge on the OH&S Program; • Impart information on the purpose of the OH&S Program; • Provide information on the related benefits; • Garner support for the OH&S Program; • Seek endorsement for the OH&S Program and its goals	• Do a presentation: An overview outlining issues, implications, business plan • Provide the entire OH&S Program document	• Before rolling it out to employees and management	Person's Name
Corporate Services & Human Resources	• Provide information on the OH&S Program; • Promote understanding of the OH&S Program and its purpose • Demonstrate how the OH&S Program aligns with company vision, mission and business strategies	Status Reports (one-page briefing notes)	Monthly	Person's Name
Board of Directors	• Knowledge of the OH&S Program to level of application; • Understanding the return on investment for the OH&S Program	• Review entire document • Provide an overview as a presentation	At a specific date	Person's Name

Stakeholder	What?	How?	When?	Account-ability?
• Senior Management • Board of Directors	Provide program outcome data (ongoing reports)	Slides (graphs) showing safety performance trends	• Annually • Quarterly	Person's Name

Through the use of similar communication plans, a plan of action for promoting the OH&S Program for each of the key stakeholder groups can be developed. Typically, these players are:

- Line Management;
- Union Executive Officers/Shop Stewards;
- Human Resources personnel; and
- External Support service providers (*e.g.*, OH&S Service Provider, EAP Service Provider, Workers' Compensation, *etc.*).

As well, an industry example of an OH&S Communication Plan is provided in Appendix 1.

3. OH&S Communication Tools

As part of the OH&S communication plan, the OH&S Program promoters must develop communication tools. These can be presentations, an OH&S Program Web page, organization/company newsletter announcements and articles, pay cheque inserts and brochures, and telecommunication screens.

The following is a discussion on OH&S communication tools that can be used to promote the OH&S Program and to explain the related services.

(a) OH&S Program: Website

In today's workplaces, one of the best methods for marketing the OH&S Program is an OH&S website. Use of an existing and accepted organization/company communication vehicle can be a powerful approach for getting the word out about the OH&S Program and its services.

(b) OH&S Program: Brochures

Brochures, a common communication tool, require planning and technical knowledge to be well developed. To begin with, it is essential to define the target audience — who is this brochure intended for? Next, an understanding of the needs of the target audience is required — what do they want and expect from an OH&S Program and its services? What information will this audience need or want regarding the OH&S Program and its services?

Once completed, attention should turn towards the information required to efficiently and effectively market the OH&S Program and its services. For example, what would the target audience need and expect to know about the OH&S Program and its services? Part of this might include performing research to find out what other companies have done in terms of marketing their OH&S Program.

In planning the brochure, the following considerations need to be taken:
- the message contained must be complete and accurate;
- the message is simple and direct;
- the tone of the brochure is positive and friendly so that the target audience recognizes that the organization/company has their best interests in mind;
- the content is informative, accurate, and worded positively;
- the contact details for key players in the OH&S Program are provided for ease of access by workers;
- bullets are used to convey important information;
- the tool is attractively presented and aligned with the corporate culture, organization/company branding, and other marketing standards and techniques;
- the graphics are appropriate to the audience and to the purpose of the brochure;
- the layout is attractive and professional looking — pay attention to the font, graphics, paper colour, and other presentation details;
- the OH&S Program and service contact personnel are included to enhance user access to the service;
- the reading level must be appropriate for the target audience — typically, this is at a grade 6-8 reading level, similar to what is used in newspapers;
- where worker literacy is an issue, the use of pictograms is recommended; and
- in workplaces with multilingual workers, the brochure should be translated.

The brochure content depends on the purpose of this communication tool, namely, what is the message being communicated and why? For example, is it to enhance stakeholder awareness and acceptance of the OH&S Program? Is it to garner stakeholder support of specific OH&S Program elements? Is it to promote timely reporting of workplace injuries/illness? Once the nature of the message is determined, then decide what information would have to be communicated to garner interest in and action on the message being sent. A sample brochure marketing the OH&S Program is contained in Appendix 2.

(c) OH&S Program: Poster

Like brochures, the development of a poster requires planning and technical expertise. However, it is an effective communication tool that can be visually displayed for employees and contractors to notice (Figure 16.4).

Figure 16.4: First Aid Response Steps

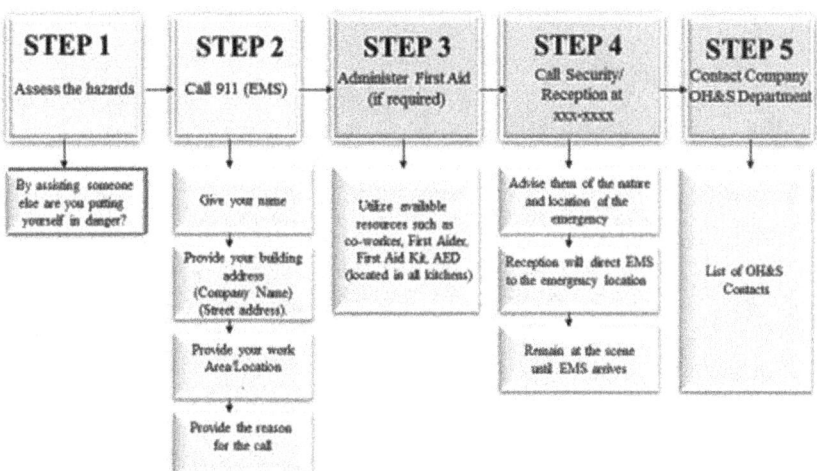

A sample poster marketing flu immunization, an aspect of the OH&S Program, is contained in Appendix 3.

(d) OH&S Program: Outcome Report

OH&S Practitioners/Professionals are expected to prepare and present program outcome reports. These reports can serve dual purposes: to describe the organization's/company's OH&S Program results, as well as to educate stakeholders on the value and benefits of the OH&S Program and its services. To do this successfully, there are several report preparations, and writing principles that need to be understood. Chapter 37, "Occupational Health and Safety: Project Management", provides guidelines for report preparation, writing, and presentation (Appendix 2).

In addition, organizations/companies can post the results of their OH&S Programs and initiatives, for example, the message issued by Unison Contracting, New Zealand (Appendix 4). This is an illustration of an excellent communication tool.

D. SUMMARY

For OH&S Programs and its services to remain visible and viable, they must be promoted within the organization/company. In return, the program promoters must demonstrate how the OH&S Program adds value to the organization/company. This information must be reported back to all the stakeholders so that they can understand that the OH&S Program is providing them with what they value. In this way, the OH&S Program is kept vibrant and valued.

Appendix 1

SAMPLE OH&S COMMUNICATION PLAN

OH&S Message	Media	Senior Management Team	Managers	Front Line Supervisors	Staff	Employees	Unions	External Stakeholders
OH&S Policy	OH&S Web site	✓	✓	✓	✓	✓	✓	
	Bulletin Board Postings	✓	✓		✓	✓	✓	
	OH&S Program Manual	✓	✓	✓	✓	✓	✓	
	Employee OH&S Handbook	✓	✓	✓	✓	✓	✓	
OH&S Program	OH&S Web site version	✓	✓	✓	✓	✓	✓	✓
	Electronic version	✓	✓	✓	✓	✓	✓	
Employee OH&S Handbook	OH&S Web site	✓	✓	✓	✓	✓	✓	✓
	Hardcopy version	✓	✓	✓	✓	✓	✓	
OH&S Action Plan	OH&S Web site	✓	✓	✓	✓	✓	✓	
OH&S Department Personnel	OH&S Web site	✓	✓	✓	✓	✓	✓	✓
	OH&S Brochure	✓	✓	✓	✓	✓	✓	
	OH&S Training Materials		✓	✓	✓	✓		

CH. 16: MARKETING OCCUPATIONAL HEALTH AND SAFETY PROGRAMS

OH&S Message	Media	Senior Management Team	Managers	Front Line Supervisors	Staff	Employees	Unions	External Stakeholders
General OH&S Information	OH&S Web site	✓	✓	✓	✓	✓	✓	
	Bulletin Board	✓	✓	✓	✓	✓	✓	
	Safety Meetings		✓	✓	✓	✓		
	Safety Training		✓	✓	✓	✓		
	Safety Posters		✓	✓	✓	✓	✓	✓
	OH&S Brochures	✓	✓	✓	✓	✓	✓	✓
Office Safety	OH&S Web site		✓	✓	✓	✓	✓	
OH&S Education	Brochures	✓	✓	✓	✓	✓	✓	✓
	Brochures	✓	✓	✓	✓	✓	✓	✓
	Employee Newsletter	✓	✓	✓	✓	✓	✓	✓
General Worker OH&S Training	New Employee Orientation	✓	✓	✓	✓	✓		
	Worker OH&S Training		✓	✓	✓	✓		
	OH&S Refresher Training		✓	✓	✓	✓		
	Public Education							✓

OH&S Message	Media	Senior Management Team	Managers	Front Line Supervisors	Staff	Employees	Unions	External Stakeholders
OH&S Program Outcomes	OH&S Web site	✓	✓	✓	✓	✓	✓	
	Reports	✓	✓			✓	✓	
	Trend analyses	✓	✓			✓	✓	
	Annual Report	✓	✓			✓	✓	✓
Incident/Investigations and Corrective Actions	OH&S Web site	✓	✓	✓	✓	✓	✓	
	Bulletin Board Postings		✓	✓	✓	✓	✓	

Appendix 2

SAMPLE OH&S PROGRAM BROCHURE

At COMPANY XYZ, our employees are our most important assets. COMPANY XYZ is committed to providing health and safety programs and services that support employee health and well-being. At Company XYZ, workplace health and safety is valued and viewed as top priority.

THE XYZ OH&S PROGRAM

COMPANY XYZ's OH&S Program is designed to:
- Ensure that employees:
    - know how to work safely
    - are able to perform their jobs safely
    - are equipped to work safely
    - are motivated to work safely
- Protect workers from workplace hazards
- Facilitate business continuity
- Lower workplace incident and injury costs
- Promote a healthy and safe work environment

OH&S TEAM

To make workplace health and safety possible, COMPANY XYZ uses a team approach to OH&S. The team includes:

- Employees
- Employer
- Union leaders
- COMPANY XYZ's OH&S Program professionals

HOW CAN YOU GET INVOLVED?

Workers

- Participate in the OH&S Program
- Comply with company safe work practices
- Participate in OH&S training programs
- Assume work duties for which you are trained and competent
- Protect the health and safety of yourself and others
- Use, care, and maintenance of your PPE
- Be positive about your work, and workplace health and safety
- Report unsafe work conditions or substandard equipment
- Refuse unsafe work
- Promote good working relationships

- Uphold the company's OH&S values and commitments

Employer
- Provide a safe and healthy workplace
- Educate workers of their OH&S and Workers' Compensation responsibilities

OH&S PROGRAM: BENEFITS

For the Employee:
- A healthy and safe work environment
- Safe equipment and work information with which to operate
- Safe work practices
- Fewer risks of injury/illness
- Ability to work safely
- Ability to report unsafe work conditions or equipment
- Recognition for a job safely done
- Regular income
- Ability to return home safely at the end of the day able to enjoy a personal life
- Know that co-workers are going home safely as well

For the Employer:
- Manages workplace hazards
- Promotes workplace health and safety, and safe work practices
- Identifies the reasons for workplace incidents through data collection and trend analysis
- Prevents employee illness/injuries
- Contains work-related injury costs
- Maintains valuable employees within the workplace
- Meets legislative obligations
- Promote a general awareness of the OH&S Program goals, benefits, and outcomes
- Develop OH&S policies and procedures
- Develop a discipline policy and clearly communicate consequences for non-compliance
- Model the desired OH&S behaviours
- Participate in monitoring and evaluation of the company's OH&S Program performance and effectiveness

- Consider safety issues when acquiring or procuring new equipment/tools, designing new facilities, or adopting new processes
- Support the efforts to mitigate losses due to worker injury by supporting modified work options
- Focus on preventative strategies, such as strengthening the safety culture and the use of leading indicators of safety

Union Leaders
- Promote workplace health and safety
- Assist with promoting member awareness of safe work practices
- Support worker training programs
- Help identify workplace hazards
- Support return-to-work efforts

CONTACTS:
OH&S Department:

Phone: (xxx) xxx-xxxx
Cell:　(xxx) xxx-xxxx
Fax:　(xxx) xxx-xxxx

NEED MORE INFORMATION?
Contact one of the following:
- Your Line Manager
- COMPANY XYZ OH&S Advisor
- COMPANY XYZ Intranet
- COMPANY XYZ OH&S Director

Appendix 3

SAMPLE OH&S POSTER[20]

[20] U.S. Department of Veteran Affairs, "Get the Flu Shot!" Reprinted with permission. Available online for public use: http://www1.va/vhapublications/ViewPublication.asp?pub_ID=1944 (date accessed: January 28, 2020).

Appendix 4

SAMPLE MANAGEMENT REPORT: UNISON CONTRACTING: GSP™ JOURNEY[21]

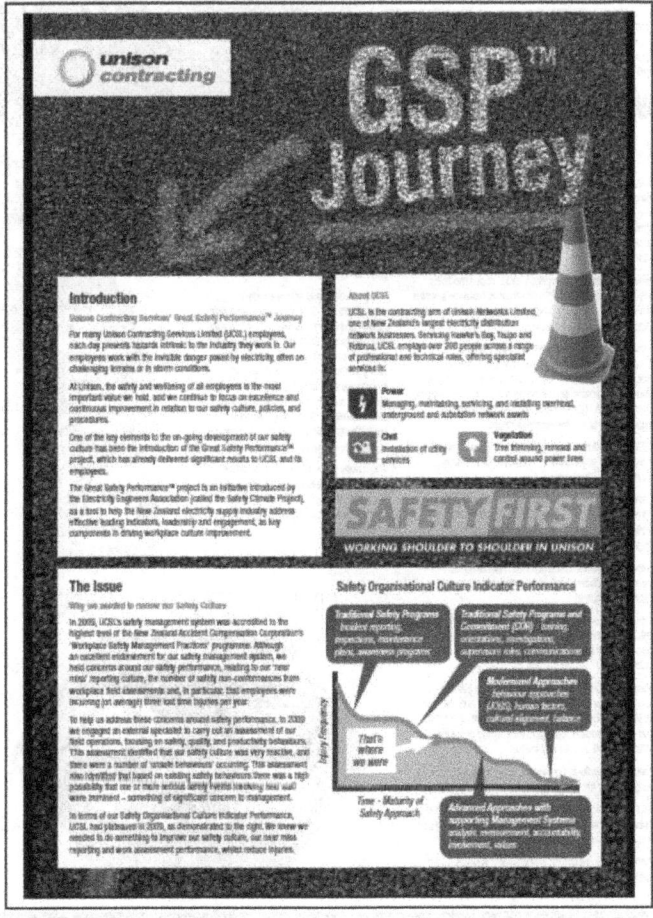

---

[21] Published with permission from Unison Contracting, "GSP Journey", Management Report (internal report, 2014), online: http://www.unison.co.nz/ (date accessed: January 28, 2020).

UNISON CONTRACTING: GSP™ JOURNEY (PAGE 2)

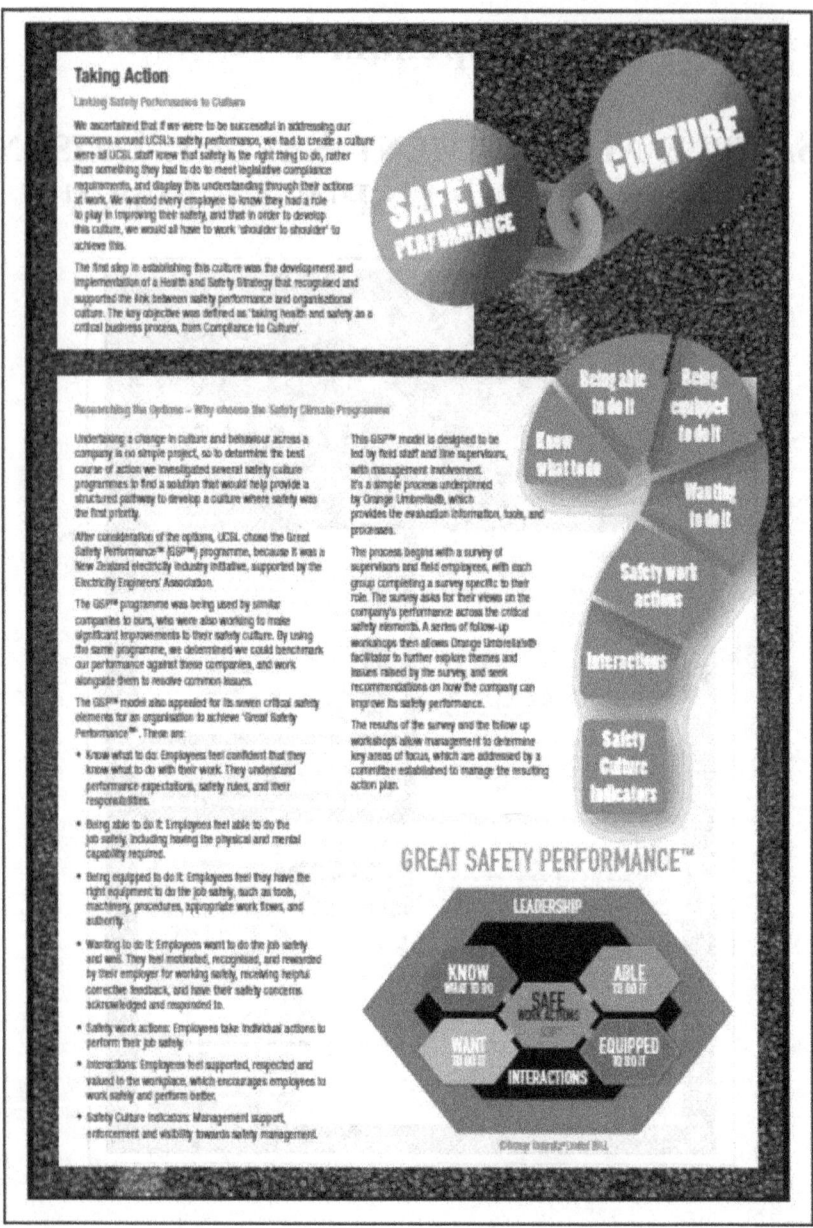

CH. 16: MARKETING OCCUPATIONAL HEALTH AND SAFETY PROGRAMS

UNISON CONTRACTING: GSP™ JOURNEY (PAGE 3)

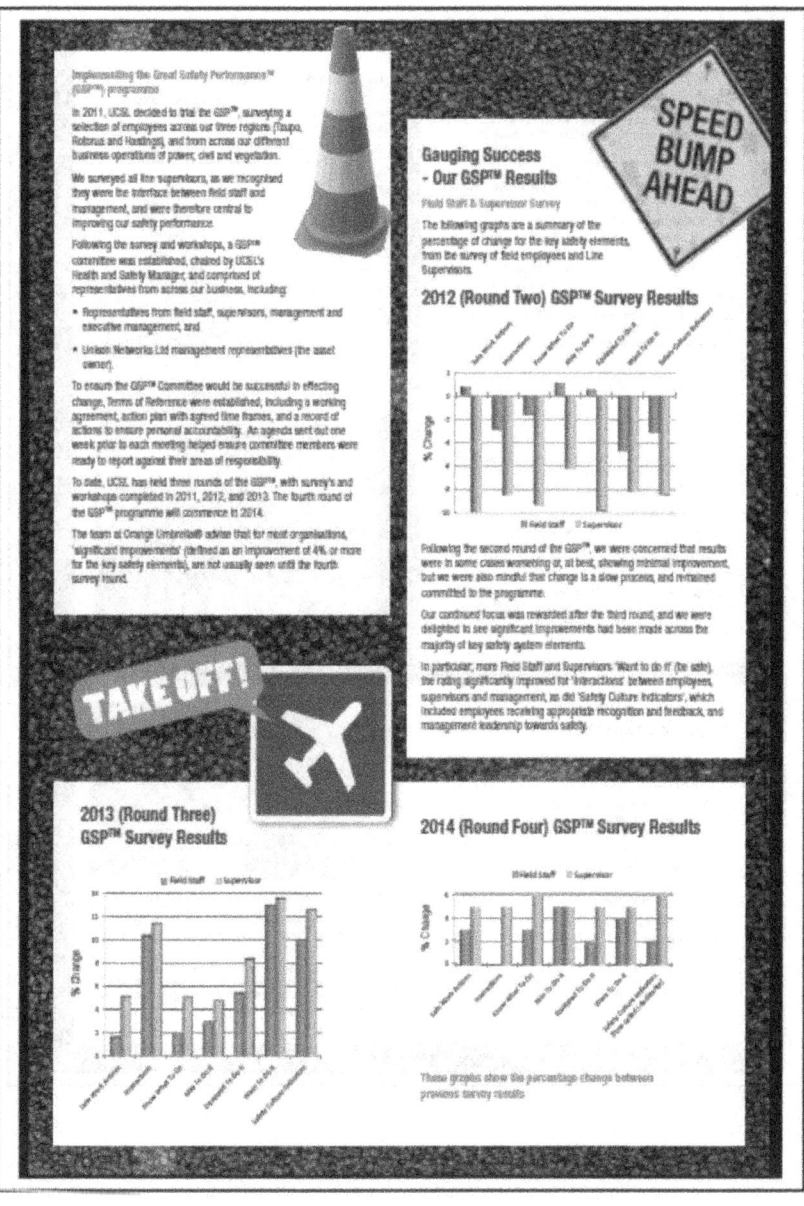

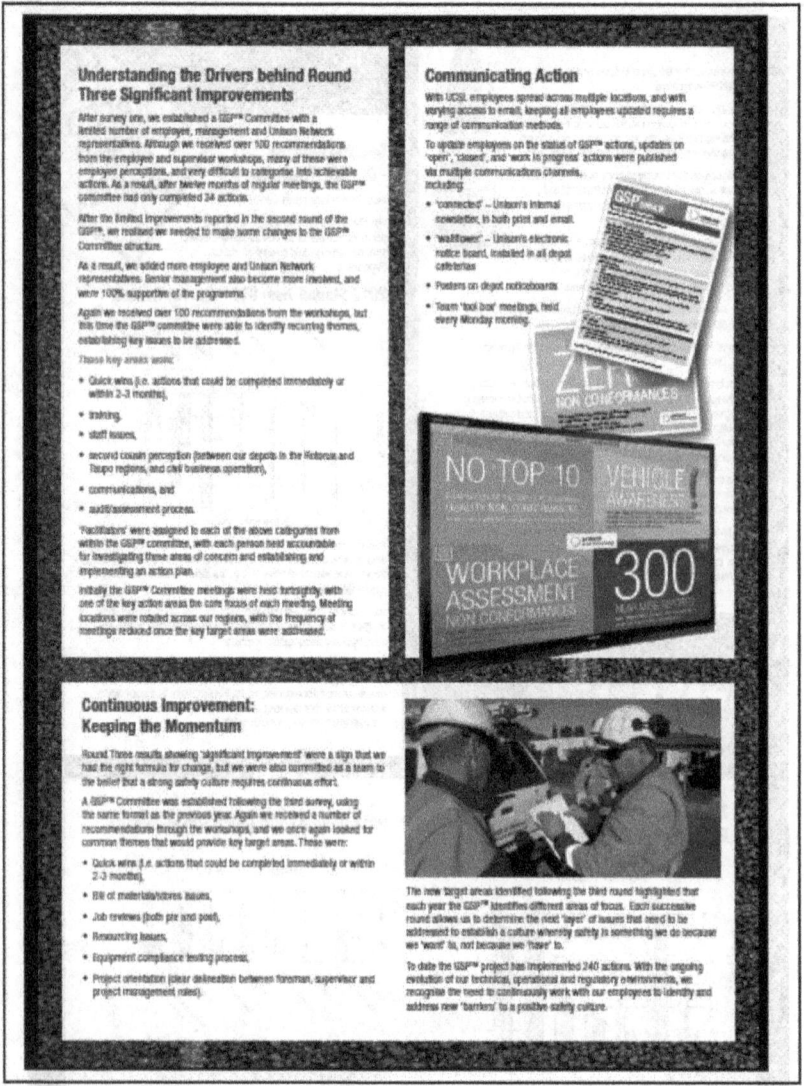

CH. 16: MARKETING OCCUPATIONAL HEALTH AND SAFETY PROGRAMS

UNISON CONTRACTING: GSP™ JOURNEY (PAGE 5)

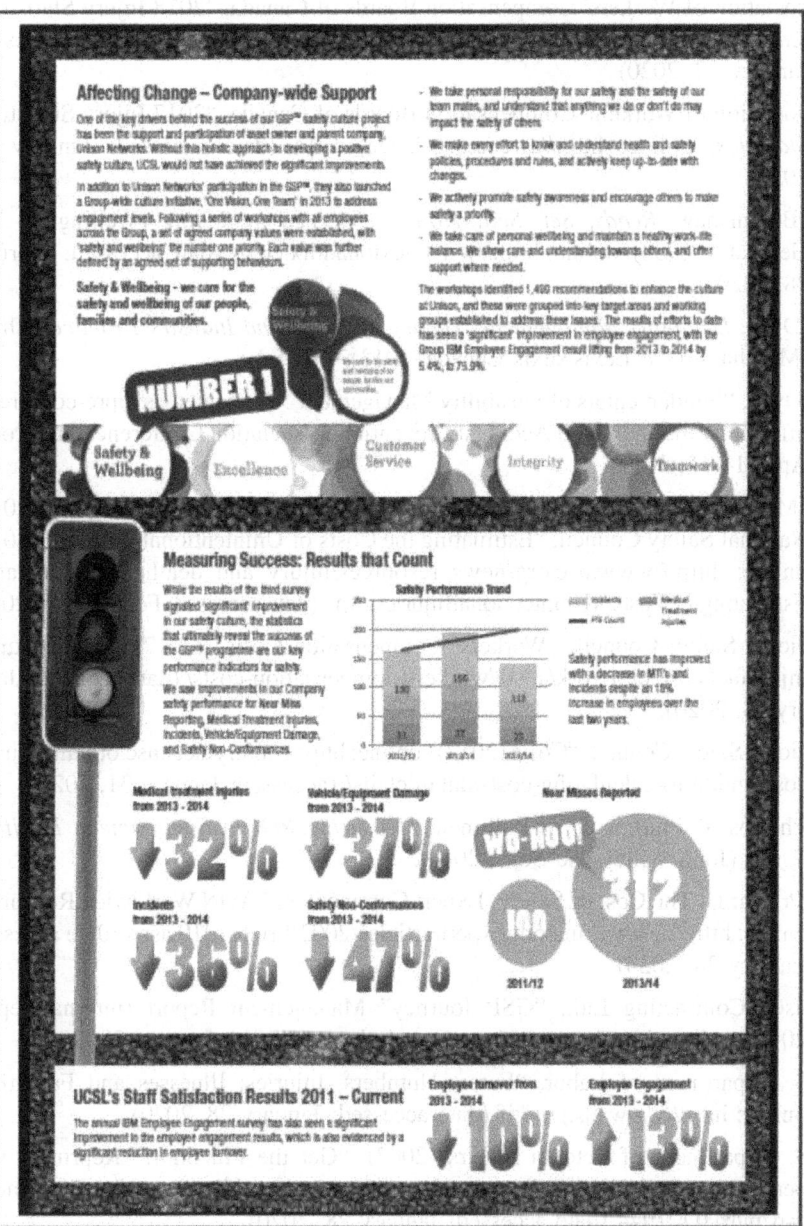

CHAPTER REFERENCES

Association of Workers' Compensation Boards of Canada, "2018 Injury Statistics", *Statistics*, online: http:// http://awcbc.org/?page_id=14#fatalities (date accessed: January 28, 2020).

Association of Workers' Compensation Boards of Canada, "2017 Injury Statistics", *Statistics*, online: http://awcbc.org/?page_id=14 (date accessed: January 28, 2020).

R. Buttenshaw, *Ready, Set, Sell: How to Succeed in Selling* (Wellington, NZ: Generator, 2008), online: http://www.acxionaddix.co.nz (date accessed: February 28, 2020).

D. Dyck, *Disability Management: Theory, Strategy and Industry Practice*, 5th ed. (Markham, ON: LexisNexis Canada, 2013).

D. Dyck, "Fundamentals of Disability Management" (Presented as a pre-conference offering at the Industrial Accident Prevention Association Conference in Toronto, April 14-15, 2007).

K. Morrison, "The ROI of Safety" *Safety & Health Magazine* (May 23, 2014). National Safety Council, "Estimating the Costs of Unintentional Injuries" (2014), online: http://www.nsc.org/news_resources/injury_and_death_statistics/Pages/EstimatingtheCostsofUnintentionalInjuries.aspx (date accessed: February 28, 2020).

National Safety Council, "Workers' Compensation Costs" (2017), online: https://injuryfacts.nsc.org/work/costs/workers-compensation-costs/ (date accessed: January 28, 2020).

National Safety Council, "Costs (2017)" online: https://injuryfacts.nsc.org/all-injuries/costs/guide-to-calculating-costs/data-details/ (accessed: January 31, 2020).

J. Phillips, P. Phillips & A. Pulliman, *Measuring ROI in Environment, Health & Safety* (John Willey and Sons, 2014).

R. Prichard, "The Cost of Safety: Expert Commentary" AON Worldwide Resources, online: http://www.irmi.com/expert/articles/2002/Prichard10.aspx (date accessed: January 28, 2020).

Unison Contracting Ltd., "GSP Journey" Management Report (internal report, 2014), online: http://www.unison.co.nz/ (date accessed: January 28, 2020).

U.S. Department of Labor, "Latest Numbers, Injuries, Illnesses and Fatalities", online: http://www.bls.gov/iif (date accessed: January 28, 2020).

U.S. Department of Veteran Affairs (2007). "Get the Flu Shot!" Reprinted with permission. Online for public use at: http://www1.vhapublications/ViewPublication.asp?pub_ID=1944 (date accessed: January 28, 2020).

Chapter 17

OCCUPATIONAL HEALTH AND SAFETY: WORKER OH&S TRAINING

A. INTRODUCTION

People learn new things every day and tend to apply what they learn to new and different situations. This is often an effortless process. Based on this premise, most government and community agencies that promote public health interventions use an educational component to elicit the desired human behavioural change. Their premise is that once people know about a given risk, they will make the recommended behavioural changes to protect themselves and others. However, from experience, "knowing" does not lead to "doing".

This chapter is dedicated to worker Occupational Health and Safety (OH&S) training and how OH&S Practitioners/Professionals can enable employers to design and deliver effective worker OH&S education.

B. WORKER OH&S TRAINING

An effective OH&S program promotes the attitude that "doing a job properly, means doing it safely the first time". To reiterate and to put Worker OH&S training into perspective, the major elements of an OH&S Program are:

- *commitment and OH&S leadership*;
- *general health and safety policies* - statements about the organization's/company's commitment to safety;
- *OH&S communication* - techniques for communicating OH&S information up and down within the organization/company;
- *hazard identification, assessment, and control* - techniques to identify and manage workplace hazards at all levels within the organization/company so that workers are protected from harm;
- *safe work practices and procedures* - instructions on how critical tasks are to be done safely;
- *rules and regulations* - work standards that must be followed by all workers;
- *personal protective equipment* - requirements and procedures on the use, maintenance, and replacement of personal protective equipment (hearing

protection, respiratory protection, hard hats, gloves, boots, aprons, goggles, *etc.*);
- *acquisition, maintenance, and repair of appropriate and adequate equipment*;
- *worker OH&S training* - workers have the right to know the risks inherent in the job and what actions they can take to manage those risks, and that they have the right to refuse unsafe work or the operation of unsafe equipment;
- *emergency preparedness and response*;
- *incident investigation and reporting*;
- *OH&S Program reporting*; and
- *program evaluation and continuous improvement.*

Developing and maintaining a strong OH&S (Safety) culture is dependent on the availability of adequate staffing levels, and **competent, well-educated employees**.

Worker OH&S training can be described as providing employees with the information, concepts, and models needed to safely do the assigned job tasks. It includes helping them to develop the requisite skills and judgement to do the work, as well as showing them what to do and how to do it safely. Worker OH&S training also involves employee assessment, determining where the employee is at, in terms of individual work performance and work performance evaluation, and then, monitoring the employee's performance and providing feedback on how well the employee is functioning. This means recognizing and praising good performance and giving constructive feedback, such as instruction on what to do differently, when shortfalls are identified.

The objective of worker OH&S training is to ease the implementation of health and safety policies into specific job practices, and to raise employee awareness and skill levels to an acceptable standard. While all employees can benefit from OH&S training, special attention should be given to the training of supervisors, trainers, and young/new employees. There is a close relationship between worker OH&S training, work direction, work instruction, and work monitoring to ensure that the work is being safely done.[1]

C. WORKER COMPETENCY VERSUS WORKPLACE COMPLACENCY

Competency is a method of ensuring that the workforce can carry out the expected work from a technical, quality, and OH&S perspective related to worker qualifications, training, and experience. **Worker competency** is defined as being "adequately qualified, suitably trained, and with sufficient [and relevant] experience to safely perform work without supervision or with only a minimal degree of

---

[1] Hatscan, *Handi-Guide to Alberta's OH&S Act, Regulation and Code* (Edmonton, AB: Hazard Alert Training (Hatscan) 2004) at CH2-7.

supervision".[2] Worker competence includes knowledge, skill, ability, training, education, relevant experience, and the authority to do the job. To meet their legal duty of protecting the health and safety of workers, employers must ensure that their workers are competent to do the assigned tasks.

Complacency occurs when workers do tasks repetitively; so much so that they tend to become bored or smug about undertaking the tasks and do the tasks in a thoughtless manner. Complacency also exists when an experienced employee takes short cuts while doing a critical task because he or she has done it so often; and as a result, believes that they are able to complete it quicker and just as safely using this alternate method. It is also present when Line Management/supervisory staff knowingly overlook complacent work practices, also believing that there is no harm in employees taking "short cuts".

Workplace Complacency Trend in incident prevention, is the theory that there is often a level of workplace complacency present in the workplace prior to an incident.[3] It can be mitigated by:

- educating Management about the dangers of complacency;
- forming an OH&S Steering Committee dedicated to keeping workplace health and safety alive and current;
- improving workplace OH&S training;
- improving the organization's/company's compliance management process;
- keeping the importance of Occupational Health and Safety foremost in employee minds;
- celebrating safety and the safety successes enjoyed by the employees and the organization/company;
- participating in community OH&S Programs; and
- positioning OH&S Practitioners/Professionals so that they are valued technical resources to all levels of Management.

There are many OH&S training measures that can be used to counter workplace complacency and promote worker competency:

1. To begin with, it is critical to provide new, or reassigned, employees with an adequate employee orientation to the job, the related critical tasks, the OH&S Program (Safety Orientation) and the OH&S culture.

---

[2] *Occupational Health and Safety Code 2006*, s. 1, made pursuant to the *Occupational Health and Safety Act*, R.S.A. 2000, c. O-2. (The 2006 Code was repealed by Alta. Reg. 87/2009 and replaced by *Occupational Health and Safety Code 2009*).

[3] D. Folk, "The Workplace Complacency Trend in Accident Prevention" *Occupational Health & Safety* (January 2007), online: http://www.ohsonline.com/articles/46371 (date accessed: February 28, 2020).

A new employee **OH&S Orientation** is an overview of the organization's/company's OH&S standards and how they are implemented. It should include an explanation of:

- the organization's/company's commitment to OH&S;
- the organization's/company's OH&S policy;
- related policies such as Substance Abuse Policy, Fit-for-Duty Policy, Non-smoking Policy, Seatbelt Use Policy, Non-use of Cell Phones When Driving, *etc.*;
- the organization's/company's OH&S rules and the rationale for each;
- worker OH&S responsibilities;
- information on organizational/company Safety Meetings, pre-job planning, and hazard identification;
- the use, maintenance, and replacement of personal protective equipment;
- the right to refuse unsafe work;
- the identification, assessment, and control of workplace hazards;
- emergency procedures and response;
- the locations of emergency and first aid supplies;
- reporting hazards, unsafe work conditions, and unsafe work practices;
- reporting work injuries/illnesses; and
- return-to-work procedures after injury or illness.

A sample new employee OH&S Orientation Checklist is provided in Appendix 1.

2. Secondly, worker OH&S training should be planned and tailored to the requirements of the job. It should also be ongoing and updated in accordance with identification of new job hazards, changes in the occupation, development of new industry practices, and/or legislative changes. One technique for designing a worker OH&S training program is to create a training matrix. (Refer to Chapter 5, "Occupational Health and Safety Program: Manual Development", Table 4.1: XYZ OH&S Training Requirements.)

Vital to worker OH&S training is to ensure that employee competency is achieved. By incorporating testing or evaluation into OH&S training programs, employers can measure the level of knowledge of the workers and therefore, the degree of effectiveness of the OH&S training activities.

3. A third method of educating workers can be achieved through **On-the-job Training (OJT)**. OJT is a hands-on explanation and demonstration of how

to do job tasks to which the employee is assigned. It should include an explanation of the potential hazards and the related control measures that the employee will be required to observe. The steps for OJT are:

- Explain the features of the job and how it relates to other job duties or other jobs.
- Discuss the various job duties and the sequence of actions required to do the job.
- Describe the potential hazards associated with the job and how to recognize and control them.
- Demonstrate each procedure, focusing on the critical tasks and ensuring that the employee completely understands each step.
- Explain the rationale for each step of the job procedure.
- Have the employee perform the job and provide feedback on the observed actions.
- Follow up periodically with the employee to determine if questions about the work exist.[4]

OJT also includes the learning initiated through Safety Meetings and Pre-job Planning sessions. These are avenues for educating employees on identifying workplace hazards and suitable control measures, as well as new industry practice.

4. **Refresher OH&S Training** is a fourth method for ensuring worker competency. It is the provision of ongoing OH&S training and growth, and a method of encouraging employees to continually improve and update their knowledge and proficiency. Bird suggests that OH&S refresher training should be taken every three years.[5] In today's world of rapid change, more often than that is advisable.

As new hazards are identified, or industry practices are enhanced, or legislative changes are issued, employee OH&S knowledge must be updated. This can be achieved through:

- OH&S training seminars such as the Canadian Centre for Occupational Health and Safety (CCOHS) online courses on Globally Harmonized System of Classification and Labelling of Chemicals (GHS) in Canada;
- *ad hoc* OH&S training such as when new threshold limit values are

---

[4] Alberta Municipal Health & Safety Association (AMHSA), *Leadership for Safety Excellence* (Nisku, AB: AMHSA, 1995).

[5] F. Bird & G. Germain, *Loss Control Management: Practical Loss Control Leadership*, revised ed. (Loganville, GA: DNV, 1996) at 263.

established, or when a new hazard is identified;
- periodic OH&S training such as routinely-scheduled confined space rescue training or First Aid Training; and
- performance-based OH&S training due to noted substandard work performance by an employee.

5. The last way to ensure worker competency is through **Performance Monitoring**. By monitoring employee performance, supervisors/ employers can identify ongoing OH&S training needs and respond to them.

Armed with the right information, employees can recognize unsafe work situations and respond accordingly. Worker OH&S training provides the knowledge, skills, and experience for employee learning.

D. WORKER OH&S TRAINING: IMPORTANCE

In the last 30 years, OH&S has "moved up" on the Management agenda because of the high cost of failures; legislation requiring employers to assess and manage risks; public opinion on risk acceptability; and the risk of high-profile disasters which negatively fuel public opinion. All OHSMS require that employees understand the organization's/company's OH&S objectives, and how to apply safety techniques for loss reduction and greater profitability.

OH&S legislation obligates employers to ensure that:
- employees are trained and clearly instructed in their job duties;
- contractors are properly trained and competent to work safely. Everyone who works for an organization/company, including self-employed people, must know how to work safely and without risks to themselves or others. So, employers need to educate them about the potential workplace hazards and risks, and how to effectively control hazards; and
- worker OH&S training should take place during working hours and paid for by the employer.

Some employee groups have specific training needs, for example, new recruits, people changing jobs or taking on extra work responsibilities, individuals in highly skilled positions, and/or young/new employees who are particularly vulnerable to incidents.[6] Employers also need to ensure that new, inexperienced, or young employees are adequately supervised.

Critical to demonstrating a duty of care, employers must keep records of worker OH&S training to prove that the relevant OH&S training has been provided to employees. It is also advisable to get employees who attend an OH&S training session, to sign a form signifying their attendance.

---

[6] Manitoba Labour and Immigration — Workplace Safety and Health Division, "Employing Young Workers: Tips for Employers and Supervisors", online: http://safemanitoba.com/ (date accessed: February 28, 2020).

Worker OH&S training is more than just instructing or showing someone how to do a task. It also involves educating employees, assessing their level of learning and modelling safe work behaviours. Given that OH&S legislation is really a "performance-based OH&S standard", it is not enough for employers to just provide training; organizations/companies must also ensure that learning has occurred, and that the OH&S knowledge/performance levels of employees reflect that learning.

Therefore, worker OH&S training/education must be thorough and effective because:

1. Employers are legally required to provide employees with training on workplace hazards, safe work practices, safe operation and maintenance of equipment, safe handling of chemicals, use of personal protective equipmen, and implementation of emergency response techniques.

2. Employers are legally required to ensure that employees are aware of the applicable OH&S legislation and their related responsibilities.

3. Employers have OH&S standards with which they want their employees to comply.

4. Some industries, because of the hazardous nature of the work, depend heavily on worker OH&S training to ensure the availability of a competent workforce.

5. Training can result in a flexible workforce capable of competently doing different jobs when required.

6. Training can improve work quality and quantity. By knowing how to use and maintain tools and equipment properly, how to efficiently complete the tasks, and how to safely do critical procedures, employee performance can be enhanced. Less time and energy is thus spent on correcting errors.

7. Training can reduce/eliminate incidents, injuries, property damage, "down time", and waste.[7] This can increase productivity and profitability.

8. Employee morale can be positively impacted by the provision of worker OH&S training: it sends the message that Management perceives its employees as valuable and worth the investment of OH&S training costs. As an added benefit, OH&S training builds employee confidence and tends to enhance job satisfaction. Job satisfaction is directly associated with employee loyalty and lessening the rate and cost of staff turnover.

9. Worker OH&S training can help to build a strong work culture that values healthy and safe work performance.

10. Proper training saves money, protects jobs, and helps to build a better workforce and organization.

---

[7] F. Bird & G. Germain, *Loss Control Management: Practical Loss Control Leadership*, revised ed. (Loganville, GA: DNV, 1996) at 249.

In short, employers should be committed to protecting the health, safety, and well-being of workers, their families, and communities.

E. WORKER OH&S TRAINING: IMPORTANT ELEMENTS

Worker OH&S training can be formal, informal, classroom instruction, online, or on-the-job instruction. It should be instructive, clear, relevant, accurate, and tailored to the intended audience's needs and learning styles. The use of adult learning principles as originally identified by Malcolm Knowles[8] is important when providing worker OH&S training. Adult learning principles include:

- recognizing that adult learners have unique learning needs;
- respecting the adult learner's previous experiences and seeking ways to incorporate those experiences into the current learning experience;[9]
- understanding that teaching/learning procedures are determined by the adult learner's needs;
- appreciating that adult learners are goal- and relevancy-oriented, and relating how the current training applies to the learner's work situation is important;[10]
- recognizing that adults are autonomous and self-directed, and need the freedom to direct themselves and be actively involved in the learning process;[11]
- using thoughtful, relevant learning situations that include the adult learner's perspectives;
- recognizing that adult learners learn from each other through class interaction;
- recognizing that adult learning is informational, and involves skill development and attitudinal change;
- understanding that adult learners are practical and need to be instructed on how the lesson will be of value to them;
- understanding that teaching should be focused on the adult learner, not the subject being taught;
- making teaching and learning a shared responsibility for the instructor and adult learner;

---

[8] National-Louis University, "Malcolm Knowles: Apostle of Andragogy", online: http://www.umsl.edu/~henschkej/henschke/malcolm_knowles_Apostle_of_andragogy.pdf (date accessed: February 28, 2020).

[9] S. Lieb, "Principles of Adult Learning", online: http://www.lindenwood.edu/education/andragogy/andragogy/2011/Lieb_1991.pdf (date accessed: February 28, 2020).

[10] *Ibid.*

[11] *Ibid.*

- continually evaluating the learning process so that both the teacher and adult learner can assess the learning achieved; and
- being willing, both teacher and adult learner, to progress from an area of comfort to a learning environment that challenges their knowledge levels.[12]

Workplace learning often occurs in relation to an employee's need to make decisions regarding workplace situations. Conditions that can impact workplace learning are:

- ***The contingent*** - the unplanned and informal nature of work that often requires impromptu decisions.
- ***The practical*** — the need to solve problems efficiently and effectively.
- ***The process*** — the need for the employee to think about past actions to learn and value professional growth.
- ***The particular*** — the need to address current situations compromising wisely as required.
- ***The affective and social domains*** — the ability to make decisions that align with the learner's professional and personal emotional and social values.[13]

Although worker OH&S training can enhance awareness on work-related topics and impart knowledge, education may not be enough to elicit the desired work behaviours. The age-old belief that by just "merely telling people what they should do" will lead to the adoption of desired behaviours has proven time and again to be faulty. Rather, behavioural change is far more complex than that and requires motivation.

From the Great Safety Performance™ model described in Chapter 6, "Occupational Health and Safety: Leadership and Commitment", the motivators for behavioural change by employees are:

- understanding, valuing, and embracing the importance of the behavioural change;
- Management rewarding the adoption and display of the desired behavioural change;
- cultural support for the desired behavioural change; and
- Management passion for and demonstration of the desired behaviours.

Additional motivators for adult learners include:

- ***External expectations***: participation to comply with instructions from

---

[12] Hatscan, "WHMIS Training" (Presented in Edmonton, AB, February 26-27, 2007).

[13] Beckett & Hager (2000), as reported in K. Sitzman, "Adult Learners Reframed" (2006) 54:6 AAOHN 292.

someone else; to fulfil the expectations or recommendations of someone with formal authority.

- *Social welfare*: participation to improve the ability to serve mankind, to prepare for service to the community, and/or to improve their ability to participate in community work.
- *Personal advancement*: participation to achieve a higher status in a job, secure professional advancement, and/or to stay ahead of competitors.
- *Escape/Stimulation*: participation to relieve boredom, to provide a break in the routine of home or work, and/or to provide a contrast to other exacting details of life.
- *Social relationships*: participation to make new friends; to meet the social need for associations and friendships.
- *Cognitive interest*: participation to learn for the sake of learning; to seek knowledge for its own sake and to satisfy an inquiring mind.[14]

Some other teaching/learning factors that impact the effectiveness of worker OH&S training are:

- *Group size*:
    - Groups of fewer than 25 attendees having similar jobs, work locations, and hazard exposures tend to result in effective learning.[15]
- *Length and frequency of the training courses*:
    - Attendance at multiple training courses that address the recognition of workplace hazards and control measures increases worker knowledge of the workplace risks.[16]
    - The use of frequent, short training sessions at the beginning of a shift resulted in favourable outcomes in terms of worker recognition of hazards and their related controls.[17]
    - Frequent training and follow-up promotes sustainable learning.[18]

---

[14] S. Lieb, "Principles of Adult Learning", online: http://honolulu.hawaii.edu/intranet/committees/FacDevCom/guidebk/teachtip/adults-2.htm (date accessed: February 28, 2020).

[15] K. Saarela, "An Intervention Program Utilizing Small Groups: A Comparative Study" (1990) 21 J. Safety Res. 149; T. Robins et al., "Implementation of the Federal Hazard Communication Standard: Does Training Work?" (1990) 32:11 J. Occup. Med. 1133.

[16] D. Parkinson et al., "Effectiveness of the United Steel Workers of America Coke Oven Intervention Program" (1989) 31:5 J. Occup. Med. 464.

[17] T. Robins et al., "Implementation of the Federal Hazard Communication Standard: Does Training Work?" (1990) 32:11 J. Occup. Med. 1133.

[18] M. Harder, "How effective is health and safety training?" *Canadian Safety Reporter* (May 2010).

- Regular Safety drills are effective.[19]
- *Mode of training*:
  - Informational campaigns, video presentations, and brochures have limited effect on evoking behavioural changes.[20]
  - Demonstrations and interactive video techniques for learning safe work practices proved to be more effective than the use of written instructions, lectures, or videos.[21]
  - The use of role play, case studies, and practice in problem-solving and decision-making tend to increase worker knowledge of workplace hazards.[22]
- *Transfer of training*:
  - When teaching basic workplace health and safety housekeeping rules, the use of contrasting illustrations of safe versus unsafe work practices is effective.[23]
  - Known safe work practices may not be used in the workplace due to physical constraints, or other conditions, that interfere with practising the desired behaviours.[24]
- *Motivational*:
  - To attain successful OH&S training results, researchers found that setting OH&S performance goals is an effective approach to use in both the training and post-training periods.[25]
- *Promotional*:
  - The use of rewards to reinforce the learning of safe work actions is

---

[19] M. Colligan & A. Cohen, "The Role of Training in Promoting Workplace Safety and Health" in J. Barling & M. Frone, eds., *The Psychology of Workplace Safety* (Washington, DC: American Psychological Association, 2004) at 223 [hereinafter "Colligan & Cohen"].

[20] Saarela *et al.* (1989), Borland *et al.* (1991) and Karmy & Martin (1980), cited in Colligan & Cohen, at 234.

[21] Leslie & Adams (1973), Rubinsky & Smith (1971), Bosco & Wagner (1988), Vaught *et al.*, (1988) and Goldrick (1989), cited in Colligan & Cohen, at 234.

[22] McQuiston *et al.* (1994), Brown & Nguyen-Scott (1992), LaMontagne *et al.* (1992), Michaels *et al.* (1992), Weiger & Lyons (1992) and Cole & Brown (1996), cited in Colligan & Cohen, at 234.

[23] Chhokar & Williams (1984), Cohen & Jensen (1984), Fox & Sulzer-Azaroff (1987), Komaki, Barwick & Scott (1978), Ray, Purswell & Schlegel (1990), Reber & Wallin (1984), Saarela (1990) and Saari & Nasanen (1989), cited in Colligan & Cohen, at 234.

[24] Carleton (1987), Scholey (1983) and St. Vincent, Tellier & Lortie (1989), cited in Colligan & Cohen, at 235.

[25] Chhokar & Williams (1984) *et al.*, cited in Colligan & Cohen, at 235.

effective during the training period and later, when applied to the workplace.[26]

- Worker OH&S training and transfer of the learning to the workplace can be enhanced by incorporating desired workplace health and safety practices into the worker's performance appraisal.[27]

- *Trainer qualifications*:
  - Having workplace health and safety training delivered by supervisors/foremen and trainers is effective in terms of positive workplace changes, raising the level of worker safety performance, and hazard recognition and control.[28]

- *Management role*:
  - Management support and endorsement of OH&S training greatly increases the impact of the training and the sustainability of worker health and safety practices.[29]
  - To reinforce and sustain positive OH&S training outcomes, Management should give hazard management high priority and make individuals accountable for doing so.[30] Indifference by Management can extinguish any potential gains targeted by worker OH&S training programs.[31] In essence, what does not interest the boss, is not viewed as important by workers.

- **Learner characteristics:**
  - Young workers have different learning needs and methods of learning.[32]
  - Supervisors are advised to use a hands-on approach in educating young workers about workplace health and safety within their operations, such as spending time to guide them through the organization's/company's OH&S materials and practices; explaining the importance of workplace health and safety for them and the

---

[26] University of Kansas (1982), Fox, Hopkins & Anger (1987) and Zohar & Fussfield (1981), cited in Colligan & Cohen, at 235.

[27] Sulzer-Azaroff (1990) and Lynch *et al.* (1990), cited in Colligan & Cohen, at 235.

[28] Maples *et al.* (1982), Lepore, Olson & Tomer (1984) and McKenzie *et al.* (1985), cited in Colligan & Cohen, at 236.

[29] Ray, Purswell & Schlegel (1990), *et al.*, cited in Colligan & Cohen, at 236.

[30] Sulzer-Azaroff *et al.* (1990) and Lynch *et al.* (1990), cited in Colligan & Cohen, at 236.

[31] Fox & Sulzer-Azaroff (1987) and Hopkins (1984), cited in Colligan & Cohen, at 236.

[32] Manitoba Labour and Immigration — Workplace Safety and Health Division, "Employing Young Workers: Tips for Employers and Supervisors", online: http://safemanitoba.com/uploads/wsh_employer_superviosr_tips.pdf (date accessed: February 28, 2020).

workplace; inviting their questions and being receptive to providing them accurate answers; modelling the desired OH&S behaviours; partnering them with an experienced and safety-conscious worker; and providing them frequent and constructive feedback on their work practices and behaviours.[33]

- **Learner retention:**
    - Learner retention is impacted by the teaching modes used. Research has shown that 72 hours post-instruction, learner retention is as follows:

Figure 17.1: The Relationship between Instructional Modes and Learner Retention

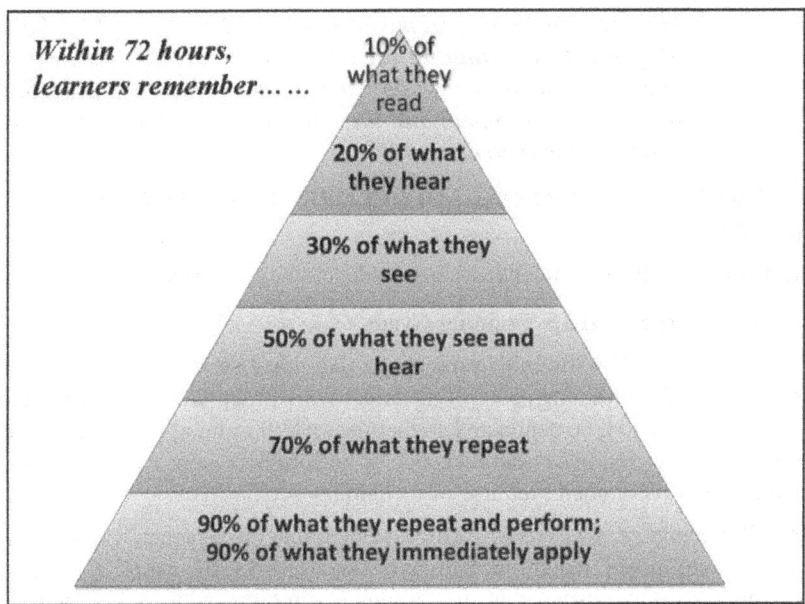

- **Follow-up Support:**
    - Learning is improved when follow-up sessions are held to enable workers to confidently apply newly learned skills, and when supervisors are coached on how to support the workers in doing so.[34]

The message is to reinforce learning and to design courses so that

---

[33] *Ibid.*

[34] Institute for Work and Health, "Study shows way to make health and safety training for office workers more effective" (January 28, 2014), online: http://www.iwh.on.ca/media/2014-jan-28 (date accessed: February 28, 2020).

learners can apply the provided content.

- **Other factors:**
  - To enhance the use of safe work practices, Management should make doing those practices the "path of least resistance" — that is, make the Safety-related materials and equipment readily available to workers.[35]
  - Although often refered to as the worker's "last line of defence", worker OH&S training is often required to support and enhance engineering and administrative hazard control methods.[36]
  - For training on safe work practices to be effective, supervisors/foremen need to model the desired work practices.
  - *Training is not a general panacea or "quick fix" for all health and safety problems within an organization. Its effectiveness will be a function of the organization's overall commitment to providing a safe work environment and the employees' perception and recognition of that commitment.*[37]

F. WORKER OH&S TRAINING: DEVELOPING AN OH&S TRAINING SESSION

The following are the recommended steps for developing an OH&S training session:

1. Identify Worker OH&S Training Needs

When planning an OH&S training program or individual session, the first step is to determine the OH&S training needs of the target work group. Educators recommend that a needs assessment be conducted so that the OH&S training is relevant, timely, and related to a desired behavioural change or performance improvement. Bird and Germain (1996) recommend the use of techniques such as:

- *Performance Analysis* — Allows the trainer to separate worker OH&S training needs from other organizational issues by using a systematic approach to ensure that the training solutions address the identified problem(s).
- *Work Analysis* — Enables the trainer to identify the requisite knowledge, skills, and abilities for specific jobs.
- *Task Observations* — Provides an evaluation of the worker performance against standard job procedures and practices. It helps to identify the specific training needs for individual employees.

---

[35] Linneman, Cannon & DeRonde & Lamphear (1991), Lynch *et al.* (1990), Seto, Ching, Chu & Fielding (1990) and Wong *et al.* (1991), cited in Colligan & Cohen, at 237.

[36] University of Kansas *et al.* (1982), cited in Colligan & Cohen, at 237.

[37] Colligan & Cohen, at 226.

- **Tests** — These can be knowledge and performance tests used to identify knowledge/performance gaps and weaknesses.
- **Worker Perception Surveys** — Bird identifies two survey methods - interviews and questionnaires. Interviews allow workers to express their learning needs to the interviewer. Questionnaires can be used to enable workers to identify problems in their jobs and areas in which they feel they need further training.[38]
- **Incident and Injury Data** — By reviewing the organization's/company's incident and injury experience, causation trends may indicate training needs.

These tools can be used by the OH&S Practitioner/Professional to define the learner's OH&S training needs, the work conditions that could impact the training, and the specific training topics to be covered.[39]

Next, answer the following questions:
- What is the teaching strategy?
- What will the teaching facility (environment) be like?
- What resources are available?
- Which jobs require specific education/training?
- Are there specific job tasks that warrant specialized training?
- What is the Safety climate like?
- What training will Management support?

2. Decide the OH&S Training Priorities

Once the OH&S training options have been identified, the next step is to decide which OH&S training topics to pursue. To do this, determine which OH&S training topics:
- are legally required;
- if not addressed, have the potential for causing the greatest harm;
- are necessary for new or inexperienced employees;
- would benefit the largest number of employees; and
- are of top priority to employees and/or the employer.

Based on these considerations, choose the suitable OH&S topic.

3. Establish the OH&S Training Objectives

Objectives are the specific aims for the OH&S training program. They are

---

[38] F. Bird & G. Germain, *Loss Control Management: Practical Loss Control Leadership*, revised ed. (Loganville, GA: DNV, 1996) at 252-255.

[39] *Ibid.*, at 255.

statements of intentions, prepared in such a way that they guide problem-solving (or training) behaviour and form the basis for measuring the results. They address two questions:

1. What must be taught?
2. What will success look like?

Objectives state the desired outcomes in a manner that is meaningful, relevant, realistic, actionable, sustainable, useful, measurable, and result oriented. Each content,[40] learning,[41] and teaching[42] objective should be written so that it is:

- **S**pecific — measures only one thing, *i.e.*, an observable behaviour;
- **M**easurable;
- **A**ttainable, but challenging;
- **R**ealistic and feasible; and
- **T**ime-oriented (**S.M.A.R.T.**).

In short, learning objectives describe the desired behavioural change anticipated as a result of the teaching-learning experience.

In terms of what must be taught and what success would look like, it is important to first determine the level of learning required. There are four levels of learning:

1. *Awareness* — the aim of the instruction is to create a level of basic understanding and knowledge about the topic. When applied to the field of disability management, this would involve providing employees, management, and other stakeholders with information on a specific OH&S Program.

2. *Knowledge* — the desired outcome is that the participants will understand and be able to recall the information provided. For example, this level of instruction would be more detailed with an expected outcome of the learners being able to recite the information provided should they be asked.

3. *Application* — in many workplace situations, the level of learning required is the participant's ability to learn and be able to apply a defined skill. In terms of OH&S, this would entail a worker knowing how to conduct a

---

[40] **Content objective** is a specific statement that speaks to the content to be covered in a training session/course. For example, *"At the end of this course, the worker will be aware of the legislative requirement for noise abatement and hearing conservation."*

[41] **Learning objective** is a specific statement that speaks to the worker learning. For example, *"At the end of this course, the worker will be able to explain how to use, maintain, and replace hearing protection."*

[42] **Teaching objective** is a specific statement that speaks to what the worker must do to demonstrate what he or she knows or how to apply a learned skill. For example, *"At the end of this course, the worker will demonstrate the correct insertion of hearing protection."*

hazards assessment and report workplace hazards. Likewise, it would involve the supervisor being able to apply the company's OH&S standards and rules.

4. *Synthesis* — the intent of the instruction is to prepare the learner to be able to take the OH&S concepts, principles and practices and apply them to a number of scenarios. As the word synthesis suggests, it is the ability to "build on" known information. An example would be the Joint Health & Safety Committee members being able to enhance a company's OH&S Program based on knowledge about OH&S industry best practices.

In writing a teaching-learning objective, an action verb must be used. Bloom's Taxonomy of measurable verbs is a useful tool for writing objectives.[43] Note that the action verb must align with the desired level of learning.

Once the desired level of learning is established, the next step is to decide what can be done to achieve the desired level of learning and what success would then look like.

There are several OH&S education measures that can be used to promote general understanding of the value and benefits of an OH&S Program:

- *General OH&S Program education provided to all stakeholders.* This measure is informational and aimed at awareness-raising.

- *Specialized education on the various stakeholder roles and responsibilities in the OH&S Program.* This measure is aimed at achieving a functional level of knowledge and application.

- *Specialist education and training for OH&S practitioners on the "best practices" in OH&S.* This measure is designed to enable the learner to apply the knowledge gained and to be able to "build on" that information.

- *A new employee Occupational Health & Safety Orientation*, designed to provide an overview of the company's OH&S standards and how it works. It should include an explanation of the:
  - company's commitment to OH&S;
  - company's OH&S Program and policy;
  - how the OH&S Program functions;
  - the employee's OH&S Program responsibilities;
  - reporting work and non-related work illness/injuries; and
  - return-to-work procedures after illness/injury.

This approach promotes a level of awareness about the OH&S Program and the new employee's role in that program.

---

[43] See https://www.utica.edu/academic/Assessment/new/Blooms%20Taxonomy%20-%20Best.pdf.

- Employee education can be reinforced through the use of ongoing OH&S presentations held at Safety Meetings or other employee group meetings.
- **Coaching** and **mentoring** are two other means of reinforcing OH&S education.
- Online training courses are an additional instructional measure. They can include videos, interactive modules, task-based exercises, *etc.*, designed for reinforcing other OH&S education measures.

4. Develop a Plan for Achieving the OH&S Training Objectives

Select content, teaching methods, media, and materials that will best meet the audience needs and enable the achievement of the stated OH&S training objectives. To begin with, research the topic, find the relevant information and select the content to be presented.

Decide on the appropriate teaching method: Will it be instructor-led teaching; self-directed learning guided by instructional materials; or a combination of both methods? Will it be delivered in a traditional teaching setting; in a worksite setting; or through the use of online learning techniques? The teaching method should align with the teaching objectives, as well as with the need for learner retention of the content (Figure 17.2).

Figure 17.2: The Learning Pyramid[44]

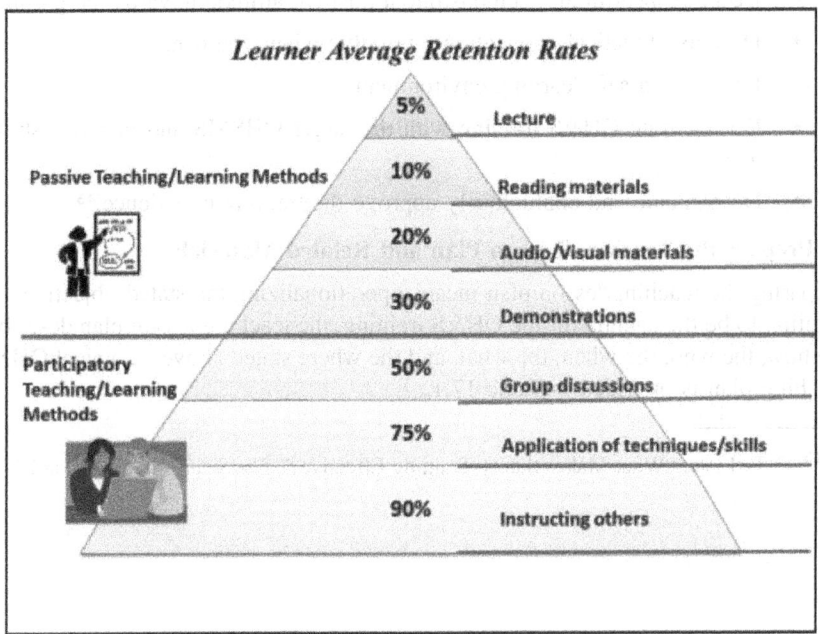

The crucial thing is to pick a training approach that best suits the target audience (the employees/management) and yields the best training outcome(s). This means deciding on details like:

- How will the OH&S training be delivered?
- Who is best suited to provide the OH&S training, to attend the training, and to assist with the training?
- When to schedule the OH&S training?
- What materials are best used to enhance the learning experience?
- Where to best host the OH&S training?

In planning the teaching/learning session consider the following tips:

- Do provide learning objectives.
- Do keep the content simple.
- Do incorporate stories.
- Do be impactful.

---

[44] Adapted from National Training Laboratories, Bethel Maine; and R. Flemmer, "The Three E's of Training" *The Safety Net* (2014) at 7.

- Do incorporate the experiences of the group.
- Do be respectful and demonstrate a positive attitude towards the group.
- Do make practical application a possible when you can.
- Do create a safe learning environment.
- Do integrate OH&S training with the larger OHSMS and business strategies.
- Do evaluate and continuously approve the training experience.[45]

5. Prepare the Teaching/Lesson Plan and Related Materials

Preparing the teaching/lesson plan means operationalizing the stated objectives. In addition to be the agenda for the OH&S training, the teaching/lesson plan describes the how, the who, the when, the what, and the where stated above. A typical OH&S teaching plan is provided in Table 17.1.

---

[45] C. DeLisle, "What Makes Safety Training Effective?" *The Safety Net* (2014) at 24-33.

Table 17.1: Worker OH&S Training: Teaching Plan
OH&S TEACHING PLAN

Title:	Great Safety Performance Model: Applying the Concepts, Methodology, and Tools			
Instructor:	Joe Black			
Type:	Half-day Workshop (3 hours) √ **Full-day Workshop (6 hours)**			
	Content (Topics)	Time Frame	Presenter	Teaching Strategy
Objectives *At the end of this workshop, the learner will be able to:*				
1. Explain Great Safety Performance (GSP) — a "breakthrough" model and methodology for maximizing the "performance system" needed to sustain great safety performance.	• Introduction of the Performance Maximizer • Description of how it was applied to workplace safety • Development of the Great Safety Performance (GSP) methodology and tools • Description of an application project and its outcomes • Explanation of the important findings	40 minutes	Joe Black	Presentation with discussion

Objectives At the end of this workshop, the learner will be able to:	Content (Topics)	Time Frame	Presenter	Teaching Strategy
2. Restate the evidence that the Great Safety Performance model generates Leading Indicator data to drive prevention.	• Discussion of the integrity of the GSP survey tool • Review of reliability and correlation data • Implication that leaders need to consider, plan for and manage the whole "performance system" in order to continuously improve to 0 incidents and injuries	20 minutes	Joe Black	Presentation with discussion
3. Implement the concepts, methodology, and tools to an actual situation: Part 1	Summarize the current situation — identify what is known and not known. Discuss the business case to support an improvement initiative. Introduce use of the GAPS Tool.	30 minutes	Joe Black	Small group exercises with report back
4. Implement the concepts, methodology, and tools to an actual situation: Part 2	Assemble the Great Safety Performance Model for the case situation: Safe Work Actions, Safety Results, Business Results, and the specific factors that make up the five Leading Indicators of the safety "performance system".	60 minutes	Joe Black	Small group exercises with report back

CH. 17: OCCUPATIONAL HEALTH AND SAFETY: WORKER OH&S TRAINING

	Objectives *At the end of this workshop, the learner will be able to:*	Content (Topics)	Time Frame	Presenter	Teaching Strategy
5.	Apply the concepts, methodology, and tools to an actual situation: Part 3	Assess gaps in safety performance using actual Leading Indicator data based on the case situation to identify root causes. Identify themes and patterns in the data — the strengths and weaknesses in the performance system.	60 minutes	Joe Black	Small group exercises with report back
6.	Use the concepts, methodology, and tools to an actual situation: Part 4	Develop an action plan of improvement solutions to strengthen the Leading Indicators and deliver improved safety results.	45 minutes	Joe Black	Small group exercises with report back
7.	Evaluate the state of the "safety culture" in their organization by rating the Leading Indicators in the "at-risk" work environments.	Participant completion of the generic GSP survey tool • Scoring and display of the survey results • Discussion of the implication of participants' results for improving safety	40 minutes	Joe Black	Individual exercise plus table discussion
8.	Use the GSP Surveys to create a "dashboard" of monitoring information.	Discuss use of periodic surveys to:	15 minutes	Joe Black	Leader-led discussion

Objectives *At the end of this workshop, the learner will be able to:*	Content (Topics)	Time Frame	Presenter	Teaching Strategy
	• measure the effectiveness of improvement initiatives • monitor Leading Indicators as preventative approach • assess the effectiveness of leadership in safety management			
9. Be able to describe/ discuss the leadership responsibilities implicit in the Great Safety Performance model.	Discussion of the expanded use of the GSP to: • improve leader's skill and align his/her behaviour • align other departments to enable worker safety • develop an energizing and supportive culture that consistently enables safe working	30 minutes	Joe Black	Leader-led discussion

Objectives At the end of this workshop, the learner will be able to:	Content (Topics)	Time Frame	Presenter	Teaching Strategy
10. Be able to describe/discuss how the GSP methodology and Leading Indicator data "raise the bar" on safety due diligence.	Discussion on what the Great Safety Performance model and methodology adds in meeting and exceeding the requirements of legislated "OH&S Duty	20 minutes	Joe Black	Leader-led discussion

6. **Deliver the OH&S Training**
    1. **Preparation:** The key to a successful OH&S training session is the preparation. To start with, make sure to know the training topic and be fully familiar with the training materials and related media equipment. Test the audio/video aids to ensure that they work properly and can be seen and heard throughout the room. Have the handouts ready and arranged for timely and easy distribution of them during the session.

        Make sure the facilities are set up to support a positive learning experience. The teaching method, media, and materials dictate how the room should be arranged. Lectures are best delivered in a classroom setting, whereas the use of "application exercises", such as case studies or role playing, requires tables for group work. Training that uses computer online training needs specialized equipment to be set up.

    2. **Rehearse the presentation:** Rehearsal allows the presenter to test the timing of the training events and to make adjustments as necessary. Take the time to "run through" the actual presentation, noting any content or time changes that need to be made. It is also an opportunity to get a feel for the room and how to best position oneself to effectively get the training messages across.

        Presenting evokes many strong emotions, some of which are not pleasant for the presenter. The important thing is to recognize your nervousness and to prepare for it. One suggestion is to have the opening statements for the presentation written out. Once underway, most presenters are comfortably able to continue.

        When presenting, there are a number of **golden rules**, namely:
        - engage the audience;
        - speak clearly and loudly enough for the participants to comfortably hear;
        - use plain, simple language: avoid the use of complex technical jargon, metaphors, and awkward examples. Rather, stick with the jargon and technical terms that are familiar to the audience;
        - use conversational hand gestures;
        - move around comfortably;
        - use concrete examples;
        - look at the audience when speaking;
        - avoid playing with a pen, pocket items, keys, coins, *etc.*;
        - only use diagrams and teaching aids that enhance the learners' understanding of the spoken materials;
        - use both text and audio materials if possible;

- create a friendly atmosphere;
- use storytelling to bring the content "alive";[46]
- make the experience enjoyable and fun;
- prepare methods to garner audience involvement (questions, show of hands);
- encourage and reward audience participation;
- stay on topic;
- seek audience feedback on the pace of the teaching, the relevancy of the materials, and the participants' satisfaction levels; *and above all*,
- **start and end the presentation on time** — it shows respect for the participants and their time.

7. Evaluate the OH&S Training

Evaluation is the only way to ensure that the OH&S training is effective and meets the stated objectives. It is also an important way to improve OH&S training programs and sessions. This topic is covered in detail in the following section.

G. WORKER OH&S TRAINING: MEASUREMENT OF EFFECTIVENESS

There are two levels of measurement available for evaluating worker OH&S training programs. One is a high-level evaluation of the entire OH&S Training Program; the other is the measurement of the effectiveness of an individual OH&S Training Course or Program.

1. Evaluation of the Entire Worker OH&S Training Program

There are many ways to measure program success. By **auditing** an OH&S training program, one can measure the program structure, processes, and outcomes against the established program objectives and/or standards. This measurement technique allows for the identification of the gaps between the current and ideal state of the program. Typically, the process includes recommendations for reaching the ideal state.

A second approach to measuring program success is to undertake a **cost-benefit analysis** projection. This involves determining the losses an organization/company would incur without the worker OH&S training program and compare those against the losses the organization/company experiences when worker OH&S training is provided. The financial differences are deemed the "benefits realized".

A third approach is to establish the **return on investment (ROI)** realized as a result of the worker OH&S training program. This is calculated by dividing the benefits realized through the OH&S training program by the cost of the training

---

[46] R. Flemmer, "The Three E's of Training" *The Safety Net* (October 2014) at 7.

program. A positive ROI indicates that a benefit exists.

Benchmarking program data against that of external parties is a fourth approach. Benchmarking is a continual and collaborative discipline that involves measuring and comparing the results of the key process with "best performers", or with one's own previous achievements. Internally, benchmarking can be used to compare against previous program outcomes, processes, practices, and performance. Externally, benchmarking involves comparing the organization's/company's OH&S training program, processes, and/or results against those of another similar organization/company.

Typically, benchmarking is undertaken to improve the quality of the service or product, address an identified problem, learn "best practices", or in response to increased pressures to improve performance. It is not an exact science; rather, a methodology designed to determine the best way to do things.

When benchmarking a worker OH&S training program, the process should be to:

1. plan what and who to benchmark;
2. collect and analyze the relevant data;
3. integrate the findings into a familiar frame of reference; and
4. develop an action plan that would implement, monitor, and evaluate the changes made.

The benefit of benchmarking is that the methodology provides a rational and objective framework for observing what others are doing, how they are doing it, and what can be done to improve the process even more. In essence, it is a quality improvement tool because the decision-making is based on facts, and on learning from the actions of others.

A fifth approach is a **client satisfaction survey** regarding the OH&S training offered. The typical approach is to ask those employees who participated in the OH&S training program for their impressions of the session, presenter(s), and learning outcomes. The questions that are usually asked are designed to measure specific elements of the program such as timeliness, accessibility, appropriateness, and universality of the program.

2. Evaluation of a Specific OH&S Training Course or Program

Training courses/programs should be designed so that evaluation is one of the delivery elements of the course/program. Evaluation can be done before the course/program begins (pre-assessments), during the course/program (process evaluations), or at the end (outcome evaluations).

The **pre-learning assessments** are a baseline measure of level of awareness/ knowledge/skills/abilities that the employee possesses prior to training. This can be done using quizzes, checklists, surveys, interviews, or demonstrations.

Process or **formative evaluations** are measures of how the learning experience is progressing. Some instructors call this a "temperature-taking" exercise. The intent is to determine if the teaching approach is aligning with the audience's expectations and needs. It is a test of relevancy. The results enable the trainer to make the appropriate adjustments to the training plan. This can be done formally using a checklist or multiple-choice questionnaire; or informally, using a show of hands to questions posed by the trainer. It is also a teaching tool that enables the trainees to test and evaluate their own levels of comprehension, and to reinforce the teaching. Thus, it increases the potential for retention of the materials learned.

Outcome evaluations are more complex. In essence, they are designed to determine if the OH&S training course/program met the participants' expectations in terms of the pre-stated OH&S training objectives. They are summative evaluations and include measurement of:

- *Participant reaction to the training*: Participant satisfaction questionnaires are the tool of choice to determine the degree to which the course/program met participant expectations and needs, and whether the participant believes that the training will make a difference in future job performance/work behaviours.

- *Change in participant awareness/knowledge levels*: Before and after quizzes/tests are used to establish if the course/program objectives have/have not been met.

- *Change in skills/behaviours/attitudes*: Through proficiency tests, direct observation of work practices/behaviours and/or self-reports on skill improvement or application of the learning concepts, trainers can measure the difference(s) made as a result of the OH&S training.

- *Change in OH&S measures*: These are objective markers that are the result of behavioural change, such as an increase in the use of hearing protection and a reduction in the number of hearing loss claims due to OH&S training on hearing conservation and use of personal protective equipment.

- *Organizational impact*: Changes to outcome measures related to worker health and safety are often used, for example, a reduction in worker injuries as they relate to the OH&S training offered. One common example is a reduction in musculoskeletal injuries as a result of ergonomic training. However, this measure is fraught with problems. For example, attempts to credit successful training outcomes with reduced injury rates can be problematic. Research indicates that only 25 per cent of the actual injury reduction could be attributed to the improved work practices that were observed following worker OH&S training.[47]

When evaluating a program/training course, controlling for confounding variables poses a huge challenge for the program designers. A **confounding variable** is

---

[47] J. Saari & M. Nasanen, "The Effect of Positive Feedback on Industrial Housekeeping and Accidents: A Long-term Study at a Shipyard" (1989) 4 Int. J. Ind. Ergon. 201.

a variable that confuses the relationship between the dependent and independent variables and that needs to be controlled through the design of the evaluation plan or via statistical procedures. The **dependent variable** is the outcome variable of interest; the variable that is hypothesized to depend on or be caused by another variable. The **independent variable** is the variable that is believed to cause or influence the dependent variable. In a research scenario, it is the variable that is manipulated.

3. OH&S Training: Measurement Model

A measurement model for evaluating an OH&S training course/program is provided in Figure 17.3. It can be used for evaluating an entire training program or a specific program/course. It illustrates the measurement of the cycle for training program/course objectives, beginning with the trainer setting the training objectives.

Figure 17.3: OH&S Training Program/Course Measurement Model

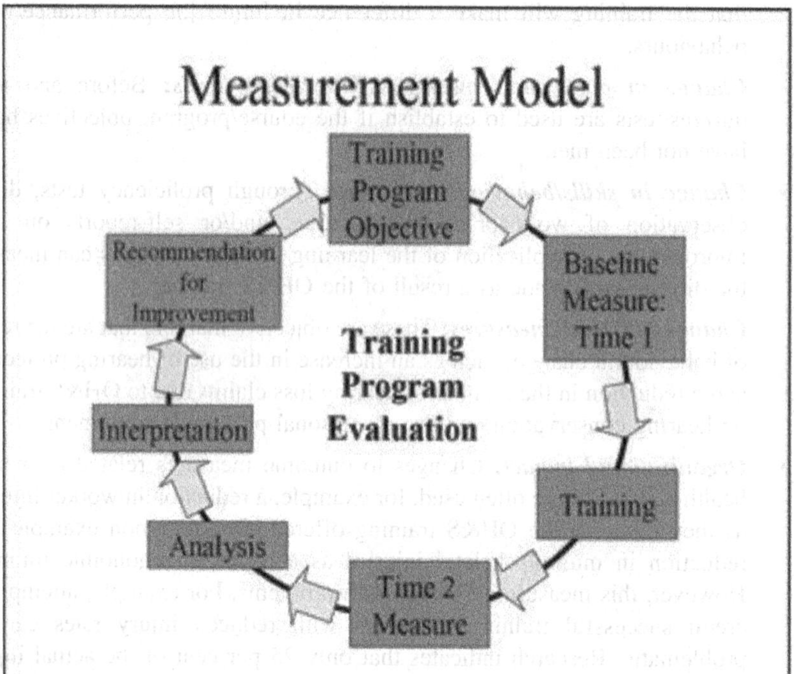

The model suggests measurement of each of the training course/program objectives prior to and following the delivery of the training. The difference in the results between Time 1 and Time 2 is determined, analyzed, and interpreted. From the outcome of that exercise, the evaluator can reach a conclusion as to whether or not the training program/course was successful, and the degree to which it did, or did not, make a difference.

H. WORKER OH&S TRAINING: MAKING IT FUN

OH&S trainers/educators constantly face the challenge of making the delivery of OH&S information exciting and interesting enough for workers to want to learn. Everyone has attended OH&S seminars and meetings that have been incredibly boring. The usual outcome is worker attendance without much sustainable learning.

Adult learning principles tell us that teaching should be self-paced, meaningful, and fun for the audience. With these principles in mind, making learning fun for workers requires innovative approaches. OH&S demonstrations, team challenges, computer-assisted learning, and the use of various audio/visual modalities can make learning inviting and interesting. For example, the "scavenger hunt technique" to teach workers about WHMIS 2015 has been successfully used by some organizations. A team challenge on OH&S awareness and knowledge has also been well received. The key is to find a method of information delivery that motivates the worker to want to learn about OH&S, and retain that knowledge.

The following are examples of industry-applied techniques for enhancing worker learning and retention:

1. An Industry Application: ENMAX: Rule Book Madness[48]

In 2001, Dale Rud, former Safety Advisor at ENMAX Corporation, Calgary, Alberta, developed just such a training technique. He devised a game that made learning about ENMAX's OH&S Program fun and challenging. Additionally, he capitalized on the principle that people remember 70 per cent of what they see, hear, and repeat.

Patterned after the well-known television show, *Jeopardy*, Rud developed a game called Rule Book Madness. Employees were provided answers to which they had to locate the appropriate question within the ENMAX Health & Safety Rule Book to earn game points. Teams of employees with the most points at the end of the game won valuable prizes.

(a) The ENMAX Health & Safety Rule Book

The ENMAX Health & Safety Rule Book (Rule Book) was developed through the joint efforts of Management and unions. The purpose was to create an everyday tool that would:

- ensure compliance with the applicable OH&S legislation and industry standards;

---

[48] Reprinted with permission from Alberta Human Resources and Employment and adapted from the article, D. Dyck & D. Rud, "ENMAX Makes Safety Training Fun" (2003) 26:3 *Occupational Health & Safety Magazine* 16. Printed with permission from ENMAX, Calgary, AB.

- set standards for participation in OH&S by all employees/contractors;
- promote commitment to OH&S by every employee/contractor;
- encourage the use of safe work practices; and
- serve as a reference tool.

When it was first published, the Rule Book was introduced in a workshop to familiarize employees with its layout and contents, and to explain the meaning and purpose of the rules it contained.

Employees and managers were enthusiastic about the new Rule Book. At the conclusion of the workshop, there was an overwhelming sense that the momentum and interest built at this workshop had to be continued. But, how?

(b) The Birth of an Idea . . .

At the monthly Safety Meetings, ENMAX supervisors urged their crews to review the Rule Book and learn the various sections, one at a time. However, this approach was received with a less than an enthusiastic response from employees. In reality, it was painful. In exasperation, one of the managers exclaimed: "We need to keep this book alive. If anyone has a better idea, please let me know!"

Rud accepted the challenge. He began dissecting the Rule Book, page-by-page, section-by-section, noting key points that related to the work that ENMAX employees perform. By turning the material into questions, Rud created an electronic game that could be presented to employees at safety meetings. The result was a game that encouraged employees to actively "thumb through" the Rule Book, in a race against time, to deliver the answers to the questions developed.

Like *Jeopardy*, Rule Book Madness used six category headings which were:
- Vehicles and transport;
- Chemical handling;
- Codes of practice;
- Electrical work;
- General rules; and
- Office safety, incidents, injuries, and First Aid.

To win, the participants had to answer the question and identify where to locate the associated rule within the Rule Book.

(c) Game Play Rules

To ensure fair play, Rud developed the following play rules:
- *Teams*:
  - Teams were chosen at random with a maximum of five players per team and a maximum of 15 teams.
  - Teams remained together for the duration of the tournament.

- Each team was assigned an identification number.
- At the end of each tournament match, the team that accumulated the most points won.
- At the end of the tournament, the points were tallied. The teams with the most points entered into a playoff challenge, which in turn, led into a championship match.

- *Game Play*:
  - Game Play began with the Game Host drawing a team number. The selected team then chose a question topic and point value.
  - Each question had to be answered using the appropriate rule, rule number, and the page on which the rule was located in the ENMAX Health & Safety Rule Book.
  - A maximum of 45 seconds was allowed for a correct question response, with a maximum of 20 seconds for each bonus question.
  - Each team had two chances to accumulate points during match play.
  - If a team asked for a higher point question and answered it incorrectly, the team was penalized that amount of points, making winning the game difficult.
  - Bonus questions that were answered incorrectly were not deducted from a team's point count.

- *Stealing Questions*:
  - If a team could not correctly answer a question, or if it went over the allotted time, the next team drawn earned the chance to "steal" that question. It had to immediately respond with the right answer to that question.
  - A team that elected to steal a question was not penalized if it could not answer a stolen question, nor did it have a chance to answer the bonus question.
  - Team members were not allowed to steal their own questions.

- *Winning the Match*:
  - Game play continued until each team had two turns.
  - The team that accumulated the most points during match play became the monthly winner.
  - Points were then carried over into the next set of matches.
  - The teams that accumulated the most points during the tournament had the chance to play off against teams from the other ENMAX Business Unit teams.
  - In order to qualify for the playoffs, a team or a player had to have

played in at least four match-play games.

- *Tiebreaker*:
  - If at the end of regulation play, a "tied score" existed between two or more teams, the Game Host would select a "tiebreaker" question. The teams then had 45 seconds in which to correctly answer. To determine which team would have the opportunity to answer first, the Game Host pulled a team number out of a hat. That team had to immediately answer the question, or forfeit the turn to the next team drawn. This process would continue until a "right" answer was obtained.

(d) Game Finals

With incentives such as cash prizes, employee interest rose. For six months, enthusiastic teams competed in games at monthly safety meetings. User awareness of the Rule Book and its contents steadily increased.

At the final round of Rule Book Madness, four teams went head-to-head. After three rounds of play, the winning team emerged. Witnessing the team win were fellow employees and Management. The electricity of the tournament was evident. Everyone was enthusiastic and geared up for a fun time. Free pizza, door prizes, and the chance to win "giveaways" added to the event.

(e) What Comes Next?

Rule Book Madness was designed to promote camaraderie, to liven up Safety Meetings, and to encourage employees to create a safety-conscious culture at ENMAX — a culture that would result in fewer injuries, enhance employee well-being, and increase employee morale and productivity. According to Rud, "[t]he problem with most company OH&S manuals is that they never get opened." Rule Book Madness however, sent employees who had lost their manuals, looking for new ones.

Not resting on his laurels, Rud developed a new game called Bullseye Safety to further promote knowledge of the Rule Book. The goal was to determine whether the information learned during Rule Book Madness was retained.

(f) Safety Can be Exciting

Learning is a dynamic process. By the same token, the delivery of safety training/education needs to be a dynamic process. What works with one workgroup may or may not work with another. The teaching approach has to be tailored to the audience — its level of knowledge, maturity, interests, and style of learning.

To keep the delivery fresh and exciting, educators/trainers need to be receptive to new ideas. A steady diet of one training/education delivery technique leads to learner apathy. Finding new approaches requires constant vigilance for innovative techniques. Working with other OH&S Practitioners/Professionals and educators in other disciplines can prove helpful. Attendance at conferences and courses on social and program marketing can be equally valuable. Lastly, many companies have communication specialists: professionals schooled in effective communication approaches. The outcome is a target audience that knows and values the information transmitted.

Know and listen to your audience, make learning about safety exciting, and make the topic relevant so that the learner can readily identify the benefits of learning for him/her. If the learning nets positive benefits at minimal cost, the audience will find the experience fun and remember the intended message.

2. An Industry Application: OH&S Crossword Puzzles Online

On the same theme of being innovative in getting OH&S messages across, the use of crossword puzzles can be implemented. One Canadian company developed a series of crossword puzzles, each with different messages, and posted them on their company intranet for employees to complete and submit back to the OH&S Department. All the correctly answered submissions received a reward that had a theme similar to the OH&S message being made. Figure 17.4 is an example of the crossword puzzle used.

Figure 17.4: OH&S — Ergonomics Crossword Puzzle

ACROSS

1. Study of fitting the task to the worker
5. Sensitive to noise greater than 85dBA
6. Skin cleaning agent
7. One eye exercise is to look _____ of the window
8. Adjust the computer _____ to slightly below eye level
9. A legal requirement for workplaces is to have a Health & _____ Program in place
11. Main type of workplace injury (plural)
13. _____ - miss: An incident that could have been an accident
14. They are required to comply with company safety standards

DOWN

1. They benefit from a Workplace Health & Safety Program
2. _____ Safety Performance: A new model for workplace safety
3. Regular monitoring of worksite conditions
4. A worker's last line of defense against a workplace hazard
8. The part of a chair that can be adjusted (plural)
10. A nuisance effect that can be eliminated from your computer screen
12. Compliance with Occupational Health & Safety requirements is the _____

OH&S — Ergonomics Crossword Answers:

```
 1E  R   2G  O   N   O   M   3I   C   S   ■
 M   ■   R   ■   ■   ■   ■   N   ■   ■   4P
 P   ■  5E   A   R   S   ■  6S   O   A   P
 L   ■   A   ■   ■   ■   ■   P   ■   ■   E
7O   U   T   ■  8S   C   R   E   E   N   ■
 Y   ■   ■   ■   E   ■   ■   C   ■   ■   ■
 E   ■   ■  9S   A   F   E   T   Y   ■  10G
 E   ■   ■   ■   T   ■   ■   I   ■   ■   L
11S  12L  I   P   S   ■   ■   O   ■   ■   A
 ■   A   ■   ■   ■   ■   ■  13N  E   A   R
 ■  14W  O   R   K   E   R   S   ■   ■   E
```

3. An Industry Application: Interactive Safety Education

The use of group participation in safety education sessions appears to be a very effective mode of worker OH&S training. For example, the use of a Power Point presentation in which various road and traffic signs were presented for worker identification, proved worthwhile. Workers who correctly identified the presented road sign were positively recognized for their knowledge. Five out of 20 road signs were pre-selected by the OH&S trainer as the signs that warranted a recognition award. This technique kept workers involved and participating.

The use of an interactive game is another way to elicit worker interest and make learning fun. One example is a WHMIS game that was designed to mimic the game, Battleship. Workers had to select a grid square that hid a WHMIS question. If correctly answered, then the worker was presented with the opportunity to fire a missile at a battleship. A certain number of missiles were programmed to "miss": but if successful at hitting the ship and it sank, the worker received a prize.[49]

The benefits of this type of approach are it:

- invited worker participation;
- enabled a self-assessment of worker knowledge level about road and traffic signs;
- promoted retention of information;

---

[49] D. Rud, industry application observed at ENMAX Corporation (Calgary, AB, 2003).

- allowed the OH&S trainer to determine the level of worker know-ledge about road and traffic signs; and
- was fun to do and generated learning energy in the session.

4. An Industry Application: Online Interactive OH&S Training

Online OH&S sessions have been produced by governments and private companies/marketers. For example, Work Safe Alberta[50] developed a number of online workplace health and safety training sessions and interactive quizzes. The topics covered include Alberta OH&S Legislation; Basic Health & Safety; Health and Safety Management System; Hazard Assessment and Control; Noise and Hearing Protection; Backs and Bums: Applying Basic Ergonomics; Fun Quizzes; Incident Investigation; Impairment and Workplace Health and Safety; *etc.*

The benefit of these sessions is that workers can self-pace their learning and suitably fit it into their workday. They can also review the material as often as they wish and show it to other workers. Lastly, it enables standardized training.

An electrical utility worked with a software developer to create an interactive, New Employee Orientation session on their substations and other unique facilities/equipment. As the workers proceed through the module, not only do they learn about the substation from the "real-life" depiction of the installation, but they are also taught the common hazards and are reminded about a number of safe work practices, such as using the appropriate steel-toed boots, hard hat, goggles, gloves, *etc.* In addition to the above-listed benefits, this approach was found to be a very cost-effective approach to employee orientation and served as proof that the employee orientation had been provided.

Several other companies use computerized training modules to teach the principles of first aid, WHMIS 2015, workplace violence, safe driving, and such.

Additionally, interactive OH&S training software programs are available commercially. These are designed to evoke decision-making by workers on common work/OH&S issues. Workers are presented with some possible decision options. Each option has a related consequence connected with it. Only one is "correct", with the others resulting in negative impacts for the worker/workplace. The value of this approach is that it forces the worker to weigh out the consequences for the decision being made.

I. THE VALUE OF OH&S TRAINING PROGRAMS

Like any other business endeavour, worker OH&S training programs must demonstrate value. Training is expensive, and the effectiveness and sustainability of learning is often questioned by Management. So how can OH&S Practitioners/Professionals demonstrate a positive return on the organization's/company's investment in worker OH&S training programs?

---

[50] Work Safe Alberta, online: http://work.alberta.ca/occupational-health-safety/268.html (date accessed: February 28, 2020).

In the past, OH&S Practitioners/Professionals tried to build a business case for training programs using the argument that training enhances worker productivity, and hence, organizational/company profitability. However, the counterargument that is often put forth is that, "[e]nterprise profitability, even productivity, depends on so many factors . . . that it is difficult [for Management] to isolate and identify. Consequently, it is not where they look for outcomes."[51]

Research has shown that what employers do value, is whether or not training programs demonstrate significant changes in work behaviours, the way employees do their jobs, and the manner in which employees think and talk about their jobs.[52] With that in mind, OH&S Practitioners/Professionals can use the measurements discussed earlier in this chapter, to explain to Management that the value of OH&S training programs can be demonstrated through:

- participant reaction to the training;
- change in participant awareness/knowledge levels;
- change in skills/behaviours/attitudes;
- change in workplace health and safety measures; and
- changes to outcome measures related to worker health and safety.

Research also indicates that organizational/company leaders are influenced by case studies illustrating the impact that training had on the behaviour of workers at other companies.[53]

On another level, an organization/company can measure the effectiveness of its training initiatives using an instrument that:

- measures leadership components such as readiness to change and available resources;
- measures the training needs and risks of the training intervention and the training action plan;
- compares the organization's/company's training practices/program against a "gold standard" for training;
- measures training design, development, and delivery techniques;
- measures and evaluates the return on investment of the training intervention; and/or
- measures the organization's/company's ability to continually improve

---

[51] Bongarde Holdings Inc., "How to Demonstrate the Value of Training Programs" (2006) 2:4 *Safety Compliance Insider* at 14, online: http://www.safetysmart.com.
[52] *Ibid.*
[53] *Ibid.*

CH. 17: OCCUPATIONAL HEALTH AND SAFETY: WORKER OH&S TRAINING

through the use of learning.[54]

J. SUMMARY

OH&S education/training does lead to changes in targeted OH&S behaviours.[55] However, the degree of impact on worker knowledge, attitudes and beliefs, or health, is yet to be determined.[56]

- ensure that employees are competent and prepared to work unharmed;
- develop a positive OH&S culture, where the use of safe and healthy work practices are the norm;
- identify ways to continually improve workplace health and safety; and
- be compliant with OH&S legislation.

Effective worker OH&S training can lead to effective learning. It is important that the learning results in:

- employees being competent in safe work practices;
- employees being capable of conducting effective hazard management thereby avoiding incidents and illness; and
- companies avoiding unnecessary human and financial costs.

Additionally, the provision of effective worker OH&S training that exceeds the minimum level of training mandated by OH&S legislation, serves to demonstrate to employees that their employer is committed to OH&S and cares about them. That message holds true for investors and the community at large.

---

[54] G. Pappas, "Effective Measurement of OSH Training Programs" (Presented at the IAPA Conference in Toronto, ON, April 2006).

[55] Institute for Work and Health, *Effectiveness of OH&S education and training*, online: http://www.iwh.on.ca/sbe/effectiveness-of-ohs-education-and-training (date accessed: February 28, 2020).

[56] *Ibid.* Despite that statement, providing OH&S information and training helps employers to:

Appendix 1

SAMPLE NEW EMPLOYEE OH&S ORIENTATION CHECKLIST

Employee Name:_____   Department:_____
Employee Number:_____

Checklist Items	Employee's Initials	Supervisor's Initials	Date
1. General Safety Orientation			
2. WHMIS Training			
3. Awareness of the Available MSDSs and How to Access Them			
4. Equipment Safety (as applicable)			
5. Ergonomic Training (as applicable)			
6. First Aid, CPR, and AED Training			
7. Respiratory Protection (as applicable)			
8. Emergency Response Plan			
9. Rescue Procedures			
10. Location of Fire Extinguishers			
11. Location of First Aid Kits			
12. Facility Security			
13. Accident/Incident Reporting Procedure			
14. PPE Training and Fitting (as applicable)			
15. Review of Company OH&S Program Manual			
16. Briefing Company Occupational Health Services			
17. OH&S Leadership and Management: Roles & Responsibilities			
18. Knowledge of Core OH&S Policies			
19. Knowledge of applicable Safe Work Procedures			

Statement of Completion:

The above items have been completed as part of the New Employee Orientation process.

Employee: _____   Supervisor: _____
Date: _____   Date: _____

Copyright D. Dyck, 2002

CHAPTER REFERENCES

Alberta Municipal Health & Safety Association (AMHSA), *Leadership for Safety Excellence* (Nisku, AB: AMHSA, 1995).

Beckett & Hager (2000), as reported in K. Sitzman, "Adult Learners Reframed" (2006) 54:6 AAOHN, 292.

F. Bird & G. Germain, *Loss Control Management: Practical Loss Control Leadership*, revised ed. (Loganville, GA: DNV, 1996).

Bloom's Taxonomy of Measurable Verbs, online: https://www.utica.edu/academic/Assessment/new/Blooms%20Taxonomy%20-%20Best.pdf (date accessed: February 28, 2020).

Bongarde Holdings Inc., "How to Demonstrate the Value of Training Programs" (2006) 2:4 *Safety Compliance Insider* at 14, online: http://www.safetysmart.com (date accessed: February 28, 2020).

Carleton (1987), Scholey (1983) and St. Vincent, Tellier & Lortie (1989), cited in M. Colligan & A. Cohen, "The Role of Training in Promoting Workplace Safety and Health" in J. Barling & M. Frone, eds., *The Psychology of Workplace Safety* (Washington, DC: American Psychological Association, 2004).

Chhokar & Williams (1984), Cohen & Jensen (1984), Fox & Sulzer-Azaroff (1987), Komaki, Barwick & Scott (1978), Ray, Purswell & Schlegel (1990), Reber & Wallin (1984), Saarela (1990) and Saari & Nasanen (1989), cited in M. Colligan & A. Cohen, "The Role of Training in Promoting Workplace Safety and Health" in J. Barling & M. Frone, eds., *The Psychology of Workplace Safety* (Washington, DC: American Psychological Association, 2004).

M. Colligan & A. Cohen, "The Role of Training in Promoting Workplace Safety and Health" in J. Barling & M. Frone, eds., *The Psychology of Workplace Safety* (Washington, DC: American Psychological Association, 2004).

C. DeLisle, "What Makes Safety Training Effective?" *The Safety Net* (2014) at 24-33.

D. Dyck, "New Employee OH&S Orientation Checklist" (Calgary, AB: Progressive Health Consulting, 2002).

D. Dyck, "Worker OH&S Training: Teaching Plan" (Calgary, AB: Progressive Health Consulting, 2003).

D. Dyck & D. Rud, "ENMAX Makes Safety Training Fun" (2003) 26:3 *Occupational Health & Safety Magazine*.

R. Flemmer, "The Three E's of Training" *The Safety Net* (October 2014) at 7.

D. Folk, "The Workplace Complacency Trend in Accident Prevention" *Occupational Health & Safety* (January 2007), online: http://www.ohsonline.com/articles/46371 (date accessed: February 28, 2020).

Fox & Sulzer-Azaroff (1987) and Hopkins (1984) cited in M. Colligan & A.

Cohen, "The Role of Training in Promoting Workplace Safety and Health" in J. Barling & M. Frone, eds., *The Psychology of Workplace Safety* (Washington, DC: American Psychological Association, 2004).

M. Harder, "How effective is health and safety training?" *Safety Reporter* (May 2010).

Hatscan, "WHMIS Training" (Presented in Edmonton, AB, February 26-27, 2007).

Hatscan, *Handi-Guide to Alberta's OH&S Act, Regulation and Code* (Edmonton, AB: Hazard Alert Training (Hatscan), 2004).

Institute for Work and Health, "Study shows way to make health and safety training for office workers more effective" (January 28, 2014), online: http://www.iwh.on.ca/media/2014-jan-28 (date accessed: February 28, 2020).

Institute for Work and Health, *Effectiveness of OH&S education and training*, online: http://www.iwh.on.ca/sbe/effectiveness-of-ohs-education-and-training (date accessed: February 28, 2020).

Leslie & Adams (1973), Rubinsky & Smith (1971), Bosco & Wagner (1988), Vaught *et al.*, (1988) and Goldrick (1989), cited in M. Colligan & A. Cohen, "The Role of Training in Promoting Workplace Safety and Health" in J. Barling & M. Frone, eds., *The Psychology of Workplace Safety* (Washington, DC: American Psychological Association, 2004).

S. Lieb, "Principles of Adult Learning", online: http://www.lindenwood.edu/education/andragogy/andragogy/2011/Lieb_1991.pdf (date accessed: February 28, 2020).

Linneman, Cannon & DeRonde & Lamphear (1991), Lynch *et al.* (1990), Seto, Ching, Chu & Fielding (1990) and Wong *et al.*, (1991) cited in M. Colligan & A. Cohen, "The Role of Training in Promoting Workplace Safety and Health" in J. Barling & M. Frone, eds., *The Psychology of Workplace Safety* (Washington, DC: American Psychological Association, 2004).

Manitoba Labour and Immigration — Workplace Safety and Health Division, "Employing Young Workers: Tips for Employers and Supervisors", online: http://safemanitoba.com. (date accessed: February 28, 2020).

Maples *et al.* (1982), Lepore, Olson & Tomer (1984) and McKenzie *et al.* (1985), cited in M. Colligan & A. Cohen, "The Role of Training in Promoting Workplace Safety and Health" in J. Barling & M. Frone, eds., *The Psychology of Workplace Safety* (Washington, DC: American Psychological Association, 2004).

McQuiston *et al.* (1994), Brown & Nguyen-Scott (1992), LaMontagne *et al.*, (1992), Michaels *et al.*, (1992), Weiger & Lyons (1992) and Cole & Brown (1996), cited in M. Colligan & A. Cohen, "The Role of Training in Promoting Workplace Safety and Health" in J. Barling & M. Frone, eds., *The Psychology of Workplace Safety* (Washington, DC: American Psychological Association, 2004).

National-Louis University, "Malcolm Knowles: Apostle of Andragogy", online: http://www.umsl.edu/~henschkej/henschke/malcolm_knowles_Apostle_of_andragogy.pdf (date accessed: February 28, 2020).

Occupational Health and Safety Code 2006 (revised 2013, 2019) (Edmonton, AB: Government of Alberta — Human Resources and Employment, 2013).

G. Pappas, "Effective Measurement of OSH Training Programs" (Presented at the IAPA Conference in Toronto, ON, April 2006).

D. Parkinson *et al.*, "Effectiveness of the United Steel Workers of America Coke Oven Intervention Program" (1989) 31:5 J. Occup. Med. 464.

Ray, Purswell & Schlegel (1990), *et al.*, cited in M. Colligan & A. Cohen, "The Role of Training in Promoting Workplace Safety and Health" in J. Barling & M. Frone, eds., *The Psychology of Workplace Safety* (Washington, DC: American Psychological Association, 2004).

T. Robins *et al.*, "Implementation of the Federal Hazard Communication Standard: Does Training Work?" (1990) 32:11 J. Occup. Med. 1133.

D. Rud, industry application observed at ENMAX Corporation, Calgary, AB (2003).

K. Saarela, "An Intervention Program Utilizing Small Groups: A Comparative Study" (1990) 21 J. Safety Res. 149.

Saarela *et al.* (1989), Borland *et al.* (1991) and Karmy & Martin (1980), cited in M. Colligan & A. Cohen, "The Role of Training in Promoting Workplace Safety and Health" in J. Barling & M. Frone, eds., *The Psychology of Workplace Safety* (Washington, DC: American Psychological Association, 2004).

J. Saari & M. Nasanen, "The Effect of Positive Feedback on Industrial Housekeeping and Accidents: A Long-term Study at a Shipyard" (1989) 4 Int. J. Ind. Ergon. 201.

Sulzer-Azaroff (1990) and Lynch *et al.* (1990), cited in M. Colligan & A. Cohen, "The Role of Training in Promoting Workplace Safety and Health" in J. Barling & M. Frone, eds., *The Psychology of Workplace Safety* (Washington, DC: American Psychological Association, 2004).

University of Kansas (1982), Fox, Hopkins & Anger (1987) and Zohar & Fussfield (1981), cited in M. Colligan & A. Cohen, "The Role of Training in Promoting Workplace Safety and Health" in J. Barling & M. Frone, *The Psychology of Workplace Safety* (Washington, DC: American Psychological Association, 2004).

Work Safe Alberta Workplace Health and Safety Interactive Quizzes, online: http://work.alberta.ca/occupational-health-safety/268.html (date accessed: February 28, 2020).

Chapter 18

OCCUPATIONAL HEALTH AND SAFETY: BEST PRACTICES

A. OH&S BEST PRACTICES: DEFINED

Best practices are a form of benchmarking that result from direct observation of clinical practices.[1,2] They are concrete solutions that work best when linked to existing problems or even to specific crises within a workplace. They are based on real examples and can be used to gradually promote system improvement. Best practices can serve as guidelines for practice and measurement of outcomes. However, changes in technology, knowledge, and practice advancements can alter any best practice. This means that benchmarks and guidelines must undergo frequent reviews and updates to remain current and credible.

B. OH&S BEST PRACTICES

The intent of this chapter is to summarize the 30 current best practices in the field of Occupational Health and Safety (OH&S). Discussion of each OH&S topic will be followed by a list of the relevant best practices.

1. Demonstrate Strong OH&S Leadership, Commitment, and Passion

For an effective OH&S system, strong OH&S leadership, commitment, and a passion for OH&S excellence are required.[3] **Commitment** can be defined as a personal thing that cannot be reduced to a formula or policy. It is a passionate "buy-in" and ownership of a belief/concept/practice.

By demonstrating a passion for and leadership that expects OH&S excellence,

---

[1] H. Bruckman & J. Harris, "Occupational Medicine Practice Guidelines" (1998) 13 Occupational Medicine: State of the Art Reviews 679.

[2] Business Dictionary, "Best Practice" (2020), online: http://www.businessdictionary.com/definition/best-practice.html (date accessed February 28, 2020).

[3] See G. Lowe & J. Wetherow, "High Trust Cultures = Health and Performance" (Presented at the Industrial Accident Prevention Association Conference, Toronto, ON, April 2007); R. Kelbert & D. Strand, "Creating a World-class Safety Culture" (Presented at the Industrial Accident Prevention Association Conference, Toronto, ON, April 2005); and AWCBC, "Knowledge Transfer for Workplace Health & Safety: An Onsite Report on the Public Forum" (Developed at the Association of Workers' Compensation Boards of Canada (AWCBC), November 19 and 20, 2001, Toronto, ON) at 4.

Management can create a strong and effective OH&S (Safety) culture, and organizational culture. In essence, "[W]hat interests the boss, fascinates the workers." The noted benefits are that "high-effectiveness companies" are associated with superior human, business, and financial business outcomes and results.[4],[5]

BEST PRACTICES

1. Begin by viewing workplace health and safety as a "core value".

2. Encourage organization/company leaders to be passionate about their individual commitments to workplace health and safety - physical and psychological health and safety.[6] Have them talk and demonstrate their interest and passion in workplace health and safety. Measure leaders on their demonstrated OH&S performance.[7] Propagate the view that "[i]f you are a poor manager of health and safety performance, you cannot be performing well at anything else", and that "[g]ood health and safety practices 'add' to the bottom line". Coming from Senior Management, these messages have a major impact.

3. Support the OH&S initiatives and research work that your industry association undertakes to enhance worker safety.[8]

4. Have a joint labour-management group serve as a steering committee for the OH&S Program. This committee would be the foundation of the OH&S Program. Its function is to advise and consult with management and union leaders, to evaluate concerns and to receive advice and suggestions for the OH&S Program from various stakeholders in the organization/company. The committee also receives reports on the OH&S Program, which assess the program's overall effectiveness. The information provided to the committee and to management is population, or aggregate data so that individual clients cannot be identified. Ideally, the steering committee membership has representatives from management and labour, and the chairperson is elected by the members for a set term.

5. OH&S responsibility should be positioned as high up in the organization/

---

[4] Towers Watson, *2013-2014 Staying@Work Report*, Canada Summary (February 2014), online: http://www.towerswatson.com/en/Insights/IC-Types/Survey-Research-Results/2014/02/2013-2014-staying-at-work-report-canada-summary (date accessed: January 31, 2020).

[5] For more details, refer to Chapter 21 "Occupational Health and Safety: Prevention of Workplace Illness and Injury".

[6] See footnote 2, above.

[7] R. Kelbert & D. Strand, "Creating a World-class Safety Culture" (Presented at the Industrial Accident Prevention Association Conference, Toronto, ON, April 2005).

[8] AWCBC, "Knowledge Transfer for Workplace Health & Safety: An Onsite Report on the Public Forum" (Developed at the Association of Workers' Compensation Boards of Canada (AWCBC), November 19 and 20, 2001, Toronto, ON) at 4.

CH. 18: OCCUPATIONAL HEALTH AND SAFETY: BEST PRACTICES

company chain of command as possible - preferably residing with the Chief Executive Officer (CEO).

6. Ensure that adequate resources (*e.g.*, Senior Management support, competent OH&S personnel, and budget) are available to the OH&S steering committee, as well as to the OH&S Program.

7. Conduct a comprehensive needs-analysis to identify specific organizational/company OH&S needs and establish baseline data. This should include an assessment of labour/management attitudes towards OH&S practices, the identification of the organization/company, or organization's/company's health and safety profile, and acknowledgement of the types of OH&S support available to the organization/company operations and employees.

8. Examine all related workplace policies and procedures in terms of their impact on the OH&S Program structure, processes, and outcomes.

9. Review and revise, where necessary, the current OH&S Program and related policies, standards, and procedures.

10. Identify and communicate the OH&S roles, responsibilities, and accountabilities of all the major stakeholders.

11. Define the available OH&S services and products.

12. Identify the current workplace hazard management process and outcomes.

13. Review and revise, as appropriate, the current OH&S strategies.

14. Develop and maintain an OH&S database.

15. Use OH&S outcome data, along with other Human Resources and employee group benefit plan outcomes, to identify and gain an understanding of OH&S issues. Target OH&S Program improvement based on OH&S results.

16. Strive for sustainability of all the above practices. Sustainability is achieved when an OH&S practice permeates all activities and processes, and when compliance is linked to manager performance and rewards. Refer to Chapter 42, "Future Challenges in Occupational Health and Safety" for details on the HS3™ Sustainability Model.

17. Implement an effective OH&S Management System in which Management ensures that:

- an OH&S philosophy, committed to by all, exists;
- passion and support for OH&S is visibly demonstrated and sustained;
- physical and psychological health and safety are of prime importance;
- Management "walks their talk";

- a belief that all incidents are preventable, exists;
- an integrated, multi-disciplinary approach is used;
- a corporate-wide OH&S framework exists;
- written policies and procedures are endorsed and supported;
- a system that ensures OH&S responsibility and accountability by all parties, exists;
- OH&S Practitioners/Professionals are in advisory capacities to all levels of management;
- OH&S Practitioners/Professionals provide technical expertise and support;
- organization/company OH&S standards go beyond "minimum standards" and are driven by dissatisfaction for the *status quo*;
- OH&S is integrated into business strategies and actions;
- a focus on the leading indicators of safety is promoted;
- OH&S data collection for analysis and evaluation occurs;
- injury management and graduated return-to-work opportunities are implemented; and
- multi-disciplinary interventions - OH&S ergonomics; Safety Engineering; toxicology; Human Resources; Employee Assistance Programs; Workplace Wellness; or medical, vocational, or occupational rehabilitation professionals - are used.

2. Strive to Develop a High-performance Work System

A high-performance work system assumes that workers can perform at high levels if they are encouraged, enabled, and permitted to do so. It proposes that workers are more committed to the organization/company and more trusting of management if they are treated with respect and viewed as being capable and intelligent individuals. The premise is that, companies using this type of human resource approach benefit in terms of better performance and greater profitability.[9]

The Great Place to Work® Institute of Canada proposes that great workplaces have strong commitment from the CEO and Senior Management; a genuine belief that people are indispensable to the success of the business; active communication between Management and employees; and the perception that they are unique and "not like the others".[10]

---

[9] A. Zacharatos & J. Barling, "High-performance Work Systems and Occupational Safety" in J. Barling & M. Frone, eds., *The Psychology of Workplace Health and Safety* (Washington, DC: American Psychological Association, 2004) at 203.

[10] G. Lowe & J. Wetherow, "High Trust Cultures = Health and Performance" (Presented

In terms of workplace health and safety, a high-performance approach aligns well with the concept of the Internal Responsibility System. It emphasizes the role of Management in promoting a safe and healthy workplace through the use of techniques, that increase worker trust in Management, that strengthen worker commitment to the organization/company and that enhance worker perceptions of a strong **OH&S (safety) work climate** — the state of a work system in terms of the perceptions of the current environment or prevailing conditions that impact safety.[11]

High employee engagement (an 80 per cent average engagement score) is associated with:

- employee support for productivity improvements;
- employee willingness to help out and make "trade-offs" to ensure success; and
- employee trust and confidence in the leaders' abilities.[12]

BEST PRACTICES

1. According to Zacharatos and Barling (2004), the top ten ways to create high-performance work systems and OH&S cultures are to:[13]

    - *Ensure employment security* for employees.
    - *Selectively hire all personnel*. Companies that have elaborate selection procedures experience lower incident and injury rates.[14]
    - *Provide extensive worker training* both in terms of the work to be done and OH&S. Education and training is crucial to ensuring that an organization/company has competent workers. Employees who have undergone extensive OH&S training experience fewer injuries than their untrained counterparts.[15]
    - *Use decentralized decision-making and self-managed teams*. The use of teams has the potential to enhance OH&S because positive team dynamics encourage responsibility for workplace health and

---

at the Industrial Accident Prevention Association Conference, Toronto, ON, April 2007).

[11] R. Lardner, M. Fleming & P. Joyner, "Towards a Mature Safety Culture" *Symposium Series No. 148* (The Keil Centre Ltd., 2001), online: http://www.keilcentre.co.uk (date accessed: January 31, 2020).

[12] N. Crawford & T. Mathers, "In Good Times and Bad" *Benefits Canada* (January 2010) at 21-23.

[13] A. Zacharatos & J. Barling, "High-performance Work Systems and Occupational Safety" in J. Barling & M. Frone, eds., *The Psychology of Workplace Health and Safety* (Washington, DC: American Psychological Association, 2004) at 203.

[14] A. Cohen, "Factors in Successful Occupational Safety Programs" (1977) 9 J. Safety Res. 168.

[15] A. Hale, "Is Safety Training Worthwhile?" (1984) 6 J. Occup. Accid. 17.

safety; ensure the sharing of ideas; enable greater control over work; and provide feedback and encouragement for working safely.[16]

- **Reduce the number of status symbols that separate workers into hierarchical levels.** Status distinctions set up barriers to open communication and promote feelings of resentment, which hinders information-sharing, the development of trust, and a mutual concern for safety issues.[17]

- **Increase the amount and quality of information-sharing** that occurs within the organization/company. Information-sharing allows employees to learn vicariously about their work and can reduce near-miss, injury, and fatality incidents.[18]

- **Provide compensation commensurable with the organization's/company's OH&S performance.**

- **Require the use of transformational leadership.** Transformational leadership provides an appropriate leadership model for:

    - demonstrating commitment to OH&S;
    - demonstrating sound safety behavioural practice;
    - "doing the right things for the right reasons";
    - inspiring others to excel;
    - promoting team work;
    - encouraging workers to "think outside the box" and challenge assumptions to come up with new solutions; and
    - demonstrating concern for the needs and interests of workers.

    Transformational leadership is associated with improved workplace health and safety; employee willingness to take the initiative on OH&S matters; the extent to which workers follow safety procedures and demonstrate good health and safety behaviours; and fewer workplace injuries.[19] More information on transformational leadership is provided in Chapter 1, "Occupational Health and Safety: Historical Perspectives", Chapter 7, "Safe Workplace = Great Workplace: Building a Sustainable Culture of Safety", and Chapter

---

[16] M. Colligan & A. Cohen, "The Role of Training in Promoting Workplace Safety and Health" in J. Barling & M. Frone, eds., *The Psychology of Workplace Health and Safety* (Washington, DC: American Psychological Association, 2004) at 223.

[17] *Ibid.*

[18] *Ibid.*

[19] *Ibid.*, at 212-13.

24, "Psychological Health and Safety: Practical Application in the Workplace".

- *Enable high-quality work practices* by assigning appropriate workloads, by encouraging great work autonomy and control, and by providing role clarity for employees (Appendix 1). All three conditions are associated with fewer workplace injuries.[20]

- *Measure the variables that are the most critical to organizational/ company success.* Measuring leading indicators of safety performance such as workplace health and safety conditions, employee attitudes toward safety, and demonstrated safety behaviours is more predictive of subsequent safety performance than measuring the lagging indicators of safety performance (past injury/illness results).[21]

Figure 18.1: Relationship between High-performance Work Practices and OH&S Performance[22]

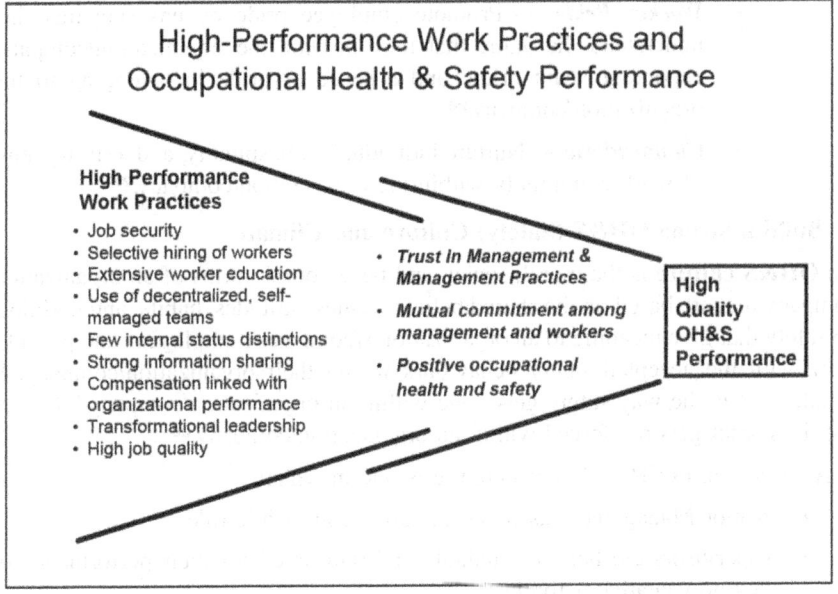

---

[20] *Ibid.*, at 213.

[21] *Ibid.*, at 214.

[22] Adapted from A. Zacharatos & J. Barling, "High-performance Work Systems and Occupational Safety" in J. Barling & M. Frone, eds., *The Psychology of Workplace Health and Safety* (Washington, DC: American Psychological Association, 2004).

2. Develop a workplace that promotes:
   - **Credibility** — This aspect encompasses open dialogue among stakeholders within the organization/company; worker competence and effective use of employees and resources; and integrity — both in terms of the managers and the organization/company modelling what they profess to believe.[23]
   - **Respect** — Make provisions for employees to receive the necessary resources and training to succeed at the job; encourage collaboration between Management and employees; and build a healthy work-life balance that enables employees to do well in these aspects of life.[24]
   - **Fairness** — Ensure that rewards are fairly distributed across the organization/company; maintain impartiality when hiring and promoting employees; and uphold "just" work decisions and non-discriminatory practices.[25]
   - **Worker Pride** — Promote employee pride by ensuring that they understand the value that they provide the organization/company; encourage team pride; and a sense of pride in belonging to this organization/company.[26]
   - **Camaraderie** — Nurture individuality, hospitality, and a strong sense of work community within the organization/company.[27]

3. Build a Strong OH&S (Safety) Culture and Climate

An **OH&S culture** is the moral, social, and behavioural norms of an organization/company that are based on the shared beliefs, values, attitudes, habits, and traditions on safety that give meaning to an organization's/company's employees and provides them with the accepted safety behaviours within their organization/company. In essence, "it is the way safety gets done within an organization/company". Simply put, it is what gets reinforced within an organization/company.[28]

A world-class OH&S (Safety) culture is one in which:
- Senior Management assumes an active and visible role;
- supervisors are held accountable and reinforced for their performance on key preventative activities;

---

[23] G. Lowe & J. Wetherow, "High Trust Cultures = Health and Performance" (Presented at the Industrial Accident Prevention Association Conference, Toronto, ON, April 2007).

[24] *Ibid.*

[25] *Ibid.*

[26] *Ibid.*

[27] *Ibid.*

[28] R. Kelbert & D. Strand, "Creating a World-class Safety Culture" (Presented at the Industrial Accident Prevention Association Conference, Toronto, ON, April 2005).

- the focus on OH&S is rigorous; and
- employees reinforce each other for demonstrating a safe work attitude and for practising safe work practices.[29]

The benefit of developing a strong workplace culture is that organizations/companies with strong corporate cultures function collaboratively and demonstrate strong corporate earnings. According to a Canadian study (2010), companies with the most-admired cultures outperformed the Toronto Stock Exchange (TSX 60) by an average of almost 600 per cent.[30]

An example of a strong OH&S culture exists at Suncor Energy Inc., formerly Petro-Canada Inc. They adopted a Zero-Harm Culture with foundation behaviours that include honesty, integrity, respect, trustworthiness, and ethical decision-making.

Culture, like a relationship, needs to be nurtured, reinforced, and shaped to suit the organization/company. This means a systematic and coordinated effort by Senior Management to promote the desired OH&S culture and workplace climate. Organizations/companies are advised to integrate the elements of the OH&S culture into the way business is done within the organization/company. This often means developing a framework that graphically depicts what is being proposed. An industry example of a graphic depiction of a strong OH&S culture is provided in Figure 18.2: Blueprints for Petro-Canada's (Suncor's) Zero-Harm Culture.[31]

BEST PRACTICES

1. Use a gap-analysis approach to determine the gaps that exist between the current and desired states of the organization's/company's OH&S culture, and suitable strategies to reduce or eliminate those gaps.
2. Determine the elements of the desired OH&S culture.
3. Design a framework depicting the elements of the OH&S culture, along with corporate standards of practice that support the desired OH&S culture (Figure 18.3).
4. Involve employees and union leaders in the design of the culture model.[32]
5. Develop an action plan for the achievement of the desired OH&S culture.

---

[29] *Ibid.*

[30] S. Klie, "Culture drives performance" Canadian HR Reporter (December 13, 2010) at 1 and 15.

[31] Petro-Canada, "Blueprint for Petro-Canada's Zero-Harm Culture" (July 2003). Reprinted with permission from Petro-Canada, Calgary, AB.

[32] R. Kelbert & D. Strand, "Creating a World-class Safety Culture" (Presented at the Industrial Accident Prevention Association Conference, Toronto, ON, April 2005).

Figure 18.2: Industry Example of a Strong OH&S Culture: Blueprints for Petro-Canada's Zero-Harm Culture

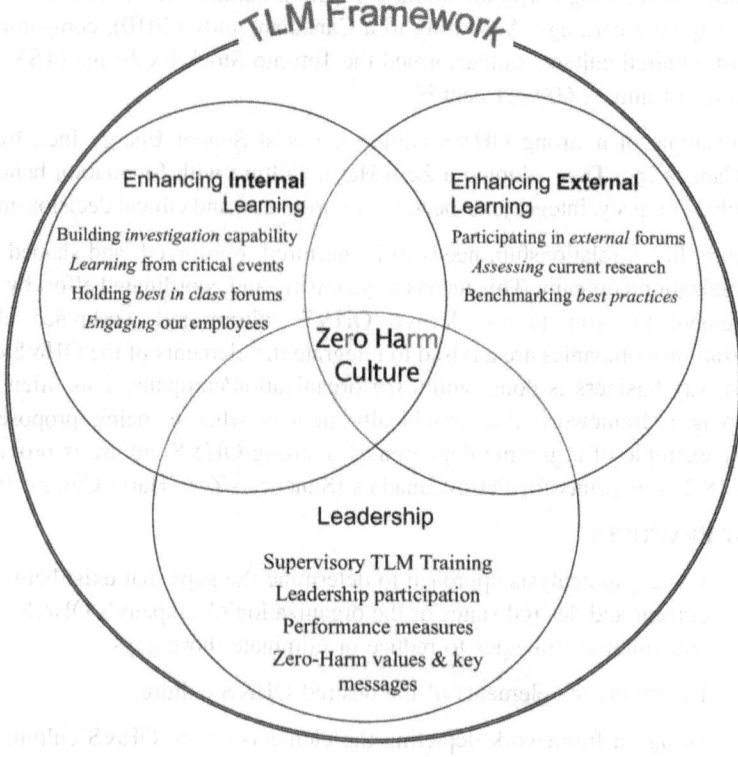

6. Ensure the selected culture model is flexible enough to accommodate individual worksite cultures.[33]
7. Regularly communicate Management messages on the importance of workplace health and safety, thereby reinforcing the OH&S culture.
8. Evaluate the OH&S culture regularly, making improvements as necessary.
9. Allot adequate time for the OH&S culture to evolve.
10. Monitor the OH&S cultural research for innovative ways to enhance the

---

[33] *Ibid.*

organization's/company's approach to culture building.

Figure 18.3: Petro-Canada's Total Loss Management (TLM) Framework[34]

Total Loss Management is a systematic approach to the management of loss, integrating reliability and quality with the reduction of risk, to people, the environment, and assets and production.

The TLM Framework is a model that demonstrates the integrated management functions that are required to develop the corporate culture desired by Petro-Canada. It is a culture that believes in the ultimate goal of "zero harm" to employees, others at our facilities, the public, the natural environment, and company assets and property.

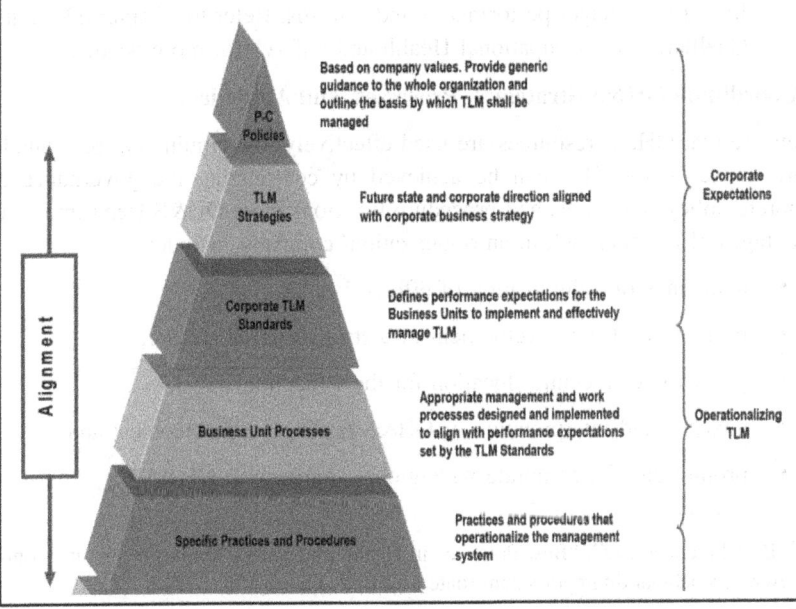

[34] Petro-Canada (now Suncor Energy Inc.), "Total Loss Management (TLM): Corporate Standards" (January 25, 2005, Calgary, AB) at 1. Reprinted with permission from Petro-Canada.

4. Incorporate Occupational Health and Safety into Corporate Business Strategies and Planning

In many companies, OH&S functions are left up to the organization's/ company's OH&S department. This approach is contra to the concept, and functioning, of an Internal Responsibility System for workplace health and safety. OH&S needs to be part of business goals, strategies, planning, and operations. OH&S is a Management and employee function and responsibility — each within their own respective scopes and level of responsibilities. An industry example is provided in Figure 18.4.

BEST PRACTICES

- Include OH&S technical expertise in business planning sessions, especially the project planning and design.
- Ensure OH&S practices are considered and factored into business goals, strategies, planning, and all appropriate business processes.
- Strive for sustainability: **Sustainability** is achieved when an OH&S practice permeates all activities and processes, and when compliance is linked to manager performance and rewards. Refer to Chapter 42, "Future Challenges in Occupational Health and Safety" for more details.

5. Coordinate OH&S Strategies, Processes, and Activities

To ensure that OH&S resources are used effectively, the elimination of redundant efforts is necessary. This can be achieved by centralizing the governance and corporate policy, standards, and procedure functions in an OH&S Department. The advantages of the approach to an organization/ company include:

- avoidance of a duplication of efforts;
- promotion of the development of a strong OH&S culture;
- provision of a central location for the collection;
- analysis and interpretation of OH&S findings and outcomes; and
- promotion of a corporate-wide/global approach to OH&S[35]

---

[35] Best Practices LLC, "Best Practices in Health and Safety Management" at 2, online: http://www.benchmarkingreports.com (date accessed: January 31, 2020).

[36] Petro-Canada (now Suncor Energy Inc.), "Total Loss Management (TLM): Corporate Standards" (January 25, 2005, Calgary, AB) at 6. Reprinted with permission from Petro-Canada.

Figure 18.4: Petro-Canada's Total Loss Management (TLM) Standards[36]

1.4 CORPORATE TLM STANDARDS

Each Organizational Unit within Petro-Canada is responsible for managing the impact of its activities and products on people, the environment, property, and corporate assets. In order to integrate TLM fully into our company's culture, this must include business units, service units, and corporate groups, who through their activities can impact the risks to which the company is exposed. This is accomplished by:

- developing a TLM system that is aligned with business strategies and plans
- understanding the risks associated with activities and products
- implementing appropriate processes, practices, and procedures to safely manage the risks
- regularly reporting performance against defined objectives and specific performance measures
- seeking input and feedback from stakeholders
- auditing the integrity and effectiveness of its TLM system
- identifying opportunities for continual improvement

To guide Organizational Units through this process, the company has developed TLM Standards in 10 Elements or subject categories. The 10 Elements are:

- **Element 1: Leadership**
- **Element 2: Health & Safety**
- **Element 3: Physical Asset System Integrity & Reliability**
- **Element 4: Contractor Management**
- **Element 5: Environmental Management Systems**
- **Element 6: Employee Practices, Capability & Development**
- **Element 7: Audits & Inspections**
- **Element 8: Stakeholder Relations**
- **Element 9: Security & Emergency Preparedness**
- **Element 10: Event Management**

In order to meet the standards within the Elements, it is expected that the Organizational Unit will have documented evidence of its activities. This is primarily to ensure that the processes are being managed properly and also to ensure that we have sufficient documentation to demonstrate due diligence over our operations.

1.5 ORGANIZATIONAL UNIT PROCESSES

The implementation of the TLM Standards is consistent with the Plan-Do-Check-Act management process cycle. By developing management processes to meet these standards, each Organizational Unit will enhance performance and continually improve through effective and productive business methods.

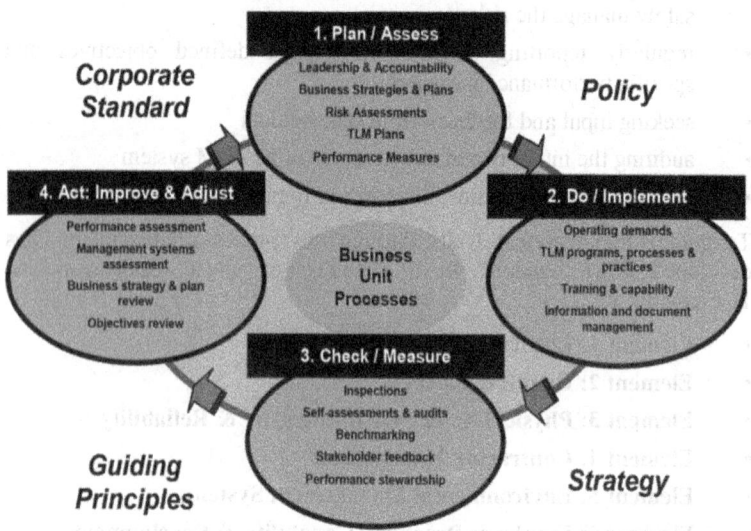

The PDCA cycle provides a dynamic scientific process of acquiring knowledge. Knowledge makes predicting the future more reliable. *If you cannot measure something, you cannot manage it. If you cannot manage it, you cannot control the outcome and the future cannot be predicted.* The result is undesired loss (*i.e.*, "incidents") and inefficiency. By gathering information, a manager can identify opportunities for continuous improvement, can use the information to plan steps for that improvement, and can measure progress.

BEST PRACTICES

1. With small and mid-sized companies, assign the oversight of the OH&S function and program to one person. For larger companies, this would translate to having a corporate OH&S Department with a Director of OH&S at the helm.[37] This approach can include decentralized OH&S roles and functions.

2. Ensure the effective functioning of the OH&S Program by having one central figure oversee the daily operations of the program. This is typically a Director/Manager of OH&S.

3. Position the OH&S Department or representative so that they can effectively drive corporate-wide OH&S strategies, and provide technical expertise on OH&S issues. Advanced companies use OH&S expertise to:

    - create the organization/company OH&S philosophy, policy, and strategy;
    - communicate specific OH&S standards;
    - design OH&S training modules;
    - use technology to make OH&S information accessible; and
    - advise business units.[38]

4. Develop an OH&S Program that supports the management of OH&S using five steps (Figure 18.5).

5. Standardize and coordinate key OH&S strategies, processes, and activities, while affording individual business units the ability to address their unique OH&S differences and needs.

6. Develop annual OH&S Action Plans that identify goals, objectives, strategies, and accountability for action (Appendix 1).

7. Track and monitor all the organization/company OH&S processes, activities, and outcomes.

8. Promote the sharing of intra-departmental learning as to OH&S processes and activities that work/do not work.

9. Research and disseminate "OH&S best practices" corporate-wide.

10. Monitor changes in legislative requirements and industry practices with a view to continuous OH&S learning and enhancement of organization/company practices.

---

[37] J. Griffin, *Ucononline Underground Construction* (April 2006) at 25.

[38] Best Practices LLC, "Best Practices in Health and Safety Management", online: http://www.benchmarkingreports.com at 2 (date accessed: January 31, 2020).

Figure 18.5: Managing Occupational Health and Safety[39]

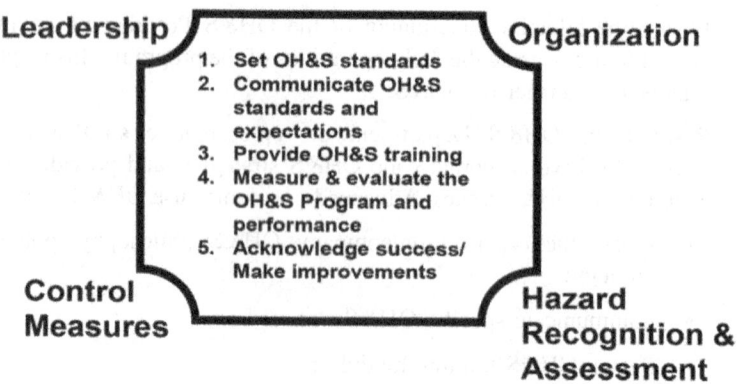

6. Create OH&S Requirements

Under Canadian and U.S., OH&S legislation, companies are required to develop an **Internal Responsibility System** for ensuring high-quality workplace health and safety. An Internal Responsibility System is the "people framework" within an organization/company. It maintains that:

- everyone within an organization/company is responsible for workplace health and safety;
- everyone must do their part to identify and solve health and safety issues/problems, and to look for ways to improve the work processes with which he/she is involved; and
- by capturing the creativity, leadership, experience, and knowledge of employees, companies can improve workplace health and safety.

When examined further, the term *Internal* means "within the workplace" and "responsibility is within the job". *Responsibility* refers to the personal duty of care that everyone has to ensure workplace health and safety. Workers have three rights:

1. the right to know about workplace hazards;
2. the right to refuse unsafe work; and

---

[39] Reprinted with permission from R. Burton, "Managing Health and Safety" slide used in "WSIB Ontario Workwell Program: Work Well Auditing for Prevention" (Presented at IAPA Conference, April 23, 2004, Toronto, ON).

3. the right to participate in health and safety decision-making activities directly or through representatives.

The term *System* indicates that the workplace is a closed system with individual parts that all work together for the good of workers and the whole organization/company. The purpose of the system is to eliminate/reduce workplace injuries/illness. Everyone operating within the Internal Responsibility System must take reasonable precautions to prevent harm to themselves and others. This is also known as due diligence.

BEST PRACTICES

1. Given the above duty of care and responsibility, companies are obliged to develop and communicate to stakeholders the OH&S standards they wish upheld. This is typically operationalized as the OH&S policy, OH&S standards, work rules, and procedures. It also means monitoring and measuring the implementation of such.

2. Develop a tracking mechanism that can provide evidence of corporate OH&S duty of care and due diligence. An industry example is the data collection that was possible to do as part of the Great Safety Performance model, Chapter 6, "Occupational Health and Safety: Leadership and Commitment". By virtue of the model's evaluation process, an organization/company can demonstrate its commitment and efforts to uphold its OH&S duty of care.

7. Create and Communicate OH&S Policies, Procedures, and Practice/Performance Standards

The intent of OH&S policies and procedures is to ensure that processes are in place and are applied fairly, equally, and consistently.[40] They should align with the organization's/company's vision, mission, and values; and the OH&S cultural framework. Their purpose is to control the way in which work is done. Procedures are defined actions that serve to standardize the OH&S Program; practice/performance standards are stated approaches to practise/perform based on recognized and accepted principles of industry practice for planned processes. They form the guidelines and rules for work practices, provide the boundaries for work activities, clarify stakeholder roles and responsibilities, and serve as a benchmark.

Once developed, the OH&S policies and procedures should be regularly communicated and enforced.

BEST PRACTICES

1. Develop and implement policies and procedures that deal with:

---

[40] Canadian Centre on Disability Studies, *Best Practices in Contemporary Occupational Health & Safety: Executive Summary, 1998*, at 8, online: http://www.disabilitystudies.ca (date accessed: January 31, 2020).

- OH&S leadership and commitment;
- the Corporate Code of Conduct;
- the respectful workplace policy;
- the substance abuse policy;
- the fit-for-duty policy;
- the anti-harassment, anti-bullying, anti-workplace violence policy;
- the non-smoking policy;
- OH&S policy, practices, and procedures;
- the personal protective equipment policy;
- safety goggle use;
- the confidentiality standard of practice;
- the documentation standard of practice;
- the Employee Assistance Program policy;
- the Disability Management Policy and Program;
- the Return-to-work Program;
- the fitness-to-work policy;
- disability leaves (*i.e.*, short-term disability, Workers' Compensation Board, long-term disability, *etc.*); and
- rehabilitation measures (*i.e.*, case/claim management, vocational rehabilitation, *etc.*)

2. Regularly review and update these policies, procedures, and practice/performance standards to assess their continued applicability.

8. Protect the Confidentiality of Employee Personal Health Information

Employee personal health and health-related data are obtained during the course of injury incident investigations, and disability claims and case management. Policies and procedures are required to deal with confidentiality and access to medical files, as well as with the retention, storage, transfer, or disposal of OH&S data.

BEST PRACTICES

1. Develop a policy that deals with the confidential management of employee personal health data. This must comply with the applicable legislation.
2. Develop a protocol for the retention, maintenance, disclosure, and disposal of personal health documentation.
3. Retain all employee personal health data in a secure and confidential manner with access by authorized personnel and only on a "need-to-know" basis.

4. Retain all employee personal health documentation for a minimum period of ten years from employee termination.[41] The exception is the legislative requirement to retain worker chemical/physical exposure data for a latent period, approximately 30 years from the last exposure date.[42] In some cases, that will be from the date of the employee's exit from the organization/company. This time frame may vary by legal jurisdiction, but these exposure records must be retained due to the latency factor in the development of exposure-related diseases.

5. Limit the dissemination of medical diagnoses to broad categories or neutral descriptors, such as occupational injury, occupational illness, non-occupational injury, or non-occupational illness. When diagnoses are used, limit them to disease classifications, or aggregate the data, so individual diagnoses cannot be determined.

9. Clarify OH&S Ownership and Responsibilities

OH&S roles, responsibilities, and accountabilities must be defined and communicated. In most Canadian and U.S. jurisdictions, employers are legally required to clarify the OH&S roles of the employees, contractors, and organization/company.

BEST PRACTICES

1. Communicate to the Board of Directors and Senior Management Team the Internal Responsibility System and their corporate duties and liabilities.
2. Promote regular reviews (at least twice a year) of the functioning and outcomes of the OH&S Program and the quality of the OHSMS by the Board of Directors and Senior Management Team.
3. Provide workshops/seminars/training on manager/employee/contractor OH&S duties, responsibilities, and accountabilities.
4. Regularly remind all stakeholders of their OH&S duties, responsibilities, and accountabilities.
5. Document the above education and communication delivered to all levels of Management and workers.
6. Evaluate the awareness and knowledge levels of the OH&S roles, responsibilities, and accountabilities of the stakeholders.

10. Promote Transformational Leadership

Transformational leadership is inspired leadership that influences the beliefs,

---

[41] Human Resources Management, *Canadian Retention Rules Summary* (2017), online: https://apps.neb-one.gc.ca/REGDOCS/File/Download/624105 (date accessed February 28, 2020).

[42] For example, the Alberta *Occupational Health and Safety Code, 2018* states in Part 4, s. 40(4) that the physician must ensure that records of the (worker) health assessment must be kept for no less than 30 years. See online: http://www.qp.alberta.ca/documents/OHS/OHS.pdf (date accessed: January 31, 2020).

values, and goals of followers so that they can perform in an extraordinary manner.[43] Leaders motivate followers through:

- providing a clear vision;
- their charisma;
- using symbolism;
- empowerment;
- intellectual stimulation; and
- integrity.

Transformational leadership has been associated with improved safety outcomes.[44] The degree to which transformational leadership is practised has been found predictive of the extent to which teams followed safety procedures and participated in safety behaviours above and beyond what was required.[45]

To promote the demonstration of transformational leadership behaviours by supervisory staff, educate them on what it takes to be a transformational leader, the advantages of transformational leadership, and then reward them for using this leadership style.[46] Refer to Chapter 1, "Occupational Health and Safety: Historical Perspectives" and Chapter 7, "Safe Workplace = Great Workplace: Building a Sustainable Culture of Safety", for additional details on the behaviours of transformational leaders.

BEST PRACTICES

1. Provide educational sessions addressing Line Management's OH&S roles, responsibilities, and accountabilities.

2. Explain the concept of transformational leadership and the practices of a transformational leader.

3. Discuss OH&S due diligence and the role of the transformational leader.

---

[43] Changing Minds, "Transformational Leadership", online: http://changingminds.org/disciplines/leadership/styles/transformational_leadership.htm (date accessed: January 31, 2020).

[44] A. O'Dea & R. Flin, "Safety Leadership in the Offshore Oil and Gas Industry" (Paper presented at the Academy of Management Annual Meeting, Toronto, ON).

[45] H. Williams, N. Turner & S. Parker, "The Compensatory Role of Transformational Leadership in Promoting Safety Behaviours" (Paper presented at the Academy of Management Annual Meeting, Toronto, ON); and J. Barling, C. Loughlin & E. Kelloway, "Development and Test of a Model Linking Safety-specific Transformational Leadership and Occupational Safety" (2002) 87:3 J. App. Psychol. 488.

[46] A. Zacharatos & J. Barling, "High Performance Work Systems and Occupational Safety" in J. Barling & M. Frone, eds., *The Psychology of Workplace Safety* (Washington, DC: American Psychological Association, 2004), 203 at 212-213.

11. Invest in Educating Upper and Line Management on Desired Management Practices

Workers involved in high-hazard work who have a supportive Line Manager experience fewer work-related incidents and injuries.[47] Companies that improve the way Line Management responds to employee health-related OH&S concerns can produce significant and sustainable reductions (nearly 50 per cent) in the number and severity of injury claims and disability-related costs.

Educated to respond, communicate, and problem-solve with employees, the workgroups led by these Line Managers were found to have fewer injury claims and disability-related costs.[48] The management-training program that was provided, focused on effective employee communication and problem-solving to help to get injured employees back to work. This training was made available to both Upper and Line Management and was found to be dramatic and sustainable.[49] Likewise, how a Line Manager responds to employee reports of a work injury, influences the speed with which the injured employee returns to work.[50]

BEST PRACTICES

1. Provide educational sessions on Line Management's OH&S roles, responsibilities, and accountabilities.
2. Focus the educational sessions to address modes of effective communication with workers; use of problem-solving skills; and effective response techniques for the management of employee OH&S concerns and reports of work injuries.
3. Incorporate Psychological Health and Safety[51] into this education.
4. Use interactive and role-playing sessions to enable Line Managers to apply the concepts provided.

---

[47] B. Yanar, M. Lay & PM Smith, "The interplay between supervisor safety support and occupational health and safety vulnerability on work injury" *Safety and Health at Work* (2018), online: doi:10.1016/j.shaw.2018.11.001.

[48] Liberty Mutual, News Release, "Reduce Future Disability Claims and Costs by nearly 50% by Improving Supervisor Response to Worker Injury and Illness" (March 6, 2006), online: http://www.libertymutualgroup.com (date accessed January 31,, 2020).

[49] Liberty Mutual, "Controlled Case Study of Supervisor Training to Optimize Injury Response in the Food Processing Industry" (2006) 26:2 WORK: A Journal of Prevention, Assessment & Rehabilitation 107.

[50] Liberty Mutual, News Release, "Reduce Future Disability Claims and Costs by nearly 50% by Improving Supervisor Response to Worker Injury and Illness" (March 6, 2006), online: http://www.libertymutualgroup.com (date accessed: January 31, 2020).

[51] UFred, *Psychological Health and Safety Certificate* (2020), online: http://www.ufred.ca/programs/school-of-occupational-health-and-safety/centre-for-psychological-health-sciences/psychological-health-and-safety-in-the-workplace-basic-level-course/ (date accessed: January 31, 2020).

5. Use mentorship as a means to coach Line Managers to use the presented concepts.
6. Offer refresher sessions or Line Manager networking sessions as a means to reinforce the learning.

12. Coordinate OH&S Staffing and OH&S Budgets

To operate successfully, the organization's/company's OH&S function must be planned, directed, coordinated, operated, and controlled.[52] In short, it must be resourced and managed.

BEST PRACTICES

1. This means:
    - having competent staff and adequate budgets;
    - developing goals and targets for the OH&S function;
    - establishing practice standards;
    - setting performance expectations; and
    - monitoring performance.
2. The practice of hiring competent OH&S Practitioners/Professionals is critical. In Canada, the Board of Canadian Registered Safety Professionals (BCRSP) is the credentialling body for Safety professionals. Its mission statement is:

    > The Board of Canadian Registered Safety Professionals (BCRSP) sets the standard for the certification of occupational health and safety professionals.[53]

With ever-increasing complexity of hazards and workplace health and safety issues, companies need highly competent OH&S personnel to assist them. The Canadian Registered Safety Professionals (CRSP) designation is widely recognized as denoting an individual who has:

- met the academic experience and examination requirements of a national registration authority (BCRSP);
- in-depth OH&S knowledge;
- the ability to use this knowledge, experience, and training to help an organization/company achieve optimum control over workplace hazards; and

---

[52] F. Bird & G. Germain, *Practical Loss Control Leadership*, revised ed. (Loganville, GA: DNV, 1996) at 27-42.

[53] Board of Canadian Registered Safety Professionals (BCRSP), "Mission, Vision, Values", BCRSP Website, online: http://www.bcrsp.ca/about-us/mission-vision-values (date accessed: January 31, 2020).

- the professional obligation to continually upgrade ones OH&S knowledge and skills.

The "new role" for the OH&S Practitioner/Professional is explained in Chapter 7, "Safe Workplace = Great Workplace: Building a Sustainable Culture of Safety".

Occupational Health Nurses (OHNs) are registered nurses who hold a certification in Occupational Health Nursing, and often possess additional training and skills in the areas of Occupational Health and Safety, relationship-building, and business management. Their mandate is to promote healthy working environments, protect the health of the worker, and prevent work-related injuries and illnesses. The value that an OHN can offer an organization/company is discussed further in Chapter 4 "Occupational Health Program".

Occupational Health Physicians (OHPs) are medical physicians who have chosen to specialize in Occupational Medicine. They are knowledgeable about medicine, the workplace, and stakeholder interests, as well as the areas of human factors, industrial hygiene, toxicology, and disability management. The value that an OHP can offer an organization/company is discussed further in Chapter 4 "Occupational Health Program".

In addition to these three specialized OH&S professionals, there are a number of other OH&S paramedical personnel whose roles are discussed in Chapter 19 "Occupational Health and Safety: Supporting Disciplines".

13. Implement an Effective OH&S Data Management System

The Liberty Mutual Research Institute recommends that Management must know their business and manage it by understanding where and why its employees get hurt, and then, focus in on those aspects.[54]

To effectively measure OH&S performance, accurate OH&S data must be collected, managed, and retained. The types of data the workplace should collect are:

- workplace hazards and controls (hazard inventory);
- management safety tour findings and corrective actions (workplace inspection logs);
- worksite inspection findings and results actions (workplace inspection logs);
- ergonomic assessments and corrective actions;
- hazard reports and corrective actions;
- worker OH&S training logs;

---

[54] Liberty Mutual, News Release, "Despite 6.2% Fall in the Number of Serious Workplace Injuries, Their Financial Impact on Employers Remains Huge" (September 20, 2005), online: http://www.libertymutualgroup.com (date accessed: January 31, 2020).

- near-miss reports and corrective actions;
- incident reports and corrective actions;
- incident costs;
- first aid reports;
- injury reports and outcomes;
- injury costs;
- modified work plans and outcomes; and
- modified work savings.

There are a number of commercial software packages available to employers.

BEST PRACTICES

1. Design and develop an effective OH&S database.
2. Make data suitable for Line Management viewing readily available.
3. Ensure the system is able to conduct trend analyses.
4. Design the system so it calculates business measures such as cost-benefits, payout ratios, return on investment, *etc*.

14. Provide, Coordinate, and Integrate Worker OH&S Education and Training

Worker OH&S education and training is vitally important to the success of any organization/company.[55] It is also a recognized management function. Management is legally obligated to inform and train workers about their OH&S duty of care and responsibilities; the safe use of equipment/tools/vehicles; safe work practices; WHMIS/TDG; working alone safety; workplace violence safety; and hazard identification, assessment, and control.

Chapter 17, "Occupational Health and Safety: Worker OH&S Training" addresses the topic of worker OH&S training in detail.

BEST PRACTICES

1. Ensure that worker OH&S training is coordinated and integrated with the general worker training programs.
2. Educate workers as opposed to merely providing worker training.
3. Ensure that new/young workers receive additional training because their

---

[55] Refer to F. Bird & G. Germain, *Practical Loss Control Leadership*, revised ed. (Loganville, GA: DNV, 1996) at 249-50; and A. Zacharatos & J. Barling, "High Performance Work Systems and Occupational Safety" in J. Barling & M. Frone, eds., *The Psychology of Workplace Safety* (Washington, DC: American Psychological Association, 2004), 203 at 207-208.

potential for workplace injuries is the greatest.[56]

4. Develop and implement a training matrix designed to address all the training needs of each worker/occupational group. An example is provided in Chapter 5, "Occupational Health and Safety Program: Manual Development", in the sample OH&S Program Manual.

5. Ensure that OH&S training includes:

    - an OH&S Orientation;
    - OH&S Training;
    - on-the-job OH&S Training; and
    - refresher OH&S Training.

6. Regularly monitor and measure the effectiveness of the OH&S Training Program. This should include obtaining participant feedback on the content, relatedness, and value to the worker; the perceived outcome value by the supervisory staff; the observed applied behavioural practices as a result of the training; and the demonstrated behaviours associated with workplace incidents.

7. Focus on continually improving the quality of the OH&S Training.

15. Measure OH&S Performance

Measuring OH&S performance is an important management function. Measurement identifies gaps between **the current state** and **the desired state** of the OH&S Program. It indicates whether the OH&S goals/objectives are being/have been met or not; and enables improvements both along the way and/or at set points in time. Continuous improvement in the area of workplace health and safety, is essential. The world of work is dynamic. To stay ahead of the competition, companies must continually improve their OH&S structures and processes if they expect the OH&S outcomes to improve: "[I]f you are not pedalling, you are going downhill."[57]

There is an adage that holds: "[I]f you can't measure it, you can't control it. If you can't control it, you can't manage it." Program measurement and evaluation is a

---

[56] WorkSafeBC, *Young Worker Injury Rate Comparison 2013-2017* (2018), online: https://www.worksafebc.com/en/health-safety/education-training-certification/young-new-worker/statistics (date accessed January 31, 2020); Alberta WCB (2018). 2017 *Workplace Injury, Disease and Fatality Statistics*, online: https://open.alberta.ca/dataset/c1c1b935-a5d5-456a-b86c-c9986fa5542d/resource/4572192b-df67-4aef-b85b-7eb21c417f65/download/2017-workplace-injury-disease-and-fatality-statistics.pdf (date accessed: January 31, 2020); and Nova Scotia Canada, "Young Worker Resources", Labour and Advanced Education (2018), online: http://novascotia.ca/lae/healthandsafety/workers.asp (date accessed: January 31, 2020).

[57] W. Rodriguez, cited in M. Paradies & L. Unger, *TapRoot®: The System for Root Cause Analysis, Problem Investigation and Proactive Improvement* (Knoxville, TN: System Improvements Inc., 2000) at 11.

management practice — a means of ascertaining whether a function is operating as planned.

When measuring an OH&S Program, it is important to carefully consider what you plan to measure. As Peter Drucker said, "[W]hat we measure gets improved. So, watch what you measure!"[58]

Likewise, "[i]t is '**what**' you measure that counts", according to Peter Urs Bender.[59] Critical to any measurement plan is determining the "what" you plan to measure. Once you establish that concept, it will then guide you as to how to conduct the measurement. The topic of program measurement is described in more detail in Chapter 2, "Occupational Health and Safety Management Systems", Chapter 6, "Occupational Health and Safety: Leadership and Commitment", and Chapter 7, "Safe Workplace = Great Workplace: Building a Sustainable Culture of Safety".

BEST PRACTICES

1. Ensure the OH&S Program has an infrastructure which includes structure, processes, and outcomes. Refer to Chapter 2, "Occupational Health and Safety Management Systems" for details on this concept.

2. Set and work towards OH&S Program goals, objectives, and targets. Refer to Chapter 2, "Occupational Health and Safety Management Systems" for details on this concept.

3. Develop a measurement plan for the OH&S Program that includes:

    a. OH&S Program goals;

    b. OH&S Program objectives;

    c. OH&S targets;

    d. a schedule/action plan for the achievement of the objectives and targets, and for the measurement of each (Table 2.1);

    e. implementation of the plan;

    f. progress monitoring (formative evaluation);

    g. measurement of the results;

    h. analysis of the results; and

    i. communication of the outcomes (Chapter 16, "Marketing Occupational Health and Safety Programs; Communicating the Results").

4. Use program measurement techniques such as auditing; benchmarking; cost-benefit analyses; determining the return on investment for having an

---

[58] P. Drucker, *Managing for Results* (New York, NY: Harper Collins Publishing, 1993).

[59] P. Urs Bender, *Leadership from Within* (Don Mills, ON: Stoddart Publishing Canada, 1997).

OH&S Program; adopting a balance score card approach; and/or using perception surveys. Refer to Chapter 2, "Occupational Health and Safety Management Systems" for details on this concept.

5. Focus on continuous improvement and tracking the evolution of the organization's/company's OH&S Program over time. Evolution of an Occupational Health and Safety Management System (OHSMS) and an OH&S Program is presented in Chapter 2, "Occupational Health and Safety Management Systems".

6. Develop a performance model for OH&S Practitioners/Professionals within the organization/company. Refer to Chapter 7, "Safe Workplace = Great Workplace: Building a Sustainable Culture of Safety", for details on how to do this.

16. Align Corporate Standards with Recognized OH&S Standards to Uniformly Measure OH&S Programs

In Canada and the U.S., OH&S standards exist in the various OH&S Acts, OHSMS standards, and industry standards. Their purpose is to establish a "yardstick for OHSMS or OH&S Programs" against which organization/company standards can be measured. Typically, the OHSMS should include:

- OH&S leadership and commitment;
- OH&S Communication;
- hazard identification, assessment and control;
- risk management;
- operational rules and controls;
- safe work practices for critical tasks;
- emergency preparedness and response plans;
- incident investigation;
- performance measurement and assessment; and
- continuous program improvement.

Companies that annually have their OHSMS audited have a lower number of workplace injury claims and related Workers' Compensation Board costs.[60]

---

[60] Alberta Workers' Compensation Board, "Safety Performance Heralds Lower Rates for 2006", *Insight* (Summer 2005), online: http://www.wcb.ab.ca/pdfs/employers/insight_archive/ summer_2005_insight_print.pdf (date accessed: January 31, 2020); Alberta Workers' Compensation Board, *Working Results: Alberta Workers' Compensation Board 2005 Annual Report* (Edmonton, AB: Workers' Compensation Board, 2005), online: http://www.wcb.ab.ca/ pdfs/public/annual_report_2005.pdf (date accessed: January 31, 2020).

BEST PRACTICES

1. Select a suitable OHSMS standard and protocol - refer to Chapter 2, "Occupational Health and Safety Management Systems" for more details.
2. Measure the organization's/company's performance against this OHSMS standard on an annual basis protocol - refer to Chapter 2, "Occupational Health and Safety Management Systems" for more details.
3. Identify the gaps between the organization's/company's OHSMS and the OHSMS standard.
4. Develop strategies to eliminate/reduce the identified gaps.
5. Prioritize the above strategies.
6. Implement the strategies and monitor their effectiveness.
7. Evaluate the corrective actions taken.

17. Develop an OH&S Communication Strategy

An essential component of any successful OH&S Program is the widespread understanding and support of stakeholders within the workplace and in the broader community. An OH&S Communication Strategy defines the corporate OH&S message, how it is to be delivered, to whom, through what means, and by when. This powerful management tool is dynamic and should be carefully monitored to ensure it aligns well with the organization's/company's OH&S standards.

Stakeholder OH&S education and open, honest communication about the OH&S Program objectives, successes, failures, and future plans are powerful tools that can alter entrenched attitudes and build trust between individuals. The reality is that the fields of Occupational Health and Safety and Disability Management are built on relationships, and these relationships need to be constantly nurtured.

BEST PRACTICES

1. Develop a communication strategy and plan to promote awareness and overcome organizational/company barriers to implementing and sustaining a strong OH&S Program. Chapter 26 "Occupational Health and Safety: Workplace Wellness Strategy", describes a number of industry-proven OH&S communication techniques.
2. Keep all key stakeholders in the information loop and part of the decision-making process.
3. Provide stakeholders with relevant outcome data and benefits realized by the OH&S Program. An excellent industry example is provided by Unison Contracting Ltd., in Chapter 16, "Marketing Occupational Health and Safety Programs: Communicating the Results", Appendix 4. This type of information garners continued support for the OH&S Program.
4. Use available communication vehicles to spread the word about the OH&S Program internally and externally to supervisors and managers, union

leaders, and employees. In this manner, the stakeholders can be part of an ongoing solution for workplace health and safety. For example, Petro-Canada has developed a brochure - *Office Workplace Leaders Tools & Tips* with topics such as *Zero-Harm in Office-based Workplaces: A Guide for Employee and Contractor Leaders and Front-Line Supervisors*.[61] In it, Line Management is reminded of the ABCs and attributes of the Petro-Canada Zero-Harm culture, along with some tips on five focus areas for them to address. Lastly, they are reminded to practise the expected leadership behaviours that Petro-Canada promotes. Petro-Canada also has a Zero-Harm Poster Campaign.

5. Ensure that stakeholders are provided tools that assist them to perform their assigned OH&S duties and responsibilities. For example, Petro-Canada has developed a Performance Diamond to support the desired high-performance culture. It is used to help Line Management to implement the Petro-Canada TLM framework (Figure 18.6).

---

[61] Petro-Canada (now Suncor Energy Inc.), *Office Workplace Leaders Tools & Tips*, "Zero-Harm in Office-based Workplaces: A Guide for Employee and Contractor Leaders and Front-Line Supervisors" (June 2006 Issue, Calgary, AB). Reprinted with permission from Petro-Canada.

Figure 18.6: The Petro-Canada Performance Diamond[62]

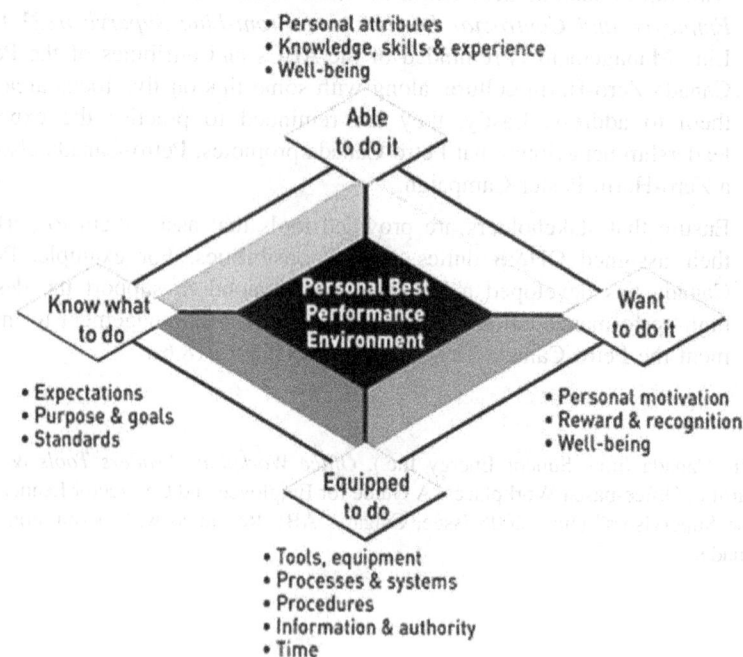

18. Strive for Technology-driven OH&S Awareness

In a work world that is technologically savvy and dependent, the use of technology to transmit messages can be very effective and efficient. Technology allows companies to push standardized information out quickly to all. This can be done using corporate intranets, electronic bulletin boards, standardized workstations, e-learning, and computerized best practices and hazard management databases. Canadian employers are "warming up to technology" for delivering health and safety information as well as boosting employee health engagement. For example, 62 per cent have online tools available at work and home; 38 per cent use a dedicated site to deliver health information; and 20 per cent endorse the use of social media tools.[63]

---

[62] Petro-Canada (now Suncor Energy Inc.), "Total Loss Management (TLM): Corporate Standards", at 9 (January 25, 2005, Calgary, AB) at 9. Reprinted with permission from Petro-Canada.

[63] Willis Towers Watson, *2015/2016 Global Staying@Work Survey, Canada* (2016) at 10,

BEST PRACTICES

1. Utilize the available technology to maintain centralized control on important OH&S activities, such as OH&S training and measurements of OH&S Program elements such as hazard assessment and management; education and training; emergency preparedness and response; operational procedures and controls; and incident investigation and reporting.
2. Use technology to efficiently communicate OH&S messages to stakeholders.
3. Garner employee engagement in OH&S efforts by using available technology.

19. Implement an Effective Disability Management Program

A Disability Management Program is a workplace program designed to facilitate the employment of persons with a disability through a coordinated effort that addresses individual needs, workplace conditions, and legal responsibilities.[64] Ideally, Disability Management Programs are proactive in nature and incorporate stakeholder involvement and accountability. Most are designed to control the personal and economic costs of employee injury or illness; convey a message that employees are valued; and demonstrate compliance with the relevant legislation.

By managing employee injury/illness and using gradual return-to-work practices, companies are able to assist recovering employees to return to work in a safe and timely manner. Both are means of secondary and tertiary prevention and have been shown to be extremely effective in retaining employees in the workplace.

Research and industry experience have supported the importance of early intervention in any absence. The Sun Life Financial study determined that through early intervention:

- employees returned to work 20% quicker than expected or 2.7 weeks sooner (an $810 per claim saving);
- 47 per cent more employees returned from STD; and
- 33 per cent fewer claims went onto LTD.[65]

Informally, many companies report more success with returning the recovering employee to the workplace if intervention begins at, or soon after, the time of injury/illness onset. By maintaining the employee in the workplace, the

---

online: https://easna.org/wp-content/uploads/2016/04/2015-2016-Global-Staying-at-Work-Survey.pdf (date accessed: February 28, 2020).

[64] NIDMAR, *Code of Practice for Disability Management* (Port Alberni, BC: NIDMAR, 2000) at 5.

[65] Sun Life Financial, "Early Intervention Programs Can Speed Disabled Employees Back to Work" (2000), online: https://www.hrhub.com/doc/early-intervention-programs-can-speed-disable-0001 (date accessed January 31, 2020).

occupational bond — the identity of the employee with the workplace — remains intact.

BEST PRACTICES

1. Design and implement a Disability Management Program (Chapter 27, "Disability Management: Overview").

2. Garner stakeholder support for the program by communicating to organization/company stakeholders the intent, value, and functioning of the Disability Management Program (Chapter 27, "Disability Management: Overview").[66]

3. Communicate stakeholder roles and responsibilities in relation to the Disability Management Program.

4. Institute early contact with the ill/injured employee. Ideally, the supervisor should do this on the first day of absence.

5. If required, implement early case management (within the first three to five days for non-occupational disabilities, and day one for occupational disabilities).

6. As appropriate, involve the Employee Assistance Program.

7. As appropriate, involve the ergonomic interventions.

8. Monitor and track the successes of the program.

9. Communicate program successes.

10. Seek employee/union/management involvement in finding solutions to identified program challenges.

11. Evaluate the Disability Management Program using auditing, benchmarking, employee perception surveys, and other measurement approaches. The intent is continuous program improvement.[67]

20. Implement a Graduated Return-to-Work Program

Graduated Return-to-Work Programs can be an effective method of systematically returning employees to health and work and can contribute to cost-containment.[68]

---

[66] D. Dyck, *Disability Management: Theory, Strategy & Industry Practice*, 6th ed. (Toronto: LexisNexis Canada, 2017).

[67] *Ibid.*

[68] Theolsongroup Benefits Group, *Do You Know the Compelling Benefits of a Return to Work Program?* (2019), online: https://theolsongroup.com/benefits-return-to-work-program/ (date accessed: January 31, 2020); and D. Lyons, "Integrated Occupational Health & Safety: Assessing Its Fit for Your Company" *Ideas at Work* (Winter 2004), online: http://www.libertymutualgroup.com (date accessed: January 15, 2015).

CH. 18: OCCUPATIONAL HEALTH AND SAFETY: BEST PRACTICES

The recommended components of a Graduated Return-to-Work Program are:[69]

- employee and supervisor submission of all the necessary claim forms;
- early intervention;
- communication with the employee and attending physician regarding the availability of modified/alternate work;
- regular follow-up with the employee and physician regarding fitness to work;
- availability of modified/alternate work for the recovering employee;
- the placement of the employee into suitable modified/alternate work;
- the monitoring of the employee's progress and fitness to work;
- a gradual return to full-time duties;
- evaluation of the case; and
- data management.

Companies that have fully functioning Graduated Return-to-Work Programs have noted significant success at returning employees to work and controlling their disability rates and costs.[70] As well, they are able to demonstrate compliance with the Canadian Human Rights legislation, the "duty to accommodate" section.

The Canadian duty to accommodate legislation varies by province. In general, this legislation indicates that the employer, employee, and unions have a tripartite responsibility to accommodate the injured/ill employee back into the workplace, up to the point of "undue hardship".[71]

The key practices of a successful Graduated Return-to-Work Program are:

- arranging acceptable practices with unions for modified/alternate work opportunities within the collective agreements;
- ensuring that the modified/alternate work offered is meaningful and gainful employment;
- having set time lines for the modified/alternate work opportunity; and
- clearly defining the differences between modified and alternate work.[72]

BEST PRACTICES

1. Develop and implement a corporate-wide Graduated Return-to-Work

---

[69] D. Dyck, *Disability Management: Theory, Strategy & Industry Practice*, 6th ed. (Toronto: LexisNexis Canada, 2017)

[70] *Ibid.*

[71] Canadian Human Rights Act, R.S.C. 1985, c. H-6, online: http://canlii.ca/t/52zkk.

[72] D. Dyck, *Disability Management: Theory, Strategy & Industry Practice*, 6th ed. (Toronto: LexisNexis Canada, 2017) at 200.

Program that has labour and Management support and participation.[73]

2. Communicate the roles, responsibilities, and accountabilities of the key stakeholders.
3. Elicit employee, union, and Line Management identification of modified/alternate work options.
4. Manage safe and timely return-to-work activities.
5. Develop flexible and creative return-to-work options.
6. Collect and manage modified/alternate work data, and the Graduated Return-to-Work Program outcomes.
7. Evaluate the program regularly.
8. Continuously improve the program.
9. Communicate the benefits, challenges, and outcomes to all key stakeholders.

21. Link the OH&S Program and Disability Management Program

OH&S Programs often operate independently of Disability Management Programs (DMPs). The irony is that the DMP data tends to identify and quantify the "failures" of an organization's/company's OH&S Program. For example, repeat back injuries within a workgroup or occupational grouping might indicate the need for industrial ergonomic interventions. Other benefits include the identification of substandard work practices, work conditions, and supervisory practices. In short, by using DMP data, companies can identify the real costs of OHSMS failures.

Likewise, actions taken/changes within an OH&S Program, or related work practice, can impact the DMP. For example, the type of work gloves purchased can have an impact on whether the employees wearing them experience hand injuries.

BEST PRACTICES
1. Link the OH&S Program with the DMP.
2. Review workplace incidents and injuries with a view to future prevention using a targeted approach.
3. Analyze workplace injuries to identify injury trends and their association with OH&S practices.

22. Link the OH&S Program and Employee Assistance Program

Workplace-focused programs can be designed to assist with the identification and resolution of personal concerns that impact employee attendance, productivity, and safety practices; and that lead to increased disability costs.

The most effective formal linkages are the ones in which there is a predetermined

---

[73] *Ibid.*

working relationship between organization/company OH&S professionals and Employee Assistance Program service providers. Appropriate consents are put in place so that the relevant issues surrounding the employee's fitness-to-work, treatment plans, and workplace accommodations can be discussed. Issues are discussed on a "need-to-know" basis and pertain to a successful re-entry to the workplace.

BEST PRACTICES

1. Assist the "troubled employee" by integrating the Employee Assistance Program services with employee attendance support and OH&S Programs.
2. Ensure that the proposed OH&S Program includes a formal linkage with the organization's/company's Employee Assistance Program.
3. Ensure that service providers attain a mutual understanding of, and respect for, their individual program goals and objectives, as well as for the overall OH&S Program goals and objectives.
4. Promote a partnership approach that allows for multi-disciplinary interventions.
5. Examine the outcome measures on the cases jointly served by the EAP and OH&S Program personnel. Knowledge of utilization rates, types of cases served, trend analyses, and success/failure rates; and anticipatory guidance for illness and injury prevention can be provided using aggregate data.
6. Seek ways to continually improve the partnership relationship(s), as well as the linkages forged. The intent is to be able to assess the value of the linkage and its contribution to the overall process. This outcome data can also be compared to those cases that were not co-managed to determine the value of the OH&S Program-Employee Assistance Program linkage.

23. Link Prevention to Workplace Illness/Injury Management

Control of workplace hazards and the prevention of workplace incidents and injuries is the mandate of OH&S Programs. However, many companies isolate the workplace injury/illness management function from its safety function. The outcome is a "silo effect" and the loss of valuable OH&S information that could enable the organization/company to identify workplace hazards. This is particularly evident with overuse injuries and psychological illness.

BEST PRACTICES

1. Develop databases that capture incident data from occurrence until the employee's successful return to work (resolution). In addition to employee demographic, occupation, and job-position data, the information collected should include date, time, and nature of the injury/illness; causal factors/actions; contributory factors/actions; days lost; interventions; modified work days; return-to-work date; duration of absence; and related costs.
2. Use aggregate data to establish associations between the nature of the

illness/injury and employee occupation, work actions, required interventions, duration of injury/illness, potential for modified work, and/or costs.
3. Undertake a gap analysis to identify problems and develop prevention strategies.
4. Monitor the implemented prevention strategies in terms of effectiveness, efficiency, and cost.
5. Implement ergonomic programs to counteract musculoskeletal injuries.
6. Embrace the Canadian Standard on Psychological Health and Safety in the Workplace[74] to promote mental health and prevent psychological disabilities.

24. Adapt to a Changing Workplace/Workforce

Over the last 60 years, the world of work and the nature/needs of the workers within the resultant workplace have dramatically changed. No longer are workers doing predominantly manual-labour jobs. The age of the knowledge-based worker and service-oriented worker has arrived. New technologies and human resource management theories/practices added to the noted changes. A large portion of the workforce consisting of contractors, consultants, or workers in part-time positions ("contingent workers").

With an aging workforce, and older and once-retired employees returning to work, employers are tasked with providing a work environment conducive to keeping the older worker healthy and challenged.

More women and immigrants are present in the workplace. Traditionally male-dominated occupations are now accepting women: unfortunately, the tools, equipment, and work processes are often poorly suited to female anthropometric measures.

The nature of workplace injuries is changing. Chronic illness, musculoskeletal injuries, and psychological illness are on the increase, as opposed to the traditional acute physical injuries due to physical trauma. These new ailments are challenging because they tend to have a slow onset, be multi-causal in nature, and require multi-discipline intervention and management. Added to this, is the threat of a corporate-wide disaster due to terrorism, workplace violence, or pandemic flu; the new work world is certainly full of unique hazards and challenges.

In addition to a changing workforce, different worker demographics, new hazards and occupational injuries/illness, globalization, and market pressures have increased

---

[74] CSA, Standard of Psychological Health and Safety in the Workplace, CAN/CSA-Z1003-13/BNQ 9700-803/2013 (2013), online: https://www.csagroup.org/documents/codes-and-standards/publications/CAN_CSA-Z1003-13_BNQ_9700-803_2013_EN.pdf (date accessed: January 31, 2020).

the pace and duration of the workday.[75] Many workers are in jobs that include shift work, mobile-work offices, business travel, 12-hour workdays, weekend work, a blurring of the lines between work and private lives, and functioning in a constant high learning curve. Emails, texting, cell phones, voice mail, and multiple device communications, add to the workday pace and stress.

BEST PRACTICES

1. Develop approaches for getting the "right OH&S information" into the "right hands" at the "right time".

2. Use audience-specific transfer of OH&S knowledge.

3. Target the OH&S knowledge-transfer messages, interventions, and products to the end-user. Each targeted group of users will have their own agenda, needs, jargon, and preferred communication medium.

4. Develop OH&S interventions/activities so that they develop confidence and user acceptance.

25. Address Workplace Health and Safety for Older Workers

Older workers have not exited the workplace as expected: in 2019, 4.2 million employed workers were aged 55 and older.[76] This poses challenges for employers namely, how to prepare for and accommodate diminishing physical capabilities, as well as how to effectively manage the increased absenteeism and disability costs due to increased medical absence rates and durations.

BEST PRACTICES

1. Assess the risks. Begin by understanding the characteristics and needs of the older worker. Compare those recognized changes that occur with human aging against the physical and psychological demands of the organization's/company's jobs. Through a gap analysis, identify the potential risks for aging employees and for the organization. Address these risks by adopting loss control strategies.

2. Offer employee benefit and support programs tailored to the needs of the older worker.

3. Structure work tasks to meet the needs of the older worker in general, and in specific, to individual workers.

4. Engineer workplaces to have safe walking surfaces, good lighting, and noise control — conditions that should be in place for all, but which are

---

[75] D. Dyck, *Disability Management: Theory, Strategy & Industry Practice*, 6th ed. (Toronto: LexisNexis Canada, 2017).

[76] Statistics Canada (2020). Table: 14-10-0017-01 *Labour force characteristics by sex and detailed age* group, monthly (2019), online: https://www150.statcan.gc.ca/t1/tbl1/en/tv.action?pid=1410001701 (date accessed: January 31, 2020).

very important to the older employee.

5. Undertake ergonomic assessment of jobs held by older employees with a view to "fitting the job to the worker" and preventing unnecessary soft tissue injuries.
6. Encourage employee fitness and the adoption of a healthy lifestyle.
7. Assist aging employees to plan and gradually transition towards retirement, if that is their career goal.

26. Address Workplace Health and Safety for Young/New Workers

Young workers continue to be "victims" of workplace incidents. They tend not to recognize workplace hazards; as well, they do not ask questions because they want to fit in and not draw attention to themselves. The outcome is that young workers are twice as likely to be injured on the job than those over the age of 25 years[77] — especially young males.[78] *For injured workers under age 25, 60 per cent were injured during their first six months on the job.*[79]

In 2018, seven of the 1,027 injury fatalities in Canada involved workers 15-19 years of age (young workers); this equated to 0.7% of the injury fatalities for that year.[80] In 2017, the U.S. experienced 84 young worker fatalities (15-19 years of age), equalling 1.6% of the injury fatalities for that year.[81] Although several strategies have been implemented to better protect young workers, the needless loss of these young workers continues.

New employees face similar work outcomes: "[w]orkers new to a job are three times more likely to be injured during the first month on the job than more experienced workers."[82]

BEST PRACTICES

---

[77] Energy Safety Canada, "Young Workers", Safety Resources (2018), online: http://www.enform.ca/media/8411/sa00_17_1.pdf (date accessed: January 31, 2020).

[78] Ontario Ministry of Labour, New and Young Workers (2016), online: https://www.labour.gov.on.ca/english/hs/sawo/pubs/fs_youngworkers.php (date accessed: January 31, 2020).

[79] Alberta Workers' Compensation Board, "Safety Awareness Campaign 2006, Alberta Workplace Injury Statistics", online: http://www.wcb.ab.ca (date accessed: January 31, 2020).

[80] AWCBC, "2018 Fatalities in Canada", Statistics (2020), online: http://awcbc.org/?page_id=14 ((date accessed: January 31, 2020).

[81] U.S. Department of Labor, Bureau of Labor Statistics, "Fatal occupational Injuries by event or exposure and age, all United States, 2017", *Injuries, Illnesses, and Fatalities*, online: https://www.bls.gov/iif/oshwc/cfoi/cftb0320.htm (date accessed: January 31, 2020).

[82] Ontario Ministry of Labour, New and Young Workers (2016), online: https://www.labour.gov.on.ca/english/hs/sawo/pubs/fs_youngworkers.php at 1 (date accessed: January 31, 2020).

CH. 18: OCCUPATIONAL HEALTH AND SAFETY: BEST PRACTICES

1. Companies need to recognize the vulnerability of the young and new employee in their specific workplace. Many industry examples exist on how an organization/company can incorporate the well-being of young workers into OH&S Programs.[83], [84], [85]

2. Develop employee training programs that provide in-depth knowledge about workplace hazards — their identification, assessment, and control. This needs to include how the young worker should report an identified hazard to his/her direct supervisor. Alberta, Jobs, Skills, Training and Labour, like other provincial agencies, offers employers tools to educate young workers.[86]

3. Use training approaches that are tailored to address young worker learning needs and to enhance young worker learning (refer to Chapter 17, "Occupational Health and Safety: Worker OH&S Training" for more details).

4. Educate the general employee population on the vulnerability of the young/new worker, and clearly state the role and responsibility it has for the protection of these workers.

5. Consider the use of various young worker information and programs such as the Passport to Safety,[87] Work Safe BC: Young Worker,[88] CCOHS

---

[83] Refer to WorkSafeBC, "Young & new workers", Education, training & certification, *Health & Safety* (2019), online: https://www.worksafebc.com/en/health-safety/education-training-certification/young-new-worker (date accessed: January 31, 2020).

[84] Saskatchewan Ministry of Labour Relations and Workplace Safety. *Young Worker Readiness Certificate Course (YWRCC)*: a mandatory test of job-readiness for all 14- to 15-year-old workers. The online test focuses on Occupational Health and Safety, employment standards and rights and responsibilities. They also require parental consent to work. Available online: https://www.saskatchewan.ca/business/hire-train-and-manage-employees/youth-in-the-workplace/take-the-young-worker-readiness-certificate-course (date accessed: January 31, 2020).

[85] Government of Alberta, *X-TREME SAFETY: Survival Guide for New and Young Workers, Cat. No. 454992*. Edmonton, AB; Government of Alberta, Human Services (2013). This booklet addresses the youth mindset of "it will never happen to me" with concrete examples of a fatality and a traumatic injury. It goes on to explain what is safe and what isn't in the workplace, as well as worker rights and how to say "no" at work when conditions are viewed as unsafe.

[86] Alberta, Jobs, Skills, Training, and Labour, Young Workers (2015), online: http://work.alberta.ca/occupational-health-safety/5369.html (date accessed: January 31, 2020).

[87] "Passport to Safety: Setting the Standard", online: http://www.passporttosafety.com (date accessed: January 31, 2020).

[88] WorkSafeBC (2011). *Young Worker Focus Report*, "Section 3: A Snapshot of Young Worker Injuries" (2015), online: https://www.worksafebc.com/en/resources/about-us/reports/protecting-young-workers-focus-report?lang=en (date accessed: January 31, 2020).

Young Worker Zone,[89] NIOSH Young Worker Health and Safety,[90] and OSHA Fact Sheet on Young Workers.[91]

6. Adopt a "zero tolerance" approach with employees who "pull pranks" on young/new employees (or on any employees for that matter). Practical jokes, horseplay, and initiation practices can result in serious injury.

27. Support/Protect the Contingent Workforce

The contingent workforce comprises contracted help, consultants, temporary workers, seasonal workers, part-time workers, students, and volunteers. This transient workforce is diverse and used to address an organization/company's temporary labour shortage and needs. Employers tend to invest very little into the OH&S training of these workers even though they are functioning in a "new environment", often doing high-risk jobs. Their work habits can impact the health and well-being of the mainstream workforce. In addition, this workforce tends to be difficult to reach.

Awareness of the need to provide OH&S education and support to this group of workers is heightened. Alberta's new OH&S Act (2018) requires employers to treat temporary workers as they their regular employees. That means clarifying the OH&S roles and responsibilities of all workers; and holding supervisors and self-employed persons accountable for workplace safety. These obligations are similarly shouldered by the temporary staffing agency — they must ensure a good person-job fit, the availability of personal protective equipment if needed, and "that the host employer is capable of looking after the worker's health and safety".[92]

Best Practices

1. Companies need to recognize the vulnerability of the new and transient worker in their specific workplace.
2. Develop orientation packages and programs that provide the relevant OH&S training to these workers in a timely manner; in essence, treat them like the corganization's/company's employees.[93,94]

---

[89] CCOHS, *Young Worker Zone* (2018), online: http://www.ccohs.ca/youngworkers/ (date accessed: January 31, 2020).

[90] NIOSH, "Young Worker Safety and Health" (2018), online: http://www.cdc.gov/niosh/topics/youth (date accessed: January 31, 2020).

[91] U.S. Department of Labor, Occupational Safety and Health, "Young Workers, OSHA Fact Sheet", online: http://www.osha.gov/Publications/young_workers.html (date accessed: January 31, 2020).

[92] Government of Alberta, *Alberta Occupational Health and Safety Act: Highlights of changes effective June 1, 2018* (2018), online: www.alberta.ca at 8 (date accessed: January 31, 2020).

[93] S. Greeves, "Planning to grow your contingent workforce in Alberta?" *Canadian Lawyer* (May 22, 2018), online: https://www.canadianlawyermag.com/author/susan-greeves/

3. Make sure the training is geared to worker jargon, needs, and understanding.
4. Develop contract wording that requires the Personnel Agencies/Service Providers that supply the contingent workforce to ensure their workers have basic OH&S knowledge.
5. Provide simple OH&S messages that are easy to access in a number of languages using a number of mediums.
6. Document the education activities undertaken.

28. Strive to be Culturally Competent and Proficient

In a world of multicultural workplaces, employers must appreciate the impact that cultural diversity and the presence of five generations in the workplace have on OH&S programming. The relevance of increased population diversity is that each cultural group has unique values, beliefs, needs, and expectations on life and work. By effectively managing diversity, organizations can enhance their competitive advantage in the Canadian labour market, and globally.

BEST PRACTICES

1. Recognize and understand the cultural groups presented in the organization.
2. Address potential cultural conflicts and noted inequities.
3. Promote inclusiveness, that is, demonstrate that employees are valued and belong in your workplace. Look for the strengths that each group offers, not their differences, and use those strengths constructively.
4. Integrate cultural competence into the organization/company's programs.
5. Provide OH&S programming and services in a manner that meets the needs of a diverse workforce.

29. Conduct More OH&S Research

According to Barling and Hutchinson, "[o]ccupational safety remains one of the least studied phenomena in organizational behaviour, with estimates suggesting it represents less than 1% of the total amount of research (conducted)."[95]

OH&S research is essential if an understanding of what causes workplace incidents is to be gained. There is a need to look at occupational injury incidents

---

planning-to-grow-your-contingent-workforce-in-alberta-15759/ (date accessed: January 31, 2020).

[91] OSHA, *Protecting Temporary Workers* (2018), online: https://www.osha.gov/temp_workers/ (date accessed: January 31, 2020).

[95] J. Barling & I. Hutchinson, "Commitment vs. Control-based Safety Practices, Safety Reputation and Perceived Safety Climate" (2000) 17:1 Canadian Journal of Administrative Science 76.

from multiple levels — the employee level, the workgroup level, and the organizational level.[96] As well, by examining near-miss and First Aid incidents, researchers can help increase the level of understanding of the factors that lead to the more serious workplace events.[97]

In terms of worker OH&S training, the factors that lead to successful worker OH&S training warrant examination. It is recognized that the implementation of OH&S training concepts and recommended behaviours depends on the worker being given the resources and support to enact those concepts and behaviours.[98] Likewise, worker motivation to learn OH&S concepts is strongly dependent on job security.[99]

Everyone agrees that worker training is critical — but few have examined what instructional techniques work and why.

Lastly, the area of leading OH&S performance indicators must be addressed. Most companies focus on their OH&S outcomes, when their energies should be placed on creating work environments that facilitate, support, and reward safe work practices and behaviours.

BEST PRACTICES

1. Initiate and support OH&S research within the workplace and industry.
2. Support government-sponsored research efforts.
3. Support the implementation of clinical research opportunities within the organization/company.

30. Get OH&S Research Ideas to Those Who Need It

According to the Association of Worker Compensation Board of Canada (AWCBC) attendees, "[r]esearch without research transfer is not full-value research." [100] Among clinical audiences, there are often lengthy gaps between the identification of new research findings and the incorporation of that learning into clinical practice. The challenge for academia is to get the OH&S clinical research out to workplaces in a timely manner.

BEST PRACTICES

1. Monitor OH&S research and seek ways to apply relevant findings to the workplace.

---

[96] J. Barling & M. Frone, *The Psychology of Workplace Health and Safety* (Washington, DC: American Psychological Association, 2004) at 300.

[97] *Ibid.*, at 301.

[98] *Ibid.*

[99] *Ibid.*

[100] AWCBC, "Knowledge Transfer for Workplace Health & Safety: An Onsite Report on the Public Forum" (Developed at the Association of Workers' Compensation Boards of Canada (AWCBC), November 19 and 20, 2001, Toronto, ON) at 4.

2. Support continuing education by the OH&S Practitioners/Professionals.
3. Encourage networking by the OH&S Practitioners/Professionals with academia and members of relevant industry groups.
4. Support the implementation of clinical research opportunities within the organization/company.

C. SUMMARY

The listing of these 30 OH&S best practices is not meant to be exhaustive in nature. Rather, this chapter merely presents some examples of currently held beliefs in OH&S best practices. As noted in the opening of this chapter, these beliefs will change as technology, knowledge, and practice advancements occur. The key is to be sensitive to the changes in OH&S practices, and to adapt them to the OH&S program and practices.

For now, the critical success factors in achieving a healthy and safe work environment and a strong OH&S culture are:

- sustained commitment and role-modelling by the organizational/ company leaders;
- use of a multi-year effort;
- use of an integrated, multi-level approach;
- adoption of a participatory, capacity-building, cultural change process;
- establishment of shared OH&S responsibilities with clear roles; and
- commitment and passion to the above as "being the way in which you operate".[101]

---

[101] G. Lowe & J. Wetherow, "High Trust Cultures = Health and Performance" (Presented at the Industrial Accident Prevention Association Conference, Toronto, ON, April 2007).

Appendix 1

OH&S Action Plan and Register

Team: *Safety Advisors and Safety Specialists*					Meeting Date: **June 1, 2020** Note: *Shaded areas are completed actions*	
ITEM #	TASK	CUSTOMER	PERSON RESPONSIBLE	PLAN DATE	ACTUAL DATE	DELIVERABLE
Goal	Develop & Implement an effective and functional OHSMS:					A functional OHSMS
1	Management Leadership Drivers: 2019 Audit Report Good Business Practice Company XYZ OH&S Policy	Company XYZ	Management and Unions OH&S Department	August 30, 2020	OH&S Department structure completed **June 11, 2020** Policy posted **May 30, 2020** The rest is underway	• Endorse the development of an OHSMS • Define and communicate stakeholder roles, responsibilities, authority, and accountability • Focus on the desired OH&S Program structure, processes, and outcomes • Encourage union involvement • Support the development of an OH&S Program Manual • Post Company XYZ's revised OH&S Policy

Team: Safety Advisors and Safety Specialists   **Meeting Date: June 1, 2020**
Note: Shaded areas are completed actions

ITEM #	TASK	CUSTOMER	PERSON RESPONSIBLE	PLAN DATE	ACTUAL DATE	DELIVERABLE
Goal			Develop & Implement an effective and functional OHSMS:		December 31, 2020	A functional OHSMS
	Comments/Update: Planning Meeting set for July 9th: Template for OH&S Program Manual obtained: Revised OH&S Policy posted on May 30, 2020.					
2	OH&S Communication Drivers: 2019 Audit Report OH&S Legislation WHMIS Legislation Company XYZ OH&S Policy	Company XYZ	Management Safety Section Communication Department	September 30, 2020	Underway	Develop an OH&S Communication Plan that includes: • OH&S Manual • OH&S Rule Book • Communiqués • Safety Meeting topics
	Comments/Update: Underway: Safety Meetings proceeding with current Safety issues. Plans underway to develop an OH&S Communication Strategy.					
3	Hazard Identification Drivers: 2019 Audit Report OH&S Legislation	Company XYZ	Management and Unions OH&S Department	June 30, 2020	Underway	• Reinforce the importance of compliance with the Company XYZ Inspection Guideline • Enhance the Company's Hazard Identification Process • Enhance the Company's Inspection Program

Team: *Safety Advisors and Safety Specialists*					Meeting Date: *June 1, 2020* Note: *Shaded areas are completed actions*	
ITEM #	TASK	CUSTOMER	PERSON RESPONSIBLE	PLAN DATE	ACTUAL DATE	DELIVERABLE
Goal		Develop & Implement an effective and functional OHSMS:				A functional OHSMS
	WHMIS Legislation 2019 Incident Analysis Report				December 31, 2020	• Enhance the Company's Incident/Incident Investigation Program • Ensure that appropriate training on Incident/Incident Investigation is provided
	Comments/Update: Work on improving the company's Incident and Investigation Form is well underway. Completion date remains June 30, 2020.					
4	Hazard Assessment & Control Drivers: 2019 Audit Report OH&S Legislation WHMIS Legislation	Company XYZ		June 30, 2020	Underway	• Develop a system for hazard assessment and control • Document control plans • Evaluate control measures
	Comments/Update: New Company Hazard Management Program completed March 29th, 2020.					
5	Documentation Drivers: Good Business Practice	Company XYZ	Information System Department Human Resources	June 30, 2020	**June 1, 2020**	Develop & implement a mechanism for recording & managing Safety data, actions and outcomes, *e.g.*: • Workplace Incident Tracking System

Team: *Safety Advisors and Safety Specialists* **Meeting Date:** *June 1, 2020*
Note: Shaded areas are completed actions

ITEM #	TASK	CUSTOMER	PERSON RESPONSIBLE	PLAN DATE	ACTUAL DATE	DELIVERABLE
Goal	**Develop & Implement an effective and functional OHSMS:**				December 31, 2020	A functional OHSMS
	OH&S Legislation		OH&S Department			
	WHMIS Legislation					
	Comments/Update: Workplace Incident Tracking System is completed and functional (February 28, 2020).					
6	Enhance the Corrective Action Process	Company XYZ	Management and Unions	June 30, 2020	Underway	• Promote timely corrective actions in response to identified problems • Enhance the Company Incident & Investigation Form to include "sign-off" re: corrective actions taken
	Drivers: Good Business Practice		OH&S Department			
	OH&S Legislation					
	Comments/Update: Work started on improving Incident & Investigation Form; Corrective Action Status Report developed, circulated and posted.					
7	Employee Education/Training	Company XYZ	Management and Unions	Ongoing	**September 2020** complete and now ongoing	• Provide OH&S Orientation to all new employees

Team: *Safety Advisors and Safety Specialists*					Meeting Date: ***June 1, 2020*** Note: *Shaded areas are completed actions*	
ITEM #	TASK	CUS-TOMER	PERSON RE-SPONSIBLE	PLAN DATE	ACTUAL DATE	DELIVERABLE
Goal			Develop & Implement an effective and functional OHSMS:		December 31, 2020	A functional OHSMS
	Drivers: 2019 Audit Report		HR			• Educate employees on their OH&S roles, responsibility, authority, and accountability
	OH&S Legislation		Training Department			• Include ERP
	TDG Legislation		OH&S Department			• Conduct regular Safety meetings with all employees
	2019 Incident Analysis Report					
	Company OH&S Policy					
Comments/Update: New employee Safety orientations introduced, February/March 2020; to be evaluated in May 2020; Regular Safety Meetings conducted.						

CH. 18: OCCUPATIONAL HEALTH AND SAFETY: BEST PRACTICES

Team: *Safety Advisors and Safety Specialists*
Meeting Date: June 1, 2020
Note: Shaded areas are completed actions

ITEM #	TASK	CUSTOMER	PERSON RESPONSIBLE	PLAN DATE	ACTUAL DATE	DELIVERABLE
Goal	Develop & Implement an effective and functional OHSMS:				December 31, 2020	A functional OHSMS
8	Safe Work Procedures	Company XYZ	Operations	Ongoing	Underway	• Development and/or update of Safe Work Procedures for critical tasks with high/medium risks
	Drivers: 2019 Audit Report		Unions		Underway	• Contribute to development of Company XYZ Safe Work Plans
	OH&S Legislation		OH&S Department & Environmental Department (technical expertise)		Ongoing	• Address new Lock Out/Tag Out procedure for a new production process
	2019 Incident Analysis Report				Completed June 30, 2020	
	Company OH&S Policy					
	Comments/Update: Developed a new Lock Out/Tag Out procedure for new production process, March 2020; Review of Safe Work Procedures upon request.					
9	Regulation Monitoring & Safety Stewardship	Company XYZ	OH&S Department	Ongoing	Ongoing all year	• Monitor OH&S regulations

Team: Safety Advisors and Safety Specialists

Meeting Date: June 1, 2020
Note: Shaded areas are completed actions

ITEM #	TASK	CUSTOMER	PERSON RESPONSIBLE	PLAN DATE	ACTUAL DATE	DELIVERABLE
Goal	Develop & Implement an effective and functional OHSMS:				December 31, 2020	A functional OHSMS
	Drivers: Good Business Practice					<ul><li>Position Company XYZ to be proactive re: OH&S practices and business initiatives</li><li>Participate in government regulatory meetings</li><li>Participate in industry/association meetings</li></ul>
	Comments/Update: Company XYZ has been represented at all the scheduled government and industry meetings.					
10	Program Evaluation	Company XYZ	OH&S Department	Start in October Complete December 15, 2020	Underway Trend Analysis Reports prepared and sent out Q1, Q2, 2020	<ul><li>Refine Safety outcome measures</li><li>Ongoing program monitoring</li></ul>
	Drivers: Good Business Practice Alberta Partnerships Program					<ul><li>Analysis of incident forms</li><li>OH&S Audit</li></ul>

Team: Safety Advisors and Safety Specialists
Meeting Date: June 1, 2020
Note: Shaded areas are completed actions

ITEM #	TASK	CUSTOMER	PERSON RESPONSIBLE	PLAN DATE	ACTUAL DATE	DELIVERABLE
Goal	Develop & Implement an effective and functional OHSMS:				December 31, 2020	A functional OHSMS
	Comments/Update: Analysis of incident/incident report data and results communicated; Monthly and quarterly Incident/Injury Reports prepared and distributed; Refining the Company's outcome measures through the development of the Great Safety Performance Model.					
11	Management Review Drivers: Good Business Practice Company OH&S Policy	Company XYZ	VP HR	Quarterly	Underway	Regular Management review of OH&S Structure, Processes, and Outcomes
	Comments/Update: Structure of OH&S Department reviewed and enhanced; Monthly and quarterly Incident/Injury Reports prepared and distributed.					
12	Develop and Implement an Office Safety Program Drivers: 2019 Audit Report OH&S Legislation Company OH&S Policy	Company XYZ	OH&S Department	March 31, 2020	March 26, 2020	A functional Office Safety Program — "user friendly" and easily accessible
	Comments/Update: Program completed in February and rolled out in March 2020; Program promotion underway; Program placed on OH&S Website.					
13	Address Company acquisitions in terms of their OH&S Programs	Company XYZ	HR	Ongoing	March 2020 N/A May 2020	• Assess the OH&S program in place at Company ABC

Team: Safety Advisors and Safety Specialists					Meeting Date: June 1, 2020 Note: Shaded areas are completed actions	
ITEM #	TASK	CUSTOMER	PERSON RESPONSIBLE	PLAN DATE	ACTUAL DATE	DELIVERABLE
Goal	**Develop & Implement an effective and functional OHSMS:**				December 31, 2020	A functional OHSMS
	Drivers: OH&S Legislation		OH&S Department			• Assess the OH&S program in place at the new gas plant
	First Aid Regulation		Management			• Support new Company XYZ acquisitions with their OH&S needs
	Company OH&S Policy					
	Comments/Update: Health & Safety Assessment of Company ABC conducted in March 2020; Safety Assessment of two other new business lines; Development of suitable OH&S action plans for all three, May 2020; New gas plant is not "on board" so far.					
14	Address Contractor Safety Management	Company XYZ	OH&S Department	Ongoing	Underway	Assess contracts/contractors re: OH&S program, procedures, and qualifications in terms of their compatibility with the Company's business interests and practices
	Drivers: Good Business Practice	Contractors				
	Company XYZ OH&S Policy					
	OH&S Act					
	Comments/Update: Underway: Participation in Contractor Program is planned to occur, June 15, 2020.					

Team: *Safety Advisors and Safety Specialists*   **Meeting Date:** *June 1, 2020*
Note: Shaded areas are completed actions

ITEM #	TASK	CUSTOMER	PERSON RESPONSIBLE	PLAN DATE	ACTUAL DATE	DELIVERABLE
Goal	Develop & Implement an effective and functional OHSMS:				December 31, 2020	A functional OHSMS
15	Develop an OH&S Action Plan for 2021 Drivers: Good Business Practice Maintain legislative compliance	Company XYZ	OH&S Department Internal Customers	January 31, 2020	January 2020	An action plan for the delivery of OH&S Services at Company XYZ in 2021
	Comments/Update: OH&S Action Plan and Register developed.					

Team: *Safety Advisors and Safety Specialists* **Meeting Date:** *June 1, 2020*
Note: Shaded areas are completed actions

ITEM #	TASK	CUSTOMER	PERSON RESPONSIBLE	PLAN DATE	ACTUAL DATE	DELIVERABLE
Goal	Develop & Implement an effective and functional OHSMS:				December 31, 2020	A functional OHSMS
16	Promote Good Safety Performance at Company XYZ Drivers: 2019 Audit Report	Company XYZ	OH&S Department		Underway	• Develop a Safety Rewards and Recognition System
			HR		Underway	• Set up a recognition/reward system for First Aid Responders/Fire Wardens
	Good Business Practice		Line Management Senior		Completed	• Address Key Performance Indicators for OH&S performance
	Company XYZ OH&S Policy		Management Unions		Underway	• Consider the recognition of safe driving as part of the proposed OH&S Reward & Recognition System
			Business Services Operations		Model developed in May 2020	• Develop the Great Safety Performance Model
Comments/Update: Input provided re: OH&S Safety Targets for 2020 established; Development of a model for Great Safety Performance May 2020 and vetted to some customer groups.						

CH. 18: OCCUPATIONAL HEALTH AND SAFETY: BEST PRACTICES

Team: Safety Advisors and Safety Specialists
Meeting Date: June 1, 2020
Note: Shaded areas are completed actions

ITEM #	TASK	CUSTOMER	PERSON RESPONSIBLE	PLAN DATE	ACTUAL DATE	DELIVERABLE
Goal	Develop & Implement an effective and functional OHSMS:				December 31, 2020	A functional OHSMS
17	OH&S Reporting Drivers: Good Business Practice	Company XYZ	OH&S Department	June 30, 2020 May 30, 2020 February 28, 2020	**June 1, 2020** Ongoing Ongoing **January 2020**	• Address file/record management • Produce regular OH&S reports • Do trend analyses of incidents • Link OH&S and Disability Management outcomes
	Comments/Update: File/record management project is complete; Monthly and quarterly OH&S Reports produced; Meetings held re: OH&S and Disability Management outcomes: Trend Analysis of OH&S Incidents for Q1, 2020 and distributed.					
18	OH&S Marketing Drivers: Contractor Report Good Business Practice	Company XYZ Public Industry	OH&S Department Operations Communication Department	• Quarterly • Ongoing	Underway **February 2020** Ongoing	• Articles in newsletter • Safety Meetings • Information in the OH&S Website

Team: *Safety Advisors and Safety Specialists*				Meeting Date: *June 1, 2020* Note: *Shaded areas are completed actions*	
ITEM #	TASK	CUS-TOMER	PERSON RE-SPONSIBLE	PLAN DATE	ACTUAL DATE	DELIVERABLE
Goal	Develop & Implement an effective and functional OHSMS:			December 31, 2020	A functional OHSMS
					• Seek opportunities to promote OH&S within Company XYZ • Nurture external relations • Promote public safety
Comments/Update: Internal newsletter article on OH&S published; OH&S Webpage developed and implemented; Articles prepared on First Aid, AED, Office Safety, and Ergonomics; Work done re: Public Safety Program.					

CHAPTER REFERENCES

Alberta *Occupational Health and Safety Code, 2018* (Alberta Queen's Printer).

Alberta WCB, 2017 *Workplace Injury, Disease and Fatality Statistics* (2018), online: https://open.alberta.ca/dataset/c1c1b935-a5d5-456a-b86c-c9986fa5542d/resource/4572192b-df67-4aef-b85b-7eb21c417f65/download/2017-workplace-injury-disease-and-fatality-statistics.pdf (date accessed: January 31, 2020).

Alberta Workers' Compensation Board, "Safety Awareness Campaign 2006, Alberta Workplace Injury Statistics", online: http://www.wcb.ab.ca (date accessed: January 31, 2020).

Alberta Workers' Compensation Board, "Safety Performance Heralds Lower Rates for 2006", online: http://www.wcb.ab.ca/pdfs/employers/insight_archive/summer_2005_insight_print.pdf (date accessed: January 31, 2020).

Alberta Workers' Compensation Board, *Working Results: Alberta Workers' Compensation Board 2005 Annual Report* (Edmonton, AB: Workers' Compensation Board, 2005), online: http://www.wcb.ab.ca/pdfs/public/annual_report_2005.pdf (date accessed: January 31, 2020).

AWCBC, "2018 Fatalities in Canada", Statistics (2020), online: http://awcbc.org/?page_id=14 ((date accessed: January 31, 2020).

AWCBC, "Knowledge Transfer for Workplace Health & Safety: An Onsite Report on the Public Forum" (Developed at the Association of Workers' Compensation Boards of Canada (AWCBC), November 19 and 20, 2001, Toronto, ON).

J. Barling & M. Frone, *The Psychology of Workplace Health & Safety* (Washington, DC: American Psychological Association, 2004).

J. Barling & I. Hutchinson, "Commitment vs. Control-based Safety Practices, Safety Reputation and Perceived Safety Climate" (2000) 17:1 Canadian Journal of Administrative Science 76.

J. Barling, C. Loughlin & E. Kelloway, "Development and Test of a Model Linking Safety-specific Transformational Leadership and Occupational Safety" (2002) 87:3 J. of Appl Psychol. 488.

Best Practices LLC, "Best Practices in Health and Safety Management", online: http://www.benchmarkingreports.com (date accessed: January 31, 2020).

F. Bird & G. Germain, *Practical Loss Control Leadership*, revised ed. (Loganville, GA: DNV, 1996).

Board of Canadian Registered Safety Professionals (BCRSP), "Mission, Vision, Values", BCRSP Website, online: http://www.bcrsp.ca/about-us/mission-vision-values (date accessed: January 31, 2020).

H. Bruckman & J. Harris, "Occupational Medicine Practice Guidelines" (1998) 13 Occupational Medicine: State of the Art Reviews 679.

R. Burton, "Managing Health and Safety" slide used in "WSIB Ontario Workwell Program: Work Well Auditing for Prevention" (Presented at IAPA Conference, April 23, 2004, Toronto, ON).

Business Dictionary, "Best Practice" (2020), online: http://www.businessdictionary.com/definition/best-practice.html (date accessed February 28, 2020).

CCOHS, *Young Worker Zone* (2018), online: http://www.ccohs.ca/youngworkers/ (date accessed: January 31, 2020).

Canadian Centre on Disability Studies, *Best Practices in Contemporary Occupational Health & Safety: Executive Summary, 1998*, online: http://www.disabilitystudies.ca (date accessed: January 31, 2020).

Canadian Standards Association, *National Standard of Canada: Psychological Health and Safety in the Workplace – CAN/CSA-Z1003-13/BNQ 9700-803/2013* (Ottawa, ON: CSA Group, 2013), online: https://www.csagroup.org/article/cancsa-z1003-13-bnq-9700-803-2013-r2018/ (date accessed: February 28, 2020).

Changing Minds, "Transformational Leadership", online: http://changingminds.org/disciplines/leadership/styles/transformational_leadership.htm (date accessed: January 31, 2020).

A. Cohen, "Factors in Successful Occupational Safety Program" (1977) 9 J. Safety Res. 168.

M. Colligan & A. Cohen, "The Role of Training in Promoting Workplace Safety and Health" in J. Barling & M. Frone, eds., *The Psychology of Workplace Health and Safety* (Washington, DC: American Psychological Association, 2004).

N. Crawford & T. Mathers, "In Good Times and Bad" (January 2010), *Benefits Canada* at 21-23.

P. Drucker, *Managing for Results* (New York, NY: Harper Collins Publishing, 1993).

D. Dyck, *Disability Management: Theory, Strategy and Industry Practice*, 6th ed. (Toronto: LexisNexis Canada, 2017).

Energy Safety Canada, "Young Workers" Safety Resources (2018), online: http://www.enform.ca/media/8411/sa00_17_1.pdf (date accessed: January 31, 2020).

Government of Alberta, *Alberta Occupational Health and Safety Act: Highlights of changes effective June 1, 2018* (2018), online: www.alberta.ca (date accessed: January 31, 2020).

Government of Alberta, *X-TREME SAFETY: Survival Guide for New and Young Workers, Cat. No. 454992*. Edmonton, AB; Government of Alberta, Human Services (2013).

S. Greeves, "Planning to grow your contingent workforce in Alberta?" *Canadian Lawyer* (May 22, 2018), online: https://www.canadianlawyermag.com/author/susan-greeves/planning-to-grow-your-contingent-workforce-in-alberta-15759/ (date accessed: January 31, 2020).

J. Griffin, *Ucononline Underground Construction* (April 2006).

A. Hale, "Is Safety Training Worthwhile?" (1984) 6 J. Occup. Accid. 17.

Human Resources Management, *Canadian Retention Rules Summary* (2017), online: https://apps.neb-one.gc.ca/REGDOCS/File/Download/624105 (date accessed February 28, 2020).

R. Kelbert & D. Strand, "Creating a World-class Safety Culture" (Presented at the Industrial Accident Prevention Association Conference, Toronto, ON, April 2005).

S. Klie, "Culture drives performance" *Canadian HR Reporter* (December 13, 2010) at 1 and 15.

R. Lardner, M. Fleming & P. Joyner, "Towards a Mature Safety Culture" (The Keil Centre Ltd., 2001), online: http://www.keilcentre.co.uk (date accessed: January 31, 2020).

Liberty Mutual, News Release, "Reduce Future Disability Claims and Costs by nearly 50% by Improving Supervisor Response to Worker Injury and Illness" (March 6, 2006), online: http://www.libertymutualgroup.com (date accessed: January 31, 2020).

G. Lowe & J. Wetherow, "High Trust Cultures = Health and Performance" (Presented at the Industrial Accident Prevention Association Conference, Toronto, ON, April 2007).

D. Lyons, "Integrated Occupational Health & Safety: Assessing Its Fit for Your Company", *Ideas at Work* (Winter 2004), online: http://www. libertymutualgroup.com (date accessed: January 31, 2020).

NIDMAR, *Code of Practice for Disability Management* (Port Alberni, BC: NIDMAR, 2000).

NIOSH, "Young Worker Health and Safety" (2018), online: http://www.cdc.gov/niosh/topics/youth (date accessed: January 31, 2020).

Nova Scotia Canada, "Young Worker Resources", Labour and Advanced Education (2018), online: http://novascotia.ca/lae/healthandsafety/workers.asp (date accessed: January 31, 2020).

A. O'Dea & R. Flin, "Safety Leadership in the Offshore Oil and Gas Industry" (Paper presented at the Academy of Management Annual Meeting, Toronto, ON).

Ontario Ministry of Labour, New and Young Workers (2016), online: https://www.labour.gov.on.ca/english/hs/sawo/pubs/fs_youngworkers.php (date accessed: January 31, 2020).

OSHA, *Protecting Temporary Workers* (2018), online: https://www.osha.gov/temp_workers/ (date accessed: January 31, 2020).

"Passport to Safety: Setting the Standard" (2004), online: http://www. passporttosafety.com/youth/FindEmployers.php (date accessed: January 31, 2020).

Petro-Canada (now Suncor Energy Inc.), "Blueprint for Petro-Canada's Zero-Harm Culture" (July 2003). Reprinted with permission from Petro-Canada, Calgary, AB.

Petro-Canada (now Suncor Energy Inc.), *Office Workplace Leaders Tools & Tips*, "Zero-Harm in Office-based Workplaces: A Guide for Employee and Contractor Leaders and Front-Line Supervisors" (June 2006 Issue, Calgary, AB). Reprinted with permission from Petro-Canada.

Petro-Canada (now Suncor Energy Inc.), "Total Loss Management (TLM): Corporate Standards" (January 25, 2005, Calgary, AB). Reprinted with permission from Petro-Canada.

W. Rodriguez, cited in M. Paradies & L. Unger, *Taproot®: The System for Root Cause Analysis, Problem Investigation and Proactive Improvement* (Knoxville, TN: System Improvements Inc., 2000).

Saskatchewan Ministry of Labour Relations and Workplace Safety. *Young Worker Readiness Certificate Course (YWRCC)*, online: https://www.saskatchewan.ca/business/hire-train-and-manage-employees/youth-in-the-workplace/take-the-young-worker-readiness-certificate-course (date accessed: January 31, 2020).

Statistics Canada (2020). Table: 14-10-0017-01 *Labour force characteristics by sex and detailed age* group, monthly (2019), online: https://www150.statcan.gc.ca/t1/tbl1/en/tv.action?pid=1410001701 (date accessed: January 31, 2020).

Theolsongroup Benefits Group, Do You Know the Compelling Benefits of a Return to Work Program? (2019), online: https://theolsongroup.com/benefits-return-to-work-program/ (date accessed: January 31, 2020)

Towers Watson, 2013-2014 Staying@Work Report, Canada Summary (2014), online: http://www.towerswatson.com/en/Insights/IC-Types/Survey-Research-Results/2014/02/2013-2014-staying-at-work-report-canada-summary (date accessed: January 31, 2020).

UFred, *Psychological Health and Safety Certificate* (2020), online: http://www.ufred.ca/programs/school-of-occupational-health-and-safety/centre-for-psychological-health-sciences/psychological-health-and-safety-in-the-workplace-basic-level-course/ (date accessed: January 31, 2020).

P. Urs Bender, *Leadership from Within* (Don Mills, ON: Stoddart Publishing Canada, 1997).

U.S. Department of Labor, Bureau of Labor Statistics, "Fatal occupational Injuries by event or exposure and age, all United States, 2017" *Injuries, Illnesses, and Fatalities*, online: https://www.bls.gov/iif/oshwc/cfoi/cftb0320.htm (date accessed: January 31, 2020).

U.S. Department of Labor, Occupational Safety and Health, "Young Workers, OSHA Fact Sheet", online: http://www.osha.gov/Publications/young_workers.html (date accessed: January 31, 2020).

H. Williams, N. Turner & S. Parker, "The Compensatory Role of Transformational Leadership in Promoting Safety Behaviours" (Paper presented at the Academy of Management Annual Meeting, Toronto, ON).

Willis Towers Watson, *2015/2016 Global Staying@Work Survey, Canada* (2016), online: https://easna.org/wp-content/uploads/2016/04/2015-2016-Global-Staying-at-Work-Survey.pdf (date accessed: February 28, 2020).

WorkSafeBC (2019). "Young & new workers", Education, training & certification, *Health & Safety*, online: https://www.worksafebc.com/en/health-safety/education-training-certification/young-new-worker (date accessed: January 31, 2020).

WorkSafeBC (2011). *Young Worker Focus Report*, "Section 3: A Snapshot of Young Worker Injuries" (2015), online: https://www.worksafebc.com/en/resources/about-us/reports/protecting-young-workers-focus-report?lang=en (date accessed: January 31, 2020).

WorkSafeBC (2018). *Young Worker Injury Rate Comparison 2013-2017*, online: https://www.worksafebc.com/en/health-safety/education-training-certification/young-new-worker/statistics (date accessed January 31, 2020).

B. Yanar, M. Lay & P.M. Smith, "The interplay between supervisor safety support and occupational health and safety vulnerability on work injury" (2019) 10:2 Safety and Health at Work, online: doi:10.1016/j.shaw.2018.11.001.

A. Zacharatos & J. Barling, "High-performance Work Systems and Occupational Safety" in J. Barling & M. Frone, eds., *The Psychology of Workplace Health and Safety* (Washington, DC: American Psychological Association, 2004).

Chapter 19

OCCUPATIONAL HEALTH AND SAFETY: SUPPORTING DISCIPLINES

A. INTRODUCTION

The field of Occupational Health and Safety (OH&S) is comprised of many players — players from a variety of disciplines. In this chapter, the roles that a number of support disciplines can assume are discussed.

B. SUPPORT FROM OTHER DISCIPLINES: VALUE AND IMPORTANCE

1. Occupational Health Professionals

Occupational Health professionals, as noted in Chapter 4, "Occupational Health Program", are nurses and physicians with advanced education in Occupational Health.

2. Occupational/Industrial Hygienists

The role of the Occupational/Industrial Hygienist is to detect, assess, and advise on the control of environmental hazards within a workplace. The field of occupational hygiene is integrated into today's OH&S systems. Occupational Hygienists focus on the nature of hazards and their potential for negatively impacting the human body systems, tissues, and organs. The management and control of health hazards that arise out of and in the course of doing work is their primary aim.

The role and responsibilities of the Occupational Hygienist are to:

- protect the health of workers;
- maintain an objective attitude towards the identification, assessment, and control of workplace/environmental hazards;
- counsel workers on health hazards, controls, and risks; and
- respect the privacy and confidences of workers and employers.[1]

---

[1] American Board of Industrial Hygiene, online: https://www.aiha.org/about-ih/Pages/default.aspx; https://www.aiha.org/about-aiha/governance/Pages/Code-of-Ethics.aspx and https://www.aiha.org/get-involved/Academy/Documents/MemberEthicalPrinciples52107.pdf (date accessed: February 28, 2020).

For OH&S professionals, a specialist like the Occupational Hygienist is an important partner in identifying, assessing, and controlling workplace hazards. As the world of work becomes more complex, this working partnership can become even more crucial. Occupational Hygienists possess specialized knowledge in the areas of noise, biological, chemical, radiation, ergonomic, and human factor hazards. Their technical expertise on air/water/soil testing, toxicology, basic human anatomy, chemicals and exposure limits, biological agents, indoor air quality and ventilation, respiratory protection, and hygiene controls are the assets that Occupational Hygienists can offer.

Some of the contributions made by Occupational Hygienists to the field of OH&S are:

- hygiene controls (refer to Chapter 9, "Occupational Health and Safety: Hazard Identification, Assessment, and Control" for details);
- hazard control methodology (refer to Chapter 15, "Occupational Health and Safety: Risk Management and Communication" for details);
- exposure standards (refer to Chapter 9, "Occupational Health and Safety: Hazard Identification, Assessment, and Control" for details); and
- exposure monitoring (refer to Chapter 9, "Occupational Health and Safety: Hazard Identification, Assessment, and Control" for details).

3. Ergonomists

Ergonomics is defined as the study of how people interact with their work environment. This involves people, machinery, and the work organization. According to the Board of Certification in Professionals Ergonomics (BCPE), an ergonomist is defined as an individual who has:

- a mastery of ergonomics knowledge;
- a command of the methodologies used by ergonomists in applying that knowledge to the design of a product, process, or environment; and
- applied his/her knowledge to the analysis, design, testing, and evaluation of products, processes, and environments.[2]

Furthermore, per the Ergonomist Formation Model (EFM), the fundamental architecture for professional competence in ergonomics, the BCPE has defined two levels of human factors/ergonomics practice:

1. *Professional*: A career problem-solver who applies and develops methodologies for analyzing, designing, testing, and evaluating systems. The professional practitioner addresses complex problems and advances ergonomics technologies and methods.

---

[2] Board of Certification in Professionals Ergonomics, Certified Professionals Ergonomists, online: http://www.bcpe.org (date accessed: February 28, 2020).

2. ***Associate***: An interventionist who applies a general breadth of knowledge of safety, health, and/or quality issues in currently operating work systems.

While the scope of practice for the professional ergonomist covers the entire breadth and depth of ergonomics knowledge, the scope of practice for the associate ergonomist is limited to the use of commonly accepted tools and techniques for the analysis and enhancement of human performance in existing systems.

Ergonomists contribute to the design and evaluation of tasks, jobs, products, environments, and systems in order to make them compatible with the needs, abilities, and limitations of people.[3] The role of the ergonomist is to advise management and workers on the design of tools, machinery, and workstations to optimize the human interface, ease of use, and work performance.

4. Toxicologists

Toxicologists investigate the possible harmful effects on living things of chemical agents (*e.g.*, drugs, pesticides, food additives), physical agents (*e.g.*, ionizing and electro-magnetic radiation), and biological agents (*e.g.*, plant and animal toxins). Although not usually employed directly by an organization, they are often contracted to investigate the hazards of chemical, physical, and biological agents at different levels of exposure. Their goal is to improve industrial safety, public, and environmental protection through a better understanding of the hazards to which living species are exposed.

In general, toxicologists:

- conduct laboratory studies on substances (*e.g.*, drugs, food additives, solvents, herbicides), or on energy (*e.g.*, radiation) to determine their effects on laboratory animals, plants, and human tissue;
- conduct research to develop new tests for use in toxicological studies;
- evaluate potential risks based on levels and periods of exposure;
- analyze and evaluate data gathered from studies and reliable scientific publications to determine appropriate controls for various chemical and physical hazards;
- develop standards or guidelines for safe levels of chemical and physical agents in workplaces, air, food, or drinking water;
- provide advice and scientific information to policy and program developers concerning the health and legal aspects of chemical use; and
- supervise and coordinate the activities of technologists and technicians.

The minimum education requirement for toxicologists is an appropriate bachelor's degree. Most toxicologists hold advanced (master's or doctoral) degrees in

---

[3] International Ergonomics Association (IEA), "What is Ergonomics", online: http://www.iea.cc (date accessed: February 28, 2020).

toxicology or a related area. A doctoral (PhD) degree is usually required to direct and administer research programs, or to teach at the college or university level.

Different specializations require different academic backgrounds. For example, veterinary toxicologists must become veterinarians first, and then take an advanced degree in toxicology. Toxicology is an interdisciplinary science which draws from diverse fields including biology, chemistry, biochemistry, pharmacology, mathematics, physiology, pathology, immunology, and genetics.[4]

5. Employee Assistance Counsellors

Employee Assistance Counsellors are registered psychologists, social workers, or nurses who function as part of an Employee Assistance Program (EAP). They are licensed within their province of residence and belong to professional associations that guide, and in some instances, govern their practice.

Employee Assistance Counsellors focus on assisting employees and their family members to deal with personal/emotional/relationship issues by:

- providing early intervention and treatment for psychological disorders;
- providing support and intervention for psychological and social aspects of a physical illness; and
- providing support as well as referrals for employees with emotional or psychological health, and/or addiction problems.

They also support the workplace by:

- assisting with return-to-work issues;
- coaching supervisor and co-workers regarding support for the returning employee;
- assisting the employee and workplace when a successful return-to-work is not feasible; and
- identifying trends in psychological illness within the workplace.

With the rise in the incidence of clinical depression and other psychological health illnesses among workers,[5] as well as an increase in workplace violence particularly in the retail, social assistance, and healthcare industries,[6] EAPs can be a valuable support to workers and their families.

---

[4] Alberta Government, Alberta Learning Information Service, OCC Info — Alberta Occupational Profiles "Toxicologist", online: http://www.alis.gov.ab.ca/occinfo (date accessed: February 28, 2020).

[5] Ipsos-Reid, "Ipsos-Reid Survey" reported by M. Shaw, "How can we . . ." *Accident Prevention Magazine* (March/April 2007) at 4.

[6] Statistics Canada, *2004 General Social Survey* reported in "In the News" *Accident Prevention Magazine* (March/April 2007) at 6.

6. Rehabilitation Therapists

Three types of Rehabilitation Therapists will be discussed — *Physical Therapists, Occupational Therapists* and *Vocational Therapists*. Although their respective focuses differ, each discipline uses a client-focused, case management approach that includes:

- early identification;
- asssessment;
- development of a rehabilitation plan;
- goal setting;
- coordination; and
- evaluation.

These services can be internally or externally located; however, most are part of an externally located agency that is providing contracted services to organizations.

(a) Physical Therapists

Physical Therapists are paramedical professionals who possess specialized training in physical therapy. They have skills in identifying physical limitations and in planning suitable recovery programs for people experiencing physical limitations. Provincially licensed and regulated, these professionals:

- provide physical treatment;
- provide information on the employee's physical capabilities and work limitations;
- can perform Functional Capacity Evaluations, ongoing functional testing and work tolerance screens; and
- can perform Job Demands Analyses (JDAs).

(b) Occupational Therapists

Also paramedical professionals, Occupational Therapists specialize in occupational therapy. Occupational therapy is a health profession concerned with promoting health and well-being through everything people do during the course of everyday life.[7] The primary goal of occupational therapy is to enable workers to participate in the occupations which give meaning and purpose to their lives.

Occupational Therapists have a broad education that provides them with the skills and knowledge to work collaboratively with people of all ages and disabilities. The challenges faced by their clients result from a change in function (*e.g.*, thinking, doing, feeling) due to illness or disability, and/or barriers in the social, institutional,

---

[7] Canadian Association of Occupational Therapists (CAOT), "Position Statement on Everyday Occupations and Health", online: http://www.caot.ca (date accessed: February 28, 2020).

or physical environment.[8] Occupational Therapists provide vocational guidance and support to recovering employees and their family members. They can perform Functional Capacity Evaluations (FCEs), ongoing functional testing, and work tolerance screens, as well as Job Demands Analyses (JDAs).

Occupational therapists use a systematic approach based on evidence and professional reasoning to enable individuals, groups, and communities to develop the means and opportunities to identify, engage in, and improve their level of work functioning.[9] The process involves assessment, intervention, and evaluation of the worker.

Occupational therapists may assume different roles such as advising on health risks in the workplace, safe driving for older adults, and programs to promote psychological health for youth. Occupational therapists also perform functions as manager, researcher, program developer, or educator in addition to the direct delivery of professional services.

Occupational therapists are generally employed in community agencies; healthcare organizations such as hospitals, chronic care facilities, rehabilitation centres and clinics; schools; social agencies; industry; or are self-employed. Some occupational therapists specialize in working with a specific age group or disability such as arthritis, developmental coordination disorder, psychological illness, or spinal cord injury.

(c) Vocational Therapists

The prime focus of Vocational Therapists is to assist the disabled person to return to work. These professionals come from a variety of educational backgrounds — occupational therapists, kinesiologists, nurses, psychologists, and social workers. Most possess professional certification such as the Canadian Certified Rehabilitation Consultant, or the American version.

7. Human Resources Professionals

Human Resources (HR) professionals specialize in human resources functions such as employee recruitment, selection and retention; employee training and development; employee career counselling; performance appraisal; and the management of organizational change, as well as the relationship of human resources functions to the organization's goals and strategic planning process.

According to Zacharatos and Barling, there are 12 Human Resources practices that organizations can use to promote workplace health and safety, namely:

- ensure employment security for employees;

---

[8] World Federation of Occupational Therapists (WFOT), "WFOT Information", online: http://www.wfot.org.au/ (date accessed: February 28, 2020).

[9] Canadian Association of Occupational Therapists (CAOT), "Definition of Occupational Therapy", online: http://www.caot.ca (date accessed: February 28, 2020).

- selectively hire when employing new personnel;
- provide extensive worker training;
- promote the use of centralized decision-making and self-managed teams;
- reduce the number of status symbols that separate employees into hierarchical levels;
- promote information-sharing throughout the organization/company;
- provide compensation increases based on the organization's performance and individual merit;
- promote the use of transformational leadership;
- educate front line leaders on how to be transformational leaders;
- recognize that high quality work measurement variables are critical to organizational success;
- adopt performance measure variables that are critical to organizational success; and
- promote and reward job quality.[10]

8. Workplace Wellness Professionals

Workplace wellness professionals tend to come from a number of educational disciplines. Their prime focus is on employee wellness through the practice of health promotion principles and practices such as those described in Chapter 21, "Occupational Health and Safety: Prevention of Workplace Illness and Injury" and Chapter 26, "Occupational Health and Safety: Workplace Wellness Strategy".

C. SUMMARY

By linking an organization's OH&S Program with its Employee Assistance Program, Disability Management Program, Human Resources Program, and Workplace Wellness Program, and by recognizing the impact that management theories can have on a workforce and workplace wellness, the opportunity to significantly reduce illness/injury incidence and impact exists. The challenge is to act on that knowledge in a proactive manner.

CHAPTER REFERENCES

Alberta Government, Alberta Learning Information Service, OCC Info — Alberta Occupational Profiles, "Toxicologist", online: http://www.alis.gov.ab.ca/occinfo (date accessed: February 28, 2020).

American Board of Industrial Hygiene, online: https://www.aiha.org/about-ih/Pages/

---

[10] A. Zacharatos & J. Barling, "High-performance Work Systems and Occupational Safety" in J. Barling & M. Frone, eds., *The Psychology of Workplace Safety* (Washington, DC: American Psychological Association, 2004), 203 at 204-205.

default.aspx; https://www.aiha.org/about-aiha/governance/Pages/Code-of-Ethics.aspx and https://www.aiha.org/get-involved/Academy/Documents/MemberEthicalPrinciples52107.pdf (date accessed: February 28, 2020).

Board of Certification in Professionals Ergonomics, Certified Professionals Ergonomists, online: http://www.bcpe.org (date accessed: February 28, 2020).

Canada Labour Code, R.S.C. 1985, c. L-2.

Canadian Association of Occupational Therapists (CAOT), "Definition of Occupational Therapy", online: http://www.caot.ca date accessed: February 28, 2020).

Canadian Association of Occupational Therapists (CAOT), "Position Statement on Everyday Occupations and Health", online: http://www.caot.ca (date accessed: February 28, 2020).

International Ergonomics Association (IEA), "What is Ergonomics", online: http://www.iea.cc (date accessed: February 28, 2020).

Ipsos-Reid, "Ipsos-Reid Survey" reported by M. Shaw, "How can we . . ." *Accident Prevention Magazine* (March/April 2007).

Statistics Canada, *2004 General Social Survey*, reported in "In the News" *Accident Prevention Magazine* (March/April 2007).

World Federation of Occupational Therapists (WFOT), "WFOT Information", online: http://www.wfot.org.au/ (date accessed: February 28, 2020).

A. Zacharatos & J. Barling, "High-performance Work Systems and Occupational Safety" in J. Barling and M. Frone, eds., *The Psychology of Workplace Safety* (Washington, DC: American Psychological Association, 2004).

Chapter 20

OCCUPATIONAL HEALTH AND SAFETY: ERGONOMICS

A. INTRODUCTION

Ergonomics is defined as the science of making workers comfortable: Applying knowledge of the physical and mental abilities of humans to the design of systems, organizations, jobs, machines, tools, and consumer products for safe, efficient, and comfortable human use. It is the study of how people interact with their work environment. The intent is to balance human capabilities with the job demands. Problems develop when the job demands exceed human capabilities. In short, ergonomics is "fitting the job to the person", as opposed to forcing the worker to adapt to the work conditions.

The focus of this chapter is on the role that ergonomics can play in the prevention and mitigation of employee illness/injury. Given that musculoskeletal injuries are difficult to rehabilitate, and the related costs are significant to employers, having an ergonomic program linked with an OH&S Program and Disability Management Program makes perfect sense.

B. MUSCULOSKELETAL DISORDERS

Although the human body adapts readily to many situations, fatigue, musculoskeletal disorders, and human error occur when the physical and psychological demands of the job exceed human capabilities.

A **musculoskeletal injury** is a disorder of the muscles or other soft tissues in the arms, legs, neck, back, or eyes, which become irritated and inflamed because of repetitive motions, excessive force, and/or extremes of motion. They are known by a variety of terms, namely:

- Repetitve strain injury (RSI);
- Musculoskeletal injuries (MSI);
- Workplace musculoskeletal disorder (WMSD);
- Cumulative trauma disorder (CTD).

Musculoskeletal disorders are typically named in association with the affected body part or activity that lead to the condition. For example, the common musculoskeletal injuries that the OH&S or Disability Management Practitioner/ Professional might encounter are:

- *Back injuries* – lumbar strain, Mechanical Back Syndrome, degenerative disc disease, *etc.*;
- *Hand and wrist problems* – Gamekeeper's Thumb, Cotton Twister's Hand, Stitcher's Wrist, Roller Ball Thumb, Carpal Tunnel Syndrome, Trigger Finger, *etc.*;
- *Elbow disorder* – Tennis Elbow, medial epicondylitis, ulnar entrapment, *etc.*;
- *Shoulder injury* – rotator cuff tendonitis, thoracic outlet syndrome, *etc.*;
- *Neck injury* - Text Neck, Tension Neck Syndrome, *etc.*

Strains and sprains of muscles/joints have been and continue to be, the leading type of workplace injury throughout Canada. Back injuries are the number one body part affected by work and the leading cause of Workers' Compensation claims, ranging from back strain to serious back damage. The most common reason for workplace back injury is improper lifting techniques.

C. MUSCULOSKELETAL DISORDERS: CAUSES

Ergonomic-related injuries don't just happen. By combining highly repetitive motions with fast, forceful movements, and awkward positions over time, an injury can develop. The basic causal elements of musculoskeletal injury are:

Repetition + Awkward Position + Force + Time + No Rest

For example, repetitive and forceful pinching and gripping is associated with carpal tunnel syndrome — the painful hand condition due to entrapment of the median nerve, and typified by tingling, numbness, burning, and weakness in the fingers and hand.[1]

An **awkward posture** is any position that is beyond the normal resting position for a joint, limb, or the entire body. It does include a stationary (static) position which is abnormal for the human body which is built to be in motion.[2]

Many of the ways in which people work — such as lifting, reaching, repeating the same motions — strain their bodies. By recognizing this hazard and the signs and symptoms of injury, workers can look for safer ways to work. Early detection also can make medical problems easier to treat and to cure.

The factors that may cause the signs or symptoms of musculoskeletal injury include:

- Sitting, standing, or holding an object in the same position for extended period;

---

[1] B. Evanoff *et al.*, "Forceful repetition a carpal tunnel risk factor" *At Work*, Issue 85, online: https://www.iwh.on.ca/at-work/85/forceful-repetition-a-carpal-tunnel-risk-factor (date accessed February 28, 2020).

[2] J. Callaghan, "Too Much Standing Hurts, Too" *At Work*, Issue 85, online: https://www.iwh.on.ca/at-work/85/too-much-standing-hurts-too (date accessed February 28, 2020).

- Making fine hand movements;
- Repeating movements, especially at high speed;
- Working with a bent wrist;
- Lifting and handling heavy objects;
- Twisting with a load;
- Working in awkward postures, such as bent over or above shoulder height;
- Reaching far forward or far behind;
- Gripping with force;
- Working with cold hands;
- Working with vibrating tools or on vibrating surfaces;
- Working with tools that put direct pressure on body parts;
- Repetitive overhead lifting or reaching;
- Doing static overhead work.

Additionally, according to Dr. Julie Côté, McGill University, "women who do the same tasks as men report pain, discomfort, and other symptoms of musculoskeletal disorders in the neck and upper limbs about twice as often as men; whereas men, are more likely to experience low-back injuries."[3] This higher risk may be due to biological (sex) differences as well as differences in social roles, activities, and behaviours (gender).[4]

People vary. Not all employees who move in these ways are affected. Also, the more of the above factors that are present in the work activity, the greater the risk of musculoskeletal injury.

D. OH&S: VALUE OF ERGONOMICS

Many employers recognize that having a robust Ergonomic Program can reduce the human, business, and financial costs associated with workplace musculoskeletal injury. Ergonomic hazards affect all industries/businesses and all workers. They relate to poor workplace design, repetitive motions, awkward body mechanics, and extreme working conditions. As well, they are extremely costly to Canadian employers.

> In Canada, musculoskeletal disorders (MSD) account for the most lost time injuries, the highest lost time claim costs, and the most lost time workdays of any type of injury. Worker's compensation board figures from various jurisdictions

---

[3] J. Côté, "Higher risk of some musculoskeletal injuries among women may be due to sex and gender differences beyond size and strength" *At Work, Issue 85*, online: https://www.iwh.on.ca/media-room/news-releases/2016-jun-23 (date accessed February 28, 2020).

[4] *Ibid.*

indicate that MSDs account for 25% to 60% of total annual compensation claims.[5]

For example:

"MSDs account for 43% of all lost-time claims, 43% of all lost-time claim costs, and 46% of all lost-time days."[6]

"The Canadian Orthopedic Care Strategy Group deems occupational and non-occupational MSDs combined to be "the most costly medical condition in Canada", again, estimating direct and indirect costs to be in the $22 billion range."[7]

"One in 15 Canadians experience repetitive strain injuries that affect normal activities."[8]

". . . a significant number of these disorders attributed to related workplace hazards.."[9]

The U.S. parallels Canada in terms of the cost of musculoskeletal injuries. According to the U.S. Bureau of Labor Statistics:

- 32% of work injury/illness cases are due to musculoskeletal disorders;[10]
- 34% of all lost work days are due to musculoskeletal disorders;[11]
- 24% of work injuries are due to overexertion, costing $13.11 billion in direct costs;[12]

---

[5] CCOHS "Product description for Musculoskeletal Disorders (MSD) Prevention Manual", *CCOHS website* (2017), online: http://www.ccohs.ca/products/publications/msd/ (date accessed February 28, 2020).

[6] Ontario Government, "Ergonomics in the Workplace" *Law and Safety* (2019), https://www.labour.gov.on.ca/english/hs/pubs/ergonomics/is_ergonomics.php (date accessed February 28, 2020).

[7] Workers Health & Safety Centre, "Making the case for MSD Prevention: The economics of ergonomics" (February 2016) at 1, online: https://www.whsc.on.ca/Files/Resources/Ergonomic-Resources/RSI-Day-2016_MSD-Case-Study_The-economics-of-ergon.aspx (date accessed February 28, 2020).

[8] WorkSafeBC, 2012. www.worksafebc.com (date accessed February 28, 2020).

[9] Workers Health & Safety Centre, "Making the case for MSD Prevention: The economics of ergonomics" (February 2016), at 1, online: https://www.whsc.on.ca/Files/Resources/Ergonomic-Resources/RSI-Day-2016_MSD-Case-Study_The-economics-of-ergon.aspx (date accessed February 28, 2020).

[10] US Bureau of Labor Statistics, News Release, "Nonfatal Occupational Injuries and Illnesses Requiring Days Away from Work" (November 2015), online: http://www.bls.gov/news.release/pdf/osh2.pdf (date accessed February 28, 2020).

[11] US Department of Labor, Occupational Safety and Health Administration, "2016. Prevention of Work-related Musculoskeletal Disorders" (2016), online: https://www.osha.gov/pls/oshaweb/owadisp.show_document?p_table=UNIFIED_AGENDA&p_id=4481 (date accessed February 28, 2020).

[12] Liberty Mutual Research Institute for Safety, "2019 Workplace Safety Index: The Top

- $1 of every $3 of Workers' Compensation costs are spent on musculoskeletal disorders;[13]
- Workers with musculoskeletal disorders missed 13 days to recover before returning to work;[14]
- Nursing assistants, laborers, and freight, stock and material movers were the prime victims;[15]
- Incidence of musculoskeletal disorders increases with age;
- Mean cost/case for upper extremity MSD injury = $8,070 versus $4,075/case for other types of work-related injury;
- The average cost per musculoskeletal injury ranges from $29,000-$32,000;[16]
- $20 billion in direct costs spent annually;[17] and
- $100-$150 billion spent annually on indirect costs.[18]

E. ERGONOMIC PROGRAM: PURPOSE

The purpose of an Ergonomic Program is to establish a systematic method for recognizing and controlling ergonomic risk factors in the workplace, whether in industrial or office settings. By providing employees with a functional and comfortable work environment, the employee remains safe, healthy, comfortable, and satisfied; this results in more employee involvement, less employee absenteeism and staff turnover, and greater employee commitment to a flexible work environment. For the organization/company, this translates to stronger employee performance and productivity, good service quality, and the ability to adapt to a changing

---

10 Causes of Disabling Injuries at Work" *Viewpoint*, online: https://viewpoint.libertymutualgroup.com/article/top-10-causes-disabling-injuries-at-work-2019/ (date accessed: January 31, 2020).

[13] US Bureau of Labor Statistics (2016). Reported by *ATI Worksite Solutions*, online: https://www.atiworksitesolutions.com/ (date accessed February 28, 2020).

[14] US Bureau of Labor Statistics, News Release, "Nonfatal Occupational Injuries and Illnesses Requiring Days Away from Work" (November 2015), online: http://www.bls.gov/news.release/pdf/osh2.pdf (date accessed February 28, 2020).

[15] *Ibid.*

[16] US Bureau of Labor Statistics (2016). Reported by *ATI Worksite Solutions*, online: https://www.atiworksitesolutions.com/ (date accessed February 28, 2020).

[17] US Department of Labor, Occupational Safety and Health Administration, "2014. Prevention of Work-related Musculoskeletal Disorders" (2014), online: https://www.osha.gov/pls/oshaweb/owadisp.show_document?p_id=4481&p_table=UNIFIED_AGENDA (date accessed February 28, 2020).

[18] US Department of Labor, Occupational Safety and Health Administration, "2014. Prevention of Work-related Musculoskeletal Disorders" (2014), online: https://www.osha.gov/pls/oshaweb/owadisp.show_document?p_id=4481&p_table=UNIFIED_AGENDA (date accessed February 28, 2020).

marketplace. It is a "win-win" for both parties.

The components of an Ergonomic Program include:

- *Management commitment* — Management commitment would include the development of an Ergonomic policy that links with the organization's/company's business strategy and demonstrates endorsement of the program. The requisite resources along with support for ergonomic improvements, must be provided. Employee training, as part of the organization's/company's OH&S education and training program, must be offered with a view to promoting employee involvement. Management commitment should be demonstrated through rewarding continuous program improvement efforts and initiatives. *"What gets rewarded, gets attention."*

- *Employee involvement* — Like Management commitment, employee commitment is critical. Commitment begins with awareness, followed by acceptance and a willingness "to try it out". It moves on to regular use, adoption of the practices into organizational procedures, and finally, incorporation of the practices into *"the way things are done around here"* — the culture.

  To operationalize program commitment, educate the workforce on the principles of ergonomics, develop ergonomic teams, conduct hazard analyses, use mutual problem solving, and learn from the findings of related incident investigations.

  Importantly, everyone in the workplace should participate in the Ergonomic Program. The use of a multidisciplinary approach to addressing ergonomic hazards is recommended.

- *Program management* — The Ergonomic Program is composed of the Ergonomic Policy and plan, stakeholder education and training, a reactive approach to ergonomic hazard management (mitigation), a proactive approach to ergonomic hazard management (prevention), medical management, and program evaluation and feedback.

- *Worksite analysis* - This can be done reactively to an identified issue, or proactively for all worksites. The intent is to identify and control ergonomic hazards.

- *Job/Position Demand Analysis* - This is another technique for identifying ergonomic hazards. For information on a Job Demands Analysis, refer to Chapter 4: "Occupational Health Program", Appendix 4.

- *Hazard prevention and control* - Ergonomic hazards in the workplace must be identified to be controlled and prevented. Line Managers, employees, and OH and OH&S Practitioners/Professionals are typically involved in evaluating the ergonomic hazards on a basic level. A thorough, expert ergonomic analysis would need to be done by an ergonomist.

- *Education and training* - All stakeholders should be offered ergonomic education and on-the-job training when the Ergonomic Program is introduced, on an ongoing basis, at new employee orientations and when new processes or products are introduced. Education and training promote ergonomic awareness, problem-solving skills, team-building skills, program management, and engineering design to continuously improve the work environment. Samples of educational materials are provided in Appendices 1, 2 and 3.

- *Medical management* - The early identification and management of musculoskeletal injuries is the ideal approach. Organizations should educate employees on the signs and symptoms of musculoskeletal injuries and how to address them. Early reporting should be encouraged. Provisions for the evaluation of musculoskeletal injuries, and referral for treatment are necessary.

To achieve this end, the following actions are required:

- Early reporting of musculoskeletal problems;
- Implementation of an evaluation and referral process;
- Medical assessment and treatment;
- Availability of a Return-to-Work (RTW) Program;
- Suitable job accommodation designed to promote recovery;
- Provide work hardening/conditioning;
- Monitor the placement and recovery;
- Documentation; and
- Ongoing case evaluation.

Having a RTW Program and job accommodation options help employees to stay in the workplace and enable them to prevent further injury. The monitoring of work accommodations and employees recovering from musculoskeletal injuries are strongly recommended. Documentation of the details of the musculoskeletal injury situation is essential for the protection of the employee and the organization. Case evaluation is advisable — it is an effective way to enhance the identification and management of musculoskeletal injuries.

F. ERGONOMIC PROGRAM: VALUE

In making a business case for the implementation of an Ergonomic Program, the OH&SPractitioner/Professional should educate Management and employees on the direct and indirect benefits that can be realized. For example:

> the purchase of a robotic palletizer – replacing a manual handling task at a warehouse – a cost of $300,000.00, reduced both labour and back injury claims, yielding a return on investment of 6 per cent in just three years.[19]

---

[19] Workers Health & Safety Centre, "Making the case for MSD Prevention: The

In the Humantech survey (2014) undertaken in the U.S., participating companies reported that musculoskeletal disorders constituted 21 to 82 per cent of their reportable injury/illness claims. By implementing an Ergonomic Program, they reduced their illness/injury rates by 4.9 to 9 per cent, and increased staff retention by 25-50 per cent. They concluded: "Return on investment ranged from 77 to 1,513 per cent, with an average 378 per cent return on the initial investment."[20]

G. ERGONOMIC PROGRAMS AND WORK ACCOMMODATION

To develop a suitable work accommodation for the recovering employee, or an employee experiencing **diminished functional capacity**[21], a systematic approach should be used to ensure a good person-job fit (Figure 20.1).

When determining a suitable work accommodation placement, begin by defining the employee's qualifications and competency to do the job. All work accommodation placements must be "safe placements". Is the employee qualified to do the job safely?

Next, determine the physical and psychological demands of the proposed work accommodation placement. Define the employee's physical and psychological capabilities and limitations. Is there a "person-job fit"? If there is a reasonable fit, then proceed to a trial placement of the employee in the proposed work accommodation assignment. If there is a "no person-job fit", seek an alternate placement; reassess the employee's capabilities and limitations to determine if there is a reasonable fit, and then proceed to a trial placement in the proposed work accommodation. If there is still a "no person-job fit", continue to seek more information and an alternate solution, leaving the employee off work until that outcome can be achieved.

---

economics of ergonomics", (February (2016) at 2, online: https://www.whsc.on.ca/Files/Resources/Ergonomic-Resources/RSI-Day-2016_MSD-Case-Study_The-economics-of-ergon.aspx (date accessed February 28, 2020).

[20] Humantech (2014), reported in Workers Health & Safety Centre, "Making the case for MSD Prevention: The economics of ergonomics" (February 2016) at 3, online: https://www.whsc.on.ca/Files/Resources/Ergonomic-Resources/RSI-Day-2016_MSD-Case-Study_The-economics-of-ergon.aspx (date accessed February 28, 2020).

[21] **Diminished functional capacity** is defined as this term is used to refer to the employee's inability to undertake or participate in the regular duties of the job.

Figure 20.1: Work Accommodation

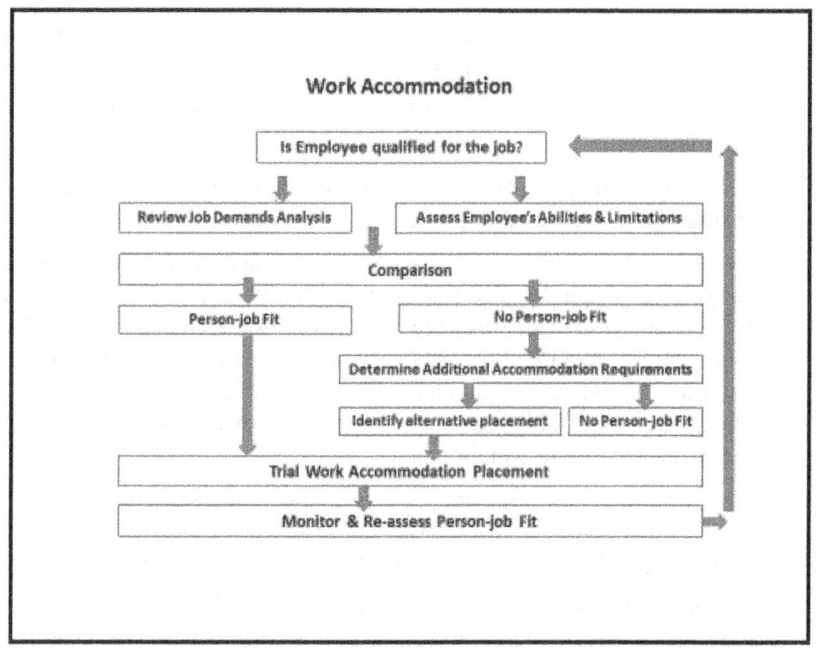

There are many ways to create a reasonable work accommodation. Typically, the OH&S/Disability Management Practitioner/Professional begins by identifying what job accommodations are warranted. Will a suitable placement require a change in job duties, job hours, or the work environment? Would workstation resign be a suitable option? Job aids and supportive devices may be needed. In the marketplace, there is a plethora of ergonomic tools, aids, and devices. Most are quite expensive. Before purchasing any ergonomic aid, it is essential to ensure that the tool/aid/device will indeed address the problem. As such, seek a "test-drive" of the device before buying.

H. ERGONOMICS AND THE AGING WORKER

Approximately 26% of today's workforce is comprised of workers 60 years or more.[22] These workers have much to offer the world of work, but to prevent injury situations, their diminishing capacities need to be recognized and accommodated. They include changes in their:

- *Muscles* — gradually lose strength, flexibility response time, and range of

---

[22] Benefits Canada, "90% of Canadians working past age 60: Stats Can" *Benefits Canada, December 2018* (2018), online: https://www.benefitscanada.com/news/90-of-canadians-working-past-age-60-stats-can-123402 (accessed February 28, 2020).

motion. At 51-55 years of age, humans possess 80% of the strength they had at aged 30 years. Body fat also diminishes.[23]

- *Bones* — lose calcium, becoming more porous. The shape of the bones changes and they tend to break easier.
- *Cartilage* — which provides cushioning in a joint, reduces and can lead to joint damage.
- *Soft tissue* — skin and other soft tissue dry out and become less elastic and flexible. Skin can become less resistant to damage.
- *Muscles* — lose their muscle mass at a rate of approximately one per cent per year after age 30. The muscle fibers shrink and if damaged, muscle tissue is slower to heal. Muscle tone decreases as does the muscle flexibility. Older people require more recovery time after doing sports and other activities. By age 65, humans have 70% of the strength that they possessed when they were 25-30 years old. Additionally, their Basal Metabolic Rate decreases, lending to weight gain.
- *Cardiovascular and respiratory systems* - breathing capacity reduces with aging. Blood vessels lose flexibility and artery walls thicken. This can lead to an increased risk of elevated blood pressure and stroke.[24]
- *Nervous system* — nerve cell mass decreases resulting in a decreased brain and spinal cord size; hence, a slowing of messages sent and received. There is a slight reduction in thought, memory, and reaction processes. The outcome is decreased pain, temperature, and pressure change perception. Inductive reasoning, selective attention, multi-tasking, and information processing, decrease with age, as do the individual's reflexes, coordination and balance.
- *Hearing* — declines with age due to changes in the inner ear. The eardrum thickens and the ability to hear higher noise frequencies decreases. The brain's ability to process sound messages becomes less. This makes it difficult for the older worker to locate the source of a sound.
- *Vision* — difficulty with reading is common after age 40 years. Problems accommodating light changes are common. Depth perception and the ability to discern colours like blue and green, change. Sensitivity to glare and the need for better lighting, increase.
- *Sensory and motor processes* — the sense of smell decreases with age which could have workplace implications. Their reaction time and response time may increase with aging.

---

[23] J. Barbour, "Ergonomics and the Aging Workforce" (2007), Presented at the CSSE Conference, Calgary, Alberta.
[24] *Ibid.*

- *Manual dexterity and feedback* — can decrease and is often associated with chronic disease states like arthritis.

As well, aging workers tend to experience the following health and lifestyle changes:

- A decrease in their maximum oxygen intake – the reduction is 10% per decade after age;[25]
- Rising systemic blood pressure – after aged 30 years, systemic blood pressure tends to increase;[26]
- Fatigue;
- Greater susceptibility to the negative effects of extreme temperatures;
- A change in shiftwork tolerance and preferences;
- A change in their training and learning styles;
- Increased disenfranchisement and disengagement with their work responsibilities;
- Increased incidence of chronic disease and illness.

The Workers' Compensation injury data indicates that older workers do have fewer workplace accidents, but when injured, they take much longer to recover. Older workers are more susceptible to musculoskeletal injuries, which by their very nature, require a lengthy recovery. Injuries are more likely if the older worker is working in a static position.[27]

The OH&S/Disability Management Practitioner/Professional can position the employer to counteract these aging effects by promoting work-design changes. The employer can assist and protect the aging employee by:

- Providing mechanical manual handling/lifting devices;
- Minimizing lifting and twisting through workstation redesign;
- Providing supportive and adjustable seating;
- Designing work so it can be done with the employee positioned in neutral postures;[28]

---

[25] S. Hawkins & R. Wiswell (2012), "Rate and Mechanism of Maximal Oxygen Consumption Decline with Aging Implications for Exercise Training" (2003) 33:12 Sports Med 877-888.

[26] E. Pinto, "Blood pressure and ageing" (2007) 83:976 Postgrad Med J. 109-114.

[27] K. Tiels, Aging Workforce – What's Ergonomics Got to Do With It? Presented at the IAPA Conference & Trade Show (Toronto April 16-18, 2007).

[28] C. Lotz, "Ergonomics Hit List: Identifying 10 Workplace Issues" (2012) 57:11 Professional Safety: Journal of the ASSE 50-51.

- Enabling work to be done from the "power zone" – that is, at the "handshake zone";[29]
- Promoting the use of a "good hand grip" – urge the use "power grips" instead of pinch grips for doing jobs and tasks;[30]
- Supplying well-designed tools to improve grip, lessen bending, and twisting;
- Allowing for frequent postural changes;
- Providing good lighting;
- Reducing visual glare;
- Requiring good housekeeping to minimize slipping, tripping, and fall risks;
- Keeping walking surfaces clean, clear, and even;
- Installing handrails and touch surfaces for stabilization of movements;
- Minimizing repetitive tasks;
- Eliminating or controlling vibration;
- Planning work assignments to avoid employee fatigue;
- Avoiding/minimizing temperature extreme exposures;
- Developing a hearing conservation program that includes an educational component;
- Reduce the amount of ambient noise;
- Speaking clearly in a quiet environment when relaying important instructions;
- Limiting the need for multi-tasking;
- Providing time to practice and reinforce work tasks;
- Using multiple warning alarm systems in the workplace;
- Minimize exposure to temperature extremes; and
- Offering ergonomic training and guidelines for all employees.

As well, the employer can encourage and support aging employees to:
- Maintain a healthy lifestyle – to exercise and eat a healthy diet;
- Add weight-bearing activities to their exercising program;
- Use material handling equipment when moving items;
- Maintain proper posture;
- Wear suitable footwear, especially when working on uneven surfaces;

---

[29] *Ibid.*

[30] *Ibid.*

- Make frequent postural changes;
- Encourage stretching before, during, and after work;
- Minimize awkward postures when working;
- Get enough sleep;
- Dress for weather extremes and drink plenty of water;
- Undergo regular vision and hearing tests and use corrective lenses and hearing aids if prescribed;
- Be aware of drug side-effects, even when taking over-the-counter medications;
- Exercise the brain and cognitive functions; and
- Welcome involvement in work challenges and decisions.

For more information on the prevention and effective management of musculoskeletal disorders refer to: the CCOHS website at: http://www.ccohs.ca/products/posters/msd/?utm_source=SilverpopMailing&utm_medium=email&utm_campaign=Liaison%20Jan%202017%20ENGLISH&utm_content=.

I. SUMMARY

By linking an organization's OH&S Program and the Disability Management Program with the Ergonomic Program, the opportunity to significantly reduce the incidence of expensive musculoskeletal injuries exists. The challenge is to act on that knowledge in a proactive manner.

Appendix 1

EDUCATIONAL MATERIALS – SAMPLE #1
General Ergonomic Safety Tips for Employees

To sustain good health:
1. ***Set up your work area to suit your body type and work.***
    - Adjust your chair so that your feet are flat on the floor with your weight shifted slightly forward. Use a footrest if your feet are not flat on the floor.
    - Adjust the chair's backrest to maintain the curve in your lower back. Add a pillow or rolled towel as an additional lumbar support if needed.
    - Tilt the chair seat pan forward to relieve thigh tension.
    - Make sure the video display terminal (VDT) screen is positioned in front of you with the top of the monitor no higher than your eye level.
    - Place the monitor screen at right angles to any windows to eliminate glare. Eliminate the source of glare whenever possible.
    - Adjust the screen brightness control to "low" and the contrast control to "high".
    - Use a document holder that allows the document and the screen to be at the same level.
    - Place the mouse at the same height as the keyboard and as close to the keyboard as possible.
    - Vary tasks throughout the day.
    - Mini-stretch breaks at 30-minute intervals throughout the day are recommended.
2. ***Avoid awkward working positions.***
    - Reposition the work or product; or adjust your position.
    - Minimize the distance between the product and yourself.
3. ***Maintain good body posturing.***
    - Work with straight wrists. Adding a foam wrist pad in front of the keyboard can help.
    - Keep elbows at a 90-degree angle to avoid strain on shoulders and damage to wrist and arm.

- Keep your elbows bent while working.
- Arm rests can be helpful for prolonged periods of typing or static arm activity. Arm rests are not needed if a wide range of movement is required.

4. **Ease standing stress.**
   - Redesign the operation/work so you can vary your position between sitting and standing.
   - Install foot rails or use footrests to allow shifting of your body weight from foot to foot.
   - Use floor coverings to cushion floors (Anti-fatigue Matting).
   - Use sit/stand stools, or sit/lean rails, to allow the use of alternate work positions.

5. **Avoid forceful movements with your hands, arms, shoulder, legs, and back.**

6. **Avoid reaching.**
   - Arrange the workstation so that the most frequently used items are positioned the closest to you.
   - Eliminate repetitive, arms-over-shoulders reaching action, and unnecessary twisting.

7. **Keep a straight wrist while working.**
   - Pick tools/equipment that allow you to grip with a straight wrist.
   - Use clamps to eliminate long periods of gripping.
   - Change grips whenever possible.

8. **Use your whole hand, or both hands whenever possible.**

9. **Use tools (jackhammers/toppers) that operate with the least vibration.**

10. **Use good lifting techniques.**
    - Move close to the load.
    - Bend your knees and keep your feet spread slightly apart.
    - Crouch down: it takes less effort to stand again.
    - Get a good grip before lifting.
    - Lift with your legs.
    - Avoid twisting while lifting.
    - Avoid pushing or pulling heavy loads: if this must be done, then push using your body weight to your advantage.
    - Limit lifting to light loads.

- Use mechanical assistance whenever possible.

11. ***Take regular rest breaks.***
    - Heed the signs that rest is needed. Feeling tired and sore can be a warning that adjustments to your work position or methods are warranted. This will ease unnecessary strain on your body.
    - For computer work, take a "mini-break" every 30 minutes, and longer breaks every 2 hours. Remember to blink your eyes while using the computer

12. ***Stretch and exercise regularly.***

    Keep your body in good physical shape by doing aerobic, strength, and stretching exercises. When you can, regularly do:
    - Head and neck stretch;
    - Body stretches;
    - Side bends;
    - Upper body twist;
    - Shoulder shrugs and circles;
    - Finger stretches;
    - Wrist stretch and rotations;
    - Thumb stretch;
    - Finger squeezes;
    - Palming your eyes.

A. Office Workstations

1. Workstation Design

Proper workstation design includes consideration of:
- Keyboards and screens;
- Brightness and contrast controls;
- Document holders;
- Task lighting;
- Proper desks, chairs, and footrests;
- Overhead lighting;
- Window location;
- Work/rest regimes.

A proper relationship between the operator, the terminal, and the environment in which the terminal is used is important to avoid fatigue and muscular discomfort, eyestrain, and irritation.

Some of the most common and troublesome symptoms of discomfort associated with VDT operation are related to postural problems. Jobs should be structured, and work schedules organized in ways that can be helpful in minimizing operator fatigue.

2. Employee Responsibility

The employee working at the screen, can do much to avoid fatigue by adjusting their posture from time to time, looking away from the screen periodically at any distant object, and blinking frequently. Lack of visual contrast can cause visual discomfort and reduce reading skills because the eye must continually adjust the brightness between the screen and the surrounding environments. Screen glare can also be a problem; glare is the reflection of light, shiny surfaces, or clothing on the screen. The glare image and the VDT image are at different optical distances and the eye attempts to focus on both images at the same time which causes visual fatigue.

Muscle fatigue may result from uninterrupted use of a VDT. This discomfort may be attributed to poor workstation design, repetitiveness of the task, postural constraints, inappropriate work/rest schedule, and the personal attributes of the employees. Employees should not perform continuous VDT work for more than two hours. Short interruptions of VDT work, such as periodic rest breaks or switching to other work activities are recommended.

---

[31] Royalty Free Image readily available on the Internet. Graphic accessed from http://www.ergonomics-info.com/ergonomic-pictures.html (date accessed February 28, 2020).

3. Computer Workstation Set-up[31]

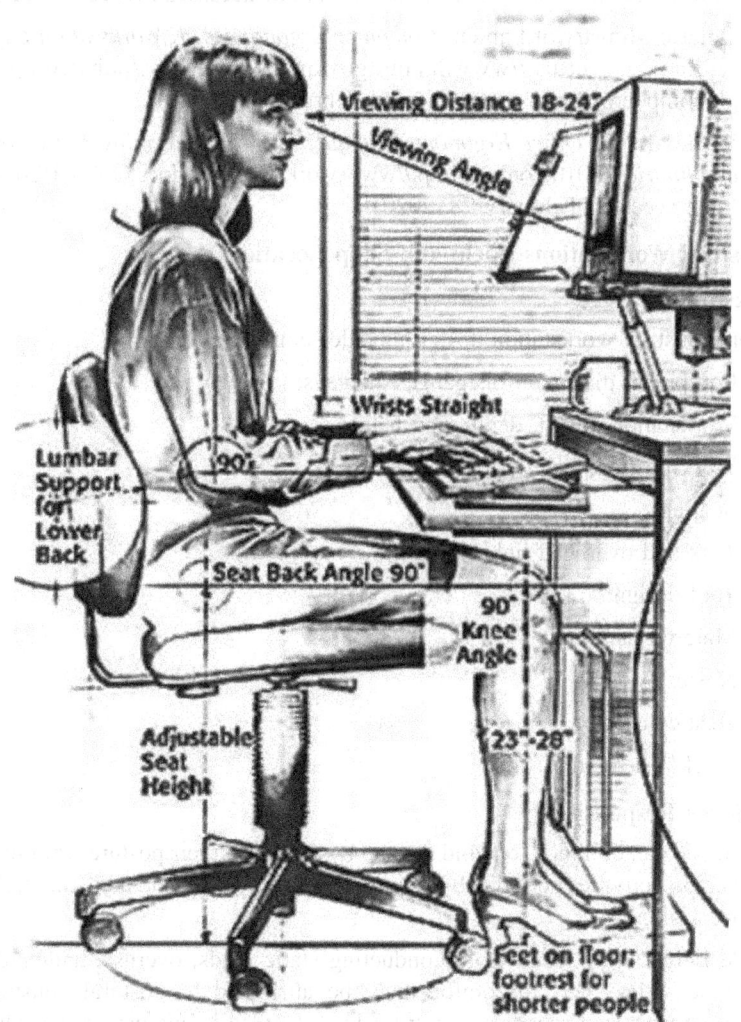

For additional information refer to:

- WorkSafeBC, *How to Make Your Computer Workstation Fit You* (2009), online: https://www.worksafebc.com/en/resources/health-safety/books-guides/how-to-make-your-computer-workstation-fit-you (date accessed February 28, 2020).

- CCOHS, *Health & Safety: Teaching Tools: Ergonomics: Workstations*, online: http://www.ccohs.ca/teach_tools/ergonomics/workstations.html (date accessed February 28, 2020).

- OSHA, *Computer Workstation eTools*, online: https://www.osha.gov/SLTC/etools/computerworkstations/index.html (date accessed February 28, 2020).
- Ontario Ministry of Labour, *Computer Ergonomics, 1. Workstation Layout*, online: https://www.labour.gov.on.ca/english/hs/pubs/comp_erg/gl_comp_erg_2.php (date accessed February 28, 2020).
- WorkSafeNB, *Office Ergonomics: Guidelines for preventing Musculoskeletal Injuries* (2010), online: http://www.worksafenb.ca/docs/OFFICEEdist.pdf (date accessed February 28, 2020).

B. Industrial Workstations (Field and Shop Locations)

1. Workstation Design

Proper industrial workstation design includes consideration of:
- Equipment dials/controls/gauges/displays;
- Workstation height, depth, and features;
- Work posture;
- Seating;
- Task and overhead lighting;
- Tool design;
- Material handling;
- Noise levels;
- Heat/cold;
- Work/rest regimes.

2. Employee Responsibility

The employee can do much to avoid fatigue by adjusting their posture from time to time, changing activities periodically, using tools as designed, and adjusting lighting to enhance vision.

Muscle fatigue may result from conducting static loads, overuse, and/or doing repetitive activities. This discomfort may be attributed to postural constraints, inappropriate work/rest schedule, repetitiveness of the task, and personal attributes. Short interruptions of work, such as periodic rest breaks or switching to other work activities are recommended to counteract muscle fatigue.

Appendix 2

EDUCATIONAL MATERIALS – SAMPLE #2
Industrial Ergonomics Program

WORKSPACE

Principle: Workspace design should be based on body dimensions. To cope with a variety of tasks, job sites, and body sizes and shapes, the use of adjustable workspaces is required.

Approach:

When designing any workspace, consider:

Head height
- Design the workspace for the tallest employee.
- The natural posture is to look slightly down when working.
- Avoid narrow viewing angles.

Shoulder height
- Position controls/gauges between shoulder and waist height.
- Avoid above-shoulder reaches.

Elbow height
- Position work surfaces slightly below elbow height.
- For precise work, work surfaces should be positioned slightly above elbow height, providing forearm rests.
- For heavy work, work surfaces should be 20 cm below elbow height.

Arm reach – horizontal and vertical
- Allow for the shortest reach when tasks require reaching up or out.
- Allow for the tallest individual when reaching down.
- Keep frequent work within forearm distance.

Knuckle height
- Keep lifting tasks between knuckle and shoulder height.

Leg length and room
- Provide workspace adjustments or footrests for various leg lengths.
- Allow for the longest leg length.

Hand size
- Allow for the smallest and largest hand sizes when dealing with gripping and access to openings.

Body bulk
- Workspace adjustments should allow for the largest and smallest body sizes.

WORKSPACE

Principle: Discomfort and fatigue arise from having to hold a posture for extended periods of time. To maintain a stationary body position, static muscle activity is required. This results in compressed blood vessels and blood flow. Fatigue and discomfort ensue.

Approach:

The longer the static muscle activity, the greater the stress on the tendons and joints. The remedy is to avoid static activity by using:

- Arm rests;
- Good seating;
- Proper manual handling techniques;
- Task variation;
- Well-designed tools/equipment.

SEATING

Principle: Seating should be aligned with the features of the job tasks. Sitting should be a dynamic activity – employees should be able to move about in their seats. Chairs/seating should be adjustable taking into account:

Approach:

Seat area
- The seat pan should be large enough to accommodate frequent movement and to prevent pressure points.
- Breathable fabric should be used.

Seat depth
- Seat pan depth should be adjustable.
- The front lip should be rounded and padded.

Seat height
- Seat height should be adjustable.

Back support
- Back support should be adjustable and designed to maintain the natural "S" curve of the back.

- It should also be adjustable horizontally and vertically.

Arm support
- If used, should be far enough apart to accommodate the largest person, and support the whole arm.
- Arm support should be adjustable in height and width.

DISPLAYS

Principle: All industrial tasks require information processing. This means good sensory input, perception, decision-making, action, and feedback. Clear communication depends on good sensory input. There must be adequate lighting and lack of masking noise.

Approach:
- Visually present complex information. Simple or urgent information can be audible.
- People have a limited capacity to perceive information. Grouping and simplifying information helps.
- To combat monotony, design the important cues demand attention.
- Keep detail to a minimum.

CONTROLS

Principle: Communication between people and machines can be made more efficient and less stressful by considering a few basic design principles.

Approach:

Field of view
- Place on/off controls and displays in the same field of view.

Frequency of use
- Position frequently used controls close to hand.
- Place frequently observed displays in the center of view.

Sequence of use
- If correct sequence of use is important, place controls and displays in sequence.

Similarity of function
- If a correct sequence of use is not important, group controls and displays according to function.

Importance of function
- Place important or emergency controls in easy reach and within normal visual view.

MANUAL HANDLING

Principle: Manual material handling controls should center on modifying the object, task, and environment.

Background:

Manual materials handling is involved in 25-30% of all industrial injuries. Related injuries include over-exertion, sprains, strains, and falls. These tend to be painful and costly workplace injuries.

Approach:

Many factors are involved, namely the object, task, work environment, and employee.

1. **The Object:**
   - Objects should be easy to grasp.
   - Use handles or handholds.
   - Avoid handling large objects that extend the arms and impede vision.
   - Lifting maximums are 25kg for men and 15kg for women.
   - Avoid lopsided weights and loose contents.
   - Mark total weight clearly.
   - Mark controlled substances (WHMIS 2015).

2. **The Task:**
   - Keep repetitive lifts between knuckle and shoulder height.
   - Use adjustable platforms where possible.
   - Keep manual materials carrying to a minimum.
   - Avoid high rates of repetitive lifting.
   - Minimize upper body (arms, shoulder, head) movements.
   - Accurate placement increases static muscle demands.

3. **The Work Environment:**
   - Avoid the extremes of heat and cold temperatures.
   - Avoid changes in surfaces such as sticky, oily, or slippery floors/surfaces.
   - Avoid congested or constricted workspaces.
   - Avoid awkward footing areas such as walking on pallets or other similar structures.

4. **The Person:**
   - Through medical/physical monitoring, avoid taxing previous injuries.
   - Teach correct lifting and manual materials handling techniques.

- Teach person limits of strength and endurance.
- Encourage physical fitness.
- Promote the use of pre-job stretching and exercising.
- Accommodate physical limits following illness/injury.
- Older workers are more vulnerable to heat and fatigue.

Most modifications can have effective outcomes. For example, by placing handholds on boxes or containers:

- They will be less likely to be dropped;
- Less static muscle activity will be required;
- Grasping will be easier and quicker; and/or
- Upper body lifting and lowering will be reduced.

MANIPULATION

Most jobs involve manipulation tasks (grasping tools, climbing, packing, assembly, writing, keypunching). Although the hand is efficient at grasping movements, it is not designed to withstand the numerous stresses placed on them by many industrial tasks. High rates of repetition, unnatural ranges of motions and angles, overloading of muscles and joints, high forces, cold temperatures, and lack of rest can lead to chronic health problems.

Approach:

The elements to consider include:

Range of motion

- Work within the neutral ranges of motion for the body.
- Never hyper-extend any joint.
- Keep elbows at right angles when applying force.
- Keep elbows at right angles when applying force.
- Keep the hand in the cupped natural position.

Length of contraction

- Keep elbows close to the body.
- Limit the amount of static muscular activities.
- Allow a rest/work ratio of 10:1.
- Avoid all muscular contractions in excess of 50% maximum.

Force of application

Never exceed 20% of the employee's maximum level of force in a prolonged or repetitive posture.

HAND TOOLS

Principle: Hand tools should be designed to be compatible with the characteristics of the human hand and arm.

Background:

Hand tools include staple guns, screwdrivers, power drills, tin-snips, pliers, and the like. If awkwardly used over time and under pressure with bent wrists, ill effects can develop.

The Considerations Required:

Muscular Effort Required: The greater the effort the more static muscular activity and control required to operate it.

Type of Muscular Activity: Static muscular activity is more fatiguing than dynamic muscular activity.

Size of Handgrip: The smaller the hand grip, the more strain on the hand and finger muscles.

Degree of Wrist Bending: The greater the bend in the wrist, the more stress on the tendons, nerves, and blood vessels passing through the wrist.

Tool Weight: The heavier the tool, the greater the muscle strain to control it.

Tool Edges: The sharper the tool edges, the more pressure on the hand and its blood flow.

Distance Between Levers: The wider the distance between the levers (lever span), the more stress on finger joints and the more fatigue.

Duration of Use: The longer the use without rest, the greater the fatigue.

HEAT

Principle: The human body functions best within its normal temperature range of 37C.

Background:

Body heat can be increased by muscular activity and environmental conditions. To keep the body in balance, both these sources of heat must be dissipated. The body's cooling mechanisms help, but prolonged attempts to cool the body can lead to adverse health effects and to accidents.

The effects of temperature extremes increase with age, level of fitness and gender. Over the age of 30 years, the person's heart rate increases more rapidly and recovers more slowly. The less physically fit the person, the more likely the negative effects of heat exposure. Women are more susceptible than men to the effects of heat.

Body Cooling:

- Perspiration and evaporation of sweat;
- Rest in a cool environment;

- Use of specialized clothing.

LIGHTING

Principle: Accurate processing of visual information requires good lighting.

Background:
- For items to be clearly seen, they must be well lit and contrast with the background.
- Lighting needs to vary from time to time, and person to person.
- Glare should be minimized.

NOISE

Principle: The human ear is sensitive. Prolonged exposure to noise over 85dBA or repeated impulse noise can cause permanent hearing loss.

Background:

Speech range is between 500 and 8000 dBA. Hearing is usually damaged within the range of 3000-4000 dBA — the range most often damaged by noise exposure. This is also the range that is important for speech perception.

Hearing Protection:

Noise control options in order of preference are:

Speech range is between 500 and 8000 dBA. Hearing is usually damaged within the range of 3000-4000 dBA — the range most often damaged by noise exposure. This is also the range that is important for speech perception.

1. Reduce the noise at the source;
2. Wear hearing protection; and
3. Limit exposure time through job rotation.

Appendix 3

EDUCATIONAL MATERIALS – SAMPLE #3
ERGONOMIC PROGRAM: PREVENTION[32]

FOR WORKERS:

HEALTHY NECK AND SHOULDERS

Set up your workstation so you can look straight ahead, with the screen about an arm's length from your face. Minimize/eliminate glare. Vary your visual activities and head/neck positions.

When using mobile devices, take breaks from constantly looking down at the screen.

HEALTHY HANDS

Use a power grip using the whole hand, to do tasks. Fatigue and discomfort tend to be associated with pinch grips, especially forceful pinch grips. Work with wrists in the neutral position to counteract these effects.

Match the hand tool with the size of your hand; avoid the use of high-vibration tools. An incompatible grip size leads to greater task effort and fatigue.

HEALTHY SHOULDERS

Work with your hands below your head; use a step stool/platform to make this possible. Choose light-weight tools then working at heights.

HEALTHY BACKS

Manual handling should be done with the use of lift-assist devices. Team-lift when manual lifting is necessary. Use good lifting techniques: bend your hips, keep the load close to the body, lift with your legs, and limit the height of the lift. Try to lift from knee height to waist; not from floor to waist.

HEALTHY BODY

Arrange your workstation so the most commonly used items are close. Avoid leaning forward, stretching, and reaching laterally. Work within close proximity to your body.

---

[32] Adapted from materials provided by Centre of Research Expertise for the Prevention of Musculoskeletal Disorders (2018). *Work Shouldn't Hurt*, Province of Ontario, online: https://www.msdprevention.com/MSD-Risk-Assessment-.htm (date accessed February 28, 2020).

Vary your work tasks and activities, making frequent postural changes and taking microbreaks to allow your body time to recover.

Assess your job tasks to identify ergonomic hazards at work and at home.

FOR THE EMPLOYER:
WORKER PAIN
- Develop an Ergonomic Program
- Undertake ergonomic hazard assessments of the major job tasks
- Implement ergonomic controls (lift-assist equipment, low-vibration hand tools, light-weight hand tools, off-the-floor storage of objects, adjustable workstations, adjustable chairs, *etc.*)
- Regularly assess the effectiveness of the ergonomic controls
- Educate workers on ergonomically safe work practices
- Design work allowing for frequent work rests to allow the body to recover
- Rotate workers who do high-hazard tasks
- Educate workers on the importance of reporting pain associated with work
- Implement a system for reporting and referring workers for medical assessment and treatment

CHAPTER REFERENCES

J. Barbour, "Ergonomics and the Aging Workforce" Presented at the CSSE Conference, Calgary, Alberta (2007).

Benefits Canada, "90% of Canadians working past age 60: Stats Can", *Benefits Canada, December 2018* (2018), online: https://www.benefitscanada.com/news/90-of-canadians-working-past-age-60-stats-can-123402 (accessed February 28, 2020).

J. Callaghan, "Too Much Standing Hurts, Too" (2016) *At Work, Issue 85*, online: https://www.iwh.on.ca/at-work/85/too-much-standing-hurts-too (date accessed February 28, 2020).

CCOHS Website (2017), online: http://www.ccohs.ca/products/posters/msd/?utm_source=SilverpopMailing&utm_medium=email&utm_campaign=Liaison%20Jan%202017%20ENGLISH&utm_content= (date accessed February 28, 2020).

CCOHS, Product description for Musculoskeletal Disorders (MSD) Prevention Manual, CCOHS website (2017), online: http://www.ccohs.ca/products/publications/msd/ (date accessed February 28, 2020).

Centre of Research Expertise for the Prevention of Musculoskeletal Disorders, *Work Shouldn't Hurt*, Province of Ontario (2018), online: https://www.msdprevention.com/MSD-Risk-Assessment-.htm (date accessed February 28, 2020).

J. Côté, "Higher risk of some musculoskeletal injuries among women may be due to sex and gender differences beyond size and strength" *Institute for Work and*

Health (2016), online: https://www.iwh.on.ca/media/2016-jun-23 (date accessed February 28, 2020).

B. Evanoff *et al.*, "Forceful repetition a carpal tunnel risk factor" *At Work, Issue 85* (2016), online: https://www.iwh.on.ca/at-work/85/forceful-repetition-a-carpal-tunnel-risk-factor (date accessed February 28, 2020).

S. Hawkins & R. Wiswell, "Rate and Mechanism of Maximal Oxygen Consumption Decline with Aging Implications for Exercise Training" (2012) 33:12 Sports Med 2003 877-888.

Humantech (2014), reported in Workers Health & Safety Centre "Making the case for MSD Prevention: The economics of ergonomics" (February 2016) at 3, online: https://www.whsc.on.ca/Files/Resources/Ergonomic-Resources/RSI-Day-2016_MSD-Case-Study_The-economics-of-ergon.aspx (date accessed February 28, 2020).

Liberty Mutual Research Institute for Safety, "2019 Workplace Safety Index: The Top 10 Causes of Disabling Injuries at Work" Viewpoint online: https://viewpoint.libertymutualgroup.com/article/top-10-causes-disabling-injuries-at-work-2019/ (date accessed: January 31, 2020).

C. Lotz, "Ergonomics Hit List: Identifying 10 Workplace Issues" (2012) 57:11 Professional Safety: Journal of the ASSE 50-51.

Ontario Government, "Ergonomics in the Workplace", *Law and Safety* (2019), online: https://www.labour.gov.on.ca/english/hs/pubs/ergonomics/is_ergonomics.php (date accessed February 28, 2020).

E. Pinto, "Blood pressure and ageing", (2007) 83:936 Postgrad Med J. 109-114.

K. Tiels, Aging Workforce – What's Ergonomics Got to Do With It? Presented at the IAPA Conference & Trade Show (Toronto April 16-18, 2007).

US Bureau of Labor Statistics (2016). Reported by ATI Worksite Solutions, online: https://www.atiworksitesolutions.com/ (date accessed February 28, 2020).

US Bureau of Labor Statistics, News Release, "Nonfatal Occupational Injuries and Illnesses Requiring Days Away from Work" (November 2015), online: http://www.bls.gov/news.release/pdf/osh2.pdf (date accessed February 28, 2020).

US Department of Labor, Occupational Safety and Health Administration (2016), "2016. Prevention of Work-related Musculoskeletal Disorders", online: https://www.osha.gov/pls/oshaweb/owadisp.show_document?p_table=UNIFIED_AGENDA&p_id=4481 (date accessed February 28, 2020).

WorkSafeBC, *How to Make Your Computer Workstation Fit You*, (2009), online: https://www.worksafebc.com/en/resources/health-safety/books-guides/how-to-make-your-computer-workstation-fit-you (date accessed February 28, 2020).

WorkSafeBC, www.worksafebc.com (2012) (date accessed February 28, 2020). Health & Safety Centre, "Making the case for MSD Prevention: The economics of ergonomics" (February 2016), online: https://www.whsc.on.ca/Files/Resources/

Ergonomic-Resources/RSI-Day-2016_MSD-Case-Study_The-economics-of-ergon.aspx (date accessed February 28, 2020).

Chapter 21

OCCUPATIONAL HEALTH AND SAFETY: PREVENTION OF WORKPLACE ILLNESS AND INJURY

A. INTRODUCTION

Organizations vary in the degree of coordination that exists between their Occupational Health & Safety (OH&S) Program and other organization/company programs such as the Employee Assistance Program (EAP), Disability Management Program (DMP), Workplace Wellness Program (WWP), and Human Resources Program (HRP). In some organizations, these programs operate independently as "silos", while in others, they are integrated to form a comprehensive workplace support system. This chapter has been written in support of the latter approach.

B. ROLE OF OCCUPATIONAL HEALTH AND SAFETY

> Workers' compensation costs are significant, and there are substantial other costs that are incurred when an injury/illness occurs. The sum of these adds directly to operational costs, and hence profits.[1]

For example, in 2018 the average Canadian workers' compensation cost for a work-related injury requiring recuperation from work was $39,534 per claim for assessable employers.[2] If an organization/company had a one per cent profit margin, it requires $8.3 million in product sales/services to cover these work-related injury costs.[3]

The message presented in ***Business Results Through Health & Safety***[4] is that

---

[1] WSIB/CSPAAT, "Business Results Through Health & Safety" (2001), online: http://www.ryerson.ca/content/dam/irm/pdfs/training/ISS_MGR/WSIB_BusinessResultsThroughHS.pdf (date accessed: February 28, 2020) at 3.

[2] Calculated using Association of Workers' Compensation Boards of Canada, "2018 Injury Statistics", *Statistics*, online: http://awcbc.org/?page_id=14 (date accessed: February 28, 2020). Involves the addition of IR5 – the Current Year Average Benefit Cost/Claim, and IR6 – Administration Cost per claim.

[3] Calculated using *Table 15.2 Calculation of Required Revenue to Offset Loss*, Chapter 15, "Occupational Health and Safety: Risk Management and Communication".

[4] WSIB/CSPAAT, "Business Results Through Health & Safety" (2001), online: http://

OH&S Programs have an important role to play in the prevention of workplace illness and injuries. Typically, prevention strategies are achieved through:
- developing OH&S policy, standards, procedures, and practices;
- focusing on hazard management;
- providing OH&S education for everyone in the workplace, particularly supervisors, managers, and employers;
- providing and maintaining personal protective equipment and fit-testing;
- enabling ergonomic accommodations to be made;
- conducting worksite walk-through assessments;
- auditing workplace practices;
- responding, reporting, and investigating occupational illness/injuries;
- tracking workplace illness/injuries; and
- enabling workplace safety to become part of the fabric of the work, as opposed to being seen as a program.

An effective OH&S Program can result in a "zero tolerance" for occupational injuries or illness. In addition, many elements of an OH&S Program are legislated, either federally or provincially. When an OH&S Program operates in conjunction with a Disability Management Program, synergies can be realized. For example, an existing Joint Health and Safety Committee can address some disability management issues, such as return-to-work practices, modified-work placements, and workplace hazard-risk reduction.

OH&S services often focus on "off-the-job" safety. This response is due to the fact that for every workplace illness or injury, seven[5] to 14[6] non-occupational illnesses/injuries occur that negatively impact the workplace. In Ontario, between 2004-2011, there were seven non-occupational injuries requiring hospital-based treatment for every work-related injury.[7] Sadly, although a 30 per cent reduction in

---

www.ryerson.ca/content/dam/irm/pdfs/training/ISS_MGR/WSIB_BusinessResultsThroughHS.pdf (date accessed: February 28, 2020) at 3. This is a valuable and free online resource that is available to OH&S professionals.

[5] A. Chambers et al., "Diverging trends in the incidence of occupational and non-occupational injury in Ontario, 2004-2011" (2015) 105:2 Am. J. Public Health 338-343, doi: 10.2105/AJPH.2014.302223, online: https://www.iwh.on.ca/summaries/ issue-briefing/divergent-trends-in-work-related-and-non-work-related-injury-in-ontario (date accessed: February 28, 2020).

[6] Based on professional knowledge gained through auditing Disability Management Programs for various organizations.

[7] A. Chambers et al., "Diverging trends in the incidence of occupational and non-occupational injury in Ontario, 2004-2011" (2015) 105:2 Am. J. Public Health 338-343, doi: 10.2105/AJPH.2014.302223, online: https://www.iwh.on.ca/summaries/ issue-briefing/divergent-

work-related injury incidents was observed, there was no reduction in the non-occupational incidents observed. As well, many of these non-occupational illnesses/injuries have proven to be difficult and costly to resolve. For example, in 2015, Saskatchewan workers experienced 28,314 unintentional off-the-job injuries. This was eight-fold more than the number of work-related injuries.[8] Reportedly, these predictable and preventable injuries cost the Saskatchewan economy CDN $180,821,918.[9]

According to the National Safety Council (NSC), the major causes of preventable injury-deaths (2017) were poisoning, falls, and drowning.[10] Some of the "off-the-job" safety initiatives that have been implemented include home safety, sport safety, water safety, cycling safety, fire prevention, and sun safety. Through prevention, non-occupational illness/injuries can be significantly reduced.

C. ROLE OF EMPLOYEE ASSISTANCE PROGRAMS IN THE PREVENTION OF WORKPLACE ILLNESS AND INJURY

Employee Assistance Program (EAP) professionals assist the employees and the organizations that they serve by focusing on attendance and disability management. Work-related injuries, illnesses, and disabilities drain millions of dollars from the Canadian economy, not to mention the toll exacted on the individual employees and their families.

In addition to the traditional EAP services of counselling and addiction intervention, an EAP can provide services such as:

- personal/relationship counselling — incidence rates for diagnosed depression of 1:10 Canadian workers and 1:7 U.S. workers, are strong drivers for the availability of corporate-sponsored EAPs;[11]
- assistance with personal and family problems that impact regular work attendance;
- critical incident stress debriefing (CISD) following a tragedy within the workplace or community;
- assistance with identifying and addressing the contributing causes of workplace incidents/injuries. Some contributing causes are employee fa-

---

trends-in-work-related-and-non-work-related-injury-in-ontario (date accessed: February 28, 2020).

[8] Safe Saskatchewan, online: http://www.safesask.com/index.cfm (date accessed: February 28, 2020).

[9] Safe Saskatchewan, online: http://www.safesask.com/html/facts/index.cfm#sthash.uGqkRItP.dpuf (date accessed February 28, 2020).

[10] National Safety Council, Home and Community Overview, *Injury Facts*, online: https://injuryfacts.nsc.org/home-and-community/home-and-community-overview/introduction/ (date accessed February 28, 2020).

[11] Ipsos-Reid Survey reported by M. Shaw, "How can we . . .", *Accident Prevention Magazine* (March/April 2007) at 4.

tigue, grief, preoccupation with personal problems or workplace discord, dysfunctional workgroups, substance abuse, *etc.*;

- development and presentation of both workplace and personal change management seminars;
- coaching clinics for supervisory staff on how to deal with the "troubled employee", and developing people skills or understanding the impact that home situations can have on employees and workgroups. Many supervisors report that they feel ill-prepared to address employee psychological health conditions;[12] and
- coaching clinics for Management on the effect that various management theories and practices can have on the workplace and employees. Based on the 2009 Ispsos Reid Survey, 29 per cent of Canadian workers report working in psychologically unsafe workplaces.

With the incidence of workplace violence on the rise, EAPs can play a significant role in assisting workers to constructively address life stressors. As well, they can help victims of workplace violence "to come to grips" with what has happened and how to move forward in their recovery process.

D. ROLE OF WORKPLACE WELLNESS IN DISABILITY MANAGEMENT IN THE PREVENTION OF WORKPLACE ILLNESS AND INJURY[13]

Canadian organizations have many of the workplace wellness components in place. For example, as early as 1997, approximately 80 per cent of Canadian companies (with 500 or more employees) had EAPs available for their employees and families.[14] The current market penetration of EAP services is estimated to be 91 per cent of the mid-sized companies[15] to 98 per cent of large-sized companies[16] that offer EAP services.

---

[12] Ipsos-Reid Survey reported by M. Shain, "Psychological Safety at Work: Legal Trends and the Implications" (August 2009) 19:5 *Benefits and Pensions Monitor* at 30–31.

[13] Adapted with permission from D. Dyck, "Workplace Wellness: What is Your Potential Return on Investment?", Human Resources Association of Calgary: A Newsletter on Human Resources Management (November 1998); and permission from D. Dyck, "Wrapping Up the Wellness Package" (January 1999) *Benefits Canada* at 16–20.

[14] Conference Board of Canada, *Compensation Planning Outlook* (Ottawa: Conference Board of Canada, 1997).

[15] Conference Board of Canada, *Healthy Brains at Work* (2016), online: http://www.sunlife.ca/static/canada/Sponsor/About%20Group%20Benefits/Focus%20Update/2016/557/HealthyBrains_Report2_EN.pdf at 20 (date accessed: February 28, 2020).

[16] Willis Towers Watson, *2015/2016 Global Staying@Work Survey, Canada* (2016), online: https://easna.org/wp-content/uploads/2016/04/2015-2016-Global-Staying-at-Work-Survey.pdf at 5 (date accessed: February 28, 2020).

CH. 21: OH&S: PREVENTION OF WORKPLACE ILLNESS AND INJURY

Most Canadian companies offer some wellness initiatives or benefits to employees, namely:

- Employee Assistance Programs (91 per cent);
- Flu immunization (76 per cent);
- Employee recognition initiatives;
- Wellness education and newsletters;
- Flexible work program;
- Health awareness/education programs;
- Smoking cessation (62 per cent);
- Nutrition education (72 per cent);
- Stress management programs (64 per cent);
- Blood pressure screening;
- Work/family balance programs;
- Time management education — 24 per cent;
- On-site fitness program — 24 per cent;
- Back care program — 21 per cent;
- Cholesterol screening — 20 per cent;
- Employee health risk assessments — 19 per cent;
- Childcare — 11 per cent; and
- Eldercare — 9 per cent.[17],[18]

It is ironic that, although most Canadian organizations/companies have many workplace wellness components in place,[19] the various programs such as the DMP, EAP, OH&S Program, Attendance Support and Assistance Program, and HRPs disjointed, operating in isolation and focusing solely on their individual program goals. There is no overall scheme in place for a comprehensive support system. The opportunity for collective synergies, and for demonstrating both individual and composite return on investment results, is missed.

---

[17] Buffett and Company Worksite Wellness Ltd., *National Wellness Survey Report 2009* (Whitby, ON: Buffett and Company Worksite Wellness Ltd., 2009) at 4, now owned by Sun Life of Canada.

[18] Willis Towers Watson, *2015/2016 Global Staying@Work Survey, Canada* (2016), online: https://easna.org/wp-content/uploads/2016/04/2015-2016-Global-Staying-at-Work-Survey.pdf at 5 (date accessed: February 28, 2020).

[19] Buffett & Company, *National Wellness Survey Report 2009*, online: http://www.buffettand company.com/NWS2009/Buffett_Company_NWS2006.pdf (date accessed: February 28, 2020).

The challenge for organizations/companies is not to find the resources to implement programs, but rather to integrate the existing components and to channel those efforts and outcomes to meet business needs. According to Dr. Catherine Connolly, McMaster University, "[w]ellness initiatives are challenging and complex to manage, so those that are not properly managed will not likely generate their anticipated benefits."[20] Hence, it is essential to identify WWP goals, objectives, and targets and to ensure that those elements are aligned with the organization's business strategy. According to Watson Wyatt:

> An integrated approach to tracking, measuring, and addressing the key determinants of workforce and organizational health is necessary to achieve and maintain a healthy organization.[21]

This sentiment is reinforced in the later releases of the Towers Watson Staying@Work Reports.[22]

For further details on how an integrated workplace health management approach can be realized, refer to Chapter 28, "Integrated Workplace Health Management".

E. POSITIONING THE OCCUPATIONAL HEALTH AND SAFETY PROGRAM WITH EMPLOYEE ASSISTANCE, WORKPLACE WELLNESS, AND DISABILITY MANAGEMENT PROGRAMS

In most companies/organizations, the OH&S Program, the EAP, and the WWP operate independently. These programs may be a mix of internally and externally resourced and operated programs. The result is a "silo" effect with each program functioning in isolation. The opportunity to provide integrated services that can offer comprehensive programming and services, and to benefit from the activities and learning of each discipline and service, is lost.

A few organizations have implemented integrated services. Historically, the University of Calgary (1998) combined the Staff Assistance Program (EAP),

---

[20] CNW, "National Wellness Survey Shows Canadian Organizations Investing in Worksite Wellness" (September 30, 2009), online: http://www.newswire.ca/en/releases/archive/September2009/ 30/c2024.html (date accessed: February 28, 2020).

[21] Watson Wyatt, *Staying@Work: Effective Presence at Work* (2008), online: http://www.watsonwyatt. com (date accessed: February 28, 2020).

[22] Watson Wyatt Worldwide, *The Health and Productivity Advantage; 2009/2010 North American Staying @ Work Report* (2010), online: http://www.towerswatson.com/en-CA/Insights/IC-Types/Survey-Research-Results/2009/12/20092010-North-American-StayingWork-Report-The-Health-and-Productivity-Advantage; Towers Watson, *Pathway to Health and Productivity: 2011/2012 Staying @ Work Survey Report* (2013), online: http://www.towerswatson.com/en-CA/Insights/IC-Types/Survey-Research-Results/2011/12/20112012-StayingWork-Survey-Report--A- Pathway-to-Employee-Health-and-Workplace-Productivity; and Willis Towers Watson, *2015/2016 Global Staying@Work Survey, Canada* (2016), online: https://easna.org/wp-content/uploads/2016/04/2015-2016-Global-Staying-at-Work-Survey.pdf (date accessed: February 28, 2020).

Managed Rehabilitation Program (Occupational Health Centre (OHC) dealing with attendance and disability management), and Human Resources support in its approach to wellness.[23] This service was integrated at a number of levels:

- *Organizational Commitment* — support of key players (*e.g.*, senior executive, leaders, unions, staff) and integration with policies and procedures.

- *Team Approach* — appropriate sharing of information and cross-referral of cases between OHC, Staff Assistance Program, Human Resources, and the healthcare community.

- *Case Management* — management of all cases associated with potential lost productivity (*e.g.*, incidental absence, short-term disability, Workers' Compensation Board, and long-term disability).

This approach resulted in considerable savings for the University. In 1997, 3,785 days were saved with a $400,000 productivity recovery (cost avoidance). In 1998, 4,563 days were saved with a productivity recovery of $477,362. The return on investment (ROI) in the first year was roughly 1:1. However, for years two and three of this program, the ROI was 2:1 — that is, for every dollar spent on the program $2 was saved.[24]

Staff responded positively to this approach, rating their level of satisfaction with the service as good-excellent (96.9 per cent). As well, 100 per cent of the assisted employees indicated they would use the service again and would make referrals to peers; and 93.4 per cent said that the presenting problem was improved.[25]

More recently, Suncor Energy Inc. (formerly Petro-Canada), uses an OH&S model that integrates Occupational Health, Disability Management, Occupational Safety, Industrial Hygiene, EAP, Workers' Compensation claim management and Human Resources practices for a number of years. The outcome is a comprehensive program that has low absenteeism and low disability costs. For example, in 2013, 88 per cent of ill/injured employees returned to work, resulting in significant savings ($2 million) in STD benefits, overtime costs, and replacement worker costs.[26] By 2016, the number of psychological claims was reduced by 15% and occupational injuries by 43%.[27] By 2018, a reduction in case duration of 4.3 days was realized for a direct savings of $4.2 million in lost time days since 2016.[28]

More companies are integrating certain elements of their DMPs. For example, in

---

[23] B. Daigle & G. Schick, "Early Intervention, Integration and Successful Resolution of Employee Health Issues" (Presented at the Health Work and Wellness Conference '98, Whistler, BC, September 27–30, 1998).

[24] *Ibid.*

[25] *Ibid.*

[26] Suncor Energy Inc, Report from Shannon Stratton, Suncor Energy Inc. (2014).

[27] Suncor Energy Inc, Report from Shannon Stratton, Suncor Energy Inc. (2017).

[28] Suncor Energy Inc., Report from Shannon Stratton, Suncor Energy Inc. (2019).

2002, 62 per cent of American employers reported using a consistent approach to managing occupational and non-occupational return-to-work programs. This was a 32 per cent increase over 2001. Fifty-one per cent had integrated their STD and LTD coverage, an increase of 39 per cent from 2001.[29] As companies recognize the financial and administrative benefits of integrated services, the expectation is that they will opt for this approach.

F. LINKS TO CORPORATE CULTURE, MANAGEMENT PRACTICES, AND HUMAN RESOURCE MANAGEMENT THEORIES

Until recently, organizations have focused on employee and program issues when trying to understand their attendance and disability management program outcomes. However, research indicates that the corporate culture, management practices, and human resource management theories have a significant impact on OH&S Programs and their outcomes.[30],[31],[32] In this section, the relevance of management practices and human resource management theories to OH&S, will be discussed.

Although the impact of corporate culture on the OH&S Management Systems and practices was discussed in several of the earlier chapters,[33] it is worth noting that 70 per cent of Canadian employers plan to focus on building a supportive health culture aimed at promoting healthy lifestyle behaviours.[34] Their priorities rest with "improving safety, enhancing productivity, raising health-risk awareness, and managing rising healthcare costs".[35] Why?

The motivating factors are the rising cost of healthcare costs and employee productivity losses that are related to employee sedentary lifestyles, obesity, and distractions caused by stress. The stress associated with absenteeism and lost productivity is perceived to be a major factor for 85 per cent of Canadian employers.

Highly stressed workers lose more than twice as many days at work and are over

---

[29] Marsh Risk Consulting, "Fourth Annual Marsh Mercer Survey on Employers' Time-Off & Disability Programs" (Workforce Risk, 2003), online: http://usa.marsh.com (date accessed: February 28, 2020).

[30] S. Dobson, "Strong link between quality of work life, patient care: Study", *Canadian HR Reporter* (November 15, 2010) at 6.

[31] Willis Watson Wyatt, *2015-2016 Staying@Work Report* (2016), online: https://www.willistowerswatson.com/-/media/WTW/PDF/Insights/2016/08/2015-2016-staying-at-work-canada-research-findings-en.pdf (date accessed, February 28, 2020).

[32] *Ibid.*, at 5.

[33] Refer to Chapter 6 "Occupational Health and Safety: Leadership and Commitment", Chapter 7 "Safe Workplace = Great Workplace: Building a Sustainable Culture of Safety and Chapter 8 "Safety Culture: Safety Climate".

[34] Willis Watson Wyatt, *2015-2016 Staying@Work Report* (2016), online: https://www.willistowerswatson.com/-/media/WTW/PDF/Insights/2016/08/2015-2016-staying-at-work-canada-research-findings-en.pdf (date accessed: February 28, 2020).

[35] *Ibid.*, at 3.

four times as likely to be disengaged [from the workplace].[36]

1. Stress Risk Management in the Workplace: Research by Dr. Martin Shain[37]

Job-related stress results from human interactions in the workplace. It needs to be managed like production and operational activity. It is up to management to address job stress at the source (*Eliminate Risk*) instead of dealing solely with the symptoms of job stress through the EAP, DMP, and OH&S Program (*Tolerate Risk*). This can best be achieved by balancing the organizational and programmatic approaches to stress risk management.

The key ingredients for good employee psychological health in the workplace are:

- mutual respect and appreciation;
- employees feeling heard and appreciated;
- open communication;
- freedom from feelings of hostility and anger; and
- a sense of self-worth and confidence.[38]

The work factors that threaten employee psychological and physical safety and contribute to workplace stress are:

- work overload and time pressures;
- lack of influence over daily work;
- too many changes within the job;
- lack of suitable education/training, and/or job preparation;
- too little or too much responsibility;
- discrimination;
- harassment;
- poor or limited communication;
- lack of quality supervision/management;
- neglect of legal and safety obligations;[39]
- inadequate staffing;
- corporate culture; and

---

[36] *Ibid.*

[37] M. Shain, "Managing Stress at its Source in the Workplace" (Presented at the Health Work and Wellness Conference '98, Whistler, BC, September 27–30, 1998).

[38] *Ibid.*

[39] *Ibid.*

- low pay.[40]

Excess workplace stress can erode employee self-efficacy and social supports — both key elements to employee well-being. In their book *Healthy Work: Stress, Productivity and the Reconstruction of Working Life*,[41] Karasek and Theorell identified the combination of high pressure (too much work in too short a period of time) and lack of influence over day-to-day work as a key contributor to cardiovascular disease. This "deadly duo" also threatens people's health by making it harder for them to take care of themselves. Getting too little exercise, smoking and drinking too much, poor nutrition, insomnia or hypersomnia and sleeping badly, and feeling generally out of control and miserable, are a few of the signs that stress is becoming disruptive.

A 24-year study of workers found that workers in low-control jobs are 43 per cent more prone to premature death; likewise, those in jobs lacking meaningful content tend to take greater risks. This study is consistent with the Whitehall studies of British civil servants. Workers with less decision-making latitude died sooner from heart disease and other similar ailments.[42]

Overwhelming pressure and lack of influence do not occur by chance. They result directly from how work is organized (*i.e.*, the allocation of work and relationship between employees) and how it is designed (the structure and content of work). The organization and design of work are dictated by both the job's technological requirements and human decisions. These two factors determine how employees relate to one another and to their jobs; they also contribute to the increase or decrease of stress levels and have an important impact on psychological and physical health (Figure 21.1). When stress from other sources — especially the home — is thrown into the equation, health is even more likely to be adversely affected.

It is not simply a matter of benefitting the "weak" by increasing their resiliency to stress. It is an issue of work conditions that can be modified to manage the risks of job-related stress.

In April 1998, the Families and Work Institute released a study that supports Dr. Shain's work.[43] It found that productivity is far more likely to be hurt by job-related stress than by family problems. Until recently, organizations tended to focus on

---

[40] Willis Towers Watson, *2015/2016 Global Staying@Work Survey, Canada* (2016), online: https://easna.org/wp-content/uploads/2016/04/2015-2016-Global-Staying-at-Work-Survey.pdf at 13 (date accessed: February 28, 2020).

[41] R. Karasek & T. Theorell, *Healthy Work: Stress, Productivity and the Reconstruction of Working Life* (Toronto: HarperCollins, 1992).

[42] Editorial, "Low-control jobs are hazardous to health" *O.H.S. Canada* (July/August 2002) at 23.

[43] M. Jackson, "Worker Squeeze: Employee Stress Cuts Productivity, Study Declares" *Calgary Herald* (April 15, 1998) at A5.

helping employees to address family problems to improve productivity. The Families and Work Institute study revealed that pay and benefits are far less important to workers than the quality of work and supportiveness of the organization.

A recent study of Canadian healthcare organizations indicates that the quality of work life — high stress, lack of time, and unreported errors — remain problematic at healthcare organizations and negatively impact patient care.[44]

---

[44] S. Dobson, "Strong link between quality of work life, patient care: Study" *Canadian HR Reporter* (November 15, 2010) at 6.

Figure 21.1: Organization and Design of Work[45]

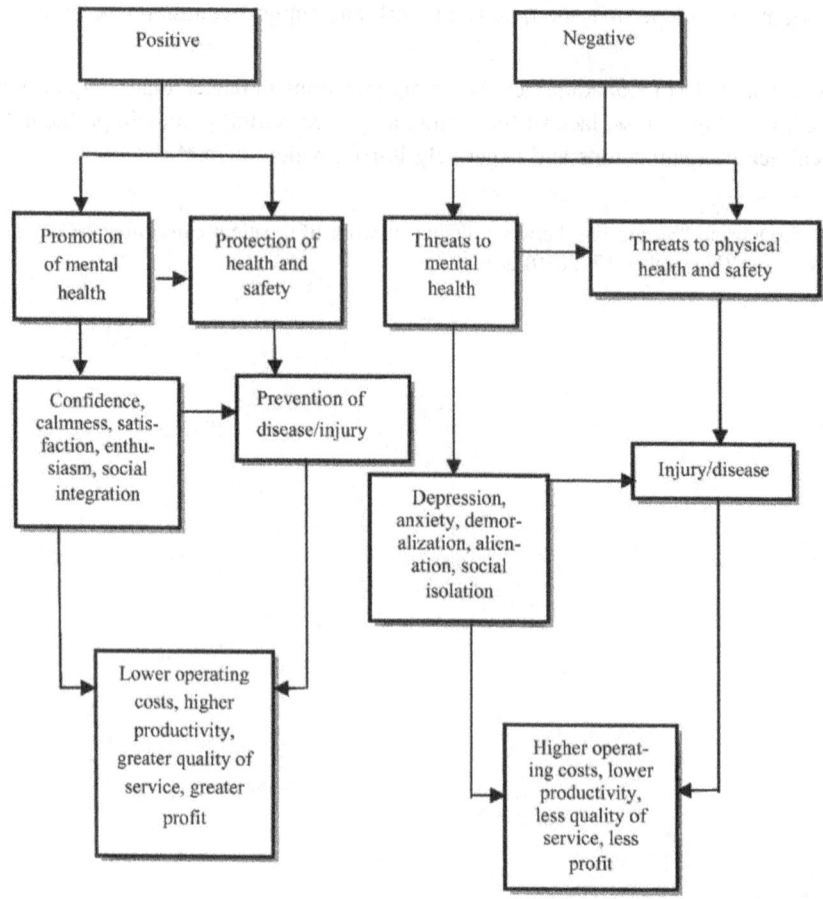

MacMillan-Bloedel, British Columbia,[46] conducted a study that showed that job stress due to high employee effort and low reward, coupled with home stress can lead to anger, and a sense of unfairness. This anger can manifest in two ways: overtly in the form of conflict (*e.g.*, workplace/home violence, road rage, sabotage) or covertly in the form of substance abuse. Anger caused by the discrepancy between effort and compensation can cause workplace injuries. Workplace injuries will occur if the employee perceives that he/she is unable to control or avoid

---

[45] M. Shain, "A New Take on Stress: Strategies that Work" Health Policy Forum (November 1997) at 12.

[46] M. Shain, "Stress, Satisfaction and Health at Work: Tuning for Performance" (Presented at the Health & Wellness Conference '99, Vancouver, BC, October 24–27, 1999).

hazards. Overt and covert anger due to work conditions end up increasing employee absenteeism and disability.

An organization can have a positive influence on the workforce and workplace outcomes by promoting good psychological health and protecting employees from injury and illness. This is described as an "upstream" approach — dealing with the root causes instead of the symptoms of the problem.

Dr. Shain notes that management can have a positive influence on the health and well-being of employees as well as on the OH&S Program outcomes.[47] Dr. Shain calls the influence of management on well-being the **Zone of Management Discretion**, which is not paternalistic towards employees, but rather focuses on prevention (Figure 21.2).

Figure 21.2: The Health Differential[48]

Technological Determinants versus Management as Influences on Employee Health

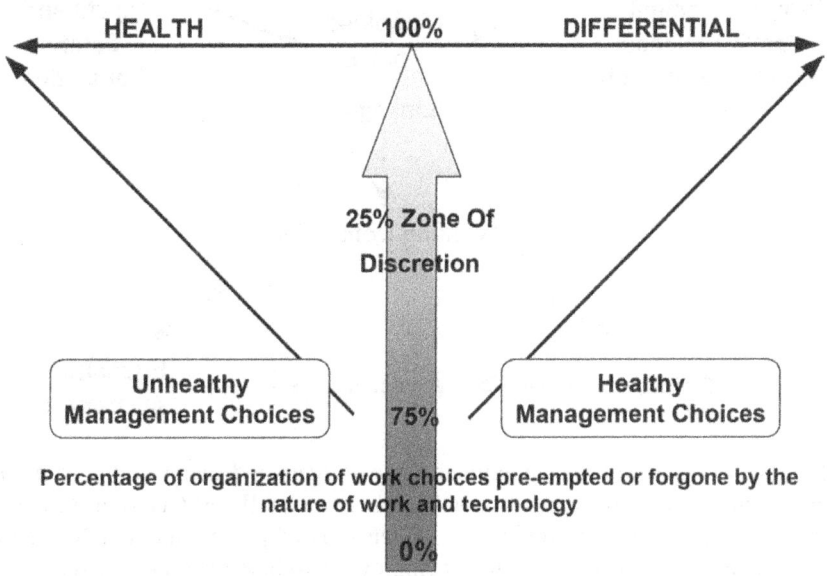

Dr. Shain also supports management recognition and control of threats to employee well-being such as high energy and effort output, high job demands, low recognition for work done, and little control over the job. He added that the last two conditions are easily changed. Employees should be provided the opportunity to

---

[47] *Ibid.*

[48] M. Shain, "Managing Stress at its Source in the Workplace" (Presented at the Health Work and Wellness Conference '98, Whistler, BC, September 27–30, 1998).

actively help the organization lower its OH&S risks and increase the resultant psychological health benefits. Being part of the problem helps people to determine workable solutions and to own the remediation when introduced.

Likewise, Dr. Shain promotes the concept of management prevention of the harmful consequences of stress. He presented a Best Practice model as early as 1998, at the Work, Health and Wellness Conference (Figure 21.3).

**Figure 21.3: Physical and Psychosocial Hazards:[49]
(A Combined Forces Approach to Hazard Abatement)**

Dr. Shain's early work has culminated into a recent endeavour produced by the Consortium for Organizational Mental Healthcare (COMH):[50] Guarding Minds @ Work — an approach that provides an evidence-based process that employers can easily and quickly implement to protect employee psychological safety and promote psychological health in the workplace.

---

[49] *Ibid.*

[50] The **Consortium for Organizational Mental Healthcare (COMH)** is a collective of mental health researchers, consultants, and practitioners who are experienced in working with a range of public and private sector organizations. Its goal is to further the creation and translation of psychological health knowledge and practice into real-world settings. To achieve this objective, the Consortium forms collaborative relationships with committed organizations, leaders, and experts to achieve specific outcomes. For more information refer to their homepage, online: http://www.comh.ca (date accessed: February 28, 2020).

According to the Consortium for Organizational Mental Healthcare (COMH):

> Guarding Minds @ Work is a response to current and emerging legal requirements in Canada for the protection of employee [psychological] health and the promotion of civility and respect at work. Legal standards increasingly require employers to develop comprehensive strategies for ensuring a psychologically safe workplace. Prudent employers need to develop policies and programs that meet these new legal standards.[51]

Most Human Resources managers are unaware of the magnitude of stress risk and hazards in the workplace. By identifying stress risks, they can have a positive impact on reducing employee absenteeism and disability. To improve employee psychological and physical health, managers must:

- encourage two-way communication with employees;
- articulate and communicate a clear vision for the future;
- train employees in new competencies and capabilities for success;
- articulate the desired organizational culture;
- re-engineer work processes to meet current demands;
- encourage employees to challenge the *status quo*;
- develop change management competencies of leaders; and
- educate managers/supervisors to be more sensitive to employee concerns.

2. Employee Engagement: Recognition of the Importance of Personal and Family Life

Employees who give their "best" to an organization and help to meet its business goals are characterized as *committed*. How are they recognized? One survey found that these employees:[52]

- work hard to improve themselves, increasing their value to the employer and make personal sacrifices to ensure the organization's success;
- recommend their organization/company as "a good place to work" and their employer's products and services as valuable; and
- believe that their workplace is "one of the best around" and intend to stay working there.

What are the drivers of employee commitment? The Aon Study found the following:

- the employees who believe that Management recognizes the importance of their personal and family lives tend to be the *committed employees*;

---

[51] *Guarding Minds @ Work: A Workplace Guide to Psychological Safety & Health*, online: http://www.guardingmindsatwork.ca/info/index (date accessed: February 28, 2020).

[52] Aon Consulting Inc., *Canada@Work* (Chicago: Aon Group, 2000).

- the most *committed employees* believe they have employee benefits that meet their needs and that they are fairly paid;
- the employers of *committed employees* communicate more effectively about a variety of topics: employee benefits, compensation, and "change initiatives"; and
- the employers of *committed employees* provide training to help employees remain current with the increased technical demands of today's jobs.[53]

This study also noted that stress plays a major role in reducing the level of employee commitment. Stress can be attributed to the difficulty in balancing the conflicting demands of the employee's work life and personal and family life. An appreciation of the importance of employee personal and family life results in increased productivity and the resulting increased profitability.

(a)   Family-friendly Policies

Attracting, retaining, and motivating the "best people" is a challenge for most organizations. For this reason, programs need to be created that recognize the needs of individuals and of diverse family situations. Employers note that flexible scheduling is one of the most important recruiting tools. Studies indicate that flexible work programs lower employee stress levels,[54] and that working from home attracts and retains staff.[55] Recent survey results indicate that 82 per cent of workers would leave their current job if they could get a job that enables them to telecommute.[56] As well, providing support for employees and families can translate into decreased absenteeism and tardiness.

Women, who are a significant part of today's workforce, have traditionally been the caregivers of society. Although today, men are assuming more of the childcare responsibilities, women remain the major caregivers. The pressures of juggling work and family tasks are strongly associated with the stress employees perceive in their lives. In general, work-family stress stems from two sources:

- *Role interference* — the roles played by individuals overlap and conflict; and
- *Role overload* — the daily work-family tasks are simply too much for one individual to handle.

---

[53] Aon Consulting Inc., *Canada@Work* (Chicago: Aon Group, 2000).

[54] Conference Board of Canada, News Release, "Eldercare Taking its Toll on Canadian Workers" (November 10, 1999), online: http://www.conferenceboard.ca (date accessed: February 28, 2020); HRM Guide, "Flexible Working Reduces Absenteeism", London, England (July 12, 2001), online: http://www.hrmguide.co.uk/flexibility/flexibility_absenteeism.htm (date accessed: February 28, 2020).

[55] S. Klie, "Working from home attracts, retains staff" *Canadian HR Reporter* (December 13, 2010) at 3.

[56] 2010 Workopolis.com Survey, reported in S. Klie, "Working from home attracts, retains staff" *Canadian HR Reporter* (December 13, 2010) at 3.

Role interference and role overload can make life very difficult for the employee and, indirectly, for the employer in terms of last-minute absences from the workplace.

When inter-role conflict occurs, these terms are referred to as Work-family Conflict (WFC) and Family-work Conflict (FWC). Note the directionality of each term. WFC occurs far more often than does FWC, with women more than men being affected. There is a strong correlation between children living in the home and the occurrence of WFC and FWC for both genders; the more children, the greater the incidence of WFC and FWC.

Work and personal life are no longer separate domains; the boundaries between each have disappeared.[57] Today, role overload is systemic, and the invasiveness of technology makes this possible. Employees may be at work shorter hours than they were in the past, but they continue to work at home. Twenty-five per cent of Canadian employees report working more than 50 hours per week, with more than 50 per cent reporting that they take work home.[58] This results in perceived high stress levels, problems balancing work and family demands, and the occurrence of a number of physical and psychological health conditions, as well as with high employee absenteeism.[59]

Obviously, a workplace that affords employees the flexibility to balance work and personal commitments can significantly reduce stress while enhancing business objectives. The ten most effective mitigation strategies in order of their effectiveness are:

1. alternative work arrangements;
2. telecommuting;
3. compressed work week;
4. flu immunization program;
5. leave for school function;
6. emergency childcare program;
7. job sharing;
8. Employee Assistance Program support;
9. Workplace Wellness Program; and
10. satellite workplaces.[60]

---

[57] L. Duxbury & C. Higgins, *Reducing Work-Life Conflict: What Works? What Doesn't?* (2012), online: http://www.hc-sc.gc.ca/ewh-semt/alt_formats/hecs-scsc/pdf/pubs/occup-travail/balancing-equilibre/full_report-rapport_complet-eng.pdf (date accessed: February 28, 2020).

[58] *Ibid.*

[59] *Ibid.*

[60] CCH, *CCH 2007 Unscheduled Absence Survey* (Riverwoods, Ill.: CCH Incorporated,

Unfortunately, the effectiveness rating of the mitigation strategy does not determine its frequency of use in the workplace. For example, the following is a comparison of the effectiveness and use of work-life programs (Figure 21.4). The most effective programs, such as the use of alternative work arrangements, telecommuting, and the compressed workweek are used less than 55 per cent of the time, while the less effective programs, such as EAPs, flu immunization programs, and WWPs are used the most.

Figure 21.4: Effectiveness and Use of Work-life Programs, 2007[61]

WORK-LIFE PROGRAM	EFFECTIVENESS RATING*	PERCENTAGE USE
Alternative Work Arrangements	3.6	54%
Telecommuting	3.5	53%
Compressed Work Week	3.3	45%
Flu Shot Programs	3.2	66%
Leave for School Functions	3.2	54%
Emergency Child Care	3.1	32%
Job Sharing	3.0	38%
Employee Assistance Plans	2.9	72%
Wellness Programs	2.9	60%
Satellite Workplaces	2.9	36%
On-site Child Care	2.9	32%
Fitness Facility	2.8	52%
On-site Health Services	2.8	33%
Work-life Seminars	2.6	43%
Career Counseling	2.6	41%
Child Care Referrals	2.6	38%
Holidays/Summer Camp	2.6	29%
Sabbaticals	2.5	35%
Concierge Services	2.4	30%
Elder Care Services	2.4	33%

*1: Not Very Effective to 5: Very Effective

2007), online: http://www.cch.com/absenteeism2007 (date accessed: February 28, 2020).

[61] *Ibid.*

(b) Options for the Family-responsive Workplace

Making the workplace family-friendly involves much more than implementing a set of programs designed to address one or two issues. Helping workers cope with work and family demands involves a strategic commitment to engage in a comprehensive transformation of the workplace culture.

This reconfiguration of the workplace involves changes in attitude, as well as clearly articulated program goals, preferably linked to the organization's strategic plan. Achieving this *is no easy task and the process takes time*. Although research shows that there are positive relationships between work and family programming and the bottom line, such evidence fails to convince leaders who do not believe in this new role for business. Adoption of this concept requires a quantum leap in management attitudes and belief systems.

In order to create a family-responsive workplace, Management must determine employee needs and tailor incentives to meet those needs. Although this sounds elementary, many Canadian employers (over 60 per cent) have not created family-responsive workplaces.

What can an employer do to help employees balance work and family life? In general, offer or promote:

- a change in the corporate culture — the attitudes, practices, values, and relationships within the organization that impact family-friendly practices. This includes belief in the legitimacy of work-family policies and practices as part of the workplace, available policies and procedures, supervisory understanding of the issues, and flexibility in dealing with situations, to name a few;
- organizational support for work-life balance;
- education for supervisors on the value of work-life balance;
- supervisor buffering of work pressures;
- flexible work arrangements;
- flexible work time — job sharing, compressed work week, shorter work week, shorter workday;
- a change in where employees work;
- programs that facilitate psychological detachment such as leisure activities, time management workshops, balancing work, and personal time, *etc.*;
- family-friendly benefit packages and programs such as:
  - childcare assistance — information and referral assistance, on-site daycare, family home care, emergency/sick-child care, daycare subsidy;
  - eldercare assistance; and
  - an EAP.

There is an incontestable relationship between work and family-related stresses, and employee productivity and absenteeism, and staff turnover. For example, a survey on workplace absence showed that absence rates fell from 8 to 6.5 days per employee when companies introduced flexible annual leave, flexible work hours, and the ability to work from home occasionally.[62] Organizations can realize improvements by making decisions on specific options and methods of implementation that help employees to alleviate some of these pressures.

The onus is on the Human Resources Professionals/Practitioners to assist in the selection of suitable employee group benefit plans, programs, and approaches that best fit the organization's/company's corporate culture and business strategies.

G. SUPPORTIVE INFRASTRUCTURE FOR THE OH&S PROGRAM

A supportive infrastructure is the system and environment within which an OH&S Program operates. It encompasses the corporate culture, organization/company policies and procedures, and linkages with other organizational resources.[63]

1. Corporate Culture

The corporate culture consists of learned values, assumptions, and behaviours that convey a sense of identity for employees and management. It acts to encourage employee commitment, organizational stability, and desired behaviours.

Depending on the nature of the corporate culture, employees, and Management may be receptive to:

- helping each other;
- looking for innovative ways to accommodate recovering employees back into the workplace;
- taking risks on certain workplace rehabilitation approaches; or
- implementing benefit plans that encourage a return to employability for the ill/injured employee.

As well, the corporate culture dictates what type of OH&S model will be adopted. For instance, a paternalistic culture tends to adopt a model that is more organizationally operated and directed. Here, the onus for worker safety is assumed by the organization. This is typical of the more traditional models of OH&S programs. On the other hand, a democratic corporate culture tends to encourage employee responsibility for workplace health and safety. In this model, the organization works

---

[62] HRM Guide, "Flexible Working Reduces Absenteeism", London, England (July 12, 2001), online: http://www.hrmguide.co.uk/flexibility/flexibility_absenteeism.htm (date accessed: February 28, 2020).

[63] Details are provided in Chapter 5, "Occupational Health and Safety Program: Manual Development", Chapter 6, "Occupational Health and Safety: Leadership and Commitment", Chapter 7, "Safe Workplace = Great Workplace: Building a Sustainable Culture of Safety", and Chapter 8, "Safety Culture: Safety Climate".

with the employee to affect a successful OH&S program leaving the employee responsible and accountable for their safety.

In a subtler fashion, the corporate culture affects occupational bonding. **Occupational bonding** is a mutually beneficial relationship between the employee, and the employer. When an organization has a corporate culture that promotes pride in belonging to the organization and that adheres to a strong work ethic, employees are more likely to see personal value in belonging to the social group and in working for the organization. They are less likely to take risks and/or to be absent from work, except for valid reasons; and more likely to return to work as soon as possible. In this instance, the occupational bond is difficult to break.

Great Place to Work Canada upholds that a great place to work is one in which employees trust their leaders, take pride in their work and value the support of co-workers (2019). The Great Place to Work® Trust Model©, from the employees' perspective, embodies five dimensions— Trust: Credibility, Trust: Respect, Trust: Fairness, Pride and Camaraderie. For more details, refer to https://www.greatplacetowork.ca/en/about-us/trust-model.

When employees view Management as trustworthy and competent, and their practices and policies are fair, they feel respected which is associated with employees' pride in their work.

Managers support and nurture "trust" behaviours through their actions, behaviours, and communications. When hiring, Managers focus on a good person-job fit. They inspire employees, thanking them for their contributions. They speak honestly and seek employee feedback. Employee growth and development is encouraged; successes are celebrated and shared equally. The workplace is inclusive, and diversity is embraced.[64] This is the type of workplace in which people want to work.

Hence, to attain a successful OH&S program, the corporate culture should value the employee and convey the message that all employees are valuable. The culture should be people-oriented in which trust between the employer and employees exists. It should also value having a safe workplace and strong safety performance (OH&S leadership), as well as adherence to the applicable OH&S legislation (Safety due diligence). All of these conditions are associated with fewer incidents and disability claims.[65]

2. Policies and Procedures

Corporate policies and procedures are designed to facilitate the achievement of the established program goals. They also reflect management's attitude and values

---

[64] Great Place to Work, *The Great Place to Work Trust Model* (2019), online: https://www.greatplacetowork.ca/en/about-us/trust-model (date accessed: February 28, 2020).

[65] R. Williams, *et al.*, *A Survey of Disability Management Approaches in Ontario Workplaces*, McMaster University and WSIB Ontario (2005), at 2, online: http://fhs.mcmaster.ca/pohem/research1.html (date accessed: February 28, 2020).

regarding workplace safety and worker well-being. For more information on Corporate policies and procedures, refer to Chapter 5, "Occupational Health and Safety Program: Manual Development", and Chapter 11, "Occupational Health and Safety: Workplace Standards and Rules".

H. WORKPLACE ILLNESS/INJURY PREVENTION STRATEGY

Watson Wyatt (2008) undertook a Staying@ Work Survey in Canada and the United States. The results speak to how organizations can attain a healthy and productive status. The secret of a healthy organization is attaining a balance between the organization's health practices and the workforce's health practices (Figure 21.5). This approach has been reinforced in the subsequent Staying@Work Reports.

Figure 21.5: Healthy Organization[66]

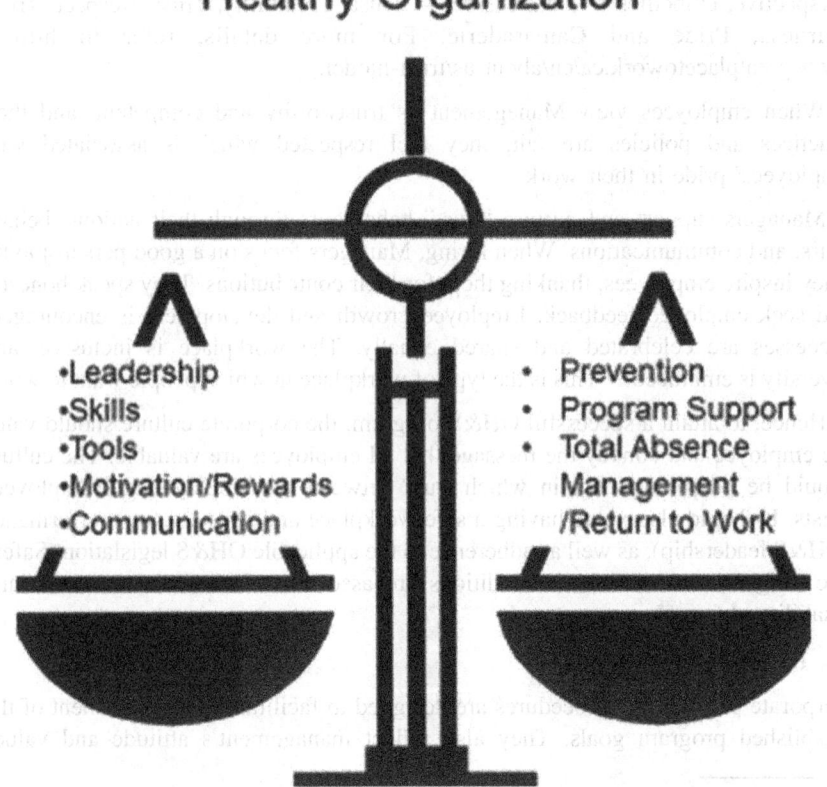

The organizational health practices include:

---

[66] A **healthy organization** achieves a balance between the organization's health practices and the workforce's health practices.

Ch. 21: OH&S: Prevention Of Workplace Illness And Injury

- *Leadership* — Providing strategic communication on the organization's/company's business goals and plans, as well as role clarity to the various stakeholders. Succinctly put, "What is to be done?", "How?" and "By whom?"
- *Skills* — Addressing the requisite employee capabilities, learning needs and application of knowledge to maximize employee performance and productivity.
- *Tools* — Providing the needed tools, data systems, information, workload, and work facilitation.
- *Employee motivation/recognition* — Inspiring employees through monetary/non-monetary rewards, performance management, job opportunities, corporate culture, and management to maximize employee performance and productivity.
- *Communication* — ensuring open and honest two-way communication.[67]

These recommended elements of work health and productivity mirror the human performance elements of the Roithmayr Performance Maximizer® described in Chapter 22, "Toxic Work Environments: Impact on Employee Illness/Injury". It advocates that the work environment should:

- enable employees to know how to do their jobs;
- ensure employees are able to do their jobs;
- equip employees to do their jobs;
- motivate employees to do their jobs;[68] and
- include workplace interactions — the "glue that brings it all together", namely the trust, honesty, respect, openness, and involvement of all stakeholders.[69]

The workforce's health practices proposed by Watson-Wyatt, 2008, include:

- *Prevention* — Through the use of OH&S Programs, Health Risk Appraisals, health management, disease management, control of employee presenteeism, conflict management, and workplace harassment and violence prevention the workforce can prevent workplace illness and injury;
- *Plan Administration/Program Support* — To effectively operate, business practices such as vendor management, plan design, and financial management are required to be in place; and

---

[67] Watson Wyatt, *Staying@Work: Effective Presence at Work* (2008), online: http://www.watsonwyatt.com (date accessed: February 28, 2020).

[68] T. Roithmayr, The Performance Maximizer®, presented in Chapter 6, "Occupational Health and Safety: Leadership and Commitment".

[69] *Ibid.*

- ***Total Absence Management and Return to Work*** — To address the human aspects of work, attendance management, integrated disability management, program administration, claim management, case management and return-to-work measures, and management are great assets to an organization/company.[70]

From the annual Towers Watson Staying@Work Survey Reports, it has been determined that organizations with highly effective health and productivity programs experience:
- fewer lost workdays due to absenteeism and disability, resulting in 1.2 fewer casual absence days and costs, and a 30% increase in employee group benefit plan savings for the company because of:
    - fewer Short Term Disability cases;
    - shorter disability durations (25 fewer workdays lost per 100 insured employees annually)
    - 4.5 per cent lower Long Term Disability rates and costs;
    - 3.5 per cent lower healthcare costs (an average of $551 loess per employee in Canada);
    - lower payroll costs for Workers' Compensation Board claims;
    - lower employee presenteeism rates and costs;
    - lower voluntary staff turnover levels — 5.3 per cent versus 10 per cent;
- integrated health management programs;
- a healthier workforce — fewer employee health risks; and
- a healthier organization and bottom-line:
    - greater financial returns and productivity improvements;
    - 40 per cent more revenue per employee (a $132,000 per employee difference);
    - 16.1 per cent higher market value;
    - 57 per cent higher shareholder returns; and
    - market premiums that are 18 points higher than those of low effectivenss organizations/companies.[71]

---

[70] Watson Wyatt, *Staying@Work: Effective Presence at Work* (2008), online: http://www.watsonwyatt.com (date accessed: February 28, 2020).

[71] Watson Wyatt, *Staying@Work: Effective Presence at Work* (2008), online: http://www.watsonwyatt. com (date accessed: February 28, 2020); Towers Watson, *2009/2010 North American Staying@Work Report: The Health and Productivity Advantage* (2010), online: http://www.towerswatson.com/en/Insights/IC-Types/Survey-Research-Results/2009/12/20092010-

Extrapolating the above information, as well as the materials presented in this chapter, a model of a Workplace Health & Productivity is provided (Figure 21.6).

In this model, the mix of an engaged workforce, effective workplace programs, program performance measurements, and evaluation, can lead to an integrated approach to workplace health; and improved workplace health behaviours, productivity and health outcomes, and costs. These health and productivity outcomes in turn, lead to superior financial performance.

Figure 21.6: Value of Workplace Health and Productivity Approaches[72]

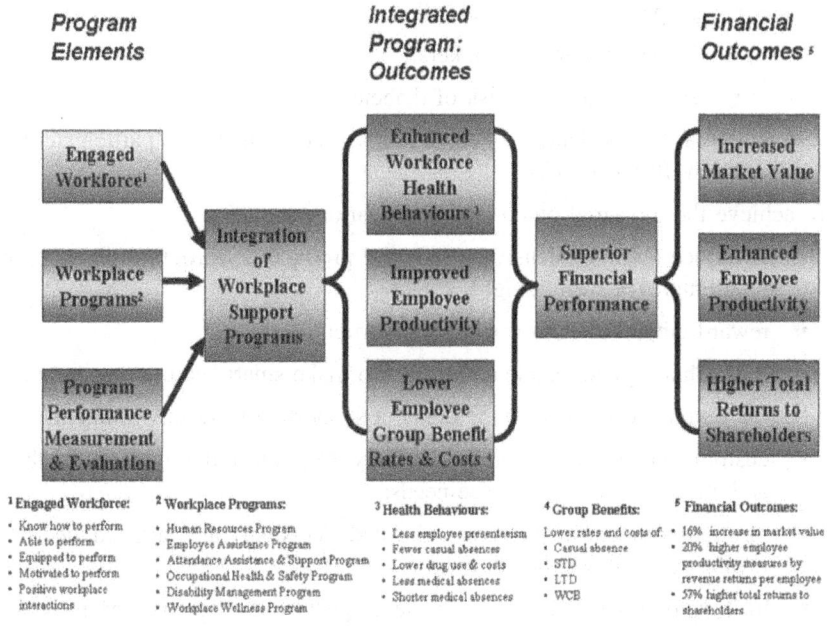

North-American-Staying Work-Report-The-Health-and-Productivity-Advantage; Towers Watson, *2011/2012 Staying@ Work Survey Report: A Pathway to Employee* Health and *Workplace Productivity* (2012), online: http://www.towerswatson.com/en/Insights/IC-Types/Survey-Research-Results/2011/12/20112012-StayingWork-Survey-Report--A-Pathway-to-Employee-Health-and-Workplace-Productivity; Towers Watson, *2013/2014 Staying@Work Report - Canada Summary* (2014), online: http://www.towerswatson.com/en-CA/Insights/IC-Types/Survey-Research-Results/2014/02/2013-2014-staying-at-work-report-canada-summary (date accessed: February 28, 2020).

[72] Adapted from T. Roithmayr, The Performance MaximizerTM (Calgary, AB: Performance by Design), online: http://www.performance-bydesign.com (date accessed: February 28, 2020); and Watson Wyatt Worldwide, *Staying@Work: Effective Presence at Work* (2008), online: http://www. watsonwyatt.com (date accessed: February 28, 2020).

Today, 60 per cent of Canadian employers offer a health and productivity program in some form.[73] However, few of those programs (only 4%) are linked with an articulated strategy that aligns with the organization's broader objectives and culture.[74] According to the 2015/2016 Staying@Work Report Canada Summary, 48 per cent of employers plan over the next three years, to create a differentiated health and productivity strategy that would allow customization of critical workforce segments. The incentives for doing so would be that effective health and productivity programs result in:

- A 50 per cent higher revenue per employee; and
- Lower employee risk factors:
  - 25 per cent less hypertension;
  - 30 per cent fewer smokers;
  - 24 per cent lower risk of diabetes; and
  - A less sedentary lifestyle: 35% more likely to participate in a wellness activity.

To achieve these desired outcomes, organizations should:

- introduce a comprehensive health and productivity program that supports total employee well-being;
- reward employee involvement and participation;
- use technology to enable employees to make smart health-related choices;
- use employee suggestions to create a supportive work environment;
- customize their Health and Productivity Program so that it offers flexibility activities that meet employee needs;
- through data analysis, make informed decisions and targeted decisions regarding employee health risks;
- measure program performance and leverage the outcomes to get Management/employee buy-in and support; and
- formalize the Health and Productivity Program.[75]

I. SUMMARY

By linking an organization's OH&S efforts with its EAP, DMP, and WWP; and by

---

[73] Willis, Tower, Watson, *2015/2016 Staying@Work Report Canada Summary* (2017), online: https://www.willistowerswatson.com/-/media/WTW/PDF/Insights/2016/08/2015-2016-staying-at-work-canada-research-findings-en.pdf (date accessed, February 28, 2020).

[74] *Ibid.*

[75] Willis Watson Wyatt, *2015-2016 Staying@Work Report* (2016), online: https://www.willistowerswatson.com/-/media/WTW/PDF/Insights/2016/08/2015-2016-staying-at-work-canada-research-findings-en.pdf (date accessed, February 28, 2020).

recognizing the impact that Management and human resource management theories can have on a workforce and workplace wellness; the opportunity to significantly reduce illness/injury incidence and impact, exists. The challenge is to act on that knowledge in a proactive manner.

CHAPTER REFERENCES

Aon Consulting Inc., *Canada@Work* (Chicago: Aon Group, 2000).

Association of Workers' Compensation Boards of Canada, "Key Statistical Measures for 2018", online: http://awcbc.org/?page_id=9759 (date accessed: February 28, 2020).

Buffett & Company, *National Wellness Survey Report 2006* (Whitby, ON: Buffett & Company Worksite Wellness Ltd., 2006), now owned by Sun Life of Canada.

Buffett and Company Worksite Wellness Ltd., *National Wellness Survey Report 2009* (Whitby, ON: Buffett and Company Worksite Wellness Ltd., 2009), now owned by Sun Life of Canada.

CCH, *CCH 2007 Unscheduled Absence Survey* (Riverwoods, Ill.: CCH Incorporated, 2007), online: http://www.cch.com/absenteeism2007/images/effUse2007/asp (date accessed: February 28, 2020).

A. Chambers, Ibrahim, S, Etches, J, & Mustard, C. Diverging trends in the incidence of occupational and non-occupational injury in Ontario, 2004-2011 (2015) 105:2 Am J Public Health 338-343, doi: 10.2105/AJPH.2014.302223, online: https://www.iwh.on.ca/summaries/issue-briefing/divergent-trends-in-work-related-and-non-work-related-injury-in-ontario (date accessed: February 28, 2020).

Conference Board of Canada, *Compensation Planning Outlook* (Ottawa: Conference Board of Canada, 1997).

Conference Board of Canada, News Release, "Eldercare Taking its Toll on Canadian Workers" (November 10, 1999), online: http://www.conferenceboard.ca (date accessed: February 28, 2020).

Conference Board of Canada, *Healthy Brains at Work* (2016), online: http://www.sunlife.ca/static/canada/Sponsor/About%20Group%20Benefits/Focus%20Update/2016/557/HealthyBrains_Report2_EN.pdf at 20 (date accessed: February 28, 2020).

Consortium for Organizational Mental Healthcare (COMH), online: http://www.comh.ca (date accessed: February 28, 2020).

CNW, "National Wellness Survey Shows Canadian Organizations Investing in Worksite Wellness" (September 30, 2009), online: http://www.newswire.ca/en/releases/archive/September2009/30/c2024.html (date accessed: February 28, 2020).

B. Daigle & G. Schick, "Early Intervention, Integration and Successful Resolution of Employee Health Issues" (Presented at the Health Work and Wellness

Conference '98, Whistler, BC, September 27–30, 1998).

S. Dobson, "Strong link between quality of work life, patient care: Study", *Canadian HR Reporter* (November 15, 2010) at 6.

L. Duxbury & C. Higgins, *Reducing Work-Life Conflict: What Works? What Doesn't?* online: http://www.hc-sc.gc.ca/ewh-semt/alt_formats/hecs-sesc/pdf/pubs/occup-travail/balancing-equilibre/full_report-rapport_complet-eng.pdf (date accessed: February 28, 2020).

D. Dyck, *Disability Management: Theory, Strategy & Industry Practice*, 3rd ed. (Markham, ON: Butterworths, 2006).

D. Dyck, "Workplace Wellness: What is Your Potential Return on Investment?", Human Resources Association of Calgary: A Newsletter on Human Resources Management (November 1998).

D. Dyck, "Wrapping Up the Wellness Package" (January 1999) *Benefits Canada*, at 16–20.

Editorial, "Low-control jobs are hazardous to health" *O.H.S. Canada* (July/August 2002) at 23.

Great Place to Work, *The Great Place to Work Trust Model* (2019), online: https://www.greatplacetowork.ca/en/about-us/trust-model(date accessed: February 28, 2020).

Guarding Minds @ Work: *A Workplace Guide to Psychological Safety & Health*, online: http://www.guardingmindsatwork.ca/info/index (date accessed: February 28, 2020).

HRM Guide, "Flexible Working Reduces Absenteeism", London, England, (July 12, 2001), online: http://www.hrmguide.co.uk/flexibility/flexibility_absenteeism.htm (date accessed: February 28, 2020).

Ipsos-Reid Survey reported by M. Shain, "Psychological Safety at Work: Legal Trends and the Implications" (August 2009) 19:5 *Benefits and Pensions Monitor* at 3–31.

Ipsos-Reid Survey reported by M. Shaw, "How can we . . .", *Accident Prevention Magazine* (March/April 2007).

M. Jackson, "Worker Squeeze: Employee Stress Cuts Productivity, Study Declares" *Calgary Herald* (April 15, 1998) at A5.

R. Karasek & T. Theorell, *Healthy Work: Stress, Productivity and the Reconstruction of Working Life* (Toronto: HarperCollins, 1992).

S. Klie, "Working from home attracts, retains staff" *Canadian HR Reporter* (December 13, 2010) at 3.

Marsh Risk Consulting, "Fourth Annual Marsh Mercer Survey on Employers' Time-Off & Disability Programs" (Workforce Risk, 2003), online: http://usa.marsh.com (date accessed: February 28, 2020).

T. Roithmayr, The Performance Maximizer®, (Calgary, AB: Performance by Design), online: http://www.performance-bydesign.com (date accessed: February 28, 2020).

Safe Saskatchewan, online: http://www.safesask.com/index.cfm (date accessed: February 28, 2020).

M. Shain, "A New Take on Stress: Strategies that Work", Health Policy Forum (November 1997).

M. Shain, "Managing Stress at its Source in the Workplace" (Presented at the Health Work and Wellness Conference '98, Whistler, BC, September 27–30, 1998).

M. Shain, "Stress, Satisfaction and Health at Work: Tuning for Performance" (Presented at the Health and Wellness Conference '99, Vancouver, BC, October 24–27, 1999).

Suncor Energy Inc., Report from Shannon Stratton, Suncor Energy Inc. (2014)

Suncor Energy Inc., Report from Shannon Stratton, Suncor Energy Inc. (2017)

Suncor Energy Inc., Report from Shannon Stratton, Suncor Energy Inc. (2019)

Towers Watson, *2009/2010 North American Staying@Work Report: The Health and Productivity Advantage* (2010), online: http://www.towerswatson.com/en/Insights/IC-Types/Survey-Research-Results/2009/12/20092010-North-American-StayingWork-Report-The-Health-and-Productivity-Advantage (date accessed: February 28, 2020).

Towers Watson, *Pathway to Health and Productivity: 2011/2012 Staying @ Work Survey Report* (2013), online: http://www.towerswatson.com/en-CA/Insights/IC-Types/Survey-Research-Results/2011/12/20112012-StayingWork-Survey-Report--A-Pathway-to-Employee-Health-and-Workplace-Productivity (date accessed: February 28, 2020).

Towers Watson, *2013/2014 Staying@Work Report - Canada Summary* (2014), online: http://www.towerswatson.com/en-CA/Insights/IC-Types/Survey-Research-Results/2014/02/2013-2014-staying-at-work-report-canada-summary (date accessed: February 28, 2020).

Watson Wyatt, *Staying@Work: Effective Presence at Work* (2008), online: http://www.watsonwyatt.com (date accessed: February 28, 2020).

Watson Wyatt Worldwide, *The Health and Productivity Advantage; 2009/2010 North American Staying @ Work Report* (2010), online: http://www.towerswatson.com/en-CA/Insights/IC-Types/Survey-Research-Results/2009/12/20092010-North-American-StayingWork-Report-The-Health-and-Productivity-Advantage (date accessed: February 28, 2020).

R. Williams, *et al.*, *A Survey of Disability Management Approaches in Ontario Workplaces* (2005), McMaster University and WSIB Ontario, at 2, online: http://fhs.mcmaster.ca/pohem/research1.html (date accessed: February 28, 2020).

Willis Tower Watson, *2015/2016 Staying@Work Report Canada Summary* (2017), online: https://www.willistowerswatson.com/-/media/WTW/PDF/Insights/2016/08/2015-2016-staying-at-work-canada-research-findings-en.pdf (date accessed, February 28, 2020).

WSIB/CSPAAT, "Business Results Through Health & Safety" (2001), online: http://www.ryerson.ca/content/dam/irm/pdfs/training/ISS_MGR/WSIB_BusinessResultsThroughHS.pdf at 3 (date accessed: February 28, 2020).

Workopolis.com Survey (2010) reported in S. Klie, "Working from home attracts, retains staff" *Canadian HR Reporter* (December 13, 2010) at 3.

Chapter 22

TOXIC WORK ENVIRONMENTS: IMPACT ON EMPLOYEE ILLNESS/INJURY[1]

By Tony Roithmayr[2]

A. INTRODUCTION

Over the years, Occupational Health and Human Resources personnel have witnessed the effects of organizational stressors on employee and manager health. Workers who perceive themselves as being "stressed":

- report making more mistakes;
- feel angry with their employers for creating the stressful situation; and
- resent their coworkers who they feel are not working as hard as they are, which can lead to a desire to leave the job.[3]

Overstressed workers also cost their employers 50 per cent more in terms of healthcare expenditures, lost workdays,[4],[5] presenteeism,[6] staff turnover costs[7] and accident costs.

---

[1] Excerpts taken from D. Dyck & T. Roithmayr, "Organizational Stressors and Health: How Occupational Health Nurses Can Help Break the Cycle" (2002) 50:5 AAOHN Journal 213. Copyright (2002) the American Association of Occupational Health Nurses, Inc. Used with permission. All rights reserved.

[2] T. Roithmayr, B.A., M.Ed., is the president of Performance by Design, online: http://www.performance-bydesign.com/ (date accessed: February 28, 2020).

[3] J. MacBride-King, "Wrestling with Workload: Organizational Strategies for Success" (Ottawa: Conference Board of Canada, 2005).

[4] Morneau Shepell, *The true picture of workplace absenteeism* (2015), online: http://www.morneaushepell.com/sites/default/files/documents/3679-true-picture-workplace-absenteeism/9933/absencemanagementreport0608-15.pdf (date accessed: February 28, 2020).

[5] Willis Towers Watson, *2015/2016 Global Staying@Work Survey, Canada* (2016) at 12, online: https://easna.org/wp-content/uploads/2016/04/2015-2016-Global-Staying-at-Work-Survey.pdf (date accessed: February 28, 2020).

[6] *Ibid.*

[7] J. MacBride-King, "Wrestling with Workload: Organizational Strategies for Success"

Typically, the response is "[w]hat can we do? Things have been this way for years." The purpose of this chapter is to encourage an understanding of the impact that a stressful, toxic work environment can have on the employee and organization; and to present a plausible argument for initiating realistic workplace changes to help "break the cycle" of workplace stress, deteriorating health, and rising cost consequences.

The bottom line is that workers vary in their tolerance of workloads. What is desirable for one worker is intolerable for another.[8] It is important for managers to understand the drivers of stress and to respect the individual differences of each employee in the workplace.

B. HAVE ORGANIZATIONS BECOME TOXIC TO HUMAN LIFE?

In too many workplaces, the following things have happened, or are happening:

- Too many and conflicting priorities
- Lack of understanding about how performance is measured
- Poor communications
- Disrespectful or bullying behaviour
- Little or no feedback and recognition

- Employees experiencing frequent headaches, workplace incidents, anxiety attacks, ulcers or high blood pressure
- Employees reporting insomnia
- Employees experiencing irritability, anger or depression
- Increased drug or alcohol abuse

- Poor employee morale
- Workplace tension
- High absenteeism
- Staff turnover
- Rising employee benefit plan costs
- Disappointing financial results

These three lists are causally related, and this relationship tends to have a destructive cycle. The cumulative effect of a multitude of organizational stressors manifests as deteriorating employee health, lost productivity, and escalating disability and employee group benefit plan costs.[9]

For example, Dr. Martin Shain reports that employees who experience high stress due to high work effort and low reward (recognition) and high strain due to high work demands and limited control over their job, suffer a threefold increase in the incidence of heart problems, back pain, work/family conflicts, substance abuse, infections, mental health problems, and injuries; and a fivefold increase in the incidence of certain cancers like colorectal cancer.[10] Other studies have shown that employees who experience chronic low-job control are more prone to premature

---

(Ottawa: Conference Board of Canada, 2005) at 1.

[8] *Ibid.*, at 1.

[9] R. Karasek & T. Theorell, *Healthy Work: Stress, Productivity, and the Reconstruction of Working Life* (New York, NY: Basic Books Inc., 1992).

[10] M. Shain, "Managing Stress at its Source in the Workplace" (Presented at the Health Work and Wellness Conference 1998, Whistler, BC, September 27–30, 1998) [unpublished].

death. The primary cause was cardiovascular in nature.[11]

The Conference Board of Canada's *Survey of Canadian Workers on Work-Life Balance* reports that high-stress levels due to the difficulty of balancing the demands of work and personal commitments are associated with health problems, work absence, and lower productivity. Respondents experiencing high stress miss twice as much work time as those who report being in low stress situations (7.2 versus 3.6 days absence).[12],[13],[14]

A 2016 study on psychological health indicates that:

- 67% of employees experiencing chronic stress indicate it negatively impacts their work;
- 48% of employees linked extreme stress and chronic illness; and
- 26% of employees with reportedly high work stress took "mental health days" from work.[15]

This finding is supported by the Willis Towers Watson 2015/2016 Global Staying@ Work Survey that indicates that highly stressed employees lose 4.2 workdays as opposed to their peers who report experiencing low stress (2.6 days).[16]

Stress situations result in the release of the hormones epinephrine (or adrenaline) and norepinephrine into the bloodstream. At the same time, the adrenal glands also release cortisol, a hormone that sends the body the message to release fatty acids for

---

[11] J. Siegrist, "Adverse Health Effects of High-Effort/Low-Reward Conditions" (1996) 1:1 Journal of Occupational Health Psychology 27; B.B. Marmot *et al.*, "Contribution of Job Control and Other Risk Factors to Social Variations in Cardiovascular Heart Disease Incidence" (1997) 350 (9073) *Lancet* 235; J. Johnson, "Long-term Psychological Work Environment and Cardiovascular Mortality among Swedish Men" (1996) 86:3 American Journal of Public Health 324; A. LaCroix, *Occupational Exposure to High Demand/Low Control Work and Coronary Heart Disease Incidence in the Framingham Cohort* (Ann Arbor, MI: University of North Carolina, UMI Dissertation Services, 1984).

[12] Conference Board of Canada, *Survey of Canadian Workers on Work-Life Balance* (Ottawa: Conference Board of Canada, 1999).

[13] J. Park, *Work Stress and Job Performance* (Ottawa: Statistics Canada, 2007), Cat. No. 75-001-XIE.

[14] L. Duxbury, *Revisiting Work-life Issues in Canada: The 2012 National Study on Balancing Work and Caregiving in Canada, 2012* (Ottawa: ON: Carleton University), online: http://newsroom.carleton.ca/wp-content/files/2012-National-Work-Long-Summary.pdf (date accessed: February 28, 2020).

[15] Morneau Shepell, *The true picture of workplace absenteeism* (2015), online: http://www.morneaushepell.com/sites/default/files/documents/3679-true-picture-workplace-absenteeism/9933/absencemanagementreport06-08-15.pdf at 2.

[16] Willis Towers Watson, *2015/2016 Global Staying@Work Survey, Canada* (2016), online: https://easna.org/wp-content/uploads/2016/04/2015-2016-Global-Staying-at-Work-Survey.pdf at 12 (date accessed: February 28, 2020).

a burst of energy. Chronically high-stress levels can negatively impact body metabolism, elevate blood pressure, damage white blood cells, worsen autoimmune disorders and inflammatory conditions (eczema and colitis), and contribute to gastrointestinal problems. Conditions like heart disease, stroke, diabetes, HIV/AIDS, and kidney failure have all been linked to chronic stress situations, as well as psychological disorders like depression and anxiety.[17]

Stress-related absences cost Canadian employers approximately $33 billion annually, plus billions in additional related-healthcare costs.[18] Workplace stress contributes to:

- 19% of absenteeism costs;
- 40% of staff turnover costs;
- 55% of Employee Assistance Program costs;
- 30% of short-term disability and long-term disability costs;
- 60% of occupational incidents; and
- 10% of prescription drug plan costs.[19]

A 2011 study indicated that each year, Canadian employers pay $6 billion in absenteeism and presenteeism costs — conditions that are strongly related to stress and psychological health conditions.[20]

So, what can be done? How can organizations eliminate or mitigate the impact of organizational stressors? Experts in the field recommend eliminating stress at its source:[21] *"to achieve high levels of employee productivity, efficiency and morale, leading executives have learned that they need to address workplace health and wellness in an integrated fashion"*.[22] This can be done effectively by focusing on the environment in which the work is being done.

---

[17] S. Cohen, D. Janicki-Deverts & G. Miller, "Psychological Stress and Diseases" (2007) 298:14 JAMA at 1685.

[18] The Canadian Encyclopedia (October 2007). *Workplace Stress Costs the Economy Millions of Dollars*, online: https://thecanadianencyclopedia.ca/en/article/workplace-stress-costs-the-economy-billions (date accessed: February 28, 2020).

[19] Chrysalis Performance Inc., reported in *IAPA: Creating Healthy Workplaces Everywhere - Healthy Workplace Week* (October 2006), online: http://www.iapa.ca/Main/articles/2006_oct_healthy_workplace_week.aspx (date accessed: February 28, 2020).

[20] P. Jacobs, (2013). Reported in Lopez-Pacheco. A. "Mental Illness adversely affecting Canada's economic potential" *Financial Post* February 5, online: http://business.financialpost.com/executive/the-economic-cost-of-mental-illness (date accessed: February 28, 2020).

[21] M. Shain, "Managing Stress at its Source in the Workplace" (Presented at the Health Work & Wellness Conference 1998, Whistler, B.C., September 27–30, 1998) [unpublished].

[22] K. Bachmann, *More Than Just Hard Hats and Safety Boots: Creating Healthier Work Environments* (Ottawa: Conference Board of Canada, 2000) at 1.

The following model provides insight into how to create a healthy and productive workforce through an integrated approach that enables employees to do their job and gain enjoyment and growth from the experience.

C. THE PERFORMANCE MAXIMIZER® MODEL

We will start with some of the fundamental aspects of human performance in the workplace. To sustain success, organizations must provide an environment that energizes, supports, and enables performance at its best. People who experience such a workplace are healthy, feel secure, and willingly hold themselves accountable for great results. The Performance Maximizer® (Figure 22.1) is a foundation concept for understanding human performance and explains what is required to enable performance at its best.

The model focuses on all the factors that shape optimal human performance. It describes, in a simple and memorable way, the conditions that exist when successful human performance occurs in the workplace.

Figure 22.1: The Performance Maximizer®[23]

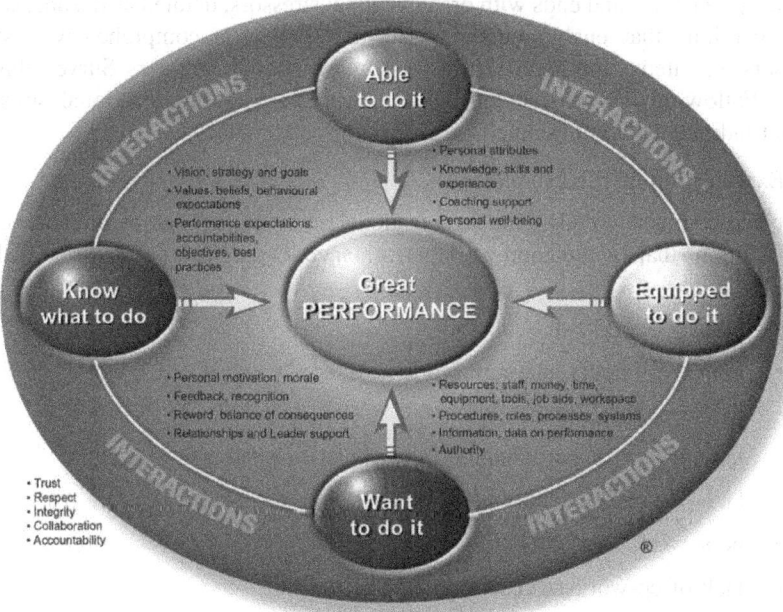

Copyright © 1998-2006 Performance! ...by Design®. All rights reserved. The Performance Maximizer ® is a registered trademark of Anthony Roithmayr. www.performance-bydesign.com

We have clustered these factors into five "Conditions for Great Performance".

---

[23] T. Roithmayr, "The Performance Maximizer®" (Calgary, AB: Performance by Design, 2000), online: http://www.performance-bydesign.com/ (date accessed: February 28, 2020).

That is, *Know What to do, Able to do it, Equipped to do it, Want to do it,* and *Interactions*,[24] that foster trust, respect, integrity, collaboration, and accountability. Expectations, best practices, skills, processes, tools, motivation, culture; all of these and much more are represented by the model.

In the absence of these conditions, "organizational stressors" develop which cause problems with employee health and on-the-job performance, and, ultimately, impact the organization's bottom line. The premise is that by focusing on the leading indicators for creating great human performance, lagging indicators such as employee absence, disability-related costs, reduced productivity, and poor profits will gradually decrease.

D. ORGANIZATIONAL STRESSORS AND HEALTH

The essential elements of Organizational Stressors and Health are presented in Figure 22.2.[25] It describes a situation that requires human performance improvement techniques to solve formidable and serious business problems. Intuitively, we recognize that a causal relationship exists among the elements presented below. But, how does this model actually work?

The cycle begins and ends with organizational stressors, defined as the absence of the conditions that enable human performance. (For a comprehensive list of stressors, see items 1 through 26 in the Organizational Stressors Survey, Figure 22.4.) Following the arrows, the model illustrates how organizational stressors impact individuals and ultimately, organization performance.

1. Explanation of the Model

The cycle begins and ends with organizational stressors, defined as the absence of the conditions that enable good work performance. Organizational stressors include exposure to:

- a heavy workload;
- long work hours and shift work;
- hectic work routine;
- lack of worker participation in workplace decision-making;
- lack of family-friendly work policies;
- poor work environment;
- lack of co-worker/supervisor support;
- conflicting or uncertain job expectations;
- job insecurity;

---

[24] **Interactions** were added to the model after the research reported in this chapter was completed.

[25] T. Roithmayr, *Performance by Design* (2000).

- lack of growth opportunity, advancement, or promotion; and
- unpleasant or dangerous physical conditions.[26]

For a comprehensive list of stressors, see items 1 through 26 in the Organizational Stressors Survey, Figure 22.4.

Following the arrows, the model illustrates how organizational stressors impact individuals and ultimately, organization performance.

---

[26] Editorial, "Stress at Work" (2007) 27:12 AAOHN News 17.

Figure 22.2: Organizational Stressors and Health

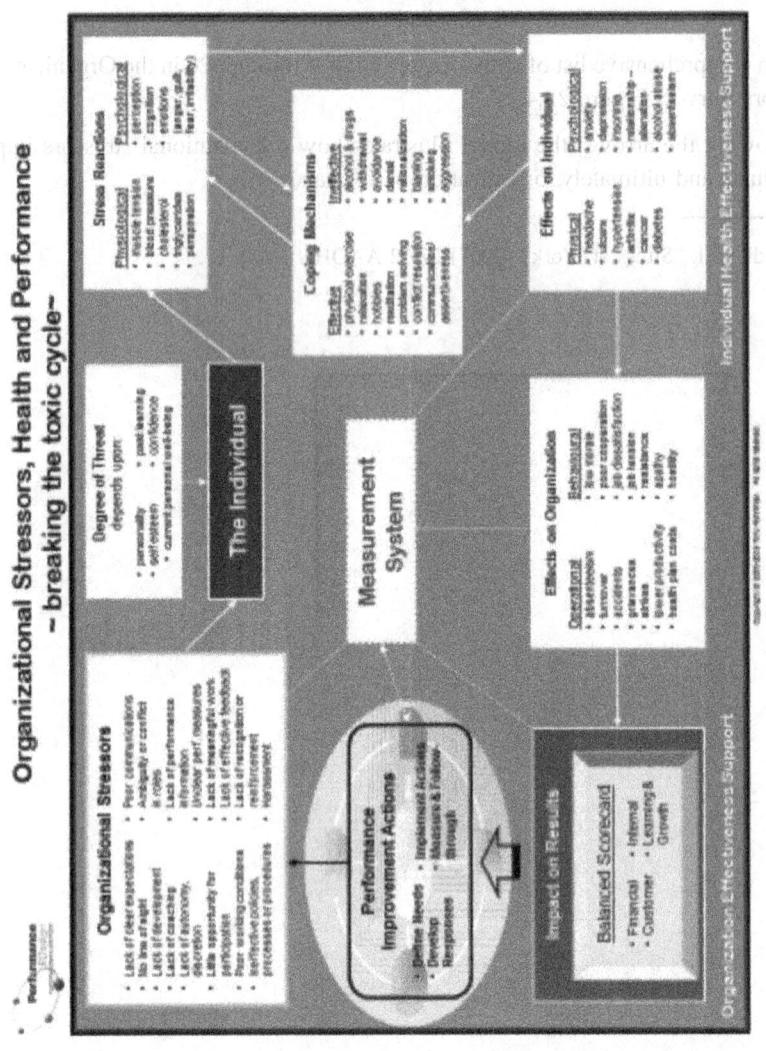

Starting with the individual (the employee), the elements include:

- **Degree of Threat**

Sustained exposure to organizational stressors can impact even the most resilient of individuals.[27] The more stressors there are and the longer the duration of the exposure, the more likely that detrimental effects will occur.

---

[27] R. Karasek & T. Theorell, *Healthy Work: Stress, Productivity and the Reconstruction of Working Life* (New York, NY: Basic Books Inc., 1990).

In general, personal impact depends on individual strengths and social supports — personality, the degree of self-confidence, level of self-esteem, emotional well-being, and personal support systems, all play a role in mitigating the effects of workplace stressors. The individual strengths and supports can heighten resilience, or, their absence can magnify the individual's vulnerability.

- **Stress Reactions**

Workplace stress produces stress reactions. Physiologically, there may be changes in blood pressure or cholesterol levels, heightened

awareness of the environment, or increases in muscle tension. Psychologically, judgement may be impaired. The person may experience irritability, anxiety, anger, an inability to concentrate, or short-term memory loss.

In 1999, the estimate was that employees process 190 messages per day (voice mail, e-mail, faxes, text messages, *etc.*). Today, they spend 33% of their workday using email.[28] This constant information processing and overloading can lead to increased stress. As well, life in general, is deemed to be more stressful than it was 30 years ago.[29]

A recent study indicates that 62 per cent of workers report experiencing "a great deal of stress" at work. For 34 per cent of workers, "stress has been so overwhelming that it has made me physically ill at times".[30] This is consistent with some previous work conducted by Karasek and Theorell, 1990.[31]

- **Coping Mechanisms**

Everyone has individual ways of dealing with everyday stressors. Good communication and problem-solving skills, regular exercise, relaxation, and social supports, as well as a variety of personal interest, can serve to lessen the effects of stress. However, when stress is distressing and prolonged, many people respond in ineffective ways: avoidance, withdrawal, panic, or aggressive behaviour — even an increased use of drugs and alcohol. These ineffective coping mechanisms can serve to deal with the stress of the moment, but can seriously impair physical and emotional health in the long run.

---

[28] Duxbury, L. and Higgins, C. (2012). *Revisiting Work-Life Issues in Canada: The 2012 National Study on Balancing Work and Caregiving in Canada*, available at. http://newsroom.carleton.ca/wp-content/files/2012-National-Work-Long-Summary.pdf (date accessed: February 28, 2020).

[29] B.A. Cryer, "Neutralizing Workplace Stress: The Physiology of Human Performance and Organizational Effectiveness" (Presented at the Psychological Disabilities in the Workplace: Prevention, Rehabilitation and Cost Control Conference, Toronto, ON, June 10 11, 1996) [unpublished].

[30] Aventis Pharma-Canada, *The Aventis Healthcare Survey* (2001), online: <http://www.sanofi-aventis.ca/1/ca/en/index.asp> (date accessed: February 28, 2020).

[31] R. Karasek & T. Theorell, *Healthy Work: Stress, Productivity and the Reconstruction of Working Life* (New York, NY: Basic Books Inc., 1990).

- **Effects on Individuals**

For years, health professionals have recognized the symptoms of workplace stress: headaches, ulcers, infections, hypertension, cardio-vascular disease, substance abuse, anxiety, hostility, and clinical depression.[32] However, society is just now beginning to appreciate the tremendous costs associated with workplace stress: productivity losses, human suffering, increased disability management and employee group benefit plan costs, and the enormous burden on our healthcare system (Figure 22.3).

Organizations are not solely responsible for the negative effects of organizational stress. However, when the average worker spends as much as two-thirds of his/her waking hours at work, or with concerns of work, it begs the question: What is the appropriate level of organizational responsibility?

Figure 22.3: Impact of Workplace Stress on Society

- Workers exposed to high job strain are twice as likely to experience psychological distress and even physical ailments, versus workers who report low job strain.[33, 34, 35, 36]
- Mental health disorders are 60% higher among working Canadians, than among non-working Canadians.[37]
- Over the previous year, 4% of Canadian employees missed 1-2

---

[32] Health Canada, *Best Advice on Stress Risk Management in the Workplace* (Ottawa: Minister of Public Works and Government Services Canada, 2000), Cat. No. H39-546/2000E; J. Occup. Health Psychol. 27.

[33] H. Orpana, L. Lemyre, & R., *Income and psychological distress: The role of the social environment* (2009), Statistics Canada Catalogue no. 82-003-X, available online at: http://www.statcan.gc.ca/pub/82-003-x/2009001/article/10772-eng.pdf (date accessed: February 28, 2020).

[34] Duxbury, L. and Higgins, C. (2012). *Revisiting Work-Life Issues in Canada: The 2012 National Study on Balancing Work and Caregiving in Canada*, available at: http://newsroom.carleton.ca/wp-content/files/2012-National-Work-Long-Summary.pdf (date accessed: February 28, 2020).

[35] Willis Towers Watson (2016). *2015/2016 Global Staying@Work Survey, Canada*, p. 12, online at: <https://easna.org/wp-content/uploads/2016/04/2015-2016-Global-Staying-at-Work-Survey.pdf> (date accessed: February 28, 2020).

[36] Willis Towers Watson (2016). *2015/2016 Global Staying@Work Survey, Canada*, p. 12, online at: <https://easna.org/wp-content/uploads/2016/04/2015-2016-Global-Staying-at-Work-Survey.pdf> (date accessed: February 28, 2020).

[37] Conference Board of Canada (2012). Healthy Brains at Work, available at: http://www.sunlife.ca/static/canada/Sponsor/About%20Group%20Benefits/Focus%20Update/2016/557/HealthyBrains_Report2_EN.pdf (date accessed: February 28, 2020).

- workdays due to depression, stress, and anxiety in 2012; 6% missed 6 days or more, and 21% missed no days.[38]
- In 2013, 47% of employed Canadians considered work to be the most stressful part of their lives.[39]
- Annually, 20% of Canadians experience stress-related illness.[40]
- 25% of Canadians have been diagnosed with depression.[41, 42]
- On any given week, at least 500,000 Canadians are unable to work due to mental illness, including 355,000 disability cases due to mental disorders, plus approximately 175,000 full-time workers absent due to psychological issues.[43]
- Only one in four cases are detected and diagnosed; as well, less than 8% are being properly treated.[44]
- The North American price tag for direct costs and productivity losses related to mental health disorders is $51 billion annually.[45]
- Of the 62% of individuals who related themselves as being "highly stressed", only 2% reported that their work environment was detri-

---

[38] IPSOS-Reid, *Mental Health in the Workplace: Largest Study Ever Conducted of Canadian Workplace Mental Health and Depression* (Toronto: IPSOS0Reid, 2007), at 13, available online at: http://www.ipsos.ca (date accessed: February 28, 2020).

[39] Partners for Mental Health (2013). Reported by Cottrell, J. (2015). "Putting stress on stress", *OH&S Canada*, April 22, 2015, available at: <http://www.ohscanada.com/features/putting-stress-on-stress/> (date accessed: February 28, 2020).

[40] Statistics Canada, 2003, cited in "Facts & Figures: Workplace Strategies for Mental Health", available online at: http://gwlcentreformentalhealth.com/display.asp?l1=2&d=2.

[41] IPSOS-Reid, (2012). *IPSOS Survey 2012 on Depression*, available online at: https://www.workplacestrategiesformentalhealth.com/pdf/GWLReleaseDeckDepressionintheWorkplace.pdf (date accessed: February 28, 2020).

[42] Centre for Addiction and Mental Health (CAMH) (2011), *Twenty-five percent of Ontarians hospitalized for depression required ER visit or readmission within 30 days, new study finds*, Newsroom, available online at: <http://www.camh.ca/en/hospital/about_camh/newsroom/news_releases_media_advisories_and_backgrounders/archives/2011/Pages/Twenty-five-percent-of-Ontarians-hospitalized-for-depression-required-ER-visit-or-readmission-within-30-days,-new-study-fin.aspx> (date accessed: February 28, 2020).

[43] Centre for Addiction and Mental Health (CAMH) (2013), "Mental Health and Addiction Statistics", CAMH homepage, available online at: http://www.camh.ca

[44] Centre for Addiction and Mental Health (CAMH) (2001), "Research", CAMH homepage, available online at: http://www.camh.net

[45] Centre for Addiction and Mental Health (CAMH) (2013), "Mental Health and Addiction Statistics", CAMH homepage, available online at: http://www.camh.ca

mental or that their co-workers/supervisors were abusive.[46]

- **Individual Health Effectiveness Support (Health and Wellness Programs)**

Organizations and institutions are not typically uncaring; nor do they choose to deliberately ignore stress-related problems. Many workplace leaders are themselves affected. In response, organizations put supportive services in place to help employees develop effective coping mechanisms, for example:

- EAPs;
- flexible employee group benefit plans;
- flexible work hours;
- DMPs;
- WWPs;
- fitness centres or subsidies;
- exercise and relaxation programs;
- Child Care Services;
- Elder Care Services; and
- health education programs.

E. ARE WE TREATING SYMPTOMS INSTEAD OF ROOT CAUSES?

No doubt, these good and very necessary employee support services do help the individual. However, as this chapter asserts, such programs do not address the "root causes" of workplace stress. A significant causal factor is the organizational environment in which the employee works. In essence, many workplace environments are "toxic to human life".

In support of this premise, examine the second part of this model, the organizational elements:

- **Effects on Organizations**

    The cost of workplace stress is not borne by employees alone. Organizations pay a huge price for the emotional mismanagement of their human capital (employees) in the form of increased operational and benefit expenses and productivity losses.[47] Low morale, lack of co-operation, workplace conflict, apathy, and hostility are among the behavioural outcomes that have occurred. The operational outcomes of this stressors cycle

---

[46] IPSOS-Reid, *Mental Health in the Workplace: Largest Study Ever Conducted of Canadian Workplace Mental Health and Depression* (Toronto: IPSOS0Reid, 2007), at 11, available online at: <http://www.ipsos.ca>

[47] M. Shain, "Stress, Satisfaction and Health at Work: Tuning for Performance" (Presented at the Health & Wellness Conference '99, Vancouver, BC, October 24-27, 1999).

include high rates of employee absenteeism and presenteeism, staff turnover, productivity losses, and increased employee group benefit plan costs. In the final analysis, the overall result impacts the organization's "bottom line".

- **Measurement**

Dealing with the causes can break this unfortunate "cycle of harm". To achieve this, an organization must truly understand what is going on. Measurement is the key.

Measurement enables an organization to make evidence-based decisions about which remedies will produce the desired improvements. This is not only about tracking EAP usage, or monitoring the drug plan usage, or counting union grievances, or determining the number of absent employees. Measurement must track all the components of the cycle illustrated in Figure 22.2 and bring the data together to form a holistic picture of the relationships active within the cycle.

The starting point for measurement should be tracking the leading indicators of safety performance that were introduced earlier. The Conditions for Performance are a bellwether for workplace-induced stress and the resulting negative impacts on organizational outcomes. It is within the leading indicators that the root causes will be found.

If you don't measure it, you can't manage it!.

If you can't manage it, you can't control it.

Measurement provides information about:

- the degree to which stressors are being experienced by employees;
- the physical and psychological effects on individuals;
- the effects on the organization:
    - behavioural outcomes,
    - operational results; and
- the financial or Key Performance Indicator (KPI) results of the organization.[48]
- Performance Improvement Actions
- Performance Improvement Actions

---

[48] A Key Performance Indicator (KPI) is defined as quantifiable measurements of the improvement, or deterioration, in the performance of activities critical to the success of a business. They must reflect the organization's/company's goals, be key to its success, and be measurable. KPIs provide the most important performance information that enables organizations/companies, or their stakeholders, to understand whether the organization is "on track" or not.

Performance and Organizational Effectiveness consultants have the "soft technology" for improving human performance in the workplace — for dealing with the "organizational stressors" as the root cause they are. OH&S Practitioners, Occupational Health Nurses and Disability Managers can assist Performance and Organizational Effectiveness consultants to:

1. conduct needs assessments;
2. analyze the "root causes" of the problems;
3. select and implement interventions; and
4. track results and evaluate the outcomes.

Through an effective organization analysis, remedies such as the following can be identified:

- *Education* — provide comprehensive briefings to top executives and Senior Management using both external and internal data to illustrate the business case associated with workplace stress and enhancing employee health and well-being;
- *Measurement system* — develop and implement measurement that tracks leading indicators of "stress toxicity" as well as individual health and operational results;
- *Promote access to* and the use of a *Workplace Wellness Program* (Individual Effectiveness Support);
- *Improve* the capacity of *management* to recognize and *respond to causes of workplace stress and employee distress*; and
- *Performance Support Practices* — become better at managing people. The most effective and enduring remedy is to implement and sustain performance support practices that focus on the "four Es" of people management:
- Establish purpose;
- Enable performance;
- Expect results; and
- Encourage success.

The importance of effective people management is not a new proposal. We know it is not just "nice to do", but essential to creating an environment that fully enables employee performance, fosters employee loyalty, and, in the long term, sustainable success.

F. USE OF GOOD PERFORMANCE SUPPORT PRACTICES IS GOOD BUSINESS

Dennis Kravetz studies the correlation between people management practices and financial success. He developed an index for rating an organization's performance in people management practices (PMP score). In 1996, he published a study that looked at the correlation of PMP scores to financial performance over a 10-year period. The following chart (Table 22.1) compares companies having low PMP scores with companies having high PMP scores:

Table 22.1: Comparison of Companies with High and Low PMP Scores[49]

Financial Factors	Companies with High PMP Scores	Companies with Low PMP Scores
Sales growth	17.1%	7.4%
Profit growth	18.2%	4.4%
Profit margin	6.4%	3.3%
Growth (earnings/share)	10.7%	4.7%
Total return (stock appreciation + dividends)	19.0%	8.8%

Organizations that consistently use good people-management practices create work environments that reduce or even eliminate a significant number of workplace stressors. Not only do these organizations enable good business results, they also foster the conditions for a healthy workplace.

A list of "best practices" for human performance support follows. Unfortunately, many organizations find it very difficult to consistently practise them:

1. ***Establish Purpose***
   - Help employees understand the organization's vision, values, goals, and business strategies.
   - Guide the development of individual performance and learning plans that will achieve organizational goals.

2. ***Enable Performance***
   - Align resource allocations with performance expectations.
   - Coach employees to overcome difficulties and build skill and knowledge.
   - Foster and maintain effective work groups.
   - Resolve performance issues and remove barriers that are beyond the control of individuals and teams.

3. ***Expect Results***
   - Facilitate the measurement of progress, contribution, and development.
   - Hold people accountable for delivering agreed-upon results.

4. ***Encourage Success***

---

[49] D. Kravetz, *People Management Practices and Financial Success: A Ten-Year Study* (Bartlet, IL: Kravetz Associates, 1996), online: <http://www.kravetz.com> (date accessed: February 28, 2020).

- Sustain communication that maintains focus, fosters commitment, and facilitates implementation.
- Recognize and celebrate progress, development, and the achieve-ment of desired results.

G. ORIGINAL RESEARCH

An informal survey (Figure 22.4) was conducted to measure the degree of organizational stress in one Canadian city. The survey looked at specific factors within the conditions we know are needed to foster and sustain successful performance.[50] The results suggest that significant organizational stressors exist in some workplaces.

In general, the number of people who responded "disagree" or "strongly disagree" is low on most items. It is a concern that few people can say "this is not a stressor" for most items.

Highlights from the survey are organized below into the "four conditions for great performance"[51] mentioned at the beginning of this chapter. In choosing the highlights, focus was placed on the percentage of combined "agree" and "strongly agree" responses and the percentage of combined "disagree" or "strongly disagree" responses. The numbers in brackets that appear below refer to the item numbers in Figure 22.4.

Know What to do

- 72 per cent of the respondents said employees are faced with conflicting priorities and demands (item 2).
- 42 per cent of the respondents reported that employees are unclear about how their performance is measured (item 5).
- Less than 25 per cent of the respondents reported goal alignment (items 1 and 4).

Able to do it

- Half the respondents reported that employees do not get the coaching and learning support they need (items 6 and 7).

Equipped to do it

- Few (10 per cent) of the respondents believed that employees get measure-ment data about their progress (item 12).
- Few (7 per cent) of the respondents believed that employees have clear and effective work processes and procedures (item 19).

---

[50] T. Roithmayr, Organizational Stressors Survey (2000), online: <http://www.kravetz.com>; D. Dyck & T. Roithmayr, "Organizational Stressors and Health: How Occupational Health Nurses Can Help Break the Cycle" (2002) 50(5) AAOHN Journal 213.

[51] Ibid.

- 63 per cent of the respondents reported that employees do not have the time to do the work required of them (item 21).

Want to do it
- Half the respondents said that employees lack the recognition they need to stay motivated while only 5 per cent indicate that they do (item 22).
- 15 per cent reported that employees get positive feedback while only 10 per cent say they get helpful corrective feedback (items 23 and 24).

This finding is consistent with research reported in 1998 by Dr. Martin Shain, University of Toronto.[52] He advocates dealing with workplace stress at the source in terms of "work design, work control, work demand, and work effort", as opposed to trying to address the negative outcomes — workplace injuries, increased operational costs, human suffering, and increased employee group benefit plan costs.

On a positive note, 83 per cent of the survey respondents believed that organizations could reduce or eliminate organizational stressors while maintaining or growing their business results (item 30).

These survey results can be compared with the *Top Ten Sources of Workplace Stress* reported by the Global Business and Economic Roundtable on Addiction and Mental Health (2001):[53]

1. *Treadmill Syndrome*: too much to do at once requiring a 24-hour workday.
2. *Random work interruptions*.
3. *Doubt*: employees unsure what is happening, or where things are headed.
4. *Mistrust*: vicious office politics that disrupt positive behaviour.
5. *Unclear company direction and policies*.
6. *Career/job ambiguity*: things happen without employees knowing why.
7. *Inconsistent performance management*: employees get raises without a performance review and positive feedback and then, get laid off.
8. *Feeling unappreciated*.
9. *Lack of two-way communication* between employees and management.
10. *Learned helplessness*: Experiencing a feeling of not contributing and having lack of control over the work and workplace.[54]

---

[52] M. Shain, "Managing Stress at its Source in the Workplace" (Presented at the Health Work & Wellness Conference '99, Vancouver BC, October 24–27, 1999).

[53] Global Business and Economic Roundtable on Addiction and Mental Health, Top 10 Sources of Workplace Stress (2001), online: http://www.mentalhealthroundtable.ca/aug_round_pdfs/Top%20Ten%20Sources%20of%20Stress.pdf (date accessed: February 28, 2020).

[54] *Ibid.*

Figure 22.4: Organizational Stressors Survey

Copyright© 2000 Tony Roithmayr. All rights reserved.

An informal survey of organizations in a Canadian city shows the degree to which Human Resource, Organizational Effectiveness and Occupational Health professionals perceive stressors within the organizations they serve. Forty-two people from about 30 organizations responded. The percentage of people who replied 0 or 1, percentage who replied 2 or 3, and percentage who replied 4 or 5 to each item are shown below.

0 = Strongly Disagree
1 = Disagree
2 = Somewhat Disagree
3 = Somewhat Agree
4 = Agree
5 = Strongly Agree

Strongly Disagree 0	1	2	3	4	5 Strongly Agree

1. are NOT clear about the specific performance expectations (results) for which they are accountable
   - 25% | 45% | 30%

2. are faced with CONFLICTING priorities and demands
   - 03% | 25% | 72%

3. are NOT clear about the behavioural expectations (personal conduct) that is consistent with the organization's values
   - 30% | 52% | 18%

4. are NOT clear about how their individual work will contribute to the goals of the organization
   - 22% | 55% | 23%

5. do NOT understand the standards for their success and how their progress and contributions will be measured
   - 13% | 45% | 42%

6. do NOT get the coaching they need to succeed in their current roles
   - 08% | 40% | 52%

7. do NOT get the learning support they need to further develop their capabilities
   - 20% | 32% | 48%

8. are in jobs that do NOT match their own personal interests and attributes; do NOT find their work meaningful
   - 28% | 65% | 07%

9. are often involved in POOR working relationships	33%	50%	17%
10. encounter workplace situations in which they feel emotionally or psychologically VULNERABLE	22%	48%	30%
11. encounter workplace situations in which they feel physically UNSAFE	68%	32%	00%
12. do NOT get measurement data about their progress	10%	63%	27%
13. do NOT get sufficient communications about developments in the organization or information that affects their job	33%	47%	20%
14. have little or NO involvement in decisions that affect their work	20%	45%	35%
15. do NOT have the necessary autonomy and discretion to successfully deliver on their accountabilities	27%	53%	20%
16. feel they have NO influence over things that happen to them at work	26%	52%	22%
17. feel their roles are UNCLEAR or are in conflict with the roles of other employees	27%	43%	30%
18. do NOT have access to the tools, equipment, job aids or other resources they need to succeed in their jobs	35%	52%	13%
19. feel encumbered with Unclear or Ineffective work practices or procedures	07%	63%	30%
20. feel their physical work environment is NOT conducive to working efficiently or effectively	22%	48%	30%
21. feel they lack sufficient time to do the work required of them	05%	32%	63%
22. do NOT get the recognition they need to stay energized and motivated in their work	05%	47%	48%
23. get little or NO positive feedback on how they do their work	15%	45%	40%

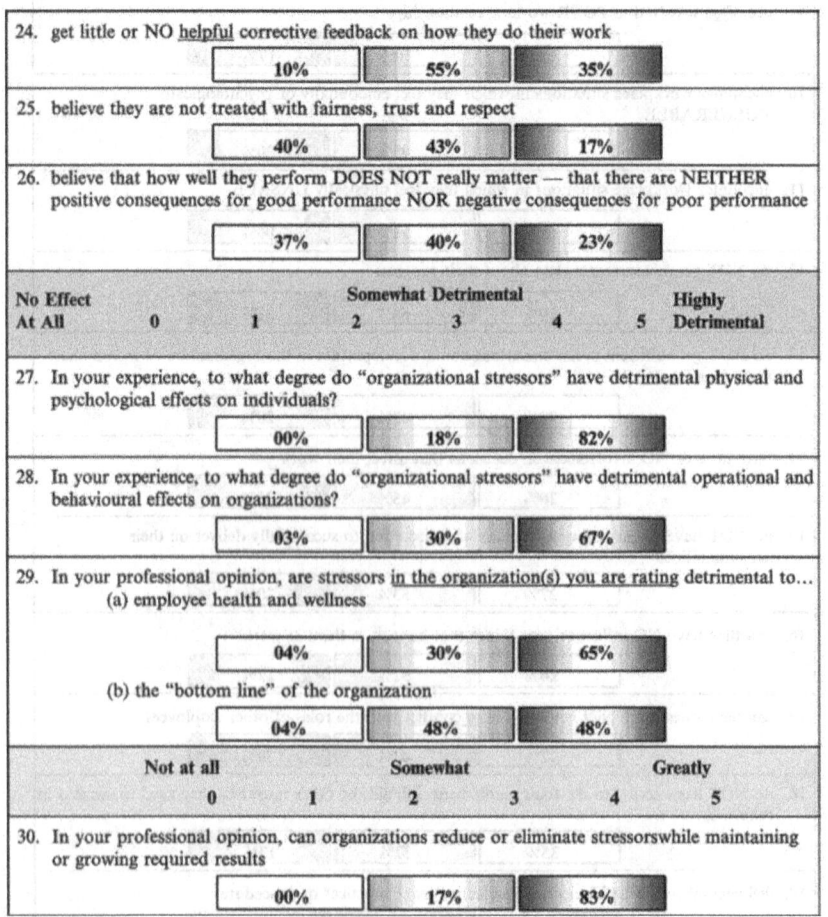

H. SUMMARY

Based on the experience of Occupational Health, OH&S, Human Resources, and Workplace Wellness Practitioners/Professionals, unmanaged organizational stress leads to increased employee stress, failing employee health, decreased productivity, increased absence and disability costs, and lowered financial results. However, the task is to prove this statement. To our knowledge, not enough data is available to tie all the elements described in this chapter into a holistic framework of cause and effect.

Organizations are interested in making evidence-based decisions regarding the allocation of their limited resources. To do this, a commitment is needed to:

- implement a measurement system to track the entire "cycle of harm";
- implement evidence-based actions to reduce or eliminate stressors; and

- track progress for three to five years and share the results publicly in order to motivate others to follow suit.

Through the use of a multi-disciplinary approach, Occupational Health, OH&S, Human Resources, and Workplace Wellness Practitioners/Professionals can act and find workable solutions to mitigate organizational stressors and the resultant employee illness/injury.

CHAPTER REFERENCES

Aventis Pharma-Canada, *The Aventis Healthcare Survey* (2001), online: http://www.sanofi-aventis.ca/1/ca/en/index.asp (date accessed: February 28, 2020).

K. Bachmann, *More Than Just Hard Hats and Safety Boots: Creating Healthier Work Environments* (Ottawa: Conference Board of Canada, 2000) at 1.

Canadian Policy Research, reported in *IAPA: Creating Healthy Workplaces Everywhere — Healthy Workplace Week* (2006), online: http://www.iapa.ca/Main/articles/2006_oct_healthy_workplace_week.aspx (date accessed: February 28, 2020).

The Canadian Encyclopedia (October 2007). *Workplace Stress Costs the Economy Millions of Dollars*, online: https://thecanadianencyclopedia.ca/en/article/workplace-stress-costs-the-economy-billions (date accessed: February 28, 2020).

CCH, *2006 CCH Unscheduled Absence Survey*, online: http://www.cch.com/absenteeism2006 (date accessed: February 28, 2020).

Centre for Addiction and Mental Health (CAMH), News Release, *Twenty-five percent of Ontarians hospitalized for depression required ER visit or readmission within 30 days, new study finds* (2011), online: http://www.camh.ca/en/hospital/about_camh/newsroom/news_releases_media_advisories_and_backgrounders/archives/2011/Pages/Twenty-five-percent-of-Ontarians-hospitalized-for-depression-required-ER-visit-or-readmission-within-30-days,-new-study-fin.aspx (date accessed: February 28, 2020).

Centre for Addiction and Mental Health (CAMH), *Newsroom*, "Statistics on Mental Health and Addictions", online: http://www.camh.ca/en/hospital/about_camh/newsroom/for_reporters/Pages/addictionmentalhealthstatistics.aspx (date accessed: February 28, 2020).

Chrysalis Performance Inc., reported in *IAPA: Creating Healthy Workplaces Everywhere – Healthy Workplace Week* (October 2006), online: http://www.iapa.ca/Main/articles/2006_oct_healthy_workplace_week.aspx (date accessed: February 28, 2020).

S. Cohen, D. Janicki-Deverts & G. Miller, "Psychological Stress and Diseases" (2007) 298:14 JAMA at 1685.

Conference Board of Canada, *Survey of Canadian Workers on Work-Life Balance* (Ottawa: Conference Board of Canada, 1999).

Conference Board of Canada, Healthy Brains at Work (2012), online: http://www.

sunlife.ca/static/canada/Sponsor/About%20Group%20Benefits/Focus%20Update/ 2016/557/HealthyBrains_Report2_EN.pdf (date accessed: February 28, 2020).

B.A. Cryer, "Neutralizing Workplace Stress: The Physiology of Human Performance and Organizational Effectiveness" (Presented at the Psychological Disabilities in the Workplace: Prevention, Rehabilitation and Cost Control Conference, Toronto, ON June 10–11, 1996) [unpublished].

L. Duxbury, *Revisiting Work-life Issues in Canada: The 2012 National Study on Balancing Work and Caregiving in Canada, 2012* (Ottawa: ON: Carleton University), online: http://newsroom.carleton.ca/wp-content/files/2012-National-Work-Long-Summary.pdf (date accessed: February 28, 2020).

L. Duxbury & C. Higgins, *Work-Life Conflict in Canada in the New Millennium: A Status Report* (prepared for Health Canada, Healthy Communities Division) (Ottawa: Health Canada, 2003), online: http://publications.gc.ca/collections/Collection/H72-21-186-2003E.pdf (date accessed: February 28, 2020).

D. Dyck & T. Roithmayr, "Organizational Stressors and Health: How Occupational Health Nurses Can Help Break the Cycle" (2002) 50:5 AAOHN Journal 213.

Editorial, "Stress at Work" (2007), 27:12 *AAOHN News* 17.

Global Business and Economic Roundtable on Addiction and Mental Health, Top 10 Sources of Workplace Stress (2001), online: http://www.mentalhealthroundtable.ca/aug_round_pdfs/Top%20Ten%20Sources%20of%20Stress.pdf (date accessed: February 28, 2020).

Health Canada, *Best Advice on Stress Risk Management in the Workplace* (Ottawa: Minister of Public Works and Government Services Canada, 2000).

Ipsos-Reid, *Mental Health in the Workplace: Largest Study Ever Conducted of Canadian Workplace Mental Health and Depression* (Toronto: Ipsos-Reid, 2007), online: http://www.ipsos.ca/ (date accessed: February 28, 2020).

P. Jacobs (2013). Reported in Lopez-Pacheco. A. "Mental Illness adversely affecting Canada's economic potential", *Financial Post*, February 5, online: http://business.financialpost.com/executive/the-economic-cost-of-mental-illness (date accessed: February 28, 2020).

J. Johnson, "Long-term Psychosocial Work Environment and Cardiovascular Mortality among Swedish Men" (1996) 86:3 American Journal of Public Health 324.

R. Karasek & T. Theorell, *Healthy Work: Stress, Productivity, and the Reconstruction of Working Life* (New York, NY: Basic Books Inc., 1992).

D. Kravetz, *People Management Practices and Financial Success: A Ten-Year Study* (Bartlet, IL: Kravetz Associates, 1996), online: http://www.kravetz.com (date accessed: February 28, 2020).

A. LaCroix, *Occupational Exposure to High Demand/Low Control Work and Coronary Heart Disease Incidence in the Framingham Cohort* (Ann Arbor, MI:

University of North Carolina, UMI Dissertation Services, 1984).

J. MacBride King, "Wrestling with Workload: Organizational Strategies for Success" (Ottawa: Conference Board of Canada, 2005).

B.B. Marmot et al., "Contribution of Job Control and Other Risk Factors to Social Variations in Cardiovascular Heart Disease Incidence" (1997) 350 (9073) *Lancet* 235.

Morneau Shepell (2015). *The true picture of workplace absenteeism*, online: http://www.morneaushepell.com/sites/default/files/documents/3679-true-picture-workplace-absenteeism/9933/absencemanagementreport0608-15.pdf at 2 (date accessed: February 28, 2020).

H. Orpana, L. Lemyre, & R. Gravel, *Income and psychological distress: The role of the social environment* (2009), Statistics Canada Catalogue no. 82-003-X, online: http://www.statcan.gc.ca/pub/82-003-x/2009001/article/10772-eng.pdf (date accessed: February 28, 2020).

J. Park, *Work Stress and Job Performance* (Ottawa: Statistics Canada, 2007), Cat. No. 75-001-XIE.

Partners for Mental Health (2013). Reported by Cottrell, J. (2015). "Putting stress on stress", *OH&S Canada* (April 22, 2015), online: http://www.ohscanada.com/features/putting-stress-on-stress/ (date accessed: February 28, 2020).

T. Roithmayr, *Performance by Design* (2000), online: http://www.performance-bydesign.com/ (date accessed: February 28, 2020).

T. Roithmayr, "The Performance Maximizer[TM]" (Calgary, AB: Performance by Design), online: http://www.performance-bydesign.com/ (date accessed: February 28, 2020).

M. Shain, "Managing Stress at its Source in the Workplace" (Presented at the Health Work & Wellness Conference 1998, Whistler, BC, September 27–30, 1998) [unpublished].

M. Shain, "Stress, Satisfaction and Health at Work: Tuning for Performance" (Presented at the Health & Wellness Conference 1999, Vancouver, BC, October 24–27, 1999) [unpublished].

J. Siegrist, "Adverse Health Effects of High-Effort/Low-Reward Conditions" (1996) 1:1 J. Occup. Health Psychol. 27.

Statistics Canada 2003, cited in Workplace Strategies for Mental Health "Mental Health Issues – Facts & Figures", online: https://www.workplacestrategiesformentalhealth.com/Mental-Health-Issues-Facts-and-Figures (date accessed: February 28, 2020).

Towers Watson, *2011/2012 Staying@Work Survey Report: Pathway to Health and Productivity* (2012), online: http://www.towerswatson.com/en-CA/ Insights/IC-Types/Survey-Research-Results/2011/12/20112012-StayingWork-Survey-Report-A-Pathway-to-Employee-Health-and-Workplace-Productivity (date accessed: February 28, 2020).

Willis Towers Watson, *2015/2016 Global Staying@Work Survey, Canada* (2016), online: https://easna.org/wp-content/uploads/2016/04/2015-2016-Global-Staying-at-Work-Survey.pdf at 10 (date accessed: February 28, 2020).

Chapter 23

PSYCHOLOGICAL HEALTH AND SAFETY IN THE WORKPLACE

By Ian Arnold,[1] Dianne Dyck and Joti Samra[2]

A. INTRODUCTION

Psychological health is identified by the World Health Organization (WHO) as being a key component of one's overall health.[3] The incidence of psychological health disorders and benefit claims is on the rise.[4],[5] The WHO estimates that 25% of the

---

[1] Dr. Ian Arnold is a Fellow of the Canadian Board of Occupational Medicine (FCBOM); a Fellow of the Royal College of Physicians and Surgeons of Canada (FRCPC – Occupational Medicine); a Certified Environmental Auditor (CEA); and a Registered Safety Professional (CRSP). He is the former Chairperson of the Workforce Advisory Committee of the Mental Health Commission of Canada. Dr. Arnold has been awarded the Canadian Workplace Wellness Pioneer Award and also received the Queen's Diamond Jubilee Medal for his work in occupational health, and workplace psychological health and safety.

[2] Dr. Joti Samra, R.Psych. is a national thought leader on issues relating to psychological health, wellness and resilience. She is the CEO and Founder of *MyWorkplaceHealth*, and *Dr. Joti Samra, R.Psych & Associates*. Dr. Samra has been involved in national initiatives that contributed to policy change in Canada, and is a Founding & Ongoing Member of the CSA Technical Committee that developed the *National Standard of Canada for Psychological Health & Safety in the Workplace (CAN/CSA-Z1003-13/BNQ9700-803/2013)*. Dr. Samra is the lead Research Scientist who created *Guarding Minds at Work: A Workplace Guide to Psychological Health & Safety*, the *Psychologically Safe Leader Assessment*, and the first online Psychological Health and Safety in the Workplace certification program, at the University of Fredericton.

[3] World Health Organization "Mental Health: Strengthening our response" (August 2014), online: http://www.who.int/mediacentre/factsheets/fs220/en/index.html (date accessed: February 28, 2020).

[4] CAMH, "Statistics on Mental Illness and Addictions", online: http://www.camh.ca/en/hospital/about_camh/newsroom/for_reporters/Pages/addictionmentalhealthstatistics.aspx (date accessed: February 28, 2020); and Marsh Risk Consulting, *Workforce Risk: Fourth Annual Marsh Mercer Survey of Employers' Time-off and Disability Programs* (2003), online: http://usa.marsh.com (date accessed: February 28, 2020).

[5] Watson Wyatt Worldwide, *Staying @ Work 2005*, online: http://www.watsonwyatt.com

population will be affected by a psychological health disorder at some time in their lives,[6] and as of 2020, depression is predicted to be the leading cause of disability worldwide.[7] The Mental Health Commission of Canada (MHCC) estimates the psychological health disorder rates to be 20% per annum, and that by age 40 years, one in two Canadians is expected to experience a mental health problem.[8] These numbers can include very serious yet uncommon conditions, such as schizophrenia, but are more likely to include conditions such as stress, burnout, depression, anxiety, or substance abuse.[9] Of note, the WHO now identifies burnout as being an occupational condition.[10]

In terms of the one-year prevalence rate of mental health disorders, estimates suggest that approximately 21.4% of workers live with a diagnosable mental health condition of some sort.[11] There is also a larger percentage of workers with sub-threshold mental health symptoms that may negatively impact their work performance, and as such, these figures increase sharply if the range of psychological distress that can be a precursor to a diagnosable disorder, is included.

---

and Marsh Risk Consulting, *Workforce Risk: Fourth Annual Marsh Mercer Survey of Employers' Time-off and Disability Programs* (2003), online: http://www.marshriskconsulting.com (date accessed February 28, 2020).

[6] World Health Organization, *Scaling up care for mental, neurological, and substance use disorders* (WHO Press: Geneva, Switzerland, 2008), online: http://www.who.int/mental_health/mhgap_final_english.pdf (date accessed: February 28, 2020).

[7] Reported in Centers for Disease Control and Prevention (CDC), Mental Health Basics (October 2013), online: http://www.cdc.gov/mentalhealth/basics.htm (date accessed: February 28, 2020); World Health Organization, *Mental Health: A Call for Action by World Health Ministers. Ministerial Round Tables, 54th World Health Assembly* (Geneva: World Health Organization, 2001), online: http://www.who.int/mental_health/advocacy/en/Call_for_Action_MoH_Intro.pdf (date accessed: February 28, 2020).

[8] Mental Health Commission of Canada (2013). *Making the Case for Investing in Mental Health in Canada*, online:https://www.mentalhealthcommission.ca/sites/default/files/2016-06/Investing_in_Mental_Health_FINAL_Version_ENG.pdf (date accessed: February 28, 2020).

[9] R.C. Kessler et al., "Lifetime and 12-Month Prevalence of DSM-III-R Psychiatric Disorders in the United States: Results from the National Comorbidity Survey" (1994) 51:1 Arch. Gen. Psychiatry 8; and R.C. Kessler et al., "Prevalence, severity, and comorbidity of 12-Month DSM-IV Disorders in the National Comorbidity Survey Replication" (2005) 62:6 Arch. Gen. Psychiatry 617.

[10] World Health Organization, "Burn-out an 'occupational phenomenon': International Classification of Diseases", Mental Health (May 23, 2019), online: https://www.who.int/mental_health/evidence/burn-out/en/ (date accessed: February 28, 2020).

[11] World Health Organization, "Burn-out an 'occupational phenomenon': International Classification of Diseases", Mental Health (May 23, 2019), online: https://www.who.int/mental_health/evidence/burn-out/en/ (date accessed: February 28, 2020).

CH. 23: PSYCHOLOGICAL HEALTH AND SAFETY IN THE WORKPLACE

1. Psychological Health Disorders: Costs

In a 2011/2012 Report by Watson Wyatt Worldwide,[12] it stated that in Canada and the U.S., psychological health conditions are the leading cause of both Short Term Disability (STD) and Long Term Disability (LTD). This finding was supported by the 2019 Deloitte Insights Report; "[o]n average mental health issues account for 30 to 40 per cent of STD claims and 30 per cent of LTD claims in Canada."[13] The current cost to employers is CDN $6 billion in lost productivity due to employee absenteeism and **staff turnover**.[14] The 2011 annual total cost of mental illness in Canada is CDN $50 billion (2.8% of Canada's gross domestic product);[15] however in 10 years, that cost is expected to reach CDN $88.8 billion.[16] By 2014, it is projected to total CDN $307 billion.[17]

In addition to impacting the direct and indirect costs to organizations, unaddressed psychological health issues can result in higher rates of workplace incidents and injuries, particularly in job positions that are safety sensitive.[18] Hence, poor mental health leads to high costs for employers.

2. Psychological Health and Performance in the Workplace

Despite all the advances made in medicine, psychological health conditions remain poorly understood and stigmatized.[19] We all, from time to time, experience sadness,

---

[12] Towers Watson, *2011-2012 Staying@Work Report: Pathway to Health and Productivity Advantage*, online: http://www.towerswatson.com/en/Insights/IC-Types/Survey-Research-Results/2011/12/20112012-StayingWork-Survey-Report--A-Pathway-to-Employee-Health-and-Workplace- Productivity (date accessed: February 28, 2020).

[13] S. Chapman *et al.*, "The ROI in workplace mental health programs: Good for people, good for business", *Deloitte's Insights* (2019), online: http://www.myworkplacehealth.com/documents-resources/ at 3 (date accessed: February 28, 2020).

[14] Mental Health Commission of Canada, *Making the Case for Investing in Mental Health in Canada* (2013), online:https://www.mentalhealthcommission.ca/sites/default/files/2016-06/Investing_in_Mental_Health_FINAL_Version_ENG.pdf (date accessed: February 28, 2020).

[15] *Ibid.*

[16] *Ibid.*

[17] *Ibid.*

[18] M. Hilton & H. Whiteford "Associations between psychological distress, workplace accidents, workplace failures and workplace successes" (December 2010) 83:8 Int. Arch. Occup. Environ. Health 923–33.

[19] Global Business and Economic Roundtable on Addiction and Mental Health, News Release, "Ground-breaking Survey on Mental Health in the Workplace" (November 19, 2007), online: http://www.gwlcentreformentalhealth.com/english/userfiles/news/pdf/s7_00456.pdf (date accessed: February 28, 2020); and Ipsos-Reid, *Mental Health in the Workplace: Largest Study Ever Conducted of Canadian Workplace Mental Health and Depression* (Toronto: Ipsos-Reid, 2007), online: https://www.workplacestrategiesformentalhealth.com/pdf/2007_factum.pdf (date accessed: February 28, 2020).

anxiety, forgetfulness, poor concentration, or mood swings. But when our psychological state becomes such that we cannot function as we normally would, a psychological health condition may exist.

Often, workers suffering from psychological health conditions tend to try to "work through their problems", which then, go unrecognized and untreated. Unfortunately, limited resources — inside and outside of the workplace — exist to help employers to deal with the impact of psychological health issues on workplace functioning. The healthcare system is also ill-equipped to deal with workplace-related psychological health issues; most physicians find it a challenge to help these workers to work productively and/or return to gainful employment.[20] Likewise, it is frustrating for employers/co-workers to understand why a worker, who may from an outsider's perspective "looks fine" yet, needs to be off work for an extended period of time, or who is demonstrating lower productivity while in the workplace. Unlike physical injuries/illnesses which tend to remain visible to others, psychological health conditions are often described as "invisible". Furthermore, unaddressed psychological health conditions do not only impact the affected individual's productivity; if one member of an integrated team is suffering, the performance of the entire team is compromised.

To gain a better understanding of this situation, consider some of the following facts. Depressive disorders rank as one of the most common reasons for visiting a Canadian physician.[21] Individuals who are dealing with depression often experience difficulty in maintaining their level of productivity at home and at work. Fewer than 20% of the people who need psychological health treatment, actually get it. Serious depression can result in a work absence of 40 or more days.[22] As well, returning to work after weeks of being on a disability leave can be a "punishing experience" for the employee. Having to face co-workers, and potentially the work situation that contributed to their illness, can be most daunting.

B. PSYCHOLOGICAL HEALTH AND SAFETY: LEGISLATIVE REQUIREMENT

According to the various provincial OH&S Acts in Canada, as well as the *Canada Labour Code Part II*, employers must provide a safe and healthy workplace for employees and other workers. This includes both physical and psychological (mental health) health and safety. Despite this interpretation, however, the WHO

---

[20] S. Calhoun & P. Strasser, "Generations at Work" (2005) 53:11 AAOHN 469–471.

[21] OMA Committee on Work and Health, *Mental Illness and Workplace Absenteeism: Exploring Risk Factors and Effective Return to Work Strategies* (April 2002), online: http://www.oma.org (date accessed: February 28, 2020).

[22] Global Business and Economic Roundtable on Addiction and Mental Health, *Roundtable Roadmap to Mental Health Disability Management in 2004–05* (Working document prepared June 25, 2004), online: http://www.mentalhealthroundtable.ca/june_2004/monitor_june2004.pdf (date accessed: February 28, 2020).

definition that clearly addresses the psychological aspect of health, is seldom found in the OH&S Acts. The provision and maintenance of a psychologically healthy and safe workplace has, however, been recognized in the legal system as an emerging legal duty, similar to the duty employers have to provide a physically healthy and safe workplace. In both realms, employers must take every reasonable precaution to protect worker safety and to demonstrate that they have done so.

A psychologically healthy and safe workplace uses all the practical and reasonable mechanisms within its control to both protect and promote the psychological health and safety of all employees. At a minimum, such a workplace guards against threats to employee psychological health and safety that are attributable to negligence, recklessness, or intentional interference.[23]

Dr. Martin Shain, in his reports to the Mental Health Commission of Canada (MHCC), states:

> We can observe seven major trends in the law becoming stronger by the year. . . . [W]e can characterize these trends as pressures building toward a perfect legal storm, where the whole is far greater than the sum of the parts.[24]

A psychologically healthy and safe workplace is no longer a "nice to do". It is now a "must do". Dr. Shain goes on to point out that five other sources of law have a bearing on the developing jurisprudence, including Workers' Compensation law, Employment contracts, Law of Torts (negligence), Human Rights legislation, and Labour Relations law.[25]

C. PSYCHOLOGICAL HEALTH AND SAFETY: DEFINITIONS[26]

Psychological health and safety pertains to safeguarding the psychological health and safety of employees and other workers (contractors, consultants, volunteers, students, *etc.*).

Psychological health is defined as a state of mental and emotional well-being; it consists of the ability to think, feel, and behave in a manner that enables employees/workers to optimally perform. To increase the psychological health of employees (and, more broadly, the workplace) includes, among other things, taking

---

[23] J. Samra, M. Gilbert, M. Shain & D. Bilsker, *Guarding Minds @ Work: A Workplace Guide to Psychological Safety and Health*, "What is Psychological Health & Safety?", online: http://www.guardingmindsatwork.ca/info/safety_what (date accessed: February 28, 2020).

[24] M. Shain, *Tracking the Perfect Legal Storm* (May 2010), online: http://www.mentalhealth commission.ca/English/system/files/private/Workforce_Tracking_the_Perfect_Legal_Storm_ENG_ 0.pdf (date accessed: February 28, 2020).

[25] *Ibid.*

[26] The following definitions are from J. Samra, M. Gilbert, M. Shain & D. Bilsker, *Guarding Minds @ Work: A Workplace Guide to Psychological Safety and Health*, "What is Psychological Health & Safety?, online: http://www.guardingmindsatwork.ca/info/safety_ what (date accessed: February 28, 2020).

steps to enhance the psychological well-being of employees.

Psychological safety is a bit different — it is the risk that a worker might experience injury to their psychological well-being. To improve workplace psychological safety, employers must act to avert injury, danger, or threats to the psychological well-being of employees. As Shain notes,

> "A **psychologically safe workplace** is . . . one in which every practical effort is made to avoid reasonably foreseeable injury to the psychological health of employees".[27]

In other words, it does not permit harm to employee psychological health in careless, negligent, reckless, or intentional ways. He further indicates that: **Psychological (mental) injury** is not the same as psychological illness. It is harm to psychological health [mental suffering] that significantly affects the ability of employees to function at work and at home.[28]

Psychological health problems must be viewed as occurring on a spectrum from mild emotional distress on one end (low mood or excessive worry), to clinically diagnosable psychological/psychiatric disorders on the other. Because milder psychological health problems are far more common in the workplace; they account for a large proportion of the negative impacts on employees and employers — often in the form of presenteeism (reduced productivity). Protecting employee/worker psychological health and safety involves both reducing the factors that contribute to milder psychological distress and problems (that, over time, may have large cumulative effects on the worker), as well as trying to reduce factors that contribute to the occurrence of clinical conditions such as depressive and anxiety disorders.

Ensuring psychological health and safety in the workplace calls for a new standard of conduct at work. This standard requires that people treat one another with fairness, civility, and respect regardless of their positional power or status within the organization. Examples of conduct that contravene this standard and may result in mental injury include bullying, harassment, discrimination, and the imposition of unreasonable work demands coupled with refusing employees the minimal levels of control over workload and work pace. For more information on civility in the workplace, refer to: CCOHS, *Civility and Respect in the Workplace*.[29]

The range of conduct at work punishable by law has expanded considerably in recent years. It is important for employers and employees to recognize that there exists a continuum of acceptable workplace behaviours and expectations that range

---

[27] M. Shain, *Tracking the Perfect Legal Storm* (May 2010), online: http://www.mentalhealth commission.ca/English/system/files/private/Workforce_Tracking_the_Perfect_Legal_Storm_ENG_ 0.pdf (date accessed: February 28, 2020).

[28] *Ibid.*

[29] CCOHS, "Civility and Respect in the Workplace" (2019) *Infographic*, online: https://www.ccohs.ca/images/products/infographics/download/Respect_Civility.png (date accessed: February 28, 2020).

from rudeness through to harassment and, on the extreme end, violence. The overall effect of the developing jurisprudence is that the types of conduct that would most likely have been tolerated in the workplace even five years ago are no longer acceptable in the eyes of the law. Given this, it is in the employer's best interest to act in a manner that prevents the risk of legal consequences.

D. PSYCHOLOGICAL HEALTH AND SAFETY: NATIONAL STANDARD

On January 16, 2013, the National Standard of Canada for Psychological Health and Safety in the Workplace was released. This voluntary standard provides organizations/companies with a systematic approach for developing and sustaining a psychologically healthy and safe workplace. Parallel to the existing occupational physical health and safety standard, this standard recommends the:

- identification of psychological hazards in the workplace;
- assessment and control of inherent risks in the workplace that are associated with hazards that cannot be eliminated (*e.g.*, stressors due to organizational change or reasonable job demands);
- implementation of practices that support and promote psychological health and safety in the workplace;
- growth of a culture that promotes psychological health and safety in the workplace; and
- implementation of measurement and review systems to ensure sustainability of the level of psychological health and safety attained.

Ideally, Canadian employers will develop a holistic approach to employee health and safety. The development of a comprehensive Occupational Health & Safety Management System that addresses employee physical and psychological health and safety is recommended. Information on the Canadian Occupational Health and Safety Management Standard, CAN/CSA-Z1000-06, and the National Standard of Canada on Psychological Health and Safety in the Workplace, 2013, CAN/CSA-Z1003-13/BNQ 9700-803/2013, are available through the Canadian Standards Association (CSA) in English and French.[30]

As of 2019, only 33 per cent of Canadian employers had developed a mental health strategy.[31] Employers, however, do not have to adopt the entire Standard on Psychological Health and Safety; rather, they can gradually implement the elements in accordance with the needs of their workforce. The 2019 Deloitte Insights Report

---

[30] Canadian Standards Association, *National Standard of Canada: Psychological Health and Safety in the Workplace – CAN/CSA-Z1003-13/BNQ 9700-803/2013* (Ottawa, ON: CSA Group, 2013), online: https://www.csagroup.org/article/cancsa-z1003-13-bnq-9700-803-2013-r2018/ (date accessed: February 28, 2020).

[31] S. Chapman *et al.*, "The ROI in workplace mental health programs: Good for people, good for business" *Deloitte's Insights* (2019) at 3, online: http://www.myworkplacehealth.com/documents-resources/ (date accessed: February 28, 2020).

recommends that a needs assessment first be undertaken to determine the baseline state of the psychological health and safety within the workplace, and then build on that foundation.[32] Their report describes the return on investment realized by 10 Canadian companies who have a mental health program.

For more information on a Psychologically Healthy and Safe Workplaces, refer to: CCOHS, *Psychologically Healthy and Safe Workplaces*.[33]

E. COMMON PSYCHOLOGICAL HEALTH CONDITIONS

Psychological health issues exist on a continuum, ranging from transient emotional distress on one end through to chronic, recurrent clinically diagnosable conditions on the other. OH&S Professionals/Practitioners who are charged with disability management responsibilities, should remain mindful of the range of psychological health issues; they all can have a significant impact on the workplace.

1. Mental Distress (Psychological Distress)

The concept of **mental distress** (psychological distress) differs from the concept of mental illness. For example, a person may experience harassment and consequentially suffer mental distress (psychological distress) but may not be clinically diagnosable as having a psychological/psychiatric disorder. Mental distress (psychological distress) can include sub-clinical symptoms of depression and anxiety, as well as severe demoralization, disengagement, and alienation.

It is important to note that in Canada, mental distress (psychological distress) may be compensable by law.

2. Mental Injury (Psychological Injury)

Another integral concept in understanding the notion of psychological safety is that of mental injury (psychological injury). The defining aspect of mental injury (psychological injury) is that it results from someone's negligent, reckless, or intentional conduct. By definition, some person or persons, is responsible for the resultant harm in whole or in part. When such conduct occurs in the workplace, the employer may be held liable, either directly or by association.

3. Post-Traumatic Stress Disorder

Post-Traumatic Stress Disorder (PTSD) is an anxiety disorder that can occur following a worker's involvement in, or observation of, a traumatic event (*e.g.*, robbery, assault, life-threatening incidence/experience). Research indicates that First Responders, *i.e.*, police, fire fighters, emergency medical services personnel, and armed forces personnel, are at particular risk for developing PTSD.

Most workers exposed to a traumatic event experience short term effects with no, or very little, time loss. However, a small percentage of workers develop PTSD.

---

[32] *Ibid.*

[33] Online: https://www.ccohs.ca/products/boutique/13factors/.

PTSD symptoms and signs include:

- *Increased Arousal*: for example, insomnia, irritability, hyper-vigilance, heightened startle response, poor concentration.
- *Avoidance*: for example, avoidance of conversation, thoughts, feelings; withdrawal from activities; avoidance of places or people that arouse recollections; recollection of problems; loss of interest and enjoyment in life; emotional restriction; and tunnel vision.
- *Intrusive Experiences*: for example, nightmares, scary thoughts, reliving the experience, flashbacks, anxiety attacks.

Translated to the workplace, workers suffering from PTSD may be absent or present and distracted. This in turn, can lead to a greater risk of incidents, lost productivity, and costly losses.

There is a greater likelihood for PTSD to occur if there is lack of social support post-incident; when there is lack of protection from repeated incidents; or when one has accumulated trauma over one's lifespan. Failure to admit that a problem exists and failure to seek timely help can both contribute to the problem.

4. Depression

Depression is a mood disorder that impacts both an individual's emotional state, as well as their physical functioning. Initial onset is often in the teen or early adulthood years. Relapse rates are quite significant: 50% of individuals with one major depressive episode experience a second episode. Two-thirds of those experiencing a second episode experience recurrent episodes throughout their lifetime.

Depression can manifest in myriad ways, including:

- feeling sad, tearful, or empty;
- depressed, flat, or irritable more often than not;
- significant loss of interest or pleasure in activities that used to be enjoyable;
- significant changes (decrease or increase) in appetite;
- significant weight loss (when not dieting) or weight gain;
- difficulty sleeping or sleeping too much;
- agitation or slowing down of thoughts and reduction of physical movements;
- fatigue or loss of energy;
- feelings of worthlessness or inappropriate guilt;
- poor concentration or having difficulty making decisions; and/or
- thinking about death or suicide.[34]

---

[34] American Psychiatric Association, *Diagnostic and Statistical Manual of Mental*

In terms of the person's functioning, depression tends to result in:
- declining home and work performance;
- poor organization and timekeeping;
- increased risk of substance abuse problems;
- frequent headaches and backaches;
- social isolation;
- demonstrations of poor judgement;
- indecisiveness;
- constant fatigue or lack of energy; and
- a range of emotional reactions, including sadness, flat affect, and irritability/anger.[35]

Depression also has high co-morbidity with a range of physical conditions, including chronic pain,[36] type-1 and type-2 diabetes,[37] hypertension, asthma, heart disease, and stroke (cerebral vascular accident).[38] For example, individuals with type 1/type 2 diabetes are twice as likely to experience depression.[39]

So, who is at *"high risk"*? According to the Global Business and Economic Roundtable on Addictions and Mental Health, the workers most vulnerable to developing a clinical depression are those who are in their prime working years; employees with 10 to 15 years with the same company; and workers new to the

---

Disorders, Fourth Edition Text Revision (Washington, DC: American Psychiatric Association, 2000).

[35] Global Business and Economic Roundtable on Addiction and Mental Health, *Roadmap to Mental Health and Excellence At Work in Canada, Summer Draft* (Presented to The Ontario Chamber of Commerce, Economic Summit on Mental Health and Productivity in Ontario, A Pilot for Canada, Sutton Place Hotel, Toronto, Ontario Canada, June 8, 2005), online: http://www.mentalhealthroundtable.ca/june_2005/RoadmapJune82005.pdf at 24 (date accessed: February 28, 2020).

[36] D. Musselman, E. Betan, H. Larsen & L. Phillips, "Relationship of depression to diabetes types 1 and 2: epidemiology, biology and treatment" (2003) 54:3 Biol. Psychiatry 317.

[37] *Ibid.*

[38] J. Schulman & P. Shapiro, "Depression and Cardiovascular Disease" (2008) 25:9 *Psychiatric Times*, online: http://www.psychiatrictimes.com/articles/depression-and-cardiovascular-disease.

[39] American Diabetes Association, "Depression" (May 2014), online: http://www.diabetes.org/living-with-diabetes/complications/mental-health/depression.html; De Groot *et al.*, (2009), noted in Workplace Strategies for Mental Health, An Initiative of the Great West Life Centre for Mental Health, Facts & Figures, online: https://www.workplacestrategiesformentalhealth.com/Mental-Health-Issues-Facts-and-Figures (date accessed: February 28, 2020).

workforce.[40] As well, women over 40 years of age tend to experience more stress and subsequent depression, than any other worker age group due to combined work and home stressors and difficulty with effectively balancing the two.[41]

Middle managers are especially vulnerable.[42] Caught between senior management demands and subordinate expectations, middle managers tend to experience the negative effects of workplace pressures. Some "tell-tale" signs and symptoms of work distress are:

- increased irritability and impatience;
- lack of concentration and an inability to stay focused;
- avoidance and distancing from work situations and social interactions;
- open displays of frustration;
- demonstration of "back stabbing" and/or passive aggressive actions;
- ever-lengthening work days;
- persistently late for meetings;
- "working at home" to avoid the negative energy at work;
- avoidance behaviours;
- missing deadlines and commitments;
- physical symptoms; and
- periodic work absences.[43]

Figure 23.1 provides a number of relevant statistics on depression and mental health.

---

[40] Public Health Agency of Canada, *A Report on Mental Illnesses in Canada* "Chapter 2 Mood Disorders" (2002), online: http://www.phac-aspc.gc.ca/publicat/miic-mmac/pdf/chap_2_e.pdf (date accessed: February 28, 2020); Global Business and Economic Roundtable on Addiction and Mental Health, *Roadmap to Mental Health and Excellence At Work in Canada, Summer Draft* (Presented to The Ontario Chamber of Commerce, Economic Summit on Mental Health and Productivity in Ontario, A Pilot for Canada, Sutton Place Hotel, Toronto, Ontario Canada, June 8, 2005), online: http://www.mentalhealthroundtable.ca/june_2005/RoadmapJune82005.pdf at 22 (date accessed: February 28, 2020).

[41] Canadian Centre for Occupational Health and Safety, "Stress Higher Among Working Women Over Forty, Study Finds" *CCOHS* (September 5, 2005) at 5.

[42] Global Business and Economic Roundtable on Addiction and Mental Health, *Roadmap to Mental Health and Excellence At Work in Canada, Summer Draft* (Presented to The Ontario Chamber of Commerce, Economic Summit on Mental Health and Productivity in Ontario, A Pilot for Canada, Sutton Place Hotel, Toronto, Ontario Canada, June 8, 2005) at 43, online: http://www.mentalhealthroundtable.ca/june_2005/RoadmapJune82005.pdf at 43 (date accessed: February 28, 2020).

[43] *Ibid.*, at 49.

Figure 23.1: Depression – Facts and Figures
Depression – Facts and Figures

The following data has been compiled from a range of sources:

- 18% of Canadians will be diagnosed with depression at some point in their lives.[44]
- 8% of Canadians will experience depression, but go undiagnosed, at some point in their lives.[45]
- In the workplace, the rates of psychological disability are the highest in the 25–44 year-old range and increase with educational status.[46]
- 66% of Canadians who experienced depression linked its onset to a non-work triggering event, such as a relationship breakdown, a death, or financial losses.[47]
- Managers estimate the average cost of employee depression to Canadian organizations to be CDN $7,000 in terms of "presenteeism" (at work, but not fully functional); and CDN $10,000 in terms of absenteeism. This is per employee per annum.[48]
- Although willing to help depressed employees, supervisors are unsure how to go about that task.[49]
- 43% of Canadians felt that acknowledging depression would prove detrimental to their careers.
- A majority (82%) of employees think that senior leaders should make helping employees experiencing depression a Human Resources priority.[50]

---

[44] Ipsos-Reid, *Mental Health in the Workplace: Largest Study Ever Conducted of Canadian Workplace Mental Health and Depression* (November 19, 2007), online: https://www.workplacestrategies formentalhealth.com/pdf/2007_factum.pdf.

[45] *Ibid.*

[46] Statistics Canada, *Mental health-related Disabilities among Canadians Aged 15 years and older*, 2012 Canadian Survey on Disability, online: http://www.statcan.gc.ca/pub/89-654-x/89-654-x2014002-eng.htm (date accessed: February 28, 2020).

[47] Ipsos-Reid, *Mental Health in the Workplace: Largest Study Ever Conducted of Canadian Workplace Mental Health and Depression* (November 19, 2007), online: https://www.work placestrategiesformentalhealth.com/pdf/2007_factum.pdf.

[48] *Ibid.*

[49] *Ibid.*

[50] *Ibid.*

- The rates of depression and suicide are higher among high-stress occupations such as physicians, nurses, soldiers, police officers, and farmers. There are also gender differences with men being four times more likely to die by suicide than women.[51]
- Employees who are diagnosed with depression and have appropriate treatment, save their employer an average of 11 days a year in prevented absenteeism.[52]

For more information on depression and other mental health conditions, refer to the *Quick Facts: Mental Illness & Addiction in Canada, Mood Disorders Society of Canada (2019)*.[53]

5. Bipolar Disorder

Bipolar disorder is a mood disorder in which the individual experiences alternating mood swings that range from being elated (manic) to being depressed. When in the manic phase, the individual feels euphoric, very upbeat, and overly optimistic. Their self-esteem is exaggerated and they tend to exhibit impulsive and reckless behaviours. Early in this phase, the worker can be very creative and productive because they need less sleep and experience increased brain activity. However, the symptoms can become disturbing because the worker experiences racing thoughts, significant agitation, and an inability to focus or concentrate. What follows is a depressive phase in which their mood is much like what is described above. The good news is that this condition is controllable through medical treatment and psychotherapy, and the person can live a productive life.

6. Schizophrenia

Schizophrenia is a psychiatric disorder that tends to first emerge in young adults, between the ages of 15 and 25 years, with males developing the disorder earlier than do females.[54] It is a condition that is characterized by delusions (false beliefs), hallucinations (hearing or seeing things that are not there), disturbances in thinking, and withdrawal from social activity. Although not totally curable, schizophrenia can be effectively treated enabling the individual to function within the workplace.

---

[51] W. Glenn, "Depression, Suicide Rate Linked to Occupations" (July/August 2008) 24:5 O.H.S. Canada 20.

[52] M. Wilson, R. Joffe & B. Wilkerson, *The Unheralded Business Crisis in Canada: Depression at Work: An Information Paper for Business, incorporating "12 Steps to A Business Plan to Defeat Depression"* (Toronto: Global Business and Economic Roundtable on Addiction and Mental Health, 2002) at 22, online: http://www.mentalhealthroundtable.ca/aug_round_pdfs/Roundtable%20report_Jul20.pdf (date accessed: February 28, 2020).

[53] Online: https://mdsc.ca/docs/MDSC_Quick_Facts_4th_Edition_EN.pdf.

[54] Schizophrenia.com, "Schizophrenia facts and statistics" (2010), online: http://www.schizophrenia.com/szfacts.htm (date accessed: February 28, 2020).

7. Alzheimer's Disease

Alzheimer's Disease (AD) may have a profound effect on the working life of the few individuals who, either for genetic reasons, are susceptible to an early onset form of AD, or who are working into their later years. Its greatest impact on the workplace will be both among those older employees who, because of changing demographics and economics, remain in the workplace and develop AD as part of the aging process; and those who may be caregivers for others with AD. For employees developing AD while still working, the challenge becomes one of early recognition of the AD symptoms because those symptoms tend to be confused with other work issues such as work performance problems or insubordination. These challenges exist not only for AD, but for all who are involved in the care of persons with mental illness.

F. AVAILABLE COMMUNITY RESOURCES

There are a number of community resources of which employers and OH&S Professionals/Practitioners need to be aware. These national resources are available in both official Canadian languages, with most being freely available.

1. A Leadership Framework for Advancing Workplace Mental Health[55]

Available from the Mental Health Commission of Canada website, this is a framework that touches on the business case, corporate social responsibility, risk management, and employee recruitment and retention. It addresses how, why, and what a senior leader can do to advance mental health safety within the workplace. It includes videos of corporate, small business, government, and union leaders discussing workplace mental health.

2. Guarding Minds at Work[56]

Guarding Minds at Work provides an evidence-based process and frame that employers can implement to protect psychological safety and promote psychological health in their workplace — and align with the tenets of the Standard.

Guarding Minds at Work is a response to current and emerging legal requirements in Canada for the protection of employee mental health and the promotion of civility and respect at work. Legal standards increasingly require employers to develop comprehensive strategies for ensuring a psychologically safe workplace. Prudent employers need to develop policies and programs that meet these new legal standards.

Guarding Minds at Work is designed to help employers address new legal requirements to maintain a psychologically safe workplace. It provides tools and information to guide employers through the process of developing comprehensive policies and programs.

---

[55] Mental Health Commission of Canada, online: http://www.mhccleadership.ca (date accessed: February 28, 2020).

[56] Online: http://www.guardingmindsatwork.ca/info (date accessed: February 28, 2020).

3. Psychologically Safe Leader Assessment

The Psychologically Safe Leadership Assessment (PSLA), provides an evidence-based assessment and action frame that allows leaders to self-assess their skills, strategies, and approaches as they pertain to psychologically safe leadership.[57] There is a leader self-assessment, as well as an employee feedback assessment. The PSLA is designed to help employers assess and align the skills of their leaders with the tenets of the National Standard.

4. Workplace Strategies for Mental Health[58]

This is a public resource that offers ideas and strategies from a number of sources — all designed to promote mental health in the workplace. It is a unique website dedicated to helping all Canadian employers who wish to address mental health issues in the workplace. It provides videos, action plans, worksheets, forms, publications, and strategies.

Two tools that may be of interest are:

a) *The Audit Tool*: The preamble to this tool states,

> This sample audit tool is an annex of the National Standard of Canada on Psychological Health and Safety in the Workplace. It is reprinted here with permission of CSA Group, Bureau de normalisation du Québec and the Mental Health Commission of Canada. Workplace Strategies for Mental Health has made this version available at no cost on their website for all organizations who wish to do a review of psychological health and safety in their workplace.[59]

b) *Supporting Employee Success*: This tool, as noted on the website, outlines a process that:

- Helps assess work-related triggers for emotional or cognitive issues;
- Supports a thoughtful approach to finding accommodations that may best support success for the employee; and
- Facilitates the employee's well-being while meeting the requirement for a safe and productive workplace.[60]

---

[57] J. Samra, *Psychologically Safe Leader Assessment* (2019), online: Psychologically-SafeLeader.com (date accessed: February 28, 2020).

[58] Workplace Strategies for Mental Health, formerly known as the GWL Centre Online: https://www.workplacestrategiesformentalhealth.com/ (date accessed: February 28, 2020).

[59] Workplace Strategies for Mental Health, formerly known as the GWL Centre Online: https://www.workplacestrategiesformentalhealth.com/ (date accessed: February 28, 2020).

[60] Supporting Employee Success, online: https://www.workplacestrategiesformentalhealth.com/pdf/Supporting_Employee_Success_EN.pdf.

c) **Working Through It**[61]

"Working Through It" provides videos and supporting handouts by and for individuals who struggle with mental health-related concerns in the workplace. Many of the tools are available for public use.

5. Mental Health Commission of Canada

A national non-profit organization created by the Canadian government in 2007, the Mental Health Commission of Canada studies mental health, mental illness, and addiction. It is a credible resource on suicide prevention, mental health, and addiction, and current population-based initiatives.

6. Health Canada[62]

Strategies and resources related to "best practices" and statistics about workplace health are available. This site includes worksheets, calculators, and publications.

7. Mental Health Works[63]

Designed for employees and employers, Mental Health Works supplies a variety of free information on mental health issues. Workshops, presentations, and consulting services on mental health in the workplace can be accessed. This site also includes e-learning, information, and other tools for employers.

8. Canadian Mental Health Association[64]

The Canadian Mental Health Association (CMHA) is a national, voluntary organization that promotes the mental health of all Canadians and supports the resilience and recovery of people experiencing mental illness. The CMHA accomplishes this mission through advocacy, education, research, and service.

9. Centre of Addiction and Mental Health[65]

The Centre of Addiction and Mental Health (CAMH) is Canada's largest mental health and addiction teaching hospital, as well as one of the world's leading research centres in the area of addiction and mental health. CAMH is fully affiliated with the University of Toronto, and is a Pan American Health Organization/World Health Organization Collaborating Centre. The CAMH website provides information, publications, and educational courses.

---

[61] Workplace Strategies for Mental Health, online: https://www.workplacestrategiesformentalhealth.com/ (date accessed: February 28, 2020).

[62] Health Canada, Occupational Health, online: http://www.hc-sc.gc.ca/ewh-semt/occup-travail/work-travail/index-eng.php (date accessed: February 28, 2020).

[63] Online: http://www.mentalhealthworks.ca/ (date accessed: February 28, 2020).

[64] More information is available online: http://www.cmha.ca/ (date accessed: February 28, 2020).

[65] Online: http://www.camh.ca/en/hospital/Pages/home.aspx (date accessed: February 28, 2020).

CH. 23: PSYCHOLOGICAL HEALTH AND SAFETY IN THE WORKPLACE

10. Mood Disorders Society of Canada[66]

According to its website description, The Mood Disorders Society of Canada (MDSC) is a national, not-for-profit, consumer-driven, voluntary health charity committed to ensuring that the voices of consumers, family members, and caregivers are heard on issues relating to mental health and mental illness and in particular with regard to depression, bipolar illness and other associated mood disorders. MDSC's mandate is to provide people with mood disorders with a strong, cohesive voice at the national level to improve access to treatment, inform research, and shape program development and government policies to improve the quality of life for people affected by mood disorders.

11. The Schizophrenia Society of Canada[67]

The Schizophrenia Society of Canada exists to improve the quality of life for those affected by schizophrenia and psychosis through education, support programs, public policy, and research.

12. The Depression Center[68]

The Depression Center offers personalized, interactive tools that have helped thousands of people challenge and overcome their depression.

13. The Panic Center[69]

The Panic Center offers personalized, interactive tools that can help people challenge and overcome their anxiety and panic states.

14. *Beyondblue*: the National Depression Initiative (Australia)[70]

Beyondblue is a national, independent, not-for-profit organization working to address issues associated with depression, anxiety, and related substance abuse disorders. The website provides valuable information on many mental health disorders and their management.

15. Mental Health Workplace Training Programs

There are two known Canadian programs, the Mental Health Workplace Training Program at Queen's University, Kingston, Ontario;[71] and the Psychological Health

---

[66] More information is available online: http://www.mooddisorderscanada.ca/ (date accessed: February 28, 2020).

[67] More information is available online: http://www.schizophrenia.ca/ (date accessed: February 28, 2020).

[68] Online: http://www.depressioncenter.net/Default.aspx (date accessed: February 28, 2020).

[69] Online: http://www.paniccenter.net/ (date accessed: February 28, 2020).

[70] Online: http://www.beyondblue.org.au (date accessed: February 28, 2020).

[71] Queen's University, online: http://www.queensu.ca/gazette/content/training-program-

& Safety in the Workplace Certificate at the University of Fredericton, New Brunswick (online training).[72] For more information, refer to the respective university websites.

16. Canadian Centre for Occupational Health and Safety (CCOHS)

The CCOHS has a number of publications and resources addressing mental health in the workplace.[73]

17. Mental Injury Tool Kit

The Mental Injury Tool Kit is a resource for workers that enables them to understand and cope with workplace stress. Posted on the Occupational Health Clinics for Ontario Workers (OHCOW) website, this guide explains workplace stress and mental distress, the common causes, actions to take, the applicable legislation, and some of the available resources. This resource is available online and via the Measure Workplace Stress App on a cellphone.[74]

18. Stress Assess Survey

The Stress Assess survey, sponsored by the Ontario Health Clinics for Ontario Workers (OHCOW), is a tool that the employee can use to identify and measure the psychosocial stress factors within their workplace. Posted online by OHCOW, this survey has two survey versions — a Personal Edition and a Workplace Edition.[75]

G. EMPLOYER SUPPORT

Today's workplaces are expected to be psychologically safe and healthy. As such, the best measures are preventive in nature and support positive mental health. CCOHS advocates that employers:

- "Encourage employee participation in decision making [concerning the workplace];
- Clearly define employees' duties and responsibilities;
- Encourage civil and respectful behaviours;
- Promote work-life balance;
- Manage workloads;
- Provide opportunities for growth and development;

---

improve- workplace-mental-health (date accessed: February 28, 2020).

[72] University of Fredericton, online: http://ufred.ca/Online-Certificate-in-Psychological-Health-and-Safety/? (date accessed: February 28, 2020).

[73] See, for example, *Steps to a Mentally Healthy Workplace* (2018), *Guarding Minds at Work* and *Psychologically Safe Leadership Assessment*. For more information, refer to www.ccohs.ca.

[74] For more information, refer to www.OHCHOW.on.ca.

[75] For more information, refer to https://stressassess.ca/#ohcow-contact-us-wrapper.

- Support employees through changes at work;
- Recognize employee contributions;
- Educate managers and JHSCs about mental health; and
- Develop policies that consider all mental and physical risk factors."[76]

Much of this approach begins with creating a work culture and environment that is respectful, and civility is "the way things are done around here." Employers should define and educate all stakeholders on civility, incorporate it into their communications and stakeholder interactions, be a role model, and address incivility with a view to neutralizing it.[77] A civil and respectful workplace is attractive and today, potential new hires are seeking this type of workplace in which to work.

H. SUMMARY

The fields of OH&S, and Disability Management continue to be fraught with interesting challenges. To practice effectively, OH&S and Disability Management Practitioners/Professionals, and Canadian employers, need to be equipped with:

- current knowledge;
- unique ways of viewing workplace relationships and practices;
- sound disability claim and case management principles and standards;
- honest and open communication skills;
- good relationship building and nurturing skills; and
- regular practice evaluation techniques.

They also need to gain an appreciation that a "systems approach" to managing mental health safety is more effective and sustainable than a disability case-by-case approach, although there is clear value in incorporating the latter as one program in the systems approach. A systems approach impacts every stakeholder in the system, is aimed at prevention, and is proactive in nature. The case-by-case approach, on the other hand, helps one employee and employee workgroup. Being reactive in nature, it is aimed at damage control and remediation. Ideally, organizations/companies use both approaches as a means of promoting good mental health and supporting the road to recovery when employees experience mental health conditions. All organizations should strive to develop a system that meets primary, secondary, and tertiary prevention goals so that, across the life continuum, it ensures an optimum level of mental illness prevention and mental health promotion.

---

[76] Canadian Centre for Occupational Health and Safety (CCOHS), *Mental Health: Tips for Employers - How to Support Positive Mental Health, Fast Fact Card* (2019), online: https://www.ccohs.ca/products/boutique/mh_employers/

[77] CCOHS, *Civility and Respect in the Workplace: Fast Facts Card* (2019), online: https://www.ccohs.ca/products/boutique/civility_ff/.

CHAPTER REFERENCES

American Psychological Association, *Stress in America 2009* (2009), online: https://www.apa.org/news/press/releases/stress/2009/stress-exec-summary.pdf (date accessed: February 28, 2020).

Beyondblue, online: http://www.beyondblue.org.au (date accessed: February 28, 2020).

S. Calhoun & P. Strasser, "Generations at Work" (2005) 53:11 AAOHN 469-471.

Canadian Centre for Occupational Health and Safety, "Stress Higher Among Working Women Over Forty, Study Finds" CCOHS (September 5, 2005).

Canadian Centre for Occupational Health and Safety (CCOHS), *Mental Health: Tips for Employers – How to Support Positive Mental Health, Fast Fact Card* (2019), online: https://www.ccohs.ca/products/boutique/mh_employers/.

CCOHS, *Civility and Respect in the Workplace: Fast Facts Card* (2019), online: https://www.ccohs.ca/products/boutique/civility_ff/.

Canadian Mental Health Association (CMHA), online: http://www.cmha.ca/ (date accessed: February 28, 2020).

CAMH, "Statistics on Mental Illness and Addictions", online: http://www.camh.ca/en/hospital/about_camh/newsroom/for_reporters/Pages/addictionmentalhealthstatistics.aspx (date accessed: February 28, 2020).

Canadian Standards Association, *National Standard of Canada: Psychological Health and Safety in the Workplace – CAN/CSA-Z1003-13/BNQ 9700-803/2013* (Ottawa, ON: CSA Group, 2013), online: https://www.csagroup.org/article/cancsa-z1003-13-bnq-9700-803-2013-r2018/ (date accessed: February 28, 2020).

Centers for Disease Control and Prevention (CDC), Mental Health Basics (October 2013), online: http://www.cdc.gov/mentalhealth/basics.htm (date accessed: February 28, 2020).

Centre of Addiction and Mental Health, online: http://www.camh.ca/en/hospital/Pages/home.aspx (date accessed: February 28, 2020).

S. Chapman, A. Kangasniemi, L. Maxwell & M. Sereneo (2019). "The ROI in workplace mental health programs: Good for people, good for business" *Deloitte's Insights* (2019), online: http://www.myworkplacehealth.com/documents-resources/ (date accessed: February 28, 2020).

CCOHS, *Civility and Respect in the Workplace* (2019), online: https://www.ccohs.ca/images/products/infographics/download/Respect_Civility.png (date accessed: February 28, 2020).

CCOHS, *Psychologically Healthy and Safe Workplaces* (2019), online: https://www.ccohs.ca/products/boutique/13factors/ (date accessed: February 28, 2020).

De Groot *et al.*, (2009), noted in Workplace Strategies for Mental Health, An Initiative of the Great West Life Centre for Mental Health, Facts & Figures, online: https://www.workplacestrategiesformentalhealth.com/Mental-Health-Issues-Facts-and-Figures (date accessed: February 28, 2020).

Depression Center, online: http://www.depressioncenter.net/Default.aspx (date accessed: February 28, 2020).

W. Glenn, "Depression, Suicide Rate Linked to Occupations" (July/August 2008) 24(5) O.H.S. Canada 20.

Global Business and Economic Roundtable on Addiction and Mental Health, News Release, "Ground-breaking Survey on Mental Health in the Workplace" (November 19, 2007), online: http://www.gwlcentreformental health.com/english/userfiles/news/pdf/s7_004756.pdf (date accessed: February 28, 2020).

Global Business and Economic Roundtable on Addiction and Mental Health, *Roadmap to Mental Health and Excellence At Work in Canada, Summer Draft* (Presented to The Ontario Chamber of Commerce, Economic Summit on Mental Health and Productivity in Ontario, A Pilot for Canada, Sutton Place Hotel, Toronto, Ontario Canada, June 8, 2005), online: http://www.mentalhealthroundtable.ca/june_2005/RoadmapJune82005.pdf (date accessed: February 28, 2020).

Global Business and Economic Roundtable on Addiction and Mental Health, *Roundtable Roadmap to Mental Health Disability Management in 2004–05* (Working document prepared June 25, 2004), online: http://www.mental healthroundtable.ca/june_2004/monitor_june2004.pdf (date accessed: February 28, 2020).

Great-West Life Centre for Mental Health in the Workplace, online: https://www.workplacestrategiesformentalhealth.com/ (date accessed: February 28, 2020).

Guarding Minds @ Work: A Workplace Guide to Psychological Health and Safety, online: http://www.guardingmindsatwork.ca/info (date accessed: February 28, 2020).

Health Canada, Occupational Health, online: http://www.hc-sc.gc.ca/ewh-semt/occup-travail/work-travail/index-eng.php.

M. Hilton & H. Whiteford, "Associations between psychological distress, workplace accidents, workplace failures and workplace successes" (December 2010) 83(8) Int. Arch. Occup. Environ. Health 923-33.

Ipsos-Reid, *Mental Health in the Workplace: Largest Study Ever Conducted of Canadian Workplace Mental Health and Depression* (Toronto: Ipsos-Reid, 2007), online: https://www.workplacestrategiesformentalhealth.com/pdf/2007_factum.pdf (date accessed: February 28, 2020).

R.C. Kessler, W.T. Chiu, O. Demler & E.E. Walters, "Prevalence, severity, and comorbidity of 12-Month DSM-IV Disorders in the National Comorbidity Survey Replication" (2005) 62:6 Arch. Gen. Psychiatry 617.

R.C. Kessler, K.A. McGonagle, S. Zhao, C.B. Nelson, M. Hughes, S. Eshleman, H. Wittchen & K. Kendler"Lifetime and 12-Month Prevalence of DSM-III-R Psychiatric Disorders in the United States: Results from the National Comorbidity

Survey" (1994) 51:1 Arch. Gen. Psychiatry 8.

Marsh Risk Consulting, *Workforce Risk: Fourth Annual Marsh Mercer Survey of Employers' Time-off and Disability Programs* (2003), online: http://usa.marsh.com (date accessed: February 28, 2020).

Mental Health Commission of Canada, *Making the Case for Investing in Mental Health in Canada* (Calgary, AB: Mental Health Commission of Canada, 2013), online: http://www.mentalhealthcommission.ca/English/node/5020 (date accessed: February 28, 2020).

Mental Health Commission of Canada, online: http://www.mhccleadership.ca.

Mental Health Works, online: http://www.mentalhealthworks.ca (date accessed: February 28, 2020).

Mood Disorders Society of Canada, online: http://www.mooddisorders canada.ca (date accessed: February 28, 2020).

Mood Disorders Society of Canada *(2019). Quick Facts: Mental Illness & Addiction in Canada, Mood Disorders Society of Canada*, online: https://mdsc.ca/docs/MDSC_Quick_Facts_4th_Edition_EN.pdf (date accessed: February 28, 2020).

D. Musselman *et al.*, "Relationship of depression to diabetes types 1 and 2: epidemiology, biology and treatment" (2003) 54:3 Biol. Psychiatry 317.

Occupational Health Clinics for Ontario Workers (OHCOW), *Mental Injury Toolkit* (2019), online: www.OHCHOW.on.ca.

Ontario Health Clinics for Ontario Workers (OHCOW), *Stress Assess Survey* (2019), online: https://stressassess.ca/#ohcow-contact-us-wrapper.

OMA Committee on Work and Health, *Mental Illness and Workplace Absenteeism: Exploring Risk Factors and Effective Return to Work Strategies* (April 2002), online: http://www.oma.org (date accessed: February 28, 2020).

The Panic Center, online: http://www.paniccenter.net (date accessed: February 28, 2020).

Public Health Agency of Canada, *A Report on Mental Illnesses in Canada* "Chapter 2 Mood Disorders" (2002), online: http://www.phac-aspc.gc.ca/publicat/miic-mmac/pdf/chap_2_e.pdf (date accessed: February 28, 2020).

Queen's University, online: http://www.queensu.ca/gazette/content/training-program-improve-workplace-mental-health (date accessed: February 28, 2020).

RiskAnalytica, T*he Life and Economic Impact of Major Mental Illnesses in Canada: 2011 to 2041*, online: http://www.riskanalytica.com/?q=node/48 (date accessed: February 28, 2020).

J. Samra, M. Gilbert, M. Shain & D. Bilsker, *Guarding Minds @ Work: A Workplace Guide to Psychological Safety and Health*, "What is Psychological Health & Safety?", online: http://www.guardingmindsat work.ca/info/safety_ what (date accessed: February 28, 2020).

Schizophrenia Society of Canada, online: http://www.schizophrenia.ca/ (date accessed: February 28, 2020).

Schizophrenia.com, "Schizophrenia facts and statistics" (2010), online: http://www.schizophrenia.com/szfacts.htm (date accessed: February 28, 2020).

J. Schulman & P. Shapiro, "Depression and Cardiovascular Disease" (2008) 25:9 *Psychiatric Times*, online: http://www.psychiatrictimes.com/articles/depression-and-cardiovascular-disease (date accessed: February 28, 2020).

M. Shain, *Tracking the Perfect Legal Storm* (May 2010), online: http://www.mentalhealthcommission.ca/English/system/files/private/Workforce_Tracking_the_Perfect_Legal_Storm_ENG_0.pdf (date accessed: February 28, 2020).

Statistics Canada, *Mental health-related disabilities among Canadians aged 15 years and older*, 2012 Canadian Survey on Disability, online: http://www.statcan.gc.ca/pub/89-654-x/89-654-x2014002-eng.htm (date accessed: February 28, 2020).

T. Stephens & N. Joubert, "The economic burden of mental health problems in Canada" (2001) 22:1 Chronic Dis. in Can. 18.

Supporting Employee Success, online: https://www.workplacestrategiesformentalhealth.com/pdf/Supporting_Employee_Success_EN.pdf (date accessed: February 28, 2020).

Towers Watson, *2011-2012 Staying @ Work Report: Pathway to Health and Productivity*, online: http://www.towerswatson.com/en-CA/Insights/IC-Types/Survey-Research-Results/2011/12/20112012-StayingWork-Survey-Report--A-Pathway-to-Employee-Health-and-Workplace-Productivity (date accessed: February 28, 2020).

University of Fredericton, online: http://ufred.ca/Online-Certificate-in-Psychological-Health-and-Safety/? (date accessed: February 28, 2020).

Watson Wyatt Worldwide, *Staying @ Work 2005*, available online at: http://www.watsonwyatt.com and Marsh Risk Consulting, *Workforce Risk: Fourth Annual Marsh Mercer Survey of Employers' Time-off and Disability Programs* (2003), available online at: http://www.marshriskconsulting.com (date accessed February 28, 2020).

M. Wilson, R. Joffe & B. Wilkerson, *The unheralded business Crisis in Canada: Depression at Work — An Information Paper for Business, Incorporating "12 Steps to a Business Plan to Defeat Depression"* (Toronto: Global Business and Economic Roundtable on Addiction and Mental Health, 2002), online: http://www.mentalhealthroundtable.ca/aug_round_pdfs/Roundtable%20report_Jul20.pdf (date accessed: February 28, 2020).

"Working Through It", available for public use from the Great-West Live Centre online: https://www.workplacestrategiesformentalhealth.com/ (date accessed: February 28, 2020).

World Health Organization, "Burn-out an 'occupational phenomenon': International

Classification of Diseases" Mental Health (May 23, 2019), online: https://www.who.int/mental_health/evidence/burn-out/en/ (date accessed: February 28, 2020).

World Health Organization, *Mental Health: A Call for Action by World Health Ministers. Ministerial Round Tables, 54th World Health Assembly* (Geneva: World Health Organization, 2001), online: http://www.who.int/mental_health/advocacy/en/Call_for_Action_MoH_Intro.pdf (date accessed: February 28, 2020).

World Health Organization "Mental Health: Strengthening our response" (August 2014), online: http://www.who.int/mediacentre/factsheets/fs220/en/index.html (date accessed: February 28, 2020).

World Health Organization, *Scaling-up care for mental, neurological, and substance use disorders* (WHO Press: Geneva, Switzerland, 2008), online: http://www.who.int/mental_health/mhgap_final_english.pdf (date accessed: February 28, 2020).

Chapter 24

PSYCHOLOGICAL HEALTH AND SAFETY: PRACTICAL APPLICATION IN THE WORKPLACE

By Mary Ann Baynton,[1] Dianne Dyck, Dan Steinke[2] and Tony Roithmayr[3]

A. INTRODUCTION

Current and evolving legislation in areas of Canadian Human Rights, Workers' Compensation Acts, labour law, tort law, and Occupational Health & Safety Acts are increasingly holding employers responsible for providing a psychologically safe workplace.[4]

Not all workplaces implicitly include psychological health and safety in their OH&S approach; but this concept is gaining impetus as a legal responsibility and a recognized business strategy. This is distinct from the duty to accommodate individuals with psychological illness which is a human rights issue. Psychological health and safety in the workplace is similar to a preventive and mitigation approach to hazards in the workplace which may impact the psychological well-being of staff.

---

[1] Mary Ann Baynton, MSW, RSW, is a workplace relations specialist who consults with organizations, governments, and institutions about addressing and improving workplace mental health including return to work, accommodation, or conflict resolution.

[2] Dan Steinke, CHSEP, CSSE, is an OHS&E Specialist who serves on the CSSE Editorial Advisory Board, and who is Vice President, Windsor Occupational Health Information Service and President, Internal Responsibility Information and Solutions.

[3] Tony Roithmayr, President, Performance by Design, Calgary, Alberta, online: http://www.performance-bydesign.com/ (date accessed: February 28, 2020).

[4] Delivered in a 2010 report by Dr. Martin Shain. Dr. Shain has a doctorate in Juridical Science and is an Adjunct Professor and Senior Consultant with the Consortium for Organizational Mental Healthcare (COMH). He also holds an appointment with the Dalla Lana School of Public Health, Faculty of Medicine, University of Toronto. He is the Co-Chair of the BC Psychologically Healthy Workplace Collaborative and serves on The Roundtable for Workplace Mental Health.

When a stakeholder in the workplace (union, Human Resources, Senior Management, Front-Line Management, Occupational Health and Safety, and employee) has a clear understanding and appreciation for the strengths as well as the responsibilities of the other stakeholders, it opens up the opportunity for more effective collaboration. This type of cooperation has the potential to benefit everyone involved in terms of fostering and maintaining psychological health and safety.

In the previous chapter, the topic of Psychological Health and Safety in the Workplace was introduced. In this chapter, the focus is on how to apply those concepts. These strategies are broken down by stakeholder group and reinforced with references to readily available national resources which are free in the public domain. Additionally, grief and grief management are addressed.

B. PSYCHOLOGICAL HEALTH AND SAFETY: APPLIED

The Performance Maximizer® model (Figure 24.1) can be used to demonstrate how psychological health and safety can be operationalized.

The Performance Maximizer® model focuses on all the factors that shape optimal human performance, and the conditions that exist when successful human performance occurs. These factors are clustered into five "Conditions for Great Performance" — that is, *Know What to do, Able to do it, Equipped to do it, Want to do it*, and the necessary *Interactions*, that foster trust, respect, integrity, collaboration, and accountability. Expectations, best practices in psychological health and safety, skills, processes, tools, motivation, culture — all of these factors and many more are represented in this model.

The Conditions for Great Performance are the leading indicators that can be used to forecast work performance outcomes. In the absence of these conditions, organizational stressors develop which cause problems with employee health and on-the-job performance, and ultimately, impact the organization's bottom line. The premise is that, by focusing on the leading indicators for creating great human performance, lagging indicators such as employee absence, disability-related costs, reduced productivity, and poor profits will gradually decrease.

---

[5] T. Roithmayr, *The Performance Maximizer®* (2000), online: http://www.performance-bydesign.com (date accessed: February 28, 2020).

Figure 24.1: The Performance Maximizer Model[5]

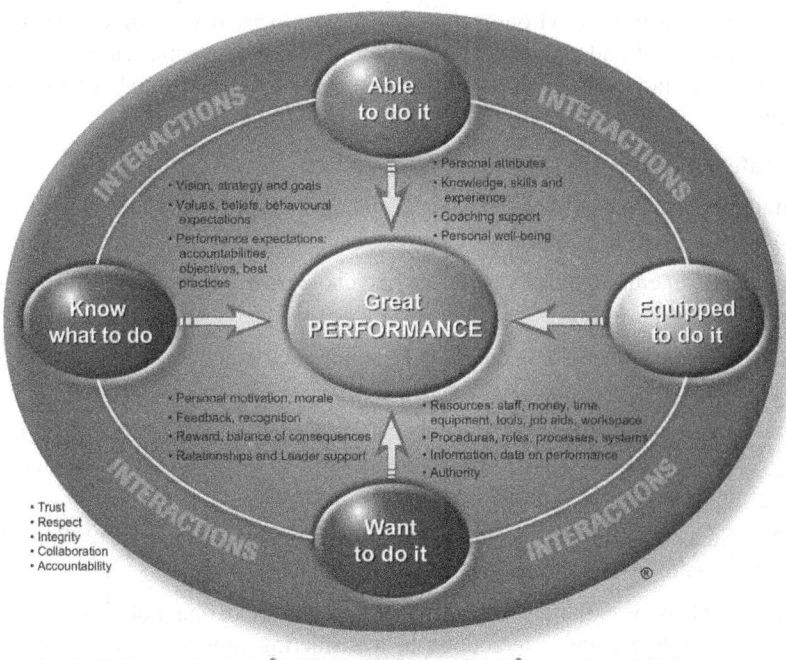

To support any workplace business function, it is critical for the organization to create a work environment in which workers:

- **Know WHAT to do** – is about clarity and alignment on what needs to be accomplished; people are crystal clear on what is expected of them;
- **ABLE to do it** – is about capability to perform; the skills and knowledge required, and the physical and emotional health necessary to be engaged and productive;
- **EQUIPPED to do it** – is about a variety of resources and organizational conditions people need to do their job: processes, procedures, systems, information, tools, workspace, *etc.*;
- **WANT to do it** – is about the factors that motivate accountability and commitment to safe, efficient, effective, work: feedback, recognition, reward, involvement, and consistent leader support;
- **INTERACTIONS** – are about the *way* people work together: human interactions that foster trust, respect, integrity, collaboration, and accountability.

Now, the question becomes, *"How can this be achieved in the workplace in terms of psychological health and safety?"* The 2013 release of the National Standard of Canada on Psychological Health and Safety in the Workplace[6] provided guidance on how to create a work environment in which employee psychological health and safety is protected and promoted. In addition, the following materials are suggestions for the practical application of psychological health and safety in the workplace.

1. Strategic Direction

To attain physical and psychological health and safety in the workplace, organizations/companies are encouraged to:

1. Make employee physical and psychological health and safety a core business strategy.
2. Engage stakeholders and clearly define their roles and responsibilities.
3. Assess the status of the organization's/company's current physical and psychological health and safety efforts. Chapter 25, "Psychological Health and Safety in the Workplace: Measurement" explains some measurement approaches that can be used for this purpose.
4. Strive to embrace the elements of the National Standard of Canada on Psychological Health and Safety in the Workplace.
5. Embed psychological health and safety in existing policies, processes, procedures, and interactions by considering how each may impact the psychological health and safety of the workforce.
6. Regularly evaluate the organization's/company's physical and psychological health and safety efforts.
7. Leverage the power of measurement to garner support for physical and psychological health and safety in the workplace, using an approach of continual improvement.
8. Involve stakeholders on the path to improvement, seeking their feedback and support.

C. STAKEHOLDER ROLES AND RESPONSIBILITIES

1. The Senior Management Role

Setting strategic direction and assigning the necessary resources to effect change in psychological health and safety are key roles of Senior Management. When a Board of Directors exists, this group would also be involved. In order for any initiative to garner support and be sustained, having a champion in Senior Management helps. This individual would be someone with both influence and decision-making

---

[6] The National Standard of Canada on Psychological Health and Safety in the Workplace is available through the Canadian Standards Association, online: http://www.csagroup.org/ca/en/services/codes-and-standards (date accessed: February 28, 2020).

capability — tenants of a Senior Management designation. He/she would understand and appreciate the value of the proposed plan and would be willing to convey such sentiments to their peers in Senior Management. When initiatives lack this type of support, and no one is at the decision-making table to champion the issue, it is much easy to dismiss it when pressures to attend to other issues arise.

Once the champion is identified, he/she may require some evidence and/or information with which to convince others of the need to address workplace psychological health. The Mental Health Commission of Canada's Workforce Advisory Committee recognized this need for a strong business case and developed a free and publicly available, online resource entitled *A Leadership Framework for Advancing Workplace Mental Health*.[7] This resource includes scientific and practice-based evidence and video testimonials from leaders in government, business, unions, and associations endorsing psychological health in the workplace as a worthwhile business issue. The framework includes the cost effectiveness of initiatives, the positive effect of improving employee psychological health on recruitment and retention, the impact of psychological health and safety on productivity, and the risk management benefits of addressing psychological health at work.

This resource assists Senior Management in understanding how all organizational departments can work towards accountability and collaboration in the initiative. Those who have used the resource indicate that including the testimonial videos in their proposal presentations to Senior Management has been helpful in exemplifying the effectiveness and plausibility of addressing workplace psychological health and safety.

Inherent in the *"psychological health and safety champion"* role is a need to advocate for a broad-based organizational focus on workplace psychological health and safety, rather than a singular project management approach. Embedding psychological health and safety strategies into various workplace policies, procedures, and departments of an organization is an effective approach to ensure a sustainable workplace psychological health and safety movement.

Psychological health and safety risk factors permeate all aspects of the employment-life span: recruitment, hiring, retention, advancement, training, performance evaluation, accommodation, return-to-work planning, retirement, and/or termination. For example, if we consider some of the stages of employment and their impact on psychological health and safety, we could ask questions such as:

(a) ***Recruiting and Hiring***

How do you recruit supervisors and managers with **emotional intelli-**

---

[7] Mental Health Commission of Canada's Workforce Advisory Committee, *A Leadership Framework for Advancing Workplace Mental Health*, online: http://www.mhccleadership.ca (date accessed: February 28, 2020).

gence?[8] How do you consider emotional capacity for a job? How can job descriptions include the psychological demands of a job? When hiring someone who is from a diverse cultural group, how can you ensure their psychological health and safety (*i.e.*, the first person of another gender, orientation, ethnicity, culture, religion, age, *etc.*, ever hired by the organization/company)? How can you make the interview and hiring process physically and psychologically healthy and safe?

(b) *Orientation and Training*

How can the orientation to processes, people, and policies help promote a psychologically healthy and safe approach? Are corporate Codes of Ethics or Conduct effective? If so, how? What type of training should supervisors and managers be required to take? How do you ensure that training results in outcomes that improve the ability to provide and sustain a psychologically healthy and safe workplace? How can employee training or lack thereof, affect workplace psychological health and safety?

(c) *Evaluation, Performance Management, Discipline, and Promotion*

Evaluation, performance management, discipline, denial of promotion, or promotion to a position of incompetence, all have a significant potential to cause psychological harm if incorrectly handled. Sometimes these issues are related to "bad-job fit", inadequate supervision, a "poisoned" work environment, personality conflicts, or insufficient training or resources. What approaches to evaluation, performance management, discipline, and promotion can prevent these problems from putting employee psychological health and safety at risk?

(d) *Intervention and Crisis Response*

Suicide, grief, violence, addiction, bullying, harassment, and conflict, all may have serious psychological impacts on the workplace. How can these issues be planned for and addressed in a psychologically healthy and safe way? Think about the organizational, managerial, union, and individual impacts.

(e) *Accommodation and Return to Work*

Accommodation and return-to-work planning can be pivotal in terms of the psychological health and safety of individual employees, especially those for whom psychological illness is a factor in their well-being. What approaches to accommodation and return-to-work can improve their psychological health and safety?

---

[8] **Emotional intelligence** is defined as the ability to perceive, control, and evaluate one's own emotions, as well as those of others, and to use that emotional information to guide thinking and behaviour.

(f) Discipline, Redeployment, and Termination

Significant changes in an organization can leave everyone unsettled. There can be **survivor guilt**[9] for those who have avoided layoffs or redeployment, as well as stress for those who are let go, and stress for those who must announce the decision to terminate. How can an organization address times of uncertainty, economic downturn, organizational restructuring, or layoffs in a way that is psychologically safe for all? For individual cases of termination or redeployment, how can the process be more psychologically healthy and safe?

Beyond these specific questions, the following are some tasks that Senior Management should consider when adopting a mandate of a psychologically healthy and safe workplace:

- Develop a work environment that supports psychological health and safety;
- Promote an Operational Approach that promotes good psychological health and safety;
- Clearly send the message that Middle and Line Management are expected to observe the psychological health and safety practices developed by the organization, and when required, to identify and support the distressed employee;
- Educate and prepare Middle and Line Management to prevent psychological injury/illness and address psychological injury/illness when they occur;
- Endorse responsible Disability Management practices to mitigate psychological injuries/illness; and
- Promote work accommodation. The cost for providing work accommodation is low (on average $500 or less) in comparison to the average cost of a psychological illness claim ($18,000).[10]

(g) Transformational Leadership

Ensuring sound leadership and clear expectations is a crucial factor in ensuring a psychologically healthy and safe workplace. If clear expectations and sound leadership are absent, employees feel disconnected from their work and unmotivated. These conditions can negatively impact

---

[9] **Survivor guilt** is the emotional condition in which the survivor believes that he/she has done something wrong in order to have survived a life-threatening event when others did not.

[10] Centre for Addiction and Mental Health (CAMH), "The numbers add up, so make your vote count", (2011) 11:3 camhconnexions at 1, online: http://www.camh.ca/en/hospital/about_camh/newsroom/connexions_newsletter/Documents/4515Connexions_Fall2011EN.pdf (date accessed: February 28, 2020).

employee psychological health and safety.

On the other hand, connection and a sense of purpose are related to good psychological health and even to recovery from psychological illnesses such as depression or anxiety-related disorders.[11] For this reason, Senior Management is advised to educate supervisory staff on transformational leadership, a well-recognized, supportive, and effective approach to employee management.

Transformational leadership is defined as inspired leadership that influences the beliefs, values, and goals of followers so that they can perform in an extraordinary manner. Transformational leaders offer a purpose that transcends short-term operational goals and focuses on higher-order intrinsic needs of the workgroup. This results in employees identifying with the leader and his/her directions.[12]

Four dimensions of transformational leadership are:

- *Charisma or Idealized Influence:* the degree to which the leader behaves in ways that followers view them as admirable and which cause followers to identify with the leader.
- *Inspirational Motivation:* the degree to which the leader articulates a vision that is appealing and inspiring to followers.
- *Intellectual Stimulation:* the degree to which the leader challenges assumptions, takes risks, and solicits the ideas of followers. Leaders with this trait stimulate and encourage follower creativity.
- *Individualized Attention:* the degree to which the leader attends to each follower's needs, acts as a mentor, or coach to the follower and listens to the follower's concerns and needs.

Transformational leaders motivate followers by providing a clear vision; their charisma; using symbolism; empowerment; intellectual stimulation; and their demonstrated integrity. It is an appropriate leadership model for managers/supervisors to show that they are:

- committed to workplace health and safety;
- demonstrating sound workplace practices and behaviours;
- doing the right things for the right reasons;
- inspiring others to excel;

---

[11] B. Schrank & M. Slade, "Psychiatry Recovery" (2007) 31 Psychiatric Bulletin 321-325.

[12] Transformational leadership is presented as a concept in Chapter 1, "Occupational Health and Safety: Historical Perspectives", Chapter 7, "Safe Workplace = Great Workplace: Building a Sustainable Culture of Safety", and Chapter 18, "Occupational Health and Safety: Best Practices". Here it is operationalized.

- promoting team work;
- encouraging employees to "think outside the box" and challenge assumptions to come up with new solutions; and
- demonstrating concern for the needs, interests, and well-being of employees.

To become a transformational leader is no easy feat, and it is a rare individual who naturally possesses all of these traits all of the time. The pressures of Front-Line Management can be enormous in terms of answering to Senior Management, regulators, Human Resources requirements, Joint Health and Safety Committees, unions, external stakeholders, peers, and especially to direct reports. When those direct reports are struggling with performance, personal or health issues, the emotional cost to Front-Line Management can be significant.

To turn an organization into one where transformational leadership is the norm requires the presence of at least three factors:

1. Senior Management has to set the standard and expectation by modelling transformational leadership themselves.

2. Those in Management roles must receive training, support, and recognition for being transformational leaders. What gets measured and recognized in the Front-Line Manager's performance review will become the focus of their attention and efforts. In organizations where managers are only held accountable or recognized for the bottom-line results, it makes sense that the majority of their attention is focused on short-term productivity. Where employee engagement and innovation are valued outcomes, Front-Line Managers are more likely to connect their job responsibilities to the impact on employees as well as with the bottom-line results.

3. Hiring and promoting individuals whose skill sets are conducive to transformational leadership is critical. There are many individuals whose skill sets are in the area of policy development, process improvement, analysis, or strategic planning. All of these skills are highly valued and in some positions, necessary; but when the position includes managing people, they are not sufficient. Senior Management should direct Human Resources Practitioners/ Professionals to recruit and select Front-Line Managers who possess emotional intelligence and the potential for transformational leadership. This approach can prevent some of the management practices that lead to stress, burnout, or health concerns for either the employees or the manager.

If these factors are not implemented, some recognized conditions that lead to employee stress, especially among managers, can occur, such as:

- a mismatch between work demands and the available resources to meet those demands;
- lack of adequate emotional intelligence;
- expectations of work productivity and deliverables without the authority to achieve the operational goals;
- work and role overload;
- unclear functional goals;
- constant "fire-fighting";
- dissonance between Senior Management demands and subordinate expectations;
- lack of support from peers/co-workers; and
- an imbalance in work and personal time.

In terms of promoting employee psychological health and safety, Roithmayr (2000) suggests that Managers must:

- clarify employee roles and responsibilities;
- be clear about performance expectations, behavioural expectations, how employee work will contribute to operational and organizational goals (line of sight), the standards against which employee performance will be judged, and how employee performance will be measured;
- support employee education and learning of needed knowledge and skills;
- seek a suitable match of job duties and employee interest and attributes, making work meaningful and rewarding;
- promote supportive work relationships;
- address work situations so that employees do not feel emotionally or psychologically vulnerable;
- address unsafe work hazards, conditions, and contribution factors;
- provide employees with information on their work performance;
- involve employees in work decisions;
- position employees so that they have the autonomy and discretion to successfully deliver on their accountabilities;
- encourage employees to play a role in what happens to them at work;
- provide employees with the tools, equipment, job aids, and other resources needed to succeed in their jobs;
- develop clear and effective work practices and procedures;

- build a work environment that is conducive to working efficiently and effectively;
- provide adequate time for employees to complete job tasks;
- provide positive feedback on employee performance, behaviours, and work contributions;
- recognize employee performance, behaviours, and work contributions;
- provide corrective feedback and coaching on how employees do their work;
- treat employees with fairness, trust, and respect; and
- communicate how important employee performance is in terms of Operational achievement — that there are positive and negative consequences associated with work performance.[13]

(h) *Summary*

While Senior Management is charged with ensuring fair and appropriate treatment of all employees, Front-Line Management is the first line of defence when ensuring that the physical and psychological health and safety of the organization's employees is protected and supported. Roithmayr indicates that managers at all levels of the organization have a role to play in ensuring the physical and psychological health and safety of its employees; and therefore, need to:

1. **Establish Purpose**
   - Help employees understand the organization's vision, values, goals, and business strategies within the context of ensuring physical and psychological health and safety.
   - Guide the development of individual performance and learning plans that will achieve organizational goals while respecting individual employee emotional well-being and their physical and psychological health and safety.

2. **Enable Performance**
   - Align resource allocations with performance expectations while making an effort to ensure employees have the resources they need to foster and maintain physical and psychological health and safety.
   - Coach employees to overcome difficulties and build skills and knowledge that can assist employees in overcoming stressful,

---

[13] T. Roithmayr, Organizational Stressors Survey (2000), online: http://www.performance-bydesign.com (date accessed: February 28, 2020).

distressful, and complex workplace situations that may impact their physical and psychological health and safety (*i.e.*, teach conflict management strategies, time management skills, *etc.*).
- Foster and maintain effective workgroups that promote the undertaking of projects that conduct projects in a manner that respects and accesses people's knowledge, supports their emotional well-being, and shares the completion of the task.
- Resolve performance issues and remove barriers that are beyond the control of individuals and teams.

3. **Expect Results**
    - Facilitate the measurement of progress, contribution, and development in a manner that invites employees' involvement in the measurement process: both in decisions about how to measure and what to measure.
    - Hold people accountable for delivering agreed-upon results. If an employee or employees are unable to deliver the agreed-upon result, make an effort to evaluate why this happened and discuss strategies to ensure this does not happen again (*i.e.*, check in if they had the needed resources, were there too many competing demands, *etc.*).

4. **Encourage Success**
    - Sustain communication that maintains focus, fosters commitment, and facilitates implementation.
    - Recognize and celebrate progress, development, and the achievement of desired results.[14]

It is important to remember that *good managers and supervisors are not "just born" — they are developed*: both on the Front-Line and in Senior Management positions. Management approaches and techniques that can be successful in the majority of cases, may prove to be ineffective or even harmful when psychological health issues are a factor. For these reasons, it is important that managers receive training and support when their job requires them to support emotionally distressed employees. Effective and responsive leadership is the crux of the physically and psychologically healthy and safe workplace; it increases employee morale, resiliency, and trust and decreases employee frustration and conflict. Even more importantly, a leader who demonstrates a commitment to workplace physical and psychological health and safety sets the tone for their colleagues, their employees, and the organization as a whole.

2. The Front-Line Manager Role

It is the task of Front-Line Management to ensure that work gets completed and the

---

[14] Ibid.

objectives of the organization are met. While every people-management situation is unique and requires good judgement, what follows are some effective people-management approaches when psychological health may be a factor.

Many managers are inadequately prepared for dealing with emotionally distressed employees and their work productivity. When faced with this type of behaviour, some Front-Line Managers resort to standard performance management approaches that include pointing out work mistakes and demanding accountability to standards of production. The reality is that when an individual is in crisis due to a health, family, financial, or relationship problem, their mindset is not receptive to what can be called the "shame or blame" approach. Shame happens when someone feels inadequate, humiliated, or condemned. Whereas guilt is often an expression of regret for behaviour; shame is an internally directed negative feeling about one's self. The reason this is relevant in the workplace is that the response to blame or shame often results in unwanted outcomes. It can be human nature for someone who feels blamed, criticized, judged, or shamed to do one of four things:

1. justify or defend their actions; or
2. withdraw; or
3. counterattack; or
4. "play the victim".

(a) Approaching Employees

By learning to express concern and to invite conversation without blaming, shaming, or judging, managers can increase their ability to solve problems, improve productivity, and reduce risk to employee psychological health and safety. What follows are some practical examples.

Upon noting substandard performance by an employee, the manager might say something like:

- *"You don't seem to be yourself. Are you feeling well?"*
- *"This anger/forgetfulness/irritability/distraction/intensity/volatility/etc. is not like you. What is happening?"*
- *"I have noticed that you seem unsettled. Is there anything troubling you?"*
- *"You used to enjoy these meetings, but lately it doesn't seem like you do. Has something changed?"*

Notice that there is an attempt in each statement to suggest change from what was once "normal" behaviour for this particular employee. It demonstrates concern while showing attentiveness to him/her. It also opens the door for the employee to "speak out" about what is going on without feeling accused or criticized.

If the employee denies that anything is wrong, or refuses to talk about it, continue to express concern in your voice, body language, and words; but refocus the conversation to the performance issues. Hence, the interaction might be:

Employee: *"There is nothing wrong with me. I am just busy at work."*

Manager: *"OK, I hear you. Let's talk about how we can make your work more manageable. Tell me what you are working on right now and we can work to prioritize the tasks so that is does not feel so overwhelming."*

Next, focus on helping the employee to express:

- What he/she is working on;
- Where the biggest challenges/stressors/pressures are;
- Which work tasks they enjoy the most;
- Which work is the easiest for him/her to do;
- Who is giving him/her work;
- Who is depending on his/her work outputs;
- Who does he/she depend on to finish his/her tasks;
- What interrupts his/her focus during the day; and
- What knowledge, skills, or resources might help him/her to complete their assigned task.

Once this is determined, work with the employee to:

- Prioritize the tasks;
- Reassign or eliminate tasks that are not a good use of this employee's time;
- Identify what can be done about the pressures and challenges; and
- Consider how to optimize or increase tasks that energize this employee, or enhance the feeling of success.

Collaborate with the employee so that he/she commits to:

- What he/she will do differently to improve his/her work situation.
- What steps he/she will take to ensure that he/she can do their tasks effectively.
- What else is required from you, the manager.
- How future success will be recognized and measured.
- When follow up will happen.
- How future issues will be addressed.

Then, **and most importantly**, follow up and follow through, as mutually agreed. As the manager and as a person, show your commitment to helping this employee. Following up and following through on these processes is imperative to ensuring that the employee really sees your understanding of their workplace needs, but also as a preventative tool — so that if the employee struggles in the future, they know that you are open to discussing their issues and working through some solutions.

Not No, But Why

When an employee is asked for their opinion or input on a work situation, be prepared to use that opinion/input. Requesting their feedback and then ignoring or discrediting the employee's contribution, sends a strong message that their opinion is not important nor valued. Yet, it would be chaos if we agreed to every suggestion, or simply said, "Yes" to every request.

There is a way to meet employee needs without actually saying, "No". It involves understanding human needs. Like a baby's cry, a single behaviour can indicate many different underlying needs/wants. For babies it could be a cue that the baby needs food, affection, warmth, a dry diaper, sleep, or is in pain. For adults, some human needs are universal and according to Maslow,[15] include:

1. **Physiological Needs**

    These needs are what human beings require to survive, *e.g.*, water, air, food, and sleep. Maslow believed that these needs are the most basic and instinctive needs in the hierarchy of needs because all other needs are secondary until these physiological needs are met.

2. **Security Needs**

    These include the need for safety and security. Security needs are important for survival, but they are not as demanding as the physiological needs. Examples of security needs include a desire for steady employment, health insurance, safe neighbourhoods, and shelter from the environment.

3. **Social Needs**

    These include needs for belonging, love, and affection. Maslow considered these needs to be less basic than physiological and security needs. Relationships such as friendships, romantic attachments, and families help fulfil this need for companionship and acceptance, as do involvement in social, community, or religious groups.

4. **Esteem Needs**

    After the first three needs have been satisfied, esteem needs become increasingly important. These include the need for things that reflect on self-esteem, personal worth, social recognition, and accomplishment.

5. **Self-actualizing Needs**

    This is the highest level of Maslow's hierarchy of needs. Self-actualizing people are self-aware, concerned with personal growth, less concerned with the opinions of others, and interested in fulfilling their potential.

---

[15] A. Maslow, Maslow's Hierarchy of Needs (1943), online: http://en.wikipedia.org/wiki/Maslow's_hierarchy_of_needs (date accessed: February 28, 2020).

Unlike these basic needs that all humans require, employment needs tend to differ from person to person. For example, one employee may have a strong need for autonomy. He/she may want you to give them a description of the desired outcomes and then to be left alone to decide how to reach those operational objectives. Other employees may find this freedom too threatening; they want more direction and coaching. They seek very specific, step-by-step instructions on how each task should be carried out. Both approaches are meeting human needs, but in a different way.

(b) Implementing Needs-based Problem Solving

All emotions, thoughts, and behaviours stem from needs. Needs are distinct from a preference or a want which is likely just one possible option for meeting a need. For example, some people have a need for full spectrum or natural light to counteract seasonal affective disorder. They may prefer or want the biggest, brightest office in the workplace, but there can be other options for meeting this need. Most problems and conflicts in the workplace (and elsewhere) result from needs that are not satisfied. To address this situation, the following steps are recommended:

1. Identify the needs of all the parties.
2. Creatively collaborate on possible strategies to ensure that the most important needs of all parties are met in a mutually agreeable way.
3. Affirm the agreement, address any misunderstandings, and discuss a process for handling the problems in the future.
4. Document the agreement, providing all parties with a copy. This approach is particularly effective when an employee's concentration, perception, or memory is compromised.

So, How Do We Say "No", Without Saying "NO"

When an employee asks for something like "the corner office with a window", rather than just telling him/her that this type of office is reserved for employees with higher employment status, you might ask, *"Why do you want the corner office?"*

The manager may discover that the employee has seasonal affective disorder (SAD) that requires a minimum amount of exposure to sunlight each day to manage it. By saying, "No", the manager is unable to find another way to meet the employee's underlying needs, such as the offer to purchase a small full spectrum light box that would have the desired effect of keeping the employee healthy at work.

Another example would be when an employee says that he/she wants to be transferred to another workgroup. As a response, the manager may ask, what the employee would want to change about the current workgroup if he/she could? Again, getting at the underlying need, rather than reacting to the request or simply saying, "No", allows for an exploration of alternate solutions.

A common Human Resources topic of debate is the value of moving an employee from workgroup to workgroup, in order to "solve" a problem. If the problem stems from the employee's lack of effective coping strategies, it is very likely that similar problems resurface after a brief "honeymoon" in the new workgroup/unit. However, if the problem was the result of how the original workgroup/unit operated or interacted, it is very likely that the employee will do well in a new work situation. Be aware though that the original problem faced will probably show up again when a new employee is placed into the original, dysfunctional workgroup. In either case, the "root" problem was not identified, nor addressed — just delayed or moved.

So, don't just say, "No"; ask, "Why?", and explore alternatives to meet employee needs.

(c) Coaching the Distressed Employee

Coaching any employee can be challenging, but when the employee is emotionally distressed, it can be even more difficult. Managers are often unsure if they are pushing too hard or not hard enough. Some will be so afraid to say or do the wrong thing that they end up doing nothing at all . . . *until a crisis happens*. At this time, the effort and fallout from the issue can end up being far worse than if it had been addressed earlier. The other response may be to begin progressive discipline, which in some cases where psychological disability such as depression or anxiety-related disorders exist, could be considered a violation of human rights. So, if doing too little or too much can both be problematic, what is a manager supposed to do? While every situation is unique and requires thoughtful consideration, what follows are some suggestions that may be helpful:

Refocus from the Problem to the Solution

Whatever we focus on tends to become the central theme of our approach. In the case of an employee's performance, when we focus on the problem, we tend to label the employee as a problem as well. This often results in an employee who is defensive and hurt by any attempt to manage performance. For example, if the problem is that the employee is not meeting deadlines and we begin to focus on not meeting deadlines. It can result in tension and frustration, especially if the employee feels the deadlines are unreasonable and the manager thinks that they are not.

When we shift that focus to the solution, we tend to open the door to an approach that encourages the employee to help us find a better way. If the approach was to ask the employee what they needed to meet a deadline and how they would approach it if they felt it was going to be a challenge as they got closer to the deadline, it can change the dynamics of the interaction. Instead of accusations and defensiveness, we can choose creative brainstorming of solutions. This approach, originally developed for employees with psychological health concerns, had the unintended consequence of forcing those who may be classified as "slackers" into either finding a solution for themselves or being recognized as not interested in success.

Use Constructive Communication Techniques[16]

Effective communication is essential when working with employees who may be

---

[16] Workplace Strategies for Mental Health, formerly, Great-West Life Centre for Mental

experiencing psychological health issues. Following are some specific strategies that have been successful in handling these often-difficult situations and conversations. You may choose to apply some of these strategies yourself or to share them with supervisors in your workplace:

- **Listening for Understanding**

    Listening for understanding takes place when you are sincerely trying to understand not just what a person says; but rather what they mean. When someone is distressed or dealing with a psychological health issue, it is not unusual for them to say things that are not really reflective of what they truly mean. By giving someone the safety and the space to articulate and then clarify or correct what they say, means you have a much better chance of understanding their perspective. Sometimes the content of what we hear will elicit an emotional response in us. As we listen to others, we may be distracted by our own "internal chatter" that can include judgements, opinions, and reactions to what is being said. When we listen for understanding, we focus on the individual and their agenda, not on our own. We listen for underlying issues and needs making us better prepared to begin a discussion about solutions.

- **Distinguishing Validation from Agreement**

    Because each person has different needs and views, there will always be some conflict in living and working with others. Effective listening can help us better problem-solve and generate solutions that meet more of everyone's needs. Listen first and acknowledge what you hear even if you don't agree with it, before expressing your point of view.

    To acknowledge that someone else's feelings are valid for them, even when we do not feel the same way, allows the person to feel heard, and therefore, better able to listen. Acknowledging another person's thoughts and feelings still leaves you with all the following options:

    - Agreeing or disagreeing with the person's point of view or actions;
    - Saying a request cannot be granted, but you are willing to explore other ways to meet the same need; and
    - Saying more about the matter being discussed.

- **Communicating Without Judgement**

    Often when we are listening to what people are saying, we will have an emotional response and make judgements/assumptions about the intent and meaning of the message. These judgements often lead us to respond in a way that fuels mistrust and conflict. To avoid a potential impasse in difficult conversations:

Health in the Workplace, online: https://www.workplacestrategiesformentalhealth.com/.

- Turn down our internal dialogue and stay focused on what is being said.
- Breathe and neutralize our emotions.
- Listen and acknowledge alternate perspectives.
- Move from judgement to curiosity by asking questions to gain better understanding.
- State your observations and experience using specific examples.
- State your perspective, needs, and desires.
- Reframe the problem into a mutual, objective statement.

Be Clear about Concerns and Expectations

During times of emotional distress, our ability to think clearly can be compromised. When working with an employee who may be struggling with psychological health issues, it is especially important to be clear about expectations. This does not mean bringing out a laundry list of problems with specific instances with dates and times. That will likely put your employee on the defensive. Nor does it mean that you should have a list of tasks to be checked off. But it does mean saying more than, "you need to have a more positive attitude".

Clarity means that the expectations are measurable, specific, and work-related. Expectations should not be focused on personality or vague values such as teamwork or a positive attitude. When someone is not well due to a psychological illness, such conversations can be confusing. Taking on issues one at a time also makes it easier for the employee to fully understand and commit to a task before moving on to others.

Offer Help

There are two parts to offering help. One part is to be aware of organizational and community resources, and to be able to offer these to your employees. Organizational resources may include an Employee Assistance Program (EAP), Disability Management Program (DMP), or Workplace Wellness Program (WWP). Community resources include clinics, crisis lines, support groups, and hospital programs. Having this information available, can save time and anguish.

The second part to offering help is to pose this question to the employee: *"What can I do to help you be successful at your job?"* This interest in the employee's well-being indicates a sign of respect for the employee's ability to consider solutions. It is an important way for the employer to get commitment to the solution created in part by the employee, rather than compliance with your solution. When this suggestion is presented, the initial response by employers is sometimes that this is a dangerous usurping of their power and control, but when they implement it, their experience is much different. Employees who are supported in finding a way to keep their job in spite of their illness, are grateful and innovative in their search for a way to express their loyalty. Moreover, the employee understands their job better than anyone else, and can come up with creative ways to accomplish the necessary outcomes without annoying co-workers or negatively impacting the bottom line.

"How can I help you be successful at work?" may seem like a simple question, but it is a powerful question that can be useful in situations that involve performance issues and psychological health concerns.

- It assumes that success at work is the goal.
- It requires the employee to consider what success at work means.
- It puts responsibility for success back with the employee.
- It puts the supervisor in the position of facilitator of success.
- It avoids focusing on the problem and instead focuses on solutions.
- It stays focused on work issues.

For this to be truly effective, it must be more than just a question. It should be the underlying intention of any supervisor to help an employee be successful at his/her job.

We need to help employees focus on a "solution"; not to contribute to the problem. This may seem challenging when the employee is experiencing a situation or condition that results in repetitively negative thoughts, such as clinical depression. Some of the words the supervisor can use to move the conversation from "problem" to the "solution" are:

- *"So, what would you like to do about that?"*
- *"What would make this better for you?"*
- *"Is there a healthier way to approach this for you?"*
- *"What is the outcome you would like to see?"*
- *"What could you do differently so that there is a different result?"*

The point here is to redirect the conversation from complaints or expressions of negativity, to realistic solutions.

Rather than pointing out what an employee is doing wrong, state what outcome, result, or behaviour is needed from the employee on the job. Be as specific as possible. Some tips are:

Avoid . . .	Instead, Try . . .
You are always "negative" in meetings.	"I want the meetings to be positive and productive for everyone." *What needs to be done for that to happen?*
You "snap" at people.	Respect in the workplace includes being able to respond to others in a helpful and professional manner. *What do we need to do so that people are not so pressured that this becomes difficult or impossible?*

Avoid ...	Instead, Try ...
You missed another deadline.	*Tell me what needs to be different to ensure that we can have our projects come in on time?*

In all of the above examples, the employee may first respond with "I don't know" or "it's someone else's fault". Redirect the conversation back to what the employee thinks the solution should be, or what the employee would do if "in charge". Once an idea is offered by the employee, begin coaching the employee on how this suggestion could be implemented and the outcome measured, as well as the employee's commitment to make that suggestion work.

Collaborate on Goals

Working with the employee to develop a realistic, achievable goals is the key to successful performance management. It is particularly important when psychological health issues are involved. Collaborating with the employee to develop goals, ensures their commitment to resolving the issue and their loyalty to the organization. If the employee has participated in setting the goals and is still unable to achieve them, it may motivate the employee to seek treatment or support if needed. In contrast, when the employer arbitrarily sets the goals and the employee fails to achieve them, this may only serve to confirm to the employee that the employer is over demanding or that their expectations are unreasonable.

This collaboration must be genuine. Ask the employee what they will do to meet the performance expectations and help to brainstorm solutions. Make sure the goals are the ones set together and agree on the completion time frames. If you set goals with, rather than for your employee, you can improve their commitment to the process.

Follow Up

Avoid letting the process end there. The employee must realize that you are committed to long-term performance solutions. Create goals that are measurable and ask the employee to do as much self-assessment as possible. Monitor the situation. Follow up on a regular basis. Have a clear way to evaluate progress and provide clarity around the consequences if the performance issues persist. Continue to be clear about your expectations. That means encouraging the employee when goals are achieved, but also making them accountable when they are not.

Agree on Next Steps

The need for collaboration continues throughout the performance management process. If the jointly set goals are not met, both parties should be involved in determining the response: that is, how the results will be documented and what the next steps will be.

It does not mean that managers should not act. Managers must hold the employee accountable for their performance and productivity. This may seem like "tough-love" towards someone who is possibly ill, by not recognizing or acknowledging their situation. In fact, it may be the impetus required for the employee to gain insight into their situation and to finally seek help.

While collaborating with employees when performance issues arise may seem difficult to managers, it actually makes the performance management process easier and particularly when there are challenges such as psychological health issues. Where the process has been truly collaborative, employees should recognize that the treatment they have received is fair. The potentially confrontational situation can be effectively diffused by clear, open dialogue.

Getting your employees' "buy-in" to a collaborative process makes performance management easier. This is true of every performance management process but is especially true when there may be psychological health concerns.

3. The Trade Union Role

In unionized workplaces, the union is the employees' certified bargaining agent. The union has a legal responsibility to protect and support employees. Over the years, the labour movement has worked to achieve safe working conditions. But the focus has primarily been on physical health and safety. Today, many unions have made great strides to support psychological health and safety in the workplace. They have engaged in training, awareness, and bargaining that includes clauses intended to have a positive impact on workplace psychological health or to support workers with psychological health disabilities.

This move by the unions is a very positive one. Of course, there will be times when the union and Management hold different perspectives and objectives. In the case of a psychologically healthy and safe workplace, however, like other OH&S concerns, the best approach is a joint one. Conflict itself can negatively impact psychological health and safety in the workplace, while collaboration and cooperation can act as protective factors. As such, it is ideal for a union to be actively engaged in creating a psychologically healthy and safe workplace.

From an individual perspective, it is also important that union stewards or representatives be supported and skilled in addressing and abating issues of psychological health and safety. These men and women are often called upon to intervene in the most complex of situations, sometimes at a cost to their own psychological well-being. Some of the training for union representatives that may be considered includes:

- Awareness of their own emotional responses and stressors;
- Recognition of psychological problems and illness;
- Learning of and how to access both organizational and community resources;
- Suicide prevention;
- Identification of hazards to psychological health and safety;

CH. 24: PSYCHOLOGICAL HEALTH AND SAFETY: PRACTICAL APPLICATION IN THE WORKPLACE

- Prevention of risk to psychological health and safety;
- Conflict resolution;
- Accommodation and return-to-work processes;
- Communication, especially at times of crisis;
- Transformational mediation techniques; and
- Non-violent communication.

There are many sources of information that help including the extensive list of accommodation ideas where psychological health is a factor, video resources to help improve understanding of psychological health issues at work, and employee resources that they can recommend. All of these free resources are available online (refer to Chapter 23, "Psychological Health and Safety in the Workplace", section F "Available Community Resources").

4. Role of Joint Health and Safety Committee

Another way in which unionized environments can contribute to psychological health and safety is through the Joint Health and Safety Committee (JHSC). The role of the JHSC does not change when it comes to psychological health and safety in the workplace. It is only necessary that they recognize these elements in their work.

In the previous chapter, the resource Guarding Minds @ Work™ was discussed. The JHSC could use this resource and their organizational audit strategies as their first step in identifying psychological health and safety concerns. Further investigation or gathering of data may be necessary depending on the audit outcomes, but in many cases, the resource itself could be sufficient.

In terms of developing a resolution to identified risks, again, the Guarding Minds @ Work™ program's suggested interventions have evidence of efficacy in reducing specified risks. This approach would be integrated into a process of continual improvement where risks are identified on a reasonably regular basis with interventions as required.

5. The Occupational Safety Professional/Practitioner Role

The Occupational Safety Professional/Practitioner should be aware of the National Standard concerning psychologically healthy and safe workplaces and support its implementation. Hazard identification and loss control need to expand to include the ways in which work is assigned and individuals interact within the work environment. Being able to spot potential for conflict, violence, or harassment and developing strategies to prevent them will be important roles for the Occupational Safety Professional/Practitioner to take on.

6. The Employee Role

From a legal perspective, employees while engaged in an occupation, must take reasonable care to protect their own health and safety, as well as that of their co-workers and the public. Employees should cooperate with their employer for the purpose of protecting the health and safety of those involved in the organization — this includes psychological health and safety. But even within a psychologically healthy and safe workplace, some employees will experience a psychological illness. For this reason, it is important to recognize and understand the difference between having a "bad day" and a psychological illness. Further, it is valuable to understand the pathway to recovery.

Individuals who have experienced illnesses such as depression, anxiety, bipolar disorders, and post-traumatic stress disorder, have shared that the journey from illness to wellness can be a "rocky road". Six themes run through many of their stories, namely that:

Theme 1. Illness often creeps up on you

Psychological illness "creeps up on you" unannounced and camouflaged as life circumstances. So, often there are physical signs that something that is not right — aches, pains, fatigue, lack of focus, but these are dismissed as just minor irritations. They are usually accompanied by stressful events. Sometimes these are significant traumas such as a relationship break-up, vehicle accident, serious illness, or loss of a loved one. Other times it is a series of events, one after another, with no time to recover balance. It is rare to be well one day and seriously ill the next. It is a gradual process, that in retrospect, they wish they had paid attention to earlier instead of telling themselves to *"suck it up and move on"*. Maybe if we learned to be a little kinder to ourselves, we could begin the healing process earlier.

Theme 2. Describing it while you are living it can be difficult

Most people are at a loss for words to describe what they are feeling when it first begins. They describe the physical symptoms and may get treated for them. They describe the life circumstances and may get some support with that. But most people seem to miss describing the ongoing negative or fearful thoughts that keep them distracted, impair their sleep, and invade their concentration. Of course, once well, many of these same individuals become articulate and knowledgeable about these earlier signs and can share this wisdom with others who may be going through something similar.

Theme 3. Finding the right treatment takes effort

It is rare for someone to get effective treatment the first time they ask for help. Many speak of being relentless in pursuit of what will make them well again. Searching for a path to recovery is in itself, a sign of strength. As one man put it, it takes more courage to reach out for help. An attitude that says, *"I am determined to find a way to feel better"* can exist even in the depths of depression or despair. While some begin their search with medications, others try talk-therapy or relaxation methods. There definitely does not seem to be just one approach to wellness; it really must be that whatever works is the right approach for that person. Being relentless in the search is the key.

Theme 4. Recovery is a process that takes time

From the time the *"right"* treatment is found until the employee is feeling well again, takes time. While a few people report remarkable overnight changes due to medication or other treatment, most report a gradual return to wellness over many weeks or months. During this time, the change can be almost imperceptible until one day the individual realizes they have not had a bad day for weeks. Most people do not *"become themselves"* again, but rather transform into a new person. Often, like after many life-changing events, people emerge with more wisdom, awareness, and insight into both their own lives and the lives of those around them. Many, report having a different perspective and appreciation for life.

Theme 5. There can be worry about relapse
Many people simultaneously embrace their newly found wellness while having a sense of dread about a reoccurrence. Even though they manage to focus on new opportunities and possibilities, that niggling worry stays with them for many years. One person said, *"I can't say I have never had a down day [since I recovered from depression], but they are not as serious, I know what to do about them and they do not last as long."* Like other chronic illnesses, it takes a commitment to lifestyle change and vigilance to stay well.

Theme 6. You should never give up
People who feel despair over ever getting better, wish their misery would end; yet they also tend to feel that it will never do so. Although they acknowledge that others have recovered, they view those outcomes as exceptions, not the rule. The message to them is *"never give up"*. Keep going for another minute, another hour or day or week or month or even year. When the veil of depression lifts or the smothering effect of anxiety is removed, the hope of a new day or a new life brings with it the realization that giving up would have meant missing out on some of their life's most glorious moments.

Thanks, is extended to the hundreds of intelligent, dynamic people who have reclaimed their well-being and are caring enough to share these insights with others. Many of these themes are included in the Workplace Strategies for Mental Health (compliments of Canada Life), formerly known as the Great West Life Centre for Mental Health in the Workplace website.[17]

For the OH&S and Disability Management Practitioners/Professionals, this information can be used to provide hope, validation, and guidance for those who appear to be struggling or resigned to feeling unwell. By understanding the experience of others, you can offer resources and support that has the potential of reducing workplace problems and improving recovery.

7. The Co-worker Role: Helping the Troubled Employee

Many forward-looking employers, in response to an emerging legal duty to provide

---

[17] Workplace Strategies for Mental Health, formerly known as the Great West Life Centre Online: https://www.workplacestrategiesformentalhealth.com/ (date accessed: February 28, 2020).Great West Life Centre for Mental Health in the Workplace, online: http://www.workplacestrategiesformentalhealth.com/wti/Home.aspx.

and maintain a physically and psychologically healthy and safe workplace, are establishing policies and procedures to attain that goal. However, recognizing and responding to individuals whose psychological health may be "at risk" is often left to supervisors and co-workers by default. How can they be supported to do this? While there could be any number of explanations for the change in a co-worker's mood, concentration, or demeanour, early intervention can reduce the chances that the current problem will become chronic or escalate. Like all health problems, it is not the responsibility of co-workers to deal with these issues. However, if someone wants to reach out or support another, they must know that each situation comes with unique elements. Here are some possible options for co-workers to consider:

- **Explore the situation**
  - Tell your co-worker that they do not seem to be him/herself lately, and specifically state what you see. *"You don't look as well as you usually do. You seem upset and distracted. Are you feeling okay?"*
  - Resist making any judgements or conclusions about what is going on. Instead, invite your co-worker to talk about what he/she is experiencing. When he/she is done, repeat what you heard, and ask him/her if that is correct.
  - Resist giving him/her advice about what to do. Instead, continue to listen and ask what you can do to help. There are two reasons for this approach. The first is that you avoid giving the wrong advice or unwanted advice, which could have unintended consequences. The second is that you are able to help your co-worker focus on what it is he/she needs. When any of us are consumed by negative or fearful thoughts, we can lose sight of what we need to move beyond them.
- **Encourage action**
  - Try to help your co-worker preserve workplace relationships and their reputation at work. This can include helping him/her to avoid unnecessary conflict or acting out when he/she is not well.
  - Encourage your co-worker to take the work breaks that are provided to go for a walk, or out for fresh air. These changes in focus and physical movement can ultimately help them to increase concentration at work.
  - Help your co-worker focus on one small step forward at a time. Trying to "fix" everything at once is overwhelming.
  - If your co-worker is overwhelmed with work, encourage him/her to write down all of the tasks that he/she is currently doing and if this seems overwhelming to him/her, consider offering to help. Encourage him/her to take the list to the manager to help them prioritize those tasks that are most important.
  - Help your co-worker to focus on solutions rather than problems. If

worried or upset about something or someone, ask what they would like to do about that, and if unsure, offer some ideas; but make sure he/she chooses their own path forward. If your co-worker is having trouble at work, look at the list of accommodation ideas on the Workplace Strategies for Mental Health website[18] and see if the suggestions for helping with concentration, stamina, workplace relationships, *etc.*, may be useful.

- **Seek additional support**
  - Look through your employee group benefits plan to see if there are any services that may be helpful such as psychological services, massage therapy, acupuncture, *etc.*
  - Suggest your co-worker speak to a healthcare professional, such as an employee health resource or family physician.
  - Based on what they say they are experiencing, look up resources in the community, online, and at the workplace. Share these with them and ask if they need anything further.
  - If your co-worker is not sure what he/she is experiencing, have them look at *Working Through It*[19] to see if any of the approaches or situations described there are helpful.
  - If your co-worker is concerned that it may be more than stress, have him/her complete, *Check Up From the Neck Up* [20] and take the results to the family doctor.
  - If your co-worker is having personal or financial issues, direct him/her to the EAP, or other community resources that help with these situations.
- **Look after yourself at the same time**
  - Remember that you are not a therapist. Refer instead to appropriate resources and just continue to be a concerned co-worker who is there for support. Do not allow your days to become filled up with discussion about problems. Help your co-worker to focus on solutions for the workplace issues.

---

[18] Workplace Strategies for Mental Health, formerly known as the Great West Life Centre Online: https://www.workplacestrategiesformentalhealth.com/ (date accessed: February 28, 2020).

[19] Workplace Strategies for Mental Health. "Working Through It", online: https://www.workplacestrategiesformentalhealth.com/employee-resources/working-through-it (date accessed: February 28, 2020).

[20] Mood Disorders Association of Ontario. "Check Up From the Neck Up", online: http://www.mooddisorders.ca/program/check-up-from-the-neck-up (date accessed: February 28, 2020).

- It is honourable that you want to help your co-worker. Remember, however, to protect your own health and well-being at the same time.

In essence, co-workers can contribute to the success/failure of a Stay-at-work program or graduated return-to-work placement.[21] It is up to Management to broker positive co-worker support by explaining what role the co-worker can play and how those efforts can promote a successful return-to-work outcome for the employee.[22]

8. The Occupational Health Nurse (OHN) Role

The occupational health nurse (OHN) can play a valuable role when a supervisor or manager has reservations about the returning employee and their ability to function in the workplace. Often, the OHN is the most qualified and available stakeholder to provide integrated OH&S services to both supervisors and employees.

One of the challenges that can occur however is that in the majority of workplaces, the OHN is in a staff position with limited to no "positional authority". As such, the supervisor may not consider turning to the OHN for advice and strategies to help address his/her needs as well as the needs of the returning employee. Many OHNs report that during this time, they feel their role becomes more focused on the administration of policies rather than on working with managers and supervisors to help employees maintain, promote, and restore health, safety, and well-being.

(a) Promote Training for Understanding

In a recent workplace situation, the OHN appeared to be suitably empowered at the beginning of a return-to-work situation for an employee who had experienced a psychological health issue. Jan, an experienced OHN, clearly understood her role in the return-to-work and accommodation process. I, Mary Ann,[23] had been brought onboard in a consulting capacity to assist Jan and was immediately impressed by her knowledge and professionalism in her interactions with the company's vice-president and Human Resources professionals. Her actions were consistent with her training, expertise, and abilities in all areas of Occupational Health and Safety.

When Jan and I met alone to discuss the return-to-work plan in more detail, she admitted she was close to resigning from her position with the company. Despite her considerable experience, Management was not wholly supportive of her efforts to help foster and support a psychologically healthy and safe workplace. Jan cited several instances where Management's actions were counterproductive to the

---

[21] Institute for Work and Health, "The undeclared stakeholders: Recognizing the role of co-workers in return to work", At Work, Issue 71, Winter 2013, online: http://www.iwh.on.ca/at-work/71/the-undeclared-stakeholders-recognizing-the-role-of-co-workers-in-return-to-work (date accessed: February 28, 2020).

[22] A. Tjulin, E. MacEachen & K. Ekberg, "Co-workers play important, but sometimes 'invisible' role in RTW" (2009) 20:3 Journal of Occupational Rehabilitation 311.

[23] Referring to Mary Ann Baynton.

process and specifically, to employees' needs. She was left feeling that "her hands were tied" and that despite all of her training, her role was primarily to "police" policy compliance and chase after employees to ensure they filled out the correct forms — actions that felt counterintuitive to her goal of helping employees whose illness made it difficult for them to comply with rules and regulations.

Jan concluded that some of the Senior Managers were not adequately trained to handle the specific needs of employees experiencing psychological health issues in the workplace. Our discussions revealed that the Management Team had never received any formal training on managing workplace psychological health and safety issues, and their reactions were consistent with their lack of knowledge and understanding. They had little idea of what they could do to foster a psychologically healthier and safer workplace, or how to help employees have a successful return-to-work experience following psychological disability.

Jan scheduled a training session with Senior Management that not only looked at the needs of the employees, but also acknowledged the stressors that are commonly experienced by managers in these situations. Jan wanted the organization's leaders to understand the emotional cost for those responsible for managing workplace psychological health issues, and the business case for making all aspects of health and wellness a priority. The training included practical strategies to workplace psychological health and safety that managers could immediately implement with their staff. Based on past history, Jan was not optimistic about the results. In the end however, she was so impressed by the fact that Senior Management wholeheartedly embraced the new strategies that she decided to stay on with the organization. When she followed up with me, she said, "*I guess it is true that when we know better, we do better.*"

(b) Consider all the Possibilities

In another situation, a Manager had reported that an employee was harassing her to the point that she felt personally threatened. The Manager confided in the Human Resources professional that she was nervous in dealing with this employee on any issue. Something wasn't adding up and the Human Resources professional consulted with the organization's OHN. Upon further investigation, it was determined that the Manager, herself, was experiencing psychological health issues — including paranoia. This condition was causing the Manager to perceive that the employee in question was "out to get her". The OHN was able to encourage the Manager to seek medical attention. Once the Manager learned her diagnosis, the OHN was able to provide her with the supports that she needed to function successfully in her role.

(c) Medicalizing Performance Issues

Another OHN was concerned about what she called the practice of "medicalizing performance issues". In these situations, employees were sent to the OHN because Managers didn't know what to do with them and didn't see any connection between the employees' lack of productivity and the current Management approaches. The OHN discovered that many of these employees lacked the knowledge, skills, or resources to properly do their jobs. She began coaching the employees to ask their Managers for the resources and support needed to enable them to be successful at work.

(d) The Rule-out-Rule

In yet another situation, the problem was that a supervisor could not identify when employees were unwell. In such cases, the supervisor would initiate disciplinary procedures when it was apparent to the OHN that the employee was struggling with a psychological health disorder.

In order to address this particular type of scenario, the Global Business and Economic Roundtable on Mental Health and Addiction introduced a concept referred to as the *Rule-out-Rule*.[24] What this means is that when an employee's attendance and/or performance becomes problematic, possible medical reasons for the problem should be considered by the supervisor prior to taking disciplinary action. Often, when an employee experiences a disruption in their psychological health, changes in behaviour, personal appearance, and work performance occur. Examples of these could include tardiness, missed deadlines, interpersonal conflict, and multiple mistakes.

In this scenario, the OHN taught supervisors that they might be observing signs and symptoms of mood disorders such as anxiety or depression. Prior to invoking progressive discipline, these supervisors were encouraged to share their observations and concerns with employees — specifically, that they were concerned about the noticed changes in behaviour, and work performance. The supervisors were taught to share specific, concrete examples, and to use objective language that avoided judgement or blame.

(e) Working for the Best Results

Some OHNs work diligently to help employees develop return-to-work plans that address their psychological health concerns, only to have the plans fall apart when employees are exposed to work stressors. Regardless of who creates the return-to-work plan, it often rests with the supervisor to ensure its successful implementation. For supervisors, this means reconciling the demands of the Operation, the co-workers, the returning employee, and their own Management responsibilities. For this reason, many OHNs view the supervisor as a crucial player in the successful return-to-work process.

Some strategies for garnering supervisor involvement in implementing a successful return-to-work plan include:

- Developing Senior Management support for supervisors to take part in facilitating successful work accommodations.
- Advocating for adequate training for supervisors in working with distressed

---

[24] Refer to Appendix 1 for details on the **Rule-out-Rule**.

employees, recognizing signs and symptoms of psychological illness in the workplace, and finding solutions that maintain productivity.
- Recommending and/or providing education so that supervisors are more self-aware in terms of their reactions to emotional behaviours such as crying, anger, frustration, and the "silent treatment". This self-awareness allows for more effective responses in times of stress.
- Involving supervisors, from the beginning, in exploring accommodation strategies that take into account the realities of the work environment and group dynamics.
- Ensuring that supervisors have support in dealing with stressful issues. Many EAPs provide that support by phone. Supervisors can be made aware of, and encouraged to access, assistance available to them through Human Resources services, Occupational Health services, or mentorship.
- Arranging for the supervisor and the employee to get together to discuss the return-to-work plan before the commencement of the return-to-work experience. This discussion can include how the plan will be implemented, if there are any anticipated challenges and the available supports.

The role of the OHN as the purveyor of health and wellness in workplaces across Canada is more important than ever before. The good news is that a growing number of resources are being developed and made publicly available to assist Canadian workplaces. Initiatives like Workplace Strategies for Mental Health are continually being advanced to respond to the changing workplace needs.

9. The Role of Employee Assistance Program

Over 91% of Canadian organizations have EAPs; however, these programs have traditionally been involved in reactive approaches to dealing with employee personal and workplace problems. Over time, some employers began viewing EAPs as the "answer to all employee problems". The result is that employees with job performance issues are asked to call the EAP rather than to work with their supervisor to find a solution. Employees who appear to be struggling with psychological health or personal issues are also asked to call the EAP. The problem is that most EAPs do not provide the services needed to diagnose or treat psychological illness. This means for the employee, further delay in accessing effective treatment, or in understanding the illness. The outcome can be a worsening of symptoms.

While it is important to put EAP services into perspective, there is a growing realization that by proactively addressing psychological health issues in the workplace, employee and organizational well-being can be enhanced. EAPs have the potential to assist organizations to move from reactively to proactively approaching psychological health and safety in the workplace. The following are approaches that EAPs may be able to assist organizations to adopt:

- **Organizational Intervention: A Systems Perspective**

From a systems perspective, EAPs can help organizations/companies establish work environments that support human performance and behaviours, as opposed to "stressing them out". It involves providing responsible leadership; work and people management systems that support human performance; clarity on workplace roles, responsibilities, and expectations; and open communication between Management and labour. The concept of maximizing human performance using a systems approach is worth adopting.[25]

Responsible workplace psychological healthy and safe leadership includes:

- a clear commitment towards workplace health and safety;
- labour-Management engagement in good psychological health and safety practices;
- a supportive workplace culture and support;
- Management accountability and incentives; and
- Management and employee education on good psychological health and safety practices and the negative effects of social stigma associated with psychological illness.[26]

From a Management perspective, the "high-risk" practices include:

- setting unreasonable work demands and timelines, *e.g.*, "*Do it at any cost*" philosophy;
- over-controlling work activities, *e.g.*, employees not allowed discretion over how work is done;
- not acknowledging employee contributions and achievements;
- operating a high-paced workplace;
- allowing unclear organization/company vision, direction, and policies to exist;
- creating perpetual doubt within the workplace;
- permitting workplace mistrust and poor employee morale to exist;
- failing to address substandard Management practices and employee performance; and

---

[25] Refer to the work by T. Roithmayr and the Performance Maximizer Model.

[26] Global Business and Economic Roundtable on Addiction and Mental Health, Roadmap to Mental Health and Excellence At Work in Canada, Summer Draft (Presented to The Ontario Chamber of Commerce, Economic Summit on Mental Health and Productivity in Ontario, A Pilot for Canada, Sutton Place Hotel, Toronto, Ontario Canada, June 8, 2005) at 17, online: http://www.mentalhealthroundtable.ca/june_2005/RoadmapJune82005.pdf.

- not enabling two-way communication and feedback to occur, especially employee feedback on their concerns about the workload and demands.[27]

Human resource management theories and practices have a direct impact on workplace illness/injury, including employee psychological health and safety.[28]

- **High-Risk Occupations**

For organizations in which "high-risk" occupations such as nursing, medicine, police, firefighting, emergency medical care, military, and penal services are provided, EAPs can help Senior Management to work with the unions and employees to identify and address work stressors that negatively impact employee psychological health and safety. The work stressors of interest are employee exposures to abnormal human situations/conditions; confrontational situations; and personal threats to life. Given the nature of the work, these psychological health and safety hazards, which are associated with these "high-risk" occupations, are somewhat inevitable. Hence, employee supports for coping with such exposures should be instituted, *e.g.*, critical incident stress debriefing, increased "down-time", psychological supports, group support, *etc.*

- **Employee-Supervisor Relationship**

Another major stressor within a workplace is the employee-supervisor relationship. A good employee-supervisor relationship enables both parties to cope with many workplace situations and demands. However, poor Management/supervisory practices can have a detrimental effect on employee psychological health, safety well-being. The signs of workgroup distress include:

- disputes and low employee morale;
- increased employee presenteeism;
- decreased productivity;
- increased staff turnover; and
- increased grievances and complaints.

EAPs can assist Senior Management to understand, identify, and address the signs of workgroup distress.

---

[27] Global Business and Economic Roundtable on Addiction and Mental Health, Roadmap to Mental Health and Excellence At Work in Canada, Summer Draft (Presented to The Ontario Chamber of Commerce, Economic Summit on Mental Health and Productivity in Ontario, A Pilot for Canada, Sutton Place Hotel, Toronto, Ontario Canada, June 8, 2005) at 27, online: http://www.mentalhealthroundtable.ca/june_2005/RoadmapJune82005.pdf.

[28] See Chapter 18, "Occupational Health and Safety: Best Practices".

- **Recognition of the "Troubled" Manager/Employee**

    Early recognition of "high stress situations" and their outcomes is another prevention approach, albeit a secondary prevention approach for organizations to adopt. In terms of Middle Managers, EAPs can coach them on the signs and symptoms of the negative impact of work stress, namely:
    - increased irritability and impatience;
    - lack of concentration and an inability to stay focused;
    - avoidance and distancing from work situations and social interactions;
    - open frustration;
    - demonstration of "back stabbing" and/or passive aggressive actions;
    - long work days;
    - persistently late for meetings;
    - "working at home" to avoid the negative energy at work;
    - avoidance behaviours;
    - missing deadlines and commitments;
    - physical symptoms; and/or
    - periodic work absences.

    Likewise, the EAP can educate the workplace on the signs and symptoms of the "troubled employee", which are:
    - slumping performance at work;
    - poor timekeeping;
    - missed deadlines;
    - a negative change in appearance;
    - increased consumption of alcohol, tobacco, or caffeine;
    - frequent headaches or backaches;
    - social withdrawal;
    - poor judgement;
    - indecisiveness;
    - tiredness/fatigue; and/or
    - unusual displays of emotion.

- **Management Intervention**

    Senior Management, in consort with the EAP service provider, can make a difference through developing an awareness of the relationship between

Management personalities, workplace practices, and the promotion of positive psychological health among employees. Intervention can exist at a "systems level", "management-practices level", and "individual-employee level". Some constructive approaches are to:

- Make clear the organization's/company's commitment to psychological health, safety, and excellence at work;
- Incorporate psychological health and safety into organizational/company workplace systems;
- Make psychological health and safety practices part of the organization's/company's vision of a healthy and safe workplace;
- Evaluate the status of the organization's/company's psychological health and safety level of awareness, practices, and status;
- Develop policy objectives that support employee psychological health and safety well-being, namely:
  - education and training;
  - primary prevention;
  - secondary prevention; and
  - graduated return-to-work offerings;
- Target common Management stress-producing practices;
- Incorporate psychological health and safety into various workplace systems, such as human resources, OH&S programs, WWPs, and DMPs;
- Guide managers/supervisors to help them evaluate work stress levels — theirs and that of employees;
- Educate supervisors/managers on how to distinguish between developing medical conditions and work performance problems, for example:
  - educate supervisors and managers to ask questions of an employee which respect the employee's privacy and help the supervisor/manager to determine if a health consultation is needed before addressing the performance issues in conventional manner;
  - encourage the employee to consult a family physician, other health professional, or the EAP service provider;
  - support the employee's success on the job while they remain at work; and
  - defer any discussion on discipline until the completion of the

health review is complete.[29]

- Promote a shared responsibility by all stakeholders for employee psychological health and safety.

D. SUPPORTING THE BEREAVED EMPLOYEE

By Dan Steinke

Grief over the loss of someone close is not something new. Workers today are no more sensitive to the pain of a loss than were their parents or grandparents. They tend to remain silent while the griever tries to work through the grief, which results in increased lost time from work, decreased concentration on the job, and an escalation of healthcare costs.[30],[31]

When a tragedy or loss occurs, employers tend to initially respond well, but struggle with providing long-term support. "Grief is a long-term process and it's a normal reaction to loss, so it is going to extend for much longer than anybody thinks."[32]

Unresolved grief consumes tremendous amounts of human energy.[33] Grief is

---

[29] Global Business and Economic Roundtable on Addiction and Mental Health, Roadmap to Mental Health and Excellence At Work in Canada, Summer Draft (Presented to The Ontario Chamber of Commerce, Economic Summit on Mental Health and Productivity in Ontario, A Pilot for Canada, Sutton Place Hotel, Toronto, Ontario Canada, June 8, 2005) at 28, online: http://www.mentalhealthroundtable.ca/june_2005/RoadmapJune82005.pdf.

[30] M. Mancini, "Employees often grappling with grief" Canadian Safety Reporter (April 2012) at 1-2.

[31] E. Morrison, "Dealing with loss and grief at work" Canadian HR Reporter (October 18, 2010) at 22-23.

[32] Z. Pedersen, "Supporting staff during a workplace tragedy" Canadian Safety Reporter (October 2012) at 1-2.

[33] E. Morrison, "Dealing with loss and grief at work" Canadian HR Reporter (October 18, 2010) at 22-23.

repressed with the symptoms going untreated. Unresolved loss can be cumulative with increasingly negative outcomes. Unfortunately, the response to grief is often a pharmacological intervention. While therapeutic involvement is at times required, there are preventative options available. Successful organizations have found that approaching grief holistically and instituting Emotional Wellness Awareness training, along with a DMP, offers more long-term benefits than any other option.

A groundbreaking, 2003 survey conducted by the Grief Recovery Institute® measured the "hidden" costs of grief in American workplaces. Grief in the workplace cost USD $75 billion, with about USD $38 billion of those grief costs being related to death of a loved one (Figure 24.2). Likewise, Canadian workplaces have similar enormous costs associated with grief.

Figure 24.2: Cost of Grief in the Workplace[34]

Grief Incident	Cost to Business
• Death of Loved One	$37.5 billion
• Divorce	$11.1 billion
• Family Crisis	$9 billion
• Financial Loss	$4.5 billion
• Death (extended family, friends, colleagues)	$7.5 billion
• Major Lifestyle Alterations	$2.4 billion
• Pet Loss	$2.4 billion
• Other Losses	$1.2 billion
Total	$75.1 billion

Ironically, Canadian and U.S. workers receive more education about basic First Aid than they do about the impacts of death, divorce, and other emotional losses. In society and at the workplace, much is learned about how to acquire things; yet, very little accurate information is available on what to do when the loss of something/someone occurs.

> While grief is normal and natural, and clearly the most powerful of all emotions, it is also the most neglected and misunderstood experience, often by both the grievers and those around them.[35]

1. What Is Grief?

The death of someone close is one of life's most stressful events. Few people can readily cope with the pain of such a loss. Learning to cope is part of the grieving process:

---

[34] J.W. James & R. Friedman, The "Hidden" Annual Costs of Grief in America's Workplace 2003 Report (2003), The Grief Recovery Institute Educational Foundation, online: hidden-annual-costs-of-grief-in-americas-workplace-2003-report/oclc/166423828.

[35] *Ibid.*

Grief is conflicting emotions caused by the end of or a change in a familiar pattern of behaviour.[36]

People need to communicate their loss and share their experiences. Some people are eager to talk about their grief, while others prefer to listen; either way, the rewards of the communication process are immense.

Conflicting feelings arise for example when someone close dies after suffering a long illness. There can be feelings of relief that your loved one is at peace, while at the same time, there can be a very painful realization that never again will one see or touch that person. Conflicting emotions of relief and pain are very typical responses to death; but they create very intense and confusing emotions for the griever and those around them.

(a) Stages of Grief

The Canadian Mental Health Association (CMHA) acknowledges that there can be many stages to grief. They summarized the three stages that most people will experience as:

Stage 1.   Numbness or shock;

Stage 2.   Disorganization; and

Stage 3.   Re-organization.[37]

There are different opinions of the number of stages of grief; while many others are adamant that stages of grief do not exist. From personal experience, the school of thought that there are no actual stages to grief seems plausible. All relationships are unique, which makes each recovery from a loss an individual act.

While there appears to be no actual grief stages, many grievers do experience some very common responses. John W. James and Russell Friedman of the Grief Recovery Institute® listed some common responses:

- *Reduced concentration* – The preoccupation with the emotions of loss and an inability to concentrate seem to be universal responses to grief.
- *A sense of numbness* – This numbness can be physical, emotional, or both.
- *Disrupted sleep patterns* – Excessive or not enough sleep is common.
- *Changed eating habits* – Eating too much, or not at all.
- *Roller coaster of emotional energy* – Grievers often feel emotionally and physically drained.[38]

---

[36] *Ibid.*

[37] Canadian Mental Health Association, Grieving, online: http://www.cmha.ca/mental-health/your-mental-health/grief/ (date accessed: February 28, 2020).

[38] J.W. James & R. Friedman, The Grief Recovery Handbook, 2d ed. (rev.) (New York, NY: Harper Paper Backs, 1998).

The griever should never be "pigeon-holed" into a stage or a particular period of time for grieving. People vacillate between the above responses or miss some stages altogether. Being aware that these responses are normal and natural is an incredible asset to the griever during the grieving process.

(b) The New Normal

The Canadian Mental Health Association uses the term "New Normal".[39] Many grievers find this term to be psychologically correct. The single most wrong idea in society today is the concept that "time heals".[40] The "steam kettle" best describes what time does. Time only moves the steam kettle closer to boiling over. Time itself does not heal. What the griever does with the time, is what helps to deal with the loss.

A grieving person can often feel that "moving forward" is somehow a lack of respect for a loved one. To socialize again is unimaginable; and to laugh again, an improbability. Guilt and resentment can immediately step forward and hold the griever back. This regressive attitude, if allowed to fester, creates illness, dependency, and in severe cases, suicide.

The "New Normal" term best describes the journey on which a griever embarks. Nothing is going to make things the way they were. It is never going to happen. A "New Normal" is an accurate and intellectual target to a tragic incident. It allows the griever to accept the pain and make a positive emotional decision to move forward. To paraphrase Eleanor Roosevelt, no one can make you feel bad about yourself without your permission to do so.[41]

To accept a new normal in one's life, the griever has accepted that their life is different. Different is not disrespectful to the person that has died, or the person grieving. In actuality, moving forward is the complete opposite. To acknowledge that life has changed, the griever moves towards acceptance that a new behaviour is a positive healthy step forward and positive is healthy.

2. Workplace Wellness Programs and Employee Assistance Programs: Link with Grief Management

Wellness is holistic! Wellness approaches life from the perspective that all human beings, function at four levels: the physical, intellectual, spiritual, and emotional levels. For optimum health, we need to look at the whole person. Just picking out an aspect of wellness that you like or feel comfortable with, and ignoring others, does not work.

---

[39] Canadian Mental Health Association, Grieving, online: http://www.cmha.ca/mental-health/your-mental-health/grief/ (date accessed: February 28, 2020).

[40] J.W. James & R. Friedman, The Grief Recovery Handbook, 2d ed. (rev.) (New York, NY: Harper Paper Backs, 1998).

[41] Actual quote is, "[n]o one can make you feel inferior without your consent." E. Roosevelt, Quotes, online: http://www.brainyquote.com/quotes/authors/e/eleanor_roosevelt_2.html (date accessed: February 28, 2020).

Being healthy is more than just the "absence of disease".[42] Often if we do not see physical evidence of a person's illness, we assume that they are healthy. Tragically, when eventually the physical signs and symptoms become apparent, we ask, "[w]*hy didn't somebody do something?*"

(a) Negative Impact of Grieving on the Workplace

If someone is experiencing a grief/loss, the person obviously will have some health and functional symptoms. Sometimes temporary medical and pharmacological intervention is required. However, if the emotional state regresses, the stress starts to affect the person's blood pressure, heart, social interaction, *etc.* Eventually the person is unable to work. Depending on the ensuing illness, the person could very well be on their way towards the development of a serious illness and perhaps, death.

If instead, the same person had knowledge of emotional wellness, the common responses to grief, and the behaviour of the people around him/her was positive and supportive, he/she would be better able to cope. The result is movement towards health and total wellness.

Grief/loss is a multifaceted stressor. It influences most parts of a person's life including physiological, psychological, sociological, and spiritual areas. The emotional level addresses issues such as stress/distress and its effects on us and our social environment, and vice-versa.

Grieving employees may be dealing with a range of physical, psychological, and social symptoms, including:

- an inability to concentrate;
- lack of motivation, apathy, decreased productivity;
- impaired decision making;
- confusion, memory lapses;
- anxiety, crying, or other emotional responses;
- social withdrawal;
- high absenteeism or presenteeism; and
- injury rate.[43]

Grieving employees often become the classic examples of "presenteeism"; because, as is expected of them, they turn up for work and try to function. However,

---

[42] J. Travis, The Wellness Continuum (1972), online: The Wellspring http://www.thewellspring.com/wellspring/introduction-to-wellness/357/key-concept-1-the-illnesswellness-continuum.cfm (date accessed: February 28, 2020).

[43] Alberta Learning Information Service, Tip Sheets, "Grieving in the Workplace: Coping with Loss" (2008), online: http://alis.alberta.ca/ep/eps/tips/tips.html?EK=11611 (date accessed: February 28, 2020).

because of their emotional distress, these employees may:

- make poor decisions;
- supervise ineffectively; and
- compromise workplace safety.

(b) Impact of Grieving on the Workplace

As already noted, many organizations have instituted an EAP, DMP, and WWP. Each of these programs has proved beneficial to employees in time of need and has met the desires of both parties. However, are these programs utilized to their optimal potential?

People deal with grief and personal tragedies in their lives differently. Grief and recovery from loss do not always conveniently fit within EAPs because they usually focus on relationship and addiction issues.

WWPs recognize the importance of balancing life issues. They assess, document, and communicate the effects of stress and lifestyle. Unfortunately, these programs rarely make the connection with the hazards associated with human emotions.

EAP and WWPs have one thing in common; they move personal issues outside the walls of the workplace. Admittedly, therapeutic intervention is necessary at times, but successful companies have come to realize that emotional wellness is a "value-added" entity of the corporation; and their most valuable resource — their employees — is their true competitive advantage in today's market.

With a DMP in place that is properly linked to the EAP and WWP, emotional issues can be managed proactively and concurrently with all of the education, OH&S training, and communications within the workplace.

(c) Workplace Response

In response to the loss of an employee, or to an employee's grief, the workplace can:

- Inform co-workers of the loss and assist them to cope with the situation;
- Enable, if appropriate, co-worker attendance at the funeral;
- Attend the funeral;
- Cover the Operational activities during this period of time;
- Become familiar with the grieving process;
- Assist the grieving party(ies) to work through their grief;
- Assess the impact that the loss will have on the affected parties and the workplace;
- Seek suitable resources to assist those involved;
- Support those grieving, recognizing that it does take time; and

- Offer specialized support, if and when warranted.[44]

3. Conclusion

Grief from loss of a family member, friend, or co-worker can have a tremendous impact on employee well-being, effectiveness, and productivity. According to The Grief Index Survey in 2003, 85% of managers indicated that their decision-making was "very poor" after their loss; 90% of those with physical jobs reported a higher incidence of injuries.[45] In addition to the emotional toll it takes, grief is the underlying cause of billions of dollars in lost revenue to businesses, large and small.

The effects of an emotional loss are as real as any other OH&S or wellness issue in the workplace. Like all OH&S incidents, if left unmanaged it leads to injured people, illness, death, neglected well-being of employees, absenteeism, division within a group, and the integrity of the entire organization. Healthcare costs, productivity loss, and quality control issues are the sum total of the enormous expense associated with unresolved/unrecognized grief.

E. SUMMARY

Psychological health and safety, although implicitly required in every Canadian OH&S Act since their inception, still remains a challenge for many organizations to implement. But knowledge and resources like this publication, are accumulating in Canada to help workplaces respond. The main thing to remember is that knowledge and effective leadership remain the "keys" to the successful implementation of a physically and psychologically healthy and safe workplace.

---

[44] E. Morrison, "Dealing with loss and grief at work" Canadian HR Reporter (October 18, 2010) at 22-23.

[45] J. Zaslow, "New Index Aims to Calculate the Annual Cost of Despair", online: http://www.wsj.com/articles/SB103773937895627388 (date accessed: February 28, 2020).

Appendix 1

AN OPERATIONAL APPROACH TO PROMOTING GOOD PSYCHOLOGICAL HEALTH

Operationally, the Roadmap to Recovery promotes the use of The Rule-Out-Rule — a management tool. The Rule-Out-Rule Approach is a means of effective management of employee performance. It enables a discussion about an employee's ability to perform in the workplace, and "teases apart" the presence of a developing medical condition from nonmedically-related work performance issues as illustrated in Figure 24.3.

Figure 24.3: Rule-Out-Rule — Performance Management[46]

Step 1: As a result of observed changes in the employee's work performance, relationships, personal affect, energy levels, and other behavioural signs, the manager/supervisor plans to privately meet with the employee to identify concern for the employee's well-being and to express concern as well as to offer support and assistance.

Note: *To effectively verify if health problems are impacting the employee's work performance, management requires the knowledge and skills to address work performance problems.*

Step 2: The manager/supervisor privately meets with the employee to express care and concern about the employee's substandard work performance. The focus centres on work performance. However, the employee is encouraged to consult with his or her family physician or another health professional to rule out any related health issues. This step typically occurs when the employee is still functioning within the workplace.

Step 3: If the employee agrees to seek a medical assessment, the manager/supervisor defers from further discussion regarding work performance until the medical review has been completed and a fitness-to-work clearance has been received.

---

[46] Global Business and Economic Roundtable on Addiction and Mental Health, Roundtable Roadmap to Mental Health Disability Management in 2004-05 (Working document prepared June 25, 2004), online: http://www.mentalhealthroundtable.ca/june_2004/monitor_june2004.pdf.

If the employee refuses, then the focus returns to work performance and what the employee plans to do about addressing the identified issues.

There may be some workplace issues that must be addressed as part of the conditions for remaining at work. For example, employees who have threatened, bullied, or physically assaulted other employees need to be informed that to remain in the workplace, this type of behaviour will not be tolerated. Under the various provincial and federal Occupational Health & Safety Acts, employers have a "duty of care" to protect all workers. However, this can be done in a constructive manner by setting the parameters around acceptable work behaviours.

In essence, it is a management tool for addressing and identifying physical and psychological health issues in the workplace. Through regular performance management combined with empathetic communication between the employee and supervisor, the symptoms of physical and psychological health problems can be identified in a timely manner.

CHAPTER REFERENCES

Alberta Learning Information Service, Tip Sheets, "Grieving in the Workplace: Coping with Loss" (2008), online: http://alis.alberta.ca/ep/eps/tips/tips.html?EK=11611 (date accessed: February 28, 2020).

Canadian Mental Health Association, Grieving, online: http://www.cmha.ca/mental-health/your-mental-health/grief/ (date accessed: February 28, 2020).

Canadian Standards Association, online: http://www.csagroup.org/ca/en/services/codes-and-standards (date accessed: February 28, 2020).

Centre for Addiction and Mental Health (CAMH), "The numbers add up, so make your vote count", *camhconnexions* (Fall 2011) 11:3 at 1, online: http://www.camh.ca/en/hospital/about_camh/newsroom/connexions_newsletter/Documents/4515Connexions_Fall2011EN.pdf (date accessed: February 28, 2020).

A. Ferrari, F. Charlson, R. Norman, S. Patten, G. Freedman, *et al.*, "Burden of Depressive Disorders by Country, Sex, Age, and Year: Findings from the Global Burden of Disease Study 2010 (2013)", 10:11 PLoS Med: e1001547. doi:10.1371/journal.pmed.1001547.

Global Business and Economic Roundtable on Addiction and Mental Health, *Roadmap to Mental Health and Excellence At Work in Canada, Summer Draft* (Presented to The Ontario Chamber of Commerce, Economic Summit on Mental Health and Productivity in Ontario, A Pilot for Canada, Sutton Place Hotel, Toronto, Ontario Canada, June 8, 2005), online: http://www.mentalhealthroundtable.ca/june_2005/RoadmapJune82005.pdf (date accessed: February 28, 2020).

Global Business and Economic Roundtable on Addiction and Mental Health, *Roundtable Roadmap to Mental Health Disability Management in 2004-05* (Working document prepared June 25, 2004), online: http://www.mentalhealthroundtable.ca/june_2004/monitor_june2004.pdf (date accessed: February 28, 2020).

Guarding Minds at Work (2010), online: http://www.guardingmindsatwork.ca/info (date accessed: February 28, 2020).

Institute for Work and Health, "The undeclared stakeholders: Recognizing the role of co-workers in return to work", *At Work*, Issue 71, Winter 2013: online: http://www.iwh.on.ca/at-work/71/the-undeclared-stakeholders-recognizing-the-role-of-co-workers-in-return-to-work (date accessed: February 28, 2020).

J.W. James & R. Friedman, *The Grief Recovery Handbook*, 2d ed. (rev.) (New York, NY: Harper Paper Backs, 1998).

J.W. James & R. Friedman, *The "Hidden" Annual Costs of Grief in America's Workplace 2003 Report* (2003), The Grief Recovery Institute Educational Foundation, online: http://www.worldcat.org/title/grief-index-the-hidden-annual-costs-of-grief-in-americas-workplace-2003-report/oclc/166423828 (date accessed: February 28, 2020).

M. Mancini, "Employees often grappling with grief" *Canadian Safety Reporter* (April 2012) at 1-2.

A. Maslow, *Maslow's Hierarchy of Needs* (1943), online: http://en.wikipedia.org/wiki/Maslow's_hierarchy_of_needs (date accessed: February 28, 2020).

Mood Disorders Association of Ontario. "Check Up From the Neck Up", online: http://www.mooddisorders.ca/program/check-up-from-the-neck-up (date accessed: February 28, 2020).

E. Morrison, "Dealing with loss and grief at work" *Canadian HR Reporter* (October 18, 2010) at 22-23.

Z. Pedersen, "Supporting staff during a workplace tragedy" *Canadian Safety Reporter* (October 2012) at 1-2.

T. Roithmayr, Organizational Stressors Survey (2000), online: http://www.performance-bydesign.com/ (date accessed: February 28, 2020).

T. Roithmayr, *The Performance Maximizer®* (2000), online: http://www.performance-bydesign.com (date accessed: February 28, 2020)..

E. Roosevelt, Quotes, online: http://www.brainyquote.com/quotes/authors/e/eleanor_roosevelt_2.html (date accessed: February 28, 2020).

B. Schrank & M. Slade, Psychiatry Recovery *Psychiatric Bulletin* 31 (2007) 321-325.

A. Tjulin, E. MacEachen & K. Ekberg, "Co-workers play important, but sometimes 'invisible' role in RTW" (2009) 20:3 Journal of Occupational Rehabilitation 311.

J. Travis, The Wellness Continuum (1972), online: The Wellspring http://www.thewellspring.com/wellspring/introduction-to-wellness/357/key-concept-1-the-illnesswellness-continuum.cfm (date accessed: February 28, 2020).

WHO, "Depression" (2010), online: World Health Organization, http://www.who.int/mental_health/management/depression/definition/en/ (date accessed: February 28, 2020).

Workplace Strategies for Mental Health, formerly, Great-West Life Centre for Mental Health in the Workplace, online: https://www.workplacestrategiesformentalhealth.com/ (date accessed: February 28, 2020).

Workplace Strategies for Mental Health. "Working Through It", online: https://www.workplacestrategiesformentalhealth.com/employee-resources/working-through-it (date accessed: February 28, 2020).

J. Zaslow, "New Index Aims to Calculate the Annual Cost of Despair", online: http://www.wsj.com/articles/SB103773937895627388 (date accessed: February 28, 2020).

Chapter 25

PSYCHOLOGICAL HEALTH AND SAFETY IN THE WORKPLACE: MEASUREMENT

By Mary Ann Baynton, Dianne Dyck and Tony Roithmayr

A. INTRODUCTION

In the previous two chapters, the concept of Psychological Health and Safety in the Workplace was introduced and applied. In this chapter, the focus is on how to measure psychological health and safety in the workplace.

B. PSYCHOLOGICAL HEALTH AND SAFETY: ASSESSMENT

The first question employers tend to ask is, "[w]here do I start?" followed by, "[w]here do I invest my limited time and resources?"

The employer's approach to psychological health and safety should be similar to that of dealing with other health and safety risk management processes. First, it is necessary to identify the factors that may pose risk to employee psychological health and safety in the workplace. Guarding Minds @ Work™ is a free public resource available to all Canadian workplaces that has identified thirteen (13) factors which have the potential to impact the psychological health and safety of employees.

The psychosocial factors (PF) identified and detailed by Guarding Minds at Work™ are:[1]

1. **Psychological Support**: Psychological Support is present in a work environment where co-workers and supervisors are *supportive of employees' psychological health concerns, and respond appropriately as needed.* Equally important are the employees' perceptions and awareness of organizational support. When employees perceive organizational support, it means they believe their organization values their contributions, is committed to ensuring their psychological well-being and provides meaningful supports if this well-being is compromised.

2. **Organizational Culture**: Organizational culture is the degree to which *a*

---

[1] Guarding Minds @ Work, *Conducting a GM@W Organizational Review: Getting Started (2013)*, online: https://www.guardingmindsatwork.ca/assets/pdfs/Organizational_Review_Getting_Started.pdf (date accessed: February 28, 2020).

work environment is characterized by trust, honesty, and fairness. In general, organizational culture has been described as "a pattern of basic assumptions invented, discovered, or developed by a given group." These assumptions are a mix of values, beliefs, meanings, and expectations that group members hold in common and that they use as behavioural and problem-solving cues. The critical task is to determine which of these assumptions enhance the psychological health and safety of the workplace and the workforce.

3. **Clear Leadership and Expectations**: Clear leadership and expectations is present in an environment where there is *effective leadership and support that helps employees know what they need to do, know how their work contributes to the organization, and know whether there are impending changes.* There are many types of leadership, each of which impact psychosocial safety and health in different ways. The most widely accepted categorizations of leadership are instrumental, transactional, and transformational. Of these, transformational leadership is considered the most powerful. Instrumental leadership focuses primarily on producing outcomes, with little attention paid to the "big picture", the psychosocial dynamics within the organization, and, unfortunately, the individual employees. Transformational leaders are seen as change agents who motivate their followers to do more than what is expected. They are concerned with long-term objectives and transmit a sense of mission, vision, and purpose. They have charisma, give individualized consideration to their employees, stimulate intellectual capabilities in others, and inspire employees.

4. **Civility and Respect**: Civility and respect is present in a work environment where employees are *respectful and considerate in their interactions* with one another, as well as with customers, clients, and the public. Civility and respect are based on showing esteem, care, and consideration for others, and acknowledging their dignity.[2]

5. **Psychological Competence and Requirements**: Psychological competence and requirements is present in a work environment where there is a *good fit between employees' interpersonal and emotional competencies, their job skills, and the position* they hold. This means that employees not only possess the technical skills and knowledge for a particular position, but they also have the psychological skills and emotional intelligence to do the job. Emotional intelligence includes self-awareness, impulse control, zeal, persistence, self-motivation, empathy, and social deftness. Of note is the fact that a subjective job fit has been found to be more important than an objective job fit, meaning it is more important for employees to feel

---

[2] CCOHS, "Civility and Respect in the Workplace, Infographic" (2019), online: https://www.ccohs.ca/products/posters/civility/ (date accessed: February 28, 2020).

they fit their job, rather than being assessed and matched to the job.

6. **Growth and Development**: Growth and development is present in a work environment where employees receive *encouragement and support in the development of their interpersonal, emotional and job skills.* Such workplaces provide a range of internal and external opportunities for employees to build their repertoire of competencies, which will not only help with their current jobs, but will also prepare them for possible future positions.

7. **Recognition and Reward**: Recognition and reward is present in a work environment where there is *appropriate acknowledgement and appreciation of employees' efforts in a fair and timely manner.* This includes appropriate and regular financial compensation as well as employee or team celebrations, recognition of years served, and/or milestones reached.

8. **Involvement and Influence**: Involvement and influence is present in a work environment where *employees are included in discussions about how their work is done and how important decisions are made.* Opportunities for involvement can relate to an employee's specific job, the activities of a team or department, or issues involving the organization as a whole.

9. **Workload Management**: Workload management is present in a work environment where *tasks and responsibilities can be accomplished successfully within the time available.* This is the psychological risk factor that many working Canadians describe as being the biggest workplace stressor (*i.e.*, having too much to do and not enough time to do it). Research has demonstrated that it is not just the amount of work that makes a difference but also the extent to which employees have the resources (time, equipment, support) to do the work well.

10. **Engagement**: Employee engagement is present in a work environment where *employees feel connected to their work, and where they feel motivated to do their job well.* Employee engagement can be physical, emotional, and/or cognitive. Physical engagement is based on the amount of exertion an employee puts into his or her job. Physically engaged employees view work as a source of energy. Emotionally engaged employees have a positive job outlook and are passionate about their work. Cognitively engaged employees devote more attention to their work and are absorbed in their job. Whatever the source, engaged employees feel connected to their work because they can relate to, and are committed to, the overall success and mission of their company. Engagement is similar to, but should not be mistaken for job satisfaction, job involvement, organizational commitment, psychological empowerment, or intrinsic motivation.

11. **Balance**: Balance is present in a work environment where there is *recognition of the need for balance between the demands of work, family, and personal life.* This psychological factor reflects the fact that everyone

has multiple roles: as professors, parents, partners, *etc.* This complexity is enriching and allows fulfilment of individual strengths and responsibilities, but conflicting responsibilities can lead to role conflict or overload.

12. **Psychological Protection**: Psychological protection is present in a work environment where employees' *psychological safety is ensured*. Workplace psychological safety is demonstrated when workers feel able to put themselves on the line, *ask questions, seek feedback, report mistakes and problems, or propose a new idea without fearing negative consequences* to themselves, their jobs or their careers. A psychologically healthy and safe workplace is one that promotes psychological well-being and actively works to prevent harm to employee psychological health due to negligent, reckless, or intentional acts.

13. **Protection of Physical Safety**: Protection of Physical Safety as present in a work environment where management takes *appropriate action to protect the physical safety of employees.* Appropriate actions may include: policies to protect employees' physical safety; training in safety-related protocols; rapid and appropriate response to physical accidents or situations identified as risky; and clearly demonstrated concern for employees' physical safety.[3]

Note that the *National Standard of Canada on Psychological Health and Safety in the Workplace* also specifies a 14th factor, namely **Other Chronic Stressors as Identified by Workers**. This is to imply that the employer must go beyond a survey assessment to engage employees in a discussion about the stressors in the workplace. One resource created to assist with this discussion is *On The Agenda*,[4] which provides slide presentations and facilitator guides to support a conversation about workplace psychological health and safety.

Validity of the Guarding Minds @ Work™ Psychosocial Factors

In March and April 2009, and again in October 2012, Ipsos Reid was commissioned to conduct the largest public opinion survey to date looking at Psychosocial Factors in Canadian workplaces. The results of these surveys provide key examples of the impact those psychosocial factors explained above have on the psychological health and safety of working Canadians. Moreover, the results from these pivotal surveys provide benchmarks that Canadian employers can compare themselves to when assessing their level of workplace psychological health and safety. For example, employees who participated in the study largely reported a moderate level of overall risk; there were also some serious areas of concern. Approximately one in three employed Canadians fall into the categories of serious or significant concerns for

---

[3] CCOHS published an infographic on "Psychologically Health and Safe Workplaces" (2019), online: https://www.ccohs.ca/products/boutique/13factors/ (date accessed: February 28, 2020).

[4] Online: https://www.workplacestrategiesformentalhealth.com/free-training-and-tools/on-the-agenda.

each of the psychosocial factors. This suggests that while Canadian employers have several strengths in fostering employee psychological health, they also have areas of needed improvement in terms of further protecting their employees' psychological safety and well-being. Table 25.1 lists the national results by Psychosocial Factor (PF) from the 2012 survey.[5]

These are the findings of a poll commissioned by the Great-West Life Centre for Mental Health in the Workplace in the fall of 2012. This online survey of 4307 employed Canadian adults was conducted via the Ipsos I-Say Online Panel, Ipsos Reid's national online panel. The results of this poll are based on a sample where quota sampling and weighting were employed to balance demographics and ensure that the sample's composition reflects that of the actual employed Canadian population according to Census data. Quota samples with weighting from the Ipsos online panel provide results that are intended to approximate a probability sample. Statistical margins of error are not applicable to online polls, however an un-weighted probability sample of this size, with a 100% response rate, would have an estimated margin of error of +/- 1.2 percentage points, 19 times out of 20, had the entire population of employed adults in Canada been polled.

While every workplace assessment will reveal results unique to their circumstance and dynamic, this chart helps when considering how an organization compares to this snapshot of a national average.

Table 25.1: Ipsos-Reid 2012 Review: Validity of Guarding Minds @ Work™ Psychosocial Factors

Psychosocial Factor (PF)	Serious Concerns	Significant Concerns	Minimal Concerns	Relative Strengths
PSR1: Psychological Support	9%	27%	37%	27%
PSR2: Organizational Culture	9%	29%	37%	25%
PSR3: Clear Leadership & Expectations	7%	30%	35%	28%
PSR4: Civility & Respect	6%	24%	41%	29%
PSR5: Psychological Competencies & Requirements	3%	22%	42%	33%

---

[5] Guarding Minds @ Work (2012), *Ipsos-Reid 2012 Review: Validity of Guarding Minds @ Work™ Psychosocial Factors*, October, online: http://www.guardingmindsatwork.ca/info/risk_factors (date accessed: February 28, 2020).

Psychosocial Factor (PF)	Serious Concerns	Significant Concerns	Minimal Concerns	Relative Strengths
PSR6: Growth & Development	9%	28%	37%	25%
PSR7: Recognition & Reward	8%	27%	35%	30%
PSR8: Involvement & Influence	5%	22%	38%	34%
PSR9: Workload Management	4%	23%	42%	32%
PSR10: Engagement	2%	12%	38%	48%
PSR11: Balance	7%	25%	39%	29%
PSR12: Psychological Protection	8%	24%	38%	30%
PSR13: Protection of Physical Safety	5%	15%	41%	38%

Psychological Health and Safety: Impact (2016)

Based on the Ipsos Reid Survey, 2016,[6] the comparison between companies that have and have not implemented the Standard, indicates a significant improvement in psychological health and safety by companies that embraced it. To summarize the results, the survey indicates that more employees:

- view their workplaces to be psychologically healthier and safe;
- are aware of mental health conditions like depression; and
- miss fewer workdays due to depression (7.4 versus 12.5 days).

To access the entire comparison, refer to the Psychological Factor Index Score.[7]

---

[6] Ipsos (2016). "Workplaces that are Implementing the National Standard of Canada for Psychological Health and Safety in the Workplace Described by Employees as Psychologically-Safer Environments", Ipsos, online: https://www.ipsos.com/en-ca/news-polls/workplaces-implementing-national-standard-canada-psychological-health-and-safety-workplace (date accessed: January 31, 2020).

[7] See online: http://ipsos-na.com/images/news-polls/media/7312-table.png.

C. PSYCHOLOGICAL HEALTH AND SAFETY: WORKPLACE EVALUATION

When evaluating a workplace program or function, it is important to assess it in terms of its structure, process, and outcomes. Using this approach, an organization/company can evaluate the status of workplace psychological health and safety.

1. Structural Evaluation

Historically, organizations/companies did not establish a workplace system in which psychological health and safety is promoted and risks prevented. To address this situation, a concerted effort to produce a National Standard of Canada on Psychological Health and Safety in the Workplace has been developed. This Standard provides employers with a description of a Psychological Health and Safety Management System (PHSMS)[8] that, incidentally, approximates an Occupational Health and Safety Management System (OHSMS). That becomes the "ideal/desired state".

Program/function evaluation identifies the gaps between *the current state* and *the desired state* of a program, indicates whether the program/function goals/objectives are met or not, and enables improvements both along the way and periodically. Hence, to evaluate an organization's/company's approach to providing for psychological health and safety in its workplace, an audit of the "current state" can be conducted.

An audit is a systematic review of an organization's/company's goals, objectives, and targets; and/or against an established or predetermined protocol or standard. In this instance, the company would compare its PHSMS against the PHSMS as defined in the National Standard.

The elements to be examined as evident/not evident would include:

- Does a PHSMS exist? Is it continually improved?
- Is the PHSMS integrated into the organization's/company's governance practices? Into other systems?
- Is the organization's/company's commitment to the development of a systematic approach for managing psychological health and safety in the workplace evident?

---

[8] Canadian Standards Association, *National Standard of Canada: Psychological Health and Safety in the Workplace – CAN/CSA-Z1003-13/BNQ 9700-803/2013* (Ottawa, ON, 2013) at 1.

- Do the organization/company leaders promote and nurture psychological health and safety in the workplace by:
  - Supporting the development and maintenance of a PHSMS?
  - Establishing an action plan for continual improvement of psychological health and safety in the workplace?
  - Creating a supportive work culture?
  - Making sure that psychological health and safety in the workplace is factored into business decision making?
  - Raising employee awareness and passion for psychological health and safety in the workplace?
- Does the organization/company engage and encourage stakeholder participation in guarding and improving psychological health and safety in their workplace?
- Does the organization/company provide mechanisms for employee participation in promoting and continuously improving psychological health and safety in their workplace?
- Has the organization/company made provisions to safeguard the confidentiality of employee personal health information?
- Does the organization/company have a system in place for:
  - Planning for the management of psychological health and safety in their workplace?
  - Assessing the psychological health and safety in their workplace?
  - Continuous improvement of the psychological health and safety in their workplace?
- Does the organization/company regularly review and evaluate its psychological health and safety system?
- Does the organization/company have a system in place for identifying psychological health and safety risks?
- Does the organization/company have a system in place for assessing psychological health and safety risks?
- Does the organization/company have a system in place for managing psychological health and safety risks?
- Does the organization/company have a system in place for collecting data on the psychological health and safety status of the workplace and its employees?
- Does the organization/company demonstrate cultural competence?
- Has the organization/company documented its psychological health and safety goals, objectives, and targets?

- Is there an action plan in place for achieving those psychological health and safety goals, objectives, and targets?
- Does the organization/company have a system in place for managing change?
- Does the organization/company have an infrastructure in place for sustaining psychological health and safety in the workplace?
- Has the organization/company allocated adequate resources (personnel, money, and time) in place for sustaining psychological health and safety in the workplace?
- Are preventative and protective measures part of the organization's/company's plan for sustaining psychological health and safety in the workplace?
- Does the organization/company have a communication strategy in place for sustaining psychological health and safety in the workplace?
- Are sponsorship, engagement, and change management measures part of the organization's/company's plan for sustaining psychological health and safety in the workplace?
- Has the organization/company clearly stated the implementation expectations it has for sustaining psychological health and safety in the workplace?
- Does the organization/company have a system in place for establishing and maintaining stakeholder competence and training regarding psychological health and safety in the workplace?
- Does the organization/company have an emergency response system in place for addressing individual psychological illness/injury in the workplace?
- Does the organization/company have an emergency response system in place for addressing organizational/company disasters/traumatic events that could lead to psychological distress in the workplace?
- Does the organization/company have a system in place for reporting and investigating psychological health and safety incidents?
- Does the organization/company instruct external parties on the particulars of its psychological health and safety efforts in the workplace?
- Does the organization/company have a system for continuous improvement of the PHSMS in place?
- Does management regularly review the performance of the PHSMS?

Although this type of audit process appears daunting in its magnitude, it can easily be incorporated into the audit process that organizations/companies undertake to evaluate their OHSMS. Both evaluate the same elements: The OHSMS audit focuses primarily on the system required to maintain the physical health and safety of workers, while the PHSMS audit focuses primarily on the system required to maintain the psychological health and safety of workers.

For a more complete description of a PHSMS audit tool, refer to the *National Standard of Canada on Psychological Health and Safety in the Workplace* (2013), Table E.1, Sample Audit Tool.[9]

2. Process Evaluation

The process aspect of the PHSMS addresses how things operate or run within a system. Measurement of stakeholder experience and perception of the quality and effectiveness of the PHSMS can be achieved through the use of a perception survey. Employees and other stakeholders are well positioned to provide feedback on the quality and effectiveness of the organization's/company's efforts to provide a psychologically healthy and safe work environment.

Guarding Minds @ Work provides organizations/companies with the tools needed to undertake an evaluation on the status of its psychological health and safety efforts in the workplace.[10] From a process evaluation perspective, organizations/companies can initially and periodically:

- Review its status in terms of the 13 psychosocial risk factors (refer to Appendix 1);
- Scan employee perception of the organization's/company's level of psychological health and safety in their workplace (refer to Appendix 2); and/or
- Survey employees to gain their perception of the organization's/company's level of psychological health and safety in their workplace (refer to http://www.guardingmindsatwork.ca/docs/dashboard/assessment/GM@W_Survey.pdf).

By obtaining this type of information, the organization/company can monitor, and track its performance, making continuous improvements as it goes along. This is known as formative evaluation.

Process improvements are a critical part of process evaluation. As such, the information, recommendations, and tools provided by Guarding Minds @ Work can assist organizations/companies to make process changes. Refer to GM@W Action Resources.[11]

---

[9] Canadian Standards Association, *National Standard of Canada: Psychological Health and Safety in the Workplace - CAN/CSA-Z1003-13/BNQ 9700-803/2013* (Ottawa, ON: CSA Group, 2013) at Table E.1, online: http://shop.csa.ca/en/canada/landing-pages/z1003-psychological-health-and-safety-in-the-workplace/page/z1003-landing-page (date accessed: February 28, 2020); or https://www.mentalhealthcommission.ca/English/what-we-do/workplace/national-standard for a free copy (date accessed: February 28, 2020).

[10] GuardingMinds@Work, *GM@W Documents and Resources* (2013), online: http://www.guardingmindsatwork.ca/resources (date accessed: February 28, 2020).

[11] GuardingMinds@Work, *GM@W Documents and Resources* (2013), online: http://www.

Some of the limitations of the Guarding Minds @ Work approach are that:

1. The focus is primarily on psychological health and safety, as opposed to a holistic approach to worker health and safety that includes worker physical health and safety;
2. It assumes that organizations/companies have the internal resources/expertise to undertake such an assessment; and
3. Although the approach involves employees, it does not include a debriefing of the survey responses post-completion. As a result the improvement benefit of the survey may be sub-optimized because:
   - employees may be left wondering what is to happen next, and if management really wanted to hear what they had to say;
   - management may miss the opportunity to get clarification on the responses received; and
   - employees may/may not get included in the action planning for change.

An alternate, and yet complimentary, approach is offered by Orange Umbrella[12] — The Great Performance Model and Approach. As described in Chapter 22, "Toxic Work Environments: Impact on Employee Illness/Injury", the Performance Maximizer Model focuses on all the factors that shape optimal human performance, and the conditions that exist when successful human performance occurs. These are clustered into five "Conditions for Great Performance" — that is, *Know What to do, Able to do it, Equipped to do it, Want to do it*, and the necessary *Interactions*,[13] which are the factors that foster trust, respect, integrity, collaboration, and accountability within the workplace.

The central focus of the "Conditions for Great Performance" is enabling its employee performance; *i.e.*, what employees do in their work and the results they achieve.

The Great Safety Performance (GSP) approach also specifically gauges the degree to which employees demonstrate "respectful workplace" behaviours in their daily interactions — behaviour that is consistent with maintaining psychological health and safety at work.

The Conditions for Great Safety Performance are leading indicators of safety that

---

guardingmindsatwork.ca/resources (date accessed: February 28, 2020).

[12] Orange Umbrella® is an organization dedicated to helping organizations to understand their performance data and implement improvement strategies. The Orange Umbrella® tools evaluate how well an organization's system is working. Factors that lead to incidents are identified before incidents happen, allowing for timely intervention. See online: http://www.orangeumbrella.co/ (date accessed: February 28, 2020).

[13] *Interactions* were added to the model after the research reported in this chapter was completed.

have a predictive relationship to work performance outcomes. In the absence of these conditions, a variety of risk factors develop which cause problems with both physical and psychological health and safety, on-the-job performance, and ultimately, impact the organization's bottom line. The premise is that by focusing on the leading indicators for creating great human performance, lagging indicators such as injuries, work absence, staff turnover, disability-related costs, and reduced productivity will gradually decrease.

Holistic in its focus, this approach addresses the physical, psychological, and environmental factors that impact employee performance. It involves the administration of an employee perception survey, along with pre- and post-survey briefings. The pre-survey introduction explains the nature of the survey, along with what is going to be done with the results and the potential benefits for the employees. The post-survey briefing addresses the findings, allows for clarification of some of the results, invites suggestions for improvement, and explains the "next steps". It is at this point that employee engagement and involvement can be operationalized. Finally, the information and suggestions provided by Guarding Minds @ Work can be incorporated into the plan for a customized process improvement.

3. Outcome Evaluation

Outcome evaluation is designed to determine if the program/function has met its stated goal(s), objectives, and targets. Guarding Minds @ Work provides organizations/companies with the tools needed to undertake an outcome evaluation on the status of its psychological health and safety efforts in the workplace.[14] They can be used as part of the outcome evaluation.

Outcome results typically involve statistical tallying of what has occurred over a set period of time. For example, the organization would establish baseline measurement and then subsequent measurement of the same performance indices at future periods in time. These measurements could include:

- Number and rate of bullying incidents;
- Number and rate of harassment incidents;
- Number and rate of workplace violence incidents;
- Number and rate of psychological injury incidents;
- Number and rate of psychological illness incidents;
- Number and rate of psychological health claims;
- Total cost of psychological health claims;
- Average cost of psychological health claims;

---

[14] Guarding Minds @ Work, *GM@W Documents and Resources* (2013), online: http://www.guardingmindsatwork.ca/resources (date accessed: February 28, 2020).

- Average cost of psychological health claims per employee.[15]

Through a comparison with the baseline measures of these indices, the magnitude and direction of change over time can be determined. This will yield a trend analysis of the psychological health and safety status of the organization/company. Actions for improvement can then be developed, ideally in collaboration with the employee population. Once again, the Guarding Minds @ Work materials can be very useful in coming up with a feasible action plan.

D. SUMMARY

To restate, psychological health and safety, although implicitly required in every Canadian OH&S Act since inception, still remains a challenge for many Canadian organizations/companies. Hence, the role of Human Resources, OH&S, and Disability Management Professionals/Practitioners is one of leadership — leadership on how psychological health and safety can be qualitatively and quantifiably measured, so that organizations/companies can develop and sustain psychological health and safety in the workplace.

---

[15] Refer to Appendix 3 for details on the calculation of these measurement indices.

Appendix 1

GM@W ORGANIZATIONAL REVIEW[16]

Conducting an Organizational Review: Getting Started

Why do an Organizational Review?

An Organizational Review of existing policies and practices related to the protection of your employees' psychological health is the first step toward determining the extent to which your workplace is psychologically safe. *A psychologically healthy and safe workplace is one that promotes employees' psychological well-being and actively works to prevent harm to employee psychological health due to negligent, reckless or intentional acts.* The Organizational Review is designed to accompany the Survey, which provides input from your employees' perspectives. Completing both the Organizational Review and the Survey allows you to generate a Comparison Report that compares management and employee perspectives.

In order to generate the Comparison Report, you must enter the results from your Organizational Review Worksheets through your Guarding Minds at Work Dashboard, **before** you close your Survey.

Workplaces may differ in the language describing various roles and positions. Guarding Minds at Work uses the terms 'employee', 'staff', 'supervisor', 'management' and 'employer'. Please use the terms appropriate for your workplace when working with Guarding Minds at Work Resources.

What is involved in the GM@W Organizational Review process?

The Organizational Review process involves the completion of up to 13 *Organizational Review Worksheets*, each corresponding to one of the Psychosocial Factors.

Each *Organizational Review Worksheet* includes:

✓ Definition of the Psychosocial Factor
✓ Benefits of addressing the factor
✓ Information that would be helpful to refer to or collect
✓ Checklist of descriptors of your workplace
✓ Short employer questionnaire
✓ Overall rating

---

[16] Reprinted with permission from Guarding Minds @ Work, Conducting a GM@W Organizational Review: Getting Started (2013), updated with the recent branding, online: https://www.workplacestrategiesformentalhealth.com/pdf/GMAW_Organizational_Review.pdf (date accessed: February 28, 2020).

What do you need to complete the Organizational Review?

The following may help you complete the *Organizational Review Worksheets*.

- Documentation: Gather and refer to reports or documents with information relevant to employee and organizational functioning.
  - *e.g., rates of and reasons for absenteeism/disability, turnover rates, benefits utilization and costs*
- Organizational and/or Market-Specific Considerations: It is important to take into account current factors or trends that may be compromising the psychological health and safety of your workplace.
  - *e.g., lack of available skilled workers, aging workforce, changes in market conditions, pending merger*
- Employee Input: Information from your staff may be helpful.
  - *e.g., employee suggestions, health and safety committee reports, employee surveys*

Who completes the Organizational Review?

The Organizational Review may be completed by one or more individuals within the organization – this may vary depending on the size and nature of the organization. For example, the Organizational Review may be completed by the owner or manager within a small-sized business or work team. In a medium-sized business, the Organizational Review may be conducted by a designated human resources professional. In a large-sized business, there may be a range of potential individuals available to undertake the process (e.g., human resources professional, occupational health & safety representative, division/department head, regional manager).

You may wish to create a subcommittee or select a small group of staff to participate in the process. To heighten the objectivity of the process, you may also consider using an external consulting group with expertise in the Guarding Minds at Work process.

Where do you go from here?

Review the 13 *Organizational Review Worksheets* corresponding to the Psychosocial Factors to determine where you might begin addressing psychological health and safety issues. We recommend that you complete the *Organizational Review Worksheets* before, or in parallel with, administering the Survey to your employees.

How to select Psychosocial Factors for the Organizational Review

It can be difficult to determine where to begin. For this reason, we recommend that you prioritize the *Organizational Review Worksheets* you will complete. Focus on <u>one</u> Psychosocial Factor at a time. This helps to streamline your efforts and increases your likelihood of success.

Begin by completing *Organizational Review Worksheets* for:

(a) Factors that disproportionately impact your organization financially.

(b) Factors that are particularly relevant to changes occurring within your organization or work unit (e.g., if there have been recent changes in leadership, you may want to select the corresponding *Organizational Review Worksheet* for PF3: Clear Leadership & Expectations).

(c) Factors that are particularly relevant to key incidents or events (e.g., legal action, suicide).

We recommend that you eventually complete all of the *Organizational Review Worksheets* in order to provide the most thorough overview of your organization's psychological health and safety. After completing each *Organizational Review Worksheet*, enter your score into the associated *Action Planning Worksheet*.

Appendix 2

GM@W INITIAL SCAN[17]

Reading the Initial Scan Results

What is the Initial Scan?

The **Initial Scan** generates two scores, the **SSOS** and the **SSIX**.

SSOS means Stress Satisfaction Offset Score. As its name implies, the SSOS treats demand and effort as "stressors" (items 3 and 4 below), offset by the "satisfiers," control and reward (items 1 and 2 below). The appeal of this approach is that the resulting metric, expressed as single number ranging from +2 to -2, sketches an immediate and compelling picture of the workforce or a unit within it in terms of job stress as offset by job satisfaction. Positive scores point toward a psychologically safer environment while more negative scores suggest the existence of more psychologically risky working environments.

The most practical advantage of the "offset" approach is that the expression of a relationship between stress and satisfaction provides clear guidance with regard to action. There is a strong and common tendency for worksite consumers of SSOS reports to want to shift the balance between stress and satisfaction as far as possible away from stress and as far as possible toward satisfaction. So, almost no matter where the score lies, there is almost always some room for improvement, and often a great deal of room.

These are the items used in the Initial Scan

1. I am satisfied with the amount of involvement I have in decisions that affect my work. **(Control as a Satisfier)**
2. I feel I am well rewarded (in terms of praise and recognition) for the level of effort I put out for my job. **(Reward as a Satisfier)**
3. In the last six months, too much time pressure at work has caused me worry, "nerves" or stress. **(Demand as a Stressor)**
4. In the last six months, I have experienced worry, "nerves" or stress from mental fatigue at work. **(Effort as a Stressor)**
5. I am satisfied with the fairness and respect I receive on the job **(Fairness and Respect as Mediators)**
6. My supervisor supports me in getting my work done **(Supervisory Support as a Mediator)**

SSIX means Stress Satisfaction Index

The SSIX is the SSOS with two extra questions added (items 5 and 6 above).

The SSOS and the SSIX provide two perspectives on risk to psychological safety and health.

As noted above, by itself, the **SSOS** provides information on how employees perceive the basic conditions of their work in terms of *demand, control, effort* and *reward*. This is important information in itself and it is often sufficient for identifying high, medium and low risk zones in your workplace with regard to psychosocial conditions of work.

---

[17] Reprinted with permission from Guarding Minds @ Work, *GM@W Initial Scan*, 2013, updated with the recent branding, online: https://www.guardingmindsatwork.ca/assets/pdfs/Reading_Initial_Scan_Results.pdf (date accessed: February 28, 2020).

With its two extra questions, the **SSIX** provides information on how SSOS scores are either raised or lowered when the role of perceived *fairness* and *supervisor support* are factored in. These two questions are called "mediators" because they mediate the impact of the basic SSO Scores by either raise or lowering them. So, for example, stressful work due to high demand and mental effort can have a greater negative impact on mental health when these conditions are made worse by perceived lack of supervisory support, fairness and respectfulness. Conversely, these same stressful conditions can have less negative impact on mental health when these conditions are made better by perceived supervisory support, fairness and respectfulness.

The additional information provided by the mediators may help you to focus on where you need to concentrate your efforts to create a psychologically safe and healthy workplace: for example, when there is little that can be done to relieve demands and effort it may still be possible to work on improving supervisory support and fairness. This is why it is useful to compare the two scores – the SSOS and the SSIX – to see where supervisory support and fairness make a positive difference and where they make a negative difference. Those environments in which they make a negative difference may recommend themselves as priority targets for workplace restoration efforts

How do I interpret Initial Scan Results?

The following quick reference guide will help you with your interpretation.

-2.5 to -0.50 = red	-0.49 to 0.0 = amber	+0.01 to +0.49 = yellow	+0.50 to +2.5 = green

Red Zone: much higher chance of mental injury, negligible chance of mental health promotion
Amber zone: elevated risk of mental injury, reduced chance of mental health promotion
Yellow zone: average risk of mental injury, better than 50:50 chance of mental health promotion
Green zone: low risk of mental injury, high potential for mental health promotion

When calculating scores, it is helpful to consider the *ranges* as well as the averages. For example, the average SSIX for a unit of 25 people might be +1.0 but the range may be between +2.5 (the maximum) and – 1.5. This is important information because it reveals significant differences in how members of a team or unit perceive and experience their psychosocial work environments. This information can be very helpful as a conversation starter with groups that are struggling, for whatever reason, with interpersonal issues that draw energy away from achieving the goals of the organization.

Remember the scores are a starting point not a destination!

Bear in mind that these scores and ranges, helpful as they are, provide only a starting or ignition point for further investigation such as conducting the **Employee Survey** as well as other methods such as interviews, feedback/focus groups, etc. It is only through these means that survey results can be validated and provide a sufficient basis for action when required.

Appendix 3

PSYCHOLOGICAL HEALTH AND SAFETY OUTCOME MEASUREMENTS

Data collection is a component of the National Standard of Canada on Psychological Health and Safety in the Workplace. However, many organizations/companies have limited capabilities in this area. The following formulas are suggested for calculating the magnitude of some of the psychological incidents that might occur within the workplace.

- **Number and Rate of Bullying Incidents**

Determined by dividing the total number of bullying incidents multiplied by 200,000 by the total number of hours worked.

$$\text{Bullying Frequency Rate} = \frac{\text{Total bullying incidents} \times 200{,}000}{\text{Total hours worked}}$$

This metric provides the incidence rate of bullying incidents per 100 workers and is similar to the Injury Frequency Rate used in the field of OH&S. For consistency, it is advisable to use calendar days when determining the total hours worked.

- **Number and Rate of Harassment Incidents**

Determined by dividing the total number of harassment incidents multiplied by 200,000 by the total number of hours worked.

$$\text{Harassment Frequency Rate} = \frac{\text{Total harassment incident} \times 200{,}000}{\text{Total hours worked}}$$

For consistency, it is advisable to use calendar days when determining the total hours worked.

- **Number and Rate of Workplace Violence Incidents**

Determined by dividing the total number of workplace violence incidents multiplied by 200,000 by the total number of hours worked.

$$\text{Violence Frequency Rate} = \frac{\text{Total violence incidents} \times 200{,}000}{\text{Total hours worked}}$$

For consistency, it is advisable to use calendar days when determining the total hours worked.

- **Number and Rate of Psychological Injury Incidents**

Determined by dividing the total number of psychological injury incidents multiplied by 200,000 by the total number of hours worked.

$$\text{Psychological Injury Frequency Rate} = \frac{\text{Total psychological injuries} \times 200{,}000}{\text{Total hours worked}}$$

For consistency, it is advisable to use calendar days when determining the total hours worked.

- **Number and Rate of Psychological Illness Incidents**

Determined by dividing the total number of psychological illness incidents multiplied by 200,000 by the total number of hours worked.

$$\text{Psychological Illness Frequency Rate} = \frac{\text{Total psychological illness} \times 200{,}000}{\text{Total hours worked}}$$

For consistency, it is advisable to use calendar days when determining the total hours worked.

- **Number and Rate of Psychological Health Claims**

Determined by dividing the total number of psychological health claims multiplied by 200,000 by the total number of hours worked.

$$\text{Psychological Health Claims Rate} = \frac{\text{Total psychological claims} \times 200{,}000}{\text{Total hours worked}}$$

For consistency, it is advisable to use calendar days when determining the total hours worked.

- **Total Cost of Psychological Health Claims**

Determined by adding all the costs related to psychological health claims for a given period of time.

- **Average Cost of Psychological Health Claims**

Determined by dividing the Total Cost of Psychological Health Claims for a given period of time by the number of psychological health claims for that specific period of time.

- **Average Cost of Psychological Health Claims per Employee**

Determined by dividing the Total Cost of Psychological Health Claims for a given period of time by the number of employees eligible to submit a psychological health claims for that specific period of time.

CHAPTER REFERENCES

CCOHS, "Civility and Respect in the Workplace, Infographic" (2019), online: https://www.ccohs.ca/products/posters/civility/ (date accessed: February 28, 2020).

CCOHS, published an infographic on "Psychologically Health and Safe Workplaces" (2019), online: https://www.ccohs.ca/products/boutique/13factors/ (date accessed: February 28, 2020).

Canadian Standards Association, *National Standard of Canada: Psychological Health and Safety in the Workplace - CAN/CSA-Z1003-13/BNQ 9700-803/2013* (Ottawa, ON: CSA Group, 2013), online: http://shop.csa.ca/en/canada/landing-pages/z1003-psychological-health-and-safety-in-the-workplace/page/z1003-landing-page (date accessed: February 28, 2020).

Guarding Minds @ Work, *Conducting a GM@W Organizational Review: Getting Started (2013)*, online: https://www.workplacestrategiesformentalhealth.com/pdf/GMAW_Organizational_Review.pdf (date accessed: February 28, 2020).

Guarding Minds @ Work, *GM@W Documents and Resources* (2013), updated with the recent branding, online: http://www.guardingmindsatwork.ca/resources (date accessed: February 28, 2020).

Guarding Minds @ Work, *GM@W Initial Scan* (2013), updated with the recent branding, online: https://www.guardingmindsatwork.ca/assets/pdfs/Reading_Initial_Scan_Results.pdf (date accessed: February 28, 2020).

Guarding Minds @ Work, *GM@W Organizational Review*, online: http://www.guardingmindsatwork.ca/resources (date accessed: February 28, 2020).

Guarding Minds @ Work (2012), *Ipsos-Reid 2012 Review: Validity of Guarding Minds @ Work™ Psychosocial Factors*, October, online: http://www.guardingmindsatwork.ca (date accessed: February 28, 2020).

Ipsos, "Workplaces that are Implementing the National Standard of Canada for Psychological Health and Safety in the Workplace Described by Employees as Psychologically-Safer Environments" (2016), online: https://www.ipsos.com/en-ca/news-polls/workplaces-implementing-national-standard-canada-psychological-health-and-safety-workplace (date accessed: February 28, 2020).

Orange Umbrella®, online: http://www.orangeumbrella.co/ (date accessed: February 28, 2020).

Chapter 26

OCCUPATIONAL HEALTH AND SAFETY: WORKPLACE WELLNESS STRATEGY

A. INTRODUCTION

Workplace Wellness Programs (WWPs) are increasingly popular in Canada and the U.S. In regard to organizations/companies with at least 50 employees, over 50 per cent of them offer a WWP. This number increases in accordance with organization/company size; for example, more than 90 per cent of larger organizations/companies offer a WWP.[1,2,3,4]

Workplace wellness from a personal perspective, can be defined as managing the psychological and physical issues in response to environmental stress, including one's work environment.[5] From an organizational perspective, **organizational wellness** is managing both business functioning and employee well-being in a manner that allows the organization to be more resistant to environmental pressures.[6]

The objectives of this chapter are to:

- define health promotion, health protection, and workplace wellness;

---

[1] A. Carrns, "Study Raises Questions for Employer Wellness Programs" in *New York Times: Your Money/Your Money Adviser* (January 6, 2014), online: http://www.nytimes.com/2014/01/07/your-money/study-raises-questions-for-employer-wellness-programs.html?

[2] Conference Board of Canada, *Compensation Planning Outlook* (Ottawa: Conference Board of Canada, 1997).

[3] Buffett and Company Worksite Wellness Ltd., *National Wellness Survey Report 2009* (Whitby, ON: Buffett and Company Worksite Wellness Ltd., 2009), now owned by Sun Life of Canada.

[4] Willis Watson Wyatt, *2015-2016 Staying@Work Report* (2016) at 5, online: https://www.willistowerswatson.com/-/media/WTW/PDF/Insights/2016/08/2015-2016-staying-at-work-canada-research-findings-en.pdf (date accessed, February 28, 2020).

[5] D. Dyck, "Workplace Wellness: A Model for Success" (Presented at the American Occupational Health Conference, New Orleans, Louisiana, April 1999) [unpublished].

[6] *Ibid*.

- explain how to design a strategy for a WWP;
- describe a model for a comprehensive WWP and its components;
- provide examples of fictitious WWPs including their development, implementation, and evaluation;
- explain how to build a business case for a WWP; and
- identify the role(s) that human resources (HR) professionals can have in the WWP and the potential value added by a WWP to the various stakeholders.

B. HEALTH PROMOTION

"**Health promotion** is the science and art of helping people change their lifestyle to move toward a state of optimal health. **Optimal health** is a balance of physical, emotional, social [workplace], spiritual, and intellectual health (Figure 26.1). Lifestyle change can be facilitated through a combination of efforts to enhance awareness, change behaviour, and create environments that support good health practices. Of the three elements, supportive environments has the greatest impact in producing lasting changes."[7]

---

[7] M.P. O'Donnell, *Health Promotion in the Workplace*, 3d ed. (Albany, NY: Delmar Thomson Learning, 2002) at 49.

Figure 26.1: Optimal Employee Health[8]

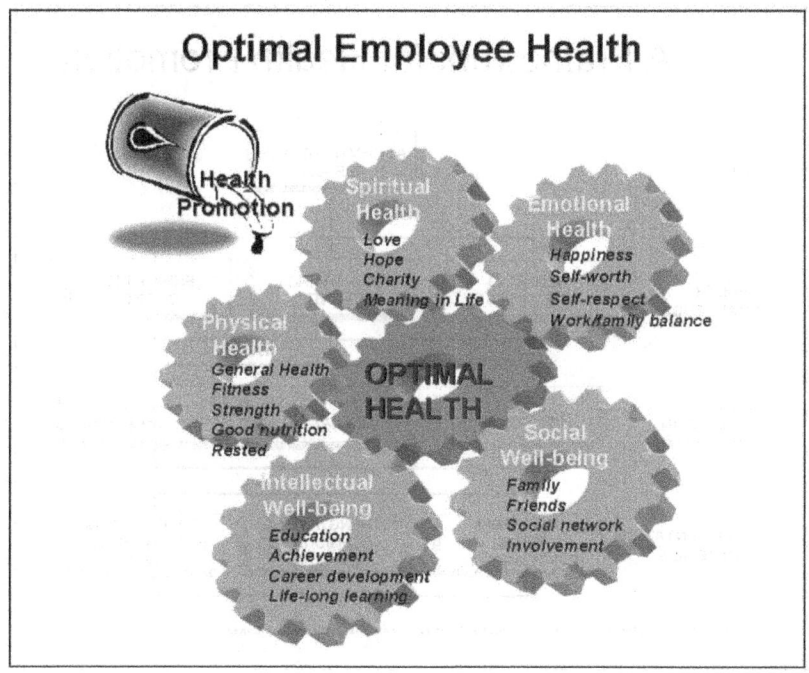

Health promotion is a process. It is important to start where people are at (people-oriented), and to work from there to enable the development or sustainability of personal and workplace control. It is action-oriented and involves a sharing of power. Health promotion involves all employees, not just those employees deemed to be "at risk". Multi-approaches aimed at effective, concrete employee participation and which focus on enabling/facilitating health, are used. As well, through the use of an ecological approach, health promotion is tailored to meet the needs of people in the context of their environment.

The Epp Model for health promotion[9] demonstrates this approach (Figure 26.2).

---

[8] Copyright D. Dyck, 2007.

[9] J. Epp, Health Promotion Model, *Achieving Health for All: A Framework for Health Promotions* (Ottawa: Health and Welfare Canada, 1986).

[10] Adapted from J. Epp, Health Promotion Model, *Achieving Health for All: A Framework for Health Promotion* (Ottawa: Public Health Agency of Canada, 1986). Reproduced with the permission of the Minister of Public Works and Government Services Canada, 2007.

Figure 26.2: A Framework for Health Promotion – The Jake Epp Model[10]

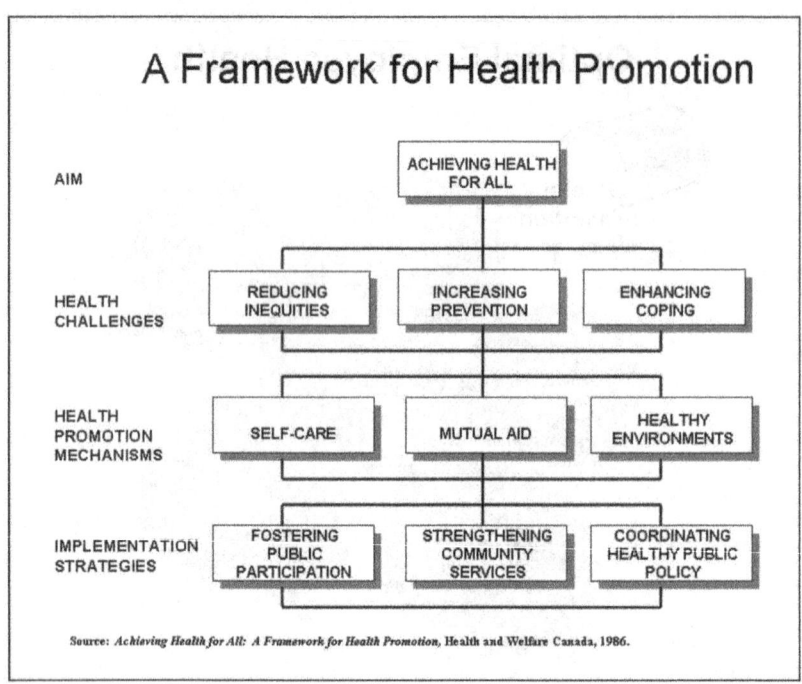

The aim is "achieving health for all". The identified health challenges are to reduce inequities, to increase prevention, and to enhance coping. The recommended health promotion mechanisms are self-care, mutual aid, and healthy environments. As pointed out by O'Donnell,[11] the latter mechanism is the most powerful approach. The implementation strategies are to foster public participation, strengthen community services, and coordinate healthy public policy.

The intent of providing this model is that it illustrates how a WWP can be designed and developed. For example, if the aim is to achieve workplace health for all employees/workers, the challenges are to reduce inequities, increase the prevention of illness/injury, and enhance employee coping abilities. This can be achieved by enabling/facilitating employee self-care and mutual aid, and by creating a safe, healthy and supportive workplace. To operationalize the approach, encourage employee participation, enhance workplace OH&S/HR services, and ensure that the related corporate policies and practices align with the ultimate goal of achieving health for all.

To take this model further, a systems approach will be used along with

---

[11] M.P. O'Donnell, *Health Promotion in the Workplace*, 3d ed. (Albany, NY: Delmar Thomson Learning, 2002) at 49.

considering the workplace as part of the larger community in which it operates (Figure 26.3). In this depiction, the workplace is encased in a circle in which the ultimate goals of workplace wellness are to enable employee health, safety, and well-being; mediate competition for scarce resources; and advocate for employee health, safety and well-being. However, there is a link with the external environment — the community/society. That link is denoted by the tail reaching outside the circle.

Figure 26.3: Workplace Wellness[12]

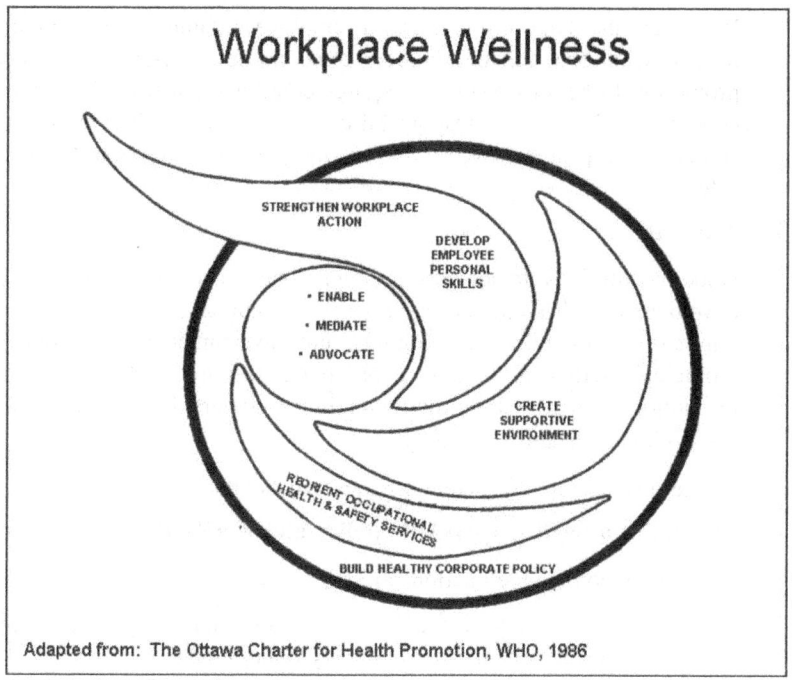

According to the Ottawa Charter for Health Promotion (1986), the goals of health promotion are to:

1. **Enable**

    Health promotion focuses on achieving equity in health. Health promotion action aims at reducing differences in current health status and ensuring equal opportunities and resources to *enable* employees to achieve their fullest potential. This includes:

---

[12] WHO, "The Ottawa Charter for Health Promotion" (1986), online: http://www.paho.org/English/AD/SDE/HS/OttawaCharterEng.pdf (date accessed: February 28, 2020).

- a supportive environment (the workplace);
- access to information;
- the requisite life skills; and
- opportunities to make healthy and safe choices.

People cannot achieve their fullest health potential unless they are able to take control of those things which determine their health.

2. **Mediate**

People are involved in this model as employees, families and community members; and they function in diverse and multiple workplaces. For health promotion to be successful, it requires coordinated actions involving all concerned. One major responsibility is to *mediate* the varying, and sometimes opposing, interests in the workplace for the pursuit of health, safety, and well-being.

3. **Advocate**

Good health is a major resource for social, economic, and personal development. It is an important dimension of human quality of life. In life, political, economic, social, cultural, and environmental behaviours and biological factors all favour or prove harmful to good health. Health promotion aims at making these conditions favourable through *advocacy* for workplace health and safety.

The mechanisms for achieving these goals are:

- **Build a healthy corporate policy and work culture:**

    The recommended actions are to:
    - put employee/organizational health and safety on the agenda of the organization/company policy-makers;
    - direct them to be aware of the health and safety consequences of their decisions;
    - direct them to accept responsibility for worker health and safety; and
    - identify the barriers to the adoption of healthy and safe corporate policies and remove them. *The aim is to make the healthy and safe choice, the easiest choice.*

- **Create a supportive environment:**

    Our society is complex and interrelated, and as a result, health, and safety cannot be separated from work and leisure. They should be a source of human health and safety. Work should help in the creation of a healthy society. In short, workplace wellness generates living and working conditions that are safe, stimulating, satisfying, and enjoyable. To achieve this end, a systematic assessment of the health impact of the workplace is required.

- **Help people to develop personal skills:**

    Health promotion (workplace wellness) supports worker personal and social development by providing information and education, and by enhancing life skills. It enables employees to learn throughout their life, to prepare themselves for various life stages, and to successfully cope with problems.

- **Strengthen workplace action:**

    Health promotion (workplace wellness) is achieved through concrete and effective workplace action in setting priorities, making decisions, planning strategies, and implementing plans to achieve better employee health. It involves empowering the work units/workplace and encouraging ownership of their identified problems and solutions through approaches like:

    - self-help;
    - social support;
    - flexible systems for employee participation;
    - full and continuous access to information/learning opportunities; and
    - adequate resources and funding.

- **Reorient Occupational Health and Safety services:**

    This mechanism suggests a shift in thought and focus from injury/illness management, to one of health and safety leadership and injury/illness prevention. It must be proactive, as opposed to being reactive, which many Canadian and U.S. companies currently are. It requires a multidisciplinary approach, not the traditional medical model for dealing with worker health and safety.

Translated into action, workplace wellness could look like the model presented in Figure 26.4. Most of this model is self-explanatory. The stakeholders all have a role to play in promoting workplace wellness. For example, the organization is charged with safe and healthy policy making, positive corporate climate (culture) development, program planning, and the provision of supportive employee benefits. However, the terms, health promotion and health protection (injury/illness prevention) in this model warrant further explanation.

---

[19] D. Dyck, "Workplace Wellness: A Model for Success" (Presented at the American

Figure 26.4: Workplace Wellness — Operational Model[19]

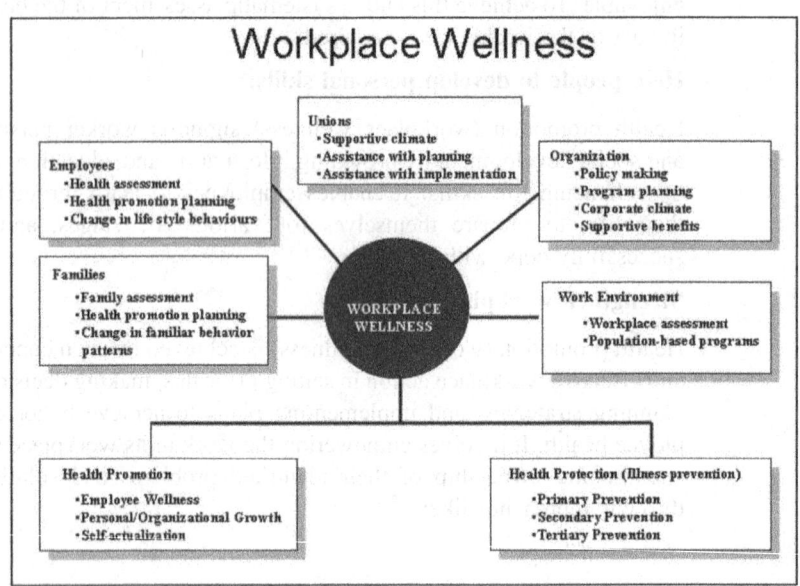

Although these terms tend to be used interchangeably, they differ in their motivational mechanism and goal orientation. To reiterate, health promotion is defined as the science and art of helping people change their lifestyles to move toward a state of optimal health. It consists of activities directed toward increasing the level of well-being and actualizing the health potential of individuals, families, communities, and society. It is not a disease- or condition-specific approach; rather it seeks to expand positive potential for health in a dynamic way.[20]

Health protection, also termed illness/injury or disease prevention, is any behaviour performed by a person, regardless of his/her perceived or actual health status, in order to protect, promote, or maintain their health, whether or not such behaviour is objectively effective toward that end.[21] The focus is to prevent a specific disease or condition by avoiding the occurrence of pathogenic insults to health and well-being. The three levels of health prevention are:

1. **Primary prevention,** which consists of activities directed toward decreasing the probability of specific illnesses, or dysfunctions in individuals, families, and communities, including active protection against unnecessary

---

Occupational Health Conference, New Orleans, Louisiana, April 1999) [unpublished].

[20] N. Pender, *Health Promotion in Nursing Practice*, 2d ed. (Norwalk, CT: Appleton & Lange, 1987) at 4.

[21] D.M. Harris & S. Guten, "Health-protective behaviour: An exploratory study" (1979) 20 *J. Health Soc. Behav.* 17.

stressors.[22] Some workplace examples are:

- immunization against influenza, hepatitis A/B, shingles, *etc.*;
- normal blood pressure control;
- normal cholesterol control;
- weight control;
- sunscreen advice; and
- use of personal protective equipment.

2. **Secondary prevention**, which emphasizes early diagnosis and treatment for health conditions thereby shortening their severity and duration, enabling individuals to regain normal functioning in a timely manner.[23] Some examples are:

- screening for vision, hearing, respiratory conditions, chemical toxicity;
- screening for cancer, diabetes, high blood pressure, elevated cholesterol; and
- learning stress management, change management, time management, and work/life balance approaches.

3. **Tertiary prevention**, which occurs once a health condition or disability becomes stable or is irreversible. The goal is to assist the individual to regain an optimal level of functioning within the constraints of the condition or disability. Some examples include:

- prescribed medication use;
- therapy compliance;
- attendance at support groups (*e.g.*, Alcoholics Anonymous, Al-Anon, Alateen, Gamblers Anonymous, and Gam-Anon).

C. WORKPLACE WELLNESS PROGRAM: STRATEGY

When designing a WWP, it is important to ensure that workplace wellness is not viewed as a separate, stand-alone program: rather, it must be an element that is aligned with the corporate vision, mission, and goals, and integrated into the very fabric of the workplace itself. That statement indicates that the aim of "achieving health for all workers" needs to be part of management and operational decision-making and planning. To explain this further, Figure 26.5 illustrates that workplace wellness is incorporated into the framework of the organization.

---

[22] N. Pender, *Health Promotion in Nursing Practice*, 2d ed. (Norwalk, CT: Appleton & Lange, 1987) at 4.

[23] *Ibid.*, at 5.

Figure 26.5: Workplace Wellness Program: Contextual Approach[24]

However, it must also be part of the consciousness of the management of the other functions within the organizations, such as Business Development, Operations, Human Resources, Total Compensation, Occupational Health and Safety, *etc.*

To develop a comprehensive and effective WWP, organizations are advised to use a systematic approach starting with the current state of organization and workers and building from there. Figure 26.6 presents a cyclical, seven-step approach.

---

[24] D. Dyck, "Workplace Wellness: A Model for Success" (Presented at the American Occupational Health Conference, New Orleans, Louisiana, April 1999) [unpublished].

[25] *Ibid.*

Figure 26.6: Workplace Wellness Strategy Preparation: Seven-Step Approach[25]

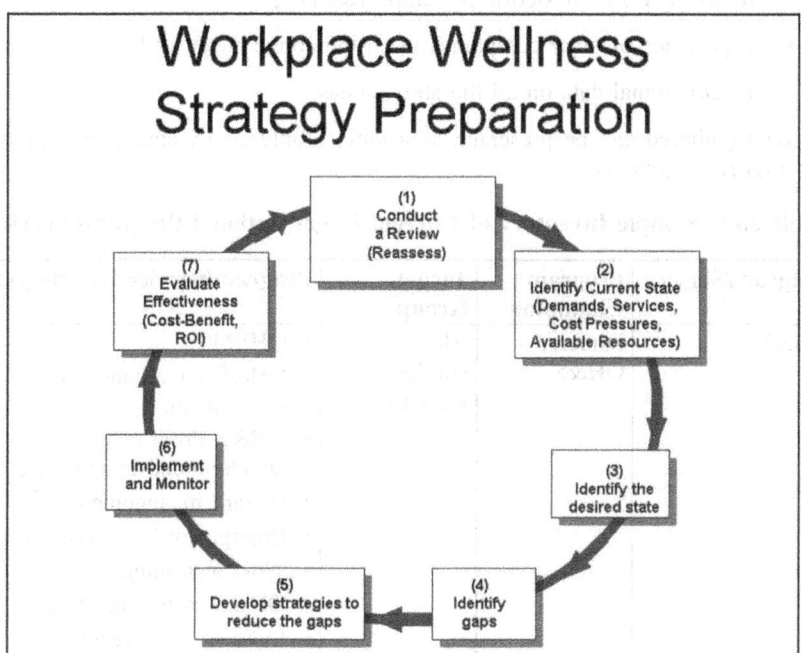

The tools that can be used to do this review include an inventory of all available programs/services and resources using a template like the one in Table 26.1 and/or a graphic depiction as in Figure 26.7.

Step 1: Conduct a Review — This step involves assessing where the organization and employees currently are in terms of workplace wellness. This means gathering data on the current:

- workplace programs/services that support worker health and safety, *e.g.*, OH&S Program, Attendance Support Program, Disability Management Program, Employee Assistance Program, Fitness Facilities, Weight Control Program, Prescription Drug Plans, Dental Plan, Vision Plan, Extended Healthcare Plan, Short-Term and Long-Term Disability Plans, Workers' Compensation Plans, Vacation Plan, Family Leave Plans, *etc.*;
- utilization rates of these programs/services;
- costs for workplace programs/services;
- other internal/external resources available;
- employee absenteeism rates/costs per employee;
- disability rates/cost per employee;

- occupational injury rates/costs per employee;
- return-to-work rates/costs per employee, *etc.*;
- experience of other companies in all the above areas; and
- local/national data on all the above areas.

The data gathered can be presented as a table (Table 26.1), and/or as a graphic depiction (Figure 26.7).

Table 26.1: Sample Inventory of Current Organizational Programs/Services

Program/Service	Program Champion	Target Group	Program/Service Offering(s)
OH&S	Director OH&S	All employees/ workers	• OHSMS • OH&S governance and stewardship • OH&S Program development/maintenance • Hazard management • Emergency Preparedness • Worker training • OH&S communication • Incident investigation • Reporting • Program evaluation • Continuous improvement
Human Resources	VP HR	All employees	• Recruitment • Selection • Compensation • Benefits • Maintenance • Career development • Labour relations • Performance management including dismissal • Policy development
Fitness Program	Director, OH&S	All employees	• Fitness benefit(s) policy • Development/maintenance of fitness facility Introduction of new fitness initiatives

Figure 26.7: EAP and its Services

Step 2: Identify the Current State — Part of the review involves the identification of baseline data on the demands, services, cost pressures, and available resources that relate to workplace wellness. This can be done by using a tool like the one depicted in Figure 26.8.

---

[39] D. Pratt, *Competitiveness and Corporate Health Promotion* (Calgary, AB: Health Business Inc. 1992).

Figure 26.8: Identification of the Total Cost of Health[39]

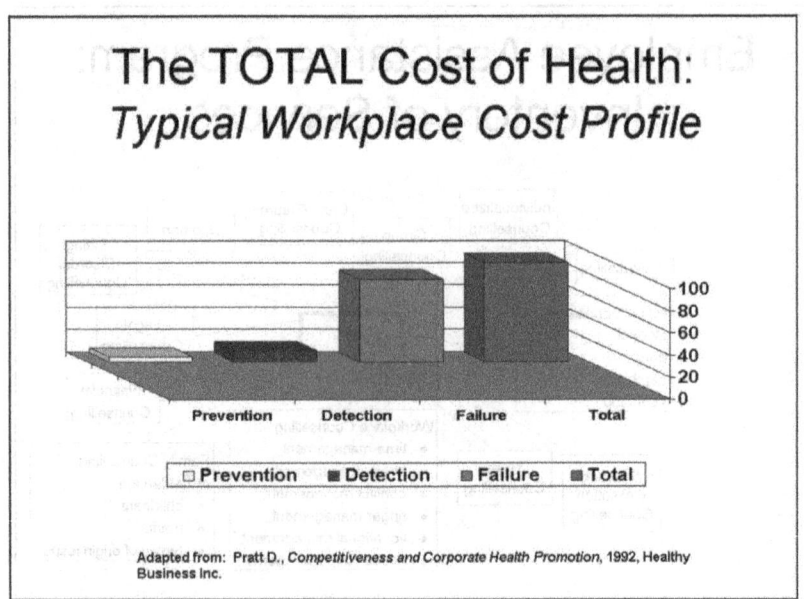

Step 3: Identify the Desired State — Part of the review involves establishing what "best practices" in the area of workplace wellness are. The best practices selected should be industry or clinically proven to be effective approaches. As well, organizational leaders must decide the desired future state for the organization/company. To do so, they must answer the question: *What does the Company want to achieve from a Workplace Wellness Program?*

Although this might seem like an easy question, it is not. However, the answer needs to be systematically arrived at. To begin with, what is the organization's/company's business strategy? A sample business strategy might be:

> *Optimize the Western Canadian Sedimentary Basin:*
> How do we best provide a range of value-added transmission, marketing and processing services to customers, and what kind of regulatory change is necessary for us to get there?
> **Integration:**
> How do we work together and combine our skills to serve customers effectively and bring them better value than the competition?
> **Value Creation:**
> How do we extract the greatest value from our efforts for both customers and shareholders?

CH. 26: OCCUPATIONAL HEALTH AND SAFETY: WORKPLACE WELLNESS STRATEGY

> **Commodity Risk:**
> *What is Oil Canada's desired mix of predictable earnings and variable earnings with commercial risks and commodity spreads? How do we manage risk successfully?*
> What is the company's vision statement? A sample could be:
> *The Company provides a healthy and safe work environment and enables the organizational structure and its employees to reach an optimal state of well-being.*
> The company will be the pre-eminent provider of high value-added integrated energy solutions worldwide.

Examining the wording of the above vision statement, certain terms provide insight into the corporate values:

Pre-eminent — delivering leadership;

High value-added — importance of shareholder and customer consideration;

Integrated — packaging services to meet customer needs;

Solutions — find value;

Energy — the future in terms of "energy"; and

Worldwide — international intent/focus.

Add to that, the corporate culture/image that the leaders wish to attain. A sample might look like:

> *The Company is:*
> - *a caring employer;*
> - *a trusting and trusted organization; and*
> - *through its actions and policy, is supportive and proactive in its employees' health, safety, and wellness.*
>
> *Employees:*
> - *value their own health, safety, and well-being;*
> - *are committed to maintaining a safe and healthy work environment for themselves and others; and*
> - *actively participate in activities that optimize their health, safety, and well-being through partnership with the Company.*

What should emerge is a workplace wellness theme: for example,

Good Health is Good Business

As well, a series of desired outcomes are identified, for example:

- *an increasingly healthy workforce;*
- *a productive workforce;*
- *efficient delivery of services;*

- *effective operations;*
- *profitability;*
- *a corporation that values and supports employees; and*
- *workplace wellness initiatives and program that are aligned with the company's business strategy.*

Now, determine what the company is willing to dedicate in terms of time, money, and other resources towards workplace wellness. For this part of the exercise, a tool like the one provided in Figure 26.9 can be used.

Figure 26.9: The Total Cost of Health: The Desired Cost Profile[40]

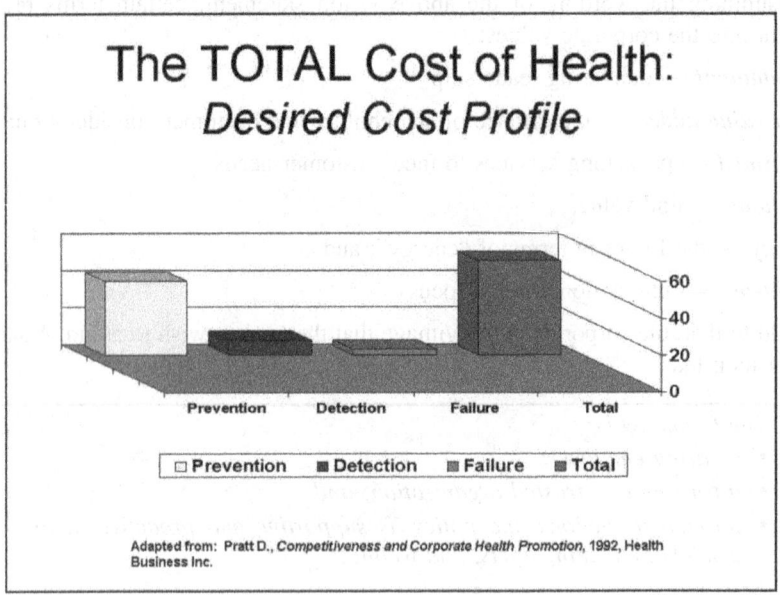

Lastly, an integrated workplace wellness approach requires involvement by many stakeholders such as:

- Senior Management for endorsement and support;
- Employees (and their families);
- Health Promotion personnel (*e.g.*, a Wellness Specialist);
- Compensation and Benefits personnel;
- Occupational Safety Practitioners/Professionals;

---

[40] *Ibid.*

- Occupational Health professionals;
- Medical/clinical services;
- EAP;
- Disability Management Program and services;
- Risk Management personnel;
- Training and Development personnel;
- Employee Association/Union leaders; and
- Finance personnel.

From this exercise, a workplace wellness vision emerges which might resemble the one portrayed in Figure 26.10.

Figure 26.10: Workplace Wellness Strategy: The Desired State[41]

To take it one step further, a company-specific workplace wellness model could be developed. Figure 26.11 is an example that was developed based on the data used to explain this strategy.

---

[41] D. Dyck, "Workplace Wellness: A Model for Success" (Presented at the American Occupational Health Conference, New Orleans, Louisiana, April 1999) [unpublished].

Step 4: Identify the gaps between the Current and Desired States — Using a gap analysis approach, determine the gaps between the two states. In this example, the gaps might be:

- *lack of employee awareness of the current services available;*
- *lack of comprehensive, consistent and observable management support to employee health, safety, and wellness;*
- *lack of awareness and measurement of employee health, safety, and wellness needs; and*
- *lack of awareness and role clarification in the delivery of health, safety, and wellness services.*

Figure 26.11: Sample Industry-Specific WWP Model[42]

Step 5: Develop strategies to eliminate/reduce the identified gaps — This step requires the prioritization and planning of the action to be taken, such as:

Goal #1: To obtain management support for workplace wellness.

Goal #2: To determine the current level of corporate and employee views on both personal and corporate wellness.

Goal #3: To increase stakeholder knowledge and use of workplace wellness concepts, programs, and services.

---

[42] *Ibid.*

CH. 26: OCCUPATIONAL HEALTH AND SAFETY: WORKPLACE WELLNESS STRATEGY

Step 6: Implementation — This is the "doing" part of the process and might include actions such as:

- *Introduce the Workplace Wellness Program strategically;*
- *Isolate the program effectiveness — quantifiably and qualitatively;*
- *Integrate the Workplace Wellness Program with other corporate strategies, such as:*
  - *Human Resource Initiatives*
  - *Benefits Plan Design*
  - *OH&S programming and services*
  - *Attendance Support programming and services*
  - *Disability Management programming and services*
  - *Communication strategies*
  - *Operational activities*
  - *Corporate business strategies.*

Step 7: Program evaluation — The last, and probably the most crucial piece, is to evaluate the actions taken. The reason is that:

If you can't measure it, you can't control it.

If you can't control it, you can't manage it.

Other reasons for program evaluation are:

- It enables the organization to function more effectively by:
  1. transforming business strategies into action-oriented performance measurements;
  2. creating an organization-wide method whereby employees can see *how* they contribute to the organization's success; and
  3. tracking performance results.
- The need to align the WWP with the organization's business strategies.

   A typical program goal for a WWP is:

   To engage every employee in the company in the Workplace Wellness Program (WWP), and to have them become a major stakeholder in the company's success.

Using that as a starting point, the next thing to do is to establish a **workplace wellness program evaluation strategy**. This strategy involves the following steps:

1. **Translate the Workplace Wellness Program vision by:**
   - identifying the critical success factors;
   - establishing an organization-wide, balanced set of performance indicators; and

- developing and communicating focused strategies.

2. **Align the organization through:**
    - cascading performance indicators throughout the organization;
    - establishing performance goals, expectations, and employee ownership;
    - linking employee day-to-day activities back to the WWP goals; and
    - creating one organization-wide agenda with specific quantifiable measures used to track progress.

3. **Provide feedback and information by:**
    - capturing and organizing relevant data;
    - documenting comments and conclusions;
    - tracking the relevant performance results; and
    - delivering timely and accurate information.

    Through the analysis of this information, continuous improvement initiatives can be developed.

4. **Continue to improve through measurement by:**
    - managing by fact;
    - creating a learning organization;
    - adjusting action plans and timeliness based on performance; and
    - linking employee compensation and performance.

    This is achieved by creating a relationship between the performance indicators and then tracking performance. When organizations/people have the right information, they do the right things for the right reasons at the right price and in a reasonable time frame.

In an attempt to further clarify a Workplace Wellness Impact Evaluation Model, Figure 26.12 is provided.

Figure 26.12: A Workplace Wellness Impact Evaluation Model[43]

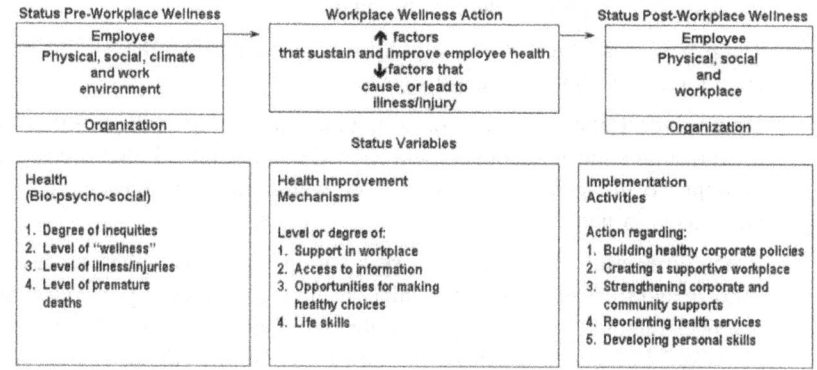

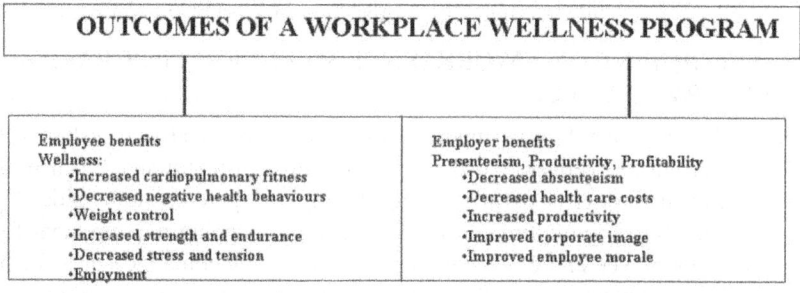

D. ORGANIZATIONAL WELLNESS: KEY ELEMENTS

Organizational wellness has a number of key elements. They include:

- **Communication** — information-sharing regarding the intent, structure, processes, and outcomes of the organization's workplace wellness initiative and program. Lack of clarity on the WWP can result in its failure.

- **Trust** — essential to any program is the belief that the program exists for the stated reasons and that Senior Management does "walk its talk" in endorsing and supporting the program. Failing to uphold the stated values can destroy the WWP.

---

[43] *Ibid.*

- **Opportunity for Personal Growth and Development** — inherent in health promotion is personal growth and development. If that element is missing or downplayed, the WWP can suffer.

- **Fairness** — since the focus of health promotion and hence, the WWP, is inclusion of all workers, and reduction inequities is one of the challenges, fairness is an important element to uphold.

- **Team Ethics** — the WWP design team must ensure that the actions taken are adopted for the right reasons.

- **Humanistic Policies** — workplace policies must address the needs of the organization's most valuable resource — the employees. Policies designed to improve the workplace and worker well-being are considered to be humanistic in nature.

- **Procedures and Practices** — organizations/companies that profess to endorse and support workplace wellness, and then introduce procedures and practices that undermine the philosophy and goals of the WWP, will doom the WWP. For example, scheduling meetings before work, at lunch or after work when employees would normally be exercising, undermines the organization's/company's talk about promoting worker fitness. Likewise, ignoring the fact that workers fail to wear their seatbelts/or continue to use cell phones while driving, flies in the face of workplace wellness.

E. DEVELOPMENT OF A WORKPLACE WELLNESS PROGRAM: AN INDUSTRY EXAMPLE

When developing a WWP, it is important to position organizational wellness, and employee and family wellness, so that they are equal in importance. Hence, they should be linked through a partnership approach and operate side-by-side. Open communication, proactivity, timely interventions, and ongoing measurement focus the WWP on employee and organizational well-being. Outcome measures for both employee and organizational wellness are geared towards productivity, effectiveness, efficiency, "value addedness" and profitability, and resource optimization. The Workplace Wellness Model (Figure 26.13) was designed to illustrate workplace wellness, what it involves, and how it functions.

Figure 26.13: Workplace Wellness Model[44]

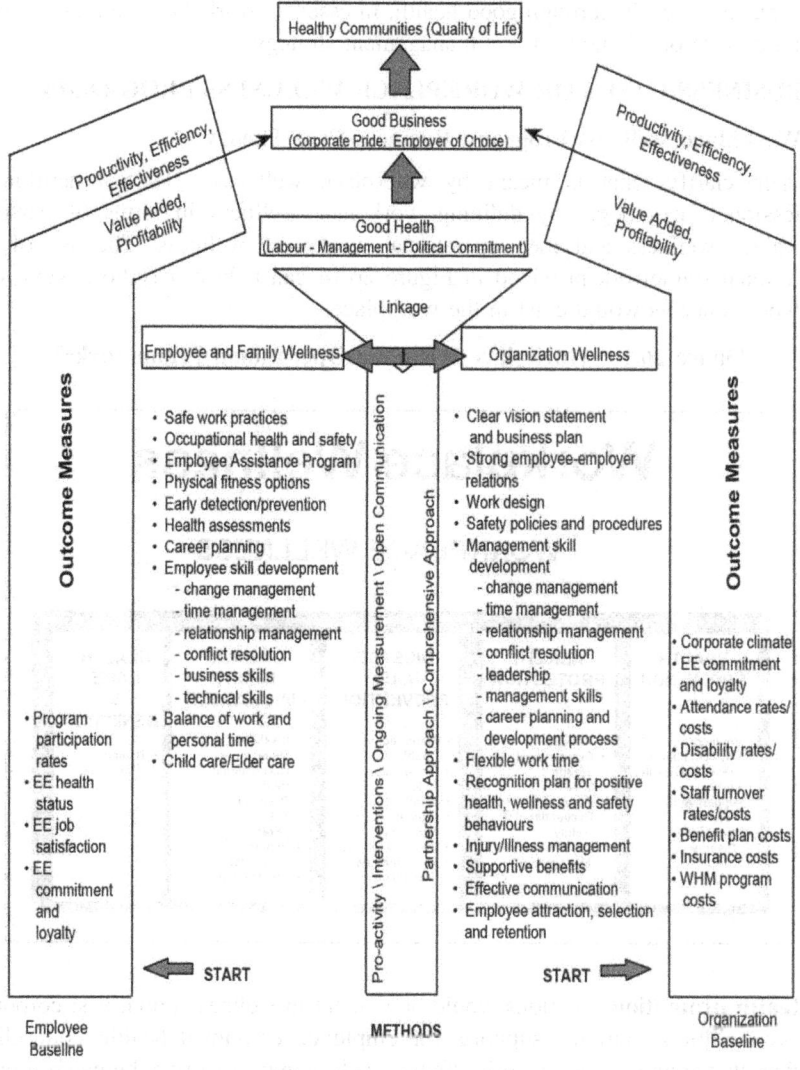

This Workplace Wellness Program Model which was originally developed by The City of Calgary, is a futuristic model. It was designed to encourage a partnership approach to creating and sustaining employee well-being. Unlike past workplace health promotion models, it is meant to encourage employee responsibility for

---

[44] D. Dyck, "Workplace Wellness: What is Your Potential Return on Investment?" (Human Resources Association of Calgary, November 1998) [unpublished].

wellness. The organization's role is to provide a receptive environment and support for positive health and well-being. However, the organizational goal is to achieve good business results through good health. In essence, workplace wellness is more than a service or a benefit, it is a management strategy.

F. BUSINESS CASE FOR WORKPLACE WELLNESS PROGRAMS

1. Workplace Wellness Program: Business Case Theory

To help clarify what is meant by workplace wellness, OH&S Practitioners/ Professionals can begin by defining workplace wellness in terms of personal workplace wellness and then operational workplace wellness. The use of the operational framework provided in Figure 26.14 can help clarify how workplace wellness concepts would exist in the workplace.

Figure 26.14: Workplace Wellness: Operational Framework[45]

Workplace Wellness

WORKPLACE WELLNESS

HEALTH PROMOTION	HEALTH PROTECTION	DISEASE/ INJURY PREVENTION	DISEASE/ INJURY MANAGEMENT	HEALTH CARE & BENEFITS
• Wellness • Physical Fitness • Emotional Health • EFAP • Supportive HR Policies • Management Theories	• Wellness • Occupational Health • EFAP • Physical Fitness • Occupational Safety • Environmental Management	• Occupational Health • Occupational Safety • EFAP • Human Resources • Physical Fitness • Environmental Management	• Disability Management • Occupational Health • Occupational Safety • EFAP • Human Resources • Physical Fitness	• Human Resources

Health promotion functions would be evident in wellness initiatives; corporate fitness facilities/programs; supports for employee emotional health; counselling services that address general worker/family well-being; supportive human resources policies; and management theories and practices (*e.g.*, compensation, labour management practices, disciplinary practices, vacation policies, *etc.*) that embrace and support worker health safety and well-being. In short, these functions enable/ facilitate the approaches of worker self-care, mutual care, and supportive work environments.

---

[45] D. Dyck, "Workplace Wellness: A Model for Success" (Presented at the American Occupational Health Conference, New Orleans, Louisiana, April 1999) [unpublished].

Health protection (injury/illnesses prevention) activities include wellness initiatives tailored towards general disease states/conditions; occupational health practices such as pre-placement screening, fitness-to-work assessments and health education; counselling services; physical fitness; occupational safety; and environmental management.

Disease/injury prevention focuses on the prevention of specific disease states/conditions and includes OH&S activities (*e.g.*, medical monitoring/surveillance, health education, *etc.*); EAP counselling; Human Resources employee benefits/plans (*e.g.*, smoking cessation programs, nutrition counselling, *etc.*); and physical fitness (*e.g.*, improve physical conditioning, weight control, increase flexibility, *etc.*).

Healthcare and benefits are designed to support employees to manage their health conditions/illnesses/injuries. Traditionally this support has resided in the area of Human Resources under the employee group benefit plans. It can also encompass Occupational Health services and supports.

The next step is to demonstrate the value that a WWP can offer an organization/company. Examples of areas impacted by a WWP are listed in Table 26.2.

Table 26.2: Areas Impacted by a Workplace Wellness Program

Area Impacted	Evidence
Productivity-related:	
• Employee morale	• Employee perception surveys
• Staff turnover	• Personnel records
• Manpower recruitment	• Manpower statistics
• Employee absenteeism	• HR Information systems (HRIS)
• Level of presenteeism	• Personnel records/supervisor observations
• Amount of physical and emotional disability	• Personnel records
• Level of motivation to work	• Employee perception surveys
• Employee loyalty and commitment to organization	• Employee perception surveys
	• Amount of staff turnover
Health-related:	
• Employee absenteeism costs	• HRIS
• Benefit plan costs	• HRIS
• Disability management costs	• OH&S data management systems
• Nature of medical absenteeism	• OH&S data management systems
• Number of Workers' Compensation claims and costs	• OH&S data management systems
• Number of medical emergencies at work and costs	• OH&S data management systems
• Number and nature of EAP cases	• EAP Service provider reports

Area Impacted	Evidence
Organization's Image:	
• Community image/reputation	• Customer/community feedback
• Workplace safety history and image	• OH&S data management systems
• History of unfair labour practices	• Legal data management systems
• Number of labour disputes	• Labour relations history
• Legal problems	• Labour relations history
• Labour disruptions	• Legal data management system
• Reputation as an employer	• Labour relations history
	• Recruitment interviews

Add to that, the benefits and costs associated with not having a WWP. For example, if it were shown from the collected data that the company was experiencing any of the following:

(1) **Employee Health:**
- 40 per cent of its employees report being "At Risk" regarding their general health and 64 per cent have significant weight problems.
- 97 per cent of the employees are interested in doing at least one thing to improve/maintain their health and 74 per cent report that they would like to exercise more.
- The identified barriers for making health/lifestyle changes are:
  - 32 per cent lack of time,
  - 20 per cent lack of energy, and
  - 15 per cent no encouragement/help from employer.

(2) **Staff Turnover:**
- 81 per cent of the staff report that they like their job.
- 52 per cent think their job offers them the right level of challenge.
- 14 per cent are strongly considering a job change within the coming year to improve their health of which 20 per cent of those respondents are in an Administrative/Clerical position.

(3) **Work Stress:**
- 57 per cent of the employees stated that the company does not make every effort to keep unnecessary stress at work to a minimum.

(4) **Employee Absenteeism Rate and Costs:**
- The employee absenteeism rate is nine per cent and costs $3M per year.

(5) **Short-Term Disability Costs:**

- The cost for Short-Term Disability (STD) was $11,297,476 with the average cost per claim being $950.
- The departments that experienced very high STD costs per employee also experienced low scores for employee morale.

(6) **Employee Feedback:**
- This company, as an enlightened employer, could:
  - communicate more openly (46 per cent);
  - train supervisors/managers to be more sensitive to employee concerns (40 per cent);
  - encourage employees to spend time improving their health (39 per cent); and
  - obtain more employee input on how work is done (35 per cent).

(7) **Employee Stressors:**
- Workplace control and influence:
  - 27 per cent of employees feel they do not have an influence over things that happen to them at work.
  - 48 per cent are dissatisfied with the amount of involvement in decisions that affect their work.
  - 35 per cent think this company could help them to improve their health by increasing the amount of input employees have with regard to how work is done.

(8) **Cost of Employee "Distress" for This Company:**

Anti-depressants cost	$156,860
Ulcer medication cost	$ 75,455
Total cost	$262,315

(9) **Strategies Cited by Employees for Coping with Stress/Worry:**
- Exercise more (58 per cent).
- Sleep more/better (35 per cent).
- Socialize more (34 per cent).
- Eat better (32 per cent).
- Spend more time with family members (27 per cent).
- Learn to relax (27 per cent).

(10) **Workplace Health and Safety Concerns:**
- Averaged 20-40 per cent; and

- Varied by the job.
(11) **Employee supports varied depending on the nature of the problem.**
(12) **The cost of prescription drug used related to lifestyle causes was on the increase.**

To put this data into business format and jargon, use a format similar to the one illustrated in Table 26.3.

Table 26.3: Sample Business Case for Workplace Wellness

Business Case for Workplace Wellness

Area	Target Action	Potential Savings
Employee Absenteeism costs	Reduce absenteeism costs by 30%	$900,000
Short Term Disability Costs	Reduce STD claim costs to an average cost of $800.00 per claim	$2,280,676
Prescription Drug Costs	Reduce prescription drug costs for lifestyle conditions ($636K) by 20%	$127,200
Staff Turnover Costs	Reduce the current staff turnover rate from 10% to 5%	$48,250,000
Workers' Compensation Costs	Obtain a rebate of 40%	$2,178,000
EAP Utilization/Costs	Maintain at 10%	$50,000
	Total =	$53,785,876

Information for building a business case can also include industry-researched facts about the effect of WWPs, such as:

- Depression in an employee population is associated with a 70 per cent increase in healthcare costs.[46] In 2016, depression cost Canadian employers $32.3B per year; anxiety states contribute to an additional $17.3B.[47]

---

[46] Goetzel *et al.* (1998) in D. Powell, "20 Characteristics of Successful Worksite Wellness Programs" (Presented at the American Occupational Health Conference, Chicago, IL, April 17, 2002) [unpublished].

[47] Conference Board of Canada, *News Release*, "Unmet Mental Health Care Needs Costing Canadian Economy Billions" (September 1, 2016), online: https://www.conferenceboard.ca/press/newsrelease/16-09-01/Unmet_Mental_Health_Care_Needs_Costing_Canadian_

- "Stressed" employees experience a 46 per cent increase in healthcare costs.[48]
- A survey of WWPs reports the benefits such as improved employee health (by 41 per cent of the WWPs), reduced healthcare costs (by 27 per cent of the WWPs), improved employee morale (by 17 per cent of the WWPs), reduced workplace incidents (by nine per cent of the WWPs), reduced employee absenteeism (by eight per cent of the WWPs) and increased productivity (by eight per cent of the WWPs).[49]

In terms of a cost-benefit analysis, identify the total cost of the proposed WWP, including the related programs, planned activities and initiatives, required resources, and facility costs. Weigh those costs against the dollar benefits quoted to determine if the benefits outweigh the cost and by how much.

As well, providing the company decision-makers with a projected return on investment (ROI) ratio is always advantageous. They understand financial language and need that type of information to make an informed decision. As well, it shows them that the WWP planners have "done their homework".

To calculate the ROI, determine the savings realized by subtracting the total cost of the various programs with a WWP from the total cost of the various programs without a WWP. Then, establish the cost associated with developing, implementing and operating the WWP. Divide the savings realized by that value. The result is the ratio of dollars saved per dollar spent up to the value of the amount spent to implement and operate the WWP. For more information on calculating the ROI refer to Chapter 2, "Occupational Health and Safety Management Systems" and Chapter 16, "Marketing Occupational Health and Safety Programs and Communicating the Results".

A final selling feature might be to graphically demonstrate to Senior Management that an effective workplace wellness strategy and WWP can result in healthy people in a healthy place (Figure 26.15). An illustration like this can provide a powerful message to organization/company leaders who value their employees and their contributions to the success of the organization/company.

Economy_Billions.aspx (date accessed February 2020).

[48] Goetzel et al. (1998) in D. Powell, "20 Characteristics of Successful Worksite Wellness Programs" (Presented at the American Occupational Health Conference, Chicago, IL, April 17, 2002) [unpublished].

[49] D. Powell, "20 Characteristics of Successful Worksite Wellness Programs" (Presented at the American Occupational Health Conference, Chicago, IL, April 17, 2002) [unpublished].

Figure 26.15: Effective Workplace Wellness Strategy

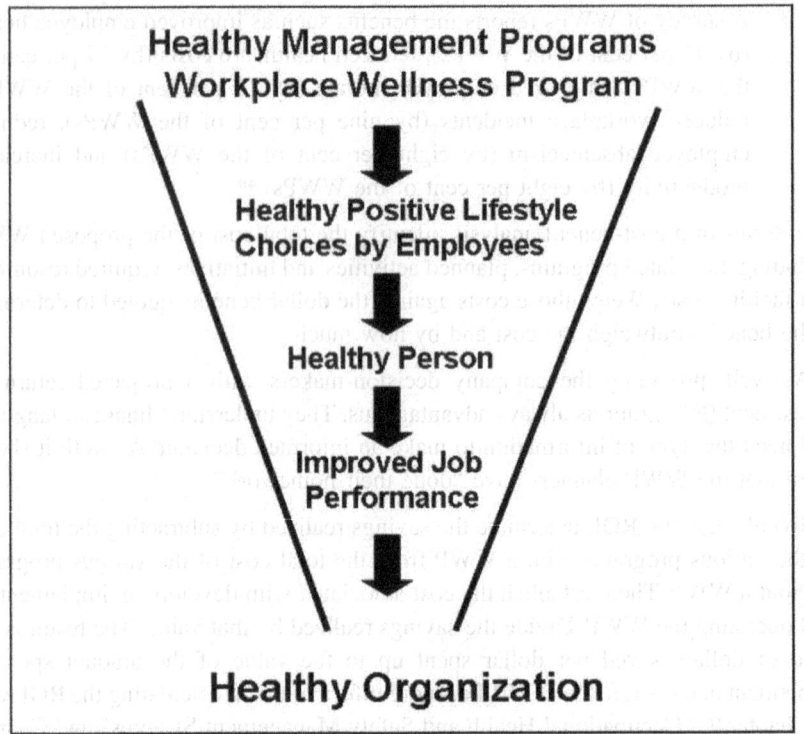

2. The Business Case: From a "Business" Perspective

Companies that focus on reactive business health culture approaches (*i.e.*, dealing with attendance and disability rates and costs; incident rates and costs; EAP utilization rates and costs; and/or staff turnover rates and costs) tend to focus their resources and energies on "failure costs". Failure costs are defined as the costs that relate to the failures of the corporate Occupational Health and Safety Management System (OHSMS).

However, detective and preventive business health culture approaches can positively impact an organization's business outcomes. By definition, *detection activities* focus on identifying workplace concerns and issues *before* they become problematic. Organizational climate evaluation, health and safety audits, employee health risk appraisals, pre-placement screenings, and incident or "near-miss" investigations are some examples of detection activities.

On the other hand, *prevention activities* focus on addressing and eliminating the identified concerns and issues so that they do not lead to health problems. Traditionally, fitness programs, nutrition counselling, smoking cessation, stress management, on and "off-the-job" safety training, immunization, and financial planning have been examples of workplace prevention initiatives. However, other approaches such as education on effective coping strategies and how to develop life coping skills could be offered to employees. Helping people to cope with their life and its challenges goes a long way towards the prevention of psychological disorders[50] — a major cause of employee medical absence. For more details, refer to Appendix 1.

A study conducted by Aon Consulting indicated that high performance leadership, communication, support of work-personal life balance, meaningful participation in and control over one's work, and development of effective interpersonal skills can help avoid workplace "failures" (*i.e.*, the failure of an employer to respond to the needs of its employees resulting in an incident).

Organizations should approach workplace health and safety as an investment, rather than a cost. Judicious investment in detection and prevention activities can significantly reduce "failure" costs. Research has shown that companies that focus on detection and prevention activities realize lower "failure" costs.[51]

Hence, a business case for a WWP should focus on demonstrating the positive impact that a WWP can have on the "bottom line". This means examining a number of outcome measures when establishing what business ROI a WWP can offer. These outcome measures include:

- the status of the corporate climate/culture;
- level of employee commitment and loyalty;
- level of employee job satisfaction;
- employee participation rate in various programs;
- employee attendance rates and costs;
- employee disability rates and costs;
- staff turnover and costs;
- employee benefit costs;
- disability and Workers' Compensation Board insurance premiums and costs; and
- the costs required to implement a WWP.

According to Danielle Pratt, President, Healthy Business Inc., formerly of

---

[50] W. Howatt, *The Coping Crisis: Discover why coping skills are required for a healthy & fulfilling life* (Toronto, ON: Morneau Shepell, Ltd., 2015).

[51] D. Pratt, "Competitiveness and Corporate Health Promotion: The Role of Management Control" (Graduate paper in Management Accounting, University of Western Ontario Business School, 1992) [unpublished].

Calgary, Alberta, the required themes in a business case for workplace wellness are the:

- impact of health, safety and wellness problems on the organization's/company's effectiveness;
- projected impact of doing nothing to address problems and how doing nothing will affect future costs, organizational resilience, employee morale, and corporate culture;
- magnitude of the potential improvement, which can be achieved by implementing workplace wellness best practices; and
- opportunity to achieve a level of excellence in employee health, safety and wellness. High-organizational performance can be achieved.

In his work on workplace health and well-being, Dr. Martin Shain, University of Toronto, reinforces the need to link workplace wellness with corporate business strategy. He has developed a *Business Health Culture Index* (BHCI) which is based on the relationship between job stressors (demand and pressure versus effort and fatigue) and job satisfiers (recognition and reward versus control).

The BHCI is a summary indicator of the extent to which the *health culture* of an organization supports its *business objectives*. It provides a simple basis for conceptualizing workplace health cultures as "business-positive", "business-neutral", or "business-negative". Business-negative health cultures actively obstruct the achievement of business objectives. Business-positive cultures facilitate the achievement of business objectives; and business-neutral cultures have no impact on business objectives.

The BHCI has the potential to yield benchmarks. Internally, it provides a baseline for action plans aimed at abating job stressors and enhancing job satisfiers. For example, a WWP objective may be to move the organization from a BHCI of 0.96 to one of 1.5, signalling an important shift in the ratio between workplace stress and job satisfaction. Externally, the index can provide a basis for comparisons within industry sectors, such as municipalities, hospitals, school settings, manufacturers, *etc*. This work serves as the basis for the "GuardingMinds @ Work" surveys, which are now readily available to Canadian employers.

The WWP business case and plan should undergo the same rigour and requirements of any other business function. To be sustainable, the WWP must speak in business terms and demonstrate targeted returns on investment.[52]

3. The Business Case: Based on Industry Findings

Many companies in Canada and the U.S. are now reporting their WWP outcomes.

---

[52] A useful resource for development of a Workplace Wellness Program is the Canadian Centre for Occupational Health & Safety (CCOHS), online: http://www.ccohs.ca/oshanswers/psychosocial/wellness_program.html (date accessed: February 28, 2020).

CH. 26: OCCUPATIONAL HEALTH AND SAFETY: WORKPLACE WELLNESS STRATEGY

The cost of workplace wellness per employee per year varies depending on whether a comprehensive approach or a single program is presented. As well, costs vary according to the formulae used and variables perceived to be program costs. Based on 13 research studies, the savings realized from WWPs were greater than their programming costs, with health cost savings averaging $3.48 per dollar, and absenteeism savings averaging $5.82 per dollar invested.[53]

For every dollar spent on WWPs, the returns have been cost savings of between $2.30 and $10.10[54] in the areas of decreased absenteeism, fewer sick days, decreased Workers' Compensation Board claim costs, lowered health and insurance costs, and improvements to employee performance and productivity.[55]

As for the benefits noted from WWPs, the following are the outcomes reported by a variety of companies:

- Canada Life developed a health promotion program in 1978 and had it independently evaluated over a 10-year period. The program showed a return of $6.85 on each corporate dollar invested:

    Reduced employee turnover, greater productivity and decreased medical claims by participating employees were primarily cited as the benefits contributing to this economic and health success.[56]

- DuPont (2002) reported an eight per cent differential in the reduction of the incidence rates of off-the-job illness between industrial worksites that offered health promotion programs and those that did not.[57]

- Providence Everett Medical Center, Everett, Washington, saved an estimated $3 million, or a cost-benefit ratio of one to 3.8, over nine years through its employee health benefit program called the Wellness Challenge®.[58]

---

[53] D. Anderson, *The Health Promotion First Act, Stay Well Health Management* (January 13, 2009), online: http://wellnessintheworkplace.net (date accessed: February 28, 2020).

[54] The average return on investment (ROI) is $3.50 in Canada: S. Pridham, "Different Strokes for Different Folks: Three Companies Take Unique Paths to Wellness", *Employee Benefit News Canada* (March-April 2008) at 16, online: http://ebn.benefitnews.com (date accessed: February 28, 2020).

[55] *Workplace Wellness Plan: ROI* (December 2, 2008), online: http://workplacewellnessprogram.com (date accessed: February 28, 2020).

[56] R. Kirby, "The ROI of Health Workplaces", *Canadian HR Reporter* (October 20, 1997) at 31.

[57] D. McReynolds, ed., *Worksite Wellness Programs: A Better Health Plan in which Everyone Benefits* (Personal Best Publications, Scott Publishing Inc., 2003) at 5.

[58] Workplace Wellness, *Workplace Wellness Program: The Bottom-Line Booster* (2009), online: http://workplacewellnessprograms.org/workplace-wellness-programs-the-bottom-line-booster/ (date accessed: February 28, 2020).

- Xerox Corp (2001) reported a 3.3 per cent differential in the decrease in frequency of WCB claims and a $2,976 differential in the average cost per claim between participants and non-participants in their workplace wellness program.[59]
- One of the longest and best-known case studies is the Johnson & Johnson Health and Wellness Program, which has been operating since the early 1980s. Reporting on data from 1990 to 1999, with more than 18,000 members, the program reported annual savings of more than $8 million, translating to savings of $225 per employee per year since 1995.[60]
- DundeeWealth, Toronto, Ontario, launched in 2006 its formal Workplace Wellness Program that embraced the slogan of "Health is Wealth". Focusing on three priorities — nutrition and weight management, physical activity and life balance — the company was already able to report positive employee participation rates. Going forward, they are tracking program participation rates, sick leave, absenteeism, and employee engagement rates.[61]
- Enbridge Inc., North York, Ontario, has operated their "Health Wise" program — an Integrated Disability Management Program for the past 20 years. In 2000, a WWP was introduced. It offered onsite physiotherapy, massage therapy, wellness seminars, nutrition counselling, and a worksite fitness facility. In 2006, Enbridge was recognized as a Top 100 Employer in Canada. In 2007, Enbridge reported long-term disability savings of $466,000 (26 per cent reduction).[62]
- Campbell Company of Canada, Listowel, Ontario, launched a "Winning Within" program in 2003. A highly interactive Workplace Wellness Program approach, the Campbell Company offered its employees an individual health assessment, followed by one-on-one professional counselling aimed at lifestyle changes. Using an integrative approach, Campbell's WWP is linked with its DMP and EAP to integrate the wellness themes. The high rate of employee involvement (more than 50 per cent) and below industry healthcare claims speaks volumes about the program's success.[63]

---

[59] D. McReynolds, ed., *Worksite Wellness Programs: A Better Health Plan in which Everyone Benefits* (Personal Best Publications, Scott Publishing Inc., 2003) at 7.

[60] R. Goetzel, R. Ozminkowski, J. Bruno, K. Rutter, F. Isaac & S. Wang, "The Long-Term Impact of Johnson & Johnson's Health & Wellness Program on Employee Health Risks" (May 2002) 44(5) *J. Occup. Environ. Med.* at 417-424.

[61] S. Pridham, "Different Strokes for Different Folks: Three Companies Take Unique Paths to Wellness", *Employee Benefit News Canada* (March-April 2008) at 16, online: http://www.ebnc.benefitnews.com (date accessed: February 28, 2020).

[62] *Ibid.*

[63] *Ibid.*

- Lighthouse Publishing, Bridgewater, N.S., reported a threefold return on its investment in its WWP.[64]
- PepsiCo and RAND Study of 64,000 people (2014) determined that WWPs designed to help individuals deal with chronic illness, disease management, yielded significant cost-savings — a monthly reduction of $30 per member in healthcare costs.[65] The lifestyle component of the WWP netted less; just $0.48 per dollar spent on the program. However, Carrns points out that WWPs do more; they project the message that the company cares about them, and help to attract people who want to work for a company that invests in employee well-being.

Some other notable research and industry workplace wellness findings include:
- British employees who feel little or no control at work have a greater risk of heart disease — 50 per cent more than those with executive positions. This study was conducted with 7,372 male and female British civil servants from 1985 to 1993.[66]
- Employees who smoke cost Canadian companies, on average, $4,256 more per year than non-smoking employees. This is due to increased absenteeism, lost productivity and increased health and life insurance premiums.[67]
- In 2002, the Centers for Disease Control and Prevention estimated that U.S. employers could save up to $12 billion annually by providing flu shots to employees. These savings were projected to come from reduced lost-time days and lower healthcare costs.[68]
- In 1996, Dr. Roy Shephard studied the effectiveness of workplace fitness programs. He found that regular participation resulted in reduced body mass by one to two per cent with a reduction of body fat of 10 to 15 per cent.

---

[64] S. Klie, "Wellness sees positive returns: Studies", *Canadian HR Reporter* (April 19, 2010) at 8.

[65] A. Carrns, "Study Raises Questions for Employer Wellness Programs" in *New York Times: Your Money/Your Money Adviser* (January 6, 2014), online: http://www.nytimes.com/2014/01/07/your-money/study-raises-questions-for-employer-wellness-programs.html?_r=0 (date accessed: February 28, 2020).

[66] R. Kirby, "The ROI of Health Workplaces", *Canadian HR Reporter* (October 20, 1997) at 31.

[67] Conference Board of Canada, *Smoking Cessation and the Workplace: Benefits of Workplace Programs* (2013), online: https://www.quitnow.ca/files/QN/files/library/Smoking_Cessation_and_the_Workplace_Briefing_3_Benefits_of_Workplace_Programs.pdf (date accessed: February 28, 2020).

[68] D. McReynolds, ed., *Worksite Wellness Programs: A Better Health Plan In Which Everyone Benefits* (Personal Best Publications, Scott Publishing Inc., 2003).

Muscle strength and aerobic capacity increased by up to 20 per cent.[69] Roy also reported on the effects of wellness on productivity as follows:

> Up to half of the burden of medical costs could be prevented by changes in personal lifestyle. Physical activity in particular has the potential to reduce both acute and chronic demands on the medical care system, with a reduction in employee turnover, an increase in productivity, a reduction in absenteeism, and a decreased risk in industrial injury.[70]

- The 1996 Wellness Survey by Buffett Taylor,[71] cited the top 10 success markers of a wellness program as depicted in the following graph (Figure 26.16).

Figure 26.16: Effectiveness of a Wellness Program: Top 10 Success Markers by Industry Sector[72]

maximum score = 2990

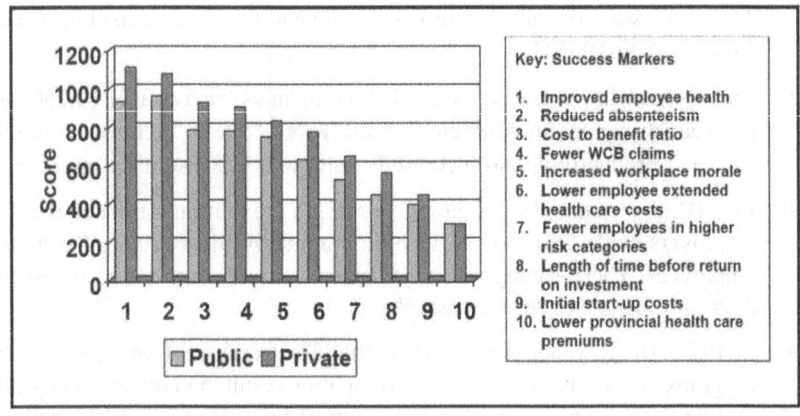

Ten years later, the participant companies in the Buffett and Company 2006 National Wellness Survey reported positive participant feedback, improved employee morale, increased worker participation, and reduced absenteeism as the top four results from their organization's wellness efforts (Figure 26.17). Interestingly, reduced employee absenteeism and improved employee morale remain high in terms of the observed results.

---

[69] R. Shephard, "Worksite fitness and exercise programs: A review of methodology and health impact" (1996) 10:6 *Am. J. Health Promot.* 436.

[70] Buffett Taylor, *The First Comprehensive Canadian Wellness Survey* (Toronto: Buffett Taylor & Associates Ltd., 1996).

[71] *Ibid.*

[72] Reprinted with permission from Buffett Taylor group of Companies, Buffett Taylor, *The First Comprehensive Canadian Wellness Survey* (Toronto: Buffett Taylor & Associates Ltd., 1996).

Figure 26.17: Buffett & Company, 2006 National Wellness Survey: Respondents' Experiences from Company Wellness Efforts[73]

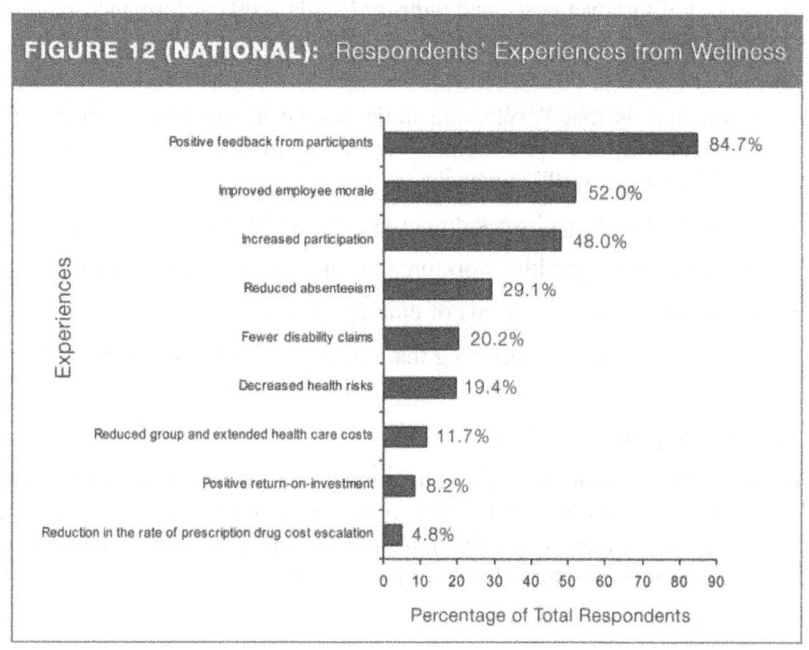

The good news is that in Canada, organizations are embracing the concept of workplace wellness. According to the Conference Board of Canada (2008), 64 per cent of organizations agree that their benefit programs focus on health promotion and disease prevention.[74] Of those surveyed, 72 per cent of organizations have a highly or moderately developed WWP; while only nine per cent of organizations report not having any workplace wellness initiatives in place.

In terms of advice, the Conference Board of Canada advocates that organizations and employees work together to address:

> . . . the socio-economic determinants of health. These determinants — income, employment and working conditions, education and literacy, housing, food security, *etc.* — have a greater impact on health outcomes than do genetics, individual choices, and the health-care delivery system.

---

[73] Buffett & Company, National Wellness Survey Report 2006 (Whitby, ON: Buffett & Company Worksite Wellness Ltd., 2006), now owned by Sun Life of Canada. Reprinted with permission from Buffett and Company Worksite Wellness Ltd.

[74] Conference Board of Canada, Healthy People, Healthy Performance, Healthy Profits: The Case for Business Action on the Socio-Economic Determinants of Health (2008), online: http://www.conferenceboard.ca/documents.aspx?did=2818 (date accessed: February 28, 2020).

By doing so, both parties can realize benefits such as *reduced* work distractions, workplace incidents, employee absenteeism, WCB claim costs, disability costs, benefit costs, staff turnover costs; and *improved* skills, work performance, employee engagement levels, employee retention rates, customer service and retention, staff recruitment and overall market competitiveness.[75]

The bottom line is that WWPs and initiatives positively impact organization/company productivity and profitability. According to the Wellness Councils of America (WELCOA), "well" companies:

- understand that employees do not get or stay healthy by chance;
- recognize that a healthy workforce means investing hard dollars;
- appreciate the diverse needs of employees; and
- understand that not everything that counts is visible on a balance sheet.[76]

(a) The Ongoing Business Case

Today, many organizations are willing to commit time and resources to workplace wellness, but need the assurance that there is a reasonable return on their investment for doing so; that is, there is a reduction in healthcare costs, increased employee health satisfaction, and increased worker productivity.[77]

For a WWP to be effective, it must:

- have "buy-in" by all stakeholders;
- be proactive in its approach;
- meet stakeholder needs;
- be aligned with the business strategy; and
- demonstrate that it adds value to the organization and employees in terms of effectiveness, efficiency, productivity, and profitability (or resource optimization).

Prior to the introduction of a WWP, baseline data should be obtained for the outcome measures that are to be used. Targets based on health promotion research can then be set for each measure. Achievements should be assessed regularly to determine whether the process is meeting its established objectives. The WWP should be aligned with, and be part of, the existing corporate business strategy. As

---

[75] *Ibid.*

[76] D. McReynolds, ed., Worksite Wellness Programs: A Better Health Plan In Which Everyone Benefits (Birmingham, AL.: Personal Best Publications, Scott Publishing Inc., 2003) at 15.

[77] Watson Wyatt, Staying@Work: Effective Presence at Work (2008) at 9, online: http://www.towerswatson.com (date accessed: February 28, 2020).

well, the WWP should have short-term, mid-range, and long-term goals and targets to justify and sustain its existence.

Measurement tools, such as corporate climate surveys, employee commitment (loyalty) surveys, rates of program utilization/participation, job satisfaction surveys, and employee group benefit rates and costs can be used to demonstrate the WWP's influence on an organization's business objectives.

How companies implement a WWP will vary depending on individual culture, business strategies, and business needs. Regardless of implementation, workplace wellness has been demonstrated to be cost-effective, and valued by employees and organizations alike. In short, "good health is good business".

G. HUMAN RESOURCES: ROLE IN WORKPLACE WELLNESS

Many stakeholders have a role to play in workplace wellness. However, the role of the Human Resources Department and personnel is crucial. They must:

- encourage management to develop a safe, healthy, and supportive workplace;
- build on the foundation of good management and business practices;
- promote the development of a positive corporate culture and climate through responsible leadership;
- develop explicit policies and procedures for individual programs and the linkages between those programs;
- ensure human rights and ethics are not violated when using human change/development technologies; and
- focus on promoting employee self-determination and self-responsibility.

H. SUMMARY

Workplace wellness is a proactive approach to workplace health and safety. Unlike the traditional OH&S and Human Resources approaches, it focuses on empowering employees to practice self-care, facilitates mutual aid within the workplace, and ensures that workplace environments are healthy, safe and supportive. An organization/company is only as strong as its weakest link: the message is: *fortify all employees and work functions, and the company, overall, will be strengthened.*

APPENDIX 1

The Globe and Mail hosts support on work life to employees and how to successfully cope. The website offers a quality of work life survey, along with a series of eight articles:

"Survey and Quality of Work Life Score (QWL): How's your life at work?"

> https://www.theglobeandmail.com/report-on-business/careers/career-advice/life-at-work/survey-hows-your-life-at-work/article16524403/

"Your Life at Work Survey: What can a manager do to help an unhappy employee?"

> https://www.theglobeandmail.com/report-on-business/careers/career-advice/life-at-work/what-can-a-manager-do-to-help-an-unhappy-employee/article17070452/

"Your Life at Work Survey: Is coping with work stress good enough?"

> https://www.theglobeandmail.com/report-on-business/careers/career-advice/life-at-work/is-coping-with-work-stress-good-enough/article17435268/

"Your Life at Work Survey: When you're unhappy, what motivates you to make a change?"

> https://www.theglobeandmail.com/report-on-business/careers/career-advice/life-at-work/when-youre-unhappy-what-motivates-you-to-make-a-change/article18370595/

"Your Life at Work Survey: Are you struggling to cope with the stress of work and life?"

> https://www.theglobeandmail.com/report-on-business/careers/career-advice/life-at-work/are-you-struggling-to-cope-with-the-stress-of-work-and-life/article18469369/

"Your Life at Work Survey: Survey says: We're stressed (and not loving it)."

> https://www.theglobeandmail.com/report-on-business/careers/career-advice/life-at-work/survey-says-were-stressed-and-not-loving-it/article22722102/

"Your Life at Work Survey: A lot of Canada's workers are stressed out. But what can be done to fix this?"

> https://www.theglobeandmail.com/report-on-business/careers/career-advice/life-at-work/video-a-lot-of-canadas-workers-are-stressed-out-but-what-can-be-done-to-fix-this/article22738814/

"Quality of Life Survey (QL). Are your satisfied with your life?"

> https://www.theglobeandmail.com/report-on-business/careers/career-advice/life-at-work/are-you-satisfied-with-your-life/article23426459/

Another source of support is the work by Dr. William Howatt, through the University of New Brunswick — Pathway to Coping Skills.

> https://www.unb.ca/cel/career/workplace-health-wellness/pathway-coping.html

Or, through Morneau Shepell:

https://www.morneaushepell.com/sites/default/files/assets/permafiles/65043/pathway-coping-online-course.pdf.

Dr. Howatt also authored "The Coping Crisis: Discover why coping skills are required for a healthy & fulfilling life" (Toronto: Morneau Sheppell, Ltd., 2015)

CHAPTER REFERENCES

D. Anderson, *The Health Promotion First Act, Stay Well Health Management* (January 13, 2009), online: http://wellnessintheworkplace.net (date accessed: February 28, 2020).

Buffett Taylor, *The First Comprehensive Canadian Wellness Survey* (Toronto: Buffett Taylor & Associates Ltd., 1996).

Buffett & Company, *National Wellness Survey Report 2006* (Whitby, ON: Buffett & Company Worksite Wellness Ltd., 2006), now owned by Sun Life of Canada.

Buffett and Company Worksite Wellness Ltd., *National Wellness Survey Report 2009* (Whitby, ON: Buffett and Company Worksite Wellness Ltd., 2009), now owned by Sun Life of Canada.

A. Carrns, "Study Raises Questions for Employer Wellness Programs" in *New York Times: Your Money/Your Money Adviser* (January 6, 2014), online: http://www.nytimes.com/2014/01/07/your-money/study-raises-questions-for-employer-wellness-programs.html?_r=0 (date accessed: February 28, 2020).

Conference Board of Canada, *Compensation Planning Outlook* (Ottawa: Conference Board of Canada, 1997).

Conference Board of Canada, *Healthy People, Healthy Performance, Healthy Profits: The Case for Business Action on the Socio-Economic Determinants of Health* (2008), online: http://www.conferenceboard.ca/documents.aspx?did=2818 (date accessed: February 28, 2020).

Conference Board of Canada, *Smoking Cessation and the Workplace: Benefits of Workplace Programs* (2013), online: https://www.quitnow.ca/files/QN/files/library/Smoking_Cessation_and_the_Workplace_Briefing_3_Benefits_of_Workplace_Programs.pdf (date accessed: February 28, 2020).

Conference Board of Canada, *News Release*, "Unmet Mental Healthcare Needs Costing Canadian Economy Billions" (September 1, 2016), online: https://www.conferenceboard.ca/press/newsrelease/16-09-01/Unmet_Mental_Health_Care_Needs_Costing_Canadian_Economy_Billions.aspx (date accessed February 2020).

D. Dyck, "Workplace Wellness: A Model for Success" (Presented at the American Occupational Health Conference, New Orleans, Louisiana, April 1999) [unpublished].

D. Dyck, "Workplace Wellness: What is Your Potential Return on Investment?" (Human Resources Association of Calgary, November 1998).

J. Epp, Health Promotion Model, *Achieving Health for All: A Framework for Health Promotion* (Ottawa: Health and Welfare Canada, 1986).

R. Goetzel, R. Ozminkowski, J. Bruno, K. Rutter, F. Isaac & S. Wang, "The Long-Term Impact of Johnson & Johnson's Health & Wellness Program on Employee Health Risks" (May 2002) 44:5 *J. Occup. Environ. Med.* 417.

Goetzel *et al.* (1998), cited in D. Powell, "20 Characteristics of Successful Worksite Wellness Programs" (Presented at the American Occupational Health Conference, Chicago, IL, April 17, 2002) [unpublished].

D.M. Harris & S. Guten, "Health-protective behaviour: An exploratory study" (1979) 20 *J. Health Soc. Behav.* 17.

Health Canada, *Smoking Cessation in the Workplace: A Guide to Helping Your Employees Quit Smoking* "Section V: Tools for Employers and Others Who Promote Health in the Workplace", *Estimating the Cost of Smoking in Your Workplace* (2008) at 7, online: http://www.hc-sc.gc.ca/hc-ps/pubs/tobac-tabac/cessation-renoncement/index-eng.php (date accessed: February 28, 2020).

W. Howatt, *The Coping Crisis: Discover why coping skills are required for a healthy & fulfilling life* (Toronto, ON: Morneau Shepell, Ltd., 2015).

R. Kirby, "The ROI of Health Workplaces", *Canadian HR Reporter* (October 20, 1997) at 31.

S. Klie, "Wellness sees positive returns: Studies", *Canadian HR Reporter* (April 19, 2010).

D. McReynolds, ed., *Worksite Wellness Programs: A Better Health Plan in which Everyone Benefits* (Birmingham, AL: Personal Best Publications, Scott Publishing Inc., 2003).

M.P. O'Donnell, *Design of Workplace Health Promotion Programs*, 2d ed. (Birmingham, MI: American Journal of Health Promotion, 1988).

M.P. O'Donnell, *Health Promotion in the Workplace*, 3d ed. (Albany, NY: Delmar Thomson Learning, 2002).

N. Pender, *Health Promotion in Nursing Practice*, 2d ed. (Norwalk, CT: Appleton & Lange, 1987).

D. Powell, "20 Characteristics of Successful Worksite Wellness Programs" (Presented at the American Occupational Health Conference, Chicago, IL, April 17, 2002) [unpublished].

D. Pratt, *Competitiveness and Corporate Health Promotion* (Calgary, AB: Health Business Inc., 1992).

D. Pratt, "Competitiveness and Corporate Health Promotion: The Role of Management Control" (Graduate paper in Management Accounting, University of Western Ontario Business School, 1992) [unpublished].

S. Pridham, "Different Strokes for Different Folks: Three Companies Take Unique

Paths to Wellness", *Employee Benefit News Canada* (March-April 2008) at 16, online: http://ebn.benefitnews.com (date accessed: February 28, 2020).

R. Shephard, "Worksite fitness and exercise programs: A review of methodology and health impact" (1996) 10:6 *Am. J. Health Promot.* 436.

Watson Wyatt, *Staying@Work: Effective Presence at Work* (2008) at 9, online: http://www.towerswatson.com (date accessed: February 28, 2020).

Willis Watson Wyatt, *2015-2016 Staying@Work Report* (2016) at 5, online: https://www.willistowerswatson.com/-/media/WTW/PDF/Insights/2016/08/2015-2016-staying-at-work-canada-research-findings-en.pdf (date accessed, February 28, 2020).

WHO, "The Ottawa Charter for Health Promotion" (1986), online: http://www.paho.org/English/AD/SDE/HS/OttawaCharterEng.pdf (date accessed: February 28, 2020).

Workplace Wellness Plan: ROI (December 2, 2008), online: http://workplacewellnessprogram.com (date accessed: February 28, 2020).

Workplace Wellness, *Workplace Wellness Program: The Bottom-Line Booster* (2009), online: http://workplacewellnessprograms.org/workplace-wellness-programs-the-bottom-line-booster/ (date accessed: February 28, 2020).

Chapter 27

DISABILITY MANAGEMENT: OVERVIEW[1]

A. WHAT IS DISABILITY MANAGEMENT?

The term "disability management" means different things to different people. For example, "disability management" has been defined as:

> . . . a collaborative partnership that involves employers, employees, unions, healthcare providers, and vocational rehabilitation professionals for the goal of minimizing the impact of injury or disability on an employee's capacity to perform his or her job.[2]

Disability management is also described as:

> . . . a workplace program that uses prevention, early intervention, and proactive return-to-work interventions to reduce the impact of injury and disability, as well as to accommodate those employees who experience functional work limitations.[3]

For the purposes of this text, **disability management** is a systematic, goal-oriented process of actively minimizing the impact of impairment on the individual's capacity to participate competitively in the work environment, and maximizing the health of employees to prevent disability, or reducing the risk of further deterioration when a disability exists.[4]

Hence, a **Disability Management Program (DMP)** is a workplace program designed to facilitate the employment of persons with a disability through a coordinated effort that addresses individual needs, workplace conditions, and legal

---

[1] Excerpts from D. Dyck, *Disability Management: Theory, Strategy & Industry Practice*, 6th ed. (Markham, ON: LexisNexis Canada, 2017).

[2] A. Ahrens & K. Mulholland, "Vocational Rehabilitation and the Evolution of Disability Management: An Organizational Case Study" (2000) 15 *Journal of Vocational Rehabilitation* 39.

[3] D. Rosenthal, N. Hursh J. Lui, R. Isom & J. Sasson, "A Survey of Current Disability Management Practice: Emerging Trends and Implications for Certification" (2007) 50 *Rehabilitation Counselling Bulletin* 76.

[4] D.G. Tate, R.V. Habeck & D.E. Galvin, "Disability Management: Origins, Concepts and Principles for Practice" (1986) 17:3 *Journal of Applied Rehabilitation Counselling* 5.

responsibilities.[5] Ideally, DMPs are proactive, as well as reactive in nature, and incorporate stakeholder involvement and accountability. Most DMPs are designed to control the human and economic costs of employee injury or illness, to convey a message that employees are valued, and to demonstrate compliance with the relevant legislation.

An **Integrated Disability Management Program (IDMP)** (Figure 27.1) is a planned and coordinated approach to facilitate and manage employee health and productivity. It is a Human Resources risk management and risk communication approach designed to integrate all organizational/company programs and resources to minimize or reduce the losses and costs associated with employee medical absence regardless of the nature of those disabilities. It is aimed at:

- assisting ill/injured employees and employees experiencing diminished work capacities;
- providing early intervention and support measures;
- facilitating a collaborative approach to managing employee disabilities;
- restoring the disabled employee's work/functional capacities to an optimal level;
- maximizing the disabled employee's capabilities;
- integrating the organization's/company's various employee support and group benefit programs;
- measuring program performance and outcomes in human, legal and business terms;
- evaluating the organization's/company's various disability management efforts and performance with a focus on continuous improvement; and
- attaining a healthy workforce through injury/illness prevention.

---

[5] National Institute of Disability Management and Research (NIDMAR), *Code of Practice for Disability Management* (Port Alberni, BC: NIDMAR, 2000) at 5.

Figure 27.1: The Umbrella of an Integrated Disability Management Program

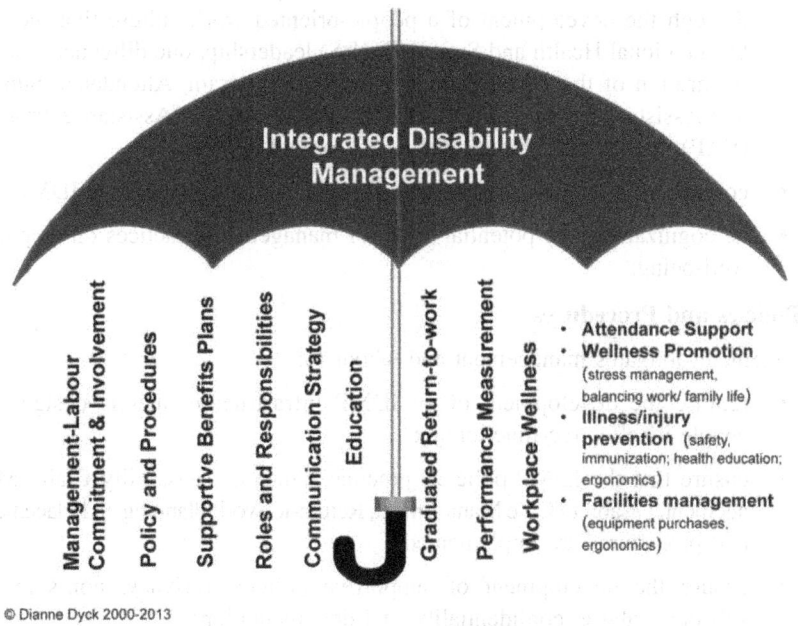

© Dianne Dyck 2000-2013

1. Management-Labour Commitment and Supportive Policies

This element compels management and labour to:

- be sensitive to the impact of disability on employees, families, work units, and the organization as a whole;
- be aware of the relevant pieces of legislation and the duty to comply (due diligence);
- promote the development of supportive policies and employee services that aim to keep employees at work;
- strive to create a people-oriented work culture that values employees when they are physically or psychologically ill/injured as well as when they are healthy;[6]
- develop management-labour agreements that protect employee employability and well-being;
- design employee benefit plans that reward a safe and timely return-to-work outcome;

---

[6] Institute for Work & Health, "Seven 'Principles' for Successful Return to Work" (Toronto: IWH, 2014) at 1, online: https://www.iwh.on.ca/sites/iwh/files/iwh/tools/seven_principles_rtw_2014.pdf (date accessed: January 31, 2020).

- support the development of flexible and creative return-to-work options;
- work together towards reducing employee presenteeism and absenteeism through the development of a people-oriented work culture that includes Occupational Health and Saftey (OH&S) leadership, due diligence, and the integration of the DMP, Human Resources Program, Attendance Support and Assistance Program, OH&S Program, Employee Assistance Program (EAP), and the Workplace Wellness Program (WWP);
- commit adequate financial and personnel resources towards the IDMP; and
- be cognizant of the potential effect of management practices on employee well-being.

2. Policies and Procedures

This element obligates management and labour to:
- require the development of an IDMP infrastructure that is designed to enable its effective functioning;
- ensure that the IDMP policies, practice standards (Disability Claim Management, Disability Case Management, Return-to-work Planning and Placement), and procedures are implemented;
- ensure the development of supportive policies (privacy, non-smoking, substance abuse, confidentiality, and documentation);
- require the development of employee support services (EAP, Attendance Control Program, WWP, OH&S Program, and Ergonomic Program, *etc.*);
- link the above employee services to function effectively under the umbrella of the IDMP; and
- ensure that the IDMP infrastructure and outcomes are regularly evaluated and updated.[7]

3. Stakeholder Education and Involvement

This element obligates management and labour to:
- establish common goals for an IDMP;
- design and develop an IDMP that addresses the organization's/company's needs;
- define and communicate stakeholder roles, responsibilities, and accountabilities in the IDMP;
- provide stakeholder education on the concepts of disability management and the various aspects of the IDMP, especially to supervisors who play an

---

[7] S. Cranston, "Integrated Disability Management: Feature Article", *Canadian Benefits and Compensation Digest* (2011) at 9-11.

influential role in getting the ill/injured employee back to work;[8]
- promote stakeholder sensitivity to the physical, psychological, social, and vocational consequences of disability;
- educate stakeholders on the importance of keeping employees at work;
- recognize and reward stakeholder participation;
- promote employee/supervisor understanding of the IDMP and their respective roles, responsibilities, and accountabilities in reducing and mitigating employee medical absenteeism; and
- motivate employees/supervisors to participate in the IDMP as well as in wellness and prevention activities aimed at illness/injury protection and prevention.

4. Supportive Benefit Programs

This element addresses the need for management and labour to:
- develop employee group benefit plans that support employee work attendance and that encourage a safe and timely return to work by ill/injured employees;
- link the organization's/company's attendance support programs with the Disability Management initiatives;
- provide EAP support that includes management and co-worker support, as well as employee and family counselling;
- integrate Workplace Wellness efforts with the organization's/company's disability management initiatives (refer to Chapter 21, "Occupational Health and Safety: Prevention of Workplace Illness and Injury");
- implement disability management plans that promote and accommodate workplace rehabilitation; and
- offer follow-up support after the employee returns to work.

5. A Coordinated Approach to Injury/Illness Management

This element requires the need for management to ensure the following actions occur, namely the:
- implementation of effective claim management processes;
- implementation of effective case management practices;
- initiation of early intervention post employee injury/illness — intervention that is viewed as considerate and caring in nature;

---

[8] IWH, *News Release*, "Effective workplace return-to-work programs incorporate health services, case coordination and work modification" (February 21. 2017), online: https://www.iwh.on.ca/media-room/news-releases/2017-feb-21 (date accessed January 28, 2020).

- provision of employee guidance towards access and use of responsible healthcare services (case management);
- implementation and use of medical/vocational fitness-to-work evaluations (*i.e.*, medical forms, job demands analyses, functional capacity assessment referrals, return-to-work clearance forms, *etc.*);
- implementation of multi-disciplinary interventions into the disability management process;
- development of alliances/linkages with external resources (*i.e.*, healthcare providers, EAPs, vocational rehabilitation, professionals, insurers, community support systems, *etc.*);
- use of early rehabilitation/retraining for the recovering employee;
- post-return-to-work follow-up to ensure sustainability of the return-to-work outcome;
- collect injury/illness data using disability management information systems; and
- regular evaluation of the claim management, case management, and return-to-work processes with a view to continuous improvement.

6. A Communication Strategy

This element deals with the need for management to ensure that:
- the needs of the key stakeholders are identified;
- the benefits of an IDMP for the key stakeholders are identified;
- the legal obligations of the employer, employee and union are recognized by those key stakeholders;
- a communication plan for the IDMP that includes a marketing component is developed;
- the IDMP's marketing strategy clearly identifies the benefits that the program can offer to the various stakeholders;
- the organization's/company's available communication vehicles are used to reach all the key stakeholders;
- an IDMP communication strategy and action plan that encourages a free flow of information to and from the key stakeholders is built; and
- the effectiveness of the implemented communication strategy and marketing plans is regularly evaluated and reported.

7. Graduated Return-to-Work

This element involves management ensuring that:

- available return-to-work options are regularly offered;[9]
- flexible and creative return-to-work options are used;
- a system of employee, union, and front-line identification of modified/alternate work options is in place;
- workplace accommodations are promoted and facilitated;
- safe and timely return-to-work practices are instituted and monitored;
- the return-to-work opportunity does not disadvantage supervisors or co-workers;[10]
- supervisor and co-worker understanding and support for the IDMP are promoted;
- supervisors are prepared to facilitate, implement, and support a timely and safe return to work;[11] and
- insurer support and participation is sought and maximized.

8. Performance Measurement

This element requires management to ensure that:

- suitable performance measures for the IDMP are established;
- strategies for measuring the desired performances/outcomes are developed;
- a plan of action for performance measurement is in place, implemented, monitored, and evaluated;
- the performance of the IDMP is evaluated as planned; and
- the return on investment realized by the IDMP is determined and reported.

9. Workplace Wellness

This element deals with the need for management to ensure that:

- disability data are analyzed;
- the jobs/positions that experience an increased incidence of injuries/illness are identified and assessed;
- the patterns of injury/illness are identified and addressed;
- prevention strategies (physical and psychological workplace health and

---

[9] S. Cranston, "Integrated Disability Management: Feature Article", *Canadian Benefits and Compensation Digest* (2011) 9-11.

[10] *Ibid.*

[11] *Ibid.*

safety, ergonomic support, positive employee-employer relationships, cultural proficiency, attendance support, *etc.*) are developed and implemented;
- incident-prevention practices are actively promoted;
- employee wellness is recognized and rewarded; and
- employee/union involvement in the workplace to increase employee job satisfaction is promoted and supported so that employee job satisfaction increases.

An IDMP addresses all aspects of employee presenteeism, absenteeism, and disability management. According to Humphreys,[12] integration of the organization's/company's resources and efforts can reduce or eliminate the factors that contribute to employee presenteeism, absenteeism, and disability issues. The result is lower rates, durations, and costs of presenteeism, medical absences, and disability issues,[13] and greater employee productivity.

B. THE CORNERSTONES OF DISABILITY MANAGEMENT PROGRAMS

In Canada, DMPs are present in 57% of organizations,[14] functioning either internally or externally. Fifty-one per cent (51%) of Canadian corporate DMPs have formal policies in place.[15] The question that now looms is:

> Do these DMPs incorporate the ten cornerstones of disability management programming?

The term **cornerstone** is defined as "something of basic importance"[16]; it is the foundation – the basis on which a concept is built. Without the presence of that component, the DMP cannot exist or function.

In the field of Disability Management, there are ten cornerstones that form a successful DMP. They are:

---

[12] B. Humphreys, "Absenteeism and Disability Management – Effective Strategies Require Confronting the Elephants in the Room", *Longwoods HR Resources Database 2010*, online: http://www.longwoods.com/content/21943 (date accessed January 28, 2020).

[13] E. Tompa, C. de Oliveria, R. Dolinschi & E. Irvin, "A systematic review of disability management interventions with economic evaluations" (2008) 18:1 *Journal of Occupational Rehabilitation* 16.

[14] Towers-Watson, *Pathway to Health & Productivity: Stay@Work Report 2011-2012* (2012) at 12, online: file:///C:/Users/Dianne/Downloads/Towers-Watson-Staying-at-Work-Report%20(4).pdf (date accessed January 28, 2020).

[15] Conference Board of Canada, *Disability Management: Opportunities for Employer Action* (Ottawa: ON: Conference Board of Canada, 2013), online: http://www.sunlife.ca/static/canada/Sponsor/About%20Group%20Benefits/Group%20benefits%20products%20and%20services/The%20Conversation/Disability/DisabilityManagement_SUNLIFE_EN.pdf (accessed January 28, 2020).

[16] *Merriam-Webster's Learner's Dictionary* (2016), *s.v.* "Cornerstone", online: http://www.merriam-webster.com/dictionary/cornerstone (accessed January 28, 2020).

1. Early Intervention

Early intervention is defined as an employer-initiated response aimed at keeping the ill/injured employee connected with the workplace; and potentially, preventing the medical absence in the first place. Early intervention tends to occur following the onset of the illness/injury. It also includes the actions taken to assist employees who are experiencing diminished functional or work capacity, to remain at work.

The intent of early intervention is to facilitate appropriate and timely treatment and rehabilitation, as well as a safe and timely return to work. Through the use of early intervention, organizations/companies are able to shorten the employee's absence duration[17,18,19] and to mitigate the potential losses associated with the disability, as well as reduce the overall human and financial costs.[20] It also preserves the status of the **occupational bond** between the employer and the ill/injured employee.

With any employee work absence, research and industry experience support the importance of early intervention. For example:

- Work-based rehabilitation begins sooner and is more successful when early intervention is employed.
- The reason for timely (prompt) attention is that after 12 weeks (three months) of absence from work, the chance of the employee returning to work is reduced to 50%. This percentage decreases significantly as the weeks pass; by 24 weeks (six months), the percentage of employees that return to work post-physical/psychological illness/injury was found to be 20%. At the 48-week point (12 months), only 2% of the disabled employees returned to work (Figure 27.2).

  Delays in recovery tend to be associated with the development of psychosocial problems that in turn, require psychological support and

---

[17] Institute for Work & Health, "Seven 'Principles' for Successful Return to Work" (Toronto, March 2007; rev. 2014) at 2, online: http://www.iwh.on.ca/seven-principles-for-rtw (accessed January 28, 2020).

[18] Institute for Work & Health, "Seven 'Principles' for Successful Return to Work" (Toronto, March 2007; rev. 2014) at 2, online: http://www.iwh.on.ca/seven-principles-for-rtw (accessed January 28, 2020).

[19] Sun Life Financial, "Early Intervention Programs Can Speed Disabled Employees Back to Work", reported in Insure.com, "Early intervention cuts disability costs makes workers happier" (2007), online: http://www.insure.com/disability-insurance/early-intervention-costs.html (accessed January 28, 2020).

[20] P. Reed, "Recent Ruling on ADA and the Value of Interventions" (2002) 27:2 *Employee Benefits Journal* 3.

treatment.[21] The longer the employee is away from work, the less likely he/she will ever return.[22] The window of opportunity for successfully bringing the employee back into the workplace appears to be within the first 30 days following the absence onset. By facilitating appropriate and timely treatments and rehabilitation, employers can assist employees to regain their health, thereby enabling them to return to work within a shorter time frame. This approach benefits the employee, family, union, and organization.

---

[21] H. Harder & L. Scott, *Comprehensive Disability Management* (Elsevier, Churchill Livingstone, Toronto: 2005) at 7.

[22] J. Curtis & L. Scott, "Integrating Disability Management into Strategic Plans: Creating Healthy Organizations" (2004) 52:7 *AAOHN Journal* 298; L. Gates, Y. Taler & S. Akabas, "Optimizing Return to Work Among Newly Disabled Workers: A New Approach to Cost Containment" (1989) 5:2 *Benefits Quarterly* 19.

- Early intervention should be tailored to the ill/injured employee's situation and needs. It must reinforce and support the importance of work as a critical aspect of human life; help to identify and promote treatment of ailments; and aim at keeping the employee in control of his/her life.[23]
- Effective early intervention and return-to-work outcomes are strongly associated with the quality of the case management offered. Bussé, 2012, demonstrated that educated, competent, and experienced case managers make a positive difference, especially when they include follow-up services once the employee returns to work.[24]

Figure 27.2: Relationship between Time Away and Return to Work[25]

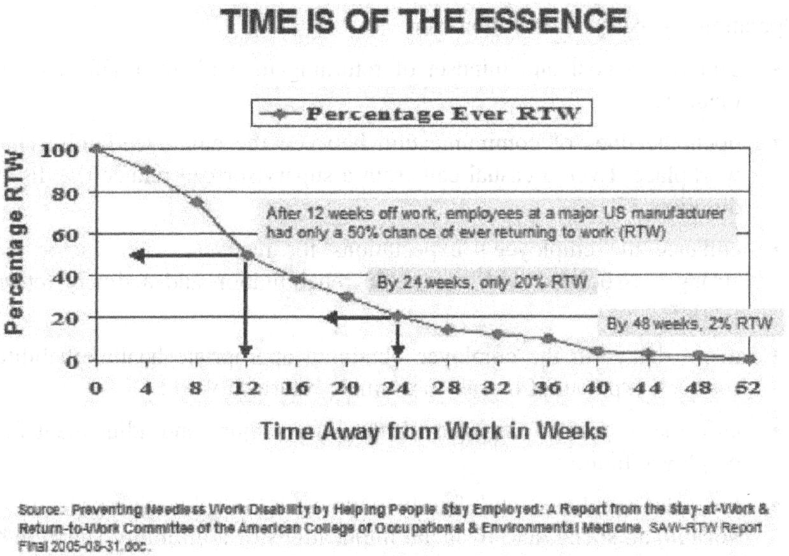
TIME IS OF THE ESSENCE
Percentage Ever RTW
Percentage RTW
100
80
60
40
20
0
After 12 weeks off work, employees at a major US manufacturer had only a 50% chance of ever returning to work (RTW)
By 24 weeks, only 20% RTW
By 48 weeks, 2% RTW
0 4 8 12 16 20 24 28 32 36 40 44 48 52
Time Away from Work in Weeks
Source: Preventing Needless Work Disability by Helping People Stay Employed: A Report from the Stay-at-Work & Return-to-Work Committee of the American College of Occupational & Environmental Medicine, SAW-RTW Report Final 2005-08-31.doc.

- Although the cost of rehabilitation services is greater when early intervention is implemented, the number of lost workdays is decreased. When

---

[23] H. Harder, "Early intervention in disability management: Factors that influence successful return to work" (2003) *International Journal of Disability, Community & Rehabilitation*, online: http://www.ijdcr.ca (accessed January 28, 2020).

[24] J. Bussé, "Case management potential area for return-to-work improvement" (2012) 68 *At Work* at 4.

[25] Partnership for Workplace Mental Health, *Assessing and Treating Psychiatric Occupational Disability*, Report released by the American Psychiatric Foundation (2005) at 21, online: http://www.workplacementalhealth.org/pdf/disabilityreportpart1.pdf (accessed January 28, 2020).

factoring in the reduced costs for time off work, the overall benefit to cost ratio was approximately seven. This meant for every dollar invested the return on the investment was $7. The intervention also had the additional benefit of identifying and eliminating workplace risks for other employees.[26]

Informally, many organizations/companies report more success with returning the recovering employee to the workplace if the intervention begins before, at, or soon after, the time of illness/injury. Typically, early intervention involves contacting the ill/injured employee and initiating case management when warranted. Organizations/companies demonstrating the best results in disability management outcomes begin early intervention by day three or five of a non-occupational medical absence, and by day one of an occupational-related medical absence.

Operationally, early intervention can:

- establish a goal and mindset of returning to work in a safe and timely manner;
- open the lines of communication between the employee/family and the workplace. Even a casual call from a supervisor can reduce the disability duration; [27]
- enhance the employee's expectations for recovery — a factor that is strongly associated with successful rehabilitation and a timely return-to-work;[28]
- avoid delays in the employee obtaining appropriate health/rehabilitation services, a potential barrier to a timely return-to-work;
- ease the process of coping with the illness/injury and adjustment for the employee/family;
- help the employee and family cope with the physical, psychological, vocational, social, and financial implications of a disability situation;[29]
- encourage family members to provide positive reinforcement and support to

---

[26] B.B. Arnetz, B. Sjögren, B. Rydéhn & R. Meisel, "Early workplace intervention for employees with musculoskeletal-related absenteeism: a prospective controlled intervention study" (2003) 45:5 *Journal of Occupational & Environmental Medicine* 499.

[27] D. Champagne, "Manage and measure disability programs", *Benefits Canada* (2016), online: http://www.benefitscanada.com/benefits/disability-management/manage-and-measure-disability-programs-32849 (accessed January 28, 2020).

[28] D. Gross & M. Battié "Work-related Recovery Expectations and the Prognosis of Chronic Low Back Pain Within a Workers Compensation Setting" (2005) 47 *Journal of Occupational and Environmental Medicine* 428, online: http://www.rtwknowledge.org/browse.php?article_id=92&view_type=research (accessed January 28, 2020).

[29] T. Riggar, D. Maki & A. Wolf, *Applied Rehabilitation Counselling* (New York: Springer Publishing Co., 1986).

the ill/injured and recovering employee;
- assist the employee and family to re-establish a sense of control of their life situation;
- reduce the negative effects of physical and psychological de-conditioning;
- prevent a break in the **occupational bond** — the mutually beneficial relationship between the employee and the employer;[30]
- enhance employee motivation to return to work;
- address both the business and human aspects that impact a timely recovery and safe return-to-work;
- minimize the separation and loss of support from co-workers;[31]
- decrease/prevent feelings of loneliness and abandonment that reduce the employee's motivation to get well;
- help prevent the development of psychological problems, such as the adoption of the "**sick role**"[32] and its related secondary gains;
- increase the likelihood of a successful rehabilitation outcome; and
- prove to be cost-effective. Research has shown a 47% return to work rate among workers referred for rehabilitation services within three months post-injury. This led to a 71% cost savings.[33] In contrast, only 33% of those referred for rehabilitation services later at the four to six-month post-injury period returned to work, and the cost savings dropped to 61%.[34]

2. Disability Claim Management

Disability claim management is the service provided to administer income loss claims through employee benefit insurance plans such as short-term disability, workers' compensation, and long-term disability. This activity includes:
- the determination of claimant eligibility to receive a benefit according to the definition of eligibility contained in the plan contract;

---

[30] D. Shrey, *Principles and Practices of Disability Management in Industry* (Winter Park, FL: GR Press Inc., 1995).

[31] T. Michalak, "Disability Management: An Assessment of Psychological Factors and Early Intervention" (2007) 6:1 *International Journal of Disability, Community & Rehabilitation*, online: http://www.ijdcr.ca/VOL06_01_CAN/articles/michalak.shtml (accessed January 28, 2020).

[32] The "sick role" is a societal-sanctioned role that an ill or injured person assumes once he/she becomes ill or injured.

[33] R. Rundle, "Move Fast if You Want to Rehabilitate the Worker" (1983) 17:18 *Business Insurance* 10.

[34] G.C. Pati, "Economics of Rehabilitation in the Workplace" (1985) 51:4 *Journal of Rehabilitation* 22.

- the facilitation of income loss replacement; and
- the processing of the claim towards a resolution or termination.

The personnel undertaking disability claim management are termed disability claim analysts, or disability claim administrators.

The disability claim management steps involve:

- *Determining claimant eligibility for income replacement benefits* – is the claimant's illness/injury work-related or not-work-related? Does the claimant meet the criteria for disability insurance coverage based on employment status? Is the illness/injury compensable or was it due to an exclusion element like participation in a riot, war, or illegal activity?
- *Gathering the information required to support a claim for income replacement* – The type of information required to support the employee's claim for income replacement varies with the disability benefit desired. For example, each Workers' Compensation Board requires at a minimum, the submission of a Worker Report of Accident, the Employer Report, and a Physician's First Report, to establish a claim. As for non-work-related claims, the required information varies, but generally involves medical validation of the illness/injury and a statement of the employee's fitness-to-work status.
- *Processing claim forms* – depending on the nature of the claim, the claim processing also varies. For example, Workers' Compensation Boards state a timeframe within which a claim has to be submitted, as well as a description of the necessary claim forms. As for non-work-related claims, the information required generally involves the completion of a report of absence of some sort. As the claim progresses, more information is needed to continue to support income replacement. The claim processing also includes how the various disability insurance plans, although mutually exclusive, can and do interface.
- *Claim adjudication* – is the process of determining whether a claim is eligible under the terms of the benefit contract or plan for benefit coverage. It involves the:

    1) receipt and review of the claim;
    2) establishment of the claimant's status and eligibility for benefit coverage;
    3) review of the eligibility requirements according to the contract/plan, and/or the collective agreement;
    4) consideration of the issue of any limitations or exclusions that may apply, consideration of the existence of any pre-existing conditions;
    5) consideration of the existence of specialized clauses that would preclude the claimant from benefit coverage (*e.g.*, risky activities, or sports);

6) determination of eligibility, acceptance or rejection of the claim; and
7) ensuring that the claimant complies with the recommended treatment plans, and to accept reasonable work accommodation. Failure to do so can result in termination of claim benefits.

- *Claim evaluation* – is the process of monitoring the claim from onset to closure, in terms of the days lost and the lost-productivity costs.

Timely and competent claim adjudication is "key" to getting the ill/injured employee access to case management and return-to-work support. That in turn, mitigates the related losses and costs for the employer. processes, making them more accurate, efficient, timely, and cost-efficient.

Claim management also includes providing the employee with the required claim submission forms and information, the details on the employee's role, information on the work accommodation available and the RTW program, and the effect of the absence on employee benefits. For the employee to responsibly and effectively manage an illness/injury situation, the disability claim administrator should ensure that the appropriate information, resources and supports are made available in a timely manner.

3. Disability Case Management

Disability case management is a collaborative process for assessing, planning, implementing, coordinating, monitoring, and evaluating the options and services available to meet an individual's health needs through communication and accessible resources to promote quality, cost-effective outcomes.[35] Disability case management promotes:

- safe and timely return-to-work efforts;
- early identification of disability claims for services and coordination of services, such as early intervention;
- maintaining contact with disabled employees;
- developing and monitoring modified/alternate work opportunities; and
- coordinating issues with the insurer and arranging for vocational rehabilitation when required.

Disability case management is intended to assist ill/injured employees to reach the highest level of medical improvement possible and to facilitate a RTW outcome in the most cost-effective manner. The Disability Case Manager is the navigator, or lynchpin, of the process.[36] According to Bussé (2012), focusing on competent

---

[35] Case Management Society of America, *Standards of Practice for Case Management* (Little Rock, AR: CMSA, 1995).

[36] J. Bussé, "Case management potential area for return-to-work improvement" (2012) 68 *At Work* at 4.

disability case management can greatly improve the organization's DMP outcomes:

> Successful case management requires skills in communication, diplomacy and relationship-building, as well as in planning, coordinating, and evaluating a rehabilitation plan.[37]

This points to the need for the disability case management to attain formalized DMP education, as well as to have the opportunity to apply the learned skills and knowledge.

Additionally, for the Disability Case Manager to competently practice, the organization needs to ensure that the role of the Disability Case Manager is clearly defined, along with the expected level of practice. This can be achieved through the development of a Disability Case Management Practice Standard,[38,39] thereby positioning the Disability Case Manager to clearly understand the role, responsibilities, and processes; to promptly reach out to the ill/injured employee; to demonstrate politeness, respect, and confidence; and to remove identified recovery and return-to-work (RTW) barriers.

4. Return-to-Work Planning

Return-to-work planning is viewed as a "socially fragile process",[40] in which the returning employee, supervisor, and co-workers face the challenge of developing new work relationships and duties. If the return-to-work placement disadvantages the supervisor/co-workers, resentment results. This outcome, in turn, can sabotage the return-to-work efforts. Hence, the return-to-work plans must anticipate and avoid negatively impacting supervisors and co-workers. Supervisor and co-worker support can positively impact the success of the inured/ill employee's return to work.[41]

A graduated return-to-work plan is designed to achieve the following objectives:

1. To ensure fair and consistent treatment for all employees returning to work;
2. To promote shared responsibility for effective graduated return-to-work

---

[37] J. Bussé, "Case management potential area for return-to-work improvement" (2012) 68 *At Work* at 4.

[38] D. Dyck, *Disability Management: Theory, Strategy and Industry Practice*, 6th ed. (Markham, ON: LexisNexis Canada Inc., 2017).

[39] COHNA-ACIIST, Disability Management Standard (2012), online: http://www.cohna-aciist.ca/assets/cohna%202012%20-%20disability%20management%20standard%20-%20electronic%20version.pdf.

[40] Institute for Work & Health, "Seven 'Principles' for Successful Return to Work" (Toronto, (2007, rev. 2014) at 3, online: http://www.iwh.on.ca/seven-principles-for-rtw (accessed January 28, 2020).

[41] IWH, "Supervisors who react with support can help injured workers return to the job" (Summer 2018) 93 *At Work*.

plans and placement among the ill/injured employee, supervisor, union, and Disability Case Manager;

3. To coordinate the graduated return-to-work plans with the disability claim management and case management services for the ill/injured employee;
4. To respect the rights and relationships present in the workplace;
5. To engage all parties in assisting the ill/injured employee to successfully return to work; and
6. To mitigate medical absence costs associated with disability claims.

Many individuals can facilitate the graduated return-to-work process — the employee, supervisor, Disability Case Manager, union, Return-to-work Coordinator, *etc*. However, regardless of who is involved, a successful graduated RTW outcome depends on a cooperative and collaborative approach between the employee, direct supervisor, union representative, management, and co-workers. Getting back to work has been shown to result in a significant improvement in the employee's health – physically and psychologically,[42] as well as with their quality of life and socioeconomic status.[43]

5. Return-to-work Placement

Graduated return-to-work and work accommodation are viewed as a "core element of disability management, leading to favourable outcomes".[44] Ideally, employees should aim to return to their own job — a position they know, and in which they can benefit from co-worker support.[45] However, work accommodation has to be mutually beneficial; that is, it must meet the employee's capabilities as well as the organization's business needs. In addition to this requirement, consideration should be given to a potential need for an ergonomic assessment to ensure a functional person-job fit.[46]

An industry example of a mutually beneficial RTW placement comes from CIBC, Canada. They accommodated a business analyst who was recovering from a mental illness. CIBC provided her:

---

[42] C. Mustard, "Getting back on the horse: Return to work has beneficial effect on health" (2012) 68 *At Work* at 5.

[43] Canadian Medical Association, *The Physician's Role in Helping Patients to Return to Work* (Ottawa: ON: CMA, 2010).

[44] Institute for Work & Health, "Seven 'Principles' for Successful Return to Work" (Toronto, 2007) at 3, online: http://www.iwh.on.ca/seven-principles-for-rtw (accessed January 28, 2020).

[45] Institute for Work & Health, "Seven 'Principles' for Successful Return to Work" (Toronto, 2007; rev. 2014) at 3, online: http://www.iwh.on.ca/seven-principles-for-rtw (accessed January 28, 2020).

[46] *Ibid.*

... with two screens for her computer so she could multitask, extra time to turn in work, and flexibility to work from home or modify hours when needed.

For every dollar spent on this work accommodation, the bank got $7.40 back in term of higher productivity and lower staff turnover.[47] As for accommodated employees, the economic benefits ranged from four to 12 times the accommodation costs.[48]

6. Confidentiality

Confidentiality is the maintenance of trust and the avoidance of invasion of privacy through accurate reporting and authorized communication. In relation to managing employee personal health information, all the individuals within the organization who collect, maintain, handle and use personal health information, are legally required to protect the confidentiality of that information.

7. Documentation

File documentation supports a well-managed disability claim and disability case, based on a well-thought-out process approach, *i.e.*, the problems are identified, actions implemented, results evaluated, and the costs and consequences considered. Documentation is crucial for effective disability claim management, disability case management, and return-to-work planning and placement. It serves to provide:

- a profile of the disability claim status and the disability claim management, disability case management, and return-to-work planning and placement services provided;
- a means of communication among members of the disability management team contributing to claim management, case management, and return-to-work planning and placement;
- a basis for planning and for continuity of claim management, case management, and return-to-work planning and placement for each disability situation;
- a basis for review, study, and evaluation of the claim, the claim management, case management, return-to-work planning and placement, and claim outcome;
- some protection for the medical and legal interests of both the employee and the organization; and
- an audit trail of activities completed which can serve as a "due diligence" tool if required.

---

[47] IWH, "Benefits outweigh costs for workplaces that accommodate people with mental illness" (Summer 2018) 93, online: https://www.iwh.on.ca/newsletters/at-work/93/benefits-outweigh-costs-for-workplaces-that-accommodate-people-with-mental-illness (accessed January 28, 2020).

[48] *Ibid.*

Documentation of the entire process is critical because it demonstrates the organization's due diligence in accommodating the employee back into the workplace. Data collection and analysis also provides reports demonstrating compliance with Canadian duty to accommodate legislation and supports the legal concept of due diligence.

8. Program Evaluation and Continuous Improvement

Program evaluation, which identifies the gaps between the current state and the desired state of a program, indicates whether the program goals/objectives are met or not, and enables program improvements. For a DMP to successfully operate and evolve, an understanding of its current state is critical, as is the recognition of the "ideal state" for a DMP.

The value obtained from conducting an evaluation of a DMP is that it:

- Creates greater stakeholder awareness of the DMP, its goals, elements, functions, and outcomes.
- Identifies opportunities for DMP improvement(s).[49,50,51]
- Provides direction for enhancement of the DMP's policy, standards, procedures, and elements.
- Increases stakeholder appreciation that "upstream" organization practices and behaviours (*leading indicators*) can positively impact "downstream" outcomes (*lagging indicators*), which are typically noted by the DMP.
- Promotes greater focus on inducing long-term behavioural and organizational culture change.[52]
- Can be used as a performance measurement for corporate incentive programs.
- Provides "real" data for organizational marketing initiatives; for enhancing the organization's image as a responsible player in Occupational Health and Safety; for leveraging system/program improvements; and for employee training programs.

---

[49] Towers Watson, *Pathway to Health and Productivity, 2011-2012 Staying@Work Survey Report* (2011) at 34, online: http://www.towerswatson.com/assets/pdf/6031/Towers-Watson-Staying-at-Work-Report.pdf (accessed January 28, 2020).

[50] S. Cranston, "Integrated Disability Management: Feature Article", *Canadian Benefits and Compensation Digest* (2011) 9.

[51] D. Champagne, "Manage and measure disability programs", *Benefits Canada* (2016), online: http://www.benefitscanada.com/benefits/disability-management/manage-and-measure-disability-programs-32849 (accessed January 28, 2020).

[52] Towers Watson, *Pathway to Health and Productivity, 2011-2012 Staying@Work Survey Report* (2011) at 34, online: http://www.towerswatson.com/assets/pdf/6031/Towers-Watson-Staying-at-Work-Report.pdf (accessed January 28, 2020).

- Demonstrates corporate due diligence in terms of managing employee health and disabilities.

Ideally, when developing a DMP, the organization establishes the desired performance measures. However, this is rarely achieved for a number of reasons.[53] Instead, DMPs are generally implemented with little forethought of what success would look like.

Once a DMP is established, the next step is to measure its actual performance by comparing the results against established DMP standards. For example, NIDMAR has established DMP standards; as well, there are industry disability management best practices that can be used as practice guidelines.[54,55,56] The effectiveness of the DMP can also be demonstrated in terms of the achievement of its stated goals, objectives, and targets.

Analyzing the data provides a measure of the DMP's cost-effectiveness, which in turn directly impacts the organization's bottom line.

9. Ethical Disability Management Practice

Ethics is defined as the science of morals, a system of principles and rules of conduct,[57] the study of standards of right and wrong, or having to do with human character, conduct, moral duty, and obligations to the community.[58] It is the moral reasoning that humans possess. In short, ethical practice is:

> Doing the right thing, at the right time, for the right person, in the right way and knowing why it is the right thing, at the right time, for the right person, in the right way.[59]

Disability management impacts corporate business plans and costs; individual/family well-being, vocational aspirations, and finances; work culture; and employee morale. As can be imagined, disability management is based on relationships and trust, and as such, ethical considerations must be addressed.

---

[53] H. Harder & L. Scott, *Comprehensive Disability Management* (Toronto, ON: Elsevier, Churchill Livingstone, 2005).

[54] D. Dyck, *Disability Management: Theory, Strategy and Industry Practice*, 6th ed. (Markham, ON: LexisNexis Canada Inc., 2017) c. 32.

[55] COHNA-ACIIST, Disability Management Standard (2012), online: http://www.cohna-aciist.ca/assets/cohna%202012%20-%20disability%20management%20standard%20-%20electronic%20version.pdf (accessed January 28, 2020).

[56] Canadian Medical Association, *The Physician's Role in Helping Patients to Return to Work* (Ottawa: ON: CMA, 2010).

[57] B. Kirkpatrick, ed., *The Cassell Concise English Dictionary* (London, England: Cassell Publishers Ltd., 1989), s.v. "ethic".

[58] *Ibid.*

[59] D. Dyck, *Disability Management* (presentation on the Ethical Aspects of Disability Management Programming to the University of Fredericton (2004-2020)) [unpublished].

Disability Management Practitioners/Professionals need to uphold ethical practices when dealing with the stakeholders involved in a disability situation. Hence, practicing in accordance with disability claim management, disability case management, and return-to-work practice standards is critical to the success of the DMP. As well, being able to manage and resolve the related goal conflicts at the individual and management level, is vital. A highly recommended approach to resolving goal conflicts includes the following steps:

1. identify and understand the underlying issues;
2. hold a case conference with the key players;
3. identify and address the issues as a group;
4. recognize that "competing agendas" exist. Unless there is objective medical rationale supporting a particular agenda, it cannot influence the return-to-work planning;
5. seek feasible solutions to rectify the situation, and using the disability management principles, select a suitable approach;
6. implement the plan;
7. monitor the return-to-work plan;
8. evaluate the outcomes; and
9. communicate the outcomes to the interested parties.

In disability claim management, disability case management, RTW planning and placement, and the program evaluation, the weighing of the ethics of a disability situation must be done in an unemotional manner so that the decision-making is rational and based on facts rather than on the emotional issues attached to the decision.

10. Legal Compliance

Disability management is a management response to Canadian legislation which upholds that:

- disabled employees cannot be discriminated against on the basis of a physical or psychological disability (Canadian human rights legislation);
- employers must provide work accommodation for workers recovering from an illness/injury (Workers' Compensation Acts, Canada Labour Code, Canadian human rights legislation);
- employees must be accommodated up to the point of undue hardship (Canadian human rights legislation); and
- employee personal health information must be respected and kept secure and confidential (Workers' Compensation Acts, privacy legislation).

As well, the disability management practices and processes are impacted by a variety of pieces of legislation. This legislation tends to vary from province to province, and from provincial to federal jurisdiction. The most important thing to note is that stakeholders involved in disability management *must* be:

- aware that specific acts and regulations are constantly changing and that they should obtain legal counsel to ensure they have the most current and up-to-date case law information when setting up programs or when dealing with specific human rights cases; and
- aware that ignorance of the law is never a valid excuse.

C. DISABILITY MANAGEMENT MODELS

There are a number of models or paradigms for disability management in today's work and marketplaces. Four examples are:

1. **Traditional Model** — This is a model in which the care plan, authorized leave, and return-to-work process are medically directed. The employer relies on the treating practitioners (primarily the employee's attending physician) to validate the illness and to help the employee to return to work. This model is often the "starting point" in disability management for many organizations, as well as for the insurer disability management models.

2. **Job Matching Model** — This is a model which involves a fitness assessment of the injured/ill employee and an analysis of the physical, social and psychological demands of the employee's job. The intent is to determine if there is a "match" or "mismatch" in terms of a safe return to work for the employee.

3. **Managed Care Model** — In a managed care model, the employee's diagnosis is referenced against standardized care plans, procedures, and diagnostic testing guidelines to determine if treatment and the physician's suggested leave duration are appropriate. This model, like the traditional model, tends to be medically driven.

4. **Direct Case Management Model** — This employee-employer approach to dealing with the employee's reduced work capacity and the employer's business needs/resources uses some of the elements of the first three models. However, it is the employee and employer who decide, based on their respective needs, the terms of the medical absence and the return-to-work plan.[60]

Although each of these four models were developed in response to different drivers, they all offer valuable contributions to the disability management process. In fact, most DMPs use some elements of each model. Typically, the Traditional

---

[60] A. Clarke, "Disability Case Management Models" (Presented at the Disability Case Management Forum, Vancouver, March 24-26, 1997) [unpublished].

Model is the starting point for a DMP,[61] and elements of the other models are then added as required, or as the organization's/company's DMP evolves.

From experience as a Disability Management Professional, and from auditing existing DMPs within various organizations, this author has learned that the best approach to disability management is one that focuses on maintaining a strong employee-employer relationship. Effective programs, such as the one depicted in the model developed by the National Institute for Disability Management and Research (Figure 27.3), maintain the employee-employer relationship, focus on the employee's capabilities versus disabilities, and are supported by a variety of technical specialists and case management approaches.

Supporting the employee and family through an illness/injury period usually promotes a "win-win" situation for all parties and related stakeholders. It also reduces the resistance to claim management and case management interventions and encourages a successful return-to-work.

---

[61] *Ibid.*

[62] Adapted from: National Institute of *Disability Management and Research (NIDMAR), Disability Management in the Workplace: A Guide to Establishing a Joint Workplace Program* (Port Alberni, BC: NIDMAR, 1995).

Figure 27.3: Employee-Employer Disability Management Model[62]

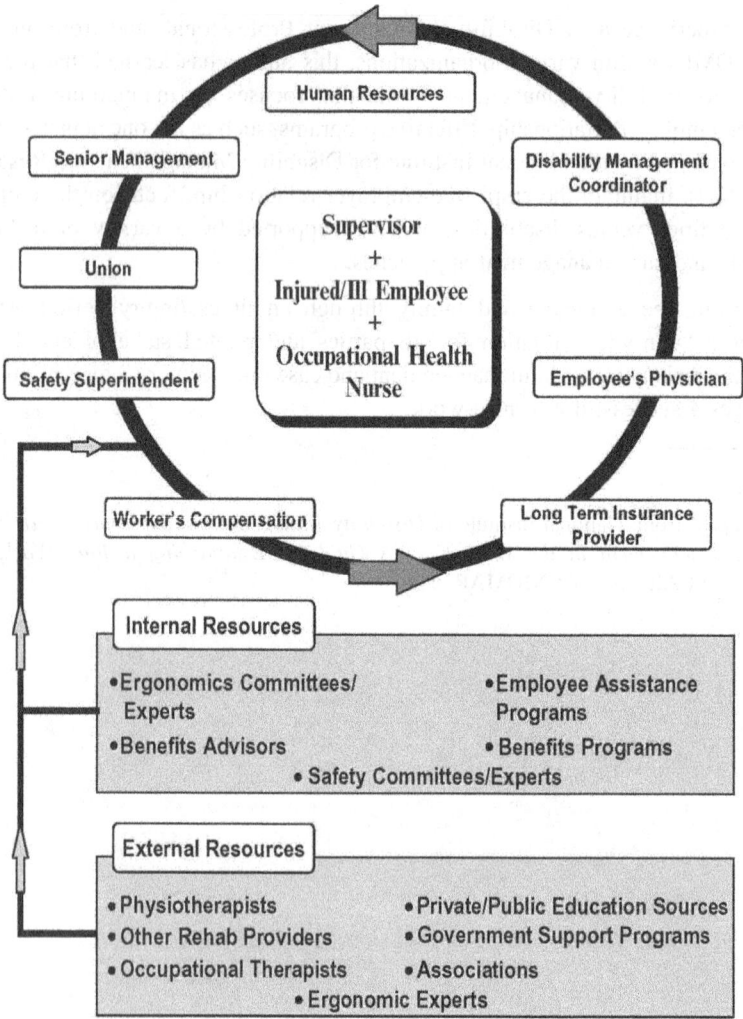

In terms of how disabilities play out in the Canadian workplace, there are many players involved, yet the employee and employer remain the key stakeholders. The nature of that relationship is critical: it is a bond — an occupational bond. When strong, both parties will move heaven and earth to keep it viable; but when shaky, it is extremely fragile and vulnerable. As well, there are a number of players who impact the occupational bond, especially during a disability period. See Figure 27.4 for a depiction of the relativeness of these players to a disability situation.

Figure 27.4: Disability Management: Key Stakeholders and Parameters

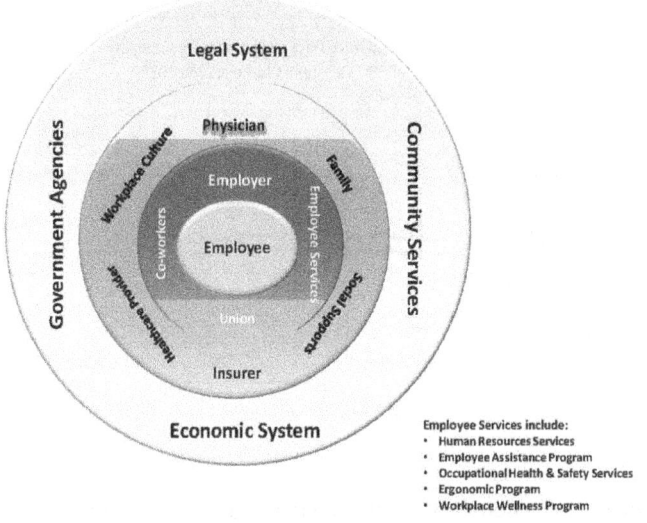

D. AN INTEGRATED DISABILITY MANAGEMENT PROGRAM: VALUE TO STAKEHOLDERS

To market and sell the merits of an IDMP, one must consider the interests and values of the key stakeholders.

1. For the Corporation

In 2019, full-time Canadian employees missed 10.3 workdays[63] for approximately 160 million lost workdays due to injury/illness and personal or family responsibilities.[64] This cost the Canadian economy about $37 billion in lost productivity.[65]

---

[63] Statistics Canada, *Table 14-10-0190-01 Work Absence of full-time employees by geography, annual* (Ottawa: Statistics Canada, 2020), online: https://www150.statcan.gc.ca/t1/tbl1/en/tv.action?pid=1410019001 (date accessed: January 31, 2020).

[64] Calculated by determining the total number of full-time employees for 2019 (15,536,200 FTEs) value and multiplying that by the average number of lost workdays for full-time employees in 2019 (10.3 days). Data obtained from Statistics Canada, *Table 14-10-0190-01 Work Absence of full-time employees by geography, annual* (Ottawa: Statistics Canada, January 31, 2020), online:https://www150.statcan.gc.ca/t1/tbl1/en/tv.action?pid=1410019001 and Statistics Canada, *Table 14-10-0287-01, Labour force characteristics, monthly adjusted*

Since 1998 when workers missed 7.8 workdays for a total of 72 million lost workdays, there has been a steady increase in employee lost-time days due to injury and illness (Figure 27.5).[66]

Figure 27.5: Average Work Absence Days and Reasons for Canadian Full-time Employees, 2000-2019

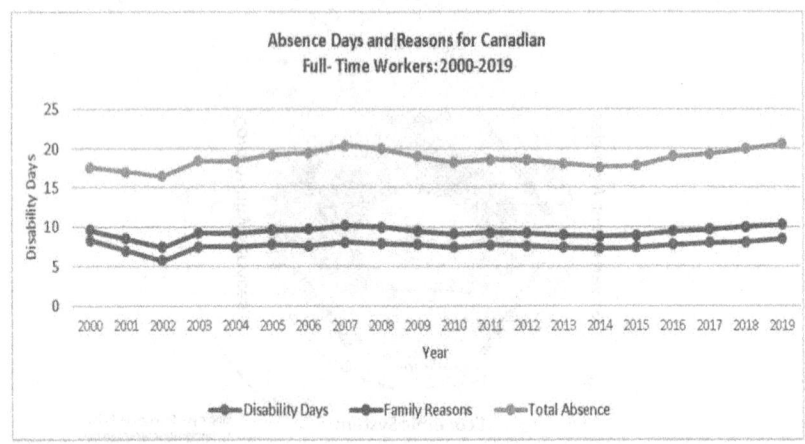

For Canadian employers, the average cost per full-time employee for workplace absence equates to $2,386 in lost productivity.[67] For an organization/company of

---

and trend-cycle, last 5 months (2019), online: https://www150.statcan.gc.ca/t1/tbl1/en/tv.action?pid=1410028701 (date accessed: January 31, 2020).

[65] Calculated by determining the daily wage from the average hourly wage for full-time Canadian employees as of October 2019 ($28.96 per hour) and multiplying that by eight hours/day and by the average number of lost time days per Canadian worker (10.3 days). That value is then multiplied by the number of full-time workers in December 2019 (15,536,200 FTEs). Data obtained from Statistics Canada, *Table 14-10-0190-01 Work Absence of full-time employees by geography, annual* (Ottawa: Statistics Canada, January 31, 2020), online: https://www150.statcan.gc.ca/t1/tbl1/en/tv.action?pid=1410019001; and Statistics Canada, *Table 11, Average Usual Hours and Wages of Employees by Selected Characteristics*, online: https://www150.statcan.gc.ca/n1/daily-quotidien/191108/t011a-eng.htm; and Statistics Canada, *Table 14-10-0287-01, Labour force characteristics, monthly adjusted and trend-cycle, last 5 months (2019)*, online: https://www150.statcan.gc.ca/t1/tbl1/en/tv.action?pid=1410028701 (date accessed: January 31, 2020).

[66] E. Akyeampong, *Work Absence Rates, 1987 to 1998* (Ottawa: Statistics Canada, 1999).

[67] Calculated by determining the daily wage from the average hourly wage for full-time Canadian employees as of December 2019 ($28.96 per hour) and multiplying that by eight hours/day and by the average number of lost time days per full-time Canadian worker (10.3 days). Data obtained from Statistics Canada, *Table 14-10-0190-01* Work Absence of full-time employees by geography, annual (Ottawa: Statistics Canada, January 31, 2020), online:

1,000 employees, this alone translates to a minimum of $2,386,000 in lost productivity costs (direct costs). According to the Hewitt Disability Absence Index 2007 Report, the *total direct* and *indirect disability absence costs* for 1,000 employees ranges from $3,500,000 to $3,675,000.[68]

Hence, employee absence rates and costs are escalating.[69] New illnesses, an aging workforce, more women in the workplace,[70] increased medical costs, decreased government support, and the lack of understanding of the total costs related to disabilities, all continue to contribute to the financial burden borne by employers. Add to this, the impact of employee presenteeism which is estimated to cost employers nine times the organization's/company's work absence costs.[71] **Employee presenteeism** is defined as the phenomenon of employees being at work, but because of wasted time, failure to concentrate, sleep deprivation, distractions, poor health, and/or lack of training, they may not be working at all.

Through an IDMP, employers can identify injury/illness causes, reduce the risk of injuries, promote employee well-being, and contain the healthcare and disability-related costs. The estimated saving is a 30 to 50 per cent reduction in lost time and related costs.[72] The net result can be a healthier workforce, reduced presenteeism and absenteeism, solid productivity, lowered disability costs, lower insurance premiums and rates, and Workers' Compensation discounts and/or rebates. As well, by helping the employee to overcome an illness/injury and to successfully return to work, the employer retains a valuable employee while decreasing disability costs and meeting legislative obligations.

2. For the Union

Unions, like management and employees, have a legal responsibility to help

---

https://www150.statcan.gc.ca/t1/tbl1/en/tv.action?pid=1410019001; and Statistics Canada, *Table 14-10-0287-01, Labour force characteristics, monthly adjusted and trend-cycle, last 5 months* (2019), online: https://www150.statcan.gc.ca/t1/tbl1/en/tv.action?pid=1410028701 (date accessed: January 31, 2020).

[68] Hewitt Associates, *Disability Absence Index Survey 2007* (2007), online: http://www.hewittassociates.com/Intl/NA/en-CA/KnowledgeCenter/ArticlesReports/DAIS_highlights.aspx (date accessed: January 31, 2020).

[69] Aon Hewitt Associates, Disability Absence Index Survey 2007 (2007), online: http://www.hewittassociates.com/Intl/NA/en-CA/KnowledgeCenter/ArticlesReports/DAIS_highlights.aspx (date accessed: January 31, 2020).

[70] B. Humphreys, "Absenteeism and Disability Management – Effective Strategies Require Confronting the Elephants in the Room" (2010), online: Longwoods HR Resources Database http://www.longwoods.com/content/21943 (date accessed: January 31, 2020).

[71] S. Aldana, *Top Five Strategies to Enhance the ROI of Worksite Wellness Programs*, Wellness Council of America (February 2009), online: http://www.absoluteadvantage.org/pdf/contentmgmt/top_5_strategies.pdf (date accessed: January 31, 2020).

[72] NIDMAR, *Occupational Standards in Disability Management: Executive Summary* (Port Alberni, BC: NIDMAR, 2000) at 2.

ill/injured employees (their members) return to work.[73] Through support and participation in an IDMP, union leaders can demonstrate their level of commitment and compliance to the Canadian human rights legislation. As well, an IDMP affords unions the opportunity to:

- promote member (employee) well-being;
- help address member "hot" health-related issues such as workplace stress, psychological illness/injury, and musculoskeletal injuries;
- maintain labour rights and principles;
- demonstrate value of the program to union members by protecting the employability of its members;
- actively participate in the return-to-work planning;[74]
- offer advice on the return-to-work placement; [75]
- support the work accommodation; [76]
- interact and build relationships;
- problem-solve in addressing areas of mutual interest and concern; and
- contribute to the organization's/company's profitability and competitiveness.

3. For the Employee

The benefits of an IDMP to the employee and family during a vulnerable period in their lives are numerous and invaluable. Some of these benefits include:

- the promotion of a speedy rehabilitation;
- the ability to maintain a sense of self-identity, self-worth, and self-respect;
- the opportunity to stay in contact with, and to gain support from, co-workers;
- the ability to remain current in their field; and
- less disruption in their normal family and workplace lives and relationships.

Graduated return-to-work (modified work) opportunities allow employees to:

- concentrate on recovery;

---

[73] B. Armstrong & S. Greckol, "Accommodation Guidelines" in *Illness and Disability Claims in the Unionized Workplace* (Winnipeg: Centre for Labour-Management Development, 1999).

[74] Canadian Human Rights Commission, *A Guide for Managing the Return-to-work* (2007), online: http://www.chrc-ccdp.ca/sites/default/files/gmrw_ggrt_en_2.pdf.

[75] *Ibid.*

[76] *Ibid.*

- keep a regular routine;
- maintain a sense of self-worth;
- make a contribution to the company;
- work at regular duties for as many hours as possible;
- keep work contacts;
- remain current with the changing work skill sets, duties, and responsibilities;
- remain current with changing technology;
- gradually adjust to full-time work;
- return to work without upgrading; and
- regain a positive health state.

In summary, an IDMP and the associated graduated return-to-work opportunities can be a "win-win" situation for everyone involved.

E. AN INTEGRATED DISABILITY MANAGEMENT PROGRAM: HYPE OR GOOD BUSINESS PRACTICE[77]

An IDMP is like any other business function. For it to survive, it must make good business sense, and offer a financial return on the resources invested. To demonstrate the impact an IDMP can have on the organization's "bottom line", first determine the costs associated with disability and the required DMP resources used to address those disability situations.

Secondly, determine the outcomes realized by other existing IDMPs, better known as **benchmarking**.[78] The value added by a DMP has been reported by a number of sources:

- An IDMP can reduce a company's benefit costs by 15 per cent to 35 per cent, depending on the benefits offered and how they are managed.[79]
- The American International Group (AIG) Claim Survey to 300 companies reported that by implementing case management procedures immediately after the occurrence of a Workers' Compensation claim, costs were reduced

---

[77] Excerpts reprinted with permission from D. Dyck, *Stating Your Case* (Ottawa: Benefits Canada, 1998) at 55-59.

[78] Benchmarking is defined as a continual and collaborative discipline that involves measuring and comparing the results of the key process with "best performers" or with one's own previous achievements.

[79] Aon Consulting, "The Case for Absence Management", *Aon Workforce Strategies* (2003), online: http://www.aon.com (date accessed: January 31, 2020).

by as much as 40 per cent.[80]

- According to the 2005 Watson Wyatt Staying@WorkTM Survey, 81 per cent of the 94 participating companies reported that they perceived documented return-to-work plans to be a key factor for managing disability-related costs, and for improving employee health, employee satisfaction, and productivity.[81] This approach is also supported in the 2011-2012 Staying@Work Report.[82] In short, managing occupational injuries makes good business sense for many Alberta companies.

- The Alberta Workers' Compensation Board (2005) reported that employers who instituted post-injury reduction services lowered their 2004 injury claim costs by 20 per cent as compared to non-participants.[83] This is a three per cent increase in savings over 10 years ago when the saving was 17 per cent.[84] This finding held true in 2011.[85] In short, managing occupational injuries makes good business sense for many Alberta companies.

- Watson-Wyatt, in the 2007 Staying@Work Survey, determined that the most effective cost-reducing health management practices include the use of:

    o disability case management with illness/injury claims;

    o documented modified work plans;

    o return-to-work plans for psychological illness/injury; and

    o supervisor/manager involvement in absence management.[86]

---

[80] AIG Claims Services Inc., "Early Intervention Cuts Workers' Compensation Costs", *Aon Commentary* (June 17, 1996).

[81] Watson Wyatt Worldwide, *Staying@Work Report 2005* (Canada, 2005), online: http://www.watsonwyatt.com/research/resrender.asp?id=w-860&page=1 (date accessed: January 31, 2020).

[82] Towers Watson, *Pathway to Health and Productivity, 2011-2012 Staying@Work Survey Report* (2011) at 8-9, online: http://www.towerswatson.com/en-CA/Insights/IC-Types/Survey-Research-Results/2011/12/20112012-StayingWork-Survey-Report--A-Pathway-to-Employee-Health-and-Workplace-Productivity (date accessed: January 31, 2020).

[83] Alberta Workers' Compensation Board, *2004 Annual Report* (Edmonton: Workers' Compensation Board Alberta, 2005) at 18, online: http://www.wcb.ab.ca/pdfs/public/annual_report_2004.pdf (date accessed: January 31, 2020).

[84] J. Cowell, "Serving Albertans Through Effective Injury Prevention and Disability Management" (Presented at the National Conference on Disability and Work: Solutions for Canadians, Sheraton Centre, Toronto, Ontario, October 7-9, 1996) [unpublished].

[85] Alberta Workers' Compensation Board, Publication no. WCB PS003, *Partnerships in Injury Reduction* (July 2011), online: http://www.wcb.ab.ca/pdfs/employers/pir_broch.pdf (date accessed: January 31, 2020).

[86] Towers Watson, *Staying @Work: Effective Presence at Work* (2007) at 10, online:

Lastly, the challenge is to present the merits of the IDMP in business language. A critical part of that language includes a cost/benefit analysis of launching such a program, the potential influence on the company's bottom line and the anticipated return on investment. The findings that can support these endeavours include:

- Since 1996, the number of companies that have implemented IDMP s has increased from 25 per cent to 51 per cent (2011-2012). The reason is simple: it is a cost-effective approach to managing worker absence and mitigating the associated costs. According to a recent Watson Wyatt Worldwide survey, savings of 0.25-1 per cent of payroll can be realized.[87]
- Shell Oil Company, Houston, Texas, implemented an "in-house" DMP to reduce non-occupational absences. The program was administered by full-time certified, corporate-based case managers and nine manufacturing location nurses. This program resulted in a 10 per cent reduction in total absence days per employee (6.9 to 6.2 days) as compared with the previous year. Business units not using this DMP had an eight per cent increase in absence days per employee (5.5 to 5.9 days). The return on investment equalled more than four to one return on investment based on direct expenditures and cost savings in terms of reduced absence days.[88]
- Suncor Energy Inc. reported that through its IDMP, 88 per cent of ill/injured employees returned to work in 2013. The notable outcomes are that the company through its work accommodation efforts, reduced the number of non-occupational disability absences by 30,000 work hours, and saved $2 million. This resulted in a significant savings in STD benefit costs and overtime/replacement worker costs.[89] In 2015 and 2016, Suncor focused on the prime reasons of employee medical absence, namely musculoskeletal and psychological disorders. By December 2016, the incidence of psychological disorders dropped from 19% to 4% of cases at the onset. Likewise, the number of occupational injury claims was reduced by 43%.[90] In 2018, Suncor reported a reduction in disability case duration of 4.3 days for a direct savings of $4.2M in lost productivity since 2016.[91]
- The direct costs of work absence and the related costs equates to between

---

http://www.towerswatson.com (date accessed: January 31, 2020).

[87] Watson Wyatt Worldwide, *Managing Health Care Costs in a New Era: 10th Annual National Business Group on Health/Watson Wyatt Survey Report, 2005*, online: http://www.watsonwyatt.com (date accessed: January 31, 2020).

[88] Shell Oil Company, Houston, "Impact of a Disability Management Program on Employee Productivity in a Petrochemical Company" (May 2006) 48:5 *J. Occup. Environ. Med.* 497.

[89] Suncor Energy Inc., *Report* (Calgary, Alberta, 2013) [unpublished].

[90] Suncor Energy Inc., *Report* (Calgary, Alberta, 2017) [unpublished].

[91] Suncor Energy Inc., *Report* (Calgary, Alberta, 2019) [unpublished].

five to six per cent of payroll (estimated to be 5.7 per cent by Watson Wyatt, 2011-2012[92]). Taking steps to address and lower worker absence just makes good business sense.

- Employers who implement at least three disability and absence management techniques have 74 per cent lower employee absence rates.[93]

F. HOW TO SELL AN INTEGRATED DISABILITY MANAGEMENT PROGRAM TO SENIOR MANAGEMENT

Many Human Resources and Occupational Health Practitioners/Professionals attend seminars, conferences, and industry focus groups on the topic of disability management. They leave these sessions convinced that an IDMP would be of value to their company or organization — however, they are unsure of how they can sell their ideas to senior management.

1. Perceived Barriers

One of the perceived barriers around initiating an IDMP is the myth that workplace accommodations are expensive. Workplace accommodation includes changes in, or reassignment of, parts of a job so that the recovering employee can return to work. This could translate into modifying existing job duties, offering transitional work, arranging for a training opportunity, providing an alternate job placement, or any combination of these.

Cantor[94] and Job Accommodation Network[95] claim that the majority of the workplace accommodations cost under $500, with 31-57 per cent of the workplace accommodations costing the employer nothing. The typical one-time expenditure was $500, with only five per cent of the workplace accommodations costing the company more than $5,000. On average, the return on investment on work

---

[92] Towers Watson, *Pathway to Health and Productivity, 2011-2012 Staying@Work Survey Report* (2011) at 10, online: http://www.towerswatson.com/en-CA/Insights/IC-Types/Survey-ResearchResults/2011/12/20112012-StayingWork-Survey-Report--A-Pathway-to-Employee-Health-and-Workplace-Productivity-at-Work-Report.pdf (date accessed: January 31, 2020).

[93] Washington Business Group on Health, *Fifth Annual Washington Business Group on Health/ Watson Wyatt Worldwide Survey on Disability Management* (Watson Wyatt Worldwide, 2004), online: http://www.towerswatson.com (date accessed: January 31, 2020).

[94] A. Cantor, "The Future of Workplace Accommodations: Containing Costs and Maximizing Effectiveness" (Presented at the National Conference on Disability and Work: Solutions for Canadians, Sheraton Centre, Toronto, Ontario, October 7-9, 1996) [unpublished].

[95] Job Accommodation Network, *Workplace Accommodations: Low Cost, High Impact, 2012* at 1, online: http://www.jan.wvu.edu/media/LowCostHighImpact.doc (date accessed: January 31, 2020).

accommodation was $10 for every dollar spent. Hence, work accommodations make good business sense.[96]

A second perceived barrier is the belief that there are limited modified work positions available within an organization for the recovering employee. This perception needs to be challenged. If all stakeholders are committed to making disability management work, modified work opportunities seem to materialize. From past experience, this author has come to appreciate that the employee population and union leaders are very resourceful at unearthing modified work positions.

A third perceived barrier is that the existing employee group benefit plans may be unsupportive of modified work. For example, group benefit plans that promote "absence with pay", or that allow the employee to earn more income by staying at home than by doing modified work, lack incentives for the employee to return to the workplace. This barrier may be real. For this reason, companies should examine their employee group benefit plans and determine the impact that each has on the employee return-to-work outcomes.

A fourth barrier is disability policies and procedures that focus on employee disability rather than capabilities. This approach promotes a "disability mindset", not an "ability mindset" for managers, Human Resources professionals, union leaders, and employees. This is one area that can negatively impact the corporate culture and that warrants serious introspection by an organization/company.

The fifth barrier centres around the belief held by some Human Resources managers that disability management has become so complex that the practice should be abandoned. This mindset tends to be reactive and akin to "throwing in the towel". Although the recent privacy legislation has made it unacceptable to access employee medical information, it has not removed the legal obligation of the employee to provide the workplace with the nature of the absence (work-related or non-work-related); the expected duration of absence; work limitations, if any; and a realistic return-to-work date. By providing Human Resources and the operations with relevant information on the employee's fitness to work in a timely manner, Disability Case Managers can overcome this barrier.

The sixth barrier is getting "stakeholder buy-in". Involving stakeholders, from the onset, in the design, development, and implementation of an IDMP can eliminate this barrier. Being part of the solution to a problem reduces the later need to market the IDMP.

The seventh barrier is gaining access to consolidated data that can be used to build a business case for an IDMP. Many organizations/companies do not possess accurate disability data, or information on their efforts to mitigate the losses associated with those disabilities. Producing historical disability data is equally impossible. Without this information, it is difficult for an organization/company to identify its current

---

[96] *Ibid.*, at 3.

situation, or to even envision what an ideal state for an IDMP would be. This is an area where human resources and occupational health and disability management practitioners can take a lead role in collecting, analyzing, and interpreting the impact of an organization's/company's disabilities and disability management initiatives.

The eighth, and perhaps most impactful, barrier is the culture of the organization/company. Management, union, and employee response towards the mitigation and prevention of employee illness/injury is reflected in the culture of the workplace — that is, in "how things are done around here". For example, in organizations/companies where management, union, and employees view illness/injury as a temporary loss and a loss that can be addressed and overcome, the response to mitigation efforts through disability management is embraced. However, there are industry examples of organizations/companies who view non-occupational injuries as "being of the employee's own making"; hence, organizational/company efforts that would normally be extended to mitigate work-related injuries/illnesses are not being offered to get the employee back to work. Since management leaders determine workplace culture, it is up to them to endorse and support the IDMP and efforts.

IDMPs return ill/injured employees to the workplace 50 per cent sooner than any other rehabilitation mode. The graduated return-to-work approach gets employees back to work and are in effect, a form of therapy. It enables the employee to recover much faster than staying at home or participating in simulated work rehabilitation programs.[97] Northwestern National Life Insurance (2002) reports that companies save $96 for every dollar spent rehabilitating an injured employee in the workplace.[98]

2. How to Move Forward

(a) Step 1: Analyze Your Situation

One way to analyze the company/organization's situation is to identify the barriers and drivers for an IDMP. Using a tool like Lewin's Force Field Analysis (Figure 27.6),[99] determine the various ways to decrease the barriers for implementing an IDMP, while increasing the program drivers.

---

[97] S. Gardner & P. Johnson, "Integrated Disability Management Programs: Good business for good organizations" (Jan-July 2004), online: http://findarticles.com (date accessed: January 31, 2020).

[98] *Ibid.*

[99] M. Brassard & D. Ritter, *The Memory Jogger: Tools for Continuous Improvement and Effective Planning* (Methuen, MA: GOAL/PC, 1994).

Figure 27.6: Lewin's Force Field for an IDMP

Driving Forces (Drivers)	Restraining Forces (Barriers)
Rising costs of disability →	← Lack of company/union awareness and "buy-in"
Disability management reduces costs →	← No early intervention
Available internal resources →	← Some non-supportive disability benefit plans
Employee Assistance Program available →	← Fear of workplace accommodation costs
Some supportive policies available →	← Perception of few graduated return-to-work opportunities
Claim management →	← Lack of rehabilitation resources
Attendance Support & Attendance Program in place →	← Mismanagement of medical claims

This Force Field Analysis helps individuals select the targets for change. By focusing on the restraining forces and looking for ways to reduce their effect, or ways to change them into driving forces, one can identify the real underlying factors preventing the implementation of an IDMP and the potential solution.

(b) Step 2: Gather Supportive Disability Data

By using research outcome findings, one can project the potential savings for the company/organization. This includes using recent survey data and the identified trends in disability costs. It also means the inclusion of any hidden costs of disability such as:

- the overtime paid for the remaining workers who assume a heavier workload while the ill/injured employee is absent;
- hiring replacement workers;
- training costs for replacement workers;
- lowered productivity due to the work-flow disruption;
- customer service disruptions;
- customer dissatisfaction;
- missed business opportunities;
- costs of employee benefits/services provided during the disability; and
- increased insurance premiums.

As well, the potential costs of "doing nothing" to manage illness/injury costs should be considered and taken into account. This can be demonstrated by showing what work absences can look like without an IDMP; and then, with an IDMP in place (Figure 27.7).

Figure 27.7: The Impact of an IDMP[100]

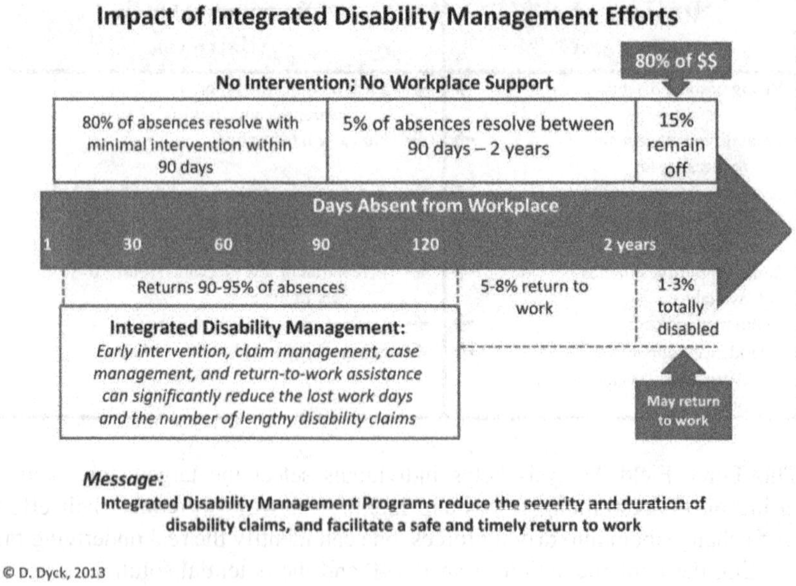

Impact of Integrated Disability Management Efforts

No Intervention; No Workplace Support

| 80% of absences resolve with minimal intervention within 90 days | 5% of absences resolve between 90 days – 2 years | 15% remain off |

80% of $$

Days Absent from Workplace

1 — 30 — 60 — 90 — 120 — 2 years

Returns 90-95% of absences | 5-8% return to work | 1-3% totally disabled

Integrated Disability Management:
Early intervention, claim management, case management, and return-to-work assistance can significantly reduce the lost work days and the number of lengthy disability claims

May return to work

Message:
Integrated Disability Management Programs reduce the severity and duration of disability claims, and facilitate a safe and timely return to work

© D. Dyck, 2013

(c) Step 3: Demonstrate the Value

Lastly, demonstrate the outcomes of whatever disability management efforts are in place, or are planned. This can be achieved by using the following principles:

- Consider the structure, process, and outcomes of the IDMP as per the established performance measures.
- Address the value that the IDMP offers to the organization. Is the program justified from a business standpoint? Is it justifiable from a financial or legal standpoint? What is the potential return on investment?
- Consider whether some of the IDMP performance measures should be valued higher than others. That is, is the impact on injury/illness severity more important than the injury/illness frequency of occurrence?
- Measure disability management data such as the frequency of claims; duration of claims; type and nature of claims; the number of long-term disability claims avoided; the return-to-work statistics; the cost of interven-

---

[100] Adapted from Kelly, Luttmer & Associates Ltd. in Hatscan, ed., *Comprehensive Disability Management* (Edmonton, AB: Hazard Alert Training & Supplies Canada Inc., 2000) c. 10, Fig. 10.1; D. Smith, "Implementing disability management: A review of basic concepts and essential components" (1997) 12:4 *Employee Assistance Quarterly* at 37-50; and S. Ritcey, "Psychological job matching" (Sept-Oct 1996) *OH&S Canada* 50.

tions; the cost of replacement workers; the degree of cost-avoidance through the graduated return-to-work opportunities; the Workers' Compensation Board costs; and any identified trends.

G. EVOLUTION OF AN INTEGRATED DISABILITY MANAGEMENT PROGRAM

Most organizations/companies become interested in disability management and establishing an IDMP as a result of increasing disability claims rates and costs. The initial DMPs tend to be very reactive in nature and then, slowly evolve towards an integrated approach to disability management (Figure 27.8).

Figure 27.8: Evolution of an Integrated Disability Management Program

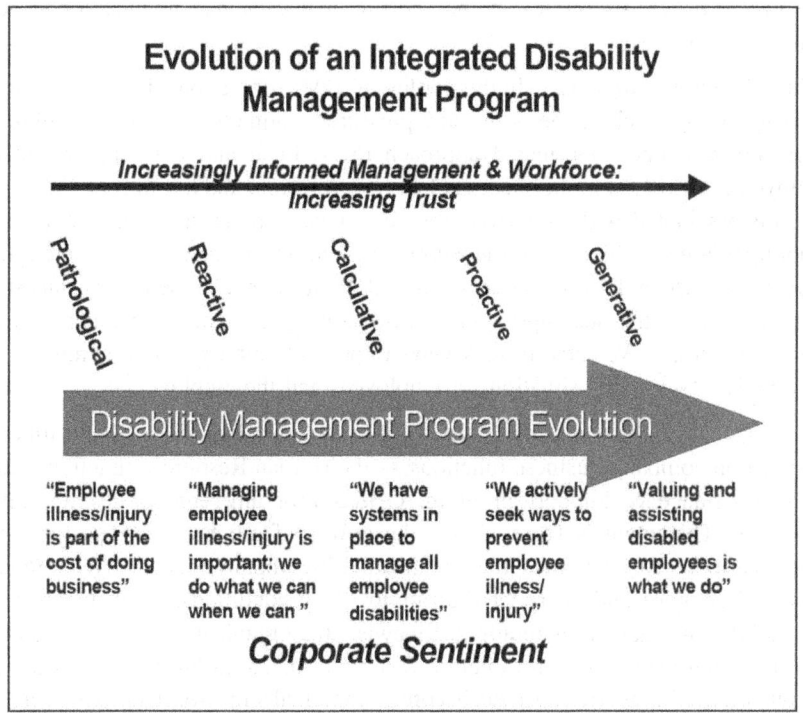

The *Pathological Stage* is where the organization/company fails to recognize that employee injury and illness are preventable occurrences. Rather, the organization/company believes that, "Employee illness and injury are part of the normal costs of doing business. After all, we are working with people. What do you expect?"

The *Reactive Stage* is typified by a realization that employee illness/injury events are costly in human and financial terms and should be addressed. "One-off attempts" at disability management are informally undertaken. The sentiment is "Managing employee illness/injury is important: we do what we can when we can." There is limited to no investment in disability management. Hence, none of the elements of a DMP exist, only a general awareness of the need to address employee illness and injury events as best they can.

The *Calculative Stage* signals a dawning of a formal DMP. The organization/company establishes a formal DMP with the necessary infrastructure, policies, procedures and data management systems. This is the onset of a formalized business function. Success with the "easy" disability claims is realized and encourages the organization/company to observe that, "We have systems in place to manage all employee disabilities." However, at this stage, many organizations/companies segregate their disability management efforts; that is, they may manage only the occupational illness/injuries, but not the non-occupational disabilities. Alternatively, they may manage disabilities in the short-term phases, but not the longer-term disability situations.

The *Proactive Stage* is a higher order of DMP operation. The organization/company "takes stock of the issues and pressures unique to each organization and comes up with one coordinated approach to facilitate and maximize continued employee contribution to the success of the organization, and to minimize or reduce the costs associated with employee absence". Occupational and non-occupational disability claims, whether in the short- or long-term stages, are treated in a coordinated manner. Disability data is tracked. Performance of the DMP is measured with a view to continuous improvement. Illness/injury prevention becomes the goal. The sentiment is: "We actively seek ways to prevent employee illness/injury." The outcome is a "win-win" situation for employees and the employer.

The final stage, the *Generative Stage*, is a true integration of the DMP with other organization/company business functions — the Human Resources function, Attendance Assistance & Support Program, OH&S Program, ergonomics initiatives, WWP, *etc*. The intent is to maximize the efforts of all these various functions/programs so as to best meet the organization's/company's needs. With today's climate of complex and interrelated disability claim conditions, a holistic approach to disability management is required. Likewise, the organization/company requires the same integrated approach to help it deal with increased business demands and complexities. The overriding goal becomes continual improvement of the organization by increasing employee satisfaction, lowering costs, and improving return-to-work times for employees on disability claims. The focus shifts from individual disability claim management to enhancing the organization such that employee disabilities are prevented. The message is then, "Valuing and assisting disabled employees is what we do."

H. SUMMARY

Employee illness and injury are costly. In Canada, employer and union efforts to manage disability and return employees have advanced to the point where 95 per cent of employers reported that they have, at a minimum, return-to-work practices in place.[101] This is encouraging because in 1999, only 23.1 per cent of Canadian employers had any disability management services.[102]

Going forward, employers and unions need to continue to proactively manage employee disabilities. An aging workforce and the presence of more women than men in the workplace – both associated with higher work absence days, will continue to present a significant challenge to Canadian employers. An IDMP is a proven risk management tool. Occupational health and human resources professionals can assist with this effort by demonstrating that disability management is good business practice, with immense possibilities for significant returns on the investment made.

CHAPTER REFERENCES

A. Ahrens & K. Mulholland, "Vocational Rehabilitation and the Evolution of Disability Management: An Organizational Case Study" (2000) 15 *Journal of Vocational Rehabilitation* 39.

AIG Claims Services Inc., "Early Intervention Cuts Workers' Compensation Costs" (*Aon Commentary*, June 17, 1996).

E. Akyeampong, *Work Absence Rates, 1987 to 1998* (Ottawa: Statistics Canada, 1999).

Alberta Workers' Compensation Board, Publication no. WCB PS003, *Partnerships in Injury Reduction* (July 2011), online: http://www.wcb.ab.ca/pdfs/employers/pir_broch.pdf (accessed January 28, 2020).

Alberta Workers' Compensation Board, *2004 Annual Report* (Edmonton: Workers' Compensation Board Alberta, 2005) at 18, online: http://www.wcb.ab.ca/pdfs/public/annual_report_2004.pdf (date accessed: January 31, 2020).

S. Aldana, *Top Five Strategies to Enhance the ROI of Worksite Wellness Programs*, Wellness Council of America (February 2009), online: http://www.absoluteadvantage.org/pdf/contentmgmt/top_5_strategies.pdf (date accessed: January 31, 2020).

Aon Consulting, "The Case for Absence Management", *Aon Workforce Strategies*, 2003, online: http://www.aon.com (date accessed: January 31, 2020).

B. Armstrong, & S. Greckol, "Accommodation Guidelines" in *Illness and Disability*

---

[101] Towers Watson, *Pathway to Health and Productivity, 2011-2012 Staying@Work Survey Report* (2011), online: http://www.towerswatson.com/en-CA/Insights/IC-Types/Survey-Research-Results/2011/12/20112012-StayingWork-Survey-Report--A-Pathway-to-Employee-Health-and-Workplace-Productivity (date accessed: January 31, 2020).

[102] Conference Board of Canada, *Supporting Workplace Health: An Exploratory Study of Stakeholder Groups* (Ottawa: Conference Board of Canada, 1999). Unpublished study reported in K. Bachmann's *More Than Just Hard Hats and Safety Boots* (Ottawa: Conference Board of Canada, 2000) at 33.

Claims in the Unionized Workplace (Winnipeg: Centre for Labour-Management Development, 1999).

B.B. Arnetz, B. Sjögren, B. Rydéhn & R. Meisel, "Early workplace intervention for employees with musculoskeletal-related absenteeism: a prospective controlled intervention study" (2003) 45:5 *J. Occup. Environ. Med.* at 499.

M. Brassard, & D. Ritter, *The Memory Jogger: Tools for Continuous Improvement and Effective Planning* (Methuen, MA: GOAL/PC, 1994).

J. Bussé, "Case management potential area for return-to-work improvement", *At Work* (2012) 68 at 4.

COHNA-ACIIST, Disability Management Standard (2012), online: http://www.cohna-aciist.ca/assets/cohna%202012%20-%20disability%20management%20standard%20-%20electronic%20version.pdf

A. Cantor, "The Future of Workplace Accommodations: Containing Costs and Maximizing Effectiveness" (Presented at the National Conference on Disability and Work: Solutions for Canadians, Sheraton Centre, Toronto, Ontario, October 7-9, 1996) [unpublished].

Case Management Society of America (CMSA), *Standards of Practice for Case Management* (Little Rock, AR: CMSA, 1995).

CCH Incorporated, "Unscheduled Employee Absenteeism Rises to Five Year High" in *2004 CCH Unscheduled Absence Survey* (Riverwoods, IL: CCH Incorporated, 2004), online: http://www.cch.com/press/news/2004/20041007h.asp (date accessed: January 31, 2020).

CCH Incorporated, *2007 CCH Unscheduled Absence Survey* (Riverwoods, IL: CCH Incorporated, 2007).

D. Champagne, "Manage and measure disability programs", *Benefits Canada* (2016), online: http://www.benefitscanada.com/benefits/disability-management/manage-and-measure-disability-programs-32849 (date accessed: January 31, 2020).

A. Clarke, "Disability Case Management Models" (Presented at the Disability Case Management Forum, Vancouver, British Columbia, March 24-26, 1997) [unpublished].

Conference Board of Canada, *Disability Management: Opportunities for Employer Action* (Ottawa: ON: Conference Board of Canada, 2013), online: http://www.sunlife.ca/static/canada/Sponsor/About%20Group%20Benefits/Group%20benefits%20products%20and%20services/The%20Conversation/Disability/DisabilityManagement_SUNLIFE_EN.pdf (date accessed: January 31, 2020).

Conference Board of Canada, *Supporting Workplace Health: An Exploratory Study of Stakeholder Groups* (Ottawa, ON: Conference Board of Canada, 1999). Unpublished study reported in K. Bachmann's *More Than Just Hard Hats and Safety Boots* (Ottawa: Conference Board of Canada, 2000) at 33.

J. Cowell, "Serving Albertans Through Effective Injury Prevention and Disability

Management" (Presented at the National Conference on Disability and Work: Solutions for Canadians, Sheraton Centre, Toronto, Ontario, 7-9 October 1996) [unpublished].

S. Cranston, "Integrated Disability Management: Feature Article", *Canadian Benefits and Compensation Digest* (2011) 9.

J. Curtis & L. Scott, "Integrating Disability Management into Strategic Plans: Creating Healthy Organizations", *AAOHN Journal* 298.

D. Dyck, *Disability Management: Theory, Strategy and Industry Practice,* 6th ed. (Markham, ON: LexisNexis Canada, 2017).

D. Dyck, *Stating Your Case* (Ottawa: Benefits Canada, 1998) at 55-59, editorial, "Disabilities Draining the B.C. Economy" (1997) 1:7 *Back to Work* 1.

S. Gardner & P. Johnson, "Integrated Disability Management Programs: Good business for good organizations" (Jan-July 2004), online: http://findarticles.com (date accessed: January 31, 2020).

L. Gates, Y. Taler & S. Akabas, "Optimizing Return to Work Among Newly Disabled Workers: A New Approach to Cost Containment" (1989) 5:2 Benefits Quarterly 19.

D. Gross & M. Battié, "Work-related Recovery Expectations and the Prognosis of Chronic Low Back Pain Within a Workers Compensation Setting" (2005) 47 *Journal of Occupational and Environmental Medicine* 428, online: http://www.rtwknowledge.org/browse.php?article_id=92&view_type=research (date accessed: January 31, 2020).

H. Harder & L. Scott, *Comprehensive Disability Management* (Toronto: Elsevier, Churchill Livingstone, 2005).

Hewitt Associates, *Disability Absence Index Survey 2007* (2007), online: http://www.hewittassociates.com/Intl/NA/en-CA/KnowledgeCenter/ArticlesReports/DAIS_highlights.aspx (date accessed: January 31, 2020).

B. Humphreys, "Absenteeism and Disability Management – Effective Strategies Require Confronting the Elephants in the Room", *Longwoods HR Resources Database 2010*, online: http://www.longwoods.com/content/21943 (date accessed: January 31, 2020).

Institute for Work & Health (IWH), "Seven 'Principles' for Successful Return to Work" (Toronto, 2014) at 1, online: https://www.iwh.on.ca/sites/iwh/files/iwh/tools/seven_principles_rtw_2014.pdf (date accessed: January 31, 2020).

IWH, "Benefits outweigh costs for workplaces that accommodate people with mental illness" (Summer 2018) 93 *At Work*, online: https://www.iwh.on.ca/newsletters/at-work/93/benefits-outweigh-costs-for-workplaces-that-accommodate-people-with-mental-illness (accessed January 28, 2020).

IWH, "Supervisors who react with support can help injured workers return to the job" (Summer 2018) 93 *At Work* (date accessed: January 31, 2020).

IWH, *News Release*, "Effective workplace return-to-work programs incorporate health services, case coordination and work modification" (February 21. 2017), online: https://www.iwh.on.ca/media-room/news-releases/2017-feb-21 (date accessed January 28, 2020).

Job Accommodation Network, *Workplace Accommodations: Low Cost, High Impact* (2007) at 3, online: http://www.jan.wvu.edu/media/LowCostHighImpact.doc (date accessed: January 31, 2020).

Job Accommodation Network, *Workplace Accommodations: Low Cost, High Impact* (2012) at 1, online: http://www.jan.wvu.edu/media/LowCostHighImpact.doc (date accessed: January 31, 2020).

Kelly, Luttmer & Associates Ltd. in Hatscan, ed., *Comprehensive Disability Management* (Edmonton, AB: Hazard Alert Training & Supplies Canada Inc., 2000) c. 10, Fig. 10.1

G. Kelly, *The Psychology of Personal Constructs* (London: Routledge, 1955; rev 1991) in association with the Centre for Personal Construct Psychology.

B. Kirkpatrick, ed., *The Cassell Concise English Dictionary* (London: Cassell Publishers Ltd., 1989) *s.v.* "ethic".

Merriam-Webster's Learner's Dictionary (2016), *s.v.* "Cornerstone", online: http://www.merriam-webster.com/dictionary/cornerstone (date accessed: January 31, 2020).

T. Michalak, "Disability Management: An Assessment of Psychological Factors and Early Intervention" (2007) 6:1 *International Journal of Disability, Community & Rehabilitation*, online: http://www.ijdcr.ca/VOL06_01_CAN/articles/michalak.shtml (date accessed: January 31, 2020).

C. Mustard, "Getting back on the horse: Return to work has beneficial effect on health" (2012) 68 *At Work* at 5.

National Institute of Disability Management and Research (NIDMAR), *Code of Practice for Disability Management* (Port Alberni, BC: NIDMAR, 2000) at 5.

National Institute of Disability Management and Research (NIDMAR), *Occupational Standards in Disability Management: Executive Summary* (Port Alberni, BC: NIDMAR, 2000) at 2.

G.C. Pati, "Economics of Rehabilitation in the Workplace" (1985) 51:4 *Journal of Rehabilitation* 22.

P. Reed, "Recent Ruling on ADA and the Value of Interventions" (2002) 27:2 *Employee Benefits Journal* 3.

J. Regan, cited in D. Thompson, "In Support of STD", *Benefits Canada* (2001), online: http://www.benefitscanada.com (date accessed: January 31, 2020).

T. Riggar, D. Maki & A. Wolf, *Applied Rehabilitation Counselling* (New York: Springer Publishing Co., 1986).

S. Ritcey, "Psychological job matching" (Sept-Oct 1996) *OH&S Canada* 50.

D. Rosenthal, N. Hursh, J. Lui, R. Isom & J. Sasson, "A Survey of Current Disability Management Practice: Emerging Trends and Implications for Certification" (2007) 50 *Rehabilitation Counselling Bulletin* 76.

R. Rundle, "Move Fast if you Want to Rehabilitate the Worker" (1983) 17:18 *Business Insurance* 10.

SAW & RTW Committee of the American College of Occupational & Environmental Medicine, "Preventing Needless Work Disability by Helping People Stay Employed: A Report from the Stay-at-Work & Return-to-Work Committee of the American College of Occupational & Environmental Medicine" (2005) at 13, online: http://cdonaldwilliamsmd.com/PrventingNeedlessDisability-finalpdf2005-11-30.pdf (date accessed: January 31, 2020).

Shell Oil Company, Houston, "Impact of a Disability Management Program on Employee Productivity in a Petrochemical Company" (2006) 48:5 *Journal of Occupational and Environmental Medicine* 497.

D. Shrey, *Principles and Practices of Disability Management in Industry* (Winter Park, FL: GR Press Inc., 1995).

D. Smith, "Implementing disability management: A review of basic concepts and essential components" (1997) 12:4 *Employee Assistance Quarterly* 37.

Statistics Canada, *Table 14-10-0190-01 Work Absence of full-time employees by geography, annual* (Ottawa: Statistics Canada, January 31, 2020), online: https://www150.statcan.gc.ca/t1/tbl1/en/tv.action?pid=1410019001.

Statistics Canada, *Table 11, Average Usual Hours and Wages of Employees by Selected Characteristics*, online: https://www150.statcan.gc.ca/n1/daily-quotidien/191108/t011a-eng.htm (date accessed: January 31, 2020).

Statistics Canada, *Table 14-10-0287-01, Labour force characteristics, monthly adjusted and trend-cycle, last 5 months (2019)*, online: https://www150.statcan.gc.ca/t1/tbl1/en/tv.action?pid=1410028701 (date accessed: January 31, 2020).

Statistics Canada, *Work Absence Rates, 2007* (Ottawa: Statistics Canada, May 2008).

Statistics Canada, V1012_04_Absence_Table1_7, *Absence rates for full time employees by sex, and province*, Labour Force Survey, 2012 (Ottawa: Statistics Canada).

Suncor Energy Inc., *Report* (Calgary, Alberta, 2013) [unpublished].

Suncor Energy Inc., *Report* (Calgary, Alberta, 2017) [unpublished].

Suncor Energy Inc., *Report* (Calgary, Alberta, 2019) [unpublished].

D.G. Tate, R.V. Habeck & D.E. Galvin, "Disability Management: Origins, Concepts and Principles for Practice" (1986) 17 *Journal of Applied Rehabilitation Counselling* 5.

E. Tompa, C. de Oliveria, R. Dolinschi & E. Irvin, "A systematic review of disability management interventions with economic evaluations" (2008) 18:1 *Journal of Occupational Rehabilitation* 16.

Towers Watson, *Pathway to Health and Productivity, 2011-2012 Staying@Work Survey Report* (2011), online: http://www.towerswatson.com/assets/pdf/6031/Towers-Watson-Staying-at-Work-Report.pdf (date accessed: January 31, 2020).

N. Vimadalal & J. Wozniak, "Best Practices to Help Employers Capture the Benefits of Integrated Disability Management" *Employee Benefits News* (March 1, 2008), online: http://ebn.benefitnews.com/news/best-practices-help-employers-capture-benefits-548491-1.html (date accessed: January 31, 2020).

Washington Business Group on Health, *Fifth Annual Washington Business Group Health/Watson Wyatt Worldwide Survey on Disability Management* (Watson Wyatt Worldwide, 2004), online: http://www.watsonwyatt.com (date accessed: January 31, 2020).

Watson Wyatt Worldwide, *Managing Healthcare Costs in a New Era: 10th Annual National Business Group on Health/Watson Wyatt Survey Report, 2005*, online: http://www.watsonwyatt.com (date accessed: January 31, 2020).

Watson Wyatt Worldwide, *Staying@Work, Report 2005* (Canada 2005) at 5, online: http://www.watsonwyatt.com/research/resrender.asp?id=w-860&page=1 (date accessed: January 31, 2020).

Chapter 28

INTEGRATED WORKPLACE HEALTH MANAGEMENT

A. INTRODUCTION

The difficult economic environment, growing epidemic of workplace stress, and ongoing challenges in changing unhealthy lifestyles are having an impact on not only direct medical costs but also on the more hidden costs associated with absence, presenteeism, overtime and replacement workers. As North American employers seek to control these rising costs, they are increasing their investment in the health and work effectiveness of their employees. What's more, employers are finding this strengthened commitment — evidenced by a philosophy and programs make employees accountable for managing and improving their own health — can have positive effects on their organization and bottom line.[1]

Organizations in Canada and the U.S. that demonstrate highly effective health and productivity strategies, have achieved significantly better business outcomes. For example:

- higher market premiums – market premiums that are 18 points higher than the market premiums of low-effectiveness organizations;[2]
- average revenues per employee that are 40 per cent higher – the difference per employee is $132,000;[3]
- an annual health care differential of $551 per employee;[4]
- lower casual absence days – 1.0 fewer unscheduled absence days;[5]

---

[1] Towers Watson, *Pathway to Health and Productivity: 2011/2012 Staying @ Work Survey Report* (2013), online: http://www.towerswatson.com/en-CA/Insights/IC-Types/Survey Research-Results/2011/12/20112012-StayingWork-Survey-Report--A-Pathway-to-Employee-Health-and-Workplace-Productivity at 2 (date accessed: January 31, 2020).

[2] *Ibid.*

[3] *Ibid.*

[4] *Ibid.*

[5] E. Gaudette, Presenting Willis Towers Watson, *2015/2016 Global Staying@Work Survey: Key Global and Canada Highlights* (2016), online: https://easna.org/wp-content/uploads/2016/04/2015-2016-Global-Staying-at-Work-Survey.pdf at 3 (date accessed: January 31, 2020).

- fewer disability days – 1.1 fewer new Long Term Disability (LTD) cases per 1,000 covered employees;[6]
- lower annual medical and prescription drug plan costs – $150 less per employee per year;[7]
- lower voluntary staff turnover rates;[8] and
- reduction in employee health risks, namely:
  - Fewer obese employees (44%);
  - Fewer employees who are at risk for diabetes (31%);
  - Fewer smokers (14%); and
  - More emplpoyees physically active (33%).[9]

In essence, there is a link between highly effective health and productivity strategies and strong business and financial results. This chapter focuses on why and how an organization can establish an Integrated Workplace Health Management (IWHM) approach and program.

B. CANADA: THE CURRENT STATE

According to the Towers Watson (2015-2016) Staying@Work Report, 82 per cent of Canadian organizations view health and productivity as a core strategy within their organization. All indications are that this degree of corporate commitment will increase. The rationale is that organizations are concerned about the rising rates of employee presenteeism, absenteeism, disability, and the related health care costs. Added to these financial drivers are the impacts of a changing workplace, changing legal drivers, changing business drivers, and changing society drivers and societal pressures. The factors that have been shown to impact workplace health, safety, and well-being include: environmental factors, social factors, economic factors, genetic/hereditary factors, and psychological factors.[10]

In response, organizations have implemented workplace programs designed to prevent illness/injuries, protect the employee from workplace hazards, and support the employee through vulnerable periods in life. For example, the typical Canadian organization offers employees support programs such as:

- Human Resources Program and supportive benefits;

---

[6] E. Gaudette, Presenting Willis Towers Watson, *2015/2016 Global Staying@Work Survey: Key Global and Canada Highlights* (2016), online: https://easna.org/wp-content/uploads/2016/04/2015-2016-Global-Staying-at-Work-Survey.pdf at 3 (date accessed: January 31, 2020).

[7] *Ibid.*

[8] *Ibid.*

[9] *Ibid.*

[10] *Ibid.*

- Occupational Health & Safety Management System (and program);
- Attendance Control Program;
- Employee Assistance Program;
- Workplace Wellness Program/initiatives/program; and
- Disability Management Program and services (*e.g.*, claim management, case management, and return-to-work planning and placement).

Coupled with that, the organization provides leadership and creates a corporate culture that provides clarity on corporate vision, goals, and strategies; performance expectations; employee individual performance levels; rewards and recognition; and career growth. In addition, Management has the legal duty to create a safe and healthy workplace — physically and psychologically safe and healthy. This means understanding the relationship between work and health; making healthy choices when designing and organizing work; monitoring the implementation of management theories and practices; and continuously improving the workplace.

C. INTEGRATED WORKPLACE HEALTH MANAGEMENT[11]

Integrated Workplace Health Management (IWHM) is defined as a management approach in which organizational resources are positioned in an integrated manner to promote workplace health, safety, and well-being for the employee and the organization. The components of the IWHM program are linked so that they can optimize their program effectiveness as well as support the functions of all the other IWHM components. The result is an enhanced approach to promote and manage workplace health, safety, and well-being.

The components of an IWHM program (Figure 28.1) function to promote workplace health, safety, and wellness; protect employees and the organization from workplace hazards; and support employees and workplace to successfully overcome various health and safety challenges that stem from a variety of sources.

These IWHM components operate on various levels: a systems level, organization level, management level, and individual level.[12] The intent is to use an integrated

---

[11] Excerpts taken with permission from D. Dyck, "Integrated Workplace Health Management: The Role of the OHN" (Spring/Summer 2014) 33:1 *OOHNA Journal*.

[12] **System Level** – an approach in which the path of maximizing workplace health, safety and well-being is the path of least resistance – the "easiest/natural way to proceed".

Organizational Level – the creation of an organization in which leadership, culture, work environment, and employee supports are aimed at workplace health, safety and well-being.

Management Level – the management practices used to address workplace health, safety and well-being.

Individual Level – the efforts that employees make towards promoting workplace health, safety and well-being for themselves and their co-workers.

and upstream approach[13] so that workplace health, safety, and well-being can be realized with the least possible risk of loss, and in a cost-effective manner.

Figure 28.1: Integrated Workplace Health Management (IWHM) Approach[14]

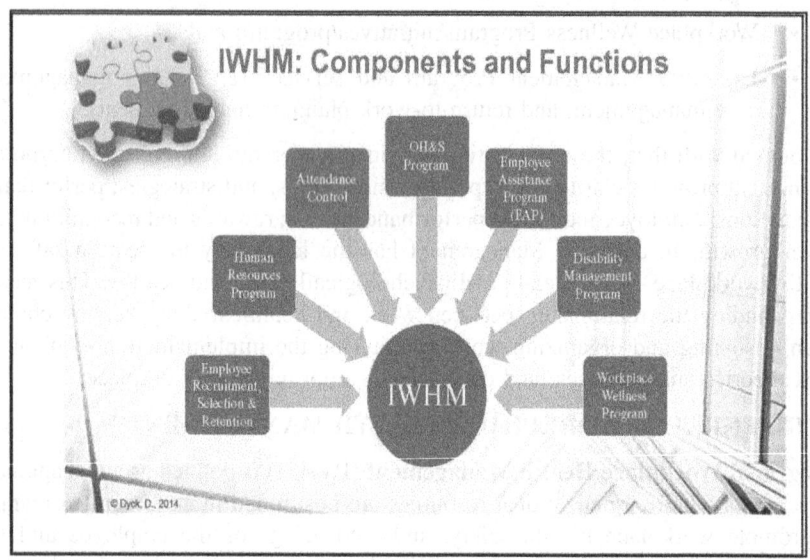

1. IWHM: Explanation

A healthy organization is one in which culture, work environment, and people-management practices are integrated and aligned to create a climate which engages, energizes, and enables employees to produce sustainable business results. Unfortunately, Canadian organizations tend to position their employee support programs and services to operate as "silos". That is, they function independently with little to no interface with each other. Their program outcomes and performance data are not combined with the intent of understanding the true and total picture of the organization's status of workplace health, safety, and well-being.

To create a healthy workplace, the organization has to set the stage. Through enlightened leadership and the creation of a supportive work culture, Management can attain operational excellence in workplace health, safety, and well-being. By

---

[13] **Upstream approach** refers to the organizational design, leadership, research, development, and production activities within an organization. In terms of the workplace, it is the actions/activities designed to promote, protect, and enhance workplace health, safety, and well-being.

[14] D. Dyck, *Integrated Workplace Health Management Course – BMC 183* (2014), prepared for the University of Fredericton and University of Calgary, online: http://www.ufred.ca/w_certificate_in_disability_management.aspx (date accessed: January 31, 2020).

understanding the value offered and the benefits afforded through the linkage of the various workplace support programs and services, Management can position the workplace to function effectively and to be highly productive. Part of the integration would be the development and implementation of a measurement strategy to collect and combine key data variables which indicate the status of the organization's health, safety, and well-being.

2. IWHM: Components and Linkages

The components of an IWHM program should interface so that they can optimize their effectiveness as well as support the function of all the **other** components. This concept (Figure 28.2), in essence, demonstrates how the contributions of each component can be used to support workplace health, safety, and well-being of the employee and the organization.

Figure 28.2: Integrated Workplace Health Management

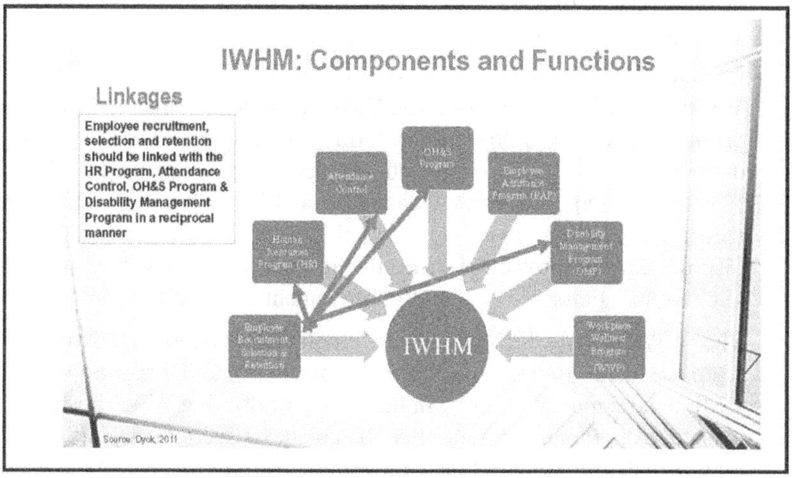

(a) Approach[15]

By way of examples, consider the following:
- Employee recruitment, selection, and retention should be linked with the Human Resources (HR) Program, Attendance Control, Occupational Health & Safety (OH&S) Program, and Disability Management Program (DMP) in a reciprocal manner. When seeking to hire a new employee, knowing the physical and psychological demands of the position is essential. The OH&S Program can play a major role in providing this type of information to HR personnel.

---

[15] *Ibid.*

- The HR Program should be linked with Attendance Control, OH&S Program, DMP, and Workplace Wellness Program (WWP) in a reciprocal manner. If HR notes that prescription drug costs are elevated due to increased use of lifestyle-factor medications, they can work with the WWP to offer targeted health education and other promotional initiatives that encourage the adoption of positive employee lifestyle practices.
- Attendance Control should be linked with the employee recruitment, selection, and retention; HR Program; Employee Assistance Program (EAP); DMP; and WWP in a reciprocal manner.[16]
- The OH&S Program should be linked with employee, recruitment, selection, and retention; HR Program; Attendance Control; EAP; DMP; and WWP in a reciprocal manner. Post-incident occurrence, the OH&S team may arrange for Critical Incident Stress Debriefing (CISD) offered by the EAP. This psychological first aid measure can prevent the serious consequences of exposure to a life-threatening event.
- The DMP should be linked with employee, recruitment, selection, and retention; HR Program; Attendance Control; OH&S Program; EAP; and WWP in a reciprocal manner. The DMP outcome data may indicate that the prime reason for employee short-term medical absences is psychological disorders. As such, the organization would be encouraged to enhance its Management practices (an HR mandate) and its EAP services. However, if the primary reason for employee work absence is due to musculoskeletal disorders, health protection measures through the OH&S Program, namely via ergonomic assessments and improvements, would be recommended.
- The WWP should be linked with employee, recruitment, selection and retention; HR Program; Attendance Control; OH&S Program; EAP; and DMP in a reciprocal manner. In the course of offering WWP initiatives, the existence of ergonomic issues may be noted. As such, the OH&S and DMP may be called upon to address the current musculoskeletal injuries and to seek ways to remedy the situation.

Although most Canadian organizations have these employee support programs and services in place, the above-suggested linkages are rarely evident.

3. Benefits of Linkages

There are numerous benefits that organizations can realize through IWHM integration, such as the occurrence of:

- Human Resources Program, Workplace Wellness Program, Employee

---

[16] See D. Dyck, *Disability Management: Theory, Strategy & Industry Practice*, 6th ed. (Markham, ON: LexisNexis Canada Inc., 2017) c. 8, which describes the operation of an Attendance Support and Assistance Program, as well as how it can link with a Disability Management Program.

Assistance Program, Occupational Health & Safety Program, Absence Control Program, and Disability Management Program and service utilization and data integration;
- a comprehensive view of the status of workplace health, safety, and wellness;
- an understanding of program savings, return on investment, and impact;
- linkages between workplace performance metrics, and employee absence and medical rates and costs;
- benchmarking of organizational data to national norms and other benchmarking indices; and
- informed decisions to be made in regards to workplace health, safety, and wellness programs and services.

Companies that use their entire corporate leverage in an integrated way, have been known to:
- retain employee talent;
- generate highly-engaged workforces positioning them to do well in the marketplace;
- deliver higher-quality products and services;
- realize greater productivity and profitability;
- deliver the greatest business value;
- enjoy stronger financial performance; and
- generate superior return on investment (ROI) to shareholders.

In terms of quantifiable financial returns, these organizations that integrate their health and productivity have been shown to realize:
- lowered medical trends by 1.2 percentage points;
- reduced disability costs by 19 to 25 per cent;
- increased productivity due to 1.8 fewer absent days per employee;
- a financial return of 11 per cent higher revenue per employee; and
- 28 per cent higher shareholder returns.[17]

These are the drivers for adopting an integrated approach to workplace health management.

---

[17] Watson Wyatt, *Staying@Work: Effective Presence at Work* (2008), online: http://www.towerswatson.com (date accessed: January 31, 2020); and E. Gaudette, presenting Willis Towers Watson, *2015/2016 Global Staying@Work Survey: Key Global and Canada Highlights* (2016), online: https://easna.org/wp-content/uploads/2016/04/2015-2016-Global-Staying-at-Work-Survey.pdf at 3 (date accessed: January 31, 2020).

D. IWHM: MANAGEMENT'S ROLE

The sections below suggest the best way to demonstrate to Management the benefits of IWHM, as well as to coach Management on how to best design and implement an IWHM approach and program. To begin with, just telling Management the company needs an IWHM is rarely enough; instead, convince Management by demonstrating how health, work productivity, and profitability relate to each other.

Start by explaining to Management the secret of achieving a healthy organization is by creating a balance between the organization's health practices and the workforce's health practices. As depicted in Figure 21.5: Healthy Organization (See Chapter 21, "Occupational Health and Safety: Prevention of Workplace Illness and Injury"), the organization's health practices should include:

- *Leadership* – provide strategic communication on the organization's/company's business goals and plans, as well as role clarity to the various stakeholders. Succinctly put, "What is to be done?", "How?" and "By whom?";
- *Skills* – address the requisite employee capabilities, learning needs and application of knowledge to maximize employee performance and productivity;
- *Tools* – provide the needed tools, data systems, information, workload, and work facilitation; and
- *Employee motivation* – inspire employees through monetary/non-monetary rewards, performance management, job opportunities, corporate culture, and management to maximize employee performance and productivity.

These recommended elements of workplace health and productivity mirror the human performance elements of the Roithmayr Performance Maximizer® (See Chapter 7, "Great Safety Performance: A Management Approach"). It advocates a work environment that:

- enables employees to know how to do their jobs;
- ensures employees are able to do their jobs;
- equips employees to do their jobs;
- motivates employees to do their jobs;[18] and
- includes workplace interactions — the "glue that brings it all together", namely the trust, honesty, respect, openness, and involvement of all stakeholders.[19]

---

[18] T. Roithmayr, *The Performance Maximizer®*, online: http://www.performance-bydesign.com (date accessed: January 31, 2020).

[19] *Ibid.*

In addition, the workforce's health practices, as proposed by Towers Watson, should include:[20]

- ***Prevention*** – through the use of Occupational Health & Safety Programs, Health Risk Appraisals, health management, disease management, control of employee presenteeism, conflict management, and workplace harassment and violence prevention, the workforce can prevent workplace illness and injury;
- ***Plan Administration*** – to effectively operate, business practices such as vendor management, plan design, and financial management are required to be in place; and
- ***Total Absence Management*** – to address the human aspects of work, attendance management, integrated disability management, program administration, claim management, case management, and return-to-work measures and management are great assets to an organization/company.

Towers Watson determined that organizations with effective health and productivity programs experience:[21]

- greater financial returns and productivity improvements, *e.g.*, 20 per cent more revenue per employee; 16.1 per cent higher market value; and 57 per cent higher shareholder returns;
- five per cent lower casual absence rates and costs;
- four per cent lower Short-term Disability rates and costs;
- 4.5 per cent lower Long-term Disability rates and costs;
- 3.5 per cent lower health care costs (American organizations);
- lower payroll costs for Workers' Compensation Board claims;
- lower presenteeism rates and costs;
- integrated health management programs;
- healthier workforce; and
- a healthier organization and bottom line.

These findings were further supported in the Towers Watson 2009/2010 and 2011/2012 Staying@Work reports.[22,23]

---

[20] Watson Wyatt, *Staying@Work: Effective Presence at Work* (2008), online: http://www.towerswatson.com (date accessed: January 31, 2020); and reinforced by Towers Watson, *Pathway to Health and Productivity: 2011/2012 Staying @ Work Survey Report* (2013), online: http://www.towerswatson.com/en-CA/Insights/IC-Types/Survey-Research-Results/2011/12/20112012-StayingWork-Survey-Report--A-Pathway-to-Employee-Health-and-Workplace-Productivity (date accessed: January 31, 2020).

[21] *Ibid.*

[22] Towers Watson, *Pathway to Health and Productivity: 2011/2012 Staying @ Work*

Next, the Integrated Workplace Health and Productivity model (Figure 28.3) should be explained to Management. In this model, the program elements — a mix of an engaged workforce, effective workplace programs, and program performance measurements and evaluation — can lead to positive integrated program outcomes and financial outcomes. These health and productivity outcomes, in turn, can lead to superior financial performance.

The program elements include an engaged workforce — workers who know how to do their job, who can do their job, who are equipped to do their job, and who are motivated to do their job. It includes workplace programs such as an HR Program and service, EAP, Attendance Support and Assistance Program, OH&S Program, DMP, and WWP. Program performance measurement and evaluation are part of this mix that can result in the integration of workplace support programs.

Figure 28.3: Integrated Workforce Health and Productivity[24]

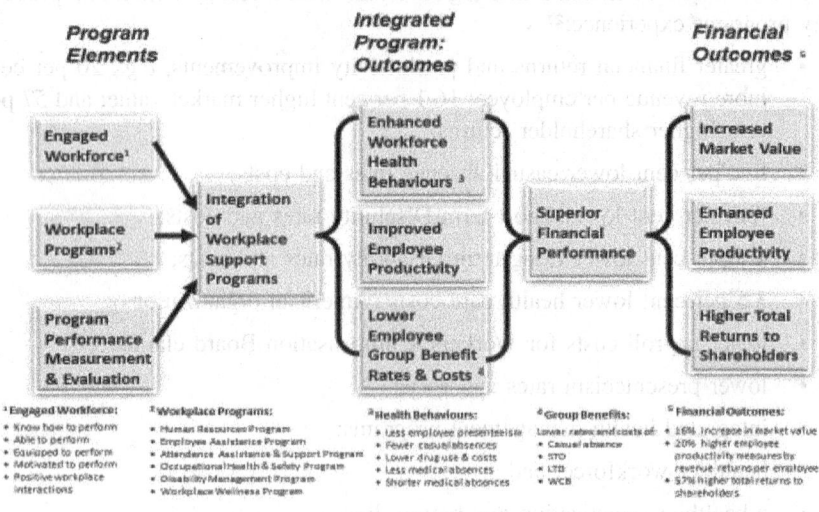

Survey Report (2013), online: http://www.towerswatson.com/en-CA/Insights/IC-Types/Survey-Research-Results/2011/12/20112012-StayingWork-Survey-Report--A-Pathway-to-Employee-Health-and-Workplace-Productivity (date accessed: January 31, 2020).

[23] Watson Wyatt, *The Health and Productivity Advantage; 2009/2010 North American Staying @ Work Report* (2010), online: http://www.towerswatson.com/en-CA/Insights/IC-Types/Survey-Research-Results/2009/12/20092010-North-American-StayingWork-Report-The-Health-and-Productivity-Advantage (date accessed: January 31, 2020).

[24] D. Dyck, *Integrated Workplace Health Management Course, BMC 183* (2014), prepared for the University of Fredericton and University of Calgary, online: http://www.ufred.ca/w_certificate_in_disability_management.aspx (date accessed: January 31, 2020).

An IWHM Program leads to enhanced workforce health behaviours, such as reduced employee presenteeism; fewer casual absences; lower drug use and costs; and fewer and shorter medical absences. It can also net improved employee productivity, and reduced employee group benefit rates and costs. The financial outcomes that can be realized have already been mentioned.

E. MARKETING THE IWHM

The benefits of workplace health management integration are great. It is incomprehensible why most companies fail to adopt such a strong strategic approach, especially since its attainment is not that difficult. The challenge then is to convey this message in a way that Management views the idea as "its own".

Hence, the mode of message delivery is important. The message has to be tailored to the specific Management personnel. Specifically, it has to speak to the benefits that workplace health management integration affords them while explaining the impacts of not embracing health management integration, *i.e.*, "the costs of doing nothing".

Using the organization's data, the costs of undesired employee health behaviours for the workplace can be demonstrated. For example, in 2014, substance use cost $1,100 per Canadian, regardless of age; totalling $38B for Canada with almost 70% of that cost being due to alcohol and tobacco use. Due to lost productivity, Canadian employers bore 40.8% ($15.7B) of those costs.[25] That translates to $770 per employee. Adding in cannabis use at $80 per person, this cost rises to approximately $850 per Canadian.[26] With the legalization of cannabis in October 2018, that cost is expected to be increased.

Work-related stress and sedentary lifestyles are linked to 24 per cent of employee illness and work absences (lost productivity). In 2018, occupational injury claims cost Canadian employers, on average, $39.5K per claim — this includes the average workers' compensation board administration costs.

In terms of non-occupational illness/injury rates and costs, the average lost workdays equalled 10.3 in 2019,[27] at an estimated average cost of $2,386.30 per full-time employee.[28] This is the direct cost; the indirect cost comes to $11,931 to

---

[25] Canadian Substance Use Costs and Harms Scientific Working Group, *Canadian substance use costs and harms (2007–2014)* (Ottawa, ON: Canadian Institute for Substance Use Research and Canadian Centre on Substance Use and Addiction, 2018), online: http://www.ccdus.ca/Eng/topics/Costs-of-Substance-Abuse-in-Canada/Pages/default.aspx (date accessed: January 31, 2020).

[26] *Ibid.*

[27] Statistics Canada, *Table 14-10-0190-01 Work Absence of full-time employees by geography, annual* (Ottawa: Statistics Canada, 2020), online: https://www150.statcan.gc.ca/t1/tbl1/en/tv.action?pid=1410019001 (date accessed: January 31, 2020).

[28] Calculated by determining the daily wage from the average hourly wage for full-time

$16,704 per employee/year, making the total disability cost $14,318 to $19,090 per employee/year.[29] Of note, these are after-tax dollars. In short, non-occupational illness/injury claims are expensive: the average total cost associated with a physical illness/injury claim is $9,000, while psychological illness claims totalled $18,000.[30]

Workplace casual absenteeism and medical absences, occupational and non-occupational in nature, cost Canadian employers $37 billion per year.[31] The top two health reasons for short-term and long-term disability insurance coverage were psychological disorders and musculoskeletal/back disorders.[32] The direct and indirect costs of workplace health-related costs have steadily increased over the years.[33] As well, employee presenteeism reportedly costs even more, in terms of lost productivity, than do casual absenteeism and medical absences.

The design of the employee group benefit plan has proven to be a major reason

---

Canadian employees as of October 2019 ($28.96 per hour) and multiplying that by eight hours/day and by the average number of lost time days per Canadian worker (10.3 days). Data obtained from Statistics Canada, *Table 14-10-0190-01 Work Absence of full-time employees by geography, annual* (Ottawa: Statistics Canada, 2020), online: https://www150.statcan.gc.ca/t1/tbl1/en/tv.action?pid=1410019001; and Statistics Canada, *Table 11, Average Usual Hours and Wages of Employees by Selected Characteristics*, online: https://www150.statcan.gc.ca/n1/daily-quotidien/191108/t011a-eng.htm.

[29] The range of indirect disability costs is calculated by multiplying the average cost of $2,386.30 per full-time employee by five and seven times. Those indirect disability costs are added to the average cost of $2,386.30 per full-time employee, to determine the range of total disability costs.

[30] Centre for Addiction and Mental Health (CAMH), "The numbers add up, so make your vote count" (Fall 2011) 11:3 *camhconnexions* at 1.

[31] Calculated by determining the daily wage from the average hourly wage for full-time Canadian employees as of October 2019 ($28.96 per hour) and multiplying that by eight hours/day and by the average number of lost time days per Canadian worker (10.3 days). That value is then multiplied by the number of full-time workers in December 2019 (15,536,200 FTEs). Data obtained from Statistics Canada, *Table 14-10-0190-01 Work Absence of full-time employees by geography, annual* (Ottawa: Statistics Canada, January 31, 2020), online: https://www150.statcan.gc.ca/t1/tbl1/en/tv.action?pid=1410019001; and Statistics Canada, *Table 11, Average Usual Hours and Wages of Employees by Selected Characteristics*, online: https://www150.statcan.gc.ca/n1/daily-quotidien/191108/t011a-eng.htm; and Statistics Canada, *Table 14-10-0287-01, Labour force characteristics, monthly adjusted and trend-cycle, last 5 months (2019)*, online: https://www150.statcan.gc.ca/t1/tbl1/en/tv.action?pid=1410028701 (date accessed: January 31, 2020).

[32] Towers Watson, *Pathway to Health and Productivity: 2011/2012 Staying @ Work Survey Report* (2013) at 10, online: http://www.towerswatson.com/en-CA/Insights/IC-Types/Survey-Research-Results/2011/12/20112012-StayingWork-Survey-Report--A-Pathway-to-Employee-Health-and-Workplace-Productivity (date accessed: January 31, 2020).

[33] D. Dyck, *Integrated Disability Management Program course IDMP 311* (University of Fredericton, 2018) [unpublished]

for lost-work time.³⁴ In addition, employee group benefits impact organizational finances. These overhead costs are paid with after-tax dollars, affecting the organization's bottom line.

By demonstrating these costs in real dollars, a strong case for an IWHM Program can be made. Additionally, point out that given most of these programs already exist within the workplace, the only cost is related to integrating their efforts. This is a relatively low-price tag.

As part of this "sales pitch", explain to Management how the performance and outcomes of the IWHM Program can be measured and monitored. IWHM performance metrics have to be established (Figure 28.4); for example, the performance of the HR Program can be measured based on the staff turnover rate, the prescription drug utilization rate, and EAP utilization rate.

Figure 28.4: Integrated Workforce Health and Productivity[35]

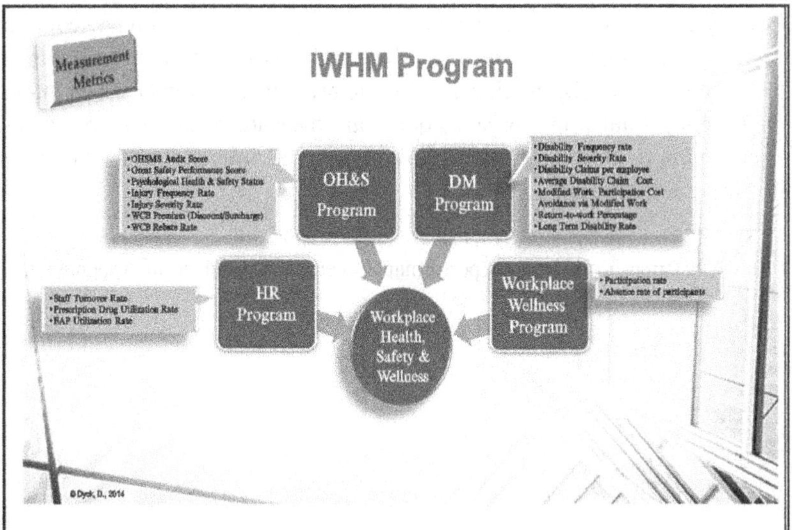

The Occupational Health Management System, OH&S Program and psychological health and safety status can be audited against a "gold standard", and the quality of the work culture and environment can be determined using the Great Performance Survey. Performance metrics such as occupational injury frequency and severity

---

[34] Watson Wyatt, *Staying@Work: Effective Presence at Work* (2008), online: http://www.towerswatson.com (date accessed: January 31, 2020).

[35] D. Dyck, *Integrated Workplace Health Management Course, BMC 183* (2014), prepared for the University of Fredericton and University of Calgary, online: http://www.ufred.ca/w_certificate_in_disability_management.aspx (date accessed: January 31, 2020).

rates, and workers' compensation insurance premium rates can be calculated and monitored.

The Disability Management Program performance metrics could include the disability frequency and disability severity rates, disability claims per employee, average disability claim costs (occupational and non-occupational in nature), modified work participation rate, cost avoidance through modified work efforts, and/or long-term disability rate.

The Workplace Wellness Program can quantify the employee participation rate, as well as impact measures such as the absence rate of WWP participants as compared to that of the non-participants.

These metrics[36] can then be used to demonstrate the effectiveness of the IWHM Program in positioning the organization to realize positive workplace health, safety, and well-being outcomes.

A balanced scorecard can then be created (Figure 28.5). A **balanced scorecard** is a performance measurement framework that can be used to develop and/or enhance a management system. By getting a "true picture" of its workplace health, safety, and well-being status, the organization can measure what matters, focus on performance improvement, prioritize efforts, make the improvements, and measure/monitor the efforts taken.

---

[36] An explanation of the various performance metrics is provided in Appendix 1.

Figure 28.5: IWHM Balanced Scorecard[37]

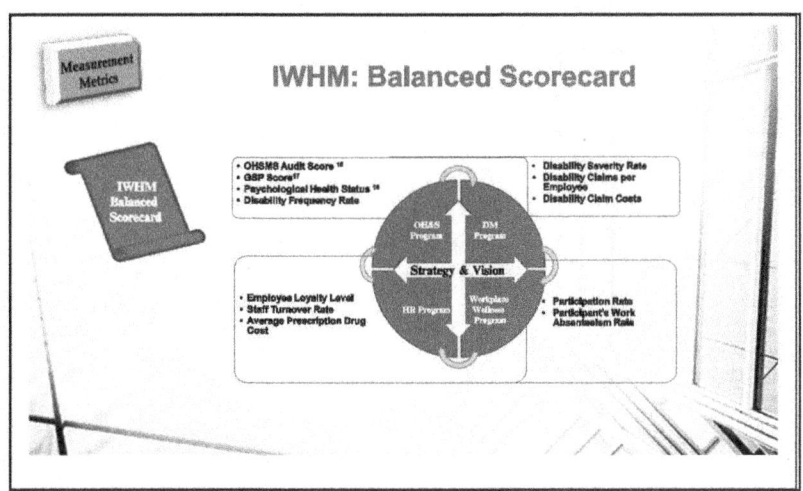

In summary, the balanced scorecard allows the organization to monitor and trend its effectiveness in creating a healthy and safe workplace. Shortcomings can be identified indicating the areas needing additional attention.

F. SUMMARY

By linking a company's/organization's Occupational Health and Safety efforts with its Human Resources Program, Employee Assistance Program, Disability Management Program, and Workplace Wellness Program; and by recognizing the impact that Management and human resource management theories can have on a workforce and workplace wellness, the opportunity to significantly reduce illness/injury incidence and impact exists. The challenge then, becomes one of acting on that knowledge in a proactive manner.

---

[37] D. Dyck, *Integrated Workplace Health Management Course, BMC 183* (2014), prepared for the University of Fredericton and University of Calgary, online: http://www.ufred.ca/w_certificate_in_disability_management.aspx (date accessed: January 31, 2020).

Appendix 1

PERFORMANCE METRICS: DEFINITIONS

Occupational Health & Safety Management System (OHSMS) Audit – measurement of the structure and outcomes of an OHSMS using an audit protocol.

Great Safety Performance Score – a measure of the conditions for great safety performance in the workplace. Developed by Roithmayr (2000-2014), this survey score measures the efforts the organization invests to create a safe workplace with safe work practices.

Psychological Health Status – a measure of the degree of psychological health, safety and well-being in a workplace. This can be measured through the use of audit tools provided in the National Standard for Psychological Health and Safety in the Workplace,[38] or from GuardingMinds @ Work.[39]

Injury Frequency Rate – the incidence rate of lost-time injury claims per 100 workers and is similar to the Injury Frequency Rate used in the field of Occupational Health & Safety. It can also be termed the **Injury Incidence Rate.**

Injury Severity Rate – the number of lost-time work hours due to injury per 100 workers.

Disability Frequency Rate – the incidence rate of disabilities per 100 workers and is similar to the Injury Frequency Rate used in the field of Occupational Health & Safety. It can also be termed the **Claim Incidence**.

Disability Severity Rate – the number disabilities per 100 workers and is similar to the Injury Severity Rate used in the field of Occupational Health & Safety. It can also be termed the **Morbidity Rate**.

Disability Claims per Employee – determined by dividing the total number of claims for a given period of time by the total number of employees for that same period of time.

Disability Claim Costs determined by dividing the total claim costs for a given period of time by the total number of employees for that same period of time. This

---

[38] Canadian Standards Association, *National Standard of Canada: Psychological Health and Safety in the Workplace – CAN/CSA-Z1003-13/BNQ 9700-803/2013* (Ottawa, ON: CSA Group, 2013), online: http://shop.csa.ca/en/canada/landing-pages/z1003-psychological-health-and-safety-in-theworkplace/page/z1003-landing-page (date accessed: January 31, 2020).

[39] GuardingMinds @ Work, *Guarding Minds @ Work: A Workplace Guide to Psychological Health & Safety* (2014), online: http://www.guardingmindsatwork.ca/ (date accessed: January 31, 2020).

can be compared against the revenue per employee to better understand the net revenue.

Modified Work Rate – is the number of modified work hours per 100 workers.

Modified Work Participation – is the percentage of ill/injured workers who participate in modified work duties.

Return-to-work Percentage – is the percentage of ill/injured workers who return to their regular work duties.

Long-term Disability Rate – is the number of ill/injured workers who are accepted on long-term disability per 100 workers.

Participants' Work Absence Rate – the absence rate for participants in a program. This can be compared against the Non-participant's Work Absence Rate to determine the magnitude of difference between the two. Costs can be assigned to these measures.

CHAPTER REFERENCES

Association of Workers' Compensation Boards of Canada, *Annual KSM Standard Report 2012*, online: https://aoc.awcbc.org/KsmReporting/KsmSubmissionReport/2 (date accessed: January 31, 2020).

Canadian Standards Association, *National Standard of Can-ada: Psychological Health and Safety in the Workplace - CAN/CSA-Z1003-13/BNQ 9700-803/2013* (Ottawa, ON: CSA Group, 2013), online: http://shop.csa.ca/en/ canada/landing-pages/z1003-psychological-health-and-safety-in-the-workplace/page/z1003-landing-page (date accessed: January 31, 2020).

Centre for Addiction and Mental Health (CAMH), "The numbers add up, so make your vote count" (Fall 2011) 11:3 *camhconnexions* at 1.

Conference Board of Canada, "Smoking Cessation and the Workplace Benefits of Workplace Programs" (June 18, 2013), online: http://www.conferenceboard.ca/e-library/abstract.aspx?did=5565 (date accessed: January 31, 2020).

D. Dyck, "Integrated Workplace Health Management: The Role of the OHN" (Spring/Summer 2014) 33:1 *OOHNA Journal*.

D. Dyck, *Integrated Disability Management Program IDMP 311* (University of Fredericton, 2018-2019) [unpublished].

D. Dyck, *Integrated Workplace Health Management Course, BMC 183* (2014), prepared for the University of Fredericton and University of Calgary, online: http://www.ufred.ca/w_certificate_in_disability_management.aspx? (date accessed: January 31, 2020).

D. Dyck, *Disability Management: Theory, Strategy & Industry Practice*, 6th ed. (Markham, ON: LexisNexis Canada Inc., 2017).

E. Gaudette, Presenting Willis Towers Watson, *2015/2016 Global Staying@Work Survey: Key Global and Canada Highlights* (2016), online: https://easna.org/wp-content/uploads/2016/04/2015-2016-Global-Staying-at-Work-Survey.pdf (date accessed: January 31, 2020).

GuardingMinds @ Work, *Guarding Minds @ Work: A Workplace Guide to Psychological Health & Safety* (2014), online: http://www.guardingmindsatwork.ca/ (date accessed: January 31, 2020).

T. Roithmayr, The Performance Maximizer®, online: http://www.performancebydesign.com/ (date accessed: January 31, 2020).

Statistics Canada, *Table 11, Average Usual Hours and Wages of Employees by Selected Characteristics*, online: https://www150.statcan.gc.ca/n1/daily-quotidien/191108/t011a-eng.htm (date accessed: January 31, 2020).

Statistics Canada, *Understanding full-time employees by sex, and province, Labour Force Survey, 2018*, CANSIM Table 279-0029 (January 2019), online: https://www150.statcan.gc.ca/t1/tbl1/en/tv.action?pid=1410019001 (date accessed: January 31, 2020).

Towers Watson, *Pathway to Health and Productivity: 2011/2012 Staying @ Work Survey Report* (2013), online: http://www.towerswatson.com/en-CA/Insights/IC-Types/Survey-Research-Results/2011/12/20112012-StayingWork-Survey-Report--A-Pathway-to-Employee-Health-and-Workplace-Productivity (date accessed: January 31, 2020).

Towers Watson, *The Health and Productivity Advantage; 2009/2010 North American Staying @ Work Report* (2010), online: http://www.towerswatson.com/en-CA/Insights/IC-Types/Survey-Research-Results/2009/12/20092010-North-American-StayingWork-Report-The-Health-and-Productivity-Advantage (date accessed: January 31, 2020).

Watson Wyatt, *Staying@Work: Effective Presence at Work* (2008), online: http://www.watsonwyatt.com (date accessed: January 31, 2020).

Willis Watson Wyatt, *2015-2016 Staying@Work Report: Canada Summary* (2016), online: https://www.willistowerswatson.com/-/media/WTW/PDF/Insights/2016/08/2015-2016-staying-at-work-canada-research-findings-en.pdf (date accessed; February 28, 2020).

Section 3

Occupational Health and Safety: Knowledge and Skills

Section 5

Occupational Health and Safety: Knowledge and Skills

Chapter 29

OCCUPATIONAL HEALTH AND SAFETY: ETHICAL PRACTICE

A. INTRODUCTION

A moral dilemma is not a choice between "the right thing to do" and "the wrong thing to do". It is a choice between two conflicting things that both seem like the right thing to do.[1]

P. Strahlendorf

Occupational Health and Safety (OH&S) Practitioners/Professionals are often faced with moral (ethical) dilemmas for which they need to reach a solution. Moral (ethical) dilemmas arise when two moral explanations leading to conflicting decisions occur. Common reactions are to:

1. arrive at a third option that satisfies both moral outcomes;
2. behave sequentially to address both options; or
3. evaluate the alternatives and select the stronger moral explanation/ solution.

The following discussion deals with the key ethical theories and their implications for OH&S Practitioners/Professionals.

B. WHY DISCUSS OCCUPATIONAL HEALTH AND SAFETY ETHICAL ISSUES?

In the course of their jobs or professional activities, OH&S Practitioners/ Professionals encounter a number of ethical dilemmas. They arise due to:

- pressures to conform to management's creeds/wishes;
- economic pressures to take the cheapest and easiest route(s);
- misrepresenting their abilities, their skills, and perhaps, their qualifications for personal gain;
- violating a client's right to confidentiality of professionally acquired information;
- failing to be objective in the application of recognized scientific methods

---

[1] P. Strahlendorf, "Professional Ethics in OH&S", *OHS Canada* (July/August, 2002) 25.

and the interpretation of the findings;
- criticizing colleagues for personal gain;[2]
- designing a program or strategy to favour a specific outcome;
- destroying or disregarding information that contradicts a desired result or outcome;
- altering or making up data to support a desired outcome;
- withholding information;
- purposely not reporting an incident;
- failing to recognize the work done by others when producing a report or presentation;
- plagiarizing;[3]
- copying electronic or audio/visual pieces of work; and/or
- neglecting to take the appropriate actions to ensure that their professional development continues.

In short, an OH&S Practitioner/Professional can be unethical without committing an illegal act.

Given that these concerns are not unique to the field of Occupational Health and Safety, various professional associations have developed codes of ethics to guide their members in dealing with ethical dilemmas. For example, physicians, nurses, disability management coordinators, safety practitioners, and psychologists all have professional codes of ethics by which they are expected and, in some instances, required to practice. The Board of Canadian Registered Safety Professionals Code of Ethics is provided in Appendix 1, and the American Association of Occupational Health Nurses Code of Ethics can be found in Appendix 2.

C. ETHICAL CONSIDERATIONS[4]

By Bonnie Rogers

Ethics is defined as the science of morals, a system of principles and rules of conduct;[5] the study of standards of right and wrong, or having to do with human character, conduct, moral duty, and obligations to the community.[6] It is the moral reasoning that human beings possess.

---

[2] *Ibid.*

[3] *Ibid.*

[4] Excerpt from D. Dyck, *Disability Management: Theory, Strategy and Industry Practice*, 3d ed. (Markham, ON: LexisNexis Canada Inc., 2006).

[5] B. Kirkpatrick, ed., *The Cassell Concise English Dictionary* (London, England: Cassell Publishers Ltd., 1989).

[6] P. Strahlendorf, "Professional Ethics in OH&S", *OHS Canada* (July/August, 2002) 25.

Ethical theories and principles guide us in making ethical decisions. The major ones include the **teleological theory — utilitarianism**, which focuses on the consequences of an action and gauges the value of that action by the end results, rather than by the means to achieve the results. It concentrates on providing the greatest good (or the least harm), for the greatest number of people. In this context, policy formation is based on cost-benefit analysis, wherein the greatest benefit is achieved for the most, but not all, for the lowest cost, is a good example of utility.

The **deontological theory** deals with action and asserts that "rightness" and "wrongness" are measured by means, rather than by consequences of an action. So, the nature of an action is more important than the outcome. By way of comparison, the deontologist would assert that confidentiality of employee health information must always be maintained, whereas the utilitarian might hold that if keeping certain kinds of health information secret would cause more harm than good, then confidentiality should be broken.

Ethical principles extend from the deontological theory, and the most widely observed principles are autonomy, nonmaleficence, beneficence, and justice. *Autonomy* is a form of personal liberty whereby the individual is regarded as having the right to self-determination. This means that the individual's values and goals must be considered in major decisions that affect his/her welfare and precludes paternalistic decision-making (when one claims to know what is best for another person), as well as requiring informed consent when decisions are made. OH&S Practitioners/Professionals who are making decisions for employees without the employee's input and consent would be in violation of this principle. The inclusion of the employee in making disability care choices is important to upholding the autonomous principles.

The second principle, **nonmaleficence**, is often referred to as the "no harm" principle. It is the foundation of most professional codes of ethics. For example, an employee with a known disability, such as a hearing loss, should not be placed in a job situation that would further compromise his/her hearing. In addition, returning an ill/injured employee to work too soon could compromise his/her continued recovery and potentially harm the employee. Here, the OH&S Practitioner/Professional must guard against having divided loyalties to the employee and organization and remain an advocate for the employee. Conducting return-to-work examinations is one way to provide protection for the health of the employee, as well as for the well-being of the organization, co-workers, and the public at large.

Beneficence, the third principle, requires that OH&S Practitioners/ Professionals act in the best interest of the employee/organization. The identification of potential health hazards through routine worksite "walk-throughs", the identification of employees at increased risk for illness or injury, and the making of recommendations for risk reduction, represent a positive occupational health intervention. Positive occupational health intervention is aimed at employee protection, thereby preventing a potential disabling event from happening. For example, the development of a back-injury prevention program for employees with a previous history of back injury would be a beneficent action. In addition, assuring the disabled employee that he/she will receive quality case management is critical to this principle.

The fourth principle, **justice**, is directed towards treating employees fairly, equally, and without discrimination. This includes providing equal opportunity for disabled persons regarding job availability and promotion; and assuring that individuals are not discriminated against because of a health condition such as HIV or another chronic disorder, when they are able to perform the job safely. This concept is embodied in Canadian human rights legislation. Another example is treating employees equally with regard to access to modified/alternate work opportunities within an organization.

In many instances, ethical principles provide us with a guide with which to weigh the risks and benefits with respect to individual health and welfare, and the development of policies and procedures to safeguard individual rights and protect health. It is incumbent upon OH&S Practitioners/Professionals to examine the situation with respect to these guiding principles and to ensure that the benefits of their actions clearly outweigh the risks.

D. ETHICS AND THE CONCEPT OF REASONABLENESS

"**Reasonableness**" is a concept on which ethics and the law are based. It stems from the rationale of what a reasonable person would do in a certain situation given the person possesses the same knowledge, training and experience. The reasonable person is therefore a hypothetical standard: "a person who is well informed, well intentioned, open-minded, calm, detached, but empathetic, unbiased, sane, and sober".[7] To solve ethical dilemmas, OH&S Practitioners/Professionals have to adopt the reasonable person state of mind.

To provide readers with an example of how ethics can be applied, a discussion on ethics and disability management follows.

E. ETHICS AND OCCUPATIONAL HEALTH & SAFETY[8]

By Jane Hall

Professional bodies and organizations establish professional codes of ethics. OH&S Practitioners/Professionals are bound to uphold the codes of the professional disciplines. Professional behaviour will be held accountable to the standards, regardless of feelings, personal beliefs, or views.

In professional practice, particularly when delivering case management services,

---

[7] *Ibid.*, at 27.

[8] Adapted from D. Dyck, *Disability Management: Theory, Strategy and Industry Practice*, 3d ed. (Markham, ON: LexisNexis Canada Inc., 2006). Reprinted with permission from LexisNexis Canada Inc. in which it was originally adapted and reprinted with permission from J. Hall, "Ethics and Case Management" (1999) 13:2 *The OEM Report* 13.

we speak of "autonomy", or the right of the employee to determine as much as possible the direction of his or her care. Autonomy is critical to keep in mind during problem solving or advocacy, especially if the employee's desires are in conflict with those of the health care providers or employers.

The other ethical principles to consider are truthfulness and justice — that is, what is "right" versus what is "wrong" in a given situation. The term "ethical dilemma" comes into play when two apparent truths are in conflict with one another. For example, an ethical dilemma may arise when a legal situation regarding a medical matter is in conflict with a religious belief. A moral wrong may not be illegal but could be unethical. A clear legal parameter may be at odds with the moral-judgement consensus. Ethics, then, is a philosophical issue rather than a scientific one. In case management, evaluation and weighing of the ethics of a case must be done in an unemotional manner, so that decision-making is rational and based on facts, rather than on the emotional issues associated with the decision.

An everyday definition of ethics could be:

> Doing the right thing, at the right time, for the right person, in the right way and knowing why it is the right thing, at the right time, for the right person, in the right way.

Dyck, 2000

1. Common Ethical OH&S Considerations

(a) Personal Beliefs versus Professional Expectations

Culture, experiences, belief systems, and academic persuasions influence OH&S Practitioners/Professionals. As human beings, we react emotionally, as well as intellectually, to situations. To deny that OH&S activities are prone to influences from our background and beliefs is to "bury your head in the sand". What is required is the, "physician know thyself" imperative. OH&S Practitioners/Professionals must take the time for introspection and examination of their beliefs, values, biases and prejudices. Only when we have acknowledged to ourselves where we stand on various issues, can we put those issues aside and deal with the factual realm, which is the professional OH&S standard.

(b) Confidentiality versus Right to Know

Confidentiality has to do with credibility and with legality. The OH&S Practitioner/Professional should be clear with employees that theirs is not a legally supported confidential relationship like the relationship between physician and patient, lawyer and client, or priest and parishioner. The OH&S Practitioner/Professional is bound to share information with other stakeholders who need to know about information pertinent to the individual case. It is strongly recommended that the OH&S Practitioner/Professional have a frank discussion with the employee at the onset of the relationship to disclose what his/her role and responsibility entails. A client who starts a sentence with, "I don't want you to tell", needs to be stopped, be counselled about the case management role, and told that confidentiality cannot be promised.

On the other hand, it is not ethical for an OH&S Practitioner/Professional to share information with anyone else who does not "need to know" the information. Shoptalk, forwarded voicemail or email messages,[9] and sharing an email with the wrong person[10] are common breaches of confidential employee information. Sharing the details of an ethical dilemma with a supervisor, however, is an appropriate method of communication.

Another example of appropriate sharing of information is when the OH&S Practitioner/Professional witnesses the employee "under the influence" on the job. The employee may request that the OH&S Practitioner/Professional not tell anyone and may promise that it will not happen again. Although alcoholism is a disease, casual drinking and violation of the company's Alcohol and Drug Policy are grounds for disciplinary action. The OH&S Practitioner/Professional is bound, as a representative of the employer, to address this unsafe situation. Hence, the individuals within the company who have a "right to know", must be apprised of this situation.

(c) Individual Wishes versus Family or Legal Constraints

Disability case management never occurs in a vacuum. Physicians and other healthcare providers, claim adjusters, employers, co-workers, and family members frequently contribute to the employee's assessment and rehabilitation plans. The manner in which the OH&S Practitioner/Professional obtains and uses the information must be carefully thought out and planned, keeping in mind the employee's best interests and wishes. The OH&S Practitioner/Professional should avoid becoming entangled in legal issues. Once again, the applicable case management standards should be used in case management practice. If a legal decision is rendered, the OH&S Practitioner/Professional is bound to obey the law. If this presents an ethical dilemma for the OH&S Practitioner/Professional, then the OH&S Practitioner/Professional's supervisor and/or lawyer should be consulted. In some cases, an OH&S Practitioner/Professional may need to withdraw from a case rather than compromise the principles of case management.

2. Tools for Determining Ethical OH&S Practices

(a) Discussion with Supervisors

Also known as the "when in doubt, check it out" philosophy, this method of dealing with ethical issues is recommended as the first step in acknowledging and confronting a problem. It is appropriate because in most situations the supervisor is accountable for the actions of individual OH&S Practitioners/Professionals that they supervise. Presumably, they have attained their supervisory position on the basis of greater knowledge and experience. Also, being more removed from the situation, they may possess more rationality and wisdom in case management practices.

---

[9] B. Morgan, "Workplace communication pitfalls" (Winter 2009) *OOHNA Journal* 32.
[10] *Ibid.*

(b) Legal Counsel

Most organizations have access to legal counsel. In ethical situations related to occupational health, legal counsel is likely to be a labour lawyer. If an OH&S Practitioner/Professional seeks legal counsel, it is appropriate to remember that there is usually some bias by the attorney(s) towards the organization's position on the situation. The OH&S Practitioner/Professional should consider the legal opinion as well as any other opinions as representing part of the picture in a specific case.

(c) Self-Appraisal

Self-appraisal is a difficult tool, but by examining our rationale, motives, and emotions as OH&S Practitioners/Professionals, we can gain an understanding of what we believe and why we believe what we do. In that way, we can gain some self-insight so that we can act on our client's behalf in an informed and non-judgemental manner.

(d) Resources

Seldom is there a situation in which the critical ethical issues have not been raised before. Thus, it is helpful to search for other similar cases and examine them for any pearls of wisdom that may apply.

F. ETHICAL DECISION-MAKING

By Dianne Dyck

There are numerous ethical conflicts and dilemmas that can arise in today's work environment, including issues such as confidentiality of employee health records, worker notification of "right to know", substance abuse, employee screening for **health indices**,[11] and "whistle-blowing". Ethical dilemmas often arise, such as cost containment versus the quality of healthcare; conflicting loyalties of the healthcare professional between the employee and the organization; and returning a worker to a safe work environment.

1. Ethical Dilemma

An **ethical dilemma** exists when two core values that the person upholds come into conflict, making it difficult for the person to decide how to move forward. In the fields of occupational health and workers' compensation, ethical issues are more likely to centre on matters of confidentiality of employee medical records, employee "right to know", action regarding potential workplace hazards, and Human Resources issues that conflict with human rights legislation and the duty to accommodate. As well, psychological issues often come to the attention of the OH&S Practitioners/Professionals, and issues of roles and responsibility may arise. OH&S Practitioners/Professionals can easily find themselves in the middle of "hot issues", and may be called upon to advocate for an employer rather than the

---

[11] **Health indices** are the measures used to establish "normal" or "abnormal" health conditions, such as body temperature, blood pressure, pulse, neurological functioning, *etc.*

employee. In such situations, the OH&S Practitioner/Professional needs to carefully assess the situation and all the issues, both medical and ethical; utilize the available resources and tools; and ultimately, make conscious, carefully thought-out decisions before taking any action.[12]

Although there is no "magic remedy" to resolve ethical dilemmas, there are a number of ethical decision-making models that can be used. The following is a discussion on two such models:

(a) Ethical Fitness™ Model[13]

Developed at the Institute for Global Ethics, this ethical decision-making model assumes that human beings have universal core values that cross religious, cultural and geographic boundaries. It serves as a decision-making guide, helping the individual to move from ethical dilemma recognition to resolution.

For example, an OH&S Practitioner/Professional becomes aware that an employee has multiple sclerosis. This employee is being considered for a promotion that will include long hours, high service demands, and extensive travel. A couple of months ago, this employee was off work for two weeks due to an exacerbation of her medical condition. She is fine now, but her manager has come to the OH&S Practitioner/Professional to ask whether there is any reason why this employee should not be considered for the new position within the company. The OH&S Practitioner/Professional has a good relationship with this manager and would like to tell him about the concerns that the employee will not be able to handle the job stress, rigours, and extensive travel demands due to health reasons. However, in the role of an OH&S Practitioner/Professional, there is an obligation to uphold the medical confidentiality entrusted to you. This is an ethical dilemma for the OH&S Practitioner/Professional.

The critical step is recognizing that an ethical dilemma exists, followed by identifying which of the core values are in conflict. Once that is achieved, the OH&S Practitioner/Professional can apply the following steps of the Ethical Fitness™ Model:

(a) *Awareness: Determine the moral issue* — First recognize that an ethical dilemma exists.

(b) *Actor: Determine the actor* — Decide who is experiencing the conflict in values.

(c) *Facts: Get the details* — Gather all the facts and assess what is learned. There may or may not be an actual conflict at hand.

---

[12] Adapted and reprinted with permission from J. Hall, "Ethics and Case Management" (1999) 13:2 *The OEM Report* 13.

[13] R.M. Kidder, *How Good People Make Tough Choices: Resolving the Dilemmas of Ethical Living* (New York: Fireside Books, 1995).

(d) *Determine if the moral issue is "right vs. wrong"* — Could this simply be an issue of "right and wrong" and not an ethical dilemma? To determine this, use five tests:

   (i) The legal test — Does the action or choice contravene any laws?
   (ii) The front page test — Would this action/decision stand the test of public scrutiny?
   (iii) The gut feeling test — What does your "gut" tell you about this decision: is it right or wrong?
   (iv) The role model test — How would you feel about someone you respect knowing you took this action or made this decision?
   (v) The professional standards test — Would your action/decision align with your professional standards?

(e) *Test for "right" by assigning one of the four dilemma paradigms* — With facts in hand and the assurance that an ethical dilemma exists, determine which core values are in conflict and why. The ethical decision-making paradigms are provided in Figure 29.1.

Figure 29.1: Ethical Decision-making Paradigms

Ethical Decision-making Paradigms[14]	
Truth versus Loyalty	Disclosure versus Confidentiality
Justice versus Mercy	Fairness versus Compassion
Short-term versus Long-term	Immediate individual needs versus Conservation for future needs
Individual versus Community	Autonomy versus Collective rights of the larger community

(f) *Apply resolution principles:* Ends-based, Rules-based or Care-based — Examine the dilemma using the three resolution principles:

   (i) *Ends-based Principle* — What is the end result of the action or decision made? Which choice would result in the greater good for the largest number of people?

   (ii) *Rules-based Principle* — What is your obligation here? If you have

---

[14] P. Edgar, "Resolving Ethical Dilemmas: Applying the Institute for Global Ethics' Ethical FitnessTM Model to Occupational and Environmental Health Practice Issues" (2002) 50:1 *AAOHN* 40.

a duty to uphold medical confidentiality, then that is what you must do.

    (iii) **Care-based Principle** — This principle is based on the Golden Rule: "Do unto others as you would have done unto you." In essence, show the level of compassion that you would like to be shown if the roles were reversed.

(g) *Determine whether a third option exists* — Is there another action or decision option that could be adopted? If so, could it allow the players to reach that "win-win" plateau that we all strive for?

(h) *Decide* — Once all the options have been considered, it is time to decide. As Edgar[15] notes, procrastination can be a decision — the decision to choose, not to decide. Usually, it is not a recommended resolution tactic.

(i) *Evaluate the decision* — Looking back on actions taken or decisions made is worthwhile. It serves to help prepare the OH&S Practitioner/Professional for future instances of ethical decision-making.

(b) Model for Ethical Decision-making in a Professional Situation[16]

For OH&S Practitioners/Professionals, ethical dilemmas abound. This is compounded by being in a position of trust: the people whom you have agreed to help trust that you will act as their advocate and provide them reasonable guidance during a vulnerable period in their lives. According to Parsons, two ethical guidelines for caregivers in a professional situation are:

- the needs of the client come first and foremost; and
- the caregiver must recognize his or her own needs and biases and avoid situations in which these might negatively impact client caregiving.

The steps that Parsons recommends for ethical decision-making are:

(a) *Parameters of the situation* — What are the facts and issues in this situation and who are all the players?

(b) *Ethical-legal issues* — Identify all the legal issues and then all the ethical ones. By separating them out, it helps to clarify the situation and what action to take.

(c) *Would legal guidelines help?* — Consult the applicable legal guidelines and determine if they would lead to the resolution of the issue.

(d) *Stakeholder rights, responsibilities and welfare* — List the stakeholders, how they are involved, their rights, responsibilities, and welfare.

---

[15] *Ibid.*

[16] R.D. Parsons, *The Ethics of Professional Practice* (Needham Heights, MA: Allyn & Bacon, 2001).

(e) *Alternate Decisions/Actions* — Develop a list of alternative decisions possible for each identified issue.

(f) *Consequences* — Assess the consequences of making each decision. Evaluate the short-term, ongoing and long-term consequences of each possible decision.

(g) *Assessment* — Present any evidence of the likelihood that the various consequences or potential benefits may occur.

(h) *Decision* — Make the decision and monitor the outcome.

G. ETHICAL DILEMMA: CASE STUDY

The following scenario is designed as an illustration of the above Model for Ethical Decision-making in a professional situation:

Hank is a long-term employee who for years has been "a thorn in management's side". Recently, as a result of a workplace incident in Area B, management took disciplinary action against two fellow employees. This exacerbated Hank's questioning of management's practices. He thinks employees and the union should be far more involved in investigating workplace incidents and questioned whether management is using these two employees as "scapegoats" for a series of "management system failures". This dialogue along with all the "heated emotions" has gone on for about three months now.

Simultaneously, the area in which Hank works has undergone significant operational changes due to a major labour shortage. Rotation through all functions of the operation is now required. This is a significant change in this company given that in the past, employees chose to work in either Area A or Area B.

Hank, a qualified tradesperson, has for years avoided working in Area B. Now, he is required to do so. Upon hearing of this operational change, Hank said he would not comply given the recent occurrence of a workplace incident in Area B. He claims that Area B is unsafe. Management tried to reason with Hank, but he cursed and swore at the Section Manager and left in a "huff".

The next day, Hank called in "sick". The following day, he produced a medical note stating that he was "unfit to work for an undetermined period of time, pending further medical investigation".

Hank was referred to the Company's Disability Management Program. This is a standard practice within this company. All the details of Hank's job (Job Demands Analysis), job performance, and the recent labour-management issues were relayed to the Disability Management Coordinator.

Upon further medical assessment of Hank's fitness to work, it was determined that Hank was indeed medically unfit to work and that he would require treatment and a lengthy recovery period.

Management questioned the above decision, believing that Hank is "bucking the system" because he does not want to work in Area B. They claim that one of Hank's co-workers just met him at the mall and that Hank told him that he was "unhappy with his current job, but at least he will never have to work in Area B". They do not believe that there is anything wrong with Hank other than he does not want to comply with the new operational change. They feel he is "playing the system", and voice strong opposition to Hank being on short-term disability.

Hank's Section Manager and the Human Resources Director approach you, the Disability Management Coordinator. They again state their concerns and want to know what is medically wrong with Hank. They tell you that management has the right to information enabling them to manage the employee and the employment relationship. They also hint that you, being in a management role, are *"one of them"*, and that you can share what you know with them. As well, you have been work-colleagues for about 20 years.

You have always had a good working relationship with both parties. You know that Hank can be a difficult person to deal with and that he is often very sarcastic. In fact, on a number of occasions, he has been impertinent to you for no apparent reason.

In discussing the current situation with the external Disability Management Service Provider, you inadvertently find out that Hank is experiencing a severe bout of mental illness compounded with a significant physical illness. This mental breakdown, although it has been coming on for a number of years, is now significant.

Using Parsons' Model,

a) *Parameters of the situation* — The facts are that Hank is medically unfit to work and will be for some time to come. There are employee-management distrust issues and ill feelings. Hank does not want to work in Area B, stating he feels it is unsafe working there. Management and his co-workers, on the other hand, are adamant that Area B is a safe place to work. All the other workers have agreed to the new operational change — they realize Management had to make this move because of a significant labour shortage.

b) *Ethical-legal issues* — The legal issues are:

- Hank is entitled to short-term disability;
- Hank is entitled to medical privacy; and
- Hank has the right to seek work accommodation if and when, he is deemed fit to work.

The ethical dilemma for the Disability Management Coordinator is:

- Keep Hank's medical information confidential at the risk of losing face with the Section Manager and Human Resources Director;

OR

- Divulge Hank's medical status and risk violating the Disability

CH. 29: OCCUPATIONAL HEALTH AND SAFETY: ETHICAL PRACTICE

Management Coordinator's Code of Ethics and the applicable terms of the privacy legislation.

c) *Would legal guidelines help* — Upon review of the applicable legal guidelines, the Disability Management Coordinator determines that under the privacy and Duty to Accommodate legislation, it is illegal to disclose what has been learned about Hank's condition. Given Hank is on a medically substantiated absence, he is eligible for benefit coverage with this company. Likewise, when and if requested, management will likely have to accommodate Hank as per the Canadian Human Rights duty to accommodate legislation.

d) *Stakeholder rights, responsibilities and welfare* — The stakeholders involved are:
   - Hank — a qualified tradesperson who is off on medical leave;
   - his Section Manager — the manager who Hank strongly dislikes and vice versa. The Section Manager has a large operation to run and has a low tolerance for "*dissention in his ranks*";
   - the Human Resources Director — the support staff member for the Section (Areas A & B) who is charged with assisting the Section Manager to effectively manage employees and their benefits;
   - co-workers — Hank's co-workers with whom he has worked for 20 years and who will now have to "pick up the slack" for Hank who is off ill despite looking well;
   - the Disability Management Coordinator — the individual whose role is to coordinate and manage disability situations within the company; and
   - the Union who has a legal responsibility to support Hank as one of its members.

e) *Alternate Decisions/Actions* — The possible decisions are:
   - #1 — Explain the privacy legislation, and advise the Section Manager and Human Resources Director that Hank is medically unfit to work and that he is obtaining the appropriate care;

   OR

   - #2 — Explain the Duty to Accommodate legislation, and counsel the Section Manager and Human Resources Director that although Hank was insubordinate, right now, he is medically unfit to work and that he is in appropriate care; or

   OR

   - #3 — Explain the Duty to Accommodate and privacy legislation, and counsel the Section Manager and Human Resources Director that although Hank was insubordinate, right now he is medically unfit to work and that he is in appropriate care;

OR

- #4 — Tell the Section Manager and Human Resources Director what you know about Hank's medical condition since this information is not documented in your information on Hank, and because you trust them to keep the information confidential. Likewise, it may help to quell some of the employee-employer tensions.

f) *Consequences* — Actions #1, #2, and #3 would all work, but #3 is probably the best option. Option #4 is unethical and illegal.

g) *Assessment* — Actions #1, #2, and #3 would all work, but #3 is probably the best option.

h) *Decision* — The decision is to select Action #3.

Although this scenario is fictitious, it is designed to demonstrate how the OH&S Professional/Practitioner can objectively formulate an ethical decision.

H. SUMMARY

To reiterate, ethics is the science of morals, a system of principles and rules of conduct, the study of standards of right and wrong, or having to do with human character, conduct, moral duty, and obligations to the community. It is the moral reasoning that humans possess.

The ethical principles covered in this chapter provide us with a guide with which to weigh the risks and benefits with respect to individual health and welfare, and the development of policies and procedures to safeguard individual rights and protect health. In the field of Occupational Health and Safety, evaluation and weighing of the ethics of a situation must be done in an unemotional manner so that decision-making is rational and based on facts rather than the emotional issues attached to the decision at hand. In short, it is: *"doing the right thing, at the right time, for the right person, in the right way and knowing why it is the right thing, at the right time, for the right person, in the right way"*.

Appendix 1

BOARD OF CANADIAN REGISTERED SAFETY PROFESSIONALS: CODE OF ETHICS[17]

Rules of Professional Conduct

Purpose

The purpose of the Canadian Registered Safety Professional (CRSP)®/ Professionnel en sécurité agréé du Canada (PSAC)® Rules of Professional Conduct (the Code) is to provide guidance to ensure that each CRSP®/PSAC® adheres to high standards of integrity and professional competence.

Competence is "the ability to perform a task, function or role up to a set of prescribed standards."

Preamble

As a condition to obtaining and maintaining certification, each CRSP®/ PSAC® commits to abide by the Code as adopted by the Board of Canadian Registered Safety Professionals (BCRSP). Each CRSP®/ PSAC® pledges to subscribe not only to the letter but also to the spirit of the Code in all their professional activities.

1. Competence

Certificants are required to:

  a. Maintain competence in carrying out professional responsibilities and provide services in an honest and diligent manner.
  b. Provide sound judgement in pursuance of their professional duties.
  c. Recognize their professional limitations and perform only those services that may be handled competently based on one's training and experience.
  d. Ensure persons working under their authority or supervision are competent to carry out the tasks assigned to them.

2. Integrity

Certificants are required to:

---

[17] Reprinted with permission from the Board of Canadian Registered Safety Professionals. Board of Canadian Registered Safety Professionals: Code of Ethics (Mississauga, ON: BCRSP, 2019).

a. Maintain honesty, integrity, and objectivity in all professional activities.
b. Protect and promote the safety and health of people, property, and the environment above any consideration of self-interest.
c. Avoid circumstances where compromise of professional conduct or conflict of interest may arise.
d. Represent their qualifications and experience accurately and not knowingly make false or misleading statements.

3. Respect in the Workplace
Certificants are required to:
a. Support, promote and apply the principles of human rights, equity, dignity, and respect in the workplace.
b. Recognize that discrimination on the basis of race, creed, colour, language, national origin, political or religious affiliation, sex, sexual orientation, age, marital status, family relationship, and disability is prohibited.

4. Professional Growth
Certificants are required to:
a. Continue professional development throughout their career and support and encourage fellow CRSPs/PSACs to develop professionally.

5. Confidentiality
Certificants are required to:
a. Protect the confidentiality of all professionally acquired information and disclose such information only when properly authorized or when legally obligated to do so.

6. Requirements
Certificants are required to:
a. Keep apprised of all relevant laws, regulations, and recognized standards of practice as it relates to their professional duties.

7. Support of the Profession and Other Professionals
Certificants are required to:
a. Uphold the honour and prestige of the profession.
b. Recognize and respect the original work, integrity and ability of their peers.

8. Support of the CRSP/PSAC Certification
Certificants are required to:
a. Comply with the relevant provisions of the CRSP®/PSAC® bylaws, policies and certification scheme.

b. Make claims regarding CRSP®/PSAC® certification only with respect to the scope for which certification has been granted.

c. Not use the certification in such a manner as to bring the certification body into disrepute, and not make any statement regarding the certification which the certification body may consider misleading or unauthorized.

d. Discontinue the use of all claims to certification that contains any reference to the certification body or certification upon suspension or withdrawal of certification, and to return any certificates issued by the certification body.

e. Not use the certificate in a misleading manner.

f. Abstain from behaviour that will cause harm to the reputation of the BCRSP and its certificants.

g. Maintain the security of the BCRSP examination information and materials, including the prevention of unauthorized disclosures of test information.

9. Accountability (Adherence)

Each certificant will rely on the BCRSP to protect the integrity of the CRSP®/PSAC®. The Professional Conduct Committee (PCC) is tasked with ensuring that responsibility is fulfilled in a fair and impartial manner. The PCC will be solely responsible for ensuring BCRSP Policy is followed to investigate complaints or allegation of misconduct against certificants.

Complaints or allegations of misconduct against certificants found to be justified by the PCC will be referred to the Discipline Committee for review.

The BCRSP may disclose any disciplinary or enforcement decision/action against a certificant along with associated information, to other organizations including without limitation, organizations related to health and safety, law enforcement agencies, and regulatory bodies.

Appendix 2

AMERICAN ASSOCIATION OF OCCUPATIONAL HEALTH NURSES: CODE OF ETHICS[18]

The American Association of Occupational Health Nurses, Inc. (AAOHN) Code of Ethics has been developed in response to the nursing profession's acceptance of its goals and values and the trust conferred upon it by society to guide the conduct and practices of the profession. As professionals, occupational and environmental health nurses accept the responsibility and inherent obligation to uphold these values.

The Code of Ethics is based on the belief that the goal of occupational and environmental health nurses is to promote the worker, worker populations and community's health and safety. This specialized practice focuses on promotion and restoration of health, prevention of illness and injury, and protection from occupational and environmental hazards. The occupational and environmental nurse has a unique role in protecting the integrity of the workplace and the work environment.

The client can be workers, workers' families/significant others, worker populations, community groups and employers. The purpose of the AAOHN Code of Ethics is to serve as a guide for licensed[19] nurses to maintain and pursue professionally recognized ethical behavior in providing occupational and environmental health and safety services.

The AAOHN Code of Ethics

1. Occupational and environmental health nurses provide health, wellness, safety and other related services to clients with regard for human dignity and rights, unrestricted by consideration of social or economic status, personal attributes, or the nature of the health status.

2. Occupational and environmental health nurses, as licensed healthcare professionals, accept obligations to society as professional and responsible members of the community.

---

[18] Copyright (2012) the American Association of Occupational Health Nurses, Inc. Used with permission. All rights reserved. Online: http://www.aaohn.org/practice/ethics.html (date accessed: January 31, 2020).

[19] The AAOHN Code of Ethics and Interpretive Statements were reviewed and revised in 2009. Based on a 2010 by-law change to include licensed practical/vocational nurses as members, a language change was made in 2012 from "registered professional nurse" to "licensed nurse".

3. Occupational and environmental health nurses strive to safeguard clients' rights to privacy by protecting confidential information and releasing information only as required or permitted by law.
4. Occupational and environmental health nurses promote collaboration with other professionals, community agencies and stakeholders in order to meet the health, wellness, safety, and other related needs of the client.
5. Occupational and environmental health nurses maintain individual competence in nursing practice, based on scientific knowledge, and recognize and accept responsibility for individual judgements and actions, while complying with appropriate laws and regulations.

CHAPTER REFERENCES

American Association of Occupational Health Nurses, Inc.: Code of Ethics, online: http://www.aaohn.org/practice/ethics.html (date accessed: January 31, 2020).

Board of Canadian Registered Safety Professionals, Board of Canadian Registered Safety Professionals: Code of Ethics (Mississauga, ON: BCRSP, 2019).

D. Dyck, *Disability Management: Theory, Strategy and Industry Practice*, 3d ed. (Markham, ON: LexisNexis Canada Inc., 2006).

P. Edgar, "Resolving Ethical Dilemmas: Applying the Institute for Global Ethics' Ethical FitnessTM Model to Occupational and Environmental Health Practice Issues" (2002) 50:1 *AAOHN*.

J. Hall, "Ethics and Case Management" (1999) 13:2 *The OEM Report*.

R.M. Kidder, *How Good People Make Tough Choices: Resolving the Dilemmas of Ethical Living* (New York: Fireside Books, 1995).

B. Kirkpatrick, ed., *The Cassell Concise English Dictionary* (London, England: Cassell Publishers Ltd., 1989).

B. Morgan, "Workplace Communication Pitfalls" (Winter 2009) *OOHNA Journal*.

R.D. Parsons, *The Ethics of Professional Practice* (Needham Heights, MA: Allyn & Bacon, 2001).

P. Strahlendorf, "Professional Ethics in OH&S" (July/August 2002) *OHS Canada*.

Chapter 30

OCCUPATIONAL HEALTH AND SAFETY: LEGAL ASPECTS

A. INTRODUCTION

An Occupational Health and Safety (OH&S) Program is a management response to legislation that upholds that:

- employers must provide, as far as is reasonably practicable, a safe, and healthy workplace for workers; and
- employers must ensure that their workers are aware of their responsibilities and duties under the OH&S legislation.

Given the amount of legal information that exists in the area of OH&S, and the differences between Canadian provincial and federal OH&S legislation, this chapter has been designed to address an overview of the pieces of Canadian legislation that impact workplaces, some recent legislative changes and their impact on the field of Occupational Health and Safety, and some relevant legal terms.

B. LEGISLATION THAT INFLUENCES WORKPLACES: AN OVERVIEW

All stakeholders in the area of Occupational Health and Safety must be aware of the relevant legislation that may affect them and their operations. Such legislation may include, but is not limited to, the provincial OH&S Acts, the Workers' Compensation Acts, *Criminal Code*,[1] section 217 (commonly referred to as Bill C-45), Workplace Hazardous Information Systems (WHMIS 2015) legislation, *Transportation of Dangerous Goods Act, 1992* (TDGA),[2] Canadian Safety Code, fire protection legislation, human rights legislation, legalization of marijuana, employee and labour relations, employment standards, and freedom of information and protection of privacy legislation.

1. Canadian OH&S Acts

In Canada, each province has its own OH&S legislation, and organizations under

---

[1] R.S.C. 1985, c. C-46.
[2] S.C. 1992, c. 34.

federal jurisdiction are under the *Canada Labour Code*,[3] Part II. Although the specific Acts and regulations vary, the general principles remain the same: employers are responsible for maintaining the health and safety of employees at their worksites. Employers are obligated to ensure that employees are aware of their OH&S responsibilities and duties. Employees also have a responsibility to work safely and protect the health and safety of themselves, their co-workers, and the general public where applicable.

Areas generally covered in the OH&S legislation include, but are not limited to:

- chemical hazards;
- physical hazards;
- electrical hazards;
- noise;
- general work safety;
- radiation;
- ventilation;
- working alone;
- workplace violence; and
- first aid.

Within these regulations there are specific requirements for hazard assessment, hazard controls (*i.e.*, engineering controls, administrative controls, and personal protective equipment), incident investigation and injury management, as well as periodic medical surveillance and follow-up depending on the exposure potential. As this legislation is reviewed and updated, employers and OH&S Practitioners/Professionals should maintain communication with government OH&S departments to ensure that they are informed of the latest changes.

2. Workers' Compensation Acts

Introduced in the late 1800s and early 1900s, the various Canadian Workers' Compensation Acts were enacted to protect and support workers injured in the workplace, as well as to protect employers from being sued. In essence, they created a "mutual gains and responsibilities system" for Canadian employers and workers.[4]

As described in Chapter 1, "Occupational Health and Safety: Historical Perspectives", workers of the era faced wretched and deplorable workplace conditions. If injured, they had little hope of getting financial compensation. The prevailing belief was that work was hazardous and workers should plan for the eventuality of getting

---

[3] R.S.C. 1985, c. L-2.

[4] AWCBC (2020). The Historic Compromise, available at: http://awcbc.org/?page_id=57 (accessed February 28, 2020).

injured. As well, employers were faced with a real risk of being sued by injured workers. These disputes had to be settled in court, tying up the court system and ending up as lengthy, costly ventures. The bottom line was that rarely did the injured worker get compensated, and should the litigation prove successful, rarely did the worker obtain compensation because often the employer would become financially insolvent.

In 1913, the modern Canadian version of a workers' compensation system was born. Sir William Meredith, Chief Justice of Ontario, delivered a report that proved to launch the first significant piece of social legislation in Canada.[5] The Meredith Report outlined a trade-off in which workers relinquish their right to sue in exchange for compensation benefits.

Although the envisioned workers' compensation system has evolved since, the five basic tenets remain:

(1) *Collective liability*: Employers as a collective group are responsible for the payment of injured workers.

(2) *No-fault insurance*: Regardless of the injury cause, the worker is covered.

(3) *Security of payment*: The worker's loss of earnings is covered regardless of the employer's financial status.

(4) *Independent administration*: The administration of this system of insurance is achieved via an independent agency. The system exists at arm's length from the government and is shielded from political influence, allowing only limited powers to the Minister responsible.

(5) *Exclusive jurisdiction*: The independent agency has the exclusive authority to make decisions on all workers' compensation matters.

In the early 1900s, provincial Workers' Compensation Boards (WCBs) were established in Canada to administer the workers' compensation system for the province.

Today, the Canadian WCB system is a third-party, government-operated insurance system designed to protect the injured/ill worker and to afford employers litigation protection. It is a no-fault, industry-funded insurance that is mandatory for certain industry groups. Employers pay all the premiums and claims costs, while employees forfeit their right to sue their employer. Claim administration includes claim submission, claim adjudication, claim appeal, if required, and claim termination.

In addition, Canadian WCBs provide injury prevention information and education; protection from legal action; financial and healthcare benefits; rate incentive

---

[5] *Ibid.*

programs; and co-ordination of the partners in the workplace safety and insurance system.

In most instances, Canadian employers are legally required to manage their WCB claims and offer work accommodation. This is best achieved through a Disability Management Program. Having an Integrated Disability Management Program in place allows the employer to offer claim management, case management, and return-to-work planning and placement in all disability, situations — occupational illness/injury and non-occupational illness/injury, short-term disability, and long-term disability. More information on Disability Management is provided in Chapter 27, "Disability Management: Overview".

3. Criminal Code, Section 217 (Commonly Referred to as Bill C-45)

On June 12, 2003, the Honourable Martin Cauchon, Minister of Justice and Attorney General of Canada, introduced Bill C-45, an *Act to amend the Criminal Code (Criminal Liability of Organizations)*,[6] which imposes criminal liability on corporations and organizations that fail to take reasonable measures to protect employees and public safety. The Act came into effect on March 31, 2004.

Bill C-45 stems from the public inquiry of Richards J. following the Westray Mine disaster of 1992. The purpose of Bill C-45 is to assign legal accountability for workplace safety to organizations and their decision-makers across Canada. Accordingly, everyone who undertakes, or has the authority to direct how another person does work or performs a task, is obligated to take reasonable steps to prevent bodily harm to that person, or any other person, arising from that work or task. This is defined as their OH&S legal duty. Failure to meet this duty is deemed a criminal offence.

OH&S criminal negligence is established where the organization or individual(s), in doing anything, or in omitting to do anything that it is their legal duty to do, shows wanton or reckless disregard for the lives of others (criminal intent). To prove criminal intent, both an unlawful act and intent must occur. Depending on the severity of the injury, the associated penalties range from 10 years to life imprisonment for individuals, and fines of $100,000 or more for organizations. Bill C-45 also provides sentencing guidelines and options for the courts.

The best defence for any organization is a sound offence.[7] By having an Occupational Health and Safety Management System (OHSMS) that contains all the requisite elements for effective functioning, organizations can demonstrate:

- that they are taking reasonable steps to prevent bodily harm to workers and the public;
- compliance with the applicable OH&S legislation; and

---

[6] S.C. 2003, c. 21.

[7] N. Keith, "Alberta OH&S Due Diligence for Managers and Supervisors" (Seminar presented in Calgary, AB: Gowlings, 2003) [unpublished].

- due diligence in meeting the intent of the OH&S legislation.

4. Workplace Hazardous Materials Information Systems Legislation

Workplace Hazardous Materials Information Systems (WHMIS-2015) is a Canadian hazard information system that is intended for safe handling, storage, use, and disposal of controlled products in the workplace. A **hazardous product** is a substance that is regulated under the *Hazardous Products Act*,[8] *Controlled Products Regulations*[9] and that meets the criteria for classifying controlled products.

The intent of WHMIS 2015 is to enable the identification of products that are hazardous to human health or the environment; to communicate the related hazards; and to describe how to safely handle, store, use, and dispose of these controlled products in Canadian workplaces. WHMIS 2015 provides hazard communication techniques such as labels, safety data sheets (SDSs), and worker training.

The responsibility for upholding WHMIS rests with four stakeholders:

(1) **The supplier of the controlled product** — Anyone who produces or imports a hazardous product must test and analyze the product for classification purposes. Once classified as a hazardous product, the product must be labelled using a WHMIS Supplier Label and providing an SDS as a condition of sale.

(2) **The employer** — The employer is anyone who employs one or more people, or is a manufacturer or importer of a hazardous product.[10] The employer is required to ensure that hazardous products are correctly and adequately labelled according to WHMIS requirements; that SDSs are readily available and accessible to workers in a format that they can understand; and that their workers are trained on the safe handling, storage, use, and disposal of hazardous products present in the worksite.

(3) **The worker** — The worker is a person employed by an employer. Workers are obligated to attend the WHMIS 2015 training provided to them, to read supplier/workplace labels and SDSs, to seek help when they do not understand a SDS, and to adhere to the WHMIS 2015 requirements such as following the precautionary measures, decanting and labelling requirements, first aid measures, and disposal procedures.

(4) **The Trade Union** — Union leaders have a tripartite obligation to make WHMIS successful.

WHMIS is federal legislation that is enforced by provincial and territorial OH&S officials. Prosecution of offenders may occur under the *Hazardous Products Act* and

---

[8] R.S.C. 1985, c. H-3.

[9] SOR/88-66.

[10] Hatscan, "WHMIS Train-the-Trainer" (Hatscan Seminar presented in Edmonton, AB, February 2007) [unpublished] 1.

provincial and territorial OH&S statutes. The aim of WHMIS is to reduce workplace incidents and injuries/illness related to chemical exposures. It applies to all Canadian workplaces where controlled products exist. For a complete description of the WHMIS legislation, refer to Chapter 5, "Occupational Health and Safety Program: Manual Development", Section 3, Related Policies and Procedures.

5. Transportation of Dangerous Goods (TDG)

The *Transportation of Dangerous Goods Act* (TDGA) and Regulations set standards for the movement of harmful chemicals to protect both the public and people moving dangerous goods from incidents, leaks, and spills during the transportation of dangerous goods.

Dangerous goods are those defined as substances that pose a risk to health, safety, property or the environment during operation and/or transportation. (In the United States, the equivalent term is **hazardous materials**.) They are divided into classes on the basis of specific chemical characteristics producing the risk. Examples are explosives, compressed gas (such as oxygen, propane, aerosols), flammable liquids (such as paint, gasoline, diesel fuel), oxidizing substances, toxic substances (formerly called poison), infectious substances, corrosive substances, and miscellaneous goods that pose enough of a risk in transport to justify regulation.

Moving dangerous goods by any means of transportation is subject to regulations. The regulations require training for those who handle, offer, or transport dangerous goods. Employers must ensure employees receive training appropriate for their level of assignment, and issue them a training certificate. Alternatively, employees may perform dangerous goods duties in the presence and under the direct supervision of a trained person. TDG also includes the identification (labelling) of dangerous goods and strict packaging standards.

Although they are mutually exclusive, two pieces of legislation tend to interface — TDG and WHMIS legislation. TDG regulations cover the loading, transportation and unloading of dangerous materials. Once in the workplace, the use, handling, storage and disposal of controlled products are regulated by WHMIS legislation.

OH&S Professionals/Practitioners must be cognizant of TDG legislation as it relates to their company's business operations. In many instances, the OH&S Professional/Practitioners will be the technical expert in this area and expected to assume a governance and stewardship role within the organization.

6. Canadian Electrical Safety Regulatory System

In Canada, there are 13 separate electrical safety regulatory authorities: Canada's 10 provinces and three territories each have legislated regulatory authorities for electrical safety. Within each provincial or territorial jurisdiction, there is a self-contained electrical regulatory infrastructure with authority to independently set and enforce electrical safety standards.

All electrical regulatory activities at the provincial or territorial level start with a statute or Act which establishes the legal framework under which the electrical safety regulatory programs operate. Acts address requirements that typically establish the scope of legislation, the authority of Inspectors and Chief Inspectors, administration provisions, offences, penalties, and the authority to make regulations and codes. A typical electrical safety regulation would cover equipment standards; qualification and licensing of installers; review of electrical designs; the issuing of permits; inspections; compliance procedures; utility connections; incident investigations; collection of fees; and other various administrative rules. Codes typically address rules and regulations that govern procedures or behaviours. In some provinces, their regulations address the items stated above and may include prescriptive or performance-based requirements.

The areas covered by the Electrical Safety regulations include:

- The roles and duties of Electrical Inspectors and the Chief Electrical Inspector operating in each province and territory.

- The adoption and enforcement of a common installation code — the CSA Standard C22.1, the Canadian Electrical Code Part I (CEC Part I).

- The acceptance of certification organizations: provincial/territorial regulations state that electrical equipment cannot be used, offered for sale, or otherwise distributed unless certified by an acceptable certification organization.

- The enforcement of regulatory requirements: in Canada, a mix of provincial/territorial governments, municipal governments, quasi-government agencies, electric utilities and, in some cases, private inspection agencies, carry out the enforcement of the various regulatory requirements. As well, through design reviews, inspections, and investigations, Electrical Inspectors and Chief Electrical Inspectors enforce the electrical safety standards. Penalties can range from fines of tens of dollars to several thousands of dollars.[11]

Given industry reliance on electricity and electrical equipment, OH&S Professionals/Practitioners must be familiar with the nature, hazards, and safe work practices associated with electricity.

7. National Fire Code of Canada

In 1956, the National Research Council of Canada (NRC) created the Associate Committee on the National Fire Code, which produced the first edition of the National Fire Code in 1963. The two Associate Committees were disbanded in October 1991 and replaced by the Canadian Commission of Building and Fire Codes (CCBFC).

---

[11] IAEI, "Ask CSA: Understanding the Canadian Electrical Safety Regulatory System. Part II: Canadian Provinces and Territories", online: http://iaeimagazine.org/magazine/2002/01/16/understanding-the-canadian-electrical-safety-regulatory-system-part-ii-canadian-provinces-and-territories/ (date accessed: January 31, 2020).

Today, the National Fire Code of Canada (NFC) sets out the technical provisions that regulate:
- activities related to the construction, use or demolition of buildings and facilities;
- the condition of specific elements of buildings and facilities;
- the design or construction of specific elements of facilities related to certain hazards; and
- protection measures for the current or intended use of buildings.

The Code's provisions are designed to achieve fire prevention and protection objectives and describe the functions that a building or facility must perform to fulfil these objectives.

The Code also regulates the ongoing safety of existing buildings in terms of:
- fire safety provisions;
- handling and storage of flammable materials in buildings;
- furnishings (where appropriate); and
- fire-related hazards associated with certain industrial processes.

With the release of the 2010 National Fire Code of Canada, OH&S professionals will note that the Code has a new organizational layout:
- *Division A:* The compliance options, objectives, and functional statements.
- *Division B:* The provisions currently termed "acceptable solutions", which relate to issues like building and occupant fire safety, indoor and outdoor storage, flammable and combustible liquids, hazardous processes and operations, fire protection equipment, and fire emergency systems in high buildings.
- *Division C:* The administrative provisions.[12]

The Code is highly relevant to all business operations, and OH&S Professionals/Practitioners must be familiar with the National Fire Code of Canada and its implications to the organization and workers.

8. Building Codes and Regulations

Under Canadian law, the regulation of buildings is a provincial responsibility and is carried out through various laws, Acts, codes, and regulations, often administered at the municipal level. Provincial legislation empowers government agencies or departments to regulate different aspects of buildings, depending on the objectives of the specific law or Act. Such legislation permits the establishment of detailed

---

[12] National Research Council of Canada, *National Fire Code of Canada, 2010*, online: http://nrc-cnrc.gc.ca/eng/publications/codes_centre/2010_national_fire_code.html (date accessed: January 31, 2020).

regulations by which the objectives of the law are to be met, or it may refer to other documents. For example, laws protecting the safety and health of building occupants usually refer to building codes for additional requirements.

Building codes generally apply to new construction and have traditionally been concerned with fire safety, structural integrity, and the health of the building's occupants. More recent codes have dealt with accessibility for handicapped persons and with energy conservation.

Zoning and planning legislation play an important role in regulating buildings by restricting the type, size, spacing, setback, and use of buildings, and by controlling general land use in a community. Its purpose is to maintain certain neighbourhood characteristics and to allow for a community's orderly development.

In addition to building codes, there are various miscellaneous Acts aimed at specific building types or at specific services within buildings. Liquor licensing, hotel, theatre, and factory Acts, for example, may affect the construction or use of specific types of buildings. Regulations under such Acts may parallel, or even conflict with, building code provisions, although the trend is to rely on building codes where practicable. Plumbing, electrical, elevator, and boiler and pressure-vessel codes are examples of standards targeted towards particular building services and may be enacted separately or combined in a single Act.

Although the regulation of buildings falls within provincial jurisdiction, past governments delegated this responsibility to municipalities. The outcome is diversified building regulations. Since municipal resources varied widely, the quality, and efficiency of building regulations also varied.

To promote uniformity, the National Building Code of Canada (NBC) published a model set of building requirements. With assistance from the NRC, the CCBFC developed the National Fire Code. The Commission also produced the Canadian Plumbing Code, Canadian Farm Building Code, and Measures for Energy Conservation in New Buildings. All are written as model legislation for adoption by authorities having jurisdiction.

Today's construction technologies and techniques differ vastly from those in use at the time the first NBC was produced in 1941. To keep pace with changes and to ensure that the latest innovations and applications are applied safely by the construction industry, a new version of the NBC is published approximately every five years. The National Research Council of Canada's (NRC) Canadian Codes Centre plays a vital role in this process by providing technical and administrative support to the CCBFC and its related committees, which are responsible for the development of the national model construction codes of Canada.

A number of standards-writing agencies produce industry standards on various aspects of building, which are referred to in building codes or in other regulations. The National Building Code, for example, refers to a total of 192 standards, including standards for construction materials, design, installation, equipment, and testing. Provincial Acts also refer to such standards or base their regulations on them. For example, the Canadian Electrical Code, produced by the Canadian Standards Association (CSA), forms the basis for electrical requirements in every Canadian province.

Other standards-writing bodies in Canada include the Canadian General Standards Board, Underwriters Laboratories of Canada, and the Canadian Gas Association. The standards they produce tend to play a major role, along with building codes, in regulating the construction of buildings in Canada.[13]

9. The Highway Traffic Acts

Each Canadian province and territory has a Highway Traffic Act. All tend to be similar in focus although how each is implemented may differ. Using the Ontario Act[14] for demonstration purposes, Canadian highway traffic Acts tend to cover:

- general administration of the Act;
- Transportation Safety Board;
- motor vehicle administration:
    - licences, registration, insurance, permits; and
    - vehicles, equipment, incidents, and removal of vehicles;
- disqualification from driving;
- general operation of vehicles:
    - speed limits;
    - traffic control devices;
    - operation of vehicles;
- off-highway vehicles;
- commercial motor transport;
- operation of commercial vehicles:
    - compliance;
    - weight;
    - general matters;
- enforcement, rights, remedies, and obligations:
    - offences;
    - evidence in prosecutions;
    - peace officers;

---

[13] A. Hansen, "Building Codes and Regulations", *The Canadian Encyclopedia*, online: http://www.thecanadianencyclopedia.com (date accessed: January 31, 2020).

[14] *Highway Traffic Act*, R.S.O. 1990, c. H-8.

- pleas, appeals, prohibition, and surrender of licences;
- rights, remedies, and obligations; and
- transitional provisions, consequential amendments, repeals, and coming into force.

Given that transportation and highway driving are key parts of today's business operations, OH&S Practitioners/Professionals need to be aware of the terms of this important Act.

10. Environmental Legislation

Environmental law refers to the laws and legal principles governing the behaviour of persons, businesses, government agencies, and other organizations. They were designed to protect the environment and health of people. The sources of environmental law are statutes and common law.

(a) Historical Background

The *Canadian Environmental Protection Act, 1999*[15] (CEPA), federal legislation, has been in place since 1988. A revised version of the Act was introduced in 1999 and enacted March 31, 2000.[16] A substantial portion of the 1988 Act dealt with the identification, control, and/or prevention of toxic substances in the environment; and the promotion of life cycle management of toxic substances. CEPA 1999 took this approach a step further, acknowledging the need to eradicate essentially all persistent and bioaccumulative substances from the environment. CEPA 1999 maintains additional legal objectives, which include reducing pollution caused by the dispersal of pollutants in Canadian waters and addressing the inconsistencies in environmental regulations applicable to federal works or undertakings on federal lands. CEPA also fulfils Canada's international and national commitments with respect to regulating air pollution and ocean dumping.[17]

The CEPA 1999 management cycle is a dynamic approach to environmental management and protection that comprises five integrated elements:

(1) risk assessment;

(2) risk management;

(3) compliance promotion and enforcement;

(4) research and monitoring; and

---

[15] S.C. 1999, c. 33.

[16] CEPA Environmental Registry™, General Information, Fact Sheet, The *Canadian Environmental Protection Act, 1999* (CEPA 1999), online: http://www.ec.gc.ca/CEPARegistry/gene_info/management.cfm (date accessed: January 31, 2020).

[17] Canada, "The *Canadian Environmental Protection Act, 1999*, Chemical Substances in Canada", online: http://www.chemicalsubstanceschimiques.gc.ca/index-eng.php (date accessed: January 31, 2020).

(5) reporting, communication and cooperation.[18]

The principles of the CEPA are:

- sustainable development;
- pollution prevention;
- precautionary principle;
- polluter-pays principle; and
- removal of threats to biological diversity.[19]

One interesting feature of CEPA 1999 is progressive enforcement. Through the use of progressive enforcement tools, such as environmental protection compliance orders, environmental protection alternative measures (*i.e.*, alternatives to court prosecution through a negotiated agreement), and ticketing for offences, compliance is encouraged.

(b) Canadian Environmental Law

In Canada, there are a number of pieces of environmental law. To assist OH&S professionals and other readers, a summary is provided in Table 30.1.

Table 30.1: The Key Canadian Environmental Laws

Governing Body	Department	Responsibility
FEDERAL	Environment Canada	• Protection and management of natural environment — air, water, soil • Administer statutes
	• Health Canada • Agriculture and Agri-Foods Canada • Dept. of Foreign Affairs and International Trade • Fisheries and Oceans Canada • Industry Canada • Justice Canada • Natural Resources Canada • Transport Canada	All deal with various environmental matters
	Standing Committee on Environmental and Sustainable Development	• Study any environmental issues submitted by House of Commons and make recommendations

---

[18] CEPA Environmental Registry, General Information, *Fact Sheet, Canadian Environmental Protection Act, 1999* (CEPA 1999), online: http://www.ec.gc.ca/CEPARegistry/gene_info/management.cfm (date accessed: January 31, 2020).

[19] Don Sayers & Associates, Environmental Practice Domain, BCRSP Study Guide (Mississauga, ON: BCRSP, 2000) at 38.

	National Round Table on the Environment and the Economy (NRTEE)	• Provide objective and accurate information on environment and economy
	Commissioner of the Environmental and Sustainable Development	• Promote sustainable development
FEDERAL AND PROVINCIAL	Canadian Council of Ministers of the Environment (CCME)	• Promote cooperation between federal and provincial authorities
	• Wildlife Ministers Council of Canada • Federal-Provincial Parks Council • Canadian Council of Forest Ministers • Federal-Provincial Agriculture Ministers Conference • Canadian Council of Energy Ministers	• Resolve issues related to environmental problems
PROVINCIAL	Ministry to deal with pollution abatement and nature conservation	• Manage Permits and licences • Ensure development projects comply with environmental standards • Inspections • Prosecutions for environmental offences
LOCAL GOVERNMENTS	Municipalities	Environmental regulation and management of: • Air • Water • Nuisance • Environmentally sensitive areas • Public health • Planning and zoning • Business licensing and regulations • Dangerous substances • Plenary powers
LAND CLAIM SETTLEMENTS	Aboriginal and government representatives	Address: • Environment assessment • Land use planning • Wildlife management • Protected areas

(c) Regulatory Instruments

The methods used by the regulators to protect the environment include:

- **Market and Economic Instruments:** By using consumer choice, environmental protection can be achieved.

- *Command and Control Instruments:* Canadian governments can protect the environment through the use of legal sanctions such as those set out below in Table 30.2.

Table 30.2: Command and Control Instruments

Section Type	Federal (F), Provincial (P), Municipal (M)	Description
Act	F P	• Statute passed in Parliament • General in nature with respect to requirements • Broad policy perspective
Regulation	F P	• Based on the Act • Means of control through application of defined rules or principles • Have specified operation details
• Permits • Certificates of Approval • Authorizations • Licences	F P	• Issued by federal/provincial departments/agencies • Usually required by an Act or Regulation
Orders	F P	Issued to specific parties due to an offence or non-compliant condition
Municipal By-laws	M	• Like regulations • Depend on and are made under authority of provincial statutes governing municipalities
• Codes of Practice • Policies • Criteria • Objectives • Guidelines	F P	• May be jointly developed • Not legally binding unless specifically referred to in an Act or Regulation • Must be reviewed and assessed as part of environmental compliance

Three other Canadian environmental initiatives include:

- ***Canada's Green Plan:*** At the end of 1990 the Canadian government announced a $3 billion five-year environmental action plan called the Green Plan. This was the federal government's first attempt to integrate environmental and economic considerations. The initial objective was to design a planning process for ecological resources that would change the structure of decision-making in government. This initiative did not succeed and the Green Plan became little more than a source of funds for "soft" policy areas such as research and public education.[20]
- ***Agenda 21:*** 1992 Earth Summit in Rio de Janeiro — Canada was one of the signing countries of Agenda 21, which called for global sustainable development by the 21st Century.[21]
- ***Kyoto Accord:*** Established in 1997, the Kyoto Protocol to the United Nations Framework Convention on Climate Change (Kyoto Accord) is an amendment to the international treaty on climate change, assigning mandatory emission limitations for the reduction of greenhouse gas (GHG) emissions to the signatory nations. The Kyoto Accord detailed the commitment of a six per cent reduction over the 1990 levels of GHGs, by the majority of developed countries by 2008-2012. The Kyoto Accord now covers more than 160 countries globally and over 55 per cent of the global GHG emissions.[22]

(d) Relevant Environmental Legislation

The relevant environmental legislation that could impact OH&S Practitioners/ Professionals includes the legislation set out below in Table 30.3.

Table 30.3: Relevant Environmental Legislation

Act	Goals	Focus	Details
Canadian Environmental Protection Act, 1999 CEPA	• Contributes to sustainable development through pollution prevention • Protects the environment, earth and its people	• Sustainable development • Pollution prevention	• Administration • Public participation

---

[20] R. Gale, "Canada's Green Plan" in *Nationale Umweltpläne in Ausgewählten Industrieländern* (Berlin: Springer-Verlag, 1997) at 97 [a study of the development of a national environmental plan with expert submissions to the Enquete Commission "Protection of People and the Environment" for the Bundestag (German Parliament)].

[21] United Nations, Agenda 21, United Nations Conference on Environment and Development (UNCED) held in Rio de Janeiro, Brazil, June 3 to 14, 1992, online: http://www.un.org/esa/sustdev/documents/agenda21/index.htm (date accessed: January 31, 2020).

[22] United Nations Framework Convention on Climate Change, "Kyoto Accord", online: http://unfccc.int/kyoto_protocol/items/2830.php (date accessed: January 31, 2020).

Act	Goals	Focus	Details
		• Precautionary principle	• Information gathering, objectives, guidelines and codes of practice
		• Polluter pays principle	• Pollution prevention
		• Removing threats to biological diversity	• Controlling toxic substances
			• Animate products biotechnology
			• Controlling pollution and managing wastes
			• Environmental matters related to emergencies
			• Government operations
			• Enforcement
			• Miscellaneous matters
			• Consequential amendments, repeal, transitional provision and coming into force
Fisheries Act[23]	Governs the fishing industry and the protection of fisheries	• Grants licences and leases	• Safe fish passage
		• Regulates methods of fishing	• Protection of fish in/near fish ways
		• Provides for habitat protection	• Fish guard and screens
		• Sanctions the powers of Fisheries officers	• Destruction of fish
			• Destruction of fish habitat
			• Pollution of fish
			• Designation of protected fish areas
Transportation of Dangerous Goods Act, 1992 **TDGA**	Promotes public safety	Addresses transportation of dangerous goods in terms of:	• Safety requirements, standards
		• Importing	• Emergency response
		• Handling	
		• Offering for transport	

---

[23] R.S.C. 1985, c. F-14.

Act	Goals	Focus	Details
		• Transporting • Reporting of accidental releases	

(e) Other Relevant Terms

An **environmental aspect** is anything that an organization does that can react with any part(s) of the environment.[24] A **high potential aspect** is one that is associated with high environmental impact.[25] An **environmental impact** is any change to the environment, good or bad, caused by, or resulting from, an organization's activities, products, and/or services.[26]

An **environmental audit** is a systematic process of objectively obtaining and evaluating evidence regarding a verifiable assertion about an environmental matter, to determine the degree of correspondence between a verifiable assertion and standards criteria, and to report to the client.[27] In essence, it measures whether the organization is doing what it says it is doing. There are three types of environmental audits:

(1) **Liability audits** — measure existing liabilities (*Liability Audit*); compliance issues (*Compliance Audit*); environmental risks (*Environmental Risk Assessment*); or acquisition risks (*Pre-acquisition Audit*);[28]

(2) **Management audits** — deal with measurement of the environmental management system (EMS) (*EMS Management Audit*); requested by a Board of Directors (*Corporate Audit*); comparison of the current EMS against a management system standard (ISO 14000) (*Management System and Control Audit*); review of the environmental policy (*Policy Audit*); or examination of a particular environmental issue (*Issue Audit*);[29] and

(3) **Activities audits** — examine all aspects of an organization's environmental performance at a particular site (*Site Audit*); identify, quantify and assess an organization's wastes (*Waste Audit*); assess a product's actual/potential environmental impact (*Product Audit*); or review activities that

---

[24] Canadian Society of Safety Engineers (CSSE), *BCRSP Exam Preparatory Course* (Toronto, ON: CSSE, 2002) at 6.

[25] *Ibid.*

[26] *Ibid.*

[27] Don Sayers & Associates, Environmental Practice Domain, *BCRSP Study Guide* (Mississauga, ON: BCRSP, 2000) at 1.

[28] *Ibid.*

[29] *Ibid.*, at 2.

cut across department/divisional boundaries (***Cross-boundary Activity Audit***).[30]

An **environmental assessment** is defined as an assessment done prior to start-up of an operation to determine the actual/potential environmental aspects and impacts.[31] However, it can also be done post-start-up because it has the ability to review historical activities to identify problems.

An **Environmental Management System (EMS)** is a management tool designed to manage an organization's environmental affairs, monitor its effect on the environment, and address the environmental "bottom line" (economics).[32] It is a structured approach consisting of a number of interrelated elements that work together to achieve the objective of effective environmental management. The key principles of an EMS are built on the cyclical concept of Plan-Do-Check-Act (**P-D-C-A**). The goal is continuous improvement. The key principles for a successful EMS include:

- Senior Management Commitment;
- An Appreciation of the Related Internal Components and External Factors;
- Compatibility with the Organizational Culture;
- Integration with the Organization's General Management System;
- A Focus on Continuous Improvement;
- Flexibility; and
- Employee Awareness and Involvement.

Some recommended EMS strategies include those set out in Table 30.4.

Table 30.4: EMS Strategies

Component	PDCA	Description	Elements	Key Phrases
ENVIRONMENTAL POLICY		• Commits to:	• Documents on the organization's overall aims, principles of action and intentions re: the environment	
		1. Continuous Improvement	• Endorsed by management	
		2. Pollution prevention	• Integral part of business strategy	

---

[30] *Ibid.*

[31] Canadian Society of Safety Engineers (CSSE), *BCRSP Exam Preparatory Course* (Toronto, ON, 2002) CSSE at 6.

[32] Don Sayers & Associates, Environmental Practice Domain, *BCRSP Study Guide* (Mississauga, ON: BCRSP, 2000) at 1-4.

Component	PDCA	Description	Elements	Key Phrases
		3. Legal compliance	• Compatible with other company policies • Framework for setting objectives and targets • Communicated to workers • Publicly available	
PLANNING	P	• Environmental aspects	Devise a plan that includes:	• *Environmental Aspects* are the elements of the organization's activities, products, services that can interact with the environment.
			• Identifying environmental aspects	• *Environmental Review* identifies the relevant Environmental Aspect(s), compares them to regulations, determines controls and establishes risks.
			• Assessing their significance through environmental review	• *Environmental Review* is the foundation upon which the rest of the EMS is built.
			• Identifying areas for cost-effective improvement	
			• Setting objectives and targets	
		• Legal and other requirements	• Organization must establish what its commitment to legal compliance will mean	
		• Objectives and Targets	• Organization must set the objective of identifying and correcting any non-compliance situations	• *Environmental Objective* is the overall goal set for achievement. It must be measurable.

Component	PD CA	Description	Elements	Key Phrases
				• ***Environmental Target*** is the detailed performance requirements that arise out of the Environmental Objective.
		• Environment-al Management Program (EMP)	• A program for achieving the Environmental Objectives and Targets • Includes time frames and the people responsible	Promotes due diligence
IMPLEMEN-TATION & OPERATION	D	• Structure and responsibility	• Define stakeholder roles, responsibilities and authority • Provide resources • Champion to ensure EMS is established	Ensures that the EMP can succeed
		• Training, awareness and competence	• Workers require appropriate training • Workers must be competent • Organization must make workers aware that they must conform; explain the importance of their work and their role and responsibility in the process	A required element of an EMP
		• Communication	Provision for sending and receiving information about EMS internally and externally	Required for successful implementation of the EMP
		• EMS documentation	• Document the EMS elements and how they interact • Maintain documentation	A required element of an EMP
		• Document control	• Procedures for controlling all documents Ready retrieval and access	A required element for due diligence reasons

Component	PDCA	Description	Elements	Key Phrases
		• Operational control	• Identify all organizational activities	A key component that includes the use of safe operating procedures (SOPs) and affirmative procurement
			• Have operating procedures for activities where environmental objectives and targets are not being met	
			• Communicate objectives to suppliers and contractors	
		• Emergency preparedness and response	Have procedures to:	• An essential element
			• Identify potential incident and emergency	• Involves spill response plans
			• Respond	
			• Prevent and mitigate environmental impacts	
			• Periodically test procedures	
CHECKING & CORRECTIVE ACTION	C	• Check results against plan	Have procedures to:	Involves the use of performance metrics
		• Determine if EMS is successful	• Monitor and measure performance	
			• Measure degree of compliance	
		• Take corrective actions	• Have procedure to define responsibility and authority for:	Includes program monitoring and use of corrective action logs or plans
			• Investigating and handling non-conformances	
			• Taking action	
			• Initiating corrective actions	
			• Recording changes made	
			• Keep records on training, audit results and management reviews	
			• Have procedures for:	

Component	PDCA	Description	Elements	Key Phrases
		• Audit regularly	• Identifying, maintaining and disposing of records • Checks conformance to standards and implementation/ maintenance of practices • Procedure should cover: • Activities and areas to be covered • Frequency of audits • Responsibilities associated with managing and conducting audits • Communication of audit results • Auditor competence • How audits are done	To ensure due diligence audits as previously described are used
MANAGEMENT REVIEW	A	• Ensure effectiveness and continued suitability	• Done periodically • Address need for change to environmental policy objectives and other elements • Based on audit results, changing circumstances, commitment to Continuous Improvement	At this stage, the use of tools like Board of Director Reports and Annual Reports is involved

The benefits of an EMS are that it can demonstrate sound management of economic, social and environmental issues — the **Triple Bottom Line**.[33] An EMS:

- is a cost-effective approach which focuses on continuous improvement through maximization of efforts, regular assessments, and identification of improvement opportunities;
- ensures targets are met;
- ensures legislative compliance;

---

[33] *Ibid.*, at 14.

- increases efficiency and reduces costs;
- improves environmental performance;
- enhances the corporate image with the public, financial sectors, and government regulators;
- can increase corporate competitiveness; and
- helps to promote retention and recruitment of employees as a result of a high state of employee pride, morale, and job satisfaction.[34]

OH&S Professionals/Practitioners are often closely involved in an organization's EMS, the principles and elements of which closely parallel that of an OHSMS. As a rule of thumb, Environmental Management tends to address issues "outside of the company fence line"; while OHSMS tend to focus primarily on issues within the fence line. Both disciplines deal with exposures and many of the contaminants are the same, just at different concentrations. Knowledge of both disciplines is advantageous.

11. Human Rights Legislation

By Sharon Chadwick

The *Canadian Human Rights Act*[35] is designed to safeguard Canadians from discriminatory practices based on race, national or ethnic origin, colour, religion, age, sex, marital status, family status, disability, sexual orientation, or conviction for an offence for which a pardon has been granted. Section 2 of the Act stipulates that every individual should have equal opportunity to make the life he/she is able to have, without being prevented from doing so by discriminatory practices.

While the Act is wide-ranging, it only deals with approximately ten specific types of discrimination, some of which include:

- communicating any message that is likely to expose a person to hatred or contempt;
- displaying or publishing in public any notice, sign, symbol, or emblem that expresses or implies discrimination or an intention to discriminate;
- refusing to employ, or to continue employing, any individual, based on a discriminatory viewpoint;
- devising employment policies, including recruitment, referral, hiring, promotion, training, or apprenticeship policies that deprive individuals of employment opportunities; and
- denying goods, services, facilities, or accommodations on discriminatory grounds.

---

[34] *Ibid.*, at 15-16.
[35] R.S.C. 1985, c. H-6.

Although some provinces are covered by their own human rights legislation, those that are not and all federal organizations are under the *Canadian Human Rights Act*, the human rights principles expressed in this legislation are fairly consistent.

In relation to OH&S, human rights legislation has a major impact on the way employers treat employees, and in particular, disability management processes.

The application and interpretation of individual human rights cases is complex and varies depending on the individual circumstances of the case. There are no black and white answers to questions in this arena — all cases must be assessed on their own merit, and much of today's awareness and understanding of human rights issues is based on case law. Even so, there is a great deal of variability, and employers are advised to obtain legal counsel when setting up policies and procedures which may have human rights implications, or when assessing difficult cases.

Generally, the principles of human rights legislation in relation to the fields of OH&S and Disability Management include the following:

(a) Discrimination

Human rights legislation prohibits discrimination in employment practices such as hiring or retaining employees. For example, the Alberta *Human Rights Act*, section 7 states:

> 7(1) No employer shall:
>
> (a) refuse to employ or refuse to continue to employ any person, or
>
> (b) discriminate against any person with regard to employment or any term or condition of employment,
>
> because of the race, religious beliefs, colour, gender, physical disability, mental disability, age, ancestry, place of origin, marital status, source of income, family status or sexual orientation of that person or of any other person.[36]

This impacts the employer for both pre-placement and return-to-work fitness requirements. Human rights legislation is based on the principle of individual assessment — persons should be evaluated on their *ability*.

The British Columbia Council of Human Rights advises employers to:

- concentrate on a person's capabilities and not disabilities;
- assess persons as individuals, not as members of a group;
- avoid making generalizations about disabilities;
- define specific employment needs according to business priorities;
- clearly state the essential components of the job; and

---

[36] R.S.A. 2000, c. A-25.5, s. 7.

- establish reasonable standards for evaluating job performance.[37]

(b) Duty to Accommodate

Employers are required to make reasonable accommodation for persons with a physical or mental disability. Duty to accommodate is a tripartite effort:

(1) *The employer* has the primary duty to originate and implement a solution as he/she is in the best position to assess how the employee can be accommodated without undue interference in the operation of the business.

(2) *The union* (if applicable) has a joint responsibility to assist in accommodating disabled employees. Unions must cooperate in the search for and implementation of accommodations as well as considering modifications or waiver of collective agreement provisions if necessary for accommodations of a particular case.

(3) *The employee* has a variety of obligations in the process of seeking and sustaining a workable solution to accommodating his/her disability. The employee has the duty to inform the employer of the need for accommodation and of the effectiveness of the measures taken to accommodate. The employee is also required to provide to the employer information regarding his/her expected return-to-work date and any limitations or restrictions. The employee has a duty to take reasonable steps to facilitate the implementation of proposed workplace accommodations.[38]

Employers and unions are required to accommodate employees with disabilities (both physical and mental) to the point of "undue hardship". This includes the return to work of ill/injured employees. Basically, the employer must be able to demonstrate that it has made reasonable attempts to accommodate the employee to the point of "undue hardship".

The development of a comprehensive Disability Management Program with clear policies and procedures, as well as clearly defined roles and responsibilities, ensures that the duty to accommodate is applied consistently for all employees, and assists stakeholders in demonstrating that appropriate steps have been taken to accommodate disabled employees.

The Centre for Labour-Management Development outlines that:

> Where the disability prevents the employee from performing some or all of the functions of a particular position, possible accommodations may include the provision of sedentary, light or modified duties, elimination of physically difficult

---

[37] British Columbia Council of Human Rights, *Disability and the Human Rights Act* (Victoria, BC: British Columbia Ministry of Labour and Consumer Service, 1998).

[38] Centre for Labour-Management Development, "Accommodation Guidelines" (Presented at the Illness and Disability: Claims in the Unionized Workplace Conference, Edmonton, AB, February 1999) [unpublished] at 69.

or hazardous duties, modification of the work environment in a manner which permits the employee to continue to carry out his or her duties, alteration of shift schedules or hours of work, *etc.*

While the nature of the employment may affect the content of the duty in a particular case, an employer is not relieved of its duty to accommodate simply because the disabled employee is a temporary employee.[39]

(c) Undue Hardship

The concept of "undue hardship" is flexible and not clearly defined. The exact interpretation and determination of "undue hardship" will depend on the individual circumstances of the case. Generally, the concept will include considerations of such factors as:

- financial costs;
- disruption of the collective agreement;
- employee morale;
- interchangeability of the work force and facilities;
- size of the employer's operation; and
- safety concerns.[40]

Regarding the safety concerns, the seriousness of risk is assessed considering four factors:

(1) *The nature of the risk* — What could happen that would be harmful?

(2) *The severity of the risk* — How serious would the harm be if it occurred?

(3) *The probability of the risk* — How likely is it that the potential harm will actually occur? Is it a real risk, or merely hypothetical or speculative?

(4) *The scope of the risk* — Who will be affected by an event if it occurs?[41]

(d) Bona Fide Occupational Requirements

Placement of employees, both on hiring and on return to work from an absence, should entail matching the tasks of the job to the person's abilities. In order to require that an employee "perform" a specific task, it must be demonstrated by the employer that the task is a *Bona Fide* Occupational Requirement (BFOR). If the employee is unable to perform a task that is a requirement, he/she may not be placed in that position. For example, subsection 7(3) of the Alberta *Human Rights Act* states:

7(3) Subsection (1) does not apply with respect to a refusal, limitation, specification

---

[39] *Ibid.*, at 59.

[40] *Ibid.*, at 73.

[41] L. McDowell, *Human Rights in the Workplace: A Practical Guide* (Toronto ON: Carswell, 1998) at 7-43.

or preference based on a *bona fide* occupational requirement.

Although the term *Bona Fide* Occupational Requirement is not defined in the legislation, basically, it means that the employer must show that the specific task is "essential" to the performance of the job.

Both the federal Charter of Rights and Freedoms and provincial human rights legislation incorporate, as a defence to the duty to accommodate, the notion of a BFOR or *bona fide* occupational qualification (BFOQ). A BFOR or qualification is an essential or critical requirement for the performance of a job. If an employer can show that an employee cannot perform all the essential requirements of a job, then the employer is not obliged to accommodate him/her.

In a 1999 case,[42] the Supreme Court of Canada set out a three-part test used to determine whether a standard (*i.e.*, an essential or critical requirement, or core characteristic, of a job) is a *bona fide* occupational requirement. The employer must demonstrate that:

1. The standard's purpose is rationally connected to the performance of the job;
2. The standard was adopted in an honest and good faith belief; and
3. The standard is reasonably necessary to the accomplishment of the legitimate work-related purpose.[43]

In *Meiorin*, the Court linked the concepts of BFOR and undue hardship. To prove that a standard is a BFOR, the employer must show that it could not accommodate the employee without undue hardship.

Shortly after the judgment came down on the *Meiorin* Case, the *Grismer* Case began.[44] Terry Grismer, a mining truck driver, suffered from Homonymous Hemianopsia due to a stroke, leaving him visually impaired. As a result, the B.C Superintendent of Motor Vehicles cancelled his driver's licence and refused, for seven years, to renew it. During this time, Grismer repeatedly passed driving tests and standard visual tests, except that he always failed the peripheral vision testing. Grismer, who was able to compensate for his poor peripheral vision by wearing special prism glasses, complained that he had been unfairly discriminated against on the basis of a physical disability.

The Supreme Court of Canada used the *Meiorin* three-part test to demonstrate that the 120-degree standard was not a *bona fide* occupational requirement. They ruled that the service provider wrongfully denied a driver's licence to a man with a

---

[42] *British Columbia (Public Service Employee Relations Commission) v. British Columbia Government and Service Employees' Union (B.C.G.S.E.U.)*, [1999] S.C.J. No. 46, 176 D.L.R. (4th) 1 (S.C.C.) [hereinafter "*Meiorin*"].

[43] *Ibid.*, at para. 54.

[44] *British Columbia (Superintendent of Motor Vehicles) v. British Columbia (Council of Human Rights)*, [1999] S.C.J. No. 73 (S.C.C.).

disability who failed a generic visual acuity test. The Supreme Court of Canada interprets the *Meiorin* three-part test to include the duty to test persons with disabilities individually and not to make assumptions based on "group abilities".

(e) Human Rights and Substance Abuse Policies

Employers who are considering the development and implementation of workplace substance abuse policies and procedures should be aware of the human rights implications related to these programs.

Alcohol and drug addictions are generally accepted as "disabilities",[45] the same as any other physical or mental disability. Employers have a duty to accommodate employees with substance abuse problems:

- *Offers of assistance*, including arranging entry into treatment, providing counselling and referral to an Employee Assistance Program;
- *Time off for treatment*, with or without weekly indemnity, use of vacation entitlement, sick leave or indefinite suspension;
- *Modified work hours* to accommodate attendance at follow-up counselling; and/or
- *Bearing the cost of monitoring the employee's compliance* with the terms of agreed-upon treatment, managing morale problems of co-workers. and perhaps, altering the employer's position to remove safety-sensitive responsibilities and duties.[46]

The issue of substance abuse testing remains controversial. In Canada, there is no specific legislation related to drug and alcohol testing in the workplace, and these cases are generally addressed under human rights legislation. Employers considering the implementation of these programs, must ensure that they research this area and work closely with legal counsel. It is imperative that these programs are not discriminatory. The need for such a program should be thoroughly researched based on the specific requirements of the company and the safety risks involved.

12. Employment Standards

Employment standards are the minimum standards of employment for employers and employees in the workplace. In each province, minimum standards of employment have been established for:

- payment of earnings;
- minimum wage;
- hours of work, rest periods, and days of rest;

---

[45] C. Sefton & B. Speigel, "Alcoholism Ruled a Handicap" (1995) 11:5 *OH&S Canada* 16.

[46] R.D. Parsons, *The Ethics of Professional Practice* (Needham Heights, MA: Allyn & Bacon, 2001) at 97.

- overtime and overtime pay;
- vacations and vacation pay;
- general holidays and general holiday pay;
- maternity and parental leave;
- termination of employment;
- employment of underage individuals;
- adolescent employment in the restaurant and food services industry.

13. Freedom of Information and Privacy Legislation[47]

By Kristine Robidoux

As with OH&S and human rights legislation, privacy legislation consists of both provincial and federal statutes that govern the collection, use, and disclosure of personal information. Again, it is important to be familiar with the legislation in the relevant jurisdiction and to know what bodies or organizations are covered under this legislation.

Public sector bodies are generally bound by provincial Freedom of Information and Protection of Privacy legislation ("FOIP"). This type of legislation is generally intended to increase government accountability by ensuring that individuals have rights of access to information in the custody or under the control of government or public bodies. FOIP legislation has been developed to grant access to public documents that may be of interest to specific groups or individuals, and to protect the privacy and limit the use of individual personal information including medical information, by public bodies.

In the private sector, there are different statutes that govern British Columbia, Alberta, and Quebec. All have private sector privacy legislation that balances the right of individuals and employees of organizations to have their personal information protected with the needs of private sector organizations to collect, use, and disclose personal information for business purposes that are reasonable. Federally regulated works, undertakings and businesses, as well as private sector businesses that are outside of British Columbia, Alberta, or Quebec, are all subject to a federal statute, the *Personal Information Protection and Electronic Documents Act*.[48] Notably, the federal statute does not apply to the personal information of employees of private sector businesses in jurisdictions without provincial private sector privacy laws, but does apply to employees of federal works, undertakings or businesses.

This analysis is critical in order to determine whether privacy law applies to the personal information relating to the disability of the employee in question, and, if so,

---

[47] K. Robidoux, LL.B., Principal (Calgary, AB: ComplianceWorks, November 2005).
[48] S.C. 2000, c. 5.

which statute applies in particular. It can be a tricky determination to make.

In 2004, Alberta and British Columbia each enacted largely similar versions of the *Personal Information Protection Act* ("PIPA").[49] These two statutes provide that private sector organizations in those provinces may only collect, use, and disclose the personal information of individuals and employees for purposes that are reasonable; and to the extent that is reasonable to carry out the purposes. An important exclusion contained in the PIPA statutes allows organizations to collect, use, and disclose "personal employee information" without the consent of the individual in some cases: *i.e.*, if the information is reasonably required by an organization and is collected, used, or disclosed solely for the purpose of establishing, managing, or terminating the employment relationship. This exclusion can be important as an organization determines its conduct and strategy with respect to an injured worker.

Assurance of confidentiality of medical information is essential in performing effective incident investigation and disability management processes, especially in the area of case management. Healthcare professionals are bound by professional ethics to maintain confidentiality, but this can be a challenge in the workplace setting, and the healthcare professional must be extremely stringent in maintaining confidentiality. These obligations are also reiterated in the privacy laws: personal information *must* be safeguarded with security that is appropriate for the relative sensitivity of the information. It must be protected against such risks as unauthorized access, collection, use, disclosure, copying, modification, disposal, or destruction. All stakeholders in the fields of OH&S and Disability Management must be aware of privacy legislation in their jurisdiction and the impact this has on the processes within their programs.

14. Personal Information Protection and Electronic Documents Act (PIPEDA)

By Kristine Robidoux and Dianne Dyck

Canada's federal private sector privacy legislation, the *Personal Information Protection and Electronic Documents Act* (PIPEDA), came into effect on January 1, 2004. This statute governs the collection, use and disclosure of personal information by federal works, undertakings, and businesses (such as federally regulated banks, telecommunications, airlines, *etc.*), as well as by private sector businesses in provinces of Canada that have not enacted their own legislation that is substantially similar to PIPEDA. Therefore, any province without its own privacy legislation must follow the federal legislation.

British Columbia, Alberta, and Quebec have enacted provincial private sector privacy legislation that governs the treatment of personal information of individuals

---

[49] *Personal Information Protection Act*, S.A. 2003, c. P-6.5; *Personal Information Protection Act*, S.B.C. 2003, c. 63.

and employees of private sector organizations in those provinces. Specifically, in British Columbia and Alberta, PIPA provides the requirements for how organizations may collect, use, disclose, and protect personal information, which may include personal health information of individuals as well as company employees. Under PIPA, individuals and employees have the right to:

- know why personal information is being collected, used, or disclosed;
- expect the organization to collect, use, or disclose personal information in a reasonable and appropriate manner;
- know who in the organization is accountable for the organization's compliance with privacy laws and practices;
- expect the organization to use the appropriate security measures to protect the information;
- expect that the information is accurate and complete;
- request corrections if required;
- complain to the organization about how it collects, uses, or discloses personal information;
- appeal to the Privacy Commissioner if a dispute over personal health information cannot be resolved using the above measures; and
- access their personal information that is in the custody or under the control of the organization, unless one or more of the exceptions under PIPA apply.

In terms of the fields of Occupational Health and Safety and Disability Management, this legislation:

- requires employers to advise employees about the nature of personal information collected, used and disclosed, along with why, how, and when, unless one of the enumerated exceptions applies;
- restricts the amount of personal information that may be collected;
- requires employee consent for the collection, use, and disclosure of personal information, unless the personal information is reasonably required by the organization and is collected, used, or disclosed solely for the purposes of establishing, managing or terminating the employment relationship;
- limits the free flow of employee personal health information between healthcare providers and the employers;
- reaffirms the need for "information firewalls" between occupational health personnel and the workplace;
- obligates OH&S Professionals/Practitioners to ensure that the employee personal information is accurate and complete; and
- requires employers to ensure that employee personal information is collected, used, retained, disclosed, and destroyed in an appropriate manner.

15. Current Legislation: Summary

It is important to stress that the field of occupational health and safety is impacted by a variety of pieces of legislation which vary from province to province and from provincial to federal jurisdiction. The most important thing to note is that stakeholders in occupational health and safety *must* be:

- aware of the current legislation in their jurisdiction; and
- aware that specific acts and regulations are constantly changing and that they should obtain legal counsel to ensure they have the most current and up-to-date case law information when setting up programs or when dealing with specific human rights cases.

Remember — *Ignorance of the law is never a valid excuse!*

C. THE IMPACT OF CHANGING LEGISLATION ON THE FIELD OF OCCUPATIONAL HEALTH AND SAFETY

The next section covers some of the current issues, considerations, and recent changes related to occupational health and safety, workers' compensation, human rights, and freedom of information and protection of privacy legislation. As these Acts vary somewhat from Canadian province to province, and from provincial to federal jurisdiction, it is imperative that stakeholders review these legislative changes as they apply to their specific jurisdictions.

1. Occupational Health and Safety Legislation

In many Canadian provinces, OH&S legislation has recently been reviewed and changed after remaining stable for years. OH&S change, although needed, has traditionally been difficult to effect due to the process involved. In the past, the public reviews and the parliamentary assent took an inordinate period of time to complete. As a result, change tended to be initiated only when an OH&S Act warranted a number of major changes. The result is that gaps between the OH&S Act and unresolved workplace issues persisted.

Alberta initiated an innovative process. By separating the OH&S Act, Regulation, and Code, provisions for timely legislative changes were made. Although the OH&S Act and Regulation remain enacted by the legislature, changes to the Code are under the jurisdiction of the Minister of Labour. Codes are easier to amend than Acts. This expediency may be necessary where there are technological advances or where an identified safety hazard needs to be corrected.

In 2018, Alberta passed a revised OH&S Act 2018 in which significant changes were introduced. For example, the requirements around the Right to Know, Right to Participate in health and safety, and Right to Refuse dangerous work, were updated. The roles and responsibilities of the various worksite parties were defined. The requirements for JHSCs, OH&S Program were introduced. OH&S compliance and enforcement were enhanced, as well as other employer duties.

A number of court decisions demonstrates the changing OH&S responsibilities of the employer, namely:

- Dominus Construction was fined $90K in 2018, for a pedestrian fatality;[50]
- J.M. Lahman Manufacturing was assessed $150K in the death of a visiting child at their facility;[51]
- *Yukon v. Yukon* found the parties liable under the OH&S Act for causing harm to public property.[52]

OH&S legislation is dynamic, and the changes introduced impact organizations. For details on recent court rulings, refer to Chapter 31, "Canadian Workplace Safety Legislation" authored by Norm Keith; and to David Corry, "Disability Management: The Law" in D. Dyck, *Disability Management: Theory Strategy & Industry Practice*, 6th ed. (Markham: LexisNexis Canada, 2017) Chapter 24. These legal experts have summarized what is currently happening in the fields of OH&S and Disability Management.

Over the years, the legal expectations for OH&S duty of care have steadily increased. Proof exists in the fact that in most Canadian provinces, the OH&S minimal standards are not enough; employers must do what it takes to protect workers and the public. For example, "The Ministry of Labour can prosecute employers under the 'general duty' clause of the *Occupational Health and Safety Act* even where the charges impose obligations that are greater than those set out in the regulations under the OHSA, the Ontario Court of Appeal has decided".[53]

Likewise, the legislative requirements have become more encompassing and prescriptive; for example, Alberta's *Occupational Health & Safety Act 2018*, now requires all workers within the workplace including the contingent workforce, to receive OH&S training, support, and equipment.[54] Another example is the national focus that has been taken regarding confined space entry and the control of the

---

[50] Ontario Ministry of Labour, *Court Bulletin*, "Construction Site Fatality Results in $90,000 fine for Construction Company" (2018), online: https://www.constructiondive.com/news/canadian-contractor-fined-ca90k-for-nonworkers-fatal-fall/529404/ (date accessed January 31, 2020).

[51] Ontario Ministry of Labour, *Court Bulletin*, "Child Fatality in Factory Results in $150,000 Fine" (2018), online: https://news.ontario.ca/mol/en/2018/11/child-fatality-in-factory-results-in-150000-fine.html (date accessed January 31, 2020).

[52] *Director of Occupational Health and Safety v. Government of Yukon, William R. Cratty and P.S. Sidhu Trucking Ltd.*, 2012 YKSC 47 (Y.T.S.C.).

[53] A. Miedema, "In important decision, Ontario appeal court says that general duty clause in OHSA can impose higher obligations than specific requirements in regulations", *Canadian Occupational Health & Safety Law* (2018) at 1, online: http://www.occupationalhealthandsafetylaw.com/in-important-decision-ontario-appeal-court-says-that-general-duty-clause-in-ohsa-can-impose-higher-obligations-than-specific-requirements-in-regulations (date accessed: January 31, 2020).

[54] Alberta, *Alberta Occupational Health and Safety Act: Highlights of changes effective June 1, 2018* (2018), online: www.alberta.ca (date accessed: January 31, 2020).

associated hazards. Recognized as a high-hazard situation, confined space entry was closely evaluated, and codes of practice developed in early 2000 so that employers could adopt them to protect workers. Given the widespread publication of these standards, employers would be hard-pressed to claim that they did not know about them. The legal requirement is for employers to implement this code of practice.

Other standards that have been enhanced are the working alone legislation, workplace violence legislation, and fall protection legislation. In each case, the duty and role of the employer is emphasized.

Canadian corporate officers/employers are expected to monitor organization/company safety efforts. It is not enough to have an organization/company OH&S policy and safe-work procedures; corporate officers/employers must ensure that they are implemented and enforced. OH&S due diligence means taking a proactive approach to workplace safety instead of adopting a reactive or "laissez-faire" approach. For Canadian corporate officers/employers, it means:

- knowledge of the OH&S law and other related legislation;
- monitoring OH&S legal developments;
- familiarity with the organization's/company's policies and procedures;
- awareness of the organization's/company's activities;
- awareness of line management's activities;
- ensuring that worker OH&S training is provided;
- knowledge of contractor activities at the worksite; and
- assurance that the organization/company OH&S Program is documented and that records on organization/company OH&S activities are retained.[55]

The challenge for organizations is to remain current with the new and upcoming pieces of legislation and to understand exactly how it will play out. In fact, organizations and industry associations are advised to get involved with new legislation before it is enacted. That way, industry input can help governments develop more applicable legislation. Likewise, OH&S Practitioners/Professionals need to remain current on the OH&S legislation as well as to be prepared to contribute to future legislative changes.

2. Workers' Compensation Legislation

A number of Canadian Workers' Compensation Boards have enacted legislation that mandates employers to return ill/injured employees to work in a safe and timely manner. For example, the Newfoundland *Workplace, Health, Safety and Compensation Act*, requires that all employers and workers are obligated under the Act to cooperate in the worker's early and safe return to suitable and available employment

---

[55] MMV Bongarde, "Selling Safety to Your CEO" (Penticton, BC: MMV Bongarde, 2006).

with the employer.⁵⁶ This may involve modified work, transitional work, full-duty regular work, a transfer to an alternative job, or trial work to assess the worker's capability.

In addition, employers in New Brunswick, Nova Scotia, Ontario, Prince Edward Island, and Quebec are obliged to re-employ injured workers unless the worker refuses the job. For more information on the different functioning and benefits offered by the various provincial Canadian Workers' Compensation Boards, refer to the Association of Workers' Compensation Boards of Canada.⁵⁷ For more information on the Canadian Workers' Compensation System, refer to Chapter 33, "OH&S: The Canadian Workers' Compensation System".

3. Human Rights Legislation: Recent Developments

By Sharon Chadwick

A recent human rights case, *British Columbia (Public Service Employee Relations Commission) v. British Columbia Government and Service Employees' Union*,⁵⁸ involved fitness-for-work testing for firefighters and the interpretation of a specific fitness standard as a BFOR.

The recommendations from the review of this case have implications for the future assessment of human rights cases. It has been recommended that a three-step test should be adopted for determining whether an employer has established, on a balance of probabilities, that a *prima facie* discriminatory standard is a BFOR.

First, the employer must show that it adopted the standard for a purpose rationally connected to the performance of the job. The focus at the first step is not on the validity of the particular standard, but rather on the validity of its more general purpose.

Second, the employer must establish that it adopted the particular standard in an honest and good faith belief that it was necessary to the fulfilment of that legitimate work-related purpose.

Third, the employer must establish that the standard is reasonably necessary to the accomplishment of that legitimate work-related purpose. To show that the standard is reasonably necessary, it must be demonstrated that it is impossible to accommodate the individual employees sharing the characteristics of the claimant without imposing undue hardship upon the employer.⁵⁹

4. Canadian Human Rights and Substance Abuse Policies

Employers who are considering the development and implementation of workplace

---

⁵⁶ R.S.N.L. 1990, c. W-11, ss. 88-89.2.
⁵⁷ Online: http://www.awcbc.org (date accessed: January 31, 2020).
⁵⁸ [1999] S.C.J. No. 46, 176 D.L.R. (4th) 1 (S.C.C.).
⁵⁹ *Ibid.*, at 2 D.L.R.

substance abuse policies and procedures should be aware of the related human rights implications. Alcohol and drug addictions are generally accepted as disabilities; as such, employers have a duty to accommodate the same as they would do for any other physical or mental disability. Employers should reference the human rights legislation and positions in their jurisdictions to access the most current information on this topic.

The issue of substance abuse testing remains controversial. In Canada, there is no specific legislation related to drug and alcohol testing in the workplace, and these cases are generally addressed under human rights legislation. Employers considering the implementation of these programs must ensure that they research this area and work closely with legal counsel. It is imperative that these programs are not discriminatory. The need for such a program should be thoroughly researched based on the specific requirements of the company and safety risks involved. For example, Suncor (2019) implemented random drug testing as a consequence of 2000 work incidents involving drug use. "Although the mandatory drug and alcohol testing program instituted by Suncor was invasive and infringed upon its employees' legal rights, the program was a *bona fide* requirement to ensure the safety of the workplace".[60,61]

OH&S Professionals/Practitioners should monitor current court decisions and the position of the human rights authorities for their jurisdiction.

5. Legalization of Marijuana

Medical marijuana has been legalized in Canada since 2001. In October 2018, recreational marijuana was legalized with edible products allowed in October 2019. For more information on the impact of these legal changes, refer to Chapter 32: "OH&S: Legalization of Marijuana".

6. Privacy Legislation: Evolving Legislation[62]

There are two areas of OH&S practice that are impacted by the privacy legislation:

(a) Request for Employee Personal Information

Organization/company representatives who try to obtain employee personal health information without informed consent from the employee, might be in violation of privacy legislation if the information was not reasonably required by the organization and collected solely for the purposes of establishing, managing, or terminating the employment relationship. If the personal information does not meet this

---

[60] T. Anandasagar & L. Hendsbee, "Legal Review: Suncor's Random Drug & Alcohol Testing Policy", *OSG* (2018) at 1, online: https://osg.ca/legal-review-suncors-random-drug-alcohol-testing-policy/ (date accessed: January 31, 2020).

[61] *Suncor Energy Inc v. Unifor Local 707A*, 2017 ABCA 313 (Alta. C.A.)

[62] D. Dyck, *Disability Management: Theory, Strategy and Industry Practice*, 4th ed. (Markham, ON: LexisNexis Canada Inc., 2009).

definition, then the consent of the individual is required.

Since the privacy laws were enacted (in Alberta, the *Health Information Act*[63] in 1998 and the *Personal Information Protection Act* in 2004), employer rights to collect employee personal health information have clearly changed. Before collecting personal information, employers should ask:

- "Is this information reasonably required?"
- "Is it fair?"
- "What will the information be used for?"
- "Is there heightened security around this information?"
- "Is there informed consent?"

To complicate matters, in certain circumstances it may not be proper to ask for consent from an employee to collect, use, or disclose the employee's personal information. For example, if the organization/company wants to access information that would disclose the presence of a psychological health condition, the privacy rights that the employee is being asked to waive are highly protected. The relationship that exists between a manager and employee is a power-relationship that can be heavily weighted in favour of the manager. It cannot be assumed that the employee knows his/her rights in this respect. If there was any coercion or duress, or if the impression was given that the employee has no choice but to consent or else face adverse employment action, asking for employee consent may be contrary to the law. Consent must be informed and may not be obtained by providing false or misleading information about the proposed collection, or by using deceptive or unlawful practices.

There is heightened security of personal health information when dealing with a psychological health condition versus a physical one. There is far less social stigma attached to a fractured hip than to mental illness. As a result, more vigilance is required to protect this information.

(b) Protection of Employee Health Information

Many organizations/companies outsource their OH&S and disability case management services. In the course of their work, the internal Occupational Health Practitioners/Professionals have greater access to employee personal health information than does the rest of the company. According to the Canadian human rights legislation:

> Only those designated people receive the information from the employee's doctor and let the employee's supervisor know how long the employee will be absent, or what specific return to work accommodation measures are reasonable.[64]

---

[63] R.S.A. 2000, c. H-5.

[64] Alberta Human Rights and Citizenship Commission, *Obtaining and Responding to Medical Information in the Workplace* (January 2009) at 6, online: http://www.albertahumanrights.

To uphold the privacy of this information, "firewalls" must be in place (Figure 30.1). In essence, there are four privacy firewalls:

(1) A major firewall between the employee, healthcare providers and the Disability Management Service Provider and the internal employer representatives (the OH&S Practitioner/Professional). The information that can be disclosed is limited to benefit eligibility, fitness-to-work (FTW), absence duration, and work limitations;

(2) A major firewall between the OH&S Practitioner/Professional and Human Resource Managers. The disclosed information is less and limited initally to fitness-to-work, duration of absence (in some instances), and anticipated work limitations with recovery;

(3) Another firewall between the Human Resource Managers and the relevant Operational managers. Here, the disclosed information centres on the modified work details and work limitations; and

(4) The last firewall is between the relevant Operational manager and the rest of the company. Here the disclosed information is scaled down to a need-to-know basis.

---

ab.ca/publications/bulletins_sheets_booklets/bulletins/obtaining_med_info_in_workplace.asp (date accessed: January 31, 2020).

Figure 30.1: Protection of Employee Health Information[65]

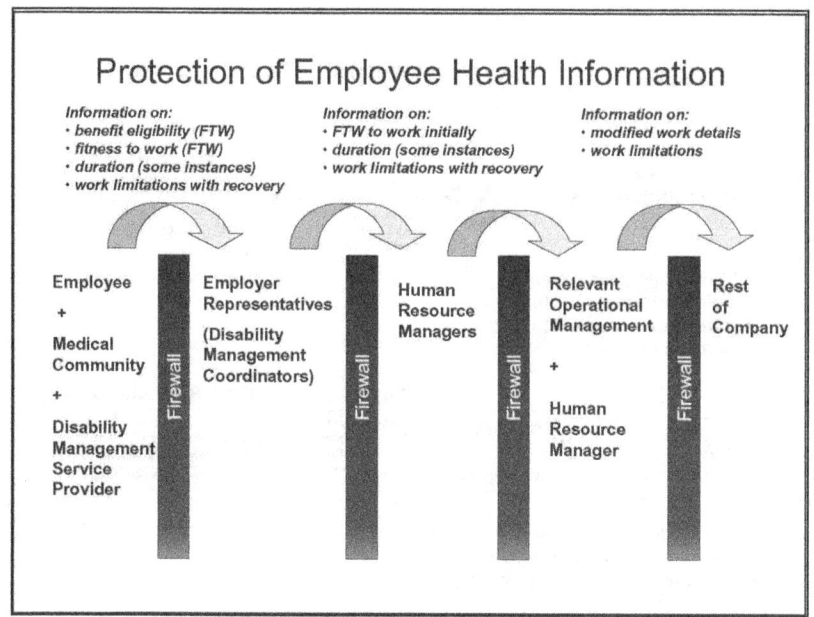

(c) Disclosure of Employee Health Information

The level of detail that OH&S Practitioners/Professionals are allowed to disclose to company managers is limited (Figure 30.2). In the acute phase of the illness/injury, only medical information exists and this cannot be provided to company representatives without the employee's informed consent. As a result, the information released is limited to the employee being unfit to work, being in a suitable treatment regimen, and to the nature of the health condition (work-related or non-work-related). This information supports the validity of the claim. In the recovery phase, details on work restrictions can be provided so that return-to-work plans can be made.

---

[65] D. Dyck, *Disability Management: Theory, Strategy and Industry Practice*, 6th ed. (Markham, ON: LexisNexis Canada Inc., 2017).

[66] *Ibid.*

Figure 30.2: Stages for the Release of Personal Health Information[66]

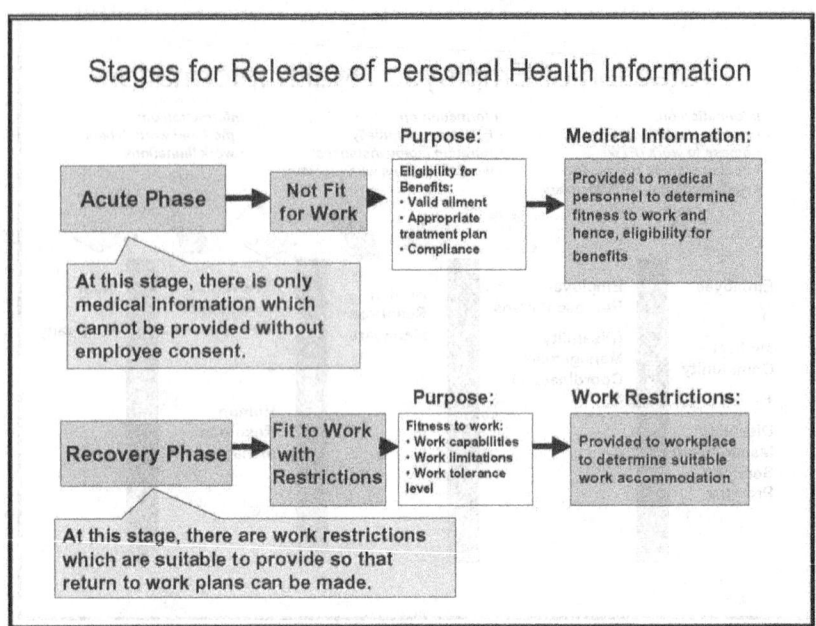

Often OH&S Practitioners/Professionals are pressured to provide more information. When this occurs, ask the question: "Is this information *reasonably required* for the sole purpose of maintaining the employment relationship?" If an employee is on short-term disability and eligible for benefit coverage as established by a Disability Management Service Provider, asking for specifics about the employee's health condition could be a violation of privacy legislation.

If the Operation Manager claims that the information is *reasonably required* because he/she wants to know what area of practice the physician is engaged in, this may also be an unlawful collection of personal information. Where the nature of the medical practice would, to a reasonable person, reveal the nature of the ailment, the manager would be able to indirectly access information that he/she could not otherwise directly access.

If it is deemed *reasonably required* because the Operational Manager is seeking to take employment action, other aspects of employment law come into play. Employment action cannot generally be taken if the employee is lawfully on medical leave.

The Occupational Health Practitioners/Professionals are permitted to access some personal health information about the employee's condition in order to assess whether or not the external Disability Management Service Provider is fulfilling its contractual commitments to the company or to ascertain the company's next steps in the work accommodation and reintegration process. Companies can challenge the external Disability Management Service Provider's findings by requesting an independent medical evaluation (IME). In such instances, company representatives are generally not entitled to medical reports, but rather to know the general nature of the employee's ailment.

Although privacy legislation is continuing to emerge, recent court decisions are fairly clear that employers are required to ensure that:

- medical information collected to administer an employee benefit plan is legitimately required;
- the medical information submitted by the employee is held in strictest confidence by qualified medical practitioners (PIPED Act, Case Summary #226);
- appropriate security safeguards are in place to protect sensitive personal medical information from unauthorized access (PIPED Act, Case Summary #226);
- employees are aware how to submit medical information so that it goes to a qualified medical professional (PIPED Act, Case Summary #226);
- employee requests to review their personal health information are honoured within 30 days of receipt of the written request (PIPED Act, Case Summary #284);
- the use of video surveillance of an employee is done only as a last resort, and in accordance with the *Privacy Act* (PIPED Act, Case Summary #269);
- the use of medical information be appropriately limited and proportionate to the operations of the organization in order to be in accordance with the *Privacy Act* (*Privacy Act*, Veterans Affairs, 2010); and
- review their information sharing agreements to ensure that the parties they share information with comply with applicable privacy legislation, including consent requirements relating to the use and disclosure of employee personal information (PIPED Act, Case Summary #2010-002).[67]

D. PSYCHOLOGICAL HEALTH SAFETY: CORPORATE DUTY OF CARE

According to the various Canadian provincial OH&S Acts, as well as the Canada Labour Code, Part II, employers must provide a safe and healthy workplace for employees and other workers. This includes both physical and psychological (mental health) safety. The provision and maintenance of a psychologically safe workplace has been recognized as an emerging legal duty, similar to the duty of care that employers have to provide a physically safe workplace. In both realms,

---

[67] For more Privacy Act and PIPEDA case summary details and decisions, see online: http://www.priv.gc.ca/cf-dc/pa/2010-11/pa_20101006_e.cfm (date accessed: January 31, 2020).

employers must take every reasonable precaution to protect worker safety and to demonstrate their performance.

A psychologically safe workplace uses all practical and reasonable mechanisms within its control to both protect and promote the mental health of all employees. At a minimum, such a workplace guards against harms to employee mental health that are attributable to negligence, recklessness, or intentional interference.[68]

A psychologically safe workplace is no longer a "nice to do". It is now a "must do". In addition to the OH&S Acts, five other sources of law also have a bearing on the developing jurisprudence. These include: Workers' Compensation law, Employment contracts, Law of Torts (negligence), Human Rights legislation, and Labour Relations law.[69] For more information on this topic, refer to Chapter 23, "Psychological Health and Safety in the Workplace".

E. IMPORTANT LEGAL TERMINOLOGY

There are a number of legal terms that OH&S Practitioners/Professionals need to recognize. This section is dedicated to providing a list of those terms and their descriptions.

Legal Terms	
Absolute discharge	A judgment in which the defendant is given no sentence and no criminal record.
Absolute liability offences	The Crown needs only to prove "beyond a reasonable doubt" that the accused committed the prohibited act. The accused would be found guilty no matter what was done to prevent the thing from happening. Liability without fault.
	A type of regulatory offence in which liability is based on proof of committing a prohibited act. Cannot use the defence of "due diligence" or "mistake of fact".
Accommodation	The process and implementation of changes to a job which enable a disabled person to perform the job productively and/or to the environment in which the job is accomplished.

---

[68] J. Samra, M. Gilbert, M. Shain & D. Bilsker, *Guarding Minds @ Work: A Workplace Guide to Psychological Safety and Health*, "What is Psychological Health & Safety?", online: http://www.guardingmindsatwork.ca/info/safety_what (date accessed: January 31, 2020).

[69] M. Shain, "Tracking the Perfect Legal Storm: Converging Systems Create Mounting Pressure to Create the Psychologically Safe Workplace" (Mental Health Commission of Canada, 2010), online: http://www.mentalhealthcommission.ca/English/system/files/private/Workforce_Tracking_the_Perfect_Legal_Storm (date accessed: January 31, 2020).

Legal Terms	
Accountability	The authority to act.
ACGIH	American Conference of Governmental Industrial Hygienists
Acquit	Release from suspicion or a legal charge.
Act	A statute, law or edict of a legislative or judicial body. It tends to be broader and less technical than a regulation or code.
Act of commission	Intentional act
Act of God	A manifestation of the forces of nature that is unpredictable and impossible to foresee.
Act of omission	An error: the reason for most incidents and injuries.
actus reus	A prohibited or wrongful act.
Administrative law	Law that governs the actions of the government and its agencies.
Administrative tribunal	A large number of special courts or bodies outside the ordinary judicial framework. They are set up under federal or provincial statutes (*e.g.*, WCB) to decide matters that may arise in the administration of some particular area of government. They interpret and apply the law.
Adverse effect discrimination	Discrimination that is neutral on its face, but it actually has negative effects on an individual or group.
Adversarial system	A system in which the Crown prosecutor and defence counsel assume antagonistic positions in debating guilt or innocence of a person(s). This system is used to reveal truths.
Agenda	Items to be covered.
Alternative measures	If the accused accepts responsibility for the crime, instead of judicial proceedings, alternate measures can be used such as community time, development of OH&S Programs and products for industry use.
Appeal	An application for judicial review by a superior court on an inferior court's decision.
Appeal officer	A federal position to which appeals about federal orders can be made.
Appeal tribunal	A court/body having jurisdiction to review the law as applied.

Legal Terms	
Appellant	One who appeals to a higher tribunal or authority.
Arbitration	The settling of a dispute by an arbitrator.
Arbitrator	Legitimate power to command or act during arbitration.
Assumption of risk — "*Volenti non Fit Injuria*"	A defence based on the plaintiff having knowledge of the dangers involved in the activity, and despite that, participated (*e.g.*, sports, work, risky activities, *etc.*).
Authority	The one who has power.
Balance of probabilities	A plaintiff's burden of proof is a balance of probabilities — more than 50 per cent certain in favour of a plaintiff.
Bargaining unit	A trade unit that acts on behalf of employees in collectively bargaining with the employer for benefits.
Bill	A draft Act which must be read three times and passed with a majority in the House of Commons, and then in the Senate, and then given Royal Assent.
Binding arbitration	The process when two parties agree to allow an independent third party to settle their dispute.
Bipartite	A two-party approach.
BNA Act	*British North American Act, 1867*, which created Canada as a country and which gave power to create courts to the Federal and Provincial legislatures.
Board of directors	Directors whose mandate is to oversee and be accountable for a government agency or business.
Bodily harm	An injury to a human body.
bona fides	Good faith.
bona fides occupational requirement (BFOR)	The employer must show that the specific task/capability is essential to the performance of the task.
Burden of Proof	In a due diligence defence, the defendant must prove that all reasonable steps were taken to avoid the particular event leading up to the charge.

	Legal Terms
Bylaw	Similar to regulations in that they depend upon and are made under the authority of provincial statutes, but are under the jurisdiction of the municipalities.
Canada Gazette	The official gazette of Canada: it publishes all items required by federal statute or regulations, statutory instruments and documents, and Public Acts of Parliament.
Canada Labour Code	Law that defines "federal work, undertakings, or business". Anything that is within the legislative authority of the Parliament of Canada, *e.g.*, telecommunications, transportation, banks, pipelines, radios, post office, *etc*.
Canadian Charter of Rights and Freedoms	Passed in 1982 and upholds the edict that the accused cannot be convicted unless four principles of criminal law are met, namely: (1) The person is presumed innocent until proven guilty. (2) The burden of proof rests with the Crown. (3) Proof of guilt is beyond a reasonable doubt (90-100 per cent); (4) Proof is made during a fair and public hearing.
Case law	Common law.
Cause of Action	A claim in law and fact sufficient enough to demand judicial attention.
Caveat emptor	"Let the buyer beware" — purchase at your own risk.
Certificate of authority	Issued by a Federal or Provincial Minister to an OH&S Officer and is to be shown to any person who asks to see it.
certiorari	A means of achieving judicial review.
Charge	In criminal law, the underlying substantive offence contained in an accusation or indictment.
Circumstantial evidence	Indirect evidence that is not based on fact and infers guilt, *i.e.*, not good enough for a conviction.

	Legal Terms
Civil law	Law that deals with private interests and personal issues such as property, financial matters, contracts and civil rights. Seeks to remedy harm.
Civil rights	The rights of an individual within a state to certain freedoms, *e.g.*, from discrimination.
Code	Collection of statutes that a body of laws or regulations systematically arranged.
Code of ethics	The standards and codes of conduct set by a professional group or another body over its members.
Collective bargaining	A method whereby the employer and employees determine the conditions of employment.
Collective liability	Joint legal responsibility.
Command and control instruments	The use of law in the form of specified prohibitions and consents that are monitored and infringement of which invites penalties.
Commencement	Beginning, first instance.
Common law	Originated in England and deals with a system of guidelines that are based on customs as opposed to statutes. Judges create the laws through precedent setting (case law).
Commutation of payments	Change of payments.
Compensation	Payment/remuneration for harm done.
Competence	Qualified, trained and experienced.
Conditional discharge	After complying with a sentencing condition, the accused is considered **not** to have a criminal record.
Conflict of interest	When an individual's perceived or actual goals/interests are in opposition to an organization/group's goals and interests.
Constitution	A country's blueprint for its government, society and laws.
Constitutional Act	Passed in 1982: gave Canada the ability to amend its own Constitution. Also provided the *Canadian Charter of Rights and Freedoms*.

Legal Terms	
Constitutional law	The law that is above all other laws. All laws must comply with the Constitution. It "rules above rules".
Contract	A legally binding agreement between two or more parties which gives rise to court enforceable obligations. To be valid, must have an offer, acceptance and considerations.
Contract law	The law that deals with contractual obligations.
Contributory negligence	See Negligence.
Convict	To prove guilty.
Coroner	An officer of the Crown whose duty is to inquire into cases of sudden or suspicious death.
Corporation	A corporate body that is legally empowered to act as an individual.
Corrective justice	Applies to situations where one person harms another. Two parties are "out of balance" in terms of one having lost something and the other having gained at the expense of the other, *e.g.*, negligence, defamation, assault and breach of contract.
Corroboration	The act of strengthening or confirming.
Court of Appeal	A provincial court which acts as a supervising body.
Court of Queen's Bench	A higher-level provincial court. Presides over civil and criminal cases.
Criminal law	Law that deals with society's prohibition against acts that threaten its very existence, *e.g.*, murder, theft, assault and terrorism.
Criminal offences	Punishment against "wicked minds". The two types are: (1) **Summary offences:** less serious offences that may result in fines and/or jail time; and (2) **Indictable offences**: serious offences with penalties ranging up to life in prison.
Cross-examination	To examine systematically for the purpose of getting at the facts that were not brought out in direct examination.
Crown	The government.

Legal Terms	
Crown agency	Agency owned by the government.
Crown corporation	Corporation owned and operated by the government.
Damages	Monetary compensation that the law awards to the one who has suffered damage, a loss or injury.
Danger	Relative exposure of people to a hazard.
Death benefits	Benefits that result from a worker's death. A lump sum payment to a surviving spouse. It amounts to $55,555.55 + $1,388.88 for each year by which the spouse's age on the death date of the worker is younger than 40 years or minus $1,388.88 for each year that the spouse's age is older than 40 years. The maximum amount = $83,333.30; the mini-mum amount = $27,777.76.
Declaratory power	The power of the federal government under the Constitution to assume jurisdiction over a local work within a federal jurisdiction by declaring it to be "for the general advantage of Canada".
Decree	A law or judgment.
Defence	A denial, answer or plea by the defendant.
Defence counsel	Speaks on behalf of the accused.
Defendant	The party responding to the claim of the plaintiff, or the accused.
Delegation	The assignment of a debt; deputation.
Deterrence	Motivation of the defendant to change substandard behaviour.
Direction	Federal orders.
Director's and Officer's liability	The liability conferred on the directors and officers of a company/organization.
Disciplinary action	Action taken as a result of being punished for a wrongful act.
Discrimination	A distinction, whether intentional or not, that is based on grounds relating to personal characteristics of the individual or group, which has the effect of imposing burdens, obligations or disadvantages.

Legal Terms	
Distributive justice	Applies to distribution of things in a society according to a "distributive criterion". Subject to goods, rights, risks, awards, *etc.* — university degrees are distributed based on merit; welfare based on need; Human Rights based on one's status as a human being.
Division of powers	Created by the 1867 Constitution and outlines the division of powers between the federal and provincial governments.
Doctrine of Common Employment	Injury caused by another worker.
Double jeopardy	A general principle in common law that no person is to be placed twice in jeopardy for the same offence/cause.
Due diligence defence	The only defence for regulatory offences (strict liability legislation). Means that every reasonable and practicable precaution was taken to prevent the event from happening. A defence based on "mistake of fact" or "reasonable precautions".
Duty	An obligation imposed by law.
Duty to re-employ	The obligation to re-employ an individual who has been injured.
Election	The exercise of choice by an unrestrained will to take or do one thing or another.
Employee	A person who has entered into an employment relationship with an employer.
Employer	Person responsible for directing work at a worksite.
Employment contract	A legally binding agreement between an employer and employee.
Employment law	Law that addresses employment issues.
Employment standards	Employment legislation for non-unionized employees that sets minimum employment standards for wages, vacation pay, hours of work, statutory holidays, overtime pay, dismissal notice, leaves, *etc.*
Enactment	Fulfillment.
ERS	External Responsibility System — government agency charged with administering OH&S legislation and enforcing OH&S standards.

Legal Terms	
Ethics	The science of morals in human conduct; moral philosophy. A broad concept of morality which includes what is good for individuals in their own lives.
Exhibit	Evidence.
Experience rating	Claim ratings over a period of time.
Expert evidence	Evidence provided by an expert witness.
Expert witness	A person possessed of a special skill or knowledge that entitles that person to provide an opinion.
Express	Intended.
Express warranty	In product liability cases, it is a written or oral promise or presentation concerning a product's quality, use or life (*e.g.*, product warranties).
Fairness	Just: the right thing to do for the right reason(s).
Federal state	A state in which the system is divided between two or more sovereign, independent authorities.
Fellow servant rule	The worker/supervisor relationship with the supervisor holding the balance of power.
Finality clause	A clause that prevents an appeal to a decision.
Garnishment of payments	Payments used to pay off debts.
General duty clause	"Every reasonable precaution" to ensure worker health and safety.
Grievance	An allegation that something imposes an illegal obligation or burden, or denies some equitable or legal rights.
Guidelines	These can be law if an Act or regulation adopts them.
Guilty	The condition of having been found by the court to have committed a crime.
Ham Royal Commission	1976: Addressed safety in mines. Introduced the concept of IRS, how it should function, and the roles and responsibilities of the stakeholders in the IRS. This concept was subsequently adopted into Canadian OH&S law.
Head of power	Legislative authority of the Parliament of Canada and provincial legislatures.

Legal Terms	
Health & Safety representative (Government Officer)	A person appointed as a Health & Safety representative under section 136 of the *Canada Labour Code* (CLC).
Healthcare benefits	Healthcare insurance coverage for healthcare services.
Healthcare practitioners	A drugless practitioner regulated under the *Drugless Practitioners Act*, or a social worker.
Healthcare professional	A member of the college of a health profession as defined in the *Regulated Health Professionals Act, 1991*.
Hearsay	A statement by someone who did not witness the events directly — not admissible.
Historic bargain	The giving up of common law rights in return for no fault benefits under an administrative regime.
Human Rights legislation	Protects Canadians from discrimination by both government and non-government sectors.
Implied	Inferred.
Implied term	Inference.
Implied warranty	In product liability cases, it relates to expectation — *i.e.*, what a product should be able to do. The mere act of marketing a product legally implies that it meets industry standards and codes and has been made with good workmanship.
Indemnify	To secure against loss or damage that may occur in the future.
Indictment	An accusation in writing of a serious offence. It sets out charges against the accused.
Inquest	An inquiry.
Inquisitorial system	A system in which a search, investigation or inquiry can be made into an event or incident.
Inspector	Government OH&S agent.
Insurance	An undertaking by one party to indemnify another party for an agreed consideration, from loss or liability in respect of an event.
Intentional harm	Injury/ill effects caused on purpose.

	Legal Terms
Internal responsibility	A Safety Management System — each employee of an organization, regardless of position or authority, is responsible for the safety of themselves, their co-workers and the general public.
intra vires	Within the powers.
IRS	Internal Responsibility System: responsibility for OH&S is internal to the workplace and internal to the job of everyone in the workplace.
IRS audit	An evaluation of the Safety Management System against set standards.
Judge	Hears the issues and acts as an impartial tribunal to arrive at a decision of guilt or innocence based on the evidence.
Judge's order	Directions issued by a judge.
Judicial review	A superior court's examination of the conduct of an inferior court.
Jurisdiction	The legal power or right of administering justice; making and enforcing laws. The power or authority to make legal decisions in a given territory, or to make decisions about a certain subject matter.
Justice	The quality of being just.
Justice of the peace	A judicial officer appointed by a provincial lieutenant-governor to carry out certain judicial or quasi-judicial duties.
Justice view	Where ethical behaviour is that which is impartial, fair and equitable in treating people.
Kantianism	A focus on the right and wrong actions, without regard to consequences, *e.g.*, modern Human Rights.
Labour law	Law associated with labour standards.
Labour market re-entry	Return to work after an illness, injury or leave.
Law	A rule of conduct imposed by authority or accepted by the community as binding.
Lawsuit	An action taken in a court of law.
Lawyer	One who practices law.
Liability	Bound or obligated by law or equity: legally responsible.
Licence	To authorize by legal permit.

Legal Terms	
Lien	A right to detain the goods of another until some claim has been settled.
Limitation period	The period of time within which an action must be brought forth and beyond which no action can be taken. For OH&S Regulations it is two years: the Statute of Limitations is seven years.
Lump sum payment	One-time payment.
mala fides	Bad faith.
Mandamus	A writ issued from a higher court and directed to a person, corporation or inferior court requiring them to do a specific thing.
Mediation	The act of intervening.
Mens rea	The guilty mind or intent.
Mens rea offence	The Crown must prove "beyond a reasonable doubt" both the *mens rea* (intent) and *actus reus* (prohibited act).
Minutes	Official records of the proceedings of a meeting, committee or society.
Modified work	A change in work duties or time to accommodate the individual currently off work due to illness/injury.
Moral rights view	Where ethical behaviour is behaviour that respects the fundamental rights shared by all human beings.
Morality	The doctrine, principles or practice of moral duties. Principles of right action regarding others. Broader than the law, but law is based on morality.
National Fire Code	A code adopted by a country.
Natural law	The belief in fundamental human goods that are in and of themselves, and not as instruments for some other objective.
Natural person	A human being who has the capacity for rights and duties.
Negligence	Failure to exercise proper care and precaution.
Negligent harm	Harm due to substandard conduct.
No fault compensation	Compensation awarded without proving negligence or fault.
Non-economic loss	A loss that is not associated with money — injury/illness, family, friends, station in society, *etc*.

Legal Terms	
Nuisances	Anything causing annoyance, inconvenience, or injury to another person.
Obstruction	Deliberate impedance or hindrance.
Occupier's liability	The liability of an occupier of land.
Offence	Illegal act.
Offence notice	Notice of an illegal act.
Office of the corporation	A corporation is a distinct legal entity, like a person. It must be treated separately from the people who own and operate it.
Officer	One holding a post or position of authority.
Order	A directive by a proper authority.
Paramountcy doctrine	The belief of superiority to others.
Partnership	A contractual agreement between numbers of people involved in a business venture.
Per cent impaired	Degree of loss of function.
Performance regulation	Requirement to fulfil one's legal obligations.
Perjury	To lie under oath.
Permanent impairment	Permanent loss of function.
Permit	A permission or warrant.
Plaintiff	The person who brought the legal action against the defendant.
Plea bargain	The practice of admitting to being guilty in return for a more lenient sentence.
Plead	The defendant's answer by matter of fact to the plaintiff's claim.
POGG clause	Peace, Order and Good Government.
Policy committees	Since 2000, large companies (more than 300 employees) under federal jurisdiction are required to have a committee to help set policies and procedures, and to undertake many of the management duties.
Preamble	A preliminary clause in a treaty, constitution or statute that states the intent, purpose and spirit of the instrument.
Precedent	A judicial decision that creates legal principles which are followed in future cases.
Prescribe	To assert a prescriptive title, *e.g.*, to prescribe by Governor in Council or determine in accordance with rules prescribed by regulation of the Governor in Council.

Legal Terms	
Presumption of innocence	Belief of "not guilty".
Prevention legislation	Statutory provisions setting out duties and rights for employers and individuals for the purpose of proactively taking steps to prevent future incident and exposures from occurring.
Principles	A sentence should always be proportionate to the crime and the degree of responsibility of the offender.
Privative clause	Prevents a person from appealing an administrative decision.
Privity of contract	The belief whereby one can enforce contractual rights against another only if one was a party to the contract.
Probation	(Suspended sentence) Defendant does not have to spend time in jail but has to report to a probation officer on a regular basis and to comply with sentencing conditions. He/ she will have a criminal record.
Procedural justice	Where a person's vital interests are to be decided by others, proceedings must be fair. Need to hear both sides of the dispute before deciding an outcome. This is in accordance with the "rules of natural justice".
Procedural regulation	A regulation that is prescriptive in its requirements.
Proclaimed	To announce publicly.
Product warranty	A contractual obligation between the seller and the buyer relating to a product's specifications, suitability for certain uses or its projected useful life.
Products liability law	Law that deals with the sale of products.
Professional	One who is characterized by or conforms to the technical and ethical standards of a profession.
Professional ethics	A subset of ethics: concerned with issues that are ethical in nature between professionals and clients, other professionals and third parties.
Property rights	A generic term that refers to any type of right to specific property whether it is personal or real property, tangible or intangible.

Legal Terms	
Prosecution	The act of exhibiting a charge against an accused person before the court.
Prosecutor	Crown counsel who presents the evidence of the accused's guilt fairly and objectively.
Provincial court	A provincial court which deals with OH&S offences, civil suits up to $4,000 in compensation and/or family disputes.
Public health	Public health is a core element of a government's attempt to improve and promote the health and welfare of citizens. It involves health promotion, health protection and illness/injury prevention activities.
Punishment	Infliction of a penalty for an offence or misdemeanour.
Quasi-judicial	The acts of persons, bodies or tribunals that are not strictly judicial, but are similar in that they have the authority to decide on important issues.
Quasi-professional	Membership in a quasi-professional association has advantages but does not govern the licence to practice. There is no exclusive right to practice.
Quorum	The minimal number of officers or members that must be present to conduct business.
Reasonable	It is a balanced, wise, prudent judgment that is understandable to others.
Reasonable mistake of fact	An honest and reasonable belief in a set of facts that were wrong.
Reasonable person	A person who is logical, intelligent, rational, sensible and who plans and takes actions in accordance with general/approved practices.
Reassessments	The WCB may request a physician to perform a second assessment of a worker if the initial assessment or the initial assessment report is deemed by the WCB to be incomplete or inaccurate.
Regulation	Law that covers a large number of specifics such as hazardous substances, JH&S Committees, First Aid and industry-specific practices.

Legal Terms	
Regulatory offences	Crimes that are not evil, but are wrong because they are prohibited.
Relevancy	To be admissible, evidence must be relevant to the case.
Remedy	The means used to enforce or redress an injury.
Reprise	A yearly rent charge or other payment out of a manor or lands.
Respondent	The person against whom a petition is presented.
Responsibility	Ability to act according to the laws of right and wrong.
Restitution	Act of making "good" or of giving the equivalent for any loss, damage, or injury.
Return to work	Re-entry into the workplace following an illness/injury.
Reverse onus of proof	The obligation of proof rests with the defendant.
Right	A claim which a person has against another or the state, which the courts will recognize.
Right to know	The "right to tell": to be informed of conditions that may cause harm.
Right to participate	The right to participate in H&S within a company.
Right to refuse	The right to refuse unsafe work without negative consequences.
Rights of action	Cause of action.
Riparian rights	Rights that accrue to owners of land on the banks of waterways.
Royal Assent	Crown approval.
Rule of law	We are subject to governance by rules and not by personal whim.
Search warrant	An order issued that authorizes a peace officer to conduct a search and seizure.
Section	Part of an Act, regulation, or code.
Sentencing	Handing down a punishment ordered by the court.
SOR	Statutory Orders and Regulations.
Specific duty	Duty of a particular stakeholder.
Specific deterrence	Deterrence of the convicted individual so that he/she will not do this wrongful act again.

	Legal Terms
Standard of care	The degree to which the organization, worker or management protect the health and safety of workers, contractors and public.
Standard of proof	Burden of proof
Stare Decisis	To stand by that which was decided. This rule is the basis of "precedent".
Statute	A law enacted by a legislative body. Sets out the broad and general duties of the workplace participants; establishes legal standards for OH&S; grants authority to government officers; sets penalties; and enables regulations to be passed.
Statutory duties & rights	Legal rights and duties.
Statutory interpretation	Interpretation of the law — usually done by a higher court.
Statutory law	Law that centres around the statutes — also called Regulatory Law. Can be passed by Federal, Provincial or Territorial governments. The intent is to deter substandard practices in a particular area of interest, *e.g.*, OH&S Law.
Statutory offences	Offences related to breaches in statutory law. Three types: (1) *Mens rea* offences (*act and intent*). (2) Absolute liability offences (*act, intent is not necessary*) (3) Strict liability offences (*act with the possibility of a due diligence defence*).
Strict liability	Liability without showing fault.
Strict liability offences	The Crown proves "beyond a reasonable doubt" that the accused committed the prohibited act. The defence must then prove on a "balance of probabilities" that the accused exercised "due diligence". These are also termed quasi-criminal offences. Allows for a due diligence defence.
Subpoena	A writ commanding a person to attend a court of justice under a penalty.
Subrogation	The substitution of one person in the place of another with succession of his or her rights to a debt.
Subsection	Part of a section.

	Legal Terms
Suits	Civil actions.
Summons	An authoritative call or citation to appear before a court or a judge.
Survivor benefits	Benefits awarded to the dependents of a worker.
Susceptible worker	An individual who by virtue of genetic make-up and/or underlying state of health is more likely to become ill/injured from a hazardous material or condition.
Summary offences	In criminal law, these are less serious offences and result in fines and/or jail time.
Test for Proof	In a criminal case. This is "guilt beyond a reasonable doubt".
Testimony	A statement under oath.
Tort law	The law that deals with civil wrongdoings to the person, including wilful or negligent damage. Injured persons may claim either financial redress or defence of their reputations.
Trespass	A wrongful act involving injury to another person or his or her property.
Trial court	A judicial examination and determination of the issues in a cause between the parties before a judge, judge and jury or referee.
Tribunal	An officer or body having the authority to adjudicate judicial or quasi-judicial matters.
ultra vires	Beyond one's legal authority or power.
Under oath	Sworn testimony.
Undertaking	The act of someone who undertakes any business.
Undue hardship	The point at which a company/organization/union is being negatively affected by the work accommodation being offered.
Unfunded liability	Liability which an agent does not have the monetary funds to cover.
Union	An organization of employees formed for the purposes that include the regulation of relations between employees and the employer.

Legal Terms	
Utilitarianism	Choosing the consequence that leads to the greatest good for the greatest number of people, *e.g.*, performing a cost/benefit analysis.
Vicarious liability	The inferred responsibility of one person for the acts of another.
Vocation rehabilitation	Coordinates rehabilitation efforts across multiple environmental systems as part of the disability management team and facilitates a return to employability.
Waiver of entitlement	Forgoing entitlement of something like WCB benefits, which is illegal for an employer to ask an employee to do.
Wantonness	Wild; mad recklessness.
Wilfully	Done from the spontaneous action of a person's will.
Witness	One who has seen or known the incident.
Work Refusal	Employee's right to refuse work is where he/she believes equipment, machinery or device to be used is likely to endanger self or others; or where the physical condition of the workplace is likely to endanger self or others or where the worker believes that any equipment, machinery or device is in contravention of the OH&S Act and likely to endanger self or others. A two-staged process: • Stage I: "Reason to Believe" by the worker that a problem exists. The claim needs to withstand the subjective test for reasonableness and requires internal investigation (most claims are addressed at this point). • Stage II: "Reasonable grounds to believe" on the part of the worker that a problem exists. Must withstand objective testing, must be reasonable, and warrants external investigation.
Workplace committees	Joint Health and Safety Committees.

F. SUMMARY

Current and accurate knowledge of the legal system and related laws and regulations is vital to successfully functioning as an OH&S Practitioner/Professional. Hence, it is important to keep a pulse on what is going on in your jurisdiction and remain active in improving OH&S legislation.

CHAPTER REFERENCES

Alberta, *Alberta Occupational Health and Safety Act: Highlights of changes effective June 1, 2018* (2018), online: www.alberta.ca (date accessed: January 31, 2020).

Alberta Human Rights and Citizenship Commission, *Obtaining and Responding to Medical Information in the Workplace* (January 2009) at 6, online: http://www.albertahumanrights.ab.ca/publications/bulletins_sheets_booklets/bulletins/obtaining_med_info_in_workplace.asp (date accessed: January 31, 2020).

T. Anandasagar & L. Hendsbee, "Legal Review: Suncor's Random Drug & Alcohol Testing Policy", *OSG* (2018), online: https://osg.ca/legal-review-suncors-random-drug-alcohol-testing-policy/ (date accessed: January 31, 2020).

Association of Workers' Compensation Boards of Canada, online: http://www.awcbc.org (date accessed: January 31, 2020).

British Columbia Council of Human Rights, *Disability and the Human Rights Act* (Victoria: British Columbia Ministry of Labour and Consumer Service, 1998).

British Columbia (Public Service Employee Relations Commission) v. British Columbia Government and Service Employees' Union (B.C.G.S.E.U.), [1999] S.C.J. No. 46, 176 D.L.R. (4th) 1 (S.C.C.).

British Columbia (Superintendent of Motor Vehicles) v. British Columbia (Council of Human Rights), [1999] S.C.J. No. 73 (S.C.C.).

Canada, "The *Canadian Environmental Protection Act, 1999*, Chemical Substances in Canada", online: http://www.chemicalsubstanceschimiques.gc.ca/index-eng.php (date accessed: January 31, 2020).

Canada Labour Code, R.S.C. 1985, c. L-2.

Canadian Environmental Protection Act, 1999, S.C. 1999, c. 33.

Canadian Human Rights Act, R.S.C. 1985, c. H-6.

Canadian Society of Safety Engineers (CSSE), *BCRSP Exam Preparatory Course* (Toronto, ON: CSSE, 2002).

Centre for Labour Management, "Accommodation Guidelines" (Presented at the Illness and Disability: Claims in the Unionized Workplace Conference, Edmonton, AB, February 1999) [unpublished].

CEPA Environmental Registry, General Information, Fact Sheet, *The Canadian Environmental Protection Act, 1999* (CEPA 1999), online: http://www.ec.gc.ca/CEPARegistry/gene_info/management.cfm (date accessed: January 31, 2020).

Criminal Code, R.S.C. 1985, c. C-46.

D. Cory, "Duty to Accommodate – Update (2005)" (Presented at the

23rd Annual Labour Arbitration Policy Conference, June 9, 2005, Toronto, ON) [unpublished].

Controlled Product Regulations, SOR/88-66.

Don Sayers & Associates, Environmental Practice Domain, *BCRSP Study Guide* (Mississauga, ON: BCRSP).

Director of Occupational Health and Safety v. Government of Yukon, William R. Cratty and P.S. Sidhu Trucking Ltd., 2012 YKSC 47 (Y.T.S.C.).

D. Dyck, *Disability Management: Theory, Strategy and Industry Practice*, 6th ed. (Markham, ON: LexisNexis, 2017).

Fisheries Act, R.S.C. 1985, c. F-14.

AWCBC (2020). The Historic Compromise, available at: http://awcbc.org/?page_id=57 (accessed February 28, 2020).

R. Gale, "Canada's Green Plan" in *Nationale Umweltpläne in Ausgewählten Industrieländern* (Berlin: Springer-Verlag, 1997).

A. Hansen, "Building Codes and Regulations", *The Canadian Encyclopedia*, online: http://www.thecanadianencyclopedia.com (date accessed: January 31, 2020).

Hatscan, *WHMIS Train-the-Trainer* (Hatscan Seminar presented in Edmonton, AB, February 2007) [unpublished].

Hazardous Products Act, R.S.C. 1985, c. H-3.

Health Information Act, R.S.A. 2000, c. H-5.

Highway Traffic Act, R.S.O. 1990, c. H-8.

Human Rights Act, R.S.A. 2000, c. A-25.5.

IAEI, "Ask CSA: Understanding the Canadian Electrical Safety Regulatory System. Part II: Canadian Provinces and Territories", online: http://www.iaeimagazine.org/magazine/2002/01/16/understanding-the-canadian-electrical-safety-regulatory-system-part-ii-canadian-provinces-and-territories/ (date accessed: January 31, 2020).

N. Keith, "Alberta OH&S Due Diligence for Managers and Supervisors" (Seminar presented in Calgary, AB: Gowlings, 2003) [unpublished].

L. McDowell, *Human Rights in the Workplace: A Practical Guide* (Toronto, ON: Carswell, 1998).

A. Miedema, "In important decision, Ontario appeal court says that general duty clause in OHSA can impose higher obligations than specific requirements in regulations" *Canadian Occupational Health & Safety Law* (2018), online: http://www.occupationalhealthandsafetylaw.com/in-important-decision-ontario-appeal-court-says-that-general-duty-clause-in-ohsa-can-impose-higher-obligations-than-specific-requirements-in-regulations (date accessed: January 31, 2020).

MMV Bongarde, *Selling Safety to Your CEO* (Penticton, BC: MMV Bongarde, 2006).

CH. 30: OCCUPATIONAL HEALTH AND SAFETY: LEGAL ASPECTS

National Research Council of Canada, *National Fire Code of Canada, 2010*, online: http://nrc-cnrc.gc.ca/eng/publications/codes_centre/2010_national_fire_code.html (date accessed: January 31, 2020).

Ontario Ministry of Labour, *Court Bulletin*, "Construction Site Fatality Results in $90,000 fine for Construction Company (2018), online: https://www.constructiondive.com/news/canadian-contractor-fined-ca90k-for-nonworkers-fatal-fall/529404/ (date accessed January 31, 2020).

Ontario Ministry of Labour, *Court Bulletin*, "Child Fatality in Factory Results in $150,000 Fine" (2018), online: https://news.ontario.ca/mol/en/2018/11/child-fatality-in-factory-results-in-150000-fine.html (date accessed January 31, 2020).

R.D. Parsons, *The Ethics of Professional Practice* (Needham Heights, MA: Allyn & Bacon, 2001) at 97.

Personal Information Protection Act, S.A. 2003, c. P-6.5.

Personal Information Protection Act, S.B.C. 2003, c. 63.

Personal Information Protection and Electronic Documents Act, S.C. 2000, c. 5.

K. Robidoux, LL.B., Principal, *ComplianceWorks* (Calgary, AB: November 2005).

J. Samra, M. Gilbert, M. Shain & D. Bilsker, *Guarding Minds @ Work: A Workplace Guide to Psychological Safety and Health*, "What is Psychological Health & Safety?", online: http://www.guardingmindsatwork.ca/info/safety_what (date accessed: January 31, 2020).

C. Sefton & B. Speigel, "Alcoholism Ruled a Handicap" (1995) 11:5 *OH&S Canada* 16.

M. Shain, "Tracking the Perfect Legal Storm: Converging Systems Create Mounting Pressure to Create the Psychologically Safe Workplace" (Mental Health Commission of Canada, 2010), online: http://www.mentalhealthcommission.ca/English/system/files/private/Workforce_Tracking_the_Perfect_Legal_Storm (date accessed: January 31, 2020).

Suncor Energy Inc v. Unifor Local 707A, 2017 ABCA 313 (Alta. C.A.)

Transportation of Dangerous Goods Act, 1992, S.C. 1992, c. 34.

United Nations, Agenda 21, United Nations Conference on Environment and Development (UNCED) held in Rio de Janeiro, Brazil, June 3 to 14, 1992, online: http://www.un.org/esa/sustdev/documents/agenda21/index.htm (date accessed: January 31, 2020).

United Nations Framework Convention on Climate Change, "Kyoto Accord", online: http://unfccc.int/kyoto_protocol/items/2830.php (date accessed: January 31, 2020).

Workplace, Health, Safety and Compensation Act, R.S.N.L. 1990, c. W-11.

Chapter 31

CANADIAN WORKPLACE SAFETY LEGISLATION[1]

by Norm Keith[1]

A. INTRODUCTION

Canadian occupational health and safety (OH&S) law developed over the last 60 years largely in reaction to workplace accidents, injury and death. Specific cases, media coverage, public inquiries and pressure, and the Canadian workers' compensation legislation, somewhat ironically, predate modern Canadian OH&S law. The Meredith Report of 1913 resulted in a no-fault insurance system for compensating workers injured on the job. At the beginning of the 20th century, it appears that governments, employers, unions, and other workplace stakeholders were more concerned with providing workers' compensation protection for injured employees and protecting employers from lawsuits, than from preventing accidents in the workplace.[2] In the second half of the 20th century, Canadian OH&S law developed.

Under the *Constitution Act, 1867*, occupational health and safety (OHS) has no specific constitutional jurisdictional designation.[3] The Act sets out the division of legislative powers between the federal and provincial governments. Occupational health and safety is not referenced in that list for either of the federal or provincial governments. Therefore, the courts were called upon to determine whether the

---

[1] D. Dyck "Internal and external consulting: Assisting clients with managing work, health and psychosocial issues" (2002) 50(3) AAOHN 111-119. Copyright (2002) the American Association of Occupational Health Nurses, Inc. Used with permission. All rights reserved.

[1] This chapter was originally published in April 2019 in N. Keith, *Workplace Health & Safety Crimes*, 4th ed. (Toronto, ON: LexisNexis Canada, 2019). It has been slightly updated regarding changes in the WHMIS 2015 legislation that have occurred since publication. It has been reproduced by kind permission of author Norm Keith and LexisNexis Canada. Norm Keith, B.A., J.D., LL.M., CRSP, is a noted author and guest speaker on OH&S law. Norm, a partner at Fasken Martineau, Toronto, Canada, can be reached at 416-868-7824 or nkeith@fasken.com.

[2] For a more complete introduction to the origins of this subject, see N.A. Keith, *Canadian Health and Safety Law* (Aurora: Canada Law Book, 2003) at 1:10.

[3] (U.K.), 30 & 31 Vict., c. 3.

provincial or the federal government has the authority to regulate the workplace through OH&S legislation.

Today, approximately eight per cent of Canadian workplaces are federally regulated and 92 per cent are provincially regulated for the purposes of labour relations, employment standards, workers' compensation, and OH&S legislation. Therefore, the vast majority of Canadian workers are regulated by provincial rather than federal OH&S statutes, regulations, and standards.

OH&S statutes set out the framework for workplace health and safety duties and responsibilities. In Canada, they are based on a blend of the **internal responsibility system** and the **external responsibility system**. The internal responsibility system, which is an overlapping system of rights and responsibilities of workplace stakeholders, is somewhat unique to Canada. A detailed discussion of the internal responsibility system appears later in this chapter.

Canadian OH&S law is also based on the external responsibility system, the lawful authority by which the applicable government regulatory body is accountable by various means. The external responsibility system has two means of enforcing OH&S requirements — standards and procedures. First is the issuance of orders or directions by inspectors or officers employed by various government regulators. The issuance of an order may be to immediately stop work or to change a work practice within a reasonable period of time. The basis for the issuance of orders or directions is an OH&S contravention. Every year, government OH&S inspectors issue thousands of orders.

Second is the laying of charges under OH&S laws as a means of enforcing the legal duties of various workplace parties. It is an OH&S offence to contravene these legal duties. The charges are regulatory in nature but must be proven just like criminal charges, beyond a reasonable doubt. This approach to establishing an offence is different than the establishment of a crime under the *Criminal Code*.[4] Under the *Criminal Code*, certain conduct is expressly designated to be a criminal offence.

Canadian OH&S law provides for enforcement, in part, by prosecutions brought as quasi-criminal, strict liability offences. There are three types of offences known to Canadian law as identified by the Supreme Court of Canada in *R. v. Sault Ste. Marie (City)*:[5]

(1) *mens rea* offences;

(2) strict liability offences; and

(3) absolute liability offences.

Since the mid-1970s, the enforcement of Canadian OH&S law with workplace

---

[4] R.S.C. 1985, c. C-46.

[5] [1978] S.C.J. No. 59, 85 D.L.R. (3d) 161 (S.C.C.) [hereinafter *"Sault Ste. Marie"*].

stakeholders that have legal duties by prosecution, has increased. Workplace stakeholders include the employer, supervisors, officers, directors, professional engineers, architects, suppliers, workers, and others. Different Canadian jurisdictions have different types of workplace stakeholders. Although many workplace stakeholders have legal duties under Canadian OH&S law, the employer is consistently the most frequently prosecuted stakeholder with OH&S offences.

The establishment of OH&S legal duties on workplace stakeholders is not new. The common law has long placed a *duty of care* on employers to provide a safe workplace free of unnecessary and foreseeable hazards for workers. Under the Anglo-Canadian common law, if an employer failed to meet a reasonable standard of worker health and safety and an injury resulted, the employer could be successfully sued for negligence. However, with the development of the Canadian workers' compensation legislation, the legal right of a worker to sue an employer for breach of this common law duty was effectively, and almost universally, terminated. However, there still are limited circumstances in which third-party lawsuits may be available for the injured party.

The *Westray Bill*[6] recognized that public safety was also an important part of worker safety. A member of the public, who is not a worker, has full right to sue the employer for negligence if injured by a workplace accident.

Under Canadian OH&S law, workplace stakeholders and employers must meet certain prescribed legislative duties and responsibilities to protect employees and workers. Some Canadian OH&S statutes focus on employees only, while others apply to the broader term "workers". For example, Part II of the *Canada Labour Code*[7] for federally regulated employers, emphasizes the duty to protect employees only. On the other hand, in the *Ontario Occupational Health and Safety Act*,[8] workplace stakeholders' duties relate to the protection of workers, not just direct employees. Therefore, the *Westray Bill*'s positive occupational safety duty in section 217.1 of the *Criminal Code*, goes farther than any current OH&S statute because it addresses public as well as worker safety.

The *Westray Bill* has introduced an expressed new legal duty regarding workplace health and safety within the criminal negligence sections of the *Criminal Code*. Theoretically, this legal duty establishes one standard for OH&S duties that is consistent across the country. The legal duty provides that those who undertake or have the authority to direct how another person does work or performs a task, are required to take reasonable steps to prevent bodily harm arising from the work to any person. The framers of the legislation clearly intended to impose this duty on a

---

[6] *An Act to Amend the Criminal Code*, S.C. 2003, c. 21, proclaimed in force March 31, 2004 (except s. 22). Introduced as Bill C-45, 2nd Sess., 37th Parl., 2003 [hereinafter "Bill C-45"].

[7] R.S.C. 1985, c. L-2 (Part II, "Occupational Health and Safety", am. S.C. 2000, c. 20).

[8] R.S.O. 1990, c. O.1.

wide range of workplace personnel. However, the duty to take reasonable steps in section 217.1 is not defined. All managerial personnel — regardless of whether they are front-line supervisors or senior officers, who have the power to direct work, are responsible for ensuring that all reasonable steps are taken to prevent bodily harm.

The duty is not confined to managers, supervisors, officers, or possibly directors; any individual employee who undertakes to direct another how to perform any task in the workplace, is also subject to this duty. As attractive as this broad duty may be from a policy perspective, it is not without problems. Imposing this obligation on employees who direct how another worker performs a task, may have the potential of creating a chilling effect on supervisors or lead hands who may become reluctant to assume such roles if they attract the potential risk of criminal liability for even minor direction or supervision of others.

The legal duty requires that those subject to the duty, take reasonable steps to prevent bodily harm to any person. Therefore, there would exist an obligation to workers and to the public at large. The public includes visitors, customers, or volunteers in the workplace. This part of the duty is also of critical importance in settings where the public is in physical attendance of where the work is performed, or where the public can be affected by adverse consequences arising from work activities.

Canadian OH&S law is enforced predominantly by issuing orders and by the prosecution of employers. Under the *Westray Bill*, it is not clear who in the workplace would most likely be prosecuted for breach of the duty. Section 217.1 established a blanket criminal offence, applying equally to all who direct the workplace activities of others. Unlike the existing Canadian OH&S law, there appears to have been no effort on the part of the drafters of the *Westray Bill* to clarify the nature or hierarchy of the duty owed. Instead, the Bill effectively states a broad and far-reaching duty, with no corresponding definitions, parameters, or guidelines as to how it will apply to various organizational decision-makers. If the current OH&S prosecutions are any indication of how OH&S criminal negligence charges will be laid, then there will likely be a much higher incidence of employers prosecuted than supervisors, and more supervisors charged than workers. Workers sometimes fail to comply as do supervisors and senior management.

The first prosecution under the *Westray Bill* in *R. v. Fantini* involved eight charges against Mr. Fantini under Ontario's *Occupational Health and Safety Act*. The matter of the criminal and regulatory charges was resolved by way of a plea of guilty by Mr. Fantini on March 3, 2005, before Gorewich J.[9] A plea arrangement between the accused counsel and counsel for the Ministry of Labour, Mr. Slansky, and counsel Mr. DeRubeis for the Crown Attorney, Mr. Dionne, was made and an agreed statement of facts provided some indication why both criminal and regulatory charges were laid. They were read onto the record by Mr. Slansky, as follows:

---

[9] *R. v. Fantini*, [2005] O.J. No. 2361 (Ont. C.J.).

MR. SLANSKY: Thank you. I will read it for the record, and I will briefly expand on paragraph one, it's the part that says the facts that set out in the information is to clarify what is required by the sections of the regulation. It's [sic] the Crown and defence agree for purposes of a plea of guilty on counts 2, 3 and 4 of the information before the court and for purposes of sentencing the following facts are true and that the court can rely on these facts without need of proof the foregoing purposes.

1. The facts as set out in counts 2, 3 and 4 of the information, as I said I will now expand on that to some extent. The regulation s. 224 of the Ont. Reg. 213/91 requires that no-one enter an excavation, or basically in common parlance, a trench unless the trench has been set up in accordance with the regulations which require under the circumstances any [sic], on the type of soil in these circumstances, that it be sloped to a 45 degree angle or that it be properly shored. Those are the requirements of that regulation. And regulation 22 which is referred to in count 3 as a requirement that the worker have appropriate headgear, protective headgear and s. 23 requires appropriate footwear as set out in the regulations.

Paragraph 2 of the Agreed Statement of Facts continues: "The defendant Domenico Fantini (hereinafter the defendant), has been a general contractor for over 25 years in his personal capacity and/or through a corporation called Vista Construction. The particular project described in the following facts, was carried out by the defendant in his personal capacity.

Paragraph 3: On Monday, April 19, 2004 a worker, Ameth Garrido (hereinafter Mr. Garrido), was found dead, as a result of what was later determined to be traumatic asphyxia, in a collapsed trench at a private residence at 15100 of the 10th Concession in King Township, in the Region of York. Mr. Garrido had been working for the defendant at this location. The defendant was getting paid $30.00 an hour for the worker's work, the worker was being paid at $18.00 an hour by the defendant.

Paragraph 4: The work was being done to install weeping tiles in a trench at the walls of the house to the footing of the basement to allow for water drainage to stop leakage into the basement. Such work inherently required a worker or workers to enter the trench. This work commenced at least two (2) days before on April 17th, 2004. The worker doing the excavation was instructed by the defendant where, how deep and how wide to make the trench. He gave no instructions to shore or properly slope the trench.

Paragraph 5: On April 19th, 2004, the defendant had been directing, "sorry", the defendant [had] been directly supervising the work of Mr. Garrido and another worker shortly before the collapse of the trench. The defendant directed the worker, Mr. Garrido and watched him work in the trench. The defendant left to get materials for the job, leaving Mr. Garrido in the trench. Shortly after the departure of the defendant, a crack formed in the soil walls of the excavation and the trench collapsed burying Mr. Garrido completely. Mr. Garrido's co-worker unsuccessfully attempted to dig out Mr. Garrido. By the time emergency personnel attended the scene and reached Mr. Garrido he was dead.

Paragraph 6: The owner of the house, was [sic] familiar with construction projects because he owned an equipment rental company, specifically directed the defendant to slope the excavation of the trench at a 45-degree angle slope to prevent the collapse.

Paragraph 7: The trench was excavated to be approximately nine feet deep, three feet wide and twenty feet long at the front of the house, against the house, where the deceased's body was found, buried under the soil. The soil was type three soil as described under s. 226(1)(4) of Regulation 213/91 under the *Occupational Health and Safety Act*. Specifically, this soil was "stiff to firm and compact to lose in consistency or is previously excavated soil" as described in the regulation. In fact, the soil had previously been excavated when the house had been constructed. Accordingly, the excavation had to be shored or sloped at a 45-degree angle to comply with s. 234-242 of Regulation 213/91. The trench was not shored or sloped at all, let alone in accordance with the regulation. This fact was the primary cause of the collapse of the trench. The collapse of the trench was the direct cause of Mr. Garrido's death."

Paragraph 8: Mr. Garrido was not wearing safety boots or protective headgear as required by regulation 213/91. The autopsy revealed that Mr. Garrido had suffered "superficial blunt force lacerations and haematomas about the head and neck". It is not known whether Mr. Garrido would have remained conscious had he been wearing protective headgear. It is not known whether this would have assisted his recovery or otherwise increased his chances of survival."

Those are the facts, the Agreed Statement of Facts.[10]

B. THE INTERNAL RESPONSIBILITY SYSTEM

As introduced above, the internal responsibility system is the underlying concept and philosophy behind the modern Canadian OH&S law. The internal responsibility system states that it is the workplace stakeholders who are best able to identify, assess, and either eliminate or control hazards in the workplace. Modern Canadian OH&S law establishes requirements, standards, and procedures to which the workplace stakeholders must comply. The internal responsibility system recognizes the limits of government resources to provide inspection, scrutiny, and enforcement of OH&S statutes and regulations in every workplace in Canada. The system is manifested by a series of overlapping legal duties, rights, and responsibilities in OH&S statutes and regulations. Workers, supervisors, and employers share legal duties for workplace safety under the internal responsibility system.

The internal responsibility system was described by Professor Swinton (1983) as follows:

> The phrase "internal responsibility system" was coined by Professor James Ham in his 1976 report on Occupational Health & Safety in Mines. Ham envisioned two types of responsibility systems: direct and contributively. The direct responsibility system would require management to define clear standards of work, assign responsibility for particular tasks, and then establish lines of accountability to ensure proper performance. This system of "direct responsibility" would be

---

[10] *Ibid.*, at paras. 14-23, *per* Gorewich J.

facilitated by a contribution system existing of an external auditing function, carried out by worker auditors and joint labour-management health and safety committees. A major flaw in regulating occupational health and safety in the past was, in Ham's opinion, the lack of worker participation.[11]

The internal responsibility system consists of a number of different elements which may vary from jurisdiction to jurisdiction across Canada. However, the core elements include the establishment of legal duties on various workplace stakeholders. Some jurisdictions, like Ontario, enumerate a long list of stakeholders that have legal duties including constructors, employers, supervisors, workers, licensees, professional engineers, architects, directors, officers, and suppliers. Others, like the federal *Canada Labour Code*, Part II, only place legal duties on employers and employees. Regardless of the list of workplace stakeholders with duties, the overriding purpose of the legal duties are to ensure the OH&S of workers.

The internal responsibility also is demonstrated by the establishment of health and safety representatives, and Joint Health and Safety Committees (JHSCs). Canadian workplaces with five or more workers and less than 20 workers, generally require health and safety representatives. The exceptions are the province of Quebec where representatives are not required, and under the federal *Canada Labour Code*, Part II, where every workplace, even one with only one worker, must have a health and safety representative. However, JHSCs may be required in Quebec based on the Minister's order or the type of industry.

JHSC members are made up of management and worker members, usually in equal number, who are designated to represent worker interests with respect to workplace health and safety. They have the legislative right to meet, inspect the workplace, make inquiries from the employer about workplace hazards, investigate workplace accidents, and participate when the regulator investigates the workplace. In many jurisdictions, JHSCs must be consulted before new OH&S programs are introduced.

Jurisdictions across Canada regulate hazardous substances and controlled products through various regulatory regimes. The purpose is to set standards that employers must follow to ensure that workers are not exposed to biological, chemical, physical, psychological, and ergonomic hazards. The Workplace Hazardous Material Information System (WHMIS 2015)[12] is a national, legislative regime that has established hazard identification, assessment, safe handling, and disposal measures for controlled products. This system provides information to workers by way of mandatory training, labelling control product containers, and providing safety data sheets (SDSs) that have additional information on how to avoid exposure to the controlled products and what to do in the event of an exposure.

---

[11] K. Swinton, "Enforcement of Occupational Health and Safety Legislation: The Role of the Internal Responsibility System" in P. Swan & E. Swinton, eds., *Studies in Labour Law* (Toronto: Butterworths, 1983) at 143.

[12] WHMIS Legislation: *Hazardous Products Regulations*, SOR/2015-17.

WHMIS 2015 is a rare Canadian example of where the federal and provincial governments cooperated on the subject of OH&S, resulting in a nationally consistent program to regulate and provide training with respect to controlled products. For example, WHMIS 2015 training in one part of the country will be valid in another part of the country. The system is an example of the internal responsibility system at its nationally consistent best.[13]

Although the internal responsibility system is discussed by legislators and policy makers when amending various Canadian OH&S statutes and regulations, for a long time a legal definition of the internal responsibility system did not exist. During the Westray Inquiry, there was a lot of discussion about the need to improve the OH&S legislation and internal responsibility system. The internal responsibility system was both lauded and criticized, but particularly criticized by the United Steelworkers Union of America at the Westray Inquiry.

After the Westray Mine Disaster, Nova Scotia made a number of significant changes to its OH&S statute, one of which was supplying a legal definition of the internal responsibility system. In section 2 of the Nova Scotia *Occupational Health and Safety Act*,[14] the following definition of the internal responsibility system is provided:

2. The foundation of this Act is the Internal Responsibility System which

    (a)   is based on the principle that

            (i)   employers, contractors, constructors, employees, and self-employed persons at a workplace, and

            (ii)   the owner of a workplace, a supplier of goods or provider of an occupational health or safety service to a workplace or an architect or professional engineer, all of whom can affect the health and safety of persons at the workplace, share the responsibility for the health and safety of persons at the workplace;

    (b)   assumes that the primary responsibility for creating and maintaining a safe and healthy workplace should be that of each of these parties, to the extent of each party's authority and ability to do so;

    (c)   includes a framework for participation, transfer of information and refusal of unsafe work, all of which are necessary for the parties to carry out their responsibilities pursuant to this Act and the regulations; and

    (d)   is supplemented by the role of the Occupational Health and Safety Division of the Department of Labour, which is not to assume responsibility for creating and maintaining safe and healthy workplaces,

---

[13] For further information on the WHMIS, refer to Chapter 5, "Occupational Health and Safety Program: Manual Development", Section 3, Part 4; Also refer to N.A. Keith, *Canadian Health and Safety Law* (Aurora: Canada Law Book, 2003) c. 4.

[14] S.N.S. 1996, c. 7.

but to establish and clarify the responsibilities of the parties under the law, to support them in carrying out their responsibilities and to intervene appropriately when those responsibilities are not carried out.

In summary, the internal responsibility system is the foundational concept of modern Canadian OH&S law. It provides the legal duties and responsibilities of workplace stakeholders. It also establishes the requirements, standards, and procedures for workplace health and safety by promoting self-management and regulation of workplace health and safety, the identification of hazards, and the mitigation of risk. The internal responsibility system reduces the need for government regulators to inspect, assess, and intervene in the workplace. Ultimately and in most cases, employers, reasonably have the highest number of duties and level of responsibility under the internal responsibility system — they hold the power and the purse strings.

C. OH&S GENERAL DUTY CLAUSES

As noted, part of the internal responsibility system is the establishment of various legal duties on workplace stakeholders. Employers and senior management have many legal duties to ensure that a workplace is healthy and safe for employees and other workers. In addition to the specific duties and regulations setting out control measures for employees, the Canadian OH&S statutes have provisions known as "general duty clauses". These general duty clauses provide a very broadly worded statement requiring employers, and on occasion, other parties to take all reasonable precautions for the health, safety, and protection of workers in the workplace. General duty clauses have similarities to new section 217.1 of the *Criminal Code* introduced by Bill C-45, and for that reason, will be briefly reviewed.

It is possible that in the judicial interpretation of the offence, as established by the *Westray Bill*, OH&S criminal negligence may refer to statutory provisions, case law, and judicial interpretation of already existing general duty clauses in the health and safety legislation. Employers are the most frequently charged party for OH&S statute violations, and a large number of those charges relate to an alleged breach of the general duty to provide a safe and healthy workplace.[15]

General duty clauses place broad OH&S responsibility on employers in their applicable jurisdictions. Interestingly, there is a relatively consistent pattern in the language of general duty clauses across various Canadian jurisdictions as will now be shown.

The federal jurisdiction, regulated by the *Canada Labour Code*, Part II, provides employers with a very broad duty to ensure the health and safety of employees. It is interesting to note that the *Canada Labour Code*, Part II does not broadly protect

---

[15] *Occupational Health and Safety Act*, R.S.O. 1990, c. O.1, s. 25(1)(a), (b), (d). In addition to general duty clauses, employers also have a specific duty with respect to safety requirements under other statutes such as the *Building Code Act, 1992*, S.O. 1992, c. 23. Specifically, an employer should obtain the opinion of a professional engineer or building code expert with respect to this obligation under the Ontario health and safety statute.

workers, as is the case with most applicable OH&S statutes across Canada, but rather restricts the employer's general duty clause to that of employees. In contrast, the *Westray Bill* protects all individuals in the workplace, including workers, volunteers, students, and members of the public, who may be at risk from hazards or activity. The federal legislation general duty clause states:

> Every employer shall ensure that the health and safety at work of every person employed by the employer is protected.[16]

In British Columbia, unlike other Canadian jurisdictions, the legislature establishes the employer and other stakeholder duties under Part III of the *Workers' Compensation Act*[17] as applied by way of the *Occupational Health and Safety Regulation*.[18] The general duty clause under the *Workers' Compensation Act* states:

> Every employer must ensure the health and safety of all workers working for that employer.[19]

In Alberta, the *Occupational Health and Safety Act*[20] provides the following general duty clause:

> **3(1)** Every employer shall ensure, as far as it is reasonably practicable for the employer to do so,
>
>   (a)   the health and safety of
>
>   (i)   workers engaged in the work of that employer, ...
>
>   (ii)   those workers not engaged in the work of that employer but present at the worksite at which that work is being carried out, and
>
>   (iii)   other persons at or in the vicinity of the worksite who may be affected by hazards originating from the worksite,

The Alberta provision moderates the extent and potential breadth of application of the general duty clause by the phrase, "as far as it is reasonably practicable for the employer to do so".

Saskatchewan has adopted a general duty clause similar to that of Alberta. The Saskatchewan *Occupational Health and Safety Act, 1993*[21] states:

> **3.** Every employer shall:
>
>   (*a*)   ensure, insofar as is reasonably practicable, the health, safety and welfare at work of all of the employer's workers;

---

[16] *Canada Labour Code*, R.S.C. 1985, c. L-2, s. 124.
[17] R.S.B.C. 1996, c. 492.
[18] B.C. Reg. 296/97.
[19] R.S.B.C. 1996, c. 492, s. 115(1).
[20] R.S.A. 2000, c. O-2.
[21] R.S.S. 1993, c. O-1.1.

In that sense, the legislature adopted similar language in its general duty clause. The Manitoba *Workplace Safety and Health Act*[22] states:

4(1) Every employer shall in accordance with the objects and purposes of this Act

(a) ensure, so far as is reasonably practicable, the safety, health and welfare at work of all his workers; . . .

The Ontario *Occupational Health and Safety Act*[23] establishes the general rights and responsibilities of government, employers, and workers, and sets minimum OH&S standards for the workplace. With the duty to take every reasonable precaution under the Ontario health and safety statute, employers have a general duty to "take every precaution reasonable in the circumstances for the protection of a worker".[24] This duty may require an employer to take precautions that include the development of an OH&S management system (OHSMS) and the suspension or discharge of workers for unsafe work practices.[25] Supervisors must also "take every precaution reasonable in the circumstances for the protection of a worker" in a similar general duty.[26]

In Quebec, the health and safety legislation states: "Every employer must take the necessary measures to protect the health and ensure the safety and physical well-being of his worker."[27]

In New Brunswick, the *Occupational Health and Safety Act*[28] requires that:

9(1) Every employer shall

(*a*) take every reasonable precaution to ensure the health and safety of his employees;

As such, "foreseeability" is a critical factor in the due diligence standard. In New Brunswick, like other provinces, an OH&S program must include activities designed to prevent the recurrence of accidents — analyzing jobs and work procedures to identify hazards and taking steps to eliminate or reduce those hazards. This concurs with the province of Prince Edward Island's *Occupational Health and Safety Act*,[29] where the general duty clause reads as follows:

12. (1) An employer shall ensure

---

[22] C.C.S.M. c. W210.

[23] R.S.O. 1990, c. O.1.

[24] *Ibid.*, s. 25(2)(h).

[25] Ministry of Labour Legal Branch, *Interpretation Opinions* (Ottawa: Ministry of Labour, November 1983) at 9, cited in N.A. Keith, *Canadian Health and Safety Law*, looseleaf (Aurora: Canada Law Book, 1997) at 3:40.2(6)(1).

[26] R.S.O. 1990, c. O.1, s. 27(2)(c).

[27] *An Act Respecting Occupational Health and Safety*, CQLR, c. S-2.1, s. 51.

[28] S.N.B. 1983, c. O-0.2.

[29] R.S.P.E.I. 1988, c. O-1.01.

(a) that every reasonable precaution is taken to protect the occupational health and safety of persons at or near the workplace;

To the same effect, in Nova Scotia, the *Occupational Health and Safety Act*,[30] amended substantially after the Westray Mine Disaster, states that:

13 (1) Every employer shall take every precaution that is reasonable in the circumstances to

(a) ensure the health and safety of persons at or near the workplace;

In the Province of Newfoundland, the *Occupational Health and Safety Act*, with more gender-inclusive language, states as follows:

An employer shall ensure, where it is reasonably practicable, the health, safety and welfare of his or her workers.[31]

In the Northwest Territories and Nunavut, the *Safety Act*[32] has an interesting and rather lengthy general duty clause, which states as follows:

4. Every employer shall . . .

. . . . .

(b) take all reasonable precautions and adopt and carry out all reasonable techniques and procedures to ensure the health and safety of every person in his or her establishment; ...

Finally, the Yukon Territory would appear to be the only Canadian jurisdiction without a general duty clause. Although the Yukon *Occupational Health and Safety Act*[33] does place legal duties on employers with respect to ensuring that workplace machinery, equipment, and processes are safe; that work techniques and procedures are used to reduce the risk of occupational illness and injury; and that other duties relating to instruction, hazard awareness, and general compliance with the Act, there is no apparent general duty clause in the Yukon Territory statute.

The wording of the OH&S general duty clauses is similar to the new section 217.1 of the *Criminal Code* as introduced by the *Westray Bill*. The phrase "reasonable steps" is arguably not as broad or as strict as the words "all", "every" and "ensure" found in many of the provincial general duty clauses. The requirement to take "reasonable steps" is similar to the phrases, "reasonable precautions" and "so far as it is reasonably practicable". Therefore, whether or not section 217.1 of the *Criminal Code* was specifically intended to be similar to a general duty clause, its purpose is clearly the same — to ensure the health and safety of workers and to ensure that the public is protected from bodily harm.

---

[30] S.N.S. 1996, c. 7.

[31] R.S.N.L. 1990, c. O-3, s. 4.

[32] R.S.N.W.T. 1988, c. S-1.

[33] R.S.Y. 2002, c. 159.

A recent decision of the Alberta Court of Appeal may have clarified the law with respect to the enforcement of general OH&S duty clause offences. It may also make it more difficult for the Crown to prove general duty offences under OH&S legislation in Alberta and across the country. On August 22, 2018, the Alberta Court of Appeal released its decision in *R. v. Precision Diversified Oilfield Services Corp.*[34] The decision clarifies the scope of the *actus reus* to be proven by the Crown where an employer is charged with failing to ensure the health and safety of a worker under the applicable OH&S legislation.

Precision was charged with the general duty clause OH&S offence after one of its workers suffered a fatal head injury while working on one of its drilling rigs. Precision was charged with failing to ensure the worker's safety "as far as it is reasonably practicable for the employer to do so", contrary to section 2(1) of Alberta's former[35] *Occupational Health and Safety Act*.[36] Precision was also charged with failing to adopt engineering or administrative controls in order to mitigate workplace hazards as required by the *Occupational Health and Safety Act*; but the focus of the appeal was on the offence in section 2(1).

Unlike true crimes, which require the Crown to prove both the prohibited act, *actus reus*, and the moral fault element, *mens rea*, the general duty clause OH&S offences are strict liability offences for which the Crown need only prove that the accused committed the prohibited act. The burden then, shifts to the accused, who can avoid conviction by establishing that all reasonable care in the circumstances was exercised — a defence commonly referred to as "due diligence".

Precision was convicted at trial on both counts — *actus reus* and *mens rea*, but the convictions were overturned by the Alberta Court of Queen's Bench. One of the issues before the Court of Appeal was whether the expression "as far as it is reasonably practicable for the employer to do so" forms part of the *actus reus* to be proven by the Crown, or part of the due diligence defence to be proven by the accused. The court questioned whether in addition to proving that the employer had failed to ensure the health and safety of the worker, the Crown also had to prove that the employer could have prevented the harm through "reasonably practicable" efforts, which the employer failed to undertake.

The Crown opposed this interpretation, arguing that it would shift the burden of disproving the accused's due diligence to the Crown. This, it argued, would be contrary to the accepted legal framework for "public welfare" offences. And, it would make OH&S prosecutions more difficult by requiring the Crown to establish what was "reasonably practicable" for an employer, rather than leaving it to the employer (who presumably has the best knowledge of what was "reasonably practicable"), to establish this in its own defence.

---

[34] [2018] A.J. No. 1005, 2018 ABCA 273 (Alta. C.A.) [hereinafter *"Precision"*].

[35] Alberta updated R.S.A. 2000, now S.A. 2017, c. O-2.1, s. 3(1).

[36] R.S.A. 2000, c. O-2 (now S.A. 2017, c. O-2.1, s. 3(1)).

The majority of the court disagreed. It held that the wording used in the legislation cast the "reasonably practicable" element, not as a codification of the "due diligence" defence, but rather as part of the offence that the Crown must prove. The majority of the court held that the Crown had to prove all of the following in order to secure a conviction of an employer under s. 2(1):

(1) the worker must have been engaged in the work of the employer;
(2) the worker's health or safety must have been threatened or compromised (*i.e.*, an unsafe condition); and
(3) it was reasonably practicable for the employer to address the unsafe condition through efforts that the employer failed to undertake.

The majority decision of the Alberta Court of Appeal downplayed the Crown's concerns about the increased difficulty of prosecution, observing that "ease of enforcement alone cannot justify disregarding the ordinary meaning of the text and adopting a strained interpretation instead". It also noted that appellate courts in both Ontario and Saskatchewan have adopted similar approaches to the Crown's burden of proof in OH&S offences.

Commentary: This decision clarifies, and potentially increases, the Crown's burden of proof in OH&S prosecutions under the general duty clause. Although the majority of the court in *Precision* listed numerous sources of evidence potentially available to the Crown in an OH&S prosecution, including:

- the circumstances of the incident;
- the results of any investigation into the cause of the incident;
- the employer's preventative efforts; and
- the employer's health and safety policy;

the reality is that the Crown often has less insight into industries and workplaces than do employers or other worksite participants. Moreover, it is not immediately clear how the Crown could satisfy its burden in situations, as in *Precision*, where the Crown may not be able to prove exactly *how* an incident occurred, and thus, to establish the specific duties that the employer breached. Requiring the Crown, instead of the employer, to lead evidence of what is "reasonably practicable" could also result in longer and more difficult prosecutions.

D. THE DEFENCE OF DUE DILIGENCE

Canadian OH&S law is made up of statutes, regulations, and industry codes and practices that are often adopted or incorporated by reference into the other statutes or the regulations. The goal of OH&S law is to prevent workplace accidents, injury, and worker death. It holds workplace stakeholders responsible for failure to comply with OH&S law. Canadian OH&S statutes focus on protecting workers and not specifically members of the public. The *Westray Bill*, on the other hand, focuses equally on both worker and public safety. Occupational health and safety law emphasizes the role of workplace stakeholders in taking responsibility to identify, assess, and control workplace hazards, while the *Westray Bill* requires compliance with the new legal duty under section 217.1 of the *Criminal Code*, failing which an individual or an organization may face prosecution for the offence of OH&S criminal negligence.

The enforcement of OH&S law is primarily by two means:

1) Government regulators have the authority to issue orders or directions to comply with the OH&S statutes and regulations. If a workplace stakeholder fails to comply with their legal duty, the OH&S regulator may issue an order or direction requiring immediate compliance, or compliance within a reasonable period of time. This is the first means of enforcement of Canadian OH&S law. Workplace stakeholders are given the right to appeal such orders or directions. Failure to comply with an order or direction, without commencing an appeal, is an offence.

2) Enforcement by way of quasi-criminal prosecution of OH&S regulatory offences: these are not criminal offences but are similar in that they result in a government prosecution.

The *Westray Bill*, on the other hand, establishes a true criminal offence for breach of the new OH&S duty under section 217.1 of the *Criminal Code*. Although the enforcement of OH&S law in Canada by way of prosecution is similar in its process to a criminal prosecution, the legal characterization of a criminal charge is different from that of a strict liability OH&S regulatory offence.

OH&S statutes across Canada establish OH&S regulatory offences. These offences, through a strict liability, give rise to the defence of due diligence. The classification of offences and the defence of due diligence are discussed below.

The Supreme Court, in *R. v. Sault Ste. Marie*,[37] indicated that strict liability offences, although not true criminal offences, are quasi-criminal in nature. Provincial and federal OH&S offences have also been defined as regulatory public welfare offences designed for the protection of public workers and interests. Provincial legislatures and the federal parliament have the power to create either true crimes or *mens rea* offences, strict liability, or absolute liability offences. Although the criminal law power is constitutionally assigned to the federal government, a provincial offence may be classified as a true crime or *mens rea* offence. These three types of offences were introduced earlier in this chapter under the heading "Introduction". However, a more complete explanation is needed to understand these types of offences and the legal defence of due diligence.

(1) *True crimes or mens rea offences*: In *mens rea* offences, the prosecution must prove beyond a reasonable doubt the prohibited act and, either as an inference from the nature of the act committed or by additional evidence, the positive state of mind on the part of the accused, such as intent,

---

[37] [1978] S.C.J. No. 59, 85 D.L.R. (3d) 161 (S.C.C.) [hereinafter "*Sault Ste. Marie*"].

knowledge, or recklessness. In a criminal offence, the prosecution has the onus of proof throughout the trial. The onus of proof never shifts to the accused.

(2) *Strict liability offences*: In strict liability offences, as in absolute liability offences, the prosecution need only prove beyond a reasonable doubt that the defendant committed the prohibited act. The prosecution need not prove a fault element; and thereafter, the accused has the defence that it reasonably believed in a mistaken set of facts that, if true, would render the act or omission innocent, or that it has taken reasonable precautions to achieve compliance; these are the two branches of the due diligence defence. Most OH&S offences are strict liability offences.

(3) *Absolute liability offences*: In absolute liability offences, the prosecution need only prove beyond a reasonable doubt that the accused committed the prohibited act, constituting the *actus reus* of the offence. There is no relevant mental element and no due diligence defence that the accused was entirely without fault. More proof of the prohibited act will lead to a conviction.

Under the *Sault Ste. Marie* decision, the Supreme Court of Canada held that in a strict liability offence, the onus of proof shifts to the defendant to establish the defence of due diligence. The shifting of the onus of proof was not seen as unfair because the defendant alone had knowledge of what was done to avoid the commission of the prohibited act. It is therefore, expected that the defendant would advance the defence of due diligence, if it was available. There are two separate branches of the due diligence defence:

(1) In the first branch, the defendant must prove that it reasonably believed in a mistaken set of facts which, if true, would render the prohibited act or omission innocent.

(2) In the second branch, the defendant must prove that it took all reasonable steps to avoid the particular prohibited event.

As such, the primary defence in the prosecution of OH&S regulatory offence is the defence of due diligence. The basis of the defence is that it would be legally and morally improper to convict a person of an offence when he/she had taken all reasonable precautions to ensure compliance with the applicable OH&S legislation. Other defences known in law, may also be available to an OH&S offence.

In the development of OH&S law, there arose an issue with respect to the availability of the due diligence defence. In *R. v. Cancoil Thermal Corp.*,[38] it was confirmed that the common law defence of due diligence, as defined in the *Sault Ste. Marie* case, is always available in a regulatory charge under Canadian OH&S law. The court dealt with the issue of the specific exclusion of the statutory defence of

---

[38] [1986] O.J. No. 290, 27 C.C.C. (3d) 295 (Ont. C.A.).

due diligence in the cases of an offence under the Ontario OH&S legislation. The law stated that certain OH&S offences were offences of absolute liability, taking away the defence of due diligence. In answering this question, the court ruled as follows:

> . . . if section 14(1)(a) [now section 25(1)(a) of the Ontario *Occupational Health and Safety Act*] were treated as creating an absolute liability offence, it would offend s. 7 of the [*Canadian*] *Charter* [*of Rights and Freedoms*], the right to life, liberty, and security of the person and the right not to be deprived thereof except in accordance with the principles of fundamental justice.[39]

The court went on to comment that since under clause 66(1)(a) of the Ontario *Occupational Health and Safety Act,* a violation of section 25(1)(a) of the applicable OH&S legislation may attract a term of imprisonment. The combination of absolute liability and the potential penalty of imprisonment was deemed a violation of section 7 of the *Canadian Charter of Rights and Freedoms*.[40] The court therefore concluded that to avoid a violation of section 7 of the Charter, clause 25(1)(a) of the applicable OH&S legislation must, at minimum, be treated as creating a strict liability offence.

Proof of the first branch of the due diligence defence is facilitated by meeting the test set down by the Supreme Court of Canada in the *Sault Ste. Marie* case. However, in order to establish the first branch of the due diligence defence, also known as the mistaken fact branch, the following elements must be proven:

(1) the accused believed in a mistaken set of facts;

(2) if the mistaken set of facts were true, they would render the act or omission innocent; and

(3) the belief by the accused was deemed reasonable by the court.

The most important legal decision interpreting the first branch of the due diligence defence in OH&S law in the last decade is prosecution in *Ontario v. London Excavators & Trucking Ltd.*[41] In that case, the accused had been hired by a general contractor to perform excavating services for the extension on a new hospital. The equipment operator came into contact with a concrete structure that was not on any of the design drawings. The equipment operators stopped the backhoe and made an inquiry to a representative of the general contractor. He was advised by the representative of the general contractor that the concrete object was part of a footing of an old nursing station and that it should be removed. However, when the backhoe operator resumed his excavating activity and dug into the concrete structure, he severed several major power cables from a local hydro utility that were connecting the hospital to the power grid. Although no one was seriously

---

[39] *Ibid.*, at 299.

[40] Part I of the *Constitution Act, 1982*, being Schedule B to the *Canada Act 1982* (U.K.), 1982, c. 11 [hereinafter the "Charter"].

[41] [1998] O.J. No. 6437, 40 O.R. (3d) 32 (Ont. C.A.).

injured, London Excavators & Trucking Company Ltd. was prosecuted by Ontario's Ministry of Labour.

At court, the company relied upon the first branch of the due diligence defence. However, the Court of Appeal for Ontario held that the defence had not been made. The Court of Appeal held that before beginning the work, more detailed and objective inquiries should have been made by the excavating contractor. Further, when the unexpected contact was made with the concrete structure, the excavating contractor could not merely rely upon the word of a supervisor of the general contractor but had to establish more reliably that there was no safety hazard. A lower court conviction was upheld by the Court of Appeal.

Proof of the second branch of the due diligence defence is facilitated by compliance with the internal responsibility system and the development of an effective OHSMS. An internal responsibility system of shared duties and responsibilities of various workplace stakeholders is manifested in an effective OHSMS. Workers, as well as employers and supervisors, have legal duties to comply with the applicable OH&S legislation and its regulations. Therefore, many courts have held that an employer is not liable when the circumstances of an accident were attributable to the inadvertence, mistake, or negligence of an employee.

One helpful example of a successful due diligence defence was that found in the prosecution of *R. v. Kenaidan Contracting Ltd.*[42] The court held that the prosecution had proven a contravention of the applicable provisions of the OH&S statute in Ontario, with which the accused had been charged. However, the court went on to assess whether or not the company had made out the second branch of the due diligence defence. Based on the following factors, the court concluded that the company had established the second branch of the due diligence defence, often referred to as the "reasonable precautions branch":

(1) There had been a pre-construction meeting in which workplace health and safety was considered as part of the overall planning of the project.

(2) A supervisor employed by the contractor was on site at all times to deal with subcontractor issues including health and safety issues.

(3) An external health and safety consultant had been retained to inspect the project, from time to time, to ensure compliance with the established health and safety programs.

(4) There were regular health and safety meetings held at a construction project with all workers.

(5) Senior representatives from the contractor attended a weekly project meeting, at which health and safety on the project was an issue.

(6) The contractor's supervisor and health and safety consultant had the clear authority to stop work if any worker was put in danger.

---

[42] (January 12, 1995), unreported decision of Justice of the Peace Kitchen (Ont. C.J.).

(7) The contractor was never made aware of the concerns relating to the issue of a subcontractor that related to the full-arrest charges against the contractor.

(8) The court determined that the health and safety program and the presence of a full-time supervisor were adequate to meet the requirements of the employer under the applicable OH&S statute and that the employer had done all that could reasonably be expected of it in the circumstances.

The two branches of the due diligence defence, according to *Sault Ste. Marie*,[43] clearly placed an onus of proof on the accused to prove the defence. That this was a departure from the long-standing Anglo-Canadian legal presumption of innocence, was significant. The Supreme Court of Canada indicated in *Sault Ste. Marie* that the accused had the onus of proof to demonstrate that it made out one of the defences in the two branches of the due diligence defence. The standard required by the accused was a civil standard of proof; that is proof on a balance of probabilities.

The validity of the reverse onus on the accused to prove the defence of due diligence, has been the subject of some legal controversy and challenge. The Charter, it was argued, should not permit an accused to have to prove its innocence. The reverse onus of the due diligence defence had exactly that effect. In *R. v. Wholesale Travel Group Inc.*,[44] after the Court of Appeal for Ontario rendered the offending provision in the *Competition Act* of no force or effect, the same issue came up in a prosecution under the Ontario *Occupational Health and Safety Act* in *Ellis-Don Ltd. v. Ontario (Labour Relations Board)*.[45] In the *Ellis-Don* case, the question of whether the reverse onus on an accused charged under the OHSA, was in violation of paragraph 11(*d*) of the Charter, was directly raised for the first time in OH&S prosecution. The *Ellis-Don* case became a test case where three cases were joined together by special order of Dubin J., then Chief Justice of Ontario. The Court of Appeal gave a split decision, 2:1, in favour of the view that the reverse onus on the accused was a violation of the presumption of innocence found in paragraph 11(*d*) of the Charter, and that it could not be saved or justified by section 1 of the Charter.

The Crown appealed both the *Wholesale Travel Group* and the *Ellis-Don* cases to the Supreme Court of Canada. The former was heard and decided first by the Supreme Court of Canada. In *Wholesale Travel Group* case, the Court reversed the Court of Appeal for Ontario decision and held that the statutory and common law reverse onus on an accused to prove on a balance of probabilities, the defence of due diligence, did not offend the Charter. The Court was divided on the question of

---

[43] [1978] S.C.J. No. 59, 85 D.L.R. (3d) 161 (S.C.C.).

[44] [1989] O.J. No. 1971, 70 O.R. (2d) 545 (Ont. C.A.), vard [1991] S.C.J. No. 79, [1991] 3 S.C.R. 154 (S.C.C.) [hereinafter "*Wholesale Travel Group*"].

[45] [1998] O.J. No. 1602, 6 Admin. L.R. (3d) 187 (Ont. C.A.), affd [2001] S.C.J. No. 5, [2001] 1 S.C.R. 221 (S.C.C.) [hereinafter "*Ellis-Don*"].

whether there was a contravention of the Charter and whether, if this was the case, the infringement was saved and justified by section 1 of the Charter. However, the arguments to ultimately justify the reverse onus on the accused in a strict liability offence, were generally accepted by the majority of the Court to include the nature of the offence and the policy reasons behind the prosecution of employers for false advertising, polluting, and OH&S offences. Placing a reverse onus on an accused in a regulatory, strict liability offence was held not to contravene the Charter. Justice Cory identified a so-called, licensing justification and a vulnerability justification, for the denial of the full presumption of innocence for an accused charged with a strict liability offence, such as an OH&S offence.

With the Supreme Court of Canada in the *Wholesale Travel Group* case permitting a reverse onus, the subsequent result in *Ellis-Don* came as no surprise. In *Ellis-Don*, the Supreme Court of Canada released a very brief decision allowing the Crown's appeal. In so doing, the Court adopted its own reasoning from the *Wholesale Travel Group* case and rejected the argument that an accused's right under paragraph 11(*d*) of the Charter is paramount over the concerns of the OH&S regulator in enforcing strict liability offences. Therefore, an accused in an OH&S offence is required to prove one or both of the branches of the due diligence defence.

Essentially, when an individual accused, or a corporate accused, are charged with an OH&S offence, they are presumed innocent until such time as the Crown has proven a *prima facie* case, beyond a reasonable doubt. Once the Crown has discharged its burden to prove the prohibited act or omission beyond a reasonable doubt, then the burden of proof shifts to the accused to prove the defence of due diligence. Although the standard of proof on the accused is a civil standard, a balance of probabilities, rather than a criminal standard, the placing of any burden of proof on the accused in the courts of its trial may be argued to be a compromise of the accused's Charter right of the presumption of innocence. However, the Supreme Court of Canada has held that is an acceptable requirement for a strict liability offence and does not infringe the Charter.

As in a criminal prosecution, enforcement of OH&S laws in Canada by way of quasi-criminal prosecution, may result in a conviction and sentencing hearing. Once an accused is convicted of an OH&S offence, then the accused is sentenced after submissions from the Crown prosecutor and the defence lawyer. In *R. v. Fantini*,[46] the plea of guilty by Mr. Fantini in the fatal construction accident where the worker was buried in a caved-in trench, is an example of resolution of a regulatory prosecution under *Occupational Health and Safety Act* without a trial. Quoted in their entirety below, are the submissions of the prosecutor, Mr. Slansky, for the Ministry of Labour of Ontario; the submissions of defence counsel, Mr. DeRubeis, counsel for Mr. Fantini; and the judgment of His Honour Judge Gorewich, accepting the joint submission of both prosecutor and defence counsel:

---

[46] [2005] O.J. No. 2361 (Ont. C.J.).

CH. 31: CANADIAN WORKPLACE SAFETY LEGISLATION

THE COURT: Would you please stand up please Mr. Fantini. On the admitted facts, I am entering a conviction on each count to which you have pleaded guilty. Thank you. Yes, Mr. Slansky?

MR. SLANSKY: Yes, what I'd like to do next is, the deceased's widow Vivian Salas, has with the assistance of someone else because her English is not great, prepared a statement which would properly be characterized as a Victim Impact Statement. She has asked me to read it on her behalf. I provided it into my friend's review prior to presenting it to the court. On her behalf, she wishes to say as follows:

> My name is Vivian Salas, Ameth Garrido was my husband. He was my companion and my best friend. Not a day goes by that I don't think of him. My heart pains every day. I see him in my dreams, but I cannot touch him or speak to him. He was my whole life and we had planned a future together that will never be. No matter what the outcome of this court proceeding, it can never bring Ameth back to me, and the only reason I am speaking today is to let the court is to let the court [sic] know that when safety precautions are disregarded and a man dies, an entire family is shattered. My future with Ameth is gone forever. I struggle every day to make a life for myself without him. A little boy will grow up without his father because a workplace was not safe. Everyone must know that this is unacceptable; Ameth did not have to die. I pray that some day I will understand why this had to happen and that my sadness and anger can be replaced by some sense of peace. The death of my husband can only have meaning if no-one else ever dies in such a senseless way, and no other family is forced to endure the suffering and pain that I still go through every day.

My submissions Your Honour are as follows. First of all, as Your Honour has [sic] already been advised, the joint submission the Crown is seeking and the defence is jointly seeking, a fine of $50,000.00 exclusive of the Victim Fine Surcharge of 25 percent. What I wish to say about the offence and the circumstances are that, excuse me, is that violations of the *Occupational Health and Safety Act* are regulatory offences. Employers are obligated to ensure that their workers are safe. Deaths of workers in the construction industry, especially in the class of trenches, are far too common. The dangers are well known, and employers must be made to understand not only that safety must be the priority for the protection of workers, but they must be made to understand that it is a priority to protect their own financial interests. As a result, the Crown is seeking a fine, as I said, in the amount of $50,000.00, to send this message to other employers. Here Mr. Garrido was killed as a result of the negligence of Mr. Fantini. This was not the mere failure to follow technical rules, this was a clear failure to follow well known, and obvious procedures to prevent the deaths of workers. Mr. Fantini knew that Mr. Garrido was going to go into the trench, he was directed to slope the trench to prevent the collapse; yet he did not do so. As a result, Mr. Garrido died. I am sure that Mr. Fantini never intended this to happen. I am sure that he feels remorse. I am sure that he is a good man with an unblemished past. However, this does not excuse his clear and seriously negligent actions. He must pay. Other employers must be made aware that should they do likewise, they will pay, otherwise senseless and tragic deaths like this will re-occur and we will have learned nothing. Subject to any questions those are my submissions.

THE COURT: Mr. DeRubeis?

MR. DERUBEIS: Thank you. Your Honour, first of all on behalf of Mr. Fantini, he would like to publicly express his remorse for what has happened. This has been a nightmare for him as well. Mr. Garrido was not just a worker, but someone that he knew and thought of as a friend. The background to this matter was that Mr. Fantini has enjoyed an impeccable reputation his entire life. He has been a very hard-working individual, working in the construction business all of his adult life. He has never been charged with any breaches of the *Occupational Health and Safety Act*, has no criminal record, is a family man, has the support of his children, his wife, he has devoted his life to them, to his work. This job in question was not, as my friend has indicated, was a slow job that was, in a sense, a favour to the homeowner. Mr. Fantini is 67 years of age. Mr. Garrido had been laid off when the company that he worked for — he found this little opportunity to assist the homeowner who had a leak in his basement. It was a job that the total value of was a few thousand dollars. There was no incentive financially to cut corners. The backhoe that was used was provided by the homeowner. The job was being paid on an hourly basis; it wasn't a tendered sum so that he was trying to cut corners to increase his profits. The homeowner was a person he had done business with for many years and he thought, in his judgment, that everything was safe and fine. The hole had been dug. He had left the site for a few minutes to go get some materials and this unfortunate tragedy occurred. When he returned the nightmare began for Mr. Fantini as well. Through his plea here today, he further expresses his sincere remorse for what has happened. He has empathy for Mr. Garrido's widow. It's obviously a tragedy. The past year has been a very difficult one for Mr. Fantini. As Your Honour will hear, under new legislation a criminal charge was laid; the Crown, the provincial Crown I anticipate, will be withdrawing the criminal charge. It has been placed. And for a man of Mr. Fantini's background, integrity, honesty and hardworking, it's been a very painful experience to have a criminal charge hanging over his head; as well and I ask Your Honour to take that into account as well. The joint submission is one that is agreeable and acceptable to both Mr. Fantini and to the Ministry of Labour; and I would ask Your Honour to impose the fine as submitted in the joint submission to Your Honour.

THE COURT: I should publicly indicate that this matter was discussed with me in a pre-trial setting at which time the facts, as the public has heard today, were laid out before me. There was some tense discussion as to the disposition of what would happen in court. This is a standard procedure; it is called the judicial pre-trial. It was clear, both in the pre-trial discussions, and obviously in the comments of counsel today, and certainly the comments of Mrs. Garrido, what a life-altering event this was for her family and of course for the deceased. His life was lost. It is a tragic, tragic event in the lives of many, many people. What has been proposed as a penalty under the *Occupational Health and Safety Act*, in my view, is appropriate. Nothing can bring back Mr. Garrido, a reality, and we feel the pain of that in the comments of Mrs. Garrido. Stand up please Mr. Fantini. On count 2, I impose a fine of $25,000.00. On count 3, I impose a fine of $20,000.00. On count 4, I impose a fine of $5,000.00. You will also be required to pay the Victim Surcharge in this matter.

I have not done the calculation, but that will be in the area of $10,000.00 and you will have the specific figure once calculated by Madam Clerk. I am hopeful that with the imposition of these fines that the message goes out to people in your profession that not only care must be taken, but extreme care must be taken, and it just does not happen to somebody else. Here it happened to Mr. Garrido, a friend of yours, a person with whom you worked in the industry. It hit very close to home. Hopefully the construction business gets the message. I am told that if this were a corporate defendant that the fines would be far greater than it is with respect to you, a small contractor. I expect that this will certainly have some impact on your financial well-being, but it is necessary, sir, in these circumstances. Mr. DeRubeis, time to pay?[47]

The following chart sets out the current penalties, including fines and jail terms, for an accused convicted of a Canadian OH&S offence.

Jurisdiction & Act	Section	Imprisonment	Maximum Fine (Individual)	Maximum Fine (Corporation)
Canada *Canada Labour Code*, R.S.C. 1985, c. L-2	148(1)(*a*)	2 years	$1,000,000	no distinction
	148(1)(*b*)	—	$100,000	no distinction
	148(2)(*a*)	2 years	$1,000,000	no distinction
	148(2)(*b*)	—	$1,000,000	no distinction
	148(3)(*a*)	2 years	$1,000,000	no distinction
	148(3)(*b*)	—	$1,000,000	no distinction
British Columbia *Workers Compensation Act*, R.S.B.C. 1996, c. 492	217(a)	6 months	$652,774.38 plus up to $32,638.75 for each day the offence continues	no distinction
	217(b)	12 months	$1,305,548.74 plus up to $65,277.44 for each day the offence continues	no distinction
	218		Any amount equal to the amount of monetary benefit arising from the offence	no distinction

---

[47] *Ibid.*, at paras. 25-30.

Jurisdiction & Act	Section	Imprisonment	Maximum Fine (Individual)	Maximum Fine (Corporation)
Alberta *Occupational Health & Safety Act*, R.S.A. 2000, c. O-2	41(1)(a)	6 months	$500,000 plus up to $30,000 for each day the offence continues for a second offence	no distinction
	41(1)(b)	12 months	$1,000,000 plus up to $60,000 for each day the offence continues for a second offence	no distinction

Jurisdiction & Act	Section	Imprisonment	Maximum Fine (Individual)	Maximum Fine (Corporation)
Saskatchewan Occupational Health & Safety Act, 1993, S.S. 1993, c. O-1.1	58(1)		$2,000	no distinction
	58(2)		$5,000 plus up to $500 for each day the offence continues	no distinction
	58(4)(a)		$10,000 plus up to $1,000 for each day the offence continues	no distinction
	58(4)(b)		$20,000 plus up to $2,000 for each day the offence continues	no distinction
	58(6)(a)		$50,000 plus up to $5,000 for each day the offence continues	no distinction
	58(6)(b)		$100,000 plus up to $10,000 for each day the offence continues	no distinction
	58(7), (8)	2 years	$300,000	no distinction
Manitoba Workplace Safety & Health Act, R.S.M. 1987, c. W210 (C.C.S.M., c. W210)	55(1)(a), (3)	6 months	$250,000 plus up to $25,000 for each day the offence continues	no distinction
	55(1)(b), (3)	6 months	$500,000 plus up to $50,000 for each day the offence continues	no distinction

Jurisdiction & Act	Section	Imprisonment	Maximum Fine (Individual)	Maximum Fine (Corporation)
Ontario *Occupational Health & Safety Act*, R.S.O. 1990, c. O.1	66	12 months	$100,000	$1,500,000
Quebec *An Act respecting Occupational health and safety*, CQLR, c. S-2.1*	236 (first offence)		$1,500	$3,000
	236 (second offence)		$3,000	$6,000
	236 (third or subsequent offence)		$6,000	$12,000
*The fines imposed under this Act will be revalorized on January 1st each year, starting in 2012, as prescribed by ss. 119 to 123 of the *Act respecting industrial accidents and Occupational diseases*, CQLR, c. A-3.001.	237 (first offence)		$3,000	$60,000
	237 (second offence)		$6,000	$150,000
	237 (third or subsequent offence)		$12,000	$300,000
New Brunswick *Occupational Health & Safety Act*, S.N.B. 1983, c. O-0.2	47	6 months	$250,000	no distinction

CH. 31: CANADIAN WORKPLACE SAFETY LEGISLATION

Jurisdiction & Act	Section	Imprisonment	Maximum Fine (Individual)	Maximum Fine (Corporation)
Nova Scotia *Occupational Health & Safety Act*, S.N.S. 1996, c. 7	74	2 years	$250,000 for a first offence plus up to $25,000 for each day the offence continues	no distinction
			$500,000 for a second or subsequent offence committed within five years of conviction for a previous offence or where an offence results in a fatality, plus up to $25,000 for each day the offence continues	
			Any amount equal to the amount of monetary benefit resulting from the offence in addition to the fine	no distinction
Prince Edward Island *Occupational Health & Safety Act*, R.S.P.-E.I. 1988, c. O-1.01	43(1), (2)	1 month	$250,000 plus up to $5,000 for each day the offence continues	no distinction

1167

Jurisdiction & Act	Section	Imprisonment	Maximum Fine (Individual)	Maximum Fine (Corporation)
Newfoundland and Labrador *Occupational Health & Safety Act*, R.S.N.L. 1990, c. O-3	67	12 months	$250,000 plus up to $25,000 for each day the offence continues	no distinction
Northwest Territories *Safety Act*, R.S.N.-W.T. 1988, c. S-1	22(2) 22(4) 22(5) 22(5.1)	1 year 6 months 1 month 1 year	$500,000 $50,000 $25,000 $500,000	no distinction no distinction no distinction no distinction

Jurisdiction & Act	Section	Imprisonment	Maximum Fine (Individual)	Maximum Fine (Corporation)
Yukon Territory *Occupational Health & Safety Act*, R.S.Y. 2002, c. 159	44(1)(a)	12 months	$150,000 plus up to $15,000 for each day the offence continues	no distinction
	44(1)(b)	24 months	$300,000 plus up to $25,000 for each day the offence continues	no distinction
	44(2)(a)	18 months	$200,000 plus up to $17,500 for each day the offence continues	no distinction
	44(2)(b)	30 months	$350,000 plus up to $27,500 for each day the offence continues	no distinction
	44(3)(a)	24 months	$250,000 plus up to $20,000 for each day the offence continues	no distinction
	44(3)(b)	36 months	$400,000 plus up to $30,000 for each day the offence continues	no distinction
	44(4)	12 months	$5,000	no distinction

E. OH&S DUE DILIGENCE AND "REASONABLE STEPS"

There is a similarity between Canadian OH&S law, OH&S general duty clauses, and the defence of due diligence on the one hand, and the new legal duty under the *Westray Bill* on the other. In the *Westray Bill* amendment to the *Criminal Code* in section 217.1, there is a duty to take reasonable steps to protect the safety of workers and the public. This duty is similar to the general duty clauses found in Canadian OH&S law and the language used by the courts in the legal defence of due diligence. There is no reverse onus in a prosecution of the new crime of OH&S criminal negligence. The question remains, however: What constitutes sufficient proof of reasonable steps in the new OH&S crime?

As already noted, due diligence is a primary defence available to an accused against charges under any OH&S statute. The second branch requires the accused to prove that it took reasonable steps or precautions to ensure compliance with OH&S statutory and regulatory standards. In a criminal prosecution of the new offence of OH&S criminal negligence, failure of the accused to take reasonable steps is part of what a prosecutor must prove to the crime.

There is every reason to suggest that if an employer complies with the new legal duty under section 217.1 of the *Criminal Code*, an accused must be compliant with the applicable OH&S laws. However, reasonable steps will be difficult to determine consistently across the country, since OH&S statutes vary by jurisdiction. In most provinces except for Quebec, for example, the OH&S legislation contains legal requirements for JHSCs. These OH&S legal requirements may serve as a yardstick in an OH&S criminal negligence prosecution; to measure whether or not reasonable steps have been taken. However, since there is no national, consistent OH&S statute, this makes the interpretation of reasonable steps more problematic. In other words, it is not exactly clear what will amount to reasonable steps under section 217.1 of the *Criminal Code*. What is clear is that there will have to be some development of the case law to establish what will constitute reasonable steps to prevent bodily harm to workers and the public.

Determining what reasonable steps an employer or manager was required to take will be a fact-specific determination. The mere failure to take reasonable steps required under an OH&S regulatory statute, or failure to exercise due diligence in an OH&S prosecution, will not necessarily result in a conviction for OH&S criminal negligence. This is due to the different onus of proof and fault elements of a criminal offence, from that of an OH&S offence. However, it is reasonable to anticipate that the courts will seek guidance from the extensive, due diligence jurisprudence arising from OH&S prosecutions that have generally imposed a high standard of care on regulated parties such as employers, constructors, prime contractors, supervisors, officers, and directors.

Commentary: What, then, does the addition of the *Criminal Code* OH&S duty mean for already regulated organizations and individuals in Canada? Aside from the obvious addition of new criminal liability for organizations, one of the most important differences between criminal and OH&S prosecutions is that a criminal standard must be breached by the individual, corporation, or organization. The criminal standard includes proof of wanton or reckless disregard for the lives or safety of others. The criminal standard also requires all elements of the offence to be proven by the prosecutor beyond a reasonable doubt.

Criminal negligence standards will continue to evolve with the development of the Bill C-45 jurisprudence. **Criminal negligence** occurs when an act or omission of an accused party shows wanton or reckless disregard for the lives or safety of others in a situation where the accused party is under a legal duty to act. In respect of organizational liability, the Crown must prove that the conduct of the senior officer of an organization represented a marked departure from the standard of care that could be expected of a reasonably prudent person in the circumstances to prevent a representative from being a party to the offence. This standard is distinct from the regulatory standard of failure to exercise all reasonable care or due diligence. Just how distinct it will be from regulatory due diligence concepts will be dependent on the case law that develops.

The second branch of the defence of due diligence, the reasonable steps branch, allows the defendant to avoid liability by proving that it took all reasonable steps to avoid a particular event. In order to successfully argue the reasonable steps branch of the due diligence defence, the Supreme Court of Canada in *Sault Ste. Marie* said:

> . . . the question will be . . . whether the accused [1] exercised all reasonable care by establishing a proper system to prevent commission of the offence and [2] by taking reasonable steps to ensure the effective operation of the system. [numbers added].[48]

Therefore, an organization can demonstrate due diligence by proving that it had an OHSMS and that it was operating effectively.

The reasonable steps branch of the due diligence defence has been applied in a number of OH&S prosecutions. Successful due diligence defences have included:

- The hazards applicable to the offence had been identified, their risks had been assessed, and controls were put in place to minimize their risk;
- Written OH&S policies and procedures were developed to communicate hazard control to workers and management;
- Personnel were provided with instruction and communication on the hazards, their risks, the controls in place to minimize those risks, and how to properly use those controls;
- The workplace was monitored and supervised and any deficiencies identified through that monitoring and the supervision process were corrected; and
- The above-noted steps were documented, reviewed and audited by senior management.

This checklist outlines the essential elements of an effective OHSMS.

An OHSMS that outlines specific responsibilities, authorities, and accountabilities for every workplace stakeholder, and that complies with the minimum

---

[48] [1978] S.C.J. No. 59 at para. 56, 85 D.L.R. (3d) 161 (S.C.C.).

requirements of OH&S law, will apply the internal responsibility system in the workplace and significantly assist in demonstrating the employer's due diligence.

Although enforcement of OH&S law in Canada, by way of orders and prosecution, is similar in process to a criminal prosecution, the legal characterization of a criminal charge is different from that of a strict liability OH&S regulatory offence. More specifically, due diligence is not a defence to the *Westray Bill* OH&S criminal negligence charges, because it is for quasi-criminal OH&S regulatory charges. One method of successfully defending an OH&S criminal negligence charge, however, is to raise a reasonable doubt in the Crown's duty to prove the elements of the offence beyond a reasonable doubt. This is applicable both to the *actus reus* of the OH&S criminal negligence offence against individuals and the *mens rea* of the OH&S criminal negligence offence. Therefore, OH&S due diligence evidence may be used to raise a doubt in the Crown's attempt to prove that representatives of an organization failed to take "reasonable steps" or to adhere to a "reasonable standard of care".

Commentary: The best method for avoiding an OH&S criminal negligence conviction is to prevent accidents that cause bodily harm. And, the most effective method of minimizing the likelihood of incidents causing bodily harm is by implementing and maintaining an effective OHSMS that demonstrates due diligence and complies with the minimum requirements of OH&S laws.

Developing, implementing and maintaining an effective OHSMS is a proven method of significantly reducing the likelihood of suffering human and the economic and legal costs associated with work-related accidents. There are several international, national, and provincial standards with respect to OHSMSs. Four excellent examples of OHSMS standards are: *BS OHSAS 18001:2007*, *CSA Z1000-14*, *CSA Z45001:19*, and *ISO 45001*.[49] These standards outline the elements of an effective OHSMS. For further details, refer to Chapter 2, "Occupational Health and Safety Management Systems", section D. OHSMS Standards: Description and Comparison.

An OHSMS standard can assist any organization, no matter the size or complexity, to realize the benefits of an OHSMS. The *Westray Bill* amplified the importance of implementing an effective OHSMS. Even in the absence of potential criminal liability, every responsible organization should strive to implement an OHSMS, which does not need to be complex nor costly.

F. AN EFFECTIVE OH&S MANAGEMENT SYSTEM (OHSMS)

The elements of an effective OHSMS, and hence, proof of the second branch of the due diligence defence, include:

---

[49] The British Standards Institution, *BS OHSAS 18001:2007* (July 31, 2007), online: http://www.bsi-global.com; Canadian Standards Association, *CSA Z-1000* (2014) and *CSA Z-45001* (2019), online: http://www.csa.ca; and ISO, *45001* (2018), online: https://www.iso.org/standard/63787.html.

1. OH&S Policy

An OH&S Policy establishes a sense of direction for an organization with respect to OH&S performance. It is usually a one-page document that outlines an organization's commitment to ensuring the health and safety of its employees and other workers by complying with the OH&S legal requirements, at minimum. The policy, which should be authorized and signed by the organization's senior management, must be appropriate to the nature and scale of the organization's OH&S risks and should include a commitment to continual OH&S improvement. Further, the policy should address the internal responsibility system and be communicated to all employees and interested parties. Employers usually post the policy in a conspicuous location at the workplace. The author recommends that senior management review its OH&S policy at least annually, for relevancy and appropriateness. In some Canadian jurisdictions, such as in Ontario, it is a legal requirement to prepare and annually review the OH&S Policy.

2. Planning

This element is the foundation of an OHSMS. Workplace hazards must be identified, their risks assessed and prioritized, and measures implemented to control (minimize) the risks of the identified hazards. This process also comprises one of the common elements of a successful due diligence defence. This is also a legal requirement in some Canadian jurisdictions. For instance, Part 2 of the Alberta *Occupational Health and Safety Code 2018*[50] requires employers to conduct a hazard assessment prior to commencing work at a worksite or prior to the construction of a new worksite. Any hazards that are identified must be either eliminated or controlled through the use of the following hierarchy: engineering controls, administrative controls, or personal protective equipment.

There are several recognized methods for carrying out the hazard identification, assessment, and control process. One method is by implementing a Job Safety Analysis (JSA) program, whereby the organization:

- identifies all jobs;
- identifies all the tasks within every identified job;
- identifies all the critical steps (those which pose a hazard to a person) within each task; and
- identifies the hazards, evaluates the risks, and recommends controls for those risks for each identified critical step.

The result of completing a detailed and thorough JSA program will be having the knowledge of foreseeable hazards associated with every job and task. This puts an organization in a position to control the risks associated with these hazards through

---

[50] Alta. Reg. 87/2009, with amendments up to and including Alta. Reg. 213/2018. Current as of January 1, 2019.

the hierarchy outlined above. Another method for carrying out the hazard identification, assessment, and control process is often referred to as a Field Level Risk Assessment, which involves a worker identifying hazards, and assessing and controlling any risks immediately prior to performing a task. Both of these methods must be incorporated into an OHSMS.

The hazard identification, assessment, and control process have several important links to other elements of the OHSMS, such as implementation, operation, and checking. The rest of the OHSMS is unable to function properly if the hazard identification, assessment, and control process is not effectively carried out.

In addition, the planning process should involve a method or procedure for identifying and complying with the OH&S legal and other requirements that apply to the organization's operations and activities. This is an excellent step in achieving compliance with the minimum requirements of the OH&S law, which is an important aspect of avoiding an OH&S criminal negligence conviction.

3. Implementation and Operation

The OHSMS must be implemented and properly operated. In that regard, the structure of the organization should be clearly defined and specific responsibilities, authorities, and accountabilities for the organization's representatives, with respect to operating the OHSMS, assigned. Further, everyone within the organization must be aware of his or her responsibilities and be competent to carry them out.

Most Canadian OH&S laws have a requirement concerning worker competency. Each jurisdiction's OH&S laws differ in how they define, apply, and use the term "competent". Generally, the term competent refers to three criteria: qualifications, training and experience. Some provinces require all supervisors to be competent. Some provinces require certain workers to be competent. It is an organization's responsibility to understand the worker competency legal requirements in the jurisdiction(s) that it carries out business in. Applying the elements of an effective OHSMS will allow an organization to demonstrate worker competency in accordance with the minimum requirements of OH&S laws. Demonstrating worker competency is an excellent example of how various elements of the OHSMS are linked. In the planning phase, the concept of a Job Safety Analysis (JSA) program was discussed. Through a JSA program, an employer will have knowledge of the reasonably foreseeable hazards and risks associated with every job as well as the controls in place to minimize those risks. This will allow for the development of safe work procedures for all known tasks. In addition, organizations can identify the necessary qualifications, training and experience required to perform a certain job or task, depending on the level of risk for each job or task.

Consultation, communication, and participation with employees is also an important aspect of the implementation and operation of the OHSMS. Workers must be advised of their legal responsibilities, the hazards they could be exposed to, the risk that those hazards pose, the controls in place to minimize those risks, and how to properly use those controls. This is a fundamental function of the internal responsibility system. All forms of instruction and communication, whether it is formal classroom training or a 10-minute safety talk, should be documented. Documentation that confirms the trainees' understanding of the information that was communicated, such as a test or quiz that the trainee completed upon conclusion of the training session, is extremely valuable. Training records that provide excellent due diligence evidence are records that are current, appropriate, documented, and demonstrate that the training was understood.

Proper consultation, communication, and participation can also be realized by the establishment of a JHSC. The legal requirements concerning the establishment, functions, and powers of a JHSC differ by province. It is an organization's responsibility to understand the JHSC requirements in the jurisdiction(s) in which it operates. JHSCs — whether established by statutory compulsion or voluntarily — fulfil employees' right to participate in the protection of their own health and safety — one of the three fundamental rights promoted by the internal responsibility system. The responsibility for implementing an OHSMS lies with the employer, but the JHSC can play a critical role in its functioning.

An effective OHSMS requires organizations to establish plans and procedures to identify the potential for and the related responses to foreseeable incidents and emergency situations. The plans should also outline how to prevent and mitigate the likelihood of illness and injury resulting from incidents and emergency situations. These emergency preparedness and response plans should be documented, tested, and periodically reviewed.

4. Evaluation: Inspections and Auditing

Once the workplace hazards have been identified, assessed, and controlled, and all workers and management have been advised of this information and received the proper qualifications, training, and experience, the OHSMS must have ongoing mechanisms to monitor and supervise the workplace in order to ensure that hazards remain controlled. These mechanisms could include, but are not limited to, regular worksite inspections, workplace walk-throughs, incident investigations, and audits.

Regular worksite inspections are a legal requirement in most Canadian jurisdictions and are a method of carrying out the element of checking and corrective action. Inspections, a method of monitoring, can be carried out in any number of frequencies, whether it be daily vehicle inspections or monthly worksite inspections. This process allows for the identification of deficiencies in the planning, implementation, and operation phases of the OHSMS. A schedule should be developed on an annual basis, along with the assignment of responsibility to conduct the inspections. The inspection should result in the assignment of responsibility to correct any noted deficiencies. This process should be documented because it allows the organization to demonstrate that the OHSMS is effectively functioning.

Whereas worksite inspections are a proactive method of monitoring the workplace, incident investigations are a reactive approach to identifying and correcting deficiencies within the OHSMS. Conducting incident investigations is another method of carrying out the element of checking and corrective action. All incidents should be investigated. The purpose of an incident investigation is to identify its root cause and make recommendations to correct that root cause to prevent recurrences.

Every Canadian jurisdiction has a legal requirement under its OH&S statute requiring employers to report certain incidents. Several other legal requirements, such as securing the scene of the incident and conducting an incident investigation, may also be triggered by certain incidents. It is an organization's responsibility to understand the incident notification requirements in the jurisdiction(s) in which it carries out business.

Every organization should also audit its OHSMS. Audits determine whether or not the OHSMS has been functioning properly, has been properly implemented and maintained, and is effective in achieving the organization's OH&S policy. Provisions for internal and external audits should be considered; the audit should cover the scope, frequency, methodologies and worker competencies, as well as the responsibilities and requirements for conducting audits and reporting results.

5. Management Review

Finally, there needs to be a process in place for management to monitor and review the successes, or to identify the weaknesses, of the OHSMS. Usually, this process is undertaken on an annual basis and includes a report to senior management on the key indicators such as conformance to the OHSMS or the measured success of its components. It involves senior management reviewing the internal or external audit results to verify that the OHSMS is properly functioning. This phase of the OHSMS allows senior management to renew its commitment to protecting the health and safety of all persons affected by its operations.

For more information on OHSMSs, their history, their value, and their structure and components, refer to Chapter 2, "Occupational Health and Safety Management Systems".

G. CONCLUSION

The use of criminal offences, such as criminal negligence and manslaughter, has changed and increased the seriousness of health and safety compliance in Canada. Section 217.1 of the *Criminal Code* now requires everyone who undertakes, or has the authority, to direct how another person does work or performs a task to take reasonable steps to prevent bodily harm to any person arising from that work or task. The term "reasonable steps" is not defined by the *Criminal Code*. Occupational health and safety laws across Canada, some of which have been in force for more than 30 years, have long required workplace stakeholders to take what can be considered to be "reasonable steps" to protect the health and safety of workers. General duty clauses on employers under the Canadian OH&S laws, have long required employers to demonstrate due diligence with respect to protecting the health and safety of workers. Due diligence is a defence to quasi-criminal regulatory charges and is available to an accused who took reasonable steps to avoid a particular event. In order to demonstrate that reasonable steps were taken, the accused must demonstrate that a system was in place to prevent commission of the offence and that the system was operating effectively.

Therefore, the legal defence of due diligence requires the development, implementation, and maintenance of an effective OHSMS. Although due diligence is not legally a defence to OH&S criminal negligence charges, due diligence evidence can be used to raise a doubt in the Crown's attempt to prove the elements of OH&S criminal negligence charges beyond a reasonable doubt.

While the offence of OH&S criminal negligence increases the potential liability for organizations and its representatives, it does not necessarily increase the level of responsibility on the part of these workplace stakeholders. The *Westray Bill* amendments to the *Criminal Code* amplify the importance of what every responsible organization should already be doing: complying with the minimum requirements of applicable OH&S laws and developing, implementing, and maintaining an effective OHSMS.

An effective OHSMS is one that conforms to an international or national OHSMS standard. An effective OHSMS reduces the likelihood of experiencing the human, economic, and legal losses and costs associated with work-related incidents, including OH&S criminal negligence convictions; while increasing the likelihood of being able to defend criminal or regulatory charges should they result from an incident. The most effective method of preventing OH&S criminal negligence convictions is to avoid having incidents that cause bodily harm.

CHAPTER REFERENCES

Alberta *Occupational Health and Safety Code 2006.*

Act respecting industrial accidents and Occupational diseases, CQLR, c. A-3.001.

An Act respecting occupational health and safety, CQLR, c. S-2.1.

An Act to Amend the Criminal Code, S.C. 2003, c. 21, proclaimed in force March 31, 2004 (except s. 22). Introduced as Bill C-45, 2d Sess., 37th Parl., 2003

British Standards Institution, *BS OHSAS 18001:2007* (July 31, 2007), online: http://www.bsi-global.com.

Building Code Act, 1992, S.O. 1992.

Canadian Charter of Rights and Freedoms, Part I of the *Constitution Act, 1982*, being Schedule B to the *Canada Act 1982* (U.K.), 1982, c. 11.

Canada Labour Code, R.S.C. 1985, c. L-2 (Part II, "Occupational Health and Safety", am. S.C. 2000, c. 20).

Canada Labour Code, R.S.C. 1985, c. C-46.

Canadian Standards Association, *CSA Z1000* (2014), online: http://www.csa.ca.

Canadian Standards Association, *CSA Z450001* (2019), online: http://www.csa.ca.

Canadian Standards Association, *CSA Z796* (1998), online: http://www.csa.ca.

Criminal Code, R.S.C. 1985, c. C-46, s. 217.1.

Ellis-Don Ltd. v. Ontario (Labour Relations Board), [1998] O.J. No. 1602, 6 Admin. L.R. (3d) 187 (Ont. C.A.), affd [2001] S.C.J. No. 5, [2001] 1 S.C.R. 221 (S.C.C.).

Hazardous Products Act, Hazardous Products Regulations, WHMIS Legislation, (SOR/2015-17).

ISO 45001 (2018), online at: https://www.iso.org/standard/63787.html

N.A. Keith, *Canadian Health and Safety Law* (Aurora: Canada Law Book, 2003).

N.A. Keith, *Workplace Health & Safety Crimes*, 4th ed. (Markham, ON: LexisNexis Inc., 2019).

Ministry of Labour Legal Branch, *Interpretation Opinions* (Ottawa: Ministry of Labour, November 1983).

Occupational Health & Safety Act, N.S. 1996, c. 7.

Occupational Health & Safety Act, R.S.A. 2000, c. 0-2, (now S.A. 2017, c. O-2.1, s. 3(1)).

Occupational Health & Safety Act, R.S.P.E.I. 1988, c. O-1.1.

Occupational Health & Safety Act, R.S.N.L. 1990, c. O-3.

Occupational Health & Safety Act, R.S.N.W.T. 1988, c. S-1.

Occupational Health & Safety Act, R.S.O. 1990, c. O.1, s. 25(1)(a), (b), (d).

Occupational Health & Safety Act, R.S.S. 1993, c. O-1.1.

Occupational Health & Safety Act, R.S.Y. 2002, c. 159.

Occupational Health & Safety Act, S.N.B. 1983, c. O-0.2.

Occupational Health & Safety Act, S.N.S. 1996, c. 7.

Occupational Health and Safety Regulation, B.C. Reg. 296/97.

Ontario v. London Excavators & Trucking Ltd., [1998] O.J. No. 6437, 40 O.R. (3d) 32 (Ont. C.A.).

R. v. Cancoil Thermal Corp., [1986] O.J. No. 290, 27 C.C.C. (3d) 295 (Ont. C.A.).

R. v. Fantini, [2005] O.J. No. 2361 (Ont. C.J.).

R. v. Kenaidan Contracting Ltd., (January 12, 1995), unreported decision of Justice of the Peace Kitchen (Ont. C.J.).

R. v. Precision Diversified Oilfield Services Corp., [2018] A.J. No. 1005, 2018 ABCA 273 (Alta. C.A.).

R. v. Sault Ste. Marie, [1978] S.C.J. No. 59, 85 D.L.R. (3d) 161 (S.C.C.).

R. v. Wholesale Travel Group Inc., [1989] O.J. No. 1971, 70 O.R. (2d) 545 (Ont. C.A.), vard [1991] S.C.J. No. 79, [1991] 3 S.C.R. 154 (S.C.C.).

Safety Act, R.S.N.W.T. (Nu) 1988, c. S-1.

K. Swinton, "Enforcement of Occupational Health and Safety Legislation: The Role of the Internal Responsibility System" in P. Swan & E. Swinton, eds., *Studies in Labour Law* (Toronto: Butterworths, 1983) at 143.

Constitution Act, 1867 (U.K.), 30 & 31 Vict., c. 3.

Workers' Compensation Act, R.S.B.C. 1996, c. 492, s. 115(1).

Workplace Safety & Health Act, C.C.S.M. c. W210.

Chapter 32

MARIJUANA LEGALIZATION: IMPACT ON THE CANADIAN WORKPLACE

A. INTRODUCTION

On October 17, 2018, Canada legalized the recreational use of marijuana under the *Cannabis Act*.[1] The legalization of medical marijuana had previously been passed in 2001[2] and was revised with the *Cannabis Act, 2018*. One year later, in October 2019, edible cannabis products were legalized. Additionally, provisions were made for each province/territory to set its laws concerning the possession limits, minimum age and location for use. Municipalities could also establish their sale and use restrictions. Consequently, this federal legislation has many ramifications for Canadian workplaces and the workforce, not to mention the societal impacts.

The purpose of this chapter is to explore the employer's response to the legalization of recreational and medical marijuana, and its impact on the workplace in terms of worker health and safety. But first, a brief description of the health effects of marijuana will be provided along with ways to detect marijuana use, and address marijuana use – medical and recreational in nature, in the workplace.

B. MARIJUANA: WHAT IS IT AND THE HEALTH EFFECTS

According to the Center for Disease Control (2018), "Marijuana is the most commonly used illegal drug in the United States, with 37.6 million users in the past year, and marijuana use may have a wide range of health effects on the body and brain".[3] In May 2019, Statistics Canada reported that 18 per cent of Canadians, 15 years and older, reported using cannabis within the last three months – a 4% increase since legalization that was primarily among males 45-64 years of age.[4] Within that time, more new cannabis users emerged (646,000); nearly twice the number from

---

[1] S.C. 2018, c. 16.

[2] *Access to Cannabis for Medical Purposes Regulations*, SOR/2016-230.

[3] Center for Disease Control (CDC), "Marijuana: How Can It Affect Your Health?", *Health Effects* (2018) at 1, online: https://www.cdc.gov/marijuana/health-effects.html (date accessed January 31, 2020).

[4] Statistics Canada, *National Cannabis Survey, first quarter 2019* (2019), online: https://www150.statcan.gc.ca/n1/daily-quotidien/190502/dq190502a-eng.htm (date accessed May 2019).

last year (327,000). Those who reported daily cannabis use remained at six per cent; these individuals believed that it was safe to drive within three hours post-use.[5]

By June 2019, the Institute for Work and Health (IWH) reported that 22 per cent of the past-year users reported using marijuana two hours before work, during work, or during work breaks.[6] Forty-four per cent felt it would be easy to use marijuana at work, or during lunch and work breaks.[7] Workers reported using marijuana primarily for recreational reasons (68%) and they reported that the main reason for doing so was to "relax". Alarmingly, 38 per cent said they used marijuana to cope with work stressors.[8]

Marijuana affects the brain resulting in memory, learning, attention, decision-making, coordination, emotional, and reaction time deficits. The degree of impairment depends on many factors: user factors such as gender, age, state of health, frequency of use, timing of use, and use with other substances; as well as the product factors such as product strength, dosage, and mode of consumption.

Marijuana has been long recognized for its medicinal properties, for example, it was used by the Chinese, in 2600 BCE, for treating cramps and pain. Despite its longevity, much remains unknown about the properties of the various cannabinoids in marijuana. So far, only three of the 100 cannabinoids have been studied.[9]

Delta-9-tetrahydrocannabinol (THC) and **Cannabinol (CBD)** are the active ingredients in marijuana (Cannabis). THC has psychoactive properties, whereas CBD does not. The potential effects of THC include analgesia, relaxation, drowsiness, euphoria, suppression of PTSD symptoms and chemotherapy-induced side-effects, and appetite stimulation. In contrast, CBD is an antidepressant, anti-convulsant, antioxidant, anti-psychotic, anti-emetic, anti-inflammatory agent, and a possible neuro-protective agent.[10]

Marijuana consumption can be via inhalation (smoking) or oral (ingestion). With inhalation, marijuana quickly gets absorbed into the blood stream, thereby maximizing the impact on the brain within 15 minutes. Oral consumption offers slower

---

[5] *Ibid.*

[6] IWH, Cannabis Use and the Canadian Workplace" (July 2019), online: https://www.iwh.on.ca/publications/cannabis-use-and-canadian-workplace-pre-legalization?utm_source=iwhnews&utm_medium=email&utm_campaign=iwhnews-2019-11 (date accessed February 28, 2020).

[7] *Ibid.*

[8] *Ibid.*

[9] G. Jackowski, "Biology of Marijuana" (presented at the Schedule 2 Employers Group Conference, October 2016, Richmond Hill, ON), online: http://www.s2egroup.com/library-2/2016-conference-presentations/ (date accessed February 28, 2020).

[10] MMG/CannTrust, *Product Monograph* (presented at the Schedule 2 Employers Group Conference, October 2016, Richmond Hill, ON), online: https://canntrust.ca/resources-toolkits/, (date accessed February 28, 2020).

and more unpredictable uptake into the blood, taking 30 to 90 minutes to take effect, with the maximum effect being reached two to three hours later, and lasting for four to eight hours.[11],[12]

Marijuana: Health Concerns

Over the years, the potency of THC in marijuana has increased,[13] affecting the brain even more. As well, depending on the method of consumption, it can result in the user getting higher levels of THC (CDC, 2018) which can have a longer effect duration (impairment of 24-48 hours[14]). Sadly, the rate of marijuana addiction is 1 in 10 users for those users over the age of 18 years; for younger users, it is 1 in 6 users. This is believed to be because the teenage brain is still developing and is more susceptible to the negative effects of THC.[15]

The mode of consumption of THC is proving problematic. Vaping has been linked with severe lung disease and death. "Most (89%) of the 27 patients [suffering with lung disease that were] interviewed in Wisconsin to date said they had used 'e-cigarettes or other vaping devices' to inhale THC products, such as waxes and oils".[16]

Marijuana poisonings have been known to happen in association with the ingestion of edibles. Edibles tend to take longer to impact the user who ends up overeating in an attempt to get a "high". Likewise, the amount of THC is difficult to quantify in edibles and the effects last longer. Sadly, accidental ingestion has led to poisonings in adults and children.

For individuals with heart conditions, smoking marijuana can increase their heart

---

[11] *Ibid.*

[12] A. Nakim, "Medical Marijuana: In the Workplace" (presented at the Schedule 2 Employers Group Conference, October 2016, Richmond Hill, ON), online: http://www.s2egroup.com/library-2/2016-conference-presentations/ (date accessed February 28, 2020).

[13] *Ibid*; M. Snider-Adler, *A Guide to Cannabis in the Workpalce, Part 1: Recreational Cannabis* (2018), online: www.cannabis@DriverCheck.ca.

[14] J. Phillips, M. Holland, D. Baldwin, L. Giffiord-Meuleveld, K. Mueller, B. Perkinson, M. Upfal & M. Dreger, "Marijuana in the Workplace: Guidelines for Occupational Health Professionals and Employers" (2015) 57:4 *JOEM* 459 at 461.

[15] C. Lopez-Quintero *et al.*, "Probability and predictors of transition from first use to dependence on nicotine, alcohol, cannabis, and cocaine: results of the National Epidemiologic Survey on Alcohol and Related Conditions (NESARC)" (2011) 115:1-2 *Drug Alcohol Depend.* 120; W. Hall, *et al.*, "Adverse health effects of non-medical cannabis use" (2009) 374:9698 *Lancet* 1383; and A.J. Budney, J.D. Sargent & D.C. Lee, "Vaping cannabis (marijuana): parallel concerns to e-cigs?" (2015) 110:11 *Addiction* 1699.

[16] Wisconsin Department of Health Services, *News Release*, "Majority of Wisconsin Lung Disease Patients Who Reported Vaping Cite THC Products: The investigation of lung disease among people who reported vaping is ongoing" (August 29, 2019), online: https://www.dhs.wisconsin.gov/news/releases/082919.htm.

rate and trigger a stroke or heart disease. As well, smoking marijuana is a vehicle for the ingestion of many toxins, irritants, and carcinogens just like with tobacco smoke.[17] As for the impact of second-hand marijuana smoke, little conclusive evidence exists as to the harmful effects.

For individuals with existing psychological health issues, the use of marijuana can cause significant and serious outcomes, such as disorientation, anxiety, paranoia, temporary psychosis, and even long-lasting mental disorders.[18]

The therapeutic value of medical marijuana remains inconclusive. It may help to treat chronic pain in some people, but is it any better than other pain-management products or therapies? It may help to counter the nausea, vomiting, and other side-effects of cancer treatment, but it does not control or cure cancer; it may even worsen the situation due to the undesired side-effects.

To obtain medical marijuana, the individual must obtain a medical document from a medical practitioner authorizing the use of cannabis for medical purposes;[19] that enables the individual to secure marijuana from licensed producers. However, medical marijuana is not an approved drug or medicine in Canada under the *Controlled Drug and Substances Act*. Because it does not have a Drug Identification Number (DIN), it is not usually covered by corporate Prescription Drug Plans in Canada.

C. MARIJUANA: DETECTION

Challenges exist in the detection of marijuana use; for instance, there is no conclusive test to identify the consumption of marijuana or its degree of impairment. In Canada, random drug testing remains illegal, except in special circumstances related to identified safety risks (settlement of *Unifor Local 707A v. Suncor Energy Inc., 2018*),[20] or post-incident. Otherwise, employers must rely on the assessment of the degree of user impairment.

The testing methods are limited to:

- Saliva testing – a swab of the mucous membranes is taken and examined. The limitation is that this mode of testing is that it detects recent use only (1 hr-12 hrs).

- Urine testing – single use THC can be detected within 2-5 hrs post-use, up to three days. Chronic or heavy THC use can be detected up to 3-30 days.

---

[17] Center for Disease Control (CDC), "Marijuana: How Can It Affect Your Health?", *Health Effects* (2018), online: https://www.cdc.gov/marijuana/health-effects.html.

[18] *Ibid*.

[19] Canada, *Cannabis for medical purposes under the Cannabis Act: information and improvements* (2019), online: https://www.canada.ca/en/health-canada/services/drugs-medication/cannabis/medical-use-cannabis.html.

[20] See online: https://scc-csc.lexum.com/scc-csc/scc-l-csc-a/en/item/17137/index.do

This is considered the "gold" standard for risk-based programs.
- Hair testing – non-invasive and detects THC between 5-7 days post-use and up to approximately 90 days.
- Blood testing – detects THC for 1-2 or more days; used by police authorities but is considered too invasive for use in the workplace.

These tests although available, do not correlate with the degree of associated employee impairment. Hence, for the employer, reliance must be on the assessment of user impairment.

When deciding to test, the employer must consider the various pieces of legislation such as the *Canadian Human Rights Act* (CHRA), OH&S Acts, Privacy laws, Labour Standards, collective agreements and regulatory requirements, along with the level of workplace supervision.[21] Substance testing is generally considered discriminatory based on disability or "perceived disability" (CHRA).

If the worker appears impaired, the employer must assess:
- If worker can do the job safely;
- If there is an impact on cognitive ability or judgment; and/or
- Are other medical conditions/treatment effects evident.

When a supervisor or manager believes that an employee is impaired, they must have the employee stop work immediately and proceed to interviewing him/her to assess for the signs or impairment. This would include interviewing other co-workers interacting with that employee.[22] The signs of impairment due to marijuana use include dizziness, drowsiness and fatigue; confusion; inability to focus; memory deficits; motor function and perception deficits; and an altered emotional state.[23]

Impairment

Impairment, defined as experiencing a state of diminished, reduced or damaged state of being,[24] is described in Chapter 9, "OH&S: Hazard Identification, Assessment and Control".

As the result of substance use, the worker can be working but be functionally incapacitated. The degree of impairment varies depending on the dosage, the properties of the substance, the gender and age of the user, the length and time since taking the substance, the worker's level of tolerance for the substance, and/or the nature of the activity. Activities involving concentration, fine motor skills, informa-

---

[21] CHR Commission, 2017.

[22] B. Kwasniewski, "Clearing the haze: Managing Cannabis in the workplace in Ontario" (Toronto, ON: Carters Professional Association, 2018).

[23] *Ibid.*

[24] Dictionary.com (2017), *s.v.* "Impairment", online: http://www.dictionary.com/browse/impairment (date accessed: January 31, 2020).

tion processing, deductive reasoning, quick response time, and memory are often the most impacted.

Some of the signs of impairment include:

A. **Cognitive**
- mood shifts (tense to relaxed)
- anxiety
- panic
- hallucinations
- increased sociability/or withdrawal
- euphoria
- perceptual changes – time & spatial
- slurred speech.

B. **Physiological**
- red eyes
- dry mouth
- poor muscle coordination
- delayed reaction time
- increased appetite
- psychomotor impairment
- increased respiratory rate
- increased blood pressure
- lack of convergence of eyes.[25]

Worker impairment can be addressed, but it takes knowledge, expertise and available resources. Regardless of the cause, impairment can be effectively identified and addressed by Occupational Health Nurses (OHNs) using the Occupational Health (OH) Nursing process and practices.

Impairment: Assessment

OHNs are qualified to undertake a systematic, rational method of planning and providing individualized nursing care. A patient-centred, goal-oriented method of "caring", the nursing process involves five major steps:

- Assessment (of company/worker's needs);
- Diagnosis (of human response needs that nursing can assist with);

---

[25] K. Hines, *Business Concern: Potential Work Limitations and Restrictions* (Fort McMurray, AB: Hines Health Services, 2019).

- Planning (of company/worker's care);
- Implementation/intervention (of care); and
- Evaluation (of the success of the implemented care).

This problem-solving process enables the OHN to determine the degree to which the substance-induced state impairs the worker's performance. Knowing the physical and cognitive demands of the worker's "own" job, the OHN can determine the degree of dissonance between the work demands and the worker's capabilities. If deemed impaired, then the OHN can activate corporate policies to eliminate the risk of having an impaired worker from the workplace. Secondly, the OHN can assist the employee to obtain appropriate medical and psychological assessment and treatment. Thirdly, when deemed recovered, the OHN can determine if the worker is indeed fit to work in his/her "own" job. With this fit-to-work information, the employer can effectively address the situation, support the employee, and uphold their due diligence.

D. MARIJUANA USE: MANAGEMENT IN THE WORKPLACE

According to Canadian OH&S law, employers have a duty to provide a safe and healthy workplace; and employees are required to protect the health and safety of themselves, co-workers, and the public, which means, they do not have the right to be impaired in the workplace. Hence, the employer has the right to prohibit the possession and use of marijuana in all forms, in the workplace by creating policies.

The purpose of marijuana management is to:

- Provide workers with ability to increase control over their physical, psychological, and emotional needs;
- Enable improvement of worker physical, psychological, and emotional health and well-being;
- Encourage activities directed at worker well-being; and
- Actualize the health potential of workers, families, and the organization.

The Employer Duty

The employer must develop an impairment policy, or add to their existing substance use policy, rules regarding the use and abuse of marijuana. As recommended by the Canadian Centre for Occupational Health & Safety (CCOHS) in 2019, the elements of an impairment policy should:

- Define impairment;
- Explain self-identification or reporting impairment;
- Address the company's confidentiality practices;
- Indicate the company's rule on possession of substance at the workplace. This is critical given that 22 per cent of employees indicated that they did

not know if a workplace policy existed or not;[26]
- Educate staff on the substance and its hazards;
- Educate workers of their legal duty to notify the employer of any impairment (British Columbia, Yukon); to not enter/stay at work if impaired (British Columbia, Yukon, Newfoundland); and/or adhere to industry-specific provisions (Saskatchewan, Manitoba, Ontario, Nunavut, Northwest Territories, Federal);
- Outline the employee support services offered (*e.g.*, Employee Assistance Program, Human Resources (HR) supports, Occupational Health and Safety (OH&S) Program);
- Address impairment testing – what it entails, and when it will be initiated;
- Explain the enforcement of this policy and the consequences of an infraction; and
- Explain how substance addiction or medical marijuana use will be managed.[27] "In most jurisdictions, human rights laws require the accommodation of employees with medical needs or disabilities (including disabilities from substance dependence)".[28] This includes reinforcing the fact that despite being prescribed medical marijuana, the employee cannot violate the company OH&S policy or safe work practices.

From a due diligence perspective, the employer should work collaboratively with the JHSC to create and implement a risk assessment and risk management plan, including worker impairment. According to Dentons (2017), the employer has the legal right to regulate consumption, possession, and trafficking of marijuana at work.[29]

Impairment due to recreational use of marijuana would be treated as a culpable action. As such, if the substance policy stated that the employees who report to work impaired for whatever reason, they may be terminated for violation of this policy.

Impairment due to the use of medicinal marijuana must be managed in accordance to how the company deals with any other disease state. The Canadian Human Rights legislation requires work accommodation if the employee asks to be

---

[26] IWH, "Cannabis Use and the Canadian Workplace" (July 2019), online: https://www.iwh.on.ca/publications/cannabis-use-and-canadian-workplace-pre-legalization?utm_source=iwhnews&utm_medium=email&utm_campaign=iwhnews-2019-11 (date accessed February 28, 2020).

[27] CCOHS, "Impairment: Cannabis in the Workplace" (2019) *CCOHS*, online: www.ccohs/ca/awareness (date accessed: January 31, 2020).

[28] *Ibid.*, at 3.

[29] Dentons, The effects of cannabis legalization on the workplace (2017), online: https://www.dentons.com/en/insights/articles/2018/february/14/the-effects-of-cannabis-legalization-on-the-workplace.

accommodated and provides proof of disability and the terms of reasonable work accommodation.

Employers must also address international travel and the use of medical marijuana. International and even cross-provincial travel prohibits the transport of medical (and recreational) marijuana.[30] Given a global marketplace, employee education is critical for business reasons.

Given that medical marijuana does not have a Drug Identification Number (DIN), prescription medication benefit coverage does not apply. That absolves the employer of paying for medical marijuana.

The Employee Duty

The employee has a legal duty to advise their employer of their functional state, and/or to not show up to work impaired, regardless of the reason – a disease state, a substance-induced state, fatigue, or a diminished capacity state. By working with the employer and in accordance with the company's substance use policy, the employee using medical marijuana can forge a workable situation to remain in the workplace.

The employee must comply with other company policies such as the non-smoking policy, respectful workplace, *etc*. Smoking medicinal marijuana in the workplace must be compliant with the company rules. Under no circumstances is it legal or permissible for the employee to enter the workplace in an impaired state, regardless of cause. Hence, to reiterate, the worker must be fit for duty, not use/possess substances in the workplace, and inform the employer of any type of impairment.

E. MARIJUANA USE: IMPACT ON THE WORKPLACE

Although the use of medical marijuana has been legal for years in Canada, experience with the legalization of recreational marijuana is new. The impact on Canada's employers is unfolding. Based on what is currently known, 13 per cent of regular cannabis users (514,000) report ingestion before and during work.[31]

Based on available literature from the U.S. and other countries that have legalized marijuana, the following observations are presented:

- **Impaired driving:** The use of marijuana leads to a "lack of concentration, impaired learning and memory, alterations in thought formation and expression, drowsiness, and sedation".[32] The result for the marijuana user

---

[30] B. Kwasniewski, *Clearing the haze: Managing Cannabis in the workplace in Ontario* (Toronto, ON: Carters Professional Association, 2018).

[31] Statistics Canada, *National Cannabis Survey, first quarter 2019* (2019), online: https://www150.statcan.gc.ca/n1/daily-quotidien/190502/dq190502a-eng.htm (date accessed May 2019).

[32] J. Phillips, M. Holland, D. Baldwin, L. Giffiord-Meuleveld, K. Mueller, B. Perkinson, M. Upfal & M. Dreger, "Marijuana in the Workplace: Guidelines for Occupational Health

is a lack of attentiveness, of perception of time and speed, and of the ability to recall past driving experiences, when operating a vehicle.[33] This was demonstrated using driving simulators and actual on-the-road driving.[34] In terms of the risk of a motor vehicle crash, "drivers who test positive for THC, were 3-6 times more likely to crash their vehicles".[35] The result is impaired performance that can result in increased risk of injury, and the related costs.

The legalization of recreational marijuana in the U.S. is strongly associated with an increased number of non-fatal motor vehicle crashes. Between 2010-2013, the rate of vehicle crashes involving marijuana increased by 300%.[36] Additionally, the National Transportation Safety Board reports that the 2018 crash of a church bus in Texas that killed 12 people, involved a 20-year old driver who was under the influence of marijuana and a sedative.[37]

In Canada, the use of marijuana has resulted in a two-fold risk of a vehicular crash and an even higher risk of a fatal crash.[38] According to Woodall, 27% of drivers who died in a vehicular crash in Ontario, 2015, tested positive for cannabis.[39]

Professionals and Employers" (2015) 57:4 *JOEM* 459 at 461.

[33] *Ibid.*

[34] *Ibid.*

[35] *Ibid.*, at 461.

[36] National Safety Council, "Marijuana at work: What employers need to know", *National Safety Council* (November 22, 2018), online: www.nsc.org/home.

[37] J. Christensen, "States that legalized recreational weed see increase in car accidents, studies say" *CNN* (October 18, 2018), online: https://www.cnn.com/2018/10/18/health/marijuana-driving-accidents-bn/index.html.

[38] D.J. Beirness, E.E. Beasley & McClafferty, *Alcohol and drug use among drivers in Ontario: findings from the 2014 roadside survey* (report prepared for the Ontario Ministry of Transportation, 2015) [unpublished]; D. Beirness & A. Porath-Waller, *Clearing the Smoke on Cannabis: Cannabis Use and Driving – An Update* (2017), online: https://www.ccsa.ca/sites/default/files/2019-04/CCSA-Cannabis-Use-Driving-Report-2017-en.pdf; S. Brown, W. Vanlaar & R. Robertson, *The Alcohol and Drug-Crash Problem in Canada 2014 – Report* (2015), online: https://www.ccmta.ca/images/publications/pdf//2014_Alcohol_and_Drug_Crash_Problem_Report.pdf; K. Woodall, B. Chow, A. Lauwers & D. Cass, "Toxicological findings in fatal motor vehicle collisions in Ontario, Canada: A one-year study" (May 2015) 60:3 *J. Forensic Sci.* 669.

[39] K. Woodall, B. Chow, A. Lauwers & D. Cass, "Toxicological findings in fatal motor vehicle collisions in Ontario, Canada: A one-year study" (May 2015) 60:3 *J. Forensic Sci.* 669.

- **Increased absenteeism and presenteeism rates and costs** were noted.[40] Employees who tested positive for marijuana missed more work than those who tested negative. Their absenteeism was 85% greater.[41] Likewise, their productivity decreased.

- **Increased work incidents:** The National Institute on Drug Abuse reports that employees who tested positive for marijuana had 55% more work accidents and experienced 85% more injuries. Why? Marijuana affects the critical functions of the human brain, four of which are essential to working safely, namely the:

    - *Basal ganglia* which is involved in motor control, planning, and initiating and terminating actions;
    - *Cerebellum* which is the centre for motor control and coordination;
    - *Hippocampus* that is critical for memory and learning facts, sequences and places;
    - *Neocortex* which is responsible for higher-cognitive functions and integrating sensory information.[42]

    Based on U.S. data (2016) where drug testing is legal, more workers came to work under the influence of marijuana in the three states in which recreational marijuana was legalized – Nevada (43%); Massachusetts (14%) and California (11%).[43]

- **Increased involuntary job loss:** The frequency of marijuana use is directly correlated with job termination,[44] a costly element of staff turnover for the employer.

- **Increased mental health:** Reportedly, marijuana can have a significant impact on the user's mental health. Health Canada indicates that for some frequent users, marijuana can increase the risk of psychosis, schizophrenia, depression, anxiety disorders, and suicide.[45] Cannabis-use disorder occurs

---

[40] J. Phillips, M. Holland, D. Baldwin, L. Giffiord-Meuleveld, K. Mueller, B. Perkinson, M. Upfal & M. Dreger, "Marijuana in the Workplace: Guidelines for Occupational Health Professionals and Employers" (2015) 57:4 *JOEM* 459.

[41] National Safety Council, "Marijuana at work: What employers need to know", *National Safety Council* (November 22, 2018), online: www.nsc.org/home.

[42] B. Demers, "Four Things employers need to know about marijuana", *WSPS* (August 2018) at 1.

[43] *Ibid.*

[44] C. Okechukwu, J. Molino & Y. Soh, "Associations between marijuana use and involuntary job loss in the United States" (2019) 61:1 *JOEM* 21.

[45] Health Canada, "Cannabis and mental health", *Health Canada* (2018), online: https://www.canada.ca/en/health-canada/services/drugs-medication/cannabis/health-effects/

in 25-50% of daily users. This addiction is associated with decreased work performance, increased absenteeism, behavioural changes (mood swings, irritability, negativity and spontaneous anger), and decreased social skills (not listening, uncooperative, argumentative, disruptive relationships), judgement errors, deteriorated personal appearance, and/or seeking more cannabis.[46]

- **Social effects:** Heavy marijuana users experience lower income, greater welfare dependence, unemployment, criminal behavior, and lower-life satisfaction.[47,48] Maternal use of marijuana is associated with low-birth weight babies and the related post-natal health risks.[49] Evidence of marijuana-triggered schizophrenia and psychosis has been documented.[50] The result is increased pressure on the Canadian healthcare system.

- **Increased need for Disability Management Services:** With increased absenteeism, Workers' Compensation claims, and psychological disabilities, Disability Case Management is needed to mitigate the related human and financial losses and costs. Some of the noted work limitations and restrictions include:

    - No flying;
    - No emergency services driving, driving in the line of duty, driving commercial transport;
    - No use of firearms or taking firearms training;
    - No defensive tactics training;
    - No operational duties;
    - No tasks where sudden incapacitation is of danger to self or others;
    - No shift/night work;
    - No situation involving physical and/or emotional confrontation;

---

mental-health.html (accessed January 31, 2020).

[46] M. Snider-Adler, *A Guide to Cannabis in the Workpalce, Part 1: Recreational Cannabis* (2018), online: from: www.cannabis@DriverCheck.ca.

[47] D. Fergusson & J. Boden, "Cannabis use and later life outcomes" (2008) 103:6 *Addict Abingdon Engl.* 969 at 977-978 (accessed January 31, 2020).

[48] J. Brook, J. Lee, S. Finch N. Seltzer & D. Brook, "Adult work commitment, financial stability, and social environment as related to trajectories of marijuana use beginning in adolescence" (2013) 34:3 *Subst. Abuse* 298.

[49] CCOHS, *Workplace Strategies: Risk of Impairment from Cannabis*, 3d ed. (2018) at 7. online: https://www.ccohs.ca/products/publications/Cannabis_pub_19.pdf.

[50] Di Forti, *et al.* (2019), in CCOHS, *et al.*, eds., *Workplace Strategies: Risk of Impairment from Cannabis*, 3d ed., (2018) at 7. online: https://www.ccohs.ca/products/publications/Cannabis_pub_19.pdf.

- No emergency response duties;
- No cognition-critical duties;
- No decision-critical duties;
- No safety-sensitive duties;
- No international travel.[51]

F. EMPLOYER RESPONSE

With the legalization of recreational and medical marijuana, came interesting challenges for Canadian employers. They need to balance the legal aspects (*e.g.*, Canadian Human Rights legislation, privacy legislation, OH&S legislation) with their business needs; as well, the employee disciplinary and supportive efforts need to be balanced. Hence, employers must position themselves to respond appropriately before faced with a marijuana-related situation. Overwhelmingly, employers are advised to implement or update their current alcohol and drug policies and procedures. Given the complexities of this tasks, legal counsel is strongly recommended.

The corporate alcohol and drug policy should accomplish the following measures:

1. It should clearly state the company's position on alcohol and drug use within the workplace.
2. The company's definition of impairment must be stated.
3. The policy should explain and set-in action procedures that are to be followed should an alcohol/drug situation occur. This would include how to deal with an employee presenting at the workplace "under-the-influence", how to respectfully approach and remove the worker from the worksite, how and when to initiate alcohol/drug testing, and the importance of upholding confidentiality throughout the process.
4. Cause testing – testing because of a serious incident or near-miss, must be explained and a procedure for doing so developed.
5. The procedure for dealing with a positive test result must be explained so that employees are clear on the consequences. Of course, employee confidentiality on the test results must be upheld.
6. The policy should extend to workforce education – education for supervisory staff, HR practitioners, OH&S practitioners, Disability Management practitioners, and the entire workforce, including contractors and part-time/seasonal personnel. They need to know about the effects of the various agents – alcohol, illicit drugs, marijuana, prescription medications, over-the-counter medication, and drug combinations; how to identify the signs

---

[51] K. Hines, *Business Concern: Potential Work Limitations and Restrictions* (Fort McMurray, AB: Hines Health Services, 2019).

of impairment; and how to respond to suspected employee impairment.
7. Provisions for ongoing education must exist.
8. A reporting procedure for notifying Management of a co-worker believed to be impaired must be established and explained. It must be emphasized that this type of reporting is each worker's duty to uphold workplace health and safety and would not result in any negative ramifications.
9. The alcohol and drug policy must be vetted with union(s), various staff positions, different levels of management and legal counsel to ensure that it is comprehensive and in accordance with the applicable legislation.
10. This policy should be reviewed annually or when necessary, by Senior Management, and updated when required.

Additionally, the employer should develop and implement a separate Fit for Duty Policy/Impairment Policy or add it to the existing alcohol and drug policy. **Fit for Duty** has been defined by the Energy Safety Canada as:

A condition in which an employee's physical, physiological and psychological state enables them to continuously perform assigned tasks safely. It encompasses physical requirements, psychological conditions, and psychological status (Figure 32.1).

Figure 32.1: Fit for Duty – Energy Safety Canada[52]

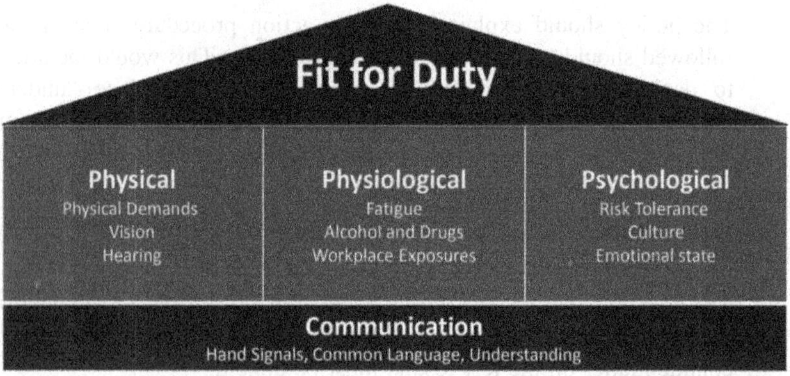

Fit for Duty		
Physical	**Physiological**	**Psychological**
Physical Demands	Fatigue	Risk Tolerance
Vision	Alcohol and Drugs	Culture
Hearing	Workplace Exposures	Emotional state
Communication Hand Signals, Common Language, Understanding		

This model addresses Management's obligations in ensuring worker fitness for duty. The worker's physical, physiological, psychological, and communication capabilities are assessed to determine fitness to safely do the job. Including this concept in the organization's/company's Fit-for-Duty requirements and policy is strongly recommended.

---

[52] Energy Safety Canada, *Fit for Duty Model* (2019), online: https://www.energysafetycanada.com/Resources/Guidelines-Reports/Fit-For-Duty. Reprinted with permission from ESC.

There are a number of Fit-for-Duty Policy templates available, for example:
- Queen's University, Fitness for Work Guideline (http://www.queensu.ca/humanresources/policies/fitness-work-guideline)
- Newgold Blackwater, Fitness for Duty Program (https://www.ceaa.gc.ca/050/documents/p80017/104417E.pdf)
- Model Policy On Fitness For Duty & Substance Abuse (Ontario Version) (https://ohsinsider.com/wp-content/uploads/2017/11/Fitness-For-Duty-Substance-Abuse-Ontario.pdf)
- TTC's Fitness for Duty: Expectations for Contractors Procedure (https://www.ttc.ca/TTC_Business/Materials_and_procurement/About_Us/Contractor_Consultant_Reference_Materials/Fitness_for_Duty.jsp).

The rationale for these policies is that the Supreme Court of Canada ruled that an employer may terminate a worker for 'just cause' when he [she] attends work under the influence of drugs and violates the corporate Fit for Duty Policy.[53]

How to Deal with an Employee Presenting as being "Under-the-influence"

CCOHS recommends that companies educate and ascertain that supervisors and employees know the signs of impairment and what actions to take. When dealing with a suspected impairment, it is critical to address the employee/worker respectfully and in an empathic, non-judgmental manner. Move to a private area and in the presence of a witness, indicate to the employee/worker the concern about his/her observed behaviours. Seek an explanation from the employee/worker as to what has been noted. If warranted and in accordance with the corporate alcohol and drug policy, notify Senior Management and the union and implement the testing procedure. Make sure the employ/worker is aware that the test results take about two days to attain. Remind the employee/worker of the consequences of a positive result and encourage them to seek help from the company's EAP and the services. Providing the employee/worker with information on the EAP is advisable.

Never let the employee/worker drive home; rather, arrange for transportation directly to their home.

A Positive Test Result

Termination of an employee using recreational substances, including marijuana, has not been readily upheld in the Canadian courts, especially if that employee can perform as expected.[54] Rather, the degree of impairment is of concern. If impairment in the workplace is evident, the employer's option is to initiate

---

[53] N. Keith, "Employee violated fitness for duty policy", *Canadian Law* (September 19, 2017), online: https://www.canadianlawyermag.com/inhouse/news/opinion/employee-violated-fitness-for-duty-policy/274504.

[54] S. Brown, "Weed in the Workplace: An Employer's Road Map", *Canadian HR Reporter* (Nov 2016) at 3, 7.

progressive discipline.[55] Having done so, and the employee continues to demonstrate impairment, then termination may be feasible.[56]

Impairment due to medical marijuana use is not acceptable and for safety reasons, must be addressed; however, because of the establishment of a disability, this situation falls under the Canadian Human Rights legislation and the duty to accommodate. Yet, it is up to the employee to advise the employer of the disability and the need for work accommodation, along with information of their medical limitations and capabilities.

As for addiction and substance dependency, they are considered to be disease states warranting work accommodation. Yet, there is no requirement to accommodate recreational substance use; that is a culpable action on the part of the employee.

For more information, refer to: *CHR Commission: Impaired at Work: Guide to Accommodating Substance Dependence*,[57] an employer tool prepared by the Canadian Human Rights Commission to help employers manage substance dependency in the workplace.

G. LEGAL RESPONSE

The Canadian legal response to substance-use infractions in the workplace is unfolding. In essence, the OH&S legislation trumps cannabis use. For example, in *Stewart v. Elk Valley Coal Corporation* (2017) the Court ruled that the company's zero-tolerance policy against substance use was not discriminatory. Elk Valley Coal Corporation's policy required employees to disclose substance dependency/addiction. If a positive drug test occurred, then termination would ensue. Stewart neglected to do so and post-incident, he tested positive for cocaine. At that point, he claimed addiction to cocaine. However, his employment was terminated.

> The Alberta Human Rights Tribunal held that S [Stewart] was terminated for breaching the policy, not because of his addiction. Its decision was affirmed by the Alberta Court of Queen's Bench and by the Alberta Court of Appeal.[58]

Although this ruling dealt with cocaine use, it sets a precedent regarding how cannabis infractions might be addressed.

The *Aitchison v. L&L Painting and Decoration Ltd.* (2018) mirrors this decision. Aitchison was determined to have unilaterally decided to self-medicate with cannabis while at work. His supervisor observed him smoking what was believed to be cannabis while working on an elevated swing-stage, untethered to the stage, and not wearing his hard hat. The Human Rights Tribunal determined that:

---

[55] *Ibid.*

[56] *Ibid.*.

[57] Canadian Human Rights Commission, Minister of Public Works and Government Services, *CHR Commission: Impaired at Work: Guide to Accommodating Substance Dependence* (2017).

[58] *Stewart v. Elk Valley Coal Corp.*, 2017 SCC 30, [2017] 1 S.C.R. 591 at para. 1 (S.C.C.).

... the applicant does not have an absolute right to smoke marijuana at work regardless of whether it is used for medicinal purposes. His actions represented a genuine health and safety risk given the safety-sensitive nature of the job site.[59]

In *French v. Selkirk Logging* (2015), a logging contractor; French was a heavy equipment operator who was diagnosed with cancer and smoking marijuana for pain management. French did not have the legal authorization to possess marijuana, but his doctor had told him that he could use it at work. French smoked marijuana regularly in the workplace. Then one day, while driving a company truck, he struck a moose. As part of the incident investigation, marijuana was found in truck. Subsequently, French's employment was terminated. The employer claimed that French quit his job and refused the invite to return to work drug-free. The Human Rights Tribunal accepted that French was disabled and used marijuana to manage pain resulting from his disability, however he did not have the required medical authorization to lawfully possess and use marijuana for medical purposes. The decision was that French was engaging in an illegal act at the workplace.

These rulings indicate that the employee does not have the absolute right to use cannabis or other substances in the workplace regardless of their medical reasons, and certainly not if doing so, violates company policies or OH&S legislation.

The *Lower Churchill v. IBEW 1620* (2018)[60] case represents a significant decision for employers, regarding safety-sensitive positions and the duty to accommodate the worker up to the point of undue hardship. In this case, the arbitrator dismissed a grievance by the unionized employee who was unsuccessful in getting hired due to potential impairment from the authorized use of medical marijuana. Based on current drug-testing technology, the employer was unable to measure the worker's degree of impairment, and therefore, was unable to measure and manage the safety risks constituting an undue hardship for the employer. The rationale used was that residual marijuana impairment lasts up to 24 hours, constituting undue hardship for safety-sensitive positions.

What does all this mean?

- Employers have a duty to accommodate employees who are legally using medical marijuana up to the point of undue hardship.
- If approached by an employee seeking accommodation relating to the use of medical marijuana, the employer should first confirm it is medically and legally authorized.

---

[59] *Aitchison v. L. & L. Painting and Decorating Ltd.*, 2018 HRTO 238 at para. 1 (H.R.T.O.), online: https://www.canlii.org/en/on/onhrt/doc/2018/2018hrto238/2018hrto238. html?searchUrlHash=AAAAAQBHQWl0Y2hpc29uIHYuIEwgJiBMIFBhaW 50aW5nIGFuZCBEZWNvcmF0aW5nIExkZC4sIDIwMTggSFJUTyAyMzggKENhbbkxJSSkAAAAAAQ&resultIn« 1 (date accessed January 31, 2020).

[60] https://www.mcinnescooper.com/wp-content/uploads/2018/06/International-Brotherhood-Lower-Churchill-Transmission-Construction-Employers%E2%80%99-Assn.-Inc.-and-IBEW-Local-1620-Tizzard-Re.pdf

- If so, then employer should seek medical information to determine if employee can safely continue performing his/her duties (from employee's doctor or an independent medical examination).
- If the information reveals that the employee would be impaired in the workplace, then the employer is likely not required to accommodate the employee's request, especially if the employee is in a safety-sensitive position.

H. SUMMARY

With the legalization of marijuana, Canadian employers faced an unprecedented challenge to workplace health and safety. Yet, many employers had in place policies and procedures to address workplace hazards; impairment had always been a hazard and marijuana just added one more dimension to hazard control.

Legally, workers must come to work fit-for-duty. A prescription for medical marijuana does not entitle substance-dosing in the workplace, endangering the safety of other workers or themselves, and/or arriving to work late or leaving early. As well, they must comply with the corporate policies and procedures.

It is the employer's duty to have and enforce a zero tolerance of marijuana use or possession policy, a substance use/abuse policy, and Fit-for-Duty policies. As well, the employer has a duty to accommodate the disabled employee up to the point of undue hardship, regardless of the nature of the disability. Yet, in the case of safety-sensitive positions, the use of medications/drugs can result in undue hardship based on safety grounds.

In essence, the legalization of marijuana in Canada brought to the forefront employee/worker Fitness-to-work, Fit-for-duty, and work accommodation issues that have existed in the shadows of the employer-employee relationship for years. Greater attention and clarity have certainly been warranted. Hopefully, as Canada moves forward with the legalization process, employer practices will make the workplace safer and healthy.

CHAPTER REFERENCES

Aitchison v. L. & L. Painting and Decorating Ltd., 2018 HRTO 238 (H.R.T.O.), online : https://www.canlii.org/en/on/onhrt/doc/2018/2018hrto238/2018hrto238. html?searchUrlHash=AAAAAQBHQWl0Y2hpc29uIHYuIEwgJiBMIFBhaW 50aW5nIGFuZCBEZWNvcmF0aW5nIEx0ZC4sIDIwMTggSFJUTyAyMzggKENhb kxJSSkAAAAAAQ&resultIndex=1(date accessed January 31, 2020).

D.J. Beirness, E.E Beasley & McClafferty, *Alcohol and drug use among drivers in Ontario: findings from the 2014 roadside survey* (report prepared for the Ontario Ministry of Transportation, 2015) [unpublished].

D. Beirness & A. Porath-Waller, *Clearing the Smoke on Cannabis: Cannabis Use and Driving - An Update* (2017), online: https://www.ccsa.ca/sites/default/files/ 2019-04/CCSA-Cannabis-Use-Driving-Report-2017-en.pdf.

J. Brook, J. Lee, S. Finch N. Seltzer & D. Brook, "Adult work commitment, financial stability, and social environment as related to trajectories of marijuana use beginning in adolescence" (2013) 34:3 *Subst. Abuse* 298.

S. Brown, W. Vanlaar & R. Robertson, *The Alcohol and Drug-Crash Problem in Canada 2014 - Report* (2015), online: https://www.ccmta.ca/images/publications/pdf//2014_Alcohol_and_Drug_Crash_Problem_Report.pdf

S. Brown, "Weed in the Workplace: An Employer's Road Map", *Canadian HR Reporter* (November 2016) at 3, 7.

A.J. Budney, J.D. Sargent & D.C. Lee, "Vaping cannabis (marijuana): parallel concerns to e-cigs?" (2015) 110:11 *Addiction* 1699.

CCOHS, *Workplace Strategies: Risk of Impairment from Cannabis*, 3d ed. (2018) at 7. online: https://www.ccohs.ca/products/publications/Cannabis_pub_19.pdf

CCOHS, "Impairment: Cannabis in the Workplace" (2019), online: www.ccohs/ca/awareness (date accessed: January 31, 2020).

Canada, "Cannabis for medical purposes under the Cannabis Act: information and improvements" (2019), online: https://www.canada.ca/en/health-canada/services/drugs-medication/cannabis/medical-use-cannabis.html.

Canadian Human Rights Commission Minister of Public Works and Government Services, *CHR Commission: Impaired at Work: Guide to Accommodating Substance Dependence* (2017).

Cann Trust/MMG, *Marketing Brochure* (Vaughan, ON: Cann Trust/MMG, 2016).

Center for Disease Control (CDC), "Marijuana: How Can It Affect Your Health?", *Health Effects* (2018), online: https://www.cdc.gov/marijuana/health-effects.html. Accessed January 31. 2020.

J. Christensen, "States that legalized recreational weed see increase in car accidents, studies say", *CNN* (October 18, 2018), online: https://www.cnn.com/2018/10/18/health/marijuana-driving-accidents-bn/index.html.

B. Demers, "Four Things employers need to know about marijuana", *WSPS* (August 2018) at 1.

Dentons, The effects of cannabis legalization on the workplace (2017), online: https://www.dentons.com/en/insights/articles/2018/february/14/the-effects-of-cannabis-legalization-on-the-workplace.

Dictionary.com (2017), *s.v.* "Impairment", online: http://www.dictionary.com/browse/impairment. Accessed January 31. 2020.

Energy Safety Canada, *Fit for Duty Model* (2019), online: https://www.energysafetycanada.com/Resources/Guidelines-Reports/Fit-For-Duty.

French v. Selkin Logging, 2015 BCHRT 101.

W. Hall, "Adverse health effects of non-medical cannabis use" (2009) 374:9698 *Lancet* 1383.

K. Hines, *Business Concern: Potential Work Limitations and Restrictions* (Fort McMurray, AB: Hines Health Services, 2019).

IWH, "Cannabis Use and the Canadian Workplace" (July 2019), online: https://www.iwh.on.ca/publications/cannabis-use-and-canadian-workplace-pre-legalization?utm_source=iwhnews&utm_medium=email&utm_campaign=iwhnews-2019-11.

G. Jackowski, "Biology of Marijuana" (presented at the Schedule 2 Employers Group Conference, October 2016, Richmond Hill, ON), online: http://www.s2egroup.com/library-2/2016-conference-presentations/ (date accessed February 28, 2020).

N. Keith, "Employee violated fitness for duty policy", *Canadian Law* (September 19, 2017), online: https://www.canadianlawyermag.com/inhouse/news/opinion/employee-violated-fitness-for-duty-policy/274504.

B. Kwasniewski, "Clearing the haze: Managing Cannabis in the workplace in Ontario" (Toronto, ON: Carters Professional Association, 2018).

C. Lopez-Quintero, *et al.*,"Probability and predictors of transition from first use to dependence on nicotine, alcohol, cannabis, and cocaine: results of the National Epidemiologic Survey on Alcohol and Related Conditions (NESARC)" (2011) 115:1-2 *Drug Alcohol Depend.* 120.

International Brotherhood of Electrical Workers, Local Union 1620 v. Lower Churchill Transmission Construction Employers' Association Inc. and Valard Construction LP (2018).

Merriam-Webster Dictionary (2017), *s.v.* "Impairment", online: https://www.merriam-webster.com/dictionary/impaired. Accessed January 31. 2020.

MMG/CannTrust, Product Monograph (presented at the Schedule 2 Employers Group Conference, October 2016, Richmond Hill, ON), online: https://canntrust.ca/resources-toolkits/ (date accessed February 28, 2020).

A. Nakim, "Medical Marijuana: In the Workplace" (presented at the Schedule 2 Employers Group Conference, October 2016, Richmond Hill, ON), online: http://www.s2egroup.com/library-2/2016-conference-presentations/ (date accessed February 28, 2020).

National Safety Council, "Marijuana at work: What employers need to know", *National Safety Council* (November 22, 2018), online: www.nsc.org/home (date accessed February 28, 2020).

C. Okechukwu, J. Molino, J. & Y. Soh, "Associations between marijuana use and involuntary job loss in the United States" (2019) 61:1 *JOEM* 21.

J. Phillips, M. Holland, D. Baldwin, L. Giffiord-Meuleveld, K. Mueller, B. Perkinson, M. Upfal & M. Dreger, "Marijuana in the Workplace: Guidelines for Occupational Health Professionals and Employers" (2015) 57:4 *JOEM* 459.

M. Snider-Adler, *A Guide to Cannabis in the Workplace, Part 1: Recreational Cannabis* (2018), online: www.cannabis@DriverCheck.ca

Statistics Canada, *National Cannabis Survey, first quarter 2019*, online: https://www150.statcan.gc.ca/n1/daily-quotidien/190502/dq190502a-eng.htm (date accessed May 2019).

Stewart v. Elk Valley Coal Corp., 2017 SCC 30, [2017] 1 S.C.R. 591 (S.C.C.)

Suncor Energy Inc. v. Unifor, Local 707A, 2018.

Cannabis Act, S.C. 2018, c. 16.

Wisconsin Department of Health Services, *News Release*, "Majority of Wisconsin Lung Disease Patients Who Reported Vaping Cite THC Products: The investigation of lung disease among people who reported vaping is ongoing" (August 29, 2019), online: https://www.dhs.wisconsin.gov/news/releases/082919.htm.

K. Woodall, B. Chow, A. Lauwers & D. Cass, "Toxicological findings in fatal motor vehicle collisions in Ontario, Canada: A one-year study" (May 2015) 60:3 *J. Forensic Sci.* 669.

World Health Organization, "Disabilities", *Health Topics, WHO* (2017), online: http://www.who.int/topics/disabilities/en/. (date accessed January 31, 2020).

Chapter 33

OCCUPATIONAL HEALTH AND SAFETY: THE CANADIAN WORKERS' COMPENSATION SYSTEM

A. INTRODUCTION

The Canadian Workers' Compensation system is designed to compensate ill/injured workers for work-related diseases and injuries. It is a no-fault insurance system funded entirely by Canadian employers. The provincial Workers' Compensation Acts govern the functioning of the Workers' Compensation Board. It is important to remember that Workers' Compensation benefits are a worker's statutory right. An effective claim management system never attempts to prevent workers from receiving benefits to which they are legitimately entitled.

Historical Perspective

Introduced in the late 1800s and early 1900s, the various Canadian Workers' Compensation Acts were enacted to protect, and support workers injured in the workplace, as well as to protect employers from being sued. They created a "mutual gains and responsibilities system" for Canadian employers and workers.

Workers of that era faced wretched and deplorable workplace conditions. If injured, they had little hope of getting financial compensation. The prevailing belief was that work was hazardous and workers should plan for the eventuality of getting injured. As well, employers were faced with a real risk of being sued by injured workers. These disputes had to be settled in court, tying up the court system and ending up as lengthy, costly ventures. The bottom line was that rarely did the injured worker get compensated, and should the litigation prove successful, rarely did the worker obtain compensation because often the employer would become financially insolvent.

In 1913, the modern Canadian version of a workers' compensation system was born. The Honourable William Meredith, Chief Justice of Ontario, delivered a report that proved to launch the first significant piece of social legislation in Canada. The Meredith Report outlined a trade-off in which workers relinquish their right to sue in exchange for compensation benefits.

B. THE WORKERS' COMPENSATION SYSTEM: INTENT

The Workers' Compensation Boards provide compensation benefits to workers who

are injured or become ill due to work. The intent is to:
- To protect workers from loss of earnings in the event of an accident or disease which arises out of, or occurs during, the course of employment; and
- To protect employers from legal action. In exchange for compensation benefits, ill/injured workers gave up the right to sue their employers (the **Statutory Bar**).

C. THE WORKERS' COMPENSATION SYSTEM: FUNDAMENTAL PRINCIPLES

Sir William Meredith (1913) recommended a compensation system that offers:
- *collective liability* – employers as a collective group are responsible for the payment of injured workers;
- *no fault insurance* – regardless of the injury/illness cause, the worker is covered if it is work-related;
- *security of payment* – the worker's loss of earnings is covered regardless of the employer's financial status;
- *exclusive jursidiction* – the provincial Workers' Compensation Board (WCB) is set up as an independent agency that has the exclusive authority to make decisions on all workers' compensation matters within the province; and
- *an independent board* – the administration of this system of insurance is achieved via an independent agency. The system exists at arm's length from the government and is shielded from political influence, allowing only limited powers to the Minister responsible.

D. THE WORKERS' COMPENSATION SYSTEM: PURPOSE

The Canadian WCB system is a third party, government-operated insurance system designed to protect the injured/ill worker and to afford employers litigation protection. It is a no-fault, industry-funded insurance that is mandatory for certain industry groups. Employers pay all the premiums and claim costs, while employees forfeit their right to take legal action against the employer. Claim administration includes claim submission, claim adjudication, claim appeal, if required, and claim termination.

In addition, Canadian WCBs provide injury prevention information and education; protection from legal action; financial and healthcare benefits; rate-incentive programs; and co-ordination of the partners in the workplace safety and insurance system.

In most instances, Canadian employers are legally required to manage their WCB claims and offer work accommodation. This is best achieved through Disability Claim Management and Disability Case Management. Having an integrated

Disability Management Program in place allows the employer to combine both of these essential elements in all disability situations — occupational illness/injury, non-occupational illness/injury, short-term disability, and long-term disability.

E. THE WORKERS' COMPENSATION SYSTEM: HOW DOES IT WORK?

Although there are differences in the provincial WCB Acts, the wording and intent are similar. What differs is how each province enacts its WCB Act and delivers the related services.

To determine if a worker is eligible for compensation benefits, each claim must be evaluated (adjudicated). **Claim adjudication** is the process of determining whether a worker's claim is eligible under the compensation terms of the WCB. This varies by province. In general, eligibility for WCB compensation coverage involves the determination of some key points:

- Whether or not the worker was in the course of employment;
- Whether or not the injury resulted from a hazard of the employer's premises;
- Whether or not the injury occurred on the employer's premises; and
- Whether the worker was being paid to do the work that caused the injury.

If the claim is accepted, the rate of WCB compensation provided to the injured/ill worker is generally less than the worker's regular salary. That rate of pay varies by province. Similarly, the waiting period for WCB coverage varies – in some provinces, there is no waiting – in others, there is a short period of time without coverage. As well, the available rehabilitation services vary.

Information on WCB Rehabilitation Services is accessible through www.awcbc.org. This website offers provincial comparisons regarding:

1) Workplace Injury Statistics;

   http://www.awcbc.org

2) Rehabilitation Services; and

   http://www.awcbc.org/common/assets/benefits/rehabilitation_services_summary.pdf

3) Law Concerning RTW and Rehabilitation.

   http://www.awcbc.org/common/assets/benefits/rehab_return_to_work.pdf

F. THE WORKERS' COMPENSATION SYSTEM: PLAYERS

Using Alberta's WCB Act as an example, an **Accident** is defined as:

   1 (1) In this Act,

   (a) "accident" means an accident that arises out of and occurs in the course of

employment in an industry to which this Act applies and includes

(i) a wilful and intentional act, not being the act of the worker who suffers the accident,

(ii) a chance event occasioned by a physical or natural cause,

(iii) disablement, and

(iv) a disabling or potentially disabling condition caused by an occupational disease; [1]

an **Employer** is defined as:

(i) an individual, firm, association, body or corporation that has, or is deemed by the Board or this Act to have, one or more workers in the individual's or its service and includes a person considered by the Board to be acting on behalf of that individual, firm, association, body or corporation,

(ii) a proprietor whose application is approved under section 15,

(iii) a corporation where the application of a director of the corporation is approved under section 15, and (iv) a partnership where the application of a partner in the partnership is approved under section 15, and includes the Crown in right of Alberta and the Crown in right of Canada insofar as the latter, in its capacity as employer, submits to the operation of this Act;[2]

A **Worker** is defined as:

"worker" means a person who enters into or works under a contract of service or apprenticeship, written or oral, express or implied, whether by way of manual labour or otherwise, and includes

(i) a learner,

(ii) a person whose application to the Board under section 15 is approved, and

(iii) any other person who, under this Act or under any direction or order of the Board, is deemed to be a worker, but does not include a person who ordinarily resides outside Canada and is employed by an employer who is based outside Canada and carries on business in Alberta on a temporary basis. [3]

Given that many Canadian companies operate nationally, the OH&S/ Disability Management Practitioner/Professional must review the definitions of a worker for each of the applicable WCB Acts.

G. THE WORKERS' COMPENSATION SYSTEM: REPORTING REQUIREMENTS

For a WCB claim to be accepted, the WCB must receive at least three reports – the Worker Report, Employer Report, and Physician Report. These reports vary by

---

[1] Alberta *Workers' Compensation Act*, R.S.A. 2000, c. W-15, s, 1(1.1a), online: http://www.qp.alberta.ca/documents/Acts/W15.pdf. (accessed January 28, 2020).

[2] *Ibid.*, s. 1(1.1.j).

[3] *Ibid.*, s. 1(1.1.z).

province but must be the official WCB forms. The data provided enables the WCB Claim Adjudicator to decide the employee's eligibility for coverage as well as the treatment and rehabilitation services that will be offered.

The Worker Report

The Worker Report must be received within a stated period of time. For example, in Alberta, the Worker Report must be submitted within two years of the injury date. This report can be submitted by the employee or dependents. A late claim is often denied; timeliness is paramount. As well, Worker Reports must be clear and understandable for WCB adjudication. Remember: The WCB adjudicator does not know the worker's job nor the industry in which the worker is employed. The more details, the better prepared the WCB adjudicator will be to decide on the worker's eligibility for WCB insurance coverage.

Insurance plans, regardless of their scope, all seek subrogation – that is, repayment of disability funds if possible, through some avenue. The typical approach is to recoup expenses for a claim it paid, when another party should have been responsible for paying at least a portion of that claim. This usually occurs when an employee is injured in a motor vehicle accident. Another example is when long-term disability (LTD) coverage is granted by a third party insurer, the employee is expected to apply for Canada Pension Plan (CPP) Disability Insurance. If granted the CPP Disability coverage, then the insurer reduces its LTD payment by that amount. For more details on the CPP Disability Benefits, refer to: http://www.hrsdc.gc.ca/eng/oas-cpp/cpp_disability/adjudframe/cppadjud.shtml.

The Employer Report

In each province, the employer has a set timeframe within which to report a workplace injury/illness. It is mandatory for the employer to file this report regardless of whether the employee chooses to report the injury or not. Failure by the employer to do so can end up in costly fines in some provinces.

For the employer, the reportable situations include:

- Time loss beyond the day of injury;
- Injuries requiring hospital treatment;
- Motor vehicle accidents;
- Injuries resulting in modified work accommodation;
- Eye glass damage;
- Dental damage;
- Injuries requiring physiotherapy, chiropractic treatment;
- Injuries requiring medication and dressings not covered by the provincial healthcare plan;
- Injuries requiring future time loss, *e.g.*, hernias.

The Physician Report

Once the employee visits the physician and it is determined that the illness/injury is work-related, then the physician has a stated period of time within which to file a Physician's First Report. It must be:

- Completed within two days after evaluating the worker if the injury will or is likely to disable him/her for more than the day of accident, or may cause complications;
- Submitted to the WCB; and
- Free of charge to the worker and employer because the WCB pays the physician.

Follow-up assessments are reported using a companion form – the Physician's Progress Report. It is:

- Completed at each follow-up appointment;
- Submitted directly to the WCB;
- Free of charge to worker or employer; and
- An assessement of the worker's fitness to work.

H. IMPACT OF THE WCB

Workers' Compensation Insurance is required in all Canadian provinces. The operation of the WCBs is based on the Meredith Principles. All WCB claims are adjudicated by the WCB; not by the employer. When a WCB claim is activated, claim management, case management, and return-to-work planning and placement are involved.

The impact that the WCB insurance system has on other disability insurance plans is that:

- All disability insurance plans are mutually exclusive;
- All types of disability insurance require proof of disability;
- The WCB is the "first payer" on claims. As the "first payer", "the WCB's responsibility is to pay medical claim costs and worker compensation for work-related injuries";[4]
- All the disability insurance plans interface. An employee may be accepted on WCB, but should that claim be terminated because the work-related portion of the claim has been paid, then the employee may be eligibile for

---

[4] WCB-Alberta, *Employer Handbook* (2015) at 3, online: https://open.alberta.ca/dataset/082f2544-bcd6-4347-a972-abbf5d7c72cf/resource/757ea160-2639-4287-88ac-f63661f9805e/download/6329973-2015-workers-compensation-board-wcb-alberta-employer-handbook-2015-01.pdf (accessed January 28, 2020).

short-term disability (STD) or LTD if the he/she is unfit to do their "own job".

- Workers on a lengthy, short-term disability can exhaust the STD elimination period, and hence, may have the option to apply for LTD insurance coverage.
- Approval for return-to-work planning rests with employer, the union, the insurer (WCB), and the employee.

To effectively work with the WCB and other disability benefit plans, the Disability Management Practitioner/Professional must determine how to work with the WCB:

- learn about the client's disability coverage, nuances and available benefits;
- determine the eligibility critria;
- recognize WCB's requirment for return-to-work planning and work accommodation; and
- identify the available WCB benefits and servcies that can support a successful return-to-work outcome.

This type of intervention is critical because research indicates that workers injured at work have shown to experience premature mortality.[5] As well, injured shift workers report poorer health outcomes than do injured day workers.[6]

I. HOW ARE WE DOING?

The Association of Workers' Compensation Boards of Canada (AWCBC) reports that in 2018, 264,438 lost-time claims were accepted. Males experienced 60% of the lost-time injury claims; young/new workers (15-24 years old) experienced 12.5% of them. Older workers (60+ years) accounted for 11% of the lost-time injury claims.

As for the industry breakdown of the 2018 lost-time injury claims, the Health and Social Services industry had the most claims – 47,014 (18%), followed by the Manufacturing industry at 14% and the Retail Trade industry at 11%.

There has not been much change in the number of new claims accepted by WorkSafeBC over the years (Figure 33.1):

---

[5] Institute for Work & Health,"Work disability puts people at risk of premature death, study finds" (winter 2015) 79 *At Work*, online: https://www.iwh.on.ca/newsletters/at-work/79/work-disability-puts-people-at-risk-of-premature-death-study-finds (date accessed: January 31, 2020).

[6] *Ibid.*

Figure 33.1: Number of New Injury Claims

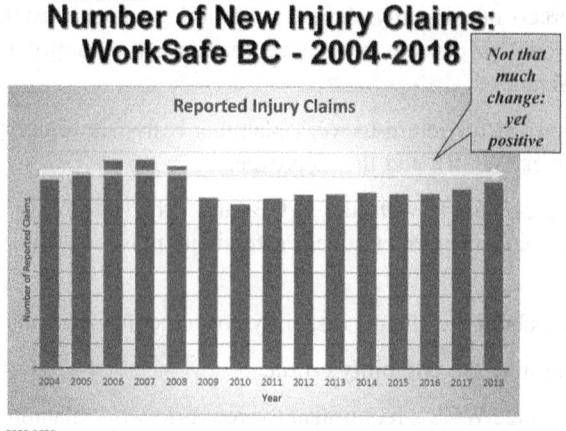

This indicates that the OH&S results have not been as effective as desired. As for the main type of workplace injury, sprain/strain injuries (musculoskeletal injuries) rank as number one; a consistent finding throughout the Canadian provinces.

In terms of worker fatalities, 2.8 Canadian workers die per calendar day (using 365 days); or 4 Canadian workers per workdays (using 250 days) (Figure 33.2).

Figure 33.2: Fatality Rates by Province: 2018

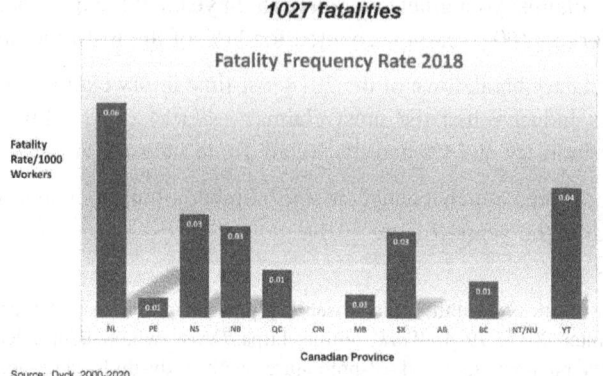

The main industries experiencing worker fatality in Canada were the construction and manufacturing industries. Having this type of information, enables industry associations to actively address their related OH&S issues. However as one can see, there is still much work to be done to make Canadian workplaces safe.

J. SUMMARY

The Canadian Workers' Compensation System has proven to be a strong insurance for ill/injured workers, as well as for employers. Like with any system, there is always room for improvement; yet, overall, it serves Canadians well.

CHAPTER REFERENCES

Alberta Municipal Health & Safety Association (AMHSA), *Leadership for Safety Excellence* (Nisku, AB: AMHSA, 1995).

Alberta *Workers' Compensation Act*, R.S.A. 2000, c. W-15, s. 1(1.1a), online: http://www.qp.alberta.ca/documents/Acts/W15.pdf. (accessed January 28, 2020).

Association of Workers' Compensation Boards of Canada, *Workplace lost-time injuries and fatalities* (2018) (2020), online: www.awcbc.org/statistics (accessed January 28, 2020).

Beckett & Hager (2000), as reported in K. Sitzman, "Adult Learners Reframed" (2006) 54:6 *AAOHN* 292.

F. Bird & G. Germain, *Loss Control Management: Practical Loss Control Leadership*, revised ed. (Loganville, GA: DNV, 1996).

Bongarde Holdings Inc., "How to Demonstrate the Value of Training Programs", *Safety Compliance Insider* (2006) 2:4 at 14, online: http://www.safetysmart.com (date accessed: February 28, 2020).

Carleton (1987), Scholey (1983) and St. Vincent, Tellier & Lortie (1989), cited in M. Colligan and A. Cohen, "The Role of Training in Promoting Workplace Safety and Health" in J. Barling & M. Frone, eds., *The Psychology of Workplace Safety* (Washington, DC: American Psychological Association, 2004).

Chhokar & Williams (1984), Cohen & Jensen (1984), Fox & Sulzer-Azaroff (1987), Komaki, Barwick & Scott (1978), Ray, Purswell & Schlegel (1990), Reber & Wallin (1984), Saarela (1990) and Saari & Nasanen (1989), cited in M. Colligan & A. Cohen, "The Role of Training in Promoting Workplace Safety and Health" in J. Barling & M. Frone, eds., *The Psychology of Workplace Safety* (Washington, DC: American Psychological Association, 2004).

M. Colligan & A. Cohen, "The Role of Training in Promoting Workplace Safety and Health" in J. Barling & M. Frone, eds., *The Psychology of Workplace Safety* (Washington, DC: American Psychological Association, 2004).

C. DeLisle, "What Makes Safety Training Effective?" *The Safety Net* (2014) 24.

D. Dyck, "New Employee OH&S Orientation Checklist" (Calgary, AB: Progressive Health Consulting, 2002).

D. Dyck, "Worker OH&S Training: Teaching Plan" (Calgary, AB: Progressive Health Consulting, 2003).

D. Dyck & D. Rud, "ENMAX Makes Safety Training Fun" (2003) 26:3 *Occupational Health & Safety Magazine*.

R. Flemmer, "The Three E's of Training", *The Safety Net* (October 2014) at 7.

D. Folk, "The Workplace Complacency Trend in Accident Prevention" (January 2007), *Occupational Health & Safety*, online: http://www.ohsonline.com/articles/46371 (date accessed: February 28, 2020).

Fox & Sulzer-Azaroff (1987) and Hopkins (1984) cited in M. Colligan and A. Cohen, "The Role of Training in Promoting Workplace Safety and Health" in J. Barling & M. Frone, eds., *The Psychology of Workplace Safety* (Washington, DC: American Psychological Association, 2004).

M. Harder, "How effective is health and safety training?" *Safety Reporter* (May 2010).

Hatscan, "WHMIS Training" (Presented in Edmonton, AB, February 26-27, 2007) [unpublished].

Hatscan, *Handi-Guide to Alberta's OH&S Act, Regulation and Code* (Edmonton, AB: Hazard Alert Training (Hatscan), 2004).

Institute for Work & Health, "Work disability puts people at risk of premature death, study finds" (winter 2015) 79 *At Work*, online: https://www.iwh.on.ca/newsletters/at-work/79/work-disability-puts-people-at-risk-of-premature-death-study-finds (date accessed: January 31, 2020).

Institute for Work and Health, "Study shows way to make health and safety training for office workers more effective" (January 28, 2014), online: http://www.iwh.on.ca/media/2014-jan-28 (date accessed: February 28, 2020).

Institute for Work and Health, *Effectiveness of OH&S education and training*, online: http://www.iwh.on.ca/sbe/effectiveness-of-ohs-education-and-training (date accessed: February 28, 2020).

Leslie & Adams (1973), Rubinsky & Smith (1971), Bosco & Wagner (1988), Vaught *et al.*, (1988) and Goldrick (1989), cited in M. Colligan and A. Cohen, "The Role of Training in Promoting Workplace Safety and Health" in J. Barling & M. Frone, eds., *The Psychology of Workplace Safety* (Washington, DC: American Psychological Association, 2004).

S. Lieb, "Principles of Adult Learning", online: http://www.lindenwood.edu/education/andragogy/andragogy/2011/Lieb_1991.pdf (date accessed: February 28, 2020).

Linneman, Cannon & DeRonde & Lamphear (1991), Lynch *et al.* (1990), Seto, Ching, Chu & Fielding (1990) and Wong *et al.*, (1991), cited in M. Colligan & A. Cohen, "The Role of Training in Promoting Workplace Safety and Health" in J. Barling & M. Frone, eds., *The Psychology of Workplace Safety* (Washington, DC: American Psychological Association, 2004).

Manitoba Labour and Immigration — Workplace Safety and Health Division, "Employing Young Workers: Tips for Employers and Supervisors", online: http://safemanitoba.com. (date accessed: February 28, 2020).

Maples *et al.* (1982), Lepore, Olson & Tomer (1984) and McKenzie *et al.* (1985), cited in M. Colligan & A. Cohen, "The Role of Training in Promoting Workplace Safety and Health" in J. Barling & M. Frone, eds., *The Psychology of Workplace Safety* (Washington, DC: American Psychological Association, 2004).

McQuiston *et al.* (1994), Brown & Nguyen-Scott (1992), LaMontagne *et al.*, (1992), Michaels *et al.*, (1992), Weiger & Lyons (1992) and Cole & Brown (1996), cited in M. Colligan & A. Cohen, "The Role of Training in Promoting Workplace Safety and Health" in J. Barling & M. Frone, eds., *The Psychology of Workplace Safety* (Washington, DC: American Psychological Association, 2004).

National-Louis University, "Malcolm Knowles: Apostle of Andragogy", online: http://www.umsl.edu/~henschkej/henschke/malcolm_knowles_Apostle_of_andragogy.pdf (date accessed: February 28, 2020).

Occupational Health and Safety Code 2006 (revised 2013) (Edmonton, AB: Government of Alberta — Human Resources and Employment, 2013).

G. Pappas, "Effective Measurement of OSH Training Programs" (presented at the IAPA Conference in Toronto, ON, April 2006) [unpublished].

D. Parkinson *et al.*, "Effectiveness of the United Steel Workers of America Coke Oven Intervention Program" (1989) 31:5 *J. Occup. Med.* 464.

Ray, Purswell & Schlegel (1990), *et al.*, cited in M. Colligan & A. Cohen, "The Role of Training in Promoting Workplace Safety and Health" in J. Barling & M. Frone, eds., *The Psychology of Workplace Safety* (Washington, DC: American Psychological Association, 2004).

T. Robins *et al.*, "Implementation of the Federal Hazard Communication Standard: Does Training Work?" (1990) 32:11 *J. Occup. Med.* 1133.

D. Rud, *Industry application observed at ENMAX Corporation* (Calgary, AB: , 2003).

K. Saarela, "An Intervention Program Utilizing Small Groups: A Comparative Study" (1990) 21 *J. Safety Res.* 149.

Saarela *et al.* (1989), Borland *et al.* (1991) and Karmy & Martin (1980), cited in M. Colligan and A. Cohen, "The Role of Training in Promoting Workplace Safety and Health" in J. Barling & M. Frone, eds., *The Psychology of Workplace Safety* (Washington, DC: American Psychological Association, 2004).

J. Saari & M. Nasanen, "The Effect of Positive Feedback on Industrial Housekeeping and Accidents: A Long-term Study at a Shipyard" (1989) 4 *Int. J. Ind. Ergon.* 201.

Sulzer-Azaroff (1990) and Lynch *et al.* (1990), cited in M. Colligan & A. Cohen, "The Role of Training in Promoting Workplace Safety and Health" in J. Barling & M. Frone, eds., *The Psychology of Workplace Safety* (Washington, DC: American Psychological Association, 2004).

University of Kansas (1982), Fox, Hopkins & Anger (1987) and Zohar & Fussfield (1981), cited in M. Colligan & A. Cohen, "The Role of Training in Promoting

Workplace Safety and Health" in J. Barling & M. Frone, *The Psychology of Workplace Safety* (Washington, DC: American Psychological Association, 2004).

WCB-Alberta *Employer Handbook* (2015) at 3, online: https://open.alberta.ca/dataset/082f2544-bcd6-4347-a972-abbf5d7c72cf/resource/757ea160-2639-4287-88ac-f63661f9805e/download/6329973-2015-workers-compensation-board-wcb-alberta-employer-handbook-2015-01.pdf (accessed January 28, 2020).

Work Safe Alberta Workplace Health and Safety Interactive Quizzes, online: http://work.alberta.ca/occupational-health-safety/268.html (date accessed: February 28, 2020).

WorkSafeBC, *Statistics 2017* (2018), online: https://www.worksafebc.com/en/resources/about-us/annual-report-statistics/2017-stats?lang=en (date accessed: January 31, 2020).

Chapter 34

OCCUPATIONAL HEALTH AND SAFETY: DIVERSITY CONSIDERATIONS

A. INTRODUCTION

Today's workforce is highly diverse. It ranges from single men and women of varying cultural backgrounds with no dependants, to those married with children and caring for elderly parents.[1]

Added to that, we are witnessing increased rates of immigration in Canada and the U.S. For example, in Canada, the current immigration for 2019 was 319,580.[2] In terms of their residence, 44% located in Ontario; 14% in Quebec and in British Columbia, and 13% in Alberta.[3]

The relevance of increased population diversity is that each cultural group has unique values, beliefs, needs, and expectations with regard to life and work. By effectively managing diversity, organizations can enhance their competitive advantage in the Canadian labour market.

B. CULTURE

Culture can be defined as the values, beliefs, customs, behaviours, and structures shared by a group of people.[4] A group may be identified by many criteria such as nationality, religion, geographic origin, language, group history, or life experiences. Although nationality often encompasses several cultures, it is the most commonly used notion of culture.

Culture is not biologically inherited: it is learned and passed on from generation to generation via **enculturation**, the repetitious and systematic inculcation of a

---

[1] T. Buller, "A Flexible Combination", *Benefits Canada* (November 2004) at 99, online: http://www.benefitscanada.ca (date accessed: January 31, 2020).

[2] Statista, "Number of Immigrants in Canada from 2000-2019" (2020), online: https://www.statista.com/statistics/443063/number-of-immigrants-in-canada/ (date accessed: January 31, 2020).

[3] *Ibid.*

[4] H.V. Ngo, *Cultural Competency: A Self-Assessment Guide for Human Service Organizations* (Calgary, AB: Cultural Diversity Institute, 2000), online: http://www.calgary.ca/csps/cns/documents/fcss/cultural_competency_self_assesment_guide.pdf?noredirect=1 (date accessed: January 31, 2020).

shared system of values, beliefs, attitudes, and learned behaviours.[5] All the beliefs, traditions, language, values, customs, rituals, manners of interacting, forms of communication, expectations for behaviours, roles, and relationships commonly shared among members of a particular group are part of that group's culture.

Culture is not static: cultures are constantly evolving and changing in response to new situations, challenges, and opportunities.

Everyone has a culture: it is like gravity, it just "is". So, how do we recognize, understand, and effectively live with our respective culture and the cultural orientations of others?

Culture is not just the group into which the person is born: people can acquire a new culture. For example, marrying into a different culture, moving to a new country, a change in economic status, or becoming disabled can lead to a cultural change.

Cultures are internally diverse: it is important to note that cultural groups are not homogeneous. There is variability between and among individuals within the same cultural and ethnic group. This variability can be the result of:

- age;
- level of education;
- family circumstances;
- rural versus urban living;
- life commitments;
- religious influences;
- level of adherence to traditional customs; and
- the degree of assimilation and acculturation into the major culture.

Culture is not determinative:[6] Individuals within the cultural group respond differently to the same cultural experiences and events. This is why making assumptions about a cultural identity typically fails.

Cultural differences are complicated by differences in status and power between

---

[5] P.G. Kittler & K.P. Sucher, *Food and Culture*, 3d ed. (Belmont, CA: Wadsworth/Thomson Learning, 2001); R. Pauly, *Cultural Diversity: Increasing Awareness* (University of Florida, Department of Medicine, 2003), reported in C. Brannon, *Cultural Competency: Values, Traditions and Effective Practices* (Nutrition Dimension Inc., 2007) at 2, online: http://www.rd411.com/ce_modules/CUC06.pdf (date accessed: January 31, 2020).

[6] L. Olsen, J. Bhattacharya & A. Scharf, *Cultural Competency: What It Is and Why It Matters* (California Tomorrow, Lucile Packard Foundation for Children's Health, 2006), online: http://www.lpfch.org/programs/culturalcompetency.pdf (date accessed: January 31, 2020).

cultures.[7] When a cultural group has most of the power, societal institutions tend to adopt the norms of the dominant culture as being the "right way to do things". For example, our workplaces tend to reflect the norms and values of the two older generations: The Veterans and Baby Boomers (for more details, please refer to Chapter 35, "The Impact of Five Generations in the Workplace on Occupational Health and Safety Programs").

Culture shapes our life experiences: it defines the way our parents disciplined us as children, the structure of family relationships, the expectations of what it means to be a boy or a girl, the values about health and approaches to healing, our body language, and what types of things do and do not get said. These learned attributes define the norms, or "how things are supposed to be", for the members of a given culture.[8] Needless to say, culture plays a significant role in the person's expectation of, and response to, the workplace.

C. OTHER RELEVANT TERMS

Acculturation: Cultural modification of an individual, group, or people by adapting to, or borrowing, traits from another culture; a merging of cultures as a result of prolonged contact. It should be noted that individuals from culturally diverse groups may desire or seek varying degrees of acculturation into the dominant culture.[9]

Assimilation: The assumption of the cultural traditions of a given people or group.

Cultural awareness: Being cognizant, observant, and conscious of similarities and differences among cultural groups.[10]

Cultural sensitivity: Understanding the needs and emotions of one's culture and the culture of others.[11]

Ethnic: Of, or relating to, large groups of people classed according to common racial, national, tribal, religious, linguistic, or cultural origins, or background.

Ethnicity: Ethnic quality or affiliation. Defined by Elliott and Fleras, ethnicity is a:

> . . . principle which explains how people are defined, differentiated, organized and entitled to group membership on the basis of certain physical or cultural characteristics. [C]an also consist of a consciously shared system of beliefs, values, loyalties and practices Ethnicity that pertain to members of a group who regard themselves as different and apart. The salient feature of ethnicity is the attachment that a person or group has a common cultural heritage.[12]

---

[7] *Ibid.*

[8] *Ibid.*

[9] H.V. Ngo, *Cultural Competency: A Self-Assessment Guide for Human Service Organizations* (Calgary, AB: Cultural Diversity Institute, 2000).

[10] *Ibid.*

[11] *Ibid.*

[12] J.L. Elliott & A. Fleras, *Unequal Relations: An Introduction to Race and Ethnic*

Ethno-cultural minorities: Includes people other than Aboriginal people who belong to cultures not generally considered part of Western society. They are also termed **cultural minorities** or **visible minorities**.[13]

Race: A tribe, people, or nation belonging to the same stock; a division of humankind, possessing traits that are transmissible by descent, and sufficient to characterize it as a distinctive human type. Race can also be described as a social construct used to separate the world's peoples.[14]

D. CULTURAL DIVERSITY

Diversity refers to "dissimilarity and variance between things and people".[15] **Cultural diversity** is the recognition that people come from a variety of gender, age, ethnic, geographic, economic, and religious backgrounds.[16] Hence, it involves an appreciation of the differences in race, ethnicity, language, nationality, or religion among various groups within a nation, community or organization. An organization is said to be "culturally diverse" if its employees include members of different groups.[17]

The main influences on the management of cultural diversity have been:[18]

1. *A Strategic Move to General Diversity Management* — Diversity management provides an enabling framework with which to address cultural diversity. To be successful at diversity management, the organization must practice openness, good communication, and flexibility. The ultimate benefit for the organization is the attainment of a diverse workforce.

2. *International Business* — Managing business internationally has had a major impact on the Western understanding of culture and how it affects

---

Dynamics in Canada (Scarborough, ON: Prentice-Hall Canada, 1992).

[13] Alberta Employment and Immigration, *Employing a Diverse Workforce: Making it Work* (2008), online: http://alis.alberta.ca/pdf/cshop/employdiverse.pdf (date accessed: January 31, 2020).

[14] H.V. Ngo, *Cultural Competency: A Self-Assessment Guide for Human Service Organizations* (Calgary, AB: Cultural Diversity Institute, 2000).

[15] C. Brannon, *Cultural Competency: Values, Traditions and Effective Practice* (Nutrition Dimension Inc., 2007) at 2, online: http://www.rd411.com/ce_modules/CUC06.pdf (date accessed: January 31, 2020).

[16] P.G. Kittler & K.P. Sucher, *Food and Culture*, 3d ed. (Belmont, CA: Wadsworth/Thomson Learning, 2001); About News, *Cultural Diversity: How it Boosts Profits*, online: http://useconomy.about.com/od/suppl1/g/Cultural-Diversity.htm (date accessed: January 31, 2020).

[17] H.V. Ngo, *Cultural Competency: A Self-Assessment Guide for Human Service Organizations* (Calgary, AB: Cultural Diversity Institute, 2000).

[18] D. Crowe & M. Hogan, *Cultural Diversity in the Workplace: Discussion Paper* (IMI Bizlab on Cultural Diversity, 2007), online: http://citeseerx.ist.psu.edu/viewdoc/summary?doi=10.1.1.126.3933 (date accessed: January 31, 2020).

individual and organizational behaviours. Likewise, Western organizations have been influenced by exposure to international management assumptions and business practices. Knowledge and understanding of cultural differences has facilitated and enhanced business negotiations, marketing, sale, and purchasing, among other activities.

3. ***The Multicultural Marketplace*** — The market segments include ethnic and language minorities; hence there are clear benefits to employing immigrant workers to address the needs of the multicultural marketplace.

4. ***Human Capital Management*** — Employees are now recognized as a source of core competence and competitive advantage. As human capital management increased in value, cultural diversity management came into vogue. Effective talent management through human resource management strategies can play a key role in achieving improved productivity and innovation.

5. ***Globalization*** — The increased ease of mobility and communications for organizations and individuals throughout the world has radically changed the world of work and the emergent workforce.

E. CULTURAL DIVERSITY: WHAT IS THE ISSUE?

Canadian and U.S. societies are changing demographically. For employers/organizations, cultural diversity is no longer a "nice thing to do". The reality is that the immigration rates in both countries have steadily increased. Why is this happening?

According to Statistics Canada:

> Immigration is becoming increasingly important to Canada's economic well-being. Roughly two-thirds of Canada's population growth comes from net international migration. Population projections show that net immi¬gration may become the only source of population growth by about 2030 and could account for virtually all net labour force growth . . . .[19]

The same situation exists in the U.S.[20]

Based on the 2016 Canadian Census, there are over 250 ethnic origins in Canada.[21] After being classified as being born in Canada (Canadian), the most frequently reported "origins" were English, French, Scottish, Irish, German, Italian,

---

[19] Statistics Canada, "Study: Canada's Immigrant Labour Market", *The Daily* (September 10, 2007), online: http://www.statcan.gc.ca/daily-quotidien/070910/dq070910a-eng.htm (date accessed: January 31, 2020).

[20] Wikipedia, "Immigration to the United States", online: http://en.wikipedia.org/wiki/Immigration_to_the_United_States#Demographics (date accessed: January 31, 2020).

[21] Statistics Canada, *Census in Brief: Ethnic and cultural origins of Canadians: Portrait of a rich heritage* (2016), online: https://www12.statcan.gc.ca/census-recensement/2016/as-sa/98-200-x/2016016/98-200-x2016016-eng.cfm (date accessed: January 31, 2020).

Chinese, North American Indian, and Ukrainian.[22] The number of individuals identified as part of a visible minority totalled 7,540,830 individuals, or 21.9 per cent of Canada's total population. This is an increase from 20.6 per cent in 2011. In 2016, Canada's visible minority equalled 22.3%; an increase of 3.2 per cent.[23] For the first time, one in five Canadians are foreign-born immigrants.[24]

In 2016, the new permanent residents who immigrated to Canada came from the following countries/areas (Table 34.1):

Table 34.1: Permanent Residents who Immigrated to Canada by Country/Area (2016)[25]

Country/Area	Percentage	Population
Philippines	15.6%	188,805
India	12.1%	147,190
People's Republic of China	10.6%	129,020
	Total Immigration	257,887

In the United States in 2017, the total U.S. population was 325.7 million, with the following ethnic distribution (Table 34.2).

Table 34.2: Race Distribution in United States (2019)[26]

Ethnicity	Percentage	Population
White Americans	76%	247.5 million
Hispanic or Latino	18%	58.6 million
Black or African Americans	13%	42.3 million
Asian	6%	19.5 million
American Indian or Alaska Native	1%	3.3 million

---

[22] *Ibid.*

[23] Wikipedia (2018), *s.v.* "Visible Minority", online: https://en.m.wikipedia.org/wiki/Visible_minority (date accessed: January 31, 2020).

[24] Statistics Canada, *Census in Brief: Ethnic and cultural origins of Canadians: Portrait of a rich heritage* (2016), online: https://www12.statcan.gc.ca/census-recensement/2016/as-sa/98-200-x/2016016/98-200-x2016016-eng.cfm (date accessed: January 31, 2020).

[25] Citizenship and Immigration Canada, *Facts and Figures: Immigration Overview — Permanent and Temporary Residents 2015* (2016), online: https://www.canada.ca/en/immigration-refugees-citizenship/corporate/reports-statistics/statistics-open-data.html (date accessed: January 31, 2020).

[26] U.S. Census Bureau, "Quick Facts: 2019", online: https://www.census.gov/quickfacts/fact/table/US/PST045219#PST04 (date accessed: January 31, 2020).

| Native Hawaiian or other Pacific Islander | 0.2% | 0.65 million |
| Two or more races | 3% | 9.8 million |

Note: These figures add up to more than 100% on this list because Hispanic and Latino Americans are distributed among all the races and are also listed as an ethnicity category, resulting in a double count.

Clearly, cultural diversity is "here to stay" for business and economic reasons. By becoming culturally competent, employers/organizations can appreciate and maximize the strengths of each of the various cultures present in their workplaces. The benefits of having a culturally diverse workforce are to:

- realize greater productivity;
- experience lower staff turnover;
- enhance their understanding of customer needs and wants;
- achieve better access to new markets;
- benefit from diverse ideas and perspectives; and
- enjoy an enhanced corporate image and reputation.[27]

Beyond the business drivers, there are legislative policies that recognize and promote racial, ethnic, and linguistic diversity. For example, in Canada, the relevant legislation includes:

- ***Canadian Charter of Rights and Freedom*** — the provision for equity rights, freedom from discrimination, and equal access to participation regardless of race, religion, national or ethnic origin, colour, sex, age, and physical or psychological disability.[28]
- ***Canada Multiculturalism Act*** — aimed at the preservation and enhancement of Canadian multiculturalism, the Act promotes full and equitable participation of individuals and communities of all cultural origins in all aspects of Canadian society. The social, cultural, economic, and political institutions of Canada have to be both respectful and inclusive and enable meaningful participation by culturally diverse people.[29]
- ***Human Rights legislation*** — prohibits practices in public services that discriminate on the basis of race, religious beliefs, colour, gender, physical

---

[27] Alberta Employment and Immigration, *Employing a Diverse Workforce: Making it Work* (2008) at 9, online: http://alis.alberta.ca/pdf/cshop/employdiverse.pdf (date accessed: January 31, 2020).

[28] H.V. Ngo, *Cultural Competency: A Self-Assessment Guide for Human Service Organizations* (Calgary, AB: Cultural Diversity Institute, 2000).

[29] *Ibid.*

or psychological disability, ancestry, place of origin, marital status, source of income, or family status.[30]

- **Employment Equity** — requires federal departments and agencies with 100 employees or more to file an annual statistical profile of the employment equity designated groups. The annual report must compare those designated groups with all other employees, in terms of such dimensions as occupational and salary levels. The principle of this law is that the employer's workforce must reflect the population from which the employer recruits.[31]

F. TRANSCULTURAL NURSING

Some of the earliest studies in terms of working with different cultures were undertaken in the field of nursing. Leininger[32] defined **transcultural nursing** as:

> . . . an essential area of study and practice focused on the cultural care beliefs, values, and life ways of people to help them maintain and/or regain their health, or to face death in meaningful ways.[33]

Transcultural nursing focuses on understanding cultures, their specific healthcare needs, and how to aid that best fits cultural lifestyles. This transcultural perspective identifies diverse and universal cultural variables found among human social groups. Today, transcultural nursing has evolved to include cultural competence — an element of nursing education programs.

G. CULTURAL COMPETENCE

In regard to the field of disability management, **cultural competence** can be described as having the ability to provide quality disability management care and services to a diverse employee population. It encompasses both systemic responses (*e.g.*, organizational policy, procedures, practices, *etc.*) as well as the delivery of OH&S and Disability Management services by skilled and sensitive practitioners/professionals. Hence, cultural competence implies a responsibility at both the organizational and individual level.

As defined by the Seattle King County Department of Public Health in 1994:

> **Cultural competency** is the ability of individuals and systems to respond respectfully and effectively to people of all cultures, classes, races, ethnic backgrounds and religions in a manner that recognizes, affirms, and values the cultural differences and similarities, and the worth of individuals, families, and communities and protects and preserves the dignity of each.[34]

---

[30] *Ibid.*

[31] *Ibid.*

[32] M. Leininger & M. McFarland, *Culture Care Diversity and Universality: A Worldwide Nursing Theory*, 2d ed. (Sudbury, MA: Jones and Bartlett, 2006).

[33] *Ibid.*

[34] Seattle King County Dept of Public Health (1994), reported in R. Fiorelli & W. Jenkins,

Operationalized, cultural competence means that OH&S and Disability Management Practitioners/Professionals must be able to identify and challenge their own cultural beliefs. In other words, they should have the "ability to see the world through different cultural lenses". Both, must be able to analyze and respond to different cultural beliefs, environments and events in ways that are meaningful — culturally and psychologically — for clients and professionals alike — and possess the ability to use that introspection so that it results in the provision of meaningful, satisfying, and competent assistance and support.

Cultural proficiency occurs along a continuum (Figure 34.1). **Cultural destructiveness** is the stage of intentional denial, rejection, or outlawing of other cultures. **Cultural incapacity** is the stage of acceptance of other cultures, but an inability to work effectively with other cultures. **Cultural blindness** is the stage when people assume that everyone is basically alike and advocates a universal approach and services for all people. **Cultural pre-competence** is the stage of awareness within systems or organizations of their strengths and areas for growth to respond effectively to culturally and linguistically diverse populations. It includes a willingness to learn and understand other cultures.[35] **Cultural competency** is the stage when people are able to effectively function in cross-cultural circumstances. It includes the development of standards, policies, practice, and attitudes that value diversity. **Cultural proficiency** is the stage where proactive promotion of cultural diversity occurs and opportunities to improve cultural relationships are sought.

"*Cultural Competency in Grief and Loss*" (2009), online: http://www.nhpco.org/sites/default/files/public/education/Cultural_Comptency_Grief-Loss_Sept_09_NL.pdf (date accessed: January 31, 2020).

[35] T. Cross, B. Bazron, K. Dennis & M. Isaacs, *Towards a Culturally Competent System of Care,* Vol. 1 (Washington, DC: CASSP Technical Assistance Center, Center for Child Health and Mental Health Policy, Georgetown University Child Development Center, 1989), online: http://www.nccccurricula.info/documents/TheContinuumRevised.doc (date accessed: January 31, 2020).

[36] Adapted from T. Cross, B. Bazron, K. Dennis & M. Isaacs, *Towards a Culturally Competent System of Care*, vol. 1 (Washington, DC: CASSP Technical Assistance Center, Center for Child Health and Mental Health Policy, Georgetown University Child Development Center, 1989), online: http://www.nccccurricula.info/documents/TheContinuumRevised.doc (date accessed: January 31, 2020); and C. Brannon, *Cultural Competency: Values, Traditions and Effective Practice* (Nutrition Dimension Inc., 2007) at 3, online: http://www.rd411.com/ce_modules/ CUC06.pdf (date accessed: January 31, 2020).

Figure 34.1: Cultural Competency Continuum[36]

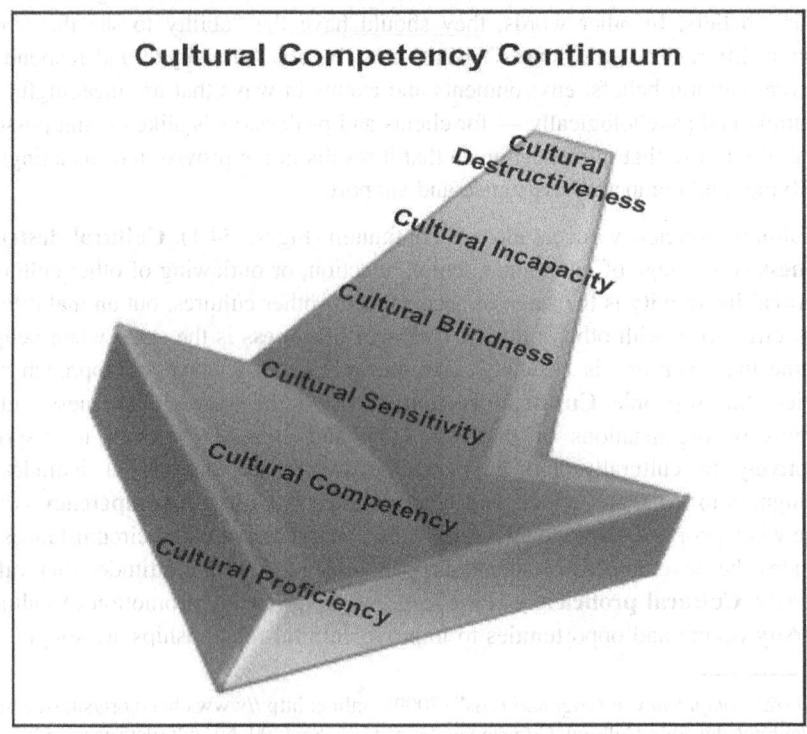

H. DEVELOPMENT OF CULTURAL COMPETENCE

The development of cultural competence goes through a series of personal and professional growth phases which include the following:

1. *Cultural awareness* — Awareness that enables the OH&S and Disability Management Practitioners/Professionals to understand how beliefs, values, and personal/political power are shaped by culture.[37] It starts with acknowledging that cultural differences exist between the Western culture and non-Western cultures. The next step is to recognize that as human beings, we view the world based on our individual culture. In essence, our culture and the socialization into that culture, colours our viewpoints. The key aspect is to realize that our individual culture is but one of many cultures, not the "only and right" culture. By identifying their personal

---

[37] P. Lister (1999) reported in *National Review of Nursing Education — Multicultural Nursing Education*, "5. Australian Nursing Education Today" (2001), online: http://www.dest.gov.au/archive/HIGHERED/nursing/pubs/multi_cultural/5.htm (date accessed: January 31, 2020).

beliefs and social customs, OH&S and Disability Management Practitioners/Professionals can develop a cultural awareness.

2. ***Cultural knowledge*** — Familiarity with the broad differences, similarities and inequalities in experience and practice among various societal groupings.[38] OH&S and Disability Management Practitioners/Professionals can develop cultural knowledge by:

   (a) researching an organization's/company's profile;

   (b) evaluating the degree to which local healthcare services meet the needs of these various minority employee groups; and

   (c) seeking information on the inequalities in health and healthcare in relation to class, gender, ethnicity and age.

3. ***Cultural understanding*** — Recognition of the problems and issues faced by individuals and groups when their values, beliefs and practices are compromised by a dominant culture.[39] OH&S and Disability Management Practitioners/Professionals can attain cultural understanding by comparing and contrasting the:

   (a) lay and professional viewpoints towards minority groups; and

   (b) experiences of employee illness/injury and the caregiving relationship. The use of "real experiences" through case studies can be valuable in achieving this end.

4. ***Cultural sensitivity*** — Regard for the employee's beliefs, values, and practices within a cultural context and showing awareness of how his/her own cultural background may be influencing professional practice.[40] OH&S and Disability Management Practitioners/Professionals can become culturally sensitive by evaluating the effectiveness of their OH&S/Disability Management Program and services in meeting the needs of various cultures. This translates to internalizing the concept of cultural understanding.

5. ***Cultural competence*** — Provision or facilitation of services which respect the values, beliefs, and practices of the employee, and which address any disadvantages arising from the employee's position in the organization/company, community and/or society.[41] OH&S and Disability Management Practitioners/Professionals demonstrate culture through a combination of knowledge, skills, and behaviours that lead to positive outcomes with culturally diverse populations.

---

[38] *Ibid.*
[39] *Ibid.*
[40] *Ibid.*
[41] *Ibid.*

I. CULTURAL COMPETENCE: ITS IMPORTANCE

Why is cultural competence important? What value does it offer to OH&S and Disability Management Practitioners/Professionals?

Studies have shown that consideration of individual cultures, preferences and needs in the provision of healthcare services:

- eliminates healthcare disparities;[42]
- reduces the misunderstandings between families and healthcare practitioners;[43]
- reduces non-compliance;
- enhances the cost-effectiveness of healthcare service provision;[44]
- increases patient/client satisfaction and retention; and
- improves health and recovery outcomes.[45]

In addition, culture has a profound impact on how people respond to preventive interventions and healthcare services, as well as in how they experience illness/injury, how they access healthcare, and how they react to recovery.

The benefits of using a culturally-competent approach to Occupational Health and Safety and Disability Management Programs include:

- the development of more appropriate return-to-work plans;
- improved quality of disability management case and claim management processes/practices;
- better OH&S services and outcomes;
- increased employee acceptance of, and compliance with, return-to-work plans;
- improved employee satisfaction with the OH&S Program and services;
- enhanced provision of EAP and support;
- increased sensitivity as to employee/workgroup needs; and

---

[42] Avesis, *Cultural Competency: 2010* (2008 revision, approved January 2009) at 5, online: http://avesis.com/pdf/cultural_competency.pdf (date accessed: January 31, 2020).

[43] L. Olsen, J. Bhattacharya & A. Scharf, *Cultural Competency: What It Is and Why It Matters* (California Tomorrow, Lucile Packard Foundation for Children's Health, 2006) at 2, online: http://www.lpfch.org/programs/culturalcompetency.pdf (date accessed: January 31, 2020).

[44] Avesis, *Cultural Competency: 2010* (2008 revision, approved January 2009) at 5, online: http://avesis.com/pdf/cultural_competency.pdf (date accessed: January 31, 2020).

[45] American Association of Colleges of Nursing, *Cultural Competency in Baccalaureate Nursing Education* (2008) at 1, online: http://www.aacn.nche.edu/education/pdf/competency.pdf (date accessed: January 31, 2020).

- the ability to work more effectively with diverse employee populations.

OH&S and Disability Management Practitioners/Professionals must be culturally aware, knowledgeable, understanding, sensitive, and competent. The outcome is the provision of Integrated OH&S/Disability Management Programs and services that positively impact the desired business and people outcomes with culturally diverse populations.

J. CULTURAL COMPETENCE: OTHER CONSIDERATIONS

Cultural competence implies a working knowledge of the various factors that impact a culture. To meet this criterion, OH&S and Disability Management Practitioners/Professionals must consider the following:

- *Role of Family* — In traditional cultures, the family plays a major role in decision-making about health issues. OH&S and Disability Management Practitioners/Professionals should take that into account and involve family members in employee meetings, case conferences, and return-to-work planning.

- *Role of Economics* — Poverty brings with it its own cultural aspects. It can be a major determinant in an employee's response to OH&S and Disability Management. Economic status can influence the employee's ability to acquire medical supplies and other resources for treatment compliance and rehabilitation. When making decisions about disability case management and return-to-work planning, OH&S and Disability Management Practitioners/Professionals should consider the different degrees of access to resources.

- *Role of Religious Beliefs* — An employee's decision about treatment, rehabilitation, and return-to-work practices can be influenced by religious beliefs. For example, because of their religious faith, employees may want to know what is wrong with them (the diagnosis), but choose not to seek the recommended medical treatment. They may also want to include traditional healing along with the scientific medical treatment modalities. Respecting these beliefs and incorporating them into the employee's care can enhance the ultimate recovery process and health outcome.

- *Role of Communication Styles* — Intercultural communication can be challenging for the OH&S and Disability Management Practitioners/Professionals as well as for the employee and family. Language barriers are well recognized and must be addressed. In addition, non-verbal communication, or body language, is an important part of how people communicate and there are differences from culture to culture. Hand, arm and head gestures; touch; and eye contact (or lack thereof) are a few of the aspects of non-verbal communication that may vary significantly depending upon cultural background.

  Facial expressions:
    - Smiling is not always an expression of happiness; it can also signify

a discussion of something sad or uncomfortable in the Chinese culture.[46]
- Winking can be viewed as a romantic or sexual invitation (in some Latin American cultures); or as a signal to children to leave the room (the Yorubas in Nigeria); or as rudeness (many Chinese cultures).[47]
- Blinking can be interpreted as a sign of disrespect and boredom in Hong Kong.[48]
- Filipinos will point to an object by shifting their eyes toward it or pursing their lips and point with their mouth, rather than using their hands.[49]
- Venezuelans may use their lips to point at something because pointing with a finger is impolite.[50]
- Crying and a public showing of emotion are culturally determined.

Head Movements:
- A signal for "yes" may be a nod of the head in Lebanon; the signal for "no" may be pointing the head sharply upward and raising the eyebrows.
- Saudis signal "yes" by swivelling their head from side to side. They may signal "no" by tipping their head backward and clicking their tongue.

Gestures:
- There are a number of gestures commonly used in Canada/U.S. that may have a different meaning and/or be offensive, to those from other cultures. One common example is the use of a finger or hand to indicate "come here please". This is the gesture used to beckon dogs in some cultures and is very offensive. Pointing with one finger is also considered to be rude in some cultures; Asians typically use their entire hand to point to something.
- In Japan, the "OK" sign is interpreted as the symbol for money. In Argentina, Belgium, France, Portugal, Italy, Greece, and Zimbabwe,

---

[46] D. Morris, *Bodytalk: The Meaning of Human Gestures* (New York: Crown Trade Paperbacks, 1994) in Management for Sciences for Health, *The Provider's Guide to Quality & Culture*, "Topic 9: Non-verbal Communication" (2009), online: http://erc.msh.org/mainpage.cfm?file=4.6.0.htm&module=provider&language=English (date accessed: January 31, 2020).

[47] *Ibid.*
[48] *Ibid.*
[49] *Ibid.*
[50] *Ibid.*

- the same sign means "zero" or "nothing". In some Eastern European countries, this is highly offensive as it indicates a bodily orifice.[51]
- Holding up crossed fingers is a gesture wishing one "good luck", while crossed fingers held behind the back negates a stated pledge or thought. In Russia, this is a way of rudely rejecting or denying something. In Argentina and Spain, this same sign is made to ward off bad luck. In China, it signifies the number 10.[52]
- In Iran, the "thumbs up" gesture is vulgar.[53]
- Tapping the underside of the elbow with the fingers of the other hand indicates that someone is stingy, in Colombia.[54]
- In Latin America, a shrug with the palms facing skyward may be interpreted as a vulgar gesture.[55]

Touch:

- While patting a child's head is considered to be a friendly or affectionate gesture in our culture, it is considered inappropriate by many Asians to touch someone on the head, which is believed to be a sacred part of the body.
- In the Middle East, the left hand is reserved for bodily hygiene and should not be used to touch another or transfer objects. In Muslim cultures, touch between opposite-gendered individuals is generally inappropriate.
- Light touching of the arm or a light kiss to the cheek is very common in some cultures, even among people who have just met. People from Latin America and Eastern Europe may be very comfortable with this kind of touching, whereas people from many Asian cultures may prefer less physical contact with acquaintances.[56]
- Physical contact, except for a handshake, early in a relationship may be uncomfortable for some Chinese. This is especially important to remember when dealing with older people or those in positions of authority.[57]
- Men in Egypt tend to be more touch-oriented; a handshake may be

---

[51] *Ibid.*
[52] *Ibid.*
[53] *Ibid.*
[54] *Ibid.*
[55] *Ibid.*
[56] *Ibid.*
[57] *Ibid.*

- accompanied by a gentle touching of the recipient's elbow with the fingers of the left hand.[58]
- A strong, warm handshake is the traditional greeting between men in Latin America. However, because most Latin Americans show affection easily, male friends, like female friends, may embrace. Women may lightly brush their cheeks together.[59]
- A Western woman should not initiate a handshake with a man in India. Many Indian women will shake hands with a foreign woman, but not a foreign man.[60]
- To many Indians, it is considered rather offensive to (even accidentally) step on someone's foot. Apologies should be made immediately.[61]

Eye contact/Gaze:

- In the mainstream Western culture, eye contact is interpreted as attentiveness and honesty? we are taught that we should "look people in the eye" when they are talking to us. In many cultures, however, including Hispanic, Asian, Middle Eastern, and Native American, eye contact is thought to be disrespectful or rude; hence for these cultures, the lack of eye contact does not mean that a person is not paying attention.[62]
- Women may especially avoid eye contact with men because it can be taken as a sign of sexual interest.[63]
- Many Middle Easterners have what North Americans and Europeans consider "languid eyes". It may appear that the person's eyes are half closed, but this does not express disinterest or disrespect.[64]
- In Ghana, young children are taught not to look adults in the eye because to do so would be considered an act of defiance.[65]
- In Latin America, good eye contact is important in both social and business situations.[66]

---

[58] *Ibid.*
[59] *Ibid.*
[60] *Ibid.*
[61] *Ibid.*
[62] *Ibid.*
[63] *Ibid.*
[64] *Ibid.*
[65] *Ibid.*
[66] *Ibid.*

Physical Contact:
- It is impolite to show the bottom of the shoe, which is often dirty in many cultures. Therefore, one should not sit with the foot resting on the opposite knee.
- Standing with the hands on the hips suggests anger, or a challenge, in Argentina.[67]
- Slouching or poor posture is considered to be disrespectful. For example, good posture is important in Taiwan, with Taiwanese men usually sitting with both feet firmly fixed to the floor.[68]

Personal Space:
- Different cultures have different tolerances for how close people position themselves when communicating. Standing "too close" may be perceived as uncomfortable or rude in some cultures, yet acceptable in others.

- *Response to Illness/Injury* — How an employee and his/her family responds to illness/injury is impacted by his/her cultural orientation. A better appreciation of the differences can be obtained by comparing the traditional and modern healthcare systems. Being part of the health maintenance for cultures, **traditional healthcare systems** have been around forever. They are part of the original culture, having filtered down, generation to generation. Traditional healthcare is the first resource the individual turns to when ill/injured.[69]

In contrast, the **modern healthcare systems** are recent social developments and tend to be the last resort when one is ill/injured. They are used when the traditional healthcare system fails. Hence, this is one reason why some people wait until their disease has reached an advanced stage before seeking care from a physician.

Other differences are that, in the traditional healthcare system, the caregiver(s) speaks the same language and upholds similar cultural beliefs and attitudes, as well as lifestyle. Since it is community-based, the traditional healthcare provider knows and understands the patient's social relationships and can enlist assistance from significant others.

In contrast, the healthcare providers in the modern healthcare systems experience professional ethnocentrisms that can negatively affect the healthcare provider-patient relationship. For example, the belief that the

---

[67] *Ibid.*

[68] *Ibid.*

[69] C.L. Edelman & C.L. Mandle, *Health Promotion Throughout the Lifespan*, 4th ed. (St. Louis, MO: Mosby, 1998).

individual's well-being surpasses the well-being of the family can be negatively viewed. Portraying the belief that the "physician knows best", can be offensive. As well, the practice of hastily handing a patient a fact sheet on a disease condition, or treatment modality, can be viewed as rude. The outcome will be non-compliance with the recommended healthcare regimens.

Lastly, the two healthcare systems have different focuses. The traditional healthcare system is ego-focused, taking cures from the client and accepting the client's symptoms at "face-value".[70] The modern healthcare systems, on the other hand, are technically oriented. An illness/injury tends not to be valid unless it can be organically defined or scientifically understood.

- **Response to Pain** — Pain is a subjective sensation and the expression of pain is culturally determined. Some cultures promote stoicism, while others accept crying and open expressions of pain and discomfort.[71]

K. CULTURAL ASSESSMENT

Cultural assessment of the ill/injured employee is required to identify his/her views and beliefs about health and illness. Beliefs about the cause, prevention and treatment of illness/injury vary among cultures. As already noted, healthcare practices can be traditional (classified as folk, spiritual, or psychic healing practices) and modern (conventional medical practices). A summary of both healthcare systems is provided in Table 34.3.

Table 34.3: Comparison of Traditional and Modern Healthcare Systems[72]

Traditional Healthcare System	Aspect	Modern Healthcare System
Existed in some format forever	*When introduced*	Developed within recent memory
Generationally promoted	*How promoted*	Commercially marketed
Initially	*When sought*	When Traditional Healthcare methods fail
Same language	*Language*	Language differences
Similar	*Cultural beliefs of caregiver(s)*	Different and varied

---

[70] *Ibid.*

[71] D. Morris, *Bodytalk: The Meaning of Human Gestures* (New York: Crown Trade Paperbacks, 1994) in Management for Sciences for Health, *The Provider's Guide to Quality & Culture*, "Topic 9: Non-verbal Communication" (2009), online: http://erc.msh.org/mainpage.cfm?file=4.6.0.htm&module=provider&language=English (date accessed: January 31, 2020).

[72] C.L. Edelman & C.L. Mandle, *Health Promotion Throughout the Lifespan*, 4th ed. (St. Louis, MO: Mosby, 1998).

CH. 34: OCCUPATIONAL HEALTH AND SAFETY: DIVERSITY CONSIDERATIONS

Traditional Healthcare System	Aspect	Modern Healthcare System
Home-based, community-based	*Location*	Clinic- or hospital-based
Good	*Knowledge of ill/ injured person*	Unknown to limited
Good	*Knowledge of available social supports*	Unknown to limited
On total family	*Focus of care*	On the individual
Ego-focused	*Focus*	Technically oriented

In addition to the usual OH&S data collected, specific cultural information can assist OH&S and Disability Management Practitioners/Professionals to gain a better understanding of the employee and his/her cultural affiliation. Areas for consideration are the individual's:

- *Place of birth* — to determine the country of origin and country(ies) of residence.

- *Length of time in Canada* — to assess the degree of acculturation to Western society.

- *First language and other languages spoken* — to obtain an understanding of the employee's level of English comprehension and how readily he/she will be able to understand the OH&S documents, forms, and materials.

- *Available support systems* — to learn who will support the employee and who makes the decision regarding healthcare.

- *Number and type of dependents* — to determine the magnitude of the employee's family responsibilities.

- *Caregiving responsibilities* — to identify the ill/injured employee's extra duties.

- *Relationships* — to find out if the ill/injured employee wants certain people to be in attendance as part of the Disability Management case management, and planning and return-to-work initiatives.

- *Response to illness/injury* — to ascertain the level of reliance on traditional and modern healthcare systems. Questions like, "What do you do when you become ill/injured?" or "How do you respond?" can be asked.

- *Traditional healthcare system* — What does that look like for this employee? Questions like, "Do you ever see a native healer or other type of practitioner when you don't feel well?", "Does this healer help?", "Do you ever take any herbs or medicines that are commonly used in your native country or by your cultural/ethnic group?", or "If so, what are they, and for what reason(s) do you take them?" can be asked.

- *Health beliefs* — Some questions to ask the ill/injured employee are:

1233

- What do you think caused your health problem(s)?
- Why do you think it started when it did?
- What does your sickness do to you? How does it affect you and your life?
- How severe is it? From your opinion, will it last a short or long period of time?
- What do you fear most about your illness/injury/disorder?
- What are the chief problems that your illness/injury has caused for you?
- What kind of treatment do you think you should receive? What are the most important results you hope to receive from that treatment?

There are a number of other factors for the OH&S and Disability Management Practitioners/Professionals to consider. For example, culturally:

- Are individuals comfortable answering personal health questions?
- When the healthcare provider asks questions, does the person or the family, perceive this inquiry as a "lack of knowledge" on the part of the healthcare provider?
- Who should be told about the illness/injury?
- Are healthcare decisions made individually or through family consensus?
- Is the gender of the healthcare provider an issue?
- Does "more medicine" equal greater severity of illness?
- Does "no medication" prescribed by the physician indicate wellness?
- Does the patient prefer to "feel" the symptoms or to "mask" them?
- Does the patient prefer one solution, or a choice of treatment options?
- Does the patient want to hear about the treatment risks?[73]

A thorough cultural assessment can assist the OH&S and Disability Management Practitioners/Professionals to better understand the cultural orientation of the ill/injured employee and the impact the employee's cultural orientation has or will have on the employee's response to the current illness/injury, the recovery process, and a safe and timely return to work.

L. RECOMMENDATIONS FOR WORKING WITH PEOPLE OF DIVERSE CULTURES

OH&S and Disability Management Practitioners/Professionals are advised to:

---

[73] A. Kleinman, *Patients and Healers in the Context of Culture* (Berkeley, CA: University of California Press, 1981).

- Understand the cultural differences that may exist.
- Be aware of one's own cultural beliefs and biases.
- Be sensitive of one's positional authority within the organization/ company, real or perceived.
- Greet employees in their own language and/or try to work with their beliefs. Acknowledge that you care and do not worry about making mistakes. Employees appreciate a sincere effort by the OH&S or Disability Management Practitioner/Professional to understand and work with them in a manner that is comfortable for them.
- Recognize that there are a number of cultural "hot buttons", namely:
    - superstitions and customs;
    - religious beliefs;
    - death and dying practices;
    - individual autonomy;
    - dietary practices;
    - physical space; and
    - body language.[74]

    Be aware of the potential for these topics to result in cultural differences and reactions. Seek ways to approach and address them effectively.
- Ask the employee, early in the OH&S/Disability Management interaction, to identify any actions deemed as offensive in his/her culture. That indicates respect and caring on the part of the OH&S and Disability Management Practitioner/Professional.
- Follow the employee's lead: if the employee moves closer or touches you in a casual manner, you may do the same.[75]
- Use hand and arm gestures with great caution. Gestures can mean very different things in different cultures.[76]
- Be careful in interpreting facial expressions. They may lead you to misinterpret the patient's feelings or to over- or underestimate the employ-

---

[74] General Healthcare Resources Inc., *Cultural Diversity* (2005) at 6, online: http://www.ghresources.com/ghr/healthcare.nsf/2010-OSHA-In-Services.pdf (date accessed: January 31, 2020).

[75] D. Morris, *Bodytalk: The Meaning of Human Gestures* (New York: Crown Trade Paperbacks, 1994) in *The Provider's Guide to Quality & Culture*, "Topic 9: Non-verbal Communication" (2009), online: http://erc.msh.org/mainpage.cfm?file=4.6.0.htm&module=provider&language=English (date accessed: January 31, 2020).

[76] *Ibid.*

ee's level of pain. This is also true of the presence or absence of crying and other expressions of pain, which are closely tied to a person's culture.[77]

- Avoid seeking eye contact from the employee/family: it may be a sign of respect by the employee's culture to not make eye contact.[78]

- Be patient establishing a rapport — the associated trust with employees from different cultures can take time.

- Thoroughly explain the OH&S/Disability Management Program process and procedures, and their rationale.

- When working with employees who have limited language proficiency, be sure to use a trained interpreter, not merely a family member or friend.

- Understand that there may be some cultural reluctance to discuss certain topics, particularly if the OH&S Practitioner/Professional, or Disability Management claim or case manager, or interpreter, is a different gender than the employee.

- Communicate effectively. As much as possible, use words (not gestures) to express the message(s). Gestures that are acceptable in our culture may be offensive or meaningless in other cultures.

- Find out how the employee views this illness/injury by asking questions such as:
    - What do you call your problem? What name does it have?
    - What do you think caused your problem?
    - Why do you think it started when it did?
    - What does your sickness do to you? How does it work?
    - How severe is it? Will it have a short or long course?
    - What do you fear most about your condition/disorder?
    - What are the chief problems that your sickness has caused for you?
    - What have you done so far to treat the sickness?
    - What kind of treatment do you think you should receive?
    - What are the most important results you hope to receive?[79]
    - What does this illness/injury mean to you? To your family?

---

[77] *Ibid.*

[78] *Ibid.*

[79] A. Kleinman, *Patients and Healers in the Context of Culture* (Berkeley, CA: University of California Press, 1980); D. Lozorik, *Pilot Study to Develop a Tool to Elicit Khmer Beliefs about the Cause of Illness* (Master's Thesis, Boston University School of Medicine, 1984) [unpublished].

- What is it like to be injured/ill within your culture?
- Identify areas of potential conflict and seek ways to address them.
- Compromise: Show respect for the employee's beliefs and be willing to work with the employee/family members to establish acceptable Disability Management claim and case management plans, and a safe and timely return to work.
- Provide clarity on the OH&S/Disability Management Program while being respectful of their culture. It is really important for workers and family members to understand the disability management process and what to expect at each stage. They need to understand the way things are done in this particular workplace. If required, they should be given access to a language and "cultural interpreter". The reality is that clinicians can be sensitive, but the whole system will not change to accommodate them. As OH&S and Disability Management Practitioners/Professionals, walk them through the disability management process and explain the options available. Help them understand their options so that they can make informed consents to treatment and return-to-work options. Advocate for them, remaining within the boundaries of the employer's and health insurance systems.

M. DIVERSITY TYPES

The first part of this chapter focuses on culture, transcultural nursing, cultural competence, and how to conduct a cultural assessment. At this point, these concepts and terms will be operationalized, beginning with a review of some types of diversity and then moving towards effective management of cultural diversity in the workplace.

1. The Older Worker (Over 60 Years of Age)

In 2017, 26 per cent of the Canadian workforce were workers aged 60 years and over.[80] Older workers have many work qualities that make them a valuable asset to any workplace. They are described as experienced, loyal, possessing strong work ethics, and willing to try a variety of roles. However, older workers do face a number of challenges, namely:

- Society as a whole tends to be **gerontophobic** (fear of the aging process) and as such, tends to be prejudiced against older workers. Some of this is perpetuated by the media and some by older workers themselves. For example, many older workers view themselves as being too old to start a new career or to tackle a new business endeavour. Having spent their entire career in one line of work, many believe that they could not do anything

---

[80] Statistics Canada, *Labour Statistics at a Glance: Reasons for Working at 60 and beyond*, 71-222-X (2018), online: https://www150.statcan.gc.ca/n1/pub/71-222-x/71-222-x2018003-eng.htm (date accessed: February 28, 2020).

else. As well, some view themselves as being overqualified to do a less demanding job.

- Older workers once they reach 65 years of age, their Long-term Disability (LTD) Insurance benefits cease and nor do LTD benefits start for any new 65-year-old worker. The rationale for this practice is that it is an industry standard: payment of Canada Pension Plan Disability Benefits stops at that age, pensions provide for retirement at aged 65, the cost to increase this age limit would be high, and any changes might lead to other problems such as administration issues in terms of how long to extend the LTD coverage for each person.[81]

- When hiring the older worker, employers question the training and development investment given that the older worker may not be with the organization/company all that long.

- Some employers hold true to the myths that older workers are physically slower, less physically able, less productive, more set in their ways, unable to adapt to new technology, and more likely to get hurt. These myths should be challenged, as the research shows them to be unfounded. For example, worker strength can improve with age. Add to that, the older worker tends to learn to work smarter and not to rely on "brawn" and "brute strength" capabilities.

However, older workers do face a number of health issues associated with the aging process. For example, 23 per cent of aging Canadian workers experience illness/injuries that negatively impact their ability to function. By age 65, the prevalence of disability increases to 42 per cent.[82] In 2019, the number of missed work days does increase with age from an average of 6.5 days for young full-time employees (aged 15 to 24 years) to 15 days for employees 65 years and older.[83] Part of this is due to the fact that older workers tend to require more than twice the recovery time when ill/injured.[84]

Occupational injury rates tend to increase up to aged 44 to 54 year cohort,

---

[81] J. Bell, "It's Time for a Rethink Around LTD Coverage for Older Employees", *Canadian Safety Reporter* (June 2014) at 4.

[82] K. Williams, "Returning to Work After Disability: What Goes Wrong?" (Presented at Canadian Human Resource Planners event, May 15, 2003, Calgary, AB) [unpublished].

[83] Statistics Canada, *Table: 14-10-0193-01 (formerly CANSIM 279-0032): Absence rates for full time employees by age group, annual, 2019* (Ottawa: Statistics Canada, January 2020), online: https://www150.statcan.gc.ca/t1/tbl1/en/tv.action?pid=1410019301 (date accessed: February 28, 2020).

[84] M. Mancini, "Aging workforce poses health, safety risks", *Canadian Safety Reporter* (July 2012) at 1-2.

and then decrease in the 55 to 64 year cohort (Table 34.4).[85]

Table 34.4: 2017 Occupational Injury and Fatality Rates

Cohort	Lost Time Injury Rate/ 100 workers	Fatality Rate/ 100 workers
15-24	1.29	0.0013
25-44	1.40	0.0019
45-54	1.54	0.0034
55-64	1.40	0.0068
65 & over	0.96	0.0891

Although the lost-time injury rates do decrease with age, older workers tend to succumb to occupational disease. Hence, the fatality rates increase with worker age. Lastly, the types of injuries differ: older workers are more susceptible to soft tissue injuries, especially back injuries.[86]

Some health conditions are associated with aging, namely, hearing loss, increased susceptibility to lengthy absences if injured, and increased prevalence of chronic health conditions. However, these human failings can be mitigated through the adoption of a supportive lifestyle and by the employer providing realistic work assignments.

- In terms of work-personal life conflict, older workers are often involved in providing care for an even older, dependant relative. Working and shouldering this type of responsibility can lead to caregiver strain — the stress of caring for an elderly dependant. Older women are particularly susceptible to this phenomenon, especially when this situation is compounded by caregiving responsibilities for grandchildren.
- Older workers tend to earn more than do younger workers, and are reluctant to reduce their work hours for less pay. This has led to resentment from younger workers, who tend to be better educated and wanting career advancement, but are held back and earning less.[87]

Today, many older workers are holding down a variety of jobs that range from entry-level to key knowledge positions. Employers appreciate the dedication, experience, and expertise that they bring and therefore, are willing to shoulder any associated risks and costs.

---

[85] AWCBC, "Lost Time Claims in Canada, 2017" and "Fatalities in Canada, 2017" (2019), online: http://awcbc.org/?page_id=14 (date accessed, February 28, 2020).

[86] T. McDonald & H. Harder, "Older Workers and Disability Management" (2004) 3:3 *International Journal of Disability, Community & Rehabilitation* 1, online: http://www.ijdcr.ca/VOL03_03_CAN/index.shtml (date accessed: January 31, 2020).

[87] Conference Board of Canada, *Young, Underpaid and Angry* (September 2014), online: http://www.conferenceboard.ca/press/speech_oped/14-09-24/young_underpaid_and_angry_the_coming_clash_over_the_income_gap.aspx.

(a) Recommendations for OH&S/Disability Management Professionals/Practitioners

OH&S and Disability Management Professionals/Practitioners are encouraged to:

- educate management and union leaders on the specific needs of the older worker.
- structure OH&S and Disability Management Program approaches so that they adequately address the needs of the older worker. For example,
    - address the decreased ability to visually adjust to changing lighting conditions;
    - appreciate that visual glare affects the older worker more;
    - be aware that small print becomes more difficult to read;
    - recognize that seeing moving objects becomes more of a challenge;
    - understand that sounds, especially high-pitched ones, become harder to hear — use a multiple of warning sources in alarm systems;
    - provide appropriate tools and equipment to compensate for weakening grip strength and debilitating conditions such as arthritis; and
    - respond to the fact that muscle and bone strength, joint flexibility, and nervous system acuity are gradually diminishing.
- recommend ergonomic assessments on the jobs held by older workers with a view to preventing unnecessary strains and sprain injuries.
- when accommodating older workers, consider garnering support from the relevant stakeholders, so that the placement proves successful.

2. Women

Women have played a major role in the workforce for many years; however, they still experience a certain degree of systemic discrimination. For example, although they hold managerial and professional positions, many earn less income than men do. Women who are trying to get into, or are employed in male-dominated occupations, continue to face many occupational barriers. Other aspects like different communication styles, relationship styles, role responsibilities in being primary caregivers, absence of mentoring programs, and the need for flexibility on work assignments set women apart from men.

Women work differently than men do; for example, they juggle business, home, and often, academic responsibilities using the following approaches:

- multi-tasking;
- delegating;
- networking to save time;
- working in a mobile fashion;

- accepting chaos and rising above it;
- being flexible;
- exercising to stay well; and
- valuing their families, placing them number one in terms of priority.[88]

Women bring to the workplace a greater level of emotional intelligence, cultural understanding, creativity and innovation, and consensus decision-making.[89]

Women are impacted by work in a manner that differs from men. For example, men report twice the number of work-related injuries as do women, except regarding the incidence of musculoskeletal injuries. Women report slightly more musculoskeletal injuries.[90] Post-injury, women take longer to return to work. The basis for this finding may be the complexity of the work, the injury, and/or the gender differences in healthcare management of the injury.[91] The effects of low job control are associated with the risk of hypertension in men, but not for women. It is also associated with diabetes in women; but not for men.[92]

In terms of well-being, women over the age of 40 years, experience more stress than do men or younger women.[93] This phenomenon is believed to be related to balancing work and family issues. From a work perspective, the higher the amount of work interference with family duties, the greater the amount of stress experienced by women. For example, business travel, which requires making alternate childcare and eldercare arrangements, is more stressful for women than for men. High job strain, for women over the age of 40, is associated with a 40 per cent increased risk of cardiovascular disease.[94] From a home perspective, the amount of caregiving that the woman is committed to provide is directly proportional to the amount of strain experienced. Additionally, as women age, they tend to be more involved in

---

[88] C. Krischer Goodman, "Women Professionals Multi-task to the Max", *McClatchy Newspapers* (April 10, 2010).

[89] H. Nash, "Women Bring Radical New Perspectives to the Board – Inspire, Harvey Nash Report", *Inspire* (March 2012), online: http://www.prnewswire.com/news-releases/women-bring-radical-new-perspectives-to-the-board---inspire--harvey-nash-report-142442725.html.

[90] P. Smith, "IWH to Explore How Work Impacts Women and Men Differently", (Fall 2014) 78 *At Work Issue* 7.

[91] *Ibid.*

[92] *Ibid.*

[93] L. Duxbury & C. Higgins, *Report Four: Who is at Risk? Predictors of Work-Life Conflict* (Ottawa: Public Health Agency of Canada, 2005) at 7, 9, 10, online: http://publications.gc.ca/site/eng/289209/publication.html (date accessed: January 31, 2020).

[94] M. Albert, cited in "Job strain takes toll on women", *Canadian HR Reporter* (November 29, 2010).

additional caregiving activities (aging parents, and grandchildren).[95]

Women are more likely to be physically inactive than are men, as well as more likely to report that they regularly experience pain and to sustain injuries. For women, the leading cause of death is cancer and heart disease.[96]

Both men and women experience more chronic health conditions as they age, however women differ from men in the nature of these disabilities.

For example, they are more likely to suffer arthritic conditions,[97] mobility problems, pain-related disabilities, and vision problems, whereas men tend to have more hearing and speech problems.[98] Some of these conditions tend to worsen with age.

Companies that have a predominantly female workforce are advised to provide family-friendly workplaces. Supportive policies and programs such as flexible work hours, flex days, childcare and eldercare information, daycare, EAP, family days, *etc.*, help women to balance their work and family commitments, thereby enabling them to meet organization/company performance expectations.

(a) *Recommendations for OH&S and Disability Management Professionals/ Practitioners*

OH&S and Disability Management Professionals/Practitioners are encouraged to:

- educate management and union leaders on the specific needs of the female worker;
- structure OH&S and Disability Management Program approaches/services so that they adequately address the needs of the female worker;
- seek opportunities for enhancing the work/life balance, especially for women;
- suggest caregiving support be included in the organization's suite of employee group benefits; and
- when accommodating female workers, consider the family responsibilities that may be impacting the worker's recovery and return to work.

3. Generations

Like age, gender, and ethnic diversity, each generation has it own beliefs, attitudes,

---

[95] *Ibid.*

[96] Statistics Canada, *Women in Canada: Gender-based Statistical Report, 2010-2011, Sixth Report*, #89-503-X (2012), online: http://www.statcan.gc.ca/pub/89-503-x/2010001/article/11543-eng.htm#a7 (date accessed: January 31, 2020).

[97] *Ibid.*

[98] Statistics Canada, *A Profile of Disability in Canada, 2001* (Ottawa: Statistics Canada, 2003), online: http://www.statcan.gc.ca/pub/89-577-x/index-eng.htm (date accessed: January 31, 2020).

values, communication styles, lifestyles, and work styles. Each has a unique set of expectations of management and the work environment. The outcome is that these generational differences can affect everything — employee recruitment, selection and retention practices; leadership; building work teams; managing and motivating employees; and maintaining and increasing productivity. This includes the management of employee disability and the return-to-work practices.

Generational diversity is a factor to consider in OH&S and Disability Management programming. This topic and the recommendations for OH&S and Disability Management Professionals/Practitioners are covered in detail in Chapter 35, "Impact of Five Generations in the Workplace on Occupational Health and Safety Programs".

4. Ethnic Groups

As already noted, immigration is on the increase in Canada and the United States. In the past five years, immigration has accounted for 70 per cent of the growth in Canada's labour force.[99] Each ethnic group brings with it a unique set of characteristics that impacts their lives, social interactions and well-being. To assist OH&S and Disability Management Professionals/Practitioners, summary descriptions of some ethnic groups are provided in Table 34.5.

Table 34.5: Summary Descriptions of a Variety of Ethnic Groups[100]

African Americans	Asians
Religions: Christian and Islam	*Religions*: Buddhism, Christianity, Hinduism, Islam, Sikh, Jain, Parsi and other traditional faiths
Family Issues: Raising children is traditionally a family/"whole village" affair: Matriarchal households	*Family Issues*: Patriarchal with ancestor reverence
Healthcare: Often uncomfortable and mistrustful of the modern healthcare system. Fear being diagnosed with a terminal disease or experiencing an invasion of privacy as a result of diagnostic procedures	*Healthcare*: Prefer healthcare practitioner of same sex. Many expect treatment that includes the administration of medication

---

[99] L. Duxbury, "Managing a Changing Workforce" (presented in Calgary, AB, 2004) [unpublished].

[100] Adapted from General Healthcare Resources Inc., *Cultural Diversity* (2005) at 3, online: http://www.ghresources.com/ghr/healthcare.nsf/2010-OSHA-In-Services.pdf (date accessed: January 31, 2020).

African Americans	Asians
Economic: Generally, members of the lower economic level of society	*Beliefs*: Numbers are important, with the lucky numbers being 3 and 8. The number 3 sounds like the word for "life"; number 8 sounds like the word for prosperity. The number 4 is very unlucky — the word sounds like "death"
Education: Lower levels of formal education	*Social*: Bowing versus shaking hands is preferable. Impolite to make and hold eye contact with elders or those in positions of power. Smiling masks emotions of anger, frustration and lack of knowledge or unhappiness
Lifestyle: Poor health status indicators	
Susceptibility to Health Problems: • Higher incidence of high blood pressure, sickle cell anemia, diabetes, obesity, cardiovascular disease, lactose intolerance • Women more likely to have HIV • Cancer death rates are about 35 per cent higher than the regular population • Diseases like prostate and breast cancer progress faster than in the regular population	*Susceptibility to Health Problems*: • For women, the common cancers are lung, breast, colon, stomach and pancreatic cancers • Higher rates of cervical cancer and mortality than in regular population • For Chinese men, common cancers are liver, colon, stomach and nasopharynx cancers • Cambodians tend to get tuberculosis, hepatitis B, and intestinal parasites • Lactose intolerance is common

Eastern Europeans	Hispanics/Latinos
Religions: Orthodox Christianity, Roman Catholicism, Islam and Judaism	*Religion*: Roman Catholicism. The Caribbean peoples have many spiritual traditions, beliefs and practices that originated in the American and African cultures
Social: Moral and physical support provided by relatives	*Social*: • Value eye contact • Friendly physical contact is common • Friendliness and treating others respectfully are valued

Eastern Europeans	Hispanics/Latinos
Family: Practice extended-family living: Multiple generations live together	***Family***: • Time with family and friends is highly valued • Children are highly valued and loved. • Parental discipline of young is mild
Healthcare: Open expression of emotions — can find personal questions uncomfortable. Note taking by healthcare providers can be viewed as suspicious. Sick individuals are encouraged to communicate suffering with others. Treatment is not complete with the issuing of a prescription	***Family***: Often workers send money "home" to extended-family members
Lifestyle: • Excessive alcohol use is common • Smoking is common among men • Limited awareness of the harmful effects of second-hand smoke • Little awareness of the importance of exercising	***Beliefs***: Education, degrees valued
Diet: • Food is culturally important: Good appetites are admired	***Diet***: Cakes and sweets are part of the regular diet
Susceptibility to Health Problems: • Higher incidence of digestive system diseases in men • Higher incidence of musculoskeletal problems in women • More smoking and obesity-related conditions	***Susceptibility to Health Problems***: • High rates of diabetes, hypertension, obesity, cervical cancer

Western Asia/Middle East	Pacific Islanders
Religions: Islam, Christianity, Jewish, Bahai, Druze, Parsi, and Zoroastrian	***Religions***: Holistic view of the world. Emphasize the interconnectedness of person, family, environment and spiritual world
Family: When outside, women are secluded from men	***Family***: Tight knit communities. Family, community and church play prominent roles

Western Asia/Middle East	Pacific Islanders
Healthcare: Prefer healthcare provider to be the same sex as the patient. Treatment expected to include the administration of pills, injections or minor procedures	*Healthcare*: Basic distrust of Western approaches to healthcare and treatment. "Scare tactics" to motivate behaviour rarely works
Diet: Most follow a strict kosher diet. Muslims fast during the holy month with no food or drink between sunrise and sunset. Fasting may include the avoidance of medications or injections. Ill people are exempt	*Social*: • Respect ancestors and elders • Interpersonal and social behaviour based on mutual respect and sharing
Lifestyle: Muslims do not consume alcohol	*Economic*: Low income and poverty are risk factors for health status
Susceptibility to Health Problems: • Egyptians have a higher incidence of parasitic diseases, blindness, typhoid fever, streptococcal disease, rheumatic fever and tuberculosis • Prone to obesity, high blood pressure, lower back pain, cardiovascular disease, diabetes, hepatitis A and B, tuberculosis, syphilis and gastric problems	*Susceptibility to Health Problems*: • Hawaiians experience high rates of breast, lung, ovarian and stomach cancer, and leukemia and non-Hodgkin's lymphoma • Hawaiians have high rates of heart disease, high blood pressure, obesity, diabetes, cancer and stroke

Sub-Saharan Africans	Native Americans/Canadian First Nations People
Religions: Wide variety of religions and languages	*Religions*: Holistic view of the world, life and health
Family Issues: Family is composed of people from the village, friends and distinct blood relatives. Some practice polygamy	*Family Issues*: Family and tribal affiliations and obligations are important
Healthcare: Males and females are circumcised in most countries	*Healthcare*: Integrate physical, social, psychological and spiritual ways of healing
Diet: Eat root crops: prefer cooked vegetable to raw ones; season foods with hot peppers	
Lifestyle: Close friends greet each other by shaking hands and asking about the health of the person and their family	*Economic*: Poverty, poor nutrition, stress and inadequate access to healthcare negatively impacts health status and well-being

Sub-Saharan Africans	Native Americans/Canadian First Nations People
Susceptibility to Health Problems: • High incidence of sickle cell anemia, lactose intolerance, malaria, dental caries, parasite diseases, post-traumatic stress and complications of female genital mutilation	***Susceptibility to Health Problems:*** • High rates of diabetes, colon and rectal cancers, and tuberculosis • Poor cervical cancer outcomes • Lactose intolerance

According to the Institute for Work and Health:

> Immigrants are exposed to a higher burden of work-related health and safety risks.[101] Immigrant men are twice as likely to require medical attention for injuries occurring at work as compared with Canadian-born men.[102]

People of all cultures share common needs in terms of personal health and safety. That is, we need to talk about our illnesses and/or health and safety concerns; we need to get competent healthcare; we need to be acknowledged, understood and valued;[103] and we need a social support system.

Based on the above descriptions of the various ethnic groups found in our society, OH&S and Disability Management Practitioners/Professionals can gain an appreciation that today's workplaces are composed of multicultural workforces with different views on life and approaches to work and life events. To complete the picture, add to this mix an awareness of one's own culture and cultural biases.

N. WESTERN CULTURE

As OH&S and Disability Management Practitioners/Professionals, we need to recognize and understand our own Western cultural beliefs and practices. In our Western society, we tend to value:

- individualism;
- personal responsibility;[104]
- a sense of personal control;
- the nuclear family as opposed to the extended family;[105]

---

[101] Institute for Work and Health, *Impact: Annual Report 2008* at 1, online: http://www.iwh.on.ca/annual-report (date accessed: January 31, 2020).

[102] *Ibid.*, at 2.

[103] R.A. Levy & J.W. Hawks, "Multicultural Medicine and Pharmacy" (1996) 7:3 *Drug Benefit Trends* 30.

[104] C. Brannon, *Cultural Competency: Values, Traditions and Effective Practice* (Nutrition Dimension Inc., 2007) at 5, online: http://www.rd411.com/ce_modules/CUC06.pdf (date accessed: January 31, 2020).

[105] *Ibid.*; see also P.G. Kittler & K.P. Sucher, *Food and Culture*, 3d ed. (Belmont, CA:

- direct, open, and honest communication;
- the "right to know";
- informality, which is considered akin to friendliness — hence, using first names is acceptable;[106]
- a future-orientation — setting short-, medium- and long-term goals is practiced;[107]
- the desire to work hard and provide a better future for our offspring;
- promptness and keeping to a set schedule;[108]
- being task-oriented;[109]
- personal accomplishments — they equate to our self-worth;
- our physical appearance — it determines "the first impression" of the person and is linked with self-esteem;[110]
- thinness — it is a societal obsession;[111]
- self-determination — fate is not a determining force;[112]
- monochronistism — we focus on and perform tasks sequentially;[113] and
- women as "partners", if not "equals".

Wadsworth/Thompson Learning, 2001); About News, *Cultural Diversity: How it Boosts Profits*, online: http://useconomy.about.com/od/suppl1/g/Cultural-Diversity.htm (date accessed: January 31, 2020).

[106] *Ibid.*

[107] C. Brannon, *Cultural Competency: Values, Traditions and Effective Practice* (Nutrition Dimension Inc., 2007) at 5, online: http://www.rd411.com/ce_modules/CUC06.pdf (date accessed: January 31, 2020).

[108] P.G. Kittler & K.P. Sucher, *Food and Culture*, 3d ed. (Belmont, CA: Wadsworth/Thompson Learning, 2001); G.G. Hall, *Culturally Competent Patient Care: A Guide for Providers and Their Staff* (Institute for Health Professions Education, October 2001), online: http://www.mercycareplan.com/ProviderManual/Appendix_2008.pdf (date accessed: January 31, 2020).

[109] C. Brannon, *Cultural Competency: Values, Traditions and Effective Practice* (Nutrition Dimension Inc., 2007), online: http://www.rd411.com/ce_modules/CUC06.pdf (date accessed: January 31, 2020).

[110] *Ibid.*

[111] J. Sobal, "Social and Cultural Influences on Obesity" in P. Björntorp, ed., *International Textbook of Obesity* (London: John Wiley and Sons, 2001) at 305.

[112] C. Brannon, *Cultural Competency: Values, Traditions and Effective Practice* (Nutrition Dimension Inc., 2007), online: http://www.rd411.com/ce_modules/CUC06.pdf (date accessed: January 31, 2020).

[113] P.G. Kittler & K.P. Sucher, *Food and Culture*, 3rd ed. (Belmont, CA: Wadsworth/Thompson Learning, 2001); About News, *Cultural Diversity: How it Boosts Profits*, online:

In contrast, the more traditional cultures tend to value:
- the welfare of the extended family over the individual;
- the extended family's involvement in decision-making;[114]
- personal relationships which take precedence over promptness and time schedules;[115]
- their belief in "fate", God, or some other supernatural factors to determine a person's health and destiny;
- politeness;
- polychronistism — being able to do many things at once — multi-tasking;[116] and
- men as typically being the head of the household.[117]

In our Western society, the modern healthcare system tends to portray Western cultural values. However, one system of healthcare is not superior to the traditional healthcare systems — just different. They can operate side-by-side to the benefit of the individual and the family unit.

O. CULTURAL VIEWPOINTS: APPROACH TO WORK AND THE WORLD OF WORK

Based on the cultural information provided, it is obvious that it is in the employer's/organization's best interest to be aware of, and to understand, how various ethnic groups approach life and how that cultural orientation impacts their world of work and work practices.

To better explain this phenomenon, a comparison of the three dominant cultures found in Canada is provided (Table 34.6).

---

http://useconomy.about.com/od/suppl1/g/Cultural-Diversity.htm (date accessed: January 31, 2020).

[114] C. Brannon, *Cultural Competency: Values, Traditions and Effective Practice* (Nutrition Dimension Inc., 2007), online: http://www.rd411.com/ce_modules/CUC06.pdf (date accessed: January 31, 2020).

[115] *Ibid.*

[116] P.G. Kittler & K.P. Sucher, *Food and Culture*, 3rd ed. (Belmont, CA: Wadsworth/Thompson Learning, 2001); About News, *Cultural Diversity: How it Boosts Profits*, online: http://useconomy.about.com/od/suppl1/g/Cultural-Diversity.htm (date accessed: January 31, 2020).

[117] *Ibid.*

Table 34.6: Cultural Viewpoints: The Approach to Work

Business Aspect	Western Perspective	First Nations Perspective	Asian Perspective
Meetings	The focus is on completing tasks and achieving goals. Time considerations are usually a driving force. Agendas, deadlines, and schedules are specifically set out. Punctuality is valued.	Interpersonal relations and affiliations are important when starting a discussion. The outcome is a prolonged discussion with the task being secondary. Meetings may not be scheduled; dealings and accomplishments are based on need, attendance and consensus. Punctuality is not expected.	The desire is to keep the conversation smooth and harmonious. Issues, circumstances and relationships are as important as the work. Meetings, agendas, schedules and punctuality are handled with regard for the individuals involved. In essence, human relations play an important role.
Individual versus Group Importance	Individual importance is valued over the importance of the group.	Group importance is valued over individual importance.	Group importance is valued over individual importance.
Competitiveness	Competition and confrontation are accepted, and individual initiative is valued. Criticism and confrontation are accepted in order to "get the job done". Criticism and opposition may be used to expose the full picture.	Co-operation brings the best results; harmony and personal humility are important in the process. Personal and group honour and dignity are valued and preserved. Criticism, disagreements or unsolicited suggestions are avoided.	Conversations are harmoniously conducted. This means refraining from open disagreements, asking difficult questions of superiors, publicly embarrassing a person, or saying things that will cause problems. Differences are best worked out quietly.

Business Aspect	Western Perspective	First Nations Perspective	Asian Perspective
Goal Achievement	Tasks are compartmentalized and considered one at a time.	Information and ideas are dealt with in the widest possible context.	Decisions are made by consensus.
	The facts directly related to the issue are presented with an emphasis on reaching a solution. Clarity is expected.	Several suggestions may be offered simultaneously, and all are considered.	The context of where and how comments are heard can be more important than what was said.
	Rational, logical, linear problem solving is valued.	This prolongs and enriches the process of problem-solving.	Communication is indirect and implicit.
	Accuracy and perfection are expected.	Problem solving is intuitive, creative and holistic.	Problem-solving is handled with regard for the individuals involved.
	Timelines and completion expectations are required to achieve the desired goals.	Inaccuracy and error are accepted.	Accuracy and perfection are expected.
	Problems are solved by the leader or when a group vote decides a course of action.	While completion is at times important, time elements are seldom considered.	Timelines and completion expectations are required to achieve the desired goals.
	Ruthless measures may be taken to attain results.	Group decision-making prevails, conflicts are resolved through consensus after divergent ideas are debated.	Decisions are made by consensus.
		Loss of dignity and disharmony is avoided.	Behaviours are directed primarily to maintaining congenial relations and affiliation within the group.
Directing Work	Direct orders and instruction are readily given.	Rather than direct orders, suggestions are better received.	Direct orders are avoided.

Business Aspect	Western Perspective	First Nations Perspective	Asian Perspective
Performance Feedback	Aversion to criticism, heeding advice, soliciting feedback may be viewed as a lack of commitment, motivation, confidence, enthusiasm, or knowledge.	Criticism, advice, confrontation and emotional outbreaks are viewed as a lack of maturity or respect.	Self-disclosure and frankness about one's emotions are viewed as inappropriate.
Leadership	Male dominance presides.	Women are expected to assume a leadership role in families and are highly regarded in tribes where they often sit as elders.	Elders are revered.
Knowledge	Knowledge is for controlling peace and order.	Knowledge is for the sake of living in harmony with nature.	Knowledge is for the sake of living in harmony with nature and man.

In essence:

> Cultures are like icebergs; some features are apparent to anyone not in a fog, while others are deeply hidden [and] are so far below the surface that they are hard to recognize.[118]

P. CULTURAL DIVERSITY MANAGEMENT

> The development of cultural competency may be best thought of not as arriving as a set of skills and knowledge, but rather as a journey and a way of being.[119]

When addressing how cultural competency can be attained, it is important to view the approach from two perspectives:

1. the Organizational Approach; and
2. the OH&S Practitioner/Professional Approach.

---

[118] E. Winters, "Cultural Issues in Communication" (2002), online: http://www.citehr.com/12441-cross-cultural-issues-communiation.html (date accessed: January 31, 2020).

[119] L. Olsen, J. Bhattacharya & A. Scharf, *Cultural Competency: What It Is and Why It Matters*, (California Tomorrow, Lucile Packard Foundation for Children's Health, 2006), online: http://www.lpfch.org/programs/culturalcompetency.pdf (date accessed: January 31, 2020).

1. The Organizational Approach

Cultural diversity is not enough; organizations must demonstrate inclusiveness.[120] **Inclusiveness** translates to employees recognizing that they are valued and belong.[121] This desired state is attained when an organization and the workplace are respectful, stable, productive, innovative, and energized.[122]

One challenge for employers/organizations is to adopt the "Platinum Rule": Treat people as they wish to be treated, not like you want to be treated.[123] This requires a change in mindset, something that many companies struggle to achieve. In essence, it involves looking for the strengths that employees bring to the workplace, not their differences.

There is a definite link between inclusiveness and cultural competency. For organizations, being culturally competent is an ongoing journey — one of intentional and continuous practices aimed at including employees from all cultural orientations in the organization's strategic planning and operations. It involves learning about, and responding to, the various cultural contexts of the community at large and the people it serves. Culturally competent organizations continuously strive to bridge the culture gaps with a view to addressing the needs of all employees.

The characteristics of culturally competent organizations are as follows:

- valuing diversity, equality, and institutionalizing these values in policy;
- being self-reflective on what they have done and what yet needs to be achieved;
- integrating cultural knowledge into the organization's work;
- encouraging and supporting staff to become culturally competent;
- addressing inequities; and
- integrating cultural competence into the organization's programs.

2. The OH&S Practitioner/Professional Approach

Given that OH&S and Disability Management Practitioners/Professionals will be providing services and assisting employees with a variety of cultural orientation, they are strongly encouraged to develop cultural competency and hopefully cultural proficiency.

In the fields of Occupational Health and Safety, and Disability Management,

---

[120] Alberta Employment and Immigration, *Employing a Diverse Workforce: Making it Work* (2008) at 9, online: http://alis.alberta.ca/pdf/cshop/employdiverse.pdf (date accessed: January 31, 2020).

[121] *Ibid.*

[122] *Ibid.*

[123] S. Calhoun & P. Strasser, "Generations at Work" (2005) 53:11 *AAOHN* 469.

instead of trying to "force fit" employees to adapt to established practices, consider redesigning those practices so that they adopt some of the employees' cultural beliefs/ practices. Create a culture of inclusion. Involve workers in the design of those programs and accommodate cultural differences where possible. Listen to the workers and show trust in them by seeking their opinions on the things that matter to them.

For example, when planning an individual employee's return-to-work placement, the Disability Management Practitioner/Professional should consider factors such as the:

- impact of cultural attitudes towards illness/injury;
- role of the family in the illness, treatment regimen, and recovery process;
- relationship development within the culture;
- preferred communication style;
- cultural aspects that could impact the return-to-work process;
- mechanisms of problem-solving and goal-setting;
- potential for cultural conflicts between the Disability Management Coordinator and worker;
- potential for the worker to freely express concerns or issues about the return-to-work process; and
- presence/absence of language barriers.

Recognize that not all diversity is visible. Be aware that disability brings many diversity issues to light. Seek understanding and offer your respect. Try to look at the situation through the eyes of the ill/injured person and determine where the challenges for recovery and return to work exist. Most of all, promote employee responsibility for the problems and solutions.

To achieve this end, the OH&S and Disability Management Practitioner/ Professional must develop good listening skills and observe without making judgements. It also means developing and practicing empathy, a comfort level with differences, self-awareness, self-reflection, flexibility, and an appreciation of multiple perspectives.

As already mentioned, cultural competency is a journey. This means that OH&S and Disability Management Practitioners/Professionals should seek ongoing opportunities to learn about cultural competency and to hone their skills. However, the first step is a personal commitment of cultural competency.

> If I waited until I knew everything about how to address diversity issues, I would never get started. The learning really begins once you get going!
>
> Julie Edwards[124]

Q. SUMMARY

Cultural competence, in terms of Occupational Health & Safety and Disability

---

[124] J. Edwards, "Looking In, Looking Out: Redefining Early Care and Education in a

Management, requires a combination of knowledge, skills, and behaviours that lead to positive return-to-work outcomes for ill/injured employees with ethnically and culturally diverse orientations. Central to cultural competency is the provision of the OH&S Program and Disability Management Program education, information, and services in a manner that meets the needs of a diverse workforce.

CHAPTER REFERENCES

About News, *Cultural Diversity: How it Boosts Profits*, online: http://useconomy.about.com/od/suppl1/g/Cultural-Diversity.htm (date accessed: January 31, 2020).

M. Albert, cited in "Job strain takes toll on women", *Canadian HR Reporter* (November 29, 2010).

Alberta Employment and Immigration, *Employing a Diverse Workforce: Making it Work* (2008), online: http://alis.alberta.ca/pdf/cshop/employdiverse.pdf (date accessed: January 31, 2020).

American Association of Colleges of Nursing, "Cultural Competency in Baccalaureate Nursing Education" (2008), online: http://www.aacn.nche.edu/education/pdf/competency.pdf (date accessed: January 31, 2020).

AWCBC, "Lost Time Claims in Canada, 2018" and "Fatalities in Canada, 2018" (2020), online at: http://awcbc.org/?page_id=14 (date accessed, February 28, 2020).

Avesis, *Cultural Competency: 2010* (2008 revision, approved January 2009), online: http://avesis.com/pdf/cultural_competency.pdf (date accessed: January 31, 2020).

J. Bell, "It's Time for a Rethink Around LTD Coverage for Older Employees", *Canadian Safety Reporter* (June 2014) at 4.

C. Brannon, *Cultural Competency: Values, Traditions and Effective Practice* (Nutrition Dimension Inc., 2007), online: http://www.rd411.com/ce_modules/CUC06.pdf (date accessed: January 31, 2020).

T. Buller, "A Flexible Combination", *Benefits Canada* (November 2004) at 99, online: http://www.benefitscanada.ca (date accessed: January 31, 2020).

S. Calhoun & P. Strasser, "Generations at Work" (2005) 53:11 *AAOHN* 469.

Citizenship and Immigration Canada, *Facts and Figures: Immigration Overview — Permanent and Temporary Residents 2015* (2016), online: https://www.canada.ca/en/immigration-refugees-citizenship/corporate/reports-statistics/statistics-open-data.html (date accessed: January 31, 2020).

Conference Board of Canada, *Young, Underpaid and Angry* (September 2014),

---

Diverse Society" (2006) in L. Olsen, J. Bhattacharya & A. Scharf, *Cultural Competency: What It Is and Why It Matters* (California Tomorrow, Lucile Packard Foundation for Children's Health, 2006), online: http://www.lpfch.org/programs/culturalcompetency.pdf (date accessed: January 31, 2020).

online: http://www.conferenceboard.ca/press/speech_oped/14-09-24/young_underpaid_and_angry_the_coming_clash_over_the_income_gap.aspx. (date accessed: February 28, 2020).

J. Cowell, "Fitness to Work" (presented at the Conference on Workers' Compensation, Calgary, 1996) [unpublished].

T. Cross, B. Bazron, K. Dennis & M. Isaacs, *Towards a Culturally Competent System of Care*, vol. 1 (Washington, DC: CASSP Technical Assistance Center, Center for Child Health and Mental Health Policy, Georgetown University Child Development Center, 1989), online: http://www.nccccurricula.info/documents/TheContinuumRevised.doc (date accessed: January 31, 2020).

D. Crowe & M. Hogan, *Cultural Diversity in the Workplace: Discussion Paper* (IMI Bizlab on Cultural Diversity, 2007), online: http://citeseerx.ist.psu.edu/viewdoc/summary?doi=10.1.1.126.3933 (date accessed: January 31, 2020).

L. Duxbury, "Managing a Changing Workforce" (Presented in Calgary, AB, 2004) [unpublished].

L. Duxbury & C. Higgins, *Report Four: Who is at Risk? Predictors of Work-Life Conflict* (Ottawa: Public Health Agency of Canada, 2005), online: http://publications.gc.ca/site/eng/289209/publication.html (date accessed: January 31, 2020).

C.L. Edelman & C.L. Mandle, *Health Promotion Throughout the Lifespan*, 4th ed. (St. Louis, MO: Mosby, 1998).

J. Edwards, "Looking In, Looking Out: Redefining Early Care and Education in a Diverse Society" (2006) in L. Olsen, J. Bhattacharya & A. Scharf, "Cultural Competency: What It Is and Why It Matters", *California Tomorrow, Lucile Packard Foundation for Children's Health* (December 7, 2006), online: http://www.lpfch.org/informed/culturalcompetency.pdf (date accessed: January 31, 2020).

J.L. Elliott & A. Fleras, *Unequal Relations: An Introduction to Race and Ethnic Dynamics in Canada* (Scarborough, ON: Prentice-Hall Canada, 1992).

General Healthcare Resources Inc., *Cultural Diversity* (2005), online: http://www.ghresources.com/ghr/healthcare.nsf/2010-OSHA-In-Services.pdf (date accessed: January 31, 2020).

Institute for Work and Health, *Impact: Annual Report 2008*, online: http://www.iwh.on.ca/annual-report (date accessed: January 31, 2020).

P.G. Kittler & K.P. Sucher, *Food and Culture*, 3d ed. (Belmont, CA: Wadsworth/Thomson Learning, 2001).

A. Kleinman, *Patients and Healers in the Context of Culture* (Berkeley, CA: University of California Press, 1981).

C. Krischer Goodman, "Women Professionals Multi-task to the Max", *McClatchy Newspapers* (April 10, 2010).

M. Leininger & M. McFarland, *Culture Care Diversity and Universality: A*

Worldwide Nursing Theory, 2d ed. (Sudbury, MA: Jones and Bartlett, 2006).

R.A. Levy & J.W. Hawks, "Multicultural Medicine and Pharmacy" (1996) 7:3 *Drug Benefit Trends* 27.

P. Lister (1999) reported in *National Review of Nursing Education — Multicultural Nursing Education*, "5. Australian Nursing Education Today" (2001), online: http://www.dest.gov.au/archive/HIGHERED/nursing/pubs/multi_cultural/5.htm (date accessed: January 31, 2020).

D. Lozorik, *Pilot Study to Develop a Tool to Elicit Khmer Beliefs about the Cause of Illness* (Master's Thesis, Boston University School of Medicine, 1984) [unpublished].

M. Mancini, "Aging workforce poses health, safety risks" *Canadian Safety Reporter* (July 2012) at 1-2.

T. McDonald & H. Harder, "Older Workers and Disability Management" (2004) 3:3 *International Journal of Disability, Community & Rehabilitation* 1, online: http://www.ijdcr.ca/VOL03_03_CAN/index.shtml (date accessed: January 31, 2020).

D. Morris, *Bodytalk: The Meaning of Human Gestures* (New York: Crown Trade Paperbacks, 1994) in *The Provider's Guide to Quality & Culture* (2009) "Topic 9: Non-verbal Communication", online: http://erc.msh.org/mainpage.cfm?file=4.6.0.htm&module=provider&language=English (date accessed: January 31, 2020).

H. Nash, "Women Bring Radical New Perspectives to the Board – Inspire, Harvey Nash Report", *Inspire* (March 2012), online: http://www.prnewswire.com/news-releases/women-bring-radical-new-perspectives-to-the-board---inspire--harvey-nash-report-142442725.html (date accessed: February 28, 2020).

H.V. Ngo, *Cultural Competency: A Self-Assessment Guide for Human Service Organizations* (Calgary, AB: Cultural Diversity Institute, 2000).

L. Olsen, J. Bhattacharya & A. Scharf, *Cultural Competency: What It Is and Why It Matters* (California Tomorrow, Lucile Packard Foundation for Children's Health, 2006), online: http://www.lpfch.org/programs/culturalcompetency.pdf (date accessed: January 31, 2020).

R. Pauly, *Cultural Diversity: Increasing Awareness* (University of Florida, Department of Medicine, 2003).

Seattle King County Dept of Public Health (1994), reported in R. Fiorelli & W. Jenkins, "*Cultural Competency in Grief and Loss*" (2009), online: http://www.nhpco.org/sites/default/files/public/education/Cultural_Comptency_Grief-Loss_Sept_09_NL.pdf (date accessed: January 31, 2020).

P. Smith, "IWH to Explore How Work Impacts Women and Men Differently" (Fall 2014) at 78 *At Work Issue* at 7.

J. Sobal, "Social and Cultural Influences on Obesity" in P. Björntorp, ed., *International Textbook of Obesity* (London: John Wiley and Sons, 2001) at 305.

Statista, "Number of Immigrants in Canada from 2000-2019" (2020), online: https://www.statista.com/statistics/443063/number-of-immigrants-in-canada/ (date accessed: January 31, 2020).

Statistics Canada, *Census in Brief: Ethnic and cultural origins of Canadians: Portrait of a rich heritage* (2016), online: https://www12.statcan.gc.ca/census-recensement/2016/as-sa/98-200-x/2016016/98-200-x2016016-eng.cfm (date accessed: January 31, 2020).

Statistics Canada, *A Profile of Disability in Canada, 2001* (Ottawa: Statistics Canada, 2003), online: http://www.statcan.gc.ca/pub/89-577-x/index-eng.htm (date accessed: January 31, 2020).

Statistics Canada, *Labour Statistics at a Glance: Reasons for Working at 60 and beyond*, 71-222-X (2018), online: https://www150.statcan.gc.ca/n1/pub/71-222-x/71-222-x2018003-eng.htm (date accessed: February 28, 2020).

Statistics Canada, "Study: Canada's Immigrant Labour Market", *The Daily* (September 10, 2007), online: http://www.statcan.gc.ca/daily-quotidien/070910/dq070910a-eng.htm (date accessed: January 31, 2020).

Statistics Canada, *Table: 14-10-0193-01 (formerly CANSIM 279-0032): Absence rates for full time employees by age group, annual, 2019* (Ottawa: Statistics Canada, January 2020), online: https://www150.statcan.gc.ca/t1/tbl1/en/tv.action?pid=1410019301 (date accessed: February 28, 2020).

Statistics Canada, *Women in Canada: Gender-based Statistical Report, 2010-2011, Sixth Report*, Cat. No. #89-503-X (2012), online: http://www.statcan.gc.ca/pub/89-503-x/2010001/article/11543-eng.htm#a7 (date accessed: January 31, 2020).

U.S. Census Bureau, "Quick Facts: 2019", online: https://www.census.gov/quickfacts/fact/table/US/PST045219#PST04 (date accessed: January 31, 2020).

Wikipedia (2018), *s.v.* "Visible Minority", online: https://en.m.wikipedia.org/wiki/Visible_minority (date accessed: January 31, 2020).

Wikipedia, "Immigration to the United States", online: http://en.wikipedia.org/wiki/Immigration_to_the_United_States#Demographics (date accessed: January 31, 2020).

K. Williams, "Returning to Work After Disability: What Goes Wrong?" (presented at Canadian Human Resource Planners event, May 15, 2003, Calgary, AB) [unpublished].

E. Winters, "Cultural Issues in Communication" (2002), online: http://www.citehr.com/12441-cross-cultural-issues-communiation.html (date accessed: January 31, 2020).

Chapter 35

IMPACT OF FIVE GENERATIONS IN THE WORKPLACE ON OCCUPATIONAL HEALTH AND SAFETY PROGRAMS

A. INTRODUCTION

For the first time in history, there are five distinct generations working in today's workplaces. According to Howland, 90 per cent of Canadian workplaces employ four generations of workers[1] and now, the fifth generation entered in 2016. Forty per cent of the surveyed Canadians report that a multi-generational workplace adds challenges to the job.[2]

The issue is that each generation has its own beliefs, attitudes, values, communication styles, and work styles. Each has a unique set of expectations of management and the work environment. The outcome is that these generational differences can affect everything: employee recruitment, selection and retention practices; leadership; building work teams; managing and motivating employees; and maintaining and increasing productivity.

This unique situation of having five generations in the workplace raises questions like:

- What impact will each of these generations have on occupational health and safety, *i.e.*, attitudes, beliefs, values, communication styles, work styles, motivations, *etc.*?
- What considerations will need to be entertained in terms of worker OH&S education/training?
- How will the inter-generational differences impact worker injury rates and processes?

OH&S Practitioners/Professionals are advised to explore and appreciate the impact that sociological and demographic changes have on workers and the world

---

[1] A. Howland, "Multi-generations Bring Challenge to Workplace", *The Vancouver Province* (May 27, 2007), online: http://www.canada.com/theprovince/news/working/story.html?id=c25b6bfa-f64a-40a5-a7d8-7de807e3ce63 (date accessed: January 31, 2020).

[2] *Ibid.*

of work. Rethinking how OH&S services are packaged and delivered is strongly recommended.

B. WHAT IS A GENERATION?

The term "generation" was traditionally defined as "the average interval of time between the birth of parents and the birth of their offspring".[3] However, this definition has changed and today, the term **generation** is viewed as a set of people (cohort) born within the same period of time (approximately a 20-year span), and whose lives and viewpoints were shaped by the events within that span of time.[4] As a result, a given generational cohort shares similar unique values, attitudes, and behaviours.

According to Strauss and Howe,[5] the 20th century generations are:

Table 35.1: The 20th Century Generations

Generation	Other Names	Period of Time
Hero Generation	GI Generation, Greatest Generation[6]	1901-1924
Veteran Generation	Traditionalists, Depression Generation, Silent Generation	1925-1942
Baby Boomers	Boom Generation, Me Generation	1943-1964
Generation X	Baby Busters, 13th Generation, Post-Boomers	1965-1981
Baby Boomlets (Gen Y)	Generation Y, Echo Boomers, Nexus, Nexters, Millennial, Generation Me	1982-1996
Generation Z	Gen 2020ers, iGeneration, Homeland	1997+

Note: Depending on the country, sociologist, researcher or author, there is a great deal of variance in terms of the actual dates of the onset and closure of the generations.

---

[3] M. McCrindle, *New Generations at Work: Attracting, Recruiting, Retraining & Training Generation Y* (Sydney, NSW, Australia: McCrindle Research, 2006) at 8, online: http://www.mccrindle.com.au/resources/whitepapers/McCrindle-Research_New-Generations-At-Work-attracting-recruiting-retaining-training-generation-y.pdf (date accessed: January 31, 2020).

[4] *Ibid.*

[5] W. Strauss & N. Howe, *Generations: The History of America's Future, 1584 to 2069* (New York: Harper Perennial, 1992) (reprint).

[6] Term coined by journalist Tom Brokaw in his book, *The Greatest Generation* (New York: Random House, 1998).

It is important to note that individuals born on the "cusp of a generation" may have a blended set of characteristics of the primary and following generation.[7] Some authors term these individuals who demonstrate the traits of either generation as "cuspers".[8]

For the purposes of this text, the following generational dates will be used:
- Veteran Generation, born 1927-1945.
- Baby Boomers, born 1946-1964.
- Generation X, born 1965-1979.
- Baby Boomlet Generation (Gen Y), 1980-1996.
- Generation Z, 1997- yet to be determined.

C. WHAT IS A GENERATION GAP?

By nature, generations tend to be skeptical of the preceding and subsequent generations. The term, **generation gap**, refers to the differences between the members of two different generations.

This phenomenon can also be referred to as the "clash of the generations", with the reasons for the differences deemed as the clash points. A **Clashpoint®**[9] is an aspect of the workplace where generational differences of perspective, attitude and opinion tend to occur. They tend to include career goals, performance feedback, performance rewards, job changes, and retirement.[10]

Generations tend to differ on many of the following:
- the social, political, and economic influencers that they experience;
- family structure and dynamics;
- educational opportunities and format;
- life values;
- attitude toward work;
- work ethics;

---

[7] D. McPhail, "Corporate Perspectives on Workplace Learning: 'Four Generations in the Workforce'" (May 9, 2006, online: http://www.wln.ualberta.ca//en/Events/Archives/Seminar%20Archives/~/media/wln/Documents/Events/Seminars_2007/Generation_Workshop.pdf (date accessed: January 31, 2020).

[8] A. Stefaniak, *Black Hole or Window of Opportunity? Understanding the Generational Gap in Today's Workplace* (Policy Brief: 04-11-07, Center for Public Policy & Administration) at 4, online: http://www.cppa.utah.edu/_documents/publications/workforce/Generations.pdf (date accessed: January 31, 2020).

[9] L. Lancaster & D. Stillman, *When Generations Collide: Who They Are, Why They Clash, How to Solve the Generational Puzzle at Work* (New York: Harper Collins, 2002).

[10] *Ibid.*

- employee loyalty;
- preferred leadership styles;[11]
- work styles;
- communication styles;
- need for performance feedback;
- preferred delivery of performance feedback, both in frequency and method of delivery;
- their relationship with technology;
- motivational drivers;
- preferred degree of recognition and type of rewards for good performance;
- job satisfaction criteria;
- career expectations;
- relationship practices;
- spending habits; and
- the preferred balance between work and personal time.

The differences between the generations affect interpersonal, team, and supervisory interactions and relationships. To put this into perspective, the Veteran and Baby Boomer generations have erroneously assumed that the newer generations entering the workforce will measure success like they do and that they will be willing to climb corporate ladders and "pay their dues" to achieve success.[12]

D. UNDERSTANDING GENERATIONAL DIFFERENCES: THE BENEFITS

In recent years, much attention has been paid to the existence of the five generations in the workplace. Is this all "hype", or is it a topic for valid concern? Many folks now know it to be the latter; the belief is supported by recent Canadian demographic statistics. In 2018, Canada's total population was 37 million;[13] the median age was

---

[11] M. McCrindle, *New Generations at Work: Attracting, Recruiting, Retraining & Training Generation Y* (Sydney, NSW, Australia: McCrindle Research, 2006), online: http://www.mccrindle.com.au/resources/whitepapers/McCrindle-Research_New-Generations-At-Work-attracting-recruiting-retaining-training-generation-y.pdf (date accessed: January 31, 2020).

[12] C. Marston, *Motivating the "What's in It for Me?" Workforce: Manage Across the Generational Divide and Increase Profits* (Hoboken, N.J.: John Wiley & Sons, 2007).

[13] Statistics Canada, "Estimates of Population, by Age Group and Sex for July 1, Canada, Provinces and Territories, Annual (Persons Unless Noted Otherwise)" (2018) Table 051-0001, online: https://www150.statcan.gc.ca/t1/tbl1/en/tv.action?pid=1710000501 (date accessed: January 31, 2020).

40.2 years of age.[14] Figure 35.1 provides a breakdown of the Canadian population by generation:

Figure 35.1: Canadian Scene, 2018: Five Generations[15]

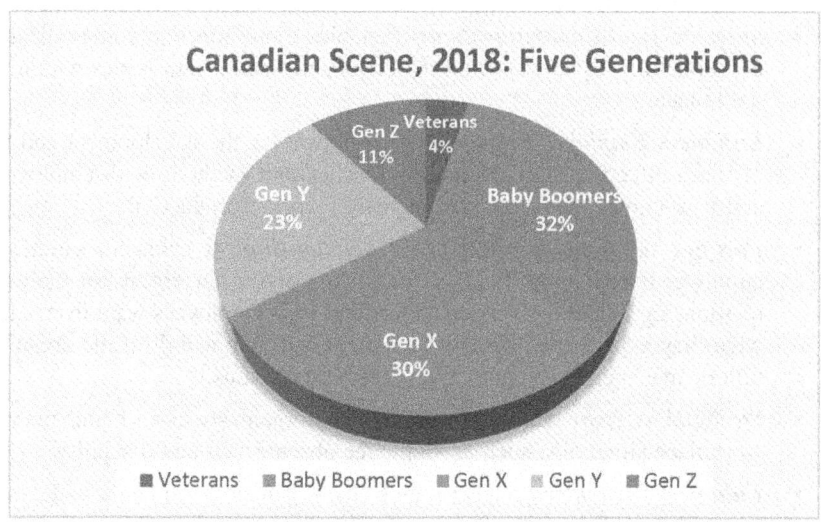

As graphically shown, in 2018, the Veterans constituted 4 per cent (1.1 million) of the workforce; Baby Boomers made up 32 per cent (9.4 million) of the workforce; Generation Xers were 30 per cent (8.9 million) of the workforce; Baby Boomlets (Gen Ys) numbered 23 per cent (6.8 million); and Gen Z totalled 11% (3.3 million). Hence, Canadian organizations and employers are advised to understand the generational differences which they will face.

The benefits of understanding the generational differences are that this approach:

- ***Supports the Corporate Culture***: By educating employees on generational differences, a milieu of acceptance, respect, and productivity can be achieved.[16,17]

---

[14] Statistics Canada, "Median age of the resident population of Canada in 2018, by province" (2019), online: https://www.statista.com/statistics/444816/canada-median-age-of-resident-population-by-province/.

[15] Calculated using Statistics Canada, "Table 051-0001: Estimates of Population, by Age Group and Sex for July 1, Canada, Provinces and Territories, Annual (Persons Unless Noted Otherwise)" (2018), online: http://www5.statcan.gc.ca/cansim/a26?lang=eng&retrLang= eng&id=0510001&paSer=&pattern=&stByVal=2&p1=-1&p2=37&tabMode= dataTable&csid= (date accessed: January 31, 2020). The workforce age was deemed 16-80 years of age. Where a generation cohort was mid-way in the 5-year age cohort, the population number was used as a percentage of that age cohort.

[16] *Ibid.*

- **Leads to Effective Recruitment, Selection and Retention**: With many organizations experiencing high staff retirement rates, there is a need to understand the major generational differences in order to both attract and keep good employees.
- **Supports Real Communication**: It enables more effective communication and hopefully, fewer misunderstandings and relationship issues within the workplace.
- **Enhances Employee Engagement**: It allows for the development and use of more effective motivational approaches and methods with employees, and it promotes teamwork and increased worker productivity.
- **Fortifies the Employer-Employee Relationship**: It enhances employer-employee expectations and hence, their respective perceptions of the work relationship. There is a need to understand what employees want from a job, what they expect from management and what they will do if the organization or management fails to meet their expectations.[18]
- **Strengthens Issue Management**: It helps organizations to better manage workplace situations, such as employee absenteeism and disability.
- **Enhances Customer Service**: It helps the organization to better understand and address customer preferences and needs across the age spectrum.

Organizations can evolve into "high-performance" workplaces by:

- understanding each generation — *e.g.*, how the generation was shaped, the key influencers, and the resultant impact on the generation's values, attitudes and behaviours;
- establishing an integrated system of policies, a supportive work environment, and work tools/technologies designed to bridge the generations and to bring out the "best" of each generation;[19] and
- creating a **"generation-neutral" workplace** — a workplace that values and utilizes the contributions of all generations.[20]

---

[17] AARP, *Leading a Multigenerational Workforce* (Washington, DC: AARP, 2007) at 4, online: http://assets.aarp.org/www.aarp.org_/cs/misc/leading_a_multigenerational_workforce.pdf (date accessed: January 31, 2020).

[18] D. McPhail, "Corporate Perspectives on Workplace Learning: 'Four Generations in the Workforce'" (May 9, 2006), online: http://www.wln.ualberta.ca//en/Events/Archives/Seminar%20Archives/~/media/wln/Documents/Events/Seminars_2007/Generation_Workshop.pdf (date accessed: January 31, 2020).

[19] K. Kirkpatrick, S. Martin & S. Warneke, *Strategies for the Intergenerational Workplace* (Gensler, July 17, 2008), online: http://www.gensler.com/uploads/documents/IntergenerationalWorkplace_07_17_2008.pdf (date accessed: January 31, 2020).

[20] AARP, *Leading a Multigenerational Workforce* (Washington, DC: AARP, 2007) at 4, online: http://assets.aarp.org/www.aarp.org_/articles/money/employers/leading_multigenerational_

E. THE FIVE GENERATIONS: DESCRIPTIONS

Although throughout the centuries generations have been identified, labelled, and described, the focus of this text will be on the five generations currently found in today's workplaces, namely:

1. Veterans and Traditionalists or the Depression Generation

Workers born between 1927 and 1945 make up the generation called The Veterans. Veterans believe in duty, honour, and country and are dedicated workers who value sacrifice, conformity, and patience. Dubbed the "Silent Generation", this generation brings a traditional and realistic attitude to the workplace. Today's business infrastructure was established by the Veterans, and their values and work ethic continue to influence modern business policies and procedures.[21]

Now averaging 75-93 years old, Veterans have retired, with some returning to the workforce due to financial reasons, or a desire to remain physically, psychologically, and socially active, as well as productive.[22]

The issue that employers face with these workers is that they are not sure how to use these very experienced, but older workers. Although Veterans are reliable workers, possess a wealth of experience, and have a strong work ethic, they are now for the most part, working in low-end service jobs. Employed in entry-level positions, they tend to make up the "shadow workforce" and absorb the "first jobs" typically held by workers entering the workplace — jobs that set the foundation for the building of work ethics of the generation entering the workplace.

Veterans, having been greatly influenced by the stock market crash and the Hindenburg Tragedy, Great Depression, World War II, VE Day, Winston Churchill, the Atomic Bomb, Korean War, and the Cold War, display traits such as:

- deep patriotic values;
- practicality;
- a prudent approach to life and spending — they tend to save their money to pad their "nest-egg";
- when spending, value quality and nationally-made products;
- a willingness to make personal sacrifices for delayed gratification;
- a black and white world view;
- strong family values — great belief in the nuclear family;
- strong religious beliefs and reliance on the church;

---

workforce.pdf (date accessed: January 31, 2020).

[21] *Ibid.*, at 9.

[22] E. Quick, "Disability Management that Works", *Benefits and Compensation Solutions Magazine* (January/February 2007) at 46.

- strong employee loyalty;
- acceptance of hierarchical leadership;
- respect of authority;
- value a command and control approach to management, and the chain of command;[23]
- do not value individualism — believe in working together towards a common good;
- participate in team work provided they lead the team;
- value work relationships;
- view work performance feedback in terms of "no news is good news";
- believe that men belong in the workplace and women in the home;
- an inability to understand the need for work/life balance — struggling with the concept of leisure and recreation time; and
- a reluctance to exercise.

Having experienced the hardships of life during World War II and the Great Depression, Veterans came into the workplace with an attitude of "head down, onward, and forward".[24] Disciplined and respectful of rules and regulations, Veterans are accustomed to a management style that gives orders and disseminates information on a "need to know" basis. For them, job satisfaction is a "job well-done", and career advancement is seniority-based, with seniority directly associated with "age".

The Veterans' work values include loyalty, dependability, persistence, hard work to get ahead, authoritarian leadership, and wisdom and experience over technical knowledge. The Veterans have been viewed as being adaptive in nature,[25] although today they remain uncomfortable with the blend of technology and age/gender/ethnic diversity in the workplace.[26]

2. Baby Boomers

Born between 1946 and 1964, these workers are the heart and soul of today's

---

[23] J. Duchscher & L. Cowen, "Multigenerational Nurses in the Workplace" (2004) 34:11 *J. Nurs. Adm.* 493.

[24] L. Lancaster & D. Stillman, *When Generations Collide: Who They Are, Why They Clash, How to Solve the Generational Puzzle at Work* (New York: Harper Collins, 2002).

[25] W. Strauss & N. Howe, *Generations: The History of America's Future, 1584 to 2069* (New York: Harper Perennial, 1992) (reprint).

[26] K. Kirkpatrick, S. Martin & S. Warneke, *Strategies for the Intergenerational Workplace* (Gensler, July 17, 2008) at 3, online: http://www.gensler.com/uploads/documents/IntergenerationalWorkplace_07_17_2008.pdf (date accessed: January 31, 2020).

workforce in many countries. This very large generation,[27] which is now between the ages of 56 and 74 years, has shaped the face of the North American work scene for a number of years. They defined "what is good and acceptable".[28]

Influenced by Civil Rights, War on Poverty, Atomic Age, Race to Space, Vietnam War, political turmoil of assassinations (John F. Kennedy and Martin Luther King), impeachment (Richard Nixon), the Beatles, Woodstock, the death of John Lennon, the sexual revolution,[29] space travel, Dr. Benjamin Spock, and the first heart transplant, this cohort has challenged the existing social institutions, conventions, and assumptions through the use of protests.

Baby Boomers were the first generation in which students were graded on school report cards for "works well with others" and "shares materials with classmates".[30] Not surprisingly, Baby Boomers prefer to work collaboratively and in team settings. Some of their contributions are business practices such as participative management, quality circles, employee involvement, and team building.[31]

Boomers seek lots of control in their lives and perceive that they do not have it. Dubbed the "whiners", today, Baby Boomers are viewed as a disenchanted group because "the promise that they bought into" was that by this time in their careers/lives, they would not be working as hard as they currently are. The result is that many Boomers go to work but are not really fully engaged.[32]

Having grown up in relative prosperity and safety, child abductions, terrorism, world pollution, and pessimism were not part of the Boomers' worldview. Hence, this generation has been described as idealistic and optimistic, believing they could change the world for the better.[33] Boomers chose the workplace as the arena in

---

[27] Baby Boomers are estimated to number 80 million in the United States of America; 10 million in Canada; and 5.3 million in Australia.

[28] A. Stefaniak, *Black Hole or Window of Opportunity? Understanding the Generational Gap in Today's Workplace* (Policy Brief: 04-11-07, Center for Public Policy & Administration) at 5, online: http://www.cppa.utah.edu/_documents/publications/workforce/Generations.pdf (date accessed: January 31, 2020).

[29] Office of Institutional Equity, Duke University, *Cross Generational Communication: Implications in the Work Environment* (April 2008), online: http://today.duke.edu/2014/02/generationaldifferences (date accessed: January 31, 2020).

[30] AARP, *Leading a Multigenerational Workforce* (Washington, DC: AARP, 2007) at 10, online: http://assets.aarp.org/www.aarp.org_/articles/money/employers/leading_multigenerational_workforce.pdf (date accessed: January 31, 2020).

[31] *Ibid.*, at 11.

[32] J. Jamrog, "The Perfect Storm" (Presented at a CHRP Seminar, Calgary, AB, September 15, 2005) [unpublished].

[33] L. Lancaster & D. Stillman, *When Generations Collide: Who They Are, Why They Clash, How to Solve the Generational Puzzle at Work* (New York: Harper Collins, 2002).

which they would prove their worth.³⁴

Generalizing, the Baby Boomers as a generation, they:
- are extremely competitive due to the enormous size of the generation;
- crave stability and dislike change;³⁵
- are strong consumers;
- buy goods partly because they like them, but mainly to impress others;
- tend to buy foreign goods for prestige reasons;
- are heavily into personal gratification;³⁶
- are known as the true "Sandwich Generation" with responsibilities to aging parents and children;
- view religion as a "hobby", although they tend to be spiritual in nature;
- are communitarians;³⁷
- prefer leadership by consensus — in fact, they were the originators of consensus-based leadership;³⁸
- value participatory leadership and coaching;³⁹
- have a "love/hate" relationship with authority;
- value being able to work collaboratively;⁴⁰
- work hard, play hard — they personify the term, "workaholic";
- value their reputation;
- tend to lack realistic views of their abilities and stamina;⁴¹
- value profitability and tend to worry a lot about money;
- prefer yearly feedback on work performance with accompanying documentation; and

---

³⁴ AARP, *Leading a Multigenerational Workforce* (Washington, DC: AARP, 2007) at 10, online: http://assets.aarp.org/www.aarp.org_/articles/money/employers/leading_multigenerational_workforce.pdf (date accessed: January 31, 2020).

³⁵ S. Calhoun & P. Strasser, "Generations at Work" (2005) 53:11 *AAOHN* 469.

³⁶ K. Kirkpatrick, S. Martin & S. Warneke, *Strategies for the Intergenerational Workplace* (Gensler, July 17, 2008), online: http://www.gensler.com/uploads/documents/IntergenerationalWorkplace_07_17_2008.pdf (date accessed: January 31, 2020).

³⁷ *Ibid.*

³⁸ *Ibid.*

³⁹ S. Calhoun & P. Strasser, "Generations at Work" (2005) 53:11 *AAOHN* 469.

⁴⁰ *Ibid.*

⁴¹ *Ibid.*

- perceive exercising as a duty.

The Baby Boomers' work values include employer loyalty, an acceptance of long work hours and stress, team-oriented, high regard for title and status symbols, demand respect and sacrifice from subordinates, "live to work", value job security, and seek to have their contributions recognized.

Facing the reality and challenges of aging, many Baby Boomers have assumed a "refuse to age" attitude. Hence, the introduction of the term, "Zoomers" to describe this generation.[42] Additionally, the Baby Boomers are opting to remain in the workplace well beyond the traditional age of retirement. However, they are assuming the role of a "free agent",[43] doing contract and consulting projects. This affords them the flexibility to do the things they want to do when they want to do them.

3. Generation X or the Baby Bust Generation

Deemed the "slackers", Generation Xers were born between 1965 and 1979. Their values differ greatly from those held by the Boomers. They grew up with self-immersed,[44] workaholic parents.[45] Being "latch-key children" who watched their parents being "laid off" after years of hard work and personal sacrifices for "the company", this cohort tends to display great skepticism, independence, resourcefulness, and "street-smarts". In addition, they were impacted by increased technology, personal computers, Bill Gates, video games, Madonna, cultural diversity, homelessness, nuclear threats and disasters (Three-Mile Island and Chernobyl), environmental disasters (Exxon Valdez Oil Spill), rap music, anti-child society, Watergate, the fall of the Berlin Wall, Tiananmen Square Uprising, and AIDS.

A relatively small generation, Generation X views itself as neglected. Mistrustful of institutions and organizations, Generation Xers readily move from job to job seeking the work knowledge and skills that they want. Their loyalties revolve around their friends and family, not their jobs.[46]

Now 41 to 55 years of age, Generation Xers, have young families to care for. Family security is important to them as well as being there for the children. They do

---

[42] M. Znaimer, online: http://www.mosesznaimer.com/ (date accessed: January 31, 2020).

[43] S. Cran, *101 Ways to Make Generations X, Y and Zoomers Happy at Work* (Vancouver, BC: Synthesis at Work Inc., 2010) at 11.

[44] J. Duchscher & L. Cowen, "Multigenerational Nurses in the Workplace" (2004) 34:11 *J. Nurs. Adm.* 493.

[45] N. Pekala, "Conquering the Generational Divide" (2001) 66:6 *Journal of Property Management* 30.

[46] R. Zemke, CEO Agenda Series, "Generations at Work" (2001), online: http://www.asaecenter.org/Resources/articledetail.cfm?ItemNumber=13053 (date accessed: January 31, 2020).

not want to emulate the Boomers who were perceived to never be there for their children.[47]

Some other characteristics of Generation Xers are that they:

- are skeptical;[48]
- are cynical;
- are pessimistic;
- are pragmatic;[49]
- are nonconformists;
- come from divorced, single parent, and/or blended families;[50]
- value self-empowerment;[51]
- are self-reliant;[52]
- are independent thinkers and actors. They are always looking for opportunities;[53]
- learned to trust themselves, not institutions[54] — hence, they tend not to be into established community groups like the Lions Club or Rotary, or religion;
- possess an attitude of "I like it and I don't care what you think";
- are reluctant to commit to a relationship: prefer trial marriages/ relationships;
- like inner-city living;

---

[47] S. Cran, *101 Ways to Make Generations X, Y and Zoomers Happy at Work* (Vancouver, BC: Synthesis at Work Inc., 2010) at 4.

[48] K. Kirkpatrick, S. Martin & S. Warneke, *Strategies for the Intergenerational Workplace* (Gensler, July 17, 2008), online: http://www.gensler.com/uploads/documents/IntergenerationalWorkplace_07_17_2008.pdf (date accessed: January 31, 2020).

[49] *Ibid.*

[50] AARP, *Leading a Multigenerational Workforce* (Washington, DC: AARP, 2007) at 12, online: http://assets.aarp.org/www.aarp.org_/articles/money/employers/leading_multigenerational_workforce.pdf (date accessed: January 31, 2020).

[51] L. Lancaster & D. Stillman, *When Generations Collide: Who They Are, Why They Clash, How to Solve the Generational Puzzle at Work* (New York: Harper Collins, 2002).

[52] K. Kirkpatrick, S. Martin & S. Warneke, *Strategies for the Intergenerational Workplace* (Gensler, July 17, 2008), online: http://www.gensler.com/uploads/documents/IntergenerationalWorkplace_07_17_2008.pdf (date accessed: January 31, 2020).

[53] S. Cran, *101 Ways to Make Generations X, Y and Zoomers Happy at Work* (Vancouver, BC: Synthesis at Work Inc., 2010).

[54] L. Lancaster & D. Stillman, *When Generations Collide: Who They Are, Why They Clash, How to Solve the Generational Puzzle at Work* (New York: Harper Collins, 2002).

- thrive on chaos and change;[55]
- are avid risk-takers;[56]
- value exercise for the well-being of their mental health;
- strongly value work/life balance and they want it *Now*;
- believe work is important[57] — they will work hard provided work does not interfere with their play plans;
- are willing to sacrifice some personal life for career advancement;
- value leader competence;
- are neither unimpressed nor intimidated by authority – rather, they prefer an informal relationship with authority figures;[58]
- seek stimulation, balance, and feedback from their work world — they tend to want regular and timely feedback on their work performance;[59]
- are action-oriented;[60]
- are results-focused[61] — enjoy achieving measurable results and streamlining systems and processes;[62]
- are techno-savvy;[63]

---

[55] AARP, *Leading a Multigenerational Workforce* (Washington, DC: AARP, 2007) at 12, online: http://assets.aarp.org/www.aarp.org_/articles/money/employers/leading_multigenerational_workforce.pdf (date accessed: January 31, 2020).

[56] K. Kirkpatrick, S. Martin & S. Warneke, *Strategies for the Intergenerational Workplace* (Gensler, July 17, 2008), online: http://www.gensler.com/uploads/documents/IntergenerationalWorkplace_07_17_2008.pdf (date accessed: January 31, 2020).

[57] L. Lancaster & D. Stillman, *When Generations Collide: Who They Are, Why They Clash, How to Solve the Generational Puzzle at Work* (New York: Harper Collins, 2002).

[58] S. Cran, *101 Ways to Make Generations X, Y and Zoomers Happy at Work* (Vancouver, BC: Synthesis at Work Inc., 2010).

[59] *Ibid.*

[60] K. Kirkpatrick, S. Martin & S. Warneke, *Strategies for the Intergenerational Workplace* (Gensler, July 17, 2008), online: http://www.gensler.com/uploads/documents/IntergenerationalWorkplace_07_17_2008.pdf (date accessed: January 31, 2020).

[61] *Ibid.*

[62] AARP, *Leading a Multigenerational Workforce* (Washington, DC: AARP, 2007) at 12, online: http://assets.aarp.org/www.aarp.org_/articles/money/employers/leading_multigenerational_workforce.pdf (date accessed: January 31, 2020).

[63] K. Kirkpatrick, S. Martin & S. Warneke, *Strategies for the Intergenerational Workplace* (Gensler, July 17, 2008), online: http://www.gensler.com/uploads/documents/IntergenerationalWorkplace_07_17_2008.pdf (date accessed: January 31, 2020).

- prefer informality;[64]
- value recognition and attention for their achievements; and
- are good at saving money.

In terms of work, they are determined to manage their own time. They value minimal supervision, a flexible work schedule, participatory leadership, and coaching.[65] Being influenced by sweeping social changes and being "the children of divorce" and familial change, Generation Xers tend to be less trusting,[66] but more adept at accepting and adapting to chaos and change. When faced with meaningful work, leaders and co-workers they respect, and a work schedule that meets their work-life balance needs, Generation Xers can be very creative and productive.[67] However, Gen Xers feel that they are overshadowed by the Boomers and hence, undervalued for their demonstrated abilities and contributions.[68]

4. Baby Boomlets (Gen Ys, Echo Boomers, Nexus, Nexters, and Millennials)

This generational group, born 1980 to 2000, is currently 20 to 40 years of age. They are in the throes of their career and entering the workforce. Baby Boomlets (Gen Ys) have lived through and been influenced by:

- the Iraq War;
- business/corporate scandals (*i.e.*, Enron);
- multiculturalism and globalism;
- global terrorism, *e.g.*, September 11th;
- rapidly expanding technology (Internet, DVD);
- the death of Princess Diana;
- Rap and Hip Hop music;
- the Oklahoma City bombing;
- the Columbine and other school shootings;
- the Dot.Com Crash;
- the proliferation of communication and mobile technologies; and
- the tsunami in the Indian Ocean.

---

[64] *Ibid.*

[65] S. Calhoun & P. Strasser, "Generations at Work" (2005) 53:11 *AAOHN* 469.

[66] K. Kirkpatrick, S. Martin & S. Warneke, *Strategies for the Intergenerational Workplace* (Gensler, July 17, 2008), online: http://www.gensler.com/uploads/documents/IntergenerationalWorkplace_07_17_2008.pdf (date accessed: January 31, 2020).

[67] *Ibid.*

[68] S. Cran, *101 Ways to Make Generations X, Y and Zoomers Happy at Work* (Vancouver, BC: Synthesis at Work Inc., 2010).

Being the largest generation of "haves and have nots", this group has been greatly impacted by environmental concerns, violence and bullying, terrorism, and being "wired to technology". To them, life without technology is inconceivable.

Baby Boomlets (Gen Ys) are the first generation to grow up without expectations of living within a nuclear family — less than 33 per cent lived in homes with two parents.[69] One estimate is that 25 per cent of the Baby Boomlets (Gen Ys) are the product of single parent families.[70] Additionally, both parents worked: one estimate is that 75 per cent of Baby Boomlets have working mothers.[71]

Interestingly, despite the breakdown of the nuclear family through divorce and single parenthood, Baby Boomlet (Gen Y) families tended to be very child-centred, with both parents involved in their children's education and activities.[72] Their parents want to give the children the "best",[73] which may be the reason Baby Boomlets (Gen Ys) have reportedly developed close parent-child relationships.[74]

Baby Boomlets (Gen Ys) were raised by parents who involved them in a whirlwind of extra-curricular activities. They were the busiest kids ever, with many of them carrying Daytimers/iPhones to track their daily activities.[75] Because their parents did not, know how to deal with their children's "hyperactivity", Baby Boomlets (Gen Ys) have the distinction of being the most medicated generation ever.[76] In fact, some sociologists feel that these children were growing up with a form of post-traumatic stress disorder.[77]

This generation tends to use online technology to forge relationships. Reportedly,

---

[69] J. Duchscher & L. Cowen, "Multigenerational Nurses in the Workplace" (2004) 34:11 J. Nurs. Adm. 493.

[70] A. Schwartz, *Generations at Work: Understanding and Influencing* (2008), online: http://www.duke.edu/web/equity/Diversity.htm (date accessed: January 31, 2020).

[71] *Ibid.*

[72] *Ibid.*

[73] K. Kirkpatrick, S. Martin & S. Warneke, *Strategies for the Intergenerational Workplace* (Gensler, July 17, 2008), online: http://www.gensler.com/uploads/documents/IntergenerationalWorkplace_07_17_2008.pdf (date accessed: January 31, 2020).

[74] A. Schwartz, *Generations at Work: Understanding and Influencing* (2008), online: http://www.duke.edu/web/equity/Diversity.htm (date accessed: January 31, 2020).

[75] A reason put forth for this phenomenon of busy children is that their parents, the Generation Xers, who got into trouble in their younger years, opted to keep their children too busy to follow in their footsteps; see A. Stefaniak, *Black Hole or Window of Opportunity? Understanding the Generational Gap in Today's Workplace* (Policy Brief: 04-11-07, Center for Public Policy & Administration) at 6, online: http://www.cppa.utah.edu/_documents/publications/workforce/Generations.pdf (date accessed: January 31, 2020).

[76] J. Jamrog, "The Perfect Storm" (presented at a CHRP Seminar, Calgary, AB, September 15, 2005) [unpublished].

[77] *Ibid.*

35 per cent of them report having found their best friends online.[78] As well, Boomlets (Gen Ys) rank high in terms of sociability, civic duty and morality.[79]

As a generation, the specific traits of the Baby Boomlets (Gen Ys) include:

- optimistic;[80]
- very pragmatic;[81]
- very realistic;[82]
- being less skeptical and angry with life and the world than are the Generation Xers;[83]
- being "Gap" shoppers;[84]
- valuing inclusive relationships and connecting with people;
- desiring flexibility in their daily routine;
- willingness to participate in community endeavours;
- judging institutions by their own merit;[85]
- strongly disliking organized groups like labour unions;
- inability to relate with religion;
- being heavily into playing sports;
- liking living with their parents;
- valuing cultural diversity in the workplace and work well in diverse workgroups;[86]

---

[78] *Ibid.*

[79] FGI, "One Workplace, Four Generations: Managing Their Conflicting Needs", *Working Well for Managers* (September 2004), online: www.mta.ca/hr/managers/workingwell_sept2004.pdf (date accessed: January 31, 2020).

[80] K. Kirkpatrick, S. Martin & S. Warneke, *Strategies for the Intergenerational Workplace* (Gensler, July 17, 2008), online: http://www.gensler.com/uploads/documents/IntergenerationalWorkplace_07_17_2008.pdf (date accessed: January 31, 2020).

[81] L. Lancaster & D. Stillman, *When Generations Collide: Who They Are, Why They Clash, How to Solve the Generational Puzzle at Work* (New York: Harper Collins, 2002).

[82] *Ibid.*

[83] *Ibid.*

[84] J. Jamrog, "The Perfect Storm" (presented at a CHRP Seminar, Calgary, AB, September 15, 2005) [unpublished].

[85] L. Lancaster & D. Stillman, *When Generations Collide: Who They Are, Why They Clash, How to Solve the Generational Puzzle at Work* (New York: Harper Collins, 2002).

[86] *Ibid.*

- possessing a global perspective on world events;[87]
- viewing leadership as "pulling together" for a cause;
- valuing collaboration and relationship-oriented leadership;[88]
- being polite towards authority figures;
- being ambitious, but not entirely focused;[89]
- desiring to be famous, and YouTube and Facebook can make that a reality;[90]
- looking to the workplace for direction and an avenue for achieving their goals;[91]
- working towards getting good grades in school;
- being extremely techno-savvy;[92]
- seeking and demanding feedback on their work performance — they are used to getting feedback "at the touch of a button";
- wanting balance in their lives; and
- valuing the saving of money.

The Baby Boomlet (Gen Y) worker is characterized as having a relatively short attention span, expecting things to happen quickly, and seeking variety in life and work. These traits are often misunderstood and negatively viewed. In actuality, Baby Boomlets (Gen Ys) have a strong sense of self-worth, are ambitious, believe that they can achieve anything, and seek a variety of tasks and work experiences to demonstrate their abilities. Their "can-do" attitude about work tasks leads them to be viewed as over-confident, self-absorbed, and demanding, looking for feedback about how they are doing frequently — even daily. Hence, they are called "Generation Me" and are viewed as being ready and willing to take on the world. Some researchers believe that they can more than measure up to their "can-do" attitudes.[93],[94]

---

[87] *Ibid.*

[88] S. Calhoun & P. Strasser, "Generations at Work" (2005) 53:11 *AAOHN* 469.

[89] A. Stefaniak, *Black Hole or Window of Opportunity? Understanding the Generational Gap in Today's Workplace* (policy Brief: 04-11-07, Center for Public Policy & Administration) at 6, online: http://www.cppa.utah.edu/_documents/publications/workforce/Generations.pdf (date accessed: January 31, 2020).

[90] S. Cran, *101 Ways to Make Generations X, Y and Zoomers Happy at Work* (Vancouver, BC: Synthesis at Work Inc., 2010).

[91] *Ibid.*

[92] L. Lancaster & D. Stillman, *When Generations Collide: Who They Are, Why They Clash, How to Solve the Generational Puzzle at Work* (New York: Harper Collins, 2002).

[93] A. Stefaniak, *Black Hole or Window of Opportunity? Understanding the Generational Gap in Today's Workplace* (Policy Brief: 04-11-07, Center for Public Policy & Administration)

Unlike other generations, Baby Boomlets (Gen Ys) have "developed work characteristics and tendencies from doting parents, structured lives, and contact with diverse people".[95] They are used to working in teams and want to make friends with people at work. Baby Boomlets (Gen Ys) tend to work well with diverse co-workers and value the knowledge possessed by older workers.[96] Their values and worldviews are expected to make significant changes to society and the work scene.

Given that approximately 75 million Baby Boomlets (Gen Ys) are preparing to enter the workplace or have done so already, it is important to understand how best to manage this generation of workers. According to Heathfield, here are some tips for working with Baby Boomlets (Gen Ys):[97]

- provide them structure;
- provide them with leadership and guidance;
- encourage their self-assuredness, "can-do" attitude, and positive self-image;
- take advantage of their comfort level with teams and teamwork by encouraging them to join in;
- listen to them;
- provide them with challenges and ever-changing tasks;
- capitalize on their ability to multi-task;
- take advantage of their computer, telephone and electronic literacy;
- benefit from their affinity to network with people;
- enable a reasonable work-life balance; and
- provide a fun, employee-centred workplace.

5. GenZ

This last generational group, born in 1997, has just reached 23 years of age and is

---

at 6, online: http://www.cppa.utah.edu/_documents/publications/workforce/Generations.pdf (date accessed: January 31, 2020); and S. Heathfield, *Managing Millennials: Eleven Tips for Managing Millennials* (2008), online: http://humanresources.about.com/od/managementtips/a/millenials.htm (date accessed: January 31, 2020).

[94] S. Cran, *101 Ways to Make Generations X, Y and Zoomers Happy at Work* (Vancouver, BC: Synthesis at Work Inc., 2010) at 7.

[95] S. Heathfield, *Managing Millennials: Eleven Tips for Managing Millennials* (2008), online: http://humanresources.about.com/od/managementtips/a/millenials.htm (date accessed: January 31, 2020).

[96] Go2 Tourism HR Society, *Capitalizing on the Generational Gap in the Workplace* (2008), online: http://www.go2hr.ca (date accessed: January 31, 2020).

[97] S. Heathfield, *Managing Millennials: Eleven Tips for Managing Millennials* (2008), online: http://humanresources.about.com/od/managementtips/a/millenials.htm (date accessed: January 31, 2020).

entering or getting ready to enter the workforce. Although this latest generation will not be explored throughout the remainder of this chapter, some current perspectives on it will be offered.

GenZers are recognized as being life-long users of the Internet. Comfortable with technology, they socialize on social media websites and use it to practice social skills that they can then use in real-life situations. They live with a cellphone in their hand; in fact, they spend more time on their cellphones than watching TV. The cellphone is how they remain in constant contact with their friends. Not only has technology strongly influenced their mode of communication, it also played a key role in their education. In fact, this generation is identified as having significantly changed how education is delivered.

GenZs are the first generation to have multi-generational parents; although mainly Generation Xers, some of their parents are Baby Boomlets.

The specific traits portrayed by this generation are that they are:

- independent;
- entrepreneurial;
- conservative;
- money-wise;
- concerned about student debt;
- competitive;
- adventuresome;
- curious;
- loyal;
- compassionate; and
- value education.

Their values are:

- freedom to get an education anywhere on Earth;
- ready, online access to medical records and other critical information;
- an ability to use their cellphones to remain "connected to friends" from wherever; and
- mobile access to organize their volunteer opportunities and corporate social responsibilities.[98]

As for their work traits, GenZ is expected to be the ideal employee of the future;

---

[98] J. Meister & K. Willyerd, *The 2020 Workplace: How Innovative Companies Attract, Develop and Keep Tomorrow's Employees Today* (New York, NY: HarperCollins Publisher, 2010) at 53.

they have all the needed skills for succeeding in a high-tech world. Additionally, they demonstrate that they are open to new ideas, and are willing to pursue new ways of doing business. They are eager to contribute to their community and want to make a difference in this world. Their willingness to take advantage of relevant professional opportunities prepares them to be better prepared for the world of work than their parents were.[99]

F. THE FOUR GENERATIONS: PERSONAL AND LIFESTYLE CHARACTERISTICS

Each generation has a unique worldview and approach to life and work. To illustrate this concept, a comparison table of the personal and lifestyle characteristics of the four generations follows (Table 35.2).

---

[99] J. Dorsey, "What do we know about the generation after millennials?", *TEDx Talks* (2015), online: TEDxHouston (retrieved 6 April 2016).

Table 35.2: Comparison of the Personal and Lifestyle Characteristics of the Four Generations[100]

Aspect	Veterans (~1927-1945)	Baby Boomers (~1946-1964)	Generation Xers (~1965-1979)	Baby Boomlets (Gen Ys) (~1980-1996)
Core Values	• Conformer • Discipline • Respect for authority • Sociability	• Optimistic • Involvement	• Skepticism • Fun • Informality	• Realism • Confidence • Extreme fun • Sociability
Outlook on life	• Adaptive	• Optimistic	• Skeptical	• Realistic

[100] Adapted from: D. McPhail, "Corporate Perspectives on Workplace Learning, 'Four Generations in the Workforce'" (May 9, 2006), online: http://www.wln.ualberta.ca/!=n/Events/Archives/Seminar%20Archives/~/media/wln/Documents/Events/Seminars_2007/Generation_Workshop.pdf (date accessed: January 31, 2020); D. Dyck, "Multi-Generational Workplaces: The OH&S Challenge" (presented at the American Occupational Health Nurses Symposium and Expo, Orlando, FL, April 2007) [unpublished]; M. McCrindle, *New Generations at Work: Attracting, Recruiting, Retraining & Training Generation Y* (Sydney, NSW, Australia: McCrindle Research, 2006), online: http://www.mccrindle.com.au/resources/whitepapers/McCrindle-Research_New-Generations-At-Work-attracting-recruiting-retaining-training-generation-y.pdf (date accessed: January 31, 2020); G. Hammill, "Mixing and Managing Four Generations of Employees", *FDU Magazine Online* (Winter/Spring 2005), online: http://www.fdu.edu/newpubs/magazine/05ws/generations.htm (date accessed: January 31, 2020); AARP, *Leading a Multigenerational Workforce* (Washington, DC: AARP, 2007), online: http://assets.aarp.org/www.aarp.org_/articles/money/employers/leading_multigenerational_workforce.pdf (date accessed: January 31, 2020); and A. Stefaniak, *Black Hole or Window of Opportunity? Understanding the Generational Gap in Today's Workplace* (Policy Brief: 04-11-07, Center for Public Policy & Administration), online: http://www.cppa.utah.edu/_documents/publications/workforce/Generations.pdf (date accessed: January 31, 2020).

Aspect	Veterans (~1927-1945)	Baby Boomers (~1946-1964)	Generation Xers (~1965-1979)	Baby Boomlets (Gen Ys) (~1980-1996)
Family	• Nuclear family with support of extended family	• Nuclear family with some disintegration occurring • Some working mothers	• "The Children of Divorce" • Single-parent families • Working parents • "Latch-key kids" • Late to marry and have children	• Single-parent families, blended families • Working parents
Education	A "dream"	A birthright	A way to get there	An incredible expense
Compelling Messages	• "Make do, or go without" • "Stay in line" • "Sacrifice" • "Consider the common good" • "Any job worth doing, is worth doing right"	• "Be anything you want to be" • "Change the world" • "Work well with others" • "Live up to expectations" • "Duck and cover"	• "Don't commit" • "Heroes don't exist" • "Get real!" • "Take care of yourself" • "Always ask, 'Why?'"	• "You are special" • "Leave no one behind" • "Connect 24/7" • "Achieve now!" • "Serve your community"

CH. 35: IMPACT OF FIVE GENERATIONS IN THE WORKPLACE

Aspect	Veterans (~1927-1945)	Baby Boomers (~1946-1964)	Generation Xers (~1965-1979)	Baby Boomlets (Gen Ys) (~1980-1996)
Communication Media	• Rotary phones • One-on-one talks • Write a memo	• Touch-tone phones • Television • Call me at any time	• Cell phone use • Call me only at work • Email messages	• Internet, email messages, chat lines, blogs • Text messages • Picture phones
Iconic Technology	• Newspaper • Telegraph • Radio	• TV • Audio Cassette • Colour TV • Fax Machines	• VCR • Walkman • Personal Computers	• Laptops/tablets/iPads • Internet/Email/SMSing • DVD • Play Station/X-Box
Music	• Frank Sinatra • Bing Crosby • Duke Ellington	• Elvis • Beatles • Rolling Stones	• INXS • Nirvana • Madonna	• Eminem • Britney Spears • Puff Daddy

Aspect	Veterans (~1927-1945)	Baby Boomers (~1946-1964)	Generation Xers (~1965-1979)	Baby Boomlets (Gen Ys) (~1980-1996)
TV and Movies	• No TV — just movies • Casablanca • All Quiet on the Western Front • Gone with the Wind	• Easy Rider • The Graduate • Jaws	• ET • Hey Hey Its Saturday • MTV	• Titanic • Pay TV • Reality TV
Popular Culture	• Jazz • Nylon stockings	• Flare jeans • Mini skirts • Barbie • Frisbee	• Rollerblades • Hyper colour • Torn jeans	• Body piercing • Tattoos • Baseball caps/hoodies • Men's cosmetics
Money Management	• Save for a rainy day • Pay cash	• Focus on long-term needs • Cash • Credit — buy now, pay later	• Focus on medium-term goals • Cautious with money • Conservative spenders • Save, save, save	• Focus on short-term wants • Earn to spend • Credit dependent
Approach to Relationships	• Uncomfortable with personal relationships • Committed to work	• Self-sacrifice	• Reluctant to commit to relationships	• Committed to success

Aspect	Veterans (~1927-1945)	Baby Boomers (~1946-1964)	Generation Xers (~1965-1979)	Baby Boomlets (Gen Ys) (~1980-1996)
Influencers	• Trial and error	• Evidential experts	• Pragmatic practitioners	• Experiential peers
Decision-making	• Reliant on commands • Individual • Trial and error	• Team-oriented • Collaborative decision-making • Consensus-based decision-making	• Self-reliant • Independent decision-making	• Focus on what is "good for all"

Aspect	Veterans (~1927-1945)	Baby Boomers (~1946-1964)	Generation Xers (~1965-1979)	Baby Boomlets (Gen Ys) (~1980-1996)
Media Images (*images that may not accurately represent reality*)	• Tired • Resent Generation X and Baby Boomlets (Gen Ys) their entitlement mentality when Traditionalists had to work so hard for what they got in life	• Workaholic • Guilt-ridden • Resent Generation X and Baby Boomlets (Gen Ys) for frequent job change and for demanding work/life balance	• Tattooed • Slacker • Resent Veterans for resisting change	• Precocious • Resent Baby Boomers for leaving the planet in a mess • Clever • "Techno-kids"
Definition of Technology	• Massive structures, *e.g.*, Hoover Dam	• Anything that makes life easier, *e.g.*, microwave oven	• Can be held in one's hand, *e.g.*, cell phone, personal digital assistant	• Can't define technology — it is just there!

CH. 35: IMPACT OF FIVE GENERATIONS IN THE WORKPLACE

Aspect	Veterans (~1927-1945)	Baby Boomers (~1946-1964)	Generation Xers (~1965-1979)	Baby Boomlets (Gen Ys) (~1980-1996)
Current Life Challenges	• Dealing with aging parents, adult children and grandchildren • Experiencing own/spousal health issues • Require help with new technologies and workplace changes • Fail to understand the "new" employees of today • Often forgotten/neglected by companies: close to retirement	• Sandwiched between demands of aging parents and children • Need flexible work schedules, especially female workers • Seeking control over own life situation • High desire to excel career-wise, but have little time left to "reach the top" • Some question if they made the right career choices	• Dislike company rules, policies: tend to question authority • Want adequate explanations for why things are done a certain way • Concerned about their "shelf-life" at work — fear obsolescence • Approach life with a fair amount of skepticism	• Don't believe that they have to "pay their dues" and "bide their time" for job advancement • Strong dislike for organized groups (*e.g.*, labour unions) • Difficult to retain as employees • Will rock the world of work

Bear in mind, this stereotyping is meant to illustrate generational differences, and is not to be taken as representative of every member of a generational group.

G. THE FOUR GENERATIONS: IMPACT ON THE WORKPLACE

Over the years, the world of work has changed. Organizational/company loyalty and commitment to employers has vanished. Being at work on time has decreased in importance. Sick leave abuse is common. Job security and high pay are not the motivators they once were. Why? Young people have watched their parents remain loyal to their employers, only to witness them experience downsizing and lay-offs.

Today, young workers are seeking jobs in which they can make a valuable contribution, work at a variety of tasks, and learn new marketable skills. They demand intellectual stimulation, team environments, transferability of work experiences, and salaries that match the rising cost of living. Interestingly, they do not believe they should have to "pay their dues" and bide their time for job advancement. Because they "work to live" and not "live to work", they refuse to be the workaholics that their parents were.

The Veterans and Baby Boomers are working their way towards retirement; however, for a variety of reasons, some are choosing not to retire as soon as was once predicted. Likewise, some older workers who have retired are back doing entry-level jobs.

With four generations in the workplace, each with their own beliefs, values, wants, and needs, employers face the daunting challenge of trying to meet worker expectations. For example, meetings are viewed by Veterans and Baby Boomers as "busy work". They prefer short meetings, speedy decisions, and only want to meet when there is an urgent need. Generation Xers and Baby Boomlets tend to value the interaction and view meetings as a way to reach a solution to an identified problem.

Veterans and Boomers value rules and regulations. Generation Xers and Baby Boomlets have been known to ignore rules, policies, and chain-of-command. Baby Boomlets live in a world of advanced technology. They are techno-wizards who embrace technology as a normal element of their lives. Veterans and Boomers had to learn to adapt to technology and for some, there remains a healthy "distrust of these new tools".

At work, generational differences can affect everything from recruiting to teamwork, to change management, to motivating workers, and to managing and increasing productivity. To better understand the generations and their impact on the workplace, it is important to first understand the work characteristics of each generation (Table 35.3).[101]

---

[101] M. Pitt-Catsouphes & M. Smyer, The Center on Aging and Work: Workplace Flexibility, "The 21st Century Multi-generational Workplace", online: Boston College http://bc.edu/research/agingandwork (date accessed: January 31, 2020).

CH. 35: IMPACT OF FIVE GENERATIONS IN THE WORKPLACE

Table 35.3: Comparison of the Work Characteristics of the Four Generations[102]

Aspect	Veterans (~1927-1945)	Baby Boomers (~1946-1964)	Generation Xers (~1965-1979)	Baby Boomlets (Gen Ys) (~1980-1996)
Work Ethic and Values	• Hard work • Respect authority • Sacrifice • "Duty" before fun • Adhere to rules • A "head down, onward and upward" approach to work	• Workaholics • Work efficiently • Crusading causes • Personal growth and fulfilment • Desire quality • Question authority	• Eliminate the task • Self-reliant • Seek structure and direction • Skeptical	• Action-oriented • Multi-tasker • Tenacious • Entre-preneurial • Tolerant • Goal-oriented
Work is . . .	a duty or obligation	an exciting adventure	a difficult challenge; a contract	a means to fulfil an end

[102] Adapted from: D. McPhail, "Corporate Perspectives on Workplace Learning 'Four Generations in the Workforce'" (May 9, 2006), online: http://www.w n.ualberta.ca/sem nar_resources/Generations_May9.ppt (date accessed: January 31, 2020); D. Dyck, "Multi-Generational Workplaces: The OH&S Challenge" (presented at the American Occupational Health Nurses Symposium and Expo, Orlando, FL, April 2007) [unpublished]; M. McCrindle, *New Generations at Work: Attracting, Recruiting, Retraining & Training Generation Y* (Sydney, NSW, Australia: McCrindle Research 2006), online: http://www.mccrindle.com.au/resources/whitepapers/McCrindle-Research_New-Generations-At-Work-attracting-recruiting-retaining-training-generation-y.pdf (date accessed: January 31, 2020); G. Hammill, "Mixing and Managing Four Generations of Employees", *FDU Magazine Online* (Winter/Spring 2005), online: http://www.fdu.edu/newspubs/magazine/05ws/generations.htm (date accessed: January 31, 2020); AARP, *Leading a Multigenerational Workforce* (Washington, DC: AARP, 2007), online: http://assets.aarp.org/www.aarp.org_/articles/money/employers/leading_multigenerational_workforce.pdf (date accessed: January 31, 2020); and A. Stefaniak, *Black Hole or Window of Opportunity? Understanding the Generational Gap in Today's Workplace* (Policy Brief: 04-11-07, Center for Public Policy & Administration), online: http://www.cppa.utah.edu/_documents/publications/workforce/Generations.pdf (date accessed: January 31, 2020).

Aspect	Veterans (~1927-1945)	Baby Boomers (~1946-1964)	Generation Xers (~1965-1979)	Baby Boomlets (Gen Ys) (~1980-1996)
Attitude towards work	• Glad to have a job — view work as a privilege • Dutiful • Personal attitudes do not factor into the equation	• Live to work • Driven • Dutiful, but focused on personal achievement and advancement	• Work to live • Balanced life • Do not want to be "tied to a desk" or even to one employer • Value achievement	• Work to live • Looking for meaning and relevance in work • Interested in life outside of work • Value achievement

Aspect	Veterans (~1927-1945)	Baby Boomers (~1946-1964)	Generation Xers (~1965-1979)	Baby Boomlets (Gen Ys) (~1980-1996)
Preferred Management Style	• Chain of Command (*Command and control*) • Prefer formality • Expect to be told what to do, or to tell others what to do • Seek explanations/logic behind management decisions • Seek job explanations clearly spelled out — dislike "napkin work orders"	• Consensus-based Command • Value having objectives and results stated in people-centred terms • Prefer to be part of work decisions — work by consensus • Originated the use of performance reviews • Like the democratic approach	• Self-command (*Self-direction*) • Respect leaders based on merit, not position or title • Prefer to be told what is to be done and not how to do it • Like multi-tasking and freedom to set own work priorities	• A mix of *Self-command*, *Consensus* and *Collaboration* • Seek continuous learning opportunities and ability to hone their work skills • Value mentoring from managers and colleagues • Respond to a supportive, positive and organized work environment • Want to be "coached", not "bossed" or micro-managed

Aspect	Veterans (~1927-1945)	Baby Boomers (~1946-1964)	Generation Xers (~1965-1979)	Baby Boomlets (Gen Ys) (~1980-1996)
	• Want to be treated fairly, consistently and with respect	• Seek to be treated as equals		• Seek motivational and achievement-oriented managers
Preferred Leaders	Commanders	Thinkers	Doers	Feelers
View of Authority	• Value and respect authority • Time = right to authority	• Love/hate relationship • Time = right to authority	• Skeptical of authority • Constantly testing authority figures	• Ambivalent about authority • Will test authority • Look to authority figures to "Help me achieve my goals"[103]
Organizational Loyalty	• Dedicated — same company for entire career	• Loyal, but also upwardly mobile — will leave for career advancement	• Cynical about organizational loyalty: tend to direct loyalty to individual supervisors/managers. • Will have a number of jobs	• Loyal to managers when they receive help to achieve self-fulfilling jobs • Mobile, looking for what they need[104]

[103] C. Marston, "Four Generations in the Workplace: A Diversity Challenge" (presentation delivered at Utah Department of Workforce Services: Roads to Success, Salt Lake City, UT, 2003) [unpublished].
[104] *Ibid.*

CH. 35: IMPACT OF FIVE GENERATIONS IN THE WORKPLACE

Aspect	Veterans (~1927-1945)	Baby Boomers (~1946-1964)	Generation Xers (~1965-1979)	Baby Boomlets (Gen Ys) (~1980-1996)
Sacrifice	Action taken for the "greater good".	Working hard for success.	Forfeiting personal time.	Do not make sacrifices: they pursue what is in their best interest.[105]
Work Assets	• Experienced • Knowledgeable • Dedication • Emotional maturity; stability • Perseverance	• Dedicated • Team player • Experienced • Knowledgeable	• Adaptable • Techno-literate • Independent • Creative • Challenge the system/status quo	• Extremely techno-savvy • Great multi-taskers • Value collaboration and teamwork • Optimistic
Work Liabilities	• Reluctant to challenge the system • Dislikes conflict • Reticent when they disagree	• Not "budget-minded" • Dislikes conflict • May focus more on the process instead of the outcome	• Skeptical • Distrust authority	• Inexperienced • Few people-management skills • Require supervision and structure

[105] C. Marston, "Four Generations in the Workplace" (presentation delivered at the Planned Parenthood Federation of America, Western Region Conference, Salt Lake City, UT, 2004) [unpublished].

Aspect	Veterans (~1927-1945)	Baby Boomers (~1946-1964)	Generation Xers (~1965-1979)	Baby Boomlets (Gen Ys) (~1980-1996)
Time at Work	• Punch in and out	• Visibility is important: long work hours (50 hours++ per week)	• Project oriented	• View work as a "gig" — in at 9 AM and out by 5 PM.
Interactive Style	• Individual work	• Collaborative • Team player • Loves to have meetings	• Strong collaborators • Entrepreneur	• Strongest collaborators • Participative
Communication Style	• Prefer formal communication — memos, letters, personal notes	• Personal contact — face-to-face meetings or phone calls	• Direct communication — corridor contact, voice mail • Immediate — email, text messages	• Email, blogs, text messages • Voice mail
Use of Technology	• "A struggle"	• "A necessary evil" • Will embrace technology if it is user friendly and makes life easier	• Comfortable with using technology • Enables them to enjoy a flexible work schedule	• "An essential part of life and work"

Aspect	Veterans (~1927-1945)	Baby Boomers (~1946-1964)	Generation Xers (~1965-1979)	Baby Boomlets (Gen Ys) (~1980-1996)
Feedback	• "No news is good news" • Having a job equals good work performance	• Do not appreciate personal feedback • Prefer an annual, written performance review • Do not want to be viewed as the "problem"	• Seek frequent feedback — "Sorry to interrupt, but how am I doing?"	• Seek frequent feedback — "Whenever I want it, at the push of a button"
Worker Expectations	• The employer will manage the employee's career	• Believe the employee's career path is a joint responsibility — employer and employee	• Aggressive regarding career development and advancement	• Career development and advancement

Aspect	Veterans (~1927-1945)	Baby Boomers (~1946-1964)	Generation Xers (~1965-1979)	Baby Boomlets (Gen Ys) (~1980-1996)
Messages and Motivators	• "Your experience is respected" • Being told that their actions directly contribute to organizational success	• "You are valued" • "You are needed" • Motivated by goals set by managers — prefer managers who get them involved and show them how to best contribute • Strive for financial benefits and compensation, title and fame	• "You can do it your way" • "We have few rules around here" • "We have new hardware and software" • Let them get the job done on their own schedule and in their own way • Like "fun", casual dress and flexible work schedules	• "You will have the chance to work with other bright, creative people" • "You and your co-workers can help turn this company around" • "You can be a hero here" • Work that has meaning • Tangible rewards

Aspect	Veterans (~1927-1945)	Baby Boomers (~1946-1964)	Generation Xers (~1965-1979)	Baby Boomlets (Gen Ys) (~1980-1996)
Rewards	• Recognition is the "pay cheque every 2 weeks" • Prefer subtle recognition • Tangible symbols of loyalty and commitment • Value plaques, certificates, pictures with top executives	• Personal appreciation • Public recognition: recognition is the "corner office with the rug", financial incentives, name in the newsletter or on the parking spot • Value money and bonuses • Promotion	• Freedom to work according to their work style and schedule • Free time is the best reward • Top-notch resources/technology • Growth and development opportunities • A focus on bottom-line results • Certification for skill and knowledge achievement	• Varied work assignments • Teamwork: fun workplaces • Recognition from their "heroes" — bosses and grandparents • Making a difference for organization and the community/society

Aspect	Veterans (~1927-1945)	Baby Boomers (~1946-1964)	Generation Xers (~1965-1979)	Baby Boomlets (Gen Ys) (~1980-1996)
Attitude about Careers	• Up to the employer to manage • One employer, one career	• Views work as a lifetime pursuit[106] • Experienced downsizing, layoffs and early retirement situations	• "Free agent" mentality • Entrepreneurial • Seek work-life balance, mutual respect and meaningful work	• "Free agent" mentality • Entrepreneurial • Seek work-life balance, mutual respect and meaningful work

[106] Research shows that 50 per cent of Baby Boomers envision working until beyond age 70 years; 40 per cent prefer a phased retirement; 67 per cent expect to do paid work after aged 60 years; 67 per cent will work to stay mentally active; and 57 per cent will work to remain physically active.

Aspect	Veterans (~1927-1945)	Baby Boomers (~1946-1964)	Generation Xers (~1965-1979)	Baby Boomlets (Gen Ys) (~1980-1996)
Training Focus	• Learning new skills benefits the company, not the individual • Prefers structured learning and demonstrations • Hands-on (on-the-job training)	• Skills needed for success, but not as important as hard work and "face-time"[107] • Prefer technical training and having their skills challenged • Seek data/evidence-based learning; workshop settings	• The more skills/knowledge the better: enhances the tool kit for future job attainment • Seek practical skills • Prefer relaxed and fun settings and use of case studies/applications	• Learning is key and will help one cope • Learn for emotional reasons • Prefer stories/participative learning, use of game playing • Needs to be fast paced
Learning Environment	• On-the-job	• Classroom style • Quiet atmosphere	• Round-table style • Relaxed ambience	• Café style • Music • Multi-modal learning

[107] C. Marston, "Four Generations in the Workplace: A Diversity Challenge" (presentation delivered at Utah Department of Workforce Services: Roads to Success, Salt Lake City, UT, 2003) [unpublished].

Aspect	Veterans (~1927-1945)	Baby Boomers (~1946-1964)	Generation Xers (~1965-1979)	Baby Boomlets (Gen Ys) (~1980-1996)
Work Environment	• Seek an office or defined space of their own • Value a workplace that communicates status and accomplishment • Decorated with tangible items or symbols of work accomplishments		• Like open workspaces; prefer closed/ secluded places inside/outside of the workplace to undertake solo work	• Comfortable working anywhere

CH. 35: IMPACT OF FIVE GENERATIONS IN THE WORKPLACE

Aspect	Veterans (~1927-1945)	Baby Boomers (~1946-1964)	Generation Xers (~1965-1979)	Baby Boomlets (Gen Ys) (~1980-1996)
Irritants	• Managers who are "touchy feely", indecisive, hesitant to make the tough decisions • Vulgar language; slang • Trendy or experimental management approaches	• Managers who are bureaucratic, reject input, are brusque, disinterested, practice "one-upmanship" and send out the sage, "It is my way or the highway" • Political incorrectness	• Being micromanaged • Managers who fail to "walk the talk" • Flashy managers: hype • Bureaucratic managers • Too much time spent on process and not enough time on outcome	• Cynicism, sarcasm • Being treated as if they are too young to understand/know • Managers threatened by their techno-savvy • Condescending managers • Inconsistency; disorganization

1299

Aspect	Veterans (~1927-1945)	Baby Boomers (~1946-1964)	Generation Xers (~1965-1979)	Baby Boomlets (Gen Ys) (~1980-1996)
Job Change	• Seek cradle to grave experience • Uncomfortable situation • Job loss is stigmatized	• Seek lengthy employment situations • Job loss is a "glitch in their career path"	• Seek career security versus job security • Job change is necessary: part of their career path	• Seek career versus job security • Expect frequent job change • Seek portable work experience that guarantees skills for whatever life offers them

Aspect	Veterans (~1927-1945)	Baby Boomers (~1946-1964)	Generation Xers (~1965-1979)	Baby Boomlets (Gen Ys) (~1980-1996)
Employee Absenteeism & Disability	• Tend to not miss much work • Work through illness/injury • If seriously injured, tend to be away longer • Can be a challenge to place in modified work • Reluctant to do "lesser jobs" • Technologically challenged • Few cross-transferrable skills	• Seeking control over life and destiny • Unrealistic life and work goals • Predisposed to mental health problems • Tend to deny psychological health problems — "struggle on" at work untreated until they "crash"	• More likely to experience sports injuries and reproductive-related absences • Willing to do modified duties • Strong transferable skills	• More likely to experience sports injuries • Willing to do modified duties • Strong transferable skills

Aspect	Veterans (~1927-1945)	Baby Boomers (~1946-1964)	Generation Xers (~1965-1979)	Baby Boomlets (Gen Ys) (~1980-1996)
Work and Family Life Balance	• In the past, there was no connection between the two lives • More comfortable with working than with leisure time • Now, seek help to shift work/life balance	• No balance • Workaholics: "Live to work" • Now, seek help to balance their work/family life	• Seek work-life balance • Want work/life balance NOW • Will *temporarily* sacrifice personal time, but only occasionally	• Strongly value work-life balance • Work is not everything • Want flexibility to balance life interests and activities • Value travel opportunities

Aspect	Veterans (~1927-1945)	Baby Boomers (~1946-1964)	Generation Xers (~1965-1979)	Baby Boomlets (Gen Ys) (~1980-1996)
Current Status	• Cautious about the changed workplace • Eager to share the work knowledge they possess	• Want to make an impact and remain relevant • Caught between the desire to succeed and the desire to slow down and enjoy their successes	• Seek balance and meritocracy • Caught between feeling that they must constantly prove themselves and being criticized for being overly ambitious, disrespectful and irreverent	• Starting out in their careers • Want to learn and contribute • Seek to be valued • Very savvy yet socially conscious • Seek to change the world of work

H. THE FOUR GENERATIONS: IMPLICATIONS FOR OH&S PRACTITIONERS/PROFESSIONALS

OH&S Practitioners/Professionals working with these four generations need to appreciate the individual characteristics of each generation. For example, the Veterans and Boomers, being older workers, tend to be away from work more days per year due to illness or personal reasons than do the younger, Generation Xers and Baby Boomlets (Gen Ys).[108] If injured, the Veterans and Boomers require much longer time off work than do younger workers to recover. However, the Veterans and Boomers have a strong work ethic and therefore are amenable to a variety of return-to-work options. Unfortunately, many of the Veterans have vocational limitations in terms of their educational background and cross-transferable skills, making some modified work options unfeasible.

Some Baby Boomers are disenchanted with their lot in life and are trying to decide what their "golden years" might look like. They tend to overestimate their abilities and stamina and are seeking to control their life and destiny. Such unrealistic goals tend to predispose them to an increased incidence of psychological health problems. However, they are a generation where admission to having psychological/nervous health problems remains "taboo", and therefore, they tend to "struggle on" without seeking professional help until they "crash". For them, recovery can be difficult because to seek help, they first have to overcome a number of long-standing values and beliefs.

The younger Generation X and Baby Boomlets (Gen Ys), on the other hand, are less likely to succumb to illness, but are more likely to get injured in sport-related incidents or other risk-taking activities. Given that the Generation Xers tend to challenge authority, organization/company rules, and policies, these workers will be looking for adequate explanations from the Disability Management Program and processes. They will be seeking innovative rehabilitation plans and modified work opportunities. They value competence and expect OH&S and Disability Management Practitioners/Professionals to provide quality, competent services.

As for the Baby Boomlets (Gen Ys), although they may also experience similar sports-related injuries, they are amenable to getting back into a social milieu. The OH&S and Disability Management Practitioners/Professionals can capitalize on this aspect and seek a suitable modified work opportunity. Luckily, these Generation X and Baby Boomlet (Gen Y) workers have strong computer skills and technological capabilities that align well with many modified work opportunities.

In addition, each generation has its unique health challenges. The Veterans and Boomers are dealing with aging, as well as the impact of years of unhealthy lifestyle habits. As such, they are experiencing cardiovascular and respiratory conditions;

---

[108] Statistics Canada, *Table: 14-10-0193-01 (formerly CANSIM 279-0032): Absence rates for full time employees by age group, annual, 2019* (Ottawa: Statistics Canada, January 2020), online: https://www150.statcan.gc.ca/t1/tbl1/en/tv.action?pid=1410019301 (date accessed: February 28, 2020).

diabetes; lung, breast and colon cancers; kidney disease; degenerative nerve and musculoskeletal conditions; osteoporosis; and stroke; to name a few. Gen Xers are still in the childbearing years, having gotten a late start on family life. They have a history of high smoking rates, which come back to haunt them in later life. Today, they tend to experience depression, anxiety, and eating disorders, which may relate back to their childhood socialization and experiences. Remember, they witnessed a number of family marital breakdowns and divorces in their formative years, and often grew up in a single-parent family. As for Gen Y, they have the benefit of youth on their side. Despite that, there are reports that this generation does experience fatigue due to lack of sleep,[109] stress, cancers, depression, and cardiovascular disease.

1. The Organizational Approach

The effective management of multi-generational workplaces can be promoted by OH&S and Disability Management Practitioners/Professionals. How can this be achieved?

OH&S and Disability Management Practitioners/Professionals can help organizations/companies to:

- understand that the phenomenon of four generations in the workplace will not "just resolve": it needs to be managed;
- appreciate that events, experiences, and conditions in the formative years helped to shape the various generation personalities;
- take the time to understand each generational group, their characteristics and why they are the way they are;
- demonstrate respect for each generational cohort;
- seek ways to bridge the gaps between the generations;
- understand the need to clearly communicate the organization's/ company's expectations of employees;
- demonstrate flexibility;
- give people the benefit of the doubt;
- assist workers to achieve their career goals; and
- develop a "cafeteria approach" to employee group benefits and support programs, such as the Employee Assistance Program, Attendance Support & Assistance Program, Disability Management Program, Occupational Health & Safety Program, Human Resources Program, and Employee Education/Training Programs.

---

[109] U.S. Department of Health & Human Services, *CDC – Health Behavior of Adults: United States, 2008-2010* (Washington, DC: US Department of Health & Human Services, 2013) at 77.

A checklist on the recommended approach to the effective management of five generations in the workplace is provided in Appendix 1.

2. Intergenerational Conflicts

When five groups with different beliefs, attitudes, values, needs and wants exist in a workplace, misunderstanding and conflict can occur. In the past, workplace policies, programs, and benefits were designed to meet the beliefs, attitudes, values, needs and "wants" of the majority. Clearly, that approach of "one size fits all" is no longer effective.

(a) Potential ClashPoints®

Some potential generational clashpoints® are:

- **Reaction to Authority**: Veterans and Boomers tend to not question or challenge authority or the status quo. This may cause confusion and resentment among the Generation Xers and Baby Boomlets (Gen Ys) who have been taught to respond to their concerns and to ask, "Why?" According to Cran, they have been told by their parents to ask questions, not to just accept the status quo.[110] As well, they have been led to believe that they can achieve whatever they set out to do.

- **Work Styles**: Baby Boomers who like to work collaboratively and in teams, can prove frustrating to Veterans and Generation Xers who prefer independent work and perceive frequent meetings as exasperating.

    Generation Xers and Baby Boomlets (Gen Ys) who have had different life experiences and use different communication styles, may fail to actively listen to Veterans and Boomers thereby missing valuable work information and key opportunities for guidance. Conversely, the older generations may ignore the Baby Boomlets (Gen Ys), deeming them too inexperienced for any meaningful contributions.

- **Performance Feedback**: Feedback styles that may appear informative and helpful to one generation might seem formal and "preachy" to another. Feedback a Generation Xer perceives as immediate and honest can seem hasty, or even inappropriate, to Veterans and Baby Boomers. Likewise, some older generations have been told that there is "a time and place" for feedback. The two younger generations may not have received that message, thereby seeking frequent and immediate feedback from colleagues and management.

    Veterans seek "no applause" for their work performance, but appreciate subtle acknowledgement that they have made or are making a difference. Baby Boomers often provide feedback to others but seldom receive the

---

[110] S. Cran, *101 Ways to Make Generations X, Y and Zoomers Happy at Work* (Vancouver, BC: Synthesis at Work Inc., 2010).

same, especially positive feedback. Hence, they can be uncomfortable being told by a Generation Xer or Baby Boomlet (Gen Y) that they are doing a great job.

Generation Xers need positive feedback to let them know that they are on the "right track". Baby Boomlets (Gen Ys), who are used to praise from doting parents and activity coaches, may mistake "silence" for disapproval. They need to know what they are doing right and what they are doing wrong.

- *Advancement*: It appears that Baby Boomers will likely remain in the workplace, especially in supervisory and management roles, longer than originally expected. For the up and coming Generation Xers, this situation creates a sort of "glass ceiling": a frustrating situation for the aspiring Generation Xers who could easily direct their resentment of the situation towards the Baby Boomers.

(b) Challenges for OH&S and Disability Management Practitioners/ Professionals

For OH&S and Disability Management Practitioners/Professionals, challenges can arise in scenarios like the following:

1. ***The Veteran employees working with Generation Xers or Baby Boomlets.***

    As noted, Veterans are used to taking orders and hence, giving orders. Information is disseminated on a "need to know" basis. Both of the younger generations want to be included and involved in the work at hand. That means open communication and the use of a collaborative and inclusive approach. The two views can be in conflict.

    Likewise, their communication styles and work styles differ. Veterans prefer formal communication or face-to-face communication; the two other generations use voice mail, email, or text messages — the latter of which the Veteran may have no idea how to receive or respond.

    The Veterans' work style is one of individual work or teamwork if they are leading the team. The Generation Xers and Baby Boomlets (Gen Ys) prefer collaborative work.

    The Veteran might also resent the younger generations' demands for work/life balance and ask, "How will all this work get done if you guys aren't willing to put in the extra time to get it done?"[111] or; "In my day, we worked until we were done."

    On the other hand, the younger generations may wonder why the

---

[111] L. Panszczyk, "Benefits and Balance in a Four-Generation Workplace", *2004 CCH Unscheduled Absence Survey* (2004), online: http://www.cch.com/absenteeism2004/excerpt.asp (date accessed: January 31, 2020).

Veterans and Baby Boomers have not been able to balance their work and personal time. Waiting until retirement to "play" seems unacceptable — they want work/personal life balance now.[112]

The difficulties noted are substantiated by a recent poll that indicates that Veterans and Baby Boomlets (Gen Ys) have difficulties working together.[113]

2. **The Baby Boomer supervising the Generation Xers.**

According to a recent poll, 50 per cent of the respondents noted that Generation Xers and Baby Boomlets (Gen Ys) have the most challenges working with Baby Boomers.[114] Looking at their respective traits, this finding is not surprising. For example, Baby Boomers "live to work". They are dedicated to the job and are prepared to work long hours to achieve success. The concept of having "fun" and "work enjoyment" is not an essential part of this process. Additionally, the Baby Boomer prefers to hold meetings, work in a team, and to reach decisions based on collaboration and consensus.

On the other hand, Generation Xers "work to live". They like to be told what needs to be done, but not how to do it. Generation Xers are independent, entrepreneurial workers. Meetings are not popular. Although they prefer collaborative decision-making, they like to multitask and to set their own work priorities. Their personal time is valued, so working long hours is not a popular practice. They do not like to be tied to a job, especially a job that fails to challenge them or to meet their learning needs. Leaving the organization is always a viable option.

The outcome is that the work motivators, work styles, and lifestyles are in conflict unless the two players can reach a common position on how they will successfully work together.

3. **The Baby Boomer Human Resources manager counselling a Baby Boomlet (Gen Y) employee.**

As already noted, research has indicated that the Baby Boomlets (Gen Ys) face challenges working with Baby Boomers. In regards to this scenario, the two generations differ on a number of key aspects, namely:

- The Baby Boomer is a "workaholic" seeking advancement and personal achievement; the Baby Boomlet (Gen Ys) is looking for

---

[112] *Ibid*.

[113] A. Howland, "Multi-Generations Bring Challenge to Workplace", *The Vancouver Province* (May 27, 2007), online: http://www.canada.com/theprovince/news/working/story.html?id=c25b6bfa-f64a-40a5-a7d8-7de807e3ce63 (date accessed: January 31, 2020).

[114] *Ibid*.

meaning and relevance in work, but also values work and the social aspects of work. The Baby Boomlet (Gen Ys) has a life outside of work.

- The Baby Boomer lives to work; whereas, Baby Boomlets (Gen Ys) work to live, with many social, community, and travel interests filling their personal time.
- The Baby Boomer is dutiful, but focused on personal achievement and advancement; whereas the Baby Boomlet (Gen Y) values work achievement.
- The Baby Boomer is motivated by goals set by his/her manager; the Baby Boomlet (Gen Y) is driven by the search for meaningful work and an enjoyable work environment.

So when the Baby Boomlet (Gen Y) comes in and announces to the Baby Boomer Human Resources manager that a lengthy trip is coming up and that he/she wants to know if the current job will be available upon return, the Human Resources manager is caught off guard. The resentment may stem from the fact that this Baby Boomlet (Gen Y) has the "nerve" to ask for a leave that the Baby Boomer had never even thought to request in the past or present.[115]

4. ***The Generation Xer or Baby Boomlet (Gen Y) working in a traditional workplace where employee loyalty, lines of authority, work rules, vertical career ladders, and linear advancement are strongly valued.***

Given the discrepancies between the employee's values, beliefs, attitudes, expectations and needs, and what this workplace offers, there is a great potential for the employee to either leave, or remain and become dysfunctional or sick.

From the generational descriptions provided, one can imagine any number of conflicts that could arise when five generations are communicating, problem solving, working together, or trying to address differences of opinions. If these conflicts cannot be effectively addressed and resolved, employee casualties can occur. Many "stress-related" illnesses have their origins in unresolved workplace issues, situations and dysfunctional relationships. These are the most difficult disabilities to manage and successfully resolve. Hence, prevention is the key!

3. Motivational and Performance Reward Differences

While no two people are motivated in the same manner, performance motivators

---

[115] L. Panszczyk, "Benefits and Balance in a Four-Generation Workplace", *2004 CCH Unscheduled Absence Survey* (2004), online: http://www.cch.com/absenteeism2004/excerpt.asp (date accessed: January 31, 2020).

tend to be generational-based. This means that managers need greater insight into age-related issues and what motivates workers. For example, Baby Boomers strive for money, a title, and the "corner office with the window". Generation Xers want freedom and job security; whereas the Baby Boomlets (Gen Ys) want a job that has meaning for them and for society. The Veteran, on the other hand, is baffled as to why these other workers need any more motivation than the pay cheque they get every two weeks.

To motivate Veterans, the manager needs to opt for formality, such as communicating face-to-face or by phone as opposed to communicating through email or voice mail. The message should be: "Your experience is respected."[116] As for recognition, the traditional forms of recognition, such as plaques, certificates, or photos with top executives are appreciated.

Baby Boomers are motivated by goals that have been set by their managers. They value having the objectives established and the desired results stated in people-centred terms. Team participation is important. The motivational message should read as: "You are valued; you are needed."[117] For them, recognition is being mentioned in the organization's/company's newsletter, having their name on a parking spot, or receiving a promotion and title.

The Generation Xers prefer to be told what needs to be done; not how to do it. They like multiple tasks and the freedom to determine their own work priorities. Frequent and frank feedback is expected. As for their work preferences, they like flexibility, dressing casually, having fun, and enjoying a healthy work/life balance. The motivational message would be: "Do it your way; forget the rules."[118]

Motivating the Baby Boomlets (Gen Ys) differs. They seek opportunities for continuous learning and enhancement of their work skills. Their managers need to appreciate the Boomlets' (Gen Ys') personal goals and respond by linking the assigned tasks to these goals. Baby Boomlets (Gen Ys) respond to a positive work environment and prefer their manager to be more of a coach and less of a boss. For them, informal communication such as casual chats in the hallway, emails or voice mail, will suffice. For them, the motivational message is: "You will work with other bright creative people."[119] Their preferred form of recognition is to be given bonus days off work.

4. Other Differences

Worker expectations and communication styles vary by generation. The Veterans do not seek much feedback. Their credo is: "No news is good news." The Boomers

---

[116] G. Hammill, "Mixing and Managing Four Generations of Employees", *FDU Magazine Online* (Winter/Spring 2005), online: http://www.fdu.edu/newspubs/magazine/05ws/generations.htm (date accessed: January 31, 2020).

[117] *Ibid.*

[118] *Ibid.*

[119] *Ibid.*

prefer to receive feedback on their work performance once a year and in writing. Both the Generation X and Baby Boomlets (Gen Ys) want immediate and frequent feedback; only the Boomlets (Gen Ys) want it daily.

In terms of work/life balance, the Veterans want help to shift the work/life balance. The Boomers seek help to balance their work and family life. Generation Xers want work/life balance now. They now have family responsibilities and prefer to spend time with their children as opposed to work. For the Boomlets (Gen Ys), work isn't everything. They want the flexibility to balance their lives and activities. Family life and travel opportunities are highly valued. They are often willing to give up a job to return to school or to travel for extensive periods of time.

The perspectives on job change radically differ. For the Veterans, a job change carries a stigma. The Boomers perceive job change as a "glitch" in their career path. To the Generation Xers, job change is necessary — it is part of their career development and path. The Boomlets (Gen Ys) feel likewise, but they expect job change to happen many times in their careers.

I. THE FIVE GENERATIONS: GENERATIONAL INTEGRATION

Successful employers are gearing up to effectively manage the phenomenon of five generations in the workplace through the use of the **generational integration** approach — that is, bringing the five generations found in today's workplaces together in such a way that the "best" of each generation is accessed and put into practice to enhance the well-being and performance of the entire workforce.[120] The intent is to enable the five generations to interact, to learn about and from each other, to bridge the gaps between the generations, and to ultimately have them "bond". The goal is to manage, motivate, and retain a multi-generational workforce.

Generational integration begins with identifying key areas of interest or areas of dissonance, such as:

- meaningful work — the opportunity to contribute to the organization and to make a difference;
- collaboration;
- learning and development;
- use of technology — a divisive force between the generations which can be bridged by engaging workers to help each other use the various technological tools; and
- flexibility.[121]

With generational integration, information flows in all directions. The most

---

[120] K. Kirkpatrick, S. Martin & S. Warneke, *Strategies for the Intergenerational Workplace* (Gensler, July 17, 2008), online: http://www.gensler.com/uploads/documents/IntergenerationalWorkplace_07_17_2008.pdf (date accessed: January 31, 2020).

[121] *Ibid.*

enlightened organizational leaders find a way to let every generation hear and be heard. They recognize that no one has all the answers. This appreciation of diversity allows each generational group to contribute and be a part of the growth and development of a department or organization.[122]

J. THE FIVE GENERATIONS: RECOMMENDATIONS FOR OH&S AND DISABILITY MANAGEMENT PRACTITIONERS/PROFESSIONALS

OH&S and Disability Management Programs should take into consideration the make-up of the corporate workforce. When developing the OH&S policy and program components and the other related policies, it is important to address the unique aspects of each generational group. The following is a discussion of some of the related issues.

OH&S and Disability Management Program *components* should reflect the needs of the employee population. For example, the safe work procedures should include the use of ergonomically designed tools, manual-handling aids, visual/audio considerations, and modifiable workstations, as well as how the safe work practice is presented to the employee work group.

OH&S and Disability Management Program *communications* must be designed to appeal to all generations. Multi-media approaches that include the traditional written documentation (brochures, newsletters, employee OH&S handbooks, *etc.*) should be combined with more interactive media approaches (webpages, electronic message boards, *etc.*). Solicit feedback about the OH&S and Disability Management Program and services using a variety of techniques, namely, hardcopy and electronic surveys, focus groups, chat boxes, blogs, *etc.*

OH&S and Disability Management Program *marketing messages* need to be tailored. The Veterans, who tend to possess strong organizational/company loyalty, value money and work relationships and have strong work ethics, respond well to the message that the OH&S and Disability Management Programs can help them and their co-workers to remain active in the workplace. The Boomers on the other hand, value evidence-based practice. As such, the marketing message must demonstrate how the OH&S and Disability Management Program have proven to be valuable to employees and the organization.

With the phenomenon of an aging workforce, the approaches used within the OH&S Program should *address the challenges faced by aging workers*. For example, aging workers experience changes in muscles, bones, cardiovascular systems, hearing, vision, mental processes, reaction times, and sensory and motor processes. Hence, OH&S Professionals/Practitioners should take this knowledge into account when designing OH&S programs, practices, and training sessions. Management education on these factors is recommended. Many work tasks can be

---

[122] Office of Institutional Equity, Duke University, *Cross Generational Communication: Implications in the Work Environment* (April 2008), online: http://www.duke.edu/web/equity/div_genera~ppt%20III.ppt (date accessed: January 31, 2020).

readily modified to reduce the wear and tear on the human body. Likewise, when counselling the aging worker, the following recommendations can be made in order to enhance the worker's well-being and productivity:

- maintain a healthy lifestyle — exercise and diet;
- get enough sleep;
- exercise the brain and cognitive functions;
- stretch before, during, and after work;
- add weight-bearing activities to their exercising program;
- dress for weather extremes – hot and cold temperatures;
- use material handling equipment to reduce lifting;
- minimize awkward postures;
- maintain proper posture;
- make frequent postural changes;
- wear suitable footwear;
- obtain regular vision and hearing tests;
- use prescribed corrective lenses and hearing aids; and
- be aware of drug side effects when taking medications.

Generation Xers perceive exercise as essential to their well-being. For them, acceptance of the concept that getting back gradually "into the game" works; however, they will expect frequent feedback on their recovery progress and recognition for their modified work efforts. The Baby Boomlets (Gen Ys) are similar on both the above issues, however, they may respond better to the message that modified work is one way of helping their work team to continue to operate through this disability period.

OH&S and Disability Management Practitioners/Professionals must use a *holistic approach* when dealing with ill/injured workers. This means understanding the characteristics of these four generations and adapting their services to meet their respective needs. Disability Management Programs, like other employee benefits, need to be "cafeteria style", providing flexibility and variability in their offerings and services.

The *choice of modified work duties* is another area for consideration. Boomers, Generation Xers, and Baby Boomlets (Gen Ys) who are technologically savvy, are well-suited for modified duties involving office work, working at home, and alternate job duties. Veterans who work in primarily labour positions pose an interesting challenge. Although knowledgeable and experienced, they may be unwilling to perform modified duties that are outside of their comfort zone or deemed to be lesser jobs than their "own" job.

Baby Boomers perceive exercise and staying physically fit as their duty. They respond well to the sports analogy that one would never put a recovering athlete back into a game situation before providing rehabilitation and a period of participation in game workouts. Modified work is the same. It provides an opportunity for work hardening aimed at enabling the employee to return to his/her regular job. Additionally, they value money and perceive modified work as worthwhile if there is a pay differential over being on short-term disability coverage.

Worker *response to support services* differs. As already noted, Veterans and Baby Boomers perceive psychological/nervous health conditions in a negative manner. To them, the stigma of psychological health disorders remains alive and well. Generation Xers and Baby Boomers are much more tolerant about human frailties. As a result, one would expect younger workers to be more receptive to using Employee Assistance Programs and other organization/company supports. Boomers, Generation Xers, and Baby Boomlets (Gen Ys) all value exercise and the benefits that participation affords. Physiotherapy and reconditioning efforts would be better received by them than by the Veterans who tend to view exercise only as a "necessary evil".

From a *prevention* point of view, Generation Xers and the Baby Boomlets (Gen Ys), who have tended to be sedentary and to consume lots of processed and fast foods, and who have been exposed to more pollution and stress, may be more prone to the development of health problems as they age.[123]

These are but a few of the generational challenges that the OH&S and Disability Management Practitioner/Professional will face. The key recommendations are to:

- *Recognize, support and build on the work aspects that all the generations seek, namely*:
    - respect from management;[124]
    - open and honest management communication;[125]
    - meaningful work;[126]
    - the ability to self-manage their work — input into work and the

---

[123] FGI (now Shepell-FGI), "One Workplace, Four Generations: Managing the Conflicting Needs", *Working Well for Managers* (September 2004), online: http://www.shepellfgi.com/EN-CA/HRFundamentals/pdf/Bridging%20the%20Generation%20Gap%20-%20CA-EN.pdf (date accessed: January 31, 2020).

[124] Ontario Restaurant Association, "Employees Today want Education, Respect and More", *Ontario Restaurant News* (January 2005), online: https://www.go2hr.ca/articles/employees-today-want-education-respect-and-more (date accessed: January 31, 2020).

[125] *Ibid.*

[126] K. Kirkpatrick, S. Martin & S. Warneke, *Strategies for the Intergenerational Workplace* (Gensler, July 17, 2008), online: http://www.gensler.com/uploads/documents/IntergenerationalWorkplace_07_17_2008.pdf (date accessed: January 31, 2020).

management of their own workloads are important to worker well-being;[127]
- work-life balance — a personal life and "down-time" are critical;[128]
- interesting work — employees want to be challenged and to feel that they are making a difference in the work that they are doing;[129]
- a good salary with benefits[130] — fair pay leads to better quality of life;[131]
- the ability to learn and grow at work;[132] and
- good performance and recognition.[133]

- *Know your clientele.*
- *Respect their unique differences.*
- *Seek modified work opportunities that not only "fit" the employee's recovery needs, but also take into consideration the generational needs of the parties involved.*
- *Be flexible and innovative in the disability management approaches.*
- *Promote the development of programs designed to support general employee health and enhance employee resilience against life stressors.*

---

[127] M. Shain, "Managing Stress at its Source in the Workplace" (presented at the Health Work and Wellness Conference '98, Whistler BC, September 27-30, 1998) [unpublished]; and D. McPhail, "Corporate Perspectives on Workplace Learning, 'Four Generations in the Workforce'" (May 9, 2006), online: http://www.wln.ualberta.ca//en/Events/ Archives/Seminar%20Archives/~/media/wln/Documents/Events/Seminars_2007/Generation_Workshop.pdf (date accessed: January 31, 2020).

[128] D. McPhail, *ibid.*

[129] *Ibid.*

[130] *Ibid.*

[131] Ontario Restaurant Association, "Employees Today want Education, Respect and More", *Ontario Restaurant News* (January 2005), online: https://www.go2hr.ca/articles/employees-today-want-education-respect-and-more (date accessed: January 31, 2020).

[132] D. McPhail, "Corporate Perspectives on Workplace Learning 'Four Generations in the Workforce'" (May 9, 2006), online: http://www.wln.ualberta.ca//en/Events/Archives/Seminar%20Archives/~/media/wln/Documents/Events/Seminars_2007/Generation_Workshop.pdf (date accessed: January 31, 2020); and Ontario Restaurant Association, "Employees Today want Education, Respect and More", *Ontario Restaurant News* (January 2005), online: https://www.go2hr.ca/articles/employees-today-want-education-respect-and-more (date accessed: January 31, 2020).

[133] D. McPhail, *ibid.*

- *Promote "ageless thinking" where employees are equal regardless of age.*[134]

K. SUMMARY

So, "Yes, there are five generations in today's workplaces." Generational differences can lead to strength and opportunity, or to stress and conflict. It all depends on how organizations/employers choose to handle the situation. Best practice solutions involve the recognition, appreciation, and respect for the traits of each generation. One option is to develop a **generational strategy** — a method of:

- understanding what makes their "employees tick";
- emphasizing the importance of teamwork;
- effective communications; and
- adopting "ageless thinking".[136]

However, it is vitally important to not stereotype employees — "the generational concept is simply 'one lens' that can be used to help us understand people".[137] Generational diversity, like gender, racial and ability diversity, is a factor to consider in managing and positioning an organization, but it is not the "whole show".[138] The answer for employers is to regularly assess the needs of the organization to determine if and how employee needs are changing, and then based on the findings, respond accordingly.

---

[134] A. Howland, "Multi-generations Bring Challenge to Workplace", *The Vancouver Province* (May 27, 2007), online: http://www.canada.com/theprovince/news/working/story.html?id=c25b6bfa-f64a-40a5-a7d8-7de807e3ce63 (date accessed: January 31, 2020).

[136] A. Howland, "Multi-Generations Bring Challenge to Workplace," *The Vancouver Province* (May 27, 2007), online: http://www.canada.com/theprovince/news/working/story.html?id=c25b6bfa-f64a-40a5-a7d8-7de807e3ce63 (date accessed: January 31, 2020).

[137] D. McPhail, "Corporate Perspectives on Workplace Learning, 'Four Generations in the Workforce'" (May 9, 2006), online: http://www.wln.ualberta.ca/en/Events/Archives/Seminar%20Archives/~/media/wln/Documents/Events/Seminars_2007/Generation_Workshop.pdf (date accessed: January 31, 2020).

[138] R. Zemke, CEO Agenda Series, "Generations at Work" (2001), online: http://www.asaecenter.org/Resources/articledetail.cfm?ItemNumber=13053 (date accessed: January 31, 2020).

Appendix 1

MULTI-GENERATIONAL WORKPLACES: EFFECTIVE MANAGEMENT

Recommended Actions:

☐ Assist the company/organization to understand that this situation will not "just resolve itself" — it needs to be managed.

☐ Help the company/organization appreciate that the various events and conditions in the generation cohorts' formative years helped to shape their generation's personalities.

☐ Encourage the company/organization to take the time to understand each generational cohort and why they are the way they are.

☐ Encourage the company/organization to demonstrate respect for each generational cohort.

☐ Promote company/organizational interest in seeking ways to bridge the gaps between the four generations.

☐ Encourage the company/organization to clearly communicate company/organizational expectations of employees.

☐ Encourage the company/organization to demonstrate flexibility in terms of how the various employee groups are supported and their needs addressed.

☐ Promote the management practice of giving workers the "benefit of the doubt" when issues/conflicts/disputes arise.

☐ Assist workers to achieve their career goals.

Copyright D. Dyck 2009-2019 — All rights reserved.

CHAPTER REFERENCES

AARP, *Leading a Multigenerational Workforce* (Washington, DC: AARP, 2007), online: http://assets.aarp.org/www.aarp.org_/articles/money/employers/leading_multigenerational_workforce.pdf (date accessed: January 31, 2020).

T. Brokaw, *The Greatest Generation* (New York: Random House, 1998).

S. Calhoun & P. Strasser, "Generations at Work" (2005) 53:11 *AAOHN* 469.

Central Intelligence Agency, "Median Age: Years", *World Factbook* (2012), online: https://www.cia.gov/library/publications/the-world-factbook/rankorder/2177rank.html (date accessed January 31, 2020)

S. Cran, *101 Ways to Make Generations X, Y and Zoomers Happy at Work* (Vancouver, BC: Synthesis at Work Inc., 2010).

M.J. Douglas, *Generation Clash! Making Generational History* (2009), online: http://content.monster.ca/7371_en-CA_pf.asp (date accessed: January 31, 2020).

J. Duchscher & L. Cowen, "Multigenerational Nurses in the Workplace" (2004) 34:11 *J. Nurs. Adm.* 493.

D. Dyck, "Multi-Generational Workplaces: The OH&S Challenge" (Presented at the American Occupational Health Nurses Symposium and Expo, Orlando, FL, April 2007) [unpublished].

FGI (now Shepell-FGI), "One Workplace, Four Generations: Managing the Conflicting Needs", *Working Well for Managers* (September 2004), online: http://www.shepellfgi.com/EN-CA/HRFundamentals/pdf/Bridging%20the%20Generation%20Gap%20-%20CA-EN.pdf (date accessed: January 31, 2020).

Go2 Tourism HR Society, *Capitalizing on the Generational Gap in the Workplace* (2008), online: http://www.go2hr.ca/ForbrEmployers/Retention/GenerationsintheWorkplace/tabid/566/Default.aspx (date accessed: January 31, 2020).

G. Hammill, "Mixing and Managing Four Generations of Employees", *FDU Magazine Online* (Winter/Spring 2005), online: http://www.fdu.edu/newspubs/magazine/05ws/generations.htm (date accessed: January 31, 2020).

S. Heathfield, *Managing Millennials: Eleven Tips for Managing Millennials* (2008), online: http://humanresources.about.com/od/managementtips/a/millenials.htm (date accessed: January 31, 2020).

A. Howland, "Multi-Generations Bring Challenge to Workplace", *The Vancouver Province* (May 27, 2007), online: http://www.canada.com/theprovince/news/working/story.html?id=c25b6bfa-f64a-40a5-a7d8-7de807e3ce63 (date accessed: January 31, 2020).

J. Jamrog, "The Perfect Storm" (Presented at a CHRP Seminar, Calgary, AB, September 15, 2005) [unpublished].

K. Kirkpatrick, S. Martin & S. Warneke, *Strategies for the Intergenerational Workplace* (Gensler, July 17, 2008), online: http://www.gensler.com/uploads/documents/IntergenerationalWorkplace_07_17_2008.pdf (date accessed: January 31, 2020).

L. Lancaster & D. Stillman, *When Generations Collide: Who They Are, Why They Clash, How to Solve the Generational Puzzle at Work* (New York: Harper Collins, 2002).

C. Marston, "Four Generations in the Workplace" (Presentation delivered at the Planned Parenthood Federation of America, Western Region Conference, Salt Lake City, UT, 2004) [unpublished].

C. Marston, "Four Generations in the Workplace: A Diversity Challenge" (presentation delivered at Utah Department of Workforce Services: Roads to Success, Salt Lake City, UT, 2003) [unpublished].

C. Marston, *Motivating the "What's in It for Me?" Workforce: Manage Across the Generational Divide and Increase Profits* (Hoboken, NJ: John Wiley & Sons, 2007).

M. McCrindle, *New Generations at Work: Attracting, Recruiting, Retraining & Training Generation Y* (Sydney, NSW, Australia: McCrindle Research, 2006), online: http://www.mccrindle.com.au/resources/whitepapers/McCrindle-Research_New-Generations-At-Work-attracting-recruiting-retaining-training-generation-y.pdf (date accessed: January 31, 2020).

T.K. McNamara, "Analysis of the March 1977 and 2007 Current Population Surveys" (unpublished raw data, 2007) in J. Dobbs, P. Healey,

K. Kane, D. Mak & T.K. McNamara, The Center on Aging & Work: Workplace Flexibility, *The Multi-Generational Workplace, Fact Sheet #9* (July 2007), online: Boston College http://agingandwork.bc.edu/documents/FS09_MultiGenWorkplace_000.pdf (date accessed: January 31, 2020).

D. McPhail, "Corporate Perspectives on Workplace Learning, 'Four Generations in the Workforce'" (May 9, 2006), online: http://www.wln.ualberta.ca//en/Events/Archives/Seminar%20Archives/~/media/wln/Documents/Events/Seminars_2007/Generation_Workshop.pdf (date accessed: January 31, 2020).

Office of Institutional Equity, Duke University, *Cross Generational Communication: Implications in the Work Environment* (April 2008), online: http://today.duke.edu/2014/02/generationaldifferences (date accessed: January 31, 2020).

Ontario Restaurant Association, "Employees Today want Education, Respect and More", *Ontario Restaurant News* (January 2005), online: https://www.go2hr.ca/articles/employees-today-want-education-respect-and-more (date accessed: January 31, 2020).

L. Panszczyk, "Benefits and Balance in a Four-Generation Workplace", *2004 CCH Unscheduled Absence Survey* (2004), online: http://www.cch.com/absenteeism2004/excerpt.asp (date accessed: January 31, 2020).

N. Pekala, "Conquering the Generational Divide" (2001) 66:6 *Journal of Property Management* 30.

M. Pitt-Catsouphes & M. Smyer, The Center on Aging & Work: Workplace Flexibility, "The 21st Century Multi-generational Workplace", online: Boston College http://bc.edu/research/agingandwork (date accessed: January 31, 2020).

E. Quick, "Disability Management that Works", *Benefits and Compensation Solutions Magazine* (January/February 2007) at 46.

A. Schwartz, *Generations at Work: Understanding and Influencing* (2008), online: http://www.duke.edu/web/equity/Diversity.htm (date accessed: January 31, 2020).

M. Shain, "Managing Stress at its Source in the Workplace" (presented at the Health

Work and Wellness Conference '98, Whistler, BC, September 27-30, 1998) [unpublished].

Statistics Canada, "Median age of the resident population of Canada in 2018, by province" (2019), online: https://www.statista.com/statistics/444816/canada-median-age-of-resident-population-by-province/.

Statistics Canada, Table 051-0001, "Estimates of Population, by Age Group and Sex for July 1, Canada, Provinces and Territories, Annual (Persons Unless Noted Otherwise)" (2018), online: https://www150.statcan.gc.ca/t1/tbl1/en/tv.action?pid=1710000501 (date accessed: January 31, 2020).

Statistics Canada, *Table: 14-10-0193-01 (formerly CANSIM 279-0032): Absence rates for full time employees by age group, annual, 2019* (Ottawa: Statistics Canada, January 2020), online: https://www150.statcan.gc.ca/t1/tbl1/en/tv.action?pid=1410019301 (date accessed: February 28, 2020).

A. Stefaniak, *Black Hole or Window of Opportunity? Understanding the Generational Gap in Today's Workplace* (Policy Brief: 04-11-07, Center for Public Policy & Administration), online: http://cppa.utah.edu/_documents/publications/workforce/generations.pdf (date accessed: January 31, 2020).

W. Strauss & N. Howe, *Generations: The History of America's Future, 1584 to 2069* (New York: Harper Perennial, 1992) (reprint).

U.S. Department of Health & Human Services, *CDC – Health Behavior of Adults: United States, 2008-2010* (Washington, DC: US Department of Health & Human Services, 2013).

R. Zemke, CEO Agenda Series, "Generations at Work" (2001), online: http://www.asaecenter.org/Resources/articledetail.cfm?ItemNumber=13053 (date accessed: January 31, 2020).

M. Znaimer, online: http://www.mosesznaimer.com/ (date accessed: January 31, 2020).

Chapter 36

INTERNAL/EXTERNAL CONSULTING: TIPS FOR OCCUPATIONAL HEALTH AND SAFETY PRACTITIONERS/ PROFESSIONALS[1]

A. INTRODUCTION

Consulting is the art of influencing people at their request.[2] In essence, it is a helpful relationship. It means reaching a conclusion of acceptable quality, of maintaining an acceptable commitment to that conclusion, of supporting the client to prevent and/or minimize any detrimental outcomes as a result of the conclusion, and of being perceived as being helpful and efficient by the client.[3]

The **Occupational Health and Safety (OH&S) consultant** by definition is a practitioner/professional who provides influence, recommendations and expertise, but has no direct power to make changes. Regardless of the degree of involvement, the consultant does not own the problem: the client does.

B. OH&S PRACTITIONERS/PROFESSIONALS AS CONSULTANTS

OH&S Practitioners/Professionals act as consultants: when someone asks them if they have five minutes to discuss an OH&S issue; when the OH&S Practitioner/ Professional advocates for an employee or champions a new idea; or when the OH&S Practitioner/Professional makes the following statements on behalf of others: "I'll look into that", "I think the real problem is . . .", "Here is what I think we should do about this situation".

In these ways, the OH&S Practitioner/Professional functions as either an internal or an external consultant. The irony is that although consulting skills are essential

---

[1] D. Dyck "Internal and external consulting: Assisting clients with managing work, health and psychosocial issues" (2002) 50:3 *AAOHN* 111. Copyright (2002) the American Association of Occupational Health Nurses, Inc. Used with permission. All rights reserved.

[2] G.M. Weinberg, *The Secrets of Consulting* (New York, NY: Dorset House Publishing, 1985).

[3] Proactive Consultants, *Effective Internal Consulting* (Calgary, AB: Proactive Consultants, 1989).

to the practice, they are not skills that OH&S Practitioners/Professionals have been well-prepared to implement. For the OH&S Practitioner/Professional, this chapter contains information that can be used to evaluate and hone internal and/or external consulting skills.

The intent of this chapter is to:

- define OH&S consulting — internal and external;
- explain how to manage any OH&S consulting request or project using a six-phase consulting model;
- explain how to build collaborative and accountable client-consulting relationships;
- learn to negotiate appropriate consulting roles with clients;
- reframe the presenting issues into working issues;
- gather relevant data while building client commitment;
- present relevant information to create the foundation for appropriate recommendations; and
- learn the top ten secrets of OH&S consulting.

While reading, think of some opportunities to apply the techniques described, and then look for work opportunities to test them.

C. CONSULTING: WHAT IS INVOLVED?

As an operational definition, **consulting** is the provision of temporary professional help to assist the client to address current or potential problems, or opportunities. It is the ability to:

- help the client discover problems and to facilitate the assessment of the client's needs and willingness to change;
- ensure clarity of roles, responsibilities, and resources through formal and/or informal contracting with the client;
- gather and present facts, observations, opinions, and feelings that assist the client to define the problem(s);
- coordinate the implementation of one or more intervention(s) (resource, expert or process) that successfully and productively address the defined problem(s) in a manner that is fully supported by the client;
- evaluate the effectiveness of the intervention and any further actions required;
- develop self-sufficiency in the client-system so as to minimize dependence on the consultant and to ensure a minimum of stress during disengagement; and
- assume any of the varying roles of a consultant as an advocate, technical

specialist, trainer/educator, collaborator in problem-solving, identifier of alternatives, fact-finder, process specialist, or as a reflector.

D. THE CONSULTING RELATIONSHIP: ITS CHARACTERISTICS

OH&S Consulting is a voluntary, temporary, and supportive relationship that involves a helper and a help-needing party (the client). It focuses on aiding the client to enable him/her to reach a solution for a potential/real problem situation. By helping clients to make use of their own knowledge, products and services, the consultant facilitates the resolution of a problem by the client.

As OH&S consultants, OH&S Practitioners/Professionals exert influence, but they do not have any power. Their leadership is non-authoritarian in nature. OH&S consultants may be involved with the problem, but they remain "outsiders". They do not own the problem or the solution. It is the client that owns the problem and makes the final decision(s).

Consulting requires discipline by both parties to recognize and play their respective roles in reaching a mutually agreed-upon solution. At each phase of the consulting process, the OH&S consultant needs to deal with three questions:

1. What do we need to accomplish?
2. What process should we use?
3. What should my role be?

There is a wide range of available consulting roles. Any role negotiated requires the use of OH&S technical expertise. This expertise will differ from role to role. The main question is: "Which role will be the most helpful in this particular situation?" To answer this question, consider:

- the skills and knowledge required to do the job;
- the time constraints, if any exist;
- the client's expectations;
- the potential for developing the client's capabilities so that they can manage a similar issue next time it arises; and
- any systems that may need to be developed and implemented for success.

The OH&S Practitioner's/Professional's role may change several times throughout an OH&S consulting project. Remember: "Help is never help unless it is perceived to be so by the client".[4] Be prepared to renegotiate any necessary role changes with the client.

There are two types of OH&S consulting relationships: internal or external. **Internal OH&S consulting** is when the consultant is a member of the organization

---

[4] D. Sousa, "Internal/External Consulting: Ten Tips for Occupational Health Nurses" (Seminar presented at the American Occupational Health Conference, San Francisco, CA, April 2001) [unpublished].

and is assuming a consulting role to assist with the resolution of a problem or the chance to capitalize on an opportunity. The client may be the whole organization for whom the consultant assumes an advocacy role or a segment of that organization. The internal OH&S consultant tends to be knowledgeable about the organization — its goals, structure, and operations. The internal OH&S consultant keeps the organization's "best-interests" in mind. Although acting as an OH&S consultant, the internal consultant does retain some power to veto proposed actions.

The **external OH&S consultant** is an "outsider" who has access to specialized services not traditionally found within the client organization. This type of consultant tends to have more than one client group, is more likely to be able to bring a wide variety of experiences to the situation and is better positioned to view the situation more objectively. Unlike the internal OH&S consultant, the external OH&S consultant has no decision-making power.

Being labelled an "OH&S consultant" does not make the OH&S Practitioner/ Professional a consultant. Rather, the knowledge possessed, and skills demonstrated, qualify the person to be called an OH&S consultant.

E. THE CONSULTING TOOL KIT

In addition to technical expertise, the OH&S consultant requires a number of specialized skills:

- *Interpersonal Skills* — building effective working relationships such as active listening skills, assertiveness skills, conflict management and issue resolution, providing support and reassurance, and giving/receiving feedback;

- *Business Skills* — linking solutions to client needs and business drivers such as strategic planning, project management, project evaluation, and cost/benefit analysis of the solution;

- *Analytical Skills* — gathering and managing the relevant information such as data gathering, data management, data analysis, data interpretation, and reporting skills;

- *Consulting Skills* — managing the consulting process such as the knowledge and ability to use the consulting model/process and adherence to professional ethics; and

- *Technical Specialist Skills* — functioning competently within the field of OH&S.

In essence, the OH&S consultant needs a tool kit (Figure 36.1).

---

[5] Adapted from L. Goode, "Consulting Successfully Within Organizations" (Seminar presented at Petro-Canada, Calgary, AB, 1995) [unpublished].

Figure 36.1: OH&S Consulting Tool Kit[5]

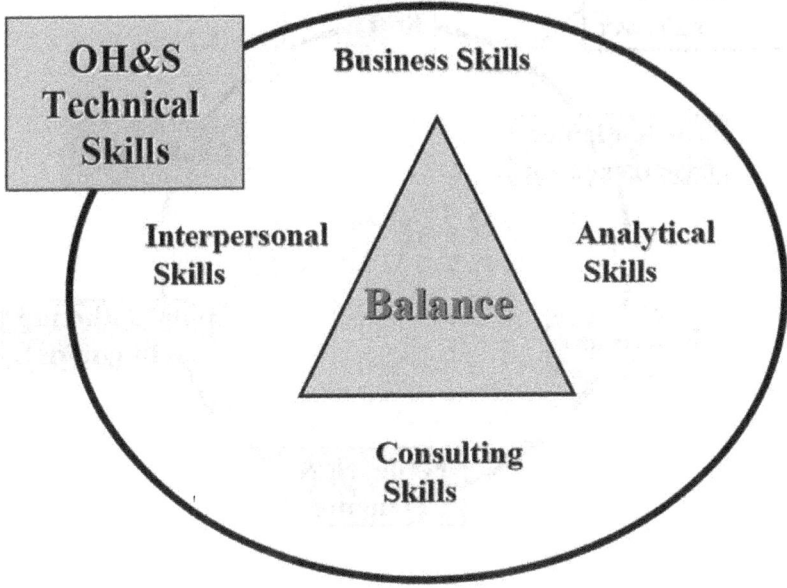

F. SIX PHASES OF AN OH&S CONSULTING PROJECT

According to Goode (1995),[6] an OH&S consulting project has six phases (Figure 36.2).

---

[6] *Ibid.*

Figure 36.2: The OH&S Consulting Process

1. Entry Phase

Entry is the first contact with a client about the concern or project. It is analogous to engaging an employee or manager to work on an issue. The purpose of the Entry Phase is to decide whether or not to work together on the project. This is a critical point: if a contract "sours"; 90 per cent of the time it goes awry in either the Entry Phase or the Contracting Phase. Remember that both parties are nervous about this phase. It is up to the consultant to manage these phases.

The criteria for the "Fit/No fit" decision include an exploration of:

- **The Task**

    Exploration of the task is critical during this stage. For example: What is the objective of the work/task? Is it possible to do? What has been done so far? Are there clear expectations for the work and a successful completion? Is it original work or routine work? What about the timing — can it be done in the allotted time frame? Are there political implications that need to be considered and/or addressed?

- **The Client-Consultant Relationship**

    Who is/are the real client(s)? Who are the real decision-makers? Who are the end-users? Is there positive chemistry between the client(s) and the consultant? Do their values and ethics align? What is the level of trust like? Can both parties "win" if this project is successful? What level of commitment does the client(s) have to the completion of this project?

- **The OH&S Practitioner/Professional's consulting experience, expertise and availability**

    What is the OH&S Practitioner's/Professional's OH&S skill level? Are the OH&S Practitioner's/Professional's personal values and ethics aligned with those of the client(s)? What are the OH&S Practitioner's/Professional's other business priorities and/or demands? Is this a "high-risk" project with "low pay-off potential"?

The recommended approach for the Entry Phase is to begin by analyzing the situation. This means clarifying the real versus the presenting problem; identifying the client/consultant's wants and expectations; identifying the client's commitment level to the project; and establishing the roles that the OH&S consultant and client will assume. Then, the OH&S consultant must reach a decision point of "Go/No Go" with the consulting assignment.

2. Contracting Phase

A **consulting contract** is an explicit agreement clarifying what the client and consultant can expect from each other and how they will work together. This can be formal or informal in nature and involves the mutual exploration and decision-making about what is to be done and how to work together. While the Entry Phase determines whether the client and consultant will work together, the Contracting Phase determines how the client and consultant will work together.

Why is contracting necessary? To begin with, it establishes a basis for working together. It is during this period that decisions such as the following are made: Will the relationship be formal or informal? Will the work be done together or independently? Who will have the final say in how things are done?

Second, if possible, it develops a 50:50 relationship. The reason for this is that the OH&S Practitioner/Professional wants the client to own the problem and the solution, so that when it comes time to exit the consulting relationship, the client will be successfully launched.

Thirdly, an OH&S consultant's maximum leverage occurs at the beginning of the working relationship, just like in a marital courtship. By clarifying expectations, establishing individual roles, making plans for the project and setting the ground rules at this point, the OH&S consultant has the greatest leverage for attaining success in the long run. *The key to a successful consulting project is to set things up right in the first place.*

Finally, most consulting "failures" can be traced back to omissions or errors in the Entry Phase or Contracting Phase. To be successful, the consulting services/products must meet the client's expectations. To achieve this, ensure that the client's expectations are clearly articulated in the Contracting Phase and *avoid over-*

promising on what can actually be delivered.

The recommended approach is to:
- confirm the client's level of commitment to the project;
- seek a balance of responsibility for the project;
- develop a clear project contract; and
- make sure that the contract is signed.

Depending on the complexity and nature of the concern/project and the client system, some or all of the following items should be clarified in the Contracting Phase:

i. **STATEMENT OF THE WORKING ISSUE AND SCOPE OF THE WORK**
   - the desired results and products;
   - the scope of the project in terms of the boundaries of the consulting process and project content; and
   - a definition of what is "success" and "failure" for this project.

ii. **CLIENT PROFILE**
   - identification of the decision-maker(s) for this project;
   - identification of all the project stakeholders and their interests/needs/wants in relation to this project;
   - identification of all people/systems affected by the project; and
   - identification of who else needs to be involved in the project and consulting arrangement.

iii. **STAKEHOLDER ROLES AND RESPONSIBILITIES**
   - establish the type and degree of involvement of stakeholders in the project; and
   - define the consultant role.

iv. **CONTRACT BY PHASE OR "ONE-SHOT" APPROACH**
   - decide on the scope of the project: Will it be treated as one large project, or a project with a number of phases each of which are perceived as an individual mini-project?

v. **COSTS AND RESOURCES**
   - agree on the fee structure, charge codes, and/or budget approvals for the project; and
   - clarify the materials, supplies, and/or resources needed.

vi. **SCHEDULE**
   - establish the start time, specific milestones, and the completion date;

- articulate the "success indicators" and when they will be reached; and
- prepare a timeline and tracking process for the project.

vii. **DATA COLLECTION**
- determine the acceptable/non-acceptable sources and methods of data collection;
- decide how feedback of data will be obtained (for whom? how? by when? at what cost?); and
- address the issue of confidentiality.

viii. **CONSTRAINTS/POTENTIAL ISSUES**
- be aware of ethical, legal, and/or political sensitivities that may impact the project; and
- identify the critical steps in the project and have contingency plans in effect.

ix. **FEEDBACK TO THE OH&S CONSULTANT**
- clarify up front, how, by whom, and when feedback on the project will be given to the OH&S consultant.

As already mentioned, critical to the success of the Contracting Phase (and ultimately, to the success of the project) is the client-consultant relationship. For this reason, it is important to explore some tips for building accountable, collaborative client-consultant relationships. To begin with, *start where "the client is at"*. Clients tend to vary in their degree of readiness to address problems/issues, as well as in their knowledge of the situation.

Support the client in his/her results and efforts. The OH&S consultant may have strong opinions about the situation at hand, but those will not help the client reach a successful resolution of the problem(s). Remember that the concern/problem is the client's concern/problem. He/she needs to "own" the concern/problem, his/her part in creating the concern/problem, the desired results, and the solution.

Work towards a partnership: ideally a 50:50 relationship concerning the consulting process. Fully explore and seek to understand the client's situation, concerns, agendas and motivations before moving into action. Be careful to stay within the client's frame of reference.

As for the interpersonal relationship with the client, there are a few things to note:
- *Before ever challenging the client on any issue, first, "earn that right".* To do so, be honest, competent, ethical, and dependable.
- *When stating that something will be done for the client, treat it as a promise.* A failure to do so can lead to distrust with this client and subsequent clients.
- *Keep the task, process and relationship visible.* This means conducting regular meetings to keep the lines of communication open, working from an

action plan, and paying attention to project successes and failures.
- *Establish clear explanations in terms of the parameters of the client-consultant relationship.*
- Use focused listening to establish a clear understanding of the goals, objectives and expected outcomes for the project. *To reach action, it requires honouring specific requests and committed promises.*

The core listening skills required in the Contracting Phase involve entirely focusing the OH&S consultant's attention on the client, deciding what needs to be understood, reflecting on the assumptions made, and questioning for understanding. Remember that the more the OH&S consultant tries to persuade a person, the less the OH&S consultant is attending to the listening aspect of the process. As well, the OH&S consultant is not listening when he/she is waiting for a turn to speak.

3. Data Gathering and Diagnosis Phase

The purpose of the Data Gathering and Diagnosis Phase is to develop a clear understanding of the situation by both the client and the consultant, to set the stage for problem solving, and to build a client commitment regarding the problem ownership and resolution. The goal of the Data Gathering Phase is to gain accurate, reliable, and complete information. Performing data gathering activities can result in consciousness raising, creating expectations, and supporting relationship building. If the client disowns or disagrees with the data, he/she will not be committed to the problem resolution and solution.

Data is operationally defined as any form of information that can be collected and is relevant to an individual, group, or organizational function. Data gathering involves systematic data collection, data analysis, and data feedback. Data collection includes devising a method to determine the nature of the problem, the kinds of information required, the number of people involved, the time available to do the project, and the cost of the method in question. A comparison of different methods of data collection is provided in Table 36.1.

Table 36.1: Comparison of Different Methods of Data Collection*

Method	Advantages	Disadvantages
INTERVIEWS	• Builds rapport • Source of broad and deep data • Adaptive — collects data on a range of topics	• Time consuming/ expensive • Can result in "interviewer response bias" • Coding/interpretation challenges • Self-reporting bias
FOCUS GROUPS	• Builds rapport • Can reach more people in less time • Works well in combination with other methods	• "Facilitator bias" • Potential for domination by some focus group participants • Can get derailed from investigation process
SURVEYS/ QUESTIONNAIRES	• Useful in describing traits of large populations • Easy to use with large samples • Inexpensive and flexible • Obtains large amount of data • Easier to probe into more sensitive areas/ issues • Responses can be easily collated, quantified and summarized	• Takes time and expertise to develop • Questions are predetermined • May appear superficial in coverage of complex topics • Seldom able to deal with the context of social life • Unable to measure social action, only recall past action • Data may be over-interpreted • "Response bias" • Responses may reflect other confounding issues
OBSERVATIONS	• Collects data on behaviour rather than reports on behaviour	• Problems coding and interpreting

---

* Adapted from: E. Babbie, *The Practice of Social Research* (Belmont, CA: Wadsworth Publishing, 1986); H. Checkoway, N. Pearce & D. Crawford-Brown, *Research Methods in Occupational Epidemiology* (Toronto, ON: Oxford Press, 1989); and Proactive Consultants, *Effective Internal Consulting* (Calgary, AB: Proactive Consultants, 1989).

Method	Advantages	Disadvantages
	• "Real-time" data, not retrospective • Can obtain large amounts of data • Good for studying subtle nuances of behaviours	• Sampling is a problem • "Observer bias/ reliability" • Hawthorne Effect • Difficult to generalize • Costly
DOCUMENTATION	• Non-reactive — no response bias • High face validity • Easily quantified	• Problems with access/ retrieval • Validity problems • Problems coding/ interpreting
LABORATORY EXPERIMENTATION	• Can control variables • Better observation • Easier to do	• Validity problems: Does this represent reality? • Results are difficult to generalize • Some situations cannot be replicated in a laboratory

As OH&S consultants, our mental filters affect both our ability to gather data and to analyze the information. Questioning is an essential information-gathering method. The quality of the data gathered depends on the quality of the questions used. In data gathering, our mental filters influence which questions we ask. Unconsciously, we can ask questions that lead us and our clients down a path to our "preferred solution" instead of to the "best solution".

In terms of an analysis of the problem, OH&S consultants can easily jump to conclusions based on their mental filters. Our mental filters make some aspects of a problem more visible to us than other aspects. For example, a financial consultant will tend to see the problem as a financial one; a business consultant will attribute the problem to business reasons; and Human Resources personnel will conclude that the issues are people-oriented. The challenge for the OH&S consultant is to be able to see an issue from many perspectives; and to identify the root cause, not to focus on the symptoms of the problem.

The secret is to use the right questions in the right order. Begin with open-ended questions to help the client describe the concern. For example, use questions like: "Can you tell me about the issue? In your opinion, what is the issue? Why has this happened? Why do you think the system is this way?" The next step is then to use content questions. Content questions involve asking for specific information: the "Whats", the "Wheres", the "Whens", the "Whos", and the "Hows". The last step is use of the binary questions. They confirm understanding and are answered with a "yes" or "no".

During this phase, the data gathering and diagnosing methods must:
- gather the relevant information in an effective and efficient manner;
- be aligned with the client's culture, norms and standards of operation; and
- be appropriate to the desired results for the project.

The recommended approach is to:
- collect the data;
- organize data into manageable issues;
- decide how to use the data;
- develop an assessment of the problem/issues;
- determine the real technical/business problem(s);
- determine the controls in place; and
- assess their effectiveness.

4. Feedback and Planning Phase

The purpose of the Feedback and Planning Phase is to ensure that the client understands the situation, owns the data/situation, has a plan of action to address the concern, and is committed to acting. It is at this point that the client needs to move out of an "understanding mode" and into an "action mode".

There are a number of feedback and recommendation issues that need to be addressed and overcome. Next to the Contracting Phase, this stage is often the most difficult with which to deal.[8] The reasons for this are as follows:

- Clients are often locked into taking action.
- There can be an overwhelming amount of data, making the selection and presentation of the most essential data difficult. However, the presentation of relevant data is important to enable the client to decide on how to move forward.
- The OH&S consultant, for a number of reasons, may not communicate the difficult realities about the project to the client or may focus only on the problems and fail to adequately support the client. When this happens, the client fails to be able to move into action and successfully solve the problem.
- The client's representative may not want other individuals within the organization to have access to the data or to be involved with the

---

[8] Proactive Consultants, *Effective Internal Consulting* (Calgary, AB: Proactive Consultants, 1989).

implementation. This can lead to implementation problems in the future.

- There may be difficulties with the feedback meeting. For example, the wrong people may be at the meeting; the people present may be unauthorized to make decisions; or there may be no chance for a face-to-face meeting with the client. This can lead to problems such as an inability to make critical decisions, a misinterpretation of the findings, and the inability to move forward.
- The consultant may prevent the client from creating his or her own solutions.

To effectively frame the recommendations, the OH&S consultant must consider the following:

- As OH&S consultants, our role is one of influence, not power. Essentially, the challenge is to influence clients to do things differently. As OH&S Practitioners/Professionals, we are in the business of helping our clients — we are not in the business of making recommendations that may or may not be used. When consulting, a shift in perspective must occur. We can only suggest and trust that we have properly prepared the client to act accordingly.
- When making recommendations, we need to remember two paradoxical points: (1) we are trying to influence the client to do things differently (to change); and (2) the client is always responsible for making the final decision to change or not to change.

Some key issues concerning the art of influence and change are:

- What are the "drivers" for the client to change?
- What recommendations are logically linked to the analysis of the problem?
- What are the broader implications of the recommendations made? What are the financial, technological, cultural, and organizational implications? Remember, the best "technical recommendation" may not be the "best recommendation".
- Change requires human energy. Decide how feasible the suggested recommendations are and how much energy is required to implement them.
- Can the client make the changes or will they need help?
- Is it feasible to hold a face-to-face meeting with the client to openly discuss the analysis and recommendations?
- Can the recommendations be framed in terms of the client's needs and business drivers?

The extent to which clients can hear and accept recommendations often depends on how we have "framed" them. To frame a recommendation, we need to understand the impact of our recommendations, the client's needs and the client's business

drivers. The recommended approach is to:
- deliver the message personally;
- plan for reaction and discussion time;
- provide the client with feedback and recommendations in a non-judgemental manner;
- be direct and simple with your explanations (avoid circling the issue(s));
- focus on process-issues versus people-issues;
- be specific and timely;
- seek reaction to the message (probe for concerns and/or resistance);
- address any client resistance;
- decide what actions to take;
- confirm that this is what the client wants to do;
- jointly develop goals and action steps for the project;
- seek client commitment;
- offer client support; and
- re-contract as required.

5. Action Phase

At this stage, someone carries out the proposed action(s). Individual responsibilities and roles have to be clarified and defined.

Implementation roles can vary between being client-directed, requiring joint participation; and being consultant-directed, requiring only consultant involvement. Depending upon which way the client chooses to proceed, the OH&S consultant can be a bystander, a mentor, or a "pair of hands".

Which role is the most appropriate for the consultant to assume? This depends on the situation, the task, the available internal resources, and the level of expertise of both parties. In general, the greater the role flexibility demonstrated by the OH&S consultant, the greater the capability to appropriately and effectively help the client.[9]

6. Evaluation and Disengagement Phase

This final stage is where the client and the OH&S consultant "part company". Success is when the client is ready, willing, and able to "go it alone". Some things to consider at closure are:
- Does the client agree that the consultant's work is done?
- Have the client's expectations been met?

---

[9] L. Goode, "Consulting Successfully Within Organizations" (Seminar presented at Petro-Canada, Calgary, AB, 1995) [unpublished].

- Who will carry the project forward?
- Is the documentation adequate?
- What has been learned from the consulting experience?
- What impact has the work had on the business?
- Is the client/consultant relationship strong?
- How can others benefit from what the OH&S consultant has learned from this project?

Failure to close the OH&S consulting project brings a cascade of undesirable outcomes, namely:

- the project continues despite the work having been completed;
- the client/consultant relationship is jeopardized;
- the consulting fees become jeopardized; and
- questions about the consultant's integrity loom.

To prevent this from happening, successful launching of the client can be accomplished by:

- evaluating the effectiveness of the Action Phase;
- seeking an extension for the OH&S consulting project if required and appropriate;
- recycling the Consulting Process if a new problem emerges;
- disengaging from the project because the original work is done, or the current work cannot be done; and
- keeping the door open for future work.

The six phases of an OH&S consulting project and relationship are summarized in terms of their actions, goals and purpose in Table 36.2.

Table 36.2: The Six Phases of an OH&S Consulting Project and Relationship: Their Actions, Goals, and Purpose

PHASE	ACTION	GOAL	PURPOSE
ENTRY	First contact to mutually determine whether to work together.	A "fit"/"no fit" decision. A reframed issue. An effective working relationship.	That the right people are doing the right work.
CONTRACTING	Mutual exploration about how to work together.	An explicit agreement.	To clarify the scope of work and the client/consultant commitments.

PHASE	ACTION	GOAL	PURPOSE
DATA GATHERING AND ANALYSIS	Information about the concern is gathered and analyzed.	A clear picture of the situation.	To create client understanding of the issue and foster intelligent action.
FEEDBACK AND RECOMMENDATIONS	Information is summarized and plans are created to address the concern.	Client understands the situation and the action plans have been created.	To create action plans that clients can realistically carry out or manage.
IMPLEMENTATION	The plan is implemented.	Sustained action by clients.	To create action in the client system because nothing changes until it changes.
CLOSURE	The consultant and client reach closure on the situation.	Clients are ready, willing and able to go it alone.	To foster long-term client success with this project/issue.

G. OH&S CONSULTING: ROLE OF THE INTERNAL CONSULTANT

The internal OH&S consultant, while encouraging client self-sufficiency, helps to solve problems so that they stay solved. This element of empowerment is crucial for a consulting assignment to be viewed as successful.

There are a number of roles that the internal OH&S consultant can play. The **Expert Role** is where the internal consultant embodies the message: "I can tell you how to fix it". This scenario is typified by the manager being a passive decision-maker, and the consultant making the decisions, as well as planning and implementing the main events. The problem with this approach is that the client or the manager does not learn the skills or gain the knowledge to deal with this type of problem in the future. Also, with less involvement, the client/manager has less commitment to the issue(s) and to the solution.

The **Pair of Hands Role** is where the OH&S consultant approaches the problem with the message: "I will fix it for you". Here, the OH&S consultant is a passive decision-maker. Decisions are made by the client/manager who specifies the desired change procedures. The problem with this approach is that the problem analysis is made solely from the client's/manager's perspective, which often results in a lack of innovation and repetition of the same solutions to the problems.

The **Collaborative Role** of "We need to work together" is characterized by joint decision-making, planning, implementation, and project evaluation. Although this approach may be viewed as duplicating efforts, it ends up being the most functional approach.

H. OH&S CONSULTING: PROJECT RECOMMENDATIONS

As OH&S consultants, OH&S Practitioners/Professionals are expected to produce recommendations for the client to consider. When making recommendations, the OH&S Practitioners/Professionals should be sure that the recommendations are:

- consistent with the data/findings;
- designed to have a positive impact;
- feasible;
- prioritized; and
- aligned with the organization's other business strategies.

CONSULTING DO'S AND DON'TS FOR OH&S CONSULTANTS	
DO:	**DON'T:**
• Act in ways that promote client trust.	• Get trapped into a "telling role".
• Be permitted to be influenced, as well as influence.	• Take advantage of the situation to show how bright, knowledgeable, and experienced you are.
• Deal with problems collaboratively.	• Meet defensiveness with pressure and arguments about "the facts".
• Communicate effectively: Listen, listen, listen!	• Confuse "helping" with "overpraising and unfounded reassurances" regarding the person being helped.
• Reframe problems/issues for clarification.	
• Remain open-minded and objective.	
• Seek to reduce "any threat" that the OH&S consultant role may generate.	
• Be non-judgemental.	
• Remember, help is only help when viewed as help.	

I. OH&S CONSULTING: ETHICAL GUIDELINES

OH&S consultants should adopt an ethical approach to consulting that includes:

- maintaining a high quality of services;
- maintaining confidentiality about client issues;
- informing the client about confidentiality limits;
- offering *ONLY* what can be competently delivered;
- not misrepresenting credentials/affiliations;
- delivering a high quality of services;
- adhering to professional standards;
- being wary of sacrificing ethics for economic/financial gains; and
- terminating the OH&S consulting project and relationship when the client is no longer benefiting.

J. CRITERIA FOR SUCCESS IN A CONSULTING SITUATION

As measures of success in an OH&S consulting situation, the following are suggested as goals for the OH&S consultant to strive towards attaining:

- a result that the client and organization can accept and successfully implement;

- a result that adds business/personal value to the client and the organization;
- an outcome that ends in having a positive (or neutral) effect on the working relationship; and
- the consultant is, and is perceived as, helpful by the client and the organization.

THE TOP TEN TIPS FOR OH&S CONSULTANTS

- Listen to the client and address his or her concerns
- Understand and work towards the client's objectives — not towards yours
- Strive to be objective and recognize when you are not
- Treat the consulting relationship as confidential unless otherwise agreed
- Work to make the client look good
- Be non-judgemental: accept the client at his or her level
- Whenever possible, give the client credit for the successes realized
- Try to understand the client's organization and the content in which people work
- At disengagement, leave the client as independent and self-reliant as possible
- Practice ethical OH&S consulting

K. FOR EXTERNAL OH&S CONSULTANTS: SUGGESTIONS FOR BUILDING YOUR OH&S CONSULTING BUSINESS

In building any OH&S consulting business, there are some basics that need to be upheld. They include the following:

- Understand and meet your client's needs. Listen! Listen! Listen!
- Do work that contributes to your business' bottom line: work that is productive, satisfying, and profitable.
- Focus on results.
- "Contract" your services upfront: avoid doing work for free.
- Have project progress meetings with your clients, particularly on large or protracted projects.
- Use a "contract" to ensure that what was promised gets accomplished and delivered.
- Set performance targets and keep track of the results.
- For standard requests and repetitive work, create efficient and standardized processes.
- Accept advice from other professionals (*e.g.*, legal, financial, business planners, *etc.*) to do your work and to achieve your business goals and targets.

- Use recognized strategic planning practices.
- Be clear about the business principles in use to run the organization/company and live by them. Remember that the characteristics of the OH&S business are strongly associated with your professional and personal reputation.
- Build your business over time, but start now.

For any business, new or established, marketing is crucial. To be effective, marketing must be targeted and focused. It should be informative and geared towards building relationships. Some marketing tips are:

- Market by attraction, rather than promotion.
- Be clear about who you are and what you offer.
- Specialize: "You can't be all things to all people".
- Know what "the competition" is doing.
- Recommend others when the services sought are out of your area of expertise.
- Be visible to decision-makers.
- Spend 5-20 per cent of your time making business contacts and doing marketing.
- Ask clients for references and referrals.
- Consider getting help from marketing professionals.
- Decide what combination of time, money and energy you want to invest into your marketing efforts.

To minimize the "feast or famine phenomenon" associated with building an OH&S consulting business, it is recommended that each day, week, month or year, make business contacts and market 5-20 per cent of the time, and prepare and do actual work 80-90 per cent of the time.

Some additional business thoughts for the external OH&S consultant are to decide how to finance your business. The options are to either infuse money into your business as needed, or to infuse money upfront. There are pros and cons to each approach. Evaluate each approach carefully.

Consider registering your business, do a business plan, use a commercial bank account, get a line of credit for the business, and evaluate the need for business insurance. Be open to advice from business associates/partners and make the appropriate legal/contractual arrangements with them. Above all, learn what you are good at and do it. Get help with what you do not do well and refer to colleagues when asked to do work outside your area of expertise. Be clear about your products and services, and then market. For more information on marketing refer to Chapter 16, "Marketing Occupational Health and Safety Programs and Communicating the Results".

L. SUMMARY

To reiterate, OH&S consulting is the art of influencing people at their request. It means reaching a conclusion of acceptable quality; maintaining an acceptable commitment to that conclusion; supporting the client to prevent and/or minimize any detrimental outcomes as a result of the conclusion; and being perceived as being helpful and efficient by the client.

OH&S Practitioners/Professionals are involved in a variety of consulting roles along with different types of consulting relationships — internal and external OH&S consulting. To consult well, OH&S Practitioners/Professionals need specialized knowledge and skills. Good OH&S consultants are not merely born; they work hard at developing the sensitivity and competency to help others and leave them self-sufficient when the project is complete.

CHAPTER REFERENCES

E. Babbie, *The Practice of Social Research* (Belmont, CA: Wadsworth Publishing, 1986).

H. Checkoway, N. Pearce & D. Crawford-Brown, *Research Methods in Occupational Epidemiology* (Toronto, ON: Oxford Press, 1989).

D. Dyck, "Internal and external consulting: Assisting clients with managing work, health and psychosocial issues" (2002) 50:3 *AAOHN* 111.

L. Goode, "Consulting Successfully Within Organizations" (Seminar presented at Petro-Canada, Calgary, AB, 1995) [unpublished].

Proactive Consultants, *Effective Internal Consulting* (Calgary, AB: Proactive Consultants, 1989).

D. Sousa, "Internal/External Consulting: Ten Tips for Occupational Health Nurses" (Seminar presented at the American Occupational Health Conference, San Francisco, CA, April 2001) [unpublished].

G.M. Weinberg, *The Secrets of Consulting* (New York, NY: Dorset House Publishing, 1985).

Chapter 37

OCCUPATIONAL HEALTH AND SAFETY: PROJECT MANAGEMENT

A. INTRODUCTION

Occupational Health and Safety (OH&S) Practitioners/Professionals often have the opportunity to bid on, or conduct project work. There are a number of relevant issues in bidding on and getting project work and managing a project to its completion. This chapter is designed to assist OH&S Practitioners/Professionals to effectively address these issues.

B. REQUEST FOR PROPOSAL: THE RESPONSE

1. Relevant Terms

Most large proposals are written in response to a **Request for Proposal** (RFP). An RFP can be issued either as an invited bid, or as an open bid. It explains what the client is interested in procuring and provides instructions on how to prepare and submit a proposal. An RFP describes the organization's:

- size, location, and characteristics;
- corporate values and beliefs;
- business interests and people needs;
- OH&S Program; and
- required OH&S services.

The current service provision arrangements might also be included, along with reasons for going to the marketplace to seek a change in their current services/products.

Government procurement is highly regulated and therefore, government RFPs tend to require submission of a more complex and comprehensive proposal in terms of the format and structure. Commercial RFPs do not have to follow the same rules and can be developed any way that the organization/company publishing the RFP wants it to be. For additional information on RFPs, refer to Chapter 40, "Outsourcing Occupational Health and Safety Services".

In addition to RFPs, some organizations, when they need information prior to issuing a solicitation, publish a **Request for Information** (RFI); others publish a **Request for Quotation** (RFQ) when all they are interested in is the price for the desired service/product.

2. Proposal Submission Information

The proposal submission information contained in the potential client's RFP describes the following to bidders:

- scope of the project;
- procedure for the bidder to acknowledge receipt of the RFP;
- format for the proposal;
- terms for proposal response submission;
- terms for proposal response rejection;
- length of time for bid acceptance;
- potential method for proposal clarification; and
- organization's policy on maintaining bidder information confidentiality.

For a bid submission to be viewed positively, it is vitally important to follow these instructions. They can "make or break" the proposal submission. In some situations, proposals that omit any of the requested elements are simply rejected.

The RFP may also include "required response letters". These letters serve a variety of purposes, such as acknowledgement of the bidder's receipt of the RFP, or confirmation of understanding of the client's preferred service agreement contract format and terms. Regardless of the purpose, be certain to include the required signed letter(s) with the proposal submission.

3. Preparation Time

Allow plenty of preparation time. RFPs can take two to three weeks to prepare, with the larger ones requiring a month or more, especially if a team of people are involved in its development. As well, time should be allowed for mailing and delivery of the RFP.

RFPs are often sent out by the client with a self-addressed, return envelope to ensure that the bid response gets to the appropriate department or person for processing. It is recommended that the bidder use this assistance whenever possible.

Lastly, know how the client wants the proposal to be submitted, including the closing time and date. The standard practice is that a late bid is considered invalid and returned unopened to the sender.

4. Proposal Details

As a prospective service provider/vendor, be sure to understand the nature of the services sought; the expected standard/quality of service; and what the client perceives that level of the service performance/product will look like. Additionally, consider whether or not there is a "fit" between the client's philosophy and business practices, and that of you and your organization/ company. If not, it may not be a "marriage" to enter into.

Given that the RFP information and questions are developed to elicit bidder responses that can be compared, it is vitally important to carefully read and understand each of the submission requirements and questions posed. Ensure the responses are clear, concise and complete. Remember they will be rated according to some predetermined value criteria set by the client and compared to the responses provided by the competition.

When reviewing RFP submissions, organizations/companies tend to factor in:

- vendor accessibility;
- fit with the corporate culture;
- service responsiveness; and
- service nature (*i.e.*, reactive versus proactive, service provider versus partnership arrangement).

Capability, adaptability, flexibility, and a willingness to work with the client to develop a workable business arrangement can be great assets. Keep that in mind and try to provide an indication of your organization's/company's strengths in these areas.

One tip is to try to indicate what it is about your service/product that makes it better qualified to meet the stated requirements of the RFP. In essence, why should the client pick you or your organization/company for this project?

Lastly and most importantly, try to find out why the organization/company is going to the marketplace. What was it about the current approach or current service provider that was not working for them? Armed with that information, position your proposal to address the identified issues.

5. Proposal Writing

Complex business arrangements must be documented in terms of a service agreement contract. The document that serves as a foundation for that service agreement is the proposal. Proposals are often produced through a complex process, usually involving many people. This is one of the many instances when it is important for the OH&S Practitioner/Professional to be able to communicate well.

It is easy to forget that the proposal process is about getting the right information on paper in the right format and written using the right tone. In fact, the "writing part" of the proposal preparation process is what scares people — especially those with limited experience in business writing. When asked to write a proposal, suddenly people with solid OH&S expertise and otherwise good problem-solving skills have no idea how to proceed. The following is some general advice for OH&S Practitioners/Professionals who are new to business proposal writing:

- **Know the client** — In business proposal writing, the only opinion that matters is that of the client. To know what to include in the proposal, or how to best format a proposal, view it from the client's perspective. Business proposal writing should answer the client's questions and explain the benefits of a service or product. To perfect the proposal writing technique, first perfect the understanding of the client's philosophy, beliefs, values, business, and current needs. The result will be the preparation of a truly effective proposal.

- **Provide the requested information** — Often people are intimidated by writing because they are focusing on the wrong things. Although the format, style, choice of words, or grammar are impotamt to successful proposal writing, what you really need to do is give the client the information they need to make an informed choice.
- **Carefully read the RFP** — Read the RFP thoroughly and know what the client is actually requesting. Prepare an outline of what is being requested and how your organization/company can deliver the requested services or product. Vet both documents with a third person to validate the alignment of both the documents.
- **Prepare an outline** — In the proposal outline, tell the client what your organization/company can do for them. Start with preparing a statement that summarizes what you or your organization/company can do, just as if an interactive discussion was occurring between you and the client. Tell them how, or why, or what, you are proposing will deliver what they want.
- **Proposal Format** — When preparing the proposal, adopt a format that portrays a professional appearance and follows an order similar to what is provided below:
  1. The *Title Page* includes the name of the project, who the proposal has been prepared for, who prepared it, and when. If using the client's organization/company logo, avoid copying it off the Internet; it may not be the approved logo. Instead, contact the client and ask for their official electronic logo.
  2. The *Executive Summary* should follow. The executive summary is, in essence, the proposal in miniature (usually one page or shorter in length). It is a brief synopsis of what the project is about, a statement of the problem, some background information, what is being proposed, the major conclusion, and you or your organization's/company's qualifications and strengths to meet the stated needs. The executive summary must contain enough information to enable readers to become acquainted with the full document without reading it. Someone reading an executive summary should get a good idea of the main points of the proposal without becoming bogged down with the details. In essence, it is a "30-second description" of the entire proposal.

     With the possible exception of the *Conclusion* and *Recommendation* sections, the executive summary is the most important section of the proposal. As such, it should be the best-written and most-polished piece; many readers will look only at the executive summary when deciding whether or not to read the entire proposal. In some organizations/companies, the proposal's executive summary is distributed so that employees are informed as to what information is available, then interested readers can request the entire proposal. Generally, the executive summary is read more frequently and by more people, than the entire proposal.

When writing the executive summary, determine if the readers of the executive summary will be the readers of the entire proposal. If dealing with two different groups of people, decide how much technical detail to include in the executive summary. If it is likely that some of the people who read only the executive summary will not have the technical background of the writer, nor the targeted reader, keep the technical information and vocabulary to a minimum.

Often, there may be three types of readers:

(a) those who want a full picture but will not check the details (*e.g.*, they might read the executive summary, some of the proposal body, the conclusions, and the recommendations);

(b) those who read everything (*e.g.*, they read the appendices, all the data, the calculations, *etc.*); and

(c) those who are in executive positions, wish to be kept informed on what is going on in the company and will say "yes" or "no" to a project (*e.g.*, they will read the executive summary, the conclusions, and the recommendations).

The executive summary must address all three of these types of readers.

Prepare the executive summary last. Since the executive summary is a condensed version of the proposal, omit any preliminaries, details and illustrative examples. Include only the main ideas, facts, and necessary background to understand the problem, options/alternatives, and major conclusions. Although brevity and conciseness are critical, do avoid taking a few sentences from key sections of the document and stringing them together.

Some recommended steps are to:

(a) review the entire proposal making notes of important concepts and elements;

(b) prepare a rough draft of the executive summary;

(c) avoid use of any introductory or transitional materials;

(d) ensure that the content is accurate and representative of the proposal. It must provide readers the essence of the information and tone contained in the entire proposal; and

    (e)   polish the content until it is smooth, seamless, and concise.

3. The *Table of Contents* enables the reader to quickly find the materials sought. If it is a lengthy proposal, include subsections as well as the major headings.

4. The *Project Description* is an in-depth explanation of the bidder's understanding of the project, what is being requested and the nature of the deliverables. In it, address the following:

   - how the client will benefit from what is being proposed;
   - how the features of the proposal align with the challenges the client currently faces. If appropriate, explain how it will position the organization/company for future business opportunities;
   - how the proposed service/product will meet, and possibly exceed, the client's request and achieve their strategic short-term and long-term goals;
   - what return on investment (ROI) the client can expect from the proposed service/product;
   - how this approach will best position the client for the future;
   - how this service or product aligns well with the client's evaluation criteria;
   - how the bidder's corporate philosophy/vision/values align with the client's; and
   - explain why the client should select the bidder. For example, what is it about the bidder's reputation, capabilities, efficiency, past performance, commitment to the project, and competitive price that distinguishes the bidder from the competition.

5. The *Project Scope* states what is and is not part of this project. It clarifies *what* the bidder proposes to undertake.

6. The *Methodology* section follows and explains *how* the project will be undertaken. If the project requires activities by both the consultant and client, be sure to identify who does what. As well, restate the project deliverables.

7. The *Timelines* provide the schedule, or *when* the project is to be done. It should be broken down into individual steps with specific time frames for onset and completion. If the project requires activities by both the consultant and client, be sure to identify who does what and in what time frame.

8. The *Project Team*, or a description of the organization/company, is

9. The section on **Professional Fees** is the price quoted for this professional OH&S work. Be sure to align it with the methodology and project timelines. There will be instances when the client wants just part of the work proposed. Breaking up the steps and pricing them separately allows the client to select what is or is not wanted, or what is or is not affordable. That is a better option than having the client turn down an entire project based on a "one-price option".

10. **References** may be requested and can be provided, or referred to, here, and then located in the Appendix. Testimonials can be used as well.

11. A **Conclusion** may or may not be needed. If the conclusion merely restates what has been said, spare the reader from having to read it again. However, if a number of details are drawn together, a summary can be very valuable.

12. In the **Appendix**, items such as RFP questionnaire responses, relevant organization/company detail, staff profiles, associated services, relevant articles, and associated service or product awards can be provided.

Marketing professional services and writing professional services proposals are very different from service/product marketing. A professional services proposal usually includes a staffing plan and the résumés of the staff who will be conducting the work. Instead of having specific service/product-line items to market, describe, and price, the writer has to develop an approach, describe it, estimate the level of effort to execute it, and then price the labour to do it. This presents the writer of the professional services proposal with a different set of challenges than those faced by other businesses. To assist the reader in how to prepare a professional proposal, a sample proposal has been included in Appendix 1.

C. FINALIST INTERVIEW

Once all the RFP submissions are analyzed and the findings reviewed, the client decision-makers select the service providers that they wish to interview. The finalists are invited to meet with the organizational representatives to explain how their services, philosophy, and plans can provide the requested OH&S services. Typically, these are formal presentations in which the service providers describe their business, service capabilities, staff qualifications, provider network, facilities, data management systems, past successes, and future plans. Presentations include a discussion of how the service provider plans to deliver the requested OH&S services, to deliver the various funding options, and to embrace the desired working relationship.

During the interview, the bidder must be prepared to respond to a series of questions that are designed to explore the possibility of a potential business relationship with the client, as well as to add rigour to the comparison process in progress. Remember that the responses given to the prepared questions will be assessed and rated against those of the other finalist(s).

Finalist interviews are the most revealing part of the service-provider selection process. On paper, many service providers sound great. However, in person, the match between the organization and service provider gets clarified.

In terms of being a finalist for a Finalist Interview, some tips to remember are:

- Come in prepared, "polished" and professional.
- If invited to do a short presentation on your company, keep it short and memorable. Speak to your strengths and be ready for questions like, "Have any organizations/companies recently discontinued your services? If so, why?"
- "Canned" approaches and presentations do not work. The client quickly picks up that they are being treated disrespectfully because you did not take the time and effort to tailor the proposed approach to meet their specific needs. It begs the question as to whether or not you will be committed to their needs, if selected.
- Know the audience and "play" to them.
- Answer the questions provided by the client — be honest, direct, and concise.
- If you do not understand what the client is asking, take the time to explore their question. It shows you are willing to listen.
- Listen to what the client is asking for — do not adopt the approach: "I am the expert, and therefore, I know what is best for you". It is easily detected and is perceived as demeaning.

D. SERVICE AGREEMENT CONTRACT DEVELOPMENT

The development of the service agreement contract begins at the onset of the project. By being aware of the value criteria, services required, desired funding arrangement, and duration of the contract, the procurement officer for the client's organization can begin to craft a service contract document.

Many organizations have standardized service contract templates. These can be used and modified to meet the needs of the proposed OH&S service contract. Regardless of the form used, the following elements should be included:

- description of services to be provided;
- expected level of service quality;
- required levels of reporting and communication;
- mutually agreed upon service performance measures;

- responsibilities of each party;
- pricing agreement;
- duration of the contract;
- payment schedule;
- legal compliance;
- hold harmless clause; and
- required business insurance coverage.

Ensure that as the selected service provider, you have a thorough understanding of each of the items listed above. It is strongly recommended that the prepared service agreement contract be reviewed by a legal counsel representing each party.

The service agreement contract should stipulate the expected levels of performance for both the hiring organization and the service provider. Once in place, the onus falls on both parties to monitor the performance levels exhibited. This includes the regular measurement of the service quality and outcomes delivered.

The measurement criteria and the techniques spelled out in the service agreement contract should be followed. A review of the results should involve both parties. Here, the partnership arrangement becomes important. Both parties should be cognizant of the issues or problems and should work together to arrive at feasible solutions.

Performance measurement is an ongoing process — not an event. Regular measurement is critical to the success of the outsourced arrangement. It allows for the identification of issues or problems, development of action plans and solutions, establishment of short- and long-term goals, trend analysis, and identification of proactive approaches to illness or injury prevention. Without regular performance measurement and open communication of performance issues, an outsourced service arrangement can fail. This is often the reason for tension between organizations and service providers, which is all too common, and can easily be prevented.

As Yogi Berra once stated, "If you don't know where you're going, you'll end up someplace else." In any service arrangement, the organization and service provider have to work together. Performance measurement is a useful tool to ensure this successfully happens.

E. PROJECT MANAGEMENT

Project management is the discipline of organizing and managing resources in such a way that these resources deliver all the work required to complete a project within defined scope, time, and cost constraints.[1] Project management processes and

---

[1] Wikipedia, "Project Management", online: http://en.wikipedia.org; Free Management Library, "All About Project Management", online: http://www.managementhelp.org/plan_

techniques are used to coordinate resources to achieve predictable results. A **project** is a temporary and one-time endeavour undertaken to create a unique product or service that brings about beneficial change or added value.[2]

Given that most projects (67 per cent), especially the large ones, do not meet their stated targets and goals, good project management skills are essential.[3] The first challenge of project management is to ensure that the project is delivered within the defined time and cost constraints. The second, more ambitious, challenge is to optimize the allocation and integration of the inputs needed to meet those pre-defined objectives. Therefore, a project is a carefully selected set of activities chosen to use resources (*e.g.*, money, people, materials, energy, space, provisions, communication, quality, risk, *etc*.) wisely to meet pre-defined objectives. The bottom line is that all projects need some degree of project management, and the best way to succeed is to follow the long-standing rule: "*Plan the work and work the plan.*"[4]

A **project manager** assumes the responsibility for the management of the project. In large projects, this individual seldom participates directly in the activities that produce the end result, but rather strives to maintain the progress and productive mutual interaction of various parties in such a way that overall risk of failure is reduced. The project manager is usually the client representative and is responsible for determining and implementing what the client needs. The ability to adapt to the various internal procedures of the contracting party, and to form close links with the nominated representatives, is essential in ensuring that the key issues of cost, time, quality, and above all, client satisfaction, can be realized. In smaller projects, the project manager may undertake a variety of roles. Regardless of project size, a successful project manager must be able to envisage the entire project from start to finish and to have the ability to ensure that this vision is realized.

1. Project Management Stages

In managing a project, many stages have to take place, some sequentially, some simultaneously. They include:

- *Project Conception Stage* — This is the formation phase, or the birth of the project idea. It can be formal, or informal in nature. The questions to be addressed are: Should we do this project? Can we do this project? If both responses are "yes", then move towards the next stage.

- *Project Initiation Stage* — This is the phase in which the project plan is developed. Project developers must determine the nature and scope of the

---

dec/project/project.htm (date accessed: January 31, 2020).

[2] Wikipedia, "Project Management", online: http://en.wikipedia.org/wiki/Project_management (date accessed: January 31, 2020).

[3] "Project Management Best Practices" (CompuCom Systems Inc., 2003).

[4] *Ibid.*

CH. 37: OCCUPATIONAL HEALTH AND SAFETY: PROJECT MANAGEMENT

proposed project. Consideration must be given to:

- Why the project is being initiated at this time? — What are the overt and covert reasons for undertaking this project?
- The assumptions involved. They should be itemized and periodically reviewed in terms of their validity and impact on the project.
- The anticipated project tasks, namely a review of the current operations; identification of the stakeholders and their interest(s) in the project; an analysis of the client's needs in measurable goals; development of a conceptual design of the final service/product; and a statement of the anticipated deliverables.
- The threats/risks that could negatively impact the project such as change in client focus, change in management interest in the project, an expanded project scope, or lack of available OH&S personnel to do the job, *etc.*
- The risk management techniques to mitigate any identified threats/risks to the project, *i.e.*, contingency planning.
- Determining the needed resources.
- The roles of the project team members.
- The roles of the client stakeholders involved in the anticipated project.
- The project timelines.
- The project costs/budget.

It is vital to understand the client's business environment and to make sure that the necessary project containment and protection controls are built into the project.

The recommended approach is to create a **project definition document** in which the following elements are articulated:

- an overview of the project;
- the project objectives and scope;
- the related assumptions and risks;
- the project methodology;
- the project team members, including their specific roles;
- documentation of client acknowledgement of understanding of the project; and
- the time/cost schedules.[5]

If this stage is not performed well, it is unlikely that the project will meet the client's

---

[5] *Ibid.*

expectations and needs. Hence, the project will fail.

- *Project Design Stage* — This is the starting phase. It involves planning the work (*i.e.*, develop a workplan); analysis and design of the project objectives; assessing and controlling foreseeable risks; estimating the needed resources; defining the services/products of the project; allocating the resources; setting up the necessary tracking systems; and announcing the project to stakeholders. The desired outcomes are that the service/product: function well; meet service/product quality standards; satisfy client needs; and can be provided within stated time and budget constraints. Project planning, although sometimes time-consuming, is extremely critical to the success of the project.

    After the initiation stage, the project service/product should be designed and tested. Testing is generally performed by a combination of testers and end users. Controls are required to ensure that the final service/product meets the client's original service/product specifications.

- *Project Execution Stage* — This is the performance phase. Activities such as organizing the work; acquiring human and material resources; assigning work tasks; doing tasks; directing activities; fixing emergent problems; keeping stakeholders informed; and controlling the project execution are part of this stage.

- *Project Monitoring and Control Stage* — This phase includes tracking and reporting the progress of the project activities; analyzing the results based on the facts achieved; forecasting future trends in the project; quality management; issue management; issue solving; and prevention of the delivery of substandard service/products.

- *Project Closure Stage* — The closing phase includes the formal acceptance of the project by the client and the completion of the project arrangement. A withdrawal of services occurs. For the project manager, the administrative activities include closing the account, debriefing the project team (post-mortem on the project), archiving the files, and documenting the lessons learned.

When bidding on a project, the OH&S consultant will likely undertake the Project Initiation Stage as part of the preparation of the project proposal. However, if successful with the bid, the OH&S consultant will have to go back to revisit the Project Initiation Stage to ensure that the plans proposed mesh with the client's expectations.

2. What Can Go Wrong?

There are many things that can go wrong with respect to project management. These are commonly called "barriers" and include:

- **Lack of due attention to project planning:** For success, project planning and working the plan are critical aspects.

- **Poor communication:** Many times a project fails because the members of the project team do not know exactly what has already been done, and hence what to do.
- **Poorly defined project goals:** Project goals need to be clear, specific, measurable, attainable, and time specific. As well, they need to be mutually agreed upon by the client and service provider, and communicated to all stakeholders involved in the project.
- **Failure to identify the risks to the project at the onset:** By identifying the project risks, the project team can develop specific plans to mitigate/avoid those risks.[6]
- **Poor project management practices:** This area is a topic in itself: suffice it to say that the project manager must have the skills required to get the project done. In general they include, planning, organizing, leading and directing, controlling (monitoring), and process improvement.
- **Failure to manage client expectations:** The project manager needs to identify, monitor, and continually manage the client's expectations of the project in terms of what is to be achieved and delivered.
- **Scope creep:** Small changes in the project scope that get added over time, and if they go unchecked, can negatively impact the entire project.[7]
- **Client fails to understand and approve requested changes to the project.**
- **Personality incompatibility:** As between the project manager and the client representative(s).
- **Disagreements:** The client and project manager must agree on numerous elements — if they do not, it can negatively impact the outcome of the project. For example, the project must meet all the elements defined in the service agreement contract to the specified service quality. If the client perceives that these expectations are not being met, and the service provider's viewpoint differs, a project-threatening disagreement can ensue.
- **Failure to comply with agreed-upon OH&S service quality standards and regulations.**
- **Labour unrests.**

3. What Leads to a Successful Project?

The five factors used to measure the degree of project success are:

1. **Scope: How much of the project was completed?**

---

[6] *Ibid.*

[7] *Ibid.*

Superior team performance is vital and can be attained by:
- planning the project from the "end to the beginning";
- developing a *Conditions of Project Acceptance* document and ensure that it is signed by the client;
- helping the team visualize and commit to what project success will look like;
- managing client expectations; and
- managing project changes/additional work.

2. **Funding: What did the project cost?**

   Did the project meet or exceed the anticipated costs? To help the client understand the costs, track the project costs itemizing each aspect and regularly communicate these financial details to the client. If project changes are requested, ensure that the client is aware upfront of the added costs to the project.

3. **Time: How long did it take?**

   Did the project meet or exceed the anticipated time frame? Time tracking is also important; as well, if project changes are requested, ensure that the client is aware upfront of the added time commitment due to this requested change.

4. **Accomplishment: Did the project meet or exceed the original project workplan?**

   Did the project meet or exceed the original project workplan? As already mentioned, working the plan is vital to project success. Deviations from the project plan can be costly in terms of time and money.

5. **Service Quality: Did the project meet or exceed the client's expectations?**

   Client expectations must be understood upfront. As well, they need to be managed: as a project evolves, client expectations tend to change and become redefined. That means their definition of success also changes. Therefore, it is vital to "establish a process to control the final outcome".[8]

F. PROJECT REPORT

As the business environment increases in complexity, the importance of skilful communication becomes increasingly important. In addition to possessing sound OH&S technical skills, the OH&S Practitioners/Professionals must develop effective communication skills. It is of little use to formulate solutions to business problems without being skilled at transmitting this information to others involved in

---

[8] E. Gonzalez, "Project Management's Fifth Discipline", *CIPSScene* (Canadian Information Processing Society, Calgary Section, 2002) at 5.

the problem-solving process. In this section, the preparation of a business project report is discussed.

1. Report Format

Written reports enable the efficient presentation of factual data. Many projects require a final report, allowing the project manager the opportunity to carefully verbalize the findings, to provide a historical record of the information being shared, and to share the same message with a wide audience in an expedient manner.

Like the project proposal, prepare the final report so that it has a professional appearance and tone. Standard report elements include:

- **Title page** — It contains the title of the project, the name of the client, the name and contact details of the project manager, date, and location of the project manager.

- **Executive Summary** — The intent of the executive summary is to provide the time-constrained reader with the important facts and findings contained in the OH&S project. It summarizes these findings and conclusions, along with any recommendations, and places them at the beginning of the study. This placement provides easy access to the more important information relevant to any decision that a business manager must make. If the business manager is interested in further details, the main body of the report is available for review.

    The executive summary should be written in a non-technical manner. It is intended for upper-level managers whose expertise often lies in business management and not in technical fields such as occupational health and safety, industrial hygiene, toxicology, ergonomics, and such. They usually have little concern for the technical aspect of the report. They primarily want to be assured that all the relevant business factors have been considered, and that appropriate procedures have been followed. If the reader decides that a more complete technical explanation is needed, that portion of the report can be reviewed.

    Although the executive summary precedes the main report when it is submitted in final form, the executive summary is written after the study has been conducted and the rest of the report is completed. The executive summary should not include new information or information that differs from the body of the report. Likewise, it should not offer conclusions based on data or information not contained in the report. As previously noted, the executive summary seldom exceeds one or two pages.

- **Table of Contents** — Presents all the relevant sections and subsections so that the business manager can readily access the desired information.

- **Project Description** — This is the body of the report which houses:
    - a brief introduction describing the nature and scope of the problem;

- any relevant history or background material essential to gaining a thorough understanding of the problem and provides clarification of the project;
- a statement which explains why the resolution of this issue is important and the critical need to formulate a course of action;
- the project goals and objectives;
- the project workplan;
- the method of project implementation (methodology) and the techniques used to attain the project results; and
- the project findings/outcomes (the relevant raw data) with any associated comments that add clarity to the data or draw attention to relevant factors that will be discussed in the next section.
- **Discussion of the Project Results** — Based on the findings from the previous section, a discussion and interpretation of the project's major implications is provided in a meaningful and non-technical manner. This section has considerable impact on the formulation of the solution to the problem as described in the beginning of the report.
- **Recommendations** — They should be presented as advice for improvement as opposed to an edict. This section tends to repeat some of the information found in the executive summary, yet allows the project manager to explain in greater detail how and why these conclusions were reached. It is important that this section be based on the results of the project findings and not other conclusions or recommendations unsupported by the contained analysis. Ideally, the recommendations should be prioritized in order of importance.
- **Conclusion** — It summarizes the main objectives and results from the perspective of the reader who has read the body of the report and is reminded what it was all about. This section should include thanks to the participating stakeholders.
- **Appendix** — This part of the report can house supporting documentation on findings, recommendations, products and such.

If OH&S project reports are prepared in this organized form, they are more useful and lend credibility and authority to the project manager. As well, the OH&S project report will command respect from those who paid for it and who are relying on its content to make important decisions.

2. Report Contents

When preparing an OH&S project report, it is important to write so that the contents are presented in a concise and factual manner. The project report should, in most cases, satisfy the needs of two somewhat different types of audiences. One type of reader could be someone with a reasonable knowledge of the OH&S area, wishing to use the findings of the report as a reference for further action. This person requires relevant theory and specific results written in a precise business style. The other type of reader is someone with a business interest in the project, but with no specific knowledge of the area. This type of person would be looking to gain, not a trivial explanation, but more of an understanding of the issues, and if necessary, be able to skip the theoretical sections without losing flow or context. The best type of project report achieves a balance between these requirements.

Consider the attitude of the target audience(s). How will they react towards the OH&S project report? If they may be somewhat hostile toward the report, offer more supporting evidence and documentation than if their anticipated reception to the report content is expected to be favourable.

The educational background and work experience of the audience(s) is a key consideration factor. In terms of style, word usage, and complexity, a report written for top executives differs considerably from the report prepared for line management. Even age, gender, and other demographic characteristics might serve to shape the report writing.[9]

The text of the report should be written in a professional format, using short precise sentences and avoiding the use of unnecessary and redundant information. If there has been some salient issue or problem with the project, note it, but do not keep repeating it throughout the main body of the text. It is quite possible that an important aspect of the project may need to be referred to in various sections of the report. If so, then describe the aspect fully in one section and reference this section where required. In short, avoid repetition.

Write the report so it does justice to the work contained within. The primary aim is clarity.[10] Minimize the use of technical jargon.[11] Be sure to remain focused on the project, the outcomes, and the results. Use the report as an opportunity to fully explain what the results mean and what is perceived as useful recommendations for the client. Be sure to clearly describe the recommended actions and the rationale for those actions.

Another consideration is the tone of the report. To the reader, is the tone of the report positive in nature? Is it respectful of the client and the client's commitment to the project? Does the content read so that the reader feels like an "equal" in the discussion process? Take time to evaluate the tone of the report because it can be

---

[9] W. Allen, "Business Report Writing", *Applied Statistics for Business and Economics* (2006) at 957-961, online: http://faculty.clayton.edu/larjoman/writing (date accessed: January 31, 2020).

[10] D. Inman, "Project Report Writing", *Project Web Guide* (London South Bank University, 2004), online: http://www.scism.lsbu.ac.uk/inmandw/projects/writing.htm (date accessed: January 31, 2020).

[11] S. Portney, *Project Management for Dummies*, 2d ed. (Hoboken, NJ: Wiley Publishing, 2007).

more influential than the content contained in the report.

In terms of a writing approach, start with a general overview and then move into coverage of the more detailed information. Consider how a person normally reads. Close attention is paid to the first page or two, and then if the gist of the content is missed, the person loses interest and skims idly through the remainder of the report trying to find significant ideas and relevant points. To avoid this terse review, engage the reader early and hold his or her attention by presenting the information in a logical and sequential manner. By the end, the reader should be well-enough informed to responsibly action the related business decisions.

Some additional tips for report writing include:
- Keep it interesting — tell people what they want/need to know.
- Ensure the writing style, grammar, punctuation, and spelling are acceptable.
- Write in the "third-person" avoiding the use of personal pronouns.
- Know and use the client's accepted report writing style where possible.
- Use sub-headings to help focus the reader's attention.
- Compare the actual project performance with the planned project performance.
- Proofread the report — one suggestion is to proof the report yourself and then to have someone who is unfamiliar with the project read and comment on it. If they can understand it and accurately explain what it contains, then the report is ready.

3. Limitations of Written Reports

Report writing is a science and an art. Learning how to write well is a critical skill for OH&S Practitioners/Professionals. However, as important as learning to write effectively is, it is just as vital to recognize the limitations associated with report writing, namely:

- Reports do not allow for verbal exchange, feedback, or other forms of interaction between the author and audience.
- The project manager cannot verify that the audience interpreted the intended message.
- The project manager never really knows if the report was read.

It is also important to realize that project reports rarely stand alone. Oral presentations are often required because the reader/audience will seek further clarification of the OH&S issues as part of their problem-solving and decision-making process. As part of the project report writing, the project manager should be prepared to develop and present an oral presentation.

G. OH&S PROJECT REPORT PRESENTATION

The oral project report presentation is a wonderful opportunity to demonstrate in detail the OH&S project from its inception to completion. Oral presentations differ from written communication in that the presenter is talking rather than writing. Although a seemingly silly statement, the point being made is that by virtue of the communication delivery, there are two critical points that influence how one prepares and delivers an oral project report presentation:

- **There is no written record.**

    Usually there is no complete written record for the audience to consult — the presenter talks, and hopefully, they listen. This means that simple, direct presentations are best: otherwise there is a risk of losing the audience's attention. Periodically, remind them of the overall structure of the presentation, and how the information fits together. In other words, do not just present the OH&S report data/ findings to them; guide the audience through the presentation and the OH&S project report.

- **Understand and use non-verbal communication techniques (body language) carefully.**

    When personally delivering a message to other people, the message sent is not just *what is being said*, it also includes *how it is said*. Attention to voice, posture, hand gestures, use of eye contact, and your overall appearance are critical because they are all sending their own messages and must align with what is being said. In fact, lots of studies show that people pay more attention to the "*hows*" of a presentation than to the "*whats*".

1. Preparing the Presentation

In preparing an oral presentation, there are six key steps:

Step 1: Planning

As with all communication tasks, before actually preparing a presentation, determine the intended audience and their anticipated reaction to the OH&S project report. Design your main idea or topic. That is your intended message. Decide on the information and evidence needed to support this message and how to structure its presentation. This forms the presentation outline or plan. Remember, you are leading the audience through the OH&S project and guiding them to the project conclusions and recommended actions. Finally, decide the presentation length. *Tip:* The higher up the organization/company hierarchy, the shorter the presentation length and the allotted time for questions.

Step 2: Preparing the introduction

The presentation introduction is critical; it lays the foundation for the entire presentation. Always connect and develop a rapport with the audience at the onset. Get their attention and show how the OH&S project relates and is important to them. Preview the main idea/topic (message) and explain the structure for the presentation. To remember these four aspects, think of the acronym **RAMP** — **R**apport, **A**ttention, **M**ain message and **P**lan.

Step 3: Preparing the body of the presentation

The body of the oral presentation is where the actual information, details, and evidence to support the main idea are provided. This part consumes the majority of the time allotted for the presentation.

There should be a number of slides in the body of the presentation, each corresponding to one of the main points in the presentation outline. Here, the argument for the main message is developed. It involves providing clear data/evidence, relevant examples, pertinent anecdotes, and supporting OH&S practice/research findings.

Since there is no written record of the oral presentation for the audience to consult, make sure that they are periodically reoriented to the structure of the presentation. Do this verbally ("now we'll move on to the second of my three main points..."), and by using overheads. Always ensure they know where they are within the presentation, and why a given section is relevant to the overall topic or idea. Otherwise the audience will lose interest, and the presentation will fall short of its intended impact.

Step 4: Preparing the conclusion

The presentation conclusion reinforces the main message. Briefly summarize the key elements and points of the OH&S project report, and if appropriate motivate the audience to act. Take advantage of the fact that an audience's attention level increases dramatically towards the end of a presentation.

Step 5: Preparing for questions

Questions are an essential part of most presentations. They allow for audience reaction and interaction, the opportunity to clarify ideas, or simply to get more information. In general, it is advisable to hold questions until the end of the presentation. This prevents repeated interruptions and provides an opportunity for the presenter to thoroughly explain the OH&S report before having to defend it. As well, the presentation tends to answer many potential questions.

If possible, prepare and practice answers to likely questions before the presentation. Anticipate and prepare for the tough questions. In particular, be able to explain and support any assumptions made during the OH&S project.

When appropriate, set and adhere to a time limit for the question period. Before finishing the question period, remind people that it is almost over by saying something like, "We're almost out of time. I can take one more quick question". If someone persistently asks questions, offer to provide more information at the end of the presentation.

Step 6: Preparing the visual aid

Visual aids must be **simple, clear, and pertinent**. Their purpose is to reinforce key presentation points and sections, not to detract from the presentation by confusing people. Accordingly, they must be carefully planned and properly used.

Computerized overheads are preferable because they are effective, inexpensive, reliable, and easy to produce and use. Overheads should contain text or a mixture of text and graphics. They include:

1. A ***visual title page*** which provides introductory information like the title of the presentation and the presenter's organization/company and name (Figure 37.2).
2. An ***overview page*** which previews your presentation's structure and main sections. It can be used several times during the course of a longer presentation to reorient the audience (Figure 37.3).
3. Charts, graphs, tables, photos, or other image-based material (Figure 37.11).

Importantly, avoid crowding material onto overheads; keep them free of visual clutter. It should go without saying that your overheads must be free of any grammatical, typographic, or spelling errors. Appendix 2 provides a sample project review presentation.

Plan which points in the overhead materials warrant emphasizing, determine which type of overhead is the best for doing this and create clear, dynamic, and colourful visual support materials that enhance the presentation message.

2. Presenting

Nervousness is part of delivering a good presentation. It enables the presenter to "get energized and up" for the experience. However, for some people, that nervousness is more detrimental than useful. As Jerry Seinfeld put it:

> People are afraid of public speaking . . . In fact, most say that it's their number one fear. Death, apparently, only comes second

The sweaty palms, the "stomach butterflies", and dry throat are all signs that the person's body is in distress.

Yet, the ability to verbally persuade or inform a group of people — in other words, to make effective oral presentations — is arguably the single most valuable skill in business. Its importance, very simply, cannot be overstated.

How can presenters overcome their fears so that they can perform to the best of their abilities? The following tips have been found to be useful:

- prepare the presentation materials yourself;
- have key notes on each slide;
- plan out the time required for each presentation and adhere to it;
- know the contents of the related OH&S report;
- know and rehearse your presentation;
- write down the opening statements so that you can get started without any verbal hesitation;

- make sure you know what the anticipated audience reaction to the report will be;
- rehearse the presentation. Try walking around, speaking each segment, and then speaking aloud the entire presentation. Rephrase ideas that are difficult to say — these will likely be hard for the audience to follow;
- be sure to time the presentation so that it does not exceed the time limit;
- keep the presentation as short as possible — avoid distractions;
- if possible, become familiar with the room where the presentation will be delivered to know how loudly to talk and how people will be seated; and
- practise and be comfortable with an effective delivery style (Table 37.1).

Table 37.1: An Effective Delivery Style

- Clearly demarcate the beginning and end of each point and segment in the presentation.
- Announce each main topic. That way, the audience knows when one topic has been completed and a new one is beginning.
- Allow a slight pause to occur after the completion of the presentation introduction then, announce the first topic.
- After finishing the final topic in the main body of the presentation, allow a slight pause before beginning the conclusion.
- Speak slowly, vigorously, and enthusiastically. Be sure you enunciate the words carefully, particularly if addressing a large group.
- Use gestures to accentuate major points. Use body movements to aid in announcing major transition points. In short, avoid standing still.
- Maintain eye contact with the audience. Doing so helps to keep the listeners involved, enhances the importance of the message, and enables the presenter to judge the audience's reaction to what is said and make any necessary adjustments.
- Avoid memorizing the presentation: one forgotten part will end up in confusion and panic. Use brief notes on a hardcopy of the slides. Highlighting key points help to refresh one's memory.
- If possible, record the rehearsed presentation. Listen objectively to what was said. Consider the main issues of audience, purpose, organization, context, content and style. Listen for tone, attitude, and clarity. Is the tone appropriate to the audience and purpose? Is each sentence easy to understand? Is the speaking pace appropriate? Are the major divisions in the presentation easy to hear? Are any sentences difficult to understand? Make any necessary adjustments to the presentation materials or delivery.

No matter what type of presentation, ultimate success as a speaker and the success of the presentation depend on the presenter's ability to establish credibility with the audience. Guidelines on planning, structuring, and delivering the presentation are important because they are designed to build your credibility with your audience. However, no amount of planning and organization will substitute for practice, which builds confidence. Practice also enhances and displays planning and the value of the presenter's ideas.

H. SUMMARY

OH&S project management and reporting are important activities for the OH&S Practitioner/Professional to conduct or oversee. Knowing the key elements and project management techniques can be invaluable to their OH&S practice.

Appendix 1

COMPANY XYZ INC.

PROPOSAL TO EVALUATE AND ENHANCE THE XYZ CASE MANAGEMENT SERVICE

Prepared by

Dianne Dyck, RN, BN, MSc, COHN(C), CRSP

Progressive Health and Safety Consulting

(address)

(contact details)

January 2020

TABLE OF CONTENTS

EXECUTIVE SUMMARY
PROJECT DESCRIPTION
    Background
    XYZ Case Management Services: Current State
    Project Scope
    Methodology
    Timelines
    Progressive Health and Safety Consulting
    Our Fees
APPENDIX A: PROFESSIONAL PROFILES

EXECUTIVE SUMMARY

It is a pleasure to have the opportunity to submit a proposal to evaluate the XYZ Case Management Service (XYZ). Progressive Health and Safety Consulting understands that XYZ wants to evaluate its current case management services that are located in City A with a view to understanding why the service has experienced limited growth, and what it would take to facilitate future growth.

The current understanding is that XYZ Case Management Service provides non-occupational case management services for clients and that it has a number of disability management resources, namely:

- a functional product offering that has been operational for three years;
- a program leader;
- qualified case management specialists;
- available marketing support, the XYZ Marketing Department; and
- access to other XYZ occupational health services for clients.

The intent of this evaluation is to assist XYZ to understand why the current case management service has not grown as expected, and what it would take to promote the desired growth. The second objective is to determine what it would take to break down the "silo effect" that is exhibited between the Case Management Service in City A and the other XYZ client services.

The proposed evaluation model is as follows:

1. **Review the Current State** of the case management service provided by the City A Case Management Group. This would include a review of stakeholder interests, market demands, work demands, supportive services, available resources and measurement, and analysis of the current processes in place and their costs;
2. **Identify the Desired State** for a case management service;
3. **Identify the Gaps** between the two states;
4. **Recommend Strategies to Reduce/Eliminate the Gaps**, which would include achievable outcome measures; and
5. **Propose a Plan of Action** to enhance the City A Case Management Service in the marketplace.

Progressive Health and Safety Consulting has expertise in the design and implementation of leading edge and proactive health, workplace wellness, occupational safety and disability management programs. Through our professional network, we can offer additional resources such as organizational design and performance maximization expertise. If warranted, XYZ may wish to embark on a strategy to maximize performance in disability management servicing by ensuring that the case management professionals know what to do, are able to do the work, are equipped to do the work, and are motivated to do the work well. That includes promoting the service in the workplace.

Our consulting fees are competitive, and we believe that Progressive Health and Safety Consulting is the logical choice to assist XYZ to enhance the functioning of its City A Case Management Service using the resources available. Our aim is to help the XYZ Case Management Service be competitive in the marketplace.

Respectfully submitted,

Dianne Dyck, RN, BN, MSc, COHN(C), CRSP
Senior Health & Safety Consultant
Progressive Health and Safety Consulting
Calgary, Alberta

PROJECT DESCRIPTION

This proposal has been prepared in response to a request for a proposal to evaluate the XYZ Case Management Service put forth by (*name of the contact*).

Background

Disability management is defined as "a systematic, goal-oriented process of actively minimizing the impact of impairment on an employee's capacity to participate competitively in the work environment, and maximizing the health of employees to prevent disability, or further deterioration when a disability exists".[12]

Ideally, Disability Management Programs are proactive in nature, and incorporate stakeholder involvement and accountability. Most are designed to control the personal and economic costs of employee injury/illness to convey the message that employees are valued and to demonstrate compliance with the relevant legislation.

The ***key elements*** of any Disability Management Program (Figure 37.1) are:

- management-staff commitment and supportive policies;
- stakeholder education and involvement;
- supportive benefits plans;
- a coordinated approach to injury/illness management with a focus on early intervention (case management);
- a communication strategy;
- a graduated return-to-work program;
- measurement of outcomes; and
- disability prevention, including workplace wellness, attendance support, and occupational health and safety initiatives.

---

[12] D.G. Tate, R.V. Habeck & G. Schwartz, "Disability management: Origins, concepts and principles for practice" (1986) 17 *Journal of Applied Rehabilitation Counseling* 5.

[13] D. Dyck, *Disability Management: Theory, Strategy and Industry Practice*, 6th ed. (Markham, ON: LexisNexis, 2017).

Figure 37.1: Umbrella of a Disability Management Program[13]

The active management and co-ordination of all non-disability claims is essential to any organization. In short, it makes good business sense. Thus, disability management services, both occupational and non-occupational in nature, need to provide:

(a) *Claim Management*

- Review all information on the disability forms to ensure it is complete.
- Forward the claim documentation to the appropriate party.
- Provide follow-up where needed.
- Notify the client of any needed information.
- Maintain claim information in a disability management database.
- Establish follow-up/communication plans and time frames with respect to the employee, supervisor, and benefit provider (*e.g.*, WCB) or adjudicator for all claims.

(b) *Monitoring Claim Status*

- Following up with the benefit provider (*e.g.*, WCB) regarding the status of the claim.
- Following up with the employee to provide support.

- Following up with the supervisor to provide status updates.
- Communicating with the employer on suspected questionable claims and pursuing appropriate action.
- Coordinating the interface of occupational and non-occupational claims.
- Communicating decisions to appropriate parties (*e.g.*, contact supervisor when a claim is denied).
- Identifying opportunities for WCB cost relief/transfer, initiating these and following up.
- Maintaining claim information on the database.
- Seeking opportunities (based on the information available) for early return to work and liaising with the supervisor and benefit provider (*e.g.*, WCB) in this regard.

(c)   Monitor Employee Return-to-Work Potential
- Support the employee in the event further medical information needs to be provided and a return to work is not possible.
- Follow-up to ensure a "Return-to-Work Authorization" from the employee's physician is provided when this point comes.
- Notify the supervisor of return-to-work dates and confirming such.

(d)   Reporting
- Provide the client with comprehensive "Disability Update Reports".
- Provide summary trend reports and statistics.
- Provide customized reports as needed.

(e)   Other Related Services
- Providing a toll-free telephone number and communicating via fax and email.
- Developing communication materials marketing the case management services to clients/employees.
- Scheduling regular client meetings to discuss claims.
- Conducting reviews/surveys with claimants to obtain feedback on disability claim management services.

XYZ Case Management Services: Current State

The current understanding is that XYZ Case Management Service provides non-occupational case management services for clients and that it has a number of disability management resources, namely:

- a functional product offering that has been operational for three years;

- a program leader;
- qualified case management specialists;
- available marketing support to the XYZ Marketing Department; and
- access to other XYZ occupational health services for clients.

Project Scope

The scope of this evaluation is:

1. To assist XYZ to understand why the current case management service has not grown as expected and what it would take to promote the desired growth.
2. To determine what it would take to break down the "silo effect" that is exhibited between the Case Management Service in City A and the other XYZ client services.

Methodology

The three areas to be evaluated are:

- case management services;
- service quality of the services provided; and
- status of the management system in maximizing employee performance.

The evaluation model is as follows:

1. **Review the Current State** of the case management service provided by XYZ. This would include: a review of stakeholder interests, market demands, work demands, supportive services, and available resources; and measurement and analysis of the current processes in place and their costs.

    This step includes:

    - assessing the structure, process, and outcomes of the current case management processes (process mapping);
    - evaluating the efficiency of these processes in terms of time, cost, and return on investment;
    - reviewing existing return-to-work practices and outcomes;
    - identifying the linkages between the case management processes offered by XYZ and other related company-client programs such as the Attendance Management Programs, Employee Assistance Programs (EAP), Occupational Health and Safety Programs (OH&S), Employee Group Benefit Programs, and Return-to-Work Programs;
    - comparing XYZ's case management processes to industry "best practices";
    - determining the service quality of the case management services provided;

- reviewing the case management services offered in the marketplace — their products and successes;
- determining the cost of XYZ's case management services as compared to that of the competition; and
- assessing the current management system in terms of maximizing professional performance — does the system enable the professional to know what to do, to be able to do the work, to be equipped to do the work, and to "want" to do the case management work?

2. ***Identify the Desired State*** for a case management service. Identify what the ideal state would be for XYZ's Case Management Service.
3. ***Identify the Gaps*** between the two states.
4. ***Recommend Strategies to Reduce/Eliminate the Gaps***, which would include achievable performance and outcome measures.
5. ***Propose a Plan of Action*** to enhance the City A Case Management Service in the marketplace.

Timelines

Project/Task	Weeks			
	1	2	3	4
Step I: Planning	■			
• Initial Planning Meeting	■			
• Collection of Documentation	■			
Step II: Documentation Review		■		
Step III: On-site Review (2 days)			■	
• On-site review			■	
• Interviews			■	
• Process mapping			■	
Step IV: Analysis & Report Writing				■
Step V: Delivery of Final Report				■

Progressive Health and Safety Consulting

Dianne Dyck is the Project Manager and principal evaluator. As such, she will be responsible for the day-to-day management and execution of the project steps and ensuring that the project is delivered on time and to your satisfaction.

Another resource that may be able to offer valuable support to XYZ and this project is (*name of an additional resource*.

Profiles for each of the above have been included in the Appendices.

Our Fees

Based on experience with similar projects, the projected cost for this work is:

Project Steps	
Planning & Document Review	$1,600
On-Site Review, Interviews & Process Mapping	$4,500
Analysis and Report Writing	$4,500
Total	**$10,600**

APPENDIX A: PROFESSIONAL PROFILE

DIANNE DYCK, RN, BN, MSc, COHN(C), CRSP SENIOR CONSULTANT

Position & Responsibilities As Senior Consultant in Health Strategies, Dianne provides health care consulting services to Canadian organizations.

Areas of Specialization Dianne has a comprehensive background in the development and management of occupational health services. Her areas of specialization include occupational health and safety; development and implementation of disability management and occupational health services; auditing of health and safety-related programs, benefit programs, disability management, ergonomics, health data information systems, employee wellness and employee assistance programs; and design and implementation of computerized occupational/health information systems. Dianne has extensive experience in the design, development, delivery, and evaluation of Disability Management and Occupational Health & Safety educational programs.

Background Dianne earned a MSc in Community Health Services from the University of Calgary. The professional designations of Specialist in Occupational Health Nursing for Canada and the United States were attained in 1995. Her undergraduate education includes a Bachelor of Nursing and a Diploma in Public Health Nursing. Prior to establishing Progressive Health and Safety Consulting, Dianne was an Advisor, Occupational Health Services at a major oil and gas company in Calgary, an OH&S Consultant for a Human Resources firm, and the Director, OH&S for an electrical utility.

Consulting Assignments	Past assignments have included provision of occupational health services for a national company; integrated auditing of environment, occupational health & safety programs; assistance in policy development and program evaluation; development and management of short term disability, long term disability and Workers' Compensation cases; design and implementation of attendance management programs; implementation of ergonomic programs; and design and implementation of a computerized occupational health information system. Recently, Dianne has facilitated the development of a workplace wellness strategy for a large corporation and a major natural gas company. Recent assignments have included auditing EAP and disability management programs for major public and private sector organizations; the design and implementation of attendance management programs; and the design of workplace wellness strategy. Some examples include a mining company, university, school board, municipality, and health authority. In addition, Dianne has conducted presentations on Disability Management, Case Management Standards, Role of EAP in Disability Management and Career Streaming, and has written articles on Managed Reha-bilitative Care, Management of Chronic Fatigue, Career Streaming, Fitness to Travel, Heart Health, Client Satisfaction Survey — Gap Analysis, Role of EAP in Disability Management and Disability Management: Hype or Good Business Practice. Dianne has recently published the sixth edition of the book, *Disability Management: Theory, Strategy and Industry Practice* and the fourth edition of *Occupational Health & Safety: Theory, Strategy and Industry Practice*.
Professional Affiliations	Dianne is a member of the College of Alberta Registered Nurses Association, Alberta Occupational Health Nurses Association, Canadian Occupational Health Nurses Association, American Board of Occupational Health Nurses and the Board of Canadian Registered Safety Professionals. She has been a member on the executive of the Alberta Association of Occupational Health Nurses and held the seat of President at the Chapter and Provincial levels.

Appendix 2

OH&S Project Presentation (Sample)

Figure 37.2: Title Slide

> **Review of Company Workplace Wellness Activities**
>
> Presented by
> Dianne Dyck
> Progressive Health Consulting
> January 2020

Figure 37.3: Slide #2

WWP Review Project

REVIEW PURPOSE:

➢ To evaluate the attendance management, disability management & workplace wellness initiatives at Company XYZ with a view to continuous improvement

Figure 37.4: Slide #3

WWP Review Project

REVIEW METHODOLOGY:

➢ Identify the Current State
➢ Identify the Desired State
➢ Identify the Gaps
➢ Recommend Strategies to Reduce Gaps

Figure 37.5: Slide #4

WWP Review Project

PROJECT FINDINGS:

- Visible company commitment to workplace wellness
- Some supportive policies, procedures & systems for an integrated Workplace Wellness Program (WWP)
- Some workplace wellness prevention activities
- Some workplace wellness detection activities
- Company Disability Management Program (DMP) is using a multi-disciplinary approach
- Good company communication practices & vehicles
- Some workplace wellness data collection

Figure 37.6: Slide #5

WWP Review Project

RECOMMENDATIONS:

1. Program Design:

- Develop an integrated approach to workplace wellness
- Enhance the position of the Wellness Advisor
- Enhance employee-employer participation in workplace wellness
- Continue to link the Employee Wellness, Occupational Health & Safety, and Employee Assistance Program services

Figure 37.7: Slide #6

WWP Review Project

RECOMMENDATIONS:

2. Communication and Training:

- Develop an effective communication strategy for the WWP
- Continue to promote workplace wellness initiatives
- Institute workplace wellness training
- Communicate WWP outcomes

Figure 37.8: Slide #7

WWP Review Project

RECOMMENDATIONS:

3. Attendance and Disability Management:

- Address attendance support & management
- Develop an integrated approach to disability management
- Develop an effective communication strategy for the integrated program
- Centralize the management & coordination of **all** disability claims

Figure 37.9: Slide #8

WWP Review Project

RECOMMENDATIONS:

3. **Attendance and Disability Management**
   (cont'd):
   - Continue to review & improve claims adjudication
   - Enhance the disability case management process
   - Standardize the documentation
   - Develop policies & procedures regarding confidentiality
   - Evaluate the attendance & disability management processes annually

Figure 37.10: Slide #9

WWP Review Project

RECOMMENDATIONS:

4. Program Measurement:

- Standardize the outcome measures for the WWP
- Develop a comprehensive method for WWP data collection & analysis
- Measure the impact of attendance & disability management initiatives
- Implement ongoing measurement, monitoring & improvement of the WWP
- Use program outcomes to set program targets & provide a focused approach to workplace wellness initiatives

Figure 37.11: Slide #10

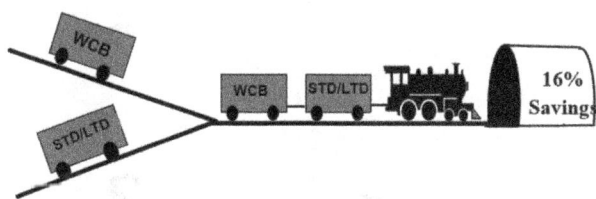

WWP Review Project

Dollar Value of An Integrated Workplace Wellness Program

Figure 37.12: Slide #11

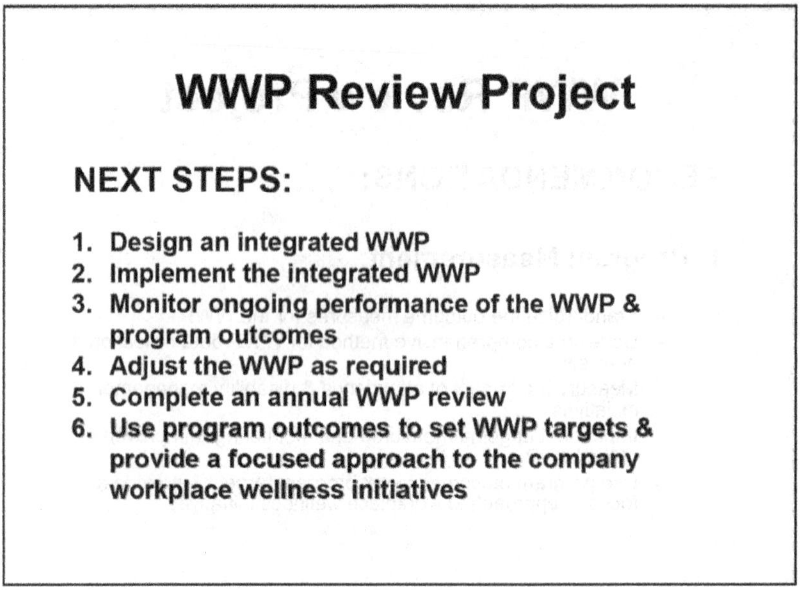

WWP Review Project

NEXT STEPS:

1. Design an integrated WWP
2. Implement the integrated WWP
3. Monitor ongoing performance of the WWP & program outcomes
4. Adjust the WWP as required
5. Complete an annual WWP review
6. Use program outcomes to set WWP targets & provide a focused approach to the company workplace wellness initiatives

Figure 37.13: Slide #12

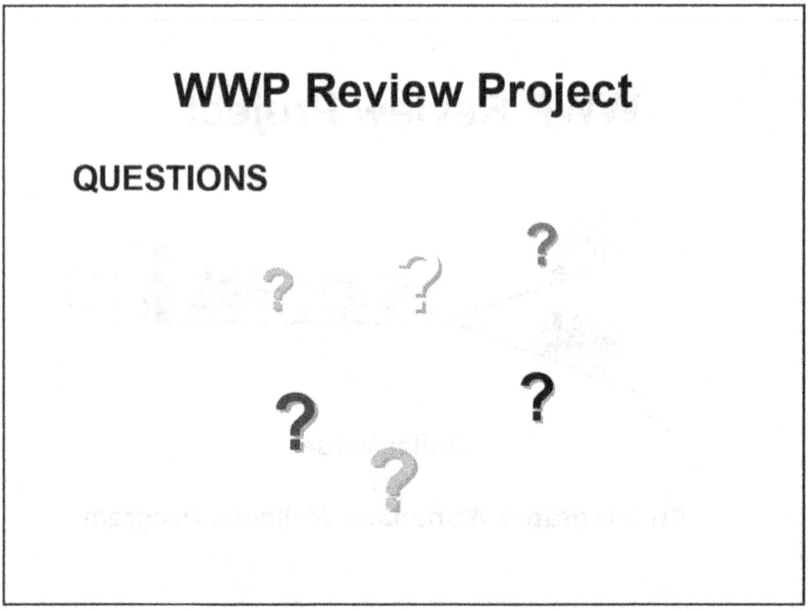

WWP Review Project

QUESTIONS

? ? ?
? ?
?

CHAPTER REFERENCES

W. Allen, "Business Report Writing" (2006), *Applied Statistics for Business and Economics*, online: http://faculty.clayton.edu/larjoman/writing (date accessed: January 31, 2020).

D. Dyck, *Disability Management: Theory, Strategy and Industry Practice*, 6th ed. (Markham, ON: LexisNexis, 2017).

Free Management Library, "All About Project Management", online: http://www.managementhelp.org/plan_dec/project/project.htm (date accessed: January 31, 2020).

E. Gonzalez, "Project Management's Fifth Discipline", *CIPSScene* (Canadian Information Processing Society, Calgary Section, 2002).

D. Inman, "Project Report Writing", *Project Web Guide* (London South Bank University, 2004), online: http://www.scism.lsbu.ac.uk/inmandw/projects/writing.htm (date accessed: January 31, 2020).

S. Portney, *Project Management for Dummies*, 2d ed. (Hoboken, NJ: Wiley Publishing, 2007).

"Project Management Best Practices" (CompuCom Systems Inc., 2003).

D. Tate, R.V. Habeck & G. Schwartz, "Disability management: Origins, concepts and principles for practice" (1986) *Journal of Applied Rehabilitation Counseling* 17.

Wikipedia, "Project Management", online: http://en.wikipedia.org (date accessed: January 31, 2020).

Chapter 38

OCCUPATIONAL HEALTH AND SAFETY: ORGANIZATIONAL BEHAVIOUR[1]

A. INTRODUCTION

The content in this chapter positions the learner to understand the basics of the field of Organizational Behaviour and the impact that an organization's philosophy, values, culture, and management theories have on workplace health and safety.

Organizational Behaviour (OB) is:

> a field of study that investigates the impact of individuals, groups, and structure on behaviour within an organization; its purpose is to apply such knowledge towards improving the organization's efficiency.[1]

OB focuses on workplace structures, lines of communication, leadership, techniques of management, relationships, and business/people issues. Interpersonal skills are critical to the success of any organization – strong interpersonal skills are needed to attract and retain high-performing employees. The technical skills needed are not enough to succeed as an organization, or an individual within an organization. Individuals today must possess strong relationship skills (people skills). That means that people must be able to form and maintain effective working relationships with individuals or groups by:

- Resolving situations effectively where emptions and attitudes of others may threaten individual or group performance;
- Demonstrating an awareness of and an understanding of people and their feelings;
- Applying a practical and effective method of coping with interpersonal differences;
- Listening effectively and be willing to entertain different points of view; and

---

[1] D. Dyck "Internal and external consulting: Assisting clients with managing work, health and psychosocial issues" (2002) 50:3 *AAOHN* 111. Copyright (2002) the American Association of Occupational Health Nurses, Inc. Used with permission. All rights reserved.

[1] N. Langton, S. Robbins, & T. Judge, *Organizational Behaviour: Concepts, Controversies, Applications*, 5th ed. (Toronto, ON: Pearson Prentice Hall, 2010) at 4.

- Communicating to convey facts, concepts or reasoning clearly to others and to receive and understand the messages sent by others.

B. ORGANIZATIONAL BEHAVIOUR: FUNDAMENTALS

The fundamentals of OB include:
- Addressing multiple levels within the organization;
- Addressing the multiple levels of functioning within an organization;
- Building on the knowledge from many disciplines;
- Applying research-based principles and practices;
- Using a systematic approach to studying organizational behaviours; and
- Employing a contingency approach to situations and making recommendations.

In essence, it functions to:
- Improve operational productivity;
- Enhance employee morale and productivity;
- Manage workforce diversity – cultural and generational;
- Reduce presenteeism, absenteeism, and staff turnover;
- Enable individuals to work together efficiently and effectively;
- Continuously improve employee and organizational performance; and
- Manage change.

C. ORGANIZATIONS: STRUCTURE AND CHARACTERISTICS

Organizations exist because a collective effort is required to accomplish many business initiatives and strategies:

> An **organization** is a consciously coordinated social unit, composed of a group of people that functions on a relatively continuous basis to achieve a common goal or set of goals.[2]

Organizations function at an:

1. *Organizational level* – they rely on work systems, human capital management and support systems, property management systems, and financial systems to succeed;
2. *Group level* – groups of employees work together as business divisions, business units, departments, workbgroups, and teams; and
3. *Individual level* – at this level, the players are the manager or supervisor/foreman, the employee, and perhaps, the union representative.

---

[2] N. Langton, S. Robbins & T. Judge, *Organizational Behaviour: Concepts, Controversies, Applications*, 5th ed. (Toronto, ON: Pearson Prentice Hall, 2010) at 4.

Organizations are comprised of people, plants (facilities and equipment), products/services, work processes/design, finances, and intelligence. They also possess distinct characteristics which might be:

- **Complex social systems** – a system made up of many parts which interrelate and are impacted by the operation of each other, as well as the functioning of the "whole".

- **Open system** – a system that operates to transform human and physical resources (inputs) into goods and/or services (outputs). An input-output interdependency exists within an organization.

- **Means-end chain** – a system in which each part of it exists to achieve a common goal ("a common end") and works together to achieve that end ("the means").

The characteristics of organizations are important for OH&S Practitioners/Professionals to recognize and understand. They explain the work environment and how it functions (Figure 38.1). As well, this is the system in which the stakeholders of an Integrated Workplace Health Management Program (refer to Chapter 28: "Integrated Workplace Health Management") operate.

Figure 38.1: Organization: Characteristics

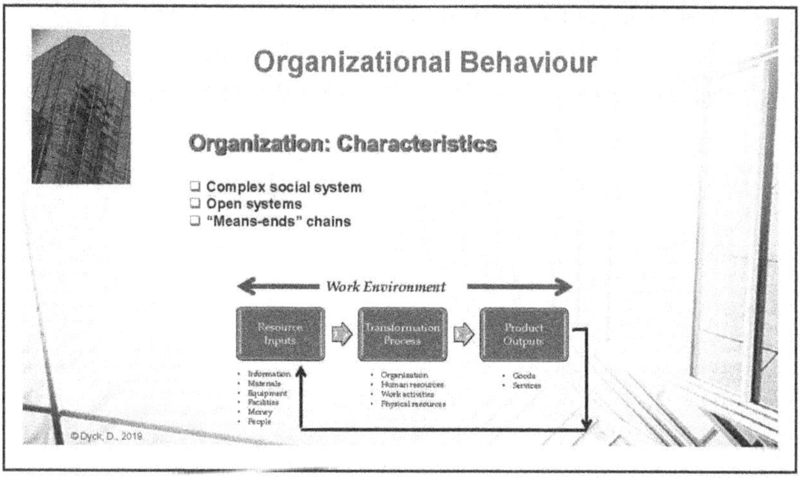

It is critical for stakeholders to recognize the input-output interdependency of an organization. Information, materials, equipment, people, facilities, and money are required for the organization to transform these inputs into products and/or services (outputs). Without these elements, an organization cannot exist. As such, it is important to support the availability of a healthy and capable workforce.

D. LEADERSHIP

Leadership is:

> ... an art that liberates people to do what is required of them in the most effective and humane way possible.[3]

Leadership is both about getting results and how those results are obtained. A **leader** is an individual who has the ability to excite, stimulate, and motivate other people to work towards a vision, making it a reality.

Leaders provide vision and strategy; they tend to press for change and can achieve extraordinary levels of performance from followers by inspiring them to go beyond their own self-interest in favour of the good of the organization.

There are many different types of leadership styles, namely: Transactional Leadership which includes Situational Leadership, Inspirational Leadership, and Charismatic Leadership; and Transformation Leadership, which tends to be the most effective. A **Transactional Leader** guides or motivates followers in the direction of established goals by clarifying the role and task requirements. A **Transformational Leader** inspires followers to transcend their own self-interest and is capable of having a profound and extraordinary effect on followers, *e.g.*, Steve Jobs, Apple Corporation, *etc*. A comparison of Transactional and Transformation Leaders is provided in Table 38.1. These leadership styles are not in opposition; rather, they can be complementary in getting things done.

---

[3] M. DePree, *Leadership is an Art* (New York, NY: Dell Publishing, 1989).

Table 38.1: Comparison of Transactional and Transformational Leaders[5]

Transactional Leaders	Transformational Leader
Guide or motivate followers in the direction of established goals by clarifying role and task requirements. Some examples are:	Inspire followers to transcend their own self-interest and are capable of having a profound and extraordinary effect on followers.
Contingent Reward Leader:	**Idealized Influencer:**
Contracts exchange of rewards for effort, promises rewards for good performance, and recognizes accomplishments	Provides vision and sense of mission, instills pride, gains respect and earns trust
Management by Exception Leader (Active):	**Inspirational Motivator:**
Watches and searches for deviations from rules and standards, and takes corrective action	Communicates high expectations, uses symbols to focus efforts, expresses important purposes in simple to understand ways
Management by Exception Leader (Passive):	**Intellectual Stimulator:**
Intervenes only if standards are not being met	Promotes intelligence, rationality, and careful problem-solving
Laissez-faire Leader:	**Leads through Individualized Consideration:**
Abdicates responsibilities, avoids making decisions	Gives personal attention, treats each follower individually, coaches, and advises

It is important for OH&S Practitioners/Professionals to understand the impact of the leaderhip styles because "organizational leaders who develop a degree of trust with their employees, see stronger work performance".[4] Over the years, there have been many leadership theories proposed; however, leaders "boil down" into two main leadership styles.

1. Leadership: Characteristics of a Leader

Strong leaders possess the:

- *Vision* to spell out clearly what they will do for followers;
- *Drive* to share their vision;

---

[5] N. Langton, S. Robbins & T. Judge, *Organizational Behaviour: Concepts, Controversies, Applications*, 5th ed. (Toronto, ON: Pearson Prentice Hall, 2010) at 426.

[4] Great Place to Work Institute Canada (2019), online: https://www.greatplacetowork.ca/en/ (date accessed: February 28, 2020).

- ***Courage*** to change what is, initiate change, and make strategic decisions to move forward;
- ***Ability*** to inspire others to achieve their goals;
- ***Foresight*** to empower others to learn new skills and to achieve their functional potential;
- ***Wisdom*** to listen, learn, and translate that knowledge into value and added performance; and
- ***Integrity*** to set a positive example and be a strong role model.

In contrast, poor leaders tend to be:

- ***Narcissists*** who are self-centred, intolerant of criticism, alienate followers, and often charm superiors;
- ***Ditherers*** who are unable to make decisions and suffer from analysis-paralysis;
- ***Panderers*** who have an excessive desire to please everyone;
- ***Avoiders*** who tend to back away from tough decisions and believe that no action equals action;
- ***Faddists*** who adopt every new thing that comes along;
- ***Tunnelers*** who tend to see the details, but never the big picture; and
- ***Fantasizers*** who are dreamers who cannot execute their ideas.[6]

An organization tends to take on the personality of its top leader(s), and that personality ripples throughout the organization's vision, philosophy, business goals, processes, and outcomes.

2. Leadership: Determinants of Leadership

Leadership is impacted by the leader's level of competence and character. To explain, what leaders are able to do (their competency) can be measured generically and specifically. For example, in generic terms, do they know about management theories, human resource management, business development, and sustainability? Specifically, what are they actually doing and how is that playing out? The leader's character – their personality and values, impact what they are likely to do.

In any given situation, leadership is impacted by both the leader's competence and character. This impacts employees in terms of employee selection, employee support, employee performance, job satisfaction, work attendance, disability management, and return-to-work attitudes.

---

[6] Jeffrey Gandz, "Talent development: The architecture of a talent pipeline that works" (2006) 1 *Ivey Business Journal* 1; and Jeffrey Gandz, *Leaders* (Ivey School of Business, Presented at Petro-Canada, Calgary, Alberta, 2006) [unpublished].

3. Management and Managers

A **manager** is the person within an organization who is responsible for performance of one or more subordinates, or functions. Managers implement the corporate vision and business strategy by working through others to achieve the desired "ends". Their management functions include:

- **Planning** – the process of setting performance objectives and identifying actions needed to accomplish them.
- **Organizing** – the process of dividing up work and coordinating the results to achieve a desired purpose.
- **Directing** – the process of directing work efforts of others to successfully accomplish assigned tasks.
- **Controlling** – the process of monitoring performance, comparing actual results to set objectives, and taking corrective action as necessary.

In essence, the manager is accountable for both task performance and human resource maintenance, as well as for the business outputs which are dependent on subordinates to attain. The manager must work with and through others to achieve the desired outputs.

Good managers seek opportunities, solve problems, analyze work situations to identify opportunities, and think systematically and scientifically. Through **Human Resource Maintenance** — the attraction, maintenance, and retention of a viable workgroup/workforce — managers procure, train, manage, and sustain employees to achieve high levels of task accomplishment.

The characteristics of a manager are focused on balancing worker maintenance and task accomplishment. To meet that end, the manager must be a/an:

- *Organizer* – able to delegate tasks;
- *Trusting* – does delegate;
- *Listener* – is caring and empathetic;
- *Interpersonal skills* – is approachable, consistent, and understands needs/skills of subordinates;
- *Visionary* – able to "see" future needs/directions;
- *Flexible* – able to adapt;
- *Communicator* – able to effectively get and receive messages;
- *Honest* – with oneself and others;
- *Coach* – able to guide and support other's endeavours;
- *Team builder* – able to be a leader and motivator; and
- *Decision-maker.*

Competency is defined as a method of ensuring that the workforce can carry out the expected work from a technical, quality, and functional perspective related to the qualifications, training, and experience of the workforce members. To function competently, managers must possess:

- *Technical Skills* – the ability to apply specialized knowledge;
- *Interpersonal Skills* – the ability to work in cooperation with other people; and
- *Conceptual Skills* – the ability to view the organization/situation as a "whole" and solve problems to benefit others.

These skills are further explained in Chapter 41, "Occupational Health and Safety Practitioners/Professionals: Career Development".

4. The Leader Role versus the Manager Role

The leader role and the manager role differ. To summarize, a comparison of these roles is offered in Table 38.2.

Table 38.2: The Leader – Manager Comparison[7]

Leader	Manager
Formulates long-term objectives for reforming the system; plans strategy and tactics	Engages in daily caretaker activities; maintains and allocates resources
Exhibits leading behaviours; acts to bring about change in others	Exhibits supervisory behaviour: acts to make others maintain standard job behaviour
Innovates for the entire organization	Administers a subsystem of an organization
Asks what and why to change standard practice	Asks how and when to engage in standard practice
Creates vision and meaning for the organization	Acts within the established culture of the organization
Uses transformational influence; induces change in values, attitudes, and behaviour using personal examples and expertise	Uses transactional influence; induces compliance in manifest behaviours using rewards, social sanctions, and formal authority
Uses empowering strategies to make followers internalize values	Relies on control strategies to get things done through others
Challenges the "status quo" and creates change	Supports and stabilizes the "status quo"

E. WORK AND WORK PERFORMANCE

Work is part of normal human activity. Most adults spend one-third to one-half of their time at work. Many people identify their work as part of "who they are" – their identity. Most people indicate that they would prefer to work even if they had no financial need to do so.[8] In essence, work is considered to be an integral component of life in a productive society.

Work is an activity that produces something of value for other people. This implies that there is a purpose to work. Monetary reward is a strong motivating factor. A **work unit** is a task-oriented group functioning within an organization that is composed of a manager and subordinates.

The effective functioning of a work unit is highly dependent on the capability and character of the manager, as well as the capabilities and character of the subordinates (employees). The level of trust and respect for each has been demonstrated as a predictor of how successful a disability management initiative and return-to-work efforts will be. Work units in which the supervisor is respected

---

[7] N. Langton, S. Robbins & T. Judge, *Organizational Behaviour: Concepts, Controversies, Applications*, 5th ed. (Toronto, ON: Pearson Prentice Hall, 2010) at 413.

[8] R.L. Kahn, *Work and Health* (New York, NY: John Wiley & Sons, 1981).

and trusted, and in which the work relationships are strong, tend to experience positive health outcomes, and positive disability management and return-to-work outcomes.

Work performance refers to *"the quality and quantity of work produced, or services provided"*. It also speaks to the quality of the relationships within the workplace. As a formula, *work performance* is:

Work Performance = quality + quality of the task accomplishments

Substandard work performance is often the outcome of illness/injury and hence is of concern to managers who are accountable for balancing task accomplishment with people management.

Work productivity is a performance measure that includes effectiveness and efficiency, with **effectiveness** being the achievement of goals, and **efficiency** being the ratio of effective work output to the input required to produce the work done.[9] A matrix of High Productivity is provided in Figure 38.2.

---

[9] N. Langton, S. Robbins & T. Judge, *Organizational Behaviour: Concepts, Controversies, Applications*, 5th ed. (Toronto, ON: Pearson Prentice Hall, 2010) at 13.

Figure 38.2: High Productivity = Performance Effectiveness + Performance Efficiency[10]

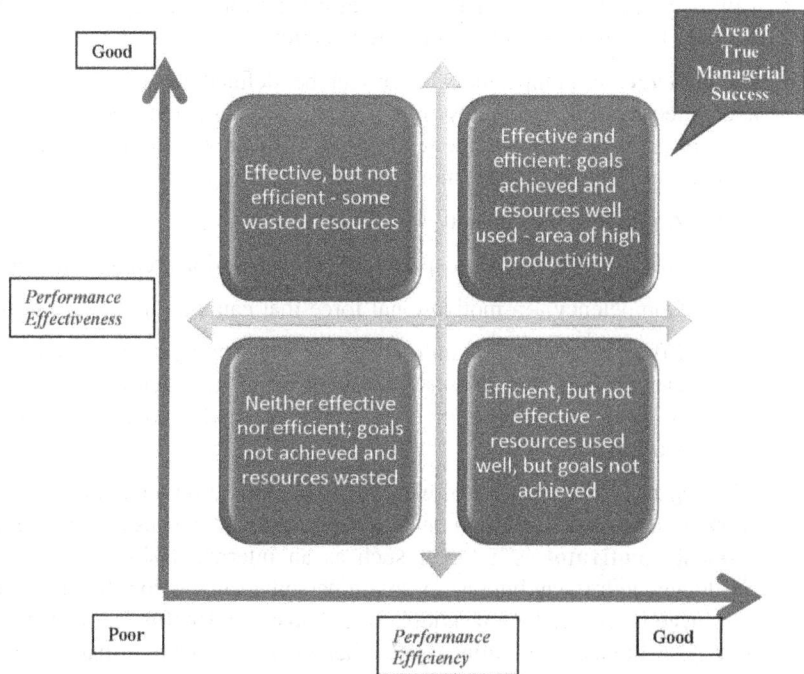

For managers who have employees frequently absent from work (high absenteeism), or non-functional when at work (presenteeism), managing the job tasks and meeting performance goals and objectives can be extremely challenging, if not impossible. By effectively managing employee performance, managers can meet their management obligations.

1. Individual Performance Factors

When examining the individual performance of leaders, managers, or employees, performance is composed of factors such as Individual Attributes (one's capacity), their Work Effort (willingness to work) and Organizational Effort (the opportunity to work).

- ***Individual attributes***: Employees possess characteristics that impact their work performance, namely their demographic factors, competencies, and psychological characteristics. Managerial influences include increasing the employee's capacity to perform the job by recruiting, selecting and training a suitable candidate for the job. Good person-job fit is critical: "[i]ndividual

---

[10] *Ibid.*

attributes must match task requirements to facilitate job performance".[11] Employee behaviour is also impacted by their culture of origin, beliefs, values, perceptions, personality, ethical orientation, life experiences, educational background, and current life situation.

- **Competency**: A competent worker can be defined as a worker that is adequately qualified, suitably trained, and with sufficient relevant experience to safely perform work without, or with only a minimal degree of supervision. The individual must possess the:
    - Aptitude – the capacity to learn.
    - Ability – the existing capacity to perform various required job tasks.
    - Competency – a motivational force that can stimulate work effort.
    - Effectance Motive – individuals who feel competent in their work can be expected to work harder at it; in this sense, competency becomes an internal force that stimulates and encourages people to work hard.
- **Work effort**: The key factor of work effort is motivation – the motivation to work, to perform. Work motivators are intrinsic and extrinsic in nature. An **intrinsic motivator** is a factor such as an internal desire to conquer a challenge, address an interest, or attain personal satisfaction that drives the employee to undertake a task/job. An **extrinsic motivator** is a factor such as pay, a bonus, recognition, and other tangible rewards, that drive the employee to undertake a task/job.

Leaders/managers are tasked with creating a work environment that motivates employees to perform.[12] **Motivation** is the intensity, direction, and persistence of effort exerted by an employee to reach a specific goal. Motivation varies by person, gender, cultural orientation, and need. Work motivation can be achieved by creating enthusiasm and allocating appropriate work-related rewards/recognition. For more information on adult motivational factors, refer to Chapter 17, "Occupational Health and Safety: Worker OH&S Training".

Job characteristics do affect work performance, but they tend to impact job satisfaction more. **Job satisfaction** is the degree to which employees perceive positively or negatively the job. Job satisfaction is higher when job motivators equal the employee's contributions to the job.

---

[11] R. French, C. Rayne, G. Rees & S. Rumbles, *Organizational Behaviour*, 2d ed. (New York, NY: John Wiley & Sons, 2011) at 78.

[12] T. Roithmayr, "Chapter 6: Occupational Health and Safety: Leadership and Commitment" in D. Dyck, ed., *Disability Management: Theory, Strategy & Industry Practice*, 6th ed. (Markham ON: LexisNexis Canada Inc., 2017).

Employee work performance is impacted by the expectations set by management and the motivators offered to them for performing at an expected level and achieving established business objectives.

Workers differ in their individual attributes, level of work effort, and the perceived level of organizational support, and that impacts their work performance. The rewards offered as well as the value that the individual employee places on those rewards impacts their motivation to do the job. Added to that comparison, employees do weigh out the rewards offered to them with what other employees are offered. If the employee believes they are being fairly treated or even "favoured", they are motivated to work hard. If an inequity is perceived, then the employee's work performance may decrease, or the employee may elect to leave the job. Well-managed rewards can lead to high levels of both individual performance and job satisfaction.

Job satisfaction is impacted by the nature of the work, the quality of supervision, relationships with co-workers, promotion opportunities, and the level of compensation (extrinsic rewards). Job satisfaction influences employee absenteeism, presenteeism, and staff turnover.

To summarize graphically these terms and their relationship, refer to Figure 38.3.

Figure 38.3: Motivation

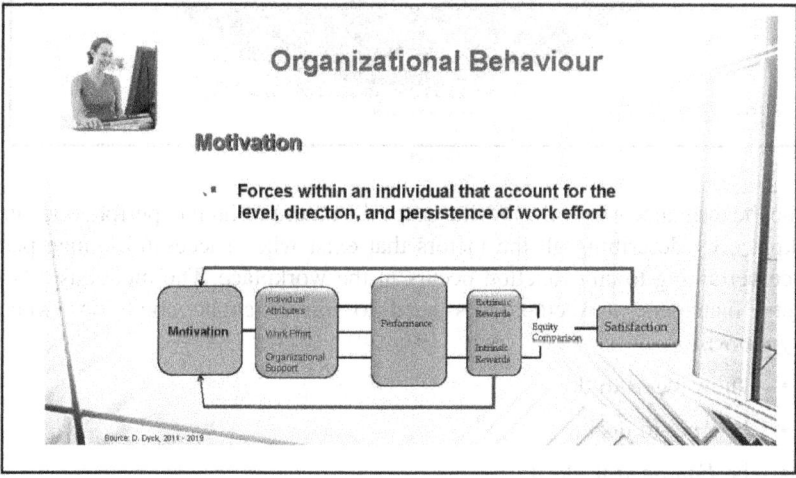

- *Organizational Support:* Successful managers recognize that perception differences exist among employees and try to understand what each employee values and what motivates the employee to perform.

Managers can motivate workers by creating a work environment in which the employee knows how to perform the tasks, is able to perform the tasks (physically, psychologically, and competently), is equipped to perform the job tasks, and is encouraged to do the job tasks because of Management's interest and involvement.

The Performance Maximizer Model (Roithmayr, 2000-2017) explains how organizations can maximize human performance within the workplace (Figure 38.4).

Figure 38.4: The Performance Maximizer Model

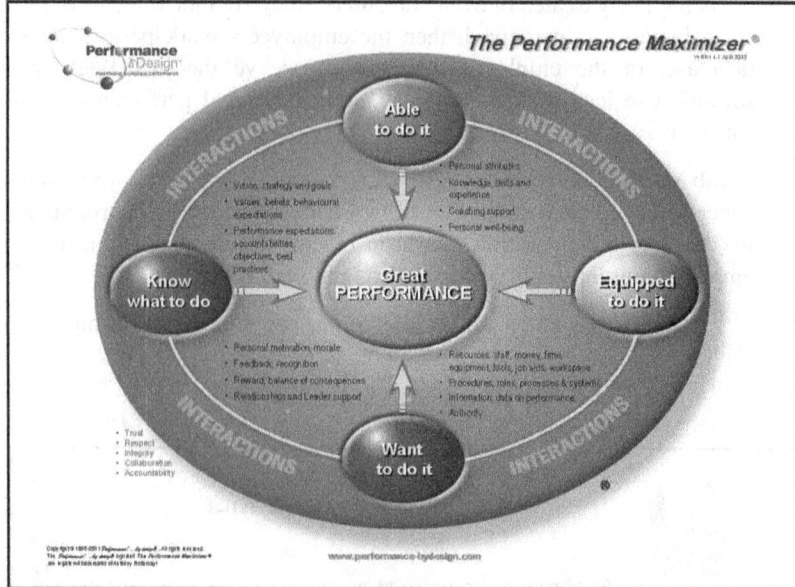

The Performance Maximizer® illustrates the nature of human performance in the workplace by describing all the factors that exist when successful human performance pertaining to any function occurs in the workplace. The model asserts that leaders, managers, and employees need to jointly create conditions whereby everyone will:

- know *What* to do;
- be *Able* to do it;
- be *Equipped* to do it;
- *want* to do it and have the experience; and
- experience *Interactions* that foster trust, respect, integrity, collaboration, and accountability.

These factors have been proven to support people in performing at their best and are referred to as the "Conditions for Great Performance". Interactions refer to the quality and effectiveness of interpersonal interactions among people in the workplace, measured as trust, respect, integrity, collaboration, and authority. The absence of these conditions constitutes obstacles and barriers to successful worker performance.

The key factors that impact employees perform are the provision of adequate resources and tools; the nature of the organizational structure and size; the technology present; the corporate culture; job design; and the group and interpersonal processes. Management can exert strong support by increasing the employees' opportunity to perform, and by doing a good job of planning, organizing, leading, and controlling the affairs of the workplace.

When faced with work performance problems, managers are advised to:

1. Determine the current state by defining their business goals and objectives and evaluating how they are meeting those expectations;
2. Define the desired state of performance;
3. Identify the gaps between the current state and the desired state; and
4. Develop strategies to eliminate or reduce those gaps. This can be achieved by using the Performance Maximizer Model (Figure 38.5).

Figure 38.5: The Performance Maximizer Model: An Approach for Addressing Work Performance Problems

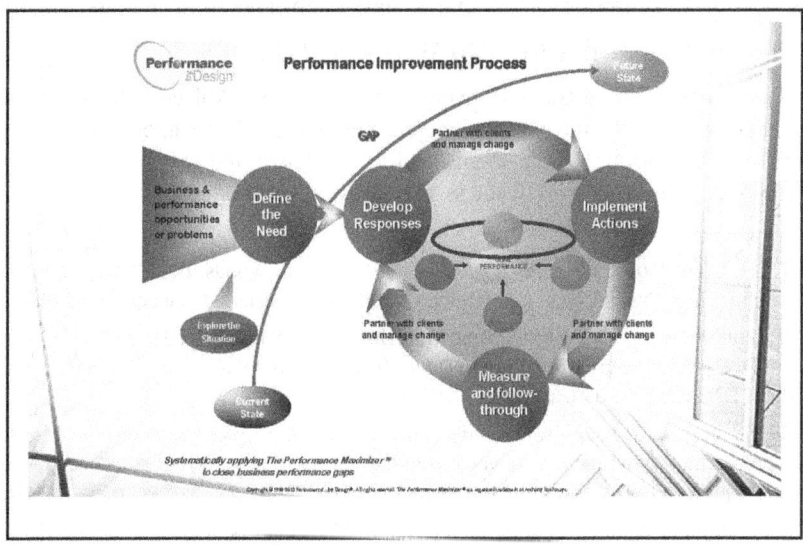

For information on maximizing human performance in the workplace, refer to Chapter 7, "Great Safety Performance: A Management Approach" for information on maximizing human performance.

In the role of an OH&S Professionals/Professional, assist organizations to identify when performance issues exist, as well as provide them with guidance on how to access appropriate help.

2. Leadership: Impact on Employees

There is a cost to substandard leadership and management. For example:

- Low employee morale results in unscheduled absence rates that are 50% higher.[13]
- Job stress is associated with high work demands, lack of role clarity, and little work control.[14]
- Job-related stress negatively impacts worker productivity more than do family problems.[15]
- Most Canadian workers (62%) report being highly stressed at work.[16]
- Personal problems that affect work performance and general health affect 15-20% of the workforce.[17]
- High job strain leads to a high risk (40% increase) of heart disease including heart attack, in women: demanding jobs with low control and opportunities for creativity, increase the risk of heart attack, stroke, or coronary bypass surgery.[18]

These are but a few facts on the losses and costs associated with the organization doing nothing to address poor leadership and/or substandard managerial practices.

F. CORPORATE CULTURE AND WORK ATTITUDES

Corporate cultures are present in every organization workplace – to quote Tony Roithmayr, cultures are like gravity, "they just are there". As such, organizations and the individuals working within them need to recognize and appreciate the impact that corporate culture has on business functions and outcomes.

---

[13] CCH, *CCH 2007 Unscheduled Absence Survey*, (Riverwoods, Ill.: CCH Incorporated, 2007), online: http://www.cch.com/absenteeism2007 (date accessed: January 31, 2015).

[14] T. Roithmayr, "Chapter 6: Occupational Health and Safety: Leadership and Commitment" in D. Dyck, ed., *Disability Management: Theory, Strategy & Industry Practice*, 6th ed. (Markham ON: LexisNexis Canada Inc. 2017).

[15] L. Duxbury & C. Higgins, *Reducing Work-Life Conflict: What Works? What Doesn't?* (2012), online: http://www.hc-sc.gc.ca/ewh-semt/alt_formats/hecs-sesc/pdf/pubs/occup-travail/balancing-equilibre/full_report-rapport_complet-eng.pdf (date accessed: February 28, 2020).

[16] S. Crompton, *What's Stressing the Stressed? Main Sources of Stress Among Workers* (Ottawa: Statistics Canada, 2011).

[17] D. Dyck, *Disability Management: Theory, Strategy & Industry Practice*, 6th ed. (Markham, ON: LexisNexis Canada Inc., 2017).

[18] *Women's Health Study* (2010) in A. Silliker, ed., (2011) *Canadian Safety Reporter* 1.

Corporate culture is the system of shared beliefs and values that develops within an organization and guides behaviour of its members.[19] It is shared by members of the organization and helps members to understand and solve things encountered internally and externally. The corporate culture helps to socialize new members and provides an identity to the organization; it influences how members perceive, think, feel, and behave.[20]

The corporate culture is man-made and dictated by:

- What Management does;
- What Management pays attention to;
- What Management condones or ignores;
- What Management measures; and
- Management controls of the resources necessary to effect change.

In essence, culture can be described as: "it is the way things are done around here".[21]

Researchers indicate that the lack of control over how one's work meets job demands and how one uses one's skills is the greatest determinant of job stress.[22] They also indicate that it is not the job demands, rather it is the structure of the organization that leads to job stress,[23] and this job stress, impacts employees with the least job control, the most.

Employee wellness is impacted by home and job stress, their individual sense of control over their lives, and their personal health profiles – biological and genetic features, physical and mental impairment, and cultural background.

Management has a choice though – a **Zone of Management Discretion**, which is the degree of influence/choice that Management has on the health, safety and well-being of employees.[24] That can lead to a 25% difference on the health of its

---

[19] T. Roithmayr, "Chapter 6: Occupational Health and Safety: Leadership and Commitment" in D. Dyck, ed., *Disability Management: Theory, Strategy & Industry Practice*, 6th ed. (Markham ON: LexisNexis Canada Inc., 2017).

[20] N. Langton, S. Robbins & T. Judge, *Organizational Behaviour: Concepts, Controversies, Applications*, 5th ed. (Toronto, ON: Pearson Prentice Hall, 2010) at 376.

[21] D. Dyck, *Disability Management: Theory, Strategy & Industry Practice*, 6th ed. (Markham, ON: LexisNexis Canada Inc., 2017) at 1311.

[22] M. Shain, "Stress, Satisfaction and Health at Work: Tuning for Performance (Presented at the Health & Wellness Conference '99, Vancouver, October 24-27, 1999) [unpublished]; and R. Karasek & T. Theorell, *Healthy Work: Stress, Productivity, and the Reconstruction of Working Life* (Toronto: Harper Collins, 1992).

[23] Willis Tower Watson, *2015/2016 Staying@Work Report Canada Summary* (2017), online: https://www.willistowerswatson.com/-/media/WTW/PDF/Insights/2016/08/2015-2016-staying-at-work-canada-research-findings-en.pdf (date accessed, February 28, 2020).

[24] M. Shain, "A New Take on Stress: Strategies that Work" (November 1997) Health

workforce. For example, Management can choose to adopt Negative Management Practices such as work factors that threaten employee psychological and physical health and safety:

- Work overload and time pressures;
- Lack of influence over daily work;
- Too many changes within the job;
- Lack of training and/or job preparation;
- Too little or too much responsibility;
- Discrimination;
- Harassment;
- Poor communication;
- Lack of quality supervision/management;
- Neglect of legal and safety obligations.

In contrast, Positive Management Practices tend to use an Upstream Approach. The organization positively influences the workforce and work outcomes by promoting good mental health and protecting employees from injury and illness. Good managers deal with the root causes instead of the symptoms of the problem. They work to develop strong organizational values by:

- Clarifying and communicating corporate values and expectations – **Role Clarity** is explaining Management expectations and desired outcomes for a particular job/position;
- Creating and maintaining a strong corporate culture;
- Adopting supportive management theories and practices such as:
    - Applying supportive policies and procedures, and
    - "Walking the Talk";
- Implementing a respectful workplace;
- Implementing a supportive infrastructure for Employee Support Programs like the Human Resources, Occupational Health & Safety, Employee Assistance, Attendance Control, Disability Management, and Workplace Wellness Programs;
- Supporting and enforcing policies and procedures;
- Recruiting and selecting suitable employees;
- Providing new employee orientation sessions;

Policy Forum 12 revised and presented in Health Canada, *Best Advice on Stress Risk Management in the Workplace* (2000) Cat. No. H39-546/2000E ISBN 0-662-29236-7 Part 2 of 2.

- Offering education/training and development programs;
- Empowering employees by giving them the responsibility for what they do and holding them accountable/answerable for the results;
- Adopting progressive recognition and reward program: **Recognition** is the practice of acknowledging work performance in terms of praise or tangible rewards. There is an adage: *"Praise in public; Reprimand in private"*;
- Recognizing the importance of personal and family life through the:
  - Use of family-friendly policies,
  - Development of a family-responsive workplace,
  - Use of a supportive versus punitive Attendance Control Program; and
- Seeking and valuing employee involvement.

G. JOB DESIGN, GOAL SETTING AND WORK SCHEDULING

Job design "means outlining the task, duties, responsibilities, qualifications, methods, and relationships required to perform the given set of a job. In other words, job design encompasses the components of the task and the interaction pattern among the employees, with the intent to satisfy both the organizational needs and the social needs of the jobholder".[25]

Job design includes four alternatives:

- *Job simplication:* it involves the standardization of work procedures and employment in clearly defined and specialized tasks.
- *Job enlargement:* this alternative offers increased task variety.
- *Job rotation:* characterized by increased task variety through the periodic rotation among the available jobs.
- *Job enrichment:* this approach has motivating factors built into job content by adding planning and evaluating duties to the job.

Job design and organization of work can have positive or negative effects on employees and their well-being as depicted in Figure 38.6.

---

[25] Business Jargons, "Job Design" (2019) at 1, online: https://businessjargons.com/job-design.html (date accessed: February 31, 2020).

Figure 38.6: The Organization and Design of Work[26]

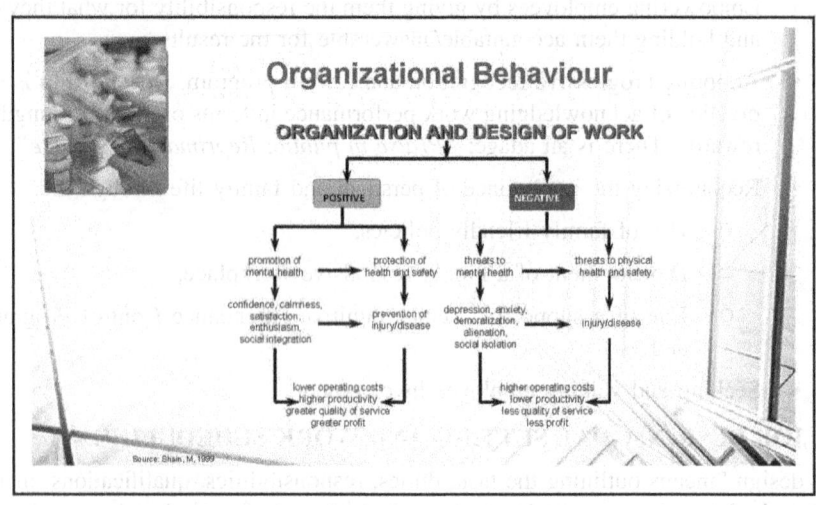

Job design and the organization of work can be supportive (Positive), or detrimental (Negative) of employee well-being, and organizational health and productivity.

1. Job Design, Goal Setting and Work Scheduling: Impact on Employees

Goal setting is the process of developing, negotiating and formalizing the targets or objectives that an employee is responsible for accomplishing. It involves stating:

- What must be done;
- How work performance will be measured;
- What is the work performance standard;
- What are the goal completion deadlines;
- How important are the goals; and
- How difficult are the goals to achieve.

Goal setting is critical to establishing and maintaining effective employee-employer relations, because it:

- Provides clarity on what is expected from the employee and why;
- Focuses on a "measurable dimension" — work performance; and

---

[26] M. Shain, "A New Take on Stress: Strategies that Work" (November 1997) Health Policy Forum 12, revised and presented in Health Canada, *Best Advice on Stress Risk Management in the Workplace* (2000) Cat. No. H39-546/2000E ISBN 0-662-29236-7, Part 2 of 2.

- Can be a powerful motivator in that it directs attention, regulates effort, increases persistence, and encourages the development of strategies and action plans.

Management by Objectives (MBO) is a process of mutual goal setting undertaken by the supervisor and employee.

Work scheduling is the assignment of work tasks and timelines. Part of goal setting is the determination of what is to be done and when. Achievable workloads should be a prime consideration. Realistic work schedules support employee control over their work and work obligations. Unrealistic ones can lead to employee stress, and if that job stress proves to be chronic in nature, ill health, and even injury. Worker fatigue is another possibility, thereby leading to the increased risk of workplace incidents and injuries.

H. GROUPS AND GROUP WORK

A **group** is a collection of people who regularly interact with one another towards a common goal – a natural social phenomenon. Its purpose is to carry out a decision/special task, to promote innovation and creativity, to result in better decisions, to enable control of group members, and to help offset the negative effects of organization growth.

1. Group Norms

Groups operate to establish group norms, which enable the group to function harmoniously. **Group norms** are the ideas or beliefs about the behaviour(s) expected from group members. Some examples are:

- Organizational and personal pride;
- Performance excellence;
- Team work;
- Open and honest communication among members;
- Leadership;
- Participation; and
- Group effectiveness.

Group norms serve to enable members to predict each other's behaviours, gain a sense of direction, and reinforce the organization's culture.

Group members tend to develop roles that support the group's performance challenges. Working in groups requires group members to form meaningful work relationships.

The performance advantages of groups are that:

- Groups make better decisions than would an individual;
- Groups are more successful at problem solving; and

- Groups are more creative and innovative because of the tendency to make more "risky" decisions.

The performance disadvantages are that groups can result in social loafing (the **Ringlemann Effect**); research indicates that group productivity drops off as more people become involved.[27]

2. Group Dynamics

Group dynamics are the forces operating in groups that affect their task performance and human resource maintenance. It involves the process through which the members work together to accomplish the assigned tasks.

Groups do not just occur; as part of their development, they pass through stages of development which correlate to their level of effectiveness. The Stages of Group Development are:

1) *Forming Stage* – This is the time for managing individual entry; it delineates who is a member and who is not.

2) *Storming Stage* – This is the time for managing group-norm development; it addresses the leadership roles, strategies for work to be done, and group dominance relationships.

3) *Norming Stage* – This is the time for managing group cohesion; it involves supporting the group-working relationships and the completion of the task at hand.

4) *Performing Stage* – This is the time for managing decision-making; it includes experimentation with new behaviours, adaptions to group norms, and doing the job.

Decision-making is the process of choosing among alternative courses of action. Decisions are made by:

- Authority without discussion;
- A Subject Matter Expert (SME);
- An average member's opinion;
- By authority after discussion;
- Majority control;
- Minority control;
- Consensus; or

---

[27] B. Latané, K. Williams & S. Harkins, "Many hands make light the work: The causes and consequences of social loafing" (1979) 37:6 *Journal of Personality and Social Psychology* 822; and K.D. Williams, S.A. Nida, L.D. Baca & B. Latané, "Social loafing and swimming: Effects of identifiability on individual and relay performance of intercollegiate swimmers" (1989) 10:1 *Basic and Applied Social Psychology* 73.

- Unanimity.

Group cohesiveness is the state where a clear alternative option/decision is selected by the majority of the group, and the opposing members feel they have been listened to and had a chance to influence the decision. All members work together to analyze the situation and decide on what, if anything, needs to be done.

3 Team Building

Team building involves a sequence of planned action steps designed to gather and analyze data on the functioning of a group and implement changes to increase organizational effectiveness. The Team Building Cycle is depicted in Figure 38.7.

Figure 38.7: The Team Building Cycle[28]

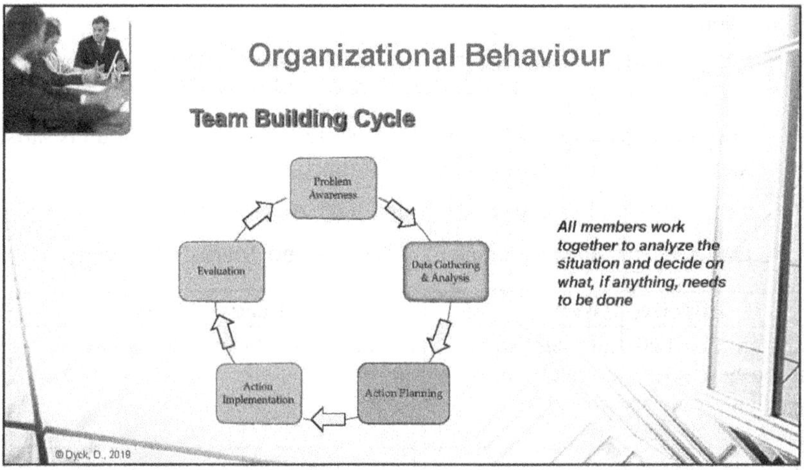

4. Group Roles and Tasks

Group roles vary with the type of group and its level of maturity. In general, group roles are the:

- **Initiator** – The member who proposes tasks or goals, defines problems, and suggests procedures or ideas for problem solving.
- **Information Seeker** – The member who requests facts, seeks information, solicits expressions of value, and seeks ideas and suggestions.
- **Information Provider** – The member who offers facts, provides information states beliefs, and gives suggestions and ideas.
- **Problem Clarifier** – The member who interprets ideas and suggestions,

---

[28] W.G. Dyer, *Team Building: Issues and Alternatives* (Reading, MA: Addison Wesley, 1977).

clears up confusion, defines terms, indicates alternatives, and gets group back on track.
- **Summarizer** – The member who pulls together ideas, restates suggestions after discussion, and offers decisions or conclusions.
- **Consensus Tester** – The member who asks to see if group is nearing decision-making, and tests possible conclusions.
- **Harmonizer** – The member who attempts to reconcile disagreements and reduce tension.
- **Gatekeeper** – The member who helps keep communication flowing and, facilitates participation by all.
- **Supporter** – The member who exudes friendliness, warmth, and responsiveness.
- **Compromiser** – The member who offers compromise, yields status, and admits error.
- **Standards Monitor** – The member who tests whether the group is satisfied with how it is proceeding and performing.

The characteristics of effective versus ineffective groups are (Table 38.3):

Table 38.3: Characteristics of Effective and Ineffective Groups

Effective Groups	Ineffective Groups
Goals are clarified and changed so that the best possible match between individual goals and the group's goals are achieved	Members accept imposed goals
Communication is two-way	Communication is one-way
Participation and leadership are distributed among group members	Leadership is delegated
Ability and information determine influence and power within the group; power is equalized and shared	Position in the group determines influence and power
Decision-making procedures are matched with the situation	Decisions are always made by the highest authority within the group
Controversy and conflict are viewed as positive and can lead to better group functioning	Controversy and conflict are ignored
Interpersonal, group and intergroup behaviour are stressed	The functions performed by members are emphasized
Problem-solving adequacy is high	Problem-solving adequacy is low
Members evaluate the effectiveness of the group and decide how to improve performance	The highest authority evaluates the effectiveness of the group and decides how to improve performance

Interpersonal effectiveness self-actualization and innovation are encouraged	"Organizational persons" who desire order, sustainability and structure are encouraged

5. Group Communication

Communication is the interpersonal process of sending and receiving symbols with meaning attached to them. It enables people, groups, or departments to interact.

Group communication can be *interactive in nature* – that is, highly focused on the task; *co-acting in nature* – individuals focus on separate parts of the task; or *counteracting in nature* – subgroups disagree with each other, impeding the group's progress and level of task completion.

I. RELEVANCE TO INTEGRATED WORKPLACE HEALTH MANAGEMENT (IWHM)

Strategically linking organizational support programs and resources maximizes the organizational efforts in providing a safe and healthy workplace. It involves linking employee support programs. It involves linking the various employee support programs:

- Employee Recruitment, Selection and Retention;
- Human Resources Program and Services;
- Attendance Control;
- OH&S Program;
- Employee Assistance Programs (EAPS);
- Disability Management Program;
- Workplace Wellness Program.

Organizational Behaviour (OB) functions to:

- Describe the system in which employee support programs operate;
- Demonstrate the relationship of leadership and corporate culture to effective integrated workplace health management (IWHM);
- Clarify stakeholder roles and responsibilities;
- Provide insight into work, work performance, job design, goal setting, and work scheduling, and how they impact IWHM;
- Describe the system in which employee support programs operate;
- Demonstrate the relationship of leadership and corporate culture to effective IWHM;
- Clarify stakeholder roles and responsibilities;
- Provide insight into work, work performance, job design, goal setting, and work scheduling, and how they impact IWHM;

- Explain groups and teamwork — Social Capital Theory[29] indicates strong workplace relationships result in successful work experiences and outcomes;
- Provide a collaborative approach that is required to effectively and efficiently manage employee support programs and services: IWHM; and
- Through collective-data management the identification of successes, shortfalls and opportunities for improvement is possible: IWHM offers that possibility.

A collaborative approach is needed to effectively and efficiently manage employee support programs and services. An IWHM Program offers organizations a way to achieve that end. Likewise, an IWHM Program enables organizations to collectively manage data that can be used to identify program successes, shortfalls, and opportunities for improvement.

J. SUMMARY

The field of Organizational Behaviour offers OH&S Practitioners/Professionals information and guidance on:

- Leadership and management;
- Work and work performance;
- Workplace culture and its impact;
- Job design, goal setting and work scheduling;
- Work groups, teams, and teamwork; and
- Their relevance to IWHM Programs.

For more information on IWHM Programs refer to Chapter 28, "Integrated Workplace Health Management".

CHAPTER REFERENCES

CCH, *CCH 2007 Unscheduled Absence Survey* (2007), (Riverwoods, Ill.: CCH Incorporated, 2007), online: http://www.cch.com/absenteeism2007 (date accessed: January 31, 2020).

M. DePree, *Leadership is an Art* (New York, NY: Dell Publishing, 1989).

L. Duxbury, & C. Higgins (2012). *Reducing Work-Life Conflict: What Works? What Doesn't?*, online: http://www.hc-sc.gc.ca/ewh-semt/alt_formats/hecs-sesc/pdf/pubs/

---

[29] **Social Capital Theory** maintains that one's willingness to help others is based on the quality of our social relationships. We are willing to help (exchange favours) only when we feel a sense of goodwill, trust, and empathy towards other members of our social group. We are also more likely to grant favours if we know that at some point in the future, the same courtesy will be returned, either directly by the recipient, or by someone else within our social group (Dr. Williams-Whitt).

occup-travail/balancing-equilibre/full_report-rapport_complet-eng.pdf (date accessed: February 28, 2020).

D. Dyck, *Disability Management: Theory, Strategy and Industry Practice*, 6th ed. (Toronto, ON: LexisNexis Canada, 2017).

D. Dyck, *Occupational Health & Safety: Theory, Strategy and Industry Practice*, 3d ed. (Toronto, ON: LexisNexis Canada, 2015).

W.G. Dyer, *Team Building: Issues and Alternatives* (Reading, MA: Addison Wesley, 1977).

R. French, C. Rayne, G. Rees & S. Rumbles, *Organizational Behaviour*, 2d ed. (New York, NY: John Wiley & Sons, 2011) at 78.

Jeffrey Gandz, "Talent development: The architecture of a talent pipeline that works" (2006) 1 *Ivey Business Journal* 1.

Jeffrey Gandz, *Leaders*, (Ivey School of Business. Presented at Petro-Canada, Calgary, Alberta, 2006) [unpublished].

Great Place to Work Institute Canada (2019), online: https://www.greatplacetowork.ca/en/ (date accessed: February 28, 2020).

Business Jargons, "Job Design" (2019) at 1, online: https://businessjargons.com/job-design.html (date accessed: February 31, 2020).

R.L. Kahn, *Work and Health* (New York, NY: John Wiley & Sons, 1981).

R. Karasek & T. Theorell, *Healthy Work: Stress, Productivity, and the Reconstruction of Working Life* (Toronto: Harper Collins, 1992).

N. Langton, S. Robbins & T. Judge, *Organizational Behaviour: Concepts, Controversies, Applications*, 5th ed. (Toronto, ON: Pearson Prentice Hall, 2010).

B. Latané, K. Williams & S. Harkins, "Many hands make light the work: The causes and consequences of social loafing" (1979) 37:6 *Journal of Personality and Social Psychology* 822.

T. Roithmayr, *Performance by Design* (Calgary, AB: , 2002-2019).

T. Roithmayr, "Chapter 6: Disability Occupational Health and Safety: Leadership and Commitment" in D. Dyck, *Management: Theory, Strategy & Industry Practice*, 6th ed. (Markham ON: LexisNexis Inc Canada, 2017).

M. Shain, "Stress, Satisfaction and Health at Work: Tuning for Performance (Presented at the Health & Wellness Conference '99, Vancouver, October 24-27, 1999) [unpublished].

M. Shain, "A New Take on Stress: Strategies that Work" (November 1997) *Health Policy Forum* 12, revised and presented in Health Canada, *Best Advice on Stress Risk Management in the Workplace*, (2000) Cat. No. H39-546/2000E ISBN 0-662-29236-7, Part 2 of 2.

K.D. Williams, S.A. Nida, L.D. Baca & B. Latané, "Social loafing and swimming:

Effects of identifiability on individual and relay performance of intercollegiate swimmers" (1989) 10(1) *Basic and Applied Social Psychology* 73.

K. Williams-Whitt, "Chapter 17 – Disability Management: The Social Capital Theory Perspective for Managing Disability Claims" in D. Dyck, *Disability Management: Theory, Strategy and Industry Practice*, 6th ed. (Toronto, ON: LexisNexis Canada, 2017).

Willis Tower Watson, *2015/2016 Staying@Work Report Canada Summary* (2017), online: https://www.willistowerswatson.com/-/media/WTW/PDF/Insights/2016/08/2015-2016-staying-at-work-canada-research-findings-en.pdf (date accessed, February 28, 2020).

Women's Health Study (2010) in A. Silliker, ed., (2011) *Canadian Safety Reporter* 1.

Chapter 39

OCCUPATIONAL HEALTH AND SAFETY: EFFECTIVE COMMUNICATION[1]

A. INTRODUCTION

OH&S Practitioners/Professionals assume a collaborative and cooperative role in which they must communicate with a variety of stakeholders. The purpose of this chapter is to explain the importance, principles, and techniques of effective communication.

B. COMMUNICATION AND ITS IMPORTANCE

Effective communication is an essential skill for OH&S Practitioners/Professionals to master. **Communication** is the ability to convey facts, concepts or reasoning clearly to others, and to receive and understand the messages sent by others. Given that OH&S documents can be legal documents, it is imperative for OH&S Practitioners/Professionals to master effective communication skills.

C. COMMUNICATION: TYPES

Oral, written, and non-verbal communication are the three main types of communication used by human beings. To effectively relay a message, the OH&S Practitioner/Professional should ensure that the *same* oral, written, and non-verbal cues are conveyed.

Some examples of the types of communication undertaken by OH&S Practitioners/ Professionals are:

1. *Oral communication* – speaking, presentations, telephoning, voice messages.
2. *Written communication* – letters, standards, reports, alerts, memos, e-mails, texts, faxes.
3. *Non-verbal communication* – facial expressions, body language, gestures, body posture, grooming, attire.

D. COMMUNICATION: ELEMENTS

When messages are sent, they must be clear and concise. The message must be

---

[1] D. Dyck "Internal and external consulting: Assisting clients with managing work, health and psychosocial issues" (2002) 50:3 *AAOHN* 111. Copyright (2002) the American Association of Occupational Health Nurses, Inc. Used with permission. All rights reserved.

transmitted without distortion. The receiver must be positioned to receive the message and comprehend it. The language and jargon used by the sender must be suitable to the receiver for understanding (Figure 39.1).

Figure 39.1 – The Elements of Communication

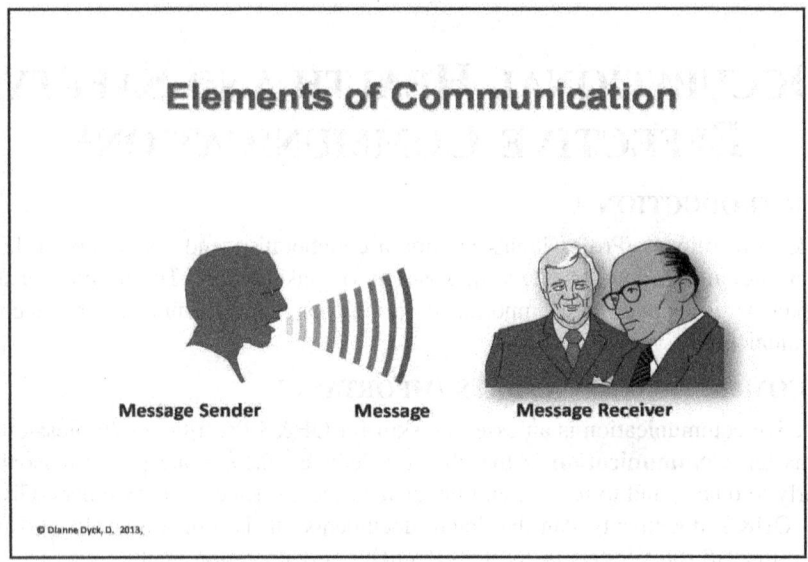

Communication problems exist when:
1. The message sender:
    - is unclear due to lack of knowledge or familiarity with the subject;
    - uses language/jargon that the receiver is unfamiliar with;
    - has a strong accent or problems speaking;
    - sends mixed messages – that is, is saying one thing, but non-verbally saying another; and/or
    - misreads the audience – the message receiver; *etc.*
2. The message becomes distorted due to background noise; transmission interruptions; competing message transmissions; *etc.*
3. The message receiver is unprepared or non-receptive to receiving the message. This may occur because the receiver is not listening, does not understand the language/jargon used, is distracted by other interests or events, is fatigued or unable to concentrate on the message, and/or does not want to hear the message, *etc.*

E. COMMUNICATION: GENERAL PRINCIPLES

There can be many reasons for communication breakdowns. It is up to the OH&S

Practitioner/Professional as the message sender, to understand the communication process and to manage the situation so that effective communication occurs. As such, knowledge and use of the following general principles of communication is recommended:

- *Know your message:* Know what you want to communicate. That sounds like a nonsensical statement — but that is where most messages get derailed.

- *Use a clear, logical delivery:* Get the message clear in your mind and then deliver it just as clearly and succinctly as you can.

- *Start with the general aspects and move to the specifics:* Messages are best conveyed when the message sender starts with generalized concepts and moves to specific details.

- *Know the receiver or target audience:* Receivers of the message have different needs, wants, expectations, and knowledge levels. To effectively communicate, the OH&S Practitioner/Professional must know the receiver's or intended audience's preferred mode of communication, terminology, and communication style.

- *Check for receipt and comprehension of the message:* Regularly check for receiver understanding. Ask questions, seek feedback, restate concepts/facts, and clarify any misconceptions.

- *Use positive words and phrases:* Avoid words like "don't, no, avoid, stop, never, *etc.*" Be sure to demonstrate empathy and understanding in your approach.

F. COMMUNICATION: ORAL COMMUNICATION

1. Speaking – Conversations

People converse to introduce themselves, to get to know people, to convey a message, to obtain information, to verify or update information, to report a decision, and to develop plans for action.

When speaking, it is important for the OH&S Practitioner/Professional to know and clearly state the message. Checking for understanding is vital. OH&S Practitioners/Professionals are advised to verify that the message is correctly getting across to the receiver. Emotions can be a barrier to getting a message across. Address them. Be empathetic. Show that you care but remain consistent with the message being delivered. At the end of the conversation, summarize what has been said and what you trust the receiver heard.

Critical to oral communication is being an effective listener. To listen well, the OH&S Practitioner/Professional needs to:

- Be able to receive incoming messages without distortion:
    - stay healthy,

- use listening aids, if required, and
  - practice listening to unfamiliar topics and speakers;
- Be attentive to the speaker;
- Focus on the message:
  - prepare to concentrate on the message,
  - focus on the central message, not the speaker's mode of delivery, and
  - either concentrate on the content, or to the relationship meaning of the message, not to both;
- Understand the message:
  - identify the thesis and main arguments, and
  - avoid assumptions – seek clarification;
- Retain the message:
  - take notes, review them after the contact, and
  - use mnemonics and other memory aids.

Listen to Understand

OH&S Professionals/Practitioners will probably spend more time using their listening skills than any other kind of skill. Like other skills, listening takes practice. To listen well, the OH&S Practitioner/Professional needs to:

- be able to receive incoming messages without distortion;
- give full attention to the speaker;
- focus on the message – listen for the main ideas;
- seek to comprehend the message; and
- be able to retain the message.

But, what does "listening to understand" actually mean? ***Real listening*** is an active process that is composed of three basic steps:

1. Hearing – hearing what is said;
2. Understanding – internalizing the message; and
3. Judging – evaluating the message.

Thoughts move about four times as fast as speech. With practice, while you are listening you will also be able to think about what you are hearing, really understand it, and give feedback to the speaker.[1]

---

[1] Homework Center, "Speaking and Listening Skills", *InfoPlease* (2015), online: http://www.infoplease.com/homework/listeningskills1.html (date accessed: February 28, 2020).

Some tips for effective listening that the OH&S Practitioner/Professional can use are:

- pay attention to the speaker;
- focus on the message – seek the main theme or idea;
- let the speaker conclude before speaking;
- finish listening before speaking;
- ask questions, probe, paraphrase, seek clarification, dialogue; and
- give feedback.

Know how to listen and you can profit even from those who talk badly.—

Plutarch

To listen well, the OH&S Practitioner/Professional must be cognizant of the two types of listening, namely:

- ***Passive Listening*** – This means allowing someone to speak, without interrupting and not doing anything else at the same time. It involves understanding perceptions; and listening to emotions, words, and body language.

  Example Case:

  Darren advises you, the OH&S Practitioner/Professional, that he was nearly struck by a pressurized hose when working in the paint shop. He reports that he is not hurt, but just wanted you to know what happened. He ends the conversation by telling you that if more information is needed, he can be reached at his work number.

- ***Active Listening*** – Active listening is reacting or doing something that demonstrates you are listening and have understood the message relayed. This means giving non-verbal cues to demonstrate that you are paying attention (nodding, making eye contact, making facial expressions appropriate to what is being said), and reflecting back the main points and summarizing what has been said. It involves encouragement, asking questions, probing, paraphrasing, seeking clarification, emphasizing, and summarizing.

  Example Case: An OH&S Situation

  Darren advises you, the OH&S Practitioner/Professional, that he was nearly struck by a pressurized hose when working in the paint shop. He reports that he is not hurt, but just wanted you to know what happened.

  The OH&S Practitioner/Professional asks Darren to explain how this incident unfolded. Darren says that he was walking around a vehicle in the paint shop and suddenly, a pressurized hose violently snaked near him. Luckily, according to Darren, he was able to avoid contact with the hose.

Although the company does not have a near-miss reporting system, he felt it important to tell you.

The OH&S Practitioner/Professional then probes: "A pressurized hose... Wow, that must have been really scary. What do you think happened?"

Darren responds that the pressurized hose broke free from the paint sprayer nozzle to which it should have been clamped. Still connected to the compressor, the hose snaked violently about. All is fixed now, but Darren was clearly shaken by the "close call".

The OH&S Practitioner/Professional then asks Darren about the equipment inspection practices in this area. He/she learns that Darren has no idea when the last formal inspection of that equipment occurred. The OH&S Practitioner/Professional probes further as to whether the worker, as part of his/her work practices, is expected to examine this equipment prior to its use. The response was, "I think so, but I am not sure."

The OH&S Practitioner/Professional, using open-ended questions, asks Darren what should be done to address this situation, and how that action might "play out". Using active listening, it is determined that:

1. Formalized near-miss reporting is not in place, and should be.
2. Regular equipment inspection is not being done, and should be.
3. Equipment inspection is not part of the work practices and should be.
4. Equipment maintenance practices should be addressed.

Being a good listener is so critical to effective communication. As also noted by Plutarch:

82% of people prefer to talk to people who are good listeners, not great speakers.

Oral Communication Styles

There are two styles of communication; however, one style tends to be more effective in certain situations.

1. **Sympathetic Communication**

    The technique of "walking in another person's shoes" — attempting to feel what another person is going through. It denotes agreement, whereas empathetic communicationin Contrast denotes understanding.

2. **Empathetic Communication (in Contrast)**

    The technique of structured listening and questioning — helps to develop and enhance relationships through a stronger understanding of what is being conveyed both intellectually and emotionally.

Example Case: A Disability Management Situation
In dealing with Susan, you learn that she is very unhappy with her medical care and

that she is seeking your advice on what to do next. She has been to a number of physicians — her attending physician and two specialists. She is unhappy with the diagnosis and treatment options received. She tells you that she is looking for another medical opinion. She is certain that the prognosis of her condition is incorrect and there must be other options other than chemotherapy.

Sympathetic Communication

Disability Management Practitioner/Professional: *"Susan, what a dilemma! What are you going to ever do? You must be so devastated."*

Susan: *"I just don't know what to do. . .. My husband is not dealing with this news very well. I have no idea how my family will ever manage with me being sick. This has to be wrong. . .."*

Disability Management Practitioner/Professional: *"Perhaps another medical opinion is the way to go. At least then, you will be assured of the accuracy of the situation and your future options."*

In this situation, the sympathetic communication approach does not help Susan. Instead, she remains stuck in a realm of grief, denial, and anger. Her reaction will likely be to seek another medical opinion.

Empathetic Communication (in Contrast)

Disability Management Practitioner/Professional: *"Susan, I certainly hear your concern and fears. Has the message from the medical practitioners been consistent?"*

Susan: *"Yes, they say that chemotherapy is my only option. If successful it could add five to 10 years to my life. But I don't want to lose my hair."*

Disability Management Practitioner/Professional: *"Chemotherapy can be an unpleasant experience; although for breast cancer the medications and approaches used today to counteract those side-effects seem to work well."*

Susan: *"They mentioned that. They also said that I could talk to some women who have gone through chemotherapy for breast cancer."*

Disability Management Practitioner/Professional: *"Did they give you some names and numbers to call? I hear that Breast Cancer survivors are excellent resource people. They are known to be most helpful to women facing cancer treatment."*

Susan: *"Yes, I have three names. But I don't even know how to go about this. Could you call them for me?"*

Disability Management Practitioner/Professional: *"If these women have offered their names, they will be most receptive to your call. Just like you called me today to tell me of your news, you can call one of them as well."*

Susan: *"What would I say?"*

Disability Management Practitioner/Professional: *"Well, let's walk through that. . ."*

At this point, the Disability Management Practitioner/Professional works with Susan to plan her call to the Cancer Survivor. It moves Susan from a place of grief, denial, and anger, to a place of action.

In certain situations, a sympathetic communication approach may prove effective; but in many more instances, use of empathetic communication is far more effective. For additional communication resources, the OH&S and Disability Management Practitioner/Professional are advised to consult:

- P. Urs Bender & R. Tracz, *Secrets of Face to Face Communications: How to communicate with power* (Toronto: Stoddart, 2001)
- Stephen Boyd's online resources:[2]
    - What Is That I Hear (January 2011)
    - You Communicate Best When You Don't Talk (December 2010)
    - Improve Your Listening Skills Today! (November 2009)
    - Win With Empathic Listening (February 2008)
    - Hear What Is Being Said (April 2007)
    - Just Someone to Listen (February 2007)
    - Listen First and Talk Second (August 31, 2005)
    - Silence Is the Beginning of Listening (May 2005)
    - Overcome Bad Listening Habits (April 2005)
    - Listen to Evaluate (September 2004)
    - You Have to Listen! (May 2004)

2. Occupational Health & Safety — Types of Oral Communication

OH&S Professionals/Practitioners, in the course of their workday, have the occasion to use oral communication on a daily basis, for example, when:

- undertaking a worksite and/or hazard assessment;
- delivering risk communication messages to stakeholders;
- providing OH&S training and education sessions;
- interviewing a worker in regards to some work aspect;
- participating in/presenting at a Safety Meeting;
- presenting an OH&S topic;
- conducting an incident investigation;
- undertaking an injury/illness case assessment;

---

[2] Resources online: http://www.speaking-tips.com/Listening-Skills/ (date accessed: February 28, 2020).

- developing a Case Management Plan;
- leading or participating in case conferences;
- conversing with the employee (family), physician, other health caregivers, insurers, workplace, lawyers, *etc.*;
- consulting with internal/external supports; and
- leaving voice mails.

The use of voice mail warrants further discussion. Voice-mail etiquette and procedures should be observed when setting up the "out-of-office" message, when leaving a voice-mail message, and when responding to received voice mails. For example, when recording the voice-mail "out-of-office" message, the recommended approach is to:

- keep it brief, informative, clear, and business-like;
- ensure the "out-of-office" message is functioning as desired;
- keep the "out-of-office" message current;
- state when the caller might expect a response;
- pick up messages frequently; and
- respond to voice mails in a timely manner.

When leaving a voice-mail message for an employee (client), supervisor, or other stakeholders, the OH&S Practitioner/Professional should:

- write down the message points before placing the call;
- speak slowly, clearly, and slightly louder than normal;
- address the "who, where and the contact information" of your call. *Remember: the message could be reviewed by anyone*;
- listen to the message – edit if needed;
- keep the message brief;
- repeat your phone number; and
- avoid leaving messages with family members (due to privacy reasons).

OH&S Professionals/Practitioners regularly use voice mail. Effective voice messaging and messages can save time and promote good customer relations. **Remember:** Your message helps to formulate an impression of you and the position/organization that you represent. Keep that fact in mind.

3. Oral Presentations

OH&S Professionals/Practitioners conduct presentations on a daily basis. Some presentations are conducted informally (*e.g.*, meetings with employees, supervisors, co-workers, union representatives) and formally (*e.g.*, Safety Meetings, case conferences, disability management meetings, project reports, program evaluation reports, marketing events). For this reason, it is critical for OH&S Professionals/Practitioners to be able to deliver effective oral presentations.

Presenting well is a valuable business skill. However, it is also one of the most challenging skills to acquire. Presenting involves more than conveying a message; it requires audience engagement and the development of trust between the presenter and the audience. Typically, the intent of presenting is to spur the audience to action — so the presentation must be informative and motivational.

Some presentation tips are as follows:
- know your target audience;
- tailor the message accordingly;
- make the message tell a story;
- be concise and succinct in your content and delivery;
- come prepared — that means, rehearse your presentation;
- dress for success;
- be prepared for things to go wrong — have a contingency plan ready;
- speak within the allotted time — overstaying your pre-arranged time is disrespectful;
- face the audience while you speak, especially when using slides;
- project your voice so everyone can hear;
- use gestures and body language for emphasis, and to give passion to your presentation;
- have legible and easy to read slides (if used);
- monitor the audience for understanding; if they appear lost, circle back to re-engage the group;
- provide concrete and feasible suggestions for future action;
- don't expect the audience to listen more than 20 minutes — schedule frequent breaks if you are doing a lengthy session; and
- the typical attention span for North Americans is seven to eight minutes — that means there is a need to present new elements within that time frame so that the audience's attention is maintained.

For additional information on preparing and delivering oral presentations, refer to Chapter 37, "Occupational Health and Safety: Project Management". Another excellent reference is P. Urs Bender, *Secrets of Power Presentations* (Willowdale, ON: Firefly Books Inc., 1990).

G. COMMUNICATION: WRITTEN COMMUNICATION

Business writing, report preparation, training courses, program development, *etc.*, are all part of the OH&S Professional's/Practitioner's role. As such, strong written communication skills are critical. Those OH&S Professionals/Practitioners that have a good command of oral and communication skills tend to stand out above the rest.

Written communication skills are learned skills. Through education and practice, the OH&S Practitioner/Professional can succeed. Some general written communication principles are:

- *Know your message:* Be clear on the information that you are trying to convey. Hence, there is a need to think through the message being sent. **Tip:** Distill your message into a single sentence. If that message can be written in one sentence, then the message sender will be able to deliver it easier.
- *Write for the receiver or target audience:* Know your audience:
    - who the intended message receiver(s) is(are);
    - their current knowledge-level on the subject;
    - what the message receiver(s) needs to know about the subject;
    - how the message receiver(s) will likely respond to the message.
- *Develop the message*: Know exactly what you want to say to the message receiver. What are the main elements of the message? Then, formulate the intent of the message — what do you want to achieve from delivering this message? What information needs to be provided? How can you motivate the receiver to act?
- *Use a clear, logical progression of thought:* Write down the main points in the most logical order. Create headings/subheadings if they clarify your thoughts. Be consistent with the message. Order the outline to align with how the receiver would think — move from general to specific information.
- *Be clear – be brief*: Write the draft and then be precise with it.
- *Be certain of your facts*: Accuracy is vital — research, verify, and critique your content. Inaccuracies discredit the message and the message sender.
- *Write at a grade 6 to 8 reading level*: This is the usual reading level used in newspapers and community publications. The use of technical jargon and acronyms should be avoided unless they are key to the communication process. Use positive words and phrases.
- *Use parallel construction of ideas/facts*: When ideas are equally weighted in their importance, give them equal sentence construction.
- *Use consistent verb tense*: Write in the present or past tense avoiding awkward future and conditional verb tense structures. Keep the verb tense consistent throughout the document.

- *Use powerful words, precise words, and active verbs*: Active verbs turn statements into action. Some action words include:

announce	inform
appoint	introduce
call	act
start	end
immediate	welcome
change	open

- *Effective writing*: Attention needs to be paid to the proper use of grammar, punctuation, and spelling. Tips on effective writing are provided in Appendix 1.

- *Eliminate wrong messages*: Ensure that what is being said aligns with what is being done. Inconsistencies indicate confusion and perhaps a hint of incompetence on the part of the message sender.

- *Proofread*: Revise. Edit.

  Tip: Not everything written is "golden". Cut out the "fluff" and extraneous words. Have an impartial person read the written product. If they understand what you are trying to say, then that is an indication that the message receiver may also comprehend it.

1. Technical Writing

Technical writing involves technical internal and external OH&S communications on behalf of the company; the results of incident/investigations; trend analysis reports; an explanation of compliance audits; presentation of OH&S Program audit results; rationale for corrective action follow-up; and, occasionally, the management of public/media relations in regards to OH&S issues.

The recommended approach is to use the general writing principles of technical writing, namely:

- Decide the content – what is this document to accomplish? Should it inform, appeal, teach, recommend, report, reply, analyze, make comparisons, or sell an idea.[3]
- Present a main point(s) or thesis.
- Make sure the content is factual and accurate.
- Use an approved format.
- Maintain focus.

---

[3] A. Shekhar, "Principles of Technical Writing", *Ezine Articles* (2010), online: http://ezinearticles.com/?Principles-of-Technical-Writing&id=4126251 (date accessed: February 28, 2020).

- Organize and develop ideas.
- Use action words.
- Use effective phrases.
- Be concise and succinct.
- Use bullets to present groups of data/ideas.
- Write grammatically correct.
- Ensure spelling and punctuation are accurate.
- Edit, edit, edit.
- Proofread.

Some examples of OH&S technical writing include the preparation of OH&S reports (Chapter 37, "Occupational Health and Safety: Project Management"); OH&S teaching/learning materials (Chapter 17, "Occupational Health and Safety: Worker OH&S Training"; and Chapter 20, "Occupational Health and Safety: Ergonomics"); OH&S writing job descriptions (Chapter 41, "Occupational Health and Safety Practitioners/Professionals: Career Development", Appendix 3); and OH&S proposals (Chapter 37, "Occupational Health and Safety: Project Management", Appendix 1).

2. Business Writing

Business writing is correspondence on behalf of an organization/company. It must reflect the philosophies and values of the organization/company, while providing the needed business information. Ensure that you understand the organization's/company's way of doing business.

The recommended approach is to use the general writing principles of business writing, namely:

- Decide what to write.
- Think about the content.
- Eliminate wrong messages.
- Consider special situations.
- Use an approved format, usually an approved letterhead and letter.
- Use positive words.
- Use effective phrases.
- Write strong sentences.
- Build effective paragraphs.
- Edit, edit, edit.

- Proofread.[4]

Letter writing is an essential aspect of the OH&S Professional/Practitioner role. The written communication principles apply, ensuring that the message, content, and language are appropriate with the intended letter receiver/audience. If possible, try to familiarize yourself with the letter receiver's/audience's terminology preferences.

In addition to all that, take into consideration special situations related to the message being sent, or to the receiver's receptivity. For example, the preparation of difficult letters, like letters that deal with rejection/refusal, a dispute/disagreement, or notices of closure/termination of a practice/service. These letters are "bad news" letters that must be delivered in an objective, factual manner without insulting or harming the organization's/company's relationship with the employee/client.

The refusal letter has two goals:

1) to say "no"; and
2) to promote goodwill.

Begin with a positive statement, followed by directly delivering the message. End with a goodwill statement.

Disagreements/disputes arise in business. Resolving them is a true test of the OH&S Practitioner's/Professional's communication skills and code of ethics, as well as the organization's/company's business practices. To reach a satisfactory resolution is simple, except when emotions are at play. Hence, the first step is to defuse the emotional component. *Tip:* Communicate calmly and clearly that the organization/company is determined to work with the letter receiver to reach a fair and equitable solution.

Individuals/company representatives naturally reject the idea of a closure/termination of a practice/service decision. However, if positioned and explained rationally and logically, they will usually understand and accept the decision.

The recommended methodology for preparing these difficult letters is to:[5]

- *Decide what to write*: Know the message that is to be sent to the receiver.
- *Determine the content*: Recognize the problem and the rationale for the refusal. Prepare a logical explanation for the decision rendered.
- *Eliminate wrong messages*: Use positive statements, not excuses. Never blame others or circumstances for the refusal/decision made. Avoid leaving the receiver believing that the refusal/decision may be overturned.
- *Consider special situations*: When a direct refusal is provided, be sure to start the letter with a neutral or warm greeting.
- *Use the approved letter format*: Business refusals should be prepared under

---

[4] S. Lamb, *How to Write It*, 2d ed. (New York, NY: Ten Speed Press, 2006).

[5] *Ibid.*

the proper letterhead. Most organizations/companies have letter formats that it uses with its customers.

- *Select and use the appropriate strong words*: Some suitable examples of strong words are:

apply	final
conclusion	decision
considered	decline
deny	refuse
evaluated	extend
unable	determination

- *Build effective phrases*: Some suitable examples include:
  - after careful consideration . . .. . . doesn't qualify
  - I must inform you that we cannot
  - our final decision
  - unquestionably, we must decline
- *Write strong sentences*: Start with a strong, positive verb and build the sentence from there. For example;
  > We have carefully considered your recommendation and determined that the information provided does not meet our company's OH&S Standard.
- *Prepare strong paragraphs*: Using strong sentences and parallel sentence construction develop paragraphs that present the difficult message in a logical manner.
- *Proofread. Revise. Edit.*

Remember: Letters are part of the official documentation of an OH&S situation, and/or disability claim/case. They, like all the other documentation, can become public in the event of a legal situation, or at the employee's/claimant's request.

3. Other Types of Written Communication

On any given day, the OH&S Practitioner/Professional may be expected to prepare a number of different types of written communication. For example, OH&S Professionals/Practitioners develop and document workplace policies, standards, safe work and other procedures, bulletins, safety alerts, training sessions/programs, audit reports, incident investigations, and OH&S program action plans and outcome reports.

In the course of managing a disability case, the OH&S Practitioner/Professional: writes case notes; responds to requests from the employee or workplace; addresses service provider or health care provider requests; prepares letters to the employee, physicians, health care providers, lawyers, *etc.*; writes internal memos and e-mail messages; prepares referral letters; develops case summaries; creates informal and formal reports; writes appeal letters; as well as documents return-to-work (RTW) planning.

Some suggestions for undertaking these varied activities follow:

a. **Case Notes**

Case notes describe all the case management activities associated with a disability situation. They are legal documents and can become public should the employee decide to divulge the information, or in the event of a legal suit. OH&S and Disability Management Professionals/Practitioners are advised to:

- Keep case notes factual – avoid assumptions or the statement of personal beliefs.
- Provide accurate accounts of what has transpired.
- Include comprehensive details of the events.
- Ensure that the data provided is relevant to the disability.
- Offer a sequential presentation of the facts.
- Spellcheck the contents.

Remember that case notes are used as a vehicle for communication among the members of the OH&S or Disability Management team. They serve as a legal representation of how the case has been managed by the organization/company and its representatives. They also speak to the OH&S and Disability Management Professional's/Practitioner's capabilities.

b. **Letters**

Writing letters is part and parcel of OH&S, disability case management, and return-to-work planning. To prepare professional letters,

- Determine the desired message to be sent in the letter.
- Avoid sending wrong messages.
- Use strong words.
- Build effective phrases, sentences, and paragraphs.
- Use an approved letter format.
- Edit, edit, edit.
- Proofread and revise, if necessary, the finished letter.

c. **Case Summaries**

Case summaries are prepared to provide an overview of what has transpired with a Workers' Compensation or other disability claim — the relevant details, the claim and case management practices, return-to-work planning and placement, and the outcomes. Hence, the OH&S and Disability Management Practitioner/Professional should:

- Be sure that the content tells a story from onset of the illness/injury to current state of the case.
- Present only the relevant facts.
- Align the actual occurrences with the planned events.
- Indicate the successes achieved and the challenges faced and to yet be faced.
- Reach a recommendation for action.

d. Use of Abbreviations

OH&S Professionals/Practitioners tend to use abbreviations as a way to speed up communication and the documentation process. However, in doing so, they must ensure that the abbreviations are standardized for ease of comprehension. This means having a list of "acceptable abbreviations" so that in the event of legal action, this list can accompany the disability claim/case file. In that way, the court will be able to accurately decipher the claim/case management notes. For further details, refer to D. Dyck, *Disability Management: Theory Strategy & Industry Practice*, 6th ed. (Markham, ON: LexisNexis Canada, 2017), Chapter 9, Appendix 4 "Approved Disability Management Abbreviations".

Other activities undertaken by the OH&S Practitioner/Professional are to prepare:

a. Informal and Formal Reports

Informal reports include program progress reports, feasibility reports, recommendations for action, interim audit reports, interim incident reports, case summaries, case conference reviews, and worksite inspection reports. They are prepared to tell a story (inform), to analyze, and to elicit action. Prepare them by using the following principles:

- tell a story;
- select the relevant content;
- indicate the successes and challenges of the situation; and
- reach a recommended course of action.

For more details, refer to Chapter 37, "Occupational Health and Safety: Project Management".

b. Online Communication – Consider

Online communication – e-mail, texting, *etc.*, require good writing skills as well as specialized online communication skills. **Tip:** Follow the principles of writing clear, brief, timely, and precise communications.[6] As well, remember that the writer is judged on his/her writing skills; this includes one's ethics, honesty, abilities, competencies, level of compassion, *etc*. Also, these types of messages can reach unintended recipients; consider how widespread the message could become. Lastly, the message is never truly deleted.

Some guidelines for online communication include:

- Know the organization's/company's rules for e-mail use.
- Use the organization/company-approved rules of etiquette.
- Use correct grammar, spelling, and punctuation, just as you would in writing a letter. The use of all capitals indicates a "shout"; so avoid using them.
- Restrict the message of the e-mail to one subject.
- State the message subject clearly and precisely in the subject line.
- Respond promptly (within 24 hours) to received e-mails.
- Ensure that attachments can be opened by the receiver.
- Keep the message to one screen-length.

c. **Discriminatory Terms**

Business communication has no room for the use of discriminatory terms. The OH&S Practitioner/Professional must be aware of the existence of discriminatory terms and avoid their use. Instead, the recommended practice is to:

- Use non-gender-specific terms.
- Use inclusive and descriptive non-gendered words.
- Use plural terms to avoid gender references.
- Use non-gendered titles.
- Use parallel or equal construction for men and women.
- Use inclusive, non-stereotypical terms.
- Avoid he/she pronouns.
- Avoid the use of gender words, such as mailman (use mailperson), policeman (use police officer), *etc*.
- Avoid the use of discriminatory words of race, nationality, or sexual preference.
- Avoid the use of age discriminatory words.

---

[6] *Ibid*.

- Avoid stereotyping people with disabilities.

H. COMMUNICATION: NON-VERBAL COMMUNICATION

Humans communicate in the spoken and written word; but most effectively, and often, they do so in a non-verbal manner. OH&S professionals/practitioners must recognize this fact and learn to effectively read and use body language.

Non-verbal communication is our strongest mode of communication. We have been using it since birth. It speaks volumes as to what we are actually thinking and feeling — often without the communicator realizing that fact. Non-verbal communication or *body language* includes facial expressions, gestures, human touch, eye contact, body posture, mode of dressing and grooming, and tone of voice. In essence:

It is not what you say, it is how you say it that matters. — Dyck, 2011

Body Language – Value

Non-verbal communication can enrich or detract from the oral and/or written communication. The key is how body language is used by the OH&S Practitioner/Professional. For example, body language offers:

- *Repetition*: It can serve to repeat the verbal message, thereby reinforcing what has been said or written.
- *Contradiction*: It can contradict a sent message. This is a situation to be avoided.
- *Substitution*: It can substitute for a verbal message — *e.g.*, a person's eyes often convey a more vivid message than do the words uttered.
- *Complementing*: It can add to, or complement, a verbal message — *e.g.*, a boss who pats a person on the back in addition to giving praise can increase the impact of the message sent.
- *Accenting*: It can accent, or underline a verbal message — *e.g.*, pounding the table can strongly underline a verbal message.
- *Facial expressions*: It can convey emotions such as happiness, sadness, anger, surprise, fear, and disgust.
- *Body movements/Posture*: It can communicate information about the person's true viewpoint and attitude on work and life.
- *Gestures*: Gestures are usually made without the person knowing. They include pointing, use of hands, beckoning, head nodding, *etc.*; actions which differ among cultures.
- *Eye contact*: In some cultures, eye contact is important to conversing; it communicates caring, affection, hostility, interest, acknowledgement, *etc.* However, in other cultures, eye contact is less critical.
- *Touch*: A firm handshake, a timid tap on the shoulder, a warm bear hug, a

reassuring pat on the back, a patronizing pat on the head, or a controlling grip, all convey a different message.

Hence, non-verbal communications are pervasive and so expressive. To be used effectively, the OH&S Practitioner/Professional must first understand their meaning and how best to use them in different cultural groups. For more information on this aspect, refer to Chapter 34, "Impact of Cultural Diversity in the Workplace".

I. SUMMARY

The purpose of this chapter was to demonstrate how important it is for the OH&S Practitioner/Professional to possess and use sound communication skills. Communication is the basis on which much of OH&S practices and relationships are based. Breakdowns in communication typically are the root cause of workplace incidents; poor workplace relations; decreased employee morale; ineffective disability claim management, disability case management, and return-to-work outcomes; and the lack of understanding of the positive impact that the OH&S function offers. As well, it can seriously damage the relationship between the employer and employee; the employer and unions; and the employee and co-workers. At times, it can result in legal actions.

To summarize, good communication embodies a number of key elements, namely:[7]

- ☑ Information is exchanged (accurate and timely).
- ☑ Sharing occurs (personal and factual).
- ☑ Non-judgemental interactions occur.
- ☑ Listening is done with understanding (empathy).
- ☑ Each other's comments, behaviours, *etc.* are checked-out to ensure that there is understanding (clarify).
- ☑ Emotion *and* reason are balanced.
- ☑ Making false/inaccurate assumptions and conclusions about the other person's motives is avoided.
- ☑ Present yourself as you *really are*; exhibit honesty and openness.
- ☑ Present non-possessive caring (caring without strings attached).
- ☑ Avoid JUDGEMENTS, EVALUATIONS, and PUTDOWNS.
- ☑ Don't hop from topic to topic without first reaching closure on the topics.

---

[7] Adapted from W. Penner, *Criteria for Good Communication* (Unpublished work, Edmonton, Alberta, 1996).

APPENDIX 1

Effective Writing: Tips

Written communication involves the proper use of grammar, punctuation and spelling. The following tips are provided to assist the OH&S Practitioner/Professional with business writing.

Use of Commas

Commas are used to assist the reader to understand your writing. Here are some rules:

Rule 1. Use commas to separate independent clauses when they are joined by any of these seven coordinating conjunctions: *and, but, for, or, nor, so, yet.*

Rule 2. Use commas after introductory a) clauses, b) phrases, or c) words that come before the main clause. For example, *"In 2015, the number of work absences exceeded the 2014 numbers."*

Rule 3. Use a pair of commas in the middle of a sentence to set off clauses, phrases, and words that are not essential to the meaning of the sentence. Use one comma before to indicate the beginning of the pause and one at the end to indicate the end of the pause. For example, *"When I finish here, and I will soon be, I'll be glad to help you."*

Rule 4. Do not use commas to set off essential elements of the sentence, such as clauses beginning with *that* (relative clauses). *That* clauses after nouns are always essential. *That* clauses following a verb expressing mental action are always essential.

Rule 5. Use commas to separate three or more words, phrases, or clauses written in a series.

Rule 6. Use commas to separate two or more coordinate adjectives that describe the same noun. Be sure never to add an extra comma between the final adjective and the noun itself, or to use commas with non-coordinate adjectives.

Rule 7. Use a comma near the end of a sentence to separate contrasted coordinate elements or to indicate a distinct pause or shift.

Rule 8. Use commas to set off phrases at the end of the sentence that refer back to the beginning or middle of the sentence. Such phrases are free modifiers that can be placed anywhere in the sentence without causing confusion.

Rule 9. Use commas to set off all geographical names, items in dates (except the month and day), addresses (except the street number and name), and titles in names.

Rule 10. Use a comma to shift between the main discourse and a quotation.

Rule 11. Use commas wherever necessary to prevent possible confusion or misreading.

Use of a Colon

A colon means "that is to say" or "here's what I mean." Colons and semicolons should never be used interchangeably.

Rule 1. Use a colon to introduce a series of items. Do not capitalize the first item after the colon (unless it's a proper noun).

Examples:

You may be required to bring many things: sleeping bags, pans, utensils, and warm clothing.

I want the following items: butter, sugar, and flour.

I need an assistant who can do the following: input data, write reports, and complete tax forms.

Rule 2. Avoid using a colon before a list when it directly follows a verb or preposition.

Incorrect: *I want: butter, sugar, and flour.*

Correct:

I want the following: butter, sugar, and flour.

or

I want butter, sugar, and flour.

Incorrect: *I've seen the greats, including: Barrymore, Guinness, and Streep.*

Correct: *I've seen the greats, including Barrymore, Guinness, and Streep.*

Rule 3. When listing items one by one, one per line, following a colon, capitalization and ending punctuation are optional when using single words or phrases preceded by letters, numbers, or bullet points. If each point is a complete sentence, capitalize the first word and end the sentence with appropriate ending punctuation. Otherwise, there are no hard and fast rules, except be consistent.

Examples:

I want an assistant who can do the following:

    a.   input data;

    b.   write reports; and

    c.   complete tax forms.

The following are requested:

- Wool sweaters for possible cold weather.
- Wet suits for snorkeling.
- Introductions to the local dignitaries.

These are the pool rules:
1. Do not run.
2. If you see unsafe behaviour, report it to the lifeguard.
3. Did you remember your towel?
4. Have fun!

Rule 4. A colon instead of a semicolon may be used between independent clauses when the second sentence explains, illustrates, paraphrases, or expands on the first sentence.

>   ***Example:*** *He got what he worked for: he really earned that promotion.*

If a complete sentence follows a colon, as in the previous example, it is up to the writer to decide whether to capitalize the first word. Capitalizing a sentence after a colon is generally a judgment call; if what follows a colon is closely related to what precedes it, there is no need for a capital.

Note: A capital letter generally does not introduce a simple phrase following a colon.

>   ***Example:*** *He got what he worked for: a promotion.*

Rule 5. Use a colon rather than a comma to follow the salutation in a business letter, even when addressing someone by his or her first name. (Never use a semicolon after a salutation.) A comma is used after the salutation in more informal correspondence.

>   ***Formal:*** *Dear Ms. Rodriguez:*
>
>   ***Informal:*** *Dear Dave,*

Use of Semicolon

A semicolon, like commas, indicate an audible pause — slightly longer than a comma's pause, but short of a period's full stop like a period. It has other uses as well:

Rule 1. A semicolon can replace a period if the writer wishes to narrow the gap between two closely linked sentences.

>   ***Examples:***
>
>   *Call me tomorrow; you can give me an answer then. We have paid our dues; we expect all the privileges listed in the contract.*

Rule 2. Use a semicolon before such words and terms as *namely, however, therefore, that is, i.e., for example, e.g., for instance, etc.*, when they introduce a complete sentence. It is also preferable to use a comma after these words and terms.

>   ***Example:*** *Bring any two items; however, sleeping bags and tents are in short supply.*

Rule 3. Use a semicolon to separate units of a series when one or more of the units contain commas.

>   ***Incorrect:*** *The conference has people who have come from Moscow, Idaho,*

Springfield, California, Alamo, Tennessee, and other places as well.

Note that with only commas, that sentence is hopeless.

Correct: *The conference has people who have come from Moscow, Idaho; Springfield, California; Alamo, Tennessee; and other places as well.* (**Note the final semicolon, rather than a comma, after *Tennessee*.**)

Rule 4. A semicolon may be used between independent clauses joined by a connector, such as *and, but, or, nor, etc.*, when one or more commas appear in the first clause.

Example: *When I finish here, and I will soon be, I'll be glad to help you; and that is a promise I will keep.*

Rule 5. Do not capitalize ordinary words after a semicolon.

Incorrect: *I am here; You are over there.*

Correct: *I am here; you are over there.*

Use of "As" and "Because"

Use "because" to give the reason of something that is important for the listener. "Because" indicates a causal connection between two elements.

"As" is used when the listener already knows the reason for the connection of two elements. "As" also denotes a similarity as opposed to a causal relationship.

Spaces between Sentences

Published work these days rarely features two spaces after a period; rather, one space is used. However, the writer can choose to use one or two spaces, but having made that choice, should then do so consistently.

Spelling

Rule 1. Spell check your work.

Rule 2. There are many words that sound the same, but that are spelled differently, and mean different things. For example:

- there, their, they're
- sum, some
- to, too, two

Rule 3. The apostrophe is used to denote ownership; not a plural. For example:

- John's phone
- The organization's business strategy

The apostrophe is also used to contract two words. For example:

- It's (it is)
- Can't (cannot)

Rule 4. Plural forms of words are created by adding an *s, es,* or *using a specific plural form of the word*. For example:

- Employees
- 1980s
- Coaches
- *Women, men, etc.*

Rule 5. When two or more words are used to describe a noun or verb, they must be joined by a hyphen. For example:

- Return-to-work program
- Fit-to-work assessment

However, when the first word ends in "y", a hyphen is not used:

- Medically unfit workers

Rule 6. Before using abbreviated forms of words, titles or phrases, you must first define the abbreviation. For example:

- Return-to-work Program (RTWP)
- Fit to work (FTW)
- Disability Management Program (DMP)

Rule 7. Single-digit numbers have to be written out; double digit numbers can be used in the numerical format. For example:

> The administration portion of a worker's compensation claim is over five times the amount that it would be in Canada.

> In Canada, one in *10* individuals will develop clinical depression at some point in their lives.

Rule 8. Long sentences should be broken down into shorter ones or bullet points used to present the various ideas. For example:

> The report, "*Management Behaviours Drive Workplace Wellness Program Results: The SMIL Model*", cites studies which indicate that to decrease healthcare expenses which are correlated with unhealthy employee lifestyles, and to increase productivity, a rising number of organizations are encouraging employees to improve their health behaviours.

Better:

> The report, "*Management Behaviours Drive Workplace Wellness Program Results: The SMIL Model*", cites studies which indicate that encouraging employees to improve their health behaviours and lifestyle practices decreases healthcare expenses.

Rule 9. A business position paper should not contain personal pronouns, like *I, you, me, we,* if possible. Rather, write in the third person, *e.g.,* the employee versus him/her, *the organization* versus *you, etc.*

Rule 10. When writing, remember that not every word is "gold". After finishing your paper, I suggest that you go through it and eliminate any redundancies and the "fluff". You are writing a business position paper and management will not read reams of words.

CHAPTER REFERENCES

S. Boyd, resources online: http://www.speaking-tips.com/Listening-Skills/ (date accessed: February 28, 2020), *Integrated Disability Management Program: Instructional Course* (Calgary, AB: 2000-2020).

D. Dyck, *Disability Management: Theory Strategy & Industry Practice*, 6th ed. (Markham: LexisNexis Canada, 2017), Chapter 9, Appendix 4 "Approved Disability Management Abbreviations".

HelpGuide.org, "Nonverbal Communication" (2011), online: http://www.helpguide.org/mental/eq6_nonverbal_communication.htm (date accessed: February 28, 2020).

Homework Center, "Speaking and Listening Skills", *InfoPlease* (2015), online: http://www.infoplease.com/homework/listeningskills1.html (date accessed: February 28, 2020).

V. Kotelnikov, "Effective Communications" (2011), online: http://www.1000ventures.com/business_guide/crosscuttings/communication_main.html (date accessed: February 28, 2020).

S. Lamb, *How to Write It*, 2d ed. (New York, NY: Ten Speed Press, 2006).

W. Penner, *Criteria for Good Communication* (Unpublished work. Edmonton, Alberta 1996).

Perdue University, Welcome to the Purdue OWL (2015), online: https://owl.english.purdue.edu/owl/owlprint/607/ (date accessed: February 28, 2020).

A. Shekhar, "Principles of Technical Writing", *Ezine Articles* (2010), online: http://ezinearticles.com/?Principles-of-Technical-Writing&id=4126251 (date accessed: February 28, 2020).

P. Urs Bender, *Secrets of Power Presentations* (Willowdale, ON: Firefly Books Inc., 1990).

P. Urs Bender & R. Tracz, *Secrets of Face to Face Communications: How to communicate with power* (Toronto: Stoddart, 2001).

A. Wolvin (2000), cited in Effective Communications, online: http://www.1000ventures.com/business_guide/crosscuttings/communication_main.html (date accessed: February 28, 2020).

Chapter 40

OUTSOURCING OCCUPATIONAL HEALTH AND SAFETY SERVICES[1]

A. INTRODUCTION

Occupational Health and Safety (OH&S) services can be provided to an organization through an in-house OH&S service, or through a contract service arrangement with an external service provider. Each option has its advantages and limitations. Organizations must decide which option is best suited to their particular OH&S and business needs. Once the decision to use an external service provider is made, the next step is to determine a "best fit" between the organization's OH&S Program, business approaches, and corporate culture, and the services offered by available external service providers.

The plethora of OH&S service providers in today's marketplace make determining a "best fit" a difficult task for organizational leaders. This chapter presents the relevant issues in outsourcing OH&S services and how to effectively address them.

B. WHY OUTSOURCE?

Companies must decide whether or not to outsource their OH&S services. After examining their business practices, many large organizations decide that internal OH&S services are outside the scope of their core business practices or competencies. For smaller companies, having a full-service, in-house OH&S service may be impractical and too costly.

However, before going to the marketplace, an organization must identify and understand their reasons for wanting to outsource OH&S services. It is paramount to know why and how the decision to outsource was made. In this manner, attention can be paid to addressing the issues that were originally identified as drivers for outsourcing the OH&S services. Without understanding why outsourcing is necessary, service providers and contracted services can be positioned in a manner that fails to address the original reasons for seeking external services.

1. Decision-making Process

According to Jane Hall (1997), a comprehensive process is needed to make sure that

---

[1] Excerpts from D. Dyck, *Disability Management: Theory, Strategy and Industry Practice*, 6th ed. (Markham, ON: LexisNexis Canada Inc., 2017).

all aspects of the decision-making process have been considered.[2] The necessary steps are listed below:

1. Identify the desired changes and the reasons behind making those changes.
2. Explore the possibility of alternative service delivery options.
3. Obtain stakeholder input into what they want from a future OH&S service.
4. Develop a communication strategy to explain the need for change, and describe the new OH&S service that is ultimately selected. According to Hall, "ineffective or inadequate communication of change is a prime reason for implementation failures".[3]
5. Establish value criteria for deciding on a suitable service provider.
6. Define the nature of the desired service provider arrangement. Is the organization seeking service/product delivery, or a partnership arrangement to deliver OH&S services?

Some of the reasons that organizations choose to outsource OH&S services include:

- *Lack of internal expertise* — By the time the reader reaches this chapter in the book, he/she will appreciate that OH&S services require a level of expertise that is lacking in many organizations. Without Occupational Health support, implementing a quality OH&S Program is difficult, if not impossible. Likewise, Disability Management expertise is required to effectively and efficiently process employee Workers' Compensation and non-occupational disability claims.

- *Lack of internal resources* — Resources, facilities, and funding are needed for an internal OH&S Program to operate. The type and amount of resources required varies depending on the industry and size of the organization. In general, office space and equipment, file cabinets, support personnel, and a budget are key to providing efficient internal OH&S services.

- *Insignificant number of services required* — If an organization is small and experiences only a few incidents per year, having an internal OH&S service can be a costly venture. This is often the major reason for outsourcing OH&S services.

- *Desired benefits of economies of scale* — This reason ties into the one provided above. By accessing OH&S services from a service provider that works with many other organizations, the organization can benefit from their infrastructure and services. Theoretically, this type of arrangement can

---

[2] J. Hall, "The Decision to Vendor and Vendor Management: A Tool for Occupational Health and Workers' Compensation Management" (1997) 11:5 *The OEM Report* at 45.

[3] *Ibid.*

provide quality OH&S services at a lower unit price than the organization could manage internally. Companies value economies of scale and often seek OH&S service arrangements that can offer savings through a contracted service arrangement.

- *Specialty expertise is required to achieve a cultural shift within the organization* — Some organizations desire extensive changes in their service practices over a short period of time. To achieve such an aggressive goal, they need the help of an external service provider. Getting there on their own would take too much time, money, and resources. This is often seen when a shift from a traditional to a more business- and people-oriented approach to OH&S, is sought.

- *Geographically-complex services are desired* — In some instances, organizations choose to keep the majority of their OH&S services in-house except for those services that are remote and better provided by a local service provider. These mini-contract arrangements are common and usually purchased as "one-off" situations.

- *Seek a method to deal with fluctuating service demands* — OH&S service demands can be unpredictable. Theoretically, service providers are better positioned to handle the "peaks and valleys of servicing".

- *Desire to purchase current skills* — Remaining current in a dynamic field like OH&S can be difficult, particularly for smaller organizations. By purchasing OH&S services from a service provider whose core business is OH&S servicing, current OH&S service practices and expertise will be obtained.

- *Desire to transfer assets and people to a third party manager* — Organizations that have in-house Occupational Health Service may wish to outsource the personnel and resources that provide OH&S services. The intent is often to keep only "core" business functions in-house and to outsource the rest.

For the above reasons, the option to outsource some aspects of an OH&S service is often sought. However, to be successful, the service provider arrangement must be well planned, implemented, and evaluated.

C. PREPARATION FOR OUTSOURCING

For organizations to attain a suitable OH&S service provider, a number of preparations must be made prior to going to the marketplace. They include:

1. Development of an OH&S Program

First, develop the type of OH&S Program that the organization wants to have in place. This important first step can be likened to house shopping. Prior to meeting with a realtor, the purchaser must decide what type of house is required. Will it be a "starter home", large family home, or a retirement residence? The same holds true for seeking an OH&S service provider. The organization must determine the extent of the OH&S services required and the framework within which they are to be delivered.

This is a vital step in a successful service provider arrangement. Having an OH&S Program in place that is known to all stakeholders, allows the participants to be more aware of their roles and areas of responsibility. As well, there is an established structure within which the service can function. Trying to operate without a framework will result in confusion. Service expectations and perceptions will become divided and poor service quality will result. This is one of the major reasons for OH&S service provider failures.

2. Determination of the Preferred Customer-Service Provider Relationship

The organization must identify the customer-service provider relationship that works for them. Some options include:

- a service that functions independently and is separated from other organizational activities (*i.e.*, service/product delivery only);
- a service that is integrated with the rest of the corporate programs and that is expected to provide and receive organizational data relevant to the OH&S Program and various prevention strategies (*i.e.*, a partnership); or
- a service that is not only integrated, but comprehensive in the approach to OH&S (*i.e.*, an enabler arrangement). This includes providing OH&S services, and participating in the long-term planning for employee/organizational health and well-being.

The continuum ranges from a completely outsourced service with minimal involvement by the purchaser, to a service provision that enables the organization to offer a comprehensive OH&S Program with an internal feel.

Before choosing the desired service provider arrangement, the organization must explore the differences between the following three options:

Option 1: The provision of a service/product is not only the least involvement that a service provider can have with the organization, but it is also the least time-consuming, risky, and costly option.

Option 2: A partnership arrangement takes more time, energy, and expertise for the service provider to deliver. It involves building relationships, knowing about the organization and its employees, and seeking ways to provide effective and efficient OH&S services.

Option 3: An enabling relationship is the most comprehensive in nature. It involves partnering, as well as enabling the organization to move towards illness/injury prevention. To achieve this type of relationship, the service provider must fully understand the organization, its people, the business strategies, service demands, work environment, and business challenges. This takes the most time and expertise and involves risk-taking. In essence, it requires making a personal investment in the organization and its issues. This is also the most costly of the three options provided.

Service gaps occur when the organization believes the service purchased is a partnership or enabling agreement, and the service provider has contracted for service delivery only. Service gaps can cause an outsourcing disaster.

3. Determination of the Nature of Services Required

The next step is to decide what parts of the OH&S Program will be maintained internally and which ones will be contracted out. Some organizations design their OH&S Program so that they keep all the program management activities in-house. For them, only some of the OH&S services are contracted out. Others seek the provision of a complete OH&S Program — for example, program management, provision of clinical Occupational Health services, Disability Management services and return-to-work responsibilities — from a service provider.

An awareness of the services required, as well as the ability to articulate them clearly, is essential to outsourcing OH&S Program services. The organization should not only list the services desired, but it should also be able to describe the nature of the services, desired service quality, required turnaround times, and desired reporting mechanisms for the services provided. Also, linkages for these services back into the organization have to be described.

The positioning of the desired client-service provider relationship is also important. Some organizations want their service provider to operate within the confines of the organization and to provide consulting services to employees, work groups, and the organization as a whole. Others prefer a completely external service with an "arm's length" relationship and very little input into the organization's business and operational practices.

4. Establishment of Service Criteria

Organizations must decide the service criteria they want from the OH&S service provider. What quality of service, level of expertise, nature of the service facility, data management, and reporting capabilities are expected? How will these be assessed and evaluated?

The organization should rate the value placed on each service criterion. For example, how does the organization prioritize the value of service quality responsiveness and cost, data management capabilities, comprehensive servicing, provision of integrated services, service provider solvency as a business entity, and the features of the service facility. This critical step influences how the bidders' response to the Request for Proposal for service provision will be evaluated.

5. Determination of the Desired Performance Criteria and Measurement Techniques

As with any service contract, the performance criteria and measurement techniques for the OH&S service provider contract must be set. These should be established in concert with the selected service provider. However, it is critical for the organization to decide on the performance criteria and to document the measurement techniques to be used.

Some typical performance measures are:
- service response time;
- service provision turnaround times;
- satisfied employees or business units;
- level of compliance with legislative requirements;
- frequency and quality of worker educational sessions;
- quality of hearing tests, vision tests, pre-placement screening, ergonomic assessments, and other employee assessments;
- rate of response to workplace incidents;
- rate of response to workplace injury incidents;
- quality of incident investigations;
- existence of non-compliance situations;
- short-term and Workers' Compensation disability durations;
- the percentage of short-term disability cases that progress to long-term disability;
- the percentage of disability cases that include modified work opportunities;
- the average duration of disability cases;
- the cost-benefit ratio per disability case;
- timely submission of invoices; and
- adherence to set service costs.

6. Establishment of Desired Funding Arrangement

Funding arrangements for the OH&S service can vary. The typical arrangements are fee-for-service, capitation, or a fee-for-service arrangement with a per capita retainer for administration services.

Companies that have set budgets and little margin for budgetary overruns, often choose a per capita option. It is also the funding option of choice for mature OH&S Programs with predictable service demands and outcomes.

The fee-for-service option can be suitable for organizations with new or changing OH&S Programs. The organization only pays for the services used. In this way, they can establish what yearly costs are incurred and then, if desired, they can establish an appropriate per capita pricing arrangement.

A hybrid mixture of fee-for-service and capitation can work well in situations where the corporate services and the care services for the organization are funded under a per capita scheme, and the business units (operational) services are paid on a fee-for-service arrangement. In this way, the combination of funding can meet an organization's various demands.

7. Establishment of a Desired Payment Arrangement

Once a desired funding arrangement is selected, the next step is to establish a payment arrangement. The options include a set fee per month, or a variable fee based on the services rendered. Again, depending on the organization's needs, available cash flows, and budgetary constraints, a suitable funding arrangement can be determined.

D. SERVICE PROVIDER MARKET SEARCH

An OH&S service provider search is a relatively new approach to determining a suitable organization-service provider match. The key is to find a suitable arrangement that meets all the organization's required criteria.

1. Steps

(a) Develop a Request for Proposal

The Request for Proposal (RFP) should describe the organization; its business and people needs; the OH&S Program; and the services sought. Typically, the approach is to provide background information on the organization; the corporate values and beliefs; and its products or services, size, and locations. A description of the organization's OH&S Program and commitment to workplace safety and employee well-being should be provided. Other available employee support services and how they link with the Employee Assistance Program also warrant explanation. The current service provision arrangements should be included, along with the reasons for seeking the current request for quotation for services.

The requested OH&S Plan design must be fully described. For example: What are the services sought? What is the expected standard of service required, and how will these be demonstrated?

RFP questions must be asked in a format that will elicit bidder responses that can be compared. Comparison of bidders is facilitated using a questionnaire. In this way, all bidders are asked the same questions and their responses can be rated according to predetermined value criteria.

Submission information must also be included in the RFP. It should describe to the bidders the scope of the project; the procedure for the bidder to acknowledge receipt of the RFP; the format for the proposal; the terms for proposal response submission; the terms for proposal response rejection; the length of time for bid acceptance; the potential method for proposal clarification; and the confidential manner in which bidder information will be handled. For ease of processing, the required response letters to be used by the bidders should be included, along with a sample service agreement contract. Bidders are asked to document their level of agreement with the contract format and terms. For more information on RFPs, refer to Chapter 37, "Occupational Health and Safety: Project Management".

(b) Select Suitable Service Providers

The organization has to decide whether to seek an invited or open bid for services. The invited bid to tender is a request for proposal from a selected number of potential service providers. The open bid is a general invitation for any service provider to respond to the RFP.

For many organizations, an open bid situation is just too cumbersome to conduct and manage. As well, the responding service providers may be unable to meet the organization's needs. More and more, organizations are choosing to pre-screen potential service providers, and then, conduct an in-depth examination of the individual service capabilities of each service provider.

(c) Distribute the RFP

The RFPs should be sent out in enough time to allow the service providers to adequately respond. RFPs can take over two weeks to prepare. As well, time should be allowed for mailing and delivery. A minimum of three weeks is recommended between issuing the RFP and the deadline date for bidder response.

It is important to clearly re-state in the accompanying cover letter the preferred method of submission and the closing time and date. As well, bidders should be reminded that any late responses will be considered invalid and returned to the sender unopened. Likewise, it is critical for organizations to abide by this statement.

RFPs should be sent out with a self-addressed, return envelope that ensures that the bid response gets to the appropriate department or person for processing. One tip is to have an identifier of some sort put on the envelope to indicate that a returned bid is enclosed. In this way, the recipient of the RFPs within the organization knows it is a returned bid and refrains from opening it until the RFP response time has elapsed. Then, all the bids can be opened at once.

(d) Data Collection and Collation

Each RFP response is reviewed in its entirety. Then, all the RFPs are dissected in terms of their responses to each of the questions asked. A standard approach is to use a spreadsheet to list the RFP questions and the individual bidder responses to each question. The spreadsheet helps compare the individual bid responses to each question.

(e) Analysis of the RFP Responses

The RFP response analysis process can take a number of formats. One is to score each question based on the value for each service criterion established by the organization. For example, if the organization predetermines that it values quality of service first, service cost second, data management capabilities third, comprehensive servicing fourth, and business solvency fifth; then scores of five points would be given to the questions that deal with service quality; four points for service cost questions; three points for data management capabilities; two points for comprehensive servicing; and one point for business solvency. In this way, each question can be objectively scored. The outcome is an overall score for each bidder. In this instance, high scores are good.

A second approach is to decide which of the RFP questions are "show-stoppers", and to rate each bidder response on those questions. The service providers with the best responses to all the critical questions are then identified as candidates for a final presentation.

With any scoring or rating system, it is important to predetermine whether the responses will be rated against each other, or whether an "all or nothing" scoring system for each question will be used. As well, plans should be made on how to deal with instances where all respondents score equally on a question. This allows for a consistent approach to the scoring technique.

Once the questionnaire responses are scored, service costs are addressed. Typically, bidders are asked to provide per capita and fee-for-service price schedules for the services requested. By applying the fee-for-service prices to the OH&S utilization rates for a former year, a comparison can be made between the fee-for-service and capitation models quoted for each bidder. Bidders are also compared respecting the services *excluded* from their respective capitation models.

Other areas for in-depth examination are:

- staffing levels;
- staff qualifications;
- service facilities;
- service provider network;
- service capabilities; and
- client references.

(f) Selection Process

Once all the responses are analyzed, the findings are reviewed by the organization's decision-makers. In addition to the items noted above, other factors like accessibility, fit with the corporate culture, service responsiveness, and service nature (reactive versus proactive, service provider versus partnership arrangement) are discussed. Based on the "fit", the decision-makers then select the service providers that they want to interview.

Finalist presentations are designed to permit the organization to meet the potential service providers and to learn more about their services, philosophies and plans to provide the OH&S services requested. Typically, these are formal presentations in which the service providers describe their business, service capabilities, staff qualifications, provider network, facilities, data management systems, past successes, and future plans. Presentations should include discussion of how the provider plans to deliver the requested OH&S services, the various funding options, and the nature of the potential working relationship.

To facilitate the interview, the organization should develop a list of questions for the service providers to answer. This allows for the further exploration of a potential business relationship, as well as adding rigour to the comparison process of the finalist presentation.

The responses to the prepared questions provided by the two or three finalists can be assessed and rated using the decision matrix tool which rates responses in order of merit. Similar rating systems can also be used. It is of prime importance to keep in mind the reasons for originally going to the marketplace, the predetermined value criteria, the desired service outcomes, and the need to find a suitable service provider.

The interview is usually the most revealing part of the selection process. On paper, many service providers sound great. However, in person, the match "fit" between the organization and service provider tends to become apparent.

(g) Response to Bidders

The successful and unsuccessful bidders expect a decision regarding the success/failure of their RFP responses. Service providers spend considerable time and resources developing RFP responses and out of professional courtesy, they deserve a response letter, regardless of the bid's success/failure.

E. SERVICE CONTRACT DEVELOPMENT

The development of the service contract begins at the onset of the project. By being aware of the value criteria, services required, desired funding arrangement, and duration of the contract, the procurement officer for the organization can begin to craft a service contract document.

Many organizations have standardized service contract templates. These can be used and modified to meet the needs of the proposed OH&S service contract. Regardless of the form used, the following elements should be included:

- description of services to be provided;
- expected level of service quality;
- required levels of reporting and communication;
- mutually agreed-upon service performance measures;
- responsibilities of each party;
- pricing agreement;
- duration of the contract;
- payment schedule;
- legal compliance;
- hold harmless clause; and

- required business insurance coverage.

F. VENDOR MANAGEMENT

Vendor management is a key aspect of a successful OH&S outsourcing arrangement. Having a comprehensive service agreement contract is only the first step in this process. Other strategies need to be in place to ensure that the outsourced service is in alignment with both organizations' needs and wants.

1. Recommended Strategies

(a) Partnering

Partnering is a method of accomplishing the mutual goals of the organization and the outsourced service in a planned and pre-described way. It involves jointly establishing service goals, objectives, and procedures. The intent is to have open communication and a solid working relationship, which foster the accomplishment of the desired goals.

(b) Quality Assurance and Continuous Improvement

Programs and OH&S services are established to address needs. As needs change, so must the programs and services. This can be successfully accomplished keeping in mind the two concepts of quality assurance and continuous improvement.

Quality assurance involves the determination that the desired quality of service is indeed being attained. There are a number of evaluation techniques that can be used. However, even before any of those can be employed, the contracting organization must establish the standard of OH&S servicing that will be expected of the service provider or vendor. These standards should be clear; valid in their rationale and intent; based on research; specific to the area of OH&S best practices; and measurable.

A number of quality assurance techniques are provided by Jane Hall,[4] and include:

- *Cat in the Corner* — visiting and observing first-hand the services provided by the service provider to employees. The intent is to evaluate the services offered.
- *The "I Don't Understand" Tool* — asking the service provider to explain a process or practice of concern.
- *Checking Out the Competition* — scanning the marketplace and comparing the various services and practices available.

Organizations should actively scrutinize the service provider's activities and make sure that the agreed-upon service demand and service quality are being met.

(c) Performance Measurement

The service agreement contract should stipulate the expected levels of performance

---

[4] *Ibid.*, at 44-47.

for the hiring organization and the service provider. Once in place, the onus falls on both parties to monitor the performance levels exhibited. This includes the regular measurement of the service quality and outcomes delivered.

The measurement criteria and the techniques spelled out in the service agreement contract should be followed. A review of the results should involve both parties. Here, the partnership arrangement becomes important. Both parties should be cognizant of the issues or problems and should work together to arrive at feasible solutions.

Performance measurement is an ongoing process — not an event. Regular measurement is key to the success of the outsourced arrangement. It allows for the identification of issues or problems; development of action plans and solutions; establishment of short- and long-term goals; trend analysis; and identification of proactive approaches to illness/injury prevention. Without regular performance measurement and open communication of performance issues, an outsourced service arrangement can fail. This is often the reason for tension between organizations and service providers, which is all too common, and can easily be prevented. In any service arrangement, the organization and service provider have to work together. Performance measurement is a useful tool to ensure this successfully happens.

(d) Service Provider Reporting

Regular reports on the services provided and the achieved outcomes are important elements to a successful contract arrangement. This feedback should be timely and should address all the requirements stated in the service agreement contract.

(e) Cost-containment

The costs associated with the outsourced OH&S services require monitoring and cost-containment. Both parties should regularly review the projected costs stated in the service agreement contract. Knowing where cost savings and cost overruns occur will help to deal with the service costs experienced or anticipated. In this way, contingency plans for the OH&S service costs can be developed, if required.

(f) Regular Meetings

Regular meetings between both parties should be held to discuss the success and limitations of the outsourced servicing. This is one way to promote open communication; to identify problems before they escalate; to determine which services are working well and point out those that require more attention; to identify any noted service trends; and to build a solid working relationship.

These six strategies are but a few examples of vendor management practices that can be implemented to ensure a successfully operated OH&S service. Partnering, open communication, regular monitoring, and measurement of performance and continuous service improvement are key aspects of a comprehensive service agreement contract.

G. VENDOR RISK MANAGEMENT

By Stacy Kirk[5]

In relation to disability management and other human service functions, there is a trend towards outsourcing functions for which the organization has limited resources or expertise. Organizations/companies recognize that a well-managed vendor arrangement can result in increased customer satisfaction, reduced costs, better quality, and better service as most organizations cannot adequately excel in every aspect of business.[6] Hewitt Associates in their *2010 HR Outsourcing: Canadian Trends and Insights Survey* noted increased interest in outsourcing benefit administration and other human resources processes.[7]

However, with the "good" comes the "bad"; there are several risks associated with outsourcing including:

- innocent data loss and fraudulent data theft;
- business knowledge loss in the element being outsourced;
- regulatory compliance concerns because vendors may need to be compliant not only where the outsourcing organization is located, but also in the location where the vendor is housed; and
- the failure of the vendor to deliver on contractual obligations.[8]

These universal risks of outsourcing are anticipated by numerous organizations; there are abundant articles addressing the vendor selection process, vendor information security risk management, and vendor management best practices.

Vendor risk management, an emerging discipline in operations management, is required when organizations/companies outsource business services and as a result, become reliant on third party vendors to achieve their business objectives.[9] **Vendor risk management** (VRM) is a comprehensive plan for identifying and decreasing potential business uncertainties and legal liabilities regarding the hiring of third party vendors for business products and services.

The introduction of the *Personal Information Protection and Electronic Docu-*

---

[5] Stacy Kirk, MA, is a human resources coordinator engaged in disability claims management, benefit administration, policy interpretation, and employment legislation compliance for a medium-sized contact centre.

[6] J. Bucki, *Vendor management: A checklist for success*, online: http://operationstech.about.com/od/vendormanagement/tp/Vendor-Management-Best-Practic.htm (date accessed: January 31, 2020).

[7] J.F. Potvin, "HR outsourcing is gaining ground", *Benefits Canada* (Toronto: Rogers Publishing Ltd., 2012).

[8] R. Shukla, "The benefits and risks of outsourcing" *Lexology* (April 6, 2010).

[9] Supplier Risk Management, Wikipedia definition, online: http://en.wikipedia.org/wiki/Supplier_Risk_Management (date accessed: January 31, 2020).

ments Act[10] (PIPEDA) in Canada, lends to employer/vendor accountability for data security — one piece of vendor risk management. The definition of accountability as interpreted by the Office of the Privacy Commissioner of Canada relies heavily on Principle 4.1.3 which declares that:

> [A]n organization is responsible for personal information in its possession or custody, including information that has been transferred to a third party for processing.[11]

PIPEDA requires organizations to undertake measures for an equivalent level of protection of personal information shared between the organization and its vendors. Several PIPEDA case summaries reflect this accountability including cases #394, #365, #333, and #313. To reiterate, Canadian organizations must be familiar with the legal requirements regarding the protection of employee personal information when using third party vendors. This is part of their vendor risk management.

Risks and accountability of the organization concerning outsourcing are features of doing business in a global economy. Specifically related to the area of disability management, organizations should not rely on decisions made by insurance companies to stop paying disability benefits as an indication that the employee is medically fit-to-return to work. The decision to cease benefit payments only means that the employee no longer meets the requirements of the insurance plan for insurance payments. In *A Guide for Managing the Return to Work*, the Canadian Human Rights Commission's stance is clear:

> . . . Do not rely on the decision of [the employee's] insurance company to stop paying benefits as an indication that he [she] can return to work safely.[12]

Disability management service providers are another example of third-party vendors with which organizations/companies deal. Organizations should be familiar with their responsibility in relation to verifying the information received from those third parties. However, responsibility is not always obvious when a third-party vendor is involved. Yet, the organization/company has the "greater obligation" to ensure that the actions of the vendor parallel the actions that the organization/company itself would have taken if it had not outsourced this aspect of the business. This onus becomes even more critical when there is no visible, or meaningful, separation between the organization and the vendor.

1. Case Law

In 2005, a nurse employed by Hamilton Health Sciences Corporation began a leave

---

[10] S.C. 2000, c. 5.

[11] Office of the Privacy Commissioner of Canada, *Interpretations* (April 17, 2012), online: http://www.priv.gc.ca/leg_c/interpretations_02_acc_e.asp (date accessed: January 31, 2020).

[12] Canadian Human Rights Commission, *A Guide for Managing the Return to Work* (2007) at 31, online: http://www.chrc-ccdp.ca/sites/default/files/gmrw_ggrt_en_2.pdf (date accessed: January 31, 2020).

of absence for medical reasons. The hospital in this matter delayed the employee's return to work after medical clearance had been provided. The medical clearance indicated some work restrictions; however, these restrictions would not have prevented the employee from returning to normal duties. A grievance was filed by the union on the employee's behalf. After a few months, a second grievance was filed because the employer still had not arranged for the employee to return to work.

The Hamilton Health Sciences Corporation outsourced their short-term disability management services to a third-party service provider, Cowan, for claim adjudication. Because this was not the first medical leave of absence taken by this employee, Cowan's stance was that the employee should be referred to the long-term disability insurance provider for claim adjudication. Furthermore, Cowan suspected the employee and attending physician of committing fraud in relation to the short-term disability claim. Hamilton Health Sciences Corporation accepted these judgments by Cowan and took no action of its own until late in the claim when it requested its own independent medical examination be done. The hospital failed to examine the details of the case even though the opinions of the third-party vendor appeared clouded by the assumption of fraud.

Arbitrator Harris found that the hospital had faulted on its obligation to accommodate their employee; instead, it relied on the beliefs and actions of its vendor even though the hospital was aware that Cowan had assumed a tainted opinion of the employee. Arbitrator Harris declared that the hospital had made the employee suffer needless financial and emotional harm to which he awarded the employee normal employment law damages as well as punitive damages.

In essence, the employer did not attempt to create a balance between management rights and employee's rights; it simply ignored its obligation to confirm the return-to-work documentation, and more importantly, ignored the employee's right to work accommodation. In his decision, Arbitrator Harris held that the:

> . . . hospital had an obligation to accommodate [the employee], but had failed to do so, relying instead on the opinions and the actions of its agent to make a decision which violated the employee's rights. It was this brazen disregard that resulted in compensatory and punitive damages being awarded to the employee.[13]

In the *Hamilton Health Sciences Corp. v. ONA (Schuster)* decision, the employer could not shed its liability by claiming that the third party vendor, Cowan, made the decision to delay the employee's return to work, and that the fault should be burdened exclusively by Cowan since there was no division in their actions: *"the Hospital simply adopted Cowan's directives"*.[14] PivotPoint Security states that regarding vendor risk management:

---

[13] E.B. Willis & W.K. Winkler, *Willis & Winkler on Leading Labour Cases* (Aurora, ON: The Cartwright Group Ltd., 2010).

[14] *Hamilton Health Sciences Corp. v. Ontario Nurses Assn. (Shuster Grievance)*, [2009] O.L.A.A. No. 493, 188 L.A.C. (4th) 327, 99 C.L.A.S. 235 at 373 (Ont. Lab. Arb.).

You can Outsource your Call Center, You can Outsource your Application Development Center. You can even Outsource your Entire IT Operation. But you CAN'T Outsource Responsibility or Liability.[15]

Vendor risk management demonstrates that organizations need to manage their vendors no matter what business aspect is outsourced — organizations are liable for their vendor's decisions and actions just as if the organization had made the decision or taken the action themselves.

2. Balance between Management Rights and Employee's Rights

With disability management, there is a need to maintain a balance between the employer's rights to operate a business and the employee's rights to equality, privacy, dignity, and physical integrity.[16] The employer must be allowed to verify return-to-work and work accommodation information; at the same time, the employee must be fairly treated regardless of the nature of the disability and accommodated up to the point of undue hardship.

Medical documentation that is unclear, or that does not provide a relevant assessment in relation to the employee's job, must be clarified by Management if they are expected to provide a safe and timely return to work for the employee, especially when a "safety-sensitive position" is part of the equation. In the case study presented in *A Guide for Managing the Return to Work*, the Canadian Human Rights Commission's standpoint is that even if a medical clearance is already obtained for a return to work:

> . . . [for] a safety-sensitive position, [the manager will] need to verify that the certificate was completed based on an understanding of what [the employee] does at work.[17]

The reasons are twofold:

1. Management needs to ensure sufficient information about the employee's capabilities and limitations, as they relate to the employee's specific job duties, has been collected in order to protect the health and safety of the employees including the returning employee; and

2. In order to show evidence of a demonstrable risk if a decision to refuse the return to work due to health or safety concerns, is to be made.

---

[15] PivotPoint Security, *Vendor (& Partner) Information Security Risk Management* (March 28, 2011) [PowerPoint slides] at 7-10, online: http://www.slideshare.net/PivotPointSecurity/third-party-vendor-risk-management (date accessed: January 31, 2020).

[16] M-C. Chartier, *Human rights and the return to work: The state of the issue* (April 28, 2006) at 1, online: http://www.chrc-ccdp.gc.ca/sites/default/files/returntowork_en_3.pdf (date accessed: January 31, 2020).

[17] Canadian Human Rights Commission, *A Guide for Managing the Return to Work* (2007) at 30, online: http://www.chrc-ccdp.ca/sites/default/files/gmrw_ggrt_en_2.pdf (date accessed: January 31, 2020).

Conversely, employees have the right to fair treatment in the workplace, regardless of the disability. The principle of dignity with risk, meaning that persons with disabilities have the ability to assume risks to their own well-being, has to be balanced with health and safety considerations for all.[18]

Canadian employees have the right to work accommodation up to the point of undue hardship. The simplified definition of duty was presented by the Ontario Human Rights Commission:

> . . . the duty to accommodate means the employer must implement whatever measures necessary to allow its employees to work to the best of their ability.[19]

Undue hardship restricts the employer's responsibility if it can show that the work accommodation would cause the employer to suffer excessive adversity. Two common grounds on which employers may attempt to claim undue hardship relate to cost and safety. The cost of a proposed accommodation would be considered unjustifiable if the costs related to the accommodation directly affect the overall survival of the organization. Cost is a "hard sell" to the Human Rights Commission because "some financial hardship is to be expected in accommodation and limited financial cost will not, on its own, result in a finding of undue hardship."[20] However, should the work accommodation create unnecessary risk to the health and safety of the employee, or to others (co-workers or customers or the general public), then an employer might be able to establish grounds for undue hardship.

In most employment circumstances, in order to establish a claim of undue hardship, the employer must demonstrate: "that all reasonable means of accommodation have been exhausted and that only unreasonable or impractical options for accommodation remain".[21] Financial cost and safety concerns are considered in determining whether undue hardship may transpire, but other factors such as "disruption of a collective agreement, negative impact on worker morale, interchangeability of work force and facilities, [and] the size of the operation" will all be examined prior to a finding of undue hardship.[22]

3. Considerations for Disability Management

As a result of the *Hamilton Health Sciences Corp. v. Ontario Nurses Assn. (Schuster*

---

[18] Ontario Human Rights Commission, *Policy and guidelines on disability and the duty to accommodate* (November 23, 2000), online: http://www.ohrc.on.ca/sites/default/files/attachments/Policy_and_guidelines_on_disability_and_the_duty_to_accommodate.pdf (date accessed: January 31, 2020).

[19] Ontario Human Rights Commission, Fact Sheet, *Duty to Accommodate* (October 6, 2010), online: http://www.chrc-ccdp.ca (date accessed: January 31, 2020).

[20] V. Dixon & N. Tymochenko, "Financial Hardship Did Not Justify Denial of Accommodation for Disability", *Labour and Employment Communiqué* (Miller Thomson LLP, 2012).

[21] *Ibid.*

[22] *Ibid.*

Grievance) decision, all organizations, and in particular the staff responsible for, or involved with workplace disability management, should pay attention to the lessons learned about return-to-work programs and disability management policies.

Use of return-to-work "best practices" could have resulted in a favourable decision for the plaintiff: for example, review any medical information submitted by the employee, verify employee fitness to return to work post-sick leave including identification of suitable work accommodation, and clarify specific details of how the condition may have an impact on the employee's job duties. If third party services are used for claim and case adjudication, then the employer needs to verify the medical clearance and understand any accommodation needs, thereby avoiding unquestioned dependence on a third-party agent.

When an organization is trying to create an equitable working environment, workplace accommodation must be a vital piece of the organization's disability management policies and other related policies. Policies are in place so that organizations can reinforce and clarify acceptable and unacceptable behaviours as well as assist managers to oversee the workforce more effectively. In summary, policies that reinforce work accommodation or the creation and maintenance of "a barrier-free work environment", can encourage consistency in decision-making and operational procedures of difficult and sensitive subject matter.[23]

In the case of *Hamilton Health Sciences Corp. v. Ontario Nurses Assn. (Schuster Grievance)*, a well-documented accommodation policy promoting the proactive approach could have avoided the interactions which led to the arbitration case and its loss; or even a well-documented work accommodation policy could have directed the Hospital to consider alternatives to its decision to follow Cowan's directives, thereby avoiding the arbitration case or at least having a substantial defence for its actions in the case.

4. Conclusion

In the *Hamilton Health Sciences Corp. v. Ontario Nurses Assn. (Schuster Grievance)* case, the decision made by Arbitrator Harris to award the employee compensatory damages and punitive damages because the employer should have known of their legal responsibilities, seems appropriate. The employer neglected its responsibility to provide a timely return-to-work plan for the employee, as well as its duty to accommodate the employee to the point of undue hardship.

Organizations/companies can learn from this decision, namely, they must:

- avoid dependence on third party vendors in relation to the disability management of their employees;
- stay connected with the disability management function; and

---

[23] Ministry of Labour, *Sample Accommodation Policy* (August 19, 2009) at 23, online: http://www.hrsdc.gc.ca/eng/labour/equality/fcp/employer_tool/docs/appendix3E.pdf (date accessed: January 31, 2020).

- manage disability management service providers in the same way that relationships with other business vendors would be managed.

The reason is simple: although many functions of business can be outsourced, liability cannot be outsourced.

H. SUMMARY

OH&S Programs and services can be outsourced in whole, or in part. Regardless of the arrangement, the sponsoring organization must remain actively involved in managing the service agreement contract. Organizations can contract out services, but not their liability and accountability for the provision of those services. This means that they must think strategically and act responsibly:

> An organization's decision to outsource or retain a particular service should be a well-thought-out process encompassing all the ramifications of the decision.[24]

CHAPTER REFERENCES

J. Bucki, *Vendor management: A checklist for success*, online: http://operationstech.about.com/od/vendormanagement/tp/Vendor-Management-Best-Practic.htm (date accessed: January 31, 2020).

Canadian Human Rights Commission, *A Guide for Managing the Return to Work* (2007), online: http://www.chrc-ccdp.ca (date accessed: January 31, 2020).

M-C. Chartier, *Human rights and the return to work: The state of the issue* (April 28, 2006) at 1, online: http://www.chrc-ccdp.gc.ca/sites/default/files/returntowork_en_3.pdf (date accessed: January 31, 2020).

V. Dixon & N. Tymochenko, "Financial Hardship Did Not Justify Denial of Accommodation for Disability", *Labour and Employment Communiqué* (Miller Thomson LLP, 2012).

D. Dyck, *Disability Management: Theory, Strategy and Industry Practice*, 6th ed. (Markham, ON: LexisNexis Canada Inc., 2017).

J. Hall, "The Decision to Vendor and Vendor Management: A Tool for Occupational Health and Workers' Compensation Management" (1997) 11:5 *The OEM Report*.

Hamilton Health Sciences Corp. v. Ontario Nurses Assn. (Shuster Grievance), [2009] O.L.A.A. No. 493, 188 L.A.C. (4th) 327, 99 C.L.A.S. 235 at 373 (Ont. Lab. Arb.).

Ministry of Labour, *Sample Accommodation Policy* (August 19, 2009) at 23, online: http://www.hrsdc.gc.ca/eng/labour/equality/fcp/employer_tool/docs/appendix3E.pdf (date accessed: January 31, 2020).

Office of the Privacy Commissioner of Canada, *Interpretations* (April 17, 2012), online: http://www.priv.gc.ca/leg_c/interpretations_02_acc_e.asp (date accessed: January 31, 2020).

---

[24] *Ibid.*

Ontario Human Rights Commission, Fact Sheet, *Duty to Accommodate* (October 6, 2010), online: http://www.chrc-ccdp.ca/preventing_discrimination/duty_obligation-eng.aspx (date accessed: January 31, 2020).

Ontario Human Rights Commission, *Policy and guidelines on disability and the duty to accommodate* (November 23, 2000), online: http://www.ohrc.on.ca/sites/default/files/attachments/Policy_and_guidelines_on_disability_and_the_duty_to_accommodate.pdf (date accessed: January 31, 2020).

PivotPoint Security, *Vendor (& Partner) Information Security Risk Management* (March 28, 2011) [PowerPoint slides] at 7-10, online: http://www.slideshare.net/PivotPointSecurity/third-party-vendor-risk-management (date accessed: January 31, 2020).

J.F. Potvin, "HR outsourcing is gaining ground", *Benefits Canada* (Toronto: Rogers Publishing Ltd, 2012).

R. Shukla, "The benefits and risks of outsourcing" *Lexology* (April 6, 2010).

Wikipedia, *s.v.* "Supplier Risk Management", online: http://en.wikipedia.org/wiki/Supplier_Risk_Management (date accessed: January 31, 2020).

E.B. Willis, and W.K. Winkler, *Willis & Winkler on Leading Labour Cases* (Aurora, ON: The Cartwright Group Ltd., 2010).

Chapter 41

OCCUPATIONAL HEALTH AND SAFETY PRACTITIONERS/PROFESSIONALS: CAREER DEVELOPMENT

A. INTRODUCTION

Few Occupational Health and Safety (OH&S) Practitioners/Professionals ever conscientiously plan out their career paths. Rather, they tend to end up in the field of Occupational Health and Safety by "happenstance". This chapter provides career planning and development information for OH&S Practitioners/Professionals, along with an appreciation of the importance of position/job descriptions.

B. CAREER STREAMING: MODEL OF CAREER DEVELOPMENT FOR OH&S PRACTITIONERS/PROFESSIONALS[1]

OH&S Practitioners/Professionals will change the type of work they do, the organizations in which they work, and the cities in which they live several times during their work lives. For this reason, they need to examine not only the first profession and job they choose, but also the series of occupations and jobs they will hold over a 40-year period.

The sequence of jobs and occupations is called a career. Hall (1976) describes a career as:

> . . . an individually perceived sequence of attitudes and behaviours associated with work-related experiences and activities over the span of the person's life.[2]

This sequence of jobs and work pursuits represents what a person does for a living. The basic assumptions underlying this definition are:

- A **career** is seen as a lifelong series of events, rather than an evaluation of how successful individuals have been in their life.
- Career success is subjective and based on personal judgement rather than peer opinions.

---

[1] Excerpts taken from: D. Dyck & M. Walker, "Career Streaming: A Model of Career Development" (1996) 44:4 *AAOHN* 177. Copyright (1996) the American Association of Occupational Health Nurses, Inc. Used with permission. All rights reserved.

[2] D.T. Hall, *Careers in Organizations* (Glenview, IL: Scott, Foreman & Company, 1976) at 96.

- A career is a series of work events made up of the things the individual feels and does over time.
- A career is best viewed as a process of work-related experiences.

A career goes through various stages: "different points of work responsibility and achievement through which people pass during the course of their work lives".[3] Schermerhorn (1991) has labelled the stages as **career entry**, **career advancement**, **career maintenance**, and **career withdrawal**, and links these stages to employee performance and age (Figure 41.1).

Figure 41.1: Career Stages, Individual Performance and Age

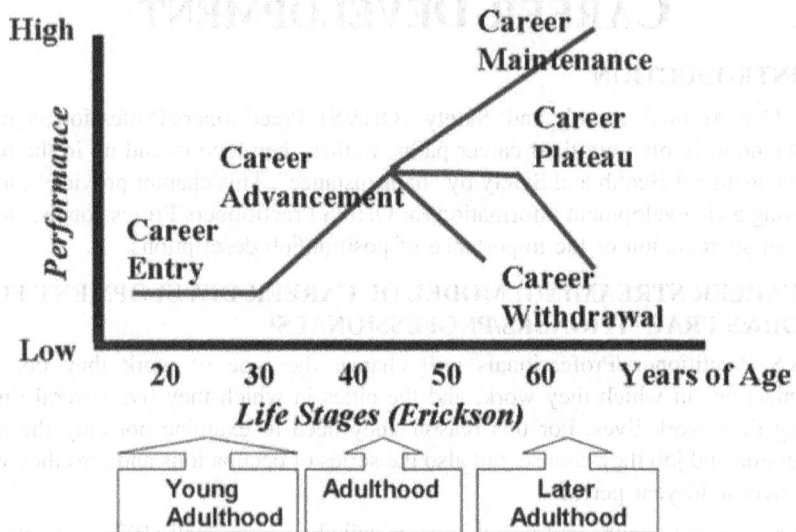

In the **career entry** stage or exploration (Hall, 1976) stage, the individual tries to match individual need, abilities and skills with organizational requirements. It is that early period when the person tries to answer the question: "Is this job for me?"

The **career advancement** stage is characterized by a steep learning curve. The person begins to better understand the work environment and organizational demands and strives to establish personal worth within the organization. It is a period of growth and acceptance of responsibility.

In the **career maintenance** stage, the individual becomes entrenched in the job. Usually, the person has strong identity ties with the organization and job, is financially bound to the position, and is unlikely to leave. It is during this stage that Hall (1976) pointed out a tendency to adopt one of three avenues:

---

[3] J. Schermerhorn, J. Hunt & R. Osborn, *Managing Organizational Behavior*, 4th ed. (New York, NY: John Wiley & Sons, 1991) at 210.

1. continued growth within the organization;
2. career plateauing; or
3. career stagnation.

During this stage, the greatest managerial challenge is to keep the employee stimulated and productive. Failure to do so ends in wasted employee time and energy.

The final stage is **career withdrawal**. Traditionally, this was associated with retirement. Now, retirement tends to occur earlier and at a point when some individuals are unready. As a result, they continue to advance their careers informally through self-employment, consulting, or re-employment with another organization.

1. Career Plateauing

Every career ultimately reaches some sort of plateau — a position from which one is unlikely to advance to a higher level of responsibility in their career.[4] A lack of career growth can occur because of personal choice, limited ability, or lack of opportunity. Today, due to both economic and demographic factors, many persons are reaching a plateau at a relatively early stage in their careers.

When an individual reaches a career plateau, there is always the danger that they will lack motivation and become apathetic towards work. They may perform poorly, become bitter, blame the organization for their situation, and export their bitterness to fellow employees. As a result, general employee morale and performance may suffer. Effectively managing the "plateaued" employee can be an important contribution to the organization. Employees can be salvaged if the organization gives the message to the individual that he/she is a valued employee and that what he/she has to offer is important.[5]

2. Career Anchors

Since World War II, people have come to expect more from their jobs than a paycheque. Among the expectations are a challenging and rewarding career, the ability to grow on the job, an opportunity to satisfy personal needs, advancement, and the chance to influence business decisions and plans. Schein (1975) termed the aspects of work that people value or need for fulfilment as **career anchors**. They include:

- *managerial competence* — the driver is the opportunity to manage people;
- *technical/functional competence* — the driver is the opportunity to use various technical abilities and special competencies;

---

[4] *Ibid.* See also R. Steers, *Introduction to Organizational Behavior*, 2d ed. (Glenview, IL: Scott, Foreman and Company, 1984).

[5] B. Moses, *Manager's Guide to Career Development* (Toronto, ON: BBM Human Resource Consultants, 1989).

- *security* — the motivation is a need for job security or stability;
- *creativity* — the motivation is a need to create or build; and
- *autonomy and independence* — the driver is the opportunity to work independently, without organizational constraints.[6]

Career planning and development activities allow employees to grow in any of these desired directions.

3. Career Planning and Development

With the entry of the "baby-boomer" generation into management positions, increased occurrence of corporate downsizings and decreased opportunities for promotion, fewer and fewer career spots are available.[7] To prevent employees from moving on, managers and Human Resource (HR) Practitioners/Professionals are challenged to redefine the philosophy, mechanics, and management of career growth.[8]

The act of putting the "humanism" back into the field of Human Resources is vitally important. By directing employees in a way that helps them realize their personal career goals, satisfy their personal needs, and be valuable assets to the corporation, managers and Human Resources Practitioners/Professionals can facilitate employee retention and reduce one source of employee dissatisfaction.[9]

Career development is the process through which individual employees, with the help of their managers, match up their skills, interests and goals with the opportunities available to them, and develop plans for achieving their goals.[10] This process should be employee-initiated and management-supported.

Schermerhorn (1991) defined **career planning** as an organization's development intervention that creates opportunities for employees to achieve long-term congruence between their individual goals and organizational career opportunities. The steps involved in career planning are:

- assessing personal strengths and weaknesses;
- analyzing opportunities;
- selecting career objectives;
- selecting and implementing the plan;

---

[6] T. Stone & N. Meltz, *Human Resource Management in Canada*, 2d ed. (Toronto, ON: Holt, Rinehart and Winston of Canada, 1992).

[7] *Ibid.*

[8] R. Goddard, "Lateral Moves Enhance Careers", 35:12 *HR Magazine* at 69–74.

[9] *Ibid.*

[10] B. Moses, *Manager's Guide to Career Development* (Toronto, ON: BBM Human Resource Consultants, 1989).

- evaluating the results and revising the plan as necessary;[11]
- increased ability to link personal career ambitions with organizational opportunities; and
- a mechanism for dealing with concerns about career plateauing.

Employees and managers benefit from career development. For employees, the process provides:

- emphasis on attaining transferable work skills versus focusing on job titles;
- assistance in making informed career decisions and developing realistic career goals;
- open communications with managers about career plans;
- a mechanism for getting feedback and improving work areas of performance; and/or
- a means of communicating training and improving weak areas of performance.

The managerial benefits include:

- improved communications with employees about their career plans;
- stimulation of employee enthusiasm and job motivation;
- a means for addressing performance problems; and
- a technique for the discovery and use of employee skills, interests and strengths.[12]

Through career development, an organization benefits from enhanced employee productivity; increased information about employee skills, experience, and interests; more productive use of Human Resources services; more realistic employee career goals; an image to employees that the organization/company cares about them; and better employee adaptation to organizational change.[13]

4. Career Planning Concepts

Three approaches to career planning are discussed. The last one, Career Streaming, is used to demonstrate a potential model for OH&S Practitioners/Professionals.

(a) Career Ladder

This is the traditional way of advancing within an organization. The employee enters the organization/company, achieves the necessary skills, and automatically progresses up the corporate ladder. It exemplifies the career path as outlined by Hall

---

[11] J. Schermerhorn, J. Hunt & R. Osborn, *Managing Organizational Behavior*, 4th ed. (New York, NY: John Wiley & Sons, 1991).

[12] *Ibid.*

[13] *Ibid.*

(1977), Steers (1984), and Schermerhorn (1991) in their Career Stages Model. This technique is popular with unions and tradespeople. Hospitals find it advantageous because the process ensures selective socialization of the workforce; that is, it maximizes the probability that supervisors and managers will be individuals whose personality, skills, and knowledge align with the institutional character.[14] Promotion from within also increases employee motivation at lower levels because they see that hard work can lead to job advancement. It is a predictable mode of advancement, but it can be quite rigid in its patterning and scope.

The major problem with this technique is that individuals can be promoted beyond their capabilities. Known as "the Peter Principle",[15] it often proves to be dysfunctional for the organization/company and the workgroup and disastrous for the individual. Jacques (1961) theorized that individuals have an accurate, unconscious awareness of the level of work they are capable of doing, the amount and quantity of work produced, and what should be equitable payment for that work. When individuals are promoted beyond their capabilities, their psychological equilibrium becomes jeopardized. The employee feels guilty and tends to behave defensively with co-workers. Ultimately, the work-pay inequity leads to feelings of insecurity and dissatisfaction.

In addition, career ladders tend to be narrow, providing few opportunities for advancement. This can lead to staff turnover by the more ambitious and assertive employees. The employees who remain, tend to feel thwarted and frustrated at being unable to attain their career goals. The usual outcome is reduced organizational effectiveness.

(b) Career Modelling

Career modelling is the development of a framework for attaining various positions in the organization. It is the assessment of the employee's current position and skills needed for that position as compared to other positions within the organization and their respective skill requirements.

This type of career planning is done at an organizational level and is a management tool designed for individual career counselling. It is used to clarify the rationale for employee development, to spell out expectations involving career growth, to assess training needs, and to establish the expectations for skill development. It communicates the opportunities for career growth within the organization and management's commitment to develop both broadly skilled managers and technical specialists.

Workforce planning and succession planning can be achieved. Career modelling can ensure that candidates with the right skills are available to meet future

---

[14] D.A. Gilles, *Nursing Management: A Systems Approach* (Toronto, ON: W.B. Saunders Company, 1989).

[15] L.J. Peter & R. Hull, *The Peter Principle* (New York, NY: W. Morrow, 1969).

organizational requirements. Also, clarifying job and skill requirements makes recruiting and employee selection easier.

The major problem with career modelling is that it still depends on upward movement to satisfy individual career aspirations. Unfortunately, with flatter organizational structures and fewer opportunities for upward advancement, new ways to keep employees interested in their jobs need to be discovered.[16]

Organizations/companies are beginning to address this career development issue in various ways. Some are using parallel career tracks, job rotation, job enrichment, or lateral and downward moves. An alternate approach is Career Streaming.

(c) Career Streaming

Organizations can provide career development for employees within a defined professional field. The challenge is to provide a range of levels of defined skills for each level. As employees progress through the "stream", they attain certain skills, and through experience and new growth opportunities, add to their "skill bank" and scope of operation within the organization.

By way of an industry example, a large Canadian petro-chemical company developed one such model for its Occupational Health Services staff. In this specific application, the entry level Occupational Health Nurse Advisor I, learns crisis counselling skills to be used with individual employees. The Occupational Health Nurse Advisor II, having acquired individual counselling skills, is then exposed to group crisis counselling (such as critical incident stress debriefing). The Occupational Health Leadership professional, on the other hand, deals with crisis planning at a corporate level as well as debriefing in a multi-workgroup situation. In summary, as one progresses through the "stream", one's skill set increases along with the scope of responsibility and accountability.

5. Career Streaming: Industry Application

This Career Streaming Model was designed to provide an opportunity for professional and career growth within the specialized area of Occupational Health and Safety, benefitting the OH&S Practitioner/Professional and the organization.

The objectives of the model are to:

- Provide a career path within the specialized area of Occupational Health and Safety services.
- Facilitate employee development by identifying key technical and core management/business skills required in each stage of the career stream.
- Provide definitions for the key technical, business and relationship skills.
- Provide direction for OH&S Practitioners/Professionals and their supervisors to plan skill development.

---

[16] R. Goddard, "Lateral Moves Enhance Careers" 35:12 *HR Magazine* at 69–74.

- Assure senior management that Occupational Health and Safety Services professional development is receiving due attention.

(a) Career Streaming Principles

Career Streaming opportunities are governed by experience, academic qualifications, demonstrated level of performance, skill development, and employee aspirations, along with the business needs and opportunities within the organization/company. For example, OH&S Practitioners/Professionals can optimize their contribution to the organization/company if they have a balance of specialist, relationship, and business skills (Table 41.1).

Table 41.1: Occupational Health Nurse: Technical Specialist, Relationship, and Business Skills[17]

Skill Set	Description	
Technical Specialist Skills	• Fitness-to-work assessments • Health surveillance • Medical monitoring • Emergency care • Hazard/risk assessment • Worksite evaluation • Emergency planning and response • Event investigation • Case management • Critical Incident Stress Debriefing (CISD) • Program development • Standard setting • Professional networking	• Training program development and delivery • Regulatory knowledge • Human factors analysis • Technical communication • Quality assurance • Environment/public health management • Social marketing • Strategic issues management • Health promotion • Risk management and communication • Program evaluation • Research

---

[17] Descriptions of these skill sets are provided in D. Dyck, *Disability Management: Theory, Strategy & Industry Practice*, 6th ed. Markham, ON: LexisNexis Canada Inc., 2017), c. 4 "Disability Management Program: Stakeholder Roles", App. 1 "Required Skill Sets for the Disability Management Coordinator".

Core Management and Business Skills	• Planning and organization • Decision-making • Problem-solving • Leadership • Financial/Business perspective • Negotiating • Information systems management • Process facilitation • Project management	• Computer skills • DM Governance • DMP Stewardship • Performance management • Employee development • Training • Change management • Presentation • Risk communication • Consulting • Strategic issues management
Relationship/People Skills	• Interpersonal skills • Communication skills • Team building/Teamwork • Coaching • Relationship building • Change management • Preceptorship	• Mentoring • Negotiating • Reputation management • Community relationship management • Mediating • Facilitation

Table 41.2: Occupational Safety Practitioner: Technical Specialist, Relationship, and Business Skills

Skill Set	Description	
Technical Specialist Skills	• Loss control • Regulatory knowledge • Hazard identification and management • Educational program development and delivery • Safety communication • Function evaluation • Risk management	• Program development, administration and management • Program evaluation • Networking • Regulatory compliance • Technical communication • Quality assurance • Social marketing • Strategic issues management

Core Management and Business Skills	- Planning and organization - Decision-making - Problem-solving - Leadership - Financial/Business perspective - Negotiating - Information systems management - Data collection and management - Process facilitation - Project management - OH&S governance	- Performance management - Employee development - Training - Change management - Presentation skills - Written communication skills - Risk management - Consulting - Strategic issues management - Computer skills - Process facilitation - DMP stewardship
Relationship/ People Skills	- Interpersonal skills - Communication skills - Team building/Teamwork - Coaching skills - Preceptorship - Change management	- Relationship management - Mentoring - Negotiating - Reputation management - Mediation

Continuing with the industry application, the technical/specialist skills are unique skills, or combinations of skills, required in each stage. For example, the Occupational Health Nurse Advisor I, is expected to enter the organization/company with two to three years supervisory and/or management experience. The core management/business skills specific to the organization/company are learned through progressive training and work experience. Occupational Health Nurses bring highly developed communication, interviewing, and teaching skills for working with individual clients to the job. New skills are group/team relationship skills.

Not all employees need to progress within their career stream to achieve job satisfaction, and not all have the same career aspirations. This is normal and expected. Also, one must recognize that all employees will eventually reach their ultimate career level as determined by their skills and abilities and the overall needs of the organization/company.

In the above example, the Occupational Health Advisors demonstrate specific skills before they move to a higher level. Determining the achievement of the practice skills remains the responsibility of supervisors and managers. Progression is not automatic: if skills are developed but are not required by the organization, there is no recognition in job compensation. However, job satisfaction likely will increase because of the contribution to the business.

(b) Development Stages

The Career Stream Model has four developmental stages. Descriptions of each developmental stage with a matrix chart to identify the applied skills at each stage are provided for Occupational Health Nurses (OHNs) in Appendix 1, and in Appendix 2 for Occupational Safety Practitioners/Professionals. Skills acquired during an earlier stage are built upon in subsequent stage experiences. The scope of the skill changes is indicated by the illustration of the OHNs Career Streaming Model as an example:

1. **OHN Advisor I** — This is the entry level for OHNs. New entrants must have an appropriate combination of education, Occupational Health experience (five to seven years) and basic supervisory skills. The primary purpose of this function is to manage the daily operations of a specific health centre. Development of new specialist skills, cross-functional relationships and understanding of the organization's/company's business endeavours is expected to require two (2) to three (3) years in this position to occur.

2. **OHN Advisor II** — At this stage of development, the role of the OHN is to influence the business of a particular operation by regularly bringing Occupational Health perspectives to management, union, and employee business. Strategic Occupational Health plans are tailored to meet operational requirements, introduced to all parties and monitored with the effects analyzed and reported. Local issues are worked cooperatively with other functions and management groups through task forces, special projects and regular contributions to Joint Health and Safety Committees. The daily operational needs of Occupational Health Services are maintained but balanced with operational influence. This influence, which focuses primarily on health, human factors, and loss control, is not always popular with management, unions, or employee groups. However, if effective, this unique perspective influences the way business is carried out in an operational setting.

3. **Occupational Health Specialist** — The role of the OHN at this stage, is to influence more than one major business unit by regularly bringing the health, human factors, and loss control perspectives to this level of decision-making and key business operations. This position provides few day-to-day, hands-on, technical Occupational Health services. Rather, the responsibilities extend to the Environmental Health perspectives by assisting the organization to address public inquiries and concerns related to health. The position takes the strategic direction plan and applies it differently in business units to achieve the same goals. The OHN is expected to identify strategic issues, to make recommendations, and to participate in developing a corporate approach. Through the use of process management, consulting, and facilitation, together with team skills, actions are taken to achieve the desired outcomes.

4. **Senior Occupational Health Specialist** — The primary purpose of this fourth stage is to:

- develop and ensure strategic direction for the Occupational Health Service for the organization;
- develop standards, policies, and audits;
- manage issues at the senior management levels, as well as with national unions, government, and community groups; and
- provide general direction and monitor other positions on health and human factor issues.

This leadership position plans cooperatively with other business groups to best serve the strategic and organizational business plan, paying particular attention to health, environment, human factors, loss control, "due diligence", and functional issues.

6. Conclusion

The Career Streaming Model has been successfully used in a number of workplace settings and with various OH&S Practitioner/Professional groups. Industry examples are provided in Appendix 1: Occupational Health Nurse Career Stream and Appendix 2: Occupational Safety Practitioner/Professional Career Stream.

C. OH&S JOB DESCRIPTIONS

Many OH&S Practitioners/Professionals function in their jobs without ever having a position/job description to explain and guide their practice. Position descriptions are important because they clarify what management expects the employee to be responsible and accountable for within a specific position. They facilitate wage and salary administration; provide a basis for manpower planning; assist with recruitment, selection, placement, orientation, and evaluation of employees; and enable job evaluation.[18] They also enable an organization to implement its purposes and ensure the efficient use of human capital. Accurate job descriptions make job performance appraisals easier by reducing the influence of subjective factors (*e.g.*, personalities, personal opinions, nature of the employee-supervisor relationship), and by focusing on objective performance measures as stated in the job description and the planned growth and development objectives.

In terms of malpractice issues, corporate insurance policies tend to cover the OH&S Practitioner/Professional provided they operate within the scope of their position. Without a job description, it would be difficult to legally establish what is the actual scope of the position in question.

A **job description** is a written record of the principal duties and scope of responsibility for a particular job/position. Usually written in a standardized format,

---

[18] D.A. Gilles, *Nursing Management: A Systems Approach* (Toronto, ON: W.B. Saunders Company, 1989) at 177–81.

it includes the required employee characteristics such as academic qualifications, work experience, skills, and aptitudes. It delineates the appropriate tasks to be performed.

Typically, position descriptions address:
- whether the position level is a management or unionized position;
- if the position is permanent, limited term, part-time, or seasonal;
- the position/job title;
- the identity of the immediate supervisor;
- the business unit in which the position resides;
- a position/job identifier (usually a code);
- the date the position was approved and by whom;
- a summary of the purpose of the position;
- the specific functions and responsibilities;
- the educational requirements;
- qualifications;
- the experience, attributes, and/or knowledge necessary;
- the span of authority;
- performance standards;
- the positions to be supervised;
- the reporting structure of the business group; and
- signatures of the position holder and immediate supervisor.

When preparing a position/job position, a good "rule of thumb" is to write it in enough detail, and with so much clarity, that an outsider, unfamiliar with the organization/company and the job, can understand it. Use a simple, direct narrative writing style and avoid the use of unfamiliar jargon and complex sentence structures. Begin each statement with an active verb that graphically describes the expected employee behaviour. For example, the position holder "conducts workplace walk through inspections on a regular basis"; "collects and analyzes incident data monthly"; and "distributes incident statistical reports to line management on a monthly basis". Samples of OH&S position descriptions are provided in Appendix 3 and Appendix 4.

D. SUMMARY

A career is an individually perceived sequence of attitudes and behaviours associated with work-related experiences and activities over the span of the individual's life. For OH&S Practitioners/Professionals, the career path can be varied and quite convoluted. For new entrants to the field of Occupational Health and Safety, hopefully the contents of this chapter will assist them to plan their careers. For seasoned OH&S Practitioners/Professionals, this chapter is designed to assist with the management of not only their own careers, but also those of the people they supervise or mentor.

Appendix 1

OCCUPATIONAL HEALTH NURSE CAREER STREAM

Position	OHN ADVISOR I	OHN ADVISOR II	OHN SPECIALIST	SENIOR OHN SPECIALIST
Qualifications/ Experience	OHN education; entry level; basic computer training; OH&S-related experience	OHN education, case management; ergonomics education; intermediate computer training; 2-3 years OH&S-related experience	OHN education & certification; case management experience, ergonomics education, advanced computer training; 5-7 years OH&S-related experience	Extensive OHN experience & certification; case management expertise; auditor training, ergonomics expertise; computer expertise; business certificate; 8-10 years OH&S-related experience
Position Scope	Narrow ⟶			Broad
Position Depth	Shallow ⟶			Deep
Application of Skill Types with Advancement	Hi / Lo — Technical/Specialist Skills		Business Skills / Relationship Skills	
Technical/ Specialist Skills	• Fitness to Work Evaluation • Health Surveillance • Medical Monitoring • Crisis Counselling • Emergency Care • Emergency Response • Worksite Evaluation • Loss Control (operational) • Regulatory Compliance • Health Promotion • Risk Management (OH) • Professional Networking	• CISD Debriefing (groups) • Human Factors Analysis • Incident Investigation • Audit • Case Management • Quality Assurance • Research • Technical Communication (individual) • Social Marketing (individual) • Program Evaluation (individual/projects)	• Standard Setting • Environmental Health Assessments • Environmental/Public Health Risk Management (operational) • Technical Communication (operational) • Social Marketing (operational) • Program Evaluation (operational)	• CISD (corporate) • Environmental/Public Health Risk Management (corporate) • Technical Communication (corporate) • Social Marketing (corporate) • Program Evaluation (national focus) • Audit Leadership
Business Skills	• OH Communication (OH advice to individual employees/workgroup; effective oral and written	• OH Communication (OH advice to one business line; technical resource to Joint Health and Safety	• OH Communication (OH advice to multiple business lines; perfect written and oral skills)	• OH Communication (OH advice company-wide; public relations; expert written and oral skills)

Position	OHN ADVISOR I	OHN ADVISOR II	OHN SPECIALIST	SENIOR OHN SPECIALIST
Business Skills	<ul><li>communication skills)</li><li>Problem-solving (basic)</li><li>Decision-making (basic)</li><li>Data Collection and Management (document and retain OH records)</li><li>Presentations (OH education sessions)</li><li>Computer Skills (basic)</li><li>Consulting (OH resource)</li></ul>	<ul><li>Committees; promote EAP; good written and oral skills</li><li>Problem-solving (intermediate)</li><li>Decision-making (intermediate)</li><li>Data Collection and Management (maintain an OH database)</li><li>Presentations (business line)</li><li>Computer Skills (intermediate)</li><li>Project Management (basic)</li></ul><ul><li>OH Performance Management (OH practices within workgroups)</li><li>OH Functional Management</li><li>Issue Management (workgroup)</li><li>Social Marketing (employee/workgroup)</li><li>Consulting (expert)</li></ul>	<ul><li>Problem-solving (strong)</li><li>Decision-making (strong)</li><li>Data Collection and Management (establish OH databases)</li><li>Presentations (multiple business lines/external)</li><li>Computer Skills (advanced)</li><li>Project Management (intermediate)</li></ul><ul><li>OH Performance Management (OH practices within a business line)</li><li>Process Facilitation</li><li>Issue Management (multiple business lines)</li><li>Social Marketing (business lines)</li></ul>	<ul><li>Problem-solving (exceptional)</li><li>Decision-making (exceptional)</li><li>Data Collection and Management (evaluate/enhance OH databases; oversee data management, channel reports accordingly)</li><li>Presentations (executive level/public/external)</li><li>Computer Skills (evaluation of computer resources)</li><li>Project Management (advanced)</li></ul><ul><li>OH Performance Management (within multiple Business Lines and the company, apply TQM principles)</li><li>Process Facilitation (corporate)</li><li>Issue Management (strategic)</li><li>Financial/Business Perspective</li><li>Social Marketing (multiple business lines)</li><li>OH Stewardship</li><li>OH Governance</li></ul>
Relationship Skills	<ul><li>Communication Skills</li><li>Interpersonal Skills</li><li>Coaching (employees)</li><li>Team Participation</li></ul>	<ul><li>Communication (business line)</li><li>Relationship Skills</li><li>Coaching (workgroups,</li></ul>	<ul><li>Communication (business line)</li><li>Community Relationship Management</li></ul>	<ul><li>Communication (company/community)</li><li>Reputation Management</li><li>Mentoring (safety</li></ul>

Position	OHN ADVISOR I	OHN ADVISOR II	OHN SPECIALIST	SENIOR OHN SPECIALIST
	■ Advocacy (employees; workgroup)	■ Team Building (business line) ■ Negotiating (conflict resolution management re: OH issues in workgroups) ■ Preceptorship ■ Change Management (individual and groups)	■ Coaching (business lines) ■ Team Building (leadership role with business lines) ■ Negotiating (conflict resolution management re: safety issues in business lines) ■ Mediation ■ Change Management (operational)	advisors) ■ Team Building (leadership role for company) ■ Negotiating (conflict resolution management re: safety issues in company) ■ Facilitation ■ Change Management (corporate)

Appendix 2

OCCUPATIONAL SAFETY PRACTITIONER/ PROFESSIONAL CAREER STREAM

Position	SAFETY ADVISOR I	SAFETY ADVISOR II	SAFETY SPECIALIST	SENIOR SAFETY SPECIALIST	DIRECTOR OH&S
Qualifications/ Experience	Entry-level position. High School Diploma or equivalent education and experience. Completed or enrolled in OH&S or another recognized certificate program. Industrial safety-related training. WHMIS training. Standard First Aid Training. 5 years related-industry experience; basic computer skills; demonstrated safe work behaviours; problem solving skills; ability to work under minimal supervision; strong communication skills; strong interpersonal skills; team player, strong appreciation and commitment for workplace safety; valid driver's licence.	Developing position. OH&S Certificate. Industrial safety-related training, such as WHMIS, Standard First Aid Training, Incident Investigation Training. 2-3 years OH&S experience; intermediate computer skills; demonstrated strong work behaviours; good problem-solving skills; ability to work independently; strong written and verbal communication skills; strong interpersonal skills; team player, organizational and administrative skills strong appreciation and commitment for workplace safety; valid driver's licence.	Performing position. A degree; or equivalent combination of education (OH&S Certificate) and experience. Professional certification – CRSP. Certified Safety Auditor. Advanced knowledge incident investigation; computer skills; ergonomics; Industrial Hygiene, Risk Communication. 4-7 years OH&S experience; demonstrated strong problem-solving skills, ability to work independently; strong written and verbal communication skills; strong interpersonal skills, team player, organizational and leadership skills; strong appreciation and commitment for workplace safety.	Advanced position. A degree or equivalent combination of education (OH&S Certificate) and experience. Professional certification – CRSP. Certified Safety Auditor, Media Relations and Risk Communications training. Business Certificate. OH&S Consulting Certificate or equivalent. Advanced Incident Investigation education and experience. Advanced computer training. Ergonomics training, Industrial Hygiene Training. 8-10 years OH&S experience; demonstrated sound problem-solving skills, self-initiative; exceptional written and verbal communication skills; exceptional interpersonal skills; team leader, sound appreciation and commitment for workplace safety.	Leadership position. A degree or equivalent combination of education (OH&S Certificate) and experience. Professional certification – CRSP, or equivalent. Business Certificate. Risk Communications training. OH&S Consulting Certificate or equivalent. Advanced computer training. Ergonomics training. Industrial Hygiene Training. Demonstrated leadership skills. 8-10 years OH&S experience; industry recognition for directing ability. Strong people and process management skills; Strong project management abilities.
Position Scope	Narrow				Broad
Position Depth	Shallow				Deep
Technical/ Specialist Skills	• Loss Control (promote among workers a level of general safety awareness, safety training, safety assessments, MSDS/ WHMIS knowledge/ training and fire prevention, EAP	• Loss Control (assist with investigating incidents that result in WCB claims; assist with ergonomics, human factor analysis, and trend analysis at workgroup level; assist with MSDS/	• Loss Control (assist with WCB claims management; undertake incident analysis and trend analysis at business line level; evaluate MSDS/WHMIS program, promote hazard control;	• Loss Control (assist with the management of general loss control at company-wide level; manage hazard control, manage incident investigation practices and processes)	• Loss Control (direct general loss control at company-wide level; direct WCB claims management and rehabilitation, direct hazard control; use OH&S outcomes to advise on

1481

Position	SAFETY ADVISOR I	SAFETY ADVISOR II	SAFETY SPECIALIST	SENIOR SAFETY SPECIALIST	DIRECTOR OH&S
	knowledge)	WHMIS management; promotion of desired safety behaviours; promote EAP awareness among workers)	promote recognition of desired safety and organizational behaviours)		related hiring and firing practices)
	• Regulatory Knowledge (develop an awareness of OH&S & WCB legislation, monitor worker compliance with OH&S and WCB legislation; promote due diligence at a worker level; assist with the interpretation of legislation at a workgroup level)	• Regulatory Knowledge (develop greater knowledge of OH&S and WCB legislation; monitor workgroup compliance; promote due diligence at a workgroup level; interpret legislation for the workgroup)	• Regulatory Knowledge (develop a level of proficiency on OH&S and WCB legislation; interpret OH&S and WCB legislation for a business line; regularly monitor the applicable legislation; assist with the development of corporate OH&S standards)	• Regulatory Knowledge (develop a level of expertise on OH&S and WCB legislation; interpret OH&S and WCB legislation at a company/industry level; assist with the provision of OH&S governance; promote responsible government/industry OH&S standards/practices)	• Regulatory Knowledge (direct company responses to emerging legislation; provide OH&S, WCB and Human Rights governance; provide stewardship in the above areas; promote responsible government/industry OH&S standards/practices)
	• Hazard Identification (understand hazard management principles; promote worksite inspections; assist with basic incident investigations and job hazard analysis at workgroup level)	• Hazard Identification (develop an in-depth understanding of hazard management principles; assist with incident investigations and job hazard analysis at workgroup level; check incident investigation reports for identified hazards and appropriate controls; prepare suitable reports and assist with trend analyses)	• Hazard Identification (prepare and analyze statistics for a business line; conduct risk analysis for business line; assist with hygiene sampling and interpretation of the results; manage the Hazard Management System and interpret the results)	• Hazard Identification (prepare and analyze statistics for a company; conduct complex risk analysis for a business line/company; link incident results with hazard management system, evaluate risk analyses)	• Hazard Identification (direct the preparation, analysis and interpretation of OH&S and Disability Management statistics company-wide in terms of identifying hazards, direct the analysis and interpretation of the Hazard Management System results; monitor the effectiveness of the hazard management system and processes)
	• Program Development (assist with the delivery of the company's emergency preparedness system, e.g. First Aid Response)	• Program Development (develop and implement emergency preparedness plans, First Aid response, monitor Industrial Hygiene; develop workgroup ergonomic solutions; monitor employee health)	• Program Development (develop ergonomic solutions for business line; monitor employee OH&S for a business line; interpret PPE testing program; address Fleet Safety)	• Program Development (oversee OH&S program development and delivery)	• Program Development (direct OH&S program development and delivery; monitor its effectiveness and efficiency)
	• Administration (assist with the documentation and retention of Safety records)	• Leadership & Administration (promote the OH&S program and how it applies to a workgroup; schedule weekly/monthly Safety activities for a workgroup; research and help group to write related work procedures)	• Leadership & Administration (promote the OH&S program and how it applies to a business line; schedule OH&S activities for a business line; assist with the preparation of related work procedures for the business line; ensure quality OH&S principles are applied; conduct project reviews; provide OH&S leadership to	• Leadership (oversee OH&S activities for a company; ensure OH&S quality principles are applied; conduct project reviews; assume budget responsibilities, assist Director, OH&S with supervisory duties)	• Leadership (oversee company Safety activities, ensure quality principles are applied; conduct project reviews, budget responsibility, supervisory skills)

CH 41: OCCUPATIONAL HEALTH AND SAFETY PRACTITIONERS/PROFESSIONALS:

Position	SAFETY ADVISOR I	SAFETY ADVISOR II	SAFETY SPECIALIST	SENIOR SAFETY SPECIALIST	DIRECTOR OH&S
	• Safety Programs (assist with Safety meetings; promote workplace safety among workers)	• Safety Programs (conduct Safety Meetings, assist with development of various OH&S Programs; manage the Worksite inspection Program; promote workplace safety within a workgroup)	• Safety Programs (standardize Safety Meetings for a business line; address OH-the-job Safety at Safety Meetings; maintain the industrial ergonomics program, promote workplace safety within a business line) other OH&S Team members)	• Safety Programs (standardize Safety Meetings for a company; assist in the maintenance of the OH&S Management System for company; maintain an OH&S reference library; promote workplace safety within a company)	• Safety Programs (direct the maintenance of the Safety Management System for company; monitor its efficiency & effectiveness, monitor company Safety performance indicators and outcomes; maintain codes of practice for Safety Advisors, direct the promotion of workplace safety; direct the Safety Recognition Programs)
	• Training (assist with conducting OH&S training; provide "new" employee Safety orientation; participate in external training opportunities)	• Training (develop and co-ordinate safety training for a work group; provide employee Safety orientation; train the trainers in OH&S; participate in external training opportunities)	• Training (develop and co-ordinate safety training for a business line; train the trainers in OH&S; participate in external training opportunities; evaluate program and trainers; evaluate employee Safety orientations)	• Training (develop and evaluate overall OH&S training initiatives)	• Training (direct and evaluate overall OH&S and Disability training initiatives)
	• Communications (promote Safety awareness among workers, explain Safety jargon/terms; participate in Safety Meetings; provide safety advice for pre-job meetings)	• Communications (provide technical safety communication to a workgroup, relate results of incident/ investigations to workgroup, present trend analysis reports to work group; explain OH&S Program audit results to work group, explain rationale for corrective action follow-up with workgroup)	• Communications (provide technical safety communication to business line; relate results of incident/ investigations to business line; present trend analysis reports to business line; present OH&S Program audit results to business line; explain rationale for corrective action follow-up with business line)	• Communications (provide technical OH&S communication to multiple business lines and the company, relate results of incident/investigations to company; present trend analysis reports, explain purpose of compliance audits; present OH&S Program audit results to company; explain rationale for corrective action follow-up with company; manage public/media relations in regards to OH&S issues)	• Communications (design and provide OH&S communication to Executive Team and Board of Directors, relate results of major incident/investigations; explain and interpret corporate trend analysis reports, present the results of the annual corporate OH&S Program audit results to Executive/Board)
	• Program Evaluation (conduct safety walkthroughs; assist with corrective action follow-up)	• Program Evaluation (conduct safety walkthroughs for a workgroup, review and assess incident investigation reports, promote corrective action follow-up)	• Program Evaluation (undertake compliance audits, safety reviews, workplace surveys)	• Program Evaluation (manage OH&S program evaluation and result interpretation)	• Program Evaluation (provide leadership and interpretation of results for company OH&S program evaluations)
	• Risk Management (promote employee/crew level of	• Risk Management (promote OH&S risk	• Risk Management (promote OH&S risk assessment and	• Risk Management (promote OH&S risk assessment &	• Risk Management (direct OH&S risk assessment

Position	SAFETY ADVISOR I	SAFETY ADVISOR II	SAFETY SPECIALIST	SENIOR SAFETY SPECIALIST	DIRECTOR OH&S
Business Skills	hazard management in terms of safety, health and environment impacts)	assessment and management for a workgroup)	management for a business line)	management for a company; communicate risk on Public/Health/Environment issues)	and management for corporation; direct risk communication on Public/Health/Environment issues)
	• Safety Communication (safety advice to individual employees, basic oral and written communication skills)	• Social Marketing (of the OH&S Program and concepts among employees)	• Social Marketing (of the OH&S Program and concepts within in a business line)	• Social Marketing (of the OH&S Program and concepts within a company)	• Social Marketing (of OH&S within the corporation)
		• Safety Communication (safety advice to workgroups; resource to on procedure development or new product/process use; promote EAP; good written and oral skills)	• Safety Communication (safety advice to a business line; technical resource on procedure development or new product/process use; promote EAP, effective written and oral skills)	• Safety Communication (safety advice to a business line; technical resource within a development or new product/process use; promote EAP, effective written and oral skill company-wide; public relations; expert written and oral skills)	• Safety Communication (safety advice company-wide, public relations; expert written and oral skills)
	• Problem-solving (basic level)	• Problem-solving (intermediate level)	• Problem-solving (strong)	• Problem-solving (exceptional)	• Problem-solving (exceptional)
	• Decision-making (basic level)	• Decision-making (intermediate level)	• Decision-making (strong)	• Decision-making (exceptional)	• Decision-making (exceptional)
	• Data Collection and Management (document and retain Safety data)	• Data Collection and Management (maintain a safety database)	• Data Collection and Management (establish safety databases)	• Data Collection and Management (evaluate/enhance safety databases, oversee data management, channel reports accordingly)	• Data Collection and Management (evaluate/enhance OH&S & Disability Management databases; direct data management, channel reports accordingly)
	• Presentations (Safety Meetings)	• Presentations (business line)	• Presentations (multiple business lines/external)	• Presentations (Executive level/public/external)	• Presentations (Executive level/public/external)
	• Computer Skills (basic level)	• Computer Skills (intermediate)	• Computer Skills (advanced)	• Computer Skills (advanced)	• Computer Skills (evaluation of computer resources, direct computer resources and OH&S Team skills)
		• Project Management (basic)	• Project Management (intermediate)	• Project Management (advanced)	• Project Management (advanced and directional)
		• Safety Performance Management (safety practices within workgroups)	• Safety Performance Management (safety practices within a business line)	• Safety Performance Management (within multiple a company, apply TQM principles)	• Safety Performance Management (direct safety performances for the corporation, evaluate safety performance, recommend related enhancements)

CH 41: OCCUPATIONAL HEALTH AND SAFETY PRACTITIONERS/PROFESSIONALS:

Position	SAFETY ADVISOR I	SAFETY ADVISOR II	SAFETY SPECIALIST	SENIOR SAFETY SPECIALIST	DIRECTOR OH&S
		<ul><li>Safety Functional Management</li><li>Issue Management (individual and workgroup)</li><li>Social Marketing (employee/workgroup)</li></ul>	<ul><li>Process Facilitation (business line)</li><li>Issue Management (business line)</li><li>Social Marketing (business lines)</li></ul>	<ul><li>Process Facilitation (company)</li><li>Issue Management (company)</li><li>Social Marketing (multiple business lines)</li><li>Financial/Business Perspective (direct)</li><li>Safety Stewardship (assist)</li><li>Safety Governance (direct)</li></ul>	<ul><li>Process Facilitation (company)</li><li>Issue Management (corporation and strategic)</li><li>Social Marketing (multiple business lines)</li><li>Financial/Business Perspective (direct)</li><li>OH&S & Disability Management Stewardship (direct)</li><li>Safety Governance (direct)</li></ul>
Relationship Skills	<ul><li>Communication (individual level, basic communication skills, basic interpersonal skills)</li><li>Interpersonal Skills (basic level)</li><li>Coaching (employees)</li><li>Team participation (team player)</li><li>Advocacy (for employees; a workgroup)</li></ul>	<ul><li>Communication (business line, intermediate communication and interpersonal skills)</li><li>Relationship Skills (developing)</li><li>Coaching (workgroups, business line)</li><li>Team Building (participatory, workgroup)</li><li>Negotiating (conflict resolution management re: safety issues in workgroups)</li><li>Perceptivity</li></ul>	<ul><li>Communication (business line, advanced communication and interpersonal skills)</li><li>Community Relationship Management (developing)</li><li>Coaching (business lines)</li><li>Team Building (leadership role with business lines)</li><li>Negotiating (conflict resolution management re: safety issues in business lines)</li><li>Mediation</li></ul>	<ul><li>Communication (company/community level, advanced communication and interpersonal skills)</li><li>Reputation Management (advanced)</li><li>Mentoring (Safety Advisors)</li><li>Team Building (leadership role for company)</li><li>Negotiating (conflict resolution management re: safety issues in company)</li><li>Facilitation</li></ul>	<ul><li>Communication (corporation, government and industry levels; expert communication and interpersonal skills)</li><li>Reputation Management (advanced)</li><li>Mentoring (Safety Advisors)</li><li>Team Building (leadership role for company)</li><li>Negotiating (conflict resolution management re: safety issues at corporation level)</li><li>Facilitation</li></ul>

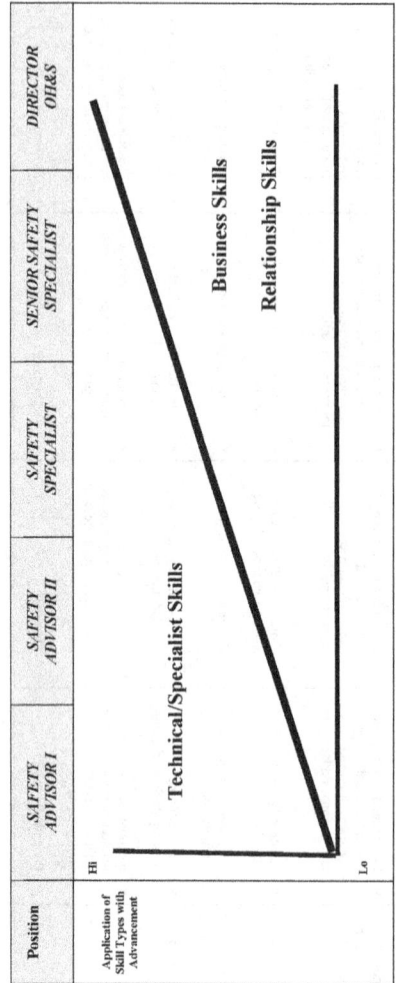

Appendix 3

Position Profile: Safety Advisor

Position Profile for:	Safety Advisor I
Position Reports to:	Manager, OH&S
Interfaces with:	*Internally*: Employees, Line Operation, Employee Training Department, Environment Department, and Human Resources Department
	Externally: Manufacturers/Suppliers and the public
	Entry-level OH&S practitioner position that focuses on individual and small group OH&S issues.
Purpose:	• Provides OH&S advice to, and programming for, employees and small workgroups.
	• Focuses on the management of individual and small group OH&S issues.
	• Models a sound OH&S attitude and behaviours.
	• Assists with the delivery of OH&S training, risk management, communication and presentations.
	• Applies basic knowledge of problem-solving, decision-making and coaching skills to the assigned tasks.
	• Works under the guidance and direction of OH&S professionals.
	• Functions as a team member to provide high-quality safety services to the various company business lines and their employees — our customers.
Accountabilities:	As a member of the OH&S Department, the specific position duties include:
	• assist with the promotion and delivery of OH&S services, including safety meetings and worker training.
	• research customer requests for safety or technical information.
	• assist with the interpretation of the applicable legislation.
	• administer existing OH&S systems such as WHMIS, Personal Protective Equipment Program, Noise Monitoring Program and other related duties.

Qualifications and Experience:	High School Diploma or equivalent education and experience. Completed or enrolled in OH&S, or another recognized certificate program. Industrial safety-related training. WHMIS training. Standard First Aid Training. Five years of related-industry experience; basic computer skills; demonstrated safe work behaviours; problem-solving skills; ability to work under minimal supervision; strong written and verbal communication skills; strong interpersonal skills; team player; strong appreciation and commitment for workplace health and safety; valid driver's license..
Special Skills and Attributes:	*Safety Technical/Specialist Skills:* • **Loss Control** (promote among workers a level of general safety awareness; safety training; safety assessments; MSDS/WHMIS knowledge/training and fire prevention; EAP knowledge) • **Regulatory Knowledge** (develop an awareness of OH&S and WCB legislation; monitor worker compliance with OH&S and WCB legislation; promote due diligence at a worker level; assist with the interpretation of legislation at a workgroup level) • **Hazard Identification** (understand hazard management principles; promote worksite inspections; and assist with basic incident investigations and job hazard analysis at workgroup level) • **Program Development** (assist with the delivery of the company's emergency preparedness system; *e.g.*, First Aid Response) • **Administration** (assist with the documentation and retention of Safety records) • **Safety Programs** (assist with Safety meetings; promote workplace safety among workers) • **Training** (assist with conducting OH&S training; provide "new" employee Safety orientation; participate in external training opportunities) • **Communications** (promote Safety awareness among workers; explain Safety jargon/terms; participate in Safety Meetings; provide safety advice for pre-job meetings) • **Program Evaluation** (conduct safety walk-throughs; assist with corrective action follow-up)

- **Risk Management** (promote employee/crew level of hazard management in terms of safety; health and environment impacts)

Business Skills:
- **Safety Communication** (safety advice to individual employees, basic oral and written communication skills)
- **Problem-solving** (basic level)
- **Decision-making** (basic level)
- **Data Collection and Management** (document and retain Safety data)
- **Presentations** (present at Safety meetings)
- **Computer Skills** (Basic level)

Relationship Skills:
- **Communication** (individual level, basic communication skills, basic interpersonal skills)
- **Interpersonal Skills** (basic level)
- **Coaching** (employees)
- **Team Participation** (team player)
- **Advocacy** (for employees; workgroup)

Date Developed: January 2020

Appendix 4

Position Profile: Senior Safety Professional

Position Profile for:	Senior Safety Professional
Position Reports to:	Manager, OH&S
Interfaces with:	*Internally*: Line Operation, Employee Training Department, Environment Department, Occupational Health Service Provider, Human Resources Department, and Legal Department
	Externally: Government agencies, industry associations, manufacturers/suppliers, and the public.
Purpose:	• Managing the Company's OH&S Program and servicing within a designated Business Line(s)
	• Functioning as a team member to provide high-quality Safety services to the various Company Business Lines and their employees — our customers
	• Providing OH&S advice and programming for our customers
	• Mentoring and coaching Safety Advisor I and IIs
	• Providing a strong OH&S leadership role within the Company and externally
	• Representing the Company's business interests on government, industry and professional committees/forums
Accountabilities:	As a member of the OH&S Department, manage and oversee the development, presentation and orientation of new workers to the Company OH&S Program. The specific position duties include:
	• **Regulatory Stewardship and Governance:** Ensure that the Company remains in compliance with the applicable legislation and industry standards.
	• **Safety Expertise:** Support the designated Business Line(s) with the development and implementation of its/their OH&S Program and practices.
	• **Hazard Management:** Play a key-support role in hazard identification, hazard assessment and hazard control within a Business Line.

- **Issue Management/Negotiating:** Assist the designated Business Line(s) with OH&S issue management and resolution.
- **Safety Performance Management:** Support the designated Business Line(s) with its/their OH&S performance management.
- **Loss Control:** Assist with the maintenance of the Company's Loss Control Data Management System.
- **Presentation:** Provide OH&S presentations in instances such as new employee orientations, safety meetings, operational meetings with Business Lines, and public safety programs.
- **Incident Investigation:** Investigate and prepare reports related to occupational and public incidents.
- **Program Evaluation:** Conduct OH&S compliance audits and prepare the resulting reports.
- **Risk Communication:** Communicate OH&S risks in terms that are understandable for the Business Lines.
- **Risk Management:** Provide OH&S training/educational seminars for employees and external third parties.
- **Program Development:** Develop customized OH&S Programs as required.
- **Communication:** Undertake effective OH&S communications that are targeted and customer focused.
- **Relationship Building:** Build effective business relationships both internally and externally.
- **Professional Growth:** Seek and avail oneself of opportunities for continued education and competency in OH&S.

Qualifications and Experience: CRSP or an equivalent OH&S Certificate; incident training and experience; ergonomics training; advanced computer training/skills; auditor training; 10–15 years Safety-related experience

Special Skills and Attributes: *Safety Technical/Specialist Skills:*

- **Loss Control** (WCB claims management; accident analysis; trend analysis at Business Line level; evaluation of MSDS/WHMIS program; engineer hazard control; recognition of desired safety and organizational behaviours)

- **Regulatory Knowledge** (interpret legislation for a Business Line; review legislation; work on Safety standards)
- **Hazard Identification** (prepare hazard analysis statistics for a Business Line; conduct risk analysis for Business Line; hygiene sampling; gas-testing)
- **Program Development** (develop ergonomic solutions for Business Line; monitor employee health for a Business Line; interpret PPE testing program; address Fleet Safety)
- **Leadership and Administration** (schedule Safety activities for a Business Line; prepare Safety policies and procedures; contract administration; ensure quality principles are applied; conduct project reviews; demonstrate supervisory skills)
- **Safety Programs** (develop and implement joint health and safety committees; develop and implement Safety Recognition Programs with operations; address Off-the-Job Safety; industrial ergonomics program)
- **Training** (develop training; train the trainers; participate in external training opportunities; evaluate program and trainers; evaluate employee Safety orientations)
- **Communications** (technical communication to Business Line; incident/investigation reports; trend analysis reports; program audit results to Business Line; corrective action follow-up with Business Line)
- **Program Evaluation** (compliance audits, safety reviews, surveys)
- **Risk Management** (safety risks for multiple Business Lines)

Business Skills:
- **Safety Communication** (safety advice to multiple Business Lines; perfect written and oral skills)
- **Problem-solving** (strong)
- **Decision-making** (strong)
- **Data Collection and Management** (establish safety databases)
- **Presentations** (multiple Business Lines/external)
- **Computer Skills** (advanced)
- **Project Management** (intermediate)
- **Safety Performance Management** (safety practices within a Business Line)

- **Process Facilitation**
- **Issue Management** (multiple Business Lines)
- **Social Marketing** (Business Lines)

Relationship Skills:
- **Communication** (Business Line)
- **Community Relationship Management**
- **Coaching** (Business Lines)
- **Team Building** (Leadership Role with Business Lines, team player)
- **Negotiating** (conflict resolution management re: safety issues in Business Lines)
- **Mediation**

Date Developed: January 2020

CHAPTER REFERENCES

D. Dyck, *Disability Management: Theory, Strategy & Industry Practice*, 6th ed. (Markham, ON: LexisNexis Canada Inc., 2017)

D. Dyck & M. Walker, "Career Streaming: A Model of Career Development" (1996) 44:4 *AAOHN* 177-182.

D.A. Gilles, *Nursing Management: A Systems Approach* (Toronto, ON: W.B. Saunders Company, 1989).

R. Goddard, "Lateral Moves Enhance Careers" 35:12 *HR Magazine*.

D.T. Hall, *Careers in Organizations* (Glenview, IL: Scott, Foreman & Company, 1979).

D.T. Hall, *Organizations: Structure and Process* (Englewood Cliffs, NJ: Prentice-Hall, 1977).

B. Moses, *Manager's Guide to Career Development* (Toronto, ON: BBM Human Resource Consultants, 1989).

L.J. Peter & R. Hull, *The Peter Principle* (New York, NY: W. Morrow, 1969).

J. Schermerhorn, J. Hunt & R. Osborn, *Managing Organizational Behavior*, 4th ed. (New York, NY: John Wiley & Sons, 1991).

R. Steers, *Introduction to Organizational Behavior*, 2d ed. (Glenview, IL: Scott, Foreman and Company, 1984).

T. Stone & N. Meltz, *Human Resource Management in Canada*, 2d ed. (Toronto, ON: Holt, Rinehart and Winston of Canada, 1992).

Section 4
Occupational Health and Safety: Future Concepts

Chapter 42

FUTURE CHALLENGES IN OCCUPATIONAL HEALTH AND SAFETY

A. INTRODUCTION

Although the field of Occupational Health and Safety (OH&S) has greatly advanced in the last 80 years, much more needs to occur in the future. This chapter is designed to raise a number of emergent issues that will impact organizations and the field of Occupational Health and Safety.

B. PROFESSIONALISM

Often groups muse over the question of whether Occupational Health and Safety is a "Profession" or not.

1. A Profession[1]

Taking the term "professional" literally, a **profession** is an occupation, vocation, or high-status career that usually involves prolonged academic training, formal qualifications and membership of a professional or regulatory body.[2] The Registered Nurses Association of Ontario (RNAO) defines profession as:

> An occupation whose core element is work based upon the mastery of a complex body of knowledge and skills. It is a vocation in which knowledge of some department of science or learning or the practice of an art founded upon it is used in the service of others. Its members possess a commitment to competence, integrity, morality, altruism and the promotion of the public good within their domain. These commitments form the basis of a social contract between a profession and society, which in return grants the profession the right to autonomy in practice and the privilege of self-regulation. Professions and their members are accountable to those serviced and to society.[3]

Some examples of professions are lawyers, accountants, engineers, teachers,

---

[1] Excerpt from D. Dyck, *Disability Management: Theory, Strategy and Industry Practice*, 6th ed. (Markham, ON: LexisNexis Canada Inc., 2017), c. 41, "Future Challenges in Disability Management".

[2] Oxford University Press, *Oxford English Dictionary*, 2d ed. (Oxford University Press, 1989).

[3] Registered Nurses Association of Ontario (RNAO), *Professionalism in Nursing, Healthy Work Environments, Best Practice Guidelines* (Toronto: RNAO, 2007) at 58, online:

professors, librarians, priests, pilots, physicians, dentists, nurses, veterinarians, pharmacists, physical therapists, and other specialized technical occupations.

A profession typically involves the application of specialized knowledge of a subject, field, or science to a customer or fee-paying clientele's situation. It is regulated by a professional body that sets competency examinations, acts as a licensing authority, and enforces adherence to an ethical code of practice. Examining these four elements further:

- **Regulation**

A profession differs from a trade group in that it regulates by statute, its members, and their participation.

- **Autonomy**

Professions have a high degree of control over their activities and practices.

- **Status and Prestige**

Society recognizes and regards professions with respect and prestige.

- **Power**

Professions have power – power to control their own members, their area(s) of expertise and their interests. According to Larkin,[4] a profession dominates, polices, and protects its area of expertise and the conduct of its members and exercises significant influence over its field.

2. Profession Characteristics: How Does the Field of OH&S Measure Up?

So what are the characteristics of a profession? That is, what are the 22 essential elements that an occupation would have to possess to be deemed a "profession" and how does the field of Occupational Health and Safety "stack up"?

- ✓ **Skills are based on theoretical knowledge:** Professionals possess extensive theoretical knowledge and related skills in a certain field or area of expertise.
- ✓ **Individual clients:** Professions have fee-paying clients/patients: payment that is direct or indirect in nature.
- ✓ **Work autonomy:** Professions retain control over their work and their theoretical knowledge. Hence, they are autonomous in their practice and can make independent judgements about their work.
- ✓ **Status and rewards:** Professions are highly respected by society and rewarded. The prestige comes from society viewing their work as having a special and valuable nature. All professions possess professional exper-

---

http://www.rnao.ca/sites/rnao-ca/files/Professionalism_in_Nursing.pdf (date accessed: January 31, 2020).

[4] G. Larkin, *Occupational Monopoly and Modern Medicine* (London, UK: Tavistock Publications, 1983).

tise — technical, specialized, and highly skilled work/abilities.

- √ **Middle-class occupations:** Professionals tend to earn more than non-professionals.
- √ **Public service and altruism:** Professions provide a valued public service and tend to do so in an unselfish way. Some are even viewed less as a profession and more as a "calling" or vocation, for example the clergy, physicians, nurses, teachers, and others.
- √ **Mobility:** Professionals own their knowledge and skills, as well as the authority to practice. Hence, they are mobile in their employment opportunities and the services they provide. This mobility is supported by the standardization of professional education and the related codes of practice.

In examining the aforementioned seven characteristics, one could say that the field of Occupational Health and Safety meets the test of being a profession as noted by the √ symbol. However, this would be a premature decision because there are more professional traits that need to be explored, namely:

- ? **Professional association:** Professions have management and administrative functions designed to control the entrance and exit requirements of members and to promote the status of their membership.
- ? **Code of professional conduct or ethics:** Professions have rules of ethical conduct which the members are required to uphold.
- ? **Extensive period of education:** Professions require their members to attain at least three years of formal university education.
- ? **Institutional training:** In addition to formalized schooling, professions usually require a lengthy period of institutional training focused on specific practical experience before being recognized as a full member within the profession.
- ? **Testing of competence:** Professions require potential candidates to successfully complete theory and practice-based examinations, as well as to provide evidence of ongoing practice competence.
- ? **Offer of reassurance:** Society looks to the professions to offer reassurance to clients/patients that things are being properly addressed and moving towards a "normal state". An example is the lawyer who assures his/her client that the proper steps are being taken to resolve an issue.
- ? **Male-dominated:** Professions have traditionally been male-dominated with the exception of nursing and teaching.
- ? **Legitimacy:** Society awards professions, based on their education and competence, legal authority to conduct specific service duties and obligations.
- ? **Indeterminacy of knowledge:** Professions deal with codes/standards of practice, not rules. They are typically faced with situations/conditions that

can only be addressed through a mix of knowledge, skills, and experience, not by following a set of prescribed rules.

A review of these nine characteristics leaves one more skeptical as to whether Occupational Health and Safety is a profession as denoted by the question marks (?). Why? Although the field of Occupational Health and Safety professes to meet some of these elements, there are a number of key elements missing, namely:

- OH&S associations exist and while much has been done to advance the field of Occupational Health and Safety, the entrance and exit requirements of members remain limited to those OH&S practitioners who voluntarily participate. Violation of the codes of ethics or acceptable OH&S practices often goes unaddressed.
- Not all practicing OH&S practitioners have undergone an extensive period of education or institutional training, or even a test of their competence. This is changing, but as of today, this situation remains.
- Many OH&S practitioners focus on the OH&S rules/legislation that need to be upheld as opposed to providing reassurance that they know how to effectively guide Management/employers towards how to work safely.
- The field of Occupational Health and Safety is changing. Once male-dominated, more and more females have entered the field. This change is not viewed as a drawback towards the attainment of professionalism, but rather as a positive aspect.
- Society has not awarded OH&S practitioners the legal right to conduct specific service duties and obligations.
- The field of Occupational Health and Safety still relies heavily on the use of rules and regulations, as opposed to codes/standards of practice. As noted, professionals face situations/conditions that can only be addressed through a mix of knowledge, skills, and experience, not by heavily relying on a set of prescribed rules.

The deciding factors to the question of whether Occupational Health and Safety is a profession are evident in the last six elements:

- ☒ **Self-regulation:** Professions are self-regulated, usually through a professional association that upholds professional standards and codes of practice. The field of OH&S has developed professional standards and a code of ethics, which are primarily upheld by CRSP-prepared practitioners. Failure to do so goes unchallenged.
- ☒ **Licenced practitioners:** Along with self-regulation comes licensure — a legally sanctioned licence to practice. OH&S practitioners in Canada can practice without any regulation or licensure.
- ☒ **Exclusion, monopoly and legal recognition:** To achieve socially sanctioned self-regulation and licensure, a profession must be legally

recognized to exclude those who do not meet the entrance criteria and to expel those members who are deemed incompetent. This is a big hurdle for the field of Occupational Health and Safety to overcome, but in some countries, this situation is changing, albeit slowly.

- ☒ **Control of remuneration and advertising:** Professions tend to set acceptable fee rates as well as to establish acceptable standards of advertising. This practice does not exist within the field of Occupational Health and Safety.

- ☒ **Ritualistic:** Professions vary in the degree and types of rituals that they uphold. For example, the clergy and legal counsel tend to uphold many rituals while physicians, nurses, and other scientifically based professions rely on research-based practices. Although within the field of Occupational Health and Safety there is a move towards evidence-based practice, the use of rituals is a long way off.

- ☒ **Inaccessible body of knowledge:** The accessibility of knowledge is usually limited or unavailable to the layperson. These topics tend not to be covered in schools, universities, or libraries; in addition, barriers may arise from the technical jargon used. Sometimes it is legally sanctioned, for example, medical/legal/religious information can be deemed privileged information, available for only those qualified to understand the information. Given the nature and focus of Occupational Health and Safety, widespread knowledge of workplace health and safety practices is a desirable outcome. Hence, this condition may never be strictly met.

Back to the original question, *Is Occupational Health & Safety a "Profession"?* The answer is "no". The field of Occupational Health and Safety is not there yet but will get there over the coming years. The rationale for this position is that the field of Occupational Health and Safety is made up of several currently recognized professions such as:

- Occupational Health Nursing;
- Occupational Medicine;
- Ergonomists;
- Safety Engineers;
- Toxicologists;
- Psychologists;
- Lawyers.

The field of Occupational Health and Safety is also composed of several quasi-professions, such as Occupational/Industrial Hygiene, Occupational Safety, Disability Case Managers, *etc*. Hence, the field is a "salad of professions and disciplines", rather than a pure science. As such, it does not currently meet the essential conditions of being a profession.

3. Professionalization

Professionalization is the process by which a profession arises from a trade or occupation. It is described as:

> ... starting with the establishment of the activity as a full-time occupation, progressing through the establishment of training schools and university links, the formation of a professional organization, and the struggle to gain legal support for exclusion, and culminating with the formation of a formal code of ethics.[5]

Professionalization involves establishing acceptable qualifications and professional norms. That requires the establishment of a professional association to oversee the conduct of members and to distinguish the "qualified" from the "unqualified amateurs".[6] This process is termed occupational closure — the division between qualified members and unqualified outsiders.[7]

Typically, trade groups/occupations "dream" of attaining the status of being a

---

[5] J. Roberts & M. Dietrich, "Conceptualizing Professionalism.: Why Economics Needs Sociology" (1999) 58:4 *The American Journal of Economics and Sociology* 977, online: http://findarticles.com/p/articles/mi_m0254/is_4_58/ai_58496769 (date accessed: January 31, 2020).

[6] Wikipedia, "Professionalization", online: http://en.wikipedia.org/wiki/professionalization (date accessed: January 31, 2020).

[7] S. Cavanagh, "The Gender of Professionalism and Occupational Closure: The Management of Tenure-related Disputes by the 'Federation of Women Teachers' Association of Ontario' 1918–1949" (2003) 15:1 *Gender and Education* 30.

"profession" — but like the fisherman who dreams of "catching that big fish" (Figure 42.1), one must ask, "Are OH&S practitioners prepared to 'land' professionalism"?

Figure 42.1: Landing Professionalism

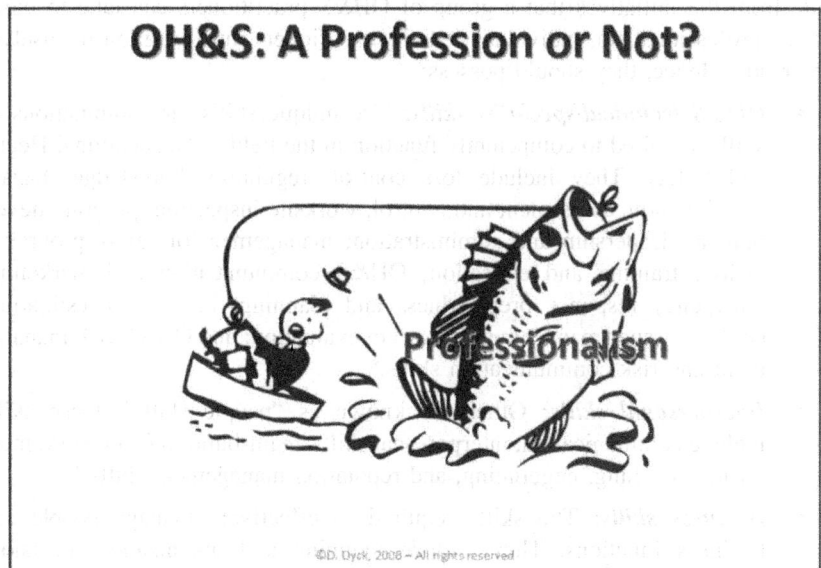

4. The Field of Occupational Health and Safety: Prepared or Not?

For the field of Occupational Health and Safety to evolve from an occupation to a profession, certain actions must be undertaken, such as:

- *Professional body/association — administration:* Establishment of a legally recognized body/association designed to set professional standards and codes of practice, administer member eligibility, and self-regulate the profession.

- *Professional standards:* Development of practice standards/codes of practice to guide members in acceptable and safe practices.

- *Professional licensure:* Establish, administer, and maintain a system to legally sanction licence to practice.

- *Educational standards and requirements:* Establish defensible (research-supported) educational standards and requirements for the profession.

- *Research and development ability:* Establish, promote, and support OH&S research and development.

- *Political actionability:* Proactively work with society, industry, and the government to promote the profession.

- **Learn from others:** Network with other newly formed professional associations to determine what worked/failed in regards to the move from an occupation to a profession.

5. The OH&S Practitioner: Ready or Not?

Aside from the initiatives that a group of OH&S practitioners can take to move towards professionalism, individual OH&S practitioners must position themselves to be ready. Hence, they should possess:

- *OH&S technical/specialist skills:* The unique skills, or combinations of skills, required to competently function in the field of Occupational Health and Safety. They include loss control; regulatory knowledge; hazard identification, assessment and control; worksite inspection; program development; leadership and administration; management of safety programs; OH&S training and education; OH&S communication and marketing; emergency response preparedness and planning; incident investigation; strategic issues management; program evaluation; and OH&S risk management and risk communication skills.[8]

- *Interpersonal skills:* Otherwise known as "people skills", these skills include communication, interpersonal skills, team-building/teamwork, mentoring, coaching, negotiating, and reputation management skills.[9]

- *Business skills:* The skills required to effectively manage people and business functions. They include planning and organization, decision-making, problem-solving, leadership, financial/business perspective, negotiating, information systems, internal consulting, process facilitation, performance management, employee development, training, change management, presentation making, and risk communications.[10]

- *Research and development skills*: The skills required to undertake defensible OH&S research and development. They involve using recognized research techniques and practices.

In short, OH&S practitioners have to position themselves towards demonstrated professionalism:

> Qualities or typical features of a profession or professional. A collection of attitudes and actions; it suggests knowledge and technical skill.[11]

---

[8] See Chapter 41, "Occupational Health and Safety Practitioners/Professionals: Career Development".

[9] *Ibid.*

[10] *Ibid.*

[11] Registered Nurses Association of Ontario (RNAO), *Professionalism in Nursing, Healthy Work Environments, Best Practice Guidelines* (Toronto, ON: RNAO, 2007) at 58, online: http://www.rnao.ca/sites/rnao-ca/files/Professionalism_in_Nursing.pdf (date accessed: January 31, 2020).

In future, the question may no longer be whether Occupational Health and Safety is a profession or not. Rather, it will be, "Are you prepared for professionalism?"

6. Summary

The road ahead towards OH&S professionalism may seem unattainable and rocky, but OH&S practitioners should "stay the course" and work collaboratively towards attaining professionalism. The key is to learn from what other professional groups have done or what OH&S groups in other countries are doing — there is no need to "re-invent the road to professionalism". Instead, move forward smartly and steadily.

C. OH&S MANAGEMENT SYSTEM: SUSTAINABILITY

The HS^{3TM} Model provides a road map for integrating health, safety, sustainability, and stewardship — all of which directly impact every organization's/company's triple bottom line: healthy people, healthy planet, and healthy profits. The HS^{3TM} Model can be used to promote healthy lifestyles, reduce workplace risk and injuries, protect the natural environment, and improve resource alignment within organizations (Figure 42.2). Synergistic HS^{3TM} planning cost-effectively links work injury management, health promotion, environmental protection, safety training and surveillance, and regulatory compliance.

Figure 42.2: The HS^{3TM} Model

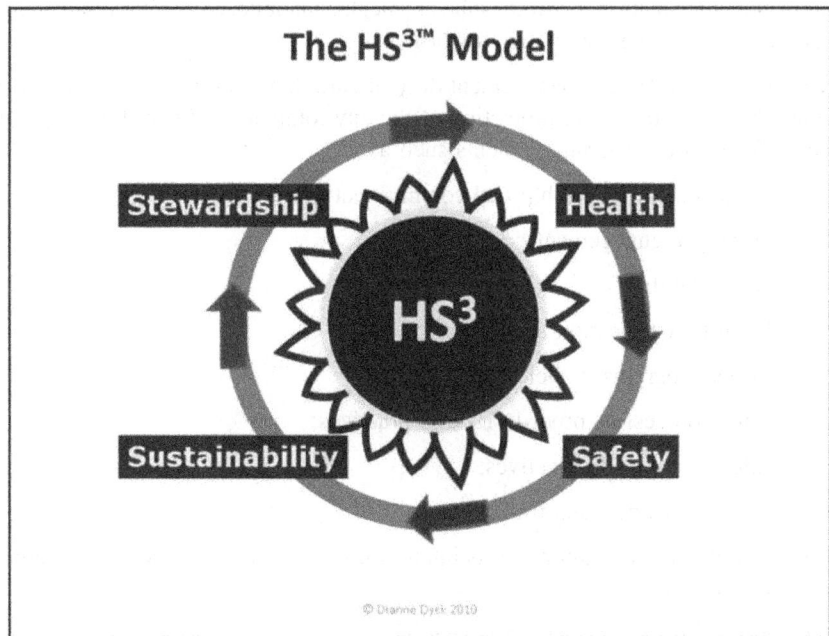

The **HS³™ Model** encompasses **H**ealth, **S**afety, **S**tewardship and **S**ustainability. It:

- promotes an integrated approach to health, safety, sustainability, and stewardship; and
- facilitates the development of a well-honed strategic plan which in turn, provides the direction needed to achieve an organization's/company's corporate vision. The strategic plan outlines how an organization/company can leverage its resources to promote healthy lifestyles, reduce risk and injuries, protect and preserve the natural environment, and provide workers the support to perform critical tasks efficiently and effectively. The strategic plan outlines how the functional integration of health, safety, sustainability, and stewardship addresses occupational health, safety, and sustainability as business issues.

Functional integration is preceded by understanding and adopting a clearly articulated definition of occupational health, safety, and sustainability and how they are linked, as well as an understanding of how they affect business.

1. Definitions

Health, in terms of employees, is the state of complete physical, mental, and social well-being and not merely the absence of disease or infirmity. In terms of an organization, Occupational Health is the physical, economic, and cultural well-being of the organization/company and its workers.

Occupational Safety is a management duty of care that requires the provision of a safe and healthy workplace, protection of the environment, and care for the public at large. It includes program elements such as:

- Management leadership and commitment;
- employee engagement;
- accountability;
- hazard identification and control;
- safety programs, policies, and plans;
- safety processes, procedures, and practices;
- safety goals and objectives;
- safety education and training;
- safety communications to maintain a high level of awareness on safety in the workplace;
- safety inspections for workplace hazards;
- safety tracking and performance metrics; and
- safety program audits.

Stewardship refers to management's responsibility to properly utilize and develop its resources, including people, property, and financial assets.

Sustainability is defined as the ability of a program/practice to endure. However, sustainability is also viewed as focusing on the "triple bottom line" of people, planet and profits. For our purposes, sustainability is achieved when an OH&S practice permeates all activities and processes and when compliance is linked to manager performance and rewards.

The HS³™ Model integrates Health, Safety, Stewardship and Sustainability, and facilitates the development of a strategic plan which, in turn, provides the direction needed to achieve the organization's vision.

HS³™ Model Planning: An Occupational Health and Safety Management System (OHSMS) ensures that the planning component includes the identification of significant health, safety, environmental aspects, and legal requirements related to the organization's/company's business. This planning focuses on establishing and implementing organizational policies that not only cover daily operations, but also emergency preparedness and response.

Ongoing operation and management; monitoring, and corrective action; ongoing improvement processes; and key support processes, such as documentation, organizational structure, communication, training, and skill assessment are all addressed in the planning process. For example, during the assessment process, as part of a core process for service delivery, the potential exists for worker exposure to workplace hazards. The plan must address the operational controls and policies in place to prevent human and environmental exposure; safety aspects, such as use of safe work practices; emergency preparedness response; and the monitoring required to assure worker health and safety.

Managers can leverage their role through synergistic planning and cost-effective actions which link health promotion, work injury management, environmental efforts, safety training and surveillance, and regulatory compliance programs.

Health, safety, sustainability and stewardship all intersect at the bottom line — healthy people, healthy planet, and healthy profits. Organizations/companies with a vision for an integrated HS3™ approach can improve the work environment, improve worker health and safety, protect and preserve the natural environment, increase worker productivity, and increase job satisfaction leading to increased employee and customer attachment levels.[12]

2. Sustainability Model Process

The functional approach to the sustainability of an OH&S Program involves the Plan, Do, Check and Act approach. Applying the performance measures contained in the OH&S Program, the approach can be graphically depicted (Figure 42.3).

---

[12] M. Weiss, "Changing the Conversation — The Occupational Health Nurse's Role in Integrated HS3" (2009) 57:7 *AAOHN* 293.

Figure 42.3: Occupational Health and Safety Sustainability Approach

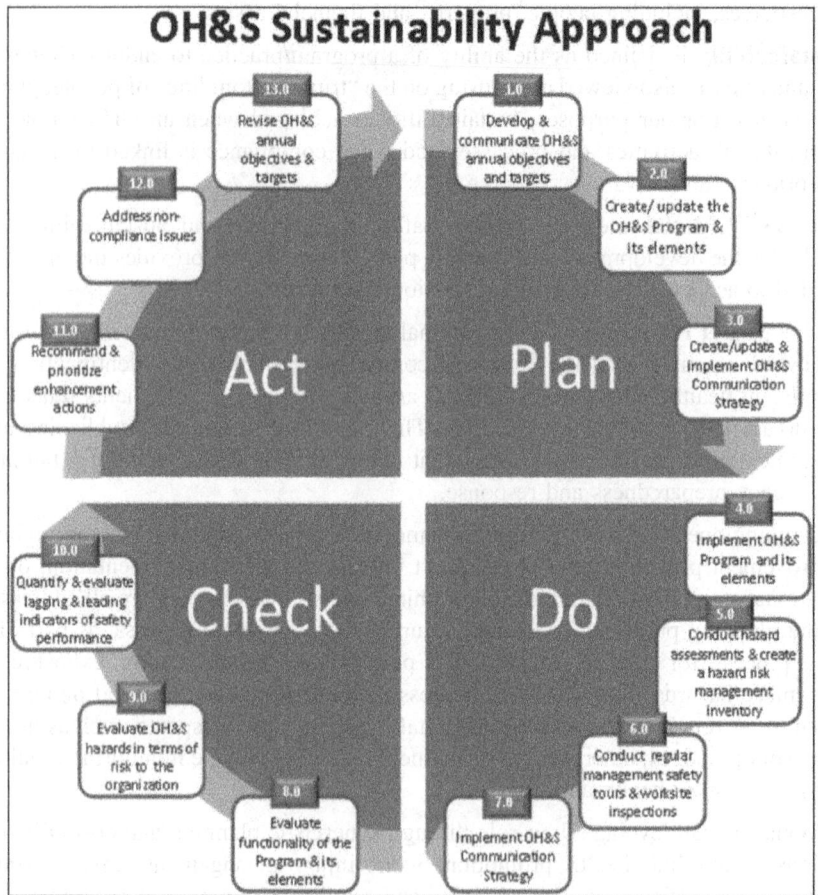

PLAN

1.0 *Develop and Communicate the OH&S Program Annual Objectives* — Establish annual OH&S Program objectives (*i.e.*, how the program will evolve) and the safety targets (*i.e.*, expected outcomes) for the OH&S Program that are based on past performance and the desired magnitude of improvement.

2.0 *Create/Update and Implement OH&S Program* — Create/update the OH&S Program and related standards, procedures, and practices, recognizing that continuous improvement of the program is required.

3.0 *Create/Update and Implement OH&S Program Communication Strategy* — The OH&S Program Communication Strategy is developed and continuously updated to promote the program, its objectives, and desired performance outcomes.

DO

4.0 *Implement the OH&S Program and its Elements* — Implement the OH&S Program and related standards, procedures, and practices as designed.

5.0 *Conduct Hazard Assessments and Create a Hazard Risk Management Inventory* — Assessment of the inherent and emergent hazards of the work is undertaken. The results are collected in terms of the magnitude of the risk, the controls implemented, the resultant risk level, and the monitoring of the effectiveness of the hazard controls. This inventory enables the organization to quantify the level of risk prior to mitigation and post-mitigation. It also serves as evidence of due diligence.

6.0 *Conduct and Report Regular Management Safety Tours and Worksite Inspections* — Regular management safety tours of office areas and worksite inspections provide performance measures related to the conditions of work areas; employee/manager compliance with the OH&S Program standards, procedures and practices; and the status of the housekeeping in the workplace.

7.0 *Execute the OH&S Program Communication Strategy* — Implement the OH&S Program Communication Strategy to engage employees in the OH&S Program and its elements, as well as to obtain user feedback on the OH&S Program elements.

CHECK

8.0 *Evaluate the Functionality of the OH&S Program and its Elements* — Examine the functionality of the OH&S Program and its elements through auditing, benchmarking, or other program evaluation measurement methods.

9.0 *Evaluate the OH&S Hazards in Terms of Risk to the Organization* — Quantify the level of risk prior to the mitigation of workplace hazards and compare that against the level of risk post-mitigation. Post-mitigation risk should be significantly reduced and at a tolerable level.

10.0 *Quantify and Evaluate the lagging and leading indicators of safety performance* — Using a variety of performance measurement techniques, measure, and evaluate the OH&S Program (structure, processes, and outcomes) in terms of effectiveness and compliance with established standards, procedures, and practices.

ACT

11.0 *Recommend and Prioritize Enhancements to the OH&S Program* — Based on the measurements, identify, recommend, and prioritize enhancements to the OH&S Program and OHSMS, with a view to continuous improvement.

12.0 *Address Non-compliance with the OH&S Program* — Once identified, instances of non-compliance with the OH&S Program are to be addressed by Management.

13.0 ***Revise Annual OH&S Program Objectives and Targets (Continuous Improvement)*** — As the OH&S Program and OHSMS evolve, so must the OH&S Program objectives.

D. THE CONTINGENT WORKFORCE

Part-time, temporary, short-term, seasonal, and contracted workers, as well as students and volunteers, are known as the **contingent workforce**. Their role is to fill in the labour gaps during periods of high work demands and during times of increased employee absenteeism. Typically, this shadow workforce enters the workplace unnoticed, performs the needed work, and leaves. Their health and safety usually go unattended.

The liability for these workers rests with the hiring (host) organization. The potential for workplace injuries and in some cases, non-occupational illness/injuries, tends to go unmanaged. Their hours of work vary with some working only a few hours a week to others who work long, unregulated hours.[13] As well, research indicates that contingent work is associated with poorer OH&S outcomes than is secure full-time work.[14]

In instances when the temporary worker is subcontracted from a staffing agency, the agency should be familiar with the host organization, the demands of the job, and the related risks. Then, the staffing agency and the host organization should ensure:

- a good person-job fit;
- the worker receives an OH&S orientation;
- the worker is provided with appropriate personal protective equipment (PPE);
- the worker knows how to do the work safely;
- the worker knows how to access OH&S support; and
- provisions for addressing worker illness/injury are in place.

As well, the two organizations need to develop worker OH&S orientations and relevant training tailored to the needs of the contingent worker. Monitoring the effectiveness of that training is strongly suggested.

Workplace rules and regulations need to include the contingent worker and their work situation. This extends from the work rules that an organization/company

---

[13] M. Quinlan & P. Bohle, "Contingent Work and Occupational Safety" in *The Psychology of Workplace Safety*, J. Barling & M. Frone, eds. (Washington, DC: American Psychological Association, 2004) at 82.

[14] *Ibid.*, at 85–87.

develops to government legislation. Regulatory agencies need to develop comprehensive compliance measures for tracking contingent work and its associated hazards. Enforcement should be part of this effort.[15]

From a prevention perspective, host organizations may elect to use only those staffing agencies that have been vetted in terms of providing OH&S education and training to temporary workers and deemed safe. They may also elect to work collaboratively with the staffing agency to orient the worker to the work, the related hazards, and the hazard controls. As well, they would jointly monitor and evaluate the worker's safe work practices.

Industry unions/associations also have a role to play in protecting the contingent worker. They should recognize the OH&S risks for the contingent worker and support organization/company and government initiatives to protect this group of workers.

Ontario's Bill 146, the *Stronger Workplaces for a Stronger Economy Act*, designed to "protect vulnerable workers and change the definition of 'worker' to include temporary and part-time contract employees", is currently at the stage of Second Reading Debate in the legislature.[16] If passed, it will change the landscape for the contingency workforce. Alberta, on the other hand, has already addressed the contingent worker in its new *Occupational Health & Safety Act* in 2018.[17]

E. INCREASED RELIANCE ON CONTRACTORS: IMPLICATIONS FOR EMPLOYERS

Many employers use contractors to undertake specialized work which the organization/company views as cost-effective to outsource. Contractor safety management is of prime importance to employers. According to OH&S legislation, employers are legally responsible for the OH&S duty of care of these workers as well as for that of their regular employees and the public.

Employers need to approach contractor safety with the same due diligence as they would use to deal with their own employees. They need to provide worker orientation and training; monitor and measure work performance; provide performance feedback; and correct contractor work performance when required. How this is achieved depends on the contract developed between the employer-company and contractor. Ultimately, however, the employer is the responsible party in ensuring workplace health and safety.

The Contractor must ensure that the work they are doing does not endanger the health and safety of all workers present on the worksite. Although the provincial

---

[15] *Ibid.*, at 100.

[16] Bill 146, *Stronger Workplaces for Stronger Economies Act, 2014*, online: http://www.ontla.on.ca/web/bills/bills_detail.do?locale=en&BillID=2916.

[17] Alberta, *Alberta Occupational Health and Safety Act: Highlights of changes effective June 1, 2018*, online: www.alberta.ca (date accessed: January 31, 2020).

legislation may vary in its requirements, the bottom-line is that the health and safety of workers and the public must be upheld.

F. USE OF OH&S PROGRAMS, ATTENDANCE SUPPORT PROGRAMS, AND DISABILITY MANAGEMENT PROGRAMS

In a time when manpower challenges exist, how can organizations/companies use tools like a sound OH&S Program, Attendance Support Program, and Disability Management Program to mitigate the shortage of skilled workers?

1. How Big is the Problem?

Over the last fifteen years in Canada, there has been a steady increase in the number of average workdays lost per full-time worker. In 1997, the employee absentee rate was 5.5 per cent with full-time workers missing, on average, 7.4 workdays per year. By 2019, Canadian full-time workers missed an average of 10.3 days for an absentee rate of 8.6 per cent.[18]

Work absences result from personal illness/injury and family responsibilities and translate into an average payroll cost of 5.7 per cent[19] and about $2,386.30 per employee per year.[20] So, the question becomes, how much absenteeism can an employer absorb before its operations are negatively impacted?

2. Why are Workers Taking Time Off Work?

The main reason is low employee morale. According to the *CCH 2006 Unscheduled Absenteeism Survey*, organizations/companies that reported low employee morale experienced an absenteeism rate of 2.9 per cent, as compared to 2.2 per cent by organizations/companies reporting high employee morale. As well, the reasons for employee absenteeism correlate with the level of employee morale. Seventy per cent of the employee absences in organizations/companies with low morale were not related to personal illness as compared with a 60 per cent rate in organizations/companies with high morale. In terms of the trending of worker absenteeism, it was

---

[18] Statistics Canada, *Table 14-10-0190-01 Work Absence of full-time employees by geography, annual* (Ottawa: Statistics Canada, 2020), online: https://www150.statcan.gc.ca/t1/tbl1/en/tv.action?pid=1410019001 (date accessed: January 31, 2020).

[19] Towers Watson, *2011/2012 Staying@Work Report* (2012) at 9, online: http://www.towerswatson.com/en/Insights/IC-Types/Survey-Research-Results/2011/12/20112012-StayingWork-Survey-Report--A-Pathway-to-Employee-Health-and-Workplace-Productivity (date accessed: January 31, 2020).

[20] Calculated by multiplying eight hours by the average number of lost time days (10.3) by the average hourly wage for Canadian workers (8 hours x 10 days x $28.96). Data obtained from Statistics Canada, *Table 14-10-0190-01 Work Absence of full-time employees by geography, annual 2019* (Ottawa: Statistics Canada, January 2020), online: https://www150.statcan.gc.ca/t1/tbl1/en/tv.action?pid=1410019001; and Statistics Canada, *Table 11, Average Usual Hours and Wages of Employees by Selected Characteristics*, online: https://www150.statcan.gc.ca/n1/daily-quotidien/191108/t011a-eng.htm (date accessed: February 28, 2020).

on the rise in organizations/companies with low employee morale and remained steady or dropping in organizations/companies with high employee morale.[21]

Other reasons for absenteeism are:

- personal illness/injury;
- personal needs;
- family issues/responsibilities including childcare and eldercare;
- entitlement mentality;
- high workplace stress;
- an aging workforce; and
- increased prevalence of generous sick plans and family-leave plans.

Based on the work absence findings from Statistics Canada, 2019, there are some additional variables that negatively impact regular work attendance. The size of an organization is one variable: large organizations experience more lost workdays (12.5 days PEPY[22]) than small ones (8.6 days PEPY).[23] Unionized workplaces have more lost workdays (14.9 days) than non-unionized workplaces (8.2 days – 1.8 times more).[24] Public sector employees miss more days (14.6 days) than private sector ones (nine days).[25] Permanent employees miss more time (10.5 days) than do non-permanent employees (7.9 days).[26] Shift workers miss slightly more workdays than do other workers. Women with preschoolers miss more (15.1 days) than men with preschoolers (7.8 days), and more than do employees without any children (10 days).[27] Long-term employees miss more days (13.5) than do employees who have

---

[21] Wolters Kluwer, *News Release* "CCH Survey Finds Most Employees Call in "Sick" for Reasons Other Than Illness" (2007), online: CCH Group.com http://news.cchgroup.com/2007/10/10/cch-survey-finds-most-employees-call-in-sick-for-reasons-other-than-illness/news/press-releases/ (date accessed: February 28, 2020).

[22] **PEPY** is an acronym for per employee per year.

[23] Statistics Canada, *Table 14-10-0198-01 Work absence of full-time employees by establishment size, annual 2019*, online: https://www150.statcan.gc.ca/t1/tbl1/en/cv.action?pid=1410019801#timeframe (date accessed: February 28, 2020).

[24] Statistics Canada, *Table 14-10-0200-01 Work absence of full-time employees by union coverage, annual 2019*, online: https://www150.statcan.gc.ca/t1/tbl1/en/cv.action?pid=1410020001#timeframe (date accessed: February 28, 2020).

[25] Statistics Canada, *Table 14-10-0196-01 Work absence of full-time employees by public and private sector, annual 2019*, online: https://www150.statcan.gc.ca/t1/tbl1/en/tv.action?pid=1410019601 (date accessed: February 28, 2020).

[26] Statistics Canada, *Table 14-10-0199-01 Work absence of full-time employees by job permanency, annual 2019*, online: https://www150.statcan.gc.ca/t1/tbl1/en/tv.action?pid=1410019901 (date accessed: February 28, 2020).

[27] Statistics Canada, *Table 14-10-0194-01 Work absence of full-time employees by sex*

been with the company less than a year (6.9 days).[28] Employees with less than grade nine education miss more workdays (14.7) than employees with university education (8.2 days).[29] Lastly, blue-collar workers miss more workdays than white-collar workers do.[30]

The Watson-Wyatt Worldwide *1997 Staying @ Work* survey indicated that factors like the employee group benefit plan design, job performance, knowledge of the benefit plan system, worker's job, wording of the collective agreement contracts,

---

and presence of children, annual 2019, online: https://www150.statcan.gc.ca/t1/tbl1/en/cv.action? pid=1410019401#timeframe (date accessed: February 28, 2020).

[28] Statistics Canada, *Table 14-10-0195-01 Work absence of full-time employees by job tenure, annual 2019*, online: https://www150.statcan.gc.ca/t1/tbl1/en/cv.action?pid=1410019501#timeframe(date accessed: February 28, 2020).

[29] Statistics Canada, *Table 14-10-0197-01 Work absence of full-time employees by educational attainment, annual 2019*, online: https://www150.statcan.gc.ca/t1/tbl1/en/cv.action?pid=1410019701#timeframe (date accessed: February 28, 2020).

[30] Statistics Canada, *Table 14-10-0285-01 Work absence of full-time employees by occupation, annual 2019*, online: https://www150.statcan.gc.ca/t1/tbl1/en/cv.action?pid=1410028501#timeframe (date accessed: February 28, 2020).

and organizational restructuring, are associated with worker absenteeism (Figure 42.4).

Figure 42.4: Other Factors Influencing Worker Absenteeism[31]

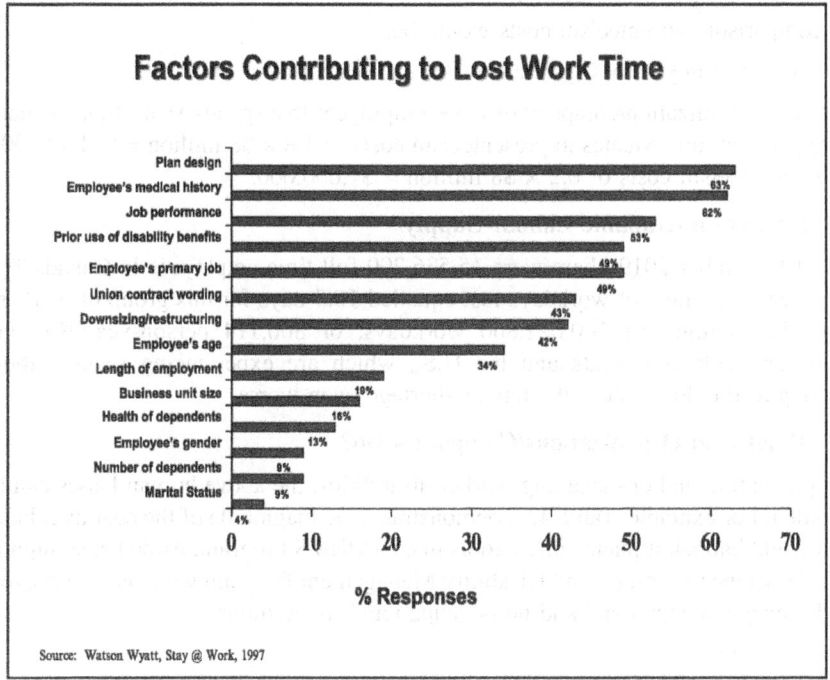

So, the reasons for worker absenteeism are multifactorial and complex. Understanding why employees are off work and the environment in which they are working, is vital if successful interventions are to be designed and implemented.

3. Presenteeism

What about workers who come to work, but because of wasted time, failure to concentrate, sleep deprivation, distractions, poor health, and/or lack of training, they may not be productive at all? Termed "presenteeism", this phenomenon is of concern to 56 per cent of the organizations/companies that participated in the *CCH 2006 Unscheduled Absenteeism Survey*.[32] Presenteeism has been found to be associated with low employee morale.

---

[31] Reprinted with permission from Watson Wyatt Worldwide, *1997 Staying @ Work Survey*, online: http://www.watsonwyatt.com (date accessed: January 31, 2020).

[32] Wolters Kluwer, *News Release*, "CCH Survey CCH Survey Finds Unscheduled Absenteeism Up in U.S. Workplaces" (2006) online: CCH Group.com http://hr.cch.com/press/releases/absenteeism/102506a.asp (date accessed: February 28, 2020).

Based on work done by the Wellness Council of America, it was determined that the costs due to lost productivity and presenteeism were nine times greater than productivity loss and costs from absenteeism.

Presenteeism = 1.8 × Organization's Annual Health Care Costs.[33]

In comparison, absenteeism costs would be:

Absenteeism = 0.2 × Organization's Annual Health Care Costs.[34]

For an organization/company of 1,000 employees that spends $8 million on health care per year, this equates to presenteeism costs of 1.8 × $8 million = $14,400,000; and absenteeism costs of 0.2 × $8 million = $1,600,000.

4. Impact on Available Labour Supply

As of December 2019, there were 15,536,200 full-time employees in Canada.[35] If the average number of workdays lost equalled 10.3 days for this group of workers, then the system lost 160,022,860 workdays, or 800,114 person-years[36],[37] For countries such as Canada and the U.S., which are experiencing severe labour shortages, this loss makes the labour shortage even worse.

5. What Can Organizations/Companies Do?

By preventing and/or managing worker disabilities, the above human losses can be avoided. For example, Table 42.1 demonstrates the magnitude of the cost-avoidance that could happen if management tools like an OH&S Program, Attendance Support and Assistance Program, and Disability Management Program were used to prevent and manage occupational and non-occupational illness/injury.

---

[33] S. Aldana, *Top Five Strategies to Enhance the ROI of Worksite Wellness Programs*, Wellness Council of America (February 2009), online: http://www.absoluteadvantage.org/pdf/contentmgmt/ top_5_strategies.pdf (date accessed: January 31, 2015).

[34] *Ibid.*

[35] Statistics Canada, *Table 14-10-0287-01 Labour force characteristics, monthly, seasonally adjusted and trend-cycle, last 5 months, 2019* (Ottawa, ON: Government of Canada, December 2019), online: https://www150.statcan.gc.ca/t1/tbl1/en/tv.action?pid=1410028701 (date accessed: January 31, 2020).

[36] A Person-year is 200 workdays.

[37] Calculated by multiplying the number of full-time employees (15,536,200) in 2019 by the average number of employee work absent days (10.3 days). Person year is 200 days and that was divided into the 160,022,860 lost workdays to determine how many person years were lost in 2019.

Table 42.1: Potential Worker Loss Avoidance

Number of FTEs Off on Disability	Rate of Reduction	Cost/Loss Avoidance (FTEs/yr)	Cost/Loss Avoidance($$)[38]
800,114	10%	80,011	$18,536,948
800,114	20%	160,023	$37,074,129
800,114	30%	240,034	$55,611,123
800,114	50%	400,057	$92,685,206

Note: This is the magnitude of saving that Canada could realize if corporate intervention strategies were widely used in 2019. (FTE=full-time employee)

To achieve these reduction rates, employers can use the following strategies:

(1) Develop a sound OH&S Program, that focuses on physical and psychological health and safety;

(2) Prevent employee presenteeism and employee absenteeism by:
- Working towards developing strong employee morale;
- Providing programs that support employees, *i.e.*, alternative work arrangements; telecommuting; compressed work week; job sharing; emergency child care; EAPs; flu immunization; on-site child care; wellness programs; on-site health services; and elder care services.
- Managing employee absenteeism in a positive manner — monitor absenteeism with a view to being supportive;[39] use programs such as Paid Leave Bank, Buy Back Program, and Bonus Program.
- Addressing employee presenteeism — educate workers; foster a culture that discourages employee presenteeism; send sick workers home; and ensure sick leave policy does not encourage employee presenteeism.

(3) Manage employee disability by implementing an integrated disability management program. Chapter 27, "Disability Management: Overview" provides an overview of an Integrated Disability Management Program which has been known to reduce employee disability by an average of 19–25 per cent. However, some organizations have seen improvements of up to 50 per cent within the first year.

---

[38] Calculated by multiplying the number of full-time employees returned to work though organizational efforts by the average daily wage ($28.96 x 8 hours= $231.68 per day).

[39] Attendance Support and Assistance Programs have demonstrated a 30 per cent improvement in worker absenteeism: see D. Dyck, *Disability Management: Theory, Strategy and Industry Practice*, 6th ed. (Markham, ON: LexisNexis, 2017).

6. What Do Organizations/Companies Need to Achieve Success?

The development of OH&S Programs, Attendance Support Programs, and Disability Management Programs requires technical expertise. This is where OH&S, Human Resources (HR), and Disability Management (DM) Practitioners/Professionals can demonstrate their value to the organizations in which they work or consult. The bottom line is that these programs will only be as good as the OH&S, DM, and HR Professionals operating them. These professionals must understand the organizational system and its challenges and have the requisite skills and experience to develop innovative solutions so that the organization is well-positioned to succeed.

G. MEASUREMENT OF LEADING VERSUS LAGGING INDICATORS OF OH&S PERFORMANCE

Over the years, employers have focused on the OH&S incident and injury results. These are "downstream indicators", or lagging indicators of workplace OH&S performance. This rear-view mirror approach of evaluating workplace OH&S performance tends to be ineffective in preventing illness/injury and affords the employer little opportunity to make a difference in a timely manner. Once measured, nothing can be done to change the data or situation. In effect, they are a *"fait accompli"*.

> The safety culture is unlikely to be positively affected when trailing [lagging] indicators are the primary focus or are the sole safety metrics an organization uses to assess performance.[40]

"Upstream indicators", or leading indicators of OH&S performance, are the conditions necessary for workers to be prepared and positioned to work safely. It translates into leaders creating a workplace culture that supports a high degree of commitment to workplace OH&S; having processes for measuring, monitoring and managing the leading indicators of OH&S performance; and demonstrating the effectiveness of the established OH&S Management System.

> Leading indicators measure actions, behaviors. and processes, the things people actually do for safety, and not simply the safety-related failures typically tracked by trailing [lagging] measures.[41]

In Chapter 6, "Occupational Health and Safety: Leadership and Commitment" and Chapter 7, "Great Safety Performance: A Management Approach", the Great Safety Performance Model is described as an industry model designed to promote a work environment, as well as a culture that promotes and facilitates Great Safety Performance within the organization. The measurement of the five leading indicators of safety is explained and demonstrated. Based on a research study conducted by Nina Novak (2006), there is a correlation between some of the leading indicators

---

[40] E. Blair & M. O'Toole, "Leading measures: Enhancing safety climate and driving safety performance" (2010) 55:8 *Professional Safety* 29.

[41] *Ibid.*

of OH&S performance and a reduction in the number of lost-time incidents. For example, the number of lost-time incidents was negatively correlated with a system in which workers:

- were equipped and knew what was expected from the work they were doing;
- were equipped and had the materials and equipment to do the assigned work;
- were equipped and worked in an environment in which the physical working conditions were safe;
- were trained and able to work safely;
- received regular and frequent recognition for the work done; and
- in which safety problems are quickly fixed by the organization/company.[42]

This is the type of evidence that indicates that employers are creating safe and healthy work environments — evidence that organization/company leaders may be required to prove if the quality of their OH&S duty comes into question.

H. SUMMARY

The field of Occupational Health and Safety will continue to be fraught with interesting challenges. To practice effectively, OH&S Practitioners/Professionals need to be equipped with current knowledge; sound OH&S principles and standards; honest and open communication skills; good relationship building and nurturing skills; and regular practice evaluation techniques.

CHAPTER REFERENCES

S. Aldana, *Top Five Strategies to Enhance the ROI of Worksite Wellness Programs*, Wellness Council of America (February 2009), online: http://www.absoluteadvantage.org/pdf/contentmgmt/top_5_strategies.pdf (date accessed: January 31, 2020).

Comments taken from discussion with I. Arnold (2006).

E. Blair & M. O'Toole, "Leading measures: Enhancing safety climate and driving safety performance" (2010) 55:8 *Professional Safety* 29.

S. Cavanagh, "The Gender of Professionalism and Occupational Closure: The Management of Tenure-related Disputes by the 'Federation of Women Teachers' Association of Ontario' 1918–1949" (2003) 15:1 *Gender and Education* 30.

CCH Inc., 2006 *CCH Annual Unscheduled Absenteeism Survey*, online: http://www.cch.com/default.asp (date accessed: January 31, 2020).

D. Dyck, *Disability Management: Theory, Strategy and Industry Practice*, 6th ed. (Markham, ON: LexisNexis, 2017).

---

[42] N. Novak, "Evaluation of Safety Indicators" (Masters Thesis, Department of Occupational Health, McGill University, September 22, 2006) [unpublished].

G. Larkin, *Occupational Monopoly and Modern Medicine* (London, UK: Tavistock, 1983).

Ontario, Bill 146, *Stronger Workplaces for Stronger Economies Act* (2014), online: http://www.ontla.on.ca/web/bills/bills_detail.do?locale=en&BillID=2916 (date accessed: January 31, 2020).

Oxford University Press, *Oxford English Dictionary*, 2d ed. (Oxford University Press, 1989).

M. Quinlan & P. Bohle, "Contingent Work and Occupational Safety" in *The Psychology of Workplace Safety*, J. Barling & M. Frone, eds. (Washington, DC: American Psychological Association, 2004).

Registered Nurses Association of Ontario (RNAO), *Professionalism in Nursing, Healthy Work Environments, Best Practice Guidelines* (Toronto, ON: RNAO, 2007) at 58, online: http://www.rnao.ca/sites/rnao-ca/files/Professionalism_in_Nursing.pdf (date accessed: January 31, 2020).

J. Roberts & M. Dietrich, "Conceptualizing Professionalism: Why Economics Needs Sociology" (1999) 58:4 *The American Journal of Economics and Sociology*, online: http://findarticles.com/p/articles/mi_m0254/is_4_58/ai_58496769 (date accessed: January 31, 2020).

Statistics Canada, *Table 11-19-1108, Average Usual Hours and Wages of Employees by Selected Characteristics*, online: https://www150.statcan.gc.ca/n1/daily-quotidien/191108/t011a-eng.htm

Statistics Canada, *Table 14-10-0190-01 Work Absence of full-time employees by geography, annual* (Ottawa: Statistics Canada, 2020), online: https://www150.statcan.gc.ca/t1/tbl1/en/tv.action?pid=1410019001 (date accessed: February 28, 2020).

Statistics Canada, *Table 14- 10-0198-01 Work absence of full-time employees by establishment size, annual 2019*, online: https://www150.statcan.gc.ca/t1/tbl1/en/cv.action?pid=1410019801#timeframe (date accessed: February 28, 2020).

Statistics Canada, *Table 14-10-0200-01, Work absence of full-time employees by union coverage, annual 2019*, online: https://www150.statcan.gc.ca/t1/tbl1/en/cv.action?pid=1410020001#timeframe (date accessed: February 28, 2020).

Statistics Canada, *Table 14-10-0196-01 Work absence of full-time employees by public and private sector, annual 2019*, online: https://www150.statcan.gc.ca/t1/tbl1/en/tv.action?pid=1410019601 (date accessed: February 28, 2020).

Statistics Canada, *Table 14-10-0199-01 Work absence of full-time employees by job permanency, annual 2019*, online: https://www150.statcan.gc.ca/t1/tbl1/en/tv.action?pid=1410019901 (date accessed: February 28, 2020).

Statistics Canada, *Table 14-10-0194-01 Work absence of full-time employees by sex and presence of children, annual 2019*, online: https://www150.statcan.gc.ca/t1/tbl1/en/cv.action?pid=1410019401#timeframe (date accessed: February 28, 2020).

Statistics Canada, *Table 14-10-0195-01 Work absence of full-time employees by job tenure, annual 2019*, online: https://www150.statcan.gc.ca/t1/tbl1/en/cv.action?pid=1410019501#timeframe (date accessed: February 28, 2020).

Statistics Canada, *Table 14-10-0197-01 Work absence of full-time employees by educational attainment, annual 2019*, online: https://www150.statcan.gc.ca/t1/tbl1/en/cv.action?pid=1410019701#timeframe (date accessed: February 28, 2020).

Statistics Canada, *Table 14-10-0285-01 Work absence of full-time employees by occupation, annual 2019*, online: https://www150.statcan.gc.ca/t1/tbl1/en/cv.action?pid=1410028501#timeframe (date accessed: February 28, 2020).

Statistics Canada, *Table 14-10-0287-01 Labour force characteristics, monthly, seasonally adjusted and trend-cycle, last 5 months, 2019* (Ottawa, ON: Government of Canada, December 2019), online: https://www150.statcan.gc.ca/t1/tbl1/en/tv.action?pid=1410028701 (date accessed: January 31, 2020).

G. Taylor, K. Easter & R. Hegney, *Enhancing Occupational Safety and Health* (Oxford: Elsevier Butterworth-Heinemann, 2004).

Towers Watson, 2011/2012 Staying@Work Report (2012) at 9, online: http://www.towerswatson.com/en/Insights/IC-Types/Survey-Research-Results/2011/12/20112012-StayingWork-Survey-Report--A-Pathway-to-Employee-Health-and-Workplace-Productivity (date accessed: January 31, 2020).

Watson-Wyatt Worldwide, 1997 Staying@Work Survey, online: http://www.watsonwyatt.com (date accessed: January 31, 2020).

Wikipedia, "Professionalization", online: http://en.wikipedia.org/wiki/professionalization (date accessed: January 31, 2020).

Wolters Kluwer, *News Release*, "CCH Survey Finds Most Employees Call in "Sick" for Reasons Other Than Illness" (2007), online: CCH Group.com http://news.cchgroup.com/2007/10/10/cch-survey-finds-most-employees-call-in-sick-for-reasons-other-than-illness/news/press-releases/ (date accessed: February 28, 2020).

Wolters Kluwer, *News Release*, "CCH Survey CCH Survey Finds Unscheduled Absenteeism Up in U.S. Workplaces" (2006), online: CCH Group.com http://hr.cch.com/press/releases/absenteeism/102506a.asp (date accessed: February 28, 2020).

Section 5
Occupational Health and Safety: Glossary

Section 5
Occupational Health and Safety Glossary

GLOSSARY

Acceptable Death Risk Rate — one death in a million to one death per 10,000.

Accident — any unplanned event that results in personal injury or property damage.

Accident Proneness Theory — which maintains that within a given set of workers, there exists a subset of workers who are more liable to be involved in incidents.

Accident Theory — the science that explains how incidents are caused. An accident theory provides categories of causes and links them together in a comprehensive manner. It helps distinguish between immediate and underlying basic or root causes.

Acculturation — cultural modification of an individual, group or people by adapting to, or borrowing, traits from another culture; a merging of cultures as a result of prolonged contact.

Actus reus — the prohibited act.

Acute illness/injury — a physical injury or sudden occurrence of an illness that results in the need for immediate temporary care.

Adaptive Organization Design — a modern organizational design in which the organization operates with minimal bureaucratic features; focuses on developing a work culture that values worker empowerment and participation; and uses team and network structures.

Administration controls — a form of hazard control that relies on management actions such as worker training; work scheduling to reduce exposure times, development and use of safe work practices; good housekeeping; product/process labelling, signage and worker warning systems; proper storage of hazardous agents; worker compliance to standards and rules; use of pre-employment screening to ensure a suitable and safe person-job fit; use of medical monitoring and medical surveillance; and provision of immunizations to workers.

Administrative Management Theory — emphasizes the flow of information in the operation of the organization with an emphasis on the economic rationality of the individual employee at work.

Alberta Partnerships in Injury Reduction Program — a voluntary program designed to create a partnership between the Alberta government and Alberta employers to promote and develop strong workplace cultures within Alberta businesses.

Annual death rate — the number of individuals from an exposed group who died from a specific cause within a given year.

ANSI/AIHA Z10 2005 — the American National Standard for Occupational Health and Safety Management Systems. This standard provides the critical management systems requirements and guidelines for the improvement of Occupational Health and Safety.

Artificial Intelligence (AI) — the theory and development of computer systems able to perform tasks that normally require human intelligence, such as visual perception, speech recognition, decision-making, and translation between languages.

ASAP — Attendance Support and Assistance Program.

Assimilation — the assumption of the cultural traditions of a given people or group.

Attendance Support and Assistance Program (ASAP) — proactive approach to promoting and supporting employee attendance at work, and to prevent "work presenteeism".

Auditing — involves documentation reviews, interviews, and observations to verify the existence and functionality of various programs, policies, and procedures.

Authority-Obedience Management — the management approach that upholds that efficiency in operations results from arranging work conditions so that the interference of human elements is kept to a minimum. This translates to a high regard for production and a low regard for relationships.

Automation — involves the use of machines or technology to automatically do tasks faster, better, and at a reduced cost, with minimal to no manpower.

Average Cost of Psychological Health Claims — is determined by dividing the Total Cost of Psychological Health Claims for a given period of time by the number of psychological health claims for that specific period of time.

Average Cost of Psychological Health Claims per Employee — is determined by dividing the Total Cost of Psychological Health Claims for a given period of time by the number of employees eligible to submit a psychological health claim for that specific period of time.

Balanced Score Card — a measurement tool that focuses on financial measures; customer service indicators; internal work efficiencies; innovation, learning and growth measures; and system measures.

Basic Needs Theory — holds that there are three levels of human needs; the need for achievement, the need for power, and the need for affiliation.

Benchmarking — continual and collaborative discipline that involves measuring and comparing the results of the key process with "best performers" or with one's own previous achievements.

Beneficence — requires that OH&S Practitioners/Professionals act in the best interest of the employee/organization.

Best practices — a form of benchmarking that results from direct observation of clinical practices.

Biased-liability Theory — is based on the view that once a worker is involved in an incident, the chances of the same worker becoming involved in future incidents are either increased or decreased as compared to the rest of workers.

Bioterrorism attack — the deliberate release of viruses, bacteria, or other germs (agents) used to cause illness or death in people, animals, or plants.

Bipolar disorder — a mood disorder in which the individual experiences alternating mood swings that range from being elated to being depressed.

Bullying Frequency Rate — is determined by dividing the total number of bullying incidents multiplied by 200,000 by the total number of hours worked.

Bureaucratic Design — an organizational design characterized by a clear-cut division of labour; strict hierarchy of authority; staffing by technical competency; the existence of formal rules and procedures; and an impersonal approach to workplace decision-making.

Bureaucratic Theory — a management theory founded by Max Weber and based on logic, order, and the concept of legitimate authority. It upholds that greater productivity can be achieved by using a proper organizational structure; a well-defined hierarchy of authority needs to exist; a clear division of labour is essential; formal rules and procedures are required; increased bureaucracy leads to greater operational efficiency and fairness for employees; career growth ought to be based on merit; and the result would be an expedient way to deal with urgent situations and multiple worksites.

Business resilience — reducing the likelihood of a business interruption occurring and reducing its impact if and when an incident occurs.

Business writing — correspondence on behalf of an organization/company. It must reflect the philosophies and values of the organization/company, while providing the needed business information.

Cannabidol (CBD) — an antidepressant, anti-convulsant, antioxidant, anti-psychotic, antiemetic, anti-inflammatory agent, and possible neuro-protective agent. It is not psychoactive.

Career — a lifelong series of events, rather than an evaluation of how successful individuals have been in their life.

Career advancement stage — characterized by a steep learning curve in which the person begins to better understand the work environment and organizational demands and strives to establish personal worth within the organization. It is a period of growth and acceptance of responsibility.

Career anchors — the aspects of work that people value or need for fulfilment.

Career development — the process through which individual employees, with the help of their managers, match up their skills, interests, and goals with the opportunities available to them and develop plans for achieving their goals.

Career entry stage — an exploration stage in which the individual tries to match individual need, abilities, and skills with organizational requirements.

Career maintenance stage — the stage at which the individual becomes entrenched in the job.

Career plateauing — a position from which one is unlikely to advance to a higher level of responsibility in their career.

Career withdrawal — the stage that was traditionally associated with retirement.

Catastrophic fault — endangers, harms, or kills a significant number of people.

CBD — cannabinol which is an antidepressant, anti-convulsant, antioxidant, anti-psychotic, anti-emetic, anti-inflammatory agent, and possible neuro-protective agent.

Charismatic Leadership — similar and complementary to Transformational Leadership, it is leadership that is grounded in the attraction that followers have for leaders who believe in themselves, possess charm and grace, and are admired.

Clarity — the company members know what is expected of them and why.

Clashpoint® — an aspect of the workplace where generational differences in perspective, attitude and opinion tend to occur. The differences include attitudes toward career goals, performance feedback, performance rewards, job changes, and retirement.

Classical Management Theory — see **Administrative Management Theory**.

Client advocacy — the activity associated with pleading or representing the cause of an employee, or organization.

Client liaison — the position or responsibility within an organization for maintaining communication links with external individuals, agencies, or organizations.

Client Satisfaction Survey — a perception survey that asks clients about their level of satisfaction with a service or product.

Close worksite — a worksite that is not more than 20-minute travel time from a healthcare facility under normal travel conditions using the available means of transportation.

Command fault — a device which appears to malfunction because it has responded as designed to a bad input.

Commitment — passionate "buy-in" and ownership of a belief/concept/practice.

Communication — the ability to convey facts, concepts, or reasoning clearly to others, and to receive and understand the messages sent by others.

Competency — a method of ensuring that the workforce is capable of carrying out the expected work from a technical, quality, and OH&S perspective related to qualifications, training, and experience.

Competent worker — a worker that is adequately qualified, suitably trained, and with sufficient, relevant experience to safely perform work without, or with, only a minimal degree of supervision.

Complacency — occurs when workers perform tasks repetitively, so much so that they tend to become bored or smug about undertaking the tasks and perform them in a thoughtless manner. Complacency also exists when an experienced employee takes shortcuts while working on a critical task because he/she has done it so often and as a result, believes that he/she is able to complete it quicker and just as safely using this alternate method. It is also present when Line Management/supervisory staff knowingly overlook complacent work practices, also believing that there is no harm in employees taking "shortcuts".

Confidentiality — maintenance of trust expressed by an individual verbally, or in writing, and the avoidance of an invasion of privacy through accurate reporting and authorized communication.

Confounding variable — a variable that confuses the relationship between the independent and dependent variables, and that needs to be controlled through the design of the evaluation plan or via statistical procedures.

Consensus — the beliefs, values, attitudes, habits, and traditions that are shared by all organizational members.

Consistency — actions and symbols within the organization that are congruent.

Consortium for Organizational Mental Healthcare (COMH) — a collective of mental health researchers, consultants, and practitioners who are experienced in working with a range of public and private sector organizations. Its goal is to further the creation and translation of mental health knowledge and practice into real world settings. To achieve this objective, the Consortium forms collaborative relationships with committed organizations, leaders, and experts to achieve specific outcomes.

Consulting — the art of influencing people at their request. It is a voluntary, temporary and supportive relationship between the helper and the party in need of help (the client). It focuses on providing temporary assistance to the client to enable him/her to reach a solution to potential (or real) problem situations or opportunities.

Consulting contract — an explicit agreement clarifying what the client and consultant can expect from each other and how they will work together.

Content objective — a specific statement that speaks to the content to be covered in a training session/course. For example, *"At the end of this course, the worker will be aware of the legislative requirement for noise abatement and hearing conservation."*

Contingency Theory — proposes that leaders adjust their leadership style based on the "readiness" of followers to perform in certain situations.

Contingent Workforce — that portion of the workforce who are contractors, consultants, or workers in part-time positions.

Continuous improvement — the process of enhancing the Occupational Health and Safety Management System to achieve improvements in overall OH&S performance in accordance with the organization's OH&S Policy.

Contributing incident causes — could be management-related factors, environment factors, and/or the physical and mental condition of the worker.

Control group — a group which is as similar as possible (in observable and unobservable dimensions) to those receiving the intervention (the study group).

Controlled products — materials that are regulated by WHMIS 2015.

Controlling — comparing results with predetermined standards of performance and taking corrective action when performance deviates from these standards.

Coping — is an attempt to overcome the obstacle by adopting a variety of human behaviours.

Core culture — composed of the corporate values, beliefs, and acceptable behaviours.

Coroner's inquests — are held, under coroner's legislation, when a workplace fatality has occurred. Coroner's juries, in a non-fault-finding exercise, review the causes of workplace fatalities and various workplace jurisdictions. They play an important role in reviewing and changing OH&S laws.

Corporate culture — the system of shared beliefs and values that develops within an organization and guides behaviour of its members. It is the way things are done within an organization.

Cost benefit — weighing costs of the OHSMS against the benefits provided.

Cost-benefit analysis — an analysis technique that weighs the costs of a program/function against the benefits provided.

Cost containment — refers to keeping costs to a minimum.

Cost effectiveness — demonstrating the results of the OHSMS from a financial perspective.

Country Club Management — a management style proposed by Blake and Mouton that is characterized by attention to worker needs for a satisfying work relationship can lead to a comfortable, friendly, organizational atmosphere, and work tempo. This translates to a low regard for production and a high regard for people and their needs/relationships.

Criminal Negligence — under Bill C-45, either results from an act or an omission relating to a legal duty, when a person shows wanton or reckless disregard for the lives or safety of other persons.

Crisis communication — encompasses those messages delivered to stakeholders during a crisis period. It is the transfer of information to help avoid a crisis situation, recover from a crisis, or maintain or enhance reputation.

Critical fault — a fault that endangers one or a few people.

Critical tasks — tasks that have the potential to produce major loss to people, property, business process, and/or environment when not performed properly.

GLOSSARY

CSA-Z1000-06 — a Canadian national standard that defines the requirements for Occupational Health and Safety Management System development and implementation.

Cultural awareness — being cognizant, observant, and conscious of similarities and differences among cultural groups.

Cultural blindness — the stage at which people assume that everyone is basically alike and advocate a universal approach and services for all people.

Cultural competence — defined as having the ability to provide quality OH&S and Disability Management care and services to a diverse employee population.

Cultural competency — the ability of individuals and systems to respond respectfully and effectively to people of all cultures, classes, races, ethnic backgrounds, and religions in a manner that recognizes, affirms and values the cultural differences, similarities, and worth of individuals, families, and communities and protects and preserves the dignity of each.

Cultural destructiveness — the stage of intentional denial, rejection, or outlawing of other cultures.

Cultural diversity — the recognition that people originate from a variety of gender, age, ethnic, geographic, economic, and religious backgrounds.

Cultural incapacity — the stage of acceptance of other cultures, but an inability to work effectively with other cultures.

Cultural pre-competence — the stage of awareness within systems or organizations of their strengths and areas for growth to respond effectively to culturally and linguistically diverse populations. It includes a willingness to learn and understand other cultures.

Cultural proficiency — the stage at which proactive promotion of cultural diversity occurs and opportunities to improve cultural relationships are sought.

Cultural sensitivity — understanding the needs and emotions of one's culture and the culture of others.

Culture — the values, beliefs, customs, behaviours, and structures shared by a group of people.

Damage incident — an incident that results in equipment or property damage only.

Dangerous goods — substances that pose risk to health, safety, property, or the environment during operation and/or transportation.

Data — any form of information that can be collected and is relevant to an individual, group, or organization function.

Decision-making — the process of choosing among alternative courses of action.

Decision Theory — focuses on the sequence of events that have to happen for an incident to be avoided, or for an incident to occur.

Deficit Principle — a satisfied need is not a motivator of behaviour: rather, people act to satisfy deprived needs.

Definitive causality — attributing observed changes to the program, while removing confounding factors.

Delta-9-tetrahydrocannabinol (THC) — a psychoactive agent that produces effects such as analgesia, relaxation, drowsiness, euphoria, suppression of PTSD symptoms and chemotherapy-induced side-effects, and appetite stimulation.

Deontological theory — deals with action and asserts that "rightness" and "wrongness" are measured by means, rather than by consequences of an action.

Departmentalization — the process whereby people and jobs are grouped into various work units or departments to create a well-functioning whole.

Dependent variable — the outcome variable of interest; the variable that is hypothesized to depend on or be caused by another variable.

Depression — a mood disorder that impacts both an individual's emotional state and their physical functioning.

Designated representative — any individual or organization to whom an employee gives written authorization to exercise a right to access personal information.

Digitalization — the use of digital technologies to change a business model and provide new revenue and value-producing opportunities; it is the process of moving to a digital business.

Digitization — the process of converting a variety of forms of information into a digital format. This a commonly done in business, for example scanning incident reports.

Direct Case Management Model — an employee-employer approach to dealing with an employee's reduced work capacity and the employer's business needs/resources. It is the employee and employer who decide, based on their respective needs, the terms of the medical absence, and the return-to-work plan.

Direct safety management costs — the costs directly identifiable in terms of management of the safety system, such as the OH&S budget, training costs, safety equipment costs, and associated program costs.

Disability Case Management — a collaborative process for assessing, planning, implementing, coordinating, monitoring, and evaluating the options and services available to meet individual health needs through communication and accessible resources. Effective case management promotes quality, cost-effective outcomes.

Disability Claim Costs — determined by dividing the total claim costs for a given period of time by the total number of employees for that same period of time.

Disability Claims per Employee — is determined by dividing the total number of claims for a given period of time by the total number of employees for that same period of time.

Glossary

Disability Management — a systematic, goal-oriented process of actively minimizing the impact of impairment on the individual's capacity to participate competitively in the work environment, and maximizing the health of employees to prevent disability or further deterioration when a disability exists.

Disability Management Program — a workplace program designed to facilitate the employment of persons with a disability through a coordinated effort that addresses individual needs, workplace conditions, and legal responsibilities.

Disability Severity Rate — the number of disabilities per 100 workers and is similar to the Injury Severity Rate used in the field of Occupational Health & Safety. It can also be termed the **Morbidity Rate**.

Distant worksite — a worksite that is more than 20 minutes, but less than 40 minutes, travel time from a healthcare facility under normal travel conditions using the available means of transportation.

Diversity — refers to "dissimilarity and variance between things and people".

Division of labour — a management approach that reduces the span of attention and/or effort of any one group and enables the development of expertise through practice and familiarity.

Divisional Structure — an organizational structure type in which workers are grouped based on similar product, or customers in same geographic area or on the same time schedule.

Divisionalized Bureaucracy — the organizational design that demonstrates the traits of:
- a hybrid form of departmentalization;
- a number of relatively autonomous internal units operating within a common organizational umbrella;
- divisions that are formed based on product/service, customers, or geographic location; and
- top management coordination assisted by a large staff complement.

Domino Theories — assume that if unsafe practices and unsafe work conditions that lead to "near miss" incidents are eliminated, then the probability of the occurrence of more serious incidents are reduced.

Downstream Safety Performance Indicators — See **Lagging indicators of safety performance**.

Due diligence — the level of judgement, care, prudence, determination, and activity that a person/organization would reasonably be expected to do under particular circumstances.

Early intervention — contacting the ill/injured employee within five days from the onset of a medical absence and initiating case management if warranted.

Effectiveness — the achievement of stated goals within the stated time frame.

Efficiency — being the ratio of effective work output to the input required to produce the work done.

Emergency First Aider — a person who holds a certificate in Emergency First Aid from a training agency.

Emergency Response Plans — provide specific instructions for responding to and dealing with emergencies.

Emerging hazards — hazards that develop with time. To determine the amount of OH&S risk within a given company system, each inherent and emerging hazard must be examined in terms of the controls in place.

Empathetic Communication — the technique of structured listening and questioning - helps to develop and enhance relationships through a stronger understanding of what is being conveyed intellectually and emotionally.

Enculturation — the repetitious and systematic inculcation of a shared system of values, beliefs, attitudes, and learned behaviours.

Energy Transfer Theory — focuses on the concept of energy release as a necessary part of incident causation.

Engagement — the mental, physical, and cognitive connection employees perceive to have with their work. It is also termed the **occupational bond**.

Engineering controls — a method of hazard control designed to address the hazard at the source.

Environmental aspect — anything that an organization does that can react with any part(s) of the environment.

Environmental assessment — an assessment done prior to start up of an operation to determine the actual/potential environmental aspects and impacts.

Environmental audit — a systematic process of objectively obtaining and evaluating evidence regarding a verifiable assertion about an environmental matter, to determine the degree of correspondence between a verifiable assertion and standards criteria, and report to the client.

Environmental impact — any change to the environment, good or bad, caused by or resulting from, an organization's activities, products, and/or services.

Environmental Management System (EMS) — a management tool designed to manage an organization's environmental affairs, monitor its effect on the environment, and address the environmental "bottom-line" (economies). It is a structured approach which consists of a number of interrelated elements that work together to achieve the objective of effective environmental management.

Equity Theory — developed by J. Stacy Adams who proposes that perceived inequity is a motivator. When a worker compares with others his/her work situation in terms of rewards, job opportunities, pay increases, and other benefits, the worker is motivated or demotivated to work.

ERG Theory — proposes that the basic human needs are:
1. Existence Needs;
2. Relatedness Needs; and
3. Growth Needs.

People do not have to satisfy a lower need before addressing a higher-order need.

Ergonomic Theory — addresses the relationship between the worker and his/her surroundings. This theory incorporates the physical and psychological factors that impact the worker internally and externally.

Ergonomics — the study of how people interact with their work environment. This involves people, machinery, and the work organization.

Ergonomist — an individual who has (1) a mastery of ergonomics knowledge; (2) a command of the methodologies used by ergonomists in applying that knowledge to the design of a product, process, or environment; and (3) has applied his/her knowledge to the analysis, design, test, and evaluation of products, processes, and environments.

Ethical dilemma — exists when two core values that the person upholds come into conflict, making it difficult for the person to decide how to move forward.

Ethics — defined as the science of morals, a system of principles and rules of conduct, the study of standards of right and wrong, or having to do with human character, conduct, moral duty, and obligations to the community.

Ethnic — of, or relating to, large groups of people classed according to common racial, national, tribal, religious, linguistic, or cultural origin or background.

Ethnicity — ethnic quality or affiliation.

Ethno-cultural minorities — includes people other than Aboriginal people who belong to cultures not generally considered part of Western society. They are also termed **cultural minorities** or **visible minorities**.

Evidence preservation — the key building block to a successful incident investigation and involves acting to keep all the position, people, parts, and paper evidence intact.

Exchange — the actual delivery of the product or service.

Expectancy — the belief that one's effort will result in attainment of desired performance goals.

Expectancy Theory — developed by V. Vroom, who proposes that worker motivation stems from factors like what employees value; the worker's perceived performance/reward ratio; and employee expectations of succeeding at work.

Exposure monitoring — the continual evaluation of the workplace and employees for potential exposure. In terms of the workplace, this involves worksite inspections and surveys, environmental monitoring, evaluation of the results, and interpretation of the findings.

External OH&S consultant — an "outsider" who has access to specialized services not traditionally found within the client organization.

External Responsibility System (ERS) — is government intervention in the form of action by a provincial or state OH&S governmental agency that comes into play when an organization's Internal Responsibility System fails.

Extrinsic motivator — a factor such as pay, a bonus, recognition, and other tangible rewards that drive the employee to undertake a task/job.

Fail-safe system — a system that cannot cause harm when it fails.

Failure — the inability of a system or component to perform its required functions within specified performance requirements.

Fatality — any death resulting from a work injury/illness regardless of the time intervening between injury and death.

Fault — a defect in a device or component, for example, a short circuit or a broken wire.

Fault-tolerant system — a system that can continue to operate with faults, though its operation will be substandard/less efficient.

First Aid — the application of accepted principles of treatment to sustain life, prevent a condition from becoming worse and to promote recovery using available equipment, supplies, facilities, and services to provide immediate and temporary care to an injured/ill worker.

First Aid incident — an occupational injury incident that is treated by a physician/nurse or other healthcare provider in which first aid is provided and in which:

- the visit is limited to observation;
- diagnostic procedures, including the use of prescription medications solely for the purpose of diagnostic purposes;
- use of non-prescription medications, including antiseptics;
- simple administration of oxygen;
- administration of tetanus or diphtheria or boosters;
- cleaning, flushing or soaking wounds on skin surface;
- use of wound coverings such as bandages and gauze pads;
- use of any hot/cold therapy except for musculoskeletal disorders;
- use of any totally non-rigid, non-immobilization means of support;
- drilling of a nail to relieve pressure;
- use of eye patches;
- removal of foreign bodies not embedded in the eye if only irrigation or removal with a cotton swab is required; and

- removal of a splinter/foreign material from areas other than eyes.

First Aid Policy — states a company's commitment to the provision of injury management.

First Aid training — education on the provision of First Aid. Companies must have individuals trained and available to provide First Aid services, if required. Many companies choose to train their workers to a basic or Standard Level of First Aid including AED training. First Aid training needs to be updated on a regular basis.

First Aider — a person who is designated by an employer to provide first aid to workers at a work site and who is an Emergency First Aider, standard First Aider, or Advanced First Aider.

Fit for Duty — a condition in which an employee's physical, physiological, and psychological state enables them to continuously perform assigned tasks safely. It encompasses physical requirements, psychological conditions, and psychological status.

Fitness to Work — the physical and psychological ability to safely undertake stated physical and psychological job demands.

Fitness-to-Work (FTW) Assessment — are "risk-based" and dependent on the demands and related hazards of the employee's assigned job.

Five Duties of Management — introduced by Fayol and proposes that a manager's duties includes:
1. foresight (Planning);
2. organization;
3. command (Leadership);
4. coordination of work; and
5. controlling.

Flexible culture — a culture in which rewards are offered for reporting "near miss" incidents much like they are given for safe work behaviours.

Formal practice/performance standards — documented company OH&S expectations, rules, and safe work practices.

Formative Evaluation — designed to validate or ensure that the goals of the program/product/training course are being achieved and to make improvements, if necessary, by means of identification and subsequent remediation of problematic aspects.

Frequency — the rate of occurrence of a loss.

Frustration-Regression Principle — according to Alderfer, a satisfied need can become a motivator when a higher order need is unmet.

Functional Structure Approach — an organizational structure that groups workers based on similar skills and common functions, *e.g.*, workers doing accounting and payment functions would be placed in a Financial Department.

Gain sharing — system of reward for increased employee productivity: a system in a company by which employees' pay is increased in line with the gains in productivity or reductions in costs achieved by the company that is the direct result of the employees' cooperation.

General Duty Clause — employers must provide employees a safe and healthy workplace, free of recognized hazards that could cause death or serious injury/illness. It requires the employer and on occasion, other stakeholders, to take all reasonable precautions for the health, safety, and protection of workers in the workplace.

Generation — a set of people (cohort) born within the same period of time (approximately a 20-year span), and whose lives and viewpoints were shaped by the events within that span of time.

Generation gap — the differences between the members of two different generations.

Generational integration approach — that is, bringing the five generations found in today's workplaces together in such a way that the "best" of each generation is accessed and put into practice to enhance the well-being and performance of the entire workforce.

Generational strategy — a method of:
- understanding what makes their "employees tick";
- emphasizing the importance of teamwork;
- effective communications; and
- adopting "ageless thinking".

Generation-neutral workplace — a workplace that values and utilizes the contributions of all generations.

Gerontophobic — fear of the aging process.

Goal — a broad statement about something you want to accomplish.

Goal setting — the process of developing, negotiating, and formalizing the targets or objectives that an employee is responsible for accomplishing.

Goal-setting Theory — developed by E. Locke who proposed that task goals can be highly motivating for workers if they are well-setup and well-managed. Likewise, worker ownership of organizational goals depends on the degree of worker participation in setting those goals.

Golden Rules of Presenting —the OH&S Professional/Practitioner should:
- engage the audience;
- speak clearly and loudly enough for the participants to comfortably hear;
- use plain, simple language: avoid the use of complex technical jargon,

metaphors, and awkward examples. Rather, stick with the jargon and technical terms that are familiar to the audience;
- use conversational hand gestures;
- move around comfortably;
- use concrete examples;
- look at the audience when speaking;
- avoid playing with a pen, pocket items, keys, coins, *etc.*;
- only use diagrams and teaching aids that enhance the learners' understanding of the spoken materials;
- use both text and audio materials if possible;
- create a friendly atmosphere;
- use storytelling to bring the content "alive";[1]
- make the experience enjoyable and fun;
- prepare methods to garner audience involvement (questions, show of hands);
- encourage and reward audience participation;
- stay on topic;
- seek audience feedback on the pace of the teaching, the relevancy of the materials, and the participants' satisfaction levels; *and above all,*
- **start and end the presentation on time** — it shows respect for the participants and their time.

Great Safety Performance Score — a measure of the conditions for great safety performance in the workplace. Developed by Roithmayr (2000-2014), this survey score measures the efforts the organization invests to create a safe workplace with safe work practices (refer to Chapters 6 and 7).

Group — a collection of people who regularly interact with one another towards a common goal - a natural social phenomenon.

Group cohesiveness — the state where a clear alternative option/decision is selected by the majority of the group, and the opposing members feel they have been listened to and had a chance to influence the decision.

Group dynamics — the forces operating in groups that affect their task performance and human resource maintenance. It involves the process through which the members work together to accomplish the assigned tasks.

Group norms — the ideas or beliefs about the behaviour(s) expected from group members.

---

[1] R. Flemmer, "The Three E's of Training" *The Safety Net* (October 2014) at 7.

Harassment Frequency Rate — is determined by dividing the total number of harassment incidents multiplied by 200,000 by the total number of hours worked.

Harm — covers injury to people, ill health, equipment/property damage, or environmental leak/spill.

Hawthorne Effect — the tendency for people who are being singled out for special attention to perform as anticipated merely because of the expectations created.

Hazard — is something with the potential to cause harm and can include substances, equipment, machines, method of work, or the work environment.

Hazard assessment — evaluation of the risks posed by these identified hazards to the people, products, plant, processes, and environment.

Hazard control — any action that reduces the risk of loss. It includes the prevention/reduction of loss exposure, reduction of loss-producing agents, and determination/avoidance of risk.

Hazard identification — is a systematic approach that involves:
- assessment of each department/area;
- preparing and reviewing a schedule of job tasks, activities, or work operations;
- discussions with management, line managers, and employees;
- reference to incident records, pre-job meetings, incident investigation reports, safety meeting minutes, and worksite inspections;
- observation of the activities under review;
- review of applicable regulations, codes of practice, and industry practices to assist with the identification of hazards associated with particular equipment or processes;
- consideration of hazards arising from reasonably foreseeable changes in processes, equipment, circumstances, or work conditions;
- documentation of identified hazards;
- identification of employees through job demands analyses (JDAs) that are potentially "at risk"; and
- identification of others (*i.e.*, visitors, contractors, the public) potentially "at risk".

Hazard severity (harm) — the extent of the injuries, ill health, and/or damage to equipment, property, process, or the environment that may be sustained if the hazard is realized.

HAZOP — an analysis technique that uses a systematic process to identify possible deviations from normal operations, and to ensure that appropriate safeguards are in place to help prevent incidents.

Health — in terms of employees, the state of complete physical, mental, and social well-being and not merely the absence of disease or infirmity. In terms of an organization, health is the physical, economic, and cultural well-being of the company.

Health indices — are the measures used to establish "normal" or "abnormal" health conditions, such as body temperature, blood pressure, pulse, neurological functioning, *etc.*

Health information — an accumulation of data relevant to the past, present, and future health status of an individual that includes all that OH&S personnel learn in the exercise of their responsibilities.

Health promotion — the science and art of helping people change their lifestyle to move toward a state of optimal health.

Health protection — also termed illness/injury or disease prevention — any behaviour performed by a person, regardless of his/her perceived or actual health status, in order to protect, promote, or maintain his/her health, whether or not such behaviour is objectively effective toward that end.

Health surveillance — the systematic collection and evaluation of employee data to identify instances of illness or health trends suggesting adverse workplace exposures coupled with actions to reduce hazardous workplace exposures.

Healthcare facility — a hospital, medical clinic or physician's office that has the capability of dispensing emergency medical treatment 24 hours a day.

Hierarchy of authority — a chain of command characterized by using a superior-subordinate or leader-follower relationship.

Hierarchy of controls — a priority of hazard controls used to manage workplace hazards that consist of:

1. engineering controls;
2. administrative and work practice controls;
3. use of personal protective equipment (PPE); and/or
4. mitigation of consequences through emergency response planning.

Hierarchy of needs — Maslow's well-known model of human needs proposes that people bring "their individual needs" to work and are driven to satisfy those human needs.

High hazard work — means work as described in Figure 11.1.

High potential aspect — an environmental aspect that is associated with high environmental impact.

Holy Trinity — a legal concept that upholds that:

1. the employer was not responsible for employee injury if the employee contributed in any way to the incident event;
2. the employer was not responsible for employee injury if a co-worker

contributed in any way to the incident that injured the employee; and/or
3. when accepting a job, an employee is also accepting the risks of the job and so they should plan for it.

HS³™ Model — encompasses **H**ealth, **S**afety, **S**tewardship and **S**ustainability, and facilitates the development of a strategic plan which, in turn, provides the direction needed to achieve the organization's vision.

Human Needs — according to Maslow, are:
- Physiological Needs — to stay alive
- Security Needs — to feel secure
- Social Needs — to belong
- Ego Needs — to be somebody
- Self-actualization — to develop potential

Human Resource Maintenance — the attraction, maintenance, and retention of a viable workgroup/workforce.

Human Resources and Behavioural Management Theory — focuses on human needs, workgroups, and social factors that impact work and upholds that people are naturally social and self-actualizing.

Hybrid Structure — an organizational structure used by larger organizations in which workers are grouped by functions and divisions.

ILO-OSH2001 — an international model that is designed to be compatible with other management system standards and guides.

Immaturity Theory — developed by Argyris, this theory proposes that if the organization's management style is in conflict with the employee's personality, it will negatively impact work performance. The greater the disparity between the individual and organizational needs, the more tension, conflict, and dissatisfaction that will result.

Immediate incident causes — unsafe acts of the worker and unsafe working conditions.

Impact Evaluation — measures the degree of change in the OH&S behaviours/well-being/attitudes of workers, workgroups, business units, or organizations that can be attributed to an OH&S initiative, program, or policy.

Impairment — experiencing a state of diminished, reduced, or damaged state of being.

Impoverished Management — proposed by Blake and Mouton, this management style is characterized by minimal effort exerted to get the work done or to sustain organizational membership. It translates to a low regard for both production and people concerns.

Incident — an event or set of circumstances that could or does result in an unintended harm or damage. It includes near-miss, damage, chemical spills, and injury incidents.

Inclusiveness — translates to employees recognizing that they are valued and belong.

Independent variable — the variable that is believed to cause or influence the dependent variable. In a research scenario, it is the variable that is manipulated.

Indirect safety management costs — the "hidden" costs such as the management time and effort to promote workplace safety, worker time to do things properly the first time, data management time, and costs, *etc.*

Individual risk — the annual probability of death for an exposed group.

Individualism-collectivism — the degree to which a society focuses on individuals or groups as resources for work and social problem solving.

Informal OH&S practice/performance standards — "what gets done in the workplace when the worker thinks no one is watching". These are the actual practices/work performances that go on within the field, worksites and offices.

Informed consent — written consent includes:
- the provision of sufficient information about the nature and consequence of the intended action to allow the employee to come to a reasoned decision;
- ensuring that the employee is mentally competent, and has the ability to understand and appreciate the nature and consequences of the procedure;
- consent being freely given;
- consent being obtained without misrepresentation or fraud;
- consent cannot be given for the performance of an illegal procedure; and
- consent is often in relation to the specific act contemplated unless the employee's life is immediately endangered and it is impractical to obtain consent.

Informed culture — an OH&S (Safety) system that collates data from all types of workplace incidents and safety audits, and then, combines that with information from proactive measures such as climate surveys. In many ways, an informed culture is an OH&S (Safety) culture.

Inherent hazards — hazards that are associated with the nature of the work being done.

Inherently safe system — a clever mechanical arrangement that cannot be made to cause harm — obviously the best arrangement, but this is not always possible.

Injury Frequency Rate — the number of lost time injury incidents per 200,000 work hours (per 100 workers).

Injury management — an important management tool for:
- reducing the negative impact(s) of workplace injury;

- reducing the potential for injury to other workers;
- controlling the related costs; and
- expediting a safe and timely return to work by the injured worker.

Injury Severity Rate — is the number of lost-time workdays due to a work injury per 200,000 work hours (per 100 workers).

Instrumentality — the belief that if one does meet performance expectations, he/she will receive a greater reward.

Insurance — an approach for legally transferring risk to a third party.

Integrated Disability Management Program — a planned and coordinated approach to facilitate and manage employee health and productivity. It is a human resources risk management and risk communication approach designed to integrate all organizational/company programs and resources to minimize or reduce the losses and costs associated with employee medical absence regardless of the nature of those disabilities.

Integrated Workplace Health Management — a management approach in which organizational resources are positioned in an integrated manner to promote workplace health, safety, and well-being for the employee and the organization.

Internal OH&S consulting — occurs when the consultant is a member of the organization and is assuming a consulting role to assist with the resolution of a problem or the chance to capitalize on an opportunity.

Internal Responsibility System — the "people framework" within an organization that maintains:
- everyone within a company is responsible for workplace health and safety;
- everyone must do their part to identify and solve health and safety issues/problems and to look for ways to improve the work processes with which they are involved; and
- by capturing the creativity, leadership, experience, and knowledge of workers, companies can improve workplace health and safety.

Intrinsic motivator — a factor such as an internal desire to conquer a challenge, address an interest, or attain personal satisfaction that drives the employee to undertake a task/job.

Isolated worksite — a worksite that is more than 40 minutes travel time from the worksite to a healthcare facility under normal travel conditions using the available means of transportation.

Job characteristics — features of the job that affect work performance, but they tend to impact job satisfaction more.

Job description — is a written record of the principal duties and scope of responsibility for a particular job/position.

Glossary

Job Matching Model — a disability management model which involves a fitness assessment of the injured/ill employee and an analysis of the physical/psychological demands of the employee's job. The intent is to determine if there is a "match" or "mismatch" in terms of a safe return-to-work for the employee.

Job Safety Analysis (JSA) — breaks a job into basic steps, identifies the hazards associated with each step and prescribes controls for each hazard.

Job satisfaction — the degree to which employees perceive positively or negatively the job.

Job/Position Demand Analysis (JDA/PDA) — is designed to identify all the physical and psychological demands of the job or position.

Joint Health and Safety Committee (JHSC) — has the mandate to identify, evaluate, and participate in the resolution of workplace health and safety issues/concerns.

Just culture — an atmosphere of trust in which response to incidents is to determine what *actually happened* as opposed to seeking to lay blame; yet, not to turn a blind eye to unsafe acts.

Justice — directed towards treating employees fairly, equally, and without discrimination.

Labelling — a WHMIS 2015 requirement for alerting workers to the identity and dangers of controlled products, and to the basic safety precautions for working with and around them.

Lagging indicators of safety performance — the outcome safety performance measures that are used by many companies, such as lost-time injury frequency and severity rates. Also known as **Downstream Safety Performance Indicators**.

Law of Effect — "states that responses that are closely followed by satisfying consequences become associated with the situation and are more likely to recur when the situation is subsequently encountered. Conversely, if the responses are followed by aversive consequences, associations to the situation become weaker."

Leaders — individuals that have the ability to excite, stimulate, and drive other people to work towards a vision, making it a reality.

Leadership — an art that liberates people to do what is required of them in the most effective and humane way possible. Leadership is both about getting results and about how those results are obtained.

Leading Indicator of Safety performance — an index designed to anticipate or forecast the Safety outcomes of current trends. They are also termed **Upstream Safety Performance Indicators**.

Learning culture — is the collection of the information needed to enhance the OH&S (Safety) performance of the organization along with the desire to make the needed changes.

Learning objective — a specific statement that speaks to the worker learning. For example, "*At the end of this course, the worker will know how to use, maintain and replace hearing protection.*"

Long-term Disability Rate — is the number of ill/injured workers who are accepted on long-term disability per 100 workers.

Loss control — the minimizing of loss due to people, property, process, plant, or profit damages/threats.

Lost time injury — an occupational injury that results in the worker losing time from work beyond the date of the incident or is likely to lose time in future.

Lost time injury incident — an incident that results in an occupational injury that results in the worker losing time from work beyond the date of the incident or is likely to lose time in future.

Low hazard work — work that is defined in Figure 11.1, as per the Alberta OH&S Code.

Machine Bureaucracy — organizational design has:
- a clear hierarchy of authority with a large middle-management group;
- highly specialized and standardized tasks;
- functional departmentalization; and
- top management decision-making.

Machine learning — an arm of computer science in which machines learn a task without being programmed to do so. It involves pattern recognition which enables computers to accomplish tasks like diagnosing medical conditions quickly and reliably.

Magnetic Resonance Imaging (MRI) — a noninvasive medical test that provides pictures of organs, soft tissues, bone, and virtually all other internal body structures through the use of a powerful magnetic field, radio frequency pulses, and a computer.

Maintenance/Motivating Factors Theory — developed by Herzberg, proposes that two types of work factors that impact worker motivation exist:

1. *Maintenance (Hygiene) Factors* (physical conditions, pay, status, benefits) — these factors prevent dissatisfaction and are neutral in regards to worker motivation.

2. *Motivating Factors* (responsibility, achievement, growth and recognition) — these factors help employees attain their ego and self-actualization needs and are therefore motivating.

Managed Care Model — a disability management model in which the employee's diagnosis is referenced against standardized care plans, procedures, and diagnostic testing guidelines to determine if treatment and the physician's suggested leave duration are appropriate. This model, like the traditional model, tends to be medically driven.

Management By Objectives (MBO) — a process of mutual-goal setting undertaken by the supervisor and employee.

Management disclosure — is having health information released to management and is limited to the following:

- report of employee fitness to work;
- determination that a medical condition exists and that the employee is under medical care;
- time that the employee has been or is expected to be off work;
- medical limitations, if any, to carry out work in a safe and timely manner; and/or
- medical restrictions, if any, regarding specific tasks.

Management theories or The Domino Theories — theories that assume, if unsafe practices and unsafe work conditions that lead to "near miss" incidents are eliminated, then the probability of the occurrence of more serious incidents are reduced. Management Theories include Heinrich's Theory, Bird's Theory and Adam's Theory.

Manager — the person within an organization who is responsible for performance of one or more subordinates, or functions.

Managerial Grid — proposed by Blake and Mouton, suggests that management is made up of one of five styles:

- Country Club Management.
- Team Management.
- Organization Management.
- Impoverished Management.
- Authority-Obedience Management.

Manual Handling — work action that includes pushing, pulling, and carrying, as well as lifting.

Market — the potential customers for whom the product or service has been designed.

Marketing — a social and managerial process through which individuals and groups obtain what they need and want by creating and exchanging products or services and value with others.

Marketing management — involves managing customer demand which, in turn, involves managing customer relationships.

Masculinity-Femininity — the degree to which a society emphasizes the so-called "masculine" traits such as assertiveness, independence, and insensitivity to human feelings.

Matrix Structure — the organizational structure that uses permanent cross-functional teams to blend the technical strengths of the functional structures with the integrating potential of divisional structures. Team members have dual alliances to their functional managers as well as to their project managers.

Mechanistic Organizational Design — an organizational structure characterized by a rigidly and tightly controlled structure that is aimed at minimizing the impact of differing human traits. Most large organizations have some elements of the mechanistic organizational design.

Medical aid injury — an occupational injury that results in medical care that is beyond first aid treatment, but that does not result in lost time from work.

Medical aid injury incident — an occupational injury incident that results in medical care that is beyond first aid treatment, but that does not result in lost time from work.

Medical monitoring — the health surveillance of employees who are exposed to specific workplace hazards.

Medium hazard work — work as described in Figure 11.1.

Mens rea — the moral fault element; the guilty mind or intent.

Mens rea **offence** — a true crimes. The Crown must prove "beyond a reasonable doubt" both the *mens rea* (intent) and *actus reus* (prohibited act).

Mental distress — is the angst that an individual feels consequent to harassment or traumatic experience but does not experience a clinically diagnosable condition. Mental distress can include sub-clinical symptoms of depression and anxiety, as well as severe demoralization, disengagement, and alienation.

Mental injury — is harm to mental health (mental suffering) that significantly affects the ability of employees to function at work and at home.

Mission statement or policy — describes Management's commitment to workplace safety and the provision of a safe work environment. It presents the high-level program objectives and describes the organization's values and beliefs towards maximizing workplace health and safety and minimizing the impact of workplace incidents on stakeholders.

Mobile devices — encompass cellphones, tablets, data-collection devices, GPS devices, *etc.*

Model — an approach created to explain a theory.

Modern healthcare systems — recent social developments which tend to be the last resort when one is ill/injured. They are used when the traditional healthcare system fails.

Modern Loss Causation Theory — maintains that incidents are the result of many causes. This model encompasses the concept of multi-linear interaction of causes and effects and involves multiple opportunities for control. It proposes that worker behaviour is motivated by a desire to reach some goal and that workers vary in their ability and motivation to do things.

Modern Management Theory — focuses on total systems thinking, contingency thinking, and an awareness of global developments in management. It views organizations as systems that function towards a common purpose. This theory also recognized that people are complex with multiple talents and abilities, and that people are variable.

Modified Work Participation — is the percentage of ill/injured workers who participate in modified work duties.

Modified Work Rate — is the number of modified work hours per 100 workers.

Motivator — the intensity, direction, and persistence of effort exerted by an employee to reach a specific goal.

Multiple Causation Theory — is an outgrowth of the Domino Theory, but it postulates that for a single incident there may be many contributory factors, causes, and sub-causes, and that certain combinations of these give rise to incidents.

Near-miss incident — an incident that has the potential to be a damage/injury incident.

Network Structure — an organizational structure approach that consists of a central core workforce that is linked through networks with external suppliers of essential business services. These networks, in the form of strategic alliances and business contracts, allow the organization to operate without having to own all its supporting functions.

Neural network — computers modelling the way the human brain problem-solves through interconnections with other computers (a network).

No-Lift patient handling — low risk lifting, pushing, pulling, lowering or restraining people or materials by using little to no minimal bodily force.

Nonmaleficence — the "no harm" principle.

Nurse — a graduate of an approved registered nursing program who maintains membership and good standing with the provincial nursing association and is an advanced first aider.

Objective — a specific aim set to achieve a desired goal.

Observable culture — what can be seen and heard within an organization. It is the way people behave, dress, talk about customers and arrange their offices. It includes the rites, rituals, norms, symbols, stories, and the heroes.

Occupational bond — the identity of the employee with the workplace.

Occupational bonding — the mutually beneficial relationship between the employee and the employer.

Occupational closure — the division between qualified members and unqualified outsiders.

Occupational disease — "a disease or ill health arising out of and directly related to an occupation".

Occupational Health — the promotion and maintenance of the highest degree of physical, mental, and social well-being of workers in all occupations; the prevention amongst workers of departures from health caused by their working conditions; the protection of workers in their employment from risks resulting from factors adverse to health; the placing and maintenance of the worker in an occupational environment adapted to his/her physiological and psychological capabilities; and, to summarize, the adaptation of work to the person, and of each person to their job.

Occupational Health and Safety (OH&S) consultant — a professional/practitioner who provides influence, recommendations, and expertise, but has no direct power to make change.

Occupational Health and Safety Management System (OHSMS) — that part of the overall management system that includes the organizational structure, planning activities, responsibilities, practices, procedures, processes, and resources for developing, implementing, achieving, reviewing, and maintaining the OH&S Policy and Program.

Occupational Health and Safety Management System (OHSMS) Audit — measurement of the structure and outcomes of an OHSMS using an audit protocol.

Occupational Health and Safety Program — See **OH&S Program**.

Occupational Health Management — the management system aimed at the promotion and maintenance of the highest degree of physical, mental, and social well-being of workers in all occupations; the prevention amongst workers of departures from health caused by their working conditions; the protection of workers in their employment from risks resulting from factors adverse to health; the placing and maintenance of the worker in an occupational environment adapted to his/her physiological and psychological capabilities; and, to summarize, the adaptation of work to the person, and of each person to their job.

Occupational Health Nurse — registered nurses who hold a certification in Occupational Health Nursing, and often, possess additional education and skills in the areas of Occupational Health and Safety, relationship building, program development, human resources management, and business management. Their mandate is to promote healthy working environments, protect the health of the worker and prevent work-related injuries/illnesses.

Occupational Health Physician — medical physicians who have chosen to specialize in the field of occupational medicine. They are knowledgeable about medicine, the workplace, and stakeholder interests, as well as the areas of ergonomics, human factors, industrial hygiene, toxicology, Workers' Compensation System, and disability management.

Occupational Health Risk — is the probability of an OH loss - typically, static risk means a loss in which there can be no hope of any gain.

Occupational Hygiene Management — elements such as hazard inventories; measurement of workplace hazards; walk-through surveys; ventilation assessments; respiratory protection; worker education; and audits.

Occupational injury — any "work-related" injury or illness suffered by an employee. It is work-related if an event or exposure in the work environment either caused or contributed to the resulting condition or aggravated a pre-existing condition.

Occupational Safety — a practice devoted to the identification, evaluation, and control of those hazards and stressors arising in, and from, the workplace which may cause losses, *i.e.*, property/equipment damage, product loss, worker injury/illness, intellectual losses, and/or financial losses.

OH&S audit — a systematic, documented verification process of objectively obtaining and evaluating audit evidence to determine whether specific OH&S activities, events, conditions, management systems, or information about these matters conform to the audit criteria, communicating the results of this process to the client.

OH&S (Safety) climate — the prevailing atmosphere of the organization, the socio-psychological environment that profoundly influences behaviour and is typically measured by employee perceptions.

OH&S consultant — a practitioner/professional who provides influence, recommendations, and expertise, but has no direct power to make changes.

OH&S consulting — a voluntary, temporary, and supportive relationship that involves a helper and a help-needing party (the client).

OH&S (Safety) culture — the moral, social, and behavioural norms of an organization that are based on the shared beliefs, values, attitudes, habits, and traditions on safety that give meaning to an organization's employees and provides them with the accepted safety behaviours within their organization.

OH&S Duty of Care (Duty) — organizations, corporations and individuals who direct others to perform work, or have the authority to do so, must take reasonable and practicable steps to provide a safe and healthy workplace, and to protect workers and the public from potential harm as a result of the work.

OH&S goal — the overall, quantifiable aim of the OH&S Program that the organization sets out to achieve.

OH&S leadership — the ability to enable and drive workplace safety.

OH&S orientation — an overview of the company's OH&S standards and how they are implemented, plus explanation of:

- company commitment to OH&S;
- company OH&S policy;
- related policies such as Substance Abuse Policy, Non-smoking Policy, Seatbelt Use Policy, Fit-For-Duty Policy, Non-use of Cell Phones When Driving, *etc.*;
- company's OH&S rules and the rationale for each rule;
- worker OH&S responsibilities;
- information on company Safety Meetings, pre-job, and hazard identification;
- use, maintenance and replacement of personal protective equipment;
- right to refuse unsafe work;
- identification, assessment, and control of workplace hazards;
- emergency procedures and response;
- locations of emergency and first aid supplies;
- reporting hazards, unsafe work conditions, and unsafe work practices;
- reporting work injuries/illnesses; and
- return-to-work procedures post-injury/-illness.

OH&S performance — the measurable results of the Occupational Health & Safety Management System, related to an organization's control of its OH&S hazards and risks, based on its OH&S Policy, goal(s), and targets.

OH&S policy — a statement made by an organization of its intentions and principles in relation to its overall OH&S performance. It provides a framework for action and for the setting of its OH&S goals, objectives, and targets.

OH&S Practitioner — a specialist in preventing workplace losses who strives to eliminate, reduce, and/or control them.

OH&S Professional — an individual who possesses a professional designation and is a specialist in preventing workplace losses who strives to eliminate, reduce, and/or control them.

OH&S Program — a complete system that ensures high safety standards throughout the company's operations and;
- reflects a strong commitment from management towards workplace health and safety;
- encourages worker commitment towards workplace health and safety;
- helps workers understand their responsibility for preventing workplace incidents;
- provides a work environment that provides the elements required to work safely, namely know how to work safely, able to work safely, equipped to

work safely, and motivated to work safely; and
- enables program evaluation and continuous improvement.

It is a defined action plan designed to prevent incidents and occupational diseases.

OH&S Program Manual — is the documented evidence of management's endorsement and commitment to Occupational Health and Safety and is designed to serve as a standard for OH&S practice within the organization.

OH&S risk communication — a science-based approach for communicating effectively in high-concern and high stress situations; emotionally charged situations; and controversial situations.

OH&S target — a detailed performance requirement, quantified where practicable, that is applicable to the organization or parts thereof, that arises from an OH&S goal, and that needs to be set and met in order to achieve that goal.

OH&S (safety) work climate — refers to the state of a system in terms of the perceptions of the current environment or prevailing conditions that impact safety.

OHSAS 18001/02 — an Occupational Health and Safety Management System specification developed through the concerted effort of a number of the world's leading national standards bodies, certification bodies, and specialist consultancies.

OHSMS — Occupational Health and Safety Management System. It is that part of the overall management system which includes organizational structure; planning activities; responsibilities; practices; procedures; processes; and resources for developing, implementing, achieving, reviewing, and maintaining the Occupational Health and Safety Policy.

On-the-job Training — a hands-on explanation and demonstration of how to do job tasks to which the worker is assigned.

Operant Conditioning Theory — "the use of consequences to modify the occurrence and form of behaviour. Operant conditioning is distinguished from Pavlovian conditioning in that operant conditioning deals with the modification of voluntary behaviour through the use of consequences, while Pavlovian conditioning deals with the conditioning of behaviour so that it occurs under new antecedent conditions".

Optimal health — a balance of physical, emotional, spiritual, intellectual, and social (workplace) health.

Option evaluation — the development of options and analysis of the feasibility of each identified option.

Organic Design Alternatives — an organizational design that operates with a decentralized authority structure in which there are few rules and procedures, little division of labour, wide spans of control, and a personal means of coordination.

Organization — is a consciously coordinated social unit, composed of a group of people that functions on a relatively continuous basis to achieve a common goal or set of goals.

Organizational Behaviour (OB) — is the study of individual and group dynamics in an organizational setting, as well as the nature of the organizations themselves. It is a field of study that investigates the impact of individuals, groups, and structure on behaviour within an organization; its purpose is to apply such knowledge towards improving the organization's efficiency.

Organizational Climate — is the manifestation of the organization's culture.

Organizational Coordination — is how work is organized and how organizations operate. It includes the structure, processes, and nature of the work relationships.

Organizational Man Management — described by Blake and Mouton, this management style is characterized by balancing the necessity to get work done with maintaining worker morale at a satisfactory level. This translates into a balance in concern for production and people.

Organizational wellness — managing both business functioning and employee well-being in a manner that allows the organization to be more resistant to environmental pressures.

Outcome evaluations — designed to determine if the OH&S training course/program met the participants' expectations in terms of the pre-stated OH&S training objectives.

Outcomes — the performance results.

Outrage — a strong human reaction to high-risk situations.

Paper evidence — includes the collection of incident report(s), safe work procedures for critical tasks, pre-job planning meeting document(s), equipment/vehicle inspection reports, equipment/plant maintenance records, loss control records, non-compliance reports, worker training logs, computer data logs, control centre communication records, relevant OH&S rules, and regulations and such.

Participants' Work Absence Rate — the absence rate for participants in a program.

Partnering — a method of accomplishing the mutual goals of the organization and the outsourced service in a planned and pre-described way. It involves jointly establishing service goals, objectives, and procedures. The intent is to have open communication and a solid working relationship, which foster the accomplishment of the desired goals.

Parts evidence — means the retrieval, labelling, and protection of each piece/part of equipment and worksite so reliable analysis can occur.

Patient handling — any task which includes the transporting or supporting of a patient by hand or bodily force (includes pushing, pulling, carrying to support, lifting, or lowering).

Patient handling Injury — an injury occurring during a patient handling task to either patients or staff.

Path-Goal Theory — developed by Robert House, this theory maintains that an effective leader is one who clarifies the paths for followers to take so they can achieve both task-related and personal work goals.

People evidence — includes witness statements; the incident report; worker/witness perceptions of what happened — how and why; technical expert opinions; and targeted interviews with key people.

PEPY — the acronym for Per Employee Per Year.

Perception Survey — measurement of worker experience and perception of the quality and effectiveness of a program, function, or corporate culture.

Personal problems — physical illness, emotional and stress-related problems, family problems, alcohol or drug abuse, or similar related matters which may adversely affect the employee's job performance and/or health.

Personal Protective Equipment (PPE) — equipment or clothing worn by a worker for protection from health or safety hazards associated with the conditions at a worksite.

Personal space — the "distance of comfort" to which we have adapted. We feel uncomfortable if others invade that space or if they are too far away for ready communication. The size of the personal space zone varies with cultures.

PHSMS — refer to the Psychological Health & Safety Management System.

Plan-Do-Check-Act Model — the Deming model used to promote program sustainability.

Policy — See **mission statement**.

Position evidence — determining where everyone and everything was prior to and at the time of the incident.

Post-Traumatic Stress Disorder (PTSD) — an anxiety disorder that can occur following a worker's involvement in, or observation of, a traumatic event (*e.g.*, robbery, assault, or life-threatening incidence/experience).

Power distance — the degree to which a society accepts a hierarchical or unequal distribution of power within organizations. An organization with a small power distance is characterized by superiors viewing subordinates as "people like me". Superiors are accessible and the general belief is that all the organizational members have equal rights.

Practice/Performance standards — stated approaches to practice/perform based on recognized and accepted principles of industry practice for planned processes. They form the guidelines and rules for work practices, provide the boundaries for work activities, clarify stakeholder roles and responsibilities, and serve as a benchmark.

Pre-learning assessments — a baseline measure of level of awareness/knowledge/skills/abilities that the participant possesses prior to training.

Presenteeism — the phenomenon of employees being at work, but because of wasted time, failure to concentrate, sleep deprivation, distractions, poor health, and/or lack of training, they may not be productive at all.

Primary failure — unexpected failure of a device that was operating within its design limits.

Primary prevention — deals with preventing problems before they exist, such as health education on heart health, smoking cessation, cancer awareness, nutrition, and off-the-job safety.

Privacy — the claim of individuals, groups, or institutions to determine for themselves when, how, and to what extent information about them is communicated to others.

Proactive due diligence — developing an OH&S Management System that ensures worker health and safety.

Probabilistically safe system — a system that has no single point of failure, and enough redundant sensors, computers and effectors so that it is very unlikely to cause harm ("very unlikely" means, on average, less than one human life lost in a billion hours of operation).

Probability — the chance of a loss occurring.

Procedures — defined actions that serve to standardize the Occupational Health and Safety Program.

Process — the way things operate within a system.

Process evaluation — See **formative evaluation**.

Process safety management — elements such as process hazard information and knowledge; process hazard analysis; process equipment integrity; process design considerations and facility set-up; pre-start-up reviews and compliance audits; sharing of process safety information; and incident findings.

Products and services — involve anything that can be offered to a market to satisfy a customer need or want.

Profession — an occupation, vocation, or high-status career that usually involves prolonged academic training, formal qualifications, and membership in a professional or regulatory body. A profession can be defined as an occupation whose core element is work based upon the mastery of a complex body of knowledge and skills. It is a vocation in which knowledge of some department of science or learning or the practice of an art founded upon it is used in the service of others. Its members possess a commitment to competence, integrity, morality, altruism, and the promotion of the public good within their domain. These commitments form the basis of a social contract between a profession and society, which in return grants the profession the right to autonomy in practice and the privilege of self-regulation. Professions and their members are accountable to those serviced and to society.

Professional Bureaucracy — an organizational design characterized by the following:
- is staffed with highly trained professionals;
- demonstrates a high level of autonomy;
- has a decentralized structure;
- possesses a large number of support staff; and
- has few middle management.

Professionalization — the process through which a profession arises from a trade or occupation.

Profit sharing — refers to various company incentive plans that provide direct or indirect payments to employees based on the company's profitability in addition to the employees' regular salary and bonuses.

Program/function evaluation — identifies the gaps between the current state and the desired state of a program, indicates whether the program/function goals/objectives are met or not, and enables improvements both along the way and periodically.

Progression Principle — according to Maslow, individuals have to satisfy a lower need before moving to a higher order of need (step-by-step progression).

Project — a temporary and one-time endeavour undertaken to create a unique product, or service that brings about beneficial change or added value.

Project definition document — a project planning tool in which the following elements are articulated:
- an overview of the project;
- the project objectives and scope;
- the related assumptions and risks;
- the project methodology;
- the project team members, including their specific roles;
- documentation of customer acknowledgement of understanding of the project; and
- the time/cost schedules.

Project management — the discipline of organizing and managing resources in such a way that these resources deliver all the work required to complete a project within defined scope, time, and cost constraints.

Project manager — the person who assumes responsibility for the management of a project.

Provincial and federal health and safety offences — regulatory public welfare offences designed for the protection of public and social interests.

Psychological health — a state of mental and emotional well-being; it consists of the ability to think, feel, and behave in a manner that enables employees/workers to optimally perform.

Psychological Health & Safety Management System (PHSMS) — approximates an Occupational Health and Safety Management System (OHSMS). It includes commitment, leadership and participation; planning; implementation; evaluation and corrective action; and management review.

Psychological Health Claims Rate — is determined by dividing the total number of psychological health claims multiplied by 200,000 by the total number of hours worked.

Psychological Health Status — a measure of the degree of psychological health, safety, and well-being in a workplace. This can be measured through the use of audit tools provided in the *National Standard for Psychological Health and Safety in the Workplace*, or from *GuardingMinds @ Work*.

Psychological Illness Frequency Rate — is determined by dividing the total number of psychological illness incidents multiplied by 200,000 by the total number of hours worked.

Psychological Injury Frequency Rate — is determined by dividing the total number of psychological injury incidents multiplied by 200,000 by the total number of hours worked.

Psychological safety — the risk that a worker might experience injury to their mental well-being. To improve workplace psychological safety, employers must act to avert injury, danger, or threats to the mental well-being of employees.

Psychological Theories — emphasize the effect of workplace social factors on employee motivation and attitude.

Psychologically safe workplace — a workplace in which every practical effort is made to avoid reasonably foreseeable injury to the mental health of employees; in other words, it does not permit harm to employee mental health in careless, negligent, reckless, or intentional ways.

Pure (static) risk — risk in which there can be no hope of gain.

Pure Chance Theory — holds that of any given set of workers, each worker has an equal chance of being involved in an incident.

Quantitative Management Approaches — focus on the use of mathematical techniques for managerial decision-making and problem solving. Their primary focus is towards decisive decision-making using economic decision-making criteria, formal mathematical models, and computer modelling.

Race — A tribe, people, or nation belonging to the same stock; a division of humankind, possessing traits that are transmissible by descent and sufficient to characterize it as a distinctive human type. Race can also be stated as a social construct used to separate the world's peoples.

Reactive due diligence — properly responding when a workplace hazard is identified.

Reasonable care — actions that would be taken by a reasonable person — one who plans and takes actions in accordance with general and approved practices.

Reasonableness — a concept on which ethics and the law are based. It stems from the rationale of what a reasonable person would do in a certain situation given the person possesses the same knowledge, training, and experience.

Recognition — is the practice of acknowledging work performance in terms of praise or tangible rewards.

Refresher training — the provision of ongoing training and growth, and a method of encouraging employees to continually improve and update their knowledge and proficiency.

Reinforcement Theory of Motivation — implies that if a person has a need, then he/she will be motivated to work hard to achieve the need desired.

Reporting culture — the active and honest workforce participation in reporting all types of incidents, completing attitude surveys, and becoming involved in how OH&S (Safety) is managed within the organization. It is characterized by an organizational climate in which the workers feel free to contribute to the informed culture.

Request for Information (RFI) — a request for needed information prior to issuing a solicitation or request for proposal.

Request for Proposals (RFP) — explains what the customer is interested in procuring and provides instructions regarding how to prepare and submit a proposal for the provision of services/products.

Request for Quotation (RFQ) — explains what a company is seeking when all they are interested in is the price.

Return on Investment (ROI) — the financial returns as a result of a financial investment.

Ringlemann Effect — refers to the phenomenon of "groups social loafing".

Risk — is the chance of loss occurring. It is a measure of the probability and the potential for severity of harm. Stated another way, it is the likelihood that harm will occur due to a hazard and the severity of its consequences.

Risk analysis — is the process of conducting a hazard identification and risk estimation for each hazard.

Risk assessment — entails:
- identifying the hazards in the work activities;
- rating the risks by determining the likelihood (probability) that harm will occur; and
- establishing the hazard severity.

Risk elimination — removal, elimination, or avoidance of a risk by not undertaking the activity, or by exiting the business or activity.

Risk financing — an approach for paying for losses through the operating budgets; by borrowing against assets; or through the use of commercial insurance.

Risk management — the process of making and implementing decisions that minimize adverse effects of accidental and business losses on an organization. As a field of endeavour, it is a practice with processes, methods, and tools for managing business risks.

Risk Management — a practice with processes, methods, and tools for managing business risks.

Risk Manager — a specialist in funding business losses who works to ensure that losses can be absorbed in normal cash flow, or buffered by reserves, or transferred to others through legal means including insurance.

Risk minimization or **Transfer** — companies transfer risk through contracts, lease, and/or insurance plans. This technique puts in place the tools necessary for controlling the severity of a concurrent loss.

Risk prevention or **Treat** — companies reduce risk by substituting the hazard agent for a safer one, by using engineering controls, by using administrative controls, and by using worker personal protection equipment.

Risk rating — a combination of the **hazard severity** and the **probability** of the occurrence.

Risk reduction or **Tolerate** — companies accept risk within the context of prescribed risk tolerances. It involves financing the identified risks through operating budgets, reserves, borrowing money, and some insurance agreements.

Robotics — an interdisciplinary branch of engineering and science that includes mechanical engineering, electronic engineering, information engineering, computer science, and others. Robotics deals with the design, construction, operation, and use of robots, as well as computer systems for their control, sensory feedback, and information processing.

Role clarity — exists due to Management explaining its expectations and desired outcomes for a particular job/position.

Rules and procedures — the work and safety standards to which workers are expected to perform.

Safety — a management duty of care that requires the provision of a safe and healthy workplace, protection of the environment, and care for the public at large.

Safety climate — See **OH&S (Safety) climate**.

Safety culture — See **OH&S (Safety) culture**.

Safety Data Sheets (SDS) — technical bulletins which provide detailed hazard and precautionary information.

Safety engineering — an applied science that assures that a life-critical system behaves as needed, even when individual system parts fail. Safety engineering refers to any act of incident prevention by a person qualified in the field.

Scientific Management Theory — focuses on universal principles and operates on the premise that people are rational and economically driven to work, and that workers will seize work opportunities that enable them to achieve economic gains.

Secondary failure — the expected failure of a component stressed beyond its design limits.

Secondary prevention — deals with the early detection of disease and the initiation of early treatment programs such as screening for vision disorders, cholesterol, diabetes, tuberculosis, and lung disorders.

Security — an approach for protecting people, property, products, business continuity, intellectual property, and assets from deliberate acts of threat.

Self-fulfilling prophecy — a prediction that, in being made, actually causes itself to become true. In terms of managing people, managers can net what they profess/demonstrate to believe about workers; workers mirror back the expectations their manager holds for them.

Severity — the extent of the loss that may be sustained if an incident/accidental loss occurs.

Sick role — a societal-sanctioned role that an ill/injured person assumes once they become ill/injured.

Situational leadership — leaders adjust their leadership style based on the "readiness" of the followers to perform in certain situations.

Social Capital Theory — suggests that our willingness to help others is based on the quality of our social relationships.

Span of Control — is the number of subordinates that reports to a superior.

Standard First Aider — means a First Aider who holds a certificate in Standard First Aid from a training agency.

Stewardship — refers to management's responsibility to properly utilize and develop its resources, including people, property, and financial assets.

Structure — the formal format of a program, *e.g.*, goals, objectives, policies, procedures, lines of authority (organizational chart), forms, roles, and responsibilities, *etc*.

Substitution — replacement of a hazardous substance, process, or piece of equipment with a less problematic option.

Supplier — the entity that produces, or imports, a product for distribution and sale in Canada.

Summative Evaluation — provides information on a program's efficacy, that is, the ability to do what it was designed to do.

Sustainability — achieved when an OH&S practice permeates all activities and processes, and when compliance is linked to manager performance and rewards.

Sympathetic Communication — the technique of "walking in another person's shoes" - attempting to feel what another person is going through. It denotes agreement, whereas empathetic communication denotes understanding.

Symptoms vs. Cause Theory — holds that unsafe acts and unsafe conditions are the symptoms of problems — the proximate causes — and not the root causes of the incident.

System of work — an interconnected set of highly specific work practices that help workers get the job done and enable organizations to fulfil their mission.

System Safety Theory (Engineering Theories) — views the organization as a closed system. It recognizes the inseparable connection between individuals, their tools and machines, and the general work environment. Changes to one part of the work system have an impact on the remaining parts and the "whole" system.

Targets — levels of OH&S performance that the company wishes to attain.

Teaching objective — a specific statement that speaks to what the worker must do to demonstrate what they know or how to apply a learned skill. For example, "*At the end of this course, the worker will demonstrate the correct insertion of hearing protection.*"

Team building — a sequence of planned action steps designed to gather and analyze data on the functioning of a group and implement changes to increase organizational effectiveness.

Team Management — a style of management in which work is accomplished by people committed to a common goal, thereby creating relationships of trust and respect. This translates to a high regard for both production and people relationships.

Team structure — an organizational structure in which permanent and temporary teams are created to improve lateral relations and to solve problems throughout the organization.

Technical writing — involves technical internal and external OH&S communications on behalf of the company; the results of incident/investigations; trend analysis reports; an explanation of compliance audits; presentation of OH&S Program audit results; rationale for corrective action follow-up; and occasionally, the management of public/media relations in regards to OH&S issues.

Teleological theory — utilitarianism — focuses on the consequences of an action and gauges the value of that action by the end results, rather than by the means to achieve the results.

Terminate (the hazard) — eliminate or avoid the risk by not undertaking the activity, or by exiting that portion of the business.

Tertiary prevention — deals with the correction of disease and/or prevention of further health deterioration as a result of disease such as rehabilitation and restoration with chronic diseases and conditions (substance abuse).

THC — delta-9-tetrahydrocannabinol which has psychoactive properties.

Theory — a set of ideas or principles that work together to explain a concept or something more tangible.

Theory X and Theory Y — holds that managers should pay more attention to the social and self-actualization needs of workers. According to this theory, managers believe and use either a Theory X or a Theory Y style, when managing their business functions (Table 1.1).

Thorndike's Law of Effect — behaviour that results in a pleasant outcome is repeated: that which does not, is not repeated.

3-D Printers — printing processes that joins or solidifies materials under computer control to create a three-dimensional object, with material being added together (such as liquid molecules or powder grains being fused together), typically layer-by-layer.

Tolerable death risk rate — one death in a million.

Total Cost of Psychological Health Claims — is determined by adding all the costs related to psychological health claims for a given period of time.

Total Quality Management (TQM) — a management theory that encompasses concepts like visionary leadership, management commitment, corporate culture, customer focus, statistical process control, benchmarking, continuous improvement, worker training, cross-functional teams, worker empowerment, self-managed teams, learning organizations, change management, and total system management.

Traditional healthcare systems — have been around forever. They are part of the original culture, having filtered down generation to generation.

Traditional Model — a disability management model in which the disabled employee's care plan, authorized leave, and return-to-work process are medically directed.

Training — a process of continually instructing workers on the skills necessary to do a job competently and safely.

Transactional leader — guides or motivates followers in the direction of established goals by clarifying the role and task requirements.

Transcultural nursing — an essential area of study and practice focused on the cultural care beliefs, values, and lifestyles of people to help them maintain and/or regain their health, or to face death in meaningful ways.

Transfer (the hazard) — move the exposure to the hazard to a third party.

Transformational Leader — inspires followers to transcend their own self-interests and they are capable of a profound and extraordinary effect on followers, *e.g.*, Steve Jobs, Apple Corporation.

Transformational Leadership — inspired leadership that influences the beliefs, values and goals of followers so that they can perform in an extraordinary manner.

Treat (the hazard) — reduce the risk by substituting the hazard agent for a safer one, by using engineering controls, by using administrative controls, and by using worker personal protection equipment. Treating the risk involves the use of safety techniques for loss control or loss prevention.

Trend analysis — the concept of collecting information and attempting to spot a pattern, or *trend*, in the information.

Triple Bottom Line — the benefits of an OH&S Environmental Management System which prove that it can demonstrate sound management of economic, social, and environmental issues.

Uncertainty avoidance — the degree to which a society perceives unequal and ambiguous situations as threatening and to be avoided. An organization with strong uncertainty avoidance is characterized by the belief that time is money, security is paramount, and documented rules and regulations are critical.

Upstream Safety Performance Indicators — See **Leading Indicators of Safety Performance**.

Valance — refers to the value the individual personally places on the rewards.

Variable — a characteristic or attribute of a person or object that varies or takes on a different value within the study population.

Violence Frequency Rate — is determined by dividing the total number of workplace violence incidents multiplied by 200,000 by the total number of hours worked.

WHMIS — stands for **W**orkplace **H**azardous **M**aterials **I**nformation **S**ystem and is Canadian legislation that addresses workers' "right-to-know" about health and safety hazards of controlled products used in the workplace.

Work — an activity that produces something of value for other people.

Work performance — the quality and quantity of work produced, or services provided. It also speaks to the quality of the relationships within the workplace. As a formula, it is: Work Performance = quality + quality of the task accomplishments.

Work "presenteeism" — attendance at work when the worker is ill/injured and unable to fully conduct the normal duties of the job.

Work productivity — a performance measure that includes effectiveness and efficiency, with effectiveness being the achievement of goals, and efficiency being the ratio of effective work output to the input required to produce the work done.

Work scheduling — the assignment of work tasks and timelines.

Work system — elements such as an environment in which workers know how to work safely, are able to work safely, are equipped to work safely, and are motivated to want to work safely. It includes employee interactions and relationship aspects such as treating each other with respect; being honest; dealing with conflicts directly and fairly; listening; adapting to diverse working and communication styles; working in unison; freely communicating; taking pride in their work; and being accountable for their own conduct and results.

Work unit — a task-oriented group functioning within an organization that is composed of a manager and subordinates.

Worker competency — being "adequately qualified, suitably trained, and with sufficient experience to safely perform work without supervision or with only a minimal degree of supervision".

Worker training — providing workers with the information, concepts, and models needed to do the assigned job tasks.

Workers' Compensation Board (WCB) — a third party, government-operated insurance system designed to protect the injured/ill worker and afford the employers litigation protection. It is a no-fault, industry-funded insurance that is mandatory for certain industry groups. Employers pay all the premiums and claim costs, while employees forfeit their right to take legal action against the employer. Claim administration includes claim submission, claim adjudication, claim appeal if required, and claim termination.

Working alone — those circumstances in which the employee is working without radio or audible contact with another worker or the worksite and/or where assistance is not readily available.

Working model — a graphic depiction of a theory. It must be general in nature and be intuitive to promote understanding.

Workplace Complacency Trend — in incident prevention is the theory that there is often a level of workplace complacency present in the workplace prior to an incident.

Workplace label — according to WHMIS 2015 legislation, a workplace label is required when decanting or replacing damaged supplier labels on controlled products.

Workplace violence — any act in which an employee is abused, threatened, intimidated, or assaulted in his/her employment.

Workplace wellness — from a personal perspective can be defined as managing both psychological and physical issues in response to environmental stress, including one's work environment.

Zone of Management Discretion — the degree of influence/choice that Management has on the health, safety, and well-being of employees.

INDEX

[References are to pages or appendices.]

A

AD (See ALZHEIMER'S DISEASE (AD))

ALZHEIMER'S DISEASE (AD)
Generally . . . 866

ASAP (See ATTENDANCE SUPPORT AND ASSISTANCE PROGRAM (ASAP))

ATTENDANCE SUPPORT AND ASSISTANCE PROGRAM (ASAP)
Generally . . . 104-105
Absence reporting: flowchart (sample) . . . 181
Labour supply, impact on available . . . 1518
Organizations/companies
    Achieving success . . . 1520
    Recommendations . . . 1518-1519
Potential worker loss avoidance (Table 42.1) . . . 1519
Presenteeism . . . 1517-1518
Problem, how big is . . . 1514
Worker absenteeism, other factors influencing (Figure 42.4) . . . 1517
Workers taking time off work, why are . . . 1514-1517

AUTOMATION
Workplace hazards, associated . . . 440-441

B

BEST PRACTICES
Action plan and register . . . 739-752
Changing workplace/workforce, adapt to . . . 730-731
Communication strategy, develop . . . 722-724
Confidentiality of employee personal health information, protect . . . 712-713
Contingent workforce, support/protect . . . 734-735
Coordinate OH&S strategies, processes, and activities
    Generally . . . 706-710
    Managing occupational health and safety (Figure 18.5) . . . 710
    Petro-Canada's Total Loss Management (TLM) standards (Figure 18.4) . . . 707-708

BEST PRACTICES—Cont.
Corporate business strategies and planning, incorporate occupational health and safety into . . . 706
Corporate standards with recognized OH&S standards to uniformly measure OH&S programs, align . . . 721-722
Create OH&S requirements . . . 710-711
Culturally competent and proficient, strive to be . . . 735
Defined . . . 695
Demonstrate strong OH&S leadership, commitment, and passion . . . 695-698
Disability management program
    Implement effective . . . 725-726
    OH&S program and, link . . . 728
Educating upper and line management on desired management practices, invest in . . . 715-716
Effective OH&S data management system, implement . . . 717-718
Employee assistance program, link OH&S program and . . . 728-729
Graduated return-to-work program, implement . . . 726-728
Health Canada . . . 868
High-performance work system
    OH&S performance and, relationship between (Figure 18.1) . . . 701
    Strive to develop . . . 698-702
OH&S policies, procedures, and practice/performance standards, create and communicate . . . 711-712
OH&S research
    Conduct more . . . 735-736
    Ideas to those who need it, get . . . 736-737
Older workers, address workplace health and safety for . . . 731-732
Ownership and responsibilities, clarify . . . 713
Performance, measure . . . 719-721
Petro-Canada Performance Diamond (Figure 18.6) . . . 724
Prevention to workplace illness/injury management, link . . . 729-730
Staffing and budgets, coordinate . . . 716-717
Strong OH&S (safety) culture and climate, build
    Generally . . . 702-705
    Blueprints for Petro-Canada's Zero-Harm Culture (Figure 18.2) . . . 704

BEST PRACTICES—Cont.
Strong OH&S (safety) culture and climate, build—Cont.
    Petro-Canada's total loss management (TLM) framework (Figure 18.3) . . . 705
Summary . . . 737
Technology-driven OH&S awareness, strive for . . . 724-725
Transformational leadership, promote . . . 713-714
Worker OH&S education and training, provide, coordinate, and integrate . . . 718-719
Young/new workers, address workplace health and safety for . . . 732-734

BEYONDBLUE: THE NATIONAL DEPRESSION INITIATIVE (AUSTRALIA)
Generally . . . 869

BFOR (See BONA FIDE OCCUPATIONAL REQUIREMENT (BFOR))

BIOTERRORISM PROTECTION
Response plan . . . 482-486

BIPOLAR DISORDER
Generally . . . 865

BOMB THREAT
Response plan . . . 480-482

BONA FIDE OCCUPATIONAL REQUIREMENT (BFOR)
Legislation . . . 1102-1104

BUILDING CODES AND REGULATION
Generally . . . 1084-1086

C

CAMH (See CENTRE OF ADDICTION AND MENTAL HEALTH (CAMH))

CANADIAN CENTRE FOR OCCUPATIONAL HEALTH AND SAFETY (CCOHS)
Generally . . . 870

CANADIAN ELECTRICAL SAFETY REGULATORY SYSTEM
Generally . . . 1082-1083

CANADIAN MENTAL HEALTH ASSOCIATION (CMHA)
Generally . . . 868

CCOHS (See CANADIAN CENTRE FOR OCCUPATIONAL HEALTH AND SAFETY (CCOHS))

CELLPHONES
Workplace hazards, associated . . . 441-442

CENTRE OF ADDICTION AND MENTAL HEALTH (CAMH)
Generally . . . 868

CHEMICAL SPILL/EMISSION RELEASE
Response plan . . . 479

CMHA (See CANADIAN MENTAL HEALTH ASSOCIATION (CMHA))

COMMUNICATION
Best practices to develop strategy . . . 722-724
Elements . . . 1417-1418
Importance . . . 1417
Integrated disability management program (IDMP) strategy . . . 996
Introduction . . . 1417
Management . . . 76
Non-verbal . . . 1435-1436
Oral
    Presentations . . . 1425-1426
    Speaking - conversations . . . 1419-1424
    Types . . . 1424-1425
Organizational behaviour, group and group work . . . 1413
Principles, general . . . 1418-1419
Program (See PROGRAM)
Risk management (See RISK MANAGEMENT)
Stakeholder incident investigation . . . 573-574
Summary . . . 1436
Types . . . 1417
Workplace standards and rules, effective techniques . . . 448-451
Written
    Generally . . . 1426-1428
    Business writing . . . 1429-1431
    Effective writing tips . . . 1437-1442
    Technical writing . . . 1428-1429
    Types, other . . . 1431-1434

CONFIDENTIALITY OF MEDICAL INFORMATION
Generally . . . 156-160

CONSULTING (See OCCUPATIONAL SAFETY PRACTITIONERS/PROFESSIONALS)

CONTRACTORS
Reliance on, increased . . . 1513-1514

CORPORATE CULTURE
Generally . . . 274
Defined . . . 274-278
Elements of (Figure 6.1) . . . 276
IAEA's Three-Level Model of Culture (Table 6.1) . . . 277
OH&S (safety) culture
    Generally . . . 281-286
    Evolution of . . . 287
    Impact on OH&S and other important elements of business (Figure 6.2) . . . 281
    Insight into organization's (Table 6.2) . . . 283-286
Subcultures . . . 278-281

CORPORATE DUTY OF CARE
Generally . . . 290-291

CULTURE EVOLUTION
DuPont Bradley Curve . . . 92-93
Fennell Model: ESSO . . . 93-95
Stages of organizational model
    Barely conscious: stage 1 . . . 95
    Consistent and collaborative: stage 4 . . . 96
    Continually improving and deeply involved: stage 5 . . . 97
    Managing: stage 3 . . . 96
    Struggling: stage 2 . . . 95-96

D

DEPRESSION
Generally . . . 861-865
Facts and figures (Figure 23.1) . . . 864-865

DISABILITY MANAGEMENT PROGRAMS
Generally . . . 991-993
Best practices
    Effective, implement . . . 725-726
    OH&S program and, link . . . 728
Cornerstones
    Generally . . . 998
    Confidentiality . . . 1008
    Disability case management . . . 1005-1006
    Disability claim management . . . 1003-1005
    Documentation . . . 1008-1009
    Early intervention . . . 999-1003
    Ethical disability management practice . . . 1010-1011
    Legal compliance . . . 1011-1012
    Program evaluation and continuous improvement . . . 1009-1010
    Return-to-work placement . . . 1007-1008

DISABILITY MANAGEMENT PROGRAMS—Cont.
Cornerstones—Cont.
    Return-to-work planning . . . 1006-1007
    Time away and return-to-work, relationship between (Figure 27.2) . . . 1001
Direct care management model . . . 1012
Disability management defined . . . 991-993
Employee-employer disability management model (Figure 27.3) . . . 1014
Injury management, effective techniques . . . 528
Integrated disability management program (IDMP)
    Generally . . . 991-993
    Evolution of
        Generally . . . 1027-1028
        Sample (Figure 27.8) . . . 1027
    Hype or good for business . . . 1019-1022
    Move forward, how to
        Analyze your situation: step 1 . . . 1024-1025
        Demonstrate the value . . . 1026-1027
        Gather supportive disability data: step 2 . . . 1025-1026
        Impact of IDMP (Figure 27.7) . . . 1026
        Lewin's Force Field for an IDMP (Figure 27.6) . . . 1025
    Organizational behaviour relevance to . . . 1413-1414
    Senior management, selling it to
        Generally . . . 1022
        Perceived barriers . . . 1022-1024
    Stakeholders (See subhead: Stakeholders)
    Umbrella of integrated program
        Communication strategy . . . 996
        Coordinated approach to injury/illness management . . . 995-996
        Graduated return-to-work . . . 997
        Management-labour commitment and supportive policies . . . 993-994
        Performance management . . . 997
        Policies and procedures . . . 994
        Sample (Figure 27.1) . . . 993
        Stakeholder education and involvement . . . 994-995
        Supportive benefit programs . . . 995
        Workplace wellness . . . 997-998
Job matching model . . . 1012
Labour supply, impact on available . . . 1518
Managed care model . . . 1012
Models . . . 1012-1015

DISABILITY MANAGEMENT PROGRAMS—Cont.
Occupational health and safety program with, positioning . . . 804-806
OH&S disability management practitioners/professionals
    Culture diversity management . . . 1253-1254
    Four generations
        Implications . . . 1304-1305
        Organizational approach . . . 1305-1306
        Recommendations . . . 1312-1316
    Intergenerational conflicts
        Generally . . . 1306
        Challenges . . . 1307-1309
        Potential ClashPoints . . . 1306-1307
    Older worker, recommendations for . . . 1240
    Women, recommendations for . . . 1242
Organizations/companies
    Achieving success . . . 1520
    Recommendations . . . 1518-1519
Potential worker loss avoidance (Table 42.1) . . . 1519
Presenteeism . . . 1517-1518
Problem, how big is . . . 1514
Stakeholders
    Average work absence days and reasons for Canadian full-time employees, 2000-2019 (Figure 27.5) . . . 1016
    Corporation, for the . . . 1015-1017
    Education and involvement . . . 994-995
    Employee, for the . . . 1018-1019
    Parameters and key (Figure 27.4) . . . 1015
    Union, for the . . . 1017-1018
Summary . . . 1028-1029
Traditional model . . . 1012
Vendor risk management . . . 1459-1460
Worker absenteeism, other factors influencing (Figure 42.4) . . . 1517
Workers taking time off work, why are . . . 1514-1517
Workplace illness and injury, role of workplace wellness in prevention of . . . 802-804

DISCRIMINATION
Legislation . . . 1100-1101

DISEASE AND INJURY PREVENTION
Generally . . . 103

DIVERSITY
Cultural assessment
    Generally . . . 1232-1234

DIVERSITY—Cont.
Cultural assessment—Cont.
    Traditional and modern healthcare systems, comparison of (Table 34.3) . . . 1232-1233
Cultural competence
    Generally . . . 1222-1224
    Considerations, other . . . 1227-1232
    Continuum (Figure 34.1) . . . 1224
    Development of . . . 1224-1225
    Importance . . . 1226-1227
Cultural diversity
    Generally . . . 1218-1219
    Issues . . . 1219-1222
    Permanent residents who immigrated to Canada by Country/area (2016) (Table 34.1) . . . 1220
    Race distribution in United States (2019) (Table 34.2) . . . 1220-1221
Cultural diversity management
    Generally . . . 1252
    OH&S practitioner/professional approach . . . 1253-1254
    Organizational approach . . . 1253
Cultural viewpoints: approach to work and world of work
    Generally . . . 1249-1252
    Table 34.6 . . . 1250-1252
Culture . . . 1215-1217
Introduction . . . 1215
People of diverse cultures, recommendations for working with . . . 1234-1237
Relevant terms, other . . . 1217-1218
Summary . . . 1254-1255
Transcultural nursing . . . 1222
Types
    Generally . . . 1237
    Ethnic groups
        Generally . . . 1243-1247
        Summary descriptions of variety of (Table 34.5) . . . 1243-1247
    Generations . . . 1242-1243
    Older worker (over 60 years of age)
        Generally . . . 1237-1239
        OH&S/disability management professionals/practitioners, recommendations for . . . 1240
        2017 occupational injury and fatality rates (Table 34.4) . . . 1239
    Women
        Generally . . . 1240-1242
        OH&S/disability management professionals/practitioners, recommendations for . . . 1242
    Western culture . . . 1247-1249

DUE DILIGENCE
Defence of . . . 1154-1169
Reasonable steps and . . . 1169-1172

DUPONT BRADLEY CURVE
Generally . . . 92-93

DUTY TO ACCOMMODATE
Legislation . . . 1101-1102

E

EAP (See EMPLOYEE ASSISTANCE PROGRAMS (EAP))

EEA (See NEW ZEALAND ELECTRICITY ENGINEERS' ASSOCIATION (EEA))

EMERGENCY PREPAREDNESS AND RESPONSE
Generally . . . 79-80
Disaster planning . . . 102
Emergency response plans (ERP)
    Generally . . . 102
    Bioterrorism protection . . . 482-486
    Bomb threat . . . 480-482
    Chemical spill/emission release . . . 479
    Company XYZ plans . . . 470-474
    Design . . . 458-459
    Elements of . . . 464-470
    Fire emergency . . . 475-478
    Major injury incident . . . 479-480
    Pandemic flu planning (See PANDEMIC FLU)
    Sample . . . 511-515
    Value of . . . 458
    Vulnerability assessments
        Generally . . . 459
        Associated risk determination . . . 460-461
        Hazards posing threat to any specific enterprise . . . 459-460
        Necessary response actions . . . 464
        Organization/company needs assessment (sample) (Table 12.1) . . . 461
        Potential impact and develop control plans . . . 461-463
        Potential impact of disaster situations (Table 12.2) . . . 463
Hierarchy of controls . . . 508-509
Introduction . . . 457-458
Measurement of effectiveness . . . 509
Programs, value of . . . 458
Summary . . . 509

EMPLOYEE ASSISTANCE COUNSELLORS
Support disciplines . . . 762

EMPLOYEE ASSISTANCE PROGRAMS (EAP)
Generally . . . 103-104
Inventory of services (Figure 26.7) . . . 959
Occupational-based assistance for workers . . . 20-21
Occupational health and safety program with, positioning . . . 804-806
OH&S program and, link; best practice . . . 728-729
Psychological health and safety practical application in workplace . . . 907-912
Workplace illness and injury, role in prevention of . . . 801-802
Workplace wellness programs and employee assistance programs: link with grief management (See WORKPLACE WELLNESS PROGRAM (WWP))

EMPLOYER
Employee-employer disability management model (Figure 27.3) . . . 1014
Marijuana, response to legalization of . . . 1193-1196
Psychological health and safety, response to . . . 870-871
Senior risk management . . . 588-589
Stakeholder (See STAKEHOLDERS)
Technology in workplace, response to . . . 443

ENVIRONMENTAL LAWS
Generally . . . 1087
Canadian environmental laws (Table 30.1) . . . 1088-1089
Command and control instruments (Table 30.2) . . . 1090
Environmental Management System (EMS) strategies (Table 30.4) . . . 1094-1098
Historical background . . . 1087-1088
Other relevant terms . . . 1093-1099
Regulatory instruments . . . 1089-1091
Relevant environmental legislation (Table 30.3) . . . 1091-1093

ERGONOMICS
Aging worker and . . . 775-779
Educational materials
    Ergonomic program: prevention - sample #3 . . . 795-796
    General ergonomic safety tips for employees - sample #1 . . . 781-786
    Industrial ergonomics program - sample #2 . . . 787-793
Introduction . . . 767

ERGONOMICS—Cont.
Musculoskeletal disorders
    Generally . . . 767-768
    Causes . . . 768-769
OH&S: value of . . . 769-771
Program
    Purpose . . . 771-773
    Value . . . 773-774
    Work accommodation and . . . 774-775
Summary . . . 779
Work accommodation and
    Generally . . . 774-775
    Example (Figure 20.1) . . . 775

ERGONOMISTS
Support disciplines . . . 760-761

ETHICAL PRACTICE
Generally . . . 1060-1061
American Association of Occupational Health Nurses: Code of Ethics . . . 1075-1076
Board of Canadian Registered Safety Professionals: Code of Ethics . . . 1071-1073
Common considerations
    Confidentiality versus right to know . . . 1061-1062
    Individual wishes versus family or legal constraints . . . 1062
    Personal beliefs versus professional expectations . . . 1061
Considerations . . . 1058-1060
Decision-making
    Generally . . . 1063
    Ethical dilemma . . . 1063-1064
    Paradigms (Figure 29.1) . . . 1065
    Professional situation, model for . . . 1066-1067
Ethical dilemma: case study . . . 1067-1070
Ethical Fitness Model . . . 1064-1066
Ethics defined . . . 1058-1060
Introduction . . . 1057
Issues . . . 1057-1058
OH&S consulting guidelines . . . 1339
Reasonableness, concept of . . . 1060
Summary . . . 1070
Tools for determining practices
    Discussion with supervisors . . . 1062
    Legal counsel . . . 1063
    Resources . . . 1063
    Self-appraisal . . . 1063

EVALUATION (See specific subject headings)

EXTERNAL RESPONSIBILITY SYSTEM
Generally . . . 67-69

F

FENNELL MODEL: ESSO
Generally . . . 93-95

FITNESS TO WORK (FTW)
Generally . . . 160-162
Medical form . . . 175
Occupation and FTW elements (Table 4.3) . . . 161

FIVE GENERATIONS IN WORKPLACE ON OH&S SAFETY PROGRAMS
Canadian scene, 2018: five generations (Figure 35.1) . . . 1263
Descriptions
    Baby Boomers . . . 1266-1269
    Baby Boomlets (Gen Y's, Echo Boomers, Nexus, Nexters, and Millennials) . . . 1272-1276
    Generation X or Baby Bust generation . . . 1269-1272
    GenZ . . . 1276-1278
    Veterans and traditionalists or depression generation . . . 1265-1266
Four generations
    Differences, other . . . 1310-1311
    Impact on workplace . . . 1286-1303
    Motivational and performance reward differences . . . 1309-1310
    OH&S practitioners/professionals, implications for (See DISABILITY MANAGEMENT PROGRAMS)
    Personal and lifestyle characteristics comparison (Table 35.2) . . . 1279-1285
    Work characteristics, comparison of (Table 35.3) . . . 1287-1303
Generational differences, benefits of understanding . . . 1262-1264
Generational integration . . . 1311-1312
Generation defined . . . 1260-1261
Generation gap defined . . . 1261-1262
Introduction . . . 1259-1260
Multi-generational workplaces: effective management . . . 1317
Summary . . . 1316
20th century generations (Table 35.1) . . . 1260

FOIP (See FREEDOM OF INFORMATION AND PROTECTION OF PRIVACY LEGISLATION (FOIP))

FREEDOM OF INFORMATION AND PROTECTION OF PRIVACY LEGISLATION (FOIP)
Generally . . . 1105-1106

FTW (See FITNESS TO WORK (FTW))

G

GUARDING MINDS AT WORK
Generally . . . 866

H

HEARING CONSERVATION PROGRAM
Generally . . . 164-169
Noise level survey results (Table 4.6) . . . 167
Occupational exposure limits (OEL) (Table 4.5) . . . 166
Selection of hearing protection (Table 4.4) . . . 165

HIGHWAY TRAFFIC ACT
Generally . . . 1086-1087

HUMAN RESOURCES PROFESSIONALS
Support disciplines . . . 764-765

HUMAN RIGHTS LEGISLATION
Generally . . . 1099-1100
Bona Fide Occupational Requirement (BFOR) . . . 1102-1104
Canadian human rights and substance abuse policies . . . 1111-1112
Discrimination . . . 1100-1101
Duty to accommodate . . . 1101-1102
Recent developments . . . 1111
Substance abuse policies . . . 1104
Undue hardship . . . 1102

I

IDMP (See INTEGRATED DISABILITY MANAGEMENT PROGRAM (IDMP))

INCIDENT INVESTIGATION
Accident theory
  Generally . . . 553-554
  Anatomy of incidents
    Generally . . . 561-563
    Chart (Figure 14.1) . . . 562
    Bird and Heinrich theories, comparison of (Table 14.1) . . . 555-556
  Causation theories . . . 554-559
  Lessons learned . . . 559-561
Alberta OHS prosecutions penalties summary 2004-2017 (Table 14.3) . . . 577
Analysis of an incident (Figure 14.2) . . . 564
Causes . . . 563-564
Corrective actions and . . . 80
Cost of incidents, calculating . . . 576-577

INCIDENT INVESTIGATION—Cont.
Definitions . . . 552
Effective techniques . . . 571-573
Importance . . . 552-553
Introduction . . . 551
Investigation . . . 563
Measurement of effectiveness . . . 574-575
Rationale for . . . 552
Sequencing
  Corrective action(s) recommended . . . 570
  Decision to investigate made . . . 565
  Evidence collected . . . 565-566
  Final report presented . . . 570-571
  Incident investigation . . . 567-569
  Incident report generated . . . 564-565
  Investigation scope defined . . . 566
  Investigative procedures . . . 569
  Leader and team selected . . . 566-567
  Protocol (Table 14.2) . . . 568-569
  Report compiled . . . 570
  Systematic root cause analysis conducted . . . 569-570
Stakeholder communication . . . 573-574
Summary . . . 577
Types of incidents . . . 553

INJURY MANAGEMENT
Generally . . . 519
Alberta Partnerships in Injury Reduction Program . . . 91-92
Annual results
  Comparison of results for four quarters, 2019 (Table 13.3) . . . 534
  Quarterly report (Table 13.4) . . . 534
  Summary statistics (Table 13.2) . . . 533
Audit questions (Table 13.8) . . . 536-537
Causes of injuries (Table 13.5) . . . 535
Causes of WCB (Table 13.7) . . . 536
Definitions . . . 520-522
Degree of hazard (Figure 13.1) . . . 522
First aid record form, sample . . . 547-549
Importance . . . 519
Incident & investigation report form . . . 539-546
Injury and illness management . . . 101-102
Measurement of effectiveness . . . 532-537
Preparedness . . . 524-527
Savings realized by modified/alternate work initiatives (Table 13.6) . . . 535
Spreadsheet (Table 13.1) . . . 533
Stakeholders roles and responsibilities . . . 523-524
Summary . . . 537
Techniques, effective
  Disability management . . . 528

INJURY MANAGEMENT—Cont.
Techniques, effective—Cont.
    Management of workplace injury incidents . . . 527-528
    Return-to-work assistance
        Generally . . . 528-531
        Hierarchy of options (Figure 13.2) . . . 531

INTEGRATED DISABILITY MANAGEMENT PROGRAM (IDMP) (See DISABILITY MANAGEMENT PROGRAMS)

INTEGRATED WORKPLACE HEALTH MANAGEMENT (IWHM)
Approach (Figure 28.1) . . . 1038
Balanced scorecard (Figure 28.5) . . . 1049
Components and linkages
    Generally . . . 1039
    Approach . . . 1039-1040
    Benefits of linkages . . . 1040-1041
    Functions and (Figure 28.2) . . . 1039
Current state in Canada . . . 1036-1037
Defined . . . 1037-1038
Explanation . . . 1038-1039
Introduction . . . 1035-1036
Management's role
    Generally . . . 1042-1045
    Integrated workforce health and productivity (Figure 28.3) . . . 1044
Marketing
    Generally . . . 1045-1049
    Integrated workforce health and productivity (Figure 28.4) . . . 1047
Performance metrics: definitions . . . 1051-1052
Summary . . . 1049

INTERNAL RESPONSIBILITY SYSTEM
Generally . . . 67-69
Structure and accountability (Figure 1.2) . . . 68
Workplace safety, Canadian legislation . . . 1146-1149

IWHM (See INTEGRATED WORKPLACE HEALTH MANAGEMENT (IWHM))

J

JDA (See JOB DEMANDS ANALYSIS (JDA))

JHSC (See JOINT HEALTH AND SAFETY COMMITTEE (JHSC))

JOB DEMANDS ANALYSIS (JDA)
Generally . . . 162-164
Form . . . 179

JOINT HEALTH AND SAFETY COMMITTEE (JHSC)
Generally . . . 128-131
Psychological health and safety practical application in workplace, role in . . . 899

L

LEADERSHIP
Generally . . . 312
Characteristics of leader . . . 1393-1394
Corporate culture (See CORPORATE CULTURE)
Defined . . . 273-274
Determinants of . . . 1394
Employees, impact on . . . 1404
Great safety performance, industry example
    Generally . . . 289-290
    Actual performance of safe work actions . . . 303
    Actual performance of system for enabling safe work actions . . . 300
    Baseline management . . . 296-299
    Conditions, importance of . . . 300-302
    "Dashboard" (Table 6.4) . . . 303
    Design of workplace safety survey (Figure 6.7) . . . 297
    Discussion . . . 305-308
    Gap analysis
        Generally . . . 299
        Develop responses and implement actions . . . 299
        Measurement and follow-through . . . 299
    Illustration of leading and lagging indicators (Figure 6.5) . . . 295
    Importance of safe work actions
        Generally . . . 303-304
        Extent to which conditions are predictive of actions . . . 304
        Extent to which conditions are predictive of outcomes . . . 304-305
        Predictive relationship . . . 304
    Mean scores
        Actual performance and importance scale (Figure 6.9) . . . 308
        Conditions (Figure 6.8) . . . 302
    Measurement of leading indicators model (Figure 6.6) . . . 296
    Model and methodology, developing . . . 294-296
    Project conclusion . . . 305
    Project results and findings . . . 299-300
    Specific improvements, sample results of (Table 6.3) . . . 300-301

LEADERSHIP—Cont.
Great safety performance, industry example—Cont.
    Workplace safety survey, reliability of . . . 299-300
Key aspects . . . 288
Leader role versus manager role
    Generally . . . 1396-1397
    Comparison (Table 38.2) . . . 1397
Managers and management . . . 1395-1396
Measurement of . . . 309-312
Organizational behaviour . . . 1392-1393
Rationale for strong . . . 288-289
Reliability and correlation tables . . . 313-314
S.O.S. Culture Change Model (Figure 6.10) . . . 311
Sustainable high-performance workplace cultures . . . 291-293
The Performance Maximizer
    Application . . . 293
    Example (Figure 6.4) . . . 292
    Project methodology . . . 293-294
Transactional leader and transformational leader, comparison of (Table 38.1) . . . 1393
Transformational leadership
    Practices . . . 327-328
    Promote, best practice . . . 713-714
    Psychological health and safety practical application in workplace . . . 883-887
    Transactional leader and, comparison of (Table 38.1) . . . 1393

LEADERSHIP FRAMEWORK FOR ADVANCING WORKPLACE MENTAL HEALTH
Generally . . . 866

LEGISLATION
Canadian workplace safety (See WORKPLACE SAFETY)
Changing, impact of . . . 1108
Governance and stewardship . . . 102-103
Introduction . . . 1077
Legal terms . . . 1118-1136
Medical marijuana, legalization of . . . 1112
Occupational health and safety legislation . . . 1108-1110
Privacy legislation (See PRIVACY)
Psychological health safety, corporate duty of care . . . 1117-1118
Summary . . . 1136-1137
Workers' compensation legislation . . . 1110-1111
Workplace, influencing
    Generally . . . 1077

LEGISLATION—Cont.
Workplace, influencing—Cont.
    Building codes and regulation . . . 1084-1086
    Canadian Electrical Safety Regulatory System . . . 1082-1083
    Canadian OH&S acts . . . 1077-1078
    Criminal Code, Section 217 (commonly referred to as Bill C-45) . . . 1080-1081
    Current legislation: summary . . . 1108
    Employment standards . . . 1104-1105
    Environmental legislation (See ENVIRONMENTAL LAWS)
    Freedom of information and privacy legislation . . . 1105-1106
    Highway Traffic Act . . . 1086-1087
    Human rights legislation (See HUMAN RIGHTS LEGISLATION)
    National Fire Code of Canada (NFC) . . . 1083-1084
    Personal Information Protection and Electronic Documents Act (PIPEDA) . . . 1106-1107
    Transportation of Dangerous Goods (TDG) . . . 1082
    Workers, compensation acts . . . 1078-1080
    Workplace Hazardous Materials Information System legislation . . . 1081-1082

M

MAGNETIC RESONANCE IMAGING (MRI)
Workplace hazards, associate . . . 443

MAJOR INJURY INCIDENT
Response plan . . . 479-480

MANAGEMENT
Administrative or classical management theory . . . 30-32
Baseline . . . 296-299
Behavioural theory and approaches . . . 33-42
Bureaucratic management theory . . . 32-33
Classical management approaches . . . 28
Corporate culture, management practices, and human resource management theories, links to . . . 806-807
Cultural diversity (See DIVERSITY)
Disability management programs (See DISABILITY MANAGEMENT PROGRAMS)
Environment Management System (EMS) strategies (Table 30.4) . . . 1094-1098
Injury (See INJURY MANAGEMENT)
Integrated workplace health management

MANAGEMENT—Cont.
(IWHM) (See INTEGRATED WORKPLACE HEALTH MANAGEMENT (IWHM))
Management strategies and their impact on modern OH&S practices (Table 1.2) . . . 48-56
Managers and . . . 1395-1396
Marijuana in workplace . . . 1187-1189
Modern management theories . . . 42-56
OH&S Management System (OHSMS) (See WORKPLACE SAFETY)
Project (See PROJECT MANAGEMENT)
Psychological health and safety, practical application in workplace (See PSYCHOLOGICAL HEALTH AND SAFETY)
Quantitative management approaches . . . 42
Risk (See RISK MANAGEMENT)
Scientific management theory . . . 28-30
Stakeholders (See STAKEHOLDERS)
System (See MANAGEMENT SYSTEM)
Theoretical background
    Motivational and modern practices, influence on
        Generally . . . 27-28
        Administrative or classical management theory . . . 30-32
        Behavioural theory and approaches . . . 33-42
        Bureaucratic management theory . . . 32-33
        Classical management approaches . . . 28
        Modern management theories . . . 42-56
        Quantitative management approaches . . . 42
        Scientific management theory . . . 28-30
Theory X or Theory Y manager beliefs and characteristics (Table 1.1) . . . 37
Vendor (See OH&S SERVICES)
Workplace hazards (See WORKPLACE HAZARDS)

MANAGEMENT SYSTEM
Generally . . . 73
Audit results example (Figure 2.9) . . . 113
Communication . . . 76
Culture evolution (See CULTURE EVOLUTION)
Cycle (Figure 2.12) . . . 118
Elements
    Generally . . . 73-74
    Basic elements (Figure 2.1) . . . 74
Emergency preparedness and response . . . 79-80

MANAGEMENT SYSTEM—Cont.
Employee qualifications, orientation and training . . . 76-77
Framework . . . 83-84
Health and safety outcomes model (Figure 2.3) . . . 82
Implementation, rationale for . . . 80-82
Improvement opportunities (Figure 2.11) . . . 114
Incident investigation and corrective actions . . . 80
Leadership
    Activities for managing control (Figure 2.2) . . . 75
    Commitment and . . . 74-76
Measurement . . . 112-118
Participant audit scores against group composite score (Figure 2.10) . . . 114
Purpose and value . . . 82-83
Review and continuous improvement . . . 80
Standards
    Generally . . . 92
    Alberta Partnerships in Injury Reduction Program . . . 91-92
    ANSI/AIHA Z10-2005 . . . 89-90
    CSA Z1000-06 . . . 90
    ILO-OSH 2001 . . . 88-89
    ISO 45001/Z45001-19 . . . 86-88
    OHSAS 18001/02 . . . 84-86
    Specification (Table 2.1) . . . 86
Structure . . . 76
Summary . . . 118-119
Sustainability, future challenges
    Generally . . . 1507-1508
    Definitions . . . 1508-1509
    HS Model (Figure 42.2) . . . 1507
    Model process (Figure 42.3) . . . 1509-1512
Workplace hazards (See WORKPLACE HAZARDS)

MARIJUANA
Detection of use . . . 1184-1187
Health effects . . . 1181-1184
Legalization of
    Employer response . . . 1193-1196
    Fit for duty - energy safe Canada (Figure 32.1) . . . 1194
    Impact on workplace . . . 1189-1193
    Introduction . . . 1181
    Legal response . . . 1196-1198
    Legislation . . . 1112
    Management in workplace . . . 1187-1189
    Summary . . . 1198

MDSC (See MOOD DISORDERS SOCIETY OF CANADA (MDSC))

MEDICAL ABSENCE REPORTING
Generally . . . 171

MEDICAL MONITORING
Generally . . . 169

MENTAL DISTRESS
Generally . . . 860

MENTAL HEALTH COMMISSION OF CANADA
Generally . . . 868

MENTAL HEALTH WORKS
Generally . . . 868

MENTAL INJURY
Generally . . . 860

MENTAL INJURY TOOL KIT
Generally . . . 870

MOOD DISORDERS SOCIETY OF CANADA (MDSC)
Generally . . . 869

MRI (See MAGNETIC RESONANCE IMAGING (MRI))

N

NATIONAL FIRE CODE OF CANADA (NFC)
Generally . . . 1083-1084

NEW ZEALAND ELECTRICITY ENGINEERS' ASSOCIATION (EEA)
Generally . . . 368-371
Safety Climate Project (SCP) (See SAFETY CLIMATE PROJECT (SCP))

NFC (See NATIONAL FIRE CODE OF CANADA (NFC))

O

OCCUPATIONAL HEALTH AND SAFETY (OH&S) (See also specific subject headings)
Generally . . . 70
Canadian and American experience
  Background history: the 1800s . . . 16-17
  Difference in systems, current . . . 26-27
  Industrial Revolution . . . 17-18
  Investigational era . . . 19
  Occupational-based assistance for workers . . . 20-21

OCCUPATIONAL HEALTH AND SAFETY (OH&S) —Cont.
Canadian and American experience—Cont.
  Summary . . . 21-22
  The 1950s and 1960s . . . 19-20
  Workers' compensation systems, birth of . . . 18-19
Defined . . . 15-16
Federal and provincial government responsibilities . . . 69-70
Future challenges
  Contingent workforce . . . 1512-1513
  Contractors, increased reliance on . . . 1513-1514
  Introduction . . . 1499
  Leading versus lagging indicators of performance, measurement of . . . 1520-1521
  Management system, sustainability (See MANAGEMENT SYSTEM)
  Professionalism
    A profession . . . 1499-1500
    Characteristics, profession . . . 1500-1503
    Field of OH&S, prepared or not . . . 1505-1506
    Landing (Figure 42.1) . . . 1505
    Practitioner, ready or not . . . 1506-1507
    Summary . . . 1507
  Professionalization . . . 1504-1505
  Summary . . . 1521
History in Canada . . . 25-26
History in United States of America . . . 22-25
Management (See MANAGEMENT)
Management system (See MANAGEMENT SYSTEM)
Organizational designs (See ORGANIZATIONAL DESIGNS)
Program (See PROGRAM)

OCCUPATIONAL HEALTH NURSES (OHN)
Generally . . . 132-133
American Association of Occupational Health Nurses: Code of Ethics . . . 1075-1076
Career stream . . . 1477-1479
Contributions made by, value of OH Service and (Table 4.2) . . . 152-155
Practice, scope of (Figure 4.1) . . . 140
Psychological health and safety practical application in workplace
  Generally . . . 904
  Consider all the possibilities . . . 905
  Medicalizing performance issues . . . 905-906

OCCUPATIONAL HEALTH NURSES (OHN)—Cont.
Psychological health and safety practical application in workplace—Cont.
  Promote training for understanding . . . 904-905
  Rule-out Rule . . . 906
  Working for the best results . . . 906-907
Qualifications . . . 139-143
Stakeholders . . . 132-133
Technical specialist, relationship, and business skills (Table 41.1) . . . 1470-1471

OCCUPATIONAL HEALTH PHYSICIANS (OHP)
Generally . . . 143-144

OCCUPATIONAL HEALTH PROGRAM
Generally . . . 137
Effects of work on health . . . 138
Health and wellness support . . . 138
Health impacts on capacity to work . . . 138
Occupational health technical support . . . 138-139
Professionals
  Occupational health nurses (OHN) . . . 139-143
  Occupational health physicians (OHP) . . . 143-144
Work and health interactions . . . 137-138

OCCUPATIONAL HEALTH RISK
Generally . . . 145-150
Going forward . . . 151
Risk reduction through OH Service (Table 4.1) . . . 150

OCCUPATIONAL HEALTH SERVICE (OH SERVICE)
Generally . . . 144
Background . . . 144-145
Contributions made by OHNs, value of OH Service and (Table 4.2) . . . 152-155
Risk reduction through OH Service (Table 4.1) . . . 150
Value . . . 151-155

OCCUPATIONAL/INDUSTRIAL HYGIENISTS
Support disciplines . . . 759-760

OCCUPATIONAL SAFETY PRACTITIONERS/PROFESSIONALS
Generally . . . 131-132
Bicycle analogy (Figure 7.3) . . . 332
Career development
  Anchors, career . . . 1465-1466

OCCUPATIONAL SAFETY PRACTITIONERS/PROFESSIONALS—Cont.
Career development—Cont.
  Introduction . . . 1463
  Model of
    Generally . . . 1463-1465
    Career stages, individual performance and age (Figure 41.1) . . . 1464
    Planning and development . . . 1466-1467
    Plateauing, career . . . 1465
    Summary . . . 1475-1476
  Career planning concepts
    Ladder, career . . . 1467-1468
    Modeling, career . . . 1468-1469
  Career streaming
    Generally . . . 1469
    Conclusion . . . 1474
    Development stages . . . 1473-1474
    Industry application . . . 1469-1470
    Occupational health nurse: technical specialist, relationship, and business skills (Table 41.1) . . . 1470-1471
    Practitioner: technical specialist, relationship, and business skills (Table 41.2) . . . 1471-1472
    Principles . . . 1470-1472
    Professional career stream . . . 1481-1486
Consulting, internal/external
  Generally . . . 1321-1322
  Building your business, suggestions for . . . 1340-1341
  Defining . . . 1322-1323
  Ethical guidelines . . . 1339
  Introduction . . . 1321
  Project recommendations . . . 1338-1339
  Relationship, characteristics of . . . 1323-1324
  Role of internal consultant . . . 1337
  Six phases of project
    Action phase . . . 1335
    Actions, goals, and purpose (Table 36.2) . . . 1336-1337
    Comparison of different methods of data collection (Table 36.1) . . . 1331-1332
    Contracting phase . . . 1327-1330
    Data gathering and diagnosis phase . . . 1330-1333
    Entry phase . . . 1326-1327
    Evaluation and disengagement phase . . . 1335-1337
    Feedback and planning phase . . . 1333-1335
    Process (Figure 36.2) . . . 1326
    Success in situation . . . 1339-1340

OCCUPATIONAL SAFETY PRACTITIONERS/PROFESSIONALS—Cont.
Consulting, internal/external—Cont.
    Summary . . . 1342
    Tool kit
        Generally . . . 1324-1325
        Chart (Figure 36.1) . . . 1325
Job descriptions . . . 1474-1475
New role for . . . 330-333
Position profile
    Safety advisor . . . 1487-1489
    Senior safety professional . . . 1491-1494
Professional career stream . . . 1481-1486
Project management (See PROJECT MANAGEMENT)
Psychological health and safety practical application in workplace, role in . . . 899
Risk manager and . . . 591-592

OHN (See OCCUPATIONAL HEALTH NURSES (OHN))

OH&S (See OCCUPATIONAL HEALTH AND SAFETY (OH&S))

OH&S (SAFETY) CULTURE
Generally . . . 357-358
Best practice to build strong
    Generally . . . 702-705
    Blueprints for Petro-Canada's Zero-Harm Culture (Figure 18.2) . . . 704
    Petro-Canada's total loss management (TLM) framework (Figure 18.3) . . . 705
Discussion . . . 384-386
New Zealand Electricity Engineers' Association (EEA)
    Generally . . . 368-371
    Safe Climate Project (SCP) (See SAFETY CLIMATE PROJECT (SCP))
Orange Umbrella Methodology (New Zealand)
    Generally . . . 364
    Approach towards assessing safety climate . . . 364-367
    Findings and learnings . . . 367
    Great Safety Performance (GSP) Model (Figure 8.2) . . . 366
    Implications of safety climate assessment process . . . 368
Public Services Health and Safety Association (PSHSA) Health and Safety Climate Assessment Project
    Generally . . . 358-359
    Findings and interpretation . . . 360-361
    Implications . . . 361-362
    Methodology . . . 359-360

OH&S (SAFETY) CULTURE—Cont.
Public Services Health and Safety Association (PSHSA) Health and Safety Climate Assessment Project—Cont.
    Observations warranting further investigation . . . 362-363
    Participation feedback, example of (Figure 8.1) . . . 361
    Transferable learning . . . 363-364
Safety Climate Project (SCP) (See SAFETY CLIMATE PROJECT (SCP))
Summary . . . 386

OH&S SERVICES
Introduction . . . 1443
Outsource (See OUTSOURCING)
Service contract development . . . 1452-1453
Service provider market search
    Generally . . . 1449
    Request for Proposal (RFP)
        Analysis of responses . . . 1450-1451
        Data collection and collation . . . 1450
        Develop . . . 1449-1450
        Distribute . . . 1450
        Response to bidders . . . 1452
        Selection process . . . 1451-1452
        Suitable service providers . . . 1450
Summary . . . 1461
Vendor management
    Generally . . . 1453
    Risk management (See RISK MANAGEMENT)
    Strategies, recommended
        Cost-containment . . . 1454
        Partnering . . . 1453
        Performance measurement . . . 1453-1454
        Quality assurance and continuous improvement . . . 1453
        Regular meetings . . . 1454
        Service provider reporting . . . 1454

ORANGE UMBRELLA METHODOLOGY (NEW ZEALAND) (See OH&S (SAFETY) CULTURE)

ORGANIZATIONAL BEHAVIOUR
Characteristics . . . 1390-1391
Corporate culture and work attitudes . . . 1404-1407
Fundamentals . . . 1390
Group and group work
    Generally . . . 1409

ORGANIZATIONAL BEHAVIOUR—Cont.
Group and group work—Cont.
    Characteristics of effective and ineffective groups (Table 38.3) . . . 1412-1413
    Communication . . . 1413
    Dynamics . . . 1410-1411
    Norms . . . 1409-1410
    Roles and tasks . . . 1411-1413
    Team building
        Generally . . . 1411
        Cycle (Figure 38.7) . . . 1411
High productivity = performance effectiveness + performance efficiency (Figure 38.2) . . . 1399
Individual performance factors . . . 1399-1404
Integrated workplace health management (IWHM), relevance to . . . 1413-1414
Introduction . . . 1389-1390
Job design, goal setting and work attitudes
    Generally . . . 1407-1408
    Employees, impact on . . . 1408-1409
    Organization and design of work (Figure 38.6) . . . 1408
Leadership (See LEADERSHIP)
Motivation (Figure 38.3) . . . 1401
Performance Maximizer Model
    Approach for addressing work performance problems (Figure 38.5) . . . 1403
    Diagram (Figure 38.4) . . . 1402
Structure . . . 1390-1391
Summary . . . 1414
Work and work performance . . . 1397-1399

ORGANIZATIONAL DESIGNS
Characteristics
    Generally . . . 57-59
    Primary organizational designs (Table 1.3) . . . 58-59
Organizational elements, relationship with . . . 60-62
Structures
    Generally . . . 63-67
    Traditional approaches to organizational structuring (Table 1.4) . . . 65-67

ORGANIZATIONAL ELEMENTS
Organizational designs, relationship with . . . 60-62

OUTSOURCING
Generally . . . 1443
Decision-making process . . . 1443-1445
Preparation for
    Desired funding arrangement, establishment of . . . 1448-1449

OUTSOURCING—Cont.
Preparation for—Cont.
    Desired payment arrangement, establishment of . . . 1449
    Desired performance criteria and measurement techniques, determination of . . . 1447-1448
    Development of program . . . 1445-1446
    Nature of services required, determination of . . . 1447
    Preferred customer-service provider relationship, determination of . . . 1446-1447
    Service criteria, establishment of . . . 1447

P

PANDEMIC FLU
Planning
    Generally . . . 486
    Business continuity
        Administrative control measures . . . 508
        Risk management . . . 492-494
        Risks . . . 488-489
    Business planning checklist (Table 12.5) . . . 502-506
    Care for employees and contract workers, provisions for . . . 507-508
    Current assumptions about outbreak . . . 486-488
    Emergency preparedness and response plan . . . 494-507
    Employee exposure at work . . . 489-491
    Occupational risk pyramid (Figure 12.1) . . . 490
    Occupational risk pyramid for pandemic influenza (Figure 12.1) . . . 490
    Prevention: recommended personal hygiene (Table 12.6) . . . 506
    Staged approaches to preparedness planning for domestic company (Table 12.4) . . . 496-500
    WHO stages (Table 12.3) . . . 495

PERSONAL INFORMATION PROTECTION AND ELECTRONIC DOCUMENTS ACT (PIPEDA)
Generally . . . 1106-1107

PIPEDA (See PERSONAL INFORMATION PROTECTION AND ELECTRONIC DOCUMENTS ACT (PIPEDA))

PLEDGE AND OATH OF CONFIDENTIALITY
Text of . . . 173

POST-TRAUMATIC STRESS DISORDER (PTSD)
Generally . . . 860-861

PRIVACY
Employee health information
    Disclosure . . . 1115-1117
    Protection of
        Generally . . . 1113-1115
        Chart (Figure 30.1) . . . 1115
        Stages of release of personal health information (Figure 30.2) . . . 1116
Employee personal information, request for . . . 1112-1113
Legislation, evolving . . . 1112

PRIVILEGED INFORMATION
Release of, authorization for . . . 177

PROGRAM
Generally . . . 16
Components . . . 97-99
Evaluation
    Generally . . . 105
    Impact . . . 108-110
    Program goals, objectives and targets . . . 106-107
    Structure, process, and outcomes . . . 106
    Summary . . . 172
    Summative and formative . . . 108
    Types, resources and techniques (Table 2.3) . . . 110
Introduction . . . 615
Marketing
    Business case for OH&S program, building . . . 622-627
    Concepts . . . 615-617
    Core concepts (Figure 16.2) . . . 618
    Cost/benefit analysis of OH&S program, model for (Figure 16.3) . . . 625
    Defined . . . 615
    Key concept . . . 617-618
    Management . . . 617
    OH&S products and services, example of some (Figure 16.1) . . . 616
    Principles, general . . . 618-622
    Relevant facts on workplace safety (Table 16.1) . . . 622
Measurement
    Generally . . . 105
    Evaluation (See subhead: Evaluation)
    Rationale . . . 110-112

PROGRAM—Cont.
Non-occupational aspects
    Attendance support and assistance program (ASAP) . . . 104-105
    Disease and injury prevention/health promotion . . . 103
    Employee assistance programs (EAP) . . . 103-104
    Stressful situations (Table 2.2) . . . 103
    Workplace wellness program (WWP) . . . 105
Occupational aspects
    Disaster planning . . . 102
    Emergency response plans (ERP) . . . 102
    Exposure monitoring and health surveillance . . . 100-101
    Injury and illness management . . . 101-102
    Legislation governance and stewardship . . . 102-103
    Workplace/employee safety and incident prevention . . . 99-100
OH&S communication plan
    Generally . . . 628
    Brochures
        Sample . . . 639-642
        Tool . . . 631-632
    Communication tools
        Generally . . . 631
    First aid response steps (Figure 16.4) . . . 633
    Goal . . . 627
    Objectives . . . 627-628
    Outcome report
        Management report, sample . . . 645-649
        Tool . . . 633
    Poster
        Sample . . . 643-644
        Tool . . . 632-633
    Program sample . . . 628-631
    Sample . . . 635-638
    Summary . . . 633
    Website . . . 631

PROGRAM MANUAL
Generally . . . 193
Administration of program . . . 196-197
Development
    Generally . . . 193
    Policies, standards and procedures . . . 193-194
Evaluation . . . 197-198
Goals and objectives . . . 196
Graphic presentation of company OH&S audit results 2017-2019 (Figure 5.1) . . . 199

PROGRAM MANUAL—Cont.
Implementation strategies . . . 197
Industry example
    Generally . . . 199-200
    Development and maintenance . . . 200-201
    Formatting . . . 201-202
    Preliminary concerns . . . 200
    Text . . . 203-270
Mission statement . . . 194-195
Policies, standards and procedures . . . 193-194
Promotion . . . 198-199
Stakeholder roles and responsibilities . . . 197
Summary . . . 202

PROJECT MANAGEMENT
Generally . . . 1351-1352
Company XYZ Inc. proposal to evaluate and enhance case management service . . . 1367-1376
Introduction . . . 1343
OH&S project presentation (sample) . . . 1379-1386
Report
    Generally . . . 1356-1357
    Contents . . . 1358-1360
    Format . . . 1357-1358
    Presentation
        Generally . . . 1360-1361
        Effective delivery style (Table 37.1) . . . 1364
        Preparing . . . 1361-1363
        Presenting . . . 1363-1364
    Written, limitations of . . . 1360
Request for proposal (RFP)
    Details, proposal . . . 1344-1345
    Finalist interview . . . 1349-1350
    Preparation time . . . 1344
    Relevant terms . . . 1343
    Submission information, proposal . . . 1344
    Writing, proposal . . . 1345-1349
Service agreement contract development . . . 1350-1351
Steps . . . 1352-1354
Successful project, factors of . . . 1355-1356
Summary . . . 1364-1365
What can go wrong . . . 1354-1355

PSYCHOLOGICAL HEALTH AND SAFETY
Common psychological health conditions
    Generally . . . 860
    Alzheimer's Disease (AD) . . . 866
    Bipolar disorder . . . 865
    Depression . . . 861-865
    Mental distress . . . 860

PSYCHOLOGICAL HEALTH AND SAFETY—Cont.
Common psychological health conditions—Cont.
    Mental injury . . . 860
    Post-Traumatic Stress Disorder (PTSD) . . . 860-861
    Schizophrenia . . . 865
Community resources, available
    Generally . . . 866
    A Leadership Framework for Advancing Workplace Mental Health . . . 866
    Beyondblue: the National Depression Initiative (Australia) . . . 869
    Canadian Centre for Occupational Health and Safety (CCOHS) . . . 870
    Canadian Mental Health Association (CMHA) . . . 868
    Centre of Addiction and Mental Health (CAMH) . . . 868
    Guarding Minds at Work . . . 866
    Health Canada . . . 868
    Mental Health Commission of Canada . . . 868
    Mental health workplace training programs . . . 869-870
    Mental Health Works . . . 868
    Mental Injury Tool Kit . . . 870
    Mood Disorders Society of Canada (MDSC) . . . 869
    Psychologically Safe Leader Assessment (PSLA) . . . 867
    Stress Assess Survey . . . 870
    The Depression Center . . . 869
    The Panic Center . . . 869
    The Schizophrenia Society of Canada . . . 869
    Workplace Strategies for Mental Health . . . 867-868
Corporate duty of care . . . 1117-1118
Definitions . . . 857-859
Disorders, costs of . . . 855
Employer support . . . 870-871
Introduction . . . 853-854
Legislative requirement . . . 856-857
Measurement in workplace
    Assessment . . . 923-928
    GM@W Initial Scan . . . 941-942
    GM@W Organizational Review . . . 937-939
    Introduction . . . 923
    Ipsos-Reid 2012 Review: Validity of Guarding Minds @ Work Psychosocial Factors (Table 25.1) . . . 927-928
    Outcome measurements . . . 943-944
    Summary . . . 935

PSYCHOLOGICAL HEALTH AND SAFETY—Cont.
National standard . . . 859-860
Operational approach to promoting good psychological health . . . 919-920
Performance in workplace and . . . 855-856
Practical application in workplace
  Applied . . . 878-880
  Co-worker role: helping troubled employee . . . 901-904
  Employee assistance program (EAP), role of . . . 907-912
  Employee role . . . 899-901
  Front-line manager role
    Generally . . . 888-889
    Approaching employees . . . 889-892
    Coaching distressed employee . . . 893-898
    Implementing needs-based problem solving . . . 892-893
  Grief, supporting bereaved employee
    Generally . . . 912-913
    Conclusion . . . 918
    Cost of grief in workplace (Figure 24.2) . . . 913
    Defined . . . 913-914
    New normal . . . 915
    Stages of . . . 914-915
  Introduction . . . 877-878
  Joint Health and Safety Committee (JHSC), role of . . . 899
  Occupational health nurse (OHN) role (See OCCUPATIONAL HEALTH NURSES (OHN))
  Occupational safety professional/practitioner role . . . 899
  Performance Maximizer Model (Figure 24.1) . . . 879
  Senior management role
    Generally . . . 880-881
    Accommodation and return-to-work . . . 882
    Discipline, redeployment, and termination . . . 883
    Evaluation, performance management, discipline, and promotion . . . 882
    Intervention and crisis response . . . 882
    Orientation and training . . . 882
    Recruiting and hiring . . . 881-882
    Summary . . . 887-888
    Transformational leadership . . . 883-887
  Strategic direction . . . 880
  Summary . . . 918

PSYCHOLOGICAL HEALTH AND SAFETY—Cont.
Practical application in workplace—Cont.
  Trade union role . . . 898-899
  Workplace wellness programs and employee assistance programs: link with grief management (See WORKPLACE WELLNESS PROGRAM (WWP))
Summary . . . 871
Workplace evaluation
  Generally . . . 929
  Outcome evaluation . . . 934-935
  Process evaluation . . . 932-934
  Structural evaluation . . . 929-932

PSYCHOLOGICALLY SAFE LEADER ASSESSMENT
Generally . . . 867

PTSD (See POST-TRAUMATIC STRESS DISORDER (PTSD))

PUBLIC SERVICES HEALTH AND SAFETY ASSOCIATION (PSHSA) HEALTH AND SAFETY ASSESSMENT CLIMATE PROJECT (See OH&S (SAFETY) CULTURE)

R

REHABILITATION THERAPISTS
Generally . . . 763
Occupational therapists . . . 763-764
Physical therapists . . . 763
Vocational therapists . . . 764

REQUEST FOR PROPOSAL (RFP)
OH&S services (See OH&S SERVICES)
Project management (See PROJECT MANAGEMENT)

RESPIRATORY CONSERVATION PROGRAM
Generally . . . 169-171

RETURN-TO-WORK
Accommodation and . . . 882
Disability management programs (See DISABILITY MANAGEMENT PROGRAMS)
Graduated program, best practices . . . 726-728
Injury management, assistance
  Generally . . . 528-531
  Hierarchy of options (Figure 13.2) . . . 531

RFP (See REQUEST FOR PROPOSAL (RFP))

RISK MANAGEMENT
Generally . . . 582-583
Cost of incident at work (Figure 15.2) . . . 584
Defined . . . 579-580
Definitions . . . 589-591
Direct to indirect cost calculation (Table 15.1) . . . 586
Due diligence checklist (Table 15.4) . . . 590
Importance . . . 580-582
Introduction . . . 579
Key roles and responsibilities . . . 588-589
OH&S
    Generally . . . 583-587
    Assessment matrix (Figure 15.6) . . . 597
    Control . . . 595-597
    Determination (Figure 15.4) . . . 594
    Estimation . . . 592-594
    Evaluation . . . 594-595
    Implement the plan . . . 597
    Management and risk management, relationship between (Figure 15.3) . . . 592
    Practitioner/professional and risk manager . . . 591-592
    Process . . . 597
    Risk communication
        Generally . . . 598-599
        Additional considerations . . . 609-610
        Common pitfalls (Table 15.7) . . . 607
        Effective techniques . . . 600-607
        Emergency: four-step model (Table 15.6) . . . 602
        Evaluation . . . 610
        Importance . . . 599-600
        Managing hostile situations . . . 607-608
        Outrage factors (Table 15.5) . . . 600
        Working with media . . . 608-609
    Techniques
        Effective . . . 592
        Four "T's" (Figure 15.5) . . . 595
Pandemic flu planning, business continuity . . . 492-494
Quantifying OH&S risk . . . 598
Required revenue to offset loss, calculation of (Table 15.2) . . . 586
Senior management (employer) . . . 588-589
Summary . . . 610
Top 10 causes of most disabling workplace injuries in 2016 (Figure 15.1) . . . 582
Vendor management
    Generally . . . 1455-1456
    Case law . . . 1456-1458
    Conclusion . . . 1460-1461

RISK MANAGEMENT—Cont.
Vendor management—Cont.
    Disability management, considerations for . . . 1459-1460
    Management rights and employee's rights, balance between . . . 1458-1459

ROBOTS
Workplace hazards, associated . . . 442-443

S

SAFETY CLIMATE PROJECT (SCP)
Generally . . . 368-371
Approach (Figure 8.3) . . . 370
Comprehensive reporting . . . 380-381
Development of positive
    Factors enhancing (Figure 8.8) . . . 380
    Factors inhibiting (Figure 8.7) . . . 379
Experience: round 1 to round 4 (Figure 8.4) . . . 371
Impact of not following through . . . 377-378
Improvement takes time . . . 381
Improvement trends for group of companies . . . 371-374
Industry involvement and support . . . 381-383
Leader influence . . . 378-380
Leadership follow-through on action plans . . . 376-377
Momentum . . . 381
Pattern of improvement
    Generally . . . 374-376
    Noted participant reactions (Figure 8.6) . . . 376
"Soft" and "traditional" OH&S skills, comparison of (Figure 8.5) . . . 374
Year 4, emerging issues
    Business continuity . . . 383
    Financial risk . . . 383
    Health and Safety at Work Act in 2014 . . . 383-384
    Public safety . . . 383

SAFETY ENGINEER
Generally . . . 133-134

SCHIZOPHRENIA
Generally . . . 865

SCP (See SAFETY CLIMATE PROJECT (SCP))

STAKEHOLDERS
Generally . . . 123-124
Disability management program (See DISABILITY MANAGEMENT PROGRAMS)
Employee . . . 126-127

STAKEHOLDERS—Cont.
Employer
    Generally . . . 125
    Front-line management . . . 126
    Upper management . . . 125-126
Incident investigation communication . . . 573-574
Joint Health and Safety Committee (JHSC) . . . 128-131
Minimizing unsafe acts, roles in (Figure 3.1) . . . 134
Occupational health nurses (OHN) . . . 132-133
Occupational safety advisor (Table 3.1) . . . 132
Occupational safety practitioners/professionals . . . 131-132
Roles and responsibilities
    Injury management . . . 523-524
    Program Manual . . . 197
Safety engineer . . . 133-134
Summary . . . 134
Trade union . . . 127-128

STRESS ASSESS SURVEY
Generally . . . 870

SUBSTANCE ABUSE POLICIES
Canadian human rights and . . . 1111-1112
Human rights and . . . 1104

SUPPORT DISCIPLINES
Employee assistance counsellors . . . 762
Ergonomists . . . 760-761
Human resources professionals . . . 764-765
Introduction . . . 759
Occupational health professionals . . . 759
Occupational/industrial hygienists . . . 759-760
Rehabilitation therapists (See REHABILITATION THERAPISTS)
Summary . . . 765
Toxicologists . . . 761-762
Workplace wellness professionals . . . 765

T

TDG (See TRANSPORTATION OF DANGEROUS GOODS (TDG))

TECHNOLOGY IN WORKPLACE
Generally . . . 433
Employer response . . . 443
Impact
    Generally . . . 438
    Big data . . . 439-440
    Business advantages . . . 440
    Corporate security issues . . . 439
    Data safeguarding issues . . . 439

TECHNOLOGY IN WORKPLACE—Cont.
Impact—Cont.
    Hours of work, change in . . . 438
    New corporate policies and procedures, need for . . . 439
    Work benefits . . . 440
    Work location, change in . . . 438
New and emerging . . . 435-437
Summary . . . 443
Technology-driven OH&S awareness, strive for; best practice . . . 724-725
Terminology . . . 433-435
Workplace hazards
    Generally . . . 440
    Associated hazards
        Automation . . . 440-441
        Cellphones . . . 441-442
        Magnetic resonance imaging (MRI) . . . 443
        Robots . . . 442-443
        3-D printers . . . 443

THE DEPRESSION CENTER
Generally . . . 869

THE PANIC CENTER
Generally . . . 869

THE SCHIZOPHRENIA SOCIETY OF CANADA
Generally . . . 869

3-D PRINTERS
Workplace hazards, associated . . . 443

TOXICOLOGISTS
Support disciplines . . . 761-762

TOXIC WORK ENVIRONMENTS
Good performance support practices as good business, use of . . . 842-844
Introduction . . . 829-830
Organizational stressors and health
    Generally . . . 834
    Model, explanation of . . . 834-840
    Research, original . . . 844-848
    Survey (Figure 22.4) . . . 846-848
Organizations as toxic to human life . . . 830-833
Performance Maximizer Model
    Generally . . . 833-834
    Sample (Figure 22.1) . . . 833
Summary . . . 848-849
Symptoms instead of root causes, treating . . . 840-842

TRADE UNION
Generally . . . 127-128
Psychological health and safety practical application in workplace . . . 898-899

TRANSPORTATION OF DANGEROUS GOODS (TDG)
Generally . . . 1082

TRIANGLE SHIRTWAIST FACTORY FIRE
History in United States of America . . . 22-23

U

UNDUE HARDSHIP
Legislation . . . 1102

V

VENDOR MANAGEMENT
OH&S Services (See OH&S SERVICES)
Risk management (See RISK MANAGEMENT)

VISITOR HEALTH AND SAFETY
Text . . . 271

W

WEBSITE
Globe and mail . . . 987-988
OH&S communication plan . . . 631

WORK ABSENCE
Medical form . . . 185
Non-medical form . . . 183

WORKER OH&S TRAINING
Elements, important . . . 658-664
Importance . . . 656-658
Instructional modes and learner retention, relationship between (Figure 17.1) . . . 663
Introduction . . . 651
Making it fun
    Generally . . . 681
    ENMAX: Rule Book Madness, industry application
        Generally . . . 681
        Birth of an idea... . . . 682
        Game finals . . . 684
        Game play rules . . . 682-684
        Health & safety rule book . . . 681-682
        Safety can be exciting . . . 684-685
        What comes next? . . . 684
    Interactive safety education, industry application . . . 686-687

WORKER OH&S TRAINING—Cont.
Making it fun—Cont.
    OH&S crossword puzzles online, industry application
        Generally . . . 685-686
        Ergonomics crossword puzzle (Figure 17.4) . . . 685-686
    Online interactive OH&S training, industry application . . . 687
Measurement of effectiveness
    Generally . . . 677
    Entire worker OH&S training program, evaluation of . . . 677-678
    Model
        Generally . . . 680-681
        Sample (Figure 17.3) . . . 680
    Specific OH&S training course or program, evaluation of . . . 678-680
New Employee OH&S orientation checklist . . . 691
OH&S training session, developing
    Deliver . . . 676-677
    Evaluate . . . 677
    Identify training needs . . . 664-665
    Learning pyramid (Figure 17.2) . . . 669
    Objectives
        Achieving plan . . . 668-670
        Establish . . . 665-668
        Priorities, decide . . . 665
    Teaching/lesson plan and related materials
        Prepare . . . 670-675
        Worker OH&S teaching plan (Table 17.1) . . . 671-675
Program . . . 651-652
Summary . . . 689
Value of programs . . . 687-689
Worker competency versus workplace complacency . . . 652-656

WORKERS' COMPENSATION
Canadian and American systems
    Birth of . . . 18-19
    Investigational era . . . 19
Canadian system
    Amount of claims . . . 1209-1210
    Fatality rates by province: 2018 (Figure 33.2) . . . 1210
    Fundamental principles . . . 1204
    Impact . . . 1208-1209
    Intent . . . 1203-1204
    Introduction . . . 1203
    New injury claims, number of (Figure 33.1) . . . 1210
    Players . . . 1205-1206
    Procedure . . . 1205

WORKERS' COMPENSATION—Cont.
Canadian system—Cont.
  Purpose . . . 1204-1205
  Reporting requirements . . . 1206-1208
  Summary . . . 1211
Legislation, impact of . . . 1110-1111
Legislation influencing workplace . . . 1078-1080
OH&S, role of . . . 799-801

WORKPLACE HAZARDS
Control . . . 78-79
Ergonomic assessment for office environment, sample . . . 427
Exposure monitoring and health surveillance . . . 100-101
Hazard
  Defined . . . 391
  Identification . . . 391-392
  Severity (Table 9.1) . . . 393
Hazard assessment and control form, sample . . . 419-421
Hazard control
  Generally . . , 395
  Documentation . . . 398
  Follow-up form (Table 9.6) . . . 398
  Mechanisms . . . 396-397
  Priority ranking for (Table 9.4) . . . 397
  Review . . . 398-399
  Risk control measures . . . 397-398
  Safeguards (Figure 9.2) . . . 407
  Time scale (Table 9.5) . . . 397
Identification and assessment . . . 77
Inspections . . . 79
Job safety analysis, sample . . . 429
Management
  Generally . . . 390-391
  Controls/safeguards (Figure 9.2) . . . 407
  Effectiveness measurement . . . 409-410
  Hazard types . . . 400-401
  Identification, assessment, and control, techniques for . . . 401-407
  Importance . . . 400
  Summary . . . 417
  Worker involvement . . . 407-409
  Worksite inspection flow chart (Figure 9.1) . . . 404
New . . . 399-400
New industry challenges . . . 410-417
Probability
  Assessment of . . . 393-394
  Hazard probability scale (Table 9.2) . . . 394
Risk assessment
  Generally . . . 392

WORKPLACE HAZARDS—Cont.
Risk assessment—Cont.
  Form, sample . . . 423-424
Risk rating
  Determination of . . . 394-395
  Table (Table 9.3) . . . 395
Technology in workplace (See TECHNOLOGY IN WORKPLACE)
Workplace Hazardous Materials Information System legislation . . . 1081-1082
Worksite inspection
  Flow chart (Figure 9.1) . . . 404
  Form, sample . . . 425

WORKPLACE ILLNESS AND INJURY
Corporate culture, link to . . . 806-807
Employee assistance programs (EAP) role in prevention of . . . 801-802
Employee engagement: recognition of importance of personal and family life
  Generally . . . 813-814
  Effectiveness and use of work-life programs, 2007 (Figure 21.4) . . . 816
  Family-friendly policies . . . 814-816
  Family-responsive workplace, options for . . . 817-818
Healthy organization (Figure 21.5) . . . 820
Human resource management theories, link to . . . 806-807
Introduction . . . 799
Management practices, link to . . . 806-807
OH&S, role of . . . 799-801
Prevention strategy . . . 820-824
Stress risk management in the workplace: research by Dr. Martin Shain
  Generally . . . 807-813
  Health differential (Figure 21.2) . . . 811
  Organization and design of work (Figure 21.1) . . . 810
  Physical and psychosocial hazards (Figure 21.3) . . . 812
Summary . . . 824-825
Supportive infrastructure for OH&S program
  Generally . . . 818
  Corporate culture . . . 818-819
  Healthy organization (Figure 21-5) . . . 820
  Policies and procedure . . . 819-820
Toxic work environments impact on (See TOXIC WORK ENVIRONMENTS)
Value of workplace health and productivity approaches (Figure 21.6) . . . 823
Workplace wellness in disability management in prevention of . . . 802-804

WORKPLACE SAFETY
Generally . . . 317
Canadian legislation
    Conclusion . . . 1176-1177
    Due diligence
        Defence of . . . 1154-1169
        Reasonable steps and . . . 1169-1172
    Internal responsibility system . . . 1146-1149
    Introduction . . . 1141-1146
    OH&S general duty clauses . . . 1149-1154
Culture evaluation, value offered . . . 355
Field safety advisor
    Business results and measures, examples of (Figure 7.5) . . . 337-338
    Overall summary of model (Figure 7.4) . . . 336
    Performance results (operational), examples of (Figure 7.6) . . . 339-341
Great safety performance, conditions for
    Able to do it . . . 325
    Equipped to do it . . . 325
    Interactions . . . 326
    Know what to do . . . 324
    Want to do it . . . 325-326
Key indicators of great/safe workplace . . . 328-329
New directions for practice of . . . 317-321
New role, defining . . . 333-341
OH&S Management System (OHSMS)
    Evaluation: inspections and planning . . . 1175-1176
    Implementation and operation . . . 1174-1175
    Management review . . . 1176
    Planning . . . 1173-1174
    Policy . . . 1173
Older workers, address workplace health and safety for; best practice . . . 731-732
Organization evaluation
    Generally . . . 342
    Evolutionary continuum (Figure 7.7) . . . 342
Performance Maximizer
    Example (Figure 7.1) . . . 322
    Guide to understanding . . . 323-324
Performance system, measuring and analyzing . . . 329-330
Performance system for building safe/great workplace
    Generally . . . 326-327
    Building great workplaces (Figure 7.2) . . . 324

WORKPLACE SAFETY—Cont.
Practitioner/professional, new role for . . . 330-333
Relevant facts on workplace safety (Table 16.1) . . . 622
Summary . . . 355
Taking a stand
    Generally . . . 343
    Current state work process (Figure 7.8) . . . 345
    Industry example . . . 343-345
    Self-reinforcing cycle of culture development
        Generally . . . 346-348
        Culture change (Figure 7.11) . . . 353-354
        Example (Figure 7.9) . . . 347
        Strategy to shift culture . . . 348-354
        Understanding unsafe actions (Figure 7.10) . . . 351
Transformational leadership practices . . . 327-328
Workplace/employee safety and incident prevention . . . 99-100
Workplace systems, understanding
    Generally . . . 321
    Introduction . . . 321-323
Young/new workers, address workplace health and safety for; best practice . . . 732-734

WORKPLACE STANDARDS AND RULES
Generally . . . 447-448
Communication techniques, effective . . . 448-451
Compliance, measurement for . . . 453-454
Importance . . . 448
Industry examples . . . 451-452
Summary . . . 454
Worker conformance, techniques for . . . 452-453

WORKPLACE STRATEGIES FOR MENTAL HEALTH
Generally . . . 867-868

WORKPLACE WELLNESS PROFESSIONALS
Support disciplines . . . 765

WORKPLACE WELLNESS PROGRAM (WWP)
Generally . . . 105
A Framework for Health Promotion - The Jake Epp Model (Figure 26.2) . . . 950
Business case for
    Areas impacted (Table 26.2) . . . 971-972

WORKPLACE WELLNESS PROGRAM (WWP)—Cont.
   Business case for—Cont.
      Buffet & Company, 2006 National Wellness Survey: Respondents' Experiences from Company Wellness Efforts (Figure 26.17) . . . 983
      Effectiveness: top ten success markers by industry sector (Figure 26.16) . . . 982
      Effective strategy (Figure 26.15) . . . 976
      Industry findings, based on . . . 978-984
      Ongoing . . . 984-985
      Operational framework (Figure 26.14) . . . 970
      Perspective, from "business" . . . 976-978
      Sample (Table 26.3) . . . 974
      Theory . . . 970-976
   Contextual approach (Figure 26.5) . . . 956
   Development of, industry example . . . 968-970
   Disability management in prevention of workplace illness and injury . . . 802-804
   Employee assistance programs and, link with grief management
      Generally . . . 915-916
      Impact of grieving on workplace . . . 917
      Negative impact of grieving on workplace . . . 916-917
      Workplace response . . . 917-918
   Globe and Mail website . . . 987-988
   Health promotion . . . 948-955
   Human resources role in . . . 985
   Impact evaluation model (Figure 26.12) . . . 967

WORKPLACE WELLNESS PROGRAM (WWP)—Cont.
   Industry-specific WWP model, sample (Figure 26.11) . . . 964
   Introduction . . . 947-948
   Inventory of current organizational program/services, sample (Table 26.1) . . . 958
   Model (Figure 26.13) . . . 969
   Occupational health and safety program with, positioning . . . 804-806
   Operational Model (Figure 26.4) . . . 954
   Optimal employee health (Figure 26.1) . . . 949
   Organizational wellness, key elements . . . 967-968
   Sample (Figure 26.3) . . . 951
   Strategy
      Generally . . . 955-967
      Desired state (Figure 26.10) . . . 963
      Preparation: Seven-Step Approach (Figure 26.6) . . . 957
   Summary . . . 985
   Total cost of health
      Desired cost profile (Figure 26.9) . . . 962
      Identification of (Figure 26.8) . . . 960
   Workplace illness and injury, disability management in prevention of . . . 802-804

WORKSITE INSPECTIONS
   Generally . . . 79
   Form, sample . . . 425

WWP (See WORKPLACE WELLNESS PROGRAM (WWP))